英汉服装服饰词汇

邢声远　主编

化学工业出版社
·北京·

本书为服装服饰专业对照性词汇工具书，共收集词目约 70000 条，主要来自服装服饰专业术语，服装、服饰名称，面料、辅料及其他织物，原料名称，款式、式样、色彩名称，服装服饰领域有关品牌、企业、组织机构、期刊杂志、设计师及其他名人词汇，以及服装、制衣业设计、研发、生产、技术、设备词汇；酌情收取与服装服饰密切相关的纺织印染、纤维、商业、管理、经贸、艺术、时尚、化妆品、美容美发、广告、洗衣等领域词汇。

本书全部词汇以英文字母顺序排列，可供服装服饰有关从业人员、院校师生及社会广大读者参考使用。

图书在版编目（CIP）数据

英汉服装服饰词汇/邢声远主编. —北京：化学工业出版社，2013.6
ISBN 978-7-122-16260-1

Ⅰ.①英… Ⅱ.①邢… Ⅲ.①服装-词汇-英、汉 ②服饰-词汇-英、汉 Ⅳ.①TS941-61

中国版本图书馆 CIP 数据核字（2013）第 003675 号

责任编辑：李晓红　徐　蔓　　　　　　　文字编辑：颜克俭
责任校对：战河红　　　　　　　　　　　　装帧设计：关　飞

出版发行：化学工业出版社（北京市东城区青年湖南街 13 号　邮政编码 100011）
印　　刷：北京永鑫印刷有限责任公司
装　　订：三河市宇新装订厂
880mm×1230mm　1/32　印张 24¾　字数 1441 千字
2014 年 7 月第 1 版第 1 次印刷

购书咨询：010-64518888（传真：010-64519686）　售后服务：010-64518899
网　　址：http://www.cip.com.cn
凡购买本书，如有缺损质量问题，本社销售中心负责调换。

定　价：128.00 元　　　　　　　　　　　　　　版权所有　违者必究

编写人员

主　　编　邢声远

副 主 编　邢宇新　张嘉秋　梁绘影

编写人员（按姓氏笔画排序）

　　　　　　马雅芳　王　红　史丽敏　邢宇东
　　　　　　邢宇新　邢声远　杨　萍　张嘉秋
　　　　　　周　硕　耿小刚　耿铭源　袁大幸
　　　　　　梁绘影　殷　娜　董奎勇　鲁彦娟
　　　　　　曾　燕　曹小红

前　言

　　我国服装服饰具有悠久的历史，并创造了光辉灿烂的中华服饰文化，在世界上一直有较大的影响。改革开放以来，我国的纺织服装业发展很快，其产量和出口量一直位居世界首位，为世界人民的衣着作出了重要贡献。纺织服装工业是一个庞大的工业体系，涉及的学科和领域较多，是国民经济中影响较大的支柱产业之一，它不仅影响到人们的生活水平和质量，影响到高科技的发展和四个现代化的建设，而且对扩大国际贸易、增加创汇和加速国民经济的发展具有重大的意义！

　　为了适应我国服装服饰产业的蓬勃发展，我们组织力量编写了这本供纺织服装服饰从业人员、工商、外贸工作者和纺织服装院校师生参考的具有新、全、准确、规范和实用性较强特点的《英汉服装服饰词汇》。编写时，本着科学性、系统性、知识性与实用性的基本原则，力争实用、查阅方便快捷。在收词方面力求实用，尽量收集出现频率高的词汇，并注意收词的层次性、典型性、先进性、科学性和实用性，特别注意近年来国际服装界出现的一些新词汇，同时也顾及国内沿用的俗称。内容以服装面料、辅料、设计、款式、色彩、裁剪、缝纫、加工工艺与设备（包括工具）、质量检测、工业管理，以及鞋、袜、帽、手套、首饰、提包、眼镜、伞、扇、美容美发等词汇为主，同时，还收集了与服装有关的计算机、营销、外贸等方面的词汇，以适应当前服装工业快速发展的需要。

　　在编写过程中，得到撒增祺、马雅琳、殷长生、张娟、王越平、刘政、萧群锋、耿灏、周湘祁、翁扬、郭凤芝、张海峰、张洪峰、张立轻、王桂英、张进武、苏蕴莲、张之泽、赵强、王喧、薛士鑫、文永奋、张长林、张文菁、张旭红、刘杰、鲁敬之、周静等同志的大力帮助，并参考了不少书刊上的文献资料。在此对被参考的文献作者和帮助过本书编写、出版的同志表示衷心的感谢和敬意！

　　由于本书涉及的学科多、内容广泛，资料来源有限，加上编者的水平和经验不足，难免有疏漏和欠妥之处，恳请纺织服装界专家、学者和读者批评指正，不胜感激！

邢声远

2014 年 3 月

使用说明

1. 本书为服装服饰专业对照性词汇工具书，共收集词目约60000条，主要来自服装服饰专业术语，服装、服饰名称，面料、辅料及其他织物、原料名称，款式、式样、色彩名称，服装服饰领域有关品牌、企业、组织机构、期刊杂志、设计师及其他名人词汇以及服装工业研发、生产、技术、设备词汇；酌情收取与服装服饰密切相关的纺织、印染、纤维、商业、管理、经贸、艺术、时尚、化妆品、美容美发、广告等领域词汇。

2. 本书全部词汇一律按英文字母顺序排列。拉丁字母以外其他字母、符号、数字、空格等不参加排序。

3. 同一词目有几个译名时，意义相同或相近的译名用","分开；意义不同的译名用";"分开。

4. 译名前面和中间的圆括号"（ ）"内的中文表示在使用时可以省略。在译名后面的圆括号"（ ）"内的中文表示对译名的注解或说明。

5. 在有的译名前面方括号"［ ］"内的中文表示该词外文词汇的原文出处，如［美］、［法］、［意］等。

6. 缩写词词目后的圆括号"（ ）"内的英文表示全称；英文全称词目后的圆括号"（ ）"内的英文表示缩写词词目。

7. 注解中的"商名"表示"商业名称"或"商标名称"。

8. 词汇中的主体名词一般以单数出现。以复数出现者仅限于pants, boots, shoes, shorts, pettipants, briefs, scissors等习惯用复数的词汇。首字母大写表示专有名词如人名、地名、商标名等。

9. 本词汇中收集的人名，在译名后分别用圆括号"（ ）"注明所在国籍和中文译名。一般按其名字的第一个字母顺序排列，少数按其姓的第一个字母顺序排列。

参 考 文 献

[1] 陈维稷主编. 中国大百科全书（纺织）[M]. 北京：中国大百科全书出版社，1984.
[2] 钱宝钧主编. 纺织词典 [M]. 上海：上海辞书出版社，1991.
[3] 周永元主编. 织物词典 [M]. 北京：中国纺织出版社，1992.
[4] 《纺织品大全》（第二版）编辑委员会编. 纺织品大全 [M]. 第 2 版. 北京：中国纺织出版社，2005.
[5] 梅自强主编. 纺织辞典 [M]. 北京：中国纺织出版社，2007.
[6] 安瑞凤主编. 现代纺织词典 [M]. 北京：中国纺织出版社，1993.
[7] 黄故主编. 新英汉纺织词汇 [M]. 北京：中国纺织出版社，2007.
[8] 王传铭主编. 现代英汉服装词汇 [M]. 北京：中国纺织出版社，1996.
[9] 王传铭主编. 汉英服装服饰词汇 [M]. 北京：中国纺织出版社，2005.
[10] 郑雄周，邢声远编著. 实用毛织物手册 [M]. 长春：吉林科学技术出版社，1987.
[11] 邢声远主编. 汉英·英汉服装服饰分类词汇 [M]. 北京：化学工业出版社，2006.
[12] 邢声远主编. 汉英·英汉纺织品与贸易分类词汇 [M]. 北京：化学工业出版社，2009.
[13] 邢声远主编. 化工产品手册 [M]. 第 4 版. 北京：化学工业出版社，2005.
[14] 邢声远主编. 纤维辞典 [M]. 北京：化学工业出版社，2007.
[15] 邢声远，郭凤芝主编. 服装面料与辅料手册 [M]. 北京：化学工业出版社，2008.
[16] 邢声远主编. 服装服饰辅料简明手册 [M]. 北京：化学工业出版社，2011.
[17] 制衣业研究所编. 英汉对照制衣业名词手册 [M]. 香港：光芒出版社，2002.
[18] 邢声远等编著. 非织造布 [M]. 北京：化学工业出版社，2003.
[19] 日中贸易用语研究会. 日英汉贸易常用词汇 [M]. 北京：商务印书馆，1985.
[20] 王友明，陈璐编. 国际经贸词典 [M]. 天津：天津大学出版社，2002.
[21] 靳顺则主编. 汉英·英汉金融外贸词汇 [M]. 北京：化学工业出版社，2005.
[22] Charlotte Mankey Calasibetta. Fairchild's Dictionary of Fashion. 郭建南等译. 北京：中国纺织出版社，2005.
[23] 靳顺则主编. 英汉·汉英商品广告词典 [M]. 北京：化学工业出版社，2007.
[24] Tortora G, Merkel S. Fairchild's Dictionary of Textiles. 黄故等译. 北京：中国纺织出版社，2004.

目 录

前言
使用说明
参考文献
正文 …………………………………………………………… 1
附录 服装服饰常用缩写词 ………………………………… 756

目 录

A

aba, abaya, abayeh, abba 阿巴粗呢,骆驼毛(或山羊毛等)织物,阿巴原色粗毛呢;阿拉伯长袍,阿拉伯大袍,回教长褂,回教罩袍
abaca, abaka 蕉麻,马尼拉麻,菲律宾麻
abacaxi fiber 菠萝叶纤维
Abai 阿巴伊外衣(土耳其男服,丝绸或仿驼毛薄呢料,袖子用粗金线饰缝)
abalone (鲍鱼)珍珠纽扣
abarcas 阿巴卡斯鞋(西班牙)
Abassi cotton 埃及阿巴西棉
abat-chauVee 低档皮板毛
abattre 凹凸花纹
abaya (=aba)
abayeh (=aba)
abb 臀和尾部粗毛,粗次毛;纬丝,纬纱
abba (=aba)
abba mantle 阿拉伯披风
abba mantle cloth 阿拉伯披风呢
abbe cape 修道院院长披风[披肩]
abbe wig, abbot's wig, abbot cape 修道院院长型假发,修道院院长假长发
abbot's cloth 粗厚方平织物
abbotsford 艾博茨福德女式呢
abbot's wig (=abbe wig)
ABC analysis ABC分析法;重点管理法
ABC art 最简单派艺术(抽象派艺术的一个变种)
ABC silk ABC绸
abdomen 腹,腹部
abdomen area 腹部(区)
abdomen girth 腹围
abdomen/thigh relationship 腹/股关系
abdominal adrt seam 腹省缝
abdominal bandage 兜肚
abdominal belt 腹带,肚带,兜肚
abdominal breadth 腹臀厚
abdominal circumference 腹围
abdominal depth 腹高
abdominal extension arc 腹围弧
abdominal extension girth 腹围
abdominal extension line 腹围线
abdominal muscles 腹肌
abdominal region 腹部(区)
abdominal slope 前腹倾角

abdominal stress 腹部压力
abdominous figure 凸腹体型,大肚子体型
abercrombie 艾伯克龙比格子呢
Aberdeen finish 阿伯丁整理
Aberdeen hose 阿伯丁短毛袜
Aberdeen socks 阿巴丁袜
abeston 古埃及石棉布
abestrine cloth 石棉织物
ABF (adhesive bonded fabric) 黏合织物,黏合法非织造布
abide by the contract 遵守合同
abilements, abillements 镶宝石的头巾(兜帽),新娘头饰
ability 能力
ability to invest 投资能力
Abis Messaline taffeta 阿比斯迈萨林塔夫绸
abito 服装(统称,意大利语)
ablaque 伊朗细生丝
abluent 洗涤的;洗涤剂
Abnakee rug 阿布纳基钩针地毯
abnet 犹太牧师的绶带;阿贝内特披巾;伊斯兰教长围巾
abnormal body type 异常体型
abnormal fiber 异形纤维,异状纤维
abnormal spoilage 非正常损坏
abnormity 缺陷
abolla [拉]阿波拉披(古罗马)
abortive design 不完全花纹
abouchouchon, aboucouchous 阿布库舒呢(法国粗纺低档呢)
aboudia 摩洛哥短白羊毛
abougedid 埃塞俄比亚进口的未漂白棉织品
above-ankle length 踝以上长度
above-calf skirt 膝盖下的裙子
above crotch (裤)袂上
above-elbow length 肘以上长度,五分长
above elbow sleeve 肘上袖
above-elbow sleeve length 五分袖长
above-knee length 过膝长度
above-the-ankle panty 及踝裤
above-the-knee pants [panties] 膝上短裤,超短裤,(短至膝上的)短裤
above-the-waist apparel 上装
abraded wool 磨损羊毛

abrang 印度纯棉平纹紫条布
abrasion-proof 耐磨
abrasion property 耐磨性能
abrasion resistance 耐磨性,耐磨牢度,抗摩擦性
abrawan 高级达卡细平布
abrawan fabric 阿布拉温织物
Abrohany 印度薄细棉布
ABS buckle 水电带扣
ABS button 水电扣
abscissa 横坐标
ABS fabrics(area bonded staple fabrics) 面黏合非织造布
absinthe (green) 苦艾绿,洋艾绿(暗黄绿色)
absinthe yellow 苦艾黄,洋艾黄(暗绿光黄色)
absorbability 吸附能力;吸湿性,吸收性
absorbable fiber 可吸收纤维
absorbable suture (人体)可吸收性缝线(外科手术用)
absorbency fiber 吸水性纤维
absorbent cotton 脱脂棉,药棉
absorbent fabric 吸湿性织物
absorbent finishing 吸湿整理
absorbent gauze 绷带纱布,医用纱布
absorbent textile 吸音纺织材料
absorb water softening agent 吸水性柔软剂
absorption 吸湿性
absorption-air permeability fiber 吸湿透气纤维
absorption fabric 吸收性织物
absorption spectrum 吸收光谱
abstract 提要
abstract art 抽象艺术
abstract cut 抽象发型
abstract design 局部花纹图案,节略花纹图案,抽象图案
abstractionism 抽象派,抽象主义
abstractism 抽象派艺术
abstract pattern 抽象图案
abstract style 抽象风格
abut 拼接,拼合缝
abutilon 苘麻,青麻,芙蓉麻,顷麻,天津麻
abutted edge 拼合缝边
abutted seam 拼合缝
abutted stitching 拼合线缝
abutted zigzag stitch 拼接锯齿线缝
abutting (缝纫中)拼接,拼合
abutting blindstitch 拼合暗缝线迹
abutting stitch 拼合线迹

AB wear 双面穿服装
abyss 深海(水)色
acacia yellow 洋槐黄
academical dress 学位礼服
academicals 大学礼服,学位礼服
academic costume, academic dress 大学礼服,学位礼服,学位服
academic gown 学位袍
academic hood 学位头巾(风帽),学位头罩,学位连领帽,学位装饰披肩
academic hood colors 学位装饰披肩的学科色带
academic style 学院派风格,学院派款式
academy blue 绿青色
academy figure 人体模特儿,人体模型
acajou 赤褐色,桃花心术色
acala cotton 阿卡拉棉,爱字棉
acanthus 毛茛饰,莨苕叶形纹样;叶板
acca 中世纪丝地金线绣织物
accent 强调
accent color 强调色
accent lighting 强调光
accentuated contrast 强化对比度
acceptable defect level(adL) 允许疵点标准
acceptable fashion 受欢迎式样
acceptable quality 可以接受的质量
acceptable quality level(aqL) 正品标准,验收合格质量标准;合格质量水平
acceptance 承兑;接受;验收
acceptance agreement 承兑协议
acceptance bill 承兑汇票
acceptance check 合格检验,验收
acceptance condition 验收条件
acceptance contract 承兑合同
acceptance fee 承兑费
acceptance for carriage 承运
acceptance letter of credit 承兑信用证
acceptance requirements 验收规格
acceptance sampling 进仓取样,收货取样,验收取样
acceptance tolerance 验收公差
accept an order 接受订货
accepting bank 承兑银行
accepting charge 承兑费
accepting procedure 承兑手续
acceptor 承兑人
accessoiriste [法]服饰搭配员;小道具管理员
accessories 服饰;辅料;附件,配件,配饰
accessories sketch 服饰画

accessorist	配件搭配专家
accessorizing	搭配服饰配件
accessory	辅料,配料;服饰品;附属品;附件;辅助设备
accessory inventory	辅料储存
accessory of lady's dress	女装服饰品
accidental color	偶生色
accident insurance	意外保险
accident lighting	事故照明
acclimate	适应环境
acclimation	环境适应性,气候适应(作用)
accommodation area	占地面积
accompanying drawing	附图
accompanying fabric	贴衬织物
accordion bag	折叠式手提包,褶饰提包
accordion band bag	折叠式提包
accordion fabric	单面提花针织物
accordion fold	蛇腹式褶皱,折叠式褶皱
accordion hand bag	折叠式手包
accordion-like wrinkles	折叠状褶皱
accordion pleated pants	襞褶裤
accordion pleated skirt	手风琴式褶裥裙
accordion pleats	手风琴褶,百褶,多褶,襞褶,多道褶裥
accordion pleat(ed) skirt	手风琴式褶裥裙,百褶裙,风箱褶裙
accordion pocket	风琴袋,立体袋,折叠袋,立体口袋,手风琴式口袋,折叠口袋
accordion stitch	单面提花组织
account	账户
accountability	会计责任
accountable	负会计责任
account current	往来账目
account day	结账日
accounting	会计
accounting documents	会计单据
accounting evidence	会计凭证
accounting item	会计科目
accounting procedure	结算手续
accounting statement	会计报表
account of disbursement	支付账
account of goods sold	销货账
account purchas credit buying	赊购
account rendered	借贷细账
account settled	账款结清
accoutrement	饰品
accoutrements	饰品,饰物;衣着;装备
accreditation	认可
accrued account	应记账
accrued taxes	应计税金
accurate wool	卷曲度正常的羊毛
acelan	腈纶短纤及斯潘德克斯氨纶弹力纱(商名,韩国)
acetal fiber	缩醛纤维
acetal resin fastener	缩醛树脂拉链
acetate crepe	醋酯绉绸
acetate crepe-back satin	绉背缎
acetate dye	醋酯染料
acetate fiber	醋酯纤维
acetate filament (yarn)	醋酯长丝
acetate horsehair	醋酯马毛
acetate jacquard lining	醋酯提花里子绸
acetate moire	醋酯人丝波纹绸
acetate moiré	醋酯丝波纹绸
acetate panne satin	醋酯丝光缎
acetate rayon	醋酯人造丝
acetate rayon staple	醋酯短纤维
acetate rayon (yarn)	醋酯人造丝
acetate satin	醋酯丝缎
acetate sharkskin	醋酯丝雪克斯丁织物,仿鲨鱼皮醋酯织物
acetate silk	醋酯人造丝
acetate/silk blends	醋酯纤维与丝的交织物
acetate spun yarn	醋酯短纤纱
acetate staple fiber	醋酯短纤,醋酯纤维
acetate stockinette	醋酯弹力织物
acetate tow	醋酯丝束,醋酯纤维束,醋酯纤维条
acetate twill lining	醋酯斜纹里子绸
acetate voile	醋酯巴里纱
acetate wool	醋酯人造毛
acetate yarn	醋酯丝
acetylated cotton	乙酰化棉(通过乙酰化作用制取部分乙酰化的改性棉纤维)
acetylated staple	乙酰化(黏胶)短纤维
acetylated viscose rayon	乙烯化黏胶纤维
acetyl fiber	乙酰化纤维,醋酯纤维
aceytuni	中世纪缎纹
acfa	埃斯帕托叶纤维
acH (arm, chest, height)	臂围、胸围、身高
achang ethnic costume	阿昌族民族装
acH-index (arm girth, chest depth, hip width index)	臂围、胸厚、臀宽指数
achiote	胭脂树红
achromatic adaptation	无色适应性
achromatic color	非彩色,无彩色;中立色;非彩色着色
achromatic color perception	无彩色感觉
achromaticity	消色
achromatism	消色

acid and alkali resistance leather shoes 耐酸碱皮鞋
acid and alkali resistance plastic boots 耐酸碱塑料靴
acid and alkali resistance rubber boots 耐酸碱胶靴
acid arving 酸剂丝鸣增光整理
acid bleaching 酸性漂白
acid blue 湖蓝色,湖蓝
acid brightening 酸性增光整理
acid chrome dye 酸性铬媒染料
acid color 色的回归
acid color dip dyeing 酸性染料浸染
acid color discharge 酸性染料拔染
acid color printing 酸性染料印花
acid color resist dyeing 酸性染料防染
acid deweighting finishing 酸减量整理
acid dye 酸性染料
acid dye for nylon 锦纶用酸性染料
acid green 酸性绿,酸绿
acid levelling color 酸性匀染染料
acid light dyestuff 酸性耐光染料
acid metal complex dye 酸性金属络合染料
acid milling dye 耐缩绒酸性染料,酸性耐缩绒染料
acid modified fiber 酸改性纤维
acid mordant dye 酸性媒染染料
Acidol dye 阿西多染料(酸性染料,商名,德国巴斯夫)
Acidol K dye 阿西多K染料(以1:2金属络合染料为基础,具有良好的配伍性、匀染性,商名,德国巴斯夫)
Acidol M dye 阿西多M染料(双磺酸基1:2金属络合染料,商名,德国巴斯夫)
acid-premetallized dye 酸性金属络合染料
acid-proof fabric 耐酸织物
acid-proof silk 防酸绸
acid resistance 抗酸,耐酸
acid-resisting 耐酸
acid soap 酸性皂
acid stain 酸渍
acid wash(ing) 酸洗;雪花洗
A-class goods 正品
Acme cotton 阿克梅棉
acock 翻缘,[帽缘]反卷
acores 法国本色亚麻布
acorn 橡果色
acorn button 橡果纽,果纽
Acri cotton 叙利亚阿克里棉
acridine dye, acridine dyestuff 吖啶染料

acridone dyestuff 吖啶酮染料
acrilan 一种聚丙烯腈纤维(商名,美国孟山都)
acrobatic shoes 杂技鞋
acrobatic slippers 杂技演员软鞋
acrobatic wear 杂技服
acromion 肩峰点
acropodian 足尖点
across back 背宽,袖隆后中点至前中心线的距离
across back shoulders 总肩(宽)
across bust point 胸点宽
across chest 胸宽,胸阔,胸围(袖隆前中点至前中心线的距离)
across front 上胸
across shoulders line 总肩宽(线)
acrylamide dye 丙烯酰胺染料
acryl bag 丙烯纤维包
acryl fiber 丙烯腈系纤维(含有不少于85％丙烯腈组分的纤维)
acrylic 聚丙烯腈纤维,腈纶;丙烯腈系纤维
acrylic blanket 腈纶毯
acrylic boa 腈纶人造毛皮
acrylic button 丙烯腈纽
acrylic coating 亚克力胶涂层,水晶胶涂层
acrylic continuous filament (聚)丙烯腈长丝
acrylic crocheted shoes 腈纶钩编鞋
acrylic embroidery thread 腈纶绣花线
acrylic fabric 腈纶织物,腈纶布
acrylic fancy suiting 腈纶花呢
acrylic fibers 聚丙烯腈系纤维;腈纶;丙烯腈系纤维
acrylic filament (yarn) 腈纶长丝,丙烯腈长丝
acrylic fleecy infant's blouse 腈纶绒娃娃衫
acrylic fleecy infant's pantihose 腈纶绒宝宝连脚裤
acrylic fleecy pyjamas 腈纶绒童睡衣
acrylic fleecy suit 腈纶绒童套装
acrylic fur scarf 腈纶毛裘围巾
acrylic gingham 腈纶格布
acrylic gloves 腈纶手套
acrylic handbag 腈纶手袋
acrylic high bulky yarn 腈纶膨体绒线
acrylic jacquard overcoating 腈纶提花大衣呢
acrylic knitting yarn 腈纶针织绒线
acrylic knitwear 腈纶衫

acrylic napped scarf　腈纶薄绒长巾
acrylic overcoating　腈纶大衣呢
acrylic plush　腈纶长毛绒
acrylic polyester knitting yarn　涤腈混纺针织绒线
acrylic rain coat　腈纶雨衣
acrylic ramie knitting yarn　腈麻混纺针织绒线
acrylic rib　腈纶罗纹
acrylic roundneck children's pullover　腈纶圆领毛针织童套衫
acrylic round-neck shirt　腈纶圆领衫
acrylic round-neck sweater　腈纶圆领肩扣绒衫
acrylic scarf　腈纶围巾
acrylic shawl　腈纶披巾
acrylic single jersey　腈纶汗布
acrylic spun yarn　丙烯腈短纤(维)纱,腈纶短纤纱
acrylic staple fiber　腈纶短纤;腈纶棉
acrylic sweater　腈纶(毛)衫
acrylic tow　腈纶丝束
acrylic trousers　腈纶裤
acrylic underwear　腈纶内裤
acrylic-vinyl fiber　丙烯腈系-乙烯系共聚纤维
acrylic-vinylon mixed raised scarf　腈维交织厚绒长围巾
acrylic/viscose fancy suitings　腈黏花呢
acrylic weather coat　腈纶雨衣
acrylic-wool blend　腈毛混纺纱
acrylic woollen type fabric　腈纶毛绒
acrylic yarn　腈纶纱
acrylic yarn gloves　腈纶手套
acrylonitrile fiber　丙烯腈系纤维
acrylonitrile-grafted rayon　丙烯腈接枝黏胶丝
acrylonitrile-protein copolymer fiber　丙烯腈-蛋白质共聚纤维
acrylonitrile-protein fiber　丙烯腈-蛋白质纤维
acrylonitrile-vinylchloride copolymer fiber　丙烯腈-氯乙烯共聚物纤维,腈氯纶
acrylonitrile-vinylidene chloride copolymer fiber　丙烯腈-偏二氧乙烯共聚纤维(偏氯腈纶;偏丙纶)
actilex　阿克蒂莱克斯(聚丙烯酸酯纤维,商名,英国考陶尔公司)
acting abilities　演技
acting area, acting zone　表演区

actinic resistance　抗光化性
actio　阿柯图(双层特殊结构吸汗快干针织物,商名,日本)
action back　活动背褶
action figurative pattern　动感图案
action gloves　竞赛手套,赛车手套,机能手套(手背上小部分镂空)
action mode　便于活动的服装
action plait, action pleat　活动褶,活络褶裥(上衣后中线对褶)
action sleeve　活动袖,活络袖
action stretch　活动性伸缩物
action stretch fabric　高弹性织物,机能弹力布料,高弹性伸缩布料,强力弹力织物
actionwear　阿克欣卫弹力耐纶织物;蒙森特公司运动服(商名,美国)
actipore fiber　活化微孔纤维
activated carbon fiber　活性炭纤维
activated fiber　活性纤维
active and fresh finish　防臭防菌整理
active beauty　动态美
active carbon　活性炭
active carbon fiber　活性碳纤维
active charcoal　活性炭
active finish　织物灭菌整理
active sportswear　运动服;排湿透气运动服(微细纤维的高密度双层织物,内外层分别为疏水性和亲水性纤维);运动服式服装;机能性运动服
active standard　现行标准
active wear　运动服,比赛服
activity chart　工作图
activity classification　操作分类
activity list　工序表
activity sampling　快速取样,工作活动取样
act of sale　销售活动
acton　衬甲衣,盔甲
acts of god　自然力
actual capital　实际资本
actual cost　实际成本
actual loss　实际损失
actual measure, actual size　实际尺寸
actuals　标准黄麻
actual size pattern　净尺寸纸样(板)
actual tare　实际皮重
actual-value　实际价格
actual weight　实际重量
actual yardage　实际用布量
A-cup　A罩杯,小号胸罩
A-cup brassiere　小号胸罩

acye line 袖窿线
ada 绣花用帆布,十字绣底布
ada canvas 刺绣用黄褐色十字布
adamant 暗石色
adamas 硬宝石
adam's apple 喉结
adams chromatic value 亚当斯色度值
adansonia fiber 猴面包树纤维
adapangia 印度阿达基亚生丝
adaptation 复制的服装;服装、服饰的仿制;服饰改制
adapter 服饰改制员
adarsa 高级印度薄纱
adassin 伊朗细生丝
addatis 印度高级棉细布
added flare skirt 加大喇叭裙
added fullness 附加量
added touch 服饰附属品,服饰配件
added value 附加值
addhi 印度手织棉刺绣底布
additional 追加;超额;附加
additional appliance 附加设备
additional catch thread 附加锁边线
additional charge 附加费
additional clause 附加条款
additional color process 加色法
additional copy 副本
additional ease 加放量
additional equipment 附属设备
additional order 追加订货(加订)
additional part 附件
additional shipment 加载
additional tax 附加税
additive color 色彩亮度
additive color mixture 加色混色
additive complementary color 补色,加法补色
additive dress 多层式女服
additive dressing 多层组合服装;附加着装
additive primary color 相加的基色
add jacket 套装夹克
add value tax 增值税
Adelaide boots 阿德雷德及踝女靴
adelaide ruby 阿得雷德红宝石
Adelaide wool 南澳洲阿德雷德羊毛
Adele Simpson 阿黛尔·辛普森(1903～,美国时装设计师)
Adeline andre 爱德琳·安德烈(法国时装设计师)
adelis(silk) 爱得利斯绸,和田绸,舒库拉绸,爱的力司绸

Adenos cotton 阿登诺棉;阿登诺棉布
adept 内行;能手
adequate sample 适当样品
adherent tissue 黏合织物
adhering fiber 黏合纤维,黏结纤维
adhesion 附着力
adhesion strength 黏附强力
adhesive 胶黏剂
adhesive backing 黏合纤维底布;黏合底布
adhesive band 黏合带
adhesive-bonded fabric 非织造布,无纺织物;无纺布
adhesive-bonded interlining 黏合衬
adhesive-bonded non woven fabric 黏合法非织造布
adhesive fabric 黏合布
adhesive failure 开胶(鞋)
adhesive fiber 黏合纤维,黏结纤维
adhesive/fusible interlining 黏衬
adhesive label 粘贴商标[标签]
adhesive lining 黏合衬
adhesive paper 胶纸
adhesive plaster 橡皮膏底布
adhesive selvage 粘合边
adhesive tape 胶粘带
adhesive woven-interlining 有纺黏合衬
adhi 印度手织棉刺绣底布
adhotar 印度手织粗平布
adia 阿得亚漂白棉布
adiathermancy 绝热性
adiathermic property 保暖性
adidas 阿迪达斯
Adire ereco 尼日利亚蓝印花布
adjacent fabric 贴衬织物
adjacent repeats 重复图案
adjective dye 间接染料
adjust 调节,调整,整理
adjustable band 袖口调节袢
adjustable belt 调节带
adjustable braces 可调节背带
adjustable clothes 可调节的衣服
adjustable form 可调尺寸人体模型
adjustable gauge 可调尺
adjustable hemmer 可调卷边器
adjustable hinged raising foot 可调活压脚
adjustable ironing board 活动烫台
adjustable knob 调节旋钮
adjustable mannequin 可调人身模型
adjustable neckband 可调领圈
adjustable ring 伸缩戒指

adjustable roller foot 可调滚轮压脚
adjustable ruler 可调尺
adjustable set square 可调节纵横直尺
adjustable side waist 可调式腰头，两侧可调腰头
adjustable slip-over 有钮套衫
adjustable split hemmer 可调开口卷边器
adjustable square 可调式角尺
adjustable suspenders 可调节背带
adjustable tab 裤腰袢；裤头袢；活动扣袢；调节带袢
adjustable-tilt ironing surface 可调整斜度的熨烫台面
adjustable vest 男子可调式背心
adjustable waist 调节式腰身
adjust armhole 调节袖窿
adjust belt 调节带
adjust hat and dress 整衣冠
adjusting stitch tongue 线迹调节杆
adjustment 调整，装配
adjustment tax 调节税
adjust needle pitch 调针距
adjust sleeve and waist line 调整袖围线与腰围线
adjust stitch 调整线迹
adjust tab 调节袢
adjuvant 辅料
Adler ribbon 艾德勒织带，阿德勒织带(商名)
administered price 管理价格
administration 管理
administrative expenses 管理费
administrative management 行政管理
administrative rules and regulations 规章制度
admiral 紫黑色，暗紫色
admiral coat 海军将官式外套，双排金纽合身型长大衣
admiralty cloth 海军呢(美国)；海军外衣(美国)
admire 赞赏
adobe 瓦色
adolescence shoes 青年鞋
adolescent 青少年
Adolfo Sardina 阿道夫·萨迪娜(1933～，美国时装设计师)
Adonis wig 阿多尼斯假发
adorn 饰物，装饰
adornment 装饰；装饰品，饰品，饰物
adorn skirt with flounce 给裙子饰荷叶边
adras, adras silk 印度阿德拉斯绸

Adri 亚迪(1930～，美国时装设计师)
Adria 亚德里亚斜纹呢
Adrian 艾德里安(美国著名戏剧服装设计师)
Adrian Cartmell 阿德里安·卡特梅尔(美国时装设计师)
adrianople red 土耳其红，鲜红
adrianople skin wool 土耳其皮板毛
adrianople twill 土耳其印花红斜纹布
Adrienne gown 背部带两个大箱形褶的宽松长裙
Adrienne Steckling-coen 阿德里安·施特克林(美国服装设计师)
adris silk 爱的力司绸
adult 成年人
adult clothing 成人服装
adult diaper 成人尿布
adult fashion 成人风格服饰
adult shoes 成年鞋
adult size, adult's size 成人尺寸，成人尺码
adumbrate 勾画
ad valorem duties 从价税
advanced 超前
advanced composite material 高级复合材料
advanced equipment 先进设备
advanced fiber nonwoven 新型纤维非织造布
advanced guard 前卫派
advanced scientific and technological achievements 科技成果
advanced technology, advanced technique 先进技术，先进工艺
advance sample 订货小样；推销样品
advancing color 前进色，近感色，近似色
advertisement charge 广告费
advertising agency(company) 广告公司
advertising cap 广告帽
advertising tape 经纱黏合带，无纬带
advertising T-shirt 广告衫
advice of acceptance 承兑通知书
adviser 顾问
aeE 爱意
aegean blue 爱琴海蓝
aegis 胸甲
aemi boots 轻便靴
aeneous (color) 青铜色
aeolian 风神绸(棉经丝纬薄织物)
aerated plastics 泡沫塑料
aerial delivery fabric 降落伞织物
aerial perspective 空间透视法，浓淡远近法
aerial port 空运港

aerobic ensemble 健身服,训练服
aerobic ensemble exercise suit 健身运动装
aerobic fashion 吸氧健身运动装
aerobics 健身运动装
aerobics brief 运动短裤
aerobics shoes 健身运动鞋,训练鞋
aerobics sportswear 健美服
aerobics wear 吸氧健身运动服
aeronautical cloth 航空用织物,气球用织物
aerophane 埃罗芬素色丝纱罗;19世纪薄绉
aeroplane fabric 飞机翼布
aerospace science 宇航科学
aeruginous color 蓝绿,铜绿色
aesthetical consciousness 审美意识
aesthetical sense 美感
aesthetic conceptions 审美观
aesthetic costume 唯美主义服式(19世纪末英国妇女的服装款式)
aesthetic dress 美学风格女装,唯美主义服式(19世纪末英国女装款式)
aesthetic education 美育
aesthetic feeling, aestheticism 美感
aesthetician 美学家,美学者,审美学家,审美学者
aestheticism 美感,审美感
aesthetic judgment 审美能力
aesthetic-oriented fiber 美观性纤维
aesthetic perception 美感
aesthetic property 美感性
aesthetic quality 美感性,美学质量(手感、外观等)
aesthetics 美学;美感性,审美观;美容
aesthetic theory 美学理论
aetz 烂花绣品,烂花刺绣
aetzstickerei 德国烂花绣品
after-skisocks 滑雪后短袜
Afgalaine 阿芙加兰平纹毛呢;阿芙加兰女式呢;阿芙加兰绉波呢
Afghan 阿富汗毛毯;阿富汗长袍;阿富汗披肩
Afghan blanket 阿富汗毛毯
afghan bokhara rug 阿富汗布哈拉地毯
Afghan coat 阿富汗(毛皮长)外套
Afghan-Dalhi 阿富汗大喜地毯
Afghanistan 阿富汗绒头地毯
Afghanistan costume 阿富汗民俗服(民族服装)
Afghanistan jacket 阿富汗皮外套
Afghanistan rug 阿富汗地毯
Afghanistan vest 阿富汗背心

Afghanistan wedding tunic 阿富汗束腰罩衫
Afghanistan wool 阿富汗羊毛
Afghan-kerki 阿富汗克尔基地毯
Afghan stitch 阿富汗毛毯缝线线迹;阿富汗钩编针法
A fiber A纤维(聚十二烷甲酰基二环己烷二胺纤维)
afifi, afifi cotton 埃及阿菲菲棉
A figure A体型,普通体型
a fine-quality writing fan 玉版扇
afiume 埃及阿菲乌姆粗亚麻
afloat 路货(运输途中的货物)
African 非洲式裙装
African bush jacket 非洲森林服
African cotton 非洲棉
African design 非洲式图案
African Development Bank 非洲开发银行
African hemp 非洲虎尾兰叶纤维
African look 非洲款式;非洲风格,非洲风
African maguey 马奎龙舌兰属叶纤维
African print 非洲风格印花
African sisal 非洲西沙尔麻
African stripe 非洲宽色条棉布
African waste 非洲阿纳菲野蚕丝
African wool 非洲羊毛,南非美利奴羊毛,非洲土种毛
Afrida lac cloth 印度虫漆深红棉布
Afridi 东印度阿夫里迪蜡防花布
Afro 仿非洲人发型,埃弗罗发型
Afro choker 乌班吉贴颈项链,非洲型贴颈项链
Afro(cut) 非洲发型,黑人发型
Afro look 非洲风貌,非洲型款式,非洲风;仿非洲人发型
Afro styles 非洲风格
afshar 喜拉斯粗织纹地毯
Afshari-khila 阿富沙里基拉地毯
after boarding 后是型
after body 服装后身
after boots 暖脚靴,冰场往返高筒靴
afterbrain 后脑
aftercare of knitted fabrics 针织物后整理
afterchrome dye 后络媒染料
after-contraction 附加收缩
after copper dye 铜盐后处理染料
after cost 后续成本
after dark 晚礼服,晚装,夜礼服
afterfive 晚礼服,夜礼服,下午五点钟以后穿的

afterfive dress 夜生活服
after flaming time 残焰时间
afterglow 晚霞色(黄红色)
afterglow time 阴燃时间
afternoon 午后装
afternoon blouse 午后女衬衫
afternoon coat 午后外套,午后大衣,午后氅(穿在午后服外面)
afternoon dress 便宴服,日礼服,半礼服,午后服装
afternoon hat 午后礼帽,半礼服女帽
afternoon pumps 午后碰拍鞋,午后潘普鞋
afternoon shift 下午班
afternoon suit 午后套装,日常套装
afterprint washing machine 印花织物后处理水洗机
after processing 后加工
after-sales service 售后服务
after-shave lotion 修脸洗剂
after shoes 滑雪后用鞋
after-shrinkage 后收缩
after six 晚礼服;夜礼服;晚宴服(下午六时后穿的晚会服);晚装
after-ski boots 暖脚靴
after-ski look 滑雪后款式
after-ski shoes 暖脚鞋
after-ski slippers 暖脚套鞋
after-ski socks 滑雪后用短袜
after-ski style 滑雪后款式
after slippers 穆克拉克套鞋
after tax 税后
aftertreatment 后处理
after-war style 战后款式
after-wash 后水洗
after welt 袜口下加固段
aga 无花果树韧皮纤维
Agabance, Agabanee 阿加班塞丝绣棉织物,叙利亚丝绣棉织物
agaguse 领围褶,领圈褶(从领口至乳部的褶子)
against all risks 保一切险
against to sample 与来样不符,不符来样
agal 阿加尔头巾(男用),羊毛绒头带,男用毛绳头箍,绳箍(阿拉伯)
agamid 无花果树韧皮纤维
agaric 毛巾布
agate 玛瑙;玛瑙棕(深棕色)
agate green 玛瑙绿,灰湖绿
agate grey 玛瑙灰
agate red 玛瑙红

agave 剑麻,西沙尔麻,龙舌兰麻,番麻
agave deciplens 类西沙尔麻
agave fiber 龙舌兰纤维,剑麻
agave rigida, agave sisalana 西沙尔麻,剑麻
age color 年龄色
aged happy 老头乐
aged metal look 旧金属感
age group 年龄群
ageing resistant finish 防老化整理
agency 代理;机构
agency agreement 代理协议
agency contract 代理合同
agenois 法国本色亚麻布
agent 代理(人),代理商,推销员
age pattern analysis 年龄类型分类
ager 蒸化机
age section 年龄组;年龄段
aggebonce 叙利亚丝绣棉织物
aggoned bunder 优级生丝,优级日本与印度生丝
aggraffe 搭扣,搭钩
aggregate 总计
aggregate scheduling 总体进度安排
aggregate value 总价值
aggressive color 挑逗色
aggressive tack 干黏性
aging resistance 抗老化性能
agis 胸甲
aglet 服饰金线,(装饰用)金线;(军装)肩带,花边帽带,帽饰花边
agnelin 阿格尼林毛
agneline 阿格尼林长绒防水粗黑呢
Agnès 阿涅斯(法国女帽设计师)
Agnès B 阿涅斯 B(法国时装设计师)
Agnès sorel bodice 阿涅斯·索雷尔上衣
Agnès sorel corsage 阿涅斯·索雷尔宽松胸衣
Agnès sorel style 阿涅斯·索雷尔公主款式
agogo look 最时髦款式
agogo style 最时髦款式
agotal 蕉麻
agpui 木槿树韧皮纤维
agra carpet 阿格拉地毯
agraf(f)e 搭扣,搭钩
agra gauze 阿格拉薄纱
agra rug 阿格拉粗厚地毯
a great gross of buttons 12 罗纽扣(即 1728 粒)
agreed 约定的

agreed quantity of trade　协议贸易量
agreement　协议;协议书
agreement of intent　意向协议
agreement price　协议价格
agreffe　搭扣
agriculture cover　农用覆盖材料
aguilles　叙利亚棉平布
Aguja　亚古哈针绣花边
agust　阿格斯特韧皮纤维
ahrami　手编条花棉布
aid　辅助手段
aida　菱形花纹平布
aida canvas　刺绣用黄褐色十字布,绣花用帆布
aida cloth　松散组织面料
aids　辅助设备
aiglet　装饰用金线;军服上肩带,帽饰花边带;绳、带两端的金属箍
aigrette　羽毛饰,羽饰;鹭鸶毛帽饰;枝装饰,仿羽毛珠宝装饰品
aigrette egret　仿羽毛的珠宝装饰品
aiguil(l)ette　金属饰带,军服肩带,穗饰
aihuku　春秋装(日本称谓)
ailanthus silk　樗蚕丝
ailesham cloth　英国中世纪细亚麻布
ailily crepe　爱丽丽纱
aimea　爱慕
aimei jacquard noil cloth　爱美提花绸绸
ai-mei jacquard suiting　爱美呢
Ainu costume　阿伊努族服装(日本)
Ainu kimono　阿伊努人短和服
air　气氛,航空
air bag fabric　安全气囊织物
airballad　艾巴拉德(聚酯及黏胶混纤丝,商名,日本尤尼吉卡)
air brush printing　喷雾印花
air-bulked yarn　喷气膨化丝,空气变形丝
airbutton clamp　气动扣夹
air cargo　空运货物
air case　航空拎包
air circulation　空气循环
air-conditioning cloth　(冬暖夏凉)空调布
air-conditioning fabric　凉爽整理织物
air container　空运集装箱
air-cooled finish　凉爽整理(使织物纱线间保持空隙)
air-covered yarn　空气包覆纱
air-cushioned needle detector　气垫验针器,气垫式验针机
air cushioned shoes　气垫鞋

air-cushioned sole　气垫鞋底
air cushion spread table　气垫式铺料台
air dome　充气帐篷
air dry　自然干燥
air dry moisture regain　标准回潮率
airer　晾衣架,烘衣架,晒衣架
air fabric　透气性织物
air fiber fabric　充气纤维织物
air filled sole　充气底(鞋)
air filtration nonwoven　空气过滤非织造布
air floating/vacuum cuting　气垫/真空裁床
air flotation cutting table　气垫裁床
air flotation/vacuum cutting table　气垫/真空裁床
airflow insole　气垫式鞋内底
air force blue　空军蓝
air force look　空军款式,空军服,空军军装风貌
air force serge　空军服哔叽
air force style　空军军装款式
air France jersey　全毛针绣乔赛,全毛针绣针织衫
air freight　航空运费
airing　晾干,晒干
air jacket　充气救生衣
air jet bulked yarn　喷气膨化变形丝
air jet interlaced yarn　空气喷射交络丝
air jet textured yarn　喷气膨化变形丝
airmen's suit　飞行服,飞行员服装
air needle cooling　气冷缝针,气动缝针冷却
air net shirt　网眼衬衫
air operated undertrimmer　气动下切线器
air-pad　充气胸罩
air permeability　透气性,通气性
air permeable fabric　透气性织物
air permeable water proofing　防水透气整理
air perviousness　透气性
air pervious waterproof fabric　透气性防雨布,防水透气织物
airplane cloth, airplane fabric　飞机用布
airplane freight　空运费
airplane luggage cloth　箱包织物,轻型行礼布
air pocket　气包,气袋
air pollution　空气污染
air-proof　不透气
air-retention capacity　空气保持性
airroll　艾洛尔(聚酯中空短纤维,商名,日本尤尼吉卡)
airship envelope fabric　飞船用气囊织物

airship fabric 气球用棉织物	aladjias 阿拉吉阿斯棉布
air silk 空心丝,气泡丝	aladjias plain cloth 阿拉吉阿斯平布
air space 空气层,空隙	alagia 印度丝棉平布
airspun acetate rayon 干[法]纺醋酯人造丝	alagoas 棉(葡萄牙称谓)
air suit 充气服,空气服	alagoas lace 巴西粗棉带
air-tangled yarn 空气变形网络丝	alamba cotton 美国阿兰巴棉
airtex 网眼夏服面料	a la mode 流行
air textile 网眼夏服面料	alamode 阿拉莫德绸
air-textured yarn (atY) 喷气变形丝,空气变形丝	à la mode, à-la-mode 穿着时髦的,流行的
	à la page 时髦的
air-textured yarn fabric 空气变形纱织物	alapat, alapat yarn 双股椰壳纤维
air the dairy 袒露胸脯	alapeen, alapine 阿勒潘毛葛
air-tight zipper 不透气拉链	alarm clock 闹钟
air transport 空运	Alaska 阿拉斯加呢;阿拉斯加棉毛混纺纱;起绒布;阿拉斯加鞋
air tucking 空气打褶;空气起裥	
air tuck stitch 空气打褶线迹	Alaskan coat 阿拉斯加外套(风帽檐上装饰毛皮的防寒外套)
air-ventilated suit 通风服	
air washer 空气洗涤室	Alaskan sealskin 阿拉斯加海狗毛皮
airway 航空	Alaska(shoes) 阿拉斯加鞋
airway bill 航空货运单	Alaska yarn 阿拉斯加棉毛混纺纱
airy 轻的;通风的,空气的	alatcha, alatchu 中亚蓝白条棉斜纹布
airy fabric 透气性织物,通气织物	alb 神父的长法衣,麻布僧衣,白麻布长袍,典礼袍
ajamis 中东棉印花布	
ajiji 阿吉古棉薄纱	albacore line 钓鱼线
ajour 网眼花纹	albahaca fiber 黄花稔属植物纤维
ajour knitted fabric 针织网眼布	Albanian embroidery 阿尔巴尼亚帆布刺绣
akashimajoofu 赤缟上布	Albanian wool 阿尔巴尼亚羊毛
akashiori 明石织	Albar 苏丹阿尔巴棉
aketon 衬甲衣,盔甲	albarazine wool 西班牙中等羊毛
Ak-hissar rug 土耳其绒头地毯	albatross 海鸟绒,细绒精纺呢,黏胶丝华达呢,仿海鸟绒棉织物,精纺绉呢,婴儿细绒布
Akiko Isomurashi 矶村晔子(日本时装设计师)	
Akira Onozuka 小野冢秋良(日本服装设计师)	Alba velvet 阿尔巴提花丝绒
	albene 法国无光醋酯纤维纱及布
akitahachijoo 秋田八丈	albernus 阿尔伯诺斯薄呢
Akura cloth 阿库拉手织粗平布,阿库拉织物	alberoni 丝金绒驼毛呢
	albert 纬面棉斜纹衬布
akwa 印度木棉	Alberta Ferretti 阿尔贝塔·费雷蒂(意大利时装设计师)
akwet-longyi 印度手织头巾布	
al 印度桑树根染料	Albert boots 阿伯特靴
alaballee 阿拉巴利薄纱	Albert Capraro 艾伯特·艾普拉罗(美国时装设计师)
Alabame cotton 美国阿拉巴马棉	
alabaster 淡黄棕色	Albert cloth 艾伯特双面大衣呢
à la capricieuse [法]阿拉卡布里新式发型;刺猬假发	Albert cord 19世纪花式羊驼呢
	Albert crepe 艾伯特绉;丝棉绉;绉线平纹绸;丧服用里绉绸
alacha 印度丝棉丝织物	
a la cheville 法国齐踝裙	Albert diagonal 艾伯特斜纹厚花呢(六页斜纹)
Alacians 法国纯棉中平布;阿拉西安棉立布	
aladja 阿拉贾织物;土库曼斯坦蓝白条子布	albert farmers satin 艾伯特缎纹棉毛呢
aladja silk/cotton fabric 阿拉贾丝棉交织绸	Albert jacket 艾伯特男式单排扣短上衣

Alberto Fabiani 阿尔贝托·法比亚尼（意大利时装设计师）
Albert overcoat 艾伯特男式宽松上衣
Albert riding coat 艾伯特骑士外套
Alberts 艾伯特斜纹织物
Albert tie 艾伯特领结（蝴蝶领结）
Albert twill 艾伯特纬面斜纹薄呢
albescent 浅白色，带白色
albesine wool 西班牙阿尔伯羊毛
albigeois 法国本色麻帆布
Albissola lace 阿比索拉小花纹花边；意大利梭结花纹花边
albumin fiber 白蛋白纤维
albumin printing 蛋白印花
alcantara 阿尔肯塔拉麂皮
alcantara wool 西班牙阿尔肯塔拉低档羊毛
alcatifa 东方风格花纹地毯
alcatquen (rug) 手工高级金线波斯裁绒地毯
alcian dye 爱尔新染料（暂溶性酞菁染料，商名，英国）
alcohol-water disperse dye 醇-水可分散性染料
alcyonne 翠鸟缎
Aldo Cipullo 阿尔多·西皮洛（法国时装设计师）
Aldo Cipullo 阿尔多·奇普洛（美国珠宝设计师）
Aldo Gucci 阿尔多·古奇（意大利时装设计师）
Alencon 阿郎松针绣花边；轻薄棉丝交织绸
Alencon bar 阿郎松针绣凸纹线条
Alencon ground 阿郎松花边底布
Alencon lace 冬季花边，阿朗松花边
Alenconnes 法国半漂白亚麻
Alepine 阿勒潘毛葛；叙利亚丝毛布；丝经毛纬丧服布；法国丝（或棉）经毛纬交织布
Aleppo cotton 叙利亚阿勒波棉
Aleppo wool 东叙利亚阿勒波羊毛
A.LERGIN 阿勒锦
Aleutian 蓝铁色貂皮，阿留申貂皮
Alexander 棉经毛纬花式斜纹里子布；埃及条纹绸
Alexander Julian 亚历山大·朱利安（美国服装设计师）
Alexander Mcqueen 亚历山大·麦奎因
Alexandra 平纹棉黑里子布
Alexandrette cotton 叙利亚亚历山大勒特棉
Alexandria 亚历山大棉毛花呢；埃及短绒棉
Alexandrine 亚历山大德里恩高支色织布
alexandrite 紫翠玉色（白天呈暗绿色，在红光线下呈红紫色）；变色宝石
Alex Gres 亚历克斯·格雷（法国服装设计师）
Alexis Kirk 亚里克斯·卡里克（美国珠宝设计师）
alfalfa 苜蓿绿灰
alforgas 阿尔福加斯厚帆布，英国出口南美的厚帆布
alforgas canvas 阿尔福加斯帆布
algaline 绷带用芦苇属纤维
Algerian eye filling stitch 阿尔及利亚贴线绣
Algerian lace 阿尔及利亚金银丝花边
Algerian silk 阿尔及利亚粗丝，阿尔及利亚绵绸
Algerian stripes 阿尔及利亚条子绸；金边棉丝条纹布；金边棉丝及衬色条纹布
Algerienne 阿尔及利亚横条纹厚呢；金边棉丝条纹布
Algerine 阿尔及平纹棉布
Algerine shawl 阿尔及利亚彩条披巾
alginate fiber 藻酸纤维，海藻纤维
alginate yarn 海藻纤维纱，藻酸纤维纱
alginic acid fiber 海藻酸纤维，藻酸纤维
algoa cord 阿尔戈阿花式灯芯绒
algodon 棉（西班牙称谓）
algodon de seda 牛角瓜籽纤维
algol dye 亚士林染料（还原染料，商名，德国）
alhambra quilt 阿尔汉布拉被单布，双股色经纱提花被单布
alibany 17世纪印度丝棉布
alice band 爱丽丝发带，弹力发带
alice in Wonderland dress 围裙装
alicienne crepe 窗帘，床单绉条布，阿里新条子泡泡纱
alicula 罗马斗篷
alicyclic aliphatic polyamide fiber 脂环族-脂肪族聚酰胺纤维，脂环族耐纶纤维，脂环族尼龙纤维
alicyclic nylon fiber 脂环族锦纶
alicyclic polyamide fiber 脂环族聚酰胺纤维
alien corporation 外国公司
alien merchant 外国商人
aligning 校直
alignment 叠边
alike dress look 相像服装风貌
alike dress style 相像服装款式
A-line A形造型，A形线条；A字形
A-line coat A形外套，A字形外套
A-line derss A形女服（窄肩，宽下摆无褶

裥), A 形连衣裙
A-line flare (skirt) A 形喇叭裙
A-line jumper A 字形马夹裙装
A-line princess dress A 字形公主连裙装
A-line silhouette A 字形轮廓
A-line skirt A 形裙,斜裙
aliphatic nylon fiber 脂肪族锦纶
aliphatic polyamide fiber 脂肪族聚酰胺纤维
Alix Gres 阿利克斯·葛雷斯(1900～,法国高级时装设计师)
Alix Gres 阿里克斯·格雷斯(德国时装设计师)
alizarin dye 茜素染料(媒染染料)
alizarin red 茜红
alizarin reds 茜草红条纹布
alizéa [法]贸易风发型
alkali containing glass fiber (中)碱玻璃纤维
alkali contracted towel 碱缩毛巾
alkali deweighting, alkali deweighting finishing 碱减量整理,涤纶仿真丝绸整理
alkali free glass fiber 无碱玻璃纤维
alkaline disperse dye 碱性分散染料(用于碱性浴染色)
alkali shrinkage single jersey 碱缩单面针织物
alkali-soluble cotton 碱溶棉
alkali-soluble fiber 海藻纤维,碱溶纤维,藻朊纤维
alkanet 红卷心菜染料
allabatis, alliabant 东印度平纹刺绣棉布
allaeanthus 阿利恩斯韧皮纤维
allahabad 印度手织毯,印度棉织物
alla haik 北非手织经向宽条头巾布
allamande corduroy silk 阿勒芒德灯芯绸
allamode, allamod silk 阿拉莫德绸
allapeen, allapine 阿勒潘毛葛;叙利亚绢毛布
all-around 多用途服式
all-around belt 全腰带
all-around jacket 通用夹克
all-around pleated skirt 百褶裙
all-around sports wear 通用运动服
all back 全后梳发型
all-bias construction 无方向结构
all blue 全蓝
all core type rayon 全芯型黏胶丝
all cotton 纯棉,全棉
all cotton stretch yarn 全棉弹力纱
all cotton yarn 纯棉纱

allegias, allejars 印度棉平布;丝棉混纺平布
allejah 18世纪色条织物
allemande 法国斜纹条子布;阿勒芒德灯芯绸
Allen cotton 美国阿伦棉
allergic dyestuff 过敏性染料
all fashion 全成形
all forward stretch stitch 全向前伸缩线迹
all gather 全折裥,全褶裥
alliabably 达卡细棉布
allibannee, allibanis 17世纪印度丝棉条纹布
allied company 联营公司
alligator calf leather 仿鳄鱼皮(牛皮革)
alligator cloth 仿鳄鱼皮漆布
alligator grained leather 仿鳄鱼皮
alligator (leather) 鳄鱼皮
alligator leather shoes 鳄鱼皮鞋
alligator print 鳄鱼皮文印花
alligator skin bag 鳄鱼皮手袋
all-in-line 整列纽扣上衣;双排纽
all-in-one 紧身女胸衣(胸罩、紧腹带、吊袜带合一),全帮胸衣,连胸紧身衣
all-in-one bikini 一片式比基尼泳衣
all-in-one body smoother 全合一内衣
all-in-one built-up neckline 一片式高领口
all-in-one collar 一体领
all-in-one cuff 一片式袖头
all-in-one flat button clamp 整套平纽扣夹紧器
all-in-one garment 上下连装
all-in-one girdle 胸罩腰带
all-in-one in-seam pocket 袋布连在面料上的内缝插袋
all-in-one petal 一片式花瓣袖
all in one piece materal 整料
all-in-one sleeve 妇女紧身袖
all-in-one vest 婴儿连体背心
all-in-one wrap skirt 一片式裹裙
all-in printing process 全料印花法
all-in vat printing process 还原染料全料印花法
all leather shoes 全皮革鞋
all parties to the contract 合同各方
alloa (wheeling) 阿洛粗绒线
alloa yarn 阿洛粗绒线
allocate order 调拨
allocation 分配
allocation of quota 配额分配
all-occasion dress 通用服

allonge（perruque） ［法］阿朗杰假发
allover 满地,全幅；满地花纹；满地花纹花边
allover apron 连身围裙,套头围裙
allover banded lace 满地花纹狭花边
allover color 满地着色
allover design 满地花纹图案,满地印花图案
allover effect 满地花纹效应
allover embroidery 满地绣
allover feeding device 综合送料装置
allover（lace） 满地花纹花边
allover length of body 身长
allover net 满地花纹网眼
allover pattern 满地花
allover print 满地印花
allowance 放缝,放余量；加工余量；折扣；容差；容许数
allowance for finish 精加工余量
alloy buckle 合金带扣
alloy button 合金扣
alloy fiber 多组分纤维,合金型纤维
all plastic sandals 全塑凉鞋
all-polyester cloth 纯涤布
all purpose 通用的
all-purpose coat 通用外套,多用途外套
all-purpose cut 多用途发型
all-purpose detergent 通用洗涤剂
all ramie sheer 爽丽纱
all risk 综合险
all rounder 狗圈式立领；狗圈式项链
all-round fiber 全能纤维
all-round hemming operation 四周缝边法
all-round pattern 满地图案；满地花
all-round performance 综合性能
all-round pleated skirt 百褶裙
all-round sewing 全面缝
all-season coat 四季外套,四季大衣,轻便大衣（欧美）
all-season dress 四季服
all-season fabric 四季料
all sheer pantyhose 透明尼龙连裤袜
all silk goods 全真丝织物
all-skin rayon fiber 全皮层黏胶纤维
all snap brim 弯檐帽
all squared 结清
all the vogue 最新流行
all weather boots 晴雨靴
all weather coat 晴雨外套,全天候外套,风雨衣

all weather fabrics 全天候织物
all weather garment 全天候服装
all weather gear 全天候军服
all weather raincoat 晴雨外套（御寒用）,全天候外套
all whit cordel handkerchief 全白粗棱手帕
all width 全幅宽,全幅
all wool 全毛,纯毛
all-wool cloth 全毛织物
all-wool fine fingering yarn 全毛细绒线
all-wool fingering yarn 全毛绒线
all wool gabardine 全毛华达呢
all-wool hose 全羊毛袜
all-wool machine knitting yarn 全毛针织绒线
all-wool serge 全毛哔叽
all-wool worsted flannel 全毛啥味呢
all-wool yarn 全羊毛纱,纯羊毛纱,纯毛纱
all year round 四季可穿
all year round coat 四季大衣
alman coat jacket （紧身衣外）男式上衣
almandite garnet 贵石榴石
almanesque 棉织品（阿根廷称谓）
alma silk 阿尔玛绸
almax 氧化铝纤维
almond 杏仁色
almond brown 杏仁棕（浅灰棕色）
almond cream 杏仁乳白色
almond green 杏叶绿（浅灰绿色）
almoner 钱包,小皮革钱袋
almuce 头巾；皮兜帽,风帽,阿尔目斯帽；兜帽斗篷
alneestloni 异面花纹斜纹毯
alnein 桤树金黄色染料
alni mayini 美国部落黑色毯
aloe fiber 芦荟属叶纤维,龙舌兰纤维
aloe lace 芦荟纤维花边
aloe thread embroidery 芦荟线绣
aloha shirt 香港衫,夏威夷衫,阿罗哈衬衫,大花纹香港衫,夏威夷开领衬衫
alongside printing 共同印花
alost lace 比利时棱结花边
alpaca 阿尔帕卡毛,羊驼毛；羊驼毛织物,驼羊呢
alpaca cloth 羊驼呢,阿尔帕卡织物
alpaca coating 羊驼毛上衣料
alpaca crepe 阿尔帕卡绉；仿羊驼毛绉；醋酯及黏胶纤维混纺绉
alpaca fabric 阿尔帕卡奥尔良呢,羊驼织物,阿尔帕卡织物,羊驼毛交织物

alpaca fancy suiting 阿尔帕卡花呢
alpaca fiber 阿尔帕卡毛
alpaca knitwear 羊驼毛衫
alpaca lining 阿尔帕卡里子呢
alpaca lustre 阿尔帕卡有光呢
alpaca mixture 阿尔帕卡交织呢
alpaca orleans 阿尔帕卡奥尔良呢
alpaca rayon 阿尔帕卡黏胶纤维绸
alpaca stitch fabric 羊驼毛双面针织物
alpaca wool 阿尔帕卡毛,驼羊毛,羊驼毛,秘鲁羊驼毛
alpaca yarn 阿尔帕卡纱
alpacianos 阿尔帕斯埃诺斯纯棉织物
alpago 19世纪中期英国厚重缎
alpargata 麻便鞋,帆布轻便鞋,西班牙便鞋,草(绳)底帆布鞋
alpen hat 高山帽
alpen stock 铁头登山杖
alphabet line 字母轮廓线
alpine 登山软帽
alpine boots 及踝登山靴
alpine cap 登山帽
alpine hat 阿尔卑斯山帽,蒂罗尔帽,登山帽
alpine jacket 登山夹克,奥地利齐腰夹克衫
alpine rose 淡玫瑰红色,阿尔卑斯玫瑰红色
alsatian madder 茜草染料
A-l silk 优质丝
alster collar 阿乐斯特领
altar cloth 比斯细薄亚麻布
altar lace 挖花花边
alter 改做(衣服);改动(设计)
alterant dye 变色染料
alteration (衣、鞋)修改;改动
alteration hand 修改员
alteration of paper pattern and marker 修改纸样和唛架
alter blouse collar 改衬衫领
alter design 修改设计
altering garment 修改衣服
alternate accordion 交错浮线提花
alternate bundle(direct) 互插排料
alternate bundle same direct 单向排料
alternate check 套格花纹
alternate half hitch knot 左右结
alternate long and short dashes line 点划线
alternate long and two short dashes line 双点划线
alternate open washer 交流冲洗式平洗机
alternate stripe 交替条纹,交互条纹,宽窄相间的直条纹,阔狭相间的直条纹

alternating-color effect 变色效应
alternating feed 交替送料,往复送料
alternating presser 交替压脚
alternating stem stitch 交替包梗绣,绕梗绣
alternating tacking stitch 交替粗缝线迹,交替疏缝线迹,交替假缝线迹
alternation 交替;修改(服装)
alternation basting 短而松的假缝(定针)
alternation hand 修改员
alternation of paper pattern 修改纸样
alternative basting stitch 交替粗缝线迹,交替疏缝线迹,交替假缝线迹
alternative fashion 另类时装,非主流时装
alternative stripe 交替条纹,互交条纹
alternative tacking stitch 交替粗缝线迹,交替疏缝线迹,交替假缝线迹
alternative type one-way fasteners 变化型单向拉链
alternative weft knitted fabric 纬编交织针织物
altessee 英国阿尔特西绉
altex 氧化铝纤维
altitude suit 高空飞行服
altobasseo 高低绒头丝绒
alumina fiber, aluminium oxide fiber 氧化铝纤维
aluminium 铝白色(银白色)
aluminized coat 镀铝上衣
aluminized coveralls 镀铝工作服
aluminized fabric 镀铝织物
aluminized Mylar 镀铝聚酯薄膜
aluminosilicate (glass) fiber 硅铝酸盐玻璃纤维
aluminum 铝;铝白色(银白色)
aluminum borosilicate fiber 硼硅酸铝纤维
aluminum fiber 铝纤维
aluminum-foil-covered asbestos suit 覆铝箔的石棉衣
aluminum foil printing 铝箔印花
aluminum graphite fiber 铝-石墨纤维
aluminum oxide fiber 氧化铝纤维
aluminum ruler (绘纸用)铝尺
aluminum silicate fiber 硅酸铝纤维,陶瓷棉
aluminum silicate refractory fiber 硅酸铝耐高温纤维
aluminum zipper 铝拉链
alumnat cloth 阿纶纳特黑呢;捷克斯洛伐克教士袍黑呢
alveolate eyelet rib fabric 蜂巢形网眼罗纹织物

alwan 西藏及印巴地区产平纹羊绒呢
always quality 质量可靠
amabouk 北爱尔兰半漂粗平纹亚麻布
Amadaure cotton, Amadowry cotton, Amadure cotton 埃及阿马杜尔棉
amadis sleeve 侠客袖,肘部钉扣的泡泡袖(19 世纪 30~50 年代)
amah's suit 妈祖装
A-Mak button 再生壳扣,仿壳扣
Amalgamated Clothing and Textile Workers Union 服装及服装工人联合会
amalgamated silver wire cloth 汞齐化银丝布
amamee 阿马米印花细平布
Aman 阿曼粗布;中东蓝色平纹织物
amanouri cotton 中东阿马奴利优级棉
amar 花莎丽(印度)
amara 阿玛拉(日本制的一种仿麂皮)
amárah 摩尔男式白色大头巾
amaranth 深紫,紫红
amaranth purple 深青莲,深紫色,苋紫色
amatorial look 色情款式
Amazon 亚马逊毛呢,亚马逊经面毛呢;妇女毛质骑马服,马裙;鹦鹉蓝
Amazon collar 亚马逊立领
Amazon corsage 亚马逊紧身胸衣
Amazones 亚马孙粗纺呢(南美洲)
amazonite 天河石,铜绿长石
Amazon look 亚马逊款式,亚马孙风貌
Amazon stone 天河石,铜绿长石
ambari hemp 熟红麻,槿麻,泽麻,印度络麻,野麻,洋麻
Ambassador 美国大使棉
ambassador style 大使服式
amber 琥珀;琥珀色,褐色
amber brown 琥珀褐
amber green 琥珀绿
ambertee 17 世纪印度印花棉布
amberty 印度印花棉布
amber yellow 琥珀色,琥珀黄
ambient air 环境空气
ambient conditions 环境条件
ambient temperature 环境温度
ambisextrous 两性服装
ambisextrous clothing 中性服装,男女不分的服装,无性别装
ambisextrous garment 两性服装
ambre 琥珀;琥珀色,褐色
ambrosia 豚草绿
amburgos 阿姆伯戈斯布;重浆白衬衫布
amekican Indian dress 印第安裙装

Ame Lia Bloomer 布卢默(美国著名戏剧服装设计师)
amens 阿曼斯精纺毛织物
American 美国时装
americana 粗制棉布床单
American aloe 美洲龙舌兰纤维
American apparel Contractors association 美国服装协会
American apparel Manufacturer association (AamA) 美国服装制造业协会
American apparel Manufacturers association 美国制衣商协会
American armhole 美式袖窿
American association for Textile Technology 美国纺织工艺协会
American association of Exporters and importers 美国进出口商协会
American association of Textile Chemists and Colorists 美国纺织化学家和染色家协会
American back 美式背样
American badger 美洲獾毛皮
American beauty 美国蔷薇红
American bottom 美式下装
American broadtail 美国大尾羔羊毛皮
American broadtail processed lamb 美国羔羊皮
American buskin 美国高筒靴,美国厚底半高筒靴
American Buyers Federation (ABF) 美国采购者联合会
American cascot 美国便装款式
American casual 美式便装
American classics 美国 20 世纪 40~50 年代时装
American Cloak and Suit Manufacturers association (acSMA) 美国外套、套装制造商协会
American cloth 仿皮布,彩色油布,漆布
American coat 美式单排扣长大衣
American College for applied arts 美国应用艺术学院
American continental 美国大陆装;美国风格
American continental style 美国风格
American costume 美国民俗服
American cotton 美国棉,美国棉花,美棉
American cotton Manufacturers' Institute 美国棉业协会
American cotton staples 美国棉花分类
American Designer Collection 美国设计

师作品展示会
American dream 美国梦
American-Egytian cotton 美国埃及种棉
American fashion 美国流行服装款式；美国时装
American Fiber,Textile,Apparel Coalition 美国纤维、纺织品、服装联盟
American Fiber Manufactures association 美国纤维制造者协会
American Flock association 美国植绒协会
American funny print 美国怪异印花，美国趣味印花
American Fur Merchants' association (AFMA) 美国毛皮商协会
American grey cloth 花旗布
American heel 美式袜跟
American heel apparatus 美式袜跟装置
American hemp 美洲龙兰舌纤维，剑麻
American Indian 美国印第安时尚
American Indian bag 印第安袋
American Indian belt 印第安皮带
American Indian blanket 美国印第安地毯
American Indian bonnet 印第安帽
American Indian design 美国印第安图案
American Indian dress 美国印第安裙装
American Indian look 美洲印第安风貌；印第安服装风格
American Indian necklace 印第安项链
American Indian print 美洲印第安印花，纳瓦霍印花
American Institute of Laundering 美国洗衣业协会
American jute 麻，青麻，芙蓉麻，天津麻
American lamb 美国羔羊皮
American long-staple upland cotton 美国长绒陆地棉
American look 美国款式，美国风，美式风格
American Marteen 美国貂皮
American merino 美国美利奴羊毛
American mink 美洲黄鼠狼皮
American mode 美国流行式样[款式]
American mohair 美国马海毛
American National Standard 美国国家标准
American natural 美国自然型西装（自然肩，整体细长）
American neckcloth 美式领带、领巾
American Needlepoint Guild 美国刺绣协会
American nostalgia 美国思乡型

Americano assilia 本色平布（东非）；美国、欧洲棉床单坯布
Americano gamti 本色粗平布（东非）；印度深色粗棉床单布
Americano marduff 本色粗斜纹布（东非）；欧美斜纹厚棉坯布
American opossum 美洲负鼠皮
American oriental 美国东方风格地毯
American oriental rug 美制东方地毯
Americano ulayiti 本色平布（东非）；美国、欧洲棉床单坯布
American peeler cotton 美国皮勒棉
American Printed Fabrics Council 美国印花协会
American Reusable Textile association 美国再生纺织品协会
American ring tail 美国产圈形尾毛
Americans 花旗布
American sable 北美黑貂皮，加拿大黑貂皮
American set 美式枕头
American sheeting 美国平纹棉布
American short-staple upland cotton 美国短绒陆地棉
American shoulder 美式肩，削肩型
American shoulders 美式加垫肩的肩部
American Society for Testing and Materials (ASTM) 美国材料试验学会；美国材料与试验协会
American Society of Knitting Technologists 美国针织工艺师学会
American spun yarn 美式精纺毛纱
American standarad (as) 美国标准
American style 美国风格；美国款式
American traditional 美国传统服装
American trench 美国堑壕外套
American trousers 美式男裤
American twill 美国二上二下斜纹布
American upland cotton 美国陆地棉
American vest 美式背心
amerikani 原色棉床单布
amertis 密织棉布
amethyst 紫水晶（色）
amethyst orchid 水晶兰花色
amethyst violet 紫晶色，青紫色
amianthus,amiantus 高级石棉，石绒
AMICA 《女友》（意）
amicable allowance 友好让价
amice 兜帽，皮帽；皮头巾，天主教披巾，阿米丝头巾，(僧侣用)长方形白麻布围

巾;(僧侣)带兜帽的衬皮披肩
amicor fiber 抗菌纤维
amidama [日]毛线球
amidated cotton 酰胺化棉
amido color 酰胺染料
amiens 亚眠斜纹细呢
amiestes,amiesties 印度棉布
amilan 阿米纶(锦纶,商名,日本)
aminized cotton 胺化棉
aminoazobenzene dye 氨基偶氮苯染料
aminoethylated cotton 氨乙基化棉
amiray 苎麻
amiray ramie 苎麻
amish costume 阿米喜服
amlikar 绣花羊绒披巾
ammonia liquor finished fabric 液氨整理织物
ammonia mercerizing 液氨丝光
amoer 图尔横棱绸
amont 海豹皮束腰外衣
amorgis 古希腊紫色亚麻细布
amorphous fiber 非晶态纤维
amorphous filament 非晶态丝
amosite 铁石棉,长纤维石棉
amosite asbestos 铁石棉,长纤维石棉
amount 全额;金额;数额
amount deposited 存款金额
amount insured 保险金额
amount of contract 合同金额
amount of deposit 存款金额
amount of fullness 鼓起度
amount of loan 贷款金额
amour 缎纹亚麻台布
amphibole (asbestos) 闪石石棉
amphimalla 粗纺双面绒布
amphimallam 粗纺双面绒布
amphitapus 粗纺双面绒布
amphoteric ion-exchange fiber 两性离子交换纤维
ample line 宽松型;宽松线形;宽松轮廓线
amplifying unit 放大机构
amplitude 幅度
amritsar 印度粗羊毛;波斯花纹粗绒头大地毯;仿开司米披巾
amsterdam 阿姆斯特丹针绣花边
amuce 中世纪披肩;毛皮里小斗篷;教士的毛皮兜帽
amulet 护身符
amundsen 阿蒙增绉纹呢
amuno 阿穆诺羊毛防蛀剂

amunzen 阿蒙增绉纹呢
amylodextrin 淀粉糊精
amy robsart satin 19世纪白地金银线花缎
anabasse 蓝白条棉毛毯;法国条子毛毯
anacasta, anacaste, anacosta, anacoste, anacostia, anacote 精纺光面哔叽,阿纳科斯特斜纹精纺呢
anadem 束发带;花冠,花环
anal groove [anal cleft] 肛门沟
analog chromatogram 模拟色谱(图)
analogous color 类似色
analogous harmony 类比调和
analogy color harmony 类似配色,相似配色
analysis of fabric 织物分析
analytical error 分析误差
analyzing 分析
ananas hemp 菠萝麻叶纤维
ananas knitting 菠萝组织编织
anapes 17世纪意大利高级纬起毛织物
anaphe silk 非洲阿纳菲野蚕丝
anap-open bottom 开裆裤
anarasi sari 阿纳拉西沙丽织物;孟加拉湾手织丝棉织物
anatolian 阿纳托利亚小毛毯;土耳其阿纳托利亚中细毛
anatolian silk 阿纳托利亚生丝
anatomical figure type 解剖用体型类型
anatomical landmarks 人体计测点
anatomy 解剖学
anaxarides 阿纳克塞利迪斯裤(伊朗)
ancelia 安萨利阿布,安萨利亚棉毛小花纹布
anchali 印度宽布带,阔带(东南亚)
anchor buckle 锚形纽环
anchor button 锚形纽扣,锚纽
anchored pants 踩脚裤,踩脚针织裤,鞋袢裤
anchored slacks 紧身裤
anchor line 缆绳,粗绳
anchor stitch 定位线迹
ancient madder 茜草印花小花纹布,仿茜草印花小花纹布
ancient style 古代风格
ancube 比利时安丘贝羊毛地毯
andalusian casaque 安达卢西亚女式傍晚束腰外衣
Andalusian fabric 安达卢西亚精纺花式毛织物;高级浮纹绸
Andalusians 安达卢西亚优质精纺衣料呢
Andalusian wool 安达卢西亚美利奴毛
Andalusian yarn 安达卢西亚四股细绒线

andalusite 红柱石
andean shift 直筒绣边安迪衬衣,安迪衬衣
anderson 苏格兰安德森高级格子布
Andes cotton 秘鲁棉
Andong moss crepe 安东料
An-dong silk suiting 安东呢
andras 安德拉斯手织细平布
Andrea Odicini 安德列亚·奥迪奇尼(意大利时装设计师)
André Courreges 安德烈·库雷热(1923~,法国时装设计师)
Andre Laug 安德烈·劳格(意大利时装设计师)
André Oliver 安德烈·奥利弗(1923~,法国时装设计师,皮尔·卡丹公司男女成衣设计师)
Andrews, Andrews cotton 安德鲁斯棉,超长海岛棉
androgynous 男女通用的服装,男女不分的服装;男性化女装
androgynous garment 两性服装
androgynous look 男女通用的服装,两性款式,无性服装风格
anga 印度伊斯兰教徒白色或有色棉长外套
angarka 北印度安嘎卡袍
Angela Cummings 安吉拉·卡明斯(美国珠宝设计师)
angel blouse 天使女衫,安琪儿女衫
angel blue 天使蓝
angel dress 天使连衣裙
angel lace 经编天使花边
Angelo tarlazzi 安杰洛·塔拉济(法国时装设计师)
angel overskirt 天使罩裙
angel's hair 金银丝(织造用);玻璃丝;假发
angel skin 天使缎
angel sleeve 安琪儿袖,天使袖,宽大衣袖
angel top 天使上衣
angelus cap 安琪儿帽,安琪勒斯帽,农夫帽
anghaka 安加克衫(印度)
angharka 印度伊斯兰男教徒左门襟及腰短外套
angiya 南印度伊斯兰女教徒短袖紧身胸衣
anglaise 素色哔叽
anglastique 安格拉斯提奎呢;毛锦弹力斜纹呢
angled bottom 有角度的裤脚,斜角裤脚
angled cuff 斜袖头,人字袖口

angled flap pocket 有袋盖的斜袋
angled needling 斜向针刺
angled pocket 斜袋
angled shawl collar 男用尖角丝瓜领
angled slit 斜向切缝
angled stitch 斜角形线迹
angle-fronted coat 晨礼服式男外套;学院外套
angle gauge 角度计;角规
angle of the mouth 口角(口唇联合)
angle pocket 斜袋,插袋
angler's vest 钓鱼背心(多口袋的背心)
anglesea hat 平檐高筒男帽
anglesey wool 安格西尔羊毛
angle sleeve 斜袖,天使袖,宽大衣袖
angle stitch 山形线迹,斜角线迹,角形线迹
anglet 帽边装饰;白色花边带;花边金属附属物
angleterre 昂格勒泰针绣花边;法国高级整理塔夫绸
angleterre edge 针绣花边的编结边
anglicanum 挖花工艺;刺绣工艺
Anglo-Greek bodice 英裔希腊女紧身胸衣
Anglo-merino 盎格鲁美利奴呢
Anglo-Saxon embroidery 盎格鲁-撒克逊刺绣
Anglo-Saxon embroidery 安格鲁-撒克逊绣
Anglo-Swiss muslin 仿瑞士彩点薄纱
angochka 印度粗棉纱手织布
angola 安哥拉棉毛呢
angola brocade 英国安哥拉花缎
angola cloth 红色斜纹布;安哥拉厚斜纹起毛大衣呢;安哥拉绣花布
angola mending 安哥拉缝补线
angora 安哥拉山羊毛;安哥拉兔毛;安哥拉披肩
angora/acrylic knitting yarn 兔毛腈纶混纺针织绒线
angora cape 安哥拉披肩
angora cashmere 马海毛斜纹薄织物
angora cloth 安哥拉毛织物,马海毛交织呢
angora fabric 安哥拉毛织物
angora gloves 兔毛手套
angora layette 安哥拉精纺毛纺
angora mohair 安哥拉马海毛
angora rabbit hair 安哥拉兔毛
angora rabbit hair scarf 安哥拉兔毛围巾
angora scarf 兔毛围巾
angora sweater 兔毛衫
angora-type core yarn 安哥拉毛型包芯纱

angora union　马海毛交织呢
angora wool　19世纪英国安哥拉毛毯用毛
angora wool muffler　羊兔毛混纺围巾
angora wool scarf　兔毛围巾
angora yarn　安哥拉兔毛(混纺)纱
angoreen　安哥拉兔毛纱
Anguilla cotton　安圭拉棉(产于西印度安圭拉岛)
angular 2 hole button　角形双眼纽扣
angular binder　角滚边器
angular check　菱形花纹
angularity　生硬难看(式样、衣着等)
angular perspective　角透视
angular stitch　角形线迹
angular stitch machine　角形线迹缝纫机
anguler twill　角度斜纹
angulus inferior scapulae point　肩胛骨下角点
anidex fiber　阿尼迪克斯纤维
anidex fiber　聚丙烯酸酯弹性纤维
aniline black　阿尼林黑
aniline black dyeing　精元染色
aniline black dye sheeting　精元布
aniline dye　苯胺染料
aniline finish　苯胺整饰(皮革)
aniline leather shoes　苯胺革鞋
aniline resist printing　精元防染印花
anilo　阿尼络韧皮纤维
animal bone button　动物骨纽扣
animal dye　动物染料
animal fiber　动物纤维
animal hair　动物毛,兽毛
animal hair fiber　动物毛纤维
animal hair-like fiber　兽毛状纤维(末端呈锥状的纤维)
animal horn button　动物角质纽扣,动物骨纽扣
animalization　纤维动物质化
animalized cotton　动物质化棉,羊毛化棉
animalized fiber　动物质化纤维,羊毛化纤维
animalized viscose rayon　动物质化黏胶纤维
animal pattern　兽皮图案,动物图案
animal print　动物图案印花
animal skin covering　狮豹披肩
animal skin design　兽皮纹图案,动物皮图案
animation button　卡通动物扣
anion-exchange fiber　阴离子交换纤维
anionic fiber　负离子纤维(电子石纤维)
anise camphor　茴香樟脑

anisole blue　茴香醚蓝
anisotropic model filament　各向异性原型丝
anisotropy　各向异性
anjengo　双股椰壳纤维纱
ankel　裤脚口贴边;罗纹袖口;翻口短袜
ankh　安克十字形
ankh necklace　T字架项链,十字架安克项链
ankh ring　T字架戒指,十字架安克戒指
ankle　踝,踝节部,脚脖,脚踝
ankle band　护踝带
ankle banded pants　束口裤,束脚裤,束踝裤;束脚袜
ankle bone　踝骨
ankle boots　高帮鞋,短筒靴,及踝短靴
ankle boot with stiletto　特细高跟女短靴
ankle bracelet　脚镯,踝环,踝镯
ankle breadth　踝宽
ankle-deep　深至踝部
ankle design　袜子踝部花纹
ankle girht　踝围
ankle guard　护踝
ankle height　踝高
ankle high socks　及踝短袜;妇女套袜;婴儿短袜
ankle-jacks　男式及踝短靴
ankle jojnt　踝关节
ankle-length　踝长,及踝长
ankle-length drawer　脚踝长内裤
ankle length garment　及踝服装
ankle level　踝节线
ankle pants　束口裤,束脚裤,束踝裤
ankle socks(hose)　齐踝短袜,及踝短袜,翻口短袜;袜套,套袜
ankle splicing　袜子高跟加固
ankle strap　系踝鞋带,脚腕鞋带;凉鞋
ankle strapped shoes　系踝鞋,踝带鞋
ankle-strap sandals　裹踝凉鞋
ankle strap shoes　系踝鞋
ankle supporter　护踝
anklet boots　短靴,高帮鞋
anklet bracelet　踝镯
ankle tied pants　束口裤,束脚裤
anklets　女短袜,及踝短袜,翻口短袜;套袜;扣带鞋;脚镯,足腰装饰物;罗纹口(做袖口、脚口)
anklet socks　套袜,翻口短袜,袜套
anklet strap shoes　扣带鞋
ankle warmer　袜套
ankle-wrap sandal　系踝凉鞋
anle batiste　安乐纺

annamese band turban 安南包头巾,越南缠裹式棉头巾
Anna Potok 安娜·波托克(意大利时装设计师)
Anna Sui 安娜·苏
annatto 胭脂树红
Ann Buck 安·巴克(英国时装设计师)
Anne Fogarty 安妮·福格蒂(1919～1981,美国时装设计师)
Anne Hall look 安妮·霍尔风貌,安妮·霍尔款式
Anne Klein 安妮·克莱茵(1923～1974,美国时装设计师)
Anne-Marie Beretta 安妮·玛丽·贝雷塔(法国时装设计师)
Ann Fogarty 安·福格蒂(美国时装设计师)
Annie Hall look 安妮·霍尔风貌
Annie Oakley costume 安妮·奥克莉装束
annotto 胭脂树红(浅橙色)
Ann Roth 安·路斯(美国著名戏剧服装设计师)
annual 年(度),年终
annual accounts 年度结算
annual balance 年终结余
annual income 岁入
annual interest 年息
annum interest rate 年利率
anomalous color vision 异常色觉
anomalous fading 反常褪色
anomalous structure 反常结构
anonymity 无个性特征
anorak 带风帽短派克大衣,皮猴;爱·诺瑞克外套;滑雪服,溜冰服,登山服;带风帽夹克衫
anorak blouse 阿诺瑞克罩衫
anorak dress 爱·诺瑞克外套;滑雪服,溜冰服;带风帽夹克衫;爱诺瑞克式连衣裙
anorak fabric 厚夹克织物,风雪大衣织物
anorak jacket 防风雪夹克,防寒短上衣,防寒夹克衫,滑雪溜冰衫
anqular stitch 角形线迹
ansate cross T字架圆环项链
anslet 男式超短紧身上衣
antelope 羚羊皮;羚羊皮包
antelope bag 羚羊皮包
antelope finished lambskin 仿羚羊皮
antelope finish(ed) suede 仿羚羊皮
antelope leather 羚羊皮革
anterior armpit point (aap) 腋窝前点

anterior bust arc 前胸弧
anterior chest width 前胸宽
anterior neck length 领前长
antery 埃及和土耳其男式及腰或膝背心
anthemion 花丛状装饰
antherea silk 蓖麻蚕丝,(印度)柞蚕丝
antherea yamamal silk 天蚕丝(日本柞蚕丝;山蚕丝)
Anthony Mascolo 安东尼·马斯科罗(意大利美容师)
anthraquinone blue 蒽醌蓝
anthraquinone dye 蒽醌染料
anthraquinone vat dye 蒽醌还原染料
anthrasole 蒽台素染料(可溶性还原染料,商名,德国)
anthrasol resist printing 溶蒽素防染印花
anthrone dye 蒽酮染料
anthropometer 身长计,高度计测器
anthropometric measurement 人体测量尺寸
anthropometric shadow scanner 人体外形扫描器
anthropometric study 人体测量学
anthropometrist 人体测量学者,人体测量学家
anthropometry 人体测量学
anthropotomy 人体解剖学
anthroquinone dyes 蒽醌染料
anti-allergic treatment 抗过敏处理
anti-art 反传统艺术,非传统艺术
anti-bacterial activated carbon fiber 抗菌性活性碳纤维
anti-bacterial deodorant fiber 抗菌防臭纤维
anti-bacterial fiber 抗菌纤维
anti-bacterial finish 抗菌整理,抑菌整理
anti-bacterial finishing agent 抗菌防臭整理剂
anti-bacterial property 抗菌性
anti-bactirial and purifying finish 抗菌防臭整理
anti-bagging finish 防拱胀整理
anti-ballistic vest 防弹背心
antibiosis 抗菌性
antibiotic fabric 抗菌织物
antibiotic finish 抗菌整理
anti-blackout suit 抗荷服
anti-cling finish 防紧贴整理
anticockle treatment 防皱整理(羊毛针织物)
anti-cold clothing 防寒服装,保暖服装
anti-conformisme 反传统裁剪的制衣法
anti-contact 绝缘布手套

anti-contact cloth gloves 绝缘布手套
anti-couture 非缝制的
anti-crease 抗皱
anti-crease cotton fabric 抗皱棉织物
anti-crease finish 防皱整理
antics 古怪姿势
anti-curl 防卷边
anti-dumping duty 反倾销税
anti-dusting 抗尘
antielectrostatic fiber 抗静电纤维
antiestablishment fashion 反体制服饰
anti-exposure suit 防曝服
anti-fashion 反流行,反时装(反欧洲传统服装风格的时装潮流)
antifelting 防毡缩整理织物;防缩绒
antifelting finishing fabric 防毡缩整理织物
antifelting treatment 防毡缩整理
antiflaming finish 阻燃整理
antiformism 反传统形式派,反形式派
anti-freezing 防冻
antifrosting pocess 防局部脱色处理
antifume finish 抗烟[气]整理,抗酸气整理
anti-fungal fiber 耐菌纤维,抗[真]菌纤维
anti-fuzz 防起毛
anti-fuzz fiber 抗起毛纤维
anti-fuzzing finish 防起毛整理
anti-G 抗超重飞行服
anti-gas cape 防毒斗篷
anti-gas clothes 防毒衣
anti-glare sewing light 遮光缝纫机灯
anti-gravity suit 抗超重飞行服
antigropelos 防水绑腿
anti-G suit 抗超重飞行衣
anti-hazard wear 安全服,抗爆服
anti-hemorrhagic fiber 止血纤维
anti-hole melt finishing 抗熔孔整理
anti-hole melt property 抗熔孔性
anti-hypothermia bag 抗低温睡袋
anti-impact property 抗冲击性能
anti-infection apparel 防感染服
anti-insect finishing agent 防虫整理剂
anti-laddering 防抽丝整理
antimacassar 防污方巾
antimagnetic 防磁
anti-matting finish 防退光整理
anti-melt finish 抗熔融整理
anti-melt finishing 抗熔整理剂
antimicrobial fiber 抗微生物纤维
antimicrobial finish 抗菌整理,抗微生物整理

antimicrobiotic finish 抗菌整理,抗微生物整理
anti-microbiral fiber 抗细菌纤维
anti-microwave fiber 防微波辐射纤维,防电磁辐射纤维
antimilling of wool fabrics 毛织物防缩整理
anti-mines shoes 防地雷鞋
antimist cloth 透明织物;防雾织物
anti-mold agent 防腐剂,防霉剂
anti-mosquito finished fabric 防蚊整理织物
anti-moth 防蛀
anti-mycotic 抗霉菌
anti-mycotic finish 抗霉菌整理
anti-napping finishing 抗起球整理
anti-neutron radiation fiber 防中子辐射纤维
anti-odour finish 防气味整理;防臭整理
anti-penetration fiber 抗渗透纤维
anti-perspiration 防汗渍
anti-perspiration finish 防汗渍整理
antipicking finish (呢面)清洁整理
anti-pill 不起球
anti-pill fabric 抗起球织物
anti-pill fiber 抗起球纤维
anti-pilling finishing 抗起球整理,防起球整理
anti-pilling tendency 抗起毛起球性
anti pollution 防污染
antique 仿古织物
antique-attic look 怀古风貌
antique bodice 女式仿古低腰紧身胸衣
antique bronze 古铜式(暗黄棕色)
antique casual 古风便装
antique cutwork 挖花
antique design 仿古设计
antique fashion 仿古款式
antique finish 仿古整理
antique finished carpet 仿古(整理)地毯
antique gold 古金色
antique ivory 古象牙色
antique lace 粗亚麻线梭结花边,仿古花边
antique leather 仿古皮革,古风皮革
antique like 仿古织物
antique look 古风款式,古代风貌
antique oriental rug 仿古东方地毯
antique rug 仿古地毯
antique satin 疙瘩双面缎(正面有纬向疙瘩效应,反面如古代绸缎)
antique table en cabochon 古穹面形(珠宝)
antique taffeta 仿古塔夫绸

antique velvet 仿古丝绒
antiquing 仿旧整理
anti-radiation fiber 防辐射纤维
anti-radio activity 抗放射性
anti-ravel spray 喷雾式防脱散剂
anti-reversing device 防倒缝纫装置
anti-rheumatic fiber 防风湿纤维
anti-rot finish 防腐整理,防菌整理
anti-rot treatment 防腐整理,防菌整理
anti-sag 防松垂
anti-septic cotton 防腐棉,清毒棉
anti-septic dressing 外科包扎布,绷带
anti-septic finish[ing] 防腐整理
anti-septic gauze 清毒纱布
anti-shrink 防缩,耐缩
anti-shrink finish[ing] 防缩整理
anti-skid shoes 防滑鞋
anti-slip finish 防滑移整理
anti-smell finish 防臭整理
anti-snag finish 防起毛整理,防擦毛整理;防钩丝整理
anti-snore mask 防打鼾面罩
anti-snore pajamas 防打鼾睡衣裤
anti-soil fiber 抗污纤维,防污纤维
anti-soiling finish 防污整理
anti-soil-redeposition 防尘污再沾着性
anti-stain finish 防污整理
anti-static 防静电性,抗静电性
anti-static agent 抗静电剂
anti-static and dust-proof clothing 抗静电防尘服
anti-static conductivity fiber 抗静电纤维
anti-static effect 抗静电效应
anti-static fabric 抗静电织物
anti-static fiber 抗静电纤维
anti-static finish 抗静电整理
anti-static finish fabric 防静电整理织物
anti-static gloves 防静电手套
anti-static nylon gloves 尼龙防静电手套
anti-static property 抗静电性
anti-static treatment 抗静电处理
anti-static uniform 防静电服
anti-static wool knitwear 抗静电毛衫
anti-sunburn 防日晒
anti-tarnish agent 防黯剂
anti-thrombocytic fiber 血小板分离用纤维,血浆分离用纤维
anti-thrombotic stockings 抗血栓袜子
anti-tickle finish 防痒整理
anti-tinea cloth 抗癣布

anti-toxic fabric 防毒织物
anti-toxic filtration nonwoven fabric 防毒过滤非织造布
anti-ultraviolet-ray finishing 抗紫外线整理
anti-X-ray fiber 抗 X 射线纤维,防 X 射线纤维
anti-yellowing 防泛黄
anti-yellowing finish 防泛黄整理
antler 鹿角色
antoinette fichu 安东尼特三角形长披肩
Antonio Canovas Del Castillo 安东尼奥·卡诺瓦斯·德尔·卡斯蒂洛(法国时装设计师)
Antonio Cerruti 安东尼奥·塞鲁蒂(意大利服装设计师)
Antonio Lopez 安东尼奥·洛佩斯(美国时装设计师)
antron 安特纶(三叶形截面的锦纶 66 纤维,商名,美国杜邦)
antwerp blue 深蓝
antwerp lace 安特卫普花边
antwerp pot lace 安特卫普瓶花花边
Ao dal, aosal 奥黛,越南民族套装
aotu cloque 凹凸绉
aoudad 羱羊皮
aoudad sheep 羱羊皮
apache 阿帕希,无赖汉领巾
apache bag 印第安袋,嬉皮士袋;阿帕希提包
apache scarf 阿帕希围巾,无赖汉领巾,无赖汉头巾
apache shirt 阿帕希 V 领套头衫
apache style 阿帕希风格,阿帕希款式,无赖汉款式
a pair of braces 一副背带
a pair of gloves 一双(副)手套
a person in charge 经办人(负责人)
aperture fabric 多孔织物
ape skin fabric 充猴皮绒织物
ape skin pattern 仿猴皮绒织物
apex 祭司帽;胸高点
apex of darts 褶尖点
apishamore 马鞍座毯
apishemean 马鞍座毯
aplied printing 直接印花
apochromatism 复消色差
apocynaceae fibers 野麻
apocynum 罗布麻,红野麻,夹竹桃麻,茶叶花,茶棵子

apodesme 古希腊束胸
apolda 德国印花粗毛披巾,印花羊毛披巾
apollo cap 阿波罗帽(毛毡做的鸭舌帽)
apollo's knots 阿波罗发结
aposematic coloration 警戒色
apou 中国夏布,苎麻布
appard 时装;服装业,服饰企业,服装产业(旧称)
apparel 衣服,衣着;穿着;服装,服饰;外观
apparel & accessories 服饰
apparel accessories 服装辅料
apparel aesthetics 服装美学
apparel and accessories 服饰
Apparel and Fashion Industry's Association 服装和时装工业协会
apparel book 服饰书籍
apparel business 服装商务,服装生意,服饰业
apparel buyer 服装采购员
apparel category 服装类目
apparel CIMS 服装计算机综合生产系统
apparel color 服装色彩;服色
apparel commerce and trade 商务
apparel computer aided design 计算机辅助服装设计
apparel computer aided manufacturing 计算机辅助服装生产
apparel computer aided planning 计算机辅助服装计划
apparel computer integral manufacture system 服装计算机综合生产系统
apparel computer integrated manufacturing 计算机综合成衣生产
apparel construction design 服装结构设计
apparel convenience 服装方便性
apparel design 服装设计
apparel designer 成衣设计师,服装设计师
apparel design scope 服装设计范围
apparel detail design 服装细部设计
apparel draper 服装商;立体裁剪师
apparel durability 服装耐久性
apparel economics 服装经济性
apparel engineering 服装工程
apparel equipment 服装设备
apparel fabric 衣料
apparel factory 服装厂;被服厂
apparel festival 服装节
apparel fiber 服装用纤维,衣着用纤维
apparel findings and notions 服装小附件
apparel firm 服装公司

apparel flammability modeling apparatus (afMA) 服装可燃烧性模拟装置
apparel focus 服装特刊
apparel form 服装形态
apparel frog 盘花扣,衣服饰扣
apparel history 服装史
apparel illustration 服饰图
apparel illustrator 服饰图画家
apparel industry 服装工业,服装行业,制衣业
apparel industry standard 服装行业标准
apparel information technology centre 服装资讯科技中心
apparel look 服装款式
apparel magazine 服装杂志
apparel maker 服装制造者;制衣业;制业主
apparel manufacturer 服装业主;服装厂,制衣厂,服装制造商(厂商)
Apparel Manufactures' Council of Canada 加拿大服装厂商联合会
apparel manufacturing 服装生产,制衣
apparel manufacturing firm 制衣公司
apparel market 服装市场
apparel market trend 服装市场趋势
apparel material calculation 服装算料
apparel material selection 服装选料
apparel modeling 服装造型
apparel news 服装快讯
apparel pattern book 服饰图样集
apparel pocket 衣袋,衣服口袋
apparel principle 服装原理
apparel promotion 服饰推广;服饰推销
apparel protection 服装防护性
apparel psychology 服饰心理学
apparel recognizability 服装标志性
apparel reporter 服饰记者
apparel sample hand 服装样品师,服装样衣师
apparel sampler 服装样品师
apparel's care 服装保管
apparel sheet 服饰目录表
apparel(shell)fabrics 服装面料
apparel show 服饰展示会
apparel size 服装号型
apparel size and style series 服装号型系列
apparel sizes and styles 服装号型
apparel sketcher 服装描画师,服饰绘图员
apparel socioeconomics 服饰社会经济学
apparel sociopsychology 服饰社会心理学

apparel speciality store 服装特色店
apparel specimen 服装标样
apparel style 服装式样
apparel stylist 服装设计师
apparel technology 服装工艺
apparel textile 服装纺织品,服用纺织品
apparel utility 服装实用性
apparel with shell sets 贝珠衣(高山族首领礼服);珠贝服装
apparel wool 衣料用羊毛
apparel (working) sketch 服装效果图
apparel yarn 衣料用纱;绒线,毛线
apparent ply yarn 花色股线
appeal 吸引购买力(指商品)
appear 呈现,显得
appearance 外貌;外观;外表;容貌;外套
appearance inspection 外观检查;外观检验
appearance of dyeing 染色外观
appearance of grounding [印花]露底
appearance quality 外观质量,外观品质
appearance rating 外观等级
appearance retention 外观保形性;外观稳定性
appearance standard 外观标准
appending label 钉上商标标签
appendix cloth (色牢度试验用)贴衬布
Appenzell enbroidery 瑞士阿本策尔刺绣
apperleen 阿珀利恩起绒呢(运动服装用)
apple 苹果,苹果红
apple blossom 苹果花色
apple green 苹果绿
apple jack cap 报童帽
apple red 苹果红
appliance circuit 电器电路
appliance embroidery 贴花绣,贴花刺绣,嵌花绣,嵌花刺绣
appliance seam 嵌花缝
applicant 开证申请人
application d'angleterre 布鲁塞尔针绣花边
application fee 申请费
application for export 出口申请
application for import 进口申请
application for patent 专利申请
application printing 直接印花
applied aesthetics 实用美学,应用美学
applied art 实用美术
applied casing 装贴的抽带管
applied design 实用设计
applied pocket 实用口袋

applied printing 直接印花
appliqué, applique 缝饰;镶饰;贴绣,贴花,补花,贴饰;缝贴花;嵌花织物,补布;嵌花;加缝刺绣
appliqué and cross-stitch 挑补花
appliqué apron 贴布围裙
appliqué carrickmacross lace 嵌花绣花边,贴花绣花边
appliquecarrickmacross lace Irlande [法] 爱尔兰贴花刺绣花边
appliqued babywear 贴花婴儿服
applique decoration 贴花,补花
appliqué embroidering 贴花
appliqué embroidery 贴花刺绣,贴花绣,嵌花绣
appliqué embroidery lace 嵌花刺绣花边
appliqué figuring 贴花花纹;刺绣花纹
appliqué handbag 贴花手提袋
appliqué honiton lace 霍尼顿贴花花边
applique jeans 贴绣牛仔裤
appliqué lace 贴花花边
appliqué lace edging 贴饰花边
appliqué net 贴花网眼纱
applique pillowcase 贴花枕(套)
applique printing (发泡式)立体印花
appliqué quilting 贴布衍缝物
appliqué seam 贴花缝
appliqué stitch 贴花线迹,嵌花针迹;贴布绣
appliqué sweater 贴布毛衣
appliqué technique 贴花技术
appliqué work 镶嵌;补花缝饰,补花制品;贴布绣,贴布绣
apply 委托
apply beaker test 委托染小样
applying interlining to collar 拼领里
apply to customs 报关
appointed surveyor 指定检验人
appointment 预约(约定)
appraisal of damage 估损
appraise 估价
appraisement 估价
appreciate 赞赏
appreciation 增值
appret 织物上浆整理
approval of import 批准进口
approvals 包退包换商品
approval sample 确认样;封样;验收样
approve 认可
approved sample 验收样
appurn 桌布,围裙

apres-guerre 战后服装款式
apres-midi 午后服,午后装
après-ski boots 暖脚靴
apres-ski clothing 滑雪后穿的衣服
apres-ski look 滑雪后着装款式
après-ski slippers 暖脚套鞋
apres swim 泳衣罩衫,泳后罩衫
apret de laine 柔软整理
apricot 杏黄色,虎黄
apricot blush 杏红
apricot brandy 杏黄白兰地酒色
apricot buff 杏黄色,浅杏黄
apricot cream 米黄色,杏色,杏奶黄
apricot nectar 杏蜜色
apricot wash 透明杏色
apron 围裙;工作裙;帘子;垂层羊毛;大盖(鞋子),皮圈
apron check 色织格子布
apron cloth 围裙布,色织条格布
apron dress 围裙装,童围裙,连衣围裙
apron dress with betise collar 贝丝领连身围裙
apron fabric 围裙织物,色织格子织物,输送带帆布
apron front 围裙鞋面,(与围裙配色呼应的)鞋面遮布
apron-like trousers 围裙裤
apron look 围裙款式
apron pocket 围裙袋,围裙兜
apron skirt 围裙式裙,连衣围裙
apron string 围裙带
apron style 围裙款式
apron swimsuit 比基尼泳衣
apron tape 围裙带
apron tongue 特长皮鞋舌
apron train 围裙长裙,围裙长摆
apron-wrap skirt 围裙式包裙
apyeil 间位芳香族聚酰胺纤维(商名,日本尤尼吉卡)
aqua 浅蓝绿色
aquagel fiber 水凝胶纤维
aqua gray 水灰色
aqua green 水绿色,浅黄绿色
aqua grey 水灰色
aqua haze 水雾色
aqualon 阿奎纶(吸水性短纤维,商名)
aquamarble 高吸湿放湿聚酯短纤维(商名,日本东洋纺)
aquamarine 蓝绿,海蓝,蓝绿色,浅绿色,水绿色;海蓝宝石,蓝晶

aqua sky 天蓝,天空蓝
aqueous dyeing 水相染色
aqueous printing ink 水性印墨(转移印花用)
ara 手织条纹棉衬衫布,手织条纹裙布,条子裙布(印度)
arab dollar 阿拉伯美元
arabesque 黄土色;阿拉伯花纹,蔓藤花纹
arabian 阿拉伯镶边网眼窗帘
arabian blue 深铁青色
Arabian carpet 阿拉伯地毯
Arabian crepe 阿拉伯绉,阿拉伯绉花绉绸
Arabian embroidery 阿拉伯刺绣,几何花纹刺绣
Arabian head wear 阿拉伯头圈
Arabian lace 阿拉伯花边
Arabian Nights look 哈伦风貌,哈伦款式
Arabian robe 阿拉伯袍
Arabian shirt 阿拉伯衬衫,阿拉伯衬衣
Arabian stripes 阿拉伯粗平纹条子布
Arabian wool 阿拉伯羊毛
Arabic cap 阿拉伯帽
Arabic design 阿拉伯式图案
arab wool 伊拉克杂色羊毛
aracaju cotton 巴西阿拉卡儒棉
arachin fiber 花生朊纤维
arachne machine 阿拉赫涅缝边机
araignee lace 蜘蛛网纹花边
arain 印度轻薄塔夫绸,轻质条格塔夫绸(东南亚)
Aralac 美国阿雷莱克酪素纤维
aramid 芳香族聚酰胺纤维(商名,俄罗斯)
aramid fiber 芳香族聚酰胺纤维;芳纶
aramid sulfone fiber 聚芳砜纤维
araneum 阿拉尼厄姆粗抽花刺绣品
Araneum lace 仿古花边,阿拉尼姆花边(锁孔大网眼)
Aran Isles(sweater) 阿兰毛衣(来源于爱尔兰的阿兰岛)
aran knit 阿兰毛衣
arasaid 苏格兰女式花连衣裙
aravzor 阿拉夫佐尔缝编织物
arbaccio 阿巴乔粗呢;撒丁岛羊毛
arba kanfoth 传统犹太男式内衣
arbascio 希腊棕色毛布
ARBITER 《男装设计》(意)
arbiter 仲裁人
arbitral 仲裁
arbitral agreement 仲裁协议
arbitral award 仲裁裁决
arbitral court 仲裁庭

arbitral procedure 仲裁程序
arbitral tribunal 仲裁庭
arbitrarily curved rule 自由曲尺
arbitrarily curved ruler 自由曲尺
arbitrary sampling 任意抽样
arbitration 仲裁
arbitration award 仲裁裁决
arbitration award on appeal 仲裁上诉裁决
arbitration by summary procedure 简易仲裁
arbitration clause 仲裁条款
arbitration expense 仲裁费
arbitration fee 仲裁费
arbitration law 仲裁法
arbitration proceeding 仲裁程序
arbitration rule 仲裁规则
arbitrator 仲裁人
arc 弧
arcadian green 田园绿
arcazabo 里昂锦缎
arch 弓形,脚弓,鞋拱
archangelsk flax 俄罗斯柔软亚麻
arched back 弓形背
arched collar 半圆形领,弓形领,拱形领
arched shape 穹面形(珠宝)
arched stitch 弓形线迹,外曲牙线迹,弓形编织线迹
arched yoke 弓形剪接布;弓形过肩,弓形育克
archetype 原始模型
archi-imperiale 意大利粗哔叽
architectural style 建筑风格;建筑风时装
architecture of textile 织物结构
arch support 脚弓垫
arckhal 浮花刺绣,凸纹刺绣
arc measurement 弧长尺寸
arcograph 圆弧规
arc preventive coveralls[overalls] 防电弧工装裤
arc styleline 弧形结构线
arctic 北极色
arctic boots 北极靴,北极长筒靴
arctic cap 防寒帽,北极帽
arctic color 北极色
arctic fox (fur) 北极狐皮
arctic ice 北极冰色
arctic mukluk 海豹皮长靴
arctics 北极狼皮;北极靴,防寒防水套鞋
arctic wolf 北极狼皮
arcus 阿克斯
ardamu silk 伊朗阿达穆生丝

ardas 伊朗重磅丝织物,阿达斯绸(伊朗)
ardasse 18世纪波斯低级生丝
ardassin(e) 伊朗细生丝
ardebil rug 阿德比波斯地毯
ardenne tapestry 阿尔当织锦;法国织锦毯
ardtoornish 阿德托尔西呢
area 面积;针织完全组织
area-bonded product 纺黏法非织造布
area-bonded staple fabric (ABS fabric) 面黏合[短纤]非织造布
area density 织物单位面积重量
area rug 小地毯
area weight 织物单位面积质量
a 10% rebate for cash payment 现金付款9折
areca fan 槟榔扇
aredas 有光薄府绸,阿里达斯府绸
are hole 臂根围
arENA 阿瑞娜
arenka 芳香族聚酰胺长丝(商名,荷兰恩卡)
areola 乳晕
areophane 素色丝纱罗
Arequipa alpaca 最优级秘鲁山羊毛
Arequipa fleece 秘鲁羊驼毛
areste 中世纪英国金丝法衣织物
Argasis 印度手工丝织品
argali wool 盘羊毛
argent 银色,银白色
Argentan lace 阿让通花边,阿根坦针绣花边(法国制六角网眼花边),针绣粗六角网眼花边
Argentan point 阿根坦六角网眼花边
Argentella lace 阿根廷拉针绣花边(意大利制网眼花边起小点子花纹),阿风太拉花边(意大利)
Argentella(point) 阿根特拉花边
argentinas 奥地利斜纹布;英国出口黑色布料;阿根廷纳斜纹布
Argentinean costume 阿根廷民族服饰,阿根廷民俗服
Argentine cloth 阿根廷单面轧光粗平布
Argentine croisée 阿根廷丝经棉纬布
Argentine merino 阿根廷美利奴羊毛
Argentine wool 阿根廷羊毛
arghan 菠萝叶纤维(商名,英国)
argoflon fiber 阿戈弗纶(聚四氟乙烯纤维,商名,意大利)
argos rug 希腊阿戈斯毛毯
argouges 漂白亚麻布,法国漂白亚麻布
argudan cotton 中国粗绒棉
argyle 菱形格子

argyle check 阿盖尔花纹,菱形花纹
argyle design 菱形图案,菱形花纹
argyle gimp 菱形花纹嵌心窄辫带
argyle gloves 菱形花纹手套
argyle knit 菱形花纹针织品
argyle pattern 菱形花纹
argyle pattern hosiery 菱形花袜
argyle pattern hosiery machine 菱形花袜机
argyle plaid 阿盖尔菱纹,菱形大方格,闪色大方格
argyle purple 阿盖尔紫
argyle socks 菱形花格短袜
argyle sweater 阿盖尔毛衣,菱形花格毛衣
argyll 菱形格子
aridas 阿里达斯绸,阿里达斯府绸
ariel 阿里尔毛织纱罗
arimatsujibori 有松绞
arimid PM 阿里米特 PM(聚酰亚胺纤维)
arim warmer 暖臂套
arisard 阿里桑特束腰式服装;苏格兰妇女长披风
aristo 阿里斯托地毯,机织割绒地毯
arithmetic mean 算术平均值
ariyular 印度横条纬面缎
Arizona cotton 美国亚利桑那埃及种棉
Arizona-Egyption cotton 亚利桑那埃及种棉
Arkansas rowden 美国阿肯色改良棉
arlequin 阿莱奎因菱形花纹
arlesienne coif 法国阿尔勒民族头饰
arm 前肢,上肢,手臂;袖子;(缝纫机)机头身;经编机针床臂
armadillolike design 犰狳花纹
armani phenomenon 阿尔玛尼现象(首创男式女装风格)
armazine 黑色塔夫绸
arm badge 臂章,袖章
arm band 臂章,袖章,臂环(衬衫吊袖子用),衬衫吊袖宽紧带
armband bracelet 上臂镯
arm bed machine 筒式底板缝纫机
arm cover 套袖,袖套
arm crook 肘窝
arm cut seam 肋省缝
arm dowel pin 机壳定位销,车台定位销
Armenian clock 亚美尼亚大氅
Armenian lace 亚美尼亚花边,亚美尼亚钩针粗花边
Armenian net 亚美尼亚斜网眼,亚美尼亚网眼花边
arm entry 袖开口

armet 头盔(15～16世纪),亚美钢盔
arm guard 护臂
arm hat 胳膊帽(可折扁后夹在腋下)
armhold press 挂肩熨烫机
armhole 袖窿,袖圈;挂肩;袖孔
armhole adjustment 袖窿调节,袖窿配置
armhole around 袖窿围长
armhole bands 挂肩边,袖窿边
armhole circumference 臂根围
armhole cowl 袖窿垂褶
armhole curve 袖窿曲线
armhole dart 袖窿省
armhole depth 袖深
armhole depth line 袖窿斜线,袖深线(袖深尺寸线),(中式服装)抬根线
armhole ease 袖深宽松量
armhole facing 袖窿贴边
armhole felling machine 袖窿镶接机
armhole gatherer 袖窿打裥机,上衣袖机
armhole girth 袖窿周长
armhole line 袖窿线(袖窿的轮廓线)
armhole narrowing 袖窿收针
armhole neatening 袖窿修整
armhole notch 袖窿对位刀眼
armhole outline 袖窿弧线,袖窿线
armhole press 挂肩熨烫机
armhole princess raglan 腋下分割连袖
armhole princess styleline 袖窿弧线分割线
armhole ridge 袖窿凸棱
armhole seam 袖窿缝
armhole seam opening press 挂肩缝熨烫机
armhole shield 腋下垫布
armhole's length of curves 袖窿弯度
armhole stay 袖窿牵条布
armhole styling 袖窿式样
armhole top press 袖山压熨机
armhole upline 袖窿翘高线
armhole width 袖窿宽,袖壮大
armiak 阿米阿克呢,驼毛斜纹织物
armilausa 阿米劳萨短披风(中世纪,欧洲)
armistice cloth 多色精纺花呢
arm length 袖长,臂长
arm-length gloves 长手套
armlet (套在上臂的)臂环,臂章;臂饰,臂钏;袖套,护臂;特短袖,小短袖
armlet sleeve suspender 袖套吊带
arm lever 机壳臂
arm-long gloves 长手套
arm oil pipe 轴臂油管;上轴油管
armoire 19世纪英国厚重棱纹绸;大型衣橱

armoire 大型衣橱
armoisin 阿姆埃辛塔夫绸,印度轻薄塔夫绸
armor 潜水服;盔甲,甲胄
armorclad 阿莫克拉德塔夫绸,法国厚塔夫绸
armored fabric 防护织物
armored stripe 阿莫德条纹
armored vest 防弹背心
armorial bearing 纹章
armo(u)r 铠甲,盔甲,甲胄;防护服
armourclad 阿莫克拉德塔夫绸
armoured vest 防弹背心
armozeen 厚黑色丝袍(教士丧服);黑色丝带(帽子饰边)
armozine 厚黑色丝袍;黑色丝带
arm pad 座椅扶手小垫子
armpit 腋下,腋窝,夹肢窝
armpit hair 腋毛
arm point 袖圈
arm rhizo sphere 臂根围
arm ring 手环镯,手环;臂环;臂钏
arms coat 饰有王徽的传令官制服;饰有纹章的外套(穿在铠甲外)
arm screw 机壳紧固螺钉
armscye 袖隆,袖孔;袖隆线
armscye girth 臂根围
armscye line 袖隆(轮廓)线
arm's eye 袖隆
arm shaft (缝纫机)主轴,缝纽机轴臂
arm side cover 后盖
armside cover thumb 顶盖压脚螺钉
arm slit 披肩前衩,伸手口
armspan 臂跨
arm spool pin 插线钉
arm-spool pin pretension guide 过线器
arms span 两臂展开宽
arm suspender 松紧袖口,衣袖吊带
arm top cover 顶盖
arm-type sewing machine 筒式(底板或底臂)缝纫机
armure 蔓(草)花绸
armure bosphore 双面卵石纹绸
armure cheviot 黑色切维奥特全毛匹染厚呢
armurelaine 经向粗棱纹丝毛交织物
armure royale 罗亚尔卵石纹绸,罗亚尔卵石纹薄呢
armure satinee 小卵石纹缎背绸
armure victoria 小花纹细薄呢
arm warmer 暖袖筒,暖臂套
arm width 袖宽;中袖

army bag 军用包
army blanket 美国军毯
army blue 蓝制服布,蓝军服布;藏青制服呢,海军制服呢
army brown 军服棕
army cap 军帽
army cloth 军服呢,军装料,灰色军服粗呢,军服布
army coating 军服呢
army duck 军用帆布
army fatigues 劳动服;军服
army greys 军用灰色衬衫布
army look 军装款式,军服款式;军装风貌;军装式时装
army oxford 军用牛津布;军用衬衫布(方平组织)
army serge 军用哗叽
army shirt 军式衬衫
army style raincoat with hood 军服式兜帽雨衣,带兜帽军服式雨衣
army sweater 军用针织衫
army-type duck 军用帆布
army uniform 军服,陆军军服
arnatto 胭脂树红
arndilly 阿地呢
arni 轻薄棉细布
Arnold Scaasi 阿诺德·斯卡西(1931～ ,美国时装设计师)
aroma microcapsule finishing 微胶囊香味整理
aroma therapy finish 芳香治疗整理
aromatic copolyamide fiber 芳香族共聚酰胺纤维
aromatic copolyester fiber 芳香族共聚酯纤维
aromatic finishing agent 香味整理剂
aromatic polyamide-benzimidazole fiber 聚芳酰胺-苯并咪唑纤维
aromatic polyamide fiber 芳香族聚酰胺纤维,芳纶
aromatic polyamideimide fiber 聚芳酰胺酰亚胺纤维
aromatic polyazomethine 芳香族聚甲亚胺纤维;聚甲亚胺纤维
aromatic polysulphonamide fiber 芳香族聚砜酰胺纤维,聚酰胺纤维,聚芳砜酰胺纤维,芳砜纶
aromatic random copolyamide fiber 芳香族无规共聚酰胺纤维
around bust 胸围

a round fan 团扇
around skirt 前开口褶叠裙
arraignee lace 蛛网花边
arrange cut piece 分片
arrange cutted piece 分片
Arrangement Regarding International Trade in Textiles 国际纺织品贸易协定
arrange sample 排列试样
arranging pieces 分片
arras 阿拉斯精纺毛织物;阿拉斯花边
arrasene 绣花丝线;绣花绒线
arrasene embroidery 挂毡刺绣,花式线刺绣
arras（lace） 阿拉斯花边
arrindy silk 阿林迪生丝,蓖麻蚕丝
arrival draft 到货汇票
arrival time 到达时间
arrow collar 可卸男衬衫领,可卸领
arrow diagram 箭头图
arrow head 箭头;箭头形刺绣;三角结
arrow head hole 三角眼疵
arrow head stitch 人字绣,人字形贴线绣;三角形线迹,箭头形线迹
arrowhead tack 箭头形加固,三角形加固
arrowhead tack stitch 三角打结线迹
arrowhead twill 人字斜纹
arrow stitch 箭头形线迹,三角形线迹;人字绣
arrow stripe 箭头形条子花纹
arrow symbol 箭头符号(纸样上表示车缝方向)
arrow wood 荚蒾色
arscot 比利时细毛哔叽
art 艺术;美术
art canvas 刺绣地布
art Deco（A.D.） 装饰艺术风格,新艺术式样,装饰派艺术
art Déco spirit 装饰艺术风格
art deco style 新样式服装风格
art design 美术图案
artepovera 观念派艺术
arterial graft 人造血管织物
art exhibition 画展
art flower 人造花,绢花
artic fox hair 北极狐毛
artichoke 草绿色
artichoke cut 洋蓟发型
artichoke green 洋蓟绿
artichoke hairstyle 洋蓟式发型
article 物品,商品,货品
article change 翻改品种

article mark 商品标志
article name 品名
article NO. 货号
article number 品号,货号
article of clothing 衣物(衣、帽、手套等)
article of haberdashery 服饰品
articulated compensating foot 铰接式补偿压脚
art ideality 艺术想象力
artide for personal adornment 饰物
artificial 人造的
artificial bristle 人造鬃丝
artificial casein fiber 人造乳酪纤维,人造牛奶纤维,人造酪素纤维
artificial corn protein fiber 人造玉米蛋白纤维
artificial cotton 人造棉,棉型黏胶纤维
artificial crinoline 人造裙撑,多箍笼状裙撑
artificial down 人造羽绒
artificial dyestuff 人造染料
artificial fiber 人造纤维,纤维素纤维（man-made fiber 的早期名称）
artificial flower 人造花,假花
artificial fur 人造毛皮,人造裘皮
artificial fur cape 人造毛皮披肩
artificial fur coat 人造毛皮大衣
artificial fur collar 人造毛皮衣领
artificial fur fabric 人造毛皮织物
artificial fur overcoat 人造毛皮大衣
artificial fur stole 人造毛皮长围巾
artificial gem 人造宝石
artificial grass 人造草坪
artificial horsehair 人造马鬃
artificial lace 仿花边
artificial lawn 化纤细布;人造草地,人造草坪
artificial leather 人造革,合成革,仿革
artificial leather carrier webs 人造革基布
artificial leather shoes 人造革鞋
artificial ligament 人工韧带
artificial light 人造光源
artificial light shade 人工光源色
artificial mineral fiber 人造矿物纤维
artificial mohair 人造马海毛
artificial peanut protein fiber 人造花生蛋白纤维
artificial protein fiber 人造蛋白质纤维
artificial schappe silk 人造绢丝
artificial silk 人造丝,普通黏胶长丝
artificial silk from collodion 硝酯人造丝,

火棉胶人造丝
artificial soybean protein fiber 人造大豆蛋白纤维
artificial suede 人造羔皮,人造麂皮
artificial tulle 仿网眼纱,充网眼纱,充花边
artificial wig 人造假发
artificial wool 人造毛,人造羊毛,毛型黏胶纤维,永久卷曲黏胶(短)纤维
artillery twill 斜纹马裤呢
artisan 手艺人
artisan sewing machine 手艺行业用缝纫机,服务行业用缝纫机
artist bow 艺术家蝴蝶结
artistical effect 艺术效果
artistic anatomy 艺术解剖学
artistic calligraphy 美术字
artistic concept 艺术概念
artistic creativity 艺术创造力
artistic design 美术设计
artistic image 艺术形象
artistic intuition 艺术直观
artistic manner 写意
artistic skill 绘画技法,艺术表现技巧
artistic style 艺术风格,艺术形象
artistic tapestry 工艺美术壁毯
artistic things 艺术品
artistic works 艺术品
ARTISTRY 雅姿
artist's canvas 油画布
artist smock 艺术家工作服
artless beauty 自然美
art lettering 美术字
art linen 装饰用亚麻平布
artlon 阿特纶(聚对苯二甲酸丁二酯纤维,商名,日本)
art muslin 高级装饰用轧光棉布;装饰用棉布
art needle work 工艺刺绣品,艺术刺绣品
Art Nouveau 新艺术风格;新艺术运动
art nouveau hose 新艺术风格袜子
art nouveau look 新艺术风格,新技术风貌
art nouveau style 新艺术服装风格
art of dressing 着装艺术
art of painting 绘画艺术
artois 阿图瓦外套
art paper 铜版纸
art piqué 变化凸棱织物
arts and crafts 工艺美术,工艺美术品
art science 艺术学
art serge 装饰用精纺哔叽

art shades (绸布的)雅致色泽,典雅色泽
art silk 刺绣丝线
art skeining threads 工艺纹线
art square 工艺美术地毯(双面提花,两端有穗)
art style 艺术风格
art-textile 艺术纺织品
art ticking 印花床单,印花枕头布
art to wear 面料艺术设计技术
artware 工艺品
art wearing 着装艺术,服装艺术,穿衣艺术
art weave 艺术花纹;缎地菱纹织物
artwork layout 样稿
arutex machine 阿鲁特克斯缝编机
asahi 旭牌黏胶丝
asalitus 西藏野山羊绒
asan 带饰边的白色祈祷毡毯
asanoha 麻叶图案
asbestiform 偏磷酸钙钠晶体纤维(商名,美国孟山都)
asbeston 石棉与棉混纺纱线
asbestos 石棉
asbestos boots 石棉靴
asbestos cloth 火烷布,石棉布,石棉织物
asbestos clothes 石棉衣,石棉服装
asbestos cord 石棉绳,石棉线
asbestos fiber 石棉纤维
asbestos float 石棉浮游纤维
asbestos gloves 石棉手套
asbestos interlining 石棉衬
asbestos lace 石棉花边,不燃性花边
asbestos thread 石棉线
asbestos uniform 石棉服
asbestos yarn 石棉纱
ascarpidor [西]艾斯卡匹德尔梳子
ascending chromatography 上行色谱
aschodur 伊朗女用黑棉布
ascot 阿司阔领带,蝉形领带,阿斯科领带(巾)
ascot collar 阿斯科领
ascoted shirt 阿斯科饰巾衬衫
ascot hat 阿斯科塞马会帽
ascot jacket 宽松男式上衣
ascot neckline 阿斯科领口
ascot scarf 阿斯科长方形披巾;阿斯科领带(巾)
ascot shirt 阿斯科衬衫(传统衬衫)
ascot tie 阔领带,领巾式领带,阿斯科领带
as-drawn fiber 初拉伸纤维

aseptic cotton 无毒棉;防腐棉
aseptic gauze 药用纱布,消毒纱布,绷带纱布
asexual look 无性别款式,无性别风貌
as-formed fiber 初纺纤维,初生纤维
asgard 阿斯加德织物(耐火织物,商名,英国)
ash 烟灰
ash blond 银灰色(或淡褐色)头发的人
ash blond hair 银灰色头发
ash (color) 烟灰色,浅灰色,淡灰色
ash gray, grey 灰色,浅灰
A-shirt (athletic shirt) A字衫(运动背心)
ashmara jute 低级黄麻
ash mota 粗硬的黄麻纤维
ashmouni cotton 埃及阿什穆尼棉
ash pink 灰粉红
ash rose 玫瑰灰
ashtamudy yarn 双股粗椰壳纤维纱
ash violet 淡白紫
Asia Minor carpet 小亚西亚长绒地毯
Asian Development Bank 亚洲开发银行
Asia textile machinery exhibition 亚洲纺织机械展览会
Asiatic cotton 亚洲棉
Asiatic silk 亚洲丝绸
asics 爱世克斯
asiglet 军服肩带;花边帽带;服饰金线
asi mani 锯齿纹丝棉织物
as-is fiber 厚态纤维
ask the price of 询价
as-melt spun fiber 初纺熔纺纤维
asonkobi silk 西非洲产阿桑科比野蚕丝
asooch 肩带;从肩至臀斜挂围巾
asota 聚丙烯短纤维(商名,奥地利林茨)
asparagus green 浅豆绿,龙须菜绿,芦笋绿
aspect ratio 长宽比,高宽比,纵横比
aspen gold 山杨黄
aspen green 山杨绿
aspen vest 麂皮背心
Aspero cotton 秘鲁产阿斯派罗棉,粗秘鲁棉
asphalt 柏油色,沥青色
asphalt felt 油毛毡
asphyxiation 窒息
assama silk 印度阿萨马丝
Assam cotton 东印度阿萨姆棉
assemblage 服装缝合;装配
assemble 法国双股绢丝;组合
assembled plastic shoes 组装塑料鞋

assembler 捆包员
assemblies 衣片
assembling 装配;缝合;组合
assembling department 装配间
assembling drawing 装配图,组合图
assembling process 装配工序
assembly (asm) 装配件,组件
assembly charts 装配图
assembly defect 缝制疵病
assembly details 缝制说明
assembly diagram 装配图
assembly line 流水作业装配线,装配线
assembly line sewing process 流水线缝纫加工
assembly man 装配工
assembly seaming system 装配线式缝纫系统
assembly workshop 组装车间
assembly-wound yarn 组合螺旋线
assessment 评估
assessor 财产估价员
assets income 资产收入
assignment of policy 保险单转让
Assili cotton 埃及产阿西利棉
assimilation effect 色泽同化效应
assisi embroidery 阿西西十字刺绣,阴纹十字绣,阿西西刺绣
assisi work 阿西西十字刺绣,阴纹十字绣
assistant designer 助理设计师
assistant engineer 助理工程师
assisting equipment 辅助设备
associated operation 联合经营
associated package 混色混码包装
association 协会
Association International polynosiques 国际波里诺西克纤维协会
Association of College Professer of Textiles and Clothing 纺织服装大学教授协会
Association of Knitted Fabrics Manufacturers 针织品生产商协会
Association of Nonwoven Fabrics Industry 非织造布工业协会
Association of South East Asian Nations 东南亚国家联盟
Association of Synthetic Yarn Manufactures 合成纤维厂联合会
Association of Yarn Distributors 纱线生产经销商协会
Assorcebunder silk 阿索斯本德丝,孟加拉最低级生丝
assorn-bund 印度丝织物

assort 拼色成包,拼件;搭配	astra work 星花刺绣
assort by color and size 分色分码	Astrazon dye 阿斯特拉松染料(腈纶用阳离子染料,商名,德国德司达)
assorted 杂色	
assorted cloth 零头布	astride overskirt 女骑装裙,妇女骑马罩裙
assorted color 杂色	astrid lace 小花纹窄花边;厚花边
assorted fiber 混杂纤维	astringent 收敛剂
assorter 服装配件工	astringent lotion 收敛性清洗剂
assorting 拣选,分级,分类	Astrocarb 阿斯特罗卡织物(黏胶基碳纤维织物,商名,美国)
assortment 分类;花包品种;(颜色、尺码)搭配;服装配件组合	
	astrolegs hose 十二宫图案袜
assortment for patterns and colors 配花配色	astronaut's cap 宇航帽
assortment for pieces 拼件	astronaut's garment 宇航服
as-spun fiber 初纺[成]纤维,初生纤维	astronaut's shoes 宇航鞋
assurance 保险	astroquartz fabric 宇航石英织物(可耐100℃高温)
Assyrian 亚述装饰纹样	
Assyrian costume 亚述服饰	A style A形,A形款式,A字形款式
Assyrian curls 亚述人鬈发	asu 印度红黄蓝色棉绣花线
Assyrian ornament 亚述装饰	a suit of ditto 同料同色的一套西服
astamudy yarn 双股椰壳纤维粗线	asylum cord 席纹高强力带织物
astar 土耳其细布	asymmetric(al) 不对称领口
astarte 阿斯塔特优质印花绸	asymmetrical bangle 不对称硬手镯
aster cloth 农用遮阳布	asymmetrical circular flounce skirt 不对称圆形荷叶边裙
asteri 伊朗印花衬里布	
aster purple 紫翠菊色,紫菀色	asymmetrical closing 偏襟,不对称门襟
astestos tape 石棉带	asymmetrical clothing 不对称服装
astestos yarn 石棉纱	asymmetrical design 不对称设计
asthenic 身材瘦长的人	asymmetrical diagonal line 不对称对角线
asticot canvas 结子花式衣料织物	asymmetrical fastening 不对称系口
asticotine 法国轻薄重缩弹力呢	asymmetrical hairstyle 不对称发型
astigmatic glasses 散光眼镜	asymmetrical inserting layout 提缝套法
astoli canvas 爱尔兰车篷防水帆布	asymmetrical neckline 不对称领口
astrachan,astrakhan 俄罗斯羔羊皮;仿羔皮织物;俄罗斯羔羊毛;经缎毛圈组织(针)	asymmetrical overlap 不对称叠门
	asymmetrical ring 不对称戒指
	asymmetrical skirt 不对称裙
astrakhan cap 阿斯特拉罕羊皮便鞋	asymmetrical straight band 不对称式立领
astrakhan cloth 仿羔羊皮织物	asymmetric balance 不对称平衡
astrakhan jacket 俄罗斯羔羊皮外套,仿羔羊皮外套	asymmetric collar 不对称领
	asymmetric curve dart 不对称曲省
astrakhan work 仿土耳其地毯绣品	asymmetric darts 不对称省
Astralene 阿斯特雷纶(涤纶弹力丝,商名,英国)	asymmetric drape 不对称垂裙
	asymmetric dress 不对称服装
Astralene-C 阿斯特雷纶-C(涤纶假捻变形丝,商名,英国)	asymmetric evening look 不对称晚装型发式
	asymmetric fiber 不对称纤维
Astralon 阿斯特拉龙(锦纶假捻变形丝,商名,英国)	asymmetric front 不对称前搭片
	asymmetric hair style 不对称发型
Astralon-C 阿斯特拉龙-C(锦纶假捻变形丝,商名,英国)	asymmetric hem 不对称下摆
	asymmetric hemline 非对称型下摆
astral work 星花刺绣	asymmetric hern 不对称下摆
astran chan 仿羔皮;羔羊毛	asymmetric hollow fiber 不对称中空纤维
a strand of pearls 一串珍珠	asymmetric line 非对称线

asymmetric neckline 不对称领口
asymmetric radiating darts 不对称放射省
asymmetric silhouette 不对称线条;不对称型;不对称轮廓
asymmetric skirt 不对称裙
asymmetric style 不对称款式
asymmetric suit 不对称套装
asymmetric swimsuit 不对称泳装
asymmetry 不对称(现象),不对称性
at 税后
ata 犹太教男士晨祷披巾
at a discount 打折(扣)
at a reduced price 减价
at buyer's option 由买方决定
at cost 按成本
atelier 时装设计室;画室;时装缝制工场
A-Tell fiber 荣辉纤维(聚对苯甲酸乙氧基[酮]酯纤维,商名,日本)
at fair price 价格公道
athlete's shoes 田径鞋
athlete's wear 田径服
athletic bra 体育胸罩,运动胸罩
athletic clothes 运动服
athletic hoses 运动(短)袜
athletic knit shirt 针织运动衫
athletic mood 运动气氛
athletic neckline 运动领口,X形领口
athletic pants 针织绒裤
athletic shirt 运动(汗)衫,运动背心,背心式运动衫
athletic shoes 运动鞋
athletic shorts 运动短裤
athletic shorts with bib 男式背带运动短裤
athletic shoulder 运动肩
athletic socks 运动袜
athletic sports look 运动竞技风貌
athletic suit 健身服
athletic supporter (男运动员)下体弹力护身,护裆,护腹
athletic tank 运动背心
athletic vest 运动背心
at-home wear 家居服
atlantic green 大西洋绿
Atlantic states cotton 美国大西洋棉
Atlantic transfer dye 阿特兰蒂克转移印花染料(商名,美国)
atlas 经纹织物,棉背缎,丝缎,经编缎纹织物
atlas tricot 经编缎纹织物
atlas fabric 经缎织物

atlas lap 经缎垫纱
atlas of fiber fracture 纤维碎裂图集
atlas paper 绘图纸,印图纸
atlas rib fabric 罗纹经缎织物
atlas silk 柞蚕丝
atlas tricot 经编缎纹织物
atlon 阿特纶醋酯纤维
atmosphere 气氛
atmosphere fading 大气(烟色)褪色
atmospheric diving suit 常压潜水服
atmospheric dryer 常温干燥机,大气干燥机
atmospheric dyeable polyester fiber 常温可染聚酯纤维
atmospheric dyeing 常压染色
atmospheric fading 烟气褪色,大气褪色
atmospheric temperature 常温
atonal coloring 无调着色
at-once-payment 立即付款
atoz yran 阿托兹纱(聚丙烯腈超高膨体纱,商名,英国)
at par 平价
atrophy 萎毛
at seller's option 由卖方决定
atta bag 阿塔麻袋
attach (缝)绱;贴,覆;钉;系
attach back crotch slay 绱大裤底
attach belt loop 绱裤带袢
attach button 缝纽扣
attach chest interlining front piece 覆衬(衣前片覆胸衬)
attach chest interlining to front piece 覆胸衬
attach collar stay 领角薄膜定位
attach collar to band 夹翻领(翻领夹进底领缝合)
attaché 密码箱
attaché case 手提箱,公文提箱;使馆包
attached collar 脱卸领,活领,活动礼服衬衫领
attached cushion (地毯)附加衬垫
attached foot 脱卸鞋(用于婴儿睡衣裤上),婴儿脱卸鞋
attach elastic 绱松紧带
attach eye to waistband 钉裤钩袢
attach facing to fly 覆挂面
attach front band 绱明门襟
attach hangtag 打吊牌
attach heel stay 绱贴脚条
attach hook and eye to waistband 钉裤钩袢
attach hook to band 手工绱领钩
attach hook to collar band 绱领构

attach hook to waistband 钉裤钩
attaching 缝合,钉合,绱,装,贴,缚,(将衣服部件)连接,缝上,钉上
attaching back crotch stay 绱大裤底(裤底装在后裆十字缝后)
attaching band 绱明门襟
attaching band to collar 上下领缝合
attaching belt loop 钉裤带,绱串裤带
attaching bodice and yoke 缝合育克和大身
attaching button 缝纽扣
attaching button and button hole tape 缝合纽扣和纽孔垫带
attaching buttonhole strip 缝纽孔垫带
attaching chest interlining to front 覆衬
attaching collar 装领
attaching collar stay 领角薄膜定位
attaching collar to band 夹翻领
attaching cuffs to sleeve 绱袖口
attaching elastic 缝松紧带
attaching eyes to waist band 钉裤钩袢
attaching facing 敷挂面
attaching fastener 钉按扣
attaching fly facing to trouser front 缝裤子左暗门襟
attaching fly shield 绱里襟
attaching heel stay 绱脚口贴条
attaching hood to garment 绱帽
attaching hook and eye to waist band 钉裤钩袢
attaching hook to collar band 装领钩
attaching hook to waist band 钉裤钩
attaching interlining 敷衬
attaching interlining to collar 缝领衬
attaching interlining to cuff 缝袖口衬里
attaching label 钉商标,缝标签
attaching lace 钉花边
attaching lining to back 敷后衣身里子
attaching lining to bodice 敷大身里子
attaching lining to hood 合帽面里
attaching lining to skirt 覆裙里
attaching pieces to body material 在大身上缝小片
attaching pocket flap 装袋盖,缝上袋盖
attaching pocket stay 敷袋口牵条
attaching pocket to garment 绱袋
attaching prefab waistband lining 缝预制腰头衬里
attaching reinforced tape 缝加固带子
attaching set-on centre plait 缝合中间褶裥
attaching sleeve 绱袖,上袖,装袖

attaching sleeve facing 上袖口滚条
attaching sleeve placket 缝制男衬衫袖衩
attaching sleeve tab 绱袖袢
attaching snap 钉按纽
attaching tag 钉吊牌
attaching tag and label 订标签
attaching tape 缝牵条,缝带,敷牵条
attaching tape to armhole 敷袖窿牵条
attaching tape to back vent 敷背衩牵条
attaching tape to front edge 敷止口牵条
attaching tape to hood brim 绱帽檐
attaching tape to lapel roll line 敷驳头牵条
attaching tape to pocket opening 敷袋口牵条
attaching tape to waist line 敷裙腰口牵条
attaching trouser curtain 绱雨水布
attaching under collar to top collar 敷领面
attaching waist band 绱腰头
attaching waist band lining 缝裤腰衬里
attaching yoke 绱过肩,上过肩,装覆肩
attaching yoke and bodice 缝合育克和大身
attaching zip fastener 装拉链
attaching zipper 缝拉链,装拉链
attach label 钉标签;绱商标
attach lining to bodice 覆大身里子
attach maker 拼合排料图
attachment 连接物,附件,附属装置;服饰附属品
attachment for pressing 整烫附属装置
attachment yoke 裁装复肩
attach rib 绱罗纹
attach right(left)fly 绱裤里(门)襟
attach rivet 钉撞钉
attach snap 钉四合扣
attach tape to armhole 覆袖窿牵条
attach tape to back vent 覆背衩牵条
attach tape to front edge 覆止口牵条
attach tape to hood brim 绱风帽檐
attach tape to lapel roll line 覆驳头牵条
attach tape to pocket opening 覆袋口牵条
attach tape to waistband of skirt 覆裙腰口牵条
attach trouser curtain 绱雨水布(缝在裤腰里下口)
attach waistband 绱腰头
attach zipper 绱拉链
attalea cloth 阿泰里水手服贴布边
attalia 阿塔利亚斜纹棉织物
attalic thread 金箔包芯线(天然纤维或化纤为芯)
attalic yarn 毛或麻芯金线

attention ability 看台能力
attic look 古雅风貌
attifet [法]阿蒂费头饰;阿提菲女帽
attifet head-dress 心形头饰
attire 衣服,服装;衣着;装束;装饰;打扮
attirement 衣服;服饰
attractive in quality and price 物美价廉
attribute 属性
attribution theory 归因理论
attushi 榆树韧皮纤维织物
A twills bag A字斜纹麻袋
A-type merino wool A型美利奴羊毛
aubergine 茄皮紫,乌紫色(紫红色)
auburn 茶褐色,金棕色,猪肝色,赭色
Auburn·56·Rex 阿勃恩·56·赖克斯棉(巴西陆地棉)
aubusson carpet 奥布松毛圈际毯
aubusson stitch 欧比松绣,针绣挂毯用十字针迹
aubusson tapèstry 法国奥布松挂毯
auction 拍卖
auction sale 拍卖
aucube 比利时地毯
audit 审计
audrey 奥黛芬
Audrey peterkin cotton 美国彼德金棉
au fuseau 网地花边(梭结花边用语)
Augsburg checks 奥古斯堡色织格布
Augusta Bernard 奥古斯塔·贝尔纳(法国时装设计师)
aulmoniere 小皮革钱袋
aumusse 奥美施头巾
au passe stitch 法国缎纹线迹
aupoz fiber 蕉麻纤维
auquili 叙利亚布袋
auramine 槐黄,碱性嫩黄
aurang-shahi 宝座盖布
aureole 光环状花边褶系带白帽;圆圈渍痕
aures 粗纺斜纹起绒织物
Aurillac lace 金银线花纹棱结花边
aurora orange 朝霞橘色(红橙色)
aurora pink 晨曦玫瑰红
aurora red 朝霞红
aurora yellow 曙黄色
ausan 白色祈祷毯
aushentic dressing 正装
auSSINO 澳西奴
Australasian wool 澳大利亚与新西兰羊毛
Australian crepe 澳洲绉
Australian merino 澳洲美利奴毛

Australian military puggaree 澳大利亚军用缠头巾
Australian nuby 澳大利亚红宝石
Australian opossum 澳大利亚负鼠毛皮
Australian padpardscha 澳洲蓝宝石
Australian ruby 澳大利亚红宝石
Australian wool 澳大利亚羊毛,澳毛,澳洲羊毛
Australian wool bale 澳洲羊毛包
Australian wool Testing authority 澳大利亚羊毛检验局
austria 奥地利斜纹伞绸
Austrian blanket 奥地利毛毯
Austrian cloth 奥地利高级毛织物
Austrian costume 奥地利民俗服,奥地利民族装
Austrian crystal 奥地利水晶饰品,奥地利水晶珠
Austrian manteau 奥地利外套
Austrian seal 奥地利兔毛
Austrian shade cloth 奥地利粗支宽条窗帘布,奥地利高级毛织物
Austrian twill 奥地利斜纹布
Austrian zephyrs 多臂提花条纹织物
Austrias 奥地利斜纹布
Austro-Hungarian lace 奥地利-匈牙利花边
Austrylon 澳斯特丽纶(锦纶66,商名,澳大利亚)
authentic dressing 正统衣着
authentic sample 可靠试样
authentic servegor 公证鉴定人
authentic style 保守型,正统式样
authority to sign 授权签字
authorized signature 授权人签字
auto coat 驾车大衣
auto-control 自动控制
auto-counter 自动计数器
auto-feed 自动送料
autoheeler 自动袜跟机,袜跟自动编结装置
auto-hydraulic balancing sole attaching machine 自动油压平衡压底机(制鞋)
autojig 自动裁边
autojig machine 自动裁边机
autolock 自动锁定
automarker 自动排料
automated controlled sewing machine of long seams 长缝轨控自动缝纫机
automated factory 自动作业工厂
automated layout 自动排料

automated line 自动线
automated production 自动化生产
automated sewing equipment 自动缝纫设备
automated sewing machine of flap pocket 自动开袋缝纫机
automated sewing machine of long seams 长缝轨控自动缝纫机
automated symbolic artwork program 自动符号工艺图解
automated system for clothing production 服装生产自动化系统
automatic air scissors （自动）气动剪刀
automatically turned welt 自动双层平针袜口
automatic backtack 自动回针
automatic bartacker 自动套结机
automatic bartacking mechanism 自动缝端加固机构
automatic baster 自动粗缝机
automatic belt loop machine 自动缝带袢机
automatic bonnet 汽车司机软帽
automatic brightness control 自动亮度调节
automatic button feeding 自动喂扣
automatic button feeding device 自动送纽装置
automatic buttonhole stitching machine 自动锁眼机，自动锁纽孔机
automatic cement drying machine 自动胶水烘干机（制鞋）
automatic clamping device 自动夹紧装置
automatic collar press 自动领子压平机
automatic collar runstitch machine 自动领圈切缝机
automatic collar turning and pressing machine 自动翻领烫领机
automatic color control 自动色彩控制
automatic condensed stitching mechanism 自动线迹密缝机构
automatic control 自动控制
automatic cropper 自动剪料头机
automatic cuffing-off and measuring machine 自动量裁机
automatic cuff pressing machine 自动压袖口机，自动压袖头机
automatic cuff runstitcher
automatic cuff topstitcher 自动袖头面缝机（裤子翻边），自动初缝及修剪袖头缝纫机
automatic cutting 自动裁剪
automatic cutting-off and measuring machine 自动量裁机
automatic cycling device 自动循环装置
automatic darning 自动织补
automatic dart sewer 自动缝省机
automatic dart sewing machine 自动缝省机
automatic decorative stitching machine 自动装饰线缝缝纫机
automatic dimensioning 自动标比尺寸
automatic display and plotting system 自动显示和绘图系统
automatic distributor 自动分料装置
automatic drafting machine 自动绘图机
automatic drawing pattern system 自动描花样系统
automatic drop loop machine 自动缝环机
automatic end trimming 自动剪头头
automatic eye(let) sewing machine 自动锁圆头纽孔机，自动圆头锁扣机
automatic fancy stitching 自动花式线缝
automatic fastening machine 自动加固缝纫机
automatic feeding mechanism 自动送料机构
automatic feed unit 自动进料装置
automatic flap pocket sewing machine 自动开袋机
automatic flat jacquard machine 自动提花横机
automatic flat knitting machine 全自动横机
automatic full fashion glove knitting machine 全自动成型手套机
automatic Gerber cutter 格柏自动裁床
automatic Gerber mover 格柏自动吊挂系统
automatic glove full-fashion knitting machine 自动全成形手套机，自动收放针手套机
automatic glove knitting machine 全自动手套针织机
automatic grading 自动放码
automatic grinding cutter 摇臂式自动磨刀裁剪机
automatic half-hose machine 自动短袜机
automatic heat forming machine 自动硫定型机（制鞋）
automatic heel seat tacking and side cement lasting machine 自动上胶中后帮打钉机（制鞋）
automatic high speed refrigerating machine 自动速冻定型机（制鞋）
automatic hook sewing machine 自动缝搭钩机
automatic hosiery knitting machine 自动

织袜机
automatic hydraulic heel-lasting machine 自动后踵结帮机(制鞋)
automatic inside shaving machine 自动鞋踵中底削边机(制鞋)
automatic intermittent positioning sewing machine 间歇定位缝纫机
automatic iron 恒温电熨斗,调温熨斗
automatic labelling machine 自动缝标签机
automatic labelling unit 自动贴商标装置
automatic line 自动生产线,自动线
automatic loader 自动装料器
automatic long seamer 自动长缝缝纫机
automatic long stitch 自动长缝线迹
automatic loom 自动织布机
automatic machine 自动缝纫机
automatic management 自动管理
automatic measuring and folding machine 自动码布折布机
automatic name setter 自动缝标牌装置
automatic needle positioner 自动缝针定位器
automatic operation 自动操作
automatic ornamental stitch 自动装饰线迹
automatic pattern 自动图案
automatic piping machine 自动滚边机
automatic placket folding machine 自动袖衩折叠机
automatic pleating machine 自动打褶机
automatic pleats making machine 自动打褶机
automatic pocket setting machine 自动钉袋机
automatic pocket welting machine 自动开袋缝纫机
automatic positioner 自动定位器
automatic press 自动烫衣机;自动印花机(印T恤衫用)
automatic presser foot lifter 自动压脚提升器
automatic pressing equipment 自动整烫设备
automatic pressing machine 自动熨烫机
automatic processor 自动洗片机
automatic production 自动化生产
automatic production line 自动生产线
automatic programming 程序自动化
automatic quilting machine 自动绗缝机
automatic rapid heat forming machine 自动快速定型加硫机(制鞋)
automatic reverse feed 自动倒向送料
automatic reverse stitching device 自动倒缝装置
automatic reverse stitching system 自动倒缝装置
automatic riveting machine 全自动铆钉机
automatic rotary type direct injection two-color sports shoes making machine 全自动轮转式双色射出结帮机
automatic rotary type four-color PVC shoes injection moulding machine 全自动回转式PVC塑胶鞋射出成型机
automatic screen print 自动筛网印花
automatic screen printing 自动网版印花
automatic seamless glove knitter 自动无缝手套机
automatic seamless hosiery machine 自动无缝圆袜机
automatic self-sharpener 自动磨刀器
automatic separation 自动单只落袜
automatic sequential buttonhole sewing machine 自动程序锁纽孔机
automatic serger 自动包边缝纫机
automatic serging machine 自动包边缝纫机,自动包缝机
automatic sewing equipment 自动缝纫设备
automatic sewing machine 自动缝纫机
automatic shirt pocket setter 自动上衬衣口袋器
automatic skip 自动跳线
automatic sleeve-placket machine 自动缝袖衩机
automatic snap fastening machine 全自动揿纽机
automatic sock knitting machine 自动织短袜机
automatic spreading machine 全自动拉布机,自动铺布机
automatic start and finish tacking 自动起始与终止加固缝
automatic stitching stop at seam end 端缝自动停缝装置
automatic stocking turner 自动翻袜装置
automatic straight-run sewing machine 自动直缝缝纫机
automatic suspension of the contract 合同自行终止
automatic system 自动化系统
automatic tacking machine 自动加固缝纫机,自动套结机
automatic tape chopper 自动切带机
automatic tape cutte 自动带切机,自动切带刀

automatic tape cutting 自动裁带
automatic tape puller 自动拉带器
automatic thread cutter 自动剪线器
automatic thread cutting device 自动剪线装置
automatic thread trimmer stitcher 自动剪线缝纫机
automatic tint control 自动色调控制
automatic toe linking 自动缝头
automatic under trimmer 自动下剪线
automatic utility press 自动万能熨烫机
automatic washing line 洗涤自动线
automatic welt tying up hosiery machine 折口袜机
automatic winding watch 自动手表
automatic wrapping machine 自动包装机,自动包缝机
automatic zigzag sewing 自动曲折形缝纫
automation 生产自动化
automation engineering 自动化工程
automobile bonnet 汽车司机软帽
automobile cap 汽车帽
automobile coat 汽车服
autonomous 自主
auto-open umbrella 自动伞
autoplotter 自动绘画仪
auto-setter 自调定形机
auto-switch 自动开关
autothread trimmer stitcher 自动剪线缝纫机
auto-tracing system 自动描图系统
autowash 自动洗涤机
autowelting 自动扎袜口
autumn 灰黄色
autumn blonde 金秋色
autumn brown 秋棕色
autumn commodity 秋季商品
autumn fair 秋季商品交易会
autumn leaf 秋叶棕(黄棕色),锈黄
autumn suits 秋服,秋装
autumn tints 秋色
autumn wear 秋装
autwerp lace 安特卫普花边,针绣扣眼线迹花边;树枝形花纹
auvergne lace 挖花梭结花边,奥韦尼花边
auxerre lace 白色亚麻梭结花边
auary drawing 辅助图
auary equipment 辅助设备
auary labor 辅助劳动
auary line 辅助线

auary material 辅料
auary staff member 辅助人员
auxochrome 助色团
auxochromic group 助色团
auxonne 法国大麻帆布
Auxtex 奥克斯特聚酯纺丝网非织造布
ava cotton 印度阿瓦棉
available time 开机时间,有效工作时间
avant-garde 前卫派,先锋派;先锋派时装
avant-garde fashion 先锋派款式,前卫流行,超前款式;先锋派时装,标新立异的款式
avant-gardism 先锋主义(前卫主义)
avant-gardistic style 前卫风格
avasca 阿瓦斯卡呢,秘鲁羊驼毛呢
ava stripe 阿瓦条子绸,缅甸阿瓦条子绸
ave maria lace 法国民间窄幅梭结花边,天主教万福玛丽亚花边
ave Maria lace 万福玛丽亚(狭幅)花边(天主教)
aventurine quartz 星彩石英
average 中间色;海损
average deposit receipt 海损保证金收据
average figure 一般体型
average policy 海损保险单
average price 平均价格
average shirting 海岛棉衬衫料
average skin temperature 皮肤平均温度
average tare 平均皮重
aviation garment 飞行服
aviation suit 航空服
aviator 飞行员服饰
aviator blouse 飞行员夹克
aviator helmet 飞行员头盔
aviator jacket 飞行员夹克
aviator's glasses 飞行员护目镜
aviator's helmet 飞行员头盔
aviator style 飞行员款式
aviator style jacket 航空式短夹克
Avicolor 阿维科洛黏胶纤维(商名,美国)
Avicron 阿维克纶潜在卷曲黏胶纤维(商名,美国)
Avignon 阿维尼昂里子绸(轻质蚕丝塔夫绸,商名,法国)
Avila wool 西班牙阿维拉羊毛
Aviloc 阿维洛克黏胶丝(商名,美国)
Avilon 阿维纶(黏胶纤维,商名,芬兰凯米拉)
Avinion 阿维尼翁金属色印花布(商名,英国)
Avisco 阿维斯科纤维(商名,美国)
Avisco acetate type 25 阿维斯科醋酯纤维

25型(变性醋酯纤维膨体纱,商名,美国)
Avisco PE 阿维斯科PE(聚乙烯纤维,商名,美国)
Avitron 阿菲特纶(聚酯长丝,商名,德国)
avocado 橄榄绿,鳄梨绿
AVON 雅芳
avonet wool 地毯用三岁绵羊毛
Avril 阿夫列尔(高湿模量黏胶线,商名,美国)
Avron 阿芙纶黏胶短纤维
Avron XL 阿芙纶XL黏胶短纤维
A/W 实际重量
award 裁决;仲裁书
award in writing 书面裁决
award sweater 字母毛衣;团队毛衣
awasai wool 伊拉克阿瓦西羊毛
awjo ［西］阿瓦霍帽
awkward jog 不雅观的凹凸面
awkward-looking garment 不雅观的服装
awl 锥子;鞋钻子;服装裁片锥子
awn 金黄色
awning 遮阳篷
awning curtain 遮阳帷幔
awnings 帆布;帐篷布;椅布
awning stripe 遮日条纹;条子帐篷布
axis 轴,轴线
Axminster carpet 阿克斯明斯特地毯
aya momen 阿亚姆门软斜纹布(商名)
ayana 手织中厚棉布
ayas 日本软斜纹布
aye rishike 日本花缎
Aylesham cloth［fabric］ 阿伊勒沙姆亚麻布,阿伊勒沙姆织物
ayrishke 日本花缎
Ayrshire blanket 埃尔郡毛毯
Ayrshire embroidery 艾尔郡刺绣,埃尔郡刺绣

ayrshire work 挖花
azalea 杜鹃花红,浅蓝光暗红色
azamgar 阿赞加缎,阿赞加尔缎(巴基斯坦),棉经丝纬缎
azara 印度棉细布
azazul 印度经条薄细布
azelon 人造蛋白质纤维,再生蛋白质纤维
azera 印度棉细布
azidine dye 叠氮染料
azidosulfonyl dye 叠氮硫酰染料
azine dye 吖嗪染料,氮杂苯染料
azizulla 孟加拉丝棉交织细布
azlon(fiber) 人造蛋白质纤维,再生蛋白质纤维(属名)
azo 偶氮染料,冰染染料
azocron dye 阿佐克朗染料
azo dye 偶氮染料
azoic disperse dye 偶氮分散染料
azoic dyes 冰染染料,不溶性偶氮染料
azoics 不溶性偶氮染料
azomethine dyestuff 偶氮甲碱染料,甲亚胺染料
AZONA 阿桑娜
azo pigment 偶氮颜料
azores 粗纺厚呢
azoton 阿佐顿变性棉
aztec design 阿兹特克图案
aztec print 阿兹特克民族花样(墨西哥)
aztec sweater 阿兹特克毛衣
aztec weave 阿兹特克织物
azure(blue) 天蓝色,蔚蓝色,晴空蓝,天青色,淡青色,中蓝
azure cerulean blue 天蓝
azure green 碧绿色,天蓝色
azurine 天青色苯胺染料
azurite blue 石青蓝(灰绿蓝色)
Azzedine alaia 阿塞丁·阿莱亚(法国时装设计师)

B

B 偏胖型(服装);B 码(鞋);中号男睡衣
babao ornament 八宝纹
babby soxes 翻口式短袜
babci 墨西哥尤卡坦半岛剑麻纤维
babies 幼儿鞋,宝宝鞋,娃娃鞋
babies' clothes[wear] 婴儿装,婴儿服
Baboosh, babouche 巴布什鞋,土耳其式拖鞋,巴布希鞋;巴布希平跟软拖鞋(伊朗北部)
babouche(＝Baboosh)
babushka [俄]婆婆头巾,俄农妇三角包头巾,三角包头巾(俄)
baby band 婴儿绑带,绑扎带(婴儿)
baby blanket 婴儿毯
baby blue 淡蓝色,婴儿蓝,宝宝蓝(柔和浅蓝色)
baby bodice 婴儿衫
baby bonnet 婴儿软帽,婴儿罩帽,娃娃帽
baby boots 柔软鞋
baby bunting 婴儿法兰绒褡裢,婴儿睡袋
baby button 圆珠纽
baby cape 婴儿披肩,婴儿短斗篷
baby clothes 婴儿服装,婴儿连衫裤
baby clothing 婴儿装
baby deer 幼鹿皮
baby diaper 婴儿尿布
baby doll 儿童玩具服饰,婴儿服饰
baby doll dress 娃娃式女服,娃娃装
baby doll look 娃娃服装款式,童装式,娃娃风貌
baby doll pajamas 娃娃式睡衣裤,童式睡衣裤
baby dolls 娃娃式鞋
baby doll shirt 娃娃式衬衣;娃娃式睡袍
baby doll shoes 娃娃鞋
baby doll silhouette 娃娃型轮廓线条,洋娃娃装轮廓
baby doll sleeve 小泡泡袖
baby-doll style 宝宝款式
baby dress 婴儿装,贝贝衫,婴儿服
baby face 娃娃脸
baby flannel 婴儿法兰绒
baby footwear 婴儿鞋,婴幼鞋,幼婴鞋
babygro 弹性连裤装
baby hood 婴儿兜帽,婴儿头巾

baby irish lace 钩编窄花边
baby jacket 幼儿夹克
baby Jane 婴儿简式连衣裙
baby jeans 儿童牛仔装
baby keeper 婴儿外套,裹婴衣
baby lace 婴儿花边,婴儿服饰花边
baby lamb 羔羊毛
baby linker 套口机
baby lock 小型包缝机
baby look shoes 婴儿式女鞋,童式女鞋
baby Louis heel 路易十五低型跟,路易斯式婴儿鞋跟
baby nap 婴儿装,婴儿服,婴儿衣(初生时)
baby napkin 婴儿尿布
baby pin 婴儿别针,婴儿别针
baby pinafore 小孩围兜
baby pink 淡粉红,宝宝粉红
baby puff steeve 婴幼儿泡泡袖
baby ribbon 婴儿缎带,儿童用细丝带
baby romper 田鸡裤,婴儿连衣裤
baby rug 婴儿毡毯
baby sash 婴儿饰带,贝贝式丝饰带
baby's bonnet 幼儿帽
baby's clothes 幼儿服装
baby set 婴儿套装
baby's footwear 婴幼鞋
baby sharkskin 雪克斯金精梳细布,精梳雪克斯金细呢
baby shoes 婴儿鞋
baby silhouette 娃娃型轮廓线条
baby's jacket 婴儿短外衣,婴儿短外套,幼儿夹克(英)
baby skirt 贝贝式短裙
baby sling 婴儿系带,娃娃系带
baby socks 宝宝袜,婴儿袜,宝宝短袜
baby's padded set 婴儿填棉套装
baby's shoes 婴儿鞋,幼儿鞋
baby's suit 婴儿套装
baby Stuart cap 斯图尔特帽,下巴系带婴儿帽
baby suit 婴儿套装
baby's vest 婴儿汗衫
baby's wear 婴儿服装
baby textile 婴儿用纺织品
baby vent 浅开衩

baby wear 婴儿服饰,婴儿服饰品(总称)
baby wool 婴儿绒线,细支团绒线
bacaya 巴卡亚阿尔帕卡毛
bachelor's gown 学士袍
bachelor's gown and hood 学士服
bachelor's hood 学士袍黑色披肩
bachlick 女式开司米三角披肩
back (B) 背,背部,后衣身,后片;(织物)背面;倒缝,倒车,退针
back accent 强调后部
back and face effect 双面花纹;织物的正反面效果
back and fore 回针
back and forth stitching 来去线缝;来回线缝
back and forward motion 往复运动
back arc 背中至袖窿底边弧长,后背宽(腋下过)
back armhole 后袖窿
back armpit point 后腋点
back bag 背包,背囊
back band 鞋后袢带,后腰带
back band shoes 踵带鞋
back bared slip 露肩衬裙
back belt 后背带,背后腰带;后空鞋
back bodice 后衣身,后衣片
back bodice block[sloper] 后衣身原型
back body 衣服后身,后衣片
back bone 脊骨;脊椎骨
back cap depth line 袖山线
back center line 后片中心线,后背中心线
back center seam 后中缝,背中缝
back center seam line 背缝(中心)线,后中缝线
back centre 后心
back centre line 后片中心线
back centre seam 后中缝
back centre seam line 背缝(中心)线
back check 背格料
back check cloth 背格条双层织物
back checked cloth 背格条双层织物,背格料
back cloth 印花衬布,底布
back-coating 背面上胶;背面涂层
back counter 后衬(鞋);后跟皮
back cowl 后背垂褶
back crease 后挺缝线
back crotch 后裆
back crotch curve 后裆弧线
back crotch length 后裆长

back crotch level 后裆线
back crotch line 后裆线,落裆线
back crotch seam 后裆缝
back crotch stay 大裤底(后裆里子)
back cuff 袖口边里料,袖头边里料
back dart 后身省,后省,后片省,腰省
back dart line 后省线
back-dip 后垂式
back drop 彩画幕布
backed cloth 二重组织,二重织物,背面加重织物,背面起绒织物
back face 织物背面,织物反面
back-fastening cardigan sweater 反穿开衫
back-fastening clothes 反穿式服装,反扣式服装,反穿衣(后扣式)
back feed dog 后送布牙
back feeding stitching 倒缝
back filled fabric 背面重浆织物,单面重浆布
back-filled white sheeting 单面浆漂布
back filling 背面上浆,单面上浆
back filling finish 背面上浆整理
back finishing 单面上浆整理
backflush chromatogram 反冲色谱
back fork 后衩
back fullness 后身宽松
back full skirt 后膨裙
back gimp 装饰用嵌芯窄辫带
back gird 后横裆
back gore 裙子后裆,三角形后裙片
back gray 印花衬布
back gray crease mark 印花衬布折痕
back gray seam 印花衬布缝头
back gray seam impression 印花衬布接缝痕(滚筒印花疵点)
back gray seam mark 印花衬布缝头印
back greige 印花衬布
back grey 印花衬布
back grey mark 衬布印
background 布(料)底;底色;背景,本底
background color 底色,背景色
background ink 背景墨水
background reflectance 背景反射
back herringbone stitch 篱笆贴线绣
back hip 后臀
back hip pocket 后裤袋
backing 衬垫;里衬,背衬;背面上胶;嵌花加固
backing cloth 衬布,底布
backing fabric 底布,衬布,背衬织物;印

花衬布
backing silk 丝衬
backing strip 衬板
backing tape 衬带
back innerlining 后夹里
back inseam line 后内缝线
back interfacing 后片衬
backlap tricot 经编手套织物
back leg 裤后片,后裤片
back length (B.L.) 后身长,背长
back length line 背长线
backless 露背装,露背女服,露背服;露背泳衣
backless bra 露背胸罩
backless brassiere 露背式胸罩
backless dress 露背裙,露背女裙
backless pumps 露跟浅口鞋,后袢带碰拍鞋
backless sandals 后带凉鞋,镂空凉鞋
backless shoes 后带鞋,露跟鞋,后空鞋
backless slippers 后空便鞋,无后帮便鞋
backless vest (配男夜礼服的)无背心,燕尾服露背式背心
backless waistcoat 燕尾服露背式背心,无背背心
back light 逆光
backlining 反面贴布,反面衬布;后里身
backlog of order 已接受订货总数
back muscle 背肌
back neck 后领圈,后领口,后颈中心点至颈肩点
back neck dart 后领省
back neck depth 后领深
back neck facing 后领贴边
back neckline (BNL) 后领圈线;后领口
back neck opening 后叠门,后开门
back neck point (B.N.P) 后颈点,颈围后中心点
back neck run 后领线
back neck width 后领宽
back notch 后定位点,后中心线刀口;后衣片刀口
back of coat rides up 后身起翘
back of fabric 织物反面
back of the coat rides up 后身起吊
back of the hand 手背
back open dart 后身通省
back opening 后开襟,后开门
back outer 后身面料
back overlap 后搭门,衣后搭门
back pack 登山背囊,双肩背包

back-pack scuba gear 背负式水下呼吸器
back pack strip 背包带
back panel 后幅,后片,后中片
back panel seam 背嵌缝
back pants 后裤片
back part 后片,后幅,后身,上衣后片
back part moulding machine 后踵定型机
back part tolerance 后容差
back patch pocket 后贴袋
back pitch 后定位点,后对位点,后凹势
back pleat 后裥
back pleated skirt 后襞褶裙
back pleat skirt 后褶裙
back pocket 后袋
back pocket opening 后袋口
back press 后身熨烫;后身熨烫机
back raised cloth 毛背织物
back rise (BR) 后裆,后直裆,后浪,落裆
back rise curve 后裆弧线
back rise line 后裆线,落裆线
back rise seam 后裆缝,落裆缝
back rise width line 后裆宽线
back robbing 簇绒织物带针疵点
back satin gabardine 缎背华达呢
back satin serge 缎背哔叽
back seam 背缝,背中缝;全成形长袜线缝;后缝(鞋子)
back seam line 背缝线
back seam opening press 背缝开熨机
back shear 背面剪毛
back shoulder dart 后肩省
back shoulder line 后肩线,肩后线
back shoulder neck 后肩颈点
back shoulder neck point 后肩颈点
back shoulder point (BSP) 后肩颈点
back shoulder run 后肩缝线
back shoulder seam 后肩缝
back side 背面,后肋,背部,后侧;臀部
back side length 后胁长;后摆缝长
backside loop 背面线圈
backsides 臀部
back side seam 背侧缝
back side thread 反面起绒纱
back-side-up 反面朝上(面斜)
backsizing 单面上浆;背面上浆
backskin 背皮,仿翻皮织物,仿鞣皮织物
backskin cloth 仿鞣皮织物
back skirt 后裙片
back skirt block 后裙片原型
back skirt yoke 后裙覆势

back sleeve 后袖,小袖
back sleeve cap 后袖山
back sleeve drape 袖后垂份
back sleeve seam 小袖缝
back sleeve yoke 后肩袖覆势,后肩袖育克
back slope 背部倾角
back socket bone 后背肩胛骨
back starching 单面上浆,背面上浆
back stay 鞋子保险皮;后撑条,背撑
back step heel 后置跟
back stitch (缝纫中)回针,倒钩针,倒缝;扣针(脚);倒针(法);倒扎针线迹,回针线迹
back stitch armhole 倒钩袖窿
back stitch at the end of opening 剪口末端处回[倒]针
back stitch back rise 倒钩后裆缝
back stitch back seam 钩后裆缝
back stitched seam 回针;倒缝缝式
back stitch embroidery 倒钩刺绣,倒缝刺绣
back stitching armhole 倒钩袖窿
back stitching back crotch 钩后裆缝
back stitch(ing) neckline 倒扎领窝(沿领窝用钩针法缝扎)
back stitch mechanism 倒针装置
back stitch sewing 回针加固缝纫
back strap 后袢带鞋;后背带扣(用以调节),背后带扣;后跟条
back strap bag 后插带提包
back strap shoes 后袢带鞋
back strings 童装后背带
back stripe 背面条纹呢,背条纹双面呢;缎面斜纹背双重织物
back swing 回针缝,复旧
back tack 回针加固缝,加固,倒回针,缝端加固
back tacker (BT) 固缝机
back tacking 加固缝,加固缝,缝端加固
back tacking device 加固缝装置,倒缝装置
back tacking guide 加固缝导向器,倒缝导向器
back tack sewing 加固缝,倒缝
back through dart 后通省
back to back 反面对反面相叠
back to back knitted fabric 双面针织物
back to face 反面对正面相叠
back to face shading 正反面色差
back to nature 回归自然
back torso 后躯干

back tracking 返回
back treatment 后背设计
back tucking 后褶裥
back tuck operation 后褶裥加工
back type 背型
back vent 背衩,后开衩
back view 后视图
back view dress 露背装,露背服
back waist arc line 后腰弧线
back waist band 腰里衬
back waist dart 后腰省
back waist length 背长,后背长
back waistline 后腰围线
back waistline point (BWP) 腰围线后中点
back waist seam line 裤后腰缝线
backward and forward tacking 正反褶裥
backward crochet 后向钩编
backward hair 后卷发型
backwardness 滞后
backward sewing 倒向缝纫
backward stitch 倒缝线迹
back washer 复洗机
back width 背宽,后背宽(腋下过)
back width line 背宽线
back winding 倒线
back-wrap dress 围裹式连衣裙,后包裹装
back yoke 后过肩;后接块布;后切替片
back yoke line 后过肩线
Bactekiller 聚酯抗菌纤维(商名,日本钟纺)
bacterial inhibition 抑菌性
bacteria-proofing 防菌整理
bacteria repellancy 抗菌性
bactericidal finish 灭菌整理
bacteriostatic fiber 抑菌性纤维
bacteriostatic finish 卫生整理
bactrian wool 双峰骆驼毛
badam khas 印度巴登细棉布
Badam lace 巴登花边(优质梭结花边)
Badam pattern 拜达姆纹,棕榈纹,火腿纹样,草履虫纹样,佩斯利旋涡纹
bad color 印色不佳
bad cover 布面覆盖不良
Baden 巴登亚麻平布
Baden hemp 巴登大麻
Baden lace 巴登梭结花边
Baden rubbers 巴登擦身浴巾
bad fit (服装)不合身
badge 章(如徽章,胸章,帽章),标志,补子
badger 獾毛皮
badger whisker 水手帽的飘带

bad hand 手感不良
bad ironing 烫工不良
badminton shoes 羽毛球鞋
badminton suit 羽毛球装
bad packing 包装不良
bad register 对花不准
bad selvage 坏边
bad selvage sewing 缝边不良
bad selvedge sewing 缝边不良
bad whites 白色不白
bad work 疵品,疵点
baeta 台面呢
baff,baft(a) 巴夫
baffeta 粗平布;平纹印花棉布;漂白重浆衬衫布;柞丝经棉纬花缎(印度、孟加拉)
baffity 未漂平布,细平布
baffs 粗平布,平纹印花棉布,漂白重浆衬衫布,柞丝经棉纬花缎(印度、孟加拉)
Bafixan dye 巴弗散染料(分散染料,商名,德国巴斯夫)
baft 粗棉布
baft(a) 粗平布;平纹印花棉布;漂白重浆衬衫布;柞丝经棉纬花缎(印度、孟加拉)
bafthowa 印度轻薄纱
baft ribbon 胶纱带
bafts 窄幅本色布,狭幅本色布
bag 包;袋;手袋;口袋;箱;衣服拱起处;小袋羊毛
bagage band 行李带
Bagalpote cotton 印度巴加尔波特棉
bagasse fiber 甘蔗渣[浆粕]纤维
bag bed sheet 被套
bag bodice 女式袋状紧身胸衣
bag bonnet 女式袋状软顶帽
bag cap 男式袋状布帽
bag closing machine 封袋机
bag cloth 口袋布,袋用织物
Bagdad lamb 巴洛达羔皮
bagdalin 巴格达林细布
bag fabric 袋织物,袋布
baggage 手提行李
bagged edge 暗线缝边(鞋子)
bagged seam 反缝(鞋)
baggie 小袋
baggies 宽松式男子游泳裤;宽翻边男长裤;萝卜裤,袋状裤
bagginess 服装起拱,局部起拱
bagging 装袋,缝袋;织物起拱;打包布;制袋材料,袋布
bagging-bag 厚黄麻袋布
bagging height 拱胀高度(服装起拱)
bagging load 拱胀负荷(服装起拱)
bagging out 翻转(贴袋缝夹里后)
bagging property 起拱性能
bagging shoes 宽松式高帮鞋
bagging tendency 起拱性,起拱倾向
bag goods 袋织物
baggy 针织服装袋状变形
baggy cloth 起拱织物,袋状织物
baggy fabric 起拱织物,袋状织物
baggy jeans 宽松牛仔裤
baggy line 袋状宽大轮廓,袋式轮廓
baggy look 袋形款式;袋状服装款式
baggy pants 袋形裤,萝卜裤,太子裤
baggys 宽身男短裤(冲浪运动员、拳师用)
baggy seat 后裆下垂
baggy shirt 袋型衬衫(浮离身体,宽松)
baggy shorts 袋形短裤,宽身男短裤(冲浪运动员、拳师用)
baggy style 袋形款式
baggy trousers 袋形长裤
baghaitloni 巴海洛尼毯子,彩花留口毛毯
bagheera 巴格希拉毛圈丝绒,匹染毛圈丝绒
bagheera velvet 毛圈天鹅绒
bag hemmer 袋卷边器
bagi pat 巴吉帕黄麻,长果种黄麻
bag making machine 缝袋机
bagnolette 带金属撑的女式兜帽
bag opening machine 开袋机
bag-out 袋缝(两块布缝合后将正面翻出),服装鼓起,服装起拱
bag overturning machine 翻袋机
bag pack 背包
bag packing machine 装袋机
bag pants 袋形裤
bagpipe sleeve 风笛宽松袖
bags 裤子,男裤(英国俚语),长裤,裤
bag-seal sewing machine 封包缝纫机
bag seam 来去缝,袋缝,法式缝,来回缝
bag seamer 缝袋机
bag seaming machine 缝袋机
bag sheeting 口袋布,袋布,袋料
bag side 贴袋边
bag sleeve 15世纪袋形袖,宽袖,袋状袖
bag-stitching machine 缝袋机
bag strapping 家具护边带;家具包装带
bag type design 袋形设计

baguette [法]条纹饰;台面呢;矩形宝石
bagwash 初洗洗衣店;初洗,一袋初洗衣服
bag wig 饰袋假发,丝囊假发,袋形假发
Bahama cotton 巴哈马棉
Bahama hemp 巴哈马剑麻
Bahama pita 巴哈马剑麻
Bahamas sisal 巴哈马剑麻
Bahia cotton 巴伊亚棉
Bahia style 巴伊亚风格
Bahmia cotton 巴米亚棉,埃及长绒棉
bah-tow neck 一字领,船形领
baidie fabric 巴叠布
Bai ethinic costume 白族服饰
baigue 单面起绒毛呢;台面呢
baihua brocade 百花缎(锦)
baihua ornament 百花纹样
bail 提链头
bailee 受委托人
bailer 委托人
Bailey cotton 美国贝利棉
bailment documents 寄存单据
baimiao 白描
bainin 爱尔兰贝宁呢,爱尔兰家纺粗毛呢
baique 比利时粗台面呢
Bairaiti cotton 印度拜拉迪棉
bair clippers 理发推子
baiye fabric 百页布(加工豆制品)
baize 台面呢
bajota 巴乔塔漂白粗平布
baju 马来西亚宽松短袖上衣,巴鹭外套(印尼)
baked print 已焙烘的印花布;焙烘印花
bakelite button 电木扣,电木纽扣
bakhtiari rug 伊朗毛毯,伊朗棉毯
bakia 巴基阿木鞋
Bakinox 不锈钢短纤维(用于抗静电织物,商名,美国)
bakisum 巴基森绸
bakrabadi 印度高级黄麻
bakshaish rug 波斯棉地毛绒头地毯
bakshis rug 波斯棉地毛绒头地毯
baku 山东巴库草帽;山东草秆编织物;东方手结地毯
bal 活领;巴尔牛津鞋
balaclava 巴拉克拉瓦盔式帽厚大衣;巴拉克拉瓦盔式帽,帽兜(只露两眼套至颈部)
balaclava cap 巴拉克拉瓦盔式帽(包头,护耳,长及肩部)
balaclava helmet 巴拉克拉瓦盔式帽(包头,护耳,长及肩部)
balaclava hood 巴拉克拉瓦盔式帽(包头,护耳,长及肩部)
bala longyi 印度手织平纹丝织物
balanac 巴兰克粗厚呢
balance 缝头;(衣服穿挂)平衡性;均衡,平衡
balance accounts 结算
balance between imports and exports 进出口平衡
balanced cloth 平衡织物(经纬密度与特数均相同的织物)
balanced crepe 平衡绉
balanced fabric 平衡织物;双面织物
balanced figure 均衡体型
balanced sided twill 双面斜纹
balanced twill 平衡斜纹
balance expenditure with income 收支平衡
balance in hand 结余
balance line 平衡线
balance mark 叠合印;剪口(打褶标记);对位记号,合印,平衡记号
balance of count 平衡织物
balance of payment 国际收支
balance of trade 贸易差额
balance sheet 资产负债表
balance wheel (缝纫机头上的)手轮
balandran [法]贝伦披裙风
balao cotton 巴罗棉
balasees 巴拉西漂白平布
balas ruby 玫红夹晶红宝石
balasse 巴拉西漂白平布
balassor 印度韧皮纤维织物
balaster 巴拉斯特金线织锦
balastre 巴拉斯特金线织锦
balata belting duck 涂巴拉塔树胶的传送带帆布
balayeuse 军装高領,关驳领;防尘裙摆折边
balbriggan 巴尔布里根内衣针织物
Balbriggan hose 巴尔布里根袜
Balbriggan pajama 巴尔布里根睡衣
bal collar 上下领,列宁装领,军装高领,关驳领,高军袋翻领
balconette bra 百克奈胸罩
baldac 宝大锦
baldachin(e), baldachino, baldaquin 宝大锦;天盖,华盖
bald head 秃顶,秃子
baldness 外观平滑
baldric, baldrik 佩带,豪华饰带,饰带,

肩带
bale 包,捆,件
bale belt 打包带
bale capacity 容积
bale carton by belt 打扎箱带
baleen corset 鲸须胸衣
BALENCIAGA 巴黎世家
Balenciaga 巴伦西亚加(西班牙时装设计师)
balernos 光柔马海毛
Balestra 巴莱斯特拉(意大利时装设计师)
bale strapping 捆扎包装
Bali dancer's costume 巴厘舞服
baline 马鬃衬,巴林织物,大麻粗布,黄麻粗布;棉经马鬃纬衬里
baling 打包;捆;扎
baling machine 打包机
baling press 打包机
baling ties 打包铁皮
baling twine 扎包绳,捆包线
baliti 菲律宾树纤维平布
balk 条格色彩或纹样不完整
Balkan blouse 巴尔干宽松上衣(至臀围,束带,宽摆),巴尔干罩衫,巴尔干外套
balk back 绒背呢
balky selvedge 不易处理的布边
ball 团(纱包装单位);纱球,线球,绒线球,团球
ballanca 巴兰卡粗厚呢
ball-and-socket fastener 凹凸式按扣
ballantine 女式小提袋
ballasor 印度韧皮纤维织物;印度亚麻平布
ball button 球形纽扣,球形按扣
ball design 团花图案
ball dress 舞会服
ball earring 球形耳环
ballerina 芭蕾舞长裙,软底低跟女便鞋
ballerina costume 长裙式芭蕾装,芭蕾舞女装
ballerina dress 短晚礼服
ballerina frock 露肩无吊带跳舞长衣
ballerina length 芭蕾舞裙长,芭蕾舞服尺寸
ballerina length skirt 芭蕾舞式宽松及小腿长裙
ballerina shoes 芭蕾舞鞋,芭蕾鞋
ballerina skirt 芭蕾舞裙
ballerina suit with trunks 长短裤与芭蕾舞衣
ballerina umbrella 芭蕾舞伞

ballerine 轻便女鞋
ballet costume 芭蕾舞服装,芭蕾舞女装
ballet dress 芭蕾舞裙装
ballet lace 芭蕾舞鞋宽缎带
ballet pump 芭蕾舞鞋
ballet shirt 舞剧团排练服;芭蕾舞层叠晚裙
ballet shoes 芭蕾舞鞋,芭蕾鞋;软底白布鞋(放射性场所用)
ballet silhouette 芭蕾舞轮廓
ballet skirt 芭蕾裙,芭蕾舞裙
ballet slippers 芭蕾舞演员鞋,芭蕾舞鞋,芭蕾鞋
ballet tights 芭蕾舞紧身衣裤;连袜紧身衣裤
ballet toe 芭蕾舞袜袜头,夹长袜头,矩形袜头
ballet toe shoes 芭蕾舞鞋,芭蕾鞋
ball flower 球心花饰,花球
ball fringe 珠形穗饰
ball gown 晚会礼服,长礼服,正规礼服,舞会服
ball heel 球状跟,球形鞋跟
ball hemmer foot 筒摆压脚
Ballindalloch 巴林德罗其
ballistic fabric 防弹布
ballistic (resistant) fabric 耐冲击织物,防弹织物
ballistic tester 冲击式强力试验机
ballistic vest 防弹背心
ball of the foot 跖球
ball of thread 线团,绒球
balloon cloth 气球用布,气球布
balloon dress 气球连衣裙,膨大气球装,灯笼形连衣裙
balloon fabric 气球织物
balloon hat 气球帽,女式气球帽
balloon line 气球线形,气球形服装款式
balloon links 链状饰边
balloon pants 灯笼裤
balloon shape 气球形
balloon shorts 宽大短裤
balloon silhouette 气球形服装轮廓
balloon skirt 气球裙,灯笼裙
balloon sleeve 气球形袖,气泡袖,灯笼袖
balloon stitch 球形针迹,起圈针迹
balloon stitch seam 起圈线迹缝
ballpoint embroidery 液体刺绣(从圆珠笔式管里挤出颜料,涂在织物上,形成刺绣效果)
ballpoint needle 圆头缝针,圆头针;圆头

刺针(非织造布针刺工艺用)
ballpoint pin 圆头大头针
ballproof clothes 防弹衣,防弹服,避弹衣
ballproof fabric 防弹织物
ballproof jacket 防弹衣,防弹夹克
ballroom dance clothing 舞厅服
ballroom neckcloth 男式舞厅领巾
ball-tipped shears 球顶剪口
ball top 毛球
ballushar 印度手织莎丽绸,巴流莎丽花绸(印度)
ball woollen knitting yarn 团绒,卷装绒线
BALLY 百丽
ballymena 巴利来纳亚麻平纹布
balmacaan 粗呢套袖翻领男式短大衣;粗呢直筒大衣
Balmacaan coat 巴尔马干大衣;连肩袖轻便外套,宽松插肩袖大衣,巴尔玛肯大衣
Balmacaan collar 巴尔玛肯领,高军装领
balmacan 巴尔玛肯大衣,粗呢直统大衣,巴尔玛肯大衣,粗呢套袖翻领男式短大衣,上下领斜肩袖外套,轻便大衣;连肩袖,插肩袖
Balmain 巴尔曼(美国时装设计师)
balmora 巴尔摩尔
balmoral 巴尔莫勒呢,巴尔莫勒尔粗纺条纹呢;巴尔(莫勒尔)牛津鞋,内耳式鞋,镶花边的靴,结带皮靴;结带鞋;斜条衬裙;巴尔莫勒尔无边圆帽,无边平顶圆帽,镶花边靴
Balmoral bodice 巴尔莫勒尔女式紧身胸衣
Balmoral boots 巴尔莫勒尔黑色短靴,巴尔莫洛靴,镶花边皮靴
Balmoral cap[bonnet] 巴尔莫勒尔帽,无边平顶圆帽,苏格兰蓝色便帽,巴尔莫洛帽
Balmoral oxford 巴尔莫勒尔内耳式系带浅帮鞋,内耳式系带浅帮鞋
Balmoral oxford shoes 巴尔(莫勒尔)牛津鞋
Balmoral petticoat 巴尔莫勒尔羊毛衬裙
balmoral (shoes) 内耳式鞋
balmoral slip 条纹呢衬裙
Balmoral tartan 巴尔莫勒尔格子花呢
balong 巴龙(菲律宾传统服装)
bal oxford 巴尔莫勒尔粗纺条纹呢;巴尔莫勒尔鞋,内耳式鞋
bal peak lapel 下斜戗驳领

balsa fiber 香[味]纤维(聚合物加香料纺丝)
balsam 蟹绿
balsam design 凤仙花纹图案
balsam green 青灰色,凤仙花绿
balso 巴尔萨术棉纤维
Baltic 波罗的海色
baltic seal 波罗的海兔毛皮
baltic tiger 波罗的海兔毛皮
baluchar 巴流莎丽花绸,印度手织莎丽绸
baluchistan carpet 羊毛驼毛制织的长毛毯
baluchon 包袱
balustre 巴拉斯特金线织锦
balzarine 棉毛轻薄纱,群青印花薄纱
balzarine brocade 匹染织花薄纱
balzorine 棉毛轻薄纱,群青印花薄纱
bambagino 意大利印花细平布
bambin 宽边女帽
bambino hat 宽边女帽
bamboo 竹黄色;竹
bamboo blind cord 编结竹帘的棉线
bamboo button 竹扣
bamboo cane 竹手杖
bamboo clappers fan 竹板扇
bamboo clip 竹夹
bamboo dice handbag 竹块编手提袋
bamboo fiber 竹纤维,竹原纤维
bamboo fiber clothing 竹子纤维服装
bamboo fiber yarn 竹原纤维纱
bamboo handbag 竹编手提袋
bamboo handle 竹柄
bamboo hangbag 竹制手提袋
bamboo hat 斗笠,斗笠帽
bamboo pole 晒衣竹
bamboo rain hat 笠帽
bamboo/ramie blended yarn 竹原苎麻纱
bamboo ruler 竹尺
bamboo/silk blended yarn 竹原纤维丝混纺纱
bamboo split hat 斗笠
bamboo tape 竹尺
bamboo (walking) stick 竹手杖
bamboo/wool blended yarn 竹原纤维毛混纺纱
bamboo (yellow) 竹黄,浅灰黄色
bamd cuff 条形克夫
bamia 巴米亚棉;秋葵纤维,刚果黄麻纤维
bamia cotton 埃及巴米亚棉
ban 高级平纹细布;香蕉茎纤维薄纱
banaati 印度粗纺毛呢;印度漂白棉布

banana 香蕉黄,浅奶油色;香蕉纤维
banana cream 香蕉奶油色
banana fiber 香蕉茎纤维
banana green 香蕉绿
banana yellow 香蕉黄
banat 印度巴纳特全毛呢
Banbury plush 班伯里长毛绒
Bancroft cotton 美国班克罗夫特棉
band 绳,带,衣带,门襟翻边,镶边,嵌条,上腰育克;袖口;领口;裙腰;扁平戒指
bandage 绷带
bandage cloth 绷带布
bandage gauze 纱布绷带
bandage scrim 绷带用平纹织物,稀薄平纹织物
bandaid 急救绷带
bandala 蕉棉,蕉麻纤维织物
bandalette 围巾;手套的宽边
bandan(n)a 印度班丹纳印花绸;班丹纳印花布;丝巾,牛仔巾,印度方巾;扎染印花手帕
bandan(n)a design 班丹纳图案
bandan(n)a print 班丹纳印花
banda stripes 班达条纹布
bandbhendi 班本迪纤维(黄麻代用品,美洲,非洲产)
bandbox 硬纸盒
band bra 束带胸罩
band brassiere 吊带乳罩,束带胸罩
band briefs 镶边三角裤,松紧口三角裤
band cloth cutter 带式裁剪机
band collar 中式立领,竖领,直领,条领,带型领
band collar shirt 立领衬衫
band conveyor 带式输送机
band cuff 条形克夫(单片对折),直筒克夫;带状袖口布
band cutter 带式裁剪刀;带式裁剪机
band cutting machine 钢带裁衣机
bandeau 束发带,细带;窄边帽带,蒙眼布条;女头带,帽圈,窄带式胸罩;绑兜
bandeau bikini top 绑兜式比基尼上装
bandeau bra 窄条胸罩
bandeau slip 胸罩式衬裙
bandeau style 班杜发型
bandeau swimsuit 无带弹力游泳装
bandeau top 绑兜式连身衬裙;绑兜式上衣
bandeau top slip 绑兜式连身衬裙
banded collar 立领,上下盘领

banded cuff 条形克夫;条形直袖口
banded jacket 有带夹克
banded lace 条形花边
banded neckline 窄条型领口(前开纽孔)
bandelet 围巾;手套的宽边
band end line 底领前宽斜线
bandero(le) 绶带;桅顶燕尾旗;丧旗
banderol 绶带
bandhana silks 印度本哈纳扎染绸
band hem 镶边下摆,镶边缘边
bandhnu 班丹纳印花布
bandhor 小亚西亚班德霍地毯;土耳其棉毛毯
banding 机织带,织带,针织带,镶边,嵌条
banditti 便帽羽饰
band knife 条刀,带式裁剪刀
band knife cloth cutting machine 带刀坯布裁剪机
band knife cutting machine 钢带裁布机;钢带裁衣机
band leg panties 镶边内裤
band leg under pants 镶边内裤
bandle linen 爱尔兰家织粗亚麻平布
bandless and faced 无腰带腰部明贴边
band neckline 立领领口,底领下口线,低级下口线
band of embroidery 绣花条
bandoleer 军用肩带,武装带,子弹带
bandoleer cloth 子弹带织物
bandoleer necklace 子弹带型项链;子弹带型项饰带
bandolier 子弹带,子弹带腰饰;军用肩带,武装带;女式肩背包;弹药宽皮带
bandolier cloth 子弹带织物
bandolier necklace 子弹带项链,弹带型项链
bandoline 胡须膏
band opening 贴边式开口
band ornament 带装装饰
band placket 贴边式开衩
band roll line 底领上口线
bands 两条胸前饰带;成形针织服的织边;鞋面加固线
band spectrum 带光谱
band strings 带流苏的线带
band tape 皮尺,卷尺
band wheel 下带轮
band width 带宽
band work 花边的填孔花纹
band wrist (衬衫的)单层袖口

bandy 废纺条子棉布
bandy stripes 班达条纹布
bang 前刘海
bangalore 印度班格洛尔手织地毯
bangara kapor 印度班格拉卡普绸(金线高级织物),班格拉卡普绸(印度)
bang hairstyle 前刘海发型
bangle 手镯；脚镯
bangle bracelet 串环,手环,串环手镯,环镯
bangra 印度荨麻手工织物
bangs 垂前发,刘海
bangtail 花缎背面残留线
bang-thro prints 压透印花布
banian 宽大的法兰绒长袍(或上衣、衬衫),宽大外套(印度),班扬衬衫(印度)
banian fiber 榕树韧皮纤维
Bani cotton 印度巴尼棉
banjo patch 垫布
bank 银行
bank deposit 银行存款
bank draft 银行汇票
banking 对齐
banking business 银行业务
banking center 金融中心
banking market 金融市场
banking the pattern 前后样片对缝
bank invoice 银行发票
bank's exchange settlement 银行结汇
bank's lien 银行留置权
bank stamp 银行戳记
bank statement 银行对账单
bankukri cotton 印度班库克里棉
bank uniform 银行制服,银行服
banne 帆布；篷布；椅布
Banner 班纳拼色短袜(袜面与袜底异色)
banner cloth 旗布
banner lap pattern sock 闪色花袜
bannerman shoes 旗鞋(中国)
banner sock 拼织短袜
bannock 面包色(鲜黄褐色)
bannockburn 班诺克本花呢,班诺克本高级花呢
Bannockburn tweed 班诺克本粗呢
Banque de France (法语)法兰西银行
bant cotton 天然转曲较多的原棉
bantis 棉织物
banyan 法兰绒衣；豪华长袍；榕树服(印度传统风格大衣)
baodai crossover 宝带绸
baodu 抱肚,围肚

baoling crepe faille 宝领绉
baoxiang flower 宝阳花纹
bar (铁、木)条；杆；棒；横线加固缝,锁眼缝迹；横档(织疵)；巴(气压单位,1巴等于 10^5 Pa)
baracan 巴拉坎厚呢；风雨大衣
baracanée 巴拉坎内经棱条色布,经棱素色织物
baracan grosgrain 巴拉坎罗缎；巴拉坎横棱纹织物
barak 驼毛外衣
Barakat cotton 苏丹巴拉卡特棉
bara-poloo silk 高级白蚕丝
baras 粗袋布
barathea 巴拉西厄领带绸；巴拉西厄毛葛,巴拉西厄军服呢,横贡呢
barawazi 彩格棉布
barb 女式花边围巾；倒钩
barbadoes cotton 巴巴多斯棉,安圭拉产海岛棉
Barbara 芭芭拉
Barbara Hulanicki 巴巴拉·胡拉尼基(英国时装设计师)
barbe [法]巴毕头饰；修女颈前折巾；长条白色孝服麻
barber 理发师
barberry 酸果红
barber's apron 理发围兜
barbershop 理发馆
barber's shop 理发馆
barbette [法]女子遮巾,(13～14世纪)女式亚麻颏下系带；修女白色头帽
barboteuse 连身短裤,连背心玩水裤
barbour 巴布尔夹克
bar button 杆式纽扣,有柄扣
bar buttonhole bar 纽孔加固缝
Barcelona 巴塞罗那丝围巾；巴塞罗那棉
Barcelona handkerchief 巴塞罗那丝头巾
Barcelona lace 巴塞罗那花边(褶裥梭结花边,西班牙制)
bar chart 条线图；条形图
barchent 低级单面绒布
bar code 条形码
bar code reader 条码识读器
Bardot style 巴多发型
bare arm 露臂
bare back 露背装,露背式,裸背
bare back blouse 女露背上衣
bare back dress 露背裙,露背装,露背服
bare backed 露背的

bare back vest 露背背心
bare bra （胸部无面料覆盖的）框架式胸罩，半胸罩
bare cloth 头眼布疵，稀布疵；稀织物
bare elastic yarn 无包覆弹性纱
bare face 不起绒织物
bare foot sandals 赤脚凉鞋
barège 巴雷格纱罗，巴雷格披巾，法国印花披巾
barége grenadine 巴雷格棉麻纱罗
bare leg stockings 无缝女袜，透明女袜
bare line 暴露型
bare look 裸露风貌，裸露款式，暴露款式
barely blue 淡蓝
bare midriff 露脐装，露腰装；露腹风貌；露腰风貌
bare-midriff dress 露腰式服装，露脐装，露腹式服装，中空式服装
bare-midriff look 露腹服装款式
bare-midriff pajamas 露腹式睡衣；露腰式睡衣
bare-midriff style 露腹款式，露腰款式，露腰式，露腹式，中空式
bare-midriff top 露腰上衣
bare neck 露颈
bare shoulder 露肩领口，裸肩
bare shouldered gown 女露肩长衣
bare shoulder neckline 露肩领口
bare side 露腋
bare silk 树皮绉绸
bare style 暴露款式，裸露式
bare top 露上装，露肩装，露胸装（露出胸围线以上，无背带）
bare top bodice 露肩式上衣，露肩式衣身
bare top dress 露胸式衣服
bare top strapless 无吊带露上跳舞长衣
barette 条状发夹
bare weight 皮重
bar fagoting 梯形开口缝，小方形开口缝（将窄带或折用四方形针迹缝牢）
bar-filling 罗纹边套口
barfoul 西非进口的棉布
bargain 廉价货；买卖
bargain away 廉价出售
barge 驳船
bargello 佛罗伦萨刺绣
barghebed 贾马瓦尔平级织物
bar graph 条线图；条形图
barhak 巴哈克驼毛厚呢
barhana 土耳其低级地毯

barillo 印度尼西亚低档丝
bark 深青褐色，鼠灰色；驼毛织物
bark cloth 树皮布（用树皮纤维织制）
bark crepe 人造丝树皮绉
bark crepe weft knitted fabric 纬编树皮皱针织物
bark fiber 红树韧皮纤维
bark silk 树皮纹绸
bar length 加固缝长度
bar lock 加固缝
bar mark 条痕疵
barme cloth 围裙
barmen lace 巴门镶边，窄花边
barmhatre 围裙
barmskin 皮围裙
Barnes cotton 美国巴恩斯棉
Barnett cotton 美国巴涅特棉
Barnsley crash 巴恩斯利窄幅棉织粗毛巾
Barnsley linen 巴恩斯利绣花用亚麻平布
barntine 中东班廷生丝
baroda carpet 巴罗达毯
baronette satin 男爵缎（作运动服用）
barong(tagalog) 网眼花衬衫（菲律宾男子穿的）
Baroque 巴洛克艺术风格
Baroque art 巴洛克艺术
Baroque art style 巴罗克服装艺术风格
Baroque color 巴罗克色
Baroque female costume 巴罗克式女服
Baroque male costume 巴罗克式男服
Baroque ornament 巴洛克装饰
Baroque pearl 巴洛克珍珠
Baroque period 15～19世纪奢华服饰时期
Baroque print 巴洛克印花；巴洛克式印花图案
Baroque style 巴洛克艺术风格
bar pin 棒形胸针
barpours 法国巴波哔叽，法国巴波哔叽
barracan 巴拉坎厚呢，巴拉坎风雨大衣呢
barrage 法国提花亚麻台布；亚麻花布；麻毛缎
barragon 纬二重单面绒布（工作服用）
barragones 斜纹裤料
barras 荷兰粗亚麻布
barratee 巴拉蒂绸
barre 横档疵，纬条痕，纬向条花，纬向条子织物
barré 纬向条子织物；纬向条花；横档疵点
barred buttonhole 加固扣眼

barred end 固结缝,加固端缝
barred herringbone stitch 人字加固线迹
barred twist 合股色节花线
barred witch stitch 人字形缝;人字形加固线迹
barre fabric 纬色条织物,纬色条针织物
barre-free fabric 无条痕织物
barrège 丝毛纱罗;丝毛披巾
barrel 无边扁平帽;筒形裙,陀螺裙
barrel bag 桶状袋,桶袋;跳伞包,筒包
barrel chest 桶胸型
barrel chested 筒胸型的
barrel coat 筒状大衣,桶状大衣
barrel cuff 筒形袖口,单袖头(男衬衫),桶形袖头
barrel curl 桶形卷发;桶形卷发发型
barrel hand bag 筒包,桶包;跳伞包
barrel heel 桶形跟,筒状鞋跟
barrel knot tie 桶状领带
barrel line 桶形线形
barrel pleating 管状褶裥
barrel-shape 琵琶桶形(珠宝)
barrel-shaped muff 筒状皮手笼
barrel silhouette 桶形廓线;桶形轮廓;樽型轮廓
barrel skirt 大筒裙,圆筒裙,樽形裙
barrel sleeve 筒状袖,桶状袖
barrel type handbag 圆筒手提袋
barre mark 纬向条痕,横条疵
barret 扁平便(或软)帽,无边扁平帽,无边平顶帽,四角帽(中世纪);毛呢雨衣
barrette 四角帽,犹太男式圆平顶无檐帽;条形发夹,条状发夹
barrier fabric 防护织物,阻挡层织物
barriness 纬向色档
barring 起棱,条痕疵,条花织疵
barring point 缝迹加固点,加固点
barring stitch 加固线迹,套结线迹
barring stitch width 加固线迹宽度,套结线迹宽度
barring tripping lever 加固缝机构开关杆
barrister's wig 律师假发
barroches 印度未漂白棉织物,未漂细布(东南亚)
barrow 无袖婴儿绒衣
barrow bag[coat] 婴儿襁褓
barry 纬向条纹,纬向条痕,横档织疵
barry dyeing 纬色档疵
Barry Kieselstein 巴里·基泽尔斯坦-考德(美国珠宝设计师)

Barrymore collar 巴里莫尔领(领边窄,领尖长)
Bar sakel cotton 苏丹塞克尔棉
bar-shaped necktie 棒形领带
bar-shaped tie 棒形领带
barshent 巴申特低级斜纹绒布
bar shoes 搭带鞋
Barsi-Natar 印度巴尔西纳塔短绒棉
bar strap shoes 搭带鞋
bar tack (缝口加固)倒回针;打结,打套结;打铆钉扣;加固缝,打结缝,加固缝纫
bartack back vent end 封背衩
bartack ends of pocket mouth 封袋口
bar tacker 打结机,套结缝纫机;套结机
bartack front rise 封小裆
bartacking 加固套结,加固针缝,封缝头
bartacking back vent 封背衩
bartacking ends of pocket mouth 封袋口
bartacking ends of sleeve slit 封袖衩
bartacking equipment 加固设备
bartacking front crotch 封小裆
bartacking front rise 封小裆
bartacking machine 套结缝纫机,套结机
bartacking sewing machine 加固缝纫机
bartacking stitch 加固线迹,套结线迹
bartack sleeve slit end 封袖衩
bartack stitch 倒针法
bartack the ends of pocket mouth 封袋口
barter 易货(贸易)
bar towel 擦玻璃用毛巾
bar tucking 加固褶裥
Baruba embroidery 巴鲁巴刺绣
barutine 伊朗低级绸
barvel 巴维尔维裙
bar warp lace machine 装饰花边机
bar width 加固缝宽度
Basacryl dye 贝西克尔染料(阳离子染料,染腈纶纤维,商名,德国巴斯夫)
basalt 黑陶色
basalt fiber 玄武岩纤维
basalt wool 玄武岩棉
basane 巴桑呢,巴萨尼粗纺毛绒斜纹织物
Ba satin 巴缎
bas de chausses [法]雪斯袜
baseball 棒球服
baseball blouson 棒球夹克
baseball boots 棒球靴
baseball cap 棒球帽
baseball clothing 棒球服

baseball dress 棒球服
baseball gloves 棒球手套
baseball jacket 棒球夹克
baseball jersey 棒球衫,针织棒球衫
baseball knit shirt 针织棒球衫
baseball shoes 棒球鞋
baseball socks 棒球袜,棒球短袜;棒球短裤
baseball uniform 棒球服
baseball wear 棒球装
base button 底扣
base cloth 地布,基布
base coat 指甲底油,底涂
base color 底色
base fabric 基布(人造革用的底布),底布
base fabric color 地色(转移印花之前织物所染的颜色)
baseless felt 无基布毡
baseless needled felt 无底布针刺毛毯
base line deviation 基线偏差
base of neck 领根,领围
base pattern 原型图
base plate 底轴
base printing 色基印花
bases 男短上衣下摆
base shade 主地色
base yarn 本色纱
bashlyk 风帽,兜帽,长尾防护头罩
basho(o)fu 芭蕉布,日本蕉麻纤维布
basic block 基本原型
basic block pattern 基型样板
basic block shape 基本原型形状
basic color 基本色彩;碱性颜料
basic color discharge 碱性染料着色拔染印花
basic color name 基本色名
basic (construction) line 基本结构线
basic curved shirt 基础型圆摆衬衫
basic design 基础图案,基调
basic dress 基本型服装;单式女服;基型连裙装
basic dye 碱性染料,阳离子染料
basic dyeing 碱性染料染色
basic fabric 基本衣料(平布、华达呢、劳动布、绒布、针织布等),基本织物,基本衣料
basic fashion (在流行中保持)基本式样
basic finish 基本整理,常规整理,一般整理(使坯布成为可售商品的整理)
basic function 基本职能,基本功能
basic item 基本商品(重点商品)

basic kimono sleeve without a gusset 无裆布和服袖
basic kimono without a gusset 无裆布和服袖
basic line 基本线,基础线(裁剪服装时首先划的基础线),基本结构线,下平线
basic line for skirt 裙子基本线
basic line for sleeve 袖子基本线
basic line for trousers 裤子基本线
basic linens 家用亚麻织物
basic line of diagram 图形基线
basic look 基本款式
basic material 主料
basic measurement 基本量身法,基本尺寸
basic motion time (BMT) 基本动作时间
basic neckline 基本领窝线,基本领口
basic one piece shawl collar 一片式基本型苹果领
basic pattern 基本式样;原型;基础图案,基础花纹;基本纸样;地组织花纹
basic pattern method 基型法
basic pattern set 整套纸样基型
basic physiological needs 基本的生理需要
basic 2 piece notched collar 两片式基本型缺角西装领
basic 1 piece shawl collar 一片式基本型青果领
basic pink 碱性桃红
basic product 基本产品
basic routine garment 日常基本型衣服
basic shade 主地色
basic shape 基本形状
basic size 基本尺寸
basic skirt 两片裙,裙子原型
basic skirt frame 基型裙衬
basic sloper 基本原型
basic specification 基本规范
basic standard 基础标准
basic stitch 基本线迹;基本针法
basic style 基本款式,基本式
basic suit 基本服装
basic surface 基础面
basic tolerance 基本公差
basic tone 明度差
basic turned up hem 基本翻折下摆
basic two-piece notched collar 两片式基本型缺角西装领
basic type 基本型号,基本类型
basic waistband 基本腰头
basified viscose rayon 动物质化黏胶纤维

basil 栎鞣羊皮革
basin 巴森布,巴森亚麻斜纹布;棉斜纹布
basine 巴辛绸
basin royal 白条亚麻褥单布
basis cotton 基础级棉
basis for claims 索赔依据
basis weight 织物单位重量
basket 席纹;板司呢,板司花呢;方平组织
basket bag 篮式编织手提包
basketball boots[shoes] 篮球鞋
basketball wear 篮球装
basket braid 方平组织饰带
basket check 席纹方格花纹,颗粒状凸纹格花纹;蓝格子
basket cloth 板丝呢,方平组织棉布,绣花用十字布
basket effect 方平效应
basket filling stitch 篮状贴线绣
basket huckaback 席纹亚麻粗布
basket knit fabric 席纹针织物
basketry 编织品
basket shirt 水手衫,海军衫,横条海军衫(水手领)
basket shoes 篮球鞋,球鞋
basket sneaker 黑色篮球鞋,篮球鞋
basket stitch 篮状绣,方平式贴线缝绣;网眼线迹,十字布刺绣线迹;席纹组织
basket stitch pattern 席纹花纹
basket wear 席纹织物,方平织物
basket weave fabric for embroidery 刺绣十字布
basket weave stitch 篮状斜纹针法
basmas 密织布(麻或棉布,土耳其用语)
BASOFIL 蜜胺纤维
basque (女)紧身短上衣,(妇女)紧身胸衣,短上衣,巴斯克衫;巴斯克连胸短裙
basque belt 束腰男服
basque beret 巴斯克贝雷帽,巴斯克软帽,[法]贝雷帽
basque bodice 巴斯克紧身上衣,巴斯克式衣身
basque costume 巴斯克装束
basque(d) hem 巴斯克衣摆
basque jacket 巴斯克夹克
basque shirt 海军衫,水手衫,海魂衫,横条海军衫(水手领)
basque silhouette 巴斯克轮廓(紧身衣加裙)
basque waistband 巴斯克腰带
basquina 华丽宽松衬裙
basquine 女式外套;紧身胸衣;华丽裙;户外夹克
Bass 柏斯
bassarisk fur 环尾猫毛皮
bass fiber 棕榈纤维,韧皮纤维
bassher 草帽
bassine 扇叶树头榈叶纤维
bassines 塔夫绸带
bast 霜色钻石
bastancini 浅蓝式亚麻细平布
bast and leaf fabric 麻织物
bastard 尺码特殊的,特殊尺码
bastard aloe 坎塔拉剑麻
bastard asbestos 变种石棉,劣质石棉
bastard jute 黄麻代用品
bastard teak 印度黄色树花染料
baste 假缝,初缝,疏缝,粗缝;中国丝绸;擦针
baste front edge 疏缝止口,止口
baste hem 疏缝底边;底边
baster 粗缝机
bastert fringe 假金丝边穗
baste stitch 擦针法;长疏缝线迹,假缝线迹
bast fabric 麻织物,麻布;韧皮织物
bast fiber 韧皮纤维,茎纤维;麻纤维
bast fiber and leaf fiber 麻纤维
bast hat 韧皮帽
basting 疏缝,假缝,粗缝,初缝,临缝;疏缝针脚
basting and tacking stitch 疏缝粗缝线迹
basting cotton 假缝线,粗缝棉线,纺缝棉线
basting foot 假缝压脚,疏缝压脚
basting front edge 止口
basting front facing 叠挂面
basting hem 底边
basting machine 假缝机,疏缝机,假缝纫机
basting margin 假缝边缘,假缝线
basting sewing machine 疏缝机,假缝机,疏缝机
basting side seam 叠摆缝
basting stitch 假缝线迹,疏缝线迹;针法
basting thread 假缝(用)线,扎线,假缝线,纫缝线
basting yarn 纫线,粗缝线,疏缝线,纫缝线
bastisseuse 帽毡成形机
bast lining 麻衬
basto 印度重浆漂白棉布
bast pulling 修呢时拔麻丝
bas tricot 编织袜子
bas tricoté [法]编织袜

bast thread 麻线
bast type feeling 麻型感
basuto 巴苏托等级马海毛
bat 垫褥,絮垫;筵棉,棉卷,棉胎;[制毡]纤维层
bataloni 麻棉交织布(大麻经,棉纬),麻经棉纬浅蓝织物
batanores 埃及进口亚麻织物
batarde 巴塔德花式斜纹呢;法国素色哔叽
batarde valenciennes 仿瓦朗西安花边
batavia 生丝经绢纬斜纹薄绸;二上二下斜纹组织
batavia silk 巴塔维亚绸,巴塔维亚斜纹绸,斜纹绸
batch bleaching 分批漂白
batch dyeing 分批染色
batched jute 油麻,软麻给乳后的黄麻
batched roll 布卷
batch finishing 分批整理
batch handling 成批处理
batching 卷布
batch number 批号
batch operation 间歇操作
batch processing 成批处理,成批加工,分批处理,分批生产
batch production 分批生产,间歇生产
batch scouring 分批煮练
batch setting 分批
batch to batch variation 批间差异
batch-type production 分批混合式生产
batchwise dyeing 间隔染色
bateau neck 一字领,船形领
bateau neckline 一字领口,船形领口
Bates' big boll cotton 美国贝氏大铃棉
Bates' favourite cotton 美国贝氏幸运棉
Bateson 背特逊
bath brussels lace 德文郡花边
bath cap 浴帽
bath cape 浴衣
bath coating 长绒法兰绒
bath curtain 浴帘
bath gown 长浴衣,浴袍,浴衣
bathing cap 游泳帽,泳帽
bathing costume 游泳衣,浴衣
bathing drawers 游泳裤
bathing dress 泳装,游泳衣
bathing headdress 游泳帽
bathing slippers 女浴鞋
bathing suit 泳装,泳衣,游泳衣;浴装
bathing suit coverups 沙滩装

bathing trunks 游泳裤
bath mat 浴室小地毯,浴室防滑垫,地巾
bathrobe 袍,晨衣;浴衣,浴袍;睡袍
bathrobe blanketing 双面绒棉毯织物
bathrobe cloth 浴衣绒布,双面起绒浴衣棉织物
bathrobe dress 浴袍式裙装
bath rug 浴室地毯
bath slippers 浴鞋
bath towel 浴巾
batia 木槿树韧皮纤维
batik 蜡染,蜡缬,蜡防印花;巴蒂克印花法,爪哇蜡脂防染印花法
batik dyeing 蜡染,蜡缬,蜡防印花,巴蒂克印花法,爪哇蜡脂防染印花法
batik effect 蜡染图案,蜡染图案花纹
batik fabric 蜡防花布
batik garment 蜡染服装
batik printing 蜡染,蜡缬,蜡防印花,巴蒂克印花法,爪哇蜡脂防染印花法
batik prints 蜡防花布
batik sarong 蜡染莎笼
batiste 法国上等细亚麻布;细棉薄布;细薄毛织物,毛高级薄绒
batiste d'ecosse 苏格兰亚麻细薄布
batiste de soie 细薄绸
batiste Hollandaise 荷兰麻布
batiste silk 巴底斯特绸,蚕丝细薄绸
batnas 印度三色印花棉布,三色平布(印度)
baton rompu 粗纺吡叽
bat,batt 絮垫,垫褥;[制毡]纤维层,毛网[针织地毯的纤维絮片];毡式织物;巴塔鞋
Battenberg braid 巴藤贝克编带(麻或棉制,有牙边)
Battenberg lace 巴藤贝克花边
batten lace 曲线粗孔花边
batter's cap 棒球帽
batter's helmet 棒球击球手头盔
battik 蜡染花布;蜡染法
battik dyeing 涂蜡印花
batting 毛絮;棉絮,棉胎,絮片
battle dress(B.D.;B/D) 军服,军制服;战地服装;卡其布军装
battle dress blouse 军装式上衣
battle jacket 战斗夹克(二重领,束腰,紧身短上衣),军服夹克,艾林豪威尔夹克;运动夹克衫
battle kit 戎装

battlemented stitching 城墙方形线迹
battlements 城墙形花边
battle pants 战斗裤
battle suit 战斗套装
batt-on-base needle felt 有底布针刺毡
batuz work 巴丘斯绣
batwing 男子蝴蝶结领带
batwing collar 蝙蝠领
batwing dolman sleeve 斗篷蝙蝠袖（介于蝙蝠袖与德尔曼袖的袖子）
batwing dress 蝙蝠衫
batwing sleeve 蝙蝠袖
batwing tie 蝙蝠结
batwing wrap 蝙蝠衫,蝙蝠形外套
bat wool 絮棉,纤维状填料
bauble 丑角权杖
baudekin, baudekyn 宝大锦；金线绣花织物
bauge 法国粗厚斜纹呢
Bauhaus 包豪斯
Bauhaus style 包豪斯风格,鲍豪斯风格
baulk finish 生坯轻缩[呢]整理
baum marten 林貂皮,欧洲林貂毛皮
baumwolle 棉（德国称谓）
baupers 17世纪英国精纺织物
bavarette 大围涎
Bavarian lace 巴伐利亚花边,巴伐利亚饰带
Bavarian leathers 巴伐利亚皮短裤
bave 茧丝
bavelot 女帽遮阳布
bavolet [法]巴伐利帽；巴伐利头饰；雨衣披肩
bawl-drick 肩带,饰带,佩带
bawneen 半漂平纹麻布；本色粗法兰绒工作服（爱尔兰农业工人穿）
baxea sandal 古代系带平跟女凉鞋
Baxer 防蛀性聚丙烯腈短纤维（商名,日本钟纺）
baxian ornament 八仙纹样
bay 台面呢；枣红色
bayadere 巴亚德横条纹绸
bayadere effect 巴亚德花纹,色泽鲜艳的横条花纹
bayadere moire 巴亚德波纹横条绸
bayadere print 巴亚德印花图案
bayadere stripe 巴亚德横条纹,巴亚德图案
bayadere stripe print 巴亚德印花横条布
bayard 彩色横条子布
bayberry 月桂果色
Bayeau lace 拜约花边

bayeta 台面呢；毛绒毯；19世纪植物染料染色法兰绒；猩红色美洲印第安毛毯
bayeta de cien hilos 斜纹长毛法兰绒
bayeta de faxuela 粗台面呢
bayeta de pellon 长毛法兰绒
bayetas 贝耶塔斯粗纺平纹毛织物
bayetones ingleses 英国粗纺大衣呢
bayette 单面法兰绒
Bayeux lace 贝叶花边,仿梭结法兰绒
Bayeux tapestry 贝叶挂毯
bayko yarn 金属花线
bayleaf 月桂叶蓝
bayutapaux 蓝白条粗棉布,红白条粗棉布
bazaar 廉价商店
bazan, bazon 棉缎纹条子布
BBB fiber BBB纤维（聚双一苯并咪唑苯并菲绕啉二酮纤维的简称）
1B cuff 单扣袖口
2B cuff 双扣袖口
B-cup 中号胸罩
B-cup brassiere 中号乳罩
beach 沙滩色（淡橄榄灰）
beach and reel 珠链饰
beachanese 醋酯丝外衣织物
beach bag 海滩包
beach bloomers 沙滩灯笼裤
beach cape 海滩披肩,沙滩披风
beach cloth 海滨薄呢
beach coat 海滨外套
beach dress 男式海滩装；男式沙滩装
beach gown 海滨袍（泳后披穿）,游泳后用外披长袍,海滨服
beach hat 海滩帽,沙滩帽
beach jacket 海滩短上衣
beach jeans 沙滩牛仔裤,海滩牛仔裤
beach pajamas 沙滩女式连衣裤,,海滩女式连衫裤
beach parasol 海滩大遮阳伞
beach poncho 沙滩披巾（用毛圈布做）,海滩披巾
beach robe 沙滩袍,海滩袍
beach sand 沙滩色
beach sandals 海滩凉鞋,沙滩凉鞋
beach shift 沙滩罩衣,海滩罩衣
beach shirt 海滩衬衫
beach shoes 海滩鞋,海滨鞋,沙滩鞋
beach shorts 百慕大短裤,沙难短裤
beach skirt 海滨裙
beach sponge cloth 海滩海绵布
beach suit 沙滩装,海滨装

beach toga 沙滩罩袍,海滩罩袍	beading leno 珠球纱罗织物
beach towel 沙滩巾	beading needle 穿珠针
beach trousers 海滩裤	beading strip 串珠饰带
beach umbrella 太阳伞	beading trim 夹缝饰边,花边镶边
beach warmer (球队)替补队员外套	bead leno 珠球纱罗织物
beach wear 海滨服(包括泳服、短裤、浴衣),沙滩装	bead netting 手套用针织物
	bead roll 一串念珠
beach wrapup 海滩披巾(正方形或长方形,毛圈布)	beads 珍珠;一串念珠;珠子项链
	beads bag 珠饰包
beacon robbing 厚双面绒布	bead seam 泡状缝
bead (装饰用)有孔小珠,念珠;珠饰	bead velvet 提花丝绒
bead and reel 珠链饰	bead work 珠球刺绣,圈纹刺绣,珠绣
bead cuff 褶裥袖头	bead yarn 珠球花线
beaded bag 串珠小包,串珠背色	beaker test 染小样试验
beaded belt 珠饰腰带,珠饰带	beak shuttle 长嘴摆梭
beaded blouse 珠饰衬衫	beam compasses 加长圆规
beaded braid 珠饰镶边	beam dyeing 卷轴染色
beaded button 珠饰扣	beam rug 混色纱地毯
beaded cuff 珠饰袖口,珠饰袖头	bean green 豆绿,豆青
beaded dress 珠饰连衣裙	beanie 小瓜皮帽;学生戴的小帽,童帽,小圆女帽;无檐小便帽,小沿便帽
beaded embroidery 珠绣,珠绣毛衫	
beaded embroidery cardigan 珠绣开襟衫	bean red 豉豆红
beaded embroidery garments 珠绣服装	bear 空头
beaded embroidery sweater 珠绣毛衫	bearcloth 粗毛呢,粗毛大衣呢,粗绒大衣呢;熊皮大衣
beaded garment 珠饰服装	beard 髯,胡须,脸须
bead edge 珠状边饰,珠边	beard hair 发毛,刚毛
beaded handbag 串珠手袋,珠饰手袋,串珠手提包	bearding 毯面绒球
	beardlet 小胡子
beaded lace edging 串珠花边	bearers 腰带,背带
beaded motif 珠饰花边	bearing 承座(饰件);风度
beaded necklace 串珠项链	bearing cloth 洗礼布,贵重装饰织物,洗礼用包裹布
beaded pattern 串珠花纹	
beaded seam 泡状缝	bear leather 熊皮
beaded shirt bib 胸前串珠饰物	bear paw 熊掌式钢甲套鞋
beaded shoes 珠饰鞋,串珠鞋	bearskin 熊毛皮,熊皮;熊皮粗绒大衣呢;熊皮高帽(美国禁卫军)
beaded sweater 珠饰毛衫,珠饰毛衣	
beaded trimming 珠饰带	bearskin cap 黑皮高帽,禁卫军帽
beaded velvet 薄地花丝绒,提花丝绒	bearskin pants 白色熊皮裤
bead embroidering 珠绣	bear's paw 熊掌鞋
bead embroidery 珍珠绣,珠绣	beating 捶布,打布
bead embroidery bag 珠绣手袋	beating finger splicing 袜子夹底加固
bead embroidery chi-pao 珠花旗袍	beating machine 拍打机(毛皮加工设备)
bead embroidery frock 穿珠绣花罩袍	Beatle boots 披头士靴
bead embroidery qi pao 珠花旗袍	Beatle cap 披头士帽
beading 网眼饰带;串珠状缘饰,小珠饰;机制镶边;双色织带技术	Beatle cut 披头士发型
	Beatles 披头士服饰
beading design 珠饰式样,串珠头花式	Beatles style 披头士乐队风格(甲壳虫队风格)
beading foot 串珠压脚	
beading galloon 串联网眼饰带	beatrice twill 纬面斜纹里子布
beading lace 珠饰花边	

beat-up　地毯每英寸绒头数
beaucatcher　卷髻
beaufort　大麻帆布;常礼服
beau ideal　理想美
beaujeu　打包麻布
beaujolais　法国麻棉布
Beaulard　皮尤勒德(法国裁缝师)
Beau sure　比尤修尔阻燃织物(商名,美国)
beaute satin　美丽缎,绉背缎
beautician　美容师,美容专家
beautician's gown　美容师工作服
beautiful lining twill　美丽绸,美丽绫,高级里子绸
beautiful packing　包装精美
beautiful pants　美感裤
beauty　美;美丽
beauty and hairdress　美容美发
beauty contest　赛美大会
beauty cream　美容霜
beauty culture　美容术,化妆术,整容术
beauty culturist　美容师,美容专家
beauty doctor　美容师,美容专家
beauty for ensemble　整体美
beauty in color　色彩美
beauty in form　形态美
beauty of modeling　造型美;形体美
beauty of nature　自然美
beauty of whole　整体美
beauty parlour　美容院
beauty patch(es)[spot]　美人斑
beauty salon　美容院
beauty shop　美容店,美容院
beauty specialist　美容师;美容专家
beauty spot　面部贴饰,美人斑
beauty treatment　美容,化妆
beauvais embroidery　刺绣挂毯
Beauvais tapestry　机织花卉挂毯
beave　半面罩
beaver　海狸毛皮;海狸绒,海狸呢,水獭呢;海狸毛色;海狸皮帽,大礼帽;海狸棕色,棕灰色;活动面甲
beaver cloth　海狸呢,水獭呢,海狸绒布
beaver color　海狸毛色
beaver dyed cony　仿海狸兔毛皮
beaverette　兔毛皮
beaver finish　海狸呢整理
beaver fur　驼灰色;海狸皮,海狸裘
beaver fustian　全棉海狸呢,纬绒海狸呢
beaver hat　海狸皮帽,水獭皮帽

beaver imitation　仿海狸皮,仿水獭皮
beaver lamb　仿海狸毛皮
beaver leather　海狸(毛)皮
beaver scarf　海狸毛围巾
beaver shawl　海狸披巾;厚羊毛围巾
beaver skin　海狸皮
beaverteen　仿獭绒,充獭绒;斜纹绒布
bebe　儿童用细丝带
bebop cap　报童帽
becca　长布条
becomingness　合适性
bed　(缝纫机)台板;床;四页精纺斜纹呢
bed bag　被套
bed cape　无袖睡衣
bedclothes　床上用品
bedcover　床罩
bedcover of warp knitted raised fabric　经编起绒床罩
bed curtain　床罩
bedding　床上用品,被褥
bedding and clothing　被服;部队衣着
bedding and clothing factory　被服厂
bedding down　(地毯绒头)厚度减低
bed filling　被褥,填絮
Bedford cloth　贝德福呢;厚实经凸条布,经条灯芯绒
Bedford cord　厚实凸条布,经条灯芯绒;经向凸条织物
Bedfordshire lace　英国贝德福德郡梭结花边
Bedford twill　贝德福凸条斜纹呢
bed gown　女睡衣
bedizen　华丽而俗气的衣服(或打扮),华而不雅的衣服
bed jacket　睡衣,短睡衣;梳头衣,罩在睡衣外的短上衣,短寝衣(上衣)
bed lace　白色滚边
bed linen　床用织物,床单(用亚麻、棉或混纺材料制成)
bed mat　床褥,床单织物
bed plate　底轴
bed point　窄幅梭结花边
bed quilt　棉被
bedraggle　裙边拖湿;裙边拖脏
bed rayle　早餐披风
bedrock price　最低价格
bed roll　铺盖
bedroom slippers　卧室用拖鞋,居家拖鞋
bed rug　床边毯
bed sacque　寝居外套,女睡衣外的宽短外衣

bed shaft （缝纫机）底轴
bed sheet 床单
bed sheeting 床单细布
bed sheet jacquard 大提花被单布
bed sheet linen 亚麻床单布
bed sheet set 被里
bed shield 床罩织物
bedside mat 床边毯
bedside towel 床沿巾
bed socks 暖袜,睡袜,床上穿的厚短袜
bed spread 床单,床罩布
bed spread fabric 床单织物
bed spread lace 床罩花边,床单花边
bed spread satin 缎纹床单布
bed spring 弹簧床垫
bedstout 四页斜纹条子或素色布
bedticking 条子褥单布
beechnut 山毛榉坚果绿色
beefcake(ry) 男性健美摄影（或照片）;男性健美展示
Beefeater's hat 英国皇家禁卫军军帽;带黑丝绸的黑海狸帽
beefy suit （呢绒制）厚实套装,厚料套装,厚套装
beege 粗呢坯,粗斜纹呢
beegum hat 大礼帽,高筒礼帽;丝质高礼帽
beehive 女子蜂窝式发型
beehive bonnet 蜂巢式罩帽;蜂窝状女草帽
beehive hairstyle 蜂巢式发型
beehive-pleats skirt 蜂巢褶裥裙
Beer 贝儿（法国时装设计师）
Beer jacket 比尔外套;大学生上衣;箱形夹克（金属纽扣,贴袋）
beeswax 蜂蜡
beeswax color 蜂蜡色
beetle 黝绿,甲虫色;捣布机,布槌
beetle color 甲虫色
beetled hemp 捣软的大麻
beetle faller 捣布槌
beetle finish 捣布整理
beetle green 黝绿
beetle lustering 捣布增光
beetling 捣布,打光（增加光泽柔软）
beetroot purple 甜菜根紫
beeveedee's（BVD's） 男装短裤
before tax 税前
before washing（B.W.） 水洗前
beggar look 乞丐风
Beggar's lace 贝格粗梭结花边
Beggar's velvet 贝格丝绒

begin color 天然色,本色
begonia pink 海棠玫瑰色
beguine cloth 原色粗呢
begum behar sari 孟加拉手织丝棉格子莎丽
behaar 印度细棉布,细布（印度）
behavior 性能,特征,性状,行为
behavioral adjustment 行为调节
behavioral science 行为科学
behind 臀部;衣服后背,衣服后襟;臀部
beibazar 土耳其二级山羊毛
beiderwand 双面提花床单布,接结双层织物
beige 本色的,未漂白的;浅灰黄色,未黄色;坯布;原色哔叽,薄斜纹呢;混色线呢
beige damas 精纺原色提花哔叽,原色提花哔叽
beige serge 原色哔叽
bei-guang brocaded damask 蓓光绸
Beijing blue 北京蓝
Beijing embroidery 京绣
Beijing Institute of Clothing Technology 北京服装学院
Beijing stitch 北京针法
beimen 美国原棉（日本称谓）
bejeans 穿牛仔装者
bejewel 饰以珠宝
belcher 贝尔彻围巾,杂色围巾,蓝白色围巾,大白点蓝色围巾
Belcher handkerchief 杂色围巾,蓝白色围巾
beldia 贝尔迪羊毛,易缩摩洛哥粗毛;中东生丝;中东低级棉
belesmes 法国大麻粗帆布
belfry cloth 粗厚方平棉织物
Belgian costume 比利时民族服
Belgian flax 比利时亚麻
Belgian lace 比利时花边
Belgian linen 比利时亚麻布
Belgian loafer 比利时懒汉鞋
Belgian-spun yarn 比利时式包芯纱
Belgian tapestry 比利时式挂毯
Belgian ticking 比利时床单布
Belgrade braid 贝尔格莱德式编带
Belica 聚酰胺织袜用弹力丝（商名,日本帝人）
Belima X 聚酯/聚酰胺裂离型超细复合纤维（商名,日本钟纺）
Belinda Bellville 贝琳达·贝尔维尔(英国时装设计师)

Belinge 比林基粗斜纹毛麻织物
bell 钟形物;降落伞衣;喇叭口
bellacosa 贝拉科萨花缎
Bellanze 聚酰胺66长丝(商名,日本钟纺)
bell bottom 喇叭脚口(裤)
bell-bottom heel 喇叭跟,凸缘鞋跟,喇叭形鞋跟
bell-bottom line 喇叭形
bell-bottom pants 喇叭裤
bell bottoms 喇叭裤,女喇叭裤
bell bottom slacks 喇叭裤
bell bottom sleeve 喇叭袖
bell bottom trousers 喇叭裤
bell-boy cap 侍者帽
bell-boy jacket 侍者外套,侍者夹克
Belle Creole 美国贝利克里棉
Belle Epoque hairstyle 贝儿发型,美好时代发型
Belleseime 贝莱萨姆(聚酯/取己内酰胺裂离型超细复合纤维制人造皮,商名,日本钟纺)
Bellflame 贝尔弗莱姆(阻燃性黏胶纤维,商名,日本)
bell-hop cap 侍者帽
bell-hop jacket 侍者外套;侍者夹克
bellied lapel 弧线形翻领,弧线翻领
bellings 毛麻呢(英国制,麻经毛纬或全毛)
bell line 钟形
Bellmanized finish 贝尔曼整理(轻薄棉布挺爽耐洗整理,商名)
bellmouth entrance 喇叭形入口
Bellock 防污聚丙烯腈短纤维(商名,日本钟纺)
bellow pocket 有褶口袋
bellows case 风琴式衣箱,折叠衣箱
bellows pleat 动作褶,动作襞,风箱式活动褶裥,风箱褶,开式褶裥(袖子后面和衣服背部中央的深度襞裥)
bellows pocket 风箱袋,老虎袋,开式对褶贴袋,胖褶袋,褶裥口袋
bellows sleeve 胖裥袖,风箱油,宽松袖
bellows tongue 风箱鞋舌,鞋舌褶,带宽舌男鞋
bells 喇叭裤
bell-shape cuff 钟形克夫
bell-shaped effect 钟形效果
bell-shaped silhouette 钟形轮廓
bell-shaped skirt 喇叭裙,钟形裙
bell-shape hat 钟形帽

bell skirt 钟形裙,喇叭裙
bell sleeve 钟形袖,喇叭袖
Belltron 抗静电聚酰胺/聚酯复合纤维(商名,日本钟纺)
bell-type 钟罩式服装款式
bell type skirt 喇叭裙,钟形裙
bell umbrella 钟形伞,圆顶形雨伞
BELLVILLES 贝拉维拉
bellwarp 英国螺旋斜纹呢
belly 腹部,肚
belly band 腰带,肚带;围腰,保暖围腰
belly button 肚脐
belly clothes 裸腹服装
belly dart 腰省
Belmont collar 伯尔莫特领(高领座,窄领边,圆领尖)
below-elbow length 肘下长度
below-elbow length sleeve 中袖,六分袖
below-elbow sleeve length 六分袖长
Below Good Ordinary 级外白棉(美国分级标准名称)
below grade cotton 等外棉
below-knee length 至膝下长度
below-knee panty 长衬裤
below par 低于(生丝)原重
below the calf 过腿肚长
below the knee panty 港裤,短裤
below-the-waist apparel 下装
belt 围绕物(带,腰带),带,皮带,线带,绳;袢;爵位绶带;女紧身胸衣
belt abdominal 腹带
belt attached to coat 连身腰带
belt backing 吊带背衬
belt-back reefer coat 背吊带双结短大衣
belt-back topper 背吊带式短大衣
belt bag 腰带包,穿带小腰包(有盖,与运动装一起穿用)
belt box 有挎带或背带的箱子
belt buckle 腰带(装饰)扣,腰带扣
belt button 带扣
belt carcass 带身,带芯
belt carrier 带,袢,蚂蝗带,蚂蝗袢
belt clamp 腰带扣,皮带扣
belt clasp 腰带扣
belt collar 带扣领,皮带领,环带领
belt conveyer[conveyor] 传送带;带式输送机
belt cover 衣裆;传送带外罩,皮带罩
belted 束腰带的
belted blouse 连腰带上衣

belted cardigan suit 束腰式卡蒂冈套装
belted coat 束腰外套
belted corset 束带紧身衣
belted fabric 胶带织物
belted garment 腰带服装
belted-in-back coat dress 背部系带女长袍
belted jacket 束带夹克
belted knit 束腰毛衣,束腰羊毛衫
belted look 束带款式,有带服装款式
belted pajamas 束腰睡衣
belted suit 腰带套装
belted topper 束带短大衣
belt fastener 腰带扣,皮带扣
belt fastening 皮带扣,皮带的扣栓物
belt feed system 皮带送料系统
belt for canvas bag 背包带
belt for gun 背枪带
belt guard 带袢,马王袢,蚂蝗袢,蚂蝗带
belt hook 腰带环,皮带钩,带钩
belting 腰带;腰衬;带料
belting ribbon 粗纹缎带,硬挺棱纹带;缎带腰衬
belt keep 腰带袢,腰带圈,带袢,马王带,蚂蝗带;蚂蝗袢
belt lacings 皮带扣,皮带卡子
beltless slacks 无腰带裤;不用腰带的裤子,不束腰带的裤子
beltless trousers 无腰带西裤
beltline 腰头;裤腰
belt loop 裤耳,裤袢,带袢,裤带袢;蚂蟥带,饰带,马王带;套环
belt-loop attaching machine 裤袢机;带子机,裤耳缝纫机,皮带环缝纫机
belt loop feeder 皮带环送料器
belt loop folder 带耳叠缝器,带袢叠缝器
belt loop sewing machine 裤耳缝纫机,带环缝纫机,缝皮带环机
belt machine 带子机;带袢机
belt of gown 衣带;袍带
belt or canvas bags 背包带
belt pocket 腰带袋
belt pouch 腰包,腰带包
belt punch 皮带打孔机
belt reefer coat 背吊带双结短大衣
belt retainer 皮带扣
belts 装饰带
belt stiffening 腰带硬衬
belt tab 腰带包头
belt through 穿引腰带
beluchistan carpet 巴罗奇地毯,土库曼地毯
belwarp 英国螺旋斜纹呢
Bemberg 铜氨纤维(商名,意大利本伯格)
bemsilk 铜氨人造丝,铜氨丝
benares 印度贝拿勒斯银丝绸
Benares hemp 贝拿勒斯麻,印度麻,菽麻
Benares work 贝拿勒斯丝绒刺绣品
Ben Bang tailor 本帮裁缝
Ben Casey collar 本·凯西领
Ben Casey shirt 本·凯西短袖衫
bench 工作台
bench coat 候车保暖外套,替补队员保暖外套,预备队员保暖外套
bench-made 定做的手工鞋
bench-made shoes 手工精制鞋
bench scale production 小规模试生产
bench warmer 预备队员外套,替补队员外套,预备队员保暖装
bench warmer jacket 候车保暖外套,预备队员保暖外套;座上保暖衣
bend 束发带,帽边缎带,帽圈;布条;条纹;背皮,最好部位的皮革
bendera 深红棉旗布
Benders cotton 美国本德棉
bending 弯曲,弯度
bending elasticity 弯曲弹性
bending energy 弯曲能
bending fatigue 弯曲疲劳
bending (flexure) property 弯曲性能
bending modulus 弯曲模量
bending resistance 抗弯性
bending rigidity 抗弯刚度,弯曲刚度
bending strain 弯曲应变
bending stress 弯曲应力
bend knife 弯刀;万能式裁剪刀
Benetton 贝纳通
benfery 编花绉
Ben Franklin glasses 本·弗兰克林眼镜
bengal 孟加拉湾生丝;印花条纹轻薄纱;丝毛交织女服面料
Bengal cloth 孟加拉呢
Bengal cotton 孟加拉湾短绒棉
Bengal hemp 菽麻
bengaline 绨,(线)绨;孟加拉呢;罗缎
bengaline de soie 丝罗缎
bengaline marquise 花罗缎
bengaline radiant 花式罗缎
bengaline velours 厚罗缎
Bengal linen 孟加拉仿亚麻布

Bengal silk　孟加拉绢丝
Bengal stripe　孟加拉蓝白条纹布
Bengal stripe brocade　孟加拉条子夹花织物
Bengal stripe shirting　孟加拉条纹衬衫布
Benglong ethnic costume　崩龙族民族服
Ben Hur sandals　本哈凉鞋,斯巴达克凉鞋
Benjamin　男式紧身大衣(19世纪)
benjy　穿边草帽
benmore　本莫
Benny　班尼大衣;男式白色工作服
Benois　比诺斯(俄国著名戏剧服装设计师)
Ben smith cotton　美国史密斯棉
bent handle dressmaker's shears　弯把裁缝剪刀
bent-handled sharp sheers　弯把尖头剪刀
bent knee girth　弯形膝围
bent trimmers　裁缝剪刀,弯把裁缝剪刀
benzaldehyde-acetalized fiber　苯甲醛缩醛化纤维
benzal green　孔雀绿(碱性三苯甲烷染料)
benzo blue　靛蓝
benzo copper dyestuff　苯并铜染料,苯并竖牢铜盐染料
benzodifuranone disperse dye　苯二呋喃酮[结构]分散染料
benzoylated cotton　苯酰化棉
Ben Zuckerman　本·朱克曼(美国时装设计师)
berams　16世纪印度印花棉布,粗布(印度)
Berar cotton　印度贝拉尔棉
Berber　贝伯轻薄缎面织物;手织毛毯;低毛圈簇绒地毯
berber-style carpet　本色地毯,天然原色地毯
berdelik　东方挂毯
beret　贝雷帽,无沿软帽,扁圆便帽
beret basque　巴斯克贝雷帽
beret hunting　贝雷帽褶,褶裥贝雷帽
beret sleeve　贝雷袖
beret snood　阔幅贝雷帽
bergama rug　伯加马地毯,土耳其优质地毯
bergamee　土耳其染纱纯毛地毯;意大利低档混纺纱地毯
bergamo　土耳其染纱纯毛地毯;意大利低档混纺纱地毯
bergamot　土耳其染纱纯毛地毯;意大利低档混纺纱地毯

bergere　浅冠宽檐大草帽
berger hat　羊倌帽
Bergerie　贝日里花毯或刺绣品
beribboned　饰以缎带
berkan　巴拉坎厚呢;风雨大衣
Berkshire heel　柏克夏袜跟,圆袜跟
Berlin blue　柏林蓝,普鲁士蓝,蓝色颜色
Berlin canvas　柏林刺绣十字绣
berlin gloves　毛织手套,棉线手套,柏林手套
Berlin silk　刺绣用柏林丝线
Berlin warehouse　毛线商店
Berlin wool　柏林绒线;细毛线
Berlin wool work　柏林绒绣
Berlin work　柏林毛线刺绣(品)
Berlin yarn　柏林绒线
Berl saddles　贝尔鞍形填料
Bermuda　贴腿短裤,百慕大式短裤
Bermuda cloth　百慕大布
Bermuda collar　百慕大领
Bermuda dress　百慕大装
Bermuda fagoting　百慕大针迹(暗花刺绣),反面绣花
Bermuda fagoting stitch　百慕大针迹
Bermuda hose　百慕大长筒袜
Bermuda jumpsuit　百慕大连衫裤
Bermuda pants　百慕大短裤
Bermudas　百慕大短裤
Bermuda shorts　(齐膝)百慕达短裤,步行短裤
Bermuda suit　百慕大套装
Bernard Perris　贝尔纳·伯里斯(法国时装设计师)
Bern Conrad　伯恩·康拉德(美国时装设计师)
Bernhardt mantle　贝因哈特斗篷
Bernhardt sleeve　贴身长袖
bernia　柏尼亚毛哔叽
be rosed look　玫瑰装饰款式
berth(a)　披肩式女服领,披巾宽领,披肩式花边领,宽圆边领;精致花边
bertha collar　披肩领,贝莎领,蓓莎领
berton hat　布列塔尼帽(前缘卷边女帽)
berundjuk　土耳其女丝绸衬衣
beryl　绿玉,绿柱石
beryl blue　水蓝色
Besfight　聚丙烯腈基碳纤维(商名,日本东邦人造丝公司)
Besfite　腈纶基碳纤维

beshir rug 土库曼地毯
Beslan 聚丙烯腈短纤维(商名,日本东邦人造丝公司)
Besloft 聚丙烯腈膨体短纤维(商名,日本东邦人造丝公司)
Beslon 聚丙烯腈短纤维(商名,日本东邦人造丝公司)
Beslon lace 贝丝纶花边
Beslon seryna lace 贝丝纶花边
besom pocket 嵌线袋
bespangled 闪光服装
bespoke line 定制服装行业
bespoke tailor 定制服装店
bespoke tailoring 定做的服装
Bessarabian lamb 比萨拉比亚羔羊毛
Bess-more 波里诺西克纤维(商名,日本东洋纺)
best bib and tucker 最好的衣服
best dresser 穿着入时者,十分讲究衣着的人,最佳穿衣者,最讲究衣者者
best-fitting 非常合身
beston droit 法国单排扣西装
betel palm fan 槟榔扇
beten 绣花袍
bethilles 贝锡勒条子布
Bethlehem head dress 伯利恒头饰;古代穆斯林头饰;伯利恒帽
betilles 菲律宾网眼细布
Betsey Johnson 贝齐·约翰逊(1942~,美国时装设计师)
betsie 硬皱领
betsie ruff 硬皱领
betweeners (女)紧身衣,紧身女服
between season wear 春秋服;换季服装
betweens needle 密缝针,手工短针
Beulon fastener 贝龙拉链
beutanol 聚乙烯涂层织物
bevel 斜角规
beveling 裁成斜边
beveling seam 分层缝
bevel-woven material 弧形织物;经纱弓曲织物
bex-fitted silhouette 贴身箱型轮廓
bezel facets 斜边小面(饰品)
B fiber 聚对苯二甲酰对苯二胺纤维(对位芳酰胺纤维;对位芳纶;芳纶 1414;芳纶-Ⅱ,凯夫拉,B 纤维)
bhagalpuri 印度手织平纹棉布
Bhangulpore cotton 班高坡棉,印度原棉
Bharua silk 巴鲁亚蚕丝

Bhatial jute 印度黄麻
bhavalpur 印度条纹绸,格子绸
Bhoga cotton 波加棉,孟加拉山区粗绒棉
bhoones 印度手织粗棉布
Bhownuggar cotton 印度波努加棉
bhurra(scarves) 布拉棉平布,美国红彩条棉布
bhyangee 西藏绵羊毛
Biagiotti 拜吉奥提
biambonnees 印度深黄或浅棕韧皮纤维织物
biancaville cotton 意大利比安卡威尔棉
Bian silk 汴绸
biarritz 双面纬向棱纹毛呢;宽袖中长手套
biarritz fantasia 西班牙丝光棉衬衫布
biarritz gloves 宽袖中长手套
bias (成衣的)斜线;斜布条;织物斜纹路,织物斜折;窄幅丝光棉细平布;斜裁,斜裁滚边料,缝纫滚条
bias binder 曲折滚边器
bias binding 斜布滚条;斜裁滚边,曲折滚边
bias binding maker 斜裁滚条器
bias check 斜格
bias collar 斜裁领
bias cuff 斜裁克夫
bias cuff collar 宽式环圈领
biascut 斜裁法
bias-cut 斜裁,斜切
bias-cut fabric 斜裁料,滚条
bias-cut garment 斜裁服装
bias-cut pattern 斜裁纸样
bias-cut skirt 斜裁裙
bias-cutter 斜条裁剪机,斜裁机
bias cutting 斜裁;斜扯织物
bias cutting device 斜裁装置
bias cutting machine 斜条裁剪机
bias draping 斜裁法
bias dress 斜裁裙装
bias edging 斜裁滚边;斜裁滚条
biased twill 急斜纹
bias fabric 斜裁布
bias-faced hem 斜条贴边,斜料贴边(用于曲线底边,喇叭裙或原料不易翻折处)
bias-faced waistline 斜裁贴边腰线
bias facing 斜料挂面;斜丝贴边
bias filling 纬料(织疵)
bias fitting 斜裁料的合身性

bias fold　斜折
bias-fold collar　斜折领
bias gored skirt　斜裁多片裙,喇叭形多片裙,斜裙
bias grain　斜丝缕
bias insertion seam　斜拼缝
bias line　斜线
bias line of front opening　大襟斜线
bias mark　斜条,斜绒
bias overhang　斜向余垂量
bias pleat　斜线褶
bias roll collar　宽式卷筒领
bias ruffle　斜条抽褶边,滚条荷叶边,斜裁抽褶边
bias sample　偏倚样品
biasse　中东生丝
bias seam　斜线缝
bias skirt　斜裙,斜纹裙,斜裁裙
bias sleeve　斜袖
bias slip　斜裁连身衬裙
bias square　斜方块
bias stretch　斜向拉伸
bias strip　滚条,斜裁布条;斜布;帮胸衬
bias tape　斜布条,斜条,滚条,斜裁滚边带
bias tape cutter　切捆条机,裁斜条机
bias tie neck　斜条打结领
bias tuck　斜褶裥
bias tunnel collar　斜裁高立领
bias turnover collar　斜裁高翻领
bias undercollar　斜裁领里
bias waistband　斜裁腰头
bias weave　斜织;三向织物
bias weft　纬斜(织疵),纬不正
bias zipper　斜拉链
biaxial fabric　双轴向织物
biaxial warp knitted fabric　双轴向经编织物
biaz　窄幅丝光棉细平布,狭幅丝光棉细平布(中亚妇女衣料)
biaza　驼毛外衣呢
bib　(童)围兜,围嘴;工装背带裤护胸;围裙的上部;衬衫上胸饰
Biba　比芭
biba dart　折线省
bib and brace　工装背带裤,背带式工装裤
bib and brace overalls　护胸工装裤,围兜背带工装裤
bib and tucker　衣服(俗称)
bib apron　连兜围裙
bib blouse　围兜式女衫(后开襟,高立领),围嘴女衫,胸兜女衫
bib collar　围兜领,胸围领,围兜式衣领
bib dress　围兜服;围嘴连衣裙,胸兜连衣裙
bibeli　土耳其丝枕头花边
bibi　比比帽
bibi bonnet　比比软帽,比比罩帽,罩形圆帽
bib jumper　围兜式背带裙
bib jump suit　工装连衫裤
biblikabad　波斯高绒头地毯
bibliography　文献目录
bib necklace　多串式项链,豪华型项链
bib overalls　带护兜的背带工装裤,工装裤
bib pants　围兜裤,背带裤
bib shorts　(有护胸的)背带短裤,有围兜的短裤
bib skirt　围兜裙
bib style　兜饰型
bib top　露背短上衣
bib top pants　工装背带裤
bib top pinafore　围兜裙
bib with arms　连袖围兜
bib with frill　褶裥胸饰;荷叶边护胸
bib yoke　围兜式剪接布;围嘴过肩
bicanere　印度骆马毛
bice　蓝色
bice blue　石青蓝(灰绿蓝色)
bice green　孔雀石绿
biceps　袖宽,袖壮;二头肌
biceps circumference　袖围,袖肥,上臂围,上臂周长
biceps line　袖宽线,袖壮线
bichon　垫子,衬垫
bichu fiber　大荨麻纤维
bico fiber　双组分纤维
bicolor　双色
bicolore　两色配色
bicolored dress　双色连衣裙
bicolor look　双拼式服装;双色服装款式
bicolor style　双拼色款式
bicomponent conductive filament　双组分导电长丝
bicomponent fiber　双组分纤维
bicomponent filament　双组分长丝,复合长丝
bicomponent film fiber　双组分薄膜纤维
bicomponent hybrid fiber　双组分混杂纤维
biconstituent filament yarn　双组分长丝
bicorne　拿破仑帽,双角帽,两角帽
bicycle bal　自行车鞋

bicycle-clip （骑车用）裤管夹
bicycle-clip hat 自行车帽,骑车夹帽（简易遮阳帽,夹在头上）
bicycle cushion's felt 自行车坐垫毡
bicycle gloves 骑车手套
bicycle helmet 骑车头盔
bicycle knickers 女式自行车裤,自行车灯笼裤
bicycle type 厚实腰部（首饰）
bid 递盘
bid bond 投标保证金
bid deadline 投标截止日期
bidder 竞买者,投标人
bid documents 投标文件
bid for 投标
bid guarantee 投标担保
Bidim 聚酯纺黏型非织造布（商名,法国罗纳-普朗克公司）
bidirectional fabric 双向加强织物
bidjar rug 长圈绒波斯地毯
Biedermeier dress 19世纪德国中产阶级服饰
biege 原色哔叽
Bielefeld 比勒费尔德上等亚麻布
bietle 鹿皮上装
bifacial embroidery 双面绣
bifiliated fiber 双组分[复合]纤维
bifocals 双光眼镜
bifurcates 分叉裤袋（裤子、裙裤等）
big bobbin 大梭芯
big Boffe cotton 大波菲棉,巴西陆地棉
big boots 宽筒靴
big buckle 大皮带扣
big business 大企业
big button 大纽扣
big cape 大披肩
big cardigan 大型卡迪根
big check 大格子
big coat 宽松长外套
big denim 宽松工装裤
big details （服装）细部扩大
big easy sweater 宽大落肩毛衣
bigeneric fiber 双组分纤维
big fashion 大款式
biggin 毕京睡帽,毕京童帽,比晶帽,儿童风帽;头套;高级律师戴的白帽
biggon 毕京睡帽,毕京童帽,比晶帽,儿童风帽;头套;高级律师戴的白帽
biggonet 毕格尼帽
big heel 大袜跟

big-hipped figure 大臀体型
big hook 大梭钩
bight 扣孔边,纽孔切口长度
big line 宽大线型,宽线型
big look 大款式,特大型
big neck 粗颈
big neckline 粗颈
big neckline type 粗脖子型
big pocket 特大口袋
big raised dish-doth 拉毛大方巾
big round neckline 大整圆领口
big shawl 大披巾
big shirt 特大型衬衫,超大型衬衫
big-shirt dress 大衬衫连衣裙;大衬衫装
big shopper bag 大购物包
big silhouette 宽松肥大型轮廓;宽松肥大型服装,特大型服装
big skirt 庞型裙,超大型裙
big stole 大型女用披肩
big style 宽松肥大款式
big swing back steamer coat 大尾轮船式外套
big tee shirt 特大超短型连袖T恤衫
big toe 拇指
big twill polyester peach 宽斜纹桃皮绒,带披肩的针织紧身衫
bijou plume 羽饰
bijouterie 小巧珠宝饰品
bike jacket 骑车夹克[衫]
biker's boots 飞车党鞋
bike shorts 自行车短裤,骑车短裤
bikini 比基尼女泳装,三点式泳装,上下分开女游泳衣;女三角裤,超短内裤;男式超短游泳裤
bikini blouse 比基尼女衫,比基尼衬衫
bikini bottom 比基尼下装,极小的三角游泳裤
bikini briefs 比基尼内裤,比基尼式女三角裤,超短三角裤
bikini chain belt 三点式泳装的金腰链
bikini clad 比基尼式女泳装
bikini clasp 比基尼式钩扣
bikini dress 比基尼服
bikini panties 比基尼超短游泳裤,游泳三角裤
bikini pants 比基尼式裤
bikini panty hose 低腰裤袜
bikinis 比基尼式裤
bikini shadow line 比基尼隐皱纹疵点
bikini straps 比基尼吊带

bikini style　比基尼服装式样；比基尼游泳衣
bikini top　比基尼上装
bi-knit　双正面针织物
bilancia　拜伦西亚发型
bilateral　双边
bilateral symmetry　左右对称调和
bilei moss crepe　碧蕾绉
biliment　女子装饰品；金花边；金线头饰
biliment lace　金丝镶珠花边
bill　票据
Bill Atkinson　比尔·阿特金森（美国时装设计师）
Bill Blass　比尔·布拉斯（1922～，美国时装设计师）
billfold　皮夹，钱夹
Bill Gibb　比尔·吉布（英国时装设计师）
billiard cloth　台球绿呢
billiard green　台球台绿，球台绿
billicock　英国小礼帽，圆顶高帽
Bill Kaiserman　比尔·凯泽曼（美国时装设计师）
bill of lading (B/L)　提（货）单
billow bag　枕头包
billowy sleeve　巨浪袖
Bill Thomas　比尔·托马斯（美国著名戏剧服装设计师）
Bill Tice　比尔·苔丝（美国女内衣设计师）
billycock　小礼帽，硬毡帽，宽边毡帽
Bi-Loft　高膨松聚丙烯腈纤维（商名，美国孟山都）
Biltex　聚酯喷气变形丝（商名，荷兰阿克佐）
bimlipatam　洋麻，槿麻
bimorphic PET filament　双形态聚酯长丝，双晶聚酯长丝
bin　裤子口袋；箱
binagacay hemp　优质马尼拉麻，优质蕉麻
binary colors　双色
binary mixture　双组分混纺
bin bagging　平纹粗黄麻布
binche lace　班什花边，比利时网眼花边
bind　叠边，滚边，镶边，滚条，带子
bind armhole　滚袖窿边，包袖笼
bind buttonhole　滚扣眼边，锁眼
bindelli　优质金银丝缎带（意大利）
binder　腹带，绷带，带子；包边带；捆扎带子；绳索；滚边器；黏合剂
binder agent　黏合剂

binder fabric　绷带织物，包扎带织物
binder feed dog　滚边送料牙
binder fiber　黏合[用]纤维，纤维状黏合剂
binder foot　滚边压脚
binder hemming　卷边
binderless cellulose nonwoven fabric　无黏合剂纤维素非织造布
binder machine　滚边缝纫机
binder plate　滚边压极
binder thread　滚边线
bind hemming　卷边，滚下摆
binding　镶边，滚边；滚条；套口
binding agent　黏合剂
binding apparatus　滚边装置
binding armhole　滚袖窿
binding button hole　滚扣眼
binding cloth　滚边布，滚条布；精装书封面用布
binding collar　滚领口
binding cord　方边带
binding cuff　滚袖口，滚克夫
binding cutter　滚边裁剪刀
binding effect　黏合效应
binding leg band　绑腿带
binding legs　滚裤脚
binding muslin　方面布
binding nature　约束力
binding neck opening　滚领口
binding-off machine　套口机，缝袜头机
binding opening　滚边式开口
binding operation　滚边操作
binding placket　滚边式开口
binding power　黏合力
binding stitch　滚边线迹
binding tape　滚边带，捆条（包边），绑带
binding tape guide　滚边导带器
bind leg band　绑腿带
bind material　黏合料
bind off　锁边；滚边，拷边；关边；收口
bingle　短发式（妇女、儿童）
binni cloth　宾尼手织平布
binocular microscope　双目显微镜
binoculars　双筒望远镜
biobarrier fabric　生物防护织物
bio-cleanroom　无菌室
biodegradable sulfur dye　可生物降解硫化染料
bioengineered fiber　生物工程纤维
bioenzyme finish　生物酶整理
biofiber　生物纤维（由微生物如细菌产生

的纤维)
biofinishing 生物整理
bioguard treatment 抗微生物处理
Biokryl 永久性抗菌聚丙烯腈纤维(商名,美国)
biological and chemical protective clothing 生化防护服
biologically absorbable fiber 生物吸收纤维
biologically active fiber 生物活性纤维
biologically compatible fiber 生物相容纤维
biological test method 生态检验方法
biological textile 生态纺织品
biomedicine 生态医学
biometeorology 生态气候学
biomimetic fiber 仿生纤维
bionics 仿生学
bionics-oriented garment designing 服装仿生设计
bionomy 生理学;生态学
bio-polishing 生物抛光整理,生物光洁整理
Biosil 毕奥西尔抗菌织物(商名,日本东洋纺)
biosteel 生物钢(蜘蛛丝蛋白质纤维)
bio-stoning 生物石磨整理
biotextiles 生物纺织品
BIOTHERM 碧欧泉
biowashing 酵素洗
biplane hat 双翼飞机帽
bipseudoindoxyl 靛蓝
birch gray 白桦灰,桦木色(浅棕灰色)
bird cage 穹顶大网眼头纱
birdcage heel 鸟笼跟
bird leg 鸟腿型
bird of paradise feather 极乐鸟羽毛
bird's back 鸟眼花纹背的织物
bird's egg green 鸟蛋绿
bird's eye 鸟眼花样,鸟眼花纹,芝麻点花纹;小菱纹织物
bird's eye design 鸟眼图案
bird's eye diaper 鸟眼花纹布
bird's eye fabric 鸟眼花纹织物
bird's eye fancy suiting 鸟跟花呢
bird's eye leno 鸟眼花纹纱罗
bird's eye linen 鸟眼花纹斜纹亚麻布
bird's eye pattern 鸟眼花纹
bird's eye pique 鸟眼凸花布
birdsnest mat 针织毛绒垫
biredshend 比雷兴地毯,紧密波斯地毯
biretta (天主教)四角帽,教土便帽
biretz 比雷兹双面毛葛

birotine 中东丝绸
birritz 宽袖中长手套
birrus 比雷斯外套(古罗马),古罗马带兜帽厚毛大衣;粗厚呢帽子;粗厚呢披肩
birthday suit 生日装
birthstone necklace 诞生石项链
birthstone ring 诞生石戒指
bis 比斯细薄亚麻布
bisage 两次染色的棉织物;两次染色的麻织物
bisaya 比萨耶麻,菲律宾槿麻
biscuit 浅棕色,饼干色
biscuit duck 旅行袋布,粗厚帆布,烤饼干帆布
Bis Dorothée 比斯·多罗西(法国针织公司)
bisect 平分线
Biserl 常压染色型聚酯纤维(商名,日本钟纺)
bisette [法]培珊花边,白亚麻枕套花边;刺绣瓣带
bisexual color 两性色
bishair wool 印度比沙尔羊毛
bishop 裙撑,裙垫(旧称)
bishop collar 主教领,法衣式领
bishop's cloth 粗厚方平织物
bishop's lawn 优质细薄棉平布
bishop sleeve 主教袖,紧口大袖,罩衫袖,法衣式宽敞袖子
bishop type sleeve 主教服式袖(上小下ণ 大的衬衫式袖)
bishrinkage mixed yarn 异收缩混纤丝
bishrinkage yarn 异收缩[变形]丝,异收缩[变形]纱
bis linen 比斯细薄亚麻布
bislint 德国细带,细狭带(德国制)
bison 深咖啡色
bisonne 法国原色里子呢
bisque 藕荷色,灰黄色,淡褐色,淡血牙色
bisso linen 比索细薄亚麻布
bissonata 比索纳塔粗纺平纹呢,法国教士粗服呢
bissonne 本色羊毛呢(法国制,作里子用)
bissuti 比萨蒂细平布
bister 棕色,深褐,枯叶棕
bistro dress 小餐馆装,小酒馆装
bi-swing 运动夹克衫,运动式外衣,运动夹克
bisymmetrical figures 双重对称花纹
bit 织物染色小样

bitlis 土耳其比特利斯地毯
bitmap 位图模式
bit moccasin 马勒摩卡辛鞋
bitre 比特尔短纤维亚麻布
bitter chocolate 深咖啡色
bittersweet 朱红色,橘红色
bivouac 比沃阿克珍珠呢
bixin 胭脂树橙色
biyu leno brocade 碧玉纱
bizarre dress 奇装异服;奇异服装
bizarre makeup 稀奇古怪的缝制
bizarre pattern 稀奇古怪的图案
bizette 棱结白亚麻帽饰花边
bizettes 粗狭网眼纱
Bjorn Borg 比欧·博格(法国时装设计师)
BL,B/L 提单
blaams linen 比利时半漂亚麻布
black 黑色,元色;黑颜料
black alkali 黑灰色
black amber 黑琥珀
black and white 黑白版,黑白照片
black and white check 黑白格子纹
black-and-white drawing 黑白画
black and white halftone 黑白网点
black-and-white pattern 黑白图形
black arm band 黑袖章
black ash 黑灰色
black basket sneakers 黑色篮球鞋
black bearskin 黑熊皮
black belt 柔道黑带
blackberry stitch fabric 黑莓组织针织物,玉米花组织针织物
Black Bird 黏胶长丝(商名,日本尤尼吉卡)
black boots 皂靴
black box calf 黑珠皮
black boxside chrome 黑粗珠皮
black brown 黑褐
blackburn printers 英国平纹棉印花坯布
black-butterfly shell button 黑蝶贝扣
black cloth 元布
black cotton plain 黑色棉平布
black cross mink 黑十字貂皮
black denim 黑色牛仔布
black dial watch 黑表面手表
black diamond 黑钻石
2-6 black drill 二六元贡
black dull 暗黑
black earth 土黑色

black embroidery 黑色刺绣
blackening 黑度
black face wool 黑面羊羊毛
black factice 黑油膏
black fiber 碳纤维;斯里兰卡黑棕榈纤维
black-figured style 黑绘风格
black-formal 黑夜礼服
black fox 黑狐皮
black gauze cap[hat] 乌纱帽
black gem 黑宝石
blackglama 美国黑貂标志
blacking 发黑
black ink 黑墨水色
blackish green 墨绿
blackish tone 近似黑色调
black jack 黑杰克鞋;方高跟夸张鞋舌女鞋
black lambskin 黑羔皮
black lenos 纱罗透孔织物;黑纬薄纱
black light 黑光
black mink 黑貂毛皮
black moss 黑苔色
black muskrat 黑麝鼠皮
black-oiled slicked jacket 黑油布夹克式雨衣
black-oiled slicker 黑油布工作外衣;黑油布宽大雨衣
black opal 黑蛋白石,黑欧珀
black organic fiber 有机碳纤维
black-out cloth 遮光布
black-out coating fabric 全黑遮光涂层织物
black-out finishing 遮光整理
black pearl button 黑珍珠贝扣
black rattler cotton 黑粹棉,改良陆地棉
black rubber apron 黑橡胶围裙
black rubber capeback coat 黑橡胶披肩大衣
black rubber firemen's coat 黑橡胶消防衣
black rubber jacket 黑橡胶夹克
black rubber leggings 黑橡胶绑腿
black rubber overalls 黑橡胶工作服
black rubber police coat 黑橡胶警察外套
black sateen 羽缎,泰西缎,棉背缎
black satin drill 二六元贡,元贡
black seed cotton 黑籽棉
black sheep-skin leather 黑绵羊皮
black shirt 黑衬衫
black silk skirt 黑绸裙

black silk yarn shoes 青丝履
black spot 黑斑,黑点疵
black suit 常礼服,黑色西装,黑色套装,半正式礼服
black superfine 黑色特细缩绒呢
black thread 油污线
black tie 黑蝶领结,黑领结;黑领带(正式场合用);带黑领结的宴会小礼服
black velvet 乌绒
black velvet leaf belt 黑天鹅绒叶饰腰带
black watch 布赖克瓦其,深色格子布
black watch tartan 深色方格呢绒
black wool 黑羊毛
black work 黑线刺绣,白底黑线刺绣
bladder green 天然绿色染料
blade 刀式;刀身,溜冰鞋冰刀;英俊潇洒男子
blade bone 肩胛骨
blade jacket 男式商务装
blade measure 肩宽
blae 灰蓝色,暗蓝色
blaireau 獾毛
blake thread 缝鞋亚麻线
blanc 漂白布(法国称谓)
blancard 优质半漂白亚麻布
blanchet 浴衣呢;睡衣呢
Blang ethnic costume 布朗族服饰
blank 袜口;坯布,毛坯,坯料;无色的
blanket 毛毯,毯;印花衬毯,厚垫布;造纸毛毯,滚筒包衬;大张毛皮
blanket and cushions 毯子和垫子
blanket bindering 毯子缝边
blanket check coat 毛毯格子外套
blanket cloth 拉绒大衣呢,毛毯大衣呢;双面棉绒布
blanket coat 拉绒外套,绒毯布外套
blanket edging machine 毛毯缝边机
blanket felt 毛毯毡
blanket finishing 呢毯整理
blanket hemming machine 毛毯缝边机
blanketing 毯料,制毯织物;保护,被覆,密封
blanket insurance 综合保险
blanket look 毛毯款式
blanket loop 方格式线袢
blanket mark 衬布痕,胶毯痕
blanket order 综合订单
blanket plaid 毛毯格子,大方格花纹;毛毯格子呢
blanket range 包袱样,试织大样

blanket ring stitch 轮形锁孔绣
blanket rug 毛毯披巾,拉绒披巾
blanket shawl 拉绒披巾,毛毯披巾
blanket sleepers 拉绒睡衣裤
blanket stitch 饰边缝线迹
blanket-stitch carrier 锁缝线迹带袢
blanket-stitch embroidery 毛毯锁边刺绣
blanket stitching 锁边线缝,饰边线缝
blanket stitch seam 锁边线缝,饰边线缝,毯子锁边缝
blanket style 毛毯款式
blanket twill 二上二下毛毯斜纹
blanket wool 毛毯料
blanking 空白
blanking die 裁剪冲模
blanking punch 裁剪冲模
blank test 空白试验
blanlc 不记名
blanquin 平纹漂白棉床单布
blarney 爱尔兰针织绒线;爱尔兰粗呢
blarney tweed 爱尔兰粗呢
blarney yarn 爱尔兰针织绒线
blashed flax 浸渍过度的亚麻
blassas 西班牙低级羊毛
blatic 湖绿[色]
blatt stitch 布拉特针迹,缎纹线迹
blazer 女上衣;运动夹克,运动上衣;布雷泽外套,软薄运动夹克衫
blazer button 外衣式大纽扣
blazer cloth 运动服条纹呢,上衣呢
blazer coat 运动西装,宽松西装,运动上衣,男式便上装(与裤子不配套)
blazer jacket 宽松外衣,便装
blazer mou 柔和的运动型西装
blazer pocket 便装口袋
blazer Sans Manche 无袖适动型西装
blazer socks 色条童袜
blazer stripe 运动夹克条纹
blazer striped shirt 丝光柳条衬衫
blazer suit 运动型套装,户外套装
blazer sweater 双排纽羊毛衫
blazon 纹章
bleach 漂白剂;漂白
bleached apricot 淡杏色
bleached aqua 漂白水色
bleached bed sheet 全白床单
bleached cloth 漂白布
bleached cotton cloth 漂白棉布
bleached denim 漂白牛仔布
bleached denims 漂白工作服;漂白工作

服色
bleached drill 漂白卡其布
bleached fabric 漂白织物
bleached goods 漂白布匹
bleached-ground ticking 漂白地经斜纹被套布
bleached jeans 漂白牛仔裤,褪色变白牛仔裤,磨白牛仔裤
bleached knit fabric 漂白针织布
bleached linen 漂白亚麻布
bleached out jeans 重漂洗牛仔裤,褪色型牛仔裤
bleached ramie sheetings 苎麻漂白细布
bleached sheeting 漂白布
bleached single jersey 漂白汗布
bleached table cloth 全白台布
bleached towel 全白毛巾
bleached tussah spun silk yarn 药水柞绢丝
bleached wool 漂白毛
bleached yarn 漂白(纱)线
bleacher 漂白坯布;漂白工厂
bleacher sheeting 漂白布
bleaching 漂白(工艺)
bleaching agent 漂白剂
bleaching fastness 漂白牢度
bleach mask 漂白美容术
bleach out 漂白款式
bleach out jeans 磨白牛仔裤
bleach out look 漂白风格
bleach wash(ing) 漂洗
blebbing 印花花纹不清,毛脚(印花布疵),色渗斑;色化斑
bleeding check 渗色法色织格子布
bleeding madras 易褪色的格子布
bleeding style 渗散印花
bleed through 渗胶
blehand,blehant 紧身女装;昂贵面料;长外袍;男窄袖服装
blemish 污点(钻石);污迹(钻石)
blend 混纺制品;混纺;修弧;圆顺
blended chemical fiber fabric 化纤混纺织物
blended color fabric 混色织物,混色布
blended dye 混合染料
blended fabric 混纺织物
blended fibre knitwear 混纺针织品
blended hand knitting yarn 混纺绒线
blended knitting yarn 混纺绒线
blended linen 亚麻混纺布
blended plush 混纺长毛绒

blended single jersey 混纺汗布
blended spun fiber 共纺纤维,混抽纤维
blended spun yarn 混纺纱
blended sweater 混纺毛衫,混纺针织衫
blended wool 混纺毛
blended yarn 混纺纱
blended yarn etched-out fabric 混纺纱烂花布
blendent 配色
blend fabric of polyester 聚酯混纺织物
blend fiber 混合纤维
blending 仿毛整理,混色整理;毛皮染色;混纺;修弧
blending-spun linen 亚麻混纺布
blend line 交接线,延伸交接线
blend off point 变更点
blends 混纺产品
bley 未漂捶光棉布;未漂捶光亚麻布
bliaud 布劳德裙;布劳德长袍,伯莱欧束腰长袍(12～13 世纪)
blighty tweed 高级粗纺呢
blimi-stitch machine 暗缝机;插边机
blind 窗帘;失光,堵塞,闭塞
blind catch stitch 隐式回三角缝,暗三角针
blind chintz 轧光棉窗帘布,横条纹窗帘布
blind cord 窗帘绳
blind edge 暗缝边
blind edge seam 暗边缝
blinders 缝边附件
blind eyelet 暗鞋孔
blind fly 缝上的口袋盖;裤子暗门襟
blind-hem 暗卷边;暗卷缝线迹,暗缝下摆
blind hem foot 暗缝缲边压脚
blind hemming 暗缝缲边,服装边缘暗卷边,黑卷边;暗缲针
blind hemming ruler 暗缝卷尺
blind hemming stitch 暗卷边线迹,暗缝线迹;暗针法,暗缲针法
blind herringbone stitch 暗人字缝线迹
blind hole 育孔(缝纫)
blind holland 窗帘亚麻布
blinding 失光
blind ladder tape 窗帘带
blind lap 暗缝下摆,搭接暗缝,暗缝叠门
blind lockstitch 暗式连锁针迹
blind looper 暗缝套口机;暗缝弯针
blind man's watch 盲人表
blind pleat 暗褶
blind roller (可以卷成筒状的)窗帘布;遮阳卷帘

blind seam 暗缝
blind stitch 缲,(缝纫中的)暗针,暗缝,暗缝线迹,手工暗切线,暗切线
blind stitch feeder 暗缝送料器
blind stitch felling machine 暗缝线迹平缝机,暗缝镶接缝纫机,缲边机
blind stitch foot 暗缝压脚
blind stitch hem 暗缝卷边
blind stitch hem guide 暗缝线迹卷边导向器
blind stitch hemmer 暗缝卷边器
blind stitch hemming 暗缝卷边
blind stitching 暗缲针缝,暗缝,暗缲线迹
blind stitching hem 缲底边
blind stitching machine 暗缝机
blind stitch latch hemmer 暗缝闩卷边器
blind stitch seam 暗缝线缝,暗缝线迹
blind stitch (sewing) machine 暗缝机
blind stitch tacker 暗缝机,暗缝加固机
blind support tape 百叶窗带
blind tacker 暗缝机
blind tacking 暗缝加固缝
blind tape 窗帘带
blind ticking 色条粗斜纹布
blind tucks 暗塔克,暗缝褶,盲塔克,盲缝褶
blind twill 暗条斜纹,暗斜纹条
bliss tweed 布利斯呢
blister 泡泡纱,泡泡织物;起泡;气泡;包泡(鞋)
blister brocade 绉纹花缎
blister cloth 双面针织绉织物,泡泡纱,泡泡呢,泡泡织物
blister crepe 泡泡纱,点纹泡泡纱,优质泡泡纱
blister design 凸纹
blistered 开衩式的;蓬松式的;泡泡点纹效应
blister fabric 凸纹(浮线)针织物
blister fabric double 双线圈凹凸织物
blister fabric single 单线圈凹凸织物
blister stitch 皱缝,褶缝;缩皱形线迹
blizzard collar (女服)防寒高立领
blob 斑渍,污渍
blobby wool 松胖羊毛
bloc 联盟
bloched and pressed 四角正方烫平
block 原型;帽楦,帽模;英国裁剪样板,定形板,烫衣板,慰烫馒头,木制假头,链内嵌段;手工印花的雕花木板

block buffer 板跑缝,由实线构成的衣片图
block check 棋盘格纹,方格花纹,棋盘纹
block construction 样板结构,原型结构,纸样结构
block-cut print graph 块割点图
block diagram 方框图,方块图,框图
blocked and pressed 定形熨烫,四角正方烫平
block felt 毡块
blockhead 木制假头;帽模
block heel 块状跟
block holder 裁剪样板架
blocking 归拔,熨烫造型,模熨;衣片与衣片在排料图上的分割线;布层粘达
blocking back piece 归拔后背
blocking board 针织物布边熨烫定形
blocking crotch 拔裆
blocking fiber 嵌段纤维
blocking front piece 推门
blocking machine 帽楦机
blocking pad 烫垫
blocking resistance 抗粘连性
blocking sleeve 归拔衣袖
blocking top collar 归拔领面
blocking under collar 归拔领里
block in sole 袜底加固
block in toe 袜头加固
block pattern 原型,裁剪样板,原型底样,服装样板,纸样
block plaid 大方格,棋盘纹
block plaid umbrella 大方格伞
block print 手工模版印法
block printing 木模(版)印花;手工模版印花
block quilting 方形绗缝,菱形绗缝
block shoulder point 原型肩点
block stripe 棒条纹,块状条纹
block to dry 边干燥边整理
block work 整只操作(皮革)
blond 浅色,亚麻色,极浅黄棕色;原色丝花边;白肤金发碧眼男人
blonde(=blond) 白肤金发碧眼女人
blonde application 机制贴花花边
blonde de fantaisie 法国机制真丝网眼花边
blonde de fil 法国六角网眼地小花纹;白麻纱梭结花边
blonde écru 梭结丝花边
blonde en persil 原色丝花边

blonde fausse　丝网眼花边
blonde lace　布隆德花边,丝花边,原色丝花边;丝带
blonde net　布隆德网地花边,梭结棉网眼纱
blond en persil　法国香菜叶纹丝花边
blond quillings　重浆网眼丝纱罗
blood fluke protective rubber boots　血防胶靴
blood red　鲜红,猩红,血红
bloodstone　血石
bloom　(面颊)红润,(毛织物)光泽整理;(丝和丝绒织物)绿灰色绒光;(染色布)表面平淡光泽;(地毯绒头)松开;喷霜(鞋)
bloomer　布鲁姆女服
bloomer dress　灯笼裤装
bloomer maillot　灯笼裤式游泳衣
bloomers　女式灯笼裤,宽大女短裤,扎口女内裤
bloomer shorts　灯笼短裤
bloomer suit　连衣泳裤
bloomer swimsuit　连裤游泳衣
bloomg lustre　鲜明光泽
blooming　增艳处理
bloom skirt　开花裙
bloomy lustre　丝绒[状]光泽
blossom　花红色,浅妃红
blossom pink　浅粉红色
blot　织物剪毛不良;疵点,污点,斑渍,污渍
blotch　印花色底,满地花纹;斑渍
blotch checks　印经蓝斑格子布(非洲)
blotch coverage　满地罩印,满地印花面积
blotch ground　满地花纹,印花色底
blotchiness　斑点外观(皮肤)
blotch printed towel　满地印花毛巾
blotch printing　满地印花,底色印花,单面印花
blotchy dyeing　斑渍染色
blouse　T恤衫,恤衬衫;女上衣,女衬衫;宽松上衣;罩衫,上衣,宽大短外套;制服上衣;军上装
blouse back　宽松后背,女衫后背
blouse coat　蓬腰外套
bloused　穿短上衣(或女衬衫)的,短上衣似的
bloused back　背色,背囊
bloused back silhouette　膨松式背部轮廓
bloused line　宽松式外形款式,上身蓬松型
blouse dress　低膨腰节衬衫套装
bloused silhouette　宽松式服装轮廓,膨松式服装轮廓,上身蓬松轮廓
bloused top　蓬松上衣
blouse fabric　衬衫织物
blouse jacket　齐腰[长]夹克
blouse lace　衬衫花边
blouse measurement　女衬衫尺寸
blouse on blouse　重叠女衫
blouse pull　女衫式毛衣
blouse sleeve　衬衫袖,紧口大袖,T恤衫式袖
blouse-slip　上衣衬裙装,连衫长衬裙
blouse steeve　紧口大袖
blouse suit　女衫式套装
blousette　女无袖上衣
blousing　上衣料,做上衣的料子,紧腰宽松式(腰部以上膨起)
blousing blouse　(女)宽松上衣,膨松上衣
blousing blouson　宽松上衣
blouson　夹克式上衣,罩衫,束摆短上衣,宽松外套;(制服上装)松紧带束腰女衫;大腰身低腰带连衣裙;布的打褶和悬垂状态
blouson blouse　膨腰女衫
blouson coat　蓬腰外套
blouson dress　低腰节膨腰上装;蓬腰连衣裙
blouson jacket　束腰短夹克,蓬腰夹克
blouson jumpsuit　膨腰式连衫裤,蓬腰式连衫裤
blouson long　法国长式宽松上衣,长宽上衣
blouson on yoke　约克船领宽松上衣
blouson pants　步行短裤,百慕大短裤
blouson silhouette　膨松型轮廓
blouson suit　短夹克套装,宽大的短上衣和裙子两件套,膨松型夹克套装;束腰女泳装
blouson swimsuit　宽松型连胸罩泳装
blouson tunic　束腰宽上衣
blouson tunicsash　女束腰宽上衣
blouzer　罩衫式运动上衣
blower　缝纫机的推布送料器,吹风送料器
blowing　(干)蒸呢
blowing apparatus　吹风装置
blown fiber　吹制纤维,喷射纤维
blowtorched hem　汽油吹管型裙摆(或衣

摆）
Bluce Oldfield　布卢斯·奥德菲尔德（英国时装设计师）
bluché　法国布谢原色细呢
blucher　布吕歇尔鞋,布吕歇尔靴,鞋面连舌系带鞋,外耳式鞋
blucher bal　布吕歇尔式皮鞋,变形布吕歇尔靴（鞋舌不与鞋面一体的系带鞋）
blucher-cut　布吕歇尔式剪裁
blucher shoes　男式缚带皮鞋,外耳式鞋
blue　蓝色,青色;蓝色染料;
blue and white stripe　蓝白条子球衣
blue and white stripe ticking　蓝布条枕套布
blue asbestos　青石棉
blue ashes　潮蓝,深灰蓝
blue atoll　珊瑚岛蓝
blue bafts　非洲蓝粗平布
blue-based color　蓝底色
bluebell　风铃草蓝色
Blue Bender cotton　美国蓝本德棉
bluebird　青岛蓝
blue black　墨蓝,蓝黑
blue black tartan　蓝黑格子呢
blue bokhara carpet　羊毛驼毛制织的长毛毯
blue bokharas　巴罗奇地毯
blue bonnet　蓝软帽,苏格兰宽顶无沿蓝色呢帽
blue clear　鲜蓝色
blue cloth　蓝布
blue coat　警察制服;大学生制服
blue-collar worker　蓝领工人
blue coral　蓝珊瑚色
blue cotton　蓝棉,极白的棉,不正常棉
blue cotton gown　蓝布大褂
blue Danube　多瑙河蓝
blue deep　暗蓝,深蓝
blue diamond　蓝钻石
blue dungaree　蓝劳动布
blue east india linen　交织缎,柞丝经棉纬花缎
blue flax　暗黑色亚麻
blue fog　雾蓝色
blue fox　蓝狐毛皮,青狐毛皮,仿制的青狐毛皮;蓝狐皮
blue glacier　冷感浅蓝色
blue glass　蓝玻璃色
blue goods　暗黑色棉布

blue grass　草地色
blue gray　蓝灰色
blue green（BG）　蓝绿,青绿
blue grey　蓝灰
blue grotto　岩洞蓝
blue haze　蓝雾色
blue helmet　维和部队蓝色头盔
blue indigo　靛蓝
blueing　上蓝;增白染料;发蓝
blueing agent　上蓝剂
blue iris　蓝虹色
blue iris mink　蓝虹貂皮
blue jay　中蓝色
blue jean　蓝斜纹布,牛仔布
blue jeans　蓝布工装裤,牛仔裤,蓝"绅士"裤,西部牛仔裤（美国）
blue jewel　珠宝蓝;蓝珠宝
blue light　浅蓝色
blue mist　雾蓝色
blue moon　月蓝色
blue mottle　蓝地白点薄纱
blue nankeen　毛蓝土布
blueness　青色,蓝色
blue pelt　青里毛皮
blue print　蓝图
blue printing paper　蓝图纸
blue purple　品蓝,蓝紫色
bluer　（色光）偏蓝
blue-red cast　蓝光红色
blue ribbon　蓝绶带
blues　美国海军蓝制服
blue sapphire　蓝宝石色
blue scale　蓝色标准（日晒牢度）
blue serge　藏青哔叽
blue shade　青光
blue shadow　阴影蓝
blue-sohn　束腰短夹克
blue spruce　蓝云杉色
blue stained cotton　灰白棉,蓝渍棉
blue steel　青钢色
blue surf　蓝浪花色
bluet　平纹蓝棉布;纬面斜纹布
blue-tinge　蓝色光
blue tint　浅蓝色
bluette　棉蓝平布;斜纹工装蓝布
blue violet　蓝紫色
blue-white　蓝白
blue-white finish　上蓝整理
blue wool　优级有光泽羊毛
blue wool fabric　蓝色毛织物标样（测耐

光色牢度用)
blue wool fabric standards 蓝色毛织物标样(评定日晒牢度用)
blue wool(light fastness)standard 蓝色毛织物标样(测耐光色牢度用)
blue wool scale 蓝色羊毛标准
bluey 蓝色包袱;旅行包;粗制外套(或衣服);澳洲丛林居民蓝色衬衫;粗制衣料;粗制外套
bluffed edges 未缝衣边
bluff finish 毛边处理
bluffing 折边装置
bluing 上蓝;增白染料;发蓝
bluish 黛青
bluish dogbane 罗布麻
bluish green (BG) 带蓝色绿
bluish white 青白,蓝白
bluish yellow 青黄,蓝黄
blumley linen 印花斜纹亚麻布
blunk 厚实印花布(英国名称,麻或棉制)
blunt 短粗针
blunting a corner 翻转衣角
blunt (pointed) needle 钝头针
blush 绒光,红光;棉的乳白色;织物有光处理
blush brush 胭脂刷
blushed leather 绒面革
blush rose 差红色(浅红色)
bluteau 网眼衬衫织物;筛绢
BN fiber 氮化硼纤维
bo 帛
boa 蟒围巾;女用长毛皮围巾,圆筒形围巾,羽毛围巾(或披肩);蟒蛇绿色
boa coat 长毛领大衣
board 烫衣垫板,台板;纸板;木板色
boarded finish 皮革搓纹整理
boarded heel 定形袜跟,热定形袜跟
boarded leather 磨面革,印痕皮,磨光皮革,搓纹革
boardness 织物硬挺度
board of directors 董事会
board rib 抽针罗纹
boardy 粗硬织物;硬挺度;手感板硬,手感粗硬
boardy feel 硬性手感
boater 康康帽;硬草帽,赛艇帽,水手帽,平边平顶硬草帽
boater tie 便装领带,船客领带
boating shirt 划船衬衫
boating shoes 船坞鞋(有防滑跟)

boating wear 功能性运动服装
boat neck 安装船领,一字领
boat neckline 船形领口,一字领领口
boat neck shirt 船形开领衬衫,一字领衬衫
boat sail drill 防风平纺棉布,船帆布
boat shoes 船鞋,帆布便鞋,休闲鞋,帆船鞋
boat shuttle 船形摆梭
bob 振子坠;短发;挂饰;博布棉;双刀滑冰鞋;短发式(妇女、儿童)
bobbed hair style 短卷发式发型
bobbie collar 小圆领,很小圆领
bobbin (缝纫机)梭心;绕线管;梭织花边梭子
bobbin case 梭匣,梭心罩,梭壳
bobbin case bouncing 缝纫机梭壳回跳
bobbin case gib plate 旋梭梭床
bobbin case holder 梭架
bobbin case holder retainer 旋梭板,月亮圈(梭心套护板)
bobbin case latch 梭匣柄;梭门盖
bobbin case opener 开梭器
bobbin case opener finger 开梭钩
bobbin case tension spring 梭壳簧
bobbinet 珠罗纱;(花边用)六角网眼纱
bobbinet lace 珠罗纱花边
bobbin fine 机制暗花细花边
bobbin fining 机制暗花凸边花边
bobbin friction ring 缝纫机用胶圈
bobbin hook 缝纫机梭心夹子
bobbin lace 梭结花边,线卷花边
bobbin lace lever 梭结花边杆
bobbinless machine 无梭缝纫机
bobbin net 六角网眼纱,珠罗纱;棉线花边带
bobbin quilling 棉线窄花边
Bobbin Show 亚特兰大制衣机械展览会
bobbin silk 绣品
bobbin tape 扁带,圆带
bobbin thread 底线,梭心线,梭线,筒卷线,轴线
bobbin thread knife 底线剪刀
bobbin thread monitor 底线控制器
bobbin thread pull off 底线拉出量
bobbin thread slipping out 底线滑脱
bobbin thread supply 供梭线
bobbin thread tension 底线张力,梭线张力
bobbin thread trimmer 底线剪刀
bobbin winder 缝纫机绕线器

bobbin winder base 缝纫机绕线座
bobbin winder bracket 缝纫机绕线架,绕线器托座
bobbin winder complete 绕线装置,绕线器组件
bobbin winder frame 缝纫机绕线架,绕线器托座
bobbin winder friction ring 绕线器摩擦轮
bobbin winder position pin 绕线器定位销
bobbin winder pulley 绕线轮
bobbin winder spindle 绕线器轴
bobbin winder stop latch 满线跳板;绕线器止动杆,绕线器满线跳杆
bobbin winder tension bracket 绕线张力架
bobbin winder tension bracket complete 绕线张力架装置
bobbin winder tension disc 绕线张力夹线板
bobbin winder tripping arm 绕线器调节板
bobbin winding 绕线
bobbin winding capacity 绕线量
bobble hat with pom-pom 束球绒线帽
bobbles 绒球
bobble skirt 蹒跚裙
bobby jeans 海滩牛仔裤
bobby pin 扁平发夹,宝贝发夹
bobby's hat 警察高顶帽
bobby socks 宝贝短袜,齐腿踝短袜,翻口式短袜
bobby soxes 妇女短袜,套袜,翻口短袜
Bob cotton 美国博布棉
bob hairstyle 短发型
Bob Mackie 鲍勃·麦奇(1940~,美国时装设计师)
bobtail 晚礼服,夜礼服
bob wig 菜花假发,短卷假发,短辫假发,短卷男假发(18世纪)
boby boots 合腿靴
boby form 上身模型
bocasin(e) 博卡辛亚麻布;优质重浆织物
boccadillos 漂白亚麻衬衫细布
boccage 亚麻花缎桌布
boccassini 博卡西尼漂白细平纱
bocking 博金粗毛呢
bocskor 尤跟女鞋
bodging 修补不良(针织物)
bodging-on 对行不对眼缝头;套口
bodg odour 汗臭
bodiasse 中国蚕丝
bodice 女装紧身上衣,紧身胸衣,紧身围腰;上衣片,宽大背心,大身(由腰到颈的服装部分)
bodice back 后片,后幅
bodice back too short 后身起吊
bodice en coeur 低领荷叶边晚装胸衣
bodice front 衣前身
bodice hug-me-tight 女用宽大背心
bodice neckline 衣身领圈线
bodice skirt 胸衣裙
bodice tuck 大身褶裥,紧身衣褶裥
bodice-type rib knit cotton vest 紧身棉罗纹背心
bodily form 体型
bodily sensation 身体感觉
bodkin 粗长针,锥子,束发针,发夹针;金缎或银缎;宝大锦
bodkin beard 锥子形胡须
bodkin-work 金线衣边
body 身体,躯干,躯体;布身,布身身骨;女紧身衣;上衣的主要部分(除领、袖外)
body adherence 服装贴身性
body armour 护身甲
body art 身体艺术,身躯艺术,人体艺术
body assists 身体的辅助动作
body back 服装后身
body bag 软包,随身包
body blank 大身衣片
body blouse 紧身衣衬衫
body boots 合腿靴,紧脚靴
body briefer 女紧身全帮内衣(胸罩式背心连短裤)
body Brussels 布鲁塞尔毛圈地毯
body build 体格,身体构造
body building 健身
body carpet 小尺寸地毯
body carpet square 拼接用地毯
body chain 体链
body chain belt 紧身链状腰带,缠身链带
body characteristic 身体特征
body clock 人体钟
body cloth 夹衣面料;马鞍用毛毯
body clothes 内衣裤,紧身衣,内衣
body coat 合身外套
body color 不透明色,体色,不透明色
body comfort 舒适性,适体性
body-conforming fit 合体
body-conscious 贴肉感,贴身感
body-conscious hugging feel 紧贴感
body consciousness 体型意识(服装紧密

吻合体型),贴肉感
body-contact 贴身
body contouring 人体外形
body core 胴体
body corporate 法人团体
body coverage 身体覆盖度
body cross-section 人体横截面
body curve 体形曲线
body cutting 立体裁剪
body demand 姿势,姿式
body dimension 人体尺寸
body fashion 贴身时装
body fitting slacks 紧身裤
body fitting trousers 紧身裤
body flex 人体屈曲度
body fluid 体液
body folded edge 大身折边
body form 上身模型
body front 前身
body garment 紧身衣
body grading 服装上身部分放样
body heat 体温
body hose 紧身连裤袜
body hugger 亵衣
body-hugging (服装)紧身,贴身
body-hugging feel 紧抱感
body-hugging suit 紧身套装
body image 人体形象
body-jewel 贴身首饰
body-jewelry 贴身首饰
body language 身势语(如耸肩、眼部表情等)
body length 身长,衣长;身高
body length knitting machine 衣坯针织机
body line 紧身轮廓线,紧身曲线,身体线条,身体轮廓线
body linen 内衣用亚麻布
body lining 衣身衬布
body lotion 洗身剂
body measurement 量体尺寸,人体尺寸;量体
body measurement system(BMS) 人体尺寸测量系统
body mechanics 健美操
body motion 全身动作
body movement 人体活动,形体动作
body nude style 裸体款式
body odour 人体气味;体臭
body of beauty 形体美
body of material 服装大身

body painting 绘身
body pants 针织紧身裤
body piece 大身衣片
body press 整烫机,大身熨烫机
body rib 下摆罗纹,大身罗纹
body rise 腰至臀高,股上
body rompers 紧身田鸡装
body satisfactory(degree) 人体满意度
body-scanning 人体扫描(电脑制衣程序)
body's contour 人体轮廓线,人体线条
body setter 袜筒定形机,无底袜定形机
body setting machine 服装成形机
body shape 体型
body shape press 大身熨烫机
body shape pressing 立体蒸汽熨烫
body shaper 整形内衣,塑身内衣
body shaping bra 贴身胸罩
body shaping slip 贴身衬裙
body shirt 紧身背心,紧身衬衫,女贴身衬衫(或背心),BS衫,扣裆紧身女上衣
body shorts 紧身女裤
body size circular-knitting machine 计件衣坯圆形针织机
body size fleece machine 计件绒布衣坯机
body slimmers 紧身减肥衣
body slip 紧身衬裙
body-smoothing shape 贴身形
body stand 人体模型,人体衣架
body stockings 紧身连裤袜,紧身连衣袜,一件式紧身女内衣
body suit(B.S.) 紧身女服,紧身连衣裤,健美服;婴儿套装;上下相连中间衣
body supporter 护身用纺织品,人体防护用品
body supporter textile 护身用纺织品
body sweater 贴身卫生衫;贴身运动衫
body tape 大身贴边
body tattoo 刺青
body temperature 人体温度
body type 体型
body warmer 絮棉(或非絮棉)背心
body wear 紧身衣裤,健美服,紧身运动装,(健身或跳舞用)形体服
body weight 体重
boehmeria nivea 苎麻
bogey 运输小车
boggy 羔皮,翻毛羔皮
bogie 运输小车
bogotana 柔软整理漂白棉细布
bogus tartan 充格子花呢,仿格子花呢

Bohemian cotton 波希米亚棉(美国改良陆地棉)
Bohemian dress 波希米亚式裙子
Bohemian flax 波希米亚亚麻
Bohemian lace 波希米亚狭花边
Bohemian look 波希米亚风貌
Bohemian ruby 波希米亚红宝石
Bohemian smock 波希米亚式罩衣
Bohemian style 波希米亚款式
Bohemian stylistic form 波希米亚款式
Bohemian ticking 波希米亚防羽绒枕芯布
Bohemian tie 波希米亚领带
bohrware 网眼刺绣
boi 博伊棉毛粗厚法兰绒
boiled fabric 煮练织物
boiled lawn 细亚麻平布,精练细亚麻平布
boiled linen 脱胶亚麻布,精练亚麻布
boiled-off silk 熟丝,脱胶蚕丝,精练蚕丝;熟绸,精练绸缎
boiled-out cotton 煮练过的棉织物;煮练过的棉纱
boiled-out fabric 煮练织物
boiled shirt 硬胸衬衫,前胸上浆的白衬衫(用于硬胸礼服)
boiler 锅炉
boiler suit 连衫裤工作服;连衫裤工作装
boiler type steam electric iron 电热式蒸汽熨斗
boiler type steam iron 电热式蒸汽熨斗
boilette 梳妆
boiling 煮呢;沸煮,沸腾
boiling off [out] 煮练
boil-off yarn 煮练纱线
boil-up 煮洗衣服
boina 扁平帽(西班牙)
boiteux 双色带
bokas 博卡布(印度蓝白条棉布)
bokhara 布哈拉绸;布哈拉地毯
bokhara cotton 布哈拉棉
bokhara rug 布哈拉地毯
bokhara wool 布哈拉羊毛
bola fiber 波拉纤维,绳用木槿属韧皮纤维
bola necktie 流星式领带
bola tie (用饰针扣住)流星式领带
bolbees 博尔皮斯漂白粗亚麻布,博尔皮斯浅蓝亚麻布
bolche 博尔奇原色平纹衬衫料

bold corduroy 粗条灯芯绒
boldekin 宝大锦(丝经、金线纬)
bold framed sunglasses 宽框边太阳镜
bold line 粗犷型;粗线条;大胆型
bold look 大胆款式(有男性的勇敢风貌),粗犷风格,大胆风貌
boldness 醒目程度
bold pattern 粗线条图案
bold strap sandals 宽带凉鞋
bold stripe 大胆条纹,醒目条纹,粗条纹
bold stripe runner 粗条纹走廊地毯
bold style 大胆风格,大胆款式
boldue [法]色(缎)带
bold wale corduroy 粗条灯芯绒;宽条灯芯绒
bold yellow 金黄色
bolero 波蕾若外套;(有袖或无袖)前胸敞开女短上衣,鲍莱罗女上衣;西班牙男短上衣
bolero blouse 波蕾若女衫
bolero cape 鲍莱罗披巾,波蕾若披风
bolero costume 波蕾若服装
bolero furskin 波蕾若毛皮短上衣
bolero hat 西班牙波蕾若男帽
bolero jacket 白来罗短夹克,波蕾若背心夹克,背心夹克
bolero kimono 和服袖高腰夹克
bolero mantle 波蕾若斗篷,波蕾若披风
bolero suit 白来罗套装(短上衣配裙),波蕾若短上衣套装,背心套装
bolero sweater 开襟短毛衫,波蕾若开襟毛衫
bolero top 波蕾诺短上衣
bolero toque 波蕾若褶饰帽
bolero type garment 波蕾若型服装
bolero vest 波蕾若无扣女背心,无扣女背心
boling pongee 薄凌纺
bolivar 博利瓦法兰绒;轻软素色法兰绒,玻利维亚条子毛绒
Bolivar county cotton 美国波利瓦棉
Bologen hemp 博洛涅大麻
Bologna crepe 波洛尼亚丝绉纱
Bologna gauze 波络尼亚丝纱罗
bolo necktie 波洛领带
bolo tie (用饰针扣住)流星式领带,博洛领带,保罗领带,饰扣式领带,波洛领带
bolsa 斜纹布袋(阿根廷制)
bolster collar 垫圈领
bolster fabric 承垫织物

bolt 一卷布；一卷缎带
Bolton coverlet 博尔顿白床单
Bolton sheeting 博尔顿斜纹粗布
Bolton thumb gloves 博尔顿拇指手套，波尔顿式手套
Bolton twill 博尔顿斜纹
bolt ring 弹簧搭扣（项链、手链上）
bolt upright posture 笔挺立姿
bombace cotton 印度孟买棉
Bombai jute 孟加拉黄麻
bombarded diamond 经轰击钻石
bombase cotton 印度孟买棉
bombasi 棉旗布；单面绒色棉布
Bombasin cotton 巴西棉
Bombasin(e) 邦巴辛斜纹绸；橡胶布底布
bombast 邦巴斯填料，棉或亚麻松软织物
bombastic shade 稻草色
bombax cotton 木棉；邦巴津织物
bombay 中厚棉床单坯布
Bombay brown 孟买褐
Bombay hair 孟买绒毛
Bombay hemp 菽麻
Bombay twill 孟买黄麻袋布
bombazet 精纺毛呢（法国制，平纹或斜纹）
bombazine 邦巴辛斜纹绸；橡胶布底布
bombe 隆起部位（服装）；军用短大衣
bomber cloth 粗纬破斜纹布；印花家具布
bomber jacket 投弹手夹克，飞行员夹克，轰炸机夹克，腰及袖口有松紧带的短夹克
bombix cotton 印度孟买棉
bombycine 薄绸，绢丝织物
Bonafil 高强聚丙烯复丝（商名，英国）
bonaid 苏格兰蓝色小呢帽
Bonan ethnic costume 保安族服饰
Bonaparte collar 拿破仑领，波拿巴领（以高高折起的上领和宽翻领为特征）
Bonaparte style 波拿巴款式
Bonaparte stylistic form 波拿巴款式
bon-bon pink 粉红色，桃红色
bon-bon sleeve 糖果式袖
bondage boots 绑带靴
bondage trousers with bum flap 用拉链和布条装饰的紧身裤
bonded 保税的
bonded area 保税区
bonded carpet 黏合地毯，黏绒地毯
bonded cushion （地毯）黏合衬垫
bonded dye 键合染料（活性染料）
bonded fabric 黏合织物

bonded-face fabric 面黏合织物
bonded fiber 黏合纤维（热溶纤维）
bonded-fiber fabric 纤维黏合非织造布
bonded knit 黏合针织物
bonded knitted fabric 黏合针织物
bonded knitting 黏合针织物
bonded lace 黏合花边
bonded mat 黏合织物；黏合纤维网
bonded nonwoven 黏合非织造布
bonded pile carpet 黏绒地毯
bonded rubber cushion 黏合橡胶衬垫
bonded seam 包缝
bonded synthetic fabric 黏合型合成纤维非织造布
bonded system 保税制度
bonded textile 黏合型纺织品
bonded thread 黏合线
bonded wed 黏合织物；黏合纤维网
bonded yarn fabric 黏合纱非织造布
bonden fabric 黏合织物
bondi beach 沙滩装，海滩装
bonding 黏合
bonding company 担保公司
bonding fabric 黏合织物，黏合衬
bonding finish 黏合整理，
bonding lace 黏合花边
bonding linings 黏合衬
bonding net 黏合网
bonding nonwoven 黏合法非织造布
bonding strength 黏合强度
bonding technique 黏合技术
bonding technology 黏合工艺
bonding treatment 胶合处理
Bond look 邦德款式（宽领西装，后身两边开衩，优雅衬衫配针织领带）
bond patch 黏合贴布
bondsman 保证人
bondyne 邦迪恩变性聚丙烯腈混纺织物（商名，美国）
bone 骨；骨制品；花边梭子（人工编结用）；发硬（缩绒制品）
bone black 骨黑色
bone brown 骨棕色（暗棕色）
bone button 骨扣
bone color 骨色
boned bodice 骨架紧身上衣，羽骨紧身上衣
boned fabric 黏合布
boned foundation 骨架式塑形内衣，骨架式整姿内衣
boned girdle 衬金属线绑肚，骨架式绑肚

boned inlay 衬衫领的插骨片
boned petticoat 带裙撑的衬裙
bone inlay 插骨片
bone lace 骨头花边,梭结花边,骨状花边
bone ornament 骨饰品,骨饰物
bone point 上等梭结花边
bone point lace 不规则网眼花边(统称)
bone ring 包纽的白环,包纽骨环
bone shoes 骨白色鞋
Bone Soeurs 布埃姐妹(法国时装设计师)
bone white 骨白色,淡灰黄色
bongra 印度平纹苧麻织物
bongrace 妇女丝绒头饰;帽檐;帽舌
boning (妇女胸衣的)羽骨;妇女胸衣用塑料带
boning and stay strip 羽骨及缝合牵条
Bonly 聚偏氯乙烯单丝(商名,日本尤尼吉卡)
Bonmouton 鲍莫顿
bonnaz embroidery 多向机绣
bonnet 软帽,系带式童帽,无边系带式女帽,(苏格兰)无边男帽,无边呢帽;(北美印第安人的)羽毛头饰
bonnet à bec [法]蝶形帽,窄檐软帽
bonnet babet 丝带装饰穆斯林女晨帽
bonnet cotton 8~16股粗棉线
bonnet de tricot [法]特里科软帽
bonnet en papillon [法]蝶形帽,窄檐软帽
bonneterle 针织品商品;针织业
bonnet lace 帽子饰带
bonnie and clyde 青年套服
Bonnie and Clyde look 邦尼·克勒代款式
Bonnie Cashin 邦尼·卡辛(1915~,美国时装设计师)
bonnie look 机智款式
bontane 腰布(红蓝色,长方形,非洲)
bonten 粗毛格子围巾;水手粗布
bonton 时髦,得体,优雅
bony chest 瘦胸型
bony neck 细颈型
bony shoulder type 瘦骨肩型
boob tube 布裹胸,直筒型弹力裹胸
boodul 阿富汗巴尔克手织红蓝黄色纬窄幅丝织物
book 存折;小包(把);账簿
book bag 书包
bookbinders' cloth 方面布
bookbinding holland 重浆方面稀平布
bookbinding leather 方面皮革

book cloth 方面布
book fold seam 往复折叠缝
book harness muslin 剪花细布
book hem 书型内折衣摆
booking 订购
booking case 打包箱
booklee 本色野蚕丝粗绸
book linen 书面织物,亚麻方面布;硬衬
book muslin 方面细布
book seam 书面缝
book value 账面价值
boomazie 俄罗斯印花哔叽绒
boondockers 野地短靴;野战军靴
boondoggle 皮辫绳(美国童子军系的)
boonnee 印度红边或黑边棉平布
boony fiber 木制纤维;麻秆纤维
boorka and mesh eye-piece 中东女式披风
boost the price 抬高价格
boots 靴,男式短筒靴;橡胶套鞋;袜筒
boota 印度手织点纹棉花
boot blow-up 鞋楦
boot bracelet 脚镯,靴镯
boot cuff 靴筒式袖口,靴筒式袖头
boot-cut pants 套靴口裤子
boot duck 制靴帆布
bootedar inalmul 印度手织点纹棉布
bootee 婴儿筒袜;编结婴儿鞋;(妇女或儿童)短筒靴;(妇女或儿童)轻便套鞋
bootery 靴鞋店(美)
boot garters 吊靴带
boot hose 带褶边长袜
boot hose tops 靴袜上端装饰边
bootie 短统靴;毛口;毛皮寝室拖鞋;轻巧鞋;小儿软鞋
booties 婴儿鞋;婴儿袜
bootiken, bootikin 小型半筒靴
boot jack V字形脱靴器
boot lace 鞋带,靴带
boot lace tie 保罗领带,饰扣式领带
boot last 鞋楦,靴楦
boot leg 靴筒,裁好的靴筒皮革
boot leg duck 制靴帆布
boot leg line 适靴裤腿线
boot length 袜筒长
boot lining 靴筒里,靴衬布
boot maker 制靴(鞋)工人,鞋匠
boot pants 小脚裤
boots 靴,长筒靴(美国),短筒靴(男用);橡胶套鞋
bootscut flared jeans 宽裤脚牛仔裤(配穿

长靴）
bootscut jeans　配靴牛仔裤
boots hose　靴型护脚
boots lace　靴带
boot sleeve　靴筒式男式克夫袖
boots leg silhouette　配靴牛仔裤型
boots length　中长裙（裙长至腿肚）
boot socks　靴袜
boots tape　藏靴带
boot stockings　长筒靴袜
boots topper　靴上筒形圈,腿套（填盖长靴与脚间空隙）
boot strap　靴袢,拔靴带
boot string　靴带
boot tree　（鞋）楦；靴楦
bop of hat　帽顶
borandjik　巴尔干地区白色棉绉
borazon　氮化硼仿宝石
Borbon cotton　波旁棉
bord　经条色织布
bordadillo　真丝提花塔夫绸
bordah　经条粗色织布
bordat　经条粗色织布
bordati　博达蒂绸,意大利棉丝交织腰带绸
borde　镶边,滚边；金银丝窄带
bordeaux　酒红,葡萄酒红（暗紫红色），酱红,枣红,紫红
border　布边；饰边花纹,色边花纹；衣襟；滚边,镶边；袜口；边镜
border checks　色边格子棉布
border design　边饰图案,裙边图案
border edging　服装饰边,滚边
bordered rug　镶边地毯,滚边地毯
border hem　镶边,饰边,贴边,裙边；边折缝
bordering　镶边,饰边
border lace　滚边,衣着饰以；装饰花边
border pattern　边花纹,边纹图案,边纹
border print　边缘印花,裙边印花图案
borders design　边饰图案
border skirt　饰边裙,花边裙
border strip　滚边,衣着饰以
border strip edging　滚边,衣着饰以
border tax　边境税
bordier point　手套绣花手背,绣花手套背
bordon　装饰用带边绳
bordthea　博德西亚大衣呢,横贡呢
bordure　镶边,滚边,包边
bore　破洞,蛛网疵

borelaps　18世纪荷兰优质平纹亚麻布
borer　穿孔器；穿孔者
borero　无纽女短上衣
borgana　伯甘纳毛图呢
borghese　贝佳斯
boron carbide fiber　碳化硼纤维
boron fiber　硼纤维
boron nitride fiber　氮化硼纤维
borre　印度棉布或荨麻布
borrego lamb　博瑞格羊羔皮
Borsalino　保萨利诺
Borsalino hat　博尔萨利诺帽（意大利）
borsehair braid　马毛织带
bortz　下等钻石
boshi crepon　博士呢
boshi jacquard crepon　花博士呢,博士花呢
boshi jacquard silk crepon　花博士呢
boshi silk crepon　博士呢
bosidian　黑曜石
bosky　博士基条子衬衫布
bosnia rug　博士尼阿直条纹窄幅地毯
bosom　胸；（衣服）胸部；胸状物；胸饰衬衫,胸襟
bosom amplifier　20世纪初的胸垫
bosom band　希腊妇女胸带
bosom blouse　胸饰衬衫
bosom flattener　束胸带
bosom friends　女士护胸
bosom knot　18世纪香味彩缎胸结
bosoms　乳房
bosom shirt　胸饰衬衫
bosom-to-shoulder area　肩胸区
bospal　衣袋沿边
bospal seam　衣袋边缝
Bosselé　博塞莱整理（拷花整理,商名,法国）
BOSSINI　堡狮龙
boss pants　老板裤（较宽松休闲裤）
Boston bag　波士顿提袋,手提包
Boston leno　波士顿纱罗织物
Boston net　波士顿网眼布；纱罗布；波士顿花边
Boston Wool Trade Association　波士顿羊毛贸易协会
bostous　博斯图呢,法国博斯图丝、毛或麻线呢
bota　棕榈花纹；西班牙男舞靴；马靴
Botah-khila　波塔基拉地毯
botanical color　植物色

botanical print 植物图案印花;植物图案印花花样
botanic design 植物图案,植物印花图案
botany 细羊毛;精纺毛纱
botany fabric 博坦尼精纺细毛织物
botany serge 博坦尼哔叽
botany suitings 博坦尼西服料
botany twill 博坦尼呢,博坦尼斜纹呢
botany wool 细羊毛,澳洲细羊毛,博坦尼羊毛
botany worsted 博坦尼精纺细毛织物
botch 粗补缀,织补
boteh 中东佩斯利花纹
both-side covering stitch 双面绷缝线迹,双面覆盖缝线迹
both-side covering stitch machine 双面绷缝机
both-side printing 双面印花
both-side raised flannelette 双面绒布
both-side raised velveteen 双面绒
both-side trousers press 裤管成型机,两侧裤管熨烫机
bottane 法国博唐绸
botte cavaliere [法]骑士长筒靴
bottekin 装饰靴,装饰半筒靴
Botticelli blue 灰蓝
Botticelli pink 明洋红
bottillon [法]半长筒靴(短筒,高筒)女靴,宽松舒适的半长靴
bottine 短筒靴,半长靴
bottle collar 瓶形领
bottled collar 瓶颈领
bottle green 深绿,瓶绿(蓝光深绿色)
bottle neck 瓶颈领
bottle neckline 瓶颈领领口(高领口)
bottle neck zone 肩颈区
bottle shaped 瓶形领带
bottle shaped tie 瓶形领带
bottom 下摆,衣裾;下装;屁股;裤脚口;底色,织物的地,布地
bottom and needle compound feed 针牙综合送料
bottom bem 底部卷边
bottom binding 贴脚条,贴脚边
bottom bra 下胸罩
bottom cloth 地织物,衬垫织物
bottom cover 起绒底面
bottom covering machine 底绷缝纫机
bottom covering stitch 底绷缝纫线迹,单面覆盖线迹

bottom covering stitch machine 单面绷缝机,底绷缝纫机
bottom covering stitch ornamenting machine 底绷线迹装饰缝纫机
bottom cover stitch 单面覆盖线迹,底绷缝线迹
bottom fabric 厚重织物
bottom facing 底贴边
bottom fashion 以裤子为中心的款式
bottom feed 底送斜,下送布,下送斜
bottom filler 鞋子填料
bottom filling 鞋子填料
bottom front 里门襟
bottom girth line 脚口围线
bottom-heavy figure 沉重型体型
bottom hem 下摆卷边,下摆折边
bottom hemmer 脚口卷边器
bottom hemming machine 裤脚卷边机
bottoming denim 先染硫化染料后套染靛蓝的牛仔布,打底套染牛仔布
bottom interlining 下摆衬
bottom leg 裤脚口,裤口
bottomless 下空装
bottom line 下摆线,脚口线,裤口线
bottom look 下装款式,以裤子为中心的款式
bottom measure 裤脚口尺寸
bottom notch 裤脚口刀口
bottom of cap 帽子底边
bottom of knee 膝下线
bottom of leg around 裤口
bottom of trousers 裤脚
bottom piece 短裤
bottom piping 裤脚口镶边
bottom price 最低价格
bottom relax 松紧下摆回缩度
bottom relaxed 下摆平度
bottom rib 大身罗纹,下摆罗纹,裤口罗纹
bottom ruffler 下打裥器
bottom ruffler binder 下打裥滚边器
bottoms 美国鞋子;短睡裤;裙裤总称;下装
bottom shaping 下摆成型
bottom stop 拉链底掣,止头
bottom strap 裤脚口袢
bottom stretch 松紧下摆拉伸度
bottom stretched 下摆拉度
bottom style 翻脚款式
bottom sweep 下摆

bottom tab　下摆袢(带),脚口扣绊
bottom tape　(裤)贴脚条,底边牵带
bottom thread　底线
bottom trimmer　下剪线器
bottom-up mode　上传模式
bottom-up theory　逆流理论(一种时装流行传播理论)
bottom view　仰视图
bottom weight fabric　厚实织物
bottom welt pocket　下摆嵌线袋
bottom width　下摆宽,裤脚宽
bottom width line　下摆宽尺寸线;下摆直线
bottom with cuff　裤卷脚,脚口卷边
botton cover stitch　单面覆盖线迹,底绷缝纫线迹
botton of trousers　裤脚
botton piece　短裤
botton piping　裤脚口镶边
boublin　鲍布林,鲍布林闪色斜纹布
boubou　西非宽大长袍,布布装(非洲民族服饰)
boucassin　布卡森里子布
bouchué　布谢原色细呢
boucle　仿羊羔皮呢,珠皮呢,结子线织物
bouclé　[法]结子捻线,毛圈花式线,结子花式线;结子线织物;珠皮呢,仿羔皮呢
bouclé fabric　毛巾布,圈圈绒织物;结子花呢,海绵呢
bouclé knitting yarn　波形绒线
bouclé lock-yarn　花式线
bouclé rayon　毛圈状人造丝,结子状人造丝
bouclé stripe　结子状条纹
boucle taffeta　韦绢塔夫绸
bouclette　球花绒;珠皮细呢
bouclette yarn　毛圈花式线,结子花式线
boucle tweed　珠皮粗花呢
bouclé twist yarn　[法]结子加捻线,毛圈花式线
bouclé yarn　毛圈花式线,波形线,结子花式线
boudoir　女更衣室
boudoir bonnet　闺房帽
boudoir cap　闺房帽,带绉边女软帽
boudoir carpet　正反异纹地毯
boudoir coat　闺房装,闺房袍
boudoir jacket　女子居家外套
boudoir shoes　闺房鞋
boudoir slippers　闺房拖鞋

bouffant　膨松型;膨展袖;膨展裙;膨展裤;膨展褶裥领巾;[法]蓬松发型
bouffant dress　抽褶膨松裙
bouffant flip　膨起外翘发式
bouffant hairdo　膨展发型
bouffant hairstyle　蓬松发型,外向蓬松式发型
bouffant neckerchief　膨展褶裥领巾
bouffant petticoat　抽褶衬裙
bouffant silhouette　膨展型;膨展型轮廓
bouffant skirt　蓬松裙,膨松裙,膨展裙
bouffant sleeve　膨松袖,膨展袖
bouffant slip　膨松套裙,蓬松套裙;向外膨起的衬裙,蓬松衬裙;膨起条状发式
bouffant style　横向展宽型发式;膨起风格
bougainvillea　鲜艳紫红色
bouge　布什精纺白呢
bought note　购货单
bouguet broche　[法]花束胸针
bouillon　金银线饰边;膨褶;鼓起的衣褶
bouillonne　[法]绉泡饰带
boulder　浅茶黄色,圆石色,卵石色
boule　梨形人造宝石
boulevardée　法国半漂大麻帆布
boulevard heel　大街鞋跟,古巴式鞋跟,布若娃跟
boullon　金银线饰边;膨褶
bouloire　法国亚麻织物
boulvardee　半漂粗大麻布
bouncy hand　弹性手感
bound　滚边
bound and faced opening　包贴式开襟
bound buttonhole　包边纽孔,加固纽孔
bound buttonhole pocket　滚边口袋,夹层袋
bound cuff　滚边袖口
bound cuff opening　滚边袖衩
bound curves　荷叶滚边
bound edge　缝边,包边
bound-edge frill　镶边皱褶
bound foot　小脚
bound-foot rain shoes　尖足雨鞋
bound hem　滚边缘边
boundle　一捆裁片
bound neckline　滚边领口
bound opening　包边式开襟
bound placket　包边式开襟
bound pocket　夹层(口)袋,滚边口袋,嵌线口袋;衣面与里子间的夹层衣袋;双

嵌线装
bound seam 包边缝,滚边缝
bound seam finish 滚边缝份处理
bound slashed pocket 嵌线插袋
bound slit 滚边开口,滚边开衩
bounty 补助金
bounty on importation 进口津贴
bouquet 花束
bouquet broche [brooch] 花束胸针
bouquet silhouette 花束轮廓
boura 布拉绸,法国布拉丝毛绸
bouracan 博拉坎厚毛织物;厚重半篷,防水外套
bourat 法国亚麻原色粗厚帆布
Bourbon cotton 波旁棉
bourdat 布拉粗布,经条粗色织布
bourdon(cord) 带边绳
bourdonette 带边绳
Bourdon lace 布尔登花边(涡形花纹)
Bourdony 德国精纺毛黏织物;布尔多内呢
bourette 绵绸;结子织物
bourette silk 䌷丝
bourette yarn 彩色丝毛结子线
bourglass line 沙漏线形
bourme 伊朗博尔梅生丝
bournous 阿拉伯式斗篷
bourrage [法]绗绣
bourras 布拉粗呢
bourré de soie 柔软绣花丝线(绢丝或经䌃丝下脚制成)
bourrelet [法]垂肉;软垫帽
bousingot [法]漆皮帽
boutique 女服店,精品店;女时装店,小型服饰专卖店,(妇女)时装用品小商店
Boutique Show 国际服装精品展(美国)
bouton D'orellé 纽扣形耳环
boutonege croisé 双排扣前襟
boutonné cloth 纬绒圈花纹织物;结子线棉织物
boutonnés [法]凹凸毛线
boutonniére [法]插孔花,别在驳领纽孔上的花;驳领眼,两袋领洞眼,插花孔眼,西装领上(插装饰品的)孔眼
boutonniere point 手工锁眼针迹花边

bouyancy bag 浮力袋
Bouyei ethnic costume 布依族服饰
Bouyei nationality's costume 布依族服饰
boven hat 牛仔帽
bovver boots 趵伐靴
bow 蝶型领结;蝴蝶结,花结;钻石领结形象
bowbacked figure 驼背体型
bow belt 蝴蝶结腰带
bow blouse 系领结女衫
bow collar 蝴蝶结领,蝶结领,领结领
bow distortion 纬弧疵
bowed cuff 蝴蝶结袖头
bowed fabric 弓纬疵点
bowed filling 纬不正,纬斜疵
Boweds cotton 美国波维斯棉
bow hat 蝴蝶帽
bowing 弓形横列;丝缕歪斜;丝缕歪斜
bowing figure O形腿体型
bowknot 蝴蝶领结,蝴蝶结
bowleg 弓形腿,膝内翻
bowler 圆顶硬呢帽,常礼帽
bowler hat 常礼帽,小礼帽,圆顶礼帽,山高帽
bowler skirt 保龄球裙
bowline knot 单套结
bowling shirt 保龄球衫
bowling shoes 保龄球鞋,保龄鞋
bow neckline 蝴蝶结领口
bow pen 小(墨线)圆规
bow ribbon 蝴蝶结
bow-shaped splicing 弓形袜底;弧形加固缝
bow string hemp 弓弦大麻,虎尾兰麻
bow tie 领结,蝴蝶领结
bow tie dress 蝴蝶领结衬衫
bow yang 扎裤脚带子
box (纸)箱,(纸)盒,包;装箱(盒);德国皮革;盒纹(鞋子)
box back 背直筒式服装;背直流式
box bag 箱形包;盒形提袋
box bottoms 箱形男裤
box calf 珠皮,纹皮;方痕小牛皮,方块粒纹小牛皮,方格粒纹小牛皮
box cape 箱形披肩
box cloth 缩绒厚

C

C 胖体型(服装体型分类代号);C 码(鞋宽,较 B 码宽,较 D 码窄或指大号男睡衣)
caadabar carpet 仿古地毯
caaporopy 荨麻属韧皮纤维
cab (裁衣余料的)碎布,碎呢
caballeros 西班牙美利奴细羊毛
caban 阿拉伯白围巾;水兵服;有袖合身上衣;律师宽大外衣
cabana cloth 薄型彩色织物,运动服织物
cabana set 卡巴拿沙滩装,卡巴拿海滨装(包括短袖上衣和短裤),男海滩套装(短袖上衣和短裤)
caban style 卡班风格
cabaret cap 卡巴莱帽
cabas 佛里吉亚帽
cabasset [法]无帽舌头盔
cabbage 碎布,裁余布料;卷心菜绿
cabbage ruff 不规则折叠的大荷叶边
cabeca 印度高档丝
cabesa wool 西班牙卡伯索原毛
cabinet press 柜式熨烫机
cabistan rug 卡比斯顿地毯
cable 凹凸花纹,绞花
cable chain stitch 绞花链状绣,钢绳链状绣;锚链式针迹
cable cords 宽条灯芯绒,阔条灯芯绒
cable fabric 绞花织物
cable hatband 缆绳状帽带
cable inquiry 来电询购
cable knit 绳编花纹
cable knitting 绞花编织
cable mattings 亚麻厚帆布
cable net 大网眼窗纱,粗棉线窗纱
cable pattern 钢绳图样,绞花图样,缆绳式花样
cable silk 饰边粗丝线
cable smocking 钢绳缩褶绣,绞花司马克
cable stitch 绞花锈;绞花组织,绳编,绳编花纹,辫子组织;链式线迹
cable stripe (毛衫)辫子花,绳索条纹
cable thread 三股缝纫线
cable webbing 粗股线斜纹带,斜纹带(粗股线制)
cabochon [法]卡波勋帽;磨光宝石,磨光穹顶宝石(珠宝)
cabot 卡博特平布,美国本色粗棉平布,[美]卡博茨平布
cabretta leather 卡伯利泰皮革,轻柔耐用的羊皮革(南美、非洲产),羊皮革(南美及非洲)
cabriolet [法]卡篷帽,篷顶罩帽
cabriolet bonnet 篷车顶式罩帽
cabriolet headdress 双轮马车式头饰
cabulla fiber 中美洲西沙尔麻,剑麻
cacao brown 可可棕毛;皮革棕色
caceres wool 中档西班牙羊毛
Cacharel 卡查雷尔(法国时装设计师)
cache-cheveux [法]隐发
cache-chignon [法]夜礼服小帽
cache-col [法]隐领围巾
cache-mesère turban 遮覆式头巾帽
cachemire 开司米羊绒,山羊绒;佩斯利涡旋花纹
cachemire de soie 羊绒披巾;高级塔夫绸
cachemire shawl 羊绒披巾
cachemirette 仿开司米斜纹绒(丝或棉经,羊毛纬,背面拉绒);女式斜纹呢(粗纺毛纱纬),斜纹呢
cache-mollet 及小腿长裙(遮盖腿肚)
cachenez 围巾
cache-poussière 防尘服
cachmerette 仿开司米斜纹绒(丝或棉经,羊毛纬,背面拉绒);女式斜纹呢(粗仿毛纱纬)
cacks 平底童鞋,婴儿软底鞋
cacomistle, cacomixie 蓬尾浣熊毛皮
cactus 仙人掌绿
cactus flower 仙人掌花色
cactus green 仙人掌绿
caddice 卡迪斯斜纹呢;精纺毛纱编带
caddie 澳式阔边帽(穿戴时倾向一侧)
caddie bag (高尔夫)球具袋
caddis 卡迪斯斜纹呢;凯蒂带,精纺毛纱编带
caddon, caddows 花式毛圈布
caddy 澳式阔边帽(穿戴时倾向一侧)
cadene rug 土耳其长毛拼条地毯
cadenette [法]卡德内发式
cadet 短指手套,卡地手套;海蓝色,灰海

色,紫灰
cadet blue 海蓝色;军官学校学生制服蓝色
cadet cloth 军校制服呢,西点军校制服呢
cadet collar 军官领
cadet grey 浅蟹灰,蓝灰,灰蓝色
Cadette 卡代特(意大利时装设计师)
cadillon 法国斜纹呢
cadis 卡迪斯斜纹呢;精纺毛纱编带
cadiz 嵌花大花边
cadiz stitch 连续锁眼针迹
cadmium blue 镉蓝
cadmium green 镉绿
cadmium orange 镉橙;海螺红;橘黄,血牙色
cadmium red 镉红
cadmium yellow 镉黄,鲜黄;鲜蓝色
cadogan wig 卡杜冈假发,卡多根假发
cadoro bra 卡多罗胸罩
caen 法国卡昂哔叽
cafa 西班牙低档棉平布
café au lait 牛奶咖啡色,淡咖啡色
cafe cream 奶咖啡色
café curtain 褶裥窗帘
café noir 浓咖啡色
caffar, caffard 卡法德呢
caffar damask 卡法德呢,纯毛制服呢,毛麻制服呢,交织条纹呢
cafry 卡法德呢
caftan 卡夫坦袍,(土耳其等国男用)束带长袖袍;阿拉伯男上衣;有腰带长袖服;阿拉伯女式宽大长袍
caftan coat 土耳其长袍;女束腰长袖衣
caftan dress 阿拉伯女式宽大长袍;类似阿拉伯男式内上衣的服装
caftan neckline 前中切口式圆领口
caftan silhouette 土耳其长袍式轮廓线条
cage (穿在衣裙上的)通花罩衫,女罩衫;薄纱罩裙,笼式加撑衬裙
cage-americaine 笼式加撑衬裙
cage dress 双层裙(里层贴身不透明,外层宽松透明或格子花纹)
cage dryer 笼式烘燥机
cage empire 笼式加撑衬裙
cage petticoat 笼式加撑衬裙
cage work 刺绣底布花纹
cagoule (登山时穿的)连帽薄防风衣;僧侣式毛衣(配有兜帽的宽松式高圆套领毛衣);盖头小披肩,覆头小披肩;卡戈童帽;卡戈头巾
caiana cotton 巴西凯阿那棉

caijing damassin 彩经缎
caiku brocade 库金;织金,彩库锦;彩库缎
cainsil 精细白亚麻及踝女式束腰外衣
caiquan 越南北部黑色长裤
caites 卡铁斯凉鞋(墨西哥)
caizhi twill damask 采芝绫
cake hat 类似登山帽的男士软毡帽
calabar skin 灰白毛皮
calabrere 意大利卡勒布雷尔棉布
calabria cotton 意大利卡勒布里亚棉
caladaris 卡勒达里条子平布
calafine 阳离子可染型聚酯纤维(商名,日本东泽纺)
calais val 圆眼瓦朗西安花边
calamacho 意大利丝缎
calamanco 卡拉曼科色条呢;卡拉曼科亚麻布
calamatta 未脱胶生丝
calami blue 菖蒲蓝色
calamine blue 淡蓝,浅蓝色
calamine violet 水锌紫色
calanca 印花棉布
calash 卡莱什头巾;篷形女头巾帽,折篷式女兜帽(18~19世纪),东篷式大兜帽
calasiris 卡拉西利斯长裙
calcao 裙裤
calcarapedes 可自动调节的男式橡胶套鞋
calceus 罗马裹腿浅帮鞋
calcia-alumina-silica fiber 氧化钙-氧化铝-氧化硅纤维
calcium alginate fiber 海藻酸钙纤维
calculation of fabric utilization and consumption 计算用料
calculator 计数器;计算器
calcutta crinkle fabric 50/50 涤棉绉皮,加尔各答绉
calcutt a hemp 黄麻
calèche 东篷式大兜帽
calecon 衬裤;外用短裤
caleeus 卡尔修斯鞋
calencart 棉平纹坯布
calencons 女式长筒袜;女式内裤
calendar watch 日历手表
calender bonded nonwoven 热轧黏合法非织造布
calender bonded nonwoven fabric 热轧黏合法非织造布
calender coater 轧辊涂布机
calendered cloth 轧光布
calendered fabric 轧光织物

calendered foxing 压延围条(鞋)
calendered half-bleached sheeting 轧光半漂布
calendered heel 压延跟
calendered moiré 波纹轧光;云纹织物
calendered nonwoven fabric 热轧法非造布
calendered sole 压延底(鞋)
calender finish 轧光整理
calender finished fabric 轧光整理织物
calender finish fabric 轧光整理织物
calendering (布料)轧光;压光(俗称油光)
calendering crease 轧光绉
calender printing 滚筒印花
calf 小腿,腓部,腿肚;小牛皮;袜腿部;腿肚围尺寸
calf bone 腓骨
calf bound 仔牛皮制的
calf circumference 腿肚围[尺寸]
calf fashioning 平袜小腿部成形
calf girth 腓围,腿肚围,小腿围
calf-high boots 中筒靴;长筒靴
calf-high rubber boots 中筒胶靴
calf-knee 膝内翻
calf leather 小牛皮,小牛皮革
calf-length 及腓高
calf-length boots 中筒靴
calf-length riding boots 中筒骑马靴
calf-length skirt 及腓长裙(遮盖腿肚)
calf line 腓围线,腿肚围线
calflong dress 卡夫龙裙装
calf measure 腓围,腿肚围度
calf skin 小牛皮,仔牛皮
calf skin dyed 染色小牛皮
calf skinners 马裤;中裤(长度到腿肚处)
calf skin silk 仿小牛皮绸
calibration 校准
calibre cut 角式切磨(首饰)
calico 平布,白布;印花布;棉布;印花粗布
calico back 平纹布背
calico design 卡利可图案
calico-like printed fabric 类印花细布
calico pattern fabric 提花细布
Calico Printer's Association 棉布印花工作者协会
calico printing 棉布印花
calico shirting 衬衫布,棉衬衫布
calicut 平纹坯布
California 加利福尼亚花呢

California blanket 加利福尼亚毛毯,细毛毯
California collar 加利福尼亚衬衫领,长尖领
California cotton 加利福尼亚棉
California embroidery 加利福尼亚编织和皮革针迹;加利福尼亚刺绣
CALIFORNIA MEN'S STYLIST 《加利福尼亚男装设计师》
Californian embroidery 加利福尼亚刺绣
Californian ruby 加利福尼亚红宝石
California pants 条格纹羊毛裤
California sports look 加利福尼亚运动款式
California sports style 西海岸常春藤款式
California wool 加利福尼亚羊毛
caliga 卡利茄袜;卡利茄靴(古罗马)
caliga sandal 古罗马克丽嘉凉鞋
caliper 厚度
calipers 卡尺,卡钳,两脚规
calisthenic costume 健美体操服
calkin 鞋底铁掌
call 对纺织品进口实施限制的通知
calla green 马蹄莲绿
call for a bid 招标
call for tenders 招标
calligraphic print 流畅线条的花纹图;手绘图案印花
calligraphic scarf 手绘围巾
calligraphy 书写;字体
callipers 测径规,两脚规,双脚规
Callot Soeurs 卡洛姐妹(法国时装设计师);卡洛特姐妹(法国高级时装店)
callotte 黑色丝绸瓜皮帽,圆顶小帽,无檐小帽
calmanco 卡利曼柯
calmande 卡尔芒德花呢
calmuc 卡尔马克长绒呢,糙面粗纺斜纹呢;中亚产地毯用毛;双色纬厚棉绒布
caloee fiber 南美苎麻
calorie 卡
calot (女、童)无檐帽;橄榄帽;瓜皮帽;(天主教用)头盖帽,教士圆帽,半球帽
calotte 瓜皮帽,教士圆帽;头盖帽(天主教),无檐小帽
calpac,calpak,calpack (近东)大羊皮帽;大毡帽;(中东)羊皮帽
calpreta fabric 防缩有光绉织物
calquier 印经塔夫绸,双股异色丝塔夫绸
calum jouree 摩擦轧光印花棉布

Calvin Klein 卡尔文·克莱恩(1942~,美国时装设计师)
calypso 卡利普索衫
calypso chemise 女式彩色棉布内裙
calypso pants 卡利普索裤
calypso shirt V字领扎结露腰衬衫
calypso style 卡利普索风格
calyptra 卡利普特拉面纱(古希腊;古罗马)
calyso shirt V字领扎结露腰衬衫
calyx-eyed needle 裂眼针
calyx-eyes 花萼针眼
cam 凸轮
camaca 卡马克厚呢(丝/骆驼毛或丝/棉交织,也有全丝,主要用于法衣或帷幔)
camail 盔甲披肩;女式及腰披肩式斗篷
camak 卡马克厚呢(丝/骆驼毛或丝/棉交织,也有全丝,主要用于法衣或帷幔)
camarines 卡马里内斯大麻;菲律宾蕉麻
camauro 罗马天主教教皇帽
camayean 卡马延布,平纹印花棉布
camayeux 同色深浅花纹
camayeux effect 同色深浅[调]效应
cambaye 东印度仿亚麻粗棉布
cambiunte 驼毛呢,仿驼毛呢
camblet 驼毛呢,仿驼毛呢
camblette 驼毛呢,仿驼毛呢
cambodia cotton 印度坎博迪亚棉
camboulas 法国棉毛半绒呢
camboys 英国厚端棉布
cambrai 机制花边
cam brake 凸轮制动器
cambrasine 法国细薄布
cambrayon 西班牙棉坯布
cambré 稀薄网状亚麻布
Cambrelle 坎布雷尔(双组分锦纶非织造布,用作鞋衬里,商名,英国)
cambresine 法国细薄布
cambric 细布,细平布,白细布,5600细布,7000细布,细薄布,细纺;单面轧光柔软平布(棉或亚麻);稀薄亚麻布
cambric finish 细薄布整理
cambric grass 苎麻
cambric muslin 漂白轧光细布;竹布
cambric shirting 衬衫细布
Cambridge 剑桥法兰绒,牛津布
Cambridge blue 淡蓝色,浅蓝色
Cambridge coat 三粒单排或双排扣的剑桥外套
Cambridge paletot 剑桥男及膝大衣

cam carrier 凸轮推杆
cam control gear 凸轮控制装置
cam drive 凸轮传动
camel 高级驼毛料,驼绒;驼色,暗棕色,浅棕色
camel back 驼背
camel-dyed mink 驼色貂皮
cameleon 织物闪色效应
cameleon fiber 变色纤维
cameleons 女式镂空鞋及靴
camel hair 驼毛,骆驼毛,驼绒毛;驼绒衣;驼绒织品
camel hair cloth 长毛骆驼绒,驼绒(布),驼绒织物
camel hair coat 驼毛大衣
camel hair knitwear 驼毛衫
camel hair over coating 驼毛大衣呢
camel hair quilted coat 驼绒袄
camel hair shawl 驼毛围巾
camel hair sweater 驼毛衫
camelina 卡默利纳粗绒呢;驼绒斜纹布;轻拉绒粗花呢,微拉绒粗花呢(方平组织)
camel leather 骆驼皮
camellia 山茶色,山茶红
camellia red 山茶红
camelot,camelott 羽纱,轻薄混纺织物;粗纬起绒织物;仿驼毛呢,驼毛呢,防雨驼毛织物
camelot baracane 法国家具厚呢;厚外套
camelote 廉价低档织物
camel's hair 驼绒毛;驼绒衣;驼绒制品
camel's hair cloth 驼绒
camel sheep hair 驼羊绒
camel suede 仿麂皮绒;仿驼毛棉布
camel teen 仿驼毛薄呢
camel wool 驼绒
camel wool fancy suiting 驼绒花呢
camel wool sweater 驼毛衫
cameo 彩色浮雕陶器;豆沙色;精细浮雕,浮雕宝石;浮雕贝壳;浮雕小徽章
cameo blue 浮雕宝石蓝
cameo brown 豆红,豆沙色;浮雕宝石棕色
cameo green 浮雕宝石绿
cameole 异色双面缎
cameo pin 浮雕宝石像章
cameo pink 浮雕宝石粉红色
cameo rose 浮雕宝石玫瑰红色
cameo silhouette 浮雕轮廓
camera 粗松亚麻织物

camera bag 相机手提袋,相机袋
camera linen 粗松亚麻织物(漂白或不漂白,漂白者通常染浅色,一般为黄色,法国制)
cameraman coat 摄影服,摄影师外套(防水、防寒,胸前有大盖袋)
camerick 轧光细薄布
cameron tartan 苏格兰格子呢
cames 女罩衫;麻布贴身长袍
cam feed decorative stitch 凸轮送料装饰线迹
cam follower 凸轮从动件
cam-handling embroidery machine 凸轮控制绣花机
cami 超短宽衬衣
cami-bookers 妇女连裤衬衣;连衫短裤
camidress 女式连裤内衣
camientries 17世纪挪威、英国毛织物
camiknickers 女连裤紧身衣,女连裤内衣(短裤背带式)
camillas cotton 粗棉布
camille 竹节条格花纹
camis 印度回教徒女式袋状长衫
camisa 衬衫,菲律宾女短衫;胸衣
camisado 宽大薄袍
camise 宽松轻薄短衫,宽松轻薄长衫;轻宽长袍;轻宽长袖衬衫,妇女宽衬衫
camisette 短小紧身胸衣
camisole 短紧身衣;超短紧身衣;吊带式长胸衣,花边胸衣;女吊带背心,贴身内衣;(旧时的)男式长袖外套,(背后可束紧的)长袖紧身外衣
camisole and tap pant set 女睡衣套装
camisole blouse 背带式无袖女罩衫
camisole bra 胸衣型胸罩
camisole dress 卡米索装;无袖女礼服裙,露肩吊带裙,花边吊带连衣裙
camisole neckline 卡米索领口,花边胸衫领口,背心领口,吊带领口,胸衣型领口
camisole skirt 背带式连衣裙
camisole slip 卡米索连身衬裙
camisole summer jump 卡米索连裤装
camisole top 内衣式衬衫,女式短袖衬衣,女式直筒衬衣
camisole top slip 卡米索连身衬裙
camisole tunic 吊带束腰外衣,棉背心式束腰外衣
cami-tap set 吊带衬衣加分喇叭口短裤套,两件套内衣
camlet 驼毛呢,仿驼毛呢,防雨驼毛织物;羽纱,轻布料,轻薄混纺织物;华丽毛料衣服;羽纱衣服
camleteen 仿驼毛呢
camlet liner 羽纱衬里
cam lever 凸轮杆
cam link 凸轮连杆
camoca, camocoa 卡马克厚呢(丝/骆驼毛或丝/棉交织,也有全丝,主要用于法衣或帷幔)
camocato 中国花缎
camoga 大麻属植物纤维
camomile 紫色
camouflage 保护色;伪装;掩饰
camouflage cap 迷彩帽,迷彩伪装帽
camouflage color 伪装色
camouflage coloring 伪装染色
camouflage fabric 伪装织物,迷彩织物
camouflage hunting 迷彩猎装
camouflage look 迷彩服式样,伪装型
camouflage net 防空网,伪装网
camouflage pants 迷彩裤
camouflage pattern 迷彩,迷彩迷彩;伪装图纹
camouflage pattern suit 迷彩服,伪装迷彩套装
camouflage print 伪装图案,掩饰性图案;伪装印花
camouflage rain suit 迷彩雨衣
camouflage shorts 迷彩短裤,掩护性的短裤
camouflage sport suit 二件套迷彩运动装
camouflage suit 伪装服
camouflage utility cap 伪装便帽
camouflage wear 迷彩服
camoyard 卡莫亚呢,山羊毛斜纹呢
camp 露营衬衫;露营短裤
campagne 金银丝流苏,窄花边
campagus 主教鞋
campaign coat 褴褛外套;长军大衣
campaigne 金银丝流苏,窄花边
campaign hat 宽边野战帽,美国战帽
campaign wig 战役假发(17～18世纪)
campanula blue 风铃草蓝色(浅紫蓝色)
campanula purple 风铃草紫色(浅红紫色)
campanula violet 风铃草紫色(浅红紫色)
campatillas 粗纺毛织物
Campbell tartan 坎贝尔苏格兰式格子呢
Campbell twill 坎贝尔斜纹
campes 康珀斯斜纹呢
camphor 樟脑

camphor ball 樟脑丸
camping 野营服
camping tentage 帐篷布
camping tent cloth 帐篷布
cam plate 凸轮盘
camp shirt 缺角西装领衬衫,野营衫
camp shorts 行军短裤
campus look 校园风貌,校园款式
campus moccasin 校园平底鞋
campus shoes 校园鞋
campus wear 校园服,学生装
cam shaft 凸轮轴
cam traverse type 凸轮往复式
cam type thread take-up lever 凸轮式挑线杆
Camyss 土耳其男式亚麻衬衫
canadaris 缎纹绸,丝棉绸
canadas 粗纺毛毯
Canadian coat 加拿大式外套
Canadian embroidery 加拿大刺绣
Canadian hemp 加拿大罗布麻
Canadian look 加拿大款式
Canadian marten 加拿大貂皮
Canadian patchwork 加拿大拼花巾;加拿大缎花床单
Canadian sable 北美黑貂皮,加拿大黑貂皮
Canadian shirt 加拿大衬衫,伐木工衬衫
Canadian snow shoes 加拿大雪地鞋
Canadian style 加拿大款式
Canadian Textile Institute 加拿大纺织学会
Canadian Textile Journal 《加拿大纺织杂志》(月刊)
Canadienne 加拿大式外套;加拿大士兵式女大衣
Canadienne coat 加拿大式外套;加拿大士兵式女大衣
canal blue 浅湖蓝;浅竹青
canale 双经棱纹绸
canalier gauntlet gloves 及腕骑士手套
canamazoo 泽麻
canamazos 未漂棉布;未漂亚麻布
canamo 黄麻
canamo de senegal 泽麻,槿麻
canapa fiber 小型大麻纤维
canapina 中国黄麻
canary 鲜黄
canary diamond 酒黄钻石
canary yellow 金丝雀黄,鲜黄色
cancan dress 康康连衣裙;康康舞衣
cancan hat 康康帽,硬草帽

cancanias 经条缎纹绸
cancan outfit 康康服装
cancan petticoat 康康衬裙
cancan slip 康康衬裙
cancelled check 注销支票
cancerogenic dyestuff 致癌性染料
candle light 烛光色
candle light peach 烛光红
candle stick fabric 烛台布
candlewick embroidery 烛芯纱穗状刺绣
candlewick fabric 烛芯纱盘花簇绒织物,仿簇绒花呢
candlewick spread 烛芯纱盘花床单
candy border 肯特条纹边
candy cane 甘蔗绿,甘蔗绿色
candy pink 糖果玫瑰红色
candys 拜占庭君主的紧身袖系带服装
candy stripe 肯特直条纹
cane 麻纤维;芦苇纤维;手杖
canebrake cotton 美国坎布拉克棉
canellee 双经棱纹绸
canepin 羔羊轻革
canescence 灰白绒毛;灰白色
canete 单丝绢丝;(菲律宾)直条凸纹细棉布;宽幅经棱细布
canezou 后背披肩,精美三角形披肩;女无袖短上衣
canfu damask satin 蚕服缎
cangan 窄幅低档棉布
caniche 卷毛呢
canile,canille 竹节条格花纹
canions 加纳斯裤,凯宁裤(16～17世纪),裤口翻边筒形紧身男裤
can-le striped crepe 蚕乐绉
can-le striped crepe damask 蚕乐绉
cannabis sativa 大麻
cannamasoz 未漂亚麻布
cannele 双经棱纹绸
cannele alternatif 交替棱条绸
cannele cord 花式灯芯绒
cannele simplete 简式棱条绸
cannelle repp 细棱条绸
cannequin 东印度漂白棉布,未漂棉布(东南亚)
cannetille 金银线(刺绣用);金银线花边;经棱条织物
cannette 单丝绢丝;菲律宾直条凸纹细棉布;阔幅经棱细布
cannon 男宽靴花边
cannon sleeve 女式加垫袖子,女式大泡

泡袖
canonicals 法衣，圣衣(牧师穿)，牧师礼服，牧师法衣
canon of taxation 课税准则
canons [法]佳衣裤饰
canopy 天篷，雨篷，遮篷；伞盖；华盖
canotier 卡诺蒂埃斜纹；康康帽，平顶有边女帽，硬草帽
canourge 粗纺毛哔叽
canques 棉衬衫布
cantai 坎泰细平布
cantala 菲律宾坎塔拉剑麻
cantaloupe 罗马甜瓜色，甜瓜色
canteen bag 水壶袋(帆布套)
canteen band 水壶背带
canteloupe 罗马甜瓜色
canteloupe red 甜瓜红
canterbury 丝经棉纺布
cantiao crepe 灿条绉
cantille 金银丝绣花线
canton 广东棉布
Canton bag 广东包
Canton cotton 广东棉绒布
Canton crepe 广绫，广东皱绸；重双绉
Canton finish 棉布平光整理
Canton flannel 广东长绒法兰绒，广东棉绒布
Canton linen 广东夏布
Canton natural steam filature 广东黄厂丝
Canton satin 绉背缎
Canton satin brocade 花广绫
Canton satin plain 素广绫
canton silk 广东生丝
Canton white steam filature 广东白厂丝
cantoon 棱面缎背厚棉布，厚急斜纹棉布，蓝色或紫色中国棉平布
canvas, canva 帆布，粗帆布，十字布，刺绣用网形粗布，帐篷布
canvas bag 帆布包
canvas belt 帆布带
canvas cloth 紧捻平纹棉织物；稀薄多股平纹毛织物
canvas cloth vamp rubber shoes 布面胶鞋
canvas coat 帆布外套
canvas cover 帆布套
canvas duck 粗帆布
canvas embroidery 垫子风格刺绣，帆布刺绣，网眼粗布棱结刺绣
canvas front 帆布衬垫
canvas gloves 帆布手套

canvas handbag 帆布手袋
canvas jumper 帆布工作服
canvas making 制衬，缝衬
canvas nurse's shoes 帆布护士鞋
canvas of cover 遮盖帆布
canvas oxford 帆布牛津鞋
canvas pants 帆布裤，珠帆裤
Canvas Products Association 国际篷帆布制品协会
canvas rubber shoes 跑鞋
canvas sandals 帆布凉鞋
canvas school bag 帆布书包
canvas shoes 帆布鞋，网球鞋
canvas sports shoes 帆布运动鞋
canvas stitch 帆布针迹
canvas tennis shoes 帆布网球鞋
canvas vamp oxford 西维喔鞋
canvas waist band 帆布裤带
canvas waist belt 帆布裤带
canvas wear 帆布服
canvas weave fabric 帆布织纹布
canvas work 垫子风格刺绣，帆布刺绣，网眼粗布棱结刺绣
canvas work stitch 网眼绣
can-wei surah 蚕维绫
canyon clay 峡谷黏土色
canyon rose 峡谷红色
cap 便帽，帽子(无檐边的)，制服帽，鸭舌帽，军帽，端帽，大学方帽；袖山
Capa 西班牙带兜帽圆披肩，斗牛士披肩
capability 能力
capability of psychological support 心理承受力
capacity of market 市场容量
capa kids 南非洲初剪马海毛
Capanaki 土耳其粗棉梭结花边
capanaki lace 卡潘纳基花边
cap and bells (宫廷丑角戴的)系铃帽，铃状帽
cap and gown 大学方顶帽和长袍，方帽长袍(大学学位服装)
caparison 华丽的衣饰
cap badge 帽徽
cap cloth 帽料，帽用织物
cap cockade 帽徽
cape 大氅，斗篷，披肩，披风
cap ease 袖山吃势
cape basuto 南非粗马海毛
cape coat 披肩男大衣，披风大衣(有披肩

cape collar 披风领,披肩领
cape cover-up 披肩,沙滩披巾
caped cloak 连披肩斗篷
caped coat 披肩大衣(无袖,呈钟形),有披肩男大衣
cape dress 披肩礼服,披肩装
caped ulster 披肩式厄尔斯特大衣
cape écharpe [法]卷领长披肩
cape first 南美洲一级马海毛
cape gloves 羊皮手套
cape hat 披巾帽
cape jacket 披肩夹克
capelet 短斗篷,短披肩(女用毛皮),小披肩;小披肩式服装
capelet blouse 披肩式女衫
capelet sleeve 短披肩袖,小披风袖
capeline 宽盆帽,平顶宽边女帽,软边宽帽,遮阳宽檐帽;(古代)武士铁盔
capella 织花布头
cape mixed 南非洲中级马海毛
cape mohair 南非洲马海毛
cape net 坚硬棉织网
cape overcoat 披肩大衣
cape piping 披风宽贴边
cape ruby 开普红宝石
capes 羊羔皮
cape sheath 披肩罩
cape shoulder 冒肩袖
capeskin 南非洲绵羊皮,好望角皮革
cape sleeve (连肩)盖袖,披肩袖,喇叭袖,短喇叭袖(童装)
cape snow-white wool 南非洲雪白美利奴毛
cape stole 披肩长围巾;披巾式披风,披巾式披肩
cape suit 披肩式上衣套装
cape ulster 披肩式厄尔斯特大衣
cape winter 南非冬季马海毛
cape wool 南非洲羊毛
cap height 袖山高
capillarity 毛细现象
capillary water 渗透水;毛细水
cap insignia 帽徽
capitonné embroidery 填塞刺绣;簇绒装饰布
cap knob 帽顶
capless wig 木制假发,无帽假发
cap of ceremony 礼仪帽;(旧时)朝冠
cap of dignity 美国加冕典礼帽
cap of estate 美国加冕典礼帽

cap of main tenance 美国加冕典礼帽
cap-of-sleeve 袖山
capot 起绒厚呢;男式宽松大衣
capote 带兜帽圆披肩,斗牛士披肩;带风帽宽厚女斗篷;带风帽长大衣(或大氅);[美]系带女帽(维多利亚时代)
capote bonnet 硬边圆帽
capothe 希腊镶边羊毛外套
cappaccino 浓咖啡色(卡普契诺)
cappa floccata 绒面圆帽
cap peak 帽舌,帽檐
capped sleeve 帽袖
cap piece 帽木
cap pompon 帽饰绒球
cappuccio 兜帽,女式软兜帽;兜帽垂物
Capri 卡布里裤(意大利)
cap ribbon 飘带(海军帽)
Capri (blue) 卡普里蓝(绿蓝色)
caprice 宽松无袖晚会女夹克
Capri girdle 卡普里紧身褡
Capri-length panty girdle 长至膝下四英寸的紧身长裤
Capri pants 卡普里裤(锥形,七分长),女式紧身长裤
Capris 卡普里裤(锥形,七分长),女式紧身长裤
caprolan 聚酰胺6纤维(商名,美国联合信号公司)
Capron 卡普纶(锦纶6纤维,商名)
cap shape 便帽,无边帽;帽型
capsicum red 辣椒红
cap sleeve 盖肩袖,蓬起袖,极短衣袖,1/8袖,帽形袖,帽袖
cap sleeve blouse 帽袖上衣,帽状袖袖口上衣
cap sleeve length 盖肩袖长
cap stretcher 帽撑
cap style 帽式发型
capsulation printing 微胶囊印花
capsule 聚酯微细短纤维(商名,日本东丽)
cap tab 帽护耳
captain's blue 船长服蓝
cap tassel 帽饰流苏
Captat Edward Molyneux 卡普塔尔·爱德华·莫利诺斯(法国时装设计师)
capuc(h)e [法]卡布什头巾
capuche 风帽,僧帽
capuchin (女用)带风帽斗篷,戴帽斗篷,连风帽大大衣
capuchin cloth 僧侣外套呢

capuchin collar 托钵僧领
capuchin hood 带风帽女斗篷
capuchon 花制小帽；连风帽女斗篷；女式及腰披肩
capucine 橘黄色；连风帽女斗篷
capucine orange 金丝雀黄(嫩黄色)
capulet 凯普莱帽，朱丽叶小帽
caputium 连帽短斗篷
cap with kissing strings 颌下系带女帽
caracal 狞猁毛皮
caracasch 伊拉克地毯用毛
caraco (18世纪后期)女式短上衣，卡拉科上衣，卡拉扣上衣(上半身合体并有腰褶，前下摆短于后下摆，后下摆起过臀部)
caraco corsage 及大腿合体短上衣
caracul 卡拉库尔羔皮，中亚羊毛皮，紫羔皮
caracul cloth 仿羔皮呢
caracule 羔皮，仿羔皮；充羔皮织物
caramel 酱色，淡褐色，黄棕色，焦糖色
caramel brown 焦糖棕
caraminian carpet 卡腊迈尼花毯，土耳其花毯
carat (CT) 克拉，公制克拉；涂24K金的薄织物
caratoe 剑麻，龙舌兰纤维
caravan 旅行帽，可折叠的小女帽
caravan bag 旅行手提包
caravan shoes 篷车鞋，胶底防水帆布鞋
caravonica cotton 墨西哥卡拉温尼卡棉
carbaso 绢丝纺，绢丝薄绸
carbasus, carbassus 西班牙亚麻；亚麻布；细平纹棉布
carbazole dye 咔唑(结构)染料
carbflex 碳纤维(商名，日本阿什兰)
carbide fiber (金属)碳化物纤维
carbochain fiber 碳链纤维
carbocyanine dye 碳花青染料
carbolan dye 卡普仑染料(酸性耐缩绒染料，商名，英国ICI)
carbolon 碳纤维(商名，日本)
carbonaceous fiber 碳纤
carbonaceous pitch fiber 碳质沥青纤维
carbon black 墨灰，炭黑色
carbon black fiber 炭黑纤维
carbon chain fiber 碳链纤维
carbon chalk 炭质划笔
carbon cloth 碳布
carbon-coated ceramic textile 镀碳陶瓷织物

carbon fiber 碳纤维
carbon fiber clothing 碳纤维服
carbon fiber mat 碳纤维编织物
carbon filament 碳长丝
carbon filament cloth 碳丝织物
carbon filament tape 碳丝狭带
carbonium color 碳鎓染料
carbonized fiber 碳化纤维
carbonized rag fiber 碳化碎呢纤维
carbonized rayon yarn 碳化人造丝
carbonized wool 碳化羊毛
carbonizing wool rags 碳化碎呢
carbon-loaded fiber 含碳纤维(导电纤维)
carbon nanofibers 纳米碳纤维
carbon paper 复写纸，碳粉纸
carbon paper marker 复写纸划样
carbon pencil 炭笔
carbon steel fiber 碳钢纤维
carboxylate acrylic fiber 羧化聚丙烯腈系纤维
carbuncle 红宝石；14世纪高领口
carcanes 东南亚棉布，棉织物
carcanet 颈珠，一串项珠；珠宝颈饰
carcassone 法国粗纺薄制服呢
carcassonnes 卡尔卡松薄呢
carcinogenic dyestuff 致癌染料
carcoat 短外套，跑车外套；短氅，短大衣，女风衣
carcuel 仿克里默羔皮织物
card board 卡纸，纸板
cardboard case 纸盒，纸箱
cardboard strip 卡纸狭条
cardboard template 卡纸板，型板
card box 纸盒
carded broadcloth 粗梳纱细平布
carded cotton fabric 粗梳棉织物
carded cotton yarn 粗梳棉纱
carded fabric 粗梳纱织物
carded filling sateen 棉纬缎
carded knitting yarn 粗梳针织物
carded percale sheeting 粗梳纱密织床单布
carded yarn 粗梳纱，普梳纱，梳棉纱
carded yarn fabric 粗梳织物
cardigan 羊毛衫,(无领)开襟毛衣，羊毛背心，(无领)开襟衫，卡迪根式开襟毛衫，卡蒂冈式开襟毛衫，卡蒂冈式夹克衫，卡迪根式夹克
cardigan bands 门襟边
cardigan bodice 卡迪根背心；卡迪根衣身

cardigan coat 卡蒂冈外套(无领,前开襟)
cardigan collar 对襟毛衫领
cardigan dress 卡蒂冈式无领前开襟上衣;无领前开襟连衣裙
cardigan front 开襟衫前片,毛衫式门襟,开门襟
cardigan jacket (无领)开襟夹克,卡迪根夹克,卡蒂冈夹克;开襟毛衫,开襟夹克衫
cardigan knitshirt 无领开襟式针织衫(多为轻薄)
cardigan knitted fabric 畦编针织物
cardigan neckline 卡迪根领口,开襟领口,开襟毛衫领口,卡帝冈领口,羊毛衫领口
cardigan silhouette 开襟式
cardigan stitch 罗纹线迹
cardigan style 开襟式
cardigan style sport coat 开襟运动服
Cardigan suit 卡帝冈套装,无领开襟毛衣裙套装,无领(开襟)套装
cardigan sweater 卡迪根毛衫,卡蒂冈毛衫,开襟式毛衫
cardigan tunic 卡迪根束腰外衣
cardigan vest 卡迪根背心,卡蒂冈背心,对襟背心
cardigan with hood 带帽开衫
cardillat 法国卡迪拉特呢
cardinal 主教服红,深红,鲜红,大红;红衣主教服;带风帽的女短外套,女式短外套;女短斗篷,女式短外套,深红色披风
cardinal cloth 主教呢
cardinal pelerine 蕾丝披肩
cardinal red 鲜红,深红
cardinal's cap (天主教)红衣主教法冠
cardinal's hat (天主教)红衣主教帽,宽边红帽(红衣主教)
cardinal's red hat (红衣主教的)宽边红帽,主教帽
cardinal stimuli 基本色刺激
carding wool 粗纺用羊毛,粗梳毛
Cardin jacket 卡丹夹克
Cardin style 卡丹风格
card of thread 纸板线
card pattern 裁剪纸样,硬纸样板
card pocket 卡片袋,名片袋
care apparel 康健服装
career apparel 职员工作服
career apparel fabric 职业装用织物

career suit 作业服(工作服、制服等)
career wear 办公服装(上班服)
career woman 职业妇女
career women's wear 职业妇女装
careett wear 职业装;职业服
care free 免烫;随意
care instruction 保管须知
care label 保养商标,洗水唛,服装使用须知标签
care labeling of textile wearing apparel 纺织品使用说明标签
careless 男式宽松披肩大衣
care of fabrics 织物的维护(包括洗涤、熨烫、晒晾、刷、防蛀、去污、储藏等)
care procedure 保养方法
care wash 洗涤小心
CARFIELD 加菲猫
cargaison 法国亚麻布
cargo 货物
cargo net 装卸网
cargo pants 船员裤,货船裤(两边有多褶的大贴袋)
cargo parachute 载重降落伞
cargo pocket 大贴袋,特大口袋
cargo pocket jeans 大贴袋牛仔裤
caribouskin 驯鹿服;驯鹿皮
carioba cotton 巴西卡里奥巴棉
carisel 平纹地毯底布
carisol 稀薄刺绣底布
carlap 帽护耳
Carlier 卡迪亚
Carlo Tivoli 卡罗·蒂沃利(意大利时装设计师)
carma 阿尔及利亚锥形高顶金属女帽
carmagnole 卡曼纽拉外套,卡曼纽拉夹克衫,卡曼纽拉服
carmagnole jacket 卡曼纽拉夹克
carmeillette 连帽女斗篷
carmeline 卡梅林中级骆马毛
carmeline wool 卡梅林中级骆马毛
carmelite 卡默利特呢,天主教修士粗重平纹呢
carmelite cloth 天主教修士粗重平纹呢
Carmen fantasy 卡门款式
Carmen outfit 卡门式服装
carmine 胭脂红,洋红,紫红,莲红
carmine rose 桃红,品红,喜蛋红,胭脂玫瑰红色
Carnaby look 卡纳比款式,坎拿比风格,坎拿比款式,坎拿比风貌

Carnaby Street 坎拿比街(20世纪60年代伦敦新时装发源地)
Carnaby tie 坎拿比领带,宽幅花领带
carnation 淡红,肉色;麝香石竹(康乃馨)
carnaval lace 婚礼服花边;针绣挖花花边
carnegie 卡尼集
carnelian 光玉髓色,红玛瑙,肉红玉髓色(番茄酱红褐色)
carnet 法国漂白亚麻布
carnival 嘉年华会;狂欢节
carnival collar 嘉年华领
carnival lace 婚礼服花边,针绣挖花花边,嘉年华花边
carnival style 嘉年华款式
carnival stylistic form 嘉年华款式
carnousie 卡鲁西
caro cloth 卡罗大衣呢
carocolillo 色织红棉布
Carol Horn 卡罗尔·霍恩(1936～,美国时装设计师)
Carolina Herrera 卡罗莱娜·赫雷拉(美国时装设计师)
Carolina Pride cotton 美国南卡罗来纳棉
Carolina Queen 叙利亚卡罗来纳皇后棉
carolinas 低档方格色织布
Caroline 卡罗琳花呢,卡罗琳花式斜纹呢,法国八页斜纹呢;西里西亚漂白亚麻布
Caroline corsage 卡罗琳紧身胸衣
Caroline hat 卡罗琳帽
Caroline plaid 卡罗琳格纹呢,19世纪英国棉毛格子花呢
Caroline Rebeux 卡罗琳·勒布(法国女帽设计师)
Caroline Sleeve 卡罗琳袖
Carolingian period costume 卡洛林王朝服饰
caroll 卡罗尔牛仔裤
Carolyne Roehm 卡罗琳·罗姆(美国时装设计师)
Caroset 法国麦尔登呢;法国双面法兰绒
carousel conveyor 转盘式传送机
carpal bone 腕骨
carpasian linen 塞浦路斯石棉布
carp collar 鲤鱼领
carpenter pants 木工裤,工装裤,木工用套裤
carpenter's apron 木匠围裙
carpenter's square 角尺
carpet 地毯,毡毯;毛毯

Carpet and Rug Institute 地毯研究院(美)
carpet backing 地毯底布,地毯背衬
carpet backing cloth 地毯底布
carpet bag 毛毯袋;毡制手提包
carpet binding 地毯滚边带
carpet blanket 厚毛毯,棉经毛纬厚毛毯
carpet cinema pile 影院用短密绒花地毯
carpet coat 毛毯外套
carpet cushion 地毯衬垫
carpet fiber 地毯用纤维
carpet linen 亚麻地毯布
carpet lining 地毯衬垫
carpet pattern 地毯图案
carpet printing 地毯印花
carpet slippers 毛毯拖鞋,绒毛拖鞋
carpet square 小方块地毯,组合地毯
carpet strip 走廊地毯
carpet sweeper 地毯清扫机
carpet thread 毛毯缝线
carpet tile 小方块地毯,组合地毯
carpet transfer printing 地毯转移印花
carpet tufted 簇绒地毯
carpet tufting fabric 地毯簇绒织物
carpet twine 缝地毯线
carpet underlay 地毯衬垫
carpet washer 地毯水洗机
carpet wool 地毯用羊毛
carpet yarn 地毯线
carpmeals 粗纺厚呢(英格兰)
carpus 腕骨
carradars 东印度细条彩格布
carranclane 英国彩格布
carre [法]方形领巾
carreau [法]小方格纹,格子花纹
carreux, carreaux 卡鲁绸,格子薄绸
carriage boots 乘车暖脚靴,防寒长靴(布制衬毛呢马车靴)
carriage cloth 车座呢
carriage dress 女式马车服
carriage lace 窄幅车用纺织品
carriage parasol 女式马车伞
carriage robes 车毯呢
carriage suit 婴儿三件套
carrick 防尘长外套,卡里克外衣(18世纪)
carrickmacross 贴花刺绣品
carrickmacross lace 卡里克马克罗斯花边,贴花刺绣花边
carrier 承运人;染色载体;带袢

carrier bag （绒或塑料制的）购物袋,手提袋
carrier dyeing process 载体染色法
carrier fiber 伴纺纤维,载体纤维
carrison cap 船形军帽
carrot orange 胡萝卜橘黄
carrot red 胡萝卜红(橙色)
Carr's melton 卡氏麦尔登呢(英国高级麦尔登呢)
carryall 大手提袋
carryall bag 万用袋
carryall clutch 女式钱包
carry bag 旅行袋
carrying cape 带裆斗篷
carrying container 手提箱
car-seat fabric 车座织物
carsey 克尔赛密绒厚呢
carsey cloth 克瑟密绒厚呢
cartab 帽护耳
car textile 汽车用织物
carthagena cotton 西印度棉
cartisane 镶金银线花边,镶重银线花边
cartography computation unit 制图计算装置
carton 纸板箱,外纸箱,卡通箱,纸板
carton cage 波纹纸盒,瓦楞纸板盒
carton mark 箱唛
cartonning 装箱
cartoon 草图,样稿;卡通,漫画
cartoon apron 卡通画围裙
cartoon look 卡通款式,漫画款式,用卡通画装饰的款式
cartoon printed handkerchief 卡通印花手帕
cartoon shirt 卡通画圆领衫,卡通衫
cartoon T-shirt 卡通画T恤衫
cartouche 卷轴装饰,涡形装饰
cartridge bag 弹药袋
cartridge belt 弹药带,子弹带腰饰
cartridge belt rib 鱼骨刺花纹
cartridge cloth 弹药绸
cartridge paper 图画纸
cartridge plaits 子弹带形褶裥
cartridge pleats 圆褶,子弹带形褶裥
cartridge pleats sleeve 弹带褶袖
cartridge pocket 弹带口袋
cartridge tucks 子弹带形褶裥
cartwheel hat 宽边圆顶女帽
cartwheel ruff 大圆盘形绉领,轮状襞领
cartwheel sleeve 二节圆形喇叭短袖

car upholstery 汽车装饰织物
carved button 雕花纽扣
carved carpet 剪花地毯
carved pile 剪花地毯,立体图案地毯
carved pile fabric 浮雕绒头织物,剪花绒头织物
carved rug 剪花地毯
Carven （Mme·Carmen Mallet)卡尔文(卡门·马利特夫人)(法国服装设计师)
caryota 棕榈纤维,棕纤维
Casa 卡撒棉布
Casablanca 卡萨布兰卡帽
Casanova blouse 卡萨诺瓦女衫,卡萨诺瓦衬衣
Casanova look 卡撒诺瓦款式
Casanova style 卡萨诺瓦款式
casaqin bodice 女式紧身胸衣
casaque 妇女合身外套(18世纪前),女孩穿的公主线裁剪的外套;骑师外套;披风领的宽松长大衣
casaquin 卡萨坤装;合体长袖及腰短上衣
casaweck 女式绗缝户外短斗篷
casban 卡斯本厚棉斜纹亮光衬里布
cascade 瀑布状花边;瀑布状物;垂瀑缘饰,瀑布装缘饰
cascade frill 瀑布式花边
cascade stripe 阶式条纹
cascade wrap 垂瀑式褶片
cascading collar 垂瀑领,阶梯式领
cascot cottons 美国陆地棉
case 包,箱,盒,柜,套,把……装箱
case brocade 盒锦
cased body 女式紧身胸衣;男式无袖短上衣
cased sleeve 嵌条女式长袖
case holder 梭芯罩架
casein button 酪素扣
casein fiber 酪蛋白纤维
casein knit fabric 牛奶针织物
casein wool 酪蛋白人造毛
case leather 箱包革
case mark 箱标,箱唛
casement 薄窗帘布
casement cloth 薄窗帘布细平布
casement fabric 窗帘布
casement repp 厚棱窗帘布
Casentino 卡森蒂诺外套(意大利),意大利卡森提洛外套

case pack label 外箱贴纸
case velvet 匣用里子丝绒
cash 款,款项;现金,现款
casha 卡沙棉法兰绒
cashambles 贵族男式筒袜
cash before delivery 交货前付款
cashemirette 仿开士米斜纹绒;女式斜纹呢
cashe-sexe 三角裤,小裤衩
cashew 腰果色
cash flow 现金流
cashgar,cashghar 粗羊绒呢
cashmere 开士米,(山)羊绒,克什米尔,紫羊绒;开士米织物(薄毛呢等);松软斜纹棉织物
Cashmere and Camel Hair Manufacturers Institute 国际羊绒与驼绒生产者协会
cashmere atlas 光亮纬面缎纹呢;开士米缎
cashmere cloeskin 开士米驼丝锦
cashmere coat 开士米羊绒大衣,开司米羊绒大衣,羊绒大衣
cashmere coating 开士米大衣呢;羊绒大衣呢
cashmere d'écosse 羊绒斜纹呢,苏格兰开士米呢
cashmere des indes 羊绒羊毛混纺呢
cashmere double 开士米斜纹面平纹底织物
cashmere fancy suiting 羊绒花呢
cashmere fiber 山羊绒,开士米羊毛
cashmere finishing 仿开士米整理
cashmere fleece 开士米起绒布,开士米线头织物
cashmere gloves 开士米手套,羊绒手套
cashmere hair 开士米山羊毛
cashmere knit goods 羊绒针织物,开士米针织物
cashmere knitted goods 羊绒针织品
cashmere knitwear 羊绒衫,开士米针织品
cashmere lapis lazuli 开士米蓝宝石
cashmerelike 仿开士米
cashmerelike acrylic machine knitting yarn 仿羊绒腈纶膨体针织绒线
cashmere-like cotton weft knit fabric 仿羊绒超柔软棉针织物
cashmere melton 开士米麦尔登
cashmere muffler 羊绒围巾
cashmere overcoating 羊绒大衣呢

cashmere rose 开士米粉红色
cashmere rug 开士米地毯
cashmere scarf 羊绒围巾
cashmere shawl 仿开士米提花披巾,开士米披巾,印度披巾,羊绒披巾
cashmere silk 开士米毛葛,开士米绸
cashmere suiting 开士米西服料
cashmere sweater 开士米毛衣,羊绒衫
cashmerette 仿开士米,仿开士米斜纹绒(丝或棉经,羊毛纬,背面拉绒);女式斜纹呢(粗纺毛纱纬)
cashmere twill 开士米斜纹
cashmere wool 羊绒,山羊绒,紫羊绒,开士米山羊毛
cashmere wool fancy suiting 羊绒花呢
Cashmere work 开士米刺绣,印度高级刺绣,羊绒刺绣品
cashmere yarn 开士米毛线,羊绒纱,开士米纱
cashmilon 聚丙烯腈纤维(商名,日本旭化成)
cash on delivery (COD) 货到付款
cashoo 阿仙药红色植物染料
cash pocket 小口袋
cash price 付现价格
casimir 仿开士米薄呢
casimir foule 山羊绒,仿开士米粗呢
casinetes 玻利维亚和秘鲁的低档棉裤料
casing 装箱,包装,装嵌式橡筋的缝入部位
casing with drawstring 衣服腰部、衣摆等处用于穿带或缝入橡筋的部位
casing with heading 用穿带装饰的面料
CASIO 卡西欧
casket cloth 寿衣呢
caspian steppes 加裆宽松裤
Caspio 聚对苯二甲酸丁二酯纤维织物(商名,日本东丽)
casque 盔状帽,盔
casquette 盔式无边帽,女式短檐草帽,鸭舌帽,法国大盖帽
casquette cache-oreilles [法]遮耳帽
cassarillos 德国平纹本色亚麻布
cassas 印度松捻细平布
cassava silk 木薯蚕丝
casscanias 色条绸(印度),印度色经条纹绸
cassette 盒,箱;卡
cassimere 毛呢,英国中厚斜纹呢,提花毛葛

cassimere glove 开士米手套
cassimerette 低级薄花呢
cassimere twill 二上二下斜纹
cassimir 法国一上二下斜纹布
cassinet 卡西奈特,英国厚斜纹棉毛织物;19世纪英国阿尔帕卡花呢
cassinetta 卡西奈塔低档全棉斜纹布
cassock （古时军人、牧人、骑手等穿用的)长斗篷或上衣;法衣,教士袍,袈裟,教士双排纽外套
cassock cloth 黑色精纺开士米呢
cassock mantle 及膝短袖女斗篷
cassure 下领折领线
cast 浅淡色调,色光;低级粗套毛;鞋楦
castalogne 卡斯塔隆毛毯
cast button 金属扣
castel branco 葡萄牙高级地毯毛
castellamare 意大利皮棉
castellated 城墙方形线迹
caste silk 中国出口的辑丝,七里丝
castilla 粗纺长毛围巾布
castle 中国白生丝
cast on 起针
cast on cloth 起口织物
castor 厚呢,海狸呢;海狸皮;仿麂皮处理的羊皮;海狸皮帽;海狸香
castor cocoon silk 蓖麻蚕丝,印度蚕丝
castor francaise 法兰西海狸毡
castor fur 海狸毛皮
castor grey 海狸皮灰色
castor leather 仿麂皮处理羊皮
castravane 土耳其生丝
casual 随便的
casual clothes 便装,便服,休闲服
casual coat 短外衣,轻便外套
casual dress 便装,女便服,生活服装
casual elegance 随意又优雅
casual garment 休闲服装
casual hand bag 轻便手提包
casual hat 便帽
casual jacket 便装夹克,轻便夹克,休闲夹克;卡曲外套
casual look 便装风貌,轻便风格,便装款式;便装型
casualness 休闲
casual overshirt 便服
casual red 卡曲红
casual sack 轻便袋装上衣
casual shirt 便服衬衫,常用衬衫,轻便衬衫

casual shoes 便鞋,卡曲鞋,低跟船鞋
casual shorts 轻便短裤
casual slacks 便裤
casual socks 便短袜,休闲袜,轻便短袜
casual sportswear 休闲运动服
casual stockings 便长袜,轻便长筒袜
casual suit 便服,休闲套装,便套装
casual trousers 便裤
casual uniform 日常制服
casual wear 便服,便装,生活服装,休闲装,轻便装
casual wear fabric 便装料,休闲衣料
casula 祭披（天主教神父做弥撒时穿的上衣）
catacaos 秘鲁木本棉
catalog selling 邮购
catalogue 商品目录
catalogue of article for sale 商品目录
catalogue retailing 目录零售（零售商提供商品目录,供消费者选购）
catalogue showroom 目录商品展示室
catalogue stylist 商品目录设计师
catalonia lace 西班牙细网眼花边
catalowne 单纱驼毛呢
catamenial 非织造卫生用品
catania 缎条装饰花绸
Catawba 卡托巴葡萄色
catawba cotton 美国南卡晚熟巴棉
catch 线缝布（毛）边
catcher ends 锁边线,压边线;地毯绒头加固线
catcher's chest protector 棒球接球手护胸器
catcher's mask 棒球接手员面具
catcher threads 锁边线,压边线;地毯绒头加固线
catching facing 钮子
catch selvage[selvedge] 假边,褶边,捕纬边
catch stitch Z形线迹,工形线迹,犬牙形线迹,曲折形线迹;三角针;环针法,三角针法
catch stitched hem Z形针法缘边,花绷缘边,回三角缝缘边
catch stitched seam 环缝,三角针撬边缝
catch stitching 曲折线缝,曲折形线缝
catch stitch seam 环缝
catch stitch tack 三角针撬边搭克
catch thread 钩边线
catch-up 番茄酱色

cat costume	猫相装束;猫样装束	cattle hide wear	牛皮服装
cate caatjes	印度棉织物	cattle leather	黄牛皮,牛皮
catechu	儿茶棕色	cattle leather dyed	染色牛皮
categories of taxes	税种	cattle leather enamelled	上釉牛皮
category	范畴	cattle leather gilded	镀金牛皮
catenary	悬链线	cattle leather silvered	镀银牛皮
cater cap	方顶帽,学士帽,学位帽	catwalk	T台;小车弄;栈桥式时装表演舞台
caterpillar	雪尼尔线	catwalking	猫步
caterpillar knitting yarn	毛虫绒线	caubeen	爱尔兰不整洁帽子
caterpillar lace	毛虫花边	Caucasian rug	高加索地毯
caterpillar point	仿毛虫花边(意大利)	caudebec	仿海狸皮毡帽
caterpillar yarn	毛虫线,毛虫状花式线	caul	女帽后部;女子头饰后部;发网
cat eye	针洞疵	cauliflower ruche	半圆褶裥饰边
cat fur	猫毛皮,猫皮	cauliflower wig	花椰菜假发
catgut gauze	全绞式亚麻纱罗织物	caul work	网
catgut linen	刺绣用亚麻布	caungeantries	16世纪英国丝毛绸
cathay	印度条子缎	cause of damage	损坏原因
cat-head knot	猫头结;双套结	causterized pattern	烂花印花花样
cathedral	(女礼服)长裙拖地的	caustic soda crepe	烧碱泡泡纱
cathedral dress	长裙拖地礼服	cauterising	烂花印花,烂花工艺
cat heel	猫跟	cauther	印度手织粗棉布,印度土布
Catherine-wheel farthingale	轮形裙环	caution	注意
Cathy Hardwick	卡西·哈德威克(1933~,美国时装设计师,出身于韩国)	caution label	警告标签
		cauto cotton	古巴考托棉
catifa	北非高级地毯	cavalier	宽边帽;骑士服;宽平花边翻领
catiol	法国婚礼及肩披巾	cavalier blouse	骑士宽短外套
cation-exchange fiber	阳离子交换纤维	cavalier boots	(骑士)马靴,骑士靴
cationic detergent	阳离子洗涤剂	cavalier cuff	骑士袖口,骑士袖头
cationic dye	阳离子染料	cavalier hat	宽檐骑士帽,骑士帽
cationic fiber-reactive dye	阳离子型活性染料	cavalier sleeve	骑士袖
cationic reactive dye	阳离子型活性染料	cavalry twill	隐斜纹厚呢,双纹呢,马裤呢
catlle leather	黄牛皮	cavella	卡维拉起绒织物
catogan	凯托根假发	cavelour	卡维劳尔耐纶丝绒
cats-and-dogs	低质量织物	caviar bag	鱼子酱包,鱼子提包
cat's-eye	金绿宝石,猫眼石;夜间闪光织物	caviar design	鱼子酱花样
		caviar dot	鱼子酱点花纹
cat stitch	Z形缝迹,Z型线迹,犬牙形线迹,Z形线迹,曲折形线迹	cavu shirt	长袖尖领男运动衫
		cawdebink	仿海狸皮毡帽
cat suit	喇叭裤;猫式套装,黑色跳跃套装,弹性紧身套服;女式长袖紧身连衣裤,女式猫样紧身连衣裤	cawdor cape	扁平折叠布帽
		caxon	开克松假发发型(18~19世纪)
		cayenne	辣椒红;南美生丝;法国本色松薄亚麻布
cattail	香蒲纤维		
cattivella	意大利薄绢绸	C.B.notch	(centre back line notch) 后心中线刀眼
cattle hair	花牛马毛		
cattle hair cloth	厚牦绒,牛毛厚呢;斜纹厚绒呢	C-cup brassiere	大号乳罩,大号胸罩
		ceara	西阿拉棉(墨西哥、巴西产)
cattle hide	大牛皮(总称),牛皮	Cebu hemp	马尼拉麻,蕉麻
cattle hide button	牛皮扣	Cebu maguey	坎塔拉剑麻
cattle hide shoes	牛皮鞋		

Cecil Beaton 塞西尔·比通(英国著名戏剧服装设计师)
cécile cut 塞西尔发型
cedar green 杉木绿,雪松绿
cedar wood 杉木色
cedilla 塞迪拉
cefiros 轻薄棉布
ceiba 木棉
ceiba fiber (丝光)木棉
ceint 腰带,胸褡
ceinture 吊袜带;腰带
ceinture bayadère [法]巴亚德阔绸腰带
ceinture drapée [法]打褶腰带
ceinture tresse [法]网状腰带
celadon 灰绿色
celadon green 青瓷绿(浅灰绿色)
celadon grey 灰绿色
celadon tint 淡灰绿色
celalinen 塞拉亚麻薄纱
celandine green 秋香色,茶绿色
celanna 醋酯里子布
celebrity jeans 名人牛仔裤
Celerina 塞勒里那提花巴里纱
celery 芹菜色
celery green 芹菜绿
celeste 天青,天蓝色
celestial blue 浅灰蓝,天空蓝,淡蓝青
Celine 塞琳
Celion 碳纤维(商名,美国赫斯特-塞拉尼斯)
Cellca 纤维素氨基甲酸酯纤维(商名,芬兰耐思特)
Cellcolor 原液染色黏胶纤维(商名,日本兴人)
cellestron silk 醋酯丝
cellophane 赛璐玢
cellophane-coated fabric 赛璐玢涂层织物
cellular blanket 多孔毛毯
cellular cloth 多孔织物,网眼织物,纱罗织物
cellular fabric 蜂窝网眼织物
cellular plastic sandals 泡沫塑料凉鞋
cellular plastic slippers 泡沫塑料拖鞋
cellular shirt 网眼衬衫
cellular vest 网眼背心
celluloid 赛璐珞
celluloid collar piece 赛璐珞片
cellulose 纤维素
cellulose acetate dye 醋纤染料
cellulose acetate rayon [纤维素]醋酯人造丝,醋酸纤维素人造丝
cellulose base fiber 纤维素纤维
cellulose bonded nonwovens 纤维素黏合法非织造布
cellulose ester fiber 纤维素酯纤维
cellulose fiber 纤维素纤维
cellulose fil(ament) 纤维素长丝
cellulose imago print 纤维素阴影印花
cellulose precursor fiber 纤维素母体纤维,纤维素[纤维]原丝(碳纤维原丝)
cellulose wadding 纤维素衬垫,纤维素填絮
cellulosic composite fiber 纤维素复合纤维
cellulosic fiber 纤维素纤维
cellulosic man-made fiber 纤维素系化学纤维
cellulosic-nylon fabric 纤维素纤维-锦纶混合织物
cellulo-silk 铜氨丝
celon 聚酰胺长丝(商名,英国考陶尔)
celtic 斜纹板司呢
Celtic ornament 凯尔特装饰
celties 斜纹板司呢,方块斜纹织物
celties twill 方块斜纹织物
cement 胶接剂(制鞋);水泥色
Cemented shoee 胶合鞋
cemented cloth shoes 胶粘布鞋
cemented process 胶粘工序(制鞋)
cemented shoes 胶合鞋,胶鞋
cement grey 水泥灰色
concealed button fly 暗纽门襟
cendal 电力纺
cengyun crepe jacquard 层云绉
cengyun jacquard crepe 层云绉
centimeter 厘米,公分
centimeter stick 公分尺,厘米尺
centimetre 公分
centipede lace 蜈蚣边
cento 零布拼料,拼布衣料
central address memory 中央地址存储器
central axis 中心轴
central bobbin shuttle (CB shuttle) 中心摆梭
central district 中心区
central European design 中欧调图案
central india cotton 印度中部棉
centrality 集中性
centralized purchasing 集中采购
centralized sale 集中销售
central knife 中间刀

central line 中心线
central opening 中心开襟
central province cotton 印度中部棉
central stitched ruffle 前中缝合的荷叶领
centre 中心,中间;中心线
centre back (CB) 后中心线
centre back dart 中背省
centre back fold (CBF) 后中心对折,后片中心对折
centre back grain 后片中心线丝缕,后中心线丝缕
centre back length 后(片)中线长,后(片)中心长(从领到腰)
centre back line (CBL) 后身中线,后中线
centre back line of hood (风)帽顶线
centre back neck 颈后中点
centre back neck point-waist (C.B.N-waist) 后颈点至腰
centre back pleat 后背中央褶,后中央褶
centre back run 后中线
centre back seam 背中缝,后中缝
centre back seam line 背缝线,后中缝线
centre back seam with bent seam 合体背中心缝,有弯曲缝的后背中央缝
centre back waist 腰后中点
centre border lace 中心花纹花边
centre color 中心色
centre crease 中折(缝),中心折缝;中折帽
centre dart 前中心褶
centre dent 领结凹部,领结涡,领结厴
centre d'Etudes Techniques des industrie de l'Habillement 法国服装工业技术研究中心
centreed zipper 中分式拉链
centre fiber 叶脉花纹;叶脉针迹
centre finding ruler 中标双向尺
centre fold 中间折叠
centre fold line 中间褶印
centre front (CF) 前中心线
centre front bust level 胸距中点
centre front depth 领驳点前中心深度
centre front fold (CFF) 前中心对折
centre front grain 前中心线丝缕
centre front length 前(片)中心长(从领到腰)
centre front line (CFL) 前(片)中心线,(门襟)搭门线,前身中线,前中线
centre front neck 颈前中点

centre front neck dart 前中颈省
centre front run 前中线
centre front waist 腰前中点
centre front waist dart 前中腰省
centre gauge 中心规
centre gimp 中心嵌芯狭辫带
centre hollowed vertical tuck 中空直线褶裥
centreing adjustment 对准中心
centre length 中线长
centre line 中心线
centre panel 中心嵌条
centre parting 平分头
centre pattern 中心花纹
centre placket length 门襟长
centre plait 中间褶裥
centre plait folder 中间打褶折边器
centre pleat 中央褶裥,中央襞裥(在夹克背中央,与背腰带相配)
centre pleat of shirt front 衬衣前门直口
centre seam 后中缝(制鞋),中缝,中央缝合线缝
centre-seamed shoe 中缝鞋
centre selvedge 中央布边
centre shoulder seam 中间缝
centre sleeve line 袖中心线
centre stitch 叶脉针迹;叶脉花纹
centre thread 芯线
centre-to-centre distance 中心距
centre vent 单衩,中心开衩,后衩,后开衩
centre waist 腰带围;中腰位
centre waist dart 前中腰省
centre waist girth 中腰围
centre web 中层织物
centric cover-core bicomponent fiber 皮芯纤维,同芯式皮芯双组分纤维
centripress drainer 离心压榨脱水机
cepken 彩色刺绣男上衣
cerafiber 硅酸铝纤维(商名,美国)
ceramic-blended fiber 掺混陶瓷[微粒]的纤维
ceramic blue 陶瓷蓝
ceramic button 陶瓷纽扣
ceramic composite yarn 含陶瓷[微粒]复合丝
ceramic fiber 陶瓷纤维,硅酸盐纤维
ceramic fiber blanket 陶瓷纤维毯
ceramic fiber glass 陶瓷玻璃纤维
ceramic finishing 陶瓷整理

ceramic powder finish 陶瓷粉整理
ceramic textile 含陶瓷[微粒]纺织品
cereal-box 饭盒肩包
cerecloth 蜡布
cerements 寿衣
ceremonial dress 礼服
ceremonial garment 庆典服
ceremonial robe 礼服,礼袍
ceremonial uniform 仪式服
ceremony costume 节日装
cerevis 塞略维斯帽
cere yarn 芯丝
cerifos checks 塞里福斯方格细布,赛里福斯细布
cerise 淡红色,樱桃红色,淡洋红色,鲜红色
cerise cloth 塞里斯条子呢
CERRUTI 1881 卓诺迪
certificate 凭证,证书,证明书
certificate in regard to hand looms 手工制纺织品产地证
certificate of competency 合格证书
certificate of delivery (C/D) 交货证明书
certificate of inspection 检验证书
certificate of insurance 保险单
certificate of manufacturer 制造厂证书
certificate of origin 产地证明书
certificate of origin textile products 纺织品产地证
certificate of patent 专利证书
certificate of quality （产品）合格证,质量证书
certificate of tare weight 皮重证明书
certification 认证
certification of approval 合格证
certification program 纤维使用须知
Certified Gemologist 合格宝石学家
certified invoice 签证发票;证实发票
certified item 鉴定项目
certified organic cotton 环保棉,未污染棉
certified public accountant 注册会计师
certified washable or dry cleanable 可洗或干洗的合格标记(国际织物防护协会的合格印记)
cerulean 淡蓝绿色,天蓝色
cerulean blue 蓝色,青色,天蓝色
ceruse white 铅白
cervical height 颈椎高
cervical point 颈椎点
cervical vertebra 颈椎

cest 女腰带,腰带
cesta （回力球）长勺手套,(古罗马)硬皮手套
cestus （古罗马拳击用）皮手套；（女用）腰带
cevennes silk 法国优质生丝(制作花边)
ceylon 塞朗棉毛交织平布
ceylonette 英国全棉衬衫布
Ceylon stitch 锡兰绣
Ceylon yellow 锡兰黄
Cézanne style 塞尚风尚
C.F waist 前中腰点
cha 中国薄印花绸
chabaori 日式宽松短上衣
chacart 印度格子花布
chaconne 恰空舞用领巾
chadar, chaddar, chadder 查达巾,高披巾,提花羊毛围巾(印度妇女用),阔幅披肩；经向蓝黑条阿拉伯棉布；高级艳绿色台球呢；山羊绒披肩料；半漂棉腰布
chaddah 印度波斯长披肩
chadidar 查迪达服
chadri 阿富汗包覆全身的女装
chadur 印度男斗篷
chafe mark 擦伤痕
chafer 子口包布
chaferconne 东印度印花亚麻布
chaffers 风帽两侧绣花垂饰物
chafing 单腿防寒裤
chaggy thread 毛芒线
chagreen 仿皮革布
chagrin 英国棉质书面用布,仿皮革绉纹绸；金线窄丝带
chagrin braid 金线窄丝带
chagrin silk 仿皮革绉纹绸
chagueta 短褂
chahar-gul 印巴高级女式呢
chaharkhana 查哈卡纳纱罗；印度棉丝平纹格绸；印巴高档羊绒纱罗
chail 印度印花棉布
chain 链；拉链；表链；链条；项圈；经纱
chain armour 锁子甲
chain belt 链带,链状饰带；链状腰带
chain bracelet 链状手镯,手镯链
chain-break effect 套格效应
chain-break worsted 套格子花呢
chain closing 金属链花边门襟
chain cotton 巴西棉
chained feather stitch 链式羽状绣

chainette 链状针迹;法国八综哔叽;法国小花纹绸;针织空心绳
chain feather stitch 羽毛状链缝
chain handle handbag 链带手袋
chain helmet 链束型帽盔
chain hole 表链孔
chaining feed dog 链状送料牙
chaining feeder 链状送料牙
chain knot 链状结
chain lace 绷子刺绣花边
chain like leno 链式罗
chain loafer 链饰鞋面的船鞋
chain lock stitch 链锁式线迹,双线链式线迹
chain loop 链式袢
chain mail 锁子甲
chain material 拉链牙原料
chain necklace 金属项链
chain of command in production 生产指挥系统
chain off 缝成长串,连续车缝裁片
chain-off thread back-tacking 辫线夹子
chain-off thread cutter 辫线剪刀
chain of responsibility 连锁责任
chainse 长及脚跟的女式白亚麻束腰内衣
chainsel 精细麻布
chain sewing 链式针迹缝合
chain size 拉链牙尺寸
chain smocking 缝缝缩褶绣,链状司马克
chain stitch 链缝线迹,锁式线迹;链式线迹;链状绣;绞花组织;绳编
chain stitcher 链式线迹缝纫机
chain stitch machine 环缝机,链式缝纫机
chain stitch seam 链缝,链式缝合,链式线缝
chain stitch seamer 链式缝合缝纫机,链式缝纫机
chain stitch selvage[selvedge] 链状针迹边
chain stitch sewing machine 链式线缝缝纫机
chain stitch smocking machine 链式装饰缝纫机
chain store 连锁店
chain twill 链形斜纹
chain twist 链式花线
chain weights 加重链
chain yarn 链式花式线,链式花线
chain zipper 链齿式拉链

chair lace 椅用花边
chair webbing 家具弹簧绷带
chaisel 细亚麻布,精细麻布
chaki, chakhi 埃及丝棉交织经面绸;恰基棉背缎(埃及)
chakmak 土耳其金丝绸
chalcedony 玉髓
chale geant [法]大披肩
chale mantille [法]西班牙披肩
chalinet 薄型平纹毛织物,印花薄型毛织物
chalk 粉笔,粉片,划粉;浅莲灰;用粉笔写、画;打……的图样
chalk blue 浅蓝色
chalk button 磁扣,白垩扣
chalked markings 划粉记号
chalk finish 增重整理,白垩粉整理
chalking (裁片间)打粉印
chalk line 画粉线
chalk mark 划粉线
chalk out 打图样
chalk paper 划粉纸
chalk pencil 划粉笔
chalk pink 浅粉红
chalk stripe 白垩条纹,粉笔条纹,细白条纹;深色地白条纹;深色地细白条子布
chalk stripes 中条子
chalk violet 浅雪青
chalky 白垩色,灰白色
chalky color 白垩色
challis 薄型平纹毛织物,印花薄型毛织物,印花薄绒
chalolas 印度高级细棉布
chalon 英国轧光斜纹薄呢
chalotas 高级细布(印度)
chalvar 阿富汗裤
chalwar 宽大裤子(土耳其)
chalys 薄型平纹毛织物,印花薄呢;印花薄型织物
chama 手织平纹棉布
chamarre 刺绣;带子;学位袍
chambers cotton 美国北卡晚熟棉
chambery 法国轻型毛葛
chambord 法国尚博德丧服呢,尚博德呢
chambord mantle 女式带兜帽七分长斗篷
chambray 青年布,自由布,桑平布;钱布雷呢,钱布雷绸(色经白纬的色织布)
chambray blue 钱布雷蓝

chambray gingham 钱布雷条格细布
chambray pink 钱布雷粉红
chambray shirt 青年布衬衫
Chambre Syndicale de al Couture Parisienne 巴黎高级服装店公会
chameleon 三色调效应
chameleon fiber 变色纤维,光敏变色纤维(光致变色纤维)
chameleonic fabric 感性变色织物,[热敏及光敏性]变色织物
chameleonic fiber 温敏性变色纤维
chameleon taffeta 三色闪光塔夫绸
chamfer 斜剪,斜裁,斜切;斜边
chammarrer 粗俗镶饰
chammer 学位袍
chamois 麂皮,羚羊皮,油鞣皮;仿羚羊皮(或仿麂皮)织物,充麂皮织物;充羚羊皮织物;羚羊皮色(黄棕色或灰黄色)
chamois cloth 仿麂皮棉布,仿麂皮织物,仿羚羊皮织物,充麂皮棉织物;充麂皮针织物,充羚羊皮棉针织物;充羚羊皮棉织物
chamois dressed 雪米羔羊皮
chamois dressed lambskin 雪米羔羊皮
chamoisette 仿麂皮棉织物;仿麂皮针织手套布
chamois fabric 仿麂皮织物
chamois garment 麂皮衣服
chamoisine 仿麂皮棉织物;仿麂皮针织手套布
chamois jacket 麂皮夹克
chamois leather 羚羊皮,麂皮革,油鞣革
chamois shorts 麂皮短裤
chamois skin 麂皮,羚羊皮,羚羊鞣皮,油鞣皮
chamois skin jacket 麂皮夹克
chamois suede 充麂皮棉织物,仿麂皮棉织物
champ 花边地,花边网眼地
champagne 黄褐色,绿黄色,香槟色
champagne color 香槟色
champagne diamond 香槟钻石
champagne rose 香槟玫瑰色
champagne yellow 香槟酒黄
champion cluster cotton 美国大铃棉
champion silk 印度及孟加拉野蚕丝
chandaha 印度手织厚重棉平布
chandar 印度进口棉布
chand-dar 织花羊绒披巾
chandelier earrings 枝形吊灯耳环

chandin 低档平纹布
chand-tara 天体图案;天体图案花缎(印度)
CHANEL 夏奈尔
Chanel bag 夏奈尔包
Chanel dress 夏奈尔服,夏奈尔式女服
Chanel handbag 夏奈尔皮包
Chanel jacket 夏奈尔夹克
Chanel length 夏奈尔长度(达膝下 5～10cm 的裙子长度)
Chanel look 夏奈尔式,夏奈尔风格
Chanel style 夏奈尔式
Chanel suit 夏奈尔套装;仙奴服(音译,开襟上衣和直线型裙的组合)
Chanel tweed 夏奈尔粗呢
changan jacquard faille 长安绸
Changchow velvet 漳绒
Changchow velvet satin 漳缎
change 更换,换去(旧鞋、帽等);换衣
changeable color 可变色
changeable earring 可变换式耳环
changeable effect 闪光效应;闪色效应
changeable silk 闪光绸
changeable taffeta 闪光塔夫绸,闪光绸;变色布;闪光布
change a collar style 更换领型
changeant 闪光织物;闪光效应
change bad quality cutted pieces 换(裁)片
change color 变色
change gear 变换牙齿
change in color 变色
change in one's overall 换上工作裤
change in shade 变色
change one's clothes 换衣服
change-over time 调换时间;更换品种时间
change pocket 零钱袋;票袋;西裤表袋;胸口袋
change purse 零钱包,零钱皮夹
change stitch 变化线迹;变形线迹
change the style of summer dresses 变更夏装款式
change time 更换品种停台时间
changhong satin striped voile 长虹绡
changing bag 暗袋
changing defective piece 更换衣片,换片
changing room 更衣室
changle jacquard sand crepe 长乐绸
chang-ot 长袄,韩式长袖丝斗篷
changshan 长衫

changtain pashmina 白色或银灰色山羊毛
changying striped voile 长缨绡
channel cardigan 无扣开襟毛衫
channel dress 无领上衣裙装
channel fashion 无领上衣裙服式
channel lip 槽边(制鞋)
channel look 无领上衣裙装,无领上衣裙装式
channel opening 开槽
channel seam 嵌条缝
channel setting 戒指沟状基座
channel suit 无扣套装
chantadi-dar 高档织花羊绒披巾
Chantal Thomas 尚塔尔·托马斯(法国时装设计师)
chantilly 尚蒂利细花花边;镶有尚蒂利细花边的服饰(如面罩、披巾等)
chantilly lace 尚蒂利细花花边(女裙和女帽用,法国制),尚蒂伊花边
chanvre 大麻(法国称谓)
chanyi georgette 蝉翼纱
Chan Yu 襜褕
Chao-fu 朝服
chaparajos 皮护腿套裤(美国牛仔穿)
chape 钩扣,皮带活动圈
chapeau 帽子
chapeau bras 三角帽(可折叠的,法国18世纪款式),折叠三角形帽
chapeau claque 折叠式高顶大礼帽
chapeau cloche 钟形女帽
chapeau de pecheur 渔夫帽
chapeau melon 瓜形帽
chapeau plumes [法]羽饰帽
chapel cap 唱诗班女帽
chapel-length train 前短后长裙裾
chapel veil 花边圆头饰(妇女在教堂穿戴)
chapeo de sol 葡萄牙女式彩色阳伞
chaperon 香片龙垂布(法官、律师);香片龙风帽,女式软兜帽,连披肩兜帽);香片龙褶皱头巾(14～15世纪);兜帽垂饰
chaperon turban 香片龙缠头巾(欧洲哥特时代)
chaperon with dagges 有扇形饰边的短帽
chapkan 宫廷卫士短上衣,佣人短上衣
chaplet 串珠,小串念珠(天主教);花冠;项圈;珠链;束发带
chaplet hair style 发辫上盘式发型,花冠式发型

chaplin look 花冠式发型
Chaplin style 卓别林风格
chapoal 聚丙烯腈单丝(商名,日本旭化成)
chaporast 克什米尔地区高档羊绒细呢
chappal 印度便鞋;印度凉鞋(皮质)
chappe 绢丝
chappe silk cloth 绢丝织物
chappe silk yarn 绢丝
chaps (美国牛仔穿的)皮护腿套裤
chapska [法]枪骑兵帽
chap's sturdy shoes 男式强力鞋
chaqueta 牛仔皮夹克;短裾
character 特性;特征
character brand 特征商标,特征品牌
character button 文字图案扣
characteristic 特色,性能,特性;规格;鉴定
characterization 品质鉴定
charadary 丝棉混纺条格布
charara cotton 埃及长绒棉
charcoal 木炭色,黑炭色
charcoal black 炭黑色
charcoal blue 深灰蓝色
charcoal brown 炭褐色
charcoal drawing 木炭素描,素描
charcoal grey 黑灰色,炭灰色
charcoal iron 炭火熨斗
charcoal sketching 木炭速写
charcoal stick 炭笔
charconnae 18世纪印度出口欧美的丝棉条格绸
chardonize rayon 无光人造丝
Chardonnay 夏敦埃酒色
Chardonnet silk 夏尔多内人造丝
chargat 波斯室内用女式三角头巾
charge 费用
charge half decay time 电荷半衰期
charka cotton 印度手轧棉
charkana 查卡纳格布
charkha cotton 手工轧花棉
charkhana 混纺格绸
char length 炭化长度(织物防燃试验项目)
Charles Creed 查尔斯·克里德(英国服装设计师)
Charles Frederick Worth 夏尔·弗雷德里克·沃斯(1826～1895,法国高级时装设计师)
Charles Greed 查尔斯·克里德(1908～

1966,英国时装设计师)
Charles James 查尔斯·詹姆斯(1906～1978,美国时装设计师)
Charles Kleibacker 查尔斯·克莱巴克尔(美国时装设计师)
Charles Poynter Redfern 夏尔·弗雷德里克·沃恩(法国时装设计师)
Charles Suppon 查尔斯·萨本(美国时装设计师)
Charleston dress 查尔斯顿装
Charleston pants 查尔斯顿裤
Charlie Chaplin coat 卓别林式外套
Charlie Chaplin suit 卓别林套装(便礼服上衣加宽肥裤)
charlock 田芥菜绿
charlock green 田芥菜绿
Charlotte 夏洛特镶花边软帽
Charlotte Coday cap 夏洛特·考狄式无边帽
charlotte Corday cap 科戴帽,夏洛特·考狄式无边帽
Charlotte Corday fichu 夏洛特·考狄长围巾
Charlton White 查尔顿白(锌钡白)
charm 魅力,魔力;小饰品,小饰物
charmanette satin 纬三重缎背斜纹绸,查门特斜纹绸
charmante satin 纬三重缎背斜纹绸
charm bracelet 护身手镯,嵌宝镯,幸福手镯,幸运手镯
charmeen 查米恩斜纹呢
charmelaine 查梅兰女式呢;一上二下棱纹布
charmelaine cloth 查梅兰女士呢
charmeuse 轻薄光面软缎,查米尤斯绉缎(柔顺女装缎料)
charmeuse cotton 棉缎
charmeuse petticoat 绉绸衬裙
charm necklace 护身项链,魔力项链
charmoy 印度夏穆艾棱条绸
charm ring 护身戒指
charms 美貌,妩媚;小饰件,小饰品
charm string 时尚项链
charro costume 墨西哥骑马装
charro pants 墨西哥牛仔裤(宽大型)
chart 图;图表
chartreuse 黄绿色
chartreuse green 卡尔特绿(黄绿色)
chartreuse yellow 卡尔特黄(浅绿黄色)
charvet 山形条纹绸

charvet silk 斜棱纹领带绸
chase boots 追猎靴
chasembles 贵族男式筒袜
chasing [布匹]叠层轧光;浮雕法,压花法(首饰)
chasselas [西非]查斯拉斯棉布
chasseur jacket 军装式及臀合身女夹克
chastity 紧身裆
chastity belt 贞操带(中世纪)
chasuble 背心装,十字裆(神父穿的无袖长袍)
chat 打包粗麻布
chateau hat 城堡帽(大帽檐麦秆帽)
chateau rose 大葡萄园红
chatee 查蒂黄麻袋布
chatelaine 查得灵链饰,短链子(女用),女用腰链,腰链(女用)
chatelaine bag 链子包
chatelaine pin 短链胸针,扣钩夹,表袋别针,短链子别针
chatelaine style 链子风格
chatelaine with etui 悬挂有女用小包的腰链
chatoyant 猫眼石;织物闪光效应
Chaumet [法]乔门
chausons 裤子
chausse 连体紧身裤袜;长袜;分离式紧身裤和长筒袜;斗篷
chaussembles 中世纪男袜
chausses [法]漏斗状滤袋;巧斯阔垂带;护腿足铁铠(中世纪)
chausses en bourse 加垫膨体裤
chausses en tonnelet 威尼斯男式马裤
chaussettes 法国白色针织袜类
chausson [法]轻便鞋;软底鞋
chausures 鞋子;靴子
chausuti 印度厚重棉平布
chautar 细棉布
chauthai 手织坚固床单布
chavonnis 印度薄细布
chayong 中国条丝绒
cheap chic 廉美装
cheap spare parts 低值易耗品
cheat 假衬衫领;男式短外套
chebha 突尼斯手编带
chechia [法]伊斯兰圆帽,朱阿夫帽,非圆帽,流苏的女毡帽,阿尔及利亚有流苏的童毡帽
chech muslin 充麻纱
check 格子花纹;格子布;格子织物;方格

图案;检验,核对;支票
check back 暗格子,隐格;背面格双层大衣料
check back fabric 背面格双层织物,背层织物
check blanket 格子毯
checkboard 棋盘花纹布;千鸟花纹布
checkboard check 棋盘格
checkboard fabric 格子织物
checkbone 颧骨
check button and snap 查纽扣(质量)
check canvas 绣花用帆布;小格刺绣十字布
check card 资料卡
check collection 支票托收
check color-and-weave effect 配色格子花纹效应
check cord 平纹条格布
check cutted pieces 验裁片
check cutting edge 检查裁片切口
check deerstalker 格子布猎鹿帽
check dress 格子花纹女服
checked bed sheet 彩格床单
checked blanket 游客毯,格子毛毯
checked board shirt 棋盘格花纹衬衫
checked button-to-hem uniform 钉纽方格制服
checked corduroy 格子灯芯绒
checked cotton suitcase 方格布箱
checked crepe 点格绉
checked crepe jacquard 格花绉
checked dobby weave 多臂小提花格子纹
checked fabric 格子布,格子织物
checked fancy 格子花呢
checked gingham 围裙式格子织物
checked hessian 黄麻垫子布;格子打包麻布
checked hose 格子长袜
checked jacquard crepe 格花绉
checked kabe crepe 格子碧绉
checked lamé taffeta 格夫绢
checked lining 格子衬里布
checked plastic raincoat with detachable hood 带可卸帽的方格塑胶雨衣
checked poplin 格子府绸
checked ruler 格子尺
checked satin 方格缎纹,阴阳缎纹
checked scarf 彩格围巾
checked taffeta 桑格绸,格绢
checked tissue 宽条银格绉

checked towel 彩格毛巾
checked trousers 格子长裤
checked twill 阴阳斜纹,阳阳小方格斜纹
check effect 格子花纹
checker 检验员;格子花,盘格子,方格花纹
checkerboard 方格图案,棋盘式图案
checkerboard check 棋盘格
checkerboard hose 格子花袜
checkerboard satin 阴阳缎纹,棋盘式缎纹
checkerboard twill 阴阳斜纹,棋盘式斜纹
checkerboard weave effect 棋盘格花纹
checkered lining 格子衬里布
checkerwork 彩格图案
checkerwork design 棋盘方格花纹
check fabric 试车布
check fancy suiting 格子花呢
check flannelette 格绒布
check flaw 查(布料)疵点
check flaxon 仿亚麻棉织物(商名)
check gingham 格子布
check grain 查(布料)纬斜
check hose 老式吊线袜
checkia 无缘圆帽(顶部有流苏)
checking 检查,检验
checking bias filling 查纬斜
checking button and snap 查纽扣
checking color and lustre 查色泽
checking color deviation 验色差
checking cut piece 验裁片
checking defect 查疵点
checking grain 查纬斜
checking interfacing 查衬布
checking lining and interlining 查里料
checking notches 查裁片刀口
checking of finished products 成品检验
checking pattern 复核划样
checking pieces 验片,验衣片
checking spot 查污渍
checking zipper 查拉链
check mark 对档
check mohair 英国彩色格子呢
check muslin 仿麻纱;凸纹格子薄纱
check number 对号码
check off 查讫,验毕
check off color 验(布料)色差
check out 检验
check pajamas 格子布睡衣裤

check pattern 格子花纹;复查花样	chelating fiber 螯合纤维
check procedure 检验步骤	chelem 西沙尔麻,剑麻
checkroom 衣帽间(美国)	chelos 印花格子细布
checks 提格布,格子布,各式格子织物;经纬条纹织物,棋盘花纹织物	Chelsea boots 切尔西靴(至踝长,两侧有弹性拼块),奇尔西宽紧靴
check sample 试验标准布样	chelsea collar 切尔西领(V字低口连翻领)
check scarf 条格围巾	
check shirt 格衬衫,格纹衬衫,格子衬衫	chelsea look 切尔西风格,切尔西风貌
check shirting 格子衬衫布	chembred cotton 美国陆地棉
check spikes 对花纹钢针(裁剪花布时用于对准各层花纹)	Chembre Syndicale de la Couture Parisienne 巴黎高级时装联合会,巴黎高级时装工联
check spot 查(布料)污渍	
check spring 挑线簧,止动弹簧	chemical agent protective clothing 防化服装,防化服装
check spring stroke 挑线簧冲程	
check spring take up 挑线簧吸收量	chemical analysis method 化学检验方法
check stitch 订缝线迹;格子针法;格子绣	chemical and oil protective gloves 防化和防油手套
check suiting silk 金格呢	
check turn-back 素色边格子布(英国)	chemical attack 化学侵蚀
check tweed 格子粗花呢	chemical-barrier suit 防化学服
check voile 缎条巴里纱	chemical-biological protective garment(CB garment) 防生化服装
check zipper 查拉链(质量)	
chedasti 印度棉条纹毯	chemical blend (fiber) 双组分复合纤维
cheddar cheese 切达干酪色	chemical bonded nonwoven fabric 化学黏合法非织造布
cheek 颊	
cheek bone 颊骨	chemical bonding machine 化学黏合机
cheek color 脸颊色	chemical boots 塑料长靴
cheeks 颊部皮革	chemical crinoline 化学处理毛硬衬;化学刺绣地布
cheeks-and-ears 衬帽;女式遮耳头巾	
cheek wrappers 法式女睡帽侧饰片	chemical environmental resistance 耐化学药品性能
cheerer 平顶男毡帽;平顶硬礼帽	
cheer girl skirt 拉拉队姑娘裙	chemical felt 化纤毡
cheer leader sweater 拉拉队长毛衫	chemical fiber 化纤,化学纤维
cheer up clothes 脱离都市服	chemical fiber fabric 化纤织物,化学纤维织物
cheese cloth (熨烫时用的)干酪包布,粗棉布	
	chemical fiber figured fabric 化纤提花织物
cheese color 干酪色,奶酪色	
cheese plate 大纽扣	chemical fiber filament yarn 化学纤维长丝纱
cheetah 猎豹皮	
cheetah skin 猎豹皮	chemical fiber garment 化纤服装
chef cap 厨师白帽,厨师帽	chemical fiber hand knitting yarn 化纤绒线
chefoo silk 烟台生丝	
chef's apron 厨师围裙	chemical fiber interlining 化纤衬
chef's cap 厨师白帽	chemical fiber super-soft knitwear 高级化纤超柔软毛衫
chef's hat 厨师帽	
chekiang lamb 浙江羔皮	chemical fiber yarn 化学纤维纱
chekmak 切克马克绸(土耳其);土耳其金线;土耳其丝棉交织绸	chemical fiber yarn-dyed wool-like palaces 色织化纤派力司
chelais 条格花纹边细平布	chemical filament 化纤长丝
chelate fiber 螯合纤维	chemical filament fiber 化纤长丝
chelating dye 螯合染料	chemical filament satin 化纤绸缎

chemical finishing 化学整理
chemical identification of fibers 纤维化学鉴定
chemical lace 烂花花边,人造花边
chemical leather 合成革,人造革,仿革
chemically-bonded nonwoven 化学[试剂]黏合的非织造布
chemically crosslinked rayon fiber 化学交联的人造丝
chemically modified cotton 化学变性棉
chemically reactive dye 活性染料,反应性染料
chemical modification 化学改性,化学变性
chemical paste 化学浆糊
chemical performance 化学性能
chemical proofing 耐化学药品性
chemical property 化学性能
chemical protective clothing 化学防护服
chemical protective helmet 防化头盔
chemical quilting 化学绗缝法
chemical relaxation 化学松弛
chemical sandals 塑料凉鞋
chemical setting 化学定形
chemical shoes 化学鞋,人造革鞋;塑料鞋
chemical softening 化学柔软处理
chemical staple fiber 化学短纤维
chemical stone washing 化学石洗
chemical stretch 化学弹力整理
chemical suits 防化服装
chemical treatment 化学处理
chemical washing 化学洗涤
chemigum 丁腈橡胶
chemiloon 女连裤内衣,女式连裤内衣
chemise 女内衣;女无袖(无领)衬衣,女吊带睡衣;衬裙;布袋装;直统连衣裙;恤米斯内衣(中世纪);男衬衫;筒式女礼服
chemise American 美式男子风格衬衫,(法)美式男衬衫
chemise de nuit 女睡衣
chemise dress 直筒形连裙装,衬裙式连裙装;布袋装
chemise frock 女式无袖连衫裙,无袖连衣裙,吊带裙
chemise gowu 珀蒂塔衬衣
chemise look 贴美身款式;无袖女衬式衣
chemise silhouette 袋形轮廓,直筒宽松轮廓
chemise skirt 无袖衬衫裙,马甲连衣裙

chemise slip 无袖衬衣式衬裙;长无袖女式内衣
chemisette 无袖衬衣;女式无袖胸衣;低领披肩
chemisette garter 吊袜带
chemise tucker 无袖衬衣
chemisier 衬衫上衣;男式女罩衫
chemism 化学机理
chemisorptive fiber 化学吸附性纤维
chemmod cotton 化学改性棉
chemsie slip 无袖衬衣式衬裙;无袖女式内衣(长及大腿)
chem-stitch 化学绗缝
chene 闪光棉布;印花平纹织物闪光绸
chengxiang kimono silk 澄香和服绉
chenille 绳绒线,雪尼尔花线;绳绒织物,仿绳绒织物
chenille axminster 绳绒毯,雪尼尔绒毯
chenille blanket 雪尼尔毛毯
chenille carpet 雪尼尔地毯
chenille cloth 绳绒织物,雪尼尔织物
chenille cord 绳绒绣花线,雪尼尔绣花线
chenille embroidery 绳绒线刺绣,雪尼尔刺绣
chenille fabric 绳绒织物,雪尼尔织物
chenille fringe 雪尼尔花线边饰
chenille knitwear 雪尼尔毛衫
chenille lace 毛虫状花边,雪尼尔花边,绳绒线花边
chenille rolio 金属芯绳绒线,金属芯雪尼尔丝线
chenille shawl 绳绒披肩,绳绒披巾,雪尼尔披巾
chenilles needle 大眼针
chenille spread 绳绒床单,雪尼尔床单
chenille sweater 雪尼尔毛衣;绳绒线毛衣
chenille towel 紧密毛巾布
chenille velvet 绳绒线平绒,雪尼尔平绒织物
chenille yarn 绳绒线,雪尼尔花线
cheongsam 长衫;旗袍,中国旗袍;中国长衫
cheongsam'changsam 旗袍
cheongsam skirt 旗袍裙
cheque 支票
chequer 方格花纹;棋盘格子;方格图案;格子花
chequer-board cut 方格棋盘式磨翻(珠宝)
chequere chain stitch 交错链状绣,魔法

链状绣
chequered muffler 格子花围巾
chequered woollens 格子(粗花)呢
chequer stitch 小型花纹图案针迹
cherconnee 混纺格绸
cheres [俄]切略斯皮带
cheripa cloth 切丽帕精纺花呢
chermeuse 休缪思绉;休缪思绢
cherolee 印度条子棉平布
cherquemolle 丝麻粗布(印度),印度丝麻混纺粗平布
cherry 鲜红,樱桃红,樱红色;红色丝带
cherry bloom 樱桃红
cherry cluster cotton 美国南卡樱桃小铃棉
cherry red 樱桃红
cherusse 上浆花边立领
cheshire printers 柴郡高级印花坯布
ches interlining 胸衬
chess board 棋台(一种熨斗烫台)
chessboard canvas 刺绣用格子白帆布
chessboard design 棋盘方格花纹
chessboard-like 棋盘花纹式样的,方格花纹的
chessboard pattern 棋盘花纹式样,方格花纹;棋盘格(电子雕刻网点排列法)
chest 前胸,胸;胸围;胸部
chest around 胸围
chest breadth 胸阔,胸宽
chest circumference 胸围
chest circumference line 胸围线
chest dart 胸省,胸褶
chester 贴身或半贴身外套
chesterfield 单(或双)排扣的软领长大衣,轻便大衣
chesterfield coat 软领长大衣,天鹅绒领暗纽长大衣
Chester Weinberg 切斯特·温博格(1930～1985,美国时装设计师)
chest girth 胸围,胸围长(第一胸围)
chest-high boots 齐胸靴
chesticore 紧身衣
chest interlining 胸衬
chest level 胸围线
chest line 胸围线
chest measurement 胸围尺寸
chestnut 栗褐色,栗色
chestnut black 栗子黑,栗壳黑
chestnut brown 栗褐,栗壳棕,栗壳色
chest piece 胸衬

chestpieces construction 胸衬结构
chest pocket 胸袋,胸前口袋
chest protector (绒布)护胸
chest-shoulder-upper-arm protector 胸肩上臂保护器
chest size 胸围尺寸,胸围大小
chest under 胸围(上围)
chest under armhole 下胸围
chest width 胸宽
chest width line 前胸宽线,胸宽线
chette 轧光印花棉布
cheval glass 穿衣镜
chevelure [法]彗星尾假发
cheverel 软山羊皮
chevilied silk 加光丝
cheville 长裙(长及小腿)
chevilled silk 拋光丝
cheviller 加光整理
cheviot 粗纺厚呢;啥味呢;精纺缩绒粗呢;条格棉衬衫布
cheviot finish 切维奥特整理(使粗纺毛织物织纹清晰)
cheviot shirting 切维奥特衬衫布,高级柔软棉衬衫料
cheviot tweed 切维奥特粗纺花呢
cheviot wool 切维奥特羊毛
Chevreau de soie 舍弗匹染罗缎
chevrette 薄山羊皮手套
chevron 山形,波浪形;(军服上的)臂章,山形军饰 V 形臂章,山型军阶标识;人字斜纹呢,山形纹呢,锯齿形装饰边
chevron stitch 山形线迹;山形绣
chevron stripe 宽条人字花纹,阔条人字花纹
chevron weave 人字斜纹,山形斜纹
chi 尺
chiadder boraals 松织印花棉布
chiaroscuro 浓淡对比
chic 漂亮,雅致,时髦(尤指妇女及其衣着),流行式样
chichen-leg sleeve 鸡腿袖
chichen skin glove 鸡皮手套
chichen yellow 鸡黄[色]
chi-chi 英国条子斜纹布
chichi rug 俄罗斯蓝毛毯
chichi style 做工精致款式,(领子、袖口等)精致褶裥款式
chicken feather 鸡羽
chicken-head knots 鸡头结
chicken-leg sleeve 鸡腿袖

chicken skin gloves　小鸡皮手套
chicken yellow　鸡黄
chicory　菊苣色
chic style　新潮款式,新潮流
chief designer　总设计师
chief engineer (CE)　总工程师
chief petty officer jacket (CPO jacket)　美国海军士官上装
chief petty officer shirt (CPO shirt)　美国海军士官衬衫(用海军蓝毛料精做的双胸袋衬衫)
chief's tartan　酋长格子呢(苏格兰)
chief value of cotton (C.V.C.)　低涤棉涤织物,棉为主的混纺织物,棉涤织物(棉高于50%)
chiffon　绡,薄纱,薄绢,薄绸,雪纺绸;女子服装装饰品;荷叶边
chiffon batiste　雪纺女式呢
chiffon crepe　雪纺绉绸
chiffonelle　高支棉粗细布,蝉翼纱
chiffon georgette　雪纺乔其纱
chiffonized　丝绒柔光整理
chiffon lace　丝绣雪纺薄绸
chiffon net　雪纺网眼黑纱
chiffons　女子服饰品
chiffon scarf　雪纺丝巾;雪纺头巾;雪纺围巾
chiffon silk scarf　雪纺丝巾
chiffon swiss organdy　瑞士蝉翼纱
chiffon taffeta　雪纺塔夫绸
chiffon twist　雪纺绳捻丝
chiffon velour　雪纺丝绒
chiffon velvet　薄丝绒,薄天鹅绒,雪纺丝绒
chiffon voile　薄绸巴里纱
chifonese　雪纺尼斯
chifton　荷叶边
chi-fu　旗服;中国清朝宫廷长袍
chignon　发髻,髻
chignon cap　发髻帽
chignon ornament　发髻装饰
chignon style　发髻发型
chignon wig　假发髻
chijimi　日本深色窗帘绸
chika　糙斑
chikan　绣花细薄布(印度)
chikan embroidered cap　印度伊斯兰男教徒绣花白布帽
chikun　印度荨麻
child figure　儿童体型,孩子体型

childish overcoat cloth　俄罗斯儿童大衣呢
childish style　儿童风格
childish suitings　俄罗斯儿童用西服呢
child patients' garment　病童服装
children badge　儿童徽章
children bag　儿童手袋
children dress factory　童装厂
children pants　童裤
children's ankle socks　齐踝童短袜
children's bag　儿童手袋
children's clothes　儿童服装
children's clothing　儿童服装,童装
children's coat　童大衣
children's footwear　童鞋
children's garment　童装
children's hat　童帽
children's jumper　童套衫
children's knitwear　针织童装
children's pants　童裤
children's printed handkerchief　儿童印花手帕
children's printed single jersey shorts　汗布印花童短袖衫裤
children's rubber boots　童雨靴
children's shoes　童鞋
children's shorts　童短裤
children's size　童装号型
children's skirt　童裙
children's slacks　童裤
children's sleepwear flammability standard　儿童睡衣燃烧性标准
children's socks　童短袜,童袜
children's stockings　儿童长袜,童长袜
children's tights　儿童连袜裤
children's towel　童巾
children's trousers　童长裤
children's T-shirt　短袖童汗衫
children('s) wear　童服,童装
children towel　儿童毛巾
child's cloth umbrella　儿童布伞
child's costume　童装
child's embroidered shoes　绣花童鞋
child's pudding　婴儿加垫小圆帽
child's sundress　幼儿太阳服
chili　干辣椒色,红棕,铁锈色
chili color　黄棕色
chilim carpet　基里姆地毯,手织光面地毯
chilkaht blanket　奇尔卡特山羊毛毯
chilkat blanket　山羊毛毛毯

chilli 英国染色粗棉布
chillo 英国染色粗棉布
chima 朝鲜女长裙
chimayó 奇马约横条纹毛毯
chimer 主教礼袍
chimere 无袖长袍，主教法衣（宽大无袖长袍）
chimney 高顶礼帽
chimney effect 烟囱效应
chimney pot 高顶毡帽
chimney pot hat 高顶礼帽
chimp jacket 津布夹克（布面有拉毛感的短夹克）
chin 颏，下巴；中国花缎
chin-chira finish 起绒整理
China artistic style 中国艺术风格
China blue 中国蓝（深蓝色），溶性蓝
China Chamber of International Commerce 中国国际商会
China checks 靛蓝格子布
China cotton 中国粗绒棉
China Council for the Promotion of International Trade 中国国际贸易促进会
China crepe 中国绉
China Entry-Exit Inspection and Quarantine Bureau (CIQ) 国家出入境检验检疫局
China Fashion Designers Association 中国服装设计师协会
China finish 陶土整理；重浆轧光整理
China Garment Association 中国服装协会
China garments 中国服装
China grass 苎麻（中国草），白苎，绿苎，苎仔，线麻，紫麻
China grass cloth 中国夏布，苎麻布
China gum cloth 印有点花的湿织丝绸
China-hemp 大麻，汉麻，火麻，魁麻，线麻
China International Garment Accessories Fair (CIGAF) 中国国际服装服饰博览会
China International Textile Machinery Exhibition 中国国际纺织机械展览会
China jute 苘麻，青麻，芙蓉麻，顷麻，天津麻
China Laboratory Accreditation Committee 中国实验室认可委员会
China lamb's wool 中国羔羊毛
China Leather Industry Association 中国皮革工业协会

China look 中国风格
China mink 中国貂皮
China mull 中国薄布；中国薄绸
China muslin 印花细布；提花细布
China National Garments Research & Design Centre 中国服装研究设计中心
China National Import & Export Commodities Inspection Corporation (CCIC) 中国商品检验公司
China nettle 苎麻；荨麻属
China Nonwoven & Industrial Textiles Association 中国非织造布与产业用纺织品行业协会
China Nonwoven Technical Association 中国非织造布技术协会
China pat 中国黄麻
China Quality Control Association 中国质量管理协会
China ribbon 中国缎带，中国丝缎带
China ribbon embroidery 中国绣花缎带
China silk 中国绸缎，中国丝绸
China size 中国鞋号
China steam filature 中国厂丝
China stripe cloth 中国条纹呢
China Textile Academy 中国纺织科学研究院
China Textile and Apparel 《中国纺织及成衣》(期刊)
China Textile Garment Technology Development centre 中国纺织服装技术开发中心
China wool 中国羊毛
chin-band 腭带，下巴处布条，下巴带子，颏带
chin bed cover 织锦床罩
chinchilla 灰鼠毛皮，灰鼠皮；栗鼠呢，珠皮呢（粗羊毛大衣呢）；银灰色
chinchilla cloth 厚重纬起绒织物
chinchilla cloth finishing 灰鼠呢整理，珠皮呢整理
chinchilla overcoating 灰鼠呢
chinchilla plush 灰鼠呢面长毛绒织物
chinchilla rabbit 栗鼠兔皮
chin cloak 围巾；头巾
chin-cloth 腭布，腰布
chin collar 颚领（高立领），托颏领，高竖领
chincum 中国生丝
chiné 印经平纹织物
chiné ribbon 印经缎带

Chinese accent 中国特征,中国味
Chinese ancient costume 中国古代服饰
Chinese ancient wedding gown 中式礼服
Chinese and Western clothing 中西服饰;华洋服饰
Chinese and Western garments 中西式服装
Chinese and Western style blouse 中西式上衣
Chinese and western style clothes 中西式服装
Chinese and western style cotton wadded coat 中西式棉袄
Chinese art 中国艺术
Chinese aspect 中国风格
Chinese attire 华服
Chinese baby wear 儿童斜襟衫,小儿斜襟衫,婴儿斜襟衫(中式)
Chinese ball button 葡萄扣,葡萄纽,中式纽扣
Chinese ball knot button 中式盘花纽
Chinese blouse 中式上装,中式上衣
Chinese blue 中国蓝,深蓝色
Chinese boots 中国靴
Chinese burr 中国类黄麻
Chinese button 花结扣,盘花扣,中式纽扣
Chinese cap 瓜皮帽,西瓜皮帽
Chinese carpet 中式地毯
Chinese collar 旗袍领,中式领,中式旗袍领,中装直领,中式直领
Chinese costume 华服,中式服装,中装
Chinese cotton 中国棉
Chinese court costume 中国宫廷服装
chinese design 中国式图案;中式设计
Chinese dog 中国狗皮
Chinese dress 长衫;中国式旗袍,旗袍
Chinese embroidery 中国刺绣
Chinese fan 蒲扇
Chinese fashion 中国式;中国时装
Chinese-foreign joint venture 中外合资企业
Chinese frog 盘花扣,胸饰扣,中式扣
Chinese garments 中国服装,中式服装
Chinese gown 长衫;旗袍;中式长袍
Chinese grass cloth 夏布
Chinese green 深绿
Chinese jacket 马褂;中式上装,中式上衣
Chinese jute 中国黄麻
Chinese knot 中式花结,中式结(用作纽扣)
Chinese knot button 中国绳结扣,花结扣
Chinese linen 夏布
Chinese local bag 单丝袋
Chinese look 中国款式,中式风格,中式风貌
Chinese lounging robe 中式居家袍
Chinese Mandarin court robes 中国清朝宫廷长袍
Chinese national costume 中国民族服装
Chinese neckline 中式领,中装领领口,中式领口
Chinese oak silk 柞蚕丝
Chinese opening 大襟,挖襟
Chinese pajama 中式睡衣
Chinese raccoon 中国浣熊毛皮
Chinese red 中国红(朱红、大红)
Chinese ribbon embroidery 中国丝带刺绣
Chinese robe (民国初时)长衫
Chinese ruler 市尺
Chinese satin 中国缎
Chinese sesban 中国野大麻
Chinese shoulder cape 中式短披肩
Chinese silhouette 中式轮廓线条
Chinese size 中国尺码
Chinese sleeve 中式袖
Chinese slippers 尖翘鞋头;高鞋面的低跟女鞋
Chinese stencil print 中国蓝印花布
Chinese stye suit 中式套装
Chinese style 中国风格,中国式样,中式
Chinese style bed sheet 中式床单
Chinese style clothes 中式服装(唐装)
Chinese style clothing 中式服装
Chinese style coat 袄,中式上衣
Chinese style cotton padded coat 中式棉袄
Chinese style cotton wadded coat 中式棉袄
Chinese style dress 旗袍;中式装
Chinese style frock 旗袍
Chinese style garment 中国式样服装
Chinese style jacket 中山装;中式上衣
Chinese-style padded coat 中式棉袄
Chinese style pajamas 中式睡衣裤
Chinese style pants (旧时)中式裤
Chinese style shirt (中式)小褂
Chinese style suit 中装,中式套装;唐装
Chinese style unlined upper garment 褂子
Chinese stylistic form 中国款式

Chinese theatrical costume 中国戏装
Chinese trousers 中式裤
Chinese tunic suit 中山装套服
Chinese turkestan 中国新疆地毯
Chinese two-piece dress 襦裙
Chinese-type jacket 襦
Chinese violet 中国紫罗蓝色
Chinese wear 中国服装
Chinese white 白色,锌白色,中国白
Chinese with felt sole 中式毡底皮靴
Chinese woolens 中国呢绒
Chinese workwear 中式工作服
Chinese yellow 中国黄
chine silk 印色丝
Chinesse yellow 中国黄(老黄色)
chine velvet 印经丝绒
chine yarn 印花纱线
chingma 青麻,芙蓉麻,天津麻,苘麻,顷麻
chini silk 印度放养桑蚕丝
chin lapel 耸领
chin line 颏线
chinlon 锦纶(尼龙)
chinlon magic tape 锦纶搭扣带;锦纶带扣
chinner 围巾;头巾
chino 斜纹棉布
chino blazer 丝光卡其夹克
chino cloth 丝光卡其军服布
chinoiserie 中国艺术风格;中国风
chinon 蛋白质接枝聚丙烯腈纤维(商名,日本东洋纺)
chinos 丝光卡其裤,丝光卡其服装
chin-piece 木刻或印版画时戴的口罩
chin quilt cover 织锦被面
Chinses style cotton-padded coat 中式棉袄
Chinses style jacket 中山装上衣
chin stays 女帽系带用的荷叶边
chin strap 腭带,颏带,领扣带;帽束带;下巴处布条,下巴带子;腰袢
chints 摩擦轧光印花棉布(印有华丽和大花型图案)
chints calico 印花棉布
chintuft 颏髭
chintz 印花棉布,油光布,轧光布,摩擦轧光印花棉布(印有华丽和大花型图案)
chintz braid 印花轧光编带,轧光印花编带

chintz finish 摩擦轧光整理
chin warmer 暖颚布,下巴罩布
Chioggia lace 契沃吉埃花边(意大利),意大利粗梭结花边带
chi-pao 旗袍
chi-pau knot 七宝结
chip bonnet 粗草帽,廉价草帽
chip hat 粗草帽
chipmuck 花栗鼠皮
chips 小碎片
chique silk 法国低档生丝
chiriman, chirimen 日本无光绉绸,无光绉绸
chirinka 俄罗斯全丝刺绣棉(或丝)方巾
chiripa 方形彩色布;方形毯子;南美毯子围裹装
chirring 褶缝
Chisato Tsumori 津森千里(日本时装设计师)
chisel 扣眼刀,纽孔凿
chitin fiber 甲壳质纤维
chiton 奇通衫;奇通袍(古希腊);古希腊露臂束腰外衣,古希腊贴身衣
chiton amphimaschalos 古希腊男常服
chiton heteromaschalos 古希腊劳动者服
chitopoly 抗菌纤维(商名,日本富士纺)
chitosan fiber 甲壳素纤维,壳聚糖纤维
chitrak 土耳其棉经丝纬织物
chits 紧密平布(东南亚),印度紧密平布
chitta 点子纹织物
chitterlings 男衬衫麻褶边
chlamys 短外套;短氅(古希腊),方形斗篷(古希腊),古希腊男短上衣(克莱梅斯),古希腊男短氅,克莱密斯短氅(古希腊);男式优质呢短斗篷
chlidema square 方形边毯
chloride bleach 氯漂
chlorinated polyvinyl chloride fiber 氯化聚氯乙烯纤维(过氯纶;PeCe)
chlorinated wool 氯化防缩羊毛
chlorination shrink proofing 氯化防缩整理
chlorine resistance 耐氯性
chlorine resist finish 抗氯树脂整理
chlorofiber 含氯纤维,聚氯乙烯纤维,氯纶
chlorofiber underwear 氯纶内衣
chocohurstle fiber 凤梨纤维
chocolate 红褐色,巧克力色,深褐色
chocolate brown 巧克力棕色,巧克力色

chocolate chip　巧克力碎屑色
chogá　乔加服(印度)
chogori　朝鲜族女式短上衣
choicest wool　最精选羊毛
choice wool　美利奴羊颈部最好毛,优选毛
choir-boy collar　诗班领(教会唱诗班用),小圆翻领
choir boy's dress　唱诗班男童服
choir robe　唱诗班长袍
choker　短项链,紧颈链,颈箍,领圈;硬高领;宽领带
choker collar　贴颈领,硬高领
choker necklace　贴颈项链
choker neckline　贴紧领领口,硬高领领口
chokli　乔克立衣(朝鲜)
chola derby hat　朱罗妇女帽
Chole　柯劳耶
cholee, choli　柯丽衫,印度传统露腰式女衫,低领口紧身胸衣
cholet　未经漂白和上浆的法国薄衣料;新奇花色手帕
chomois suede　仿麂皮棉织物
chongkwen　中国伞绸;轻薄塔夫绸
chong-sam　长衫
choori-dars　丘里-达斯裤(印度),印度宽大裤(裤脚紧,超长式)
choose　选择,挑选
choose and purchase　选购
choose dress materials　挑选衣料
choose samples　选样
chop　乔拍商标,生丝牌号
chopine　(16～17世纪妇女穿的)软木高底鞋,花盆鞋,乔平高底女鞋(16～17世纪)
chop mark　生丝牌号,丝绸牌号,织物牌号
chop pants　剪边裤,剪切型裤子,截筒裤,截筒型裤子
choppat　乔帕特轻塔夫绸(印度),印度薄塔夫绸
chopped sleeve　光袖,平袖
chopper　划样裁剪师
chopper bar fabric　经编压纱织物
chopping　裁剪
chorantine fast dyestuff　氯冉享耐光染料(商名,瑞士)
chord　弦;着浅色
chore jacket　劳动夹克
chorus shirt　排练用服装

chou　(妇女衣帽上)球结,花结
chou hat　卷心菜帽,卷心帽
chouse　大花结
chou silk　绸
chowtar　印度细棉布
chrisom　初生婴儿白色洗礼巾(或洗礼服),洗礼巾
christening bonnet and shoes　洗礼鞋帽
christening dress　圣洗服,受洗服,洗礼装
christening robe　圣洗服,受洗服,洗礼袍
christening shawl　洗礼披巾
Christian Dior　克里斯汀·迪奥(1905～1957,法国时装设计师)
Christian Lacroix　克里斯汀·拉克鲁瓦(1951～,法国时装设计师)
Christina Aujard　克里斯汀·奥加尔德(法国时装设计师)
Christina Claus　克里丝蒂娜·克劳斯(加拿大时装设计师)
Christopher Lebourg　克里斯托夫·勒堡(法国时装设计师)
chroma　(色彩的)浓度,色度,彩度
chroma contrast　彩度对比
chroma scale　彩度标
chromatic　颜色的,色彩的
chromatic aberration　色相差,色视差
chromatic characteristic　彩色特性
chromatic circle　色环,色圆
chromatic color　彩色色调,有彩色
chromatic difference　色差,色度差
chromatic difference of magnification　放大色差
chromatic difference of rotation　旋转色差
chromatic dispersion　色散
chromaticism　彩色学,色彩学
chromaticity　色度,色品;染色性
chromaticity chart　色品图
chromaticity diagram　色品图,色度图
chromaticity display computer　(CDC) 色度显示计算机
chromaticity index　色度指数
chromaticity spacing　色度间距
chromaticness　色品度,知觉色度
chromatic parallax　色视差;色适应性
chromatics　色彩学,颜色学
chromatic sensation　色彩感觉
chromatic sensitivity　色感度,感色性
chromatic spectrum　彩色光谱
chromatic transference scale　彩色沾色样卡

chromatic value 色彩值,色度值
chromatist 色彩学家
chromatogram 色谱,色谱图
chromatograph 色谱仪
chromatology 色彩学
chrome collagen fiber 含铬的胶质纤维,铬胶纤维
chrome complex dye 含铬染料,铬络合染料
chrome dye 铬媒染料
chrome green 铬绿
chrome leather 铬鞣革
chrome mordant azo dye 铬媒染偶氮染料
chrome mordant dyeing process 预铬媒染法
chrome printing 铬媒染料印花
chrome red 铬红
chrome yellow 铬黄
chromic-alumina silica fiber 含铬硅酸铝纤维
chromiferous dye 含铬络合染料
chrominance 色度,彩矢量
chromo embroidery 多彩印版刺绣,覆纸刺绣
chromogene dye 铬精染料
chronograph 计时器,秒表
chrysanthemum design 菊花纹
chrysoberyl 金绿宝石
chrysoidine crystals 碱性橙
chrysolite aquamarine 金绿玉海水蓝宝
chrysotile 温石棉
chrysotile asbestos 优质石棉
chshion-sole socks 弹性袜底短袜
Chuan embroidery 蜀绣,川绣
Chuan embroidery fabric 川绣织物
chubbie sizes 丰满女孩尺码,胖女孩尺码
chubbily sizes 超重女孩尺寸
chubby 无领短宽式毛皮外套,圆胖服
chubby coat 丰满外套(短宽式外套)
chubby fur coat 中长裘皮大衣
chubut wool 阿根廷优质羊毛
chucumci 西沙尔麻
chuddar 经向蓝黑条阿拉伯棉布;高级艳绿色台球呢;山羊绒披肩料;棉半漂腰布
chuddar, chudder 披肩布(印度北部妇女用)
chukka 皮靴,查卡靴
chukka boots 查卡靴,马球靴式高帮皮靴
chukka hat 查卡帽,马球帽

chukka knit shirt 马球针织衫
chukker shirt 开领短袖马球衬衫,扣子衫
chullo and Montera 盔状绒线帽
chumese hemp 印度麻,菽麻
chunbo habotai 春波纺
chundari 印度扎染棉布;印度扎染绸
chunderkana 印度中档棉手帕布
chunfeng jacquard poult 春风葛
chung-shan chou silk 南京中山绸
chung-shan suit 中山装
chunguang faille 春光葛
chunhua brocade 春花缎
chunky 厚而膨松的服装,厚实服装;矮胖的,结实的
chunky heel 粗短跟,超宽型高中鞋跟
chunky shoes 粗短跟鞋,琼琦鞋(超重型款式),厚实鞋
chunmei jacquard sand crepe 春美绸
chunri 印度扎染棉布;印度扎染绸
chunxiang brocade 春香缎
chunyi jacquard taffeta (chine) 春艺绸
chuquelas 条子丝棉交织塔夫绸(印度制)
church embroidery 教堂刺绣
Churchill suit 丘吉尔装
church lace braid 教堂花式辫带
church linen 比斯细薄亚麻布
churumbezi 土耳其细布
chute 降落伞
chute fabric 降落伞织物
chutney 酸辣酱色
chuyu cap 丘乌帽(南美)
chymer 学位袍
ciciclia 土耳其花缎
ciclaton 盔甲上穿上的外衣
cidaris 波斯男式截头锥形帽
cider cloth 稀平布
cienture 饰带;皮带
CIF (cost, insurance and freight) 到岸价,包括成本保险费及运费
cigarette jeans 筒式细棉布衬裤
cigarette mitt 雪茄露指手套
cigarette pants 细长直线型香烟裤,香烟裤(直筒裤)
cilana 泡泡效应;永久绉纹效应
cilaprel 瑞士泡泡纱
cilasilk 西拉泡泡纱
cilatex 西拉花布
cilice 粗毛布;粗毛布衣服(僧人和忏悔者所穿),粗毛布内衣

cilicium 山羊毛织物；粗帆布
cinch 束腰宽带，紧腰饰带
cinch belt 合腰饰带，马腹带式腰带
cinch buckle 腰带扣
cinch closing 双环系带扣合
cinched waist 束紧腰围
cinched waistline 紧腰式腰线
cincher 紧腰饰带，合腰饰带，女腰带
cincture 线带，绳，袢；围绕物（带，腰带）；绣饰束带
cinder grey 灰渣色
cinema pile carpet 影院用短密绒花地毯
cine-mode 电影时装款式，银幕时装式样
cinetique art 活动艺术
cingulum 腰带；妇女胸褡
cinnabar 朱红，朱砂，黄褐，肉桂色
cinnamon 肉桂色
cinnamon brown 肉桂棕色（黄褐色）
cinqtrous 法国五角网眼花边
cintas 带子
Cinzia Ruggeri 钦齐亚·鲁杰里（意大利时装设计师）
Ciporovica 佛加利亚手织簇绒地毯
circassian 锡卡西昂棉毛交织呢
circassian rug 土耳其短绒地毯
circassina wrapper 宽松围裹装
circle 围裙
3-circle 三围
circle buckle 圆形带扣
circle cape 圆斗篷
circle coat 圆裁外套
circle measuring tool[device] 测图工具[装置]
circle pocket 圆形贴袋
circle shirt 圆裙
circle-shoulder jacket 圆肩夹克
circle skirt 圆裙
circle sleeve 二节圆形喇叭短袖
circle stitch 圆形线迹
circlet 小圈，环形饰物
circle template 画圆样板，圆形模板
circle tent dress 圆篷帐连衣裙
circuit diagram 线路图
circular 长披肩或斗篷；无袖外衣，女用无袖外套；喇叭式衣裙下缘
circular auto-control heating oven 立体热风对流烘干机
circular bags 圆筒布
circular beard 满脸大胡子
circular blade （裁剪）圆刀；（制鞋）皮刀

circular cape 圆斗篷，波浪形披风，圆形披风
circular cape sleeve （波浪褶）披肩短袖，，披肩袖，披风袖，圆披肩袖
circular cap sleeve 盖肩袖，圆盖袖
circular cuff 喇叭袖头，圆形袖头，圆袖口，喇叭袖口
circular cut 圆形裁剪，圆形剪裁
circular cut sleeve 圆形袖
circular cutting knife 圆形裁剪刀，圆刀
circular eyelet knitting machine 圆形网眼针织机
circular hem 圆形下摆，圆裙裙摆
circular hemline sleeve 圆摆[线]袖
circular hollow tape 套带
circular hood 风兜，圆头巾
circular hose 圆筒袜，圆袜；筒状物
circular hosiery knitting machine 圆袜机
circular knife 圆刀（裁剪机）
circular knife cutter 圆刀裁剪机，圆刀电剪
circular knit 圆筒形针织物
circular knit hose 圆袜
circular knit hosiery 圆袜
circular knitted fabric 圆形针织物
circular knitted ribbed stocking 罗纹圆袜
circular knitting 圆机针织
circular knitting machine 圆筒针织机,圆形纬编针织机,圆形针织机,针织圆筒机
circular letter of credit 流通信用证
circular linker 圆形套口机
circular milanese knitting machine 圆形米兰尼斯经编机
circular rib fabric 圆筒形罗纹织物
circular ruffle 无褶荷叶边，波浪边
circular sewing 圆周缝纫
circular shoes 波兰鞋
circular skirt 圆裙，圆形喇叭裙
circular stockings 圆袜
circular template 圆形模板
circular-type garment 圆形服装
circular web 圆筒形针织物
circular wrinkle below back neckline 后领窝起涌
circular yoke 圆形过肩；圆形抵肩
circulater layout 循环套料法
circulation channel 流通渠道
circum ference 周边
circumference 周长；围长

circumference ease 围度放松量
circumference measurements 围度尺寸
circumference operation 圆周作业(制衣)
circumferences 围度
circumferential seam 包缝,环缝
circumfolding hat 男士低顶可折叠礼帽
circus band cloth 乐队制服
ciré (仿漆皮)蜡光整理(丝织物经上蜡热轧后产生高度光泽)
ciré finishing 蜡光整理,光泽处理
ciréing 蜡光整理
cire satin 蜡光缎
cire silk 轧光绸,蜡光绸
cirus band cloth 乐队制服
ciselé velvet 凸绒刻花天鹅绒
Cisilian embroidery 西西里刺绣
citamci 墨西哥低级剑麻,西沙尔麻
CITIZEN 西铁城
citrine 柠檬色,柠檬色石英晶体
citrine citron 柠檬黄
citron 香橼黄
citron yellow 香橼黄
citrus 柑橘黄色(嫩黄色)
city casual 城市便装,城市卡曲
city classic 城市传统服饰
city clothing 都市服装
city cowboy jeans 都市式牛仔裤
city crops 旧棉布样品
city cut jeans 都市式截断牛仔裤
city dandy look 城市花花公子款式
city girl 都市少女
city jeans 都市式牛仔裤,上街牛仔裤
city pants 都市裤,街市裤
city shorts 都市短裤,女西装短裤,热裤(与运动上衣配套)
city shorty 都市短上衣,上街短上衣
city sport look 都市轻便款式,都市轻快款式
city sports look 都市轻快款式
city style 城市款式
city western style 城市西部款式
city western stylistic form 城市西部款式
civet 灵猫香,香猫香,麝猫香;鼬猫毛
civet cat 麝猫皮
civet cat fur 灵猫毛皮,香猫毛皮,麝猫毛皮
civetet 香猫皮
civet skin 灵猫皮,香猫皮

civies (俚语)便服,便装
civil aviation uniform 民航服
civil clothes 便服,休闲装,民用服装,西服便装
civil garments 民用服装
civilian cap 便帽
civilian clothes 便服,便装,民用服装,西服便装
civilian wear 便服,便装
civilized clothes 都市服装
civilized fashion 都市款式,文明款式
civilized look 文明风貌,文明款式
civil textiles 民用纺织品
civrine cameleon 多色棱纹织物,双色棱纹织物
clad fiber 包层纤维,涂层纤维
claim 索赔
claimant 索赔人
claim clause 索赔条款
claim for inferion quality 质量差索赔
claim indemnity 索赔
claims accepted 接受索赔
claims settlement 理赔
Claire Borrat 克莱尔·博拉(法国时装设计师)
Claire Mccardell 克莱尔·麦克卡德尔(1906～1958,美国时装设计师)
claires 法国上等细亚麻布
claith 织物(苏格兰称谓)
clam 夹子
clam closing lever 钳口闭合杆
clam diggers 捞哈裤,半长裤(长到膝下)
clam lifting arm 钳口提升杆
clamminess 粘湿性
clammy 冷湿感
clammy handle 滑腻手感
clamp 夹子,钳子
clamp check 夹紧压脚
clamped printing 夹缬
clamp fastener 扣夹
clamp screw 夹紧螺钉
clamp stop motion 夹紧自停装置,离合装置
clamp washer 夹持垫圈
clam-shaped dart 蛤形褶,菱形褶,鱼形褶
clandian 克兰丁条子薄呢
clan green shade 蓝绿色
clan plaid 苏格兰氏族格子花呢
clan tartan 苏格兰氏族格子花呢

clapper 压板
clarence 男式及踝系带靴
Clare Potter 克莱尔·波特(美国时装设计师)
claret 紫红
claret red 葡萄酒红,暗紫红色,红酒色,深紫红色
claret violet 葡萄酒紫
clarines 法国上等细亚麻布
clarino 克拉尼诺(特种合成皮革)
CLARINS 娇韵诗
clarity 净度(钻石),清晰度,透明度,透明感
clarity grade 瑕疵等级(钻石)
clarté 织物表面清洁整理
clash 抵触(颜色等)
clasp 带扣,扣环,钩子,钩搭,别针,夹子,对抱扣811,扣子,扣紧物
clasp buckle 对抱环扣
clasp buckle closure 对抱环扣扣合件
clasp coat 抱合式外套;无纽扣大衣
clasping (手套腕口)镶扣或镶带(工艺),腕口镶扣工艺
class fashion 上流时装
classic 经典服装,妇女的传统服装,(不受时尚影响的)传统服式(通常指1925～1942年间制造的)
classic accessory 经典服饰
classical 二等生丝;经典/古典风格
classical coat 古女装
classical costume 古典服装
classical method 传统方法
classical silk 次优级生丝
classical style 古典风格
classic balley dress 古典芭蕾舞服
classic blue 古典蓝
classic box jacket 典型的箱型外套
classic clothes 传统服装,古典服装
classic coat 传统外套
classic costume 传统服装,古典服装,经典服装
classic denim 古典牛仔布
classic elegance 古典而优雅
classic ensemble 典雅套装
classic floral print 古典花卉印花
classicicist style 古典主义风格
classicism 古典主义
classicism costume style 古典主义服装风格
classicist style 古典主义风格

classic jacket 正统夹克衫
classic look 传统风貌,传统款式,传统风格;传统型
classic model 古典型
classic model suit 古典型西装
classic notched collar 典型的平驳领(缺角西装领)
classic pants 传统式样裤子
classic pattern 标准纸样;传统花样,古典花样,典型花样;传统服装
classic print 古典印花
classic pumps 典型浅口无带鞋
classic shirt 传统式衬衫
classic shirt dress 典型衬衫裙
classic ski pants with knit ankle 针织布裤脚管的古典式滑雪裤
classic skirt 传统式裙子
classic sleeve placket 大小袖衩,琵琶头老式袖衩
classic straight shoulder 直肩
classic style 古典风格,古典主义
classic suit 正式套装;正统男西装;古朴传统服装
classic sweater 传统运动衫,普通运动衫
classic synthetic fiber 经典合成纤维(早期生产低速纺丝,再经过拉伸制成的合成纤维)
classic tailoring 传统缝制法
classic travel wardrobe 典型的旅游服装
classicue 二等生丝
classic walking shoes 传统步行鞋
classic wedding dress 传统的婚礼服
classic yoke-topper 古典覆肩式短大衣
classification 类别
classify from form 形状上分类
classify from techniques 工艺分类
classique 法国粗平布;厚棉床单布
class number 缝针针号,针号
classy cotton 上等棉
Claude Montana 克劳德·蒙塔纳(1949～,法国时装设计师)
Claude Saint Cyr 克劳德·圣·西尔(法国女帽设计师)
Claudio la Viola 克劳迪奥·拉维奥拉(意大利时装设计师)
clause 条款
clavi 克拉维服饰(古罗马)
clavicle 锁骨
clavshoot's vest 飞靶射手背心
claw button 羊角扣

claw hammer 燕尾服
clay 泥土色,黏土色,浅棕色,软木棕,土褐
clay diagonal 克莱精纺毛哔叽,克莱精纺斜纹呢
clays 克莱精纺毛哔叽,克莱精纺斜纹呢
clay serge 克莱精纺毛哔叽,克莱精纺斜纹呢
clayshoot's vest 飞靶射手背心
clay twill 链形斜纹
clay worsted 克莱精纺毛哔叽,克莱精纺斜纹呢(英国)
cleanability 清洗性
clean area garment 无尘无菌服,净化室工作服
clean ball 清洁熨烫台
clean bill 发票
clean B/L 清洁提单
clean color 纯洁色,清色,透明色
cleaner production 清洁生产
clean finish 光洁整理,光面整理;净加工
clean finish binder 光边滚边器
clean-finished seam 卷边缝
clean finishing 毛织物显纹整理
clean garment 净化服
clean gun 去污枪,(熨烫用)喷水枪
cleaning 净化,清洁,清洗,洗涤
cleaning cloth 百洁布
cleaning gun 去污喷枪
cleaning in cleanroom (CIC) 无尘衣超级净化处理
cleaning naphtha 去渍油
cleaning table 清除衣物污垢台
clean L/C 光票信用证
clean-limbed figure 四肢匀称体型
clean look 干净朴素风貌,清洁款式,整洁风貌
cleanness clothing 无尘服
clean print 轮廓光洁印花花纹
clean room 净化室
clean room garment 净化室服装
clean run 平滑,圆顺
clean shave 胡须剃净
cleansing cream 洁面膏
clean stitch seam 光缝,整洁缝
clean width 净宽
clear 透明
clearance 间隙
clearance of inventory 出口存货
clearance on shoulder 斜肩间隙

clear blue 鲜蓝
clear cerulean 鲜青
clear color 清色,透明色
clear-cut 轮廓清晰,轮廓鲜明
clear-cut finish 剪净整理,光洁整理
clear design 明显花纹
cleared goods 成形针织品
clear finish 光洁整理,光面整理
clear finished worsted 缩绒剪毛精纺呢
clear foundation lace 透明细白花边
clear green 鲜绿
clearing cloth 擦布,揩布
clearing form 结算方式
clear-knee power stand 带动力L型机架(不撞膝盖)
clear muslin 细薄平布
clear orange 浅橘黄,浅橙
clear plastic ruler 透明尺;透明方格塑料尺
clear starch 无色上浆
clear white 纯白色
clear woolen finish 毛呢光面整理
cleat 防滑钉;防滑革片,防滑金属片,防滑栓
cleats (装有防滑钉的)防滑鞋
cleavage 露胸服装;妇女上衣胸槽;乳沟,胸槽
cleavage area 胸槽区
clematis 铁线莲紫(红紫色)
clematis blue 铁线莲蓝色
Cleopatra 克娄巴特拉蓝(鲜蓝色)
Cleopatra style 克娄巴特拉款式
Cleopatra stylistic form 克娄巴特拉款式
clerical cape 牧师披风,牧师披肩
clerical collar 牧师领,短竖领
clerical dress 牧师礼服
clerical frock 牧师礼服;神职人员衬衫前身
clerical garment 牧师服,僧侣服装
clerical gown 牧师礼服,僧侣长袍
clerical grey 深灰色
clerical robe 牧师罩袍
clericals 牧师服装
clerical shirt 牧师衬衫,教士衬衫
clerical vest 牧师背心
cleric shirt 牧师衬衫,教士衬衫
cleveland cotton 美国克利夫兰棉
clevyl 聚氯乙烯纤维(商名,法国罗纳-普朗克)
clew 线团;缚牢吊床的绳

cliche 手工印花模板
clicker 针织品冲裁机
clicker press machine 冲裁机
climate control fabric 可调节[人体]环境温湿度的织物
climatic chamber 模拟气候室,空调箱,人工气候室,人工环境室
climatized garment 调温服
climbing breeches 登山裤
climbing harness 胸腰保护带
climbing pants 登山裤
climbing shoes 登山鞋
climbing shorts 登山短裤(多口袋)
clinchamps 大麻粗厚帆布
cling 服装紧贴感
clinginess 服装紧贴感,紧贴感
clinging clothes 紧身服装
clinging property 缠合性,贴身性,紧身性
clinging skirt 贴体裙装
cling property 衣服贴身性,衣服紧身性,衣服缠合性
cling-test 织物紧贴性试验;静电吸附试验
clingy 缠身衣,紧身衣
clingy look 紧身款式,紧贴款式
clingy style 紧身款式
CLINIQUE 倩碧
clinquant 金编带,金丝编带
clip 夹子;回形针(曲别针);剪,修剪;别针,首饰别针;(骑自行车时用的)裤腿夹;布屑;发夹
clip-back earring 背夹式耳环
clip brooch 装饰物用别针
clip closing 金属搭扣门襟
clip cord handkerchief 剪花手帕
clip cutted pieces 修剪裁片
clip earring 夹式耳环
clip-front knit blouse 夹扣胸襟针织上衣
clip hat 夹式帽
clip off 剪掉
clip-on bow tie 夹扣蝴蝶领结,夹式蝶结
clip-on earring 夹式耳环
clip-on rigid plastic shield 硬塑料夹式眼睛罩
clip-on sunglasses 夹式太阳眼镜
clip-on tie 夹式领带,夹式领结
clip-on wire mesh shield 夹紧线网眼罩
clipped dot voile 点纹巴里纱
clipped figure 浮点花纹

clipped pattern 剪贴花纹
clipped spot 浮点花纹
clipped spot gingham 剪花格子布
clipper hooks 卡子,夹扣
clippers 剪刀
clipper seam 卡子接缝,夹扣接缝
clipper technique 轧剪技术
clip pin 对夹式别针
clipping and carving 地毯剪花,剪花
clippings 低级羊毛;零头碎呢,零头
clipping underarm seam allowance 修剪袖下缝缝头
clips 剪花带
clip spot lawn 高级小点子剪花细平布
clip spots 浮点花纹
clip spots fabric 浮点花纹织物
clip thread residue 剪线头
clip underarm's seam allowance 抬根缝剪口
clisse [法]杯套;瓶套
clo 克罗,克罗值
cloak 披风,斗篷,大氅,无袖外套;大衣(总称)
cloak-and-suiter 现成服装店
cloak bag breeches 斗篷袋式裤
cloak cord 斗篷系带
cloaking 外套料(粗梳毛织物),大衣料
cloakroom 衣帽间,寄存衣物处
clobber 衣服(英国俗称)
cloche (圆顶狭边)钟形女帽;吊钟帽
cloche hat 钟形女帽
cloche with feathers 羽饰钟形女帽
clock 袜统刺绣;网眼花纹,绣花花纹,袜子上的装饰花纹;吊线边花袜子
clock-cover line 钟罩型
clocked hose 吊线边花袜子
clock for check attendance 考勤钟
clock for work and rest 作息钟
clock hose 吊线边花袜子
clock installation 子母钟
clock pattern 绣花花纹
clocks and watches 钟表
Clodagh O'kennedy 克洛达·奥肯尼迪(爱尔兰时装设计师)
clogs 屐,木屐,木鞋,木套鞋(木底鞋);木拖鞋;松糕鞋
clogué 泡泡组织织物;泡泡纱布
cloisonné 景泰蓝
cloisonné necklace 景泰蓝项链
cloister cloth 厚窗帘布

cloister fabric 轻薄织物
cloky 泡泡点纹;泡泡组织织物
cloky organdy 泡泡蝉翼绸
clolured goods 色布
clomino design 多米诺印花
cloque 泡泡点纹;泡泡组织织物
cloqué [法]泡泡点纹
cloqué effect 泡泡纱效应
cloqué fabric 凸纹(浮线)针织物
cloque organdy 泡泡蝉翼绉(经耐久坚硬化处理,部分透明)
close 关闭;缝合
close centre seam of collar interlining 缝合领衬
close centre seam of top collar 缝合领面
close clipping 紧贴剪毛
close coat 合身上身,合身上衣,有纽扣的上衣
close coat
close collar jacket 关门领
close crotch 缝合裤裆
closed and zipper 双封尾型拉链
closed basket stitch 竹篮十字绣
closed buttonhole stitch 密锁眼绣
closed cam lace 封闭式镶条花边
closed clothes 无缝衣
closed collar 立领,关门领
closed collar jacket 立领夹克
closed cretan stitch 细密克里特针法,密克里特绣
closed culet 小底尖(钻石)
closed dart 尖省
closed door discount house 批发俱乐部
closed end fastener 闭口式拉链,双封尾型拉链
closed end zipper 密尾拉链(尾部封死),双封尾型拉链
closed feather stitch 密羽状绣,密羽毛绣
closed garment 无缝服装,无缝衣服
closed herringbone stitch 泡绣
closed principle 封闭性原则
closed seam 包缝,封头缝,暗缝
closed sleeve 封闭式袖
closed stitch 密针脚
closed toe 封闭袜头,无缝袜头
closed toe knitting 无缝袜头的缝织,缝袜头系统
close fabric 厚重织物
close face fabric 呢面织物
close face finish 密实呢面整理

close fitting 紧贴,贴体;紧身尺码
close-fitting garment 紧身服
close-fitting gown 紧身服
close herringbone stitch 密人字缝
close hood seam 缝合风帽缝,合帽缝
close joint 合缝
close-knit 细网眼针织物
close leaf 编带平纹针迹
close leaf motifs 密叶花纹
close leaf stitch 填叶线迹,填叶针迹,闭圈式线迹
closely bound goods 紧密织物
closely woven fabric 紧密织物
close-mesh 细网眼
close-open (C.O.) 关-开,合-断,启-闭
close saddle skirt 女用骑马裙裤
closes bids 密封递价
close side and sleeve 缝合侧边和袖底
close sleeve 缝袖子
close stitch 密缝;密缝线迹,密针脚
closet case 壁橱衣架
close texture 质地紧密
close the centre seam of collar interlining 合领衬
close the centre seam of top collar 合领面
close weave 紧密织物
close zippers 封住拉链洗涤
closing 扣子,缝合;腰头门襟;填空整理,硬挺上浆整理
closing line 止口线
closings 封闭服装开口,门襟
closing thread 缝线
closing wire 闭口针
closure 服装闭合体(拉链、纽扣、搭扣等总称),服装扣合件
closure of trousers 腰头门襟
closures 封套物,封闭服装开口,扣合物
closure seam 封头缝,封闭缝
clot 劳动鞋
cloth 布,织物,布料,呢绒,衣料
cloth and silk 布帛(棉布和丝绸)
cloth appliqué 贴花缝绣织物
cloth band 下领
cloth belt 布腰带;成卷状料子
cloth blanket 平纹薄绒毯
cloth-both-sides 夹胶布
cloth braid (饰边用)窄花边,狭花边
cloth buyer 衣料商
cloth cap 布帽
cloth clip 布夹,裁布夹(固定布片),裁

片夹
cloth coat 布大衣
cloth construction 织物结构
cloth constructor 织物花纹设计师,织物设计师
cloth contraction 织物收缩,织物缩率
cloth conveyor system 送布系统
cloth cover 布面;熨烫罩布;织物丰满
cloth cutter 裁布机,裁布刀,裁剪机
cloth cutting machine 裁衣机
cloth defect 织物疵点
cloth design 织物设计
cloth drill 钻布针;定位针
clothe 穿衣
clothed net 机织渔网
cloth embroidery 百纳刺绣,丝线嵌花刺绣,金银线嵌花刺绣
clothes 服装,衣着,衣服,衣,衣裳;(总称)被褥
clothes and hat 衣冠
clothes and ornaments of every dynasty 历代服饰
clothes bag 换洗衣服袋
clothes basket 换洗衣服篮
clothes brush 衣刷
clothes-conscious 衣着意识
clothes drawing 服装效果图
clothes dryer 干衣机,烘衣机
clothes for surgical utilization 外科用服装
clothes hanger 衣架,挂衣钩
clothes hook 挂衣钩
clothes horse 晒衣架;讲究穿着的人
clothes keeping 服装保养
clothes length line 衣长线
clothes line 晾衣绳,晒衣绳
clothes machine 服装机械
clothes man 旧衣商
clothes mender 补衣工人
clothes moth 蠹虫,衣服蛀虫,衣鱼,衣娥
clothes parts 衣片
clothes peg 晾衣架,晒衣夹,晒衣架(英国)
clothes pin 木衣架,晒衣夹,衣夹
clothes press 衣柜,衣橱
clothes prop 晒衣绳支架
clothes repair 补衣
clothes shop 服装店
clothes sizes 服装尺寸;服装号型
clothes stand 衣帽架,衣架
clothes tree (柱式)衣帽架

clothes-wadding 填絮,棉絮,服装垫料
clothes yard 晒衣场
cloth examination 验布,织物检验
cloth examiner 验布工;验布员
cloth fastener 布搭扣
cloth feeder 给布装置,送布器
cloth feeding 送布;送料
cloth finish 呢面整理;人造丝织物起绒整理
cloth folding 折布机
cloth folding machine 折布机
cloth gard 布码
cloth hat 布帽,轻便帽
cloth holder 布架,衣架
clothier 布料商,服装商,布匹商;织布者;制衣者;裁缝
clothing (总称)服装,衣服,衣裳;服饰,被服;制衣
clothing accessories 衣饰
clothing advertisements 时装广告
clothing articles 衣物
clothing basic theory 服装基础理论
clothing behaviour 衣着行为
clothing botton 衣摆
clothing cabinet 衣柜,衣橱
clothing care 服装保养
clothing comfort 服装舒适性,衣着舒适性
clothing comment 服装评论
clothing components 服饰配件
clothing computer aided design 服装计算机辅助设计(服装 CAD)
clothing computer aided manufacture 服装计算机辅助制造(服装 CAM)
clothing construction 服装结构
clothing counter 服装柜
clothing course 服装课程
clothing cues 服装暗示
clothing curriculum 服装课程
clothing department 服装部
clothing design 服装设计;服饰图案
clothing designer 服装设计师
clothing design system 服装款式设计系统
clothing expenditure 服装耗料
clothing factory 服装厂,被服厂
clothing for cold weather 防寒服
clothing fork 衣叉
clothing form 服装模型,服装外形轮廓
clothing headgear and footwear 服装鞋

帽;衣着
clothing heat-moisture transfer 服装湿热传递
clothing hem 衣摆
clothing higher education 服装高等教育
clothing hygiene and safety property 服装卫生安全性能
clothing hygienist 服装卫生学家
clothing illustration 服装画
clothing industry 服装工业,衣着工业
clothing industry training authority 制衣业训练局,制衣培训班
clothing information 服装信息
clothing institute 服装研究所
clothing insulation 衣服保暖性;衣服绝缘性
clothing insulation value 服装隔热值
clothing label 服装标识
clothing leather 服装革,衣着皮革
clothing length 衣长
clothing looking machine 看布机
clothing machinery 服装机械
clothing machinery & tools 服装缝纫设备与工具
clothing manufacture 成衣制作
Clothing Manufacturers Association of the USA 美国制衣者协会
clothing mark 服装标识
clothing materials 服装材料,衣料
clothing micro-climate volume 服装小气候容量
clothing mirror 穿衣镜
clothing operation 服装制作
clothing ornament 服饰
clothing personality 服装个性
clothing personalization 服装个性化
clothing physiologist 服装生理学家
clothing physiology 服装生理学,衣着生理学
clothingpress 衣柜
clothing pressure 服装压力
clothing principle 服装原理
clothing program 服装课程
clothing psychology 服装心理学
clothing quota 服装配额
clothing relic 服装文物
clothing sanitation 衣服卫生
clothing science 服装科学
clothing shop 服装商店
clothing socialization 服装社会化

clothing social psychology 服装社会心理学
clothing stand 衣帽架
clothing store 服装店
clothing subject 服装学科
clothing syllabus 服装教学大纲
clothing symbolism 服装象征意义
clothing technological operation 服装制作与工艺
clothing technology 服装工艺
clothing technology demonstration centre 服装工艺示范中心
clothing test 衣着试验
clothing textile 服装纺织品,衣着用纺织品
clothing theme 服装主题
clothing thermal insulating value 服装隔热值
clothing toxin 衣橱毒素,衣服毒素(衣服上含有游离甲醛等毒素)
clothing twill 斜布
clothing value 克罗值(纺织品和服装隔热性的国际通用单位)
clothing waist 衣腰
clothing wool 粗梳毛
cloth in rope form 绳状织物,绳状布
cloth insertion sheet 双面胶布
cloth inspecting machine 验布机
cloth inspecting table 看布台
cloth inspection 验布
cloth laying machine 摊布机
cloth-lined paper 衬布纸
cloth loop 布环
cloth marker 表层布上划样;漏画;画皮
cloth measure 布尺
cloth measuring and cutting machine 量裁(布)机
cloth measuring machine 量布机
cloth mechanics 织物力学
cloth moths 衣服蛀虫
cloth of acca 金蓝色闪光绸
cloth of areste 中世纪金线花纹织物
cloth of bruges 中世纪英国金丝花缎
cloth of cologne 中世纪德国金花绸
cloth of gold[silver] 金[银]线织物
cloth of gold 金线织物
cloth of gold tissue 金地绸
cloth of pall 中世纪梅红马甲绸
cloth of raynes 中世纪细亚麻平布
cloth of silver 银线织物

cloth of tars 丝绸;丝毛绸
clothology 服装学
cloth pattern cutting machine 切布样机
cloth plate 缝制台
cloth plate complete set 整套缝台
cloth plate extension 缝制台加长板
cloth pore 织物孔隙
cloth presser foot 送料压脚
cloth printed cretonne 大型花印花装饰布
cloth proofing 布长涂胶
cloth puller device 滚轮送料装置
cloth quality 织物密度;织物品级,织物质量,织物品质
cloth-rack 衣帽架,衣架
cloth roll 布卷
cloth rolling and inspecting machine 卷布验布机
cloth rolling machine 卷布机
cloth sample 布样
cloth sandals 布凉鞋
cloth scissor 布剪
cloth scouring 织物洗涤
2-cloth seaming machine 缝两层布的缝纫机
cloth sewing 缝
cloth shank 布扣柄
cloth shoes 布鞋
cloth slippers 布拖鞋
cloth slitting machine 切布机
cloth sole 布(鞋)底
cloth specimen 布样
cloth spreader 铺料机
cloth spreading 拉布铺料,铺布,平布,扩布
cloth spreading machine 拉布机,坯布扩幅机,平布机
cloth stockings 布袜
cloth stop line 晒衣绳
cloth tape[rule] 布卷尺
cloth tape 布卷尺
cloth tape rule 布卷尺
cloth temperature probe 布温探头
cloth thickness gauge 织物测厚规,织物厚度仪
cloth-top boots 布顶靴(1920～1930年)
cloth trunk 堆布车
cloth turning 翻布
cloth turning machine 翻布机
cloth umbrella 布伞
cloth warehouse 布库

cloth waste 碎布
cloth weight 织物重量
cloth winding inspection machine 卷布验布机
cloth winding machine 卷布机
cloth with geometrical figure 印有几何图案的布料
cloth-wrapper 包袱
clothy 手感有身骨
clothy appearance 布面纹路清晰
cloth yard 布码,量布长度
clothy feel 轻软手感
clothy hand 布质坚实、柔软而有身骨
clotted soap 块皂,凝块皂,皂块
clotted wool 花式粗纺呢,膨松毛呢
cloud 污迹,污点,云斑疵;宽松女披肩衣;轻柔女围巾
cloud band design 云带花纹
cloud blue 蓝云色
cloud collar 云肩
cloud cream 云白色
cloud design 云彩图案
clouded ribbon 波纹条带
cloud grey 云灰色
cloudiness 云状花纹;无光泽
cloud pink 淡藕红色,淡妃色
cloud-scrolls design 云雷纹
cloud-shaped rule 云尺(云状弧线尺)
cloud-yarn 多色云彩花色丝,竹节花线,云彩花线
cloudy dyeing 云斑染疵
cloudy goods 云斑织疵;云斑织物
cloudy patch 云斑疵,云点
cloudy patterned suiting 云纹呢
cloudy patterned suiting silk 云纹呢
cloudy print 云斑印疵
clout 零布,余料,破布;给(衣服)打补丁;给(鞋底)钉铁掌
clouties 亚麻布(英国制),英国亚麻布
clouty 亚麻织物;用布制作
clo value 克罗值(纺织品隔热单位)
clove 丁香色,红灰色
clover 三叶草色
clover collar 苜蓿叶形西装领
clover darning stitch 苜蓿缀纹绣
clover green 苜蓿叶绿
clover lapel 苜蓿叶形翻领,圆驳领
clover leaf knot 苜蓿草结
clover leaf lapel 苜蓿叶形翻领,苜蓿叶形下领片

clover leaf lapel collar 苜蓿叶形西装领
clover pink 苜蓿花紫
clover revers 苜蓿叶形西装领,苜蓿叶形翻领
clown hat 小丑帽
clowning around effect 小丑诙谐效果
clown look 小丑款式
clown pantaloons 丑角裤子,小丑裤
clown style 小丑款式
clown suit 小丑服,小丑套装
club bow 棒状蝶领结
club bow tie 俱乐部蝶领结,平结领结
club check 两色小格子花纹,双色小格子花纹
club cutting 直剪(美容)
club foot 畸形足
club hand 畸形手
club stripe 多色直条花纹(用于领巾,小外套等),俱乐部条纹,克拉勃(俱乐部)多色条纹
club tie 多色斜条丝领带
clump 特厚鞋底
Cluny guipure 克伦尼网眼花边,克伦尼亚麻粗梭结花边,手工编结克伦尼网眼花边
Cluny lace 克吕尼花边,克伦尼粗机织花边,有凸纹的亚麻机织花边,克伦尼粗梭结花边
cluny tapestry 英国毛经丝纬挂毯
cluster earring 串珠耳环
cluster fringe 锯齿形流苏
cluster gather 集中式碎褶
cluster pleated skirt 群褶裙
cluster pleats 成群褶裥,丛状褶
cluster ring 大小宝石串珠戒指
cluster stripe 丛状条纹,束条纹
cluster tucks 成群缝褶,簇状活褶
clutch 离合器;无带女提包,无提手的夹包,女用无带提包
clutch bag 女用无带提包,手握公务包,手握包
clutch coat 裹襟式大衣,围裹式大衣
clutch purse 手握包,手握提包,女用无带提包
clydella 克莱德拉棉毛混纺呢
CMYK color CMYK 模式
coach coat 观赛外套
coach lace 装饰车辆用窄花边
coachman coat 马车夫大衣,收腰长大衣
coachman hat 马车夫帽

coachman's cloth 密绒厚呢
coachman's coat 车夫大衣,车夫外套,马车夫外套
coachman's frock 马车夫福乐克
coagulated fiber 凝固纤维
coal black 煤黑
coal scuttle bonnet 煤筐女帽,煤桶形罩帽
coarse adjustment 粗调
coarse bobs 粗棉线大网眼窗纱
coarse calico 粗棉布,粗平布
coarse cloth 粗布,粗织物
coarse-cord 粗凸纹
coarse cut 低机号
coarse denier 粗旦
coarse drill 粗斜纹
coarse fiber 粗纤维
coarse gauge 粗隔距,粗针距;粗号,低机号
coarse gauge knit 粗针距针织物
coarse hand knitting yarn 粗绒线
coarse knit 粗针距针织物
coarse pick 粗纬疵
coarse plain cloth 粗(平)布
coarse pulp felt 抄浆毛毯
coarse sheeting 粗布,白粗布,粗平布
coarse stitch 疏缝,疏针脚
coarse thread 粗线
coarse touch 粗糙感
coarse under fleece 粗绒毛
coarse wool 粗梳毛,粗羊毛
coarse yarn 粗支纱
coarse yarn fabric 粗支纱织物,特粗织物
coast ivy look[style] 西海岸常春藤款式
coast ship insurance 沿海船舶保险
coat 袄;大衣;男式上衣,外套,妇女、孩童短大衣;(动物)皮毛;衫;女西装;衬裙
coat-and-jacket press board 大衣和外套烫衣板
coat and pants (C. & P.) 男式上衣和短裤
coat and skirt (一套)上衣和裙子,妇女衣裙套装;妇女外出服装
coat armor 纹章
coat belt 大衣腰带
coat button 大衣纽扣,衣衫扣
coat clothing 外套式服装
coat collar 大衣领
coat-core fiber 皮芯纤维

coat cuff　外套袖口,外翻袖口,外套袖头
coat dress　外套,大衣(宽腰,直筒型);外套式连衣裙;(从领口到下摆有一排扣子的)女式紧式外衣
coated cloth　涂层织物;涂层布料
coated collar　胶领
coated diamond　涂层钻石
coated fabric　上胶织物,涂层织物,涂层衣料
coated finished fabric　涂层整理织物
coated glass fabric　涂层玻璃纤维织物
coated gloves　涂塑手套
coated nylon jacket　涂层尼龙雨衣
coated nylon palace　涂层尼丝纺
coatee　紧身短上衣(妇女、孩童用)
coat ensemble　女式大衣套装
coat fabric　上衣面料
coat front　上衣前襟,上衣前身;上衣前胸硬衬
coat front padding　上衣前身垫衬
coat hanger　衣架,挂衣钩,衣帽架
coating　上衣料,大衣料;(织物)涂层;上胶
coating alpaca　羊驼毛上衣料
coating cloth　外套衣料
coating finishing　涂层整理
coating lace　上衣绣花花边
coatings　涂层产品
coating velvet　上衣丝绒,外衣丝绒,棉背厚丝绒
coat jumper　无袖单(或双)排纽连裙装
coat length　上衣长度,衣长
coatless　没有(或不穿)外衣的
coatless fashion　不穿外衣式样
coat lining　上衣衬;大衣里
coat matching　大衣套装;外套配套
coat of arms　饰有王徽的传令官制服,(穿在铠甲外的)饰有纹章的外套
coat on coat　外层大衣(指大衣、斗篷等),大外套
coat on sweater　卡迪根加外套
coat pocket　大衣袋,衣袋
coat rack　衣帽架
coat room　衣帽间
coat set　儿童外套组合
coat shirt　外套衬衫(无钮长衬衫),西服衬衫
coat style pajamas　外套型睡衣
coat style sweater　外套式毛衫;外套式毛衣
coat suit　外套套装(外套与西裤同料)
coat sweater　外套式毛衣(长V领,有扣开襟式,有或无罗纹下摆)
coattail　(男上衣)后摆;燕尾服下翼;女长外衣下摆
coat tail　男上衣后摆,女长外衣下摆,(燕尾服的)燕尾
coat tree　衣帽架,柱式衣架
cobalt blue　钴蓝色
cobalt blue king's blue　钴蓝
cobalt green　钴绿色
cobalt ultramarine　钴绀青色
cobalt violet　钴紫色
cobalt violet deep　深钴紫色
cobalt violet pale　浅钴紫色
cobalt yellow　钴黄色
cobbler　退修的染整织物;鞋匠
cobbler's apron　鞋匠围裙
cobbler's twine　制鞋麻线
cobbler's wax　(制鞋用)线蜡,鞋线蜡
cobcab　科布开凉鞋
Cobourg, coburg　科伯呢,科伯斜纹呢,棉毛交织仿开士米斜纹呢
cobweb cotton　密西西比晚熟棉
cobweb lawn　薄亚麻织物
cobweb skirt　蛛网裙
coc　线绠
cocarde　帽上的花结,线带花结;帽章
coccyx　骶骨
cochineal　虫红,胭脂红
cochineal coral handkerchief　红色丝手帕,红色丝围巾
cochran cotton　美国科克伦棉
cock　翘帽檐
cockade　帽徽,帽上的花结,线带花结,帽章,帽花结
cocked hat　三角帽,(海军用)卷边帽,两端尖的帽子(二角帽),折叠三角帽(18世纪款式)
cocked shape　二角帽,卷边帽
cocked suit　二角帽
cocker boots　科克靴
cockers　半高筒靴;绑腿
cocking　成衣检验
cockled bar　皱条痕,皱状条痕
cockled edge　皱边
cockled effect　褶裥效果
cockled fabric　起皱织物
cockle hat　贝壳帽
cockliness　织物起皱,纱线起皱

cockling 皱面外观,织物起皱;抽褶接缝;纱线皱缩
cockscomb 鸡冠花型
cockscomb red 鸡冠红
cockscrew curl 鸡冠花型
cocktail 鸡尾酒会服,鸡尾羽
cocktail apron 装饰小围裙
cocktail bag 鸡尾酒会包,鸡尾酒会手提包
cocktail coat 燕尾服;晚会外套,晚会大衣
cocktail dress 燕尾服,晚礼服,女式常用礼服,鸡尾酒会服
cocktail hat 鸡尾酒会帽,晚礼服帽
cocktail ring 鸡尾戒指,鸡尾酒会戒指,晚宴戒指
cocktail shoes 晚礼服鞋
cocktail skirt 鸡尾裙
cocktail suit 鸡尾酒会套装,酒会礼服,燕尾服
cocktail umbrella 鸡尾酒会伞,鸡尾伞
coco 科科,椰子纤维
cocoa 可可色,黄棕色
cocoa brown 可可棕色(黄棕色)
cocoa matting 椰壳纤维席纹布
cocoanut fiber 椰子绒,椰子纤维
cocoa sacks 可可袋
Coco Chanel 可可·夏奈尔(1883~1971,法国时装设计师)
coconada cotton 印度科科拿大棉
coconada hemp 菽麻
coconada rug 印度科科纳达地毯,马德拉斯地毯
coconada stripes 印度红色棉平布
coconut fiber 椰子绒,椰壳纤维
coconut shell button 椰壳扣
cocoon 蚕茧色;茧形装(宽肩,蝙蝠袖,立领),及膝宽松外套
cocoon coat 茧形大衣,茧形外套
cocoon dress 茧形女装
cocoon jacket 茧形夹克
cocoon silhouette 蚕茧轮廓
cocoon stripping spun yarn 茧衣绢丝
cocus 绑腿;粗制高筒靴;渔民靴
coddle-free carpet 不需保养的地毯
codovec 海狸皮帽
cod piece (15~16世纪)男裤前褶裥,男紧身裤下体盖片(15~16世纪),阴囊护片(15~16世纪)
cod placket 裤前扣合件

codrington 男式宽松外套
coed dressing 两性装,女学生装(男女同校)
coed hairdo 女学生发式(男女同校)
coeur fleuri 法国漂白小几何花纹被褥麻布
coffee 咖啡色
coffee bean print 咖啡豆印花花样
coffee-brown 咖啡棕
coffer headdress 女式箱形小头巾
coffin cloth 寿衣呢
coffo suede 绵羔皮
cofiber 共黏合纤维,相互黏合纤维
coggers 绑腿
cognac 科涅克白兰地酒色
cognitive 认知
cohras prints 科勒斯蜡防印花布
coif 考福帽,贴头帽,压发帽,衬帽;(尼姑或农妇用)头巾;短发,修饰头发
coiffitte 科费特帽(14世纪)
coiffure 头巾;女便帽;头饰;妇女花边头饰;妇女发型
coiled choker 螺旋项链
coiled plaits 盘绕的发辫
coiling soup 亮油影条(鞋)
coil wire boning 重绕螺旋羽骨
coil zipper 环扣拉链
coimbatore cotton 萨勒姆棉
coin 袜子上的装饰花纹
coin bracelet 金币手链镯,硬币手镯
coin de feu 室内穿高领短外套
coin dot 大圆点花纹,硬币花纹
coin knot 金钱结
coin laundry 自动洗衣店
coin necklace 古币项链
coin-op cleaner 投币自动干洗机
coin-operated self-service drycleaning machine 投币自动干洗机
coin-op laundry 投币自动洗涤
coin-puff 双面织物
coin purse 零钱包
coin setting 金币镶嵌座
coin shoes 懒汉鞋,硬币鞋,硬布鞋
cointise 头盔垂饰;剪花头饰
coin-vertible 双面织物
coir 椰壳纤维
coir fabric 椰壳纤维织物
coir fiber 椰壳纤维
coir mat 椰壳纤维垫

coir matting 椰壳纤维席纹布
coir raincoat 蓑衣
coke 科克圆顶礼帽
coker cotton 美国东南陆地棉
cokers 绑腿；粗制高筒靴；渔民靴
colax 科拉克斯（吸水性聚丙烯腈纤维，商名，日本）
colbertan 方形网地花边
colbertan lace 法国机织花边
colberteen lace 法国机织花边
colbert embroidery 柯尔贝尔刺绣,库尔伯特彩色刺绣,绣地针绣
colbertine 方形网地花边
colbertine lace 法国机织花边
col chale 披肩领
col claudine 小圆领
colcothar 英国红,氧化铁红
cold 寒冷威
cold calendering 冷轧光
cold color 寒色,冷色（蓝、绿等）
cold cream 冷霜,润肤膏
cold-embossing 冷轧纹
cold pad-batch dyeing 冷轧堆染色
cold pad-batch dyeing process 冷轧堆染色法
cold permanent 冷定型（美发）
cold proof coat 防寒服
cold proof material 防寒材料
cold shortness 冷脆性
cold sweat 冷汗
cold tone 寒调,冷色调
cold water shrinkage 冷水收缩率,呢绒冷水预缩
cold wave 冷烫,化学烫
cold weather fabric 冷天织物
cold weather gear 防寒服
cold weather permeable cap 冷天透气帽
cold white fluorescence 冷白荧光
coleraines 优质半漂白亚麻布
coleta 泽麻,槿麻
colid 可立特（苎麻、涤纶和黏胶丝混纺色织布）
collage effect 拼贴效果,粘贴画效果
collagen fiber 骨胶原纤维
collant 紧身衣,紧身裤,考兰紧身裤,紧身舞服
collapsible modelform 可拆卸的人体模型,可折叠的人体模型
collapsible umbrella 折叠伞
collar 衣领,硬领,假领；领饰；项饰；领圈,颈圈；鞋口缘饰

collar and cuff set 女式可拆卸领和袖口
collar and sleeve 领袖部位
collar attached shirt 装领衬衫（指不结领带衬衫或极正统衬衫）
collar band 领脚,领角；领座,底领,衬衫衣领的下盘
collar band end line 底领前宽斜线
collar banding sewing machine 滚领机
collar band inside line 底领上口线
collar band interfacing 下领衬,下领夹里
collar band is longer[shorter] than collar 底领伸出[缩进]
collar band lean out of collar 底领外露
collar band lining 下领衬；下领夹里
collar band outside line 底领下口线
collar band tab 领舌
collar band too long 底领挥出
collar band too short 底领缩进
collar binding sewing machine 滚领机
collar blocking machine 热压领机,压领机
collar bone （使衬衫领子保持硬挺的）插骨片,领衬,锁骨插骨片
collar bootee 毛口
collar box 衣领盒
collar button 领扣
collar buttoner 衣领钉扣器
collar circumference 领围,领长
collar clasp 领钩
collar cloth 衣领布,半硬领布
collar contour trimmer 修领器,修领脚机
collar controller 挖领机
collar corner 领角
collar creasing press with heat and cool head 冷热双头领型机
collar-cuff-flap former 领、袖口、袋盖熨烫机
collar detail 衣领片
collar deviates from front centre line 绱领偏斜
collar edge 领边,领外口
collar edge appears loose 领外口松
collar edge appears tight 领外口紧
collaret 滚边；针织窄花边领圈；女用衣领,装卸式领；女用披肩,褶边立领(16世纪)
collaret'collarette 女用领饰；披肩；围巾
collarette 滚边；针织窄花边领圈；女用衣领,装卸式领,褶边立领(16世纪)；女

用披肩
collarette guide 导带器,缝纫机的导带器
collarette tape 滚带;领圈
collarette with décolleté 18世纪妇女低胸饰领
collar face 领面
collar facing 领面
collar fall 衣领翻下部分
collar felt cloth 领底呢衬里
collar fly 衬衫领口蝴蝶片,领口蝴蝶插片,衣领蝴蝶片
collar fold line 领上口
collar former 领型机,烫领机
collar forming 拔领脚
collar guard 护领
collar hand 领座宽
collar heat notcher 电热点领机
collar height 后领高,领高
collar hook 风纪扣
collaring 毛口
collar inlay 领插角片,领口片,衣领插骨片
collar interfacing 上领衬
collar interlining 领衬
collar keeper 领托,领条,支领
collar lace 衣领用花边
collar leaf 领角衬,领口片,领头片
collar leaf lapel 苜蓿叶形下领片
collar length 领长,领围,领大
collarless blouse 无领衬衫,无领女衬衫
collarless collar 无领领型,无面领的领型(保留底领)
collarless high button style pajamas 无领高襟式宽大睡衣
collarless jacket 无领夹克衫,无领短外衣
collarless robe 无领袍
collarless short-sleeved dress 无领短袖连衣裙
collar lining 领里,领衬,领夹里
collar marking machine 点领机
collar melton 领夹里粗呢,衬领呢
collar modeling 领型,领下口
collar neckline 领下口
collar notch 领豁口
collar on collar 超重衣款式
collar opening 领口,领口大小
collar outer edge too loose 领外口松
collar outer edge too tight 领外口紧
collar outside line 翻领外口线
collar patch 领章

collar pin 领针
collar point 领尖,领尖点线,领角
collar point length 领尖长
collar point line 领尖线
collar point trimming machine 切领嘴机
collar point turning machine 翻领(角)机
collar point width 领尖宽
collar press 领子压平机
collar presser 压领机
collar press pad 烫领馒头
collar riding up 爬领疵
collar roll allowance 翻领座势;领子翻折所需的放松量
collar roll line 翻领上口线
collar run stitcher 衣领初缝机
collar run stitch machine 领圈缝纫机
collar scoring set 衣领划线装置
collar seam 领缝,领缝线
collar separate shirt 活领衬衫,脱卸领衬衫
collar shape 领型
collar shape design 领型设计
collar shaping device 领子成形器
collar shaping press 领型机
collar slip 长礼服背心的白色添加领
collar spread 两领尖间距离,两领尖距离,领尖距离
collar stand 领高,领脚,领座,领脚长
collar stand away from neck 领离脖(领子不贴脖)
collar stay 领座,领角,底领,领尖撑,衬衫领插骨片,领口薄膜,领插角片,领口片
collar stiffener 衣领硬挺剂
collar stop 领止点
collar strap 领带袢,领条,领托
collar stud 领扣,领纽
collar style 领型
collar style knit shirt 连领式针织衬衫
collar style line 领款轮廓线,领外口线,领款造型线
collar support 领衬,撑领圈
collar supporter 领条,领托
collar tab 领袢
collar terminology 领子术语
collar top 领面
collar top line 鞋沿口
collar turner 翻领机,翻领器,翻领撞针
collar turning and pressing machine 翻领压领机

collar turning machine　翻领机
collar type　领型;领子类型
collar underline　领下口;领下口线
collar velvet　衣领绒
collar vest　西装背心,有领背心
collar width　领面宽
collar width line　翻领前宽斜线
collar width with stand　后领高
collecting rag　碎布回收
collection　服装精品;设计师留下的精品;服饰系列;新装发表会,发布会,时装系列表演,时装展览
COLLECTION FEMME CHIC　《漂亮女生》(法)
collection of the instrument　票据托收
Collections Konfektionsware Dusseldorf　杜塞尔多夫成衣展(德国)
Collections Premieres Dusseldorf　杜塞尔多夫先导博览会(德国)
college cap　大学方帽,方顶帽,学士帽,大学帽
colleger　学位帽,学士帽
college style　大学服装款式,大学生款式;校园风格
college wear　学生服
collegians　男式黑色短靴,牛津靴
collegienne style　女中学生风格
collerate　上领圈
collet　领座;小型领;(戒指)宝石镶座,戒座
collette　本色帆布
colley west on ward　斜穿侧缝开衩夹克
collier　[法]颈饰,项圈;络腮胡子
collier de fleurs　[法]花型颈饰
collie shoe　苦力鞋(无后跟)
colobium　内衬衣;束腰内衣;圣餐服
colofficier　官服领
cologne　科隆香水(德国)
colombiana　匹染厚斜纹呢
colombian pita　凤梨叶纤维
colonial look　殖民地款式
colonial nostalgia look　美国怀旧风貌(或款式)
colonial pants　美式肥短裤
colonial shoes　可乐利鞋(中跟,低鞋缘,鞋舌硬挺且延伸至脚背并饰大型扣环),殖民地鞋(17~18世纪)
colonial style　殖民地风格
colonial velour　棉麻交织绒
colonial wool　英联邦羊毛

colonial yellow　淡棕黄色
color,colour　颜色,色彩,彩色;气色,面色;颜料,染料;(色彩标记的)授带,徽章;衣帽
colorability　染色性,上染性
color accent　缀色增亮
color advance-retreat feeling　色彩进退感
color analysis display computer　(CADC)颜色分析显示计算机
color and luster　色泽
color and weave effect　色织效应
colorant　颜料
colorant mixture computer　(COMIC)配色计算机
colorant staining　沾色
color appearance system　显色系
color area　色彩区域;色块
color associate　色联想
coloration　染色法,着色法;天然色泽
color atlas　色度表,色卡本,色卡簿
color attractive feeling　色彩醒目感
color attribute　色彩属性
color balance　色彩平衡,色平衡
color balance feeling　色平衡感
color-ball pin　彩头针
color band　彩色带
color bar　色档疵,颜色条
color base printing method　色基印花法
color bleeding　串色(鞋),渗色
color blindness　色盲
color block　色块;色轮
color box　颜料盒
color brocade spun silk　彩花绢纺
color camera　彩色摄像机
color card　色卡,染色样本
color cast　色光
color change　变色,换色
color-changeable cloth　(颜色可变化的)变色布
color-changeable glasses　变色镜
color change printing　变色印花
color chart　色彩表
color check　彩格
color checked handkerchief　格子手帕
color checked printed face cloth　彩格印花方巾
color checked scarf　彩格围巾
color checked shioze　彩格纺
color chip　色点,色片
color circle　色环

color clarity 色泽鲜明度；色泽清晰度
color class 色组
color coating 彩色涂层
color code 色标，色码；色彩分类
color combination 配色，色彩的组合
color conditioning 色彩调节
color connection 色彩关系
color consistency 色泽一致性
color continuity 色彩连续性
color contrast 色彩对比，色彩反差，色对比，颜色反差
color coordinate 色彩配套
color coordination 色彩调和，色彩调配
color counterbalance 色彩平衡
color crocking 色剥落
color degradation 变色，褪色
color denim jeans 彩色牛仔裤
color design 色彩设计
color deterioration 变色，褪色
color deviation 色差
color difference 色差，色光不符
color display 彩色显示
color distortion 色彩失真
colored 色，彩色，有色
colored band 色谱带
colored belt 有色带
colored check 彩色条格布
colored checked handkerchief 格子手帕
colored check fabric 花格子织物
colored cloth 花布；染色布，色布
colored corduroy skirt 十色灯芯绒裙
colored cotton 天然有色棉，天然彩色棉
colored denim 彩色牛仔布
colored diaper 菱形格色布
colored discharge printing 色拔印花，着色拔染印花
colored dress 彩衣
colored embroidery 彩色刺绣，色线刺绣
colored fabric 有色织物
colored female rubber boots 彩色轻便靴
colored fiber 有色纤维
colored flat tape 扁花带
colored goods 花布，染色布，色布
colored greys 混色灰呢
colored hair 色发
colored heel 花色袜跟
colored jacquard knitted fabric 色织提花针织物
colored jeans 彩色牛仔裤，非蓝色牛仔裤，花色牛仔裤

colored knops 彩色结子纱
colored knob yarn 彩色结子纱
colored light 色彩亮度
colored list 有色布边
colored melton 彩色麦尔登
colored pattern 配色样板
colored pencil 彩色铅笔
colored plaiting 花色添纱
colored polyacrylic fiber 有色聚丙烯腈纤维
colored rags 杂色呢片
colored resist printing 着色防染印花
colored shirt 花衬衫，有色西装衬衫
colored silk 彩绸
colored spot 色花疵
colored spots shirt 花点衬衫
colored spun yarn 混色纺纱线
colored stockings 有色长袜
colored stripe 彩条，花色条带
colored stripe band 彩条带
colored striped 彩条
colored striped single jersey 彩横条汗布
colored thread 彩色线，染色线
colored twill 杂色斜纹棉布
colored weft 色纬
colored woven 色织
colored woven cloth 色织布
colored woven flannel 色织法兰绒，色织条格法兰绒
colored woven goods 色织物
colored yarn shirting chambray 色织细平布
color effect 彩色效应，配色效应，色效应
color effect yarn 色彩效应花色纱
Color-Eighteen 十八色
color enlarger 彩色放大机
color factor 色彩要素
color fading 褪色
color fast cloth 不褪色的布料
color fastness 染料坚牢度，染色牢度，色牢度
color fastness rating 染色牢度等级
color feeling 色彩感觉，色感，色彩感情，色彩敏感性
color field 色彩派
color field printing 色彩派绘画（大色域绘画）
color figured knitting 多花色编织
color film 彩色胶片
colorful-checked face cloth 彩格方巾

colorful hair cap 彩色假发套,现代色彩假发帽
colorful lace 彩色花边
colorful thread 彩线
color gamut 全色域,色域
color-glittering fabric 闪色织物
color glow 色彩鲜艳夺目
color gradation 色调层次
color-graded glasses 变色片眼镜
color grading 色彩分级
color graph 彩图,色图
color guide 色样
color harmony 配色协调,色彩调和,色和谐,色谐,色泽和谐
color harmony chart 颜色谐调卡
colorhold 不褪色,保色
color hue 色调
color illusion 色彩错觉
colorimetric purity 色纯度
colorimetric specification 色度规格
colorimetry 比色法,色度学
color in color pattern socks 嵌色吊线花袜,镶色绣花短袜
color in color wrap hosiery machine 嵌色吊线花袜机
color index 颜色指数
color index number of hand knitting yarn 绒线色号
coloring 着色;着色法
coloring material 色素
color intensity 色彩强度
colorist 配色师,色彩设计师,着色师,印染工作者;染发师
coloristic sensation 色彩感觉
color joy-sad feeling 色悲欢感,色彩悲欢感
color key 色标,色码
color kitchen 电子配色间
color lacking uniformity 色不匀
colorless 无色
color light-dark feeling 色彩明暗感
color light-weight feeling 色彩轻重感,色轻重感
color look 多彩款式
color matching 比色,仿色,配色,色彩搭配
color matching computer 配色电子计算机
Color Matching System（CMS） 配色系统
color measurement 测色

color mechanics 色彩力学
color meter 色泽仪
color mixing 调色,混色
color mixing system 混色系
color mode 色彩模式
color model 色彩模型
color motif 色彩基调
color move-quiet feeling 色彩动静感,色动静感
color name 色名
color No 色号
color number 色数,色号
color of ground 底色
color on color 重色配色法
color overlapping 搭色
color painting 彩色画
color paste 色糊
color patch 色渍疵
color pattern 色彩花样
color pattern card silk 黏胶丝色卡绸
color-patterned fabric 配色花纹织物
color perception 色感,色知觉
color permutation 颜色排列
color photography 彩色摄影术
color picker 颜色拾取器
color planning 配色方案,色彩设计
color powder 颜色粉
color prediction 色彩预测
color printing 彩印
color properties 颜色性质
color psychology 色彩心理学
color purity 色彩纯度,色纯度
color quality 色质
color quality control system 颜色品质控制系统
color reaction 显色反应
color registration 彩色重合
color relation 色彩关系
color rendering 颜色性
color-rendering index 颜色性指数
color reproduction 彩色再现,色重演
color-resist printing 色防印花
color response 色反应
color retention 色泽稳定性,保色性
color reversal film 彩色反转胶片
color reversion 变色
color rich-poor feeling 色彩贫富感,色贫富感
color rinse 洗涤染发剂
color rotation 色顺

color rubber boots 彩色胶靴
colors 彩带；徽章，绶带（团体标志）；彩色衣服；彩色衣料
color salt printing method 色盐印花法
color sample 色样
color scale 色标，颜色标准级别
color scales lab 颜色标度
color scheme 配色方案，色彩设计
color screen 滤色镜
color selective mirror 分色镜
color sensation 色感觉，色觉
color sense 色彩感觉，色感
color sensitivity 感色灵敏度，感色性，色感度
color separation 分色，色分离
color separation film 分色描样片
color separation system 分色系统
color shade 色光，色泽，色调
color/shade matching 配色，色彩匹配
color sheep wool 彩色羊毛
color silk 彩色蚕丝
color soft-hard feeling 色彩软硬感，色软硬感
color solid 色立体，颜色立体
color space 色空间
color specification system 色标系
color speck 色斑疵
color square coordination 色块组合
color staining 搭色，色斑，色泽，色渍
color standard 色度标准
color's three attributes 色的三属性
color stick 棒状染发剂
color stimulus 色刺激
color stockings 彩色长袜
color strength 颜色力分，颜色深度
color stretch-draw feeling 色彩伸缩感，色伸缩感
color striped crepe twill 彩条双绉（香乐绉）
color striped fabric 彩条织物
color striped jersey 彩条衫，针织彩条衫
color striped single jersey 彩横条汗布
color style 多彩款式
color substance 色彩实体
color symbol 色彩象征，色象征
color symmetry 色对称
color system 表色系，色彩系统
color temperature 色温
color terminology 颜色术语
color theory 色彩理论

color three-dimensional 颜色立体
color tolerance 色公差
color tonality 色调
color tone 色调
color trend 色彩流行趋势
color triangle 原色三角形
color unevenness 色泽不匀
color value 给色量；色彩明暗程度
color variety 色彩种类
color video screen 彩色影像荧光屏
color vision 色视觉
color warp velour 彩经绒
color wash fastness 耐洗色牢度
color wave 色波
colorway 色彩配合，色彩设计（英），配色色位
color ways 色位
color weaving 色织
color wheel 色轮，色相环，颜色旋转盘
color-woven fabric 色织物
color-woven taffeta 色织塔夫绸
color yield 给色量
colour 面色
Co Ltd. 有限公司
colter 聚丙烯腈异形截面纤维（商名，日本东邦人造丝公司）
colth folder 折布机
colthorp pride cotton 美国科托普棉
columbia cotton 美国哥伦比亚棉
column 柱
columnar heel 柱形细高跟
column dress 圆柱形女服
column line 圆柱形
column skirt 筒裙
column spacing 柱间跨距
comb 梳
combat boots 军靴，军鞋，半统军靴，战斗靴，作战靴
combat cloth 竞赛呢
combat collar 战斗领，衬衫领（领前有三条宽的装饰缝）
comb design 蜂窝图案
combed broad cloth 精梳细平布
combed cotton fabric 精梳棉织物
combed duck 精梳帆布，精梳棉帆布，薄棉帆布
combed percale 精梳密织细布
combed poplin 精梳府绸
combed yarn 精梳棉纱
combed yarn cotton white thread 精梳全

棉白线
combed yarn fabric 精梳织物
combinaison 法国连裤衬衫,上下相连的套装
combinaison short'combinshort 短裤型跳跃套装
combination 上下相连的内衣,上下相连的套装,连衫裤;糅合,组合
combination acetate yarn 醋酯合成线
combination apparel 组合服装
combination button 混合扣
combination corset cover and knicker-bockers 连衣灯笼裤;紧身衣外的连裤内衣
combination design 组合设计
combination fabric 混纺织物,混合织物
combination felt 复合毛毯
combination gloves 镶皮手套,镶拼手套
combination gusset 缝合式腋下插角布,连袖腋下插角布
combination lace 混合花边
combination layout 组合排料
combination neck and armhole facing 领口连袖窿贴边
combination neck and placket band 领连口襟
combination of technology and trade 技贸结合
combination printing 同浆印花
combination ruler and curve 组合尺
combinations 连衬裤内衣,连衬裙内衣;上下相连的服装
combination shades 拼色色泽;拼色样卡
combination shirt 连裤衬衣
combination shoes 组合鞋(不同颜色或材料制成),不同布料制成的鞋
combination shorts 连短裤装
combination stitch 组合线迹,复合线迹,反面加固线迹
combination tanning 结合鞣
combination textured yarn 组合变形丝,混合变形丝
combination twills 复合斜纹
combination type carpet 拼装式地毯
combination yarn 混色纱
combination yarn fabric 混合股线织物
combined fabric 胶合织物;夹层布
combined fiber 复合纤维
combined filament yarn 混纤丝
combined garments 套装,组合服装

combined motion 联合动作
combined package 拼件成包,拼装包装
combined piece 拼匹
combined pressing stand 组合烫台
combined rib fabric 复合罗纹布
combined shipment 混合装载
combined stitch 复合线迹,组合线迹
combined twill 复合斜纹
combined weaving and knitting fabric 科威尼特织物,机织针织联合织物
combined yarn 混纤复丝,混纤丝
combing jacket 梳妆夹克
combing wool 精梳毛
combining production with sell 产销结合
combishorts 连短裤装
combi-trend 繁复风格
combi-zigzag 复合曲折线迹
combi-zigzag stitch 复合曲折线迹
combourg ordinaire 法国粗亚麻布
comb-out 发型梳理,梳发定型
comboy 英国印花长裹裙;条格披巾;英国厚端棉布
combric 细纺
comburg ordinaire 法国粗亚麻布
combustible fiber 可燃纤维
combustion behaviour 燃烧特性
comeback wool 归宗羊毛
come clean 克姆克林防污织物(商名)
come in 当今,到成熟季节,开始流行,上市
come into vogue 开始流行
come off 褪色
come up to advanced world standard 达到世界先进水平
comformability 贴身性
comformity 一致性
comfort 舒适,(织物的)舒适性,穿着舒适,盖被(美国名称)
comfortability 舒适性,贴身性,顺应性
comfortable 舒适的;盖被(美国名称)
comfortable finishing 舒适整理
comfortable fit 服装的穿着舒适合体性,舒适合身
comfortable handling 手感舒适
comfort durability 舒适耐久性
comforter 棉被,盖被;绒织头巾,细长毛围巾
comfort fabric 舒适性织物
comfort factor 舒适因素
comfort fiber 舒适纤维(具有棉纤维般吸

湿性的聚酯短纤维,商名,美国)
comfort hose　舒适长袜
comfort of wear　穿着舒适性
comfort property　舒适性
comfort shoes　舒适低跟鞋(老人,病人用)
comfort stretch　舒适伸缩
comfort stretch apparel　舒适弹力服装
comfort stretch fabric　舒适弹力织物
comfort wear　穿着舒适
comfy　绒线围巾;镶皮针织手套;袜套
comical　古怪衣着
comical attire　怪里怪气的衣着
comingled yarn　混纤交络丝
comingling yarn　混纤交络丝
comled pique　精梳凹凸织物
comma heel　逗号跟
commander's cap　宇航帽
commercial braid　商用宣传绶带
commercial color　商用色;直接染料
commercial dye　直接染料
commercial fastness properties　商业[通用]色牢度
commercial fiber　商品纤维
commercial finish　工业整理,商品化整理
commercial firm　贸易公司
commercial garment　大量出售的服装,商品化服装
commercial garment design　成衣设计
commercial illustrator　商业画家
commercialized dye　商品化染料
commercial laundering　商业洗涤
commercial matching　容差配色
commercial moisture content　公定含水率
commercial pattern　裁片样板,商业样板,商业纸样
commercial pleating　打褶布料,百褶布料
commercial print　广告花样,广告花样印花
commercial printing　直接印花
commercial shrinkage　商业公定缩水率
commercial trade mark　商标
commericial print　商业印花花样
commission　佣金
commission agency　佣金代理
commission agent　代理商
commission business　委托贸易
commission charges　代理手续费
commission designing　代客设计
commission drading　代客推档

commissioning　投产,运转,正式使用,委托加工
commission knitter　佣金针织商
Commission Sericole Internationale　国际蚕丝业协会
commission sewing　代客缝制
commissure　缝口,合缝处,接合处
Committee for Implementation of Textile Agreement　纺织协定执行委员会
Committee of the Wool Textile Industry in the European Union　欧盟毛纺织工业委员会
commode　克莫德头巾(17～18世纪);衣橱,衣柜
commode petal　克莫德头巾
commodious quantity　宽裕量
commodity　商品
commodity exchange　商品交易所
commodity inspection　商品检验,商检
commodity inspection certification　商检证明书
commodore　海军卡其,英国海军卡其
commodore cap　海军女式帽
commodore dress　海军式裙装
common and braid wool　粗羊毛
common chain stitch seam　普通链式线迹缝
common gingham　普通条格布
common inspection　共同检验
common market　共同市场
common parts　通用零件
common pin　普通别针
common-sense heel　大众跟,大众鞋跟
common silk　桑蚕丝,普通天然丝
common twill　正则斜纹,普通斜纹
common velvet　普通天鹅绒
common wool　普通羊毛
communion dress　女孩圣餐礼服(白色)
communion veil　女孩圣餐面纱(白色透明)
commuter look　持月票者款式
compact　连镜小粉盒
compact blouse　简洁女衫
compact cloth　紧密织物
compact face powder　粉扑
compact feel　坚实手感
compact handle　结实手感
compacting　织物热压预缩
compactness　紧密性
companion fabric　组合使用织物(两种或

多种织物,设计成一起使用)
companion pleat 相配褶裥
company culture 企业文化
company's nankeen 宽幅棉布
company standard 企业标准
comparable price 可比价格
comparative clothing 比较衣着学
comparative enlargement 比拟放大
compass cloak 法式圆形斗篷
compasses （制图用）圆规
compatibility 可混用性,配伍性,相容性
compatible shrinkage 相容收缩(面料、辅料间)
compensating mechanism 补偿装置
compensating presser foot 补偿压脚
compensating unit 补偿装置
compensation trade 补偿贸易
compenzine 花式三股线
competition-striped knit shirt 二色[相间]条纹运动衫
competition-striped shirt 比赛条格服
competition strips 双色条纹运动衫
competitive 竞争
complement 定额装备
complementary color 补色,余色,互补色
complete cost 完全成本
complete economic integration 完全经济一体化
complete in specification 规格齐全
complete inspection 全数检查
completely washable fabric 机可洗织物,可充分洗涤的织物(用71.7℃热水机洗而不过分缩水或变形变色)
complete mechanization 全盘机械化
complete repeat 花样循环
complete set 成套,整套设备
complete set of equipment 成套设备,整套设备
complex color harmony 不调和的配色方法,复式配色
complex dye 复合染料
complex dyestuff 络合染料
complex fiber 复合纤维
complex French curve 法式多功能曲线板
complexion 肤色,面色,气色
complex yarn 复丝,复合纱,复合线,花式线
compliance 柔量;柔顺性
component 成分
component analysis 成分分析

component color 组分色
component fiber 组分纤维,复合纤维
component part 组件
composite color 复合色
composite design 结构设计
composite dye 复合染料,拼合染料
composite fabric 复合织物
composite fiber 复合纤维
composite fiber fabric 双组分纤维织物
composite filament 复合[长]丝
composite green 混合绿
composite guide 复合导缝
composite laying 组合式铺叠
composite materials 复合材料
composite seam 结构缝(各式线缝统称)
composite synthetic fiber 复合合成纤维
composite yarn 复合纱,组合纱
composite yarn etched out fabric 全包芯纱烂花布
composite yellow 混合黄
composition 成分;构成,合成,组成;构图;合成物
composition button 树脂纽扣
composition cloth 防水帆布,多层复合布
composition design 构图设计
composition dot 植绒花纹,植绒点子
composition fiber 复合纤维
compound button 复合扣
compound cloth 多层织物
compound color 混合色
compound construction 复合织物
compound fabric 多层织物,复合织物
compound feed 复合送布
compound pattern 复合花纹,复合花样
compound shade 复色,拼色
compound sole 复合底(鞋)
compound twill 复合斜纹
compound yarn 花色线,股线
comprehensive most-favored-nation clause 全面最惠国条款
comprehensive planning [program] 全面规划,总体规划
compressional resistance 耐压性
compression bandage 压缩性绷带
compression elasticity 压缩弹性
compression packing 压缩包装法
compressive shrink finish 机械预缩整理
compressive shrinking 预缩
compressive strength 抗压强度
comptah cotton 印度暗棕棉

compulsory standard of footwear 强制性鞋类标准
computed value 计算价格
computer 电脑,电子计算机
computer aided design（CAD） 计算机辅助设计
computer-aided design and drafting（CADD） 计算机辅助设计与制图
computer aided design/computer aided manufacture（CAD/CAM） 计算机辅助设计与制造
computer aided design system 电脑辅助设计系统
computer aided engineering（CAE） 计算机辅助工程
computer aided garment design 计算机辅助服装设计
computer aided garment instruction system（CAIS） 计算机辅助服装教学系统
computer aided garment manufacturing system 服装CAM系统
computer aided garment process planning system（CAPP system） 计算机辅助服装工艺设计系统
computer aided layout（CAL） 计算机辅助排料
computer-aided machine shop operating system（CAMOS） 计算机辅助车间操作系统
computer aided management（CAM） 计算机辅助管理
computer aided manufacture（CAM） 计算机辅助生产(或制造)
computer aided pattern（CAP） 计算机辅助纸样设计
computer aided phototypesetting 计算机辅助照相排版
computer aided production planning（CAPP） 计算机辅助生产计划
computer aided testing（CAT） 计算机辅助测试
computer architecture 计算机体系结构
computer art 计算机艺术
computer automatic cutting system 计算机自动裁剪系统
computer color matching（C.C.M） 计算机配色,电子配色,计算机拼色
computer color matching system 计算机配色系统,电脑配色系统
computer control 电脑控制
computer control apread system 电脑铺料系统
computer controlled programmed sewing machine 电脑程控缝纫机
computer cutting machine 计算机自动裁片机
computer data processing 计算机数据处理
computer dress 计算机服
computer dress pattern design 计算机服装纸样设计
computer embroidery design and punching system 电脑绣花设计打板系统
computer embroidery design editing system 电脑刺绣设计系统
computer embroidery machine 电脑绣花机
computer grading 电脑放码,计算机推档
computer grading and marking system 计算机放码和排码系统
computer grading system 电脑放码系统
computer graphics 计算机绘图,计算机图形学
computer information management system（CIMS） 电子信息管理系统
computer integrated clothing system 计算机综合成衣系统
computer integrated manufacture（CIM） 计算机一体化生产
computer integrated manufacturing（CIM） 电脑综合制造
computer integrated manufacturing system（CIMS） 计算机集成制造系统
computerized automatic cutting table 电脑自动裁床
computerized automatic knitting machine 电脑自动针织机
computerized color measure system 计算机测色系统
computerized color separation system 计算机分色系统
computerized embroidering 电脑绣花
computerized embroidery 电脑绣花
computerized embroidery machine 电脑绣花机
computerized engraving button 电脑雕刻纽扣
computerized fabric designing 计算机花纹设计
computerized monitoring system 计算机

监测系统
computerized numerical control（CNC） 计算机数控
computerized pattern grading 计算机辅助样板［放缩］推档
computerized pattern grading & marker making machine 计算机放样排板机
computerized pattern grading marking and cutting system 计算机辅助样板推档、划样和裁剪系统
computerized prediction 计算机预测
computerized pressing 电脑熨烫
computerized searching system 计算机检索系统
computerized service routine 计算机服务程序
computerized sole attaching conveyor 计算机贴鞋底传送机
computerized woven label 电脑织唛
computerize engraving button 电脑雕刻扣
computer jacquard 电脑提花
computer management/information system for apparel industry 制衣业电脑管理信息系统
computer marker planning 电脑码克排料，电脑排唛架
computer match prediction 计算机配色预测
computer network 计算机网络
computer pattern 电子计算机花纹
computer pattern preparation system 计算机纹样准备系统
computer pattern socks 电脑花袜
computer pattern system 计算机提花系统
computer peripheral device 计算机外围设备
computer process control（CPC） 计算机过程控制
computor aided garment process planning system（CAPPS） 计算机辅助服装工艺计划系统
concatenation of thread 串线
concave dart line 瘪形省线
concaved shoulder 凹肩型，凹线肩线
concave stitching line 凹形缝线，瘪形省线
concealed 暗襟
concealed button fly 暗纽扣门襟
concealed check fancy suiting 隐格花呢
concealed fastener 暗拉链，隐形拉链

concealed hood in collar 后领内风帽
concealed hook buckle 隐钩环扣，隐钩式环扣
concealed pocket 暗袋
concealed seam 暗缝，隐蔽缝
concealed sheet 暗襟
concealed stripe fancy suiting 隐条花呢
concealed wiring 暗布线
concealed zip-fastener 暗缝拉链
concealed zipper 暗缝拉链，隐形拉链，暗拉链
conceal fastener 隐形拉链
conceal seam 暗缝，隐蔽缝
concentrative marketing strategy 密集性市场策略
concentric sheath-core fiber 同心皮芯纤维
conception of design 设计构思
concept of product as a whole 产品整体概念
concept-production time 概念设想到投产的时间
concertina 束腰
conch, concha 海螺壳（服饰），海螺式女长服
concha belt 贝壳饰链
conch shell 海螺壳红，深藕红
conch stitch 贝壳形线迹
concierge hairstyle 女管家发型
conclude an agreement 达成协议
concord grape 和谐紫色
concrete grey 水泥灰色
condeaux 半漂白大麻帆布
condense dyes 缩聚染料
condensing the stitches 缝迹加密
conditioned atmosphere 公定温湿度
conditioning fiber and textile 调温纤维和纺织品
condor 兀鹰
condor pattern 兀鹰图案
conductance nonwoven fabric 导电非织造布
conducting-core heterofilament 导电芯型双组分长丝
conductive fabric 导电织物
conductive fiber 导电纤维
conductivity 传导性
conduct of the arbitration 仲裁进行方法
conductor's cap 售票员帽，列车员帽
conductors'uniform 列车员制服

cone 锥体,锥形,宝塔筒子
cone cap 圆锥帽
cone heel 锥形鞋跟
cone of thread 宝塔线
cone pattern 圆锥形花纹
cone raised pattern 圆锥形凸花
cone shaped skirt 锥型裙
cone skirt 圆锥形裙,锥形裙
coney 家兔毛皮,兔毛皮;兔皮大衣
coney shin coat 兔皮大衣
confection 时髦女装,女用成衣,女用服饰品
confederate dye 天然染料
confetti 五彩碎纸,五彩纸屑红色
confetti dot 彩色点纹,彩色点子图案
confidence 信用
confidence interval 置信区间
confident 大胆发型(17世纪)
configuretion 形状
confirmation dress 坚信礼服(天主教等)
confirmation of order 订货确认
confirming sample 确认样
conflex P 导电性聚丙烯纤维(商名,英国考陶尔)
conflex V 导电性黏胶纤维(商名,英国考陶尔)
conformability 适合性,顺应性
conformatear 头部测量仪
conformation 形态
conformity certification 合格认证
confused color fabric 迷彩织物
conghua taffeta faconne 丛花绢
conglomerate 跨行业公司
congo hemp 刚果大麻
congo jute 刚果黄麻
congo red 刚果红(直接红色染料),刚果红色
congress canvas 刺绣十字布
congress cap 国大党白色党帽(印度)
congress gaiter 康格里斯靴(19世纪),议会及踝松紧靴
congress shoes 高帮男靴,康格里斯靴(19世纪)
congruent points 叠合点
conical cap 圆锥形帽
conical hat 圆锥形帽
conical shank button 锥柄纽扣
conical skirt 圆锥形裙
conjugated fiber 复合纤维,共轭纤维(尤指双组分并列型复合纤维)

conjugated filament 复合丝,共轭丝
conjugated micro fiber 共轭型超细纤维,共轭型微纤维
conjugated yarn 复合丝,共轭丝
conjugate fiber 热塑性双组分纤维
conjugate filament hosiery 复合丝袜
conjugate-spun fiber 复合丝,共轭丝
conjulame 聚丙烯腈复合纤维(商名,日本)
conk 头发拉直
connaught 刺绣十字布
connaught cloth 康纳特布
connaught yarn 爱尔兰针织绒线
connecting stitch 钉扣线迹
connemara tweed 科纳马拉粗花呢(英国)
connoisseur 行家
conrréges flower socks 库雷热大绣花女袜
conserva-rich 高级保守时装
conservative color 保守色
conservative fashion 保守时装;保守款式
conservative look 保守风貌,保守款式
conservative pattern 传统图案
conservative suit 老式衣服
considerations 轻撑裙
consignment 托运
consignment note 托运单
consignor 托运人
consistent pattern 相一致的样板
consistent stitch formation 连续线迹成形
consolidation shrinkage 织物的洗后收缩;服装的洗后收缩
conspicious figure 露头角
conspicuousness 显眼
constance bag 积琪莲手袋
constant interval scale 等间隔刻度
constant pitch cross winding 等螺距交叉卷绕
constant ruffler 均匀打褶器
constitute 构成
constitution cord 阔棱厚灯芯绒,宽棱厚灯芯绒
constitution corduroy 阔棱厚灯芯绒
constraint pattern layout 强制性样板排料
constructed garment 结构服装,做工精致服装(很多部位有黏合衬,大身和袖有里子)
constructed tailoring 制作严密的缝制法

constructing fabric 复合针织物
construction 组织结构,(面料)组织,织物结构;(使服装)调和;服装缝制
construction boots 工地靴,建筑靴,建工靴
construction building 工业厂房
construction design 结构设计
construction fabric 建筑用织物
constructionist 结构主义者
construction line 服装结构线
construction pressing 半制品加工整烫,中间缝制整烫
construction seam 结构缝
constructivism fashion 构成主义流行款式
construct textile 建筑用织物
consultant 顾问
consultation call 纺织品设限通知
consulting gown 白色大褂
consulting service 咨询服务
consumable parts 易耗件
consumables 消费性商品
consumer 消费者
consumer appeal 顾客吸收力(指商品)
consumer behaviour 消费者行为
consumer fashion acceptance 消费者对时装的接受力
consumer goods 生活用织物;消费品
consumer orientation 消费者导向
consumer-satisfying product 消费者满意产品
consumer-satisfying service 消费者满意服务
consumer taste 消费品位
consumer trends 消费者倾向
consumption 费用,消费,消耗,耗料
consumption currency 消费流行
consumption impulse 消费冲动
consumption level 消费层
consumption psychology 消费心理;消费心理学
consumption quota 消耗定额
contact lens 隐形眼镜
contact printing method 压印法(码克复制)
contailles 法国低档绸
container 集装箱
container bag 尼龙袋,装物袋
container board 盒纸板,硬纸板
container conveyor 输送带

container freight station (CFS) 集装箱货运站
container loader 装箱器
container package 集装箱包装
container sealer 封箱器
container shaper 容器成形器
container shipping 集装箱货运
contance 法国大麻床罩
contemporary heel 现代式鞋跟
contemporary model 现代流行西装型,现代流行型
contemporary styling 当今流行款式;当今时尚
contemporary traditional style 现代传统款式
contemporary traditional stylistic form 现代传统款式
contemporary woman 现代妇女
content label 成分唛
contexture 结构,织物,交织
continental collar 欧式领,欧洲大陆领
continental count 米制支数
continental cut 欧式裁剪,欧洲大陆式裁剪
continental cut pants 欧式长裤,欧洲大陆式长裤
continental cut shirt 欧洲型衬衫,西装衬衫,西服衬衫
continental cut suit 大陆式西装,欧式西装
continental hat 瑞士军帽,欧式三角帽
continental heel 欧式鞋跟,欧洲大陆鞋跟
continental look(or mode) 欧式款式,欧洲大陆式风貌,欧洲大陆型
continental mode 欧洲大陆式款式
continental model 欧式(西装模型),欧洲大陆型
continental necktie 欧洲大陆型领带
continental pants 欧式长裤(裤腰合身,不系腰带的窄管男裤,前身口袋多半采用水平或曲线状开口)
continental pocket 欧式袋(弧形袋口),欧洲大陆式袋
continental sack coat 欧式男便装夹克
continental slacks 欧式长裤
continental style 欧式,欧洲大陆式
continental stylistic form 欧洲大陆款式
continental suit 欧式套装(自然肩线,配以宽下窄裤子,不系腰带,斜插袋),欧洲大陆式套装

continental tie 欧式领带,欧洲大陆领结
contingency 不可预见费
continuous band 连续式提花带
continuous bound placket 包连式开衩
continuous brocade 通纬花缎
continuous buttonholing machine 连续锁眼机
continuous dyeing 连续染色
continuous feed 连续送料
continuous filament rayon yarn 黏胶复丝,黏胶长丝
continuous finishing 连续整理
continuous lap 包连式开衩
continuous lap opening 连续开衩
continuous lap placket 连续开衩,直袖衩
continuous line 实线
continuous loops 连续纽圈式
continuous operation 持续运行
continuous pattern 连续花纹
continuous placket 连续式服装开口,连续性开口
continuous seam 连续缝
continuous-thread dart 省尖不断线的省
continuous-thread tuck 端点不断线的塔克褶
continuous tone 连续浓淡色调
continuous type 无裤腰型,无裤腰的西装裤型
continuous unit-pattern 连续纹样
contouche 女袍,带悬袖的袍
contour 轮廓
contour belt 合腰形腰带,体形带;轮廓饰带,曲线饰带
contour bra 造型胸罩
contour clutch 法式钱包
contour cut 按体型裁剪
contour dart 长腰省,曲线省,两头尖省
contour design 合身设计
contoured-fit jeans 贴体牛仔裤
contoured front 服装前片分割
contoured jet with reinforcement pocket 加固口袋;缘边双嵌线
contoured waistband 合腰形腰带
contour embroidery 轮廓刺绣
contour fabric 成形织物
contour final 最终轮廓线
contour gauge 仿形样板,靠模样板
contour heel 小方块高袜跟
contour hood 合形风兜
contouring color 勾画轮廓色

contour line 等高线
contour machining 成形加工,仿形加工
contour outline 勾草图
contours 女子身体曲线;外形,结构,特征
contour seamer 无导轨缝纫机
contour sewing machine 靠模缝纫机
contour sheets 成形床单
contour stiching 廓线针迹,廓线线迹
contour stitch 廓线针迹,轮廓针法
contour stitcher 仿形缝纫机,轮廓合缝机
contour waistband 合腰形腰带
contract 承包;合同;公共场合用纺织品
contract carpet 配套地毯,订货地毯
contract carpeting 配套地毯料,订货地毯料
contract for processing with customer's materials 来料加工合同
contraction finishing 收缩整理
contractive color 收缩色
contract law 合同法
contrapuntal pattern 对位组合花样,组合花样
contrary design 逆反设计
contrast 对比,对照;反差
contrast binding 错色滚条,错色镶边
contrast color 对比色,衬色,对照色,反衬色
contrast color harmony 对比色调和
contrast color scheme 对比色配色
contrast color seam 拼色缝,镶色缝,镶色线缝
contrast gloss 对比光泽,反衬光泽
contrast grade 对比度,反差度
contrast harmony 对比调和,协调对比
contrasting moods in complementary shades 补色对比风格
contrast stitching 拼色线缝,镶色线缝
contro 康屈洛松紧线(商名);康屈洛松紧带
control 控制
control briefs 弹力内裤,吊袜三角裤
control chart 控制图
control color 美容控制色
control cycling pants 自行车弹力裤,自行车束腹裤
controlled programmed sewing 程控缝纫
control panties 束腹型短内裤,束腹型内裤,弹力针织内裤
control panty hose 束腹型裤袜
control sample 对照样品

control seam 试样调节缝,合体控制缝
control sequence 控制时序;控制顺序;控制序列
control signal 控制信号
control system 控制系统
control technique 检查技术,控制技术
control test 对照试验
control-top panty hose 束腹型裤袜
control undercover garment 对照标准服装
convas making 缝衬
convection 对流
convective heat transfer 对流传热
convenient ruler 简便尺
convent cloth 粗纺平纹女衣呢,修女绉呢
convention 习俗
conventional art 常规技术,传统技术
conventional design 传统纹样
conventional flat iron 火烙铁,火烙熨斗
conventional method 传统方法
conventional stockings 普通无缝长筒女袜
conventional symbol 习惯用符号
conventional tariff 协定税则
conventional type 常规类型,普通型号
conventional zipper 封尾型拉链,普通拉链
conversational design[pattern] 风俗画图案
conversational design 风俗画图案
conversational pattern 风俗画图案
conversational print 风俗画印花布
conversation bonnet 带红绿丝带的翻边女帽
converse 匡威
conversion effect 转色效应
converted fabric 被整理的织物;净坯
converted flannel 被整理的棉法兰绒
converted goods 被整理的织物
converted products 复制产品
converted ticking 漂白印花被套布
convertibility 两用性
convertible collar 开关领,两用领,换形领,开襟领
convertible cuff 活袖口,可换的袖口,活袖头,两用袖头,可变袖头
convertible garment 可变换式服装
convertible open 两用开门领
convertible shorts 可变换短裤
convex and concave crepe 凹凸绉

convex method crease elasticity testing 凸形法折皱弹性试验
convex point pattern 凸点纹
convex stitching line 凸形缝线,胖形省线
conveyer belt press 传送带式粘合机
conveyer feed 传送带送料
conveyer line system 传送带流水[作业]线系统
conveyer press 传送式熨烫机
conveyer pressing system 履带烫衣机
conveyer system 流水作业
conveying equipment 输送设备
conveyor pressing system 履带烫衣机
convict stripes 黑色横条子白地棉布;苦役服
convoy 缘饰条格披巾
cony, coney 兔毛皮,家兔毛皮,兔皮大衣
cony hair 兔毛
cony hat 兔毛帽
coogan 儿童背带裤(澳、新)
cook cotton 美国库克棉
cooket hat 卷边帽
cooking cap 厨师帽,炊事员帽
cooking coat 炊事衣
cooking uniform 炊事服
cooking wear 炊事服
cook jacket 厨师外套
cool 酷,新潮的
cool beauty 素静美
cool color 冷色
cool color scheme 冷色配色
cool hat 凉帽
coolie 苦力外套,劳工型外套(立领,插肩袖,宽松型,长至腰下);对襟绣花外套
coolie coat 苦力外套,劳工型外套(立领,插肩袖,宽松型,长至腰下);对襟绣花外套
coolie hat 苦力帽,斗笠帽
coolie jacket 苦力外套
coolie shirt 劳工衫
coolie shoes 劳工鞋,苦力鞋
coolie straw hat 斗笠草帽
cool iron 低温熨烫
cool ironing 低温熨烫
cool-keeping fabrics 凉爽织物
cool lamp 冷光源
coolmax 六角凹槽形合成纤维(商名,美国杜邦)
coolmax bra top (背心式)凉爽胸衣
coolnice 导湿涤纶(商品名,中国)

coolplus 吸湿排汗纤维
cool skin tone 冷皮肤色调
cool suit 冷却服
cool tailoring suit 轻凉西装
cool tone 冷色调
cool-warm contrast 冷暖对比
cool water 冷水
coomptah cotton 印度库姆塔棉
coonskin 浣熊皮
coonskin cap 浣熊皮帽,大卫帽
coonskin coat 浣熊毛皮外套
cooperation 协作
cooperation program 合作项目
coordinate 配套
coordinate digitizer 坐标数字化仪
coordinate look 配套式;协调型;整体感
coordinate pattern 搭配色图案
coordinates （颜色、原料、式样等）配合协调的衣服,调和套装,协调的衣服,套装
coordinates look 成套式
coordinate suit 调和套装,西服套装
coordination 调和,协调;配套商品;组合
coordination color 搭配色
Coordination Committee for the Textile Industries in the European Union 欧洲纺织工业联合会
coordination style 调和风格
coordinative button 组合纽
coordinator 服装搭配设计师,形象设计师
coosong 黑色棉平布
coothay 库赛彩条缎
cop-and-cop doubled yarn 混合双股线
copang 科平厚花呢
cope 披风,护肩,斗篷,斗篷式长袍,蔻普斗篷,主教披风,剑桥大学博士所穿的斗篷
copees 印度花格布
cope line 拱顶型
Copen blue 哥本哈根蓝
Copenhogen blue 哥本哈根蓝,灰光蓝色
cope piping 披风宽贴边
copied pattern 复制纸样
coping 厚花呢,厚重毛织物
cop of thread 纸蕊收
copolyamide fiber 共聚酰胺纤维
copolyether ester fiber 共聚醚酯纤维
copolymer fiber 共聚物纤维
copotain 锥形高帽
copou 中国薄布

copped shoes 细尖头鞋
copper ammonia fiber 铜氨纤维
copper brown 紫铜棕
copper color 古铜色(暗黄棕色),铜色
copper green 铜绿色
copper lace 古铜色花边,铜色花边
copper-plated zipper 镀铜拉链
copper rayon 铜氨人造丝
copper red 铜红(棕橘色)
copper rust 红锈色
coppery 铜色
co-production 合作生产
copy 仿形,副本,复制品,复印件,临摹
copy design 描图样
copying box 拷贝箱
copying motion 仿形运动
copy of contract 合同幅本
copy pattern 复制图
copy rule 仿形尺
coq plumage 欧洲军官帽子羽饰
coquelicot 虞美人红(橙红色)
coquille 贝壳形花边,(击剑用的)护剑盘
cora 科拉绸
corah silk 科拉绸
coral 珊瑚;珊瑚红,珊瑚色
coral blush 珊瑚红
coral button 珊瑚扣
coral cloud 珊瑚云色
coral haze 朦胧珊瑚色
coraline 厚针绣花边
coralline 珊瑚色
coralline point 珊瑚纹手工针绣花边
coral necklace 珊瑚项链
coral pink 浅珊瑚红,浅橘红色
coral reef 珊瑚礁色
coral rose 珊瑚玫瑰色
coral sands 珊瑚沙色
coral stitch 珊瑚针迹,粗线紧密线迹
coram 德国漂白粗亚麻布
coravia 菠萝属叶纤维
corazza 后开扣、紧身袖男衬衫
corbeau 乌鸦黑
cord 带,绳,线,绳带;服饰辫线;棱,凸纹;灯芯绒,棉条绒,趟绒;双丝绸;花色粗针法;经绒组织
cordal 法国韧皮纤维粗帆布
cordaleen 科达林垂直棱纹女式呢
cordat 粗绒哔叽,亚麻帆布
cordat serge 科达粗厚哔叽
cord belt 绳形腰带

cord braid 实芯细圆辫带
cord carpet 起圈地毯,凸条纹地毯
cord de chine 凸纹直条薄呢
cordé bag 嵌带提花手提包
corded 起棱纹的,起凸线的
corded alpaca 阿尔帕卡经凸条纹布
corded baft 英国经向凸条坯布
corded band trimming 镶条边饰
corded belt 镶边
corded button 编结纽
corded buttonhole 衬线纽孔,镶线纽孔
corded cuff 绳饰袖口,绳饰袖头
corded edge 滚绳饰边
corded edge braid 绳索编带
corded elastic 线绳松紧带
corded fabric 起棱织物.直棱织物,横棱织物,凸条纹织物
corded faced edge 嵌线镶边
corded gather 嵌线抽褶
cord edge braid 绳索编带(织带)
corded knitted fabric 灯芯条针织物
corded machine buttonhole 嵌线机锁纽孔
corded necklace 绳状项链
corded pants 棱条织物裤
corded piping 衬线滚边
corded plush 棱条长毛绒
corded seam 夹心嵌条缝
corded self-finished edge 本布嵌线贴边
corded shirring 嵌线抽褶
corded silk ribbon 绳纹丝带,灯芯缎带
corded tubing 串带管
corded tuck 衬线褶裥,嵌芯裥饰
corded velveteen 灯芯绒,棉条绒,趟绒,条绒
corded voile 经向凸条巴里纱
cord effect 经向凸纹效应
cordelan 科迪纶(聚乙烯醇和聚氯乙烯双组分阻燃纤维,商名,日本)
cord elastic shirring 松紧绳抽褶
cordelat 科迪拉绒面斜纹厚呢,科迪拉长绒厚呢
cordeliere 绳索编制端头打结腰带
cordella lace 网地粗线勾边花纹花边
cord embroidering 嵌绣,嵌线绣花
cord embroidery 嵌线绣花,嵌绣
corder 缝饰带工,制绳线工
cordettes 大麻布(做头巾用,法国制)
cord fabric 凸条纹织物,棱纹织物
cord handbag 绳编袋
cord hanger 挂衣领袢,领袢

cordillas 粗毛呢
cordillat 科迪拉绒面斜纹厚呢,科迪拉长绒厚呢
cording 花边,饰边,镶边;嵌线,包梗,凸条浮雕,金银饰绦,凸线刺绣;镶嵌线
cording embroidery 凸线刺绣
cording foot 滚边压脚,嵌线压脚
cording lace 绳饰花边
cording seam 凸线缝,嵌缝
cording stitch 嵌线线迹;凸线刺绣;凸线针法
cordington 绒厚呢
cord lace 粗线花边,条绒花边,绳状花边
cordlane 细棱纹织物
cordless cutter 电池式裁剪机
cord loop 纽襻,线环
cordoba wool 阿根廷科尔多巴羊毛
cordon 绶带,绶章,细绳,装饰带
cordon braid 绶带,绶章带
cordoncillas 科唐西拉斯低级平布
cordoncillos 粗平纹棉布
cordonnet 粗丝线,花边外缘饰线
cordonnet en laine 毛制装饰带绳
cordonnet silk 粗丝线
cordons 金银编带
cordoroy 条绒织物,灯芯绒
cordovan 哥多华皮革,科尔多瓦皮革(西班牙),马尻革),西班牙革;皮革色;
cordova wool 科尔多瓦羊毛
cord piping 滚边编带,滚边带
cord quilting 意大利绗缝
cordrein 化纤夏服料
cords 灯芯绒裤
cord sash 嵌芯圆线带
cord scalloping machine 扇形嵌线缝纫机
cord seam 凸缝
cords of knitting 针织饰带
cord stitch 凸纹线迹,盘线针法,绳纹线迹,嵌线线迹;绕绣
cord stripe 直条凸纹,直条凸纹织物
cord suspender 挂衣服用领袢,挂衣领袢,领袢
cord tie 缎带领带,警长领带,团长领带,西部领带,细绳领带,鞋带型领带
cord trousering 裤料条花呢
cord trousers 棱条花布裤
cordura 科杜拉(聚酰胺66长丝,商名,美国杜邦)
cordurette 横棱纹呢
corduroy 灯芯绒,棉条绒,趟绒,条绒

corduroy cap 灯芯绒便帽,灯芯绒帽
corduroy gingham printed bed sheet 灯芯绒印花床单
corduroy jeans 灯芯绒牛仔裤
corduroy pants 灯芯绒裤,工装裤
corduroys 灯芯绒裤
corduroy shirt 灯芯绒衬衫
corduroy shoes 灯芯绒鞋
corduroy trousers 灯芯绒裤
cord velours 凸纹天鹅绒
cord velvet 棱纹丝绒,凸条丝绒
cord weave 棱纹
cordyback hat 仿海狸皮毡帽
coreded silk 绳纹丝织物
core design 核心设计,中心设计
coredon 二上一下粗纺呢
core sample 核心衣样,中心样衣
cores button 象牙果扣
core-sheath compound fiber 皮芯型复合纤维
core-sheath type 皮芯型化纤
core-skin fiber 皮芯纤维
co-respondents 观赛鞋
core-spun elastic yarn 包芯弹力纱
core-spun polyester/cotton sewing thread 涤/棉包芯缝纫线
core-spun sewing thread 包芯缝纫线
core-spun spandex 包芯氨纶丝
core-spun T/C sewing thread 涤棉包芯缝纫线
core-spun yarn 包芯线;包芯花线
core thread[yarn] 包芯线,包芯花线
core yarn fabric 包芯纱织物
corfam 科芬皮革
Corfam shoes 科芬鞋
corfu lace 希腊粗花边
corgach 科加其
corinch pink 桃红
corinth 酱红色
corinth pink 桃红色
cork 软木,豆沙色,浅棕
cork edge 锁口边,拷边
cork jacket 浮水衣,软木救生衣
cork lace 爱尔兰考克花边
cork rug 软木衬地毯
corkscrew 螺旋纹精纺呢,螺旋斜纹呢
corkscrew cloth 璧绉,螺旋斜纹呢
corkscrew curl 螺旋发型,螺旋卷发型
corkscrew repp 螺旋纬纱横棱织物
corkscrew thread 螺旋花线

corkscrew twill 螺旋斜纹织物
corkscrew twill fabric 螺旋斜纹织物
corkscrew warp weave 经面螺旋纹织物
corkscrew weave 经面斜纹,螺旋斜纹
corkscrew weft weave 纬面螺旋纹织物
corkscrew yarn 螺旋花线
cork wig 软木假发(18世纪)
corn 玉米黄色
corncrowing hairstyle 辫子发型
corned shoe 宽头鞋
corned-spun 包芯
corner blunted 翻转衣角
corner cap 角帽(16～17世纪),三角丝绒帽,四角丝绒帽;学位帽
cornered style line 有转角的款式线条
corner lace 角用花边,手帕花边
corner pressing 转角压烫
corner seam 转角缝
corner shape 女式卷边帽
corner stitch 转角线迹
corner stitching 缝角
corner turning 修剪转角
cornet 白布帽,大白帽(慈善修女戴),上浆薄纱女帽(15世纪),圆锥头巾(14～18世纪)
cornet cap 修女大白帽
cornet sleeve 喇叭荷叶边紧身袖,喇叭袖
corn fiber 玉米蛋白质纤维,玉蜀黍蛋白质纤维
cornflower blue 矢车菊蓝色(紫蓝色)
cornhusk 玉米穗壳色
cornisilk 佛手黄色(淡黄色)
corn protein fiber 玉米蛋白质纤维
corn rowing hairstyle 辫子头发型
corn rows 多辫发型
corn sacks 玉蜀黍麻袋
cornsilk 玉米穗丝色
cornstalk 玉米秆色
corolla 花冠
coromandel 英国出口非洲的棉粗布
corona 冠,圆顶(光头部分),科罗纳冠(罗马);科罗纳绉;黏胶短纤维(商名,日本)
corona-crepe corona 科罗纳绉
corona discharge 电晕放电
coronal 花冠,花环
coronation braid 变直径棉编带,加冕编带,加冕装饰带(华贵织带)
coronation cloth 加冕布,加冕呢
coronation color 加冕七色(英国)

coronation gimp 加冕带,冠状花式带
coronation robe 加冕典礼用袍
coronet 冠,冕,束发带冠,冠状头饰
coronet blue 冠冕蓝
corotte 印度克罗特印花粗棉布
corozo button 象牙果纽扣
corpilon 抗起球聚丙烯腈纤维(商名,日本)
corporate (CI) 企业标识
corporate behavior 企业行为
corporate clothing 劳动服,工作服
corporate culture 企业文化
corporate identity 企业特征
corporate image 公司形象,企业形象
corporate responsibility 共同责任
corporation duty[tax] 公司税
corps 紧身胸衣,女式紧身上衣
corps de fer 铁质紧身胸衣
corps pique 绗缝紧身胸衣
corrdinator 服装搭配设计师
corrected grain leather 修饰面革
correction to program 程序校正
correct measurements 正确量身(法)
correct sample 修正样
correlation 关联性,相关性
correlation analysis 相关分析
correlation diagram 相关图
correlogram 相关图
correspondents 双色低跟镂花皮鞋
correspondent shirt 记者衬衫
correspondent shoes 双色低跟镂花皮鞋,双色男鞋
corridor rug 走廊地毯
corriedale wool 考力代羊毛
corrientes 阿根廷科连特斯羊毛
corrosion resistant fiber 耐腐蚀纤维
corrugate 起波纹
corrugated board 瓦楞纸板
corrugated card board 波效纸板,瓦楞纸板
corrugated pleat 瓦楞褶
corrugation 瓦楞形
corsage 女服胸部;女胸衣,紧身衣;女服胸饰,花饰
corsage en corset 紧身晚装内衣
corsaire 海盗裤(紧贴于腿肚)
corsair pants 海盗裤(齐腿肚的紧身裤)
corselet 胸衣,束腹,胸甲,束衣
corselet belt 束身宽腰带
corselet cloth 提花胸罩布

corselet'corselette (女)胸衣
corselet sash 束身宽腰带
corselet skirt 紧身连衫裙
corselette 胸衣,束腹,胸甲,束衣,紧身褡
corselette cloth 提花紧身织物,提花胸罩布
corselet top 束身上衣
corselet waistline 束宽腰带的腰线
corsery 印度棉布
corses 丝带
corset 束衣,胸衣,紧身胸衣,束腹带,(硬)围腰,紧身褡,紧腰衣
corset batiste 厚重织物,胸罩类织物
corset belt 束身腰带
corset bodice 紧身胴衣
corset brocade 胸衣锦缎
corset broché 胸衣织物
corset busks 紧身上衣
corset cloth 紧身衣料
corset coutil 内衣、胸罩用布
corset cover 紧身衣罩衫(19世纪),胸罩,(妇女套在紧身褡外面的)背心
corset crepe 匹染胸衣绉,胸衣绉
corset dress 紧身裙装
corseted waist 细束腰
corset elastic 弹力胸衣
corset fabric 胸罩织物,胸衣织物
corset frock 紧身胸衣式上衣
corseting 穿胸衣,紧身褡试样
corset jean 斜纹胸衣布
corset lace 胸衣花边
corset net 胸衣花边网
corset over bloomers 穿在灯笼袖衬衫外的紧身胸衣
corsetry 束腹;胸罩;吊袜带
corsetry net 胸衣花边网
corsets 束腹
corset top 紧身褡上衣
corset trimming 胸衣饰带
corset waistline 紧腰式腰线
corsicaine 法国印经小方格绸
Corsican blue 科西嘉蓝色
Corsican tie 科西嘉领带,紫色窄颈巾
corsican wool 科西嘉羊毛
corslet 胸甲,胸衣,束腹,束衣,护身盔甲
corso skirt 鞍形裙
cortex 皮层
corundum 刚玉(首饰)
corunna stripes 科伦纳棉条子细布

cosin 曙红
cosmetic 化妆品；化妆用的
cosmetic case 化妆用小镜匣
cosmetic gauze 化妆面纱
cosmetician 美容师,化妆师；化妆品制造商；化妆品经销商
cosmetic lens 美容眼镜
cosmetics bag 化妆手袋
cosmetic support stockings 医用整形袜
cosmetique 发蜡条
cosmetologist 美容师
cosmetology 美容,美容术；美容学；美容业,整容术；化妆品制造术
cosmic look 太空服式,宇宙服式
cosmic print 宇宙印花
cosmic rays 宇宙射线
cosmic style 宇宙款式
cosmic wear 太空服,宇宙服
cosmocorps look[style] 宇航服款式,宇航风貌,宇航服风貌
cospinning fiber 共纺纤维,混抽纤维
cossa 印度棉平布
Cossack 哥萨克服装(立领,束带长衬衫)
Cossack blouse 哥萨克女衫(立领,膨松袖,偏襟,镶有绣花饰边,有腰带)
Cossack boots 哥萨克靴
Cossack cap 哥萨克帽
Cossack collar 哥萨克领(偏襟立领,也有绣花)
Cossack costume 哥萨克民族服饰
Cossack forage cap 哥萨克军便帽
Cossack fur cap 哥萨克皮帽
Cossack hat 哥萨克皮帽(无帽檐,上宽下窄)
Cossack jacket 哥萨克夹克
Cossack look 哥萨克风貌,哥萨克款式,哥萨克型
Cossack midi 哥萨克半长外套
Cossack neckline 哥萨克领口
Cossack officer 哥萨克军服
Cossack pajamas 哥萨克睡衣裤
Cossack rainboots 哥萨克雨靴
Cossacks 哥萨克裤(裤腰打褶,脚踝处裤管口穿松紧带),肥大式灯笼裤
Cossack shirt 哥萨克衬衫(偏襟立领)
Cossack style 哥萨克款式
Cossack stylistic form 哥萨克款式
Cossack trousers 哥萨克裤
cossaes 印度松捻细平布
cossr 东南亚粗印花布,印度粗印花布

cossas 细平布
cosse green 豆荚绿(鲜黄绿色)
cost and freight (C&F) 离岸加运费价
cost-benefit analysis 成本效益分析
cost control 成本管理
costive 不透气织物,不透水织物
costo 肋骨
cost of goods sold 产品销售成本
cost of production 生产成本
costume 服装,装束,女服,全套服饰,职业装,套装,戏服；服装式样
costume art 服装美术,服装艺术
costume babies 服饰人型(17～18世纪)
costume cambric 杂色低档轧光平布
costume conception 服装意识
costume consultant 服饰顾问
costume contest 服装比赛,时装比赛
costume copy writer 服饰文案撰写者
costume culture 服饰文化
costume design 服装设计
costume designer 服装设计师,舞台服装设计师
costume doll 服饰人型(17～18世纪)
costume editor 服饰编辑
costume future 未来时装
costume history 服装史
costume identical principle 着装一致性原则
costume jacket 女式夹克,短外衣
costume jewelry 廉价珠宝；人造珠宝饰物(总称)
costume language 服饰语言
costume look 舞台服装款式
costume master 服装管理员
costume mistress 服装管理员
costume museum 服饰博物馆
costume of national minorities 少数民族服装
costume of the Zang nationality 藏族服装
costume play (cosplay) 服装秀
costume pleasure 服装快感
costume popularization 服装大众化
costume prop 服饰道具
costume property 服饰道具
costumer (为化妆舞会、剧团等服务的)服装制作人；衣商,服装商；衣帽架
costume ratailing 服饰零售业
costume rehearsal 试装排练
costumery 服装(总称),服饰,衣饰；服装设计技艺

costume security 着装安全感
costume shop 服装间
costume show 服装秀
costume show style 服装表现风格
costume sketch 服装设计
costume skirt 西服裙,直裙,直筒裙;直身裙
costume slip 定制连身衬裙
costume style 舞台服装款式
costume stylist 服饰搭配师
costume subculture 服饰亚文化
costume suit 连裙套装
costume taste 服装趣味
costume tweed 粗花呢
costume velvet 阔幅高级棉绒(统称)
costumey 服装的,戏装似的,(衣服等)过于华丽的
costumey clothes 华丽服装
costumier (为化妆舞会、剧团等服务的)服装制作人;服装衣料;服装配饰;服装设计;衣商,服装商,服装出租商;衣帽架
cosy 壶套,保暖罩
cot 手指套,脚趾套;护套
cote 凸条的,罗纹的;外套
cote de cheval 法国经条骑装布
cotehardi (14～15世纪欧洲男女服式的)柯特阿第外衣;紧身对襟长袍
cotehardie 柯特阿第外衣(14～15世纪欧洲男女服式),柯特哈迪外衣
cotel 科坦尔皱缩凸条织物
cotelaine 漂白棱纹细布
cotêlé 科特莱毛葛,凸条毛葛
coteline 经棱条细平布
cote menue 细棱纹哔叽
cote onglaise faconnée 素色法国哔叽
cote piqué 棱纹哔叽
cote satinée 单色法国哔叽,人造丝光亮硬缎
cote syrienne 法国素哔叽
cotha more 爱尔兰粗呢大衣料
cothurnus (古希腊悲剧演员穿的)半高筒厚底靴
coti americano 床罩
cotillion 彩色条纹呢裙料,黑白条粗纺呢,黑白条粗纺毛料
coton azul 色织斜纹布
coton de maquillage [法]分妆用棉
cotonette 棉毛混纺织物
cotonine 棉麻厚帆布

cotonis 丝经棉纬花式布(印度),意大利丝经棉纬花布
coton peigné [法]细棉线
coton pierre 巴西棉
cotorinas 墨西哥式背心(采用鲜明横条手织布)
cotta 短袖或无袖白法衣,外衣,束腰外衣
cottage bonnet 农舍帽,夏日草帽,考蒂姬帽
cottage cloak 农舍带兜帽头篷
cottage dress 农舍服
cottage front 乡村紧身衣
cottage style bonnet 乡村帽
cotte 考特内衣(13～14世纪),男式紧身衣,宽松束腰外衣,紧身外衣夹克;接角布
100% cotton 全棉制品
cotton 棉,棉花,棉纱,棉布,棉织品
Cotton, Silk and Man-made Fibers Research Association 棉、丝、化学纤维研究协会(英国)
cottonade 仿呢料斜纹厚棉布
cotton and wool mixtures 棉毛混纺织物,棉毛交织物
cotton baby blanket 棉童毯
cotton back satin 棉背缎
cotton bag sheeting 棉袋布,袋用织物
cotton bathrobe flannel 双面棉绒浴衣织物
cotton batting 棉絮,棉绒胎
cotton beaver 充海狸皮棉绒布
cotton Bedford 厚实凸条棉布
cotton bed sheet 纯棉床单
cotton blanket 棉毯,线毯
cotton blanket bed cover 线毯床罩
cotton blanket cloth 双面棉绒浴衣织物
cotton blanketing 棉毯料
cotton boots 棉靴
cotton broad cloth 宽幅棉布;各色细平棉布;棉绒面呢
cotton brocade 棉锦缎
cotton cambric 棉细纺布
cotton canvas 棉帆布
cotton canvas bands 帆布鞋带
cotton cashmere 棉印花斜纹;全棉仿毛开司米织物
cotton chamois-color cloth 仿麂皮棉布
cotton check cord fabric 纬昌呢
cotton checks 蓝白条格棉布

cotton chenille rug 雪尼尔棉毯
cotton chiffonvoile 棉制薄绸纱
cotton chintz 擦光印花棉布
cotton cloth 棉布
cotton clothes 棉布服装,全棉服装,棉布衣服
cotton coatray 小仓雪花布(日本)
cotton cord 棉绳;棉细棱纹织物
Cotton Council International 国际棉花委员会
cotton covert 棉芝麻呢
cotton crepe 绉布,棉绉纱,泡泡纱
cotton dacron 棉涤纶,棉的确良
cotton damask 棉锦缎
cotton diaper 菱形花纹棉布,棉尿布
cotton dobby fabric 小提花棉织物
cotton down-proof satin drill 纯棉防绒缎
cotton drawers 男内裤,男汗裤
cotton dress 棉衣
cotton drill 斜纹棉布
cotton duck 棉粗布,棉帆布
cotton dull thread 棉无光缝纫线
cotton duvetyn 棉起绒织物,棉仿麂皮绒织物
cotton dyed down-proof cloth 全棉防绒布
cottonee 土耳其棉背缎
cotton embroidery thread 棉绣花线
cotton enduit (有头巾的)防雨棉夹克
cottonette 棉毛混纺织物,棉针织泳衣布
cotton fabric 棉布,棉织物
cotton felt 双面棉绒布
cotton ferret 棉带
cotton fiber 棉纤维
cotton-filled label 充棉唛头
cotton flannel 棉法兰绒,绒布
cotton fleecy gloves 针织绒手套
cotton floater 棉包橡皮包布,棉包油包布
cotton floral print pajamas 印花棉布睡衣
cotton foulard 软薄棉绸
Cotton Foundation 棉花发展促进会
cotton gabardine 棉华达呢
cotton garment 棉布服装
cotton georgette 纯棉乔其纱,棉乔其纱
cotton glazed thread 棉蜡光缝纫线
cotton gloves 棉纱手套,纱手套,线手套
cotton goods 棉织品,棉制品
cotton hooked rug 长绒刺绣棉毯,长绒棉毯
cotton hosiery 棉袜

Cotton Incorporated 棉业联合会
cotton interlock gloves 棉毛手套,棉纱手套
cotton interlock jersey 棉毛衫
cotton interlock singlet 棉毛衫
cotton interlock singlet and trousers 棉毛衫裤
cotton interlock trousers (针织)棉毛裤
cotton interlock underwear 棉毛衫裤
cottonized chemical fiber fabric 棉型化纤织物
cottonized flax 棉化亚麻纤维
cotton jacquard fabric 大提花棉织物
cotton jersey gloves 汗布手套
cotton knit petticoat 针织棉衬裙
cotton knitted lining 棉针织里布
cotton knitted underpants 线裤
cotton knitwear 棉针织品,线衣线裤
cotton lace 纯棉花边,棉花边
cotton lasting 厚实棉斜纹织物
cotton-like fabric 仿棉织物,棉型织物
cotton-like fiber 棉型纤维
cotton-like finishing 仿棉整理
cotton lined clothes 棉夹衣,絮棉衣
cotton linen 充亚麻棉布,仿亚麻棉布
cotton lining 棉衬里布,弱捻纱表强捻纱里的经编织物
cotton lycra 全棉拉架,全棉弹力布
cotton manufactured goods 棉织品
cotton mark 纯棉标志
cotton maternity frock 孕妇棉布袍
cotton mercerized thread 棉丝光缝纫线
cotton molleton 棉莫列顿绒布
cottonopolis 棉天鹅绒
cotton-padded 棉袄
cotton-padded cap 棉帽
cotton-padded clothes 棉衣
cotton-padded coat 棉大衣,棉衣
cotton-padded jacket 棉袄
cotton-padded leather shoes 棉皮鞋
cotton-padded pants 棉裤
cotton-padded rubber shoes 棉胶鞋
cotton-padded shoes 棉鞋
cotton-padded trousers 棉裤
cotton-padded vest 棉背心
cotton pants 棉布裤
cotton piece goods 棉布
cotton pillow 棉枕头
cotton pique 凸花棉布
cotton plain fabric 平纹棉织物

cotton plains 纯棉平布
cotton plush 棉长毛绒
cotton/polyester fabric 棉涤纶织物
cotton-polyglycidyl methacrylate fabric 棉-聚甲基丙烯酸缩水甘油酯织物
cotton/polypropylene plain cloth 涤/丙平布,棉/聚丙烯平布
cotton pongee 绵茧绸,丝光棉平纹布
cotton poplin 棉府绸,棉衬衫布
cotton poplinette 纯棉府绸,纯棉全纱府绸
cotton prayer towel 全棉朝圣布
cotton press cloth 打包棉布
cotton print 印花棉布,花布
cotton print sweatshirt 全棉印花圆领衫
cotton quilted coat 绗缝棉衣,棉大衣
cotton quilted jacket 绗缝棉夹克
cotton rag 旧棉絮
cotton rag rug 碎布地毯
cotton rags 棉絮
cotton/rayon lining 黏棉里子绸
cotton rayon mixed bed blanket 线绨被面
cotton reel 棉线轴
cotton rep 棱纹棉布,纬棱条厚棉布
cotton rib 全棉罗纹
cotton rope 线绳
cotton rope in color 染色棉绳
cotton rope in nature 原色棉绳
cotton rope in white 漂白棉绳
cottons 棉制品
cotton 棉缎
cotton satin[sateen] 棉缎
cotton satin fabric 缎纹棉织物
cotton scarf (棉)纱围巾,棉绒围巾
cotton serge 棉哔叽
cotton sewing 缝棉布
cotton sewing thread 棉缝纫线
cotton sheeting 床单布
cotton shirt 全棉衬衫
cotton shoes 布鞋,棉鞋
cotton shorts 棉布短裤
cotton socks (棉)纱袜,纱短袜
cotton spandex knit fabric 棉/氨纶针织物
cotton spandex woven fabric 棉/氨纶机织物
Cotton's patent frame 自动成型平机
cotton-spun acrylic 棉型丙烯腈系纤维
cotton staples, American 美国陆地棉
cotton stockinette gloves 汗布手套

cotton stretch 弹力棉织物
cotton striped shorts 条纹棉短裤
cotton suide 绵羔皮,绵羊羔皮
cotton suiting for men 男线呢
cotton suiting for women 女线呢
cotton suitings 充毛纯棉织物,线呢
cotton suitings for man 男线呢;袍料
cotton suitings for woman 女线呢
cotton sweater 棉毛运动衫,卫生衫,卫生衣,卫衣
cotton sweater and trousers 卫生衫裤
cotton sweat shirt and sweat pant combined 套式棉汗衫裤
cotton system shirt 柯登式针织衬衫
cotton taffeta 棉塔夫绸
cotton tape 棉布带
cotton tapestry 棉织花毯
cotton terry robe 毛巾布长袍
cotton textile 棉纺织品
cotton textile agreement 棉纺织品协定
cotton thick thread 粗棉线
cotton thread 棉缝纫线,棉线
cotton thread blanket 线毯
cotton thread socks (棉)线袜
cotton thread sweater 棉线衫
cotton thread tape 线带
cotton/tinsel mixed velvet 锦萤绒
cotton towelling 全棉毛巾布
cotton towelling apron 全棉毛巾围裙
cotton tropical 充毛薄型棉织物,仿毛薄型棉织物
cotton trousering 仿毛棉裤料,厚实织物,色织绒呢
cotton trousers 棉毛裤,卫生裤
cotton tuscany laces 棉线网扣
cotton tweed 仿粗花呢色织棉布
cotton twill 全棉斜纹布
cotton twill fabric 斜纹棉织物
cotton twill shorts 全棉斜纹短裤
cotton twill trousers 纯棉斜纹裤,棉斜纹裤
cotton type (C type) 棉型,棉型化纤
cotton type chemical fiber fabric 棉型化纤织物
cotton type fabric 棉型织物
cotton type fiber 棉型化纤,棉型纤维
cotton type viscose staple fiber 人造棉
cotton underwear 棉内衣
cotton union quilt covering 线绨被面
cotton unlined clothes 棉单衣

cotton velvet 棉绒,棉天鹅绒
cotton velveteen 棉绒,纬绒,平绒,棉绒织物
cotton venetian 元贡呢;泰西缎
cotton voile 棉巴里纱
cotton wadded clothes 棉衣
cotton wadded coat 棉大衣
cotton wadded garment 棉衣
cotton wadded jacket 棉袄
cotton wadded legging 棉绑腿
cotton wadded robe 棉袍
cotton wadded short gown 短棉袄
cotton wadded trousers 棉裤
cotton wadded vest 棉袄背心
cotton wadding 棉絮,棉胎
cotton waffle pique skirt 格子凸花棉裙
cotton waist band 棉腰带
cotton wannamaker 万那梅克棉
cotton warp 棉经交织物,棉毛交织呢
cotton warp and nylon weft fabric 棉经锦纶纬织物
cotton warp linen 棉亚麻交织布,棉麻交织物
cotton warp union 棉经交织物,棉毛交织呢
cotton wax 棉蜡线
cotton wool 原棉,脱脂棉,药棉
cotton worsted 充精纺毛料棉织物
cotton wrapped core spun thread 混合包芯线
cotton yarn 棉纱
cotton yarn and cotton piece goods 棉纱棉布
cotton yarn-dyed jacquard chambray 色织纯棉提花青年布
cotton yarn-dyed preshrunk checked flannelette 全棉防缩色织棉绒布
cotton yarn gloves (棉)纱手套
cotton yarn hosiery 棉纱线袜
cotton yarn sweater 棉线衫
cotton yarn tape 纱带
cottony finish 仿棉整理
cotts 粗硬毛;粗毛毯
couche 丝平绒
couched embroidery 贴线缝刺绣
couching 挑线,贴线缝
couching embroidery 贴线刺绣
couching stitch 挑绣针迹,贴线缝绣针迹
couel 女式头巾
couleur de rose 玫瑰色,粉红色

coulisse 法国束带
Council for Textile Recycling 美国再生纺织品协会
Council of American Embroiderers 美国刺绣者协会
Council of Fashion Designer of America (CFDA) 美国时装设计饰协会
council of Fashion Designers (CFD) 东京时装设计者协会
count (布料的)纱支,支数;计数
countel lining 后跟里
countenance 小皮暖手筒
counter 后帮(制鞋),鞋后踵,鞋帮后跟部的坚硬部分
counter-change check 交错格子织物,棋盘花纹织物
counter-change pattern 交错变换花纹,交错变色花纹
counter-clockwise twist 反手捻
counter-current 对流
counter-current airing 对流通风
counter-draw 摹图,拓画,映绘
counterfoil 票据存根
counter lining 后跟里
countermeasure 对策,对付措施
counterpane 床罩
counter pattern 对称花样;对称式样
counter purchase 互购
counter revolution 反转
counter-rotation 倒转
counter sample 对等样品,回样,试销样
counter stitch 对称线迹
count-lea-strength product 品质指标
countrified ways 土气服装
country 乡村风格
country cloth 乡下布
country hide 农家皮
country line 乡村型
country look 乡村风貌,乡村式样,乡村装;乡村疗养装
country plaid 乡村格子纹
country quotas 国别配额
country shoes 乡村鞋
country stripe 乡村条纹
country stylistic form 乡村款式
country suit 乡村式套装,郊外套装,(男子)郊游装
country taste 乡村风格
country tweed 乡村花呢
country wear 郊外装,郊游装

country-western look 西部风貌(美国)
60 count yarn 60支纱
county shoes 乡村鞋
coupe 匹长
coupe-vent 风衣
coupling dye 偶合染料
coupure 开司米斜纹织物
Courréges 库雷热(法国时装设计师)
Courréges boots 库雷热靴,带内衬的长筒皮靴
Courréges flower socks 库雷热大绣花女袜
courréges glasses 库雷热太阳镜
courréges helmet 库雷热帽盔(类似宇航帽)
Courréges look[style] 库雷热风貌,库雷热款式
Courreges stylistic form 库雷热款式
courrier dummies 人台,服装展示用模型
course 横列,线圈横列
course length 列环长
course mark 横条疵
courses per inch (C.P.L) 每英寸横列数
coursing joint 成行缝
courtaille 法国大麻帆布
court breeches 大礼服裤
court dress 大礼服,朝服,宫廷礼服
Courtek M 抗菌聚丙烯腈纤维(商名,英国考陶尔)
Courtelle 聚丙烯腈纤维(商名,英国考陶尔)
Courtelle Neocrome 聚丙烯腈微细短纤维(商名,英国考陶尔)
courte pointe 凸花或印花被褥料
courtepy 粗布短外衣
courtepye 短外衣
courtex M 防菌纤维
court hat 宫廷帽
courtier boots 朝靴
court plaster 薄橡皮膏布
courtrai finish 考特赖整理
courtrai flax 比利时考特赖亚麻
courtrai lace 比利时考特赖花边
court shoes 船形高跟浅帮鞋(与宫廷礼服配穿),宫廷鞋;晚宴鞋,后带扣鞋,后袢带鞋,踵带鞋
court tie 宫廷鞋,男式低帮牛津鞋;女式系带牛津鞋
coutance 法国大麻床罩布
couteline 亚麻条纹床罩布;蓝白条棉床罩布
couter 铠甲肘盖
coutil 人字斜纹布;内衣、乳罩用布;床垫用布
coutil facon de bruxelles 法国窄条纹床垫
coutille 内衣、乳罩用布;床垫用布
coutils de brin 法国粗被褥布
couton 棉花
coutrai lace 考特赖花边(比利时)
couture 女式时装,高级时装;美国高级设计师作品;女式时装设计师和裁缝师(总称);制衣,裁缝;女式时装设计与缝制;女式时装业,高级时装店
couture clothes 定制服装
couture design 女式时装设计
couture house 时装设计室,特制时装屋,女式时装店
couture lace 高级时装花边
couture society 时髦界(西方国家称呼对时装潮流有决定性影响的人物)
couturier 男时装设计师,时装男裁缝;时装店男店主;女式时装店
couturiere 女时装设计师,时装女裁缝;时装店女店主
coututier method (花边)镶贴法
couvrechief 盎格鲁撒克逊头巾,诺尔曼头巾
covele 考维尔(聚丙烯腈超细纤维,商名,日本)
coventry bule 考文垂靛蓝绣花毛线
coventry ware 考文垂精纺毛织物
cover 盖,罩,套,(布)面,覆盖层;织物紧度,布面丰满
coverage area 覆盖区
coveralls 连衫裤装,连衣裤工作服
coveralls automatic press 自动熨烫工作服机
coveralls rain suit 连身雨衣裤
coverchief 妇女头巾,女头巾,小披巾
covercloth 帆布,帐篷布
cover coat 短外套,短大衣,短氅
cover-core 皮芯型,包芯型
covered buckle 包扣环
covered button 包纽,包扣
covered double-rim button 双圈包纽,双圈包扣
covered edge 包边
covered elastic 包线橡筋
covered elastic yarn 包覆弹力纱
covered heel 包跟,覆皮跟

covered hem 包覆式下摆
covered hooks and eyes 钩状扣子,饰缝钩眼扣
covered mold button 包布模压纽,包布环扣
covered rubber thread 包覆橡筋芯线
covered rubber yarn 包覆橡筋芯纱
covered seam 绷缝,包边缝,覆盖缝
covered shoulder pad 布包肩垫
covered snap 包按扣,包布按扣
covered sole shoes 包底鞋
covered stitch 绷缝,包边缝,覆盖缝
covered woollen-spun yarn 粗纺毛包覆纱
covered worsted 缩绒精纺毛织物
covered yarn 包缠花式纱线,包线
covered zipper 隐形拉链
cover factor (cf) 织物覆盖系数
covering chain stitch 覆盖链式线迹,绷缝线迹
covering chain stitch, top and bottom 双面覆盖链式线迹
covering kerchief 幂䍡;幂䍠(中国古代)
covering looper 绷缝弯针
covering property 被覆性,覆盖性
covering seam 绷缝,覆盖缝
covering spreader 绷缝拨线钩
covering stitch 绷缝线迹,覆盖线迹
covering stitch machine 绷缝机
covering stitch seam 绷缝,覆盖缝
covering thread 绷缝线,包覆线,面线
covering velvet 密绒天鹅绒
covering work 绷缝加工
covering yarn 包芯纱
coverlet 被单,床罩
coverlid 被单,床罩
coverlight 涂层锦纶工业用织物
cover placket 暗门襟
covers 家具套子
cover seam 覆盖缝
cover seam machine 覆盖缝缝纫机,绷缝机
covers old 包装残旧
cover spun yarn 包绕纱,长丝包缠纱
cover stitch 包缝线迹
coverstock 包布(卫生巾,尿布等),包覆料
cover stock 吸湿非织造布
covert 芝麻呢,中厚急斜纹外套料,薄平纹运动服棉布;棉斜纹工装布;毛防雨布

covert cloth 斜纹大衣呢,芝麻粗纺花呢,芝麻花呢
covert coat (骑马射击时穿的)轻皮短外套,芝麻呢轻便短大衣
covert coating 精纺花呢,宽幅线呢
covert dress (狩猎、骑马穿)轻便短外套;避尘衣
covert green 暗绿色
cover thread 绷缝线,覆盖线
covert topcoat 轻便大衣
covert twill coatings 二上二下斜纹宽幅线呢
covert weave 经面斜纹
cover-up 罩袍,罩衣,罩衫,有头巾的长袖海滨装
cover-up look[style] 兜帽长袖海滨装款式
cover up theory 遮羞说
cow 母牛皮
cowbeck 帽用毛料
cowboy belt 牛仔腰带
cowboy boots 牛仔长靴,牛仔靴
cowboy costume 牛仔服
cowboy dress 牛仔装
cowboy hat 宽边高顶帽,牛仔帽
cowboy jacket 牛仔夹克
cowboy jodhpurs 牛仔马裤
cowboy outfits 牧童外套
cowboy pants 牛仔裤
cowboy shirt 牛仔衬衫
cowboy's jacket 牛仔夹克
cowboy suit 牛仔套装,牛仔装
cowboy vest 牛仔背心
cowboy wear 牛仔服
cowboy yoke 牛仔装式覆肩
Co-We-Nit 科韦尼特经编织物
cow hair 牛毛
cowhide 母牛皮,母牛革,牛皮,12千克以上皮革
cowhide belt 牛皮带
cowhide garments 牛皮服装
cowhide upper leather 牛面皮
cowichan sweater 科维昌式厚毛衣(加拿大),印第安人厚毛衣
cowl 僧袍,道袍;苏格兰睡帽,兜帽;荡领,垂褶领
cowl collar 考尔领,垂褶领,围巾领
cowl drape 考尔式柔褶(造型)
cowl dress 带风帽衣服
cow leather 熟牛皮

cowl hood 修士风帽(连在外套上的兜帽)
cowlick 翘发
cowling 帽,套
cowl inset 嵌入式斜裁垂褶领
cowl leather 熟牛皮
cowl neck 考尔领,荡领,披帽式领,垂褶领
cowl neck dress 垂褶领女装
cowl-necked sweater 垂褶领毛衣;大翻领衫
cowl neckline 考尔领口,荡领口,垂褶领口,胸前褶领(口)
cowl neck sweater 垂褶领大衣
cowl skirt 荡裙,荡褶楔形裙
cowl sleeve 考尔袖,荡袖,垂褶袖
cowoven fabric 纤维增强树脂用交织布
cowsoong 黑色棉平布
cow split 牛二层革
cow split boots 二层革牛皮靴
cow split gloves 牛(二层)皮手套
coxcomb 鸡冠花发型,鸡冠花型,鸡冠帽(古时丑角戴);线连花边;小丑头巾
cox royal arch cotton 美国喀莱阿棉
coyote 郊狼毛皮
coyote fur 郊狼毛皮
coyote skin 小狼皮
cozy 保温套
CPO (chief petty officer) shirt 美国士官衬衫
crab apple 沙果橙
crabapple 酸苹果色
crab-back bathing suit 蟹背状男泳装
crab green 浅蟹绿
crab grey 浅蟹灰
crack 松裆,稀弄
crack dyeing 裂纹防染
cracked selvage 破边
crack grain 裂面(皮革)
cracking 龟裂(鞋)
cracking grain 裂面(皮革)(鞋)
crackle 冰裂纹;机制网眼布
crackle effect 裂纹效果
crackle finish 裂纹整理,龟裂效应
crackle net 裂纹形网眼织物,裂纹形网织物,细裂纹型
crackling effect 裂纹效应,龟裂效应,冰花效应;皱纹整理,起霜整理涂层
crackly net 机制网眼布
crack marks 稀路疵(针织物)
crack pattern 裂纹图形

crack printing 裂纹防染
cracow 尖长翘头鞋
cradle 托架,支架
cradle feature 元宝袜底
cradle sole 元宝袜底
craft 工艺,手艺
craft bag 工艺包
craft cap 工艺帽
craft design 工艺设计
craft dress 工艺服装
crafts 手工艺
craft shoes 工艺鞋
craft(s) manship 技艺,手艺
craftsmen's machine 手工缝纫机
Craft Yarn Council of American 美国手工编织委员会
Craiganputtach 苏格兰粗纺呢
crakow 尖长翘头鞋
crammed stripe 密经条纹,密纬条纹
crammed striped fabric 密条纹织物
crammed stripe fabric 稀密织物
cramoisie, cramoisy 深红色的;红布,深红色布
cranberry 酸果曼红(深红色)
cranette serge 精纺细哔叽
crank connecting rod complete 曲柄连杆组
crank driving 曲柄传动
crank opening 曲襟
crank range 曲柄动程
cranky 小方块亚麻布,棉麻褥套布
cranky checks 英国蓝白方格棉布
cranston 克兰斯顿彩色条格呢
crape 黑纱;绉呢,绉绸,绉布;针织薄纱
crape cloth 绉纱毛织品,绉织物
crape de chine 双绉
crape hair 人造毛发(演员用)
crape moirett 细经粗纬松结构呢
crape sole 凹凸鞋底,凸鞋底
craping defect 绉疵
craping machine 起绉缝纫机
crapyness 折皱性,皱折性
craquele net 机制网眼布
crash 粗布,白粗布,粗平布
crash effect 粗麻布效应
crash helmet 安全帽,防护帽,滑雪头盔
crash towel 粗毛巾
crash toweling 粗毛巾布
crass 粗厚的麻布

crate 柳条箱
cravat （旧式）领带，克拉瓦特领带；围巾，女用围巾，克拉瓦特围巾；领饰，领巾
cravate étroit ［法］细窄领带
cravate façonnée 法国提花领带绸
cravat with pin 带领夹式饰针的领带；带饰针的领巾
cravenette 大花型瑰丽印花装饰布；克莱文特防雨布，克莱文特防雨呢
cravenette worsted cloth 克莱文特呢
crawcaw 长头鞋
crawford cotton 美国克劳福德棉
crawlers 童裤，儿童背带裤，田鸡装，田鸡服，田鸡连衫裤（童装）；婴儿罩衣裤
crayon 粉笔，蜡笔，颜色笔；蜡笔画
crayon chalk 细白条纹
crayon dyeing 色笔描绘染色
crayon stripe 粉笔条纹，细白条纹
crayon touch 蜡笔画风格
craze 狂热流行
crazing 裂浆（鞋）
crazy cloth 无规则花纹棉布，自由纹布
crazy Madras 碎马德拉斯狭条布
crazy quilt 碎布缝成的被褥，百衲被褥
crazy shirt 新潮印花运动汗衫
crea 硬挺漂白棉布
cream 奶黄色，奶油色；润肤油质制品
cream blush 奶油红
cream color 淡黄色，奶油色
cream damask 奶油色棉，亚麻花缎
creames 奶油色华达呢
cream gold 奶油金黄色
cream pearl 奶白珍珠；奶白珍珠色
cream pink 奶油玫瑰红
cream puff 奶油泡芙色
cream tan 奶油咖啡色
cream twill linen 刺绣用亚麻布
creamy white 奶白色
crea para sabanas 漂白棉衬衫布
creas 漂白亚麻帆布
creas anchas 波希米亚粗麻布
crease 打折痕，弄皱，起皱，折痕，（衣服等）折缝，褶子，皱，皱痕，皱折，褶裥，死折痕（织物疵点）
crease and shrink resistant finish 防皱防缩整理
crease angle method 折皱角度试验法
crease bar 皱档
crease finished fabric 轧绉织物

crease finishing 折皱整理
crease-fold dart leg 烫折后省柱
crease folding 皱折折叠
crease line 裤缝线，烫迹线；折缝，折线
crease line flare 挺缝线喇叭造型
crease line inward 烫迹线内撇
crease line leans to inside 裤烫折线内撇，烫迹线内撇
crease line leans to outside 裤烫折线外撇，烫迹线外撇
crease line out ward 烫迹线外撇
crease mark 印花皱印（疵病），折痕疵，折印
crease pressed-in 熨烫裤线，熨烫折痕
crease proof finish 防皱整理
creaser 碎褶压脚
crease recovery 回折性，折皱弹性，折皱回复性，折痕回复性
crease recovery angle 折皱回复角；折痕回复角
crease resistance 防皱性，抗皱性，耐折皱性
crease resistance finish fabric 防皱整理织物
crease resistant fabric 防皱织物
crease resistant finish 防皱整理
crease resistant finish fabric 防皱整理织物
crease resist finish 防皱整理
crease resisting cloth 防皱布，防皱织物
crease retention 褶裥保持性，折缝耐久性
creases at right facing 底襟里起皱，里襟里起皱
creases at underarm 腋窝起绺
creases below neckline 领窝不平
crease sensitivity 易折皱性，折皱灵敏性
crease setter 裤线定形器，褶印定形器
crease sharpness 褶裥清晰度
crease sharpness index 折痕清晰度指数
crease-shedding 除皱
creases in trousers 裤腿折痕，裤线
creases on two ends of pocket mouth 袋口角起皱
crease stitching 起皱线缝
crease streak 皱折条花疵
crease turner 翻角熨平器
creasing 起皱
creasing length 折幅
creasing machine 槽型压边机，折幅机，褶裥机

creasing property 折皱性
creating meaning storehouse 服装创意库
creation 定制服装；服装作品；高级时装创作
creative design 创意设计
creative designer 创意设计师
creative idea 创意
creative imagination 创造力
creative work 创作
creator 时装设计师
credibility 可信度
credit 信用；信用证
creedmore 劳动鞋，劳动靴
cree moccasin 克雷软底鞋
creep 蠕变
creepalong set 两件套婴幼儿服
creeper （婴儿）连衫裤，爬行服，小孩罩衣，田鸡装，蛤蟆装
creeper pajamas 婴儿连裤袜睡衣
creepers 儿童背带裤，蛤蟆装，（婴儿）爬行衣裤，田鸡装，田鸡服，田鸡连衫裤（童装）
creep-free hemming 无滑移卷边
creep-free seam 无潜伸缝
creeping apron 婴儿连衫装
crefeld velvet 克拉菲尔德丝绒
cremes 奶油色华达呢
cremona 聚乙烯醇系纤维（商名，日本可乐丽）
crenelated hem 城垛形衣边，方齿形衣边
crenelle 城垛形衣边，方齿形衣边
crenulated 异形丝
Creole 克里奥耳鞋
Creole evening dress 古法式晚礼服，克里奥耳晚礼服
creole wool 阿根廷克雷奥尔羊毛
creoulo cotton 仙乐棉
crepaline 绉边薄织物
crepe 绉，绉呢，绉绸，绉布，绉纱，绉纹布，绉纹呢
crepe algerian 印花细棉绉布（商名）
crepe alpaca 羊驼绉
crepe antique 重双绉
crepe back satin 绉背缎
crepe bar 绉档疵
crepe beatrice 经向隐条绉向织物（商名）
crepe berber 匹染起绉茧绸（商名）
crepe brocade 叠花绉
crepe charmeuse 查来尤斯绉缎，夏缪斯绉缎

crepe chenette 塞内特绉
crepe chiffon 雪纺绉，薄绉（棉织物），俄罗斯雪纺绸
crepe corduroy 绉纹灯芯绒
crepe crepe 重捻绉
crepe damask satin 花绉缎
crepe de chine （俄罗斯）双绉，丝经毛纬绉，精梳棉线绉
crepe de chine faconne 小花双绉
crepe de chine travers 横棱双绉
crepe de dante 棉毛丝交织绉
crepe de laine 薄型绉呢
crepe de santé 法国卫生内衣绉，毛经丝纬桑特绉
crepe d'espagne 西班牙绉，丝经毛纬稀绉
crepe diana 棉丝交织绉缎
creped-soled shoes 软橡胶底鞋
crepe effect 绉纹效应
crepe elizabeth 伊丽莎白绉
crepe embossing 拷花绉，绉纹拷花
crepe fabric 胡桃呢，绉纹呢，绉织物
crepe faille 罗缎绉
crepe faille sublime 重磅罗纹绉绸
crepe fault 裙子皱
crepe finish 绉缩整理
crepe flannel 绉拉法兰绒
crepe georgette 乔其纱，乔其绉
crepe gown 绉纱女袍
crepe jacquard 提花绉，花绉
crepe janigor 杰尼贡绉，棱绉绉
crepe jeans 皱纹牛仔裤
crepe jersey 乔赛绉
crepe knit 绉纹针织物，绉纹针织品
crepe knitted fabric 起绉针织物
crepe lady's dress 绉纹女衣呢
crepe lady's dress cloth 绉纹女衣呢
crepe lady's dress worsted 绉纹女衣呢
crepe lease 亮光绉
crepe-like staple 高卷曲羊毛
crepe-like texture 绉状结构
crepe line 充双绉
crepe lisse 亮光绉
crepella 强捻平纹呢
crepe marocain 马罗坎横棱纹，马罗坎绉
crepe meteor 流星绉
crepe mohair 马海绉（蚕丝与马海毛纱交织）
crepe morette 莫雷特绉呢
crepe mosseux 股线防缩绉纱
crepe motette 强捻经纱薄绉

crepenette 绉纹棉府绸,绉纹丝府绸
crepe nodule 波纹绉
crepe olive lapel 皱纹青果驳领
crepe ondese 粗黏胶绉,翁迪斯绉
crepe ondor 翁多绉
crepe ondule 波纹绉丝织物
crepe outsole 皱胶底(鞋)
crepe plissé 泡泡纱
crepe poplin 府绸绉
crepe rachel 棉经毛纬印花绉
crepe radio 棱纹绉,棱纹绉丝织物
crepe rayon 人丝绉,人造丝绉
crepe-resisting cloth 防皱织物
crepes 绉,绉织物
crepe satin 贝绉缎,绉缎,绉背缎
crepe satin brocade 花绉缎
crepe satin plain 素绉缎
crepe setting 绉纱定型,绉织物定型
crepe shirt 绉绸衬衫
crepe silk 绉丝线
crepe-smoothing steamer 蒸汽除皱器,蒸汽除皱熨机
crepe-soled shoes 鸡皮绉底鞋
crepe suzette 顺纤乔其
crepe tape 绉带
crepe twalle 人丝绉绸,人造丝绉绸
crepe twist 紧捻,绉捻
crepe twist yarn 绉捻纱
crepe weave fabric 绉纹织物
crepe weave like wrinkle 绉纱型皱纹
crepe woolen 皱纹毛织物
crepe wool fabric 绉纹毛织物
crepe yarn 绉捻线,绉纱
crepe zephyr 色织条子棉绉纹布
crepide 库雷匹达鞋(古罗马)
crepine 小点绉绸;饰边
creping 起绉,起皱整理(强捻织物用水或碱液等处理,使织物表面形成皱纹)
crepoline 棱纹绉
crepon 古立波呢,树皮绉织物,厚绉纱,重皱纹织物
crepon beorgette 顺纤乔其纱
crepon effect 绉缩效应
creponette 绉府绸
crepon finishing 绉缩整理
crepon georgette 乔其绉,顺纤乔其纱
crepon givre 斜纹绉
creponne 家具布
crepon style 绉纹款式
creppella 绉纹布,绉纹呢

crequillas 轻薄棉布
crescent 粗丝线;月牙形首饰;月牙形花边
crescentin 绢纺绸
crescent pocket 新月形口袋,月牙形袋,月牙形口袋
crescent-shaped bag 新月形手袋
crescent-shaped box 月牙形提袋
crescent-shaped pad 新月形肩垫,月牙形肩垫
crescent shape lace 蜈蚣边
crescent sleeve 新月袖,月牙袖
creseau 刺绣用丝织物,双面起毛斜纹呢
creshi 克勒希丝交织物
cresida sandal 露趾凉鞋
creslan 克雷斯纶(聚丙烯腈纤维,商名,美国氰胺公司)
crespinette 丝发网
crespolina 棉织物
cress green 芹菜绿,水芹绿
crest 顶饰,前饰,羽毛饰;盾形纹章,饰章;头盔
crestfil 克雷斯特菲尔仿麂皮织物(商名,日本)
crests (服装上的)纹饰
cretan lace 棱结花边
cretan stitch 克兰顿针迹,叶状绣花针迹,克里特绣
crete 提花窗帘布;环边凸纹花编带
crete braid 克里特编带,环边凸纹花编带
crete lace 克里特花边,克里特棱结花边,希腊式挖花花边
cretes 大提花细平布
cretona 棉牛津布
cretonne 白麻布,大花型印花装饰布,大花型瑰丽印花装饰布
creva 巴西抽绣
cré Ve-coeur [法]伤心发式
crevell 棉底真丝纬绒织物
crevenette 色织华达呢,杂色花线制织的华达呢,华达呢类织物,克赖文内特防水处理,雨衣料,防雨卡
crew cut 平头,水手发型
crewel 刺绣用松捻双股细绒线,松捻双股细绒线
crewel lace 松捻绒线镶边,松捻绒线饰边
crewel needle (长眼)绣花针
crewel silk 绣花线
crewel stitch 刺绣线迹,绒线刺绣线迹
crewel work 绒线刺绣,绒绣

crewel yarn 刺绣用细绒线
crew look 水手款式
crew neck 船员领,水手领,圆领
crew neckline 海员领口,圆口紧领,水手领口
crew neck pullover 圆领套头衫,水兵领套头衫
crew-neck sweater 水手领毛衫,圆领套头毛衣
crew socks 2+2罗口短袜,船员短袜,水手袜,毛线短袜
crew sweater 船员毛衣
criade 克瑞得裙撑,防水布裙撑
crib blanket 粗毛毯
crib cloth 粗筛布
crib mosquito net 罩式蚊帐
cricket-cap 鸭舌帽
cricketing 板球呢
cricket shirt 男孩两用衫,运动衫,板球衫
cricket sweater 板球针织衫,板球套衫
crimp 皮革弯折器,屈曲
crimpability 卷曲性
crimp cloth 泡泡纱
crimp collar band 底领起皱起绉
crimp crepe 卷缩丝绉织物
crimp developing fiber 可卷曲纤维,潜在卷曲纤维
crimped fiber 卷曲纤维
crimped loop ruche 卷曲毛边,褶裥饰边
crimped rayon staple 卷面黏胶短纤维
crimped-set yarn 卷曲定形丝
crimped staple 卷曲短纤维
crimped viscose 卷曲黏胶纤维
crimp fabric 泡泡纱,绉纹织物
crimping 翻边,卷曲,卷边
crimping fringe 绉丝穗
crimple 皱缩,皱褶
crimp-proof finish 防皱整理
crimp retention 卷曲保持性
crimps 卷发,卷曲头发;绉纹织物
crimp stripe 泡泡纱条纹,皱条纹
crimpy top collar 领面起泡
crimpy wool 卷曲羊毛
crimson 绯红,梅红,深红
crimson red 深红,艳红色
crin 马毛;粗丝,蛹丝
crinkle 皱,皱纹,折皱,波状,卷曲,绉纹;条子泡泡纱
crinkle cloth 泡泡绉,绉布
crinkle cotton 棉泡泡纱

crinkled crepe 泡泡纱
crinkled effect 起绉效应,印花皱缩效应
crinkled fabric 绉条织物
crinkled fiber 卷缩纤维
crinkled textured yarn 波曲丝,卷缩变形丝,波曲变形丝
crinkled wool 卷曲处理羊毛
crinkle fabric 顺纤绉,绉条织物
crinkle finish 泡泡纱整理;皱缩整理
crinkle georgette 顺纬乔其,顺纤乔其纱
crinkle patent leather 皱漆皮
crinkle retention 绉缩保持性
crinkles 走样;皮革变形
crinkle type yarn 卷缩变形丝
crinkle yarn 编拆法变形丝,假编法变形丝,卷缩纱
crinkling 卷曲,起皱,皱缩
crinkly cloth 绉纹织物
crinol 人造马鬃毛
crinolette （旧时的）箍形裙架,箍形裙衬,女式腰后小裙撑
crinoline 硬衬布;裙撑,硬衬布衬裙,克丽诺琳裙撑,(旧时)裙衬架,衬架裙;传统结构的裙衫
crinoline era 裙撑时期
crinoline hoop 衬裙箍,裙撑箍
crinoline muslin 硬挺细布,硬挺细布衬
Crinoline period costume 克里诺琳风格服装
crinoline petticoat 硬衬衣衬裙,衬架式女裙
crinoline skirt 膨鼓裙,硬衬布衬裙
crinoline sleeve 克丽诺琳袖
crinoline style 膨鼓式,裙撑式
crin silk 马毛;钓鱼丝
crin vegetal 人造鬃丝,植物性鬃丝
criolla wool 阿根廷克雷奥尔羊毛
criollo cotton 巴西克里奥洛棉
cripy 65 芳香味聚酯中空纤维(商名,日本)
crisp 挺括,线条利落;挺爽手感,织物挺括感;英国优质细亚麻纤维;绉布
crisped crepe 绉线;绉绸
crisp handle 挺爽手感
crisp-high tech apparel 劲健型高技术服装,挺爽型高技术服装
crispin 无领斗篷;女式短斗篷;男式晚间斗篷;[法]皮护手
crispinette 遮面网,面纱
crispness 干爽性,挺爽性

crisp sheer 挺爽的透明薄织物
crisp style 劲健风格,挺爽风格
crisp-white 纯白色的
crispy handle 坚挺手感,爽脆手感
criss-cross back 交叉式后背
criss-cross girdle 交叉式紧身褡
criss-cross halter top 垂瀑式吊带交叠襟上装
criss-cross lines 交叉线花型
criss-cross stitch 十字缝迹,十字交叉线迹
criss-cross strap 交叉成十字形的皮革带,十字形带子鞋
cristalline 高支双经双纬席纹布
Cristobal Balenciaga 克里斯特巴尔·巴伦夏加(1890~1972,法国籍西班牙时装设计师)
criterion 准则
critical operation 关键操作
critical outlets 精密织物,要求高的纺织产品
critical pressing 临界压烫
croc bag 鳄鱼皮包
croceate color 藏红花色
crochet 钩针;钩编花边;钩针编织品
Crochet Association International 国际钩针编织协会
crochet bag 钩编手袋
crochet beading 小珠饰缀缝
crochet blouse 抽纱衫;通花衣;钩编衫
crochet braid 钩编用棉辫带
crochet cotton 钩编棉线
crochet cotton braid 钩编用棉辫带
crocheted bag 钩编手袋
crocheted blouse 通花衬衫
crocheted button 钩编纽扣,钩针编织纽扣
crocheted cap 钩编无檐帽
crocheted dress 钩编服装,钩编针织服装
crochet edging 钩编滚边,钩编饰边
crocheted hat 钩编无檐帽
crocheted jute bag 黄麻钩织包
crocheted lace 钩针编织花边
crocheted scarf 钩编围巾
crocheted shawl 钩针披风
crocheted shoes 钩编鞋
crocheted vest 钩编背心
crochet elastic cord 钩编松紧带
crochet garment 手编钩针衫
crochet gauze 钩花纱罗
crochet gloves 钩编手套

crochet hook 钩编针,钩针,手工钩针
crocheting 钩编;钩编工艺品
crochet knit 钩编,钩编领带布
crochet knitwear 钩针衫
crochet lace 钩编花边,钩针编织花边,绦子花边
crochet look 仿钩编织物外观,类钩编织物外观
crochet look fabric 仿钩编织物
crochet machine 钩编机,钩针针织机
crochet needle 钩编针,链缝针
crochet quilt 提花棉床单
crochet shawl 钩编围巾
crochet silk 钩编丝线
crochet stitch 钩编线迹
crochet sweater 钩针衫
crochette 钩编织品
crochet twist 钩编丝线
crochet twist silk 钩编用丝线
crochet wear 钩编服装
crochet work 钩编花边
crochet yarn 钩编纱线
crocidolite asbestos 青石棉
crocking 摩擦褪色
crocking fastness 耐摩擦色牢度
crocodile 鳄鱼,鳄鱼皮
crocodile cloth 类鳄鱼皮凸条纹毛织物
crocodile design 鳄鱼皮花纹,鳄鱼纹
Crocodile kids 小鳄鱼
crocodile leather 鳄鱼(皮)革
crocodile leather skin 鳄鱼皮
crocodile skin 鳄鱼皮
crocodile spanner 鳄头扳手
crocodiling 鳄纹
crocus 藏红花紫,深玫瑰紫;橘黄色
crocus cloth 粗袋布
croded tuck 衬线褶裥
croisé 斜纹织物;织物背面斜纹;棉经毛纬薄呢
croisé silk 三页斜纹里子绸
croisé velvet 斜纹地平绒
croix de chevalier 模板印花装饰布;歇瓦利埃印花装饰呢
Crombie 克龙比式;克龙比式大衣(或上衣、夹克等)
Crompton Axminster carpet 克朗普顿阿克斯明斯特地毯
cromwell collar 克伦威尔宽翻领
cromwell shoes 克伦威尔女鞋
crook back 驼背

crooked cloth 走样印花布,起伏不平的布
crooked selvage 曲边
crooked shoe 特制适合左或右脚的鞋
crooked stitch 线迹歪斜
croos bred wool 半细毛
crop 剪平头;整张鞣革
cropped pants 长度在膝至踝关节内的裤子,截短裤
cropped sweater 及腰毛衫;及腰毛线上衣,短毛线上衣,短羊毛衫(长至腰围)
cropped terry 针织绒
cropped terry pile 割圈丝绒
cropped top 腰节以上的短上衣,剪短上衣
cropped wading 防水短夹克
cropped worsted 剪毛精纺织物
cropper jacket 剪短夹克
cropper shirt 剪短衬衫
cropping 剪毛
crop top 短上衣,腰节以上的短上衣,剪短上衣
croquet [法]锯齿形花边
croquis 时装画稿,草图,速写,素描
cross 交叉,十字架项饰,十字形
cross back width 背宽
cross bandage 十字绷带
cross bar 短横线,横档疵;开口止点;棱格布
crossbar check 纵横格纹
cross bar dimity 棱格细布
cross basket stitch 编篮式线架
cross belt 和服挂袖带,斜挂皮带
cross bobbin winder 十字形梭芯绕线器
cross border 横边花,花式横边,织物纬饰条
cross bred fabric 杂交种毛织物
cross bred wool 杂交种羊毛
cross bred wool fancy suiting 半细毛花呢
cross breeding flax 杂交亚麻
cross check 相互校验
cross cloth 发带,扎头巾
cross coloring 经纬异色花纹
cross coordination 交叉协调
cross corded fabric 纬向凸条布
cross country gear 越野滑雪服
cross country skating shoes 越野滑雪鞋
cross country ski boots 越野滑雪鞋
cross cultural look 交叉文化风貌
cross cut 横纹裁剪,横裁
cross cutting 横裁

cross dress 穿异性服装
cross dyed cloth 交染织物
cross dyed effect 交染效应
crossed-band neckline 交叠式V字领口
crossed blanket stitch 交叉毛毯绣,十字锁边绣
crossed dyeing 交染染色法
crossed front edge 止口搅
crossed seam 十字缝,交叉缝
crossed twill 芦席斜纹
cross-elasticity 交叉弹性
crosses 小块毛皮
cross-eyes 斗鸡眼
cross fiber 十字形截面纤维
cross fox 十字狐毛皮,十字狐皮
cross front 交叉前襟
cross gartering 交叉式绑袜;交叉式绑带
cross gauze 纱罗织物
cross-girdled belt 希腊绑扎饰带,希腊饰带
cross hat 十字帽
cross hatch 断面线,网纹,网状影线
crossing at the back vent 背衩搅
crossing at the front edge 止口搅
crossing bulky seam 十字形凹凸缝,大十字形松散缝
cross joint 节点
cross knot 十字结
cross lacing 交叉系带
cross-laid yarn fabric 交叉铺纱黏合非织造布
cross land cotton 美国克罗斯兰棉
cross laser marker 十字镭射灯
cross-laying 交叉叠铺
cross liabilities 相对责任
cross line pattern 十字格花纹
crosslinked cellulose fiber 交联纤维素纤维
crosslinked polyacrylate fiber 交联聚丙烯酸酯纤维(高吸水性纤维)
crosslinked polymer fiber 交联型聚合物纤维
crosslinked rayon 交联黏胶纤维
crosslinked rayon staple 交联黏胶短纤维
crosslinked viscose rayon 交联黏胶丝
crosslink finishing 交联整理
crossmark 十字标记
cross muffler collar 交叉领巾领(毛衣领)
crossover 交叉,横纹襟纽,色纬横条布;异性融合

crossover bra　交叉系带式胸罩
crossover button　横纹襻纽
crossover collar　叠领，横纹领，交叉领
crossover design　十字形花纹，横条花纹
crossover dress　前叠襟衣裙
crossover fabric　横条花纹织物
crossover fashion　交叉型款式，融合款式
crossover fichu with sash　有饰带的交叠型肩巾
crossover front closing　叠门襟
crossover in fashion　交叉型款式，新音响款式(1977年日本男用时装设计主题)
crossover leno　窗帘纱
crossover neck　横交领形
crossover neckline　叠领，叠领口，横纹领形
crossover sleeve　交叉型袖
crossover stitch　十字交叉线迹
crossover stitching　十字交叉线缝
crossover strap　交叉搭带鞋
crossover style　横条花纹
crossover thong sandal　夹带凉鞋
crossover tie　交叉领带(在脖前交叉扣住)
crossover V-neck　V形叠领
cross piece　十字块
cross-plating knit fabric　交换添纱针织物
cross pocket　横开口袋
cross pointing motif　交换添纱花纹；十字形针绣图案
cross-rib　纽花
cross ribs　横棱纹
cross sacing　十字花纽带
cross seam　十字缝，交叉缝
cross section　横截面，截面
cross-shaped fastening　十字形加固缝
cross-shaped fastening seam　十字形加固缝
cross shawl collar　交叉围巾领
cross shoulder　肩宽
cross spreading　双合铺布
cross stitch　十字线迹，十字交叉线迹；挑缝，十字缝；十字绣
cross stitch canvas　刺绣十字布
cross stitch crotch　绷十字缝，花绷十字缝
cross stitch embroidery　十字刺绣
cross stitch fagoting　十字缝针法接缝
cross stitching　十字线缝，十字线迹，十字刺绣
cross stitching crotch　花绷十字缝
cross stitching facing　十字缝贴边
cross stitch tack　十字缝加固
cross stitch work　挑花
cross strap shoes　交叉绊带鞋
cross striation　横条纹
cross stripe　横条纹，横格条纹；色纬横条子布
cross tape　窄布条
cross thong sandals　夹带凉鞋
cross thread carrier　过线钩
cross triangle　三角形装饰缝褶；珠地网眼织物
cross tuck　交错活褶；珠地网眼织物
cross tuck knit fabric　交叉集圈针织物
cross tucks　十字缝褶，十字塔克
cross ways grain　横向纹路，横向丝绺
cross-winding　交叉卷绕
crosswise　横向
crosswise fold　横向折皱印
crosswise fold layout　对折排料
crosswise grain　横丝绺
crosswise grainline　横向布纹
crosswise relaxation　横向松弛
crosswise shrinkage　横向收缩
cross woven fabric　纱罗织物，绞经织物
crotalaria juncea　印度麻，菽麻
crotal color　红棕色，金褐色
crotch　大腿根处，裤裆，两腿分叉处，胯部，横裆；裤衩；裤浪底(十字骨)；T形拐杖
crotch area　裤裆区
crotch base　股长上点，直裆底，直裆基点
crotch depth　立裆，直裆，上裆，股上，裤裆深，直裆深
crotch depth ease　直裆松量
crotch depth line　直裆线
crotched shoes　树叉形鞋，丫巴鞋
crotchet　钩针编织品
crotchet cap　钩编帽
crotchet chain　钩编链
crotchet shawl　钩编围巾，钩编方形披巾
crotch fitting　裤裆合身度
crotch height　裤裆高，胯高
crotch length　裤裆长，裆长，股上，直裆围
crotch length adjustment　裤裆长度调节
crotch level　裤裆线
crotch line　横裆线，裤裆线
crotch lining　裤裆里布，裤裆里布，直衬里
crotch measurement　裆位测量，裆位尺寸
crotch panel　裤裆垫布

crotch pants 裆裤
crotch piece 裤子叉裆片料,并裆料,拼裆
crotch placket 直裆门襟
crotch point 裆点
crotch press （裤）浪底熨烫机,裤裆压烫机
crotch reinforcement 裆部加固
crotch seam 裆缝
crotch stay 裤裆加固条
crotohet chain 钩编链
crotohet edging 钩编织品饰边
crow black 乌鸦黑
crowded stitch 线迹过密
crowfoot 四页破斜纹
crowfoot satin 土耳其缎,四枚缎
crow foot twill 四页破斜纹
crow hat with barbette 13世纪中叶妇女带缠头巾的冠状帽
crowise fold 横向折叠
crown 帽顶,帽身,冕,皇冠,王冠,花冠;头部;头纱;蜂腰帽
crown ease 袖山吃势
crown height 袖山高
crown jewels 御宝
crown knot 冠结
crown lining 帽顶衬布
crown sable 冠貂皮
crowon height 袖山高
crow's feet 鸦爪印疵,皱纹印,爪纹,鸡爪印
crow's foot 鸦爪固定缝,松叶固定缝
crow's foot tack 鸦爪饰缝
crow twill 三上一下斜纹
crow weave 三上一下经面斜纹,一上三下纬面斜纹
croydon 克劳伊登布
croydon finish 克劳伊登整理（整理后棉布纹路清晰）
crozier 主教权杖
crudillo corona 未漂亚麻布
crudillo gallo 西班牙亚麻布
cruiser 无袖麦基诺夹克
cruising blazer 大型快艇用的运动型西装,巡航运动夹克,运动型西装（乘快艇用）
cruising wear 艇用服装,巡航服,远洋巡航服
crumb cloth 彩线绣厚花缎,亚麻厚花缎
crumping 起皱
crumple 压皱,折皱,起绉;皱纹

crumples at dart point 省尖起泡
crumples at the collar 领面起泡
crumpling resistance 抗皱性
crunching feel 嘎吱声手感
crus 胫,小腿
crusade 18世纪英国粗纺呢
crusader hood 十字军连领帽,十字军头巾
crusader's cross 马耳其十字架挂饰,十字军徽项链,十字军十字架挂饰
crush 亚麻细平布
crushable clothe 可折叠钟形帽
crushable pouch 可揉袋
crush and crease resistant finish 防皱整理
crushed berry 赤豆红
crushed fedora 可折叠费多拉帽
crushed grape 榨葡萄色
crushed leather 粗糙皮革,未加工皮革;压光皮革
crushed plush 拷花长毛绒,压花长毛绒
crushed strawberry 桃红色,深粉红色
crushed style 插入型（用于胸袋装饰手绢）,花瓣式饰巾,塞入式饰巾
crushed velour 压花丝绒
crushed velvet 压花丝绒,拷花丝绒
crushed violets 碎紫罗蓝色
crusher hat 可折叠的帽子,软毡帽
crush finish 折皱整理
crush hat 可折叠的帽子,软毡帽,摺叠帽
crush mark 皱痕,压痕,揉皱痕
crush proofing 防皱整理
crush proof pile 抗压绒头
crush resistance 抗压回复性
crush resistant velvet 防倒绒绒织物
crush towel 压花毛巾料,拷花毛巾料
cru silk 本色丝,天然色丝
crutch 裤裆,两脚分叉处,胯部;裤衩;裤底十字骨;T形拐杖
crutch depth 裤裆深
crutch gusset 裤衩;裤衩裆布,裤浪底拼布,裤衩布
crutch hat 可卷式男呢帽（便于装入口袋）
crutch lining 裤裆衬里,裤裆里子
crylor 聚丙烯腈短纤维（商名,法国罗纳-普朗克）
cryptic color 保护色,隐藏色
crystal 水晶;水晶饰物;水晶色
crystal blue 水晶蓝
crystal brocaded velvet 水晶绒

crystal button 水晶扣
crystal cream 奶白色
crystal crepe 水晶绉
crystal embroidery 水晶刺绣
crystal fiber 晶体纤维
crystal glasses 水晶片眼镜
crystal green 水晶绿
crystal grey 水晶灰色
crystalline 丝毛棱纹呢
crystalline fiber 晶态纤维
crystalline muslin 棉-人丝交织棱条细布
crystallized glass fiber 微晶玻璃纤维,晶化玻璃纤维
crystal millennium dress 水晶千禧裙
crystal ornament 水晶饰物
crystal pleated 水晶褶裙,有细小的热定形褶裥的褶裙
crystal pleats 水晶褶
crystal pleat skirt 明褶裙,水晶褶裙
crystal rose 水晶玫瑰红
crystal shoes 水晶鞋
crystal transfer 水晶烫贴
crystal wash 水晶洗(牛仔布用)
C/T corduroy 棉/涤灯芯绒
C/T fabric 棉涤纶织物
CTN. No. 箱号
c type 偏瘦体型(服装体型分类代号)
C-type merino C型美利奴羊毛
C-type merino wool C型美利奴羊毛
cub 幼狐色
cuban heel 半高跟鞋,元宝形中跟鞋;古巴跟,古巴式鞋跟,直形鞋跟
cuban hemp 古巴叶纤维
cuban jute 古巴黄麻韧皮纤维
cuban sand 古巴砂色
cubavera jacket 有四个明贴袋的白色棉运动夹克
cube 长方体
cube heel 方形跟
cubica 英国库比卡薄呢
cubic skirt press 立体烫裙机,女装裙整烫机,立体裙子整烫机
cubist design 立体派图案
cubital 袖套
cubitus 前臂
cuboid 矩形体,骰骨
cucullus 库库勒斯帽(古罗马)
cucumber braid 狭条梭结花边
cucumber green 爪皮绿
cuddly look coat 库特立式外套

cue 发辫
cueitl 彩色围裹式长棉裙
cuero de diablo 厚斜纹棉布
cuff 袖口(宽),袖头(宽),卡夫;衣袖卷边,裤脚翻边;套袜罗口;长手套的腕部;翻口皮;装袖头;袖口布
cuff band 假翻袖头,袖口带状物,袖口搭袢,假反袖口,袖口狭饰边
cuff basting 袖头粗缝
cuff binding patch 袖口滚条贴片
cuff bottom line 袖头止口线
cuff bracelet 椭圆手镯,袖头式手镯
cuff brim 卡夫帽檐,袖头式帽檐
cuff button 链扣,袖扣,袖纽
cuff cloth 袖头布
cuff cover 套袖,袖套
cuff ease 袖头宽松量
cuffed boat neckline 船形翻领口
cuff edge line 袖头上口线
cuffed hem 翻裤脚
cuffed leg 翻边裤管,反折式裤管
cuffed shorts 翻边短裤
cuffed sleeve 带袖头衣袖,翻边袖,翻袖口
cuffed socks 翻脚口袜,折边袜子
cuffed trousers 翻贴边裤
cuff-entry placement 袖克夫定位
cuff extended (松紧)袖头拉度
cuff guards 护腕
cuff height 袖头高
cuff hem 袖口折边
cuff interfacing 克夫里衬
cuff interlining 袖口衬,袖头衬
cuff jacket with rib 给夹克衫绸罗纹袖头
cuff join 袖头缝合
cuff lace 袖口饰边
cuff length 袖口长,袖罗纹长,袖头长
cuffless 没有折回裤脚边的,不翻边裤子;无袖头,无卷边袖头;无脚口卷边裤,无裤脚卷边
cuffless bottom 不翻裤脚口,无翻脚口脚口
cuffless trousers 不翻边男裤
cuff line 克夫线,袖口线,袖头线
cuff lining 袖口衬里,袖头里子
cuff link 袖扣,袖口链扣,袖纽
cuff-link hole 衬衫袖扣孔,袖扣孔
cuff open 袖口全长
cuff opening 袖口,袖头,袖衩,袖克夫开衩

cuff pants 翻贴边裤
cuff panty 平口裤
cuff pressing machine 袖口压烫机,袖头压烫机,压袖头机
cuff relaxed (松紧)袖头平度
cuff rib 袖口罗纹
cuff rib length 袖口罗纹长
cuff rib width 袖口罗纹宽
cuff roll line 袖头上口线,翻边的翻折线
cuff ruff 襞褶袖口
cuffs button 袖头装饰纽扣
cuff slit 袖口开衩,袖头衩
cuffs rib length 袖头罗纹长
cuffs rib width 袖头罗纹宽
cuff stitch 平罗纹线圈
cuff strap 假翻袖头,假反袖口,袖头搭袢,袖头狭饰边
cuff tacking 袖口加固缝,袖头加固缝
cuff top girdle 高腰紧身褡
cuff top line 袖头上口线
cuff top stitcher 袖口面缝机,袖头面缝机
cuff trimming silk 袖边绸
cuff turning machine 翻袖头机
cuff type 袖口型,袖口类型,袖头型,袖头类型
cuff vent 袖头衩,袖衩
cuff width (裤脚)翻边宽,克夫宽,袖头宽,裤卷脚宽
cuirass 妇女胸衣;(包括胸甲和背甲的)上半身铠甲,胸甲
cuirboulli 浸蜡硬革
cuir laine 法国结子厚花呢
cuir savage look 光泽外观
cuish 护腿甲
cuissardes [法]及股长靴,时装高筒靴
cuisse 护腿甲,腿甲
cuizhu matellasse 翠竹锦
cul [法]可发型
cul de crin 法国裙撑
cul de paris [法]巴黎股垫,法国裙撑
culet 护臀甲
culmination 时装全盛期
culotte dress 连衣裙裤
culotte jumpsuit 裙裤式连身裤
culotte pajamas 长至脚跟的睡衣(大裤脚),及跟睡衣
culotte pants 裙裤
culottes 裙裤,裤裙,开衩裙
culottes anglaise [法]紧身短裤

culottes de cheval 马裤
culottes dress 裙裤装,套装裙裤
culotte skirt 裤裙,裙裤,叠折裙
culotte slip 长至膝或以上的宽摆衬裙
culotte suit 裙裤套装
culpepper cotton 卡波波棉
cultivated 家蚕丝绸
cultivated silk 桑蚕丝,家蚕丝
cultrured pearl necklace 养殖珍珠项链
cult sandals 偶像凉鞋
culture change 文化变迁
cultured history 文化史
cultured pearl bracelet 养殖珍珠手镯
cultured pearl button 养殖珍珠扣
cultured taste 文化品位
culture pearl 养殖珍珠
cumbi 孔比呢,羊驼毛呢
cumbliers 彩色条格棉披巾
cummerbund 腹带,绶带,徽带,宽腰带,斜胸带,印度腰带
cummerbund skirt 宽腰带裙
cummervest (宽带绕成的)卡马背心,肯马背心
cumulative error 累积误差
cumulative frequency polygon 累积频率直方图
cun 寸
cup 帽,乳罩窝,罩杯,胸杯
cup apex 乳罩窝顶,胸罩窝顶
cup bust 杯形胸罩
cup collar 杯式领
cup feed overseaming machine 圆盘送料包缝机
cup form 托胸罩,托乳罩
cup heel 酒杯跟
cupioni 花式结子线
cupped pattern 杯形纸样
cupping 撮拢
cup radius 胸杯半径
cuprammonium fiber 铜氨纤维
cuprammonium rayon 铜氨丝,铜氨人造丝
cuprammonium rayon fiber 铜氨丝,铜氨纤维
cuprammonium silk 铜氨丝,铜氨人造丝
cuprammonium staple fiber 铜氨短纤维
cuprammonium yarn 铜氨丝
cupra rayon 铜氨人造丝
cuprate silk 铜氨人造丝
cuprel 库普勒尔(铜氨纤维,商名,美国)

cuprene fiber 铜氨纤维
cupro fiber 铜氨丝,铜氨人造纤维,铜氨纤维
cuprous-ion dyeing process 亚铜离子染色法
cup seamer 包缝机
cup seaming 包缝,包缝缝合
cup seaming [sewing] machine 包缝机
cup-shape collar 杯形领
cup-shape sleeve 杯形袖
cup sole 盘形底(鞋)
curatin satin brocade 窗帘缎
curb chain 锁链
curch(ef) 克切帽,妇女头巾
curéhat 神父帽
curl 卷发,鬈毛
curl clip 发卷,卷发夹,卷发器,卷发筒
curled cap 带绲边女软帽
curled fiber 卷曲纤维
curled hair 鬈发;鬈发型
curled list 卷边疵
curled pile 卷曲绒头
curled selvage 翻边,卷边,翻边疵
curler 卷发夹,卷发器,卷发筒
curling 卷帽边工艺,卷曲
curling elasticity 卷曲弹性
curling irons 卷发钳,烫发钳
curling pin 卷发发夹
curling songs 烫发钳
curling tendency 卷曲性
curling tongs 卷发火剪,烫发钳
curl neck 波纹领
curl pile 卷曲绒头;毛圈织物
curl pile fabric 仿羔皮织物
curl rayon staple 卷曲黏胶短纤维
curls 发卷,卷发,卷曲式发型
curls cut 鬈发型
curl-surfaced texture 表面具有圈绒的织物
curl weave 耐压机织地毯
curl yarn 卷毛纱,起圈花线
curly cord 卷毛灯芯绒
curly edges 卷边缘
curly hair 卷发,鬈毛,鬈发型
curragh 爱尔兰针绣花边
curragh lace 爱尔兰针绣花边
currant red 茶燕子红(深红色),醋栗红
currency of settlement 结算货币
current fashion 流行式样
current fashion apparel 流行时装,流行式样的服装
current fashion trends 当前流行趋势
current price 现行价格
current repair 小修
current shade 流行色泽
current standard 现行标准
curricle coat 箱形男外套;女式翻领合身长外套
curriery 皮革整理,鞣制;制革工场,鞣革工场;制革业,鞣革业
curry 咖喱(粉)色
curtain 幕,帷幔,窗帘;窗帘布;窗帘绳,帷幔绳;帽口条
curtain drapery 臀垫
curtained hat 帷帽
curtain fabric 窗帘布,窗帘织物
curtain gauze 窗帘纱
curtain grenadine 提花纱罗窗帘
curtaining 彩色提花窗帘布
curtain lace 花边窗帘,花边窗纱
curtain madras 浮纬织花窗帘纱
curtain marquisette 窗帘纱;平纹薄纱罗
curtain net 网眼窗帘
curtain serge 帷幕厚哔叽
curtain silk 窗帘绸
curtain tassel 窗帘穗子
curtain waistband 西裤裤腰
curvaciousness 曲线美(女性)
curvature 弓形布面,曲率,弯曲
curve 弧线,曲线,弯曲,弯曲部分
curve belt 低腰裙腰带
curve board 曲线板
curve bottom 弧形下摆,曲裾
curve cut 曲线切割
curved belt 曲线腰带,曲线合身腰带,曲线形腰带
curved bottom 弧形下摆
curved bound pocket 弧形嵌线袋
curved button pocket skirt 钉纽弯形袋裙子
curved cuff 圆角袖口
curved dart 弧形省,弯曲省
curved hemline 弧形底摆
curved inset pocket 弧形镶嵌口袋
curved jet with reinforcement pocket 弧形加固双嵌线袋
curved lapped seam 弯曲的压缉缝
curved line of front opening 大襟弧线
curved line of front rise 前裆弧线
curved measure 曲线尺,弯尺

curved needle 弧形针
curved over-lapping neckline 弧形叠领口
curved petersham tape 彼得沙姆曲衬带
curved pile 地毯剪花绒头
curved plain seam 弧形平缝
curved pleats 弧形褶裥,曲线褶
curved pocket 服装弧形袋,弧形袋
curved rule(r) 曲线尺,触角计测器
curved seam 弧形缝,弧形接缝,曲线缝
curved sewing 弧形缝纫
curved shirttail 圆口衬衫后摆
curved skirt 曲下摆裙
curved slit 曲面缝,弧形缝
curved style line 曲线造型线
curved tuck 缉弧形细褶
curved tuck seam 缉弧形细褶缝
curved twill 波形斜纹,曲线斜纹
curved twill weave 波形斜纹,曲线斜纹
curved yoke 曲绕托肩
curve gauge 曲线板
curve graduation 曲线修匀
curve inserted layout 互套法
curve line 曲线,弧线
curve line of back rise (裤)后裆弧线
curve line of front opening (中装)大襟弧线
curve line of front rise (裤)前裆弧线
curve line of underarm (中装)裉缝弧线
curve modeling 曲线建模
curve of beauty 曲线美
curve plotter 绘图器
curve ruler 曲尺,弯尺,曲线板,曲线规
curves 曲线板;曲线轮廓(女性健美体型)
curve stick 曲尺
curvilinear motion 曲线运动
cushion 衬垫,垫子,(女子衬在头发下的)发垫,插针垫
cushion back carpet 衬垫毯
cushion dot 浮点花纹
cushion dot fabric 浮点花纹织物
cushion foot 毛圈袜底
cushion heel 弹性袜跟
cushioning material 垫料,衬垫
cushion lace 棱结花边
cushion material 垫料,衬垫
cushion nonwoven 靠垫布,靠垫用非织造布
cushion nonwoven fabric 非织造靠垫布
cushion sole 绒头袜底,弹性袜底,绒袜底

cushion sole socks 垫底短袜,绒头袜底短袜,弹性袜底短袜
cushion-stitch 短直针迹,斜向平行针迹,仿机织纹针迹,双面针迹
cushion style embroidery 柏林毛线刺绣品,椅垫刺绣,帆布满地绣
cusier silk 优级缝纫丝线
cusir 缝纫丝线
cussidah 东印度细布
custom 定做的,定制的;风俗
customary packing 习惯包装
custom clothes 定做服装,定制服装
custom cut 融合美发法
custom designer 定制服装设计师
customer's accessory 客供辅料
customer's preference 顾客品味
customization 服装定制,定做
custom-made 定制服装,定做服装;定做的,定制的
custom-made clothes 定制服装
custom-made garment 定做服装,定制服装
custom-made slip 定制连身衬裙
custom method 定做方法,定做缝制方法
custom of trade 贸易惯例
custom-order 定制服
custom-order clothing 定做服装,定制服装
customs 海关
customs debenture 报关单
customs declaration 海关申报单
customs pass 海关通行证
custom suit 定制的衣服
customs uniform 海关服,海关制服
custom tailor 定制服装裁缝;定制西服店;定制,定做
custom-tailored garment 定做精制服装
custom-tailoring 定制服装
custom tufted 手工簇绒
custom tufted carpet 手工簇绒地毯
custume director 服装指导
custumer 服装制作人(为化妆舞会、剧团等服务)
custume storage 服装贮藏室
custumier 服装制作人(为化妆舞会、剧团等服务)
cut 裁,剪,切,开剪;(裁剪)式样;发型;剪刀;克特毛纱支数制,克特亚麻纱支数制
cut across 疵布开匹,开剪

cut along this line　裁开线
cut and open　分割切开线
cut and sew　裁剪成形
cut and sewn　裁剪与缝制
cut and sew sweatshirt　裁剪和缝制汗衫
cut and try method　试探法
cut away　常礼服,燕尾服,晨礼服;剪掉;男上衣下摆裁成圆角
cutaway coat　常礼服
cutaway collar　圆角领
cutaway front　晨礼服剪掉前摆的设计
cutaway jacket　礼服夹克
cutaway model　剖面模型
cutaway neckline　大圆形领口
cutaway shoulder　挖肩
cutaway view　剖视图;剖面图
cut-before-stitching　先裁后缝
cut buttonhole　开扣眼
cut canvas work　起绒十字布刺绣
cut cashmere　开士米斜纹呢
cutch　儿茶棕色
cut chenille　绳绒线,雪尼尔花线
cut chinchilla　灰鼠条纹呢,直条纹灰鼠呢;毛织物经条纹效应
cut dart　剪省缝,开省线
cut details　(裁剪)配零料
cut double cloths　沟纹双层呢
cut edge　切边
cut edge fabric　切割边织物
cute look　伶俐风貌(13~18岁女孩),伶俐款式
cut eye　洞眼
cut fabric　针洞疵
cut-fashioned　裁剪成形(用于毛衫成衣,有别于全成形毛衣法)
cut-figured velour　割绒花纹绒
cut-fit-trim　缝衣全进程,服装加工全过程
cut from end　从布头上裁零
cut fur skin　割过的毛皮
cut gloves　裁制手套
cut goods　开剪织物
cut hose [hosiery]　裁制袜
cuticle cream　增软脂
cuticle pusher　老皮剪
cuticle rem over　角质层除去剂
cuticolor　肉色
cut-in-one　连裁(如衣连领、衣连袖)
cut-in-one sleeve　一片袖
cut-in pocket　开缝口袋

cut into　裁(剪)成
cut lace　镂空花边,剪下的花边花纹
cutless　无缝的
cut line　优质长亚麻;强韧细绳
cut lines sole　切纹底(鞋)
cut listing　破边
cut-make-trim　缝衣全进程,服装加工全过程
cut mark　分匹墨印
cut mark length　墨印长度
cutni　意大利手织丝棉绸
cut off　裁(剪)下;毛边裤脚
cut-off a piece for bust modelling　撇门
cut-off a piece for modelling　撇势
cut-off jacket　截下摆夹克
cut-off jeans　毛边脚口的牛仔短裤,毛边牛仔短裤
cut-off pants　截短裤,截筒裤
cut-offs　剪截式短裤,截短裤(毛边蓝色牛仔短裤)
cut-on centre plait　裁剪中央裥
cut on fold　折叠裁剪,对折裁剪
cut-on shirt front　裁剪衬衫前片
cut-on the bias　斜裁
cut open　裁剪开,裁开
cut-open state　剪开图
cut open to layout　剖缝套法(圆筒针织物)
cut out　裁剪出,裁出;镂空,挖剪工艺;剪下图样
cut-out armhole　挖袖窿
cut-out back　镂空后身,(服装)露背
cut-out dart　剪开的省道
cut-out design　挖剪设计
cut-out dress　剪去部分的服装
cut-out gloves　赛车手套,机能手套,竞赛手套(手背上镂空,增加屈挠性能)
cut-out lace motif　挖剪花边图案
cut-out neck hole line　挖领窿线
cut-out neckline　挖剪式领口,挖领口(低领),挖领圈
cut-outs　挖剪部分,挖去部分
cut-out slit　装饰切缝
cut-out styleline　挖剪结构线
cut-out swimsuit　紧身镂空网眼泳衣
cut parts [pieces]　裁片,片料
cut pick　缺纬;断纬
cut piece　衣片,裁片;片料
cut pile　割绒
cut pile carpet　割绒地毯
cut pile fabric　割绒织物

cut pile towel 割绒毛巾
cut pile weave 丝绒织物
cut pile weft knit fabric 纬编割绒织物
cut plush fabric 割绒织物
cut pocket 挖袋
cut pocket mouth [opening] 开袋口
cut pressed design 花压板花纹
cut presser fabric 经编花压板提花织物
cut-proof fabric 防切断织物,防切割织物
cut ribbon 裁制带子
cut ruche [法]褶带,袖口的饰边,袖头饰边;女装领口;中开匹(双幅布)
cuts 裁片,裁剪,片料,剪刀;(英国足球运动员的)短裤
cut & sew 裁缝
cut shape 裁形
cut silk brocade 缂丝(克丝,刻丝)
cut size 裁剪毛头
cut staple metal fiber 金属短纤维
cut stockings 裁制袜子
cut symbol 裁剪符号
cuttanee 坎坦斯棉背织条条纹绸,麻棉混纺交织布;丝棉混纺交织布
cut tape 裁制带子,剪裁带条
cutted pieces 裁片,片料
cutter 裁刀,裁剪机,电刀,切刀;排板师,排版师;裁剪工
cutter path 刀具轨迹
cutter shirt 活领衬衫,敞领衬衫
cutter shoes 平底鞋,平跟女式船鞋,平跟女式无带鞋
cutter's workmanship 裁剪师的手艺
cut the garment according to the figure 量体裁衣
cut through 开匹,开剪,疵布开匹
cutting 裁,剪,裁剪,开裁,开剪
cutting, making up, trimming (CMT) 裁剪、缝制、整装(简称裁、缝、整)
cutting away 燕尾摆
cutting bed 裁床
cutting before stitching 先裁后缝
cutting board 裁剪台
cutting button hole 开扣眼
cutting by each color 分色裁剪
cutting by each roll 分卷匹裁剪
cutting by each size 分尺寸裁剪
cutting by one-piece 一件裁剪
cutting dart 剪省缝
cutting dart seam 剪省缝

cutting details 配零料
cutting die 啤刀
cutting disc 圆裁剪刀,圆盘切割刀,圆刀
cutting drawing 裁剪图
cutting equipment 裁剪设备
cutting from ends 布头裁片,碎料裁片
cutting guide line 裁标线
cutting illustration 裁剪图
cutting in 缩刀(毛皮、皮革工艺)
cutting knife 裁布刀,裁刀,切刀
cutting layout 裁剪排料
cutting line 裁剪划线,裁剪线
cutting line mark 毛样符号,毛样号(裁片上缝头标记)
cutting loss 裁剪废料,裁剪损耗
cutting machine 裁剪机,裁布机,裁剪电刀,开剪机
cutting margin 裁剪划线
cutting mark 裁剪标记
cutting marker 裁剪划线器;裁剪排料
cutting mat 裁剪垫报
cutting measurements 裁剪尺码,裁剪尺寸
cutting motif fabric 剪花织物
cutting operation 裁剪操作
cutting out garment 裁剪衣服
cutting over 放刀(毛皮、皮革工艺)
cutting parts 裁零料
cutting pattern 毛尺寸样板
cutting pattern hollow handkerchief 镂空剪花手帕
cutting plain line 缝纫切割线,切割线
cutting pocket mouth 开袋口
cutting process 裁剪工序
cutting quantity 裁剪数量
cutting rate 裁剪率
cutting remnants 零头布
cutting room 裁剪车间
cutting scissors 理发剪刀
cutting shears 裁衣大剪
cutting shop 裁片间
cutting space 挖纽孔部位,挖纽孔长度,纽孔间距
cutting system 裁剪方式
cutting table 裁剪台,裁剪案板,裁床,裁剪桌
cutting technique 裁剪技术
cutting technology 裁剪技术
cutting template 裁剪模板,裁剪样板
cutting trade 服饰业

cutting up board　裁切板
cutting up trade　裁剪行业(美国)
cutting waste　裁剪废料
cutting width　裁剪宽度
cutting work　裁剪工作
cuttle　折叠,码布长度,折布长度
cuttle fold　卷折
cuttler　折布机;折叠机
cuttler machine　裁布机,裁剪机
cuttng edge　切边
cut tracking　裁剪轨迹,裁剪跟踪
cutty sark　女式短内衣,苏格兰女式短内衣,短衣服(衬衣裙等)
cut up　裁得出
cut up into　裁得成
cut velvet　满地花丝绒,配饰丝绒
cut voile　剪绒巴里纱,绒头纱
cut way coat　大衣
cut wig　花椰菜假发
cut work　挖花花边;雕绣
cut work lace　挖花花边;雕绣
CVC blended yarn　CVC 混纺纱
C.V.C.cloth　棉涤混纺布料(含棉量 50% 以上)
C/V plain cloth　棉/维平布
C/V poplin　涤/维府绸
cyan　青色,蓝绿色
cyan-blue　天蓝,湖蓝,深蓝,翠蓝,晴空蓝
cyanean　青色的,蓝绿色的
cyanine　花青
cyanine blue　深蓝,深紫蓝,原色蓝
cyanine dyes　菁类染料
cyclamen　仙客来红紫色(涂暗红紫色)
cyclas　山克勒斯外衣(古希腊)
cycle jacket　骑车夹克(紧身,黑皮齐腰,拉链或纽扣扣合)
cyclic stress　反复应力
cyclic three-dimensional stretching　反复三向拉伸
cycling helmet　自行车头盔

cycling jersey　自行车紧身运动衫
cycling mitten　自行车手套
cycling shoe cover　自行车鞋套
cycling shoes　骑自行车鞋,骑自行车运动鞋
cycling shorts　自行车(运动)短裤
cycling sportswear　自行车运动服
cycling suit　骑自行车(运动)套装
cycling wear　骑自行车服
cyclist attire　自行车装束
cyclist pants　自行车裤
cyclist's cape　自行车斗篷,摩托车驾驶员斗篷
cyclized fiber　环化纤维
cycloid　旋轮线
cylinder-bed　缝纫机筒形底板,底臂
cylinder-bed machine　悬臂式缝纫机,圆筒式缝纫机
cylinder cuff-attaching machine　圆筒形缝袖机
cylinder printing　滚筒印花
cylinder-shaped torso　圆筒形身段
cylinder silhouette　圆柱形轮廓
cylindrical goods　圆筒形针织坯布
cymar　宽松女上衣,宽大轻便无袖衣,西玛外衣(17～18 世纪)
cynara　黏胶丝绉布
cynara crepe　黏胶丝绉布
cypress　黑纱
cypress green　柏树绿(暗黄绿色)
cyprus lace　塞浦路斯金银线花边,挖花边
cyprus wool　塞浦路斯羊毛
Czech and Slovakian embroidery　捷克斯洛伐克刺绣
Czechoslovakian costume　捷克斯洛伐克民俗服
Czechoslovakian embroidery　亚麻布彩绣
Czechoslovakian ticking　床罩布
czerkeska　哥萨克军用及小腿外套

D

D　D码(鞋宽较C码宽而较E码窄,亦指男用睡衣的特大号);打褶符号
da　洋麻,槿麻
dab　能手
dabbakhis　印度条纹棉细布
dabbidar lungi　伊斯兰头巾;腰布(伊斯兰教)
dabbis　斜纹棉布;平纹棉布;棉制仿甘布龙布
dab hand　能工巧匠
dabins　达宾高支漂白细棉平布
da brocade　大锦
Dacca　达卡牌黄麻
Dacca cotton　达卡棉;仿达卡细平布
Dacca muslin　达卡细平布
Dacca silk　达卡无捻绣花丝线
Dacca twist　达卡细布
dacey　印度达赛粗野蚕丝
dacian cloth　浮雕布,拷花布,压花布
Dacre　得克瑞
Dacron　的确良,涤纶(俗称的确良),大可纶(聚酯长丝和短纤维,商名,美国杜邦)
dado　意大利双色棉印花装饰布
daffodil, daffodile　水仙黄
dag　衣服的装饰边;剪边装饰
dagbezi　土耳其细布
dagged　花瓣形饰边
dagged sleeve　剑形袖
dagges　剪边装饰
dagging　齿边缘饰(14～15世纪)
daggings　剪边装饰
daghestan rug　高加索地毯
dagswain　粗毛呢,粗呢
daharra　摩洛哥宽式长袍
dahlia mauve　大丽花紫色
dahlia purple [violet]　大丽花紫,浅紫红色
dahua (rayon) satin　大华绸
Dai bracade　傣锦
Dai ethnic costume　傣族民族服
daily inspection　日常检查
Daily News Record　每日新闻(介绍男装和纺织业情况的日报,由美国仙童出版公司出版)
daily output　日产量
daily suit　日常套装
daily-use goods　日用品
Dai nationality brocade　傣锦
Dai nationality's costume　傣族服饰
Dainimer　高强度聚乙烯纤维(商名,日本东洋纺)
dairygirl shirt　挤奶女工衬衫 牛奶妹衬衫
daisy stitch　代西绣 链式针迹
Daiwabo　聚丙烯和聚乙烯双组分纤维纺黏法非织造布(商名,日本大和纺)
daka　土库曼斯坦达克原色平布
dalecarlian lace　瑞典打褶花边
Dalhousie　达尔豪斯 达尔豪斯粗格呢(苏格兰)
Da Lian International Fashion Festival　大连国际服装节
Dalkeith Eureka cotton　英国达尔科斯尤里卡棉
Dalmatian　达尔马提亚服(克罗地亚)
dalmatian lace　南斯拉夫粗棱结花边
dalmatian sleeve　南斯拉夫达尔马提亚大袖
dalmatic　罗马天主教主教的法衣,达尔马提亚袍;英国国王加冕服;美国新娘服
dalmatica with clavi　带克拉维彩饰条的达尔马提亚袍
dalmation sleeve　喇叭袖
dalmiyan　印度和巴基斯坦丝网
daltonism　色盲
damaged fabric　有损伤的织物
damage to fabric　戳毛
damaging-out　修补衣服
damaras　薄质花塔夫绸(印度)
damas　花缎,锦缎
damas caffart　法国仿东方花缎
Damascene lace　大马士革花边,手制小花纹花边
damascene ring　大马士革戒指
damascening　波形花纹;金银丝镶嵌
damas chine　印经花缎
damas de caux　条纹亚麻布
damas de lyon　丝织装饰布
damas de Naples　高级装饰用花缎
damas des indes　法国细花缎

damas en dorure 法国金丝花卉锦缎
damask 红玫瑰色,紫红色,蔷薇花色,淡红色;花缎,织锦缎,缎子
damask darning stitch 花缎缀纹绣
damask diaper 经缎相间的条子花纹
damask figured satin 花缎
damask flower 花缎,织锦缎
damask gauze 花式丝纱罗
damask linen 织花麻布
damask lisère 金银线花缎
damask pointille 方格点子花缎
damask printing 仿锦缎印花,印花仿锦缎
damask rose 蔷薇花色
damask satin 缎,缎纹织锦
damask silk 绫
damask swiss organdy 仿花缎印花织物(透明底纹上用涂料或一般印花作出不透明花纹)
damask ticking 锦缎床垫布
damask twill 斜纹地提花锦缎
damask velour 织锦丝绒
damasse 锦缎,花缎,大花彩色丝缎
damassé arabesque 阿拉伯式图案花缎(用花、叶、茎等花纹)
damassé brocat 金银丝晚装花缎
damassé jardiniere 大花彩色丝缎
damassé moiré 波纹花缎
damassé pointille 方格点子花缎
damassé raye 彩色经条纹花缎
damas serge 印度斜纹地花缎
damassin 金银线织锦缎,金银线轧花缎
dambrod 黑白格子棉布,棋盘格棉布
dambrod pattern 棋盘格子纹,素色大方格纹
dame joan ground 梭结花边的六角形连线;六角形连线针绣
dames en dorure 法国金花丝锦缎
damier 素色大方格花纹
dammesek 丝绸
damp 潮湿,潮湿感,干中带潮;阻尼,衰减
damp dry 半干,带潮
damping 打湿,给湿,喷雾
dampness 潮湿
damp pressing 汽蒸压烫,汽蒸熨烫
damp-proof package cloth 防湿包装布
damp-proof packing 防潮包
damp-proof shoes 防潮鞋,防湿鞋
damp rag (熨烫用)水布,熨烫湿布
damp setting 湿[热]定型

damson 暗紫色
Dan Beranger 唐·贝朗热(法国时装设计师)
dance bag 舞包
dance dress 女跳舞服,舞蹈服
dance mask 跳舞面具
dance set 舞蹈套服;女内衣组合
dance shoes 舞蹈鞋
dance skirt 舞裙
dance wear 舞蹈装,舞蹈服
dancing clog 跳舞木鞋
dancing dress 舞蹈装,舞蹈服
dancing hemline 舞裙裙边
dancing shoes 跳舞鞋,舞鞋
dancing skirt 舞裙
dancing slippers 舞鞋
Dancomb 丹贡布,精梳斜纹织物
Dancool [美]丹库尔,但库耳涤棉耐久定形织物(商名)
dandelion 蒲公英黄
dandiacal clothes 浮华服装
dandizette 纨绔女子风格
dandruff 头垢,头皮屑
dandy 时髦男士,时髦人
dandy blouse 褶边装饰女衫
dandy collar 纨绔领
dandy hat 华丽女帽,纨绔帽
dandyism 时髦
dandy look 花花公子风格,纨绔风格,男式女装风格
dandy shirt 克夫、门襟有褶边装饰的衬衫
dandy shoes 花花公子鞋
dandy style 登迪款式
Danel 丹内耳色织涤棉织物
Danfit process 丹菲特防缩整理(用于针织物的防缩整理)
Danflair 丹弗赖尔色织条格织物
dangaris 巴基斯坦粗棉布
dangerous goods 危险货物
dangline earing 垂耳领
dangling part 悬垂部件(袖子、领带等)
Daniel Hechter 丹尼尔·埃切特(1938~,法国时装设计师)
Daniel line 丹尼尔线型
Daniel Tribouillard 丹尼尔·特里布亚(法国服装设计师)
Danish cloth 丹麦棉平布
Danish costume 丹麦民族服饰
Danish embroidery 丹麦抽花刺绣,丹麦抽绣

Danish lace 丹麦花边
danish trousers 丹麦裤
dan-mo 弹墨
Danny Noble 丹尼·诺布尔(美国时装设计师)
dan-roug (tussah) boucle 丹绒绸
danton 可折叠高领
Dantrel [美]丹垂尔
Dantwill [美]丹忝尔
Danufil 黏胶短纤维(商名,德国赫斯特)
DAPHNE 达芙妮
daphne blue 月桂蓝
daphne cobalt 钴蓝色
dapperpy 杂色呢绒
dapple 斑纹色
dapple grey 深灰色
dapplin 赫或都普林
darab 达勒布,印度红边粗棉布
darale 印度未漂红边棉布
Dardanelles canvas 土耳其达达尼尔粗帆布
daresh 印染织物
dari 达里,达里棉拜毯;达里粗斜纹布
dariabanis 印度达里阿伯尼漂白棉布
darida 植物纤维花布
darish 毛蓝布
darish blue 毛蓝
dark (dk) 暗色,深色,黝黑
dark beaver 海狸深棕
dark blue 暗蓝,暗青,深蓝,藏青,深蓝,凡拉明蓝
dark bluish color 鳝鱼青
dark bronze 黑古铜色,茶色,深褐
dark brown 深棕
dark brown skin 暗棕色皮肤,深棕色皮肤
dark camel 暗驼色
dark citron 暗香橼色
dark color (dc) 暗色,深色;暗色调
dark-colored blouse 深[浅]色衬衫
dark-colored denim 深色牛仔布
dark-colored shirt 深色衬衫
dark denium 深蓝色
dark earth 暗土色
darkening 变暗,加深
dark glasses 墨镜
dark green (dkg) 暗绿,墨绿,青绿,深绿,苍翠
dark grey (dkg) 深灰色
dark greyish (dg) 暗灰色调
dark gull grey 深鸥灰色
dark heavy pattern 色花纹

dark indigo 靛蓝(洗衣时上蓝用)
dark ivy 深常春藤色
dark mink 养殖黑貂皮
dark navy 深藏青
darkness 暗度
dark olive green 灰茶绿色
dark purple 深紫色,绀青色
dark red 暗红,深红
dark reddish purple 绛紫,酱紫
dark reddish violet 暗红光紫
dark region 暗区
dark room 暗室
darkroom equipment 暗室设备
dark rose purple 暗紫玫瑰色
dark rose taupe 古铜色(暗黄棕色)
dark selvage 印花糊边(疵病)
dark shade 深色,暗色泽
dark skin 黑皮肤
dark suit 黑礼服,暗色套服,暗色套装,男便礼服,办公室套装,办公套装
dark suit subfusc 黑礼服(牛津大学师生穿)
dark tone 深色调
dark-tone plaid frock 深色格子长外套
dark value 暗调,暗调子
dark yellow 深黄
Darlexx Superskin 达利克斯泳衣布
darn 织补
darnamas 达纳马白布,土耳其优级漂白棉布
darned embroidery 缝花花边;织补刺绣,缀纹刺绣
darned filet lace 织补针迹方网眼花边
darned lace 补缀花边,绣花网眼花边
darned netting 绣花网眼花边,绣花网眼纱,缀纹刺绣
darned wheel 缀纹轮状绣
darned work 绣花网眼纱,缀纹刺绣,绣花网眼花边
darner 织补针
darning 织补,缝补;织补物;缝花花边
darning and embroidering hoop 织补和绣花绷架
darning and embroidery foot 织补和绣花压脚
darning arm 织补式机头
darning ball 球形织补衬托架
darning clover stitch 苜蓿缀纹绣
darning damask stitch 花缎缀纹绣
darning double stitch 双缀纹绣

darning edge　织补边
darning egg　馒头烫垫；球形织补托架
darning foot　织补压脚
darning for woolens　织补毛料衣服
darning Japanese stitch　日本式缀纹绣
darning machine　织补机
darning mushroom　织补蘑菇
darning needle　织补针
darning plate　织补针板
darning rip　织补刀
darning rosette stitch　玫瑰缀纹绣
darning seam　织补缝
darning socks　织补短袜
darning stitch　（绣花）缀纹绣，织补绣；；织补线迹
darning surface stitch　表面织补绣，表面缀纹绣
darning thread　织补纱线，织补线
darning work　织补
darning yarn　织补纱线
dart　省（衣片上缝去的部分俗称），省道，褶；裥；短省，缝裥；开省（俗称开省道），缉省
dart angle　省角量
dart apex　省尖点，省尖，褶尖
dart bubble　省尖起泡
dart cluster　组合省
dart curve　省（道）弧
darted collar　缉省领，短缝衣领
darted neck　收省颈
darted neckline　短缝领口，收省领口，褶缝领口
darted sleeve　收省袖
dart end　省止点
dart equivalent　省道等量变换；省道等效变化
dart excess　省余量
dart flange　开省落肩边
dart fullness　省缝丰满度，省缝丰满性
dart intake　省份量
dart leg　省柱（省的两条边）
dart length　省长
dartless foundation　无省服装基础纸样
dartless knit　无省针织衫
dartless pattern　无省纸样
dartless sleeve　无省袖
dart line　省道线，褶线
dart manipulation　省道变化，省道处理，省的移位，省缝处理
dart marking　省缝记号

dart placement　省位
dart point　省端点，省尖点，省尖
dart position mark　开省号，开省记
dart pressing　省道熨烫
darts apex　褶尖
dart seam　档缝，省缝，褶缝
dart seamer　短缝缝纫机，辑省缝纫机
dart seam opening press　省缝分开熨烫机
dart sewer　短缝缝纫机，辑省缝纫机，省缝机，缝省机，短缝机
dart sewing machine　缝省机，省道缝合机
dart sewing unit　辑省装置，短缝装置
dart site　省位
dart slash　褶形切线
dart space　省量
dart tuck　半活褶，半褶，塔克省，开花省
dart width　省大
darya　印度原色野蚕丝绸
daryai　印度黄色手织绸，印度浅绿色手织绸
das　洋麻，槿麻
Dashiki　达西奇套衫，非洲颜色鲜艳的无领和服袖衬衫，大稀奇装
dasija　巴基斯坦黄条丝罗
dasuti　股线棉布
data　资料；数据
data bank　数据库，数据中心，资料库
data base　数据库
data graduation　数据修匀法
datails　设计细节
date dress　约会服装
date red　枣红
datong brocaded poplin　大同绸
da-tong brocaded rayon poplin　大同绸
datum line　基准线
daub printing　涂抹印花
dauphine　杜法因呢，装饰用薄毛呢，丝绒，毛绒
Daur ethnic costume　达斡尔族服饰
davao hemp　蕉麻
davetyn　毛绒，丝绒
David Dameron　戴维·达默龙（美国时装设计师）
David Webb　戴维·韦布（美国时装设计师）
Davy Crockett hat　大卫帽（浣熊毛皮，帽后有尾巴），浣熊皮帽
dawei jacquard crepon　大伟呢
dawei suiting silk　大伟呢
dawn　黎明色

dawn blue 黎明蓝
dawn grey 黎明灰
dawn rayon brocade 曙光绉
Dawoer ethnic costume 达斡尔族民族服
day break 拂晓色
day frock 日间福乐克装
daylight color 日光色；日光荧光灯
day shift 日班
daytime bag 外出手提包
daytime dress 白日服,外出服
daytime leather bag 白天用皮革包
daytime shift 常日班
daytime wear 白日服
Da Yun shoes 大云鞋
day worker 计日工,计时工,临时工
dazhong brocaded poplin 大众绸
dazzling coat 眩惑外套
D.B.lapel 双排纽驳领
3D body scanning 人体三维扫描
D.B.suit 双排纽套装
2D computer aided garment design 二维计算机辅助服装设计
deacon 初生牛犊皮（小牛皮）
dead appearance 黯淡外观
deadbeat clothing 过时服装
dead color 呆板色,死色,萎色
dead fold 死裥
deadline 限期
dead match 符合来样（染色）
dead stock 滞销品
Deang ethnic costume 德昂族民族服
Deanna Littell 迪娜·利特尔（美国时装设计师）
Deauville scarf 多维尔丝头巾（法国）
debage 德巴热呢,法国色纬毛呢,棉经色毛纬呢
debai-rumi 土耳其红色平纹绸
debeige 法国色纬毛呢
debenture 退税凭单
de bevoise brassière 无袖低领胸衣
debuani 东非穆斯林彩条头巾布
debut 首次表演,首次发表会,首批作品
debutant look 雏儿风貌
decagon 十边形
decal 贴花纸；贴花转印图案
decalcomania 贴花纸,移画印花法,贴花转印图案
decal printing 转移印花
decan hemp 洋麻,槿麻
decating blanket 蒸呢包布

decating knit goods 蒸煮的针织品
decating mark 搭头印（布疵）,皱痕
decatising 罐蒸,全蒸,蒸煮,蒸呢
decatizing wrapper 蒸呢包布
deccan hemp 槿麻,洋麻
deccan rug 印度风格地毯
decency 得体,端庄,高雅,合乎礼仪,正派
deceptive cut 背道发型
dechine 双绉类织物
deci 迪西（测皮革面积的单位）
decibel（dB） 分贝
decimeter 分米
decitex（dt；dtex） 分号；分特
deck 覆盖物；航海服饰
deck chair canvas 窄幅躺椅帆布
deck chair fabric 躺椅用布
deck jacket 甲板短夹克
deckle edged 有毛边的,未裁齐的
deck pants 甲板短裤（长于膝的贴身短裤）
deck shoes 船鞋,甲板鞋；校园软鞋,帆布便鞋
decky 胸垫；礼服衬（包括胸背和领）
declination 倾斜
declining skirt 前短后长裙,斜摆裙
décolletage 低领袒胸服,露肩服,袒胸露肩衣服；（袒胸衣服的）低领
décolleté 露肩的,穿露肩衣服的
décolleté bra(ssiere) 低胸式胸罩,袒露式胸罩
décolleté evening gown 袒胸夜礼服
décolleté neckline 低胸式领口,袒露领口
décolleté oxford 低口牛津鞋
décolleté sandals 袒露鞋
decolor 去色,褪色,脱色
decolorant 脱色剂
decoloration 脱色,剥色
decoloring 脱色,剥色
decoloring assistant 脱色助剂
decoloring by oxidation 氧化脱色
decolorise 脱色,剥色
decolorization 脱色（作用）
decolorize 脱色,剥色
decolorizing 退色剂
decolorizing agent 去色剂,脱色剂
decolorizing carbon 脱色碳
deconstructed style 解构主义色彩风格
deconstruction 解构
deconstructivism 解构主义
decorate 装饰

decorate cuffs and flaps with plush 用长毛绒饰袖头和袋盖
decorated pump 装饰碰拍鞋,装饰潘普鞋
decorating color 点缀色
decoration （衣服的）装饰,边饰;勋章;装饰品,饰物
decoration color 点缀色
decoration drawing 图案设计
decoration hole 假扣眼,装饰扣眼,假纽孔
decoration ribbon 帽墙带
decoration ribbon on cap 帽墙带
decoration theory 装饰说
decorative 装饰的
decorative art 装饰艺术
decorative banding 装饰性镶边
decorative bar 装饰性加固缝
decorative bias facing 装饰性斜贴边
decorative button 装饰扣,花扣,装饰纽扣
decorative cables 花式绞花
decorative chain scallop finish 装饰性链式荷叶边（扇形边,月牙边）
decorative cover 装饰顶盖
decorative design 装饰图案;装缀设计
decorative double seam 装饰用双排缝
decorative effect 装饰效应
decorative elastic band 装饰性松紧带
decorative elastic double seam 装饰伸缩双线缝
decorative emphasis 装饰强调
decorative enlargement 图案放大
decorative fabric 装饰布,装饰织物
decorative facing 装饰镶边
decorative feature 装饰特征
decorative felt 装饰毡
decorative fringe 装饰流苏
decorative grain 装饰褶（鞋子）,装饰皱纹（鞋子）
decorative hem 装饰卷边,装饰贴边
decorative hole 假纽孔（装饰用）
decorative jacquard raised fabric 大提花装饰绒布
decorative lace 装饰花边
decorative lapped seam 贴布叠缝,装饰叠缝
decorative line 装饰线
decorative linen 亚麻装饰布
decorative motif 装饰花边,装饰花纹,装饰图案
decorative narrow rolled edging 装饰滚边,装饰性狭滚边
decorative overedging 装饰包边

decorative patch stitch 装饰补缀线迹
decorative pattern 装饰图案
decorative picot finish 装饰性锯齿处理,装饰性锯齿光边;装饰性小环边
decorative pieces 饰件,装饰件
decorative pins 饰针
decorative plastic double seam 装饰伸缩双线缝
decorative plate coversheeting 装饰板贴面
decorative plush 长毛绒装饰布,装饰性长毛绒
decorative seam 装饰缝
decorative seamer 装饰线迹缝纫机
decorative seaming 装饰缝纫
decorative sewing 装饰缝纫
decorative shirring 装饰抽褶
decorative single-needle stitch 装饰单针线迹
decorative stitch 装饰线迹
decorative stitching 装饰缝纫;装饰线缝,装饰针脚
decorative stitching machine 装饰线迹缝纫机
decorative stitch sewing machine 装饰缝缝纫机
decorative stockings 花式长袜,装饰长袜
decorative style 装饰款式
decorative tacking machine 装饰线迹缝纫机
decorative tape attaching foot 装饰带缝纫压脚
decorative thread 装饰线
decorative touch 装饰性
decorative tucks 装饰塔克,装饰缝褶
decorative zipper 树脂拉链,塑霸拉链,塑钢拉链,装饰拉链
decor fabric 装饰织物
découpage 剪贴艺术
dedicated computer 专用计算机
deed tax 契据税
deep (dp) 深色的,深色调
deep armhole 深袖窿
deep azalea 深杜鹃色
deep back vent 后背长开衩
deep blue 暗蓝,深蓝
deep brimmed hat 低檐女帽
deep brown 暗褐,深棕色
deep-buff 二层皮（椰皮）
deep cerulean 暗青
deep claret 深紫色

deep cobalt 深蓝色
deep color 深色,浓色
deep colored shirt 深色衬衫
deep cowl 低位垂褶,底位垂褶
deep curve 深凹势,深弯势
deep-cut armhole 挖低袖窿
deep-cut armhole design 深袖窿设计
deep-cut square sleeve 方形低袖窿
deepening color 深色,浓色
deepening the reform 深化改革
deeper bust cup 深型胸罩
deep green 暗绿,墨绿
deep grey 暗灰,深灰
deep jungle 丛林色
deep layer stand collar 深叠襟主领
deep mahogany 深红木色
deep olive 深橄榄色
deep orange 深橘黄,深橙
deep periwinkle 深海螺色
deep pile 长毛绒,长毛绒织物,仿毛皮织物
deep pile fabric 长毛绒,长毛绒织物,仿毛皮织物
deep pink 深桃红色,深粉红色
deep pinning 深针痕
deep placket 长开衩,长开襟,宽前襟,深叠合
deep plaited skirt 深褶裙
deep pleated cowl 深悬褶,深垂褶
deep plum 深梅红色
deep pocket 深口袋
deep processing 深加工
deep processing product 深加工产品
deep prussian collar 深普鲁士领
deep red,deep brown 酱红
deep red 酱红,深红
deep ruffle 深褶荷叶边
deep sea 深海色
deep-sea coral 深海珊瑚
deep-sea diver 深海潜水服
deep-sea diving suit 深海潜水服
deep shade 深色,浓色
deep sky blue 深天蓝
deep sleeve 大袖窿,深袖窿袖子
deep teal 深凫色
deep-textured fabric 立体感织物
deep turban 深缠头巾帽
deep ultramarine 深群青
deep vees V形大袒胸,深V形领口
deep violet lotus 深紫莲色
deep water 海蓝色

deep welt 长袜口
deep wistaria 深紫藤色
deep wool 长壮羊毛
deep yellow 深黄,暗黄
deep yoke 深低肩,深过肩
deer 鹿棕(淡褐色)
deer leather shoes 鹿皮鞋
deerskin 鹿皮,鹿皮革;鹿皮服装
deerskin flying suit 鹿皮飞行服
deerstalker 福尔摩斯帽,前后翘猎帽,猎鹿帽
deerstalker hat 福尔摩斯帽,前后翘猎帽,猎鹿帽
defeat 缺陷,故障
defective color vision 视觉缺陷,异常色觉
defective edge 毛边
defective fraction 疵品率,不合格率
defective goods 疵布,次品
defective product 疵品
defective selvadge 边不良,毛边疵
defects of end products 成品缺陷
defence apparel 防护服
deferred charge 延期费
defilade property 遮蔽性
defined bust 定型胸罩
deformation 变形,走样
deformed pocket 袋形走样
défriser [法]弄直
dégagé 法国轻便低领女服
degage neckline 宽松领口,露颈(离颈)领口
degage neckline, off neckline 宽开领口
dégagé neekline 离颈领口,露颈领口
degradation-resistant fiber 抗降解纤维
degrading 降等;降级;降解
degrains 去纹皮革
degreased wool 脱脂羊毛
degree (deg.) 度,级
degree gown 学位服
degree of whiteness 白度
degskin gloves 狗皮手套
degummed silk 精练绸缎;熟丝
degummed silk fabric 精练丝织物,精炼丝绸
degumming 脱胶
dehairing 去除戗毛,去除抢毛,分绒工艺,羊绒分梳
dehumidifier machine 除湿机
dehydrate weight 干重
dehydration 脱水
deicing property 防冰性能

deigwara twill 斜纹薄花呢
de joinville 镶边宽领结
de joinville teck 预先系好的丝绸领结
Delacoule 德拉库尔
delaine 高级精梳羊毛;印花毛纱;高级薄花呢
delaine merino C型美利奴羊毛
delaine merino wool C型美利奴羊毛
delainette 仿薄花呢
delamination 层离,脱层
delay 延期
delayed elastic recovery 缓弹性恢复
delay elastic deformation 缓弹性变形
Delcerro 秘鲁德塞罗棉
Delcerro cotton 秘鲁德塞罗棉
delfina 西班牙薄黄麻帆布
Delfino 聚酯异收缩混纤丝(商名,日本东洋纺)
Delfos 美国德字棉,美国德尔福斯棉
Delfos cotton 美国德字棉,美国德尔福斯棉
delft blue 东方蓝(暗蓝色)
Delhi work 德里绣,缎地链线迹刺绣,德里刺绣
delicacy 娇弱,清秀,纤弱,雅致
delicate 鲜嫩
delicate color 娇嫩色泽,嫩娇色,柔和色
delicate cycle 轻洗涤程序
delicate fabric 娇嫩织物,宜轻洗的织物,易损织物,易变形织物,轻薄织物
delicate fiber 细纤维
delicate lace 纤薄花边
delightful dressing (D.D.) 欢愉的装饰
delineate 勾图,描绘;描轮廓;描外形
delineation 勾画轮廓;画像
delineator 描画者;描画器;裁剪图样;时装报道员
delinere 中档漂白亚麻布
deliquescent effect 吸湿效应
delivered duty-paid (DDP) 完税后交货
deliver the goods in batches 分批交货
delivery 支付
delivery eye (of bobbin case) 梭心出线孔
delphi method 德尔菲法
delphinium blue 靛蓝色,翠雀草蓝色
Delphos 特尔斐夜礼服
Delphos dress 特尔斐裙;德尔弗斯服饰
Delta cotton 美国三角洲棉
delta-E 色差
Deltapine cotton 美国岱字棉,岱塔派棉
deltatype webber cotton 美国角形早熟棉

deltoid muscle 三角肌
delustered fiber 无光纤维,消光纤维
delustered rayon yarn 无光人造丝
delustered yarn 无光丝,消光丝
delustering 消光
delusting 除光,退光,消光
deluxe 高光泽电光整理布
deluxe adjustable tape foot 高级可调缝带压脚
deluxe fabric 高光泽电光整理布
deluxe goods 顶级商品,豪华商品
deluxe ready-to-wear 豪华名牌服装
deluxe utility sewing machine 高级实用缝纫机
delver 上光棉平布
delvi 阿拉伯包头条子布
demand draft (d/d) 即期汇票
demand exceeds supply 供不应求
demarara jute 阿卡西黄麻
dememrara cotton 南非德梅拉拉棉
demi-bas [法]半筒袜
demi-boots 短靴,矮腰靴,轻便靴
demi-bosom 短胸衣
demi-bra(ssiere) 半罩式胸罩(露部分乳房),短式胸罩
demi-buff mink 半浅黄色貂皮
demi-castor 矮海狸皮帽
demi-cornal 短冕状头饰
demi-corset 短紧身褡
demi-cup bra 半罩式胸罩(露部分乳房),短式胸罩
demi-drap 轻缩绒薄呢
demi-gigot sleeve 半羊腿袖
demi-gross weight 半毛重
demi-londres 轻缩绒粗纺松软呢
demi-luster 半有光羊毛;半有光羊毛纱
demi-luster yarn 半有光丝;半有光纱
demi-maunch 中袖
demi-mini 超超短连裙装,超超短裙,超超短装
demi-mousseline 细薄平布;细薄条子布
demiostage 苏格兰有光呢
demirdji rug 土耳其中长绒头松软地毯
demitint 晕色(介于深,浅色之间)
demi-toe reinforcement 袜头半加固
demi-toilette 半正式礼服
demittons 厚重棉织物(英国),重磅棉布
demiwedge 半楔形鞋跟
demiwedge heel 半楔形鞋跟
demi-wig 半顶假发,半假发

demo 示范,展销产品,示范表演
democrat hat 民主主义者帽
demonstration 论证;示范,表演
denatured goat hair 变性山羊毛
denatured wool 变性羊毛
dence 浓重
Dendara 埃及丹达拉棉
denes blanket 兔毛毯
denier (D,d) 旦尼尔(旦)
denier number 旦数
denier per filament 单丝旦数
denim 粗斜纹布,粗斜纹棉布,劳动布,牛仔布;坚固呢(蓝斜纹布)工作服,工装裤
denim abrasion 牛仔布仿旧整理,牛仔布磨洗
denim bag 牛仔袋
denim blue 蓝色斜纹工作服
denim cap 牛仔布帽
denim cotton 粗斜纹棉布,劳动布
denim garments 牛仔服装
denim garments collection 牛仔服装系列
denim hat 牛仔布帽
denim jacket 粗斜纹布短外衣,劳动布夹克,牛仔夹克
denim jersey 毛织斜纹运动衫
denim-like knitted fabric 牛仔针织物
denim look 劳动布风格,牛仔布风格
denim overalls 牛仔工作服
denim padded jacket 牛仔填棉夹克
denim pants 牛仔裤,劳动布裤,粗斜纹布裤
denims 工装裤,劳动布工作服
denim shirt 牛仔(布)衬衫
denim shorts 牛仔短裤
denim skirt 牛仔裙
denim suit 粗斜纹布服装,蓝粗斜纹布装,牛仔(布)套装
denim tops and bottoms 牛仔套裤,牛仔套装
denim top-stitched pocket 牛仔双明线贴袋
denim trousers 牛仔裤
denim vest 牛仔(布)背心
denim waistcoat 粗斜纹背心,牛仔背心
denim washing 牛仔布仿旧整理,牛仔布水洗
Denmark cock 丹麦男式三角帽
Denmark satin 丹麦缎纹呢
dense hollow fiber 厚壁中空纤维,致密中空纤维

dense pattern 密纹图案
dense pile velvet 高级密乔绒
dense texture fabric 高密织物,紧密织物
dense upholstery silk 密纹装饰绸
densimeter, densitometer 密度计;(测布料用的)经纬密度尺
density 密度
dentelle 荷叶边
dentelle àcartisane [法]金银凸花花边
dentelle a la vierge 法国梭结花边
dentelle au fuseau 筒子梭结花边
dentelle au lacet 花边条
dentelle d'application [法]镶贴花边,梭结花边
dentelle de 爱尔兰花边
dentelle de fil 单纱花边
dentelle de la chasse 狩猎花纹花边
dentelle de la sorciere 丝网眼纱
dentelle de la vierge [法]处女花边,锯齿边宽幅梭结花边
dentelle de liege 比利时网眼花边
dentelle de lin 法国亚麻花边
dentelle de moresse 摩洛哥花边条
dentelle des indes 抽绣
dentelle lace 梭结花边;锯齿边阔帽
dentelle net 花边图案粗网眼窗纱
dentelle redin 网眼地花边
dentelle renaissance 手工钩编花边
dentelliere 花边编结机
dents 花边边饰
dents de rat 金银线锯齿边
deny 粗纺毛织物
deo cotton 印度迪欧棉
deodorant 除臭剂,防臭剂,脱臭剂
deodorant fiber 除臭纤维
deodorant shampoo 灭菌除臭洗发剂
deodorizing finish 除臭整理
deo kapas 印度迪欧棉
déoordonné 不调和
de-ozonized fabric 除恶臭织物
department stores 大百货商店
departure 启运
DéPéCHE MODE 《最新服式》(法国)
DéPéCHE MODE SPECIAL PRêTAPORTER 《最新服式专题》(法国)
depiction 雕刻图案
depilatory 脱毛剂
depth 深度;色泽深度,色泽浓度
depth of hip 臀长
depth of scye 袖隆深

derbent rug 山羊毛机织长绒地毯
derby 常礼帽,圆顶礼帽;德比(贝)领带,四步活结领带;低跟运动鞋
derby bartack 帽状套结
Derby boot 德比靴
derby hat 常礼帽,德比帽,圆顶高帽
Derby red 德比红,深橙红色
Derby rib 德比式罗纹组织
Derby tie 德贝领带,德比领带,四步活结领带
Derby top 德比式(6+3)螺纹袜口
derh pati 粗制腰布
deriband 印度漂白棉布
derivative fiber 衍生纤维
derivative gauze 变化纱罗织物
derma-respiration 皮肤呼吸
dermatitis 皮炎
dernier cri 最新款式的服装
dernier mode 服装等最新样式
derrafendi cloth 高档手织棉布
derriere 臀部
derries 印度色织棉布
derris 印度色织布
derry 德里布,印度色织棉服料
Derung ethnic costume 独龙族服饰
dervish skirt 回教托钵僧裙
dervish tulle 粗制印花网织物,伊斯兰教托钵僧薄纱
descending 色调递降
Desco bag 迪斯可袋
des 199 cotton 美国商业陆地棉
description of goods 货物分类目录
desert boots 绒面革皮靴,沙漠长靴,沙漠深靴
desert color 沙漠色
desert dust 浅黄咖色
desert fatigue cap 沙漠工作帽
desert flower 沙漠花色
desert mist 沙漠雾色
desert palm 沙漠棕榈色
desert rose 沙漠玫瑰色
desert sand 沙漠色
desert shoes 沙漠鞋
déshabillé 法国女便服
déshabillé style 裸露式
deshi cotton 印度德希棉
deshi jute 德希黄麻
desi 印度黑色黄麻;德希棉
design 设计,构思;图案,花纹,款式;花样,图样,样式

design ability 设计能力
design and color 花色
design area 花型面积,花型范围
designated line 指定线
designation strip 标条,名牌
design boutique 设计师精品店
design brushing 刷花
design capacity 设计能力
design chart 设计图
design collection 设计师时装发布会
design comprehensive studies 设计综合研究
design concept 设计理念,设计思想
design conception 设计构思;设计观念
design data book 设计资料集
design department 设计部
design detail 设计详图
design detailing 设计细部
design details input 花纹详图输入
design drawing 设计图;服装款式造型图;服装效果图
design dress in smart shape 设计时新女装
designed organization 归口单位
designed ribbon 图案饰带
designed tape 花色带
design element 设计元素,设计基础
designer 设计员;(流行服饰)设计师;制图(打样)人,图案师
designer brand 设计师商标
designer-cutter 设计裁剪师
designer jeans 标名牛仔裤,名家牛仔裤,设计师牛仔裤
designer-label style 名师标牌款式,设计师标牌服装款式
designer-name style 名设计师命名款式,设计师命名款式
designer scarf 名仕饰巾,签字饰巾,印有设计师名字的真丝围巾
designer's clothes 标名服装,命名服装,设计师标名服装
designer's dart (省尖到胸高点的)设计师省道
designer's house 设计师标名厂商号
designer's idea 设计师思想
designer's inspiration 服装设计灵感
designer's intent 设计师意图
designer store 名师商店
design feature 设计特征
design flocking 图案植绒
design focus 设计焦点

design garment 设计衣服
design graph 设计图样,花纹图样
design hosiery 花式连裤袜
design idea 设计概念,设计思想
designing thought 设计思想
design inspiration 设计灵感,设计说明书
design judgement 设计评审
design matrix 花纹矩阵设计
design motif 花样单元
design number 花号,印花图案号
design of hexagon pattern 六边形花纹设计
design of rectangular pattern 矩形花纹设计
design package 设计包装
design paper 方格纸,图案纸,意匠纸
design parameter 设计参数
design patent 设计专利
design pattern 设计纸样
design performance 设计性能
design philosophy 设计理念
design points stitch 花色线圈
design process 设计流程
design processing 设计处理
design program 设计方案
design project 设计规划
design proposals 设计方案
design reduction 图案缩小
design register 印度对花
design ribbon 提花带
design room 打样间,打样室,花样间,设计室
design sample 花样样品
design scheme 设计方案
design screen 设计图案显示展
design source 设计源
design specification sheet 设计明细单;设计明细表
design stitch 花样线迹,花纹线迹
design style 设计风格
design sweater 花色针织套衫
design system 设计分类
design test 鉴定试验
design trademark 设计商标
design variable 设计变数,设计变量
desirable hand 合适手感,手感舒适
desire 棉丝混纺织物
desized cotton 退浆棉布
desizing 去浆糊
desizing cotton 退浆棉布
desk clock 台钟,座钟
de-skilling 非技术操作,降低作业的技术水平
desooksoy 印度棉布,印度棉织物
despaissis silk 铜铵人造丝
despeissis rayon 德斯佩息斯人造丝(铜铵人造丝旧称)
dessinateur [法]素描或样式画家
dessinateur modéliste [法]服装设计画家,服装效果图设计师
dessin caviar 小格子花样,鱼子酱花样
destaticizer 除静电剂
destatic property 除静电性
destrados 西班牙粗地毯
detachable 活动袖口
detachable belt 活动腰带
detachable coat 脱卸外套
detachable collar 活动领
detachable cuff 活动袖头,可卸袖口,可卸克夫
detachable gloves 可拆卸手套
detachable hood 拆卸式兜帽,可卸兜帽,脱卸头巾
detachable jacket 活动上衣;可拆卸夹克
detachable lining 活衣里,活里子
detachable pantyhose 三段组合式裤袜,脱卸式裤袜
detachable part 可卸件,可装拆件
detachable shank button 可卸有柄纽
detachable skirt 可拆卸裙,可卸裙
detachable sleeve 活络袖
detachable work plate 缝纫机可卸式工作台板
detachable zip-in lining 脱卸式里子,活里子
detached chain stitch 平式花瓣线迹
detached coat 脱卸外套
detail (服装或资料的)细目,细节;(设计)详作;零(配)件
detail design 细部设计
detail drawing 工笔画,明细图,细部图,详图
detailed jacket (拼接)花式夹克
detailed list 明细表
detail of design 设计详图
detail operation schedule 详尽作业程序表
detail schedule 进度计划明细表
detecting period 检测周期
detergency 去污力,洗净力
detergency builder[promoter] 助洗剂
detergent 洗涤剂,清洁剂,去垢剂
detex 分号;分特

detil 希腊式锯齿图案
Dettigen cock 帽檐上翻的男式三角帽
deuteranopic chromaticity confusion 绿色盲错乱
developed area 发达地区
deviation in shade 色差
device （有特定用途的）设备，装置，器件
devil suit 恶魔装
Devonia ground 花边的曲折线迹
Devonia lace 德文郡凸花花边
Devon long-wool 德文郡长羊毛
Devonshire 德文郡色织条子斜纹棉布
Devonshire lace 德文郡花边
devorant 仿烂花织物
dew 露水色
dewaxed cotton 脱脂棉
dewdrops 露珠布
deweighting 减量处理，碱减量处理
deweighting finish 减量整理
deweighting finish fabric 减量整理织物
dexterous craft 灵巧手艺
Dexter wipes 狄士打湿擦布，狄士打白洁布
2D-fashion design 二维服装设计系统
D figure D体型，肥胖体型
dha 德哈大麻
dhaka 印度印花布
dharwar-American cotton 美国达瓦尔棉
Dharwar cotton 印度达瓦尔棉
Dharwar No.1 cotton 印度达瓦尔一号棉
dhildren's dress 儿童服装
Dhollerah cotton 印度暗白棉
Dhoosa 印度杜沙棉
dhoosa cotton 印度杜沙棉
dhoosootie 重磅帆布
dhootie, dhooty, dhotee, dhoti, dhotie, dhoty 色织提花细布，印度腰布，缠腰布；道蒂服（印度）
dhooty 色织提花细布，印度腰布
dhotar 土耳其粗棉布
dhotee 色织提花细布，印度腰布
dhoti 缠腰布
dhotie 缠腰布
dhoty 色织提花细布，印度腰布
dhour 披巾中央边缘花纹（印度）；印度开司米围巾图案
dhourdar 开士米头巾
3D human body model 三维人体模型
dhupatti 扎染莎丽，扎染丝莎丽
dhupchan 日影布，日影绸（红绿闪色绸）
dhurrie 棱纹原棉布；印度原棉毯，印度

厚毛毯
di 滴（服装加工专业用语）
diable fuerte 灯芯绒，经向灯芯布
diablement fort 麻经棉纬织物
diacetate fiber 二醋酯纤维
diadem 冕，王冠
diadem cap 王冠状浴帽
diadem fanchon bonnet 蕾丝和无鹅绒光环形檐帽
diagonal 对角线，斜线符号；贡斜纹，斜纹花纹；斜纹织物
diagonal basting 扳针法，斜线疏缝
diagonal basting stitch 斜向疏缝线迹，斜边粗缝线迹
diagonal basty 斜线疏缝
diagonal basty stitch 斜向疏缝线迹，斜边粗缝线迹
diagonal braid 斜编带
diagonal buttonhole 斜式纽孔
diagonal cloth 斜纹毛织物
diagonal connecting stitch 斜向连接（钉纽）线迹
diagonal crease 斜向折印（皱纹，褶，裥）
diagonal cross-over stitch 斜向交叉线迹，对角线线迹
diagonal cut(ting) 斜裁
diagonal dart 斜省，对角省
diagonal displacement 对角线形纬斜
diagonal English leather 英国斜纹仿麂皮经编织物
diagonal eyelet 斜网眼
diagonal fold rib effect 对角交叉凹凸效应
diagonal interrib 针织斜纹呢
diagonal nap cloth 波纹珠呢，波纹卷毛绒
diagonal pattern 斜纹花型，斜纹图案，斜纹式图案
diagonal print 斜纹图案印花，斜纹花型印花
diagonals 斜纹棉坯布
diagonal slip stitch 扳针
diagonal stay 斜撑条，对角拉撑
diagonal stitch 扳针法；对角缎纹绣；对角条纹；对角棱纹线迹，对角线线迹
diagonal stitching 斜扎线，对角线缝
diagonal stripe 对角条纹
diagonal strut 斜撑
diagonal tacking 斜线加固
diagonal weave 贡斜纹
diagonal worsted 精纺粗斜纹呢
diagonal wrinkles at sleeve cap 绱袖不圆

顺,袖山起皱
diagonal wrinkles at sleeve lining 袖里拧(袖里、面错位)
diagonal yoke 交叉式育克,斜抵肩
diagram 曲线图,图,图表,图解,图形
diagrammatic sketch 草图,示意图
diagrammatic view 简图,图示
diagram patterns 服装衣片图
dial control drop feed 刻度盘控制送布
dial looper 圆盘式套口机
dial thread tension regulator 圆盘式线张力调节器
dial transfer jack 袜口钩子
diamanté 闪光珠饰礼服
diamanté dress 闪光珠饰礼服
diamantee 法国线经马甲绸
diamanté top 闪光珠饰短上衣
diamantine 英国粗纺呢,英国轧光斜纹呢
diamond 金刚石,钻石,菱形花,菱形花纹
diamond barring 菱形条花,钻石条花
diamond bracelet 钻石手镯
diamond braid 菱花辫线;菱花细编带
diamond check 菱形格子
Diamond cotton 美国钻石棉
diamond dart 双向褶,鱼形褶;菱形绸
diamond draught diaper 菱形花纹亚麻布
diamond earrings 钻石耳饰
diamond effect 菱形效应
diamond fabric 经编缎纹织物
diamond filling stitch 菱形编结绣,菱形贴线绣
diamond heel 菱形袜跟
diamond hip type 菱型臀型
diamond hose 菱形袜
diamond knee 加厚菱形袜膝段
diamondlerx 钻石光(用于首饰照明)
diamond linen 小菱纹亚麻织物
diamond neck 菱形领
diamond necklace 钻石项链
diamond neckline 菱形领圈,钻石领口
diamond net 菱形网眼
diamond ornament 菱形装饰
diamond pattern 菱形图案
diamond pattern nylon blouse 菱形尼龙上衣
diamond petal 菱形花瓣
diamond pin 钻石别针
diamond point 菱形图案;金刚钻针
diamond point toe 菱形袜头
diamond printing 钻石印花

diamond ring with pearl 镶珠钻戒,珍珠钻石戒指
diamonds by the yard 镶钻石的14k按码金链
diamond-shaped dart 菱形褶,双向褶,鱼形褶
diamond-shaped marking 菱形记号
diamond shoes 钻石鞋
diamond smocking 菱形缩褶绣,菱形司马克,钻菱缩褶绣
diamond snap 宝石按扣;宝石按钮
diamond stitch 菱形编结绣;菱形刺绣针迹
diamond twill 菱形斜纹
Diana fashion 戴安娜时装
Diane Pernet 戴安娜·珀内特(美国时装设计师)
Diane Von Furstenberg 戴安娜·冯·弗斯滕伯格(1947~,美国时装设计师)
diaper 菱纹图案;菱纹织物,细亚麻毛巾布,彩色格子斜纹棉布;尿布,强吸湿性织物
diaper backsheet 尿布后片
diaper bathing suit 菱纹浴衣
diaper-bottom bikini 菱裆比基尼
diaper cloth 小菱形织物,鸟眼花纹织物;婴儿用布
diaper covering 尿布包裤
diaper flannel 平纹棉法兰绒
diaper linen 亚麻尿布
diaper liner 尿布垫
diaper pad 尿布垫
diaper panty 尿布包裤
diaphalene 浅色精梳丝光细布
diaphane 白色或印花透明薄纱
diaphaneity 透明度;透明性
diaphanous gauze 透明薄纱
diaphanous tabbies 薄丝织物
diaphone 白色或印花透明薄纱
diaphragm 横隔膜
diaphragm slit 膜片缝,光阑缝
diapistus 中世纪高级织物
diathermancy 透热性
diazo dye 重氮染料,偶氮染料
diazotize color 重氮化染料
diazotized and developed dye 重氮显色染料
dice 阴阳菱形花纹,阴阳小方格花纹
dice check 阴阳菱形花纹,阴阳小方格花纹
Dicel 醋酯长丝(商名,英国)

dice pattern 阴阳菱形图案,阴阳小方格图案
dicer 钢盔,头盔;硬帽,男用大礼帽
dice rib 阴阳小方格罗纹
dice twill 阴阳斜纹,阴阳小方格斜纹
dice venetian 色子贡,骰子贡
dichroic mirror 二向色反射镜
dichroism 二向色性
dick 皮围裙(英国)
dicker 10 张皮革;V 形装饰布(女服胸前),胸衿,胸前 V 形装饰布(女服)
dickey, dickie, dicky 装饰衬领,假衬衫;领结(英国);围涎;(女服胸前)V 形装饰布
dickey-front shirt 上浆胸衬衫;乌贼胸衬衫
dickie 装饰衬领,假衬衫;领结(英国);围涎;(女服胸前)V 形装饰布
Dickson cotton 美国迪克森棉
dicky 假衬衫
dicky dress 假前胸装
Didier Lecoanet 迪迪埃·莱科特(法国著名时装设计师)
die 刀模,型,模子;裁剪型板;冲裁刀具
die clicker 冲裁机;冲模裁剪机;冲压剪机
die cloth cutting machine 冲模裁剪机,冲压裁剪机;打眼机,冲压编带机
die cut 下料,冲裁
die-cut label 模压唛头
die cutter 冲裁机,冲模裁片机,冲压裁机,冲切裁剪机,钢芯雕刻机
die cutting 冲切;冲压裁剪
die cutting machine 冲裁机,冲模裁床,冲压裁剪机
diehua cloque 叠花绉
dielectric boots 绝缘靴
dielectric helmet 绝缘头盔
Dieppe lace 迪埃普梭结针绣花边
Dieppe point lace 法式梭结花边
Diepper point lace 小鸡花边
die press 模型熨烫法
die pressing 压翻领袖口机;压模熨烫机
Diesel 迪泽儿
Dietmer Sterling 迪特玛·斯特林(法国时装设计师)
difference of stitch length 线迹长度差
differential arm 差动机头
differential belt feed 差动皮带送料
differential bottom feed 差动下送料
differential cloth feeding 差动喂布

differential dyeing 差异染色
differential dyeing fiber 差异染色性纤维
differential feed 差动送料
differential feed flat bed machine 差动送料平底板缝纫机
differential feed type 差动送料式
differential feed type lower feeding 差动送料式下送料
differential fiber 改性纤维,差别化纤维
differentially dyeable fiber 改性可染纤维
differential sampling 差别抽样法
differential stretching feed 差动伸缩送料
differential top belt feed 差动皮带上送料
differential top feed 差动上送料
differentiated filament yarn 变异长丝
different material lapel 异布驳领
different parts of the body 身体各部分
different twist yarn fabric 异捻丝织物
different types of bodies 各种体型
different warp fabric 异经织物
difficult-to-cut fabric 难裁剪织物.难裁织物
difficult-to-cut material 难裁缝料
difficulty combustible fabric 难燃织物,阻燃织物
diffraction streak 衍射条纹
diffuse shading 漫射明暗度
diffusion 扩散
Digby Morton 迪格比·莫腾(英国时装设计师)
digit 数字
digital ink jet printing 数码喷墨印花
digital jet printing machine 数码喷墨印花机
digital switch 数字式开关
digital-tone 数字色调
dignity 端庄
dignity look 端庄风貌
dihua damask satin 涤花缎
Dik Brandsma 迪克·布兰斯马(法国时装设计师)
Di lazzaro 迪·拉扎罗(意大利时装设计师)
dilutee 非熟练工
dima 叙利亚窄幅棉布
dimakso 生丝(阿拉伯语)
dimantino 斜纹呢
dim color 暗色
dim effect 朦胧[印花]效果
dim effect printing 朦胧印花,迷彩印花,

暗淡印花
dimei satin brocade 涤美缎
dimension 尺寸,尺码,尺度
2-dimensional block pattern 平面服装样板,二维服装样板
dimensional change 尺寸变化
dimensional construction 服装立体构成
3-dimensional cutting 立体裁剪
dimensional drawing 尺寸图
3-dimensional effect 立体效应,三维效应
3-dimensional fashion 立体成形
3-dimensional handling 立体加工
dimensional instability 尺寸不稳定性
dimensional interchange ability 尺寸互换性
dimensional metrology 尺寸测量法
3-dimensional printing 立体印花
3-dimensional printing effect 立体印花效应
dimensional restorability 形状复原性
2-dimensional shape 二维形态
3-dimensional shape 立体形态
dimensional stability 尺寸稳定性,形稳性
dimensional stability finishing 尺寸稳定性整理
dimensional stability index 尺寸稳定性指数
dimensional stability of knitted fabric 针织物尺寸稳定性
dimensional stability tester 尺寸稳定试验仪
dimensional tolerance 尺寸公差
dimensioning 量尺寸
dimension line 尺寸线
dimension scale 尺寸比例
dimidje 光亮丝绸宽松土耳其裤
dimity 麻纱(棉织物),条格麻纱
dimity binding 凸纹纱带
dimity check 格子麻纱
dimity cord 棱条麻纱
dimity crossbar 棱格麻纱,格条麻纱,棉格条
dimity hair cords 麻纱
dimity ruffling 狭幅棉绉布,窄幅棉绉布
dimity stripe 绉条细布,泡泡条子床罩布
dimity wool 无光羊毛
dimple 酒窝,领结涡,波纹,领结屑
dimple pants 侧褶裤
dimple-sleeve jacket 侧褶袖夹克
dim style 朦胧式样
dinginess 黯淡色,暗淡色,深黑色;衣衫褴褛

dingxiang brocade 丁香缎
dining clothes 大餐衣
dink 丁克帽(瓜皮帽款式)
dinky 丁克帽(瓜皮帽款式)
dinner cloth 正餐桌巾
dinner clothes 餐服
dinner coat 小礼服,晚会便服
dinner dress 女餐服,晚礼服,夜礼服,晚宴服,晚餐服,女式小礼服
dinner jacket 小礼服,无尾礼服,晚宴夹克礼服,晚礼服,男子餐服
dinner jeans 正式牛仔裤,宴会牛仔裤
dinner ring 鸡尾酒会戒指,晚宴戒指
dinner suit 晚礼服,夜便礼服(男用套装),晚宴套装
diobiris 粗厚五枚缎绒里丝毛呢
Diolen 聚酯纤维(商名,荷兰阿克佐)
Dior new look 迪奥新风貌(1947),迪奥新款式
dip 前低后高的腰围线点;染色试样
dip belt 曲线饰带
dip-dyed hose 浸染裤袜,浸染袜子
dip-dyed hosiery 浸染袜类
dip dyeing 浸染
diploidion 古希腊长衣
diplois 迪拍劳依丝外衣(古希腊)
dipped fabric 浸渍织物,浸染织物
dipped hose 浸染袜子
dip-top boots 长牛仔靴
direct azo dye 直接偶氮染料
direct blending dye 直接混纺染料
direct bordeaux 直接耐酸枣红
direct closing method 直接结算方法
direct color 直接染料
direct copper grey 直接铜盐灰
direct cotton dye 染棉直接染料,直接染棉染料
direct dye 直接染料
direct fast dye 直接耐晒染料(直接L型)
direct fast yellow 直接耐晒嫩黄
direct indigo blue 直接靛蓝
directional fabric 方向性面料,方向性织物,顺丝绺面料
directional layout 按丝绺排料
directional line 方向线,裁剪线
directional pressing 按方向熨烫
directional stability 方向稳定性
directional stitching 顺丝绺缝线
directional structure fabric 定向结构织物
direction of slippage 织物中纱线易滑移

的方向
direction of twill 斜纹方向;捻向
direction of twist 捻向
direct liability 直接责任
direct mark 顺向符号,顺向号
direct marketing 直接销售,直销
direct measurement system 直接测量尺寸法,直接测量系统
Directoire bonnet 执政内阁式帽
Directoire coat 执政内阁式大衣
Directoire gown 执政内阁式女装
Directoire jacket 执政内阁式夹克
Directoire skirt 执政内阁式七片裙
Directoire style 督政府发型,执政风格
Directoire waistline 执政内阁式腰线,督政府腰线(18~19世纪)
director's suit 白天准礼服,董事长套装,干部西服套装,男式交际服
direct port 直达港
direct printing 直接印花
direct printing with mordant dyes 媒染料直接印花
direct printing with substantive color 直接染料直接印花
direct tax 直接税
direct trade 直接贸易
direct worker 直接生产工人
dirndl 旦多尔连衣裙,旦多尔装,紧身腰裥服装,阿尔卑斯村姑装;简单细裥裙,抽褶裙
dirndl apron 阿尔卑斯村姑式围裙,旦多尔围裙,装饰用围裙
dirndl blouse 阿尔卑斯村姑式连衣裙,旦多尔连衣裙
dirndl dress 紧身腰裥服装;阿尔卑斯村姑装;简单细裥裙,抽褶裙;旦多尔装
dirndl line 阿尔卑斯村姑款式,村姑风格式样,村姑型
dirndl necklace 阿尔卑斯村姑式颈饰,旦多尔颈饰
dirndl pants 阿尔卑斯村姑裤,旦多尔裤(抽裥)
dirndl-peasant skirt 村姑-农妇裙
dirndl petticoat 抽裥衬裙
dirndl shorts 村姑短裤
dirndl silhouette 旦多尔型,紧身连衣裙轮廓
dirndl skirt 阿尔卑斯村姑裙,旦多尔裙;抽褶裙
dirndl suit 阿尔卑斯套装

dirt 污垢
dirt and oil 油污
dirt and oil patch 油污斑
dirt bike steel toe 钢头摩托靴
dirt repellent finish 防污整理,拒污整理
dirt repellent treatment 防污处理
dirt resistance 抗污性
dirt stain 灰渍,污渍
dirty 暗淡的,褐色的,带灰的;织物暗淡的外观
dirty buck 脏鹿皮鞋;脏鹿皮
dirty green 暗绿色
dirtying 沾污,沾色
dirty red 暗红色
disadvantage 疵点
disappearing arms dress 缺袖窿装
disappearing fiber 溶解性纤维
disarray 不整齐的衣服
disazo dye 双偶氮染料,双重氮染料
discarded section 舍弃部分(纸样设计)
disc feed overseaming machine 圆盘送料包缝机
discharge 拔染;放电,静电放电
dischargeable color 拔染染料
discharge design 拔印花样
discharge printed 拔染印花
discharge printed fabric 拔染印花织物
discharge printed handkerchief 拔染印花手帕
discharge printing 拔染印花;雕印
discharge prints 拔染印花织物
discharge-resistant dye 防拔染[用]染料,防拔染[着色]染料
discharge-resist printing 防拔染印花
discharge style 拔染印花
discharges with mordant dyes 媒染染料拔染
discharge zone 放电区
discharging fee 卸货费
discharging of tannin mordant 单宁媒染拔染
disclaimer liability 否认责任
disco bag 迪斯科包,迪斯科袋
disco clothes 迪斯科装
disco fashion 迪斯科装
discolor 褪色膏,脱色
discoloration 脱色,褪色,变色
discolored pick 错色纬疵
discoloring 变色,脱色
discolorization 脱色,褪色,变色
discomfort (DI) 不快指数;不舒服,不舒

适感
disconstruction 无结构主义
discontinuous brocade 挖花花缎
discontinuous metallic filament 金属短纤维,不连续金属丝
discord 色彩不调和,色彩不一致
discordance 不调和
discotheque dress 迪斯科装
discotheque sandals 夜总会跳舞凉鞋
discotheque style 夜总会款式(低领,短下摆)
discount 打折,贴现
discounter 减价商店
discount store 折扣商店
discreet shade 保安色彩
discrimination 鉴别
discrimination level 判别水平
disfigured design 对花不准
disguise 化装,假扮,伪装
dishabille (古代)家便服,便服,穿着便服(或睡衣)
disharmony 不调和;不和谐
dish cloth 茶巾;揩碗布;抹布
dishdasha 伊拉克长至脚踝的衬衫
dishonour 拒付
dishrag shirt 男式针织宽松运动衫;无扣低腰衫
dish towel 擦盘碟的中粗毛巾
dishy 好看的,漂亮的
disinfection 灭菌.消毒
disk feed 圆盘送料
di-song polyester twill 涤松绫
disong twill 涤松绫
disorderly period 杂乱无章时期
dispatcher's office 调度室
disperse disazo dye 分散重氮染料,双偶氮分散染料
disperse dye 醋酸染料,分散染料
disperse dyeing 分散染料染色
disperse metallic dye 分散金属络合染料
disperse pink red 蓝光桃红
disperse premetallized dye 分散金属络合染料
disperse red 分散蓝光红
dispersibility 分散性
dispersing reactive dyes 活性分散染料
dispersive cationic dyes 分散性阳离子染料
displacement-free seam 无位移缝;无位移缝纫
displacement-free sewing 无位移缝纫

displacement in sewing direction 缝纫方向的位移
displacement printing 防染印花;置换印花
display 陈列.展览
display cabine [case] 陈列橱
display of fashion dress 时装展示
disposable 用即弃产品,用可弃
disposable diaper 用即弃尿布
disposable fabric 一次性织物,用即弃织物
disposable nonwoven 用即弃非织造布
disposable nonwoven fabric 用即弃非织造布
disposable pants 纸裤
disposable PE raincoat 一次性雨衣
disposables 一次性使用制品,用即弃制品
disposable suit 一次性服装
disposable swim suit 一次性泳装
disproportion 不相称
disproportionate figure 不匀称体型
disputes 争议
disque hat 宽檐帽
disrobing 脱衣
dissonant colors 不协调色
dissymmétry 非对称性
distance between two teats 乳峰距;乳[间]宽
distance line 距离线
distinctive line 纹路清晰
distorted heel and toe 袜头跟歪角
distorted loops 三角眼疵
distorted selvage[selvedge] 布边歪斜
distortion 扭变,失真,歪边疵
distortion-free seam 无扭变缝
distortion-free sewing 无扭曲缝纫
distortion of threads 经纬歪曲
distributing dyestuff 匀染染料
distributing property 匀染性能
distribution curve 分布曲线
distributive law 分配律
district check 地区格子呢(苏格兰)
disturbed traverse 花纹错乱
ditan polyester twill 涤弹绫
ditsosi 单面长圈毛毯
dittos 同料同色衣服,同面料的男式套装
ditto suit 同料同色的一套西服
divergence angle 分叉角
diver's coverall 潜水员工作服
diversification of management 经营多样化
diversified twill 变化斜纹
diversity suit 多样性套装

diver's submarine armor 潜水衣
diver's suit 潜水服
divi coir 椰壳纤维
divided-adjuster 分规调节器
divided culotte style 裙裤样式
divided satin stitch 分茎绣
divided skirt 分衩裙;裙裤
dividers 分线规,两脚规
dividing collar 大前开门领
diving calf 暗褐色小牛皮
diving clothing 潜水服
diving collar 大前开门领
diving dress 潜水服,潜水衣
diving helmet 潜水头盔
diving suit 潜水服
divinity calf 暗褐色小牛皮
division 分割
divisional color 分界色
divisional system 事业部制
division of costs 费用划分
dixian oxford 涤纤绸
dixia satin brocade 涤霞缎
Dixie cotton 美国迪克西棉
DIY(do it yourself) 自己动手
diyogi 膨体平纹毯
dizzying slit (服装)高开衩
djellaba(h) 洁露芭袍,摩洛哥宽长袍
djiba 洁露芭袍,摩洛哥宽长袍
djule 伊朗背面起毛毛毯
D.Lecoanet 德·勒科纳(法国时装设计师)
doaria bufta 印度手工纺纱织布
dobby 小提花织物
dobby-bordered fabric 多臂边花织物
dobby check 多臂提花格子
dobby cloth 多臂提花布,多臂提花织物
dobby fabric 多臂提花织物
dobby fancy corduroy 提花灯芯绒
dobby lining 小提花衬里
dobby poplin 小提花府绸
dobby stripe 多臂提花条纹,小提花条
dōbuko 旅行风衣;日本双排扣
docket 标志,签条,标签;打印
doctor cuff 医生袖头
doctor Martens 马丁博士气势靴
doctor's bag 铰合式手提旅行包;医用皮箱
Doctor's gown 博士服
Doctor's gown and hood 博士服
doctor wear 医生服
docuwa 粗犷柔软格子布

document against payment(d/p) 付款交单
documentary design 仿古艺术图案
documentary print 仿古印花纹织物
document retrieval 文献检索
dodot 蜡防印花
Dody Brode 多迪·布罗达(法国时装设计师)
doe 母鹿色,浅棕灰,驼灰
doe fur 雌鹿毛皮
doek (南非土著妇女用的)白布头巾
doeskin 羚羊皮;仿麂皮,仿麂皮织物,驼丝锦,礼服呢
doe skin 母鹿皮革;母山羊皮绵羊皮,羔羊皮,驼丝锦,克罗丁
doeskin 驼丝锦(紧密缎纹毛织物),礼服呢,仿麂皮织物;羚羊皮,仿麂皮,小牛绒面革,羊羔绒面革
doeskin fabric 仿麂皮织物
doeskin finish 仿麂皮整理
doeskin gloves 麂皮手套,仿麂皮手套
doeskin pile 仿麂皮绒
doesootjes 东印度漂白细棉布
dogaline 中世纪肥袖直裁宽松袍
dog collar 狗项圈,宽带项圈;后开门立领,牧师领,狗圈小立领,用宝石装饰的领子
dog collar necklace 宽宝石项链,宽带贴颈短项链;珠饰项链,贴颈饰项圈
dog collar neckline 项圈式领口
dog collar scarf 项圈式围巾
dog-ear pocket 垂耳袋,狗耳形袋
dog fur 狗毛皮,狗皮
dog-leash belt 狗领带
dog-leash fastener 狗链型带扣
dog-legged edge 弯曲布边,荷叶边布疵
dog-legged selvedge 弯曲布边,荷叶边布疵
dog line 直条轧痕
dogs 劣等棉
dog's-ear collar 垂耳领
dog skin 仿海豹绒,狗皮绒,狗皮,狗皮革
dog skin gloves 狗皮手套
dog's tooth 犬牙花纹粗纺呢绒
dog's tooth check 犬牙格子花纹
dog streaks 经向条花
dogul 多古羊绒斜纹呢
dohar 印度毛披巾
doilly lace 垫布花边
doily 英国毛织物;小揩布,小垫布
Do it yourself (DIY) 自己动手做
do-it-yourself style 随你便款式,自由服

装款式
dokanni 印度双丝条子棉布
dolara 印度多拉腊布
Dolce&Gabbana/D&G 多尔切和加巴纳
dollar-round toe 银元形鞋头
dollerah cotton 印度道洛拉棉
doll fedora 娃娃式浅顶软帽
doll hat 德尔帽,娃娃帽,20世纪30年代小巧的蜂鸟形帽
doll's dress 娃娃衣服
doll's hair stitching machine 假发缝纫机,植发机
doll's miniature hat 玩偶式小帽
doll stitch 缝泽娃娃头发线迹,植发线迹
doll twist style 娃娃卷发型,娃娃髫发型
dolly 捣衣棒,捣衣杆;家用洗衣机,洗呢机,捶打洗涤机;捆痕(捆布卷造成)
dolly bird 漂亮的摩登女郎(英)
dolly look 娃娃风貌,洋娃娃风格,洋妹妹风貌
dolly varden (19世纪)多莉瓦登印花女服;花束布,花束绸(此花束为纹样)
dolly varden hat 多莉瓦登花饰女帽
dolman 土耳其式长袍;(宽大袖)女外套,女短斗篷,袖窄腋宽女外衣,腋部宽大的女外衣
dolman cardigan 多尔门对襟毛衣
dolmanette 钩编女式短斗篷
dolman sleeve 德尔曼袖(音译,袖口窄,袖襬宽),斗篷袖,土耳曼袖,腋部宽大连袖
dolman sleeve blouse 连肩袖衬衫
dolman sweater 德尔曼毛衣,腋部宽大连袖毛衣(蝙蝠袖,双翻领或船领,罗纹下摆)
dolman with gusset 有袖裆的连身袖服装
dolphin skin 海豚毛皮
domarji 达勒布
dome button 拱形纽扣
domeck 低档棉缎
dome coat 半球形外套
domed cricket 板球帽
dome fastener (用于手套等的)按扣,揿纽,摁纽,揿扣
dome-front style 揿纽对襟式
dome hat 拱顶帽
dome line 圆顶型
dome ring 拱顶环
dome-side style 单侧揿纽式
dome silhouette 拱顶型

dome skirt 撑开的细裥裙,拱形裙,穹形裙
domestic 国产(的);手织棉织物
domestic content 纤维含量
domestic design 国内设计
domestic duck flank plumage 白花鸭毛
domestic flat knitting machine 家庭针织横机
domestic laundering 家庭洗涤法(国际标准化组织规定的耐洗试验法)
domestic market 国内市场
domestico crudo [liso] 本色棉床单布
domestic oriental (rug) 仿东方地毯,有光地毯
domestics 本地产品,国货;衬衫料;家用棉织物,室内纺织用品
domestic sewing machine 家用缝纫机
domestic steam iron 家用蒸汽熨斗
domestic textiles 家用纺织品;内销纺织品
domestic type 国内型
domestic washing machine 家用洗衣机
domestic wool 本地羊毛,国毛,本土羊毛;美国西部羊毛
domet 双面厚绒布
domet flannel 双面厚绒布;盖肩衬
domet[t] 双面厚绒布;盖肩衬
domette 厚绒布,双面厚绒布;盖肩衬,肩头衬
domette flannel 双面厚绒布
dome umbrella 拱顶伞
dominance 统一化;支配色,主调色
dominant color 支配色,主调色
dominant harmony 主色彩调和
dominant hue 主色调
dominant shade 主色调
domination 支配
domingo henap 剑麻,西沙尔麻
domino 连帽化装斗篷;半截面具,黑色小面具;穿连帽化装斗篷的人,戴半截面具的人
domino color 多米诺色
domino design 多米诺印花
domino effect 多米诺效应;骨牌花纹,西洋骨牌花纹
domino stitch 齿形线迹
Dom pedro shoe 及踝男式厚重工作靴
Donald Brooks 唐纳德·布鲁克斯(1928~,美国时装设计师)
Donald Davies 唐纳德·戴维斯(爱尔兰时装设计师)
donat 杜纳特鞋跟

donau linen　奥地利彩边亚麻桌布
donchery　法国厚毛哔叽
donegal　多内加耳粗呢
Donegal carpet　爱尔兰多尼盖尔高级手结厚毯
Donegal tweed　爱尔兰多尼盖尔粗花呢,仿爱尔兰多尼盖尔粗花呢,英国粗花呢,多尼盖尔粗花呢
Donegal wool tweed　多尼盖尔粗花呢
dongaree　深蓝或褐色斜纹工装布
Dong brocade　侗锦
dongery　斜纹工装布,双经单纬布
Dong ethnic costume　侗族服饰
dongfang silk　东方绸
dong-feng gauze　东风纱
Dong Guo shoes　东郭履
Dong Hua University　东华大学
dong-li peach-skin fabric　东丽丝
Dong nationality brocade　侗锦
Dongxiang ethnic costume　东乡族服饰
donkey coat　驴子外套(斜切丝爪领,大口袋,皮纽,多为褐色,系美国东海岸传统外套)
donkey jacket　(野外作业)风雨衣,野外风雨衣;女式防风厚上衣
donkey skin　驴皮
donkey stitching machine　简易缝头机,坯布缝头机
Donna Karan(DKNY)　唐娜·卡伦(1948~,美国时装设计师)
Donna Maria　八经十二纬绸布,宗教用薄纱
donna Maria sleeve　多娜·玛丽亚袖
donn 400 cotton　达恩陆地棉
donnilette　女式绗缝羊毛披风
do not bleach　禁止漂白
do not dry clean　禁止干洗
do not iron　禁止熨烫
do not tumble　禁止滚筒烘干
Donskol wool　顿河羊毛
donsu　日本缎子,日本锦缎
don't mentions　男裤
dooklee　印度原色粗制平纹绸
doorea　高级达卡细平布,优质达卡细布
dooriah　密经缎条漂白棉布
door knocker earring　扁平圈耳环,门环形耳环
dopata　印度高级面纱棉细布,印度优质细布
dopatta　多帕塔巾,尤帕纳布(印度)
dopatti　多帕铁巾(印度)

dope dyeing　纺前染色法
dora　横条纹基里姆地毯
doram　杜伦粉
dori　印度帐篷绳
doriah　密经缎条漂白棉布,密织缎条细布
doria stripes　多里厄条子细布(美国)
Doric chiton　杜瑞克奇通衫,多利克式奇通衫
dorino　波斯尼亚女式外出服
dorma mats　孟加拉国和印度手编垫子
dormeuse　法式睡帽
dormick　花缎式锦缎;多尼克地毯
dorm shirt　长至膝上的衬衫式睡袍
dornock　菱形格子粗亚麻布
Dorothee Bis　多罗泰·比斯(法国时装设计师)
Dorothy bag　多萝西提包(抽带束口)(英国)
dorsal view　背视图
D'Orsay　侧空便鞋
D'Orsay coat　德奥赛大衣(19世纪欧洲)
D'Orsay cut　德奥赛款式
D'Orsay pumps　德奥赛鞋,多尔赛鞋
D'Orsay shoes　侧空鞋,陶乐赛鞋
D'Orsay slippers　德奥赛拖鞋
dorset　道塞呢,英国多赛特棉粗平布
Dorset cloth　英国多赛特低档平纹呢
dorsetteen　多尔赛廷丝毛呢(英国),无经丝纬织物
Dorset wool　英国多尔塞特羊毛
doru　横条纹基里姆地毯
do rukha　织物表面
dorure　法国式金银饰带,法式金银饰带
dorure fausse　镀金纸条纬织物
doschella　高档开司米围巾
dosefitting　贴体
dosia　中国毛袜
dossal　垂帘,幔布;吊账;靠背饰布
dosuti　印度金线棉布
dot　点,点纹,水珠花样,圆点图案
dot-and-blot print　斑点印花
dot button　打点纽扣,圆点纽扣
dot-coated adhesive　点状涂胶黏剂
dot dash line　点划线
dot dash stitch　点划线迹
dot design　水滴图案,小点图案
dotera　长袍用优质条纹绸
dotis　印度印花棉布
dot-mark　圆点标记,纸样净端点
dotohi　红蓝边棉床单布

dot pattern 水珠花样
dots 圆点花样印花
dots dashes line 双点划线
dot stitch 单线结粒绣,点状缀纹绣;点纹线迹
dotted stripe 点子条
dotted crepe 点点绉
dotted effect 点缀效果
dotted fancy yarn 珠球花线
dotted line 反面轮廓线,虚线
dotted muslin 点子花薄呢,点子花薄纱
dotted pattern 针点花样,点缀花样
dotted stripe 点缀条纹,针点条纹,针头条纹
dotted swiss 瑞士点纹薄纱
dotted weft motif 纬点子花纹
dotted yarn 结子线
dotted yarn suiting 结子花呢
double 头跟加固;双排纽扣;股线
double action stitch 顺逆交替(双动式,复式)线迹
double alpaca 阿尔帕卡双面呢
double and galloon 丝鞋带,丝鞋结
double and tacking machine 双排加固缝纫机
double and twist denim (D and T denim) 双色纬纱劳动布;双色纬纱粗斜纹布
double and twist yarn 双色螺旋花线,仿混色螺旋花线,混色仿螺旋花线
double angle buttonhole knife 双角钮孔切刀
double atlas 双经缎织物
double Atlas fabric 双梳栉经缎织物
double backing 地毯的第二层底布
double back stitch 曲折线迹,枕套花边的松针迹,双回线迹;双回针绣
double back tape 双面粘合带
double belt 双层皮带,双层腰带
double bias facing 双层斜贴边
double blade ruffler 双刀折边器
double blanket 双幅毛毯
double blister fabric 双到凸纹浮线织物
double-blister jacquard knitted fabric 胖花提花针织物,双胖提花针织物
double border 花纹里外边线
double border lace 双面边纹花边
double braid 双层空心绳;双面编带
double breadth 双幅
double breast 双排纽式(对襟)
double breast blouson 双排纽短夹克
double-breasted 双排纽;双排扣

double-breasted box o'coat 双排纽箱形大衣
double-breasted closing 双排纽门襟
double-breasted coat 双排扣外套;双排纽箱形大衣
double-breasted collar and rever 双排纽驳折领
double-breasted front 双排扣前襟
double-breasted jacket 双襟式上衣,双排纽夹克,双排纽上衣
double-breasted job 双排纽上衣
double-breasted lapel (DB lapel) 剑领
double-breasted lapel 双襟驳头,双排纽翻领,剑领
double-breasted overcoat 双排扣大衣
double-breasted pinwale 双排纽条纹短大衣
double-breasted sack 双排扣袋型上衣,双排纽短夹克
double-breasted six-button jacket 双襟式六扣上衣
double-breasted suit 双排扣外衣,双排纽套装,双排扣西服
double-breasted vest 双襟背心,双排纽背心(马甲)
double-breast suit 双排纽套装
double broadrib 2+2双面宽罗纹
double brocade 双层锦
double button 双纽式,双排纽
double buttoned pyramid coat 角锥形双纽大衣
double buttonhole stitch 双重锁眼绣
double carpet 双层地毯
double cashmere 开士米双层织物,开司米双层织物
double cassinet 混纬缎
double chaining looper 双链式弯针
double chain stitch 双链式线迹;双链绣
double chain stitcher 双链式线迹缝纫机
double chain stitch seam 双线链式线缝
double chain stitch sewing 双链式线迹缝纫机
double chain stitch sewing machine 双线链式线迹缝纫机
double check 双格子花纹,双重格子
double chemical lace 立体烂花花边,双面烂花花边
double chesterfield 双排扣契司达有腰身外套,双排扣有腰身外套
double circular skirt 外观双层圆裙
double cloth 两面呢,双层织物
double cloth stitch 双层针绣;双人字绣花

针迹
double coat 双棱哔叽(德国)
double coin knot 双线结
double collar 双折领,双褶领,双层折领,双重领,双层翻领
double-colored twist yarn 双色花线
double-color jacket 双色夹克
double computer jacquard knitting machine 电脑提花双面针织机
double connection knot 双联结
double cord seam 压双线缝
double cote 法国10经10纬哔叽
double cream 重奶油色
double crepe 双层绉
double cross leno 全绞式纱罗
double cross stitch 双人字形线迹,双十字形线迹;双十字绣,星状绣
double cuff 反折裤管,双卡夫;双克夫,双袖口(两折袖头)
double curve cut 双重弧形下摆
double curved needle overlock 双弯针包缝
double cut 重叠修剪发型
double cut W[M] shape W形[M形]双重切割
double cylinder automatic rib machine 双针筒自动罗纹机
double cylinder hosiery machine 双针筒圆袜机
double cylinder jacquard machine 双针筒提花针织机,双针筒提花织机
double cylinder weft knitting machine 双针筒圆形纬编针织机
double damask 八枚花缎
double-darning stitch 双缀纹绣
double dart 对称身双向褶,对称省,双向褶,鱼形褶,菱形褶;对称身短褶
double decker pocket skirt 双叠袋裙
doubled edge 双边
double diagonal stitch 十字线迹,锯齿形线迹
double disc looper thread take-up 双盘式弯针挑线
double dot 双重点纹
double duty dress 两用装,女式两用套装
double duty marking paper 两用排板纸(兼顾对格对条排料)
double edge 全边(鞋子)
double edge automatic sewing machine 双边自动缝纫机
double-edged ruffle 双褶端褶边

double-ended dart 菱形褶,双向褶,鱼形褶
double-ended needle 双头针
double ends 双经疵,双纱疵;双丝头疵;双头疵
double ends fabric 双经布
double end tacking machine 双排加固缝纫机
double entry pocket 双向袋口口袋
double extension cuff 双重延伸袖头
double fabric 双层织物
double fabric feed 双面送料
double fabric sewing 层缝,双面送料缝纫
double faced 双面织物;两面穿上衣
double faced adhesive interlining 双面黏合衬
double faced and color cloth 双面双色呢
double faced fabric 双面织物
double faced flocking 双面植绒
double faced fusible interlining 双面黏合衬
double faced jacket 双面夹克
double faced jacquard 双面提花
double faced material 双面布料
double faced pile 双面起毛,双面起绒
double faced satin 双面缎纹织物
double faced tape 双面胶带
double faced terry cloth 双面毛巾布
double faced terry weft knitted fabric 双面毛圈针织物
double faced twill 双面斜纹
double face fabric 同面织物
double face knitted fabric 两面针织物,双面针织物
double face lady's dress 双面女衣呢
double face lady's dress worsted 双面女衣呢
double face printed label 双面印唛
double face printing 双面印花,双面印色
double face rayon/acetate damask 正反花绸
double face ribbon 双面缎带
double face satin 双面缎
double face weft and warp 双面经纬
double fagot stitch 双面装饰线迹
double fancy knitted fabric 双面花色针织物
double fancy knitwear 双面花色毛衫
double fancy stitch sewing machine 双道装饰线迹缝纫机,双行装饰线迹缝纫机
double feather stitch 双羽毛缝,双羽毛饰边绣

double feed machine 双系统针织机
double felled seam 双折边叠缝
double filling duck 双纬帆布
double filling flat duck 双纬帆布
double fleece weft knit 双面绒纬编针织物
double float foot bottom 双吃虚线袜底
double fly front skirt 前身中间双暗裥女裙
double fold 双折边
double fold bias binding 双折斜裁滚边
double fold hem 双折下摆,双重折边
double fold layout 双折排料
double fold plain binder 双折边普通滚边器
double folds 双折边
double fold tape 折叠带子
double footage 双幅
double French darts 双法式省
double French piping foot 法式双滚边压脚
double frill 双层花边
double full cardigan knit 双面畦编针织物
double galloon 丝鞋带
double genoa 横条棉绒布(英),英国棱条棉绒布
double-girdled 束胸腰带款式
double gloves 双层手套
double gown 两面穿长袍,两面穿厚长袍
double green 甲基绿
double group presser 复式压脚
double guide bar stitch-bonding 双梳栉缝编法
double gumming 双面涂胶
double head boxing machine 双头箱式缝纫机
double head sewing machine 双头缝纫机
double heater yarn 双加热器变形纱
double-hem stitch 双抽丝线绣,双花饰线绣,意大利抽绣
double herringbone stitch 双人字绣,双人字缝
double hose 头跟加固袜子
double hosiery 头跟加固袜子
double ikat 纱线扎染织物
double insurance policy 双重保险单
double interlining 双层里衬
double inverted pleat 双阴裥
double jacquard 双面提花
double jean 哔叽织物(旧称)
double jersey 双面乔赛,双面针织物
double jersey with polyester face and cotton back 涤盖棉针织物

double joint 宽缝,双接缝
double knee 双层膝
double knit fabric 双面针织布
double knit machine 双面针织机;双面圆形针织机
double knit(s) 双面针织物
double knitted finger 加固手指手套
double knitted plaited fabric 双面编结添纱针织物
double knitted terry 双面[针织]毛巾布
double knitting 双面针织品;双面织
double knitting fabric 双面针织物
double knitting jacquard 双面针织提花
double knitting yarn 四股手工织毛线
double knot 双头结,双层领带结
double knot stitch 双点结绣,双回针穿线,双扎结绣;双接缝针法
double laid-in weft-knitted fabric 双面纬编衬纬针织物
double lapped double-stitched seam 双线搭接双排缝,双折边叠缝
double lapped felling 双线叠缝折边,双圈缝卷边
double lapped seam 双搭缝,双线叠缝
double lap seam felling 双线搭缝折边,双线折边,双线搭接平缝
double layer 双层
double layer bias binding 双层斜镶边
double layered quilted knit fabric 夹层绗缝针织物
double layered upholstery fabric 双层袋组织装饰织物
double lazy daisy stitch 双层套针绣;双重雏菊线绣
double leviathan stitch 复式大十字绣花线迹
double line 双线线缝
double line twill 双线条斜纹
double linked cuff 双层袖口,双层袖头
double lock chainstitch 双线链式线迹
double locked chainstitch 双线链式线迹
double locked chainstitch seam 双线链式线迹缝
double locked machine 双线锁式线迹缝纫机
double locked stitch 锁式针迹双线线迹缝
double locked stitcher 双线锁式缝纫机
double lock flat knitting machine 双系统平机,双系统横机

double lock machine 双针床横机,V型横机
double lock seam 双线锁式线迹缝
double lock stitch 双线锁式缝线迹
double lock stitching 双线锁式线缝
double lock zigzag machine 双线锁式线迹缝纫机
double London 双股线经斜纹带
double longs 5/10～9号缝针
double look 双重款式
double looped felling 双圈缝折边
double looped seam 双圈缝
double loop towel 双面毛圈毛巾
double mantle 节日斗篷
double marl yarn 斑点花式线
double mercerized fabric 双丝光布(纱、布先后丝光)
double mercerized finish 金丝光,双丝光(纱线与织物均进行丝光)
double mercerized T-shirt 双丝光T恤衫(双丝光布制)
double mercerized weft knit fabric 双丝光针织物
double milanese 双面细米兰尼斯经编丝织物
double milled fabric 双缩绒织物
double moquette 双面毛圈或毛绒织物
double napped 双面起绒的
double neckline 二重领型领口,二重领口
double needle 双针
double needle bar fabric 双面针织物
double needle flat sewing machine 双针平缝机
double needle lockstitch machine 双针平车,双针平缝缝纫机
double needle machine 双针缝纫机
double needle sewing 双针缝
double needle sewing machine 双针缝纫机
double needle sewing quilter 双针摆缝机
double needle stitched piping (双针)压条
double ombré 双色深浅条纹
double overcoating cloth 双重厚大衣呢
double overlock 双线包缝
double overlock stitch 双线包缝线迹,双线包缝
double pattern rayon poult 双花绸
double pekinese stitch 双回针穿线绣
double picot 双锯齿边
double picot stitch 双环边线迹,双锯齿边线迹;编织线迹

double piece moquette 双层天鹅绒
double pile 双面绒头织物
double pile velveteen 双层立绒,双层丝绒,双层天鹅绒
double piped 双滚边,双嵌线
double piped foot 双滚边压脚
double piped pocket 双滚边袋,双唇袋,双滚边口袋
double piped pocket with flap 双滚边带盖口袋,双嵌线带盖口袋
double piped zip pocket 双滚边拉链口袋,双嵌线拉链口袋
double piping foot 双滚边压脚,双嵌线压脚
double piping pocket 双滚边口袋,双嵌线袋
double piqué 双面凹凸织物
double piqué fabric 点纹罗纹布
double piqué rib fabric 点纹罗纹布
double plain 牙签条花呢
double plain cloth 双层平纹织物
double plains 双层平纹织物
double pleats 双褶,叠褶
double plover stitch 双锁边缝
double plush 针织绒布,双面长毛绒
double plush fabric 双面毛绒织物
double ply 双层,双股,补强层
double ply handkerchief 双层手帕
double ply hosiery 双层袜
double pocket 双口袋贴袋
double pointed dart 橄榄省
double pointed flap 双尖角袋盖
double pointed waistline dart 双尖角腰省
double poplin 厚毛葛
double print effect 叠色效应
double printing 叠印疵
double puffed sleeve 双泡袖
double purl 锁纽眼针法,套结针法
double purl stitch 双绣边线迹
doubler 衬布(鞋子);棉毡
double raised jacquard design 双面凸纹提花花纹
double rib 双罗纹,双罗纹针织物;双面经编织物
double ribbon 双面织花带
double rib raschel fabric 双罗纹拉舍尔织物
double rib top 双罗纹口
double rib vamp shoes 双梁鞋
double rib warp loom 双针床经编机

doublerie 法国印花帆布
double rim button 双圈扣
double ring closing 双环系带扣合
double rivet button 双铆扣
double riveted 双重铆结
double roller presser foot 双滚柱压脚
double rotating transverse shuttle 双横向旋转梭
double rotation hook 双旋转梭
double round collar 双圆领
double row differential feed 双排差动送料
double rowed openwork stitch 双线抽纱
double row stitch 双行线迹
double ruffle 双层褶边,双层皱边
double running embroidery 双面绣花,双平针刺绣
double running stitch 双平针法,双平针线迹;双面花式刺绣
doubles 双股棉线;黑丝鞋带
double safety stitch 双行加固线迹,双排安全线迹
double safety stitch over edging 双排安全线迹包缝
double sample inspection plan 复式抽验方案
double sampling 复式抽样
double satin 双面缎
double satin de Lyons 里昂双面有光缎
double satin ribbon 双面有光缎带
double screen 双帘网
double seam 双行线迹,双折边线缝,双行线缝
double seaming 双折边线缝,双行线缝
double seated trousers 臀部双层加固裤
double selvage [selvedge] 卷边,翻边(织疵),翻边疵
double serge 双经双纬哔叽
double sewing head curtain hemming machine 双头窗帘缝纫机
double sewing line 双行线缝,重叠线迹,双行线迹
double sewing stitch 重叠线迹
double sewn seam 两次缝纫线缝
double shirt 双重衬衫
double shot 双经双纬哔叽
double shoulder press 双肩压烫机
double shrink proof fabric 双防缩织物
double-sided 双面的
double-sided flannel 双面法兰绒
double-sided imitation lace effect 双面防花边效应
double-sided loop pile fabric 双面毛圈起绒织物
double-sided plush 双面绒,双面长毛绒
double-sided stitching 双面缝
double-sided twill 双面斜纹织物
double-side loop pile fabric 双面毛圈起绒织物
double-side singeing 双面烧毛
double-side terry fabric knitting machine 双面毛巾布针织机,双面毛圈织物针织机
double skirt 双层裙
double sleeve 双层袖,双重袖,双层泡泡袖,双袖口
double sleeve press 双袖压烫机
double sliders zipper 双头拉链
double sole 加固袜底,夹底
double sole cutting 袜底加固部分的剪线
double spot 双点(黏合衬)
double star 双星
double stick tape 双面粘牵带
double stitch 双道明线,双行线迹,双排线迹,双缝;双轨装饰绣
double-stitched edge 双排针迹缝边;双排针迹缝
double-stitched flat seam 双道缝迹平缝
double stitch edge 双边,双排线迹缝边
double-stitched hem 双压线下摆
double-stitched overcasting seam 双行滚边缝
double-stitched seam 双次缝,双线缝
double-stitched seam finish 双次缝缝份处理
double-stitched welt seam 双排贴边缝,双重边缝
double-stitching 绱双线,双线缝,双线缝纫
double stripe 双条纹,双重条纹
double style 双重款式
doublet 大布利特上衣,(旧时)男式紧身上衣,紧身短上衣;苏格兰夹克;骑装上衣;紧身背心,马甲;双重衣
double tack 双加固缝
double take collar 两用领
double take dress 两用装
doublet and hose 男装
double tariff system 双重税制
doublete 双色花塔绸,双色花塔夫绸
double tent line 喇叭式套裙式样

double texture fabric 夹胶布,双层胶布
double thickness sash 双片饰带,双片腰带
double threaded back stitch 穿线针法
double thread overcasting seam 双线包缝
double thread overlock stitch 双线包缝线迹
double thread shoe border stitcher 双线鞋沿缝边机
double toe 加固袜头
double toe even hinge foot 双趾铰链压脚
double tone color 双色调色
double top 双口,双层袜口
double top stitched seam 双排明缝,分缉缝
double tricot 双面经编针织物
double tricot machine 双针床经编机
double turn hem 双翻边下摆
double twilled merino 复式斜纹类美利奴毛织物
double twill weave 双斜纹,左右交叉斜纹
double Vandyke 双梳栉经缎织物
double velvet 双层丝绒
double vent 西装的双开衩
double veviathan stitch 复式大十字绣花线迹
double-V twill 双山形斜纹窄幅布
double-V yoke snowsuit 双V形覆肩式雪衫裤
double wall 双层瓦楞纸板
double wall fabric 双层织物
double warp 双股线经平纹织物
double warp bagging 双经袋布
double warp flannelette 双经彩花条子棉法兰绒
double warp lining 双股线经里子布,线经里子布
double warp machine 双针床经编机
double warps 线经织物
double warp tricot 双梳栉经编织物
double wearing 双重穿着式
double weft 双层袜口,双口
double weft brocade 双纬花绸
double wefted cloth 纬二重织物
double weft fabric 双纬布
double weft flannelette 双纬绒
double weft seam 双线贴缝
double welt 双层袜口,双口;双贴边
double welt back hip pocket 双嵌线后裤袋
double welt inside pocket 滚边里袋,双嵌里袋
double welt pocket 双嵌线袋,双嵌线口袋,双贴边口袋
double welt pocket with flap 有盖双嵌线口袋
double welt seam 双线贴缝,双贴边缝
double whip stitch 包边线缝,双褡线缝
double whip stitching 包边线缝,双搭线缝
double whip stitching with gimp 双搭接夹线缝
double whip stitch with gimp 双褡接夹线缝
double width 双幅
double width fabric 双幅织物
double with counterchange design 14世纪印花上衣,有交错花型的大布利特上衣
double work 双纱针织物
double worsted 中世纪英国重磅精纺呢
double woven blankets 双层织造(的)毯子
double woven fabric 双层织物
double woven pile fabric 双层织造的长毛绒织物
double woven tapestry velvet 万紫绒
double yarn fabric 双纱布
double yoke 双覆肩
double-zipper foundation 便于穿脱的双拉链妇女整姿内衣
doubling 服装衬里
doubling and tacking machine 折幅缝筒机,套结缝纫机,折边缝纫机
doubling foxing 贴合围条(鞋)
doubling machine 织物对折机
doublings 棉里子布
doublure 衣服里子;漂白军服粗呢
doughboy jacket 步兵夹克(美军上衣,立领,肩章,四个有盖的袋)
dough face 假面具,面具
doughnut fiber 中空纤维,环形截面纤维
doughnut print 炸面饼圈印花
douillette 女式绗缝羊毛披风
do up 扣上(衣服);把头发向上盘;洗烫(衣服等)
doupioni, doupion, douppioni 双宫丝;双宫绸
doupioni foulard 双宫斜纹绸
doupioni gingham 条格双宫绸
doupioni habutai 双宫纺绸
doupioni pongee 双宫绸,双宫塔夫绸
doupioni silk 桑蚕双宫丝;双宫绸
doupioni silk crepe 双宫绉

doupioni silk taffeta 双宫丝塔夫绸
doupioni taffeta 双宫塔夫绸
doupioni twill 双宫斜纹绸
dove 鸽灰(浅灰色)
dove color 鸽灰色(浅红灰色),暖灰色,淡红灰色
dove grey 鸽灰色(略带紫红的浅灰色),鸽子灰,紫灰色
dovetail (斜裙裁片时的)倒顺排料;燕尾服,夜礼服
dovetailed tapestry 手织燕尾式挂毯
dovetailed twill 燕尾形斜纹织物
doweaves 三轴向经纬交织物
dowel pin 暗销
dowlas 道拉斯粗亚麻布;道拉斯粗棉布;粗被单布;环状毛巾
down 鸭绒,羽绒,绒毛;向下推档
Downbell 羽绒型聚酯短纤维(商名,日本钟纺)
down cloth 鸭绒织物,水鸟绒毛织物
down clothing coat 羽绒大衣
down coat 羽绒服
down content 含绒量
down costume 羽绒服
down fiber 羽绒纤维
down garment 羽绒服
down garment factory 羽绒服厂
down hair 茸毛,绒毛;下髦发型
down hill pants 滑雪裤
down jacket 羽绒夹克,羽绒外套
down-like 仿羽绒
down-like fiber 仿羽绒纤维
down padded trousers 羽绒裤
down products 羽绒制品
down proof 防羽绒刺出性
down-proof cloth 防羽布,防绒布
down-proof fabric 防绒布,防绒织物,防羽绒布,防绒布
down-proof finish 防羽绒整理
down-proof finished fabric 防羽绒整理织物
down-proofness test 防羽绒刺出性试验
down quilt 羽绒服
down quilting machine 羽绒绗缝机
down roof fabric 防绒布
downs 粗纺厚呢,塘斯呢,粗纺棉经厚呢
down-ski 羽毛衣
down-ski wear 滑雪服
downs-sealing 羽绒毛密封作用
down style 梳辫发型

downton lace 精细棱结花边
down trousers 羽绒裤
down-turn folder 向下折边器
down-turn hemming device 向下卷边器
down vest 羽绒背心
down wear 羽绒服,羽绒服装
down wool 塘种羊毛,英国塘种羊毛
downy calves 假腿肚垫
downy wool 绒毛,软毛
dowrah 印度多拉黄麻
dowrah jute 印度多拉黄麻
dozen 打(量词,意谓十二)
drab 黄褐色,淡褐色,灰黄色;黄褐色厚呢;褐色布,灰色布,褐色三页斜纹布,原色厚斜纹布
drabbet 粗斜纹亚麻布
drab printing fabric 抓毛印花布
draft 草案,草图,底稿,图样,打样;汇票
drafted pattern 手绘样板
drafter 制图机
drafting 制图;打样
drafting board 绘图板,制图板
drafting design 结构设计
drafting machine 打样机,绘图机,制图机
drafting method 打样方法,画样方法
drafting paper 绘图纸,制图纸
drafting room 绘图室
drafting square 制图尺
drafting table 绘图桌
Draft International standard (DIS) 《国际标准草案》
draftman 绘图员
Draft Proposal (DP) 建议草案(国际标准化组织 ISO 的工作用语)
Draft Revisions of an International Standard (DRIS) 《国际标准修改草案》
draftsman 制图员;打样人
Draft Standard 标准草案(国际标准化组织的工作用语)
draft technical report (DTR) 技术报告草案
drag 男子穿的女子服装(俚语)
draggle-tail 拖地长裙;浅地衫褂,曳地衫褂
dragon design 龙纹
dragon fly 蟹青;蜻蜓绿;浅蟹绿
dragon lame 金银龙缎
dragon long robe 长龙袍
dragon-phoenix 龙凤呈祥(图案)

dragon robe 龙袍	draped collar 垂坠领,褶裥型领
drainage canvas 排水帆布	draped cowl neckline 挂帽领口
drainage fabric 排水帆布	draped elbow sleeve 垂肘袖
drain mark 水纹疵;水印疵;水渍	draped heel 垂褶跟,褶饰鞋跟
drain pipe pants 瘦腿紧身裤	draped neck 褶裥型领
drain pipes 瘦腿紧身裤,瘦腿裤	draped neckline 垂坠领口,垂缀领口,自然皱领口,褶型领口
drainpipe trousers 瘦腿裤	
Drake cluster cotton 美国德雷克棉	draped pattern 立体裁片
Drake cotton 美国商业陆地棉	drap edredon 重缩绒大衣呢
drak gull gray 深鸥灰色	draped silhouette 悬垂型,垂坠型
Dralon 聚丙烯腈纤维(商名,德国拜耳)	draped skirt 垂饰裙,垂褶裙,褶皱裙,坠纹裙,自然皱裙
dramatic belt 戏装上饰带(腰带)	
dramatic costume 戏装,戏剧服装	draped sleeve 垂褶袖
dramatic curve 明显起伏的曲线	draped toque 垂褶豆蔻帽
dramatic dress 戏装,戏剧服装	draped twist neckline 褶绉扭结领口
dramatic effect 戏剧性效果	draped waistband 垂饰腰头;褶裥腰头
dramatic garment 戏剧服装	draped wrap 垂饰裙,垂褶裙,坠纹裙,自然皱裙
drap 棉织物;毛织物;缩绒呢	
drapability (衣服、布料的)悬垂性	drape effect 褶裥效果,褶裥效应
drapable fabric 悬垂性织物	drape-flex test 织物悬垂弯曲试验
drapable nonwoven 悬垂性非织造布	drape meter 悬垂仪
drapade 桑米尔哔叽	drape model 褶裥式
drap bresilienne 斜纹丝毛交织呢	drape natte 缩绒毛呢
drap chats 夏兹女式黑呢	drapeometer 织物悬垂性试验仪
drap croisé 法国斜纹布	draper 立体裁剪师,女服装裁剪师;卖布者,布料商;织布者
drap d'alma 双列斜纹精梳毛织物;阿尔马斜纹薄绸	
	draper's goods 呢绒布匹,呢绒匹料,细绒布匹
drap d'argent 法国银丝缎	
drap de beaucamp 法国纯毛麻经毛纬斜纹呢	drapery 布匹,匹头(布匹、呢绒的总称);布店,布业;服装,打折的服装;服装业;装饰织物,悬挂织物,帷幕,毛织物
drap de berry 法国粗纺呢绒	
drap de chasse 丝经棉纬横纹呢	
drap de dame 软薄绒呢	drapery damask 双面提花厚锦缎
drap de gobelin 粗纺红呢绒	drapery fabric 装饰布,装饰织物
drap de gros bureau 粗纺大衣呢	drapery rep 棱条装饰绸
drap de Lyons 里昂华丽绸	drapery skirt 围式裙
drap de milord 小花纹哔叽	drape sack coat 褶裥宽松大衣
drap de prince 八经三纬哔叽	drape shoulder 垂肩
drap de silesia 法国薄呢	drape sleeve 垂褶袖
drap de soie 法国蚕丝织物,全丝斜纹硬挺绸	drape style 垂褶风格
	drape suit 长上装瘦裤腿男套装
drap d'Ete 法国夏用斜纹薄毛织物	drape test 悬垂性试验
drap d'or 法国金丝缎	drape tester 织物悬垂性试验仪
drape 褶裥自然皱,褶皱;服装式样;(布料的)悬垂性;垂褶;垂坠;精纺缩绒呢,(常用复)窗帘;立体裁剪	drape trousers 垂褶裤
	drap geraldine 粗纺暗色毛呢
	drap imperial 法国棉毛平纹呢
	draping 悬垂性;立体裁剪
drape allowance 褶裥缝份	draping cutting 立体裁剪
drape and wrap-over 垂褶缠绕裙	draping property 悬垂性
drape coefficient 悬重系数	draping quality 悬垂性
draped cap 垂饰帽	draping steps 立体裁剪步骤

drap ling 悬垂线
drap piqué 马甲花绸
drappo 丝绸(意大利称谓)
drap royal 印花马甲呢,缩呢;小凸条斜纹绸
drap sanglier 松结构纯毛丧服呢
drap satin 粗纺光呢
draps croises 法国斜纹毛织物
drap soleil 宽横条光呢
drap zephir 仿开司士米呢
draughting paper 打样设计纸,样板纸
draw 划,画,绘制,绘图;划线笔;划线刀
draw blue print (裁剪)打样;划样
drawboy 手工提花织物;鞋面呢
draw cord (衣服腰部或下摆等处的)拉绳,系绳;松紧带;松紧绳
draw design (裁剪)打样,划样
drawee 付款人
drawer 制图工具;制图员
drawer bottom 翻袖口,衬裤罗口
drawers 衬裤,内裤,三角裤,长内裤,汗裤;小衣(方言)
drawers with separate legs 分腿式长腿袜
drawing 绘画,绘图,制图,素描;图,图样,草图,图纸
drawing board 绘图板,制图板;制图桌
drawing board adjustment 图板调整装置
drawing coat 系带外套
drawing compasses 制图圆规
drawing easel 画凳
drawing gloves 射手手套
drawing head 可调节纵横直尺
drawing-in 收拢(首饰加工)
drawing ink 绘图墨水
drawing in pencil 铅笔画
drawing instrument 绘图仪器,制图仪器
drawing master 绘画描师
drawing office 绘图室,设计室,制图室
drawing paper 绘图纸
drawing pattern 划样
drawing pen 绘图笔,鸭嘴笔
drawing pin 图钉
drawing quilting line 划绗缝线
drawing scale 制图比例尺
drawing silk 画绢
drawing stitch 拼缝线迹,并缝线迹;平整缝
drawing string 束带,抽带

drawing table 画图板,绘图板,制图板;绘图桌
drawing up 抽出来
drawing work lace 抽绣花边
draw knot 抽纱线迹;自紧结
draw loom 手工提花绵缎
drawn fabric embroidery 扎线刺绣
drawn fabric stitch 抽纱针迹
drawn fabric work 抽纱刺绣
drawn line 实线
drawn piece 弓斜疵
drawn selvage 紧边疵
drawn thread work 抽纱刺绣品
drawn work 抽纱,抽绣;抽纱品,抽绣品,抽花刺绣品
drawn work blouse 抽绣衬衫
drawn work embroidery handkerchief 抽绣手帕
drawn work garments 抽纱服装
drawn work handkerchief 抽绣手帕,抽纱手帕
drawn work lace 抽绣花边
drawn work stitch 抽丝绣
draw pattern (裁剪)表层划样;画皮
draw point 手套背面梗条纹饰缝
draw quilting 划绗棉线
drawstring (衣、裤的)束带,兜帽系带;拉绳,抽绳,抽带,系绳
drawstring anorak 束带滑雪上衣
drawstring bag 束带提包
drawstring belt 抽带
drawstring blouse 束带领口女衫
drawstring blouson 束带式蓬腰女衫
drawstring bra 抽带胸衣
drawstring closure 束带领口
drawstring coat 束带外套
drawstring collar 束带领
drawstring cord 抽绳
drawstring cuff 抽带袖口
drawstring drawers 束带长内裤;束带长衬裤
drawstring handbag (袋口)束带手袋,抽绳手提袋
drawstring neck 抽带领
drawstring neckline 伸缩形领口,束带领口,串领领口
drawstring pants 束带紧腰裤
drawstring puffed sleeve 抽带袖口泡泡袖
drawstring shirt 束带式衬衫
drawstring shorts (拉绳)束带短裤

drawstring sleeve 伸缩式袖,束带袖,束带袖口
drawstring snood 束发带帽;束发网
drawstring string 抽绳
drawstring sweater (下摆)束带毛衫,下摆束带开襟衫
drawstring top 束带式上装
drawstring waist 抽绳腰部,束腰款式
drawstring waistline 抽带腰节线
draw textured yarn (DTY) 拉伸变形丝;拉伸变形纱
draw thread work 抽绣;抽线制品
draw vacuum 抽真空
draw work 抽绣
draw yarn 抽纱
Dr.Denton sleepers 丹顿睡衣裤
dreaded ring 横路疵,横影条疵
dreadnaught 厚呢大衣;仿熊皮粗绒大衣呢,厚呢
dreadnought 厚呢大衣;仿熊皮粗绒大衣呢,厚呢
Dreater 聚酯絮棉(商名,日本东丽)
Drécoll 德雷科尔(法国高级时装店)
dredge sleeve duck 水龙带帆布
Dresden 德累斯顿印经小花纹;经浮花呢
Dresden point 德累斯顿刺绣
Dresden point lace 德累斯顿抽绣花边,粗枕套花边;德累斯顿抽花刺绣品
Dresden ribbon 德累斯顿彩色带
Dresden silk 德累斯顿绸(小花绸),德累斯顿印经绸
dress 服饰,服装(统指外穿),衣服,礼服,连衣裙,(特定场合穿的)套裙服装;装束,穿衣
dress making 女服裁制
dress accessories 服装辅料
dress àla constitution 三色旗服
dress alikes look 相似服装款式
dress and accessory 服饰
dress and bloomers 20世纪初的灯笼短裤;连衣裙式游泳衣
dress and jacket combination 女服和短外衣的组合装
dress and make up 梳妆
dress and pant set 婴儿罩衣及尿裤套装
dress and personal adornment 服饰
dress and personal adornment taboo 服饰禁忌
dress art 服装艺术
dress beauty 服饰美

dress boots 文雅长靴,礼服皮靴,高雅长靴,优雅高筒靴
dress braid 镶边,边缘饰带
dress canvas 服用染色帆布
dress cape 礼服斗篷,礼服披风
dress career apparel fabric 职业装织物
dress chambray 钱布富花式布(丝光,漂白,预缩)
dress chesterfield 切斯特菲尔德礼用大衣
dress clip 裙夹
dress clothes 正式场合服装
dress clothing 礼服,西装外套,燕尾服
dress coat 节日服,燕尾服,晚宴外套,礼服,社交礼服,晚礼服,夜礼服
dress coating 礼服料
dress cover 礼服外套
dress culture 服饰文化
dress data 服装资料
dress design 服装设计
dress designer 服装设计师
dress down 简装,便服;穿着轻便,放松的穿衣方法
dressed fabric 起绒圈针织物;上浆整理织物
dressed fox fur 已硝狐毛皮
dressed manikin 穿衣假人
dressed ribbon 上浆整理的带子
dressed skin 熟皮
dress ehadgear 礼帽
dress elevator 系裙腰带(藏在裙下)
dresser 穿衣者,穿着讲究者;梳头师;服装(管理)员,剧团服装员;梳妆台,化妆台
dresser line 匹印
dresses 女便装
dress exporter 服装出口商
dress fabric 妇孺衣料,服用织物
dress face finish 绒面整理,重缩绒整理(不露底纹)
dress fastener 衣扣
dress flannel 服装法兰绒,冬季全毛织物
dress form 服型,款型;服装人体模型(女装),(半身)模型架,胸架
dress frock coat 男式双排扣礼服
dress gingham 方格色织布,优质格子布
dress goods (女、童)外衣料,妇孺衣料,服用织物
dress guard 护衣装置(女式自行车上),衣挡
dress handkerchief 礼服装饰手套

dress hanger 衣架
dress headgear 礼帽
dress holder 裙托
dress improver 妇女托裙腰垫,腰垫
dressiness 讲究穿着,时髦服装,时装
dressing 穿戴,穿衣,穿着;装饰,服饰,修饰;化妆;装饰品
dressing art 穿着艺术,着衣艺术,着装艺术
dressing booth 试衣室
dressing cape 梳妆披巾,梳妆用披肩
dressing case 化妆盒
dressing equipage 全套化妆用品
dressing fur 硝革过的毛皮
dressing gauze 药用纱布,绷带纱布
dressing glass 穿衣镜,梳妆镜
dressing gown 装束便袍,家居便服,室内服装,室内衣,化妆衣,晨衣,宽衣
dressing jacket 室内夹克
dressing mirror 穿衣镜
dressing mull 药用纱布,绷带纱布
dressing of cut parts 裁片修边
dressing robe 晨衣,便袍,装束便袍(化妆衣,宽衣,晨衣,室内长袍,长浴衣,家居服)
dressing room 化妆室
dressing room mirror 化妆室镜子
dressing sack(or sacque) 化妆外套,理发罩衫
dressing satisfactory degree 服装穿着满意度
dressing selvage lace 饰边花边,装饰花边
dressing smock 化妆外套,化装罩衫
dressing table 化妆台,梳妆台
dressing time 更衣时间
dress inverness 披肩式男用外套
dress length 衣长;裙长;一件衣服布料
dress linen 亚麻平布,亚麻夏服布,亚麻外衣布
dress liner linen 亚麻服装衬布
dress lining linen 亚麻服装衬布
dress lounge 半正式晚宴夹克
dressmaker 服装工,裁缝工,女服裁剪师;女式服装店
dressmaker basting 裁缝假缝,缝纫定针
dressmaker coat 分割、褶裥等装饰的女式大衣,装饰女式大衣
dressmakered 定制女服,女服制作
dressmaker hat 装饰性妇女帽
dressmaker pin 女装别针,细别针

dressmaker's bent trimmers 弯形裁缝剪刀
dressmaker's brim 有明线帽边
dressmaker's carbon 裁缝用炭笔
dressmaker's carbon paper 裁缝用复写纸
dressmaker's chalk 裁缝用划粉
dressmaker's dart 女裁缝师缝褶省(不缝到胸高点的省)
dressmaker's dummy 木制人体模型,女装的模型,人台
dressmaker's gauge 弧形定规
dressmaker's model 女装的模型
dressmaker's pattern 女(童)服装纸样
dressmaker's shears 裁缝剪刀,裁剪刀
dressmaker's straight trimmers 直形裁剪刀
dressmaker's style book 突出女性美的服装,柔性服装;女装款式样本
dressmaker's tracing paper 裁缝用描图纸
dressmaker style 女装款式
dressmaker suit 定做的女套装;柔性套装;突出女性美的套装
dressmaker's weights 裁缝用小重锤
dressmaker's workroom 裁缝工作室,女装缝制工场间
dressmaker-type garment 女装类服装
dressmaking 服饰制作(总称),制衣,女服裁制,服装裁制;成衣
dressmaking fastenings 成衣扣件
dressmaking patterns 制版
dressmaking skirt 柔美女裙
dressmaking suit 柔美套装
dressmaking technique 成衣技术,制衣技术
dress match 服装协调,服装选配,衣服配穿
dress material 面料,衣料
dress museum 服饰博物馆
dress necklace 礼服用项链
dress net 服装用网眼纱,衣用网眼纱
dress of beauty 服饰美
dress-off 服装比赛
dress opening 连裙装式开襟或开口,连衣裙装饰开口
dress parade 时装展览,时装表演
dress pattern 服装纸样,服装样板
dress pieces 衣服片料
dress pin 大头针
dress placket 连裙装式开襟或开口
dress preserver 护衣汗垫(妇女腋下)
dress rehearsal 着装彩排

dress room 化妆室
dress set 连衣裙套
dress shield 吸汗垫布(衣袖腋处);防护外衣;防护斗篷
dress shirt 传统前通襟衬衫,礼服衬衫,西服衬衫,礼宴衬衫,上浆衬衫
dress shoes 便鞋,西装鞋,礼服用鞋
dress shop 女服店,女装店
dress skirt 两用裙
dress slacks 半正式场合穿着的裤子,礼服西裤
dress smock 化妆外套,理发罩衫,披风
dress socks 薄型男短丝袜
dress stable show 服装静态展示
dress stand 人体模型,胸架,形体架
dress style 服装款式
dress style-book 服装款式书;服装样本
dress suit (男)夜礼服,燕尾服,晚礼服,晚会服(男)
dress sword 礼服佩剑,装饰佩剑
dress taffeta 外衣用塔夫绸
dress tartan 女装格子呢,苏格兰格子呢服装
dress trousers 礼服裤子
dress turtle 礼宴龟颈领衬衣
dress uniform 军服,军礼服(美国);仪仗队制服
dress up 穿上盛装,打扮,妆饰,装扮,乔装打扮;须穿盛装的场合,须穿礼服的;盛装
dress up clothes 高雅服装
dress up jeans 宴会牛仔裤,正式牛仔裤
dress vest(or waistcoat) 礼服背心
dress wear 妇孺服装
dress wellington 晚宴用连鞋袜
dress with cape 披肩连衣裙
dress with pantaloons 帝国风格礼服及马裤
dress wool fabric 俄罗斯精梳西服呢
dress worsted 精纺女式呢
dressy 时髦的,衣着讲究的,打扮漂亮的,打扮入时的;时髦服装
dressy clothes 高雅服装;时髦服装;讲究穿着
dressy coat 高雅大衣
dressy evening slippers 时新夜礼鞋
dressy garment 裙装
dressy look 端丽款式,高雅款式,时髦款式,时新风貌
dressy shoes 精巧的鞋子,考究的鞋子
dressy suit 高雅套装
dress zipper 双封尾形拉链
dribblet 细小裁片,小裁片
dried herb 干草色
dried moss 干苔藓色
drier 烘干机
drift 动向
drill 本白斜纹布,卡其,斜纹(布);钻孔,钻孔器;平纹棉布;平纹裤料布
drill dress 体操服
drillette 纬面缎纹布
drill hole 打定位眼,钻眼
drilling (在裁片上)钻孔,钻孔位;卡其布,斜纹布
drilling mark 钻眼符号
drill negro firme 匹染棉布
drills florentine 厚卡其
drill ticking 经面斜纹床垫布
drill yarn-twisted 全线卡其;线卡
D-ring D形环;D形扣,巴黎扣
D-ring belt D形环窄皮带(穿戴时皮带端穿过两个D形环)
D-ring closing 双环系带扣合
drip cup 油杯
drip dry 滴干
drip-dry garment 滴干免烫服装
drip-drying 快干性
drive belt 传动电动机
drive motor 传动电动机
driver fork 传动牙叉
driver gloves 驾驶手套,驾车手套,司机手套
drive shoes 套筒靴
driving arm 传动臂
driving cape 马夫披肩,驭马斗篷
driving coat 驾驶员大衣
driving crank link 传动曲柄连杆
driving gloves 驾驶员手套
driving overcoat 驾车大衣
driving shoes 驾驶鞋
driving wheel 下带轮,主动轮
drizzle 雾雨色
drizzle jacket 毛毛雨夹克(防雨上衣),雾雨夹克
Dr.Martens 马丁博士气垫鞋,马丁博士鞋
Dr.Martens boots 马丁博士气垫底靴,马丁博士靴
Dr.Martens footweer 马丁博士气垫鞋
drochell 细绢网

droguet 18世纪法国花缎;粗毯;特经绸
droguet lisere 双色经双色纬织物
droguet lustrine 金银丝纬换纹织物
droop 垂份,宽松飘垂的连衣裙
droop full sleeve 落肩袖
drooping grain 垂悬的丝缕(暗示要加省或开刀)
droopy cloche 垂式钟形帽
drop 胸腰落差(衣服尺码)
drop armhole 大袖窿(腋部宽大)
drop a veil 放下面罩
drop cloth 家具用罩布
drop compass 小圈圆规
drop cut 切小条子(皮革)
drop design 嵌立图案
drop earring 垂耳环,垂挂式耳环
drop flap 下垂口袋盖
drop-front 带纽扣开衩长裤
drop full sleeve 落肩袖
drop head 翻斗,缝纫机翻斗,卧斗式
droping-in 坐跟(鞋)
drop lea 地毯粗底布;地毯无绒头部分
drop loop 将裤子皮带圈装低
drop loop machine 缝环缝纫机
dropped armhole 低袖窿
dropped back 后垂式下摆,后长前短式下摆
dropped buttocks 低臀
dropped cuff 垂式袖口,垂式袖头
dropped lapel jacket 低驳领夹克
dropped neckline 低垂领口
dropped puff sleeve 下垂式灯笼短袖,垂式膨胀袖
dropped shoulder 落肩
dropped shoulder sleeve 落肩袖
dropped skirt 低腰裙
dropped sleeve 垂袖
dropped stitch 漏针,脱散线圈,脱圈组织
dropped torso dress 低腰女服
dropped waistline 低腰线
dropped waistline dress 低腰连衣裙
dropping 毛皮拉伸
droppings 落毛
drop shoulder 垂肩,落肩;低肩袖;落肩式,过宽肩式;肩线下倾
drop shoulder sleeve 低肩袖,落肩袖,露肩袖
drop shoulder sweater 袒肩式运动衫
drop size 差数尺寸(胸围与腰围的差数)
drop stitch 空针,漏针疵;水滴状绖带贴绣;脱散线圈
drop stitch knitting 漏圈花纹针织物;漏针花纹针织物
drop stitch pattern 漏圈花纹,漏针花纹
drop torso 短厚躯干
drop type belt loop 下垂式套环(饰带)
drop & variable feed 上下差动送布
drop waist dress 落腰女装
drop waisted skirt 低腰裙
drop waist line 低腰线型,降腰线型
drosin 荷兰绢丝外衣绸
drugget 粗毛地毯;棉毛混纺地毯;印花地毡;地毯罩布;粗而耐用的织物
drug shoes 药物鞋
druid 棉帆布
druid's cloth 纯棉帆布;僧侣服布
drum 耳鼓,转鼓(毛皮加工设备)
drum farthingale 英国式裙撑
drum handbag 鼓形手提袋
drum major's hat 鼓手长帽
drummond worsted 德朗蒙德灰色斜纹呢
drum plotter 滚筒式绘图器
drum washing machine 滚筒式洗涤机
Drung nationality's costume 独龙族服饰
dry 干燥,烘干
dry-bonded fabric 干式黏合织物(非织造布)
dry bulb temperature 干球温度
dry clean 干洗
dry cleanability 耐干洗性
dry cleanable garment 干洗服装
dry cleaner 干洗工人;干洗机;干洗剂;干洗商,干洗商店
dry cleaning 干洗;干洗的衣物
dry cleaning agent 干洗剂
dry cleaning and dyeing shop 干洗洗染店
dry cleaning equipment 干洗设备
dry cleaning intensifier 干洗增效剂
dry cleaning machine 干洗机
dry cleaning resistance 耐干洗性
dry cleaning resistant 耐干洗
dry cleaning resistant fabric 耐干洗织物
dry cleaning resistant finish 耐干洗整理
dry cleaning shop 干洗店
dry cleaning solvent 干洗溶剂
dry clean machine 干洗机
dry clean only 只可干洗
dry clean or hand wash with care 干洗或小心手洗
dry crease 干折皱

dry decatizing （干）蒸呢
dryer 烘干机
dryer felt 烘缸毛毯
dryer felt duck 造纸毛毯
dry fabric 织物干燥
dry-finishing 干整理
dry-formed fabric 干法成形非织造布
dry friction 干摩擦
dry goods 匹头，纺织品；现成衣服
dry goods store 布店
dry handle 干燥手感
dry head type 无油润滑型
dry heat setting 干热定型
dry in air 晾晒
drying in the open 自然干燥
drying of wool fabrics 羊毛织物烘干
drying temperature 烘干温度
drying tumbler 干衣机
dry ironing 干熨烫（不喷水雾）
dry pelt 干毛皮
dry rose 陈旧玫瑰色
dry skin 干毛皮
dry soil resistance 干抗污性
dry spun yarn 干纺纱
dry strength 干强度
dry suit 防水服
dry touch 滑爽感
dry touch and skin friendly fabric 干爽舒适织物
dry tumble 干式转筒整理
dry wash 干洗，洗过而未烫的干衣服（美国）
dry weight 干重
dry wrinkle recovery 干折皱回复
dsedim rug 东方缝合地毯
dsujnabe rug 中亚手工结绒地毯
dual bobbin winder 双梭芯绕线器
dual coat 两用衫
dual feed system 双送料装置
dual-opening zipper 双头拉链
dual-purpose fastener 两用拉链
dual-purpose gloves 两用手套
dual silhouette 两种外轮廓并存造型
dual underfront shirt hemmer 衬衫用内向双卷边器
Du Barry costume 杜巴里服
Du Barry sleeve 杜巴里袖
dubbin 皮革保护油
ducal bonnet 公爵帽，公爵夫人帽
duchesse 高密缎

duchesse lace 白色凸纹枕套带
duchesse mousseline 花式斜纹绸
duchesse pleat 裙后箱形褶
duchesse satin 高密缎，丝硬缎
duchess lace 细网眼凸纹梭结花边，镶嵌花边
duchester 杜奇斯特丝绒，美国天鹅绒
duck 粗布，帆布；棉布衬（西装用）；帆布衣裤
duck arse 鸭屁股式男发型
duck belting 帆布带
duck bill bonnet 鸭舌帽，鸭舌软帽
duck bill cap[hat] 法国革命时期长舌女帽
duck bills 鸭嘴鞋
duck down 鸭绒
duck-egg blue 鸭蛋青色
duck fabric 防水布；帆布
duck feather 鸭羽毛
duck green 鸭绿色
duck-hunter 英国侍者用条纹亚麻夹克
ducking 粗布；帆布
ducks 白色帆布裤子，帆布衣服，帆布裤子
duck's arse 鸭屁股式男发型
duck's egg green 浅青色，鸭蛋绿
duck shoes 帆布鞋
duck's tail hair 鸭尾发型
duck suit 帆布装
ducktail hairstyle 鸭尾式发型
ductility 可延展性，延性
dude jeans 度假牛仔裤（美国）
Dudley diamond 多得利钻石
duds 非正式服装，破衣，旧衣，破旧衣服
due bill 到期票据
dueling blouse 击剑衬衫
dueling shirt 击剑衬衫
dues 应付款
duet pin 双别针
duffel 低级毛毯；粗厚起绒地毯；粗厚起绒呢料；外衣用粗呢；粗呢大衣
duffel，duffle 粗厚呢料
duffel bag 露营袋，筒式包，行李袋，圆筒旅行袋
duffel blanket 低级毛毯
duffel cloth 起绒粗呢
duffel clothing 粗呢大衣[上衣]
duffel coat （用纽袢条的）连帽粗呢雪大衣，粗呢大衣
duffer 暖和短外套

Duff-Gorden 达夫·戈顿(英国时装设计师)
duffle coat 粗呢大衣,连帽粗呢风雪大衣(用纽袢条)
duffle front 浮标型系合前襟
duffle sweater 迪弗尔毛衣
Dufy touch 杜飞点缀
dugebia 英国丝棉交织条格绸
duhua brocaded silk table cloth 独花织锦台毯
duhua Yajiang bourette 独花鸭江绸
dull (d) 色光暗淡,浊色,消光,无光
dull appearance 无光外观,暗淡外观
dull black 暗黑
dull blue 暗蓝色
dull color 暗淡色,灰暗色
dull coloring 色彩单调
dull cotton thread 无光棉缝纫线
dull decatizing 罐蒸,全蒸,加入蒸呢
dull fabric 消光织物
dull fiber 无光纤维,消光纤维
dull finish 平光柔软整理,消光整理;失光(鞋)
dull grey 深灰
dull hose 暗淡袜子
dulling effect 暗晦效应,灰暗现象,无光效应
dull-lustered rayon yarn 无光人造丝
dullness 晦暗,消光;无光度,消光度
dull purple 暗紫色
dull rayon 无光人造丝
dull rayon shioze habotai 无光纺
dull red 暗红
dull taffeta 无光塔夫绸
dull thread 无光丝,无光线
Dulong ethnic costume 独龙族民族服
dum 编织用棕榈叶纤维
dumba 大尾羊羊毛
dumb jacket 短皮夹克,皮制短夹克(总称)
dumb look 菱黯外观
dummy 人体模型,假人,人台
dummy buttonhole 假纽孔
dummy fitting method 立体裁剪法
dummy-papiermaché 胶纸人体模型
dummy pocket 假口袋
dummy selvage [selvedge] 假边
dummy try-on （在模特上穿)试身,服装试样,疏缝试衣装
Dumobin 当默宾格子呢(苏格兰),高级苏络兰格呢
dumont blue 大青色
dump 倾销
dumping （商品)倾销
dumping duty 倾销税
dumpling-like shoes 烧卖鞋
dun 暗褐,焦茶色,鼠灰,灰褐色,灰兔褐色
Duncan cotton 美国邓肯棉
dunce('s) cap 笨人帽,(旧时学校给成绩差学生戴的)圆锥形毡帽
dunchee hemp 印度敦奇大麻
dunchese 丝硬缎
dundas tartan 彩色套格呢
Dundee 敦提女式呢,光面女式呢;打包麻布
dune yellow 沙丘黄(淡黄褐色)
dungaree 粗蓝布,劳动布,粗棉布;粗布工作服,牛仔裤,粗蓝布工作裤
dungarees 劳动裤,牛仔裤
dunrobin 敦罗宾格子细呢
duo-length coat 双长度大衣(底边装绳带用以改变长短或由拉链控制,常为毛皮大衣)
Duon 杜恩
duos （不同布料组合的)调和套装,异料套装(用不同衣料制作而十分调和的套装)
duotone 双色版;双色调
duotone curve 双色调调整曲线
duotonic pattern 同色浓淡图案
dupatta 杜帕梯披风(印度)
dupatti chunni 杜帕塔披风(印度)
dup-chan 红经绿纬平纹绸
dupion fabric 仿丝织物,双面织物;双层织物
duplex board 白纸板
duplex damask 双面同花锦缎
duplex fabric 双面织物,双层织物,双面手套布
duplex fabric in different colors 双面异色布
duplex fiber 皮芯型复合纤维
duplex pressing board 两段式熨烫台
duplex prints 双面复合印花织物,双面印花织物
duplex sheeting 双面斜纹绒布,双面绒布
duplex soft-filled sheeting 双面绒布
duplicate 复制
duplicate copy 副本

duplicate cutter 复样裁剪师
duplicate hand 复制样衣师
duplicate maker 复样缝制员
duplicate printing 双面印花,双面复合印花
duplicate sample 复样,留底样本,存档样本
DUPONT 杜邦
duprene 氯丁橡胶
durability 耐久性,耐用性
durability of wear 耐穿性
durability to washing 耐洗涤性
durability under repeated washing 耐重复洗涤性
durable anti bacterial finishing 耐久抗菌整理
durable antistatic finish 持久性抗静电整理
durable antistatic finishing 耐久抗静电整理
durable crease 耐久折痕,耐久褶裥
durable finish 耐久性整理
durable finishing 耐久性加工,耐久性整理
durable flame retardant finishing 耐久阻燃整理
durable glaze finish (DG finish) 耐久性上光整理
durable goods 耐久织物
durableness 耐久性
durable nonwoven 耐用型非织造布
durable pleating 耐久褶裥,耐久折痕
durable press (D.P) DP 整理衣料,DP 整理;定型熨烫,耐久熨烫,耐久压烫
durable press finish 耐久压烫整理
durable press level 耐久压烫程度
durable product 耐久性制品
durable setting 耐久定形
durable softening and water-repellent finish 耐久柔软拒水整理
durable soil-release 持久去污性,耐久防污
Durango cotton 美国杜兰戈棉
durant 厚毡呢
duration 有效期
duration of afterglow 阴燃期(阻燃试验)
duration of flame 有焰燃烧期
duree quilt 大花床单
Durene Association America 美国丝光棉线协会
duria 手织经条棉丝绒
duroelasts 硬弹体
durois 杜鲁瓦精纺厚呢

duro-plasts 硬塑料
durreeaee 条花绸
durrie 棱纹厚棉布;印度厚毯
durris rug 印度手织棉毯
durry 棱纹厚棉布;印度厚毯
dusk 微暗黑色,黄昏色,暗色
dusky orchid 暗紫色
dusky green blue 深青灰
dust 粉末,灰尘;淡褐色
dust cap 防尘帽,防尘罩
dust cloak 防尘长外套,防尘罩衣
dust cloth 布罩,防尘套
dust coat, dustcoat 防尘罩衣,防尘外衣,罩衣,罩衫;风衣
dust color 灰暗色
dusted interlining 撒粉衬
duster 防尘布;避尘衣,风衣,宽松便服,旅行用遮灰罩衫;抹布
duster cloak 防尘罩衣,防尘长外套
duster clothing coat 防尘外套
duster coat 防尘罩衣,防尘长外套
dust free coat 防尘服,无尘服
dust gown 防尘外套,罩袍
dust grey 土灰
dust jacket 家具防尘布套
dust-laden air 含尘空气
dustman uniform 环卫服
dust mask 防尘面罩,防尘口罩
dust-proof effect 防尘效应
dust resistant finish 防尘整理
dust ruffle 裙摆折边,裙摆褶边,防尘褶边
dusty cedar 雪松色
dusty coral 模糊珊瑚色
dusty grey 土灰色
dusty jade green 浅灰绿,豆绿
dusty lavender 朦胧紫
dusty orange 灰橙色
dusty pink 带灰粉红色,灰粉红
dusty rose 朦胧玫瑰色
dusty wrapper 家具防尘套
Dutch blue 浅紫蓝
Dutch bob 巴斯特·布朗发式;荷兰短发,荷兰齐耳短发式
Dutch bonnet 荷兰帽,荷兰罩帽
Dutch bort 荷兰劣钻
Dutch boy cap 荷兰帽
Dutch boy heel 荷兰跟,荷兰鞋跟
Dutch boy pants 荷兰式少年裤(腰部圆阔,裤脚自然变细);荷兰式牛仔裤

Dutch boy skirt	荷兰童裙
Dutch breeches	荷兰裤,荷兰式马裤
Dutch camlet	荷兰羽纱;素色棱格交织呢
Dutch cap	荷兰帽,荷兰式夹顶女帽
Dutch carpeting	荷兰走廊地毯
Dutch coat	荷兰男式短上衣
Dutch collar	荷兰领
Dutch costume	荷兰民族服饰
Dutch girl skirt	荷兰青年女裙,荷兰少女裙
Dutch hat	荷兰帽
Dutch heel	荷兰鞋跟,荷兰青年鞋
Dutch Klompen	荷兰木鞋
Dutch linen	荷兰强力细实亚麻布,细实亚麻布(荷兰)
Dutchman's breeches	荷兰宽大裤子
Dutch neckline	荷兰领领口(圆形领口)
Dutch pink	荷兰粉红
Dutch tape	亚麻带
Dutch waist	荷兰式女紧身胸衣
Dutch white	荷兰白
duty	税
duty free	免税
duty on buyer's account	关税由买方负担
duty shoes	工作鞋
duvet	鸭绒垫子;羽绒马甲;羽绒被
duvetine	起绒织物
duvetyn	起绒织物
duvetyne de soie	绢丝绒;绢丝呢
dux collar	男式窄立领
dwarf costume	小矮人服装
dwarf gusto	矮人乐
dyads	补色配色(二色配色)
dye	染色;染料
dyeability	可染性,染色性
dyeable fiber	易染纤维
dyeables	与衣服颜色匹配的鞋子
dyeable shoes	与衣服颜色匹配的鞋子
dye compatibility	染色配伍性
dyed/bright rayon brocade	争春绸
dyed cattle leather	染色牛皮
dyed cloth	染色布,色布
dyed corduroy	染色灯芯绒
dyed cotton	染色棉花
dyed drill	染色卡其布
dyed fabric	染色织物
dyed figured cotton Italians	花色平纹棉布
dyed fox fur skin	染色狐皮
dyed goods	染色布,色布
dyed goods handkerchief	匹染手帕
dyed interlock fabric	染色棉毛布
dyed knit fabric	染色针织布
dyed knits	针织色布
dyed nylonpalace	染色尼丝纺
dyed oxford	染色牛津布
dyed polyester viscose blended serge	什色涤黏哔叽
dyed poplin	染色府绸,什色棉府绸
dyed rayon brocade	织闪缎
dyed rayon/polyamid mixed brocatelle	蓓花绸
dyed sheeting	染色细布
dyed shirtings	染色平布,染色细布,染色细纺
dyed shirtings (sulfur black) schreinered	电光硫化元布
dyed silk	染色绸
dyed style printing	染地印花
dyed T/C fabric	染色涤棉布
dyed union brocaded velvet	锦绣绒
dyed washing	染色水洗布
dyed yarn	染色纱,色纱;色织
dyeing	染色(工艺)
dyeing cloth	色布(染色布)
dyeing defect	色花疵
dyeing speck	染斑
dye lot	染色批量
dye mark	染渍
dyer's greenweed	植物染料
dyes	染料
dyes for polypropylene fibre	丙纶染料
dyestain	染渍
dye streak	染色条花疵,色绺疵,色柳疵
dyestuff	染料,颜料
dyestuff printing	染料印花
dye up-take	上染
dynacurl	仿波斯羔皮,代纳克尔羔牛绒
dynamic abrasion	动态磨;动态磨损;动态耐磨÷
dynamic balance	动平衡;动态平衡
dynamic behavior	动态性状
dynamic bending	动态弯曲
dynamic deformation	动态变形
dynamic 3D graphics	动态三维图形显示
dynamic drape	动态悬垂性
dynamic elasticity	动态弹性
dynamic equilibrium	动态平衡
dynamic fatigue	动态疲劳
dynamic handling	动态处理
dynamic measurement	动态测定
dynamic ornamental stitch	可变装饰线迹

dynamic piecing 动态衣片设计
dynamic recovery 动态回复
dynamic relaxation 动态弛豫
dynamic resilience 动态回弹
dynamic state 动态
dynamic symmetry 力的对称；动态的对称
dynamic test 动态试验
dynamic torque 动态转矩
dynamic viscoelastic 动态黏弹性
dynamited silk 增重丝绸
dynasty 年代
dzoul 原色山羊毛地毯

E

E E码(鞋宽,较D码宽而较EE码窄)
eagle ornament 鹰纹样
ear 耳廓
ear band 夹环式耳坠
ear bob 耳环;耳坠
ear button 纽扣型耳环
ear cap (御寒用)耳套,护耳罩,防寒耳套
ear drop 耳环;耳坠
ear flap 耳扇,帽子耳扇,帽瓣
Earl-glo 尔格卢里子绸
ear lobe 耳垂
ear lock 垂鬓,耳际鬓发
early adopter 流行先驱
early American look 美国初期(开拓时代)服装风格,美国早期款式
early Carolina cotton 南卡罗来纳棉
early flax 石棉
early majority 先驱追随者
ear muffs 耳套,护耳罩,耳罩
ear muffs hairstyle 耳髻发型
earning foreign exchange 出口创汇
ear pendant 耳环
ear phone 耳机;耳机式发结,耳机式发型
ear piece (帽上的)护耳片,护耳罩,眼镜脚
ear protector 护耳器
earring 耳环,耳饰
ear shell 石决明
ear shield 护耳罩
ear string 男式穿左耳黑丝线
ear tab 耳扇,护耳罩
eartern and western style cotton wadded jacket 中西式棉袄
earth black 土黑
earth color 土黄色,土色(深棕色或茶色)
earth colors 矿物颜料
earth flax 高级石棉
earthquake gown 女式地震保暖装
earth red 土红色
earth shoes 大地鞋
earth yellow 土黄
earthy fashion 乡土装
ear warmer (御寒用)耳套,耳罩
ease 放宽(尺寸);松份,松量,松裕;舒适
ease and gather control 松份与褶量的控制
ease and natural 穿着舒展自然
ease bottom 放宽下摆
ease-care 易保管
ease control notch 吃势量[控制]刀眼
ease distribution 吃势分配
eased seam 缩缝;缩缝缝合
easel 画架
ease neckline 放领线
ease of care 随便穿,免烫
ease of clothes 服装松度
ease-of-handing 易加工处理
ease of ignition 易点燃性
ease stitch 大针距线迹
ease waistband 放腰头
easing (尺寸,衣服)宽松;放宽,吃势,润势;放松缝线张力
easing mark 吃势记号,放宽记号
easing sleeve cap 收袖山
Easter bonnet(or hat) 带花饰春天女帽,复活节软帽
Easter egg 复活节彩蛋色
eastern and western style coat 中西式上衣
Eastern and western style cotton wadded jacket 中西式棉袄
eastern brocade 东方花锦绸
eastern carpet 东方地毯
Eastern charm 东方神韵
Eastern civilization 东方文明
Eastern flavour 东方情调
easternization 东方化
Eastern wool 美国本土羊毛
Easter Parade 复活节罩帽
east improved cotton 佐治亚晚熟棉
east India wool 东南亚羊毛
easy bag 随意袋;随意包
easy cap 便帽
easy care 免烫,洗可穿;易保管
easy care and minimum care 免烫,洗可穿;易处理加工
easy care clothes 快干免烫衣服
easy care cotton 免烫棉布
easy care finish 免烫整理,洗可穿整理,随便穿整理
easy care performance 免烫性能
easy care property 免烫性能,洗可穿性能,随便穿性能

easy care suit 便套装	会刺绣,宗教型纹样绣
easy casual 便服	ecclesiastical lace 教堂式花边
easy coat 宽舒外套	ecclesiastical robe 法衣,弥撒祭服
easy dress 便装	ecclesiastical vestments 教堂法衣
easy fit 宽适型	ecclesiastic embroidery 比索细薄亚麻布;
easy-fitting 穿着舒适的	教堂式刺绣,教会刺绣,宗教型纹样绣
easy fitting coat 宽蔽外套	ECCO 爱步
easy-iron 易熨烫	Ecco shoes(or sneaks) 爱可运动鞋
easy-made 看样定做	EC eco audit 欧共体生态审核
easy of maintenance 易于保养	EC eco label 欧共体生态标志(生态标
easy-on-blouse 穿着方便上衣(多半指衬衫)	签)
	échantillon 布样,包袱样
easy order 简易定做	echarpe 肩带,绶带;披巾
easy pants (宽松舒适的)便裤,松紧裤	echarpe-anneau 环形披肩
easy servicing 小修	echelle 艾柴尔胸衣
easy sleeper 宽蔽睡长袍	echigo chizmi 日本粗苎麻布
easy sleeve 宽蔽袖	echizen 日本优质绸
easy strok 飞云走笔	echo 回声发型
easy style 宽蔽风格	Eclipse Tie 单排眼尖舌女鞋
easy suit 休闲套装,便套装	eco-design 绿色环保设计
easy threader 简便穿线器	eco-diaper 生态尿布(可被生物降解)
easy-to-dye polyester fiber 易染绦纶纤维	eco-dyestuff 生态染料
easy-to-follow 易于仿效	eco-friendry dye 生态染料
easy-to-get-into 易穿入	ecolier(ere)look 小学生款式
easy-to-iron 易熨烫	ecological accessories 环保辅料
easy-to-make dress 简做服装	ecological button 环保纽扣
easy-to-move 容易操作	ecological clothing (天然材料制)生态服装,环保服装
easy-to-wear outlook 便于穿着式样	
eau de cologne [法]科隆水	ecological cotton 生态棉布,环保棉布
eau de Nil 深绿色	ecological cotton garment 环保棉布服装
E.Bergere E.贝热尔(法国时装设计师)	ecological design 绿色环保设计
ebonite 硬质橡胶	ecological dyestuff 生态染料,环保型染料
ebony 深蟹青,暗蓝色,乌木黑,檀黑色	ecological fabric 生态纺织品
ebony brown 茶褐色,乌木棕(色)	ecological fashion 自然式服装,生态时装
ébredon 三页斜纹呢	ecological fiber 无污染纤维,生态纤维
ébredon vegetable cotton 印度本土棉	ecological finish 生态整理
ecaille 鱼鳞绸;鱼鳞花边;鱼鳞饰片	ecological print 生态图案;生态学印花
ecaille de poisson 鱼鳞纹花边网地	ecological textile 生态纺织品
ecaille work 金银片缀饰	ecology 生态学
Eccelide 丝毛条纹呢	ecology fashion 自然式服装,生态时装
eccentric clothes 奇异服装,奇装异服	ecology look 生态风貌
eccentricity 古怪	ecology stylistic form 生态学款式
eccentric look 古怪款式	E-commerce 电子商务
eccentric sheath-core bicomponent fiber 偏心皮芯型双组分纤维	economic 经济
	economic accounting system 经济核算制
eccentric sheath-core type hosiery monofils 偏心皮芯型针织单丝	economical marking 经济排料,拼裁
	economic association 经济联合体
eccentric style 古怪款式,与众不同风格	economic batch quantity 经济批量
eccentric yarn 螺旋花线	economic benefit 经济效益
ecclesiastical embroidery 教堂式刺绣,教	economic globalization 经济全球化

economic growth rate 经济增长率
economic lay 经济辅料
economic layout 套料排料
economic layout of different patterns 供套法
economic responsibility 经济责任
economic system reform 经济体制改革
economic technical norms 经济技术指标
economy pinking shears 花边剪刀,实用锯齿边剪刀
economy price 经济实惠的价格
economy size 经济尺寸
economy type 经济实惠型,实用型
ecossaise 苏格兰丝织格子绸
ecossaise ribbon 苏格兰格子丝带
ecosse 双色棱纹织物
Eco-Tex, eco-tex(tiles) 生态纺织;生态服装,生态纺织品,环保纺织品
eco-textile label 生态纺织品标签
ecrase 绉纹效应,压花效应;压花皮革
ecru 本色织物,淡褐色织物;本色,米黄色,淡灰褐色,未漂白的,生麻色
ecru chiffon habotai 生纺
ecru damask 本色花缎
ecru drab 淡褐色
ecru lace 本色花边,米色花边
ecru silk cloth 本色生丝织物
ecsaine 人造革;轻便易洗的非织造布
ectomorph(y) 瘦型体质
edafuhri cloth 手织棉平布
Edardian coat 爱德华大衣
Edardian jacket 爱德华式夹克
Edardian look 爱德华风格
Edardian puff sleeve 上膨下紧式袖
Edardian style 爱德华式
Edardian suit 爱德华装
edelweiss lace 德国烂花花边
edestin fiber 麻仁蛋白纤维
edge 边饰,布边,边沿,衣边;镶边;(鞋子)边、棱
edge abrasion 布边磨损
edge and point presser 马凳
edge baster 衣边粗缝工
edge bump 碰破边疵
edge cementing and folding machine 加固折边缝纫机
edge control 布边控制
edge control seamer 边缝自控缝纫机
edge control seaming machine 边缝自控缝纫机,控边缝纫机
edge cover 包边

edge covering 包边
edge cover stitching 包边线缝
edge curl 料边卷曲
edge cutting knife 切边刀
edge cutting machine 切边机
edge decorating stitch 饰边线迹
edged with lace 镶边
edge finish （毛巾）布边整理;缝份边缘加工处理
edge finishing stitching 布边线迹,原身包边(卷边)线缝,布边线缝
edge folder (machine) 布边折叠器;皮革折边机
edge folding machine 折边机
edge following 自动缝边
edge foot 靠边压脚
edge fusing 热熔成边
edge guide 边沿导向脚器,导布器,吸边器
edge guide mechanism 导布机构
edge guiding hemming 导布卷边压脚
edge gumming 浆边
edge hemming ruler 卷边尺
edge line of left fly 裤门襟止口线
edge machine 缝边机,锁边机
edge marks 边色不匀,布边色差
edge missed stitch 边漏缝,未缝住边
edge neatening 修边
edge nick 边裂口
edge notcher 缝料边切口器
edge-on-edge 缝料相叠
edge pick stitching machine 边缝缝合机
edge piping 镶边,滚边
edge piping and edge trimming machine 切边滚边机
edge piping blazer 滚边运动夹克
edge piping & edge trimming machine 切边滚边机
edge press 衣边压烫机
edge press block 分烫烫台
edger 切边机,修边机
edge recycling machine 布边回收机
edge roll 卷边,针织物卷边
edge rolling property 针织物的卷边性
edge seamer 边缝机
edge seaming and trimming machine 缝边机,锁边机
edge sewing operation 缝边操作
edge skirt with lace 给裙镶花边
edge slack 荷叶边,宽急边(织疵),荷叶

边疵
edge stitch 缝边线迹,止口线迹
edge stitched hem 光边边缘,缝边折边的（裙子或衣）下摆
edge stitched seam 缝边折边缝
edge stitcher 锁边机,缝边器
edge stitching 缝边
edge stitch pleats 有边缝线的裥
edge stop 缝边限位器
edge straightener 布边伸直器
edge tape 镶边带,贴边带
edge taping 镶边
edge taping machine 镶边缝纫机,镶边机
edge-to-edge cardigan 对比开衫,对襟开衫
edge-to-edge dart 双接省
edge-to-edge front opening 对襟
edge-to-edge lining 对接里子
edge-to-edge printing 边到边印花,全幅印花(针织品)
edge-to-edge ruffler 边对边皱褶器
edge-to-edge seam 对拉链
edge treatment 布边处理
edge trimming （衣服）饰边,修边,剪边,整边
edge trimming machine 修边机
edge warmer 防寒耳套
edging 饰边,滚边,缘饰
edging finishing 剪边
edging lace 镶边花边
edging stitch 锁边线迹,锁边花边
edging tape 滚边带
edging trim 缘饰
Edinburgh jacket 爱丁堡夹克
Edinburgh tweed 爱丁堡粗呢
Edisto cotton 埃迪斯托海岛棉
Edith Head 伊迪丝·黑德（美国时装设计师）
edozuma 日本已婚女性和服
edredon 粗纺斜纹呢
Edward Henri Molyneux 爱德华·亨利·莫利纳斯（法国时装设计师）
Edwardian coat 爱德华式大衣
Edwardian costume 爱德华服式（英王爱德华七世时欧美流行服式）
Edwardian jacket 爱德华夹克（紧身,两侧开衩,高折V领和宽翻领）
edwardian puff steeve 上膨下紧式袖
Edwardian style 爱德华风格,爱德华式
Edwardian suit 爱德华装
Edward Molyneux 爱德华·莫利纳斯

（1891～1974,法国时装设计师）
eel bluish 鳗青色
eel green 鳗青色,鳝鱼青
eel skin 鳗鱼皮
eelskin masher trousere 鳗鱼皮式紧身裤
eelskin sleeve 鳗鱼皮式袖
eel skirt 贴身多片裙
eel yellow 鳝鱼黄
effece side 织物正面
effect 效应;效果
effect drawing （服装设计）效果图
effective face 织物有效面,效应面
effective feed in motion 有效送布量
effectiveness 有效性
effective period 有效期
effective pile 有效绒头
effective pile thickness 有效绒头厚度
effective range 使用范围
effective side 织物正面
effective width 有效门幅
effect thread 特色效应嵌线,花色线,主调色线
effect yarn 花色线,饰纱,花式线的装饰纱,花式纱线,花式线
efficiency-related wages 效益工资
effile 布边完美的布;17世纪法国丧服流苏
egg albumin fiber 蛋白纤维
egg blue 蛋青
egge 装饰边
egg line 蛋壳轮廓
egg pad 馒头,馒头烫垫
egg plant 茄皮紫,茄紫色
egg plant grey 茄灰
egg shape 椭圆形
eggshape dress O形装;蛋形装
eggshell 蛋壳黄
eggshell color 蛋壳色
eggshell finish 暗光整理,消光整理
eggshell line 蛋壳形
eggshell silhouette 蛋壳轮廓
eggshell skirt 蛋壳裙
egg toe 蛋形鞋头
egg yellow 蛋黄色
Egipto cotton 埃吉托秘鲁棉
Egon von Furstenberg 埃贡·冯·弗斯滕伯格（美国服装设计师）
egret 冠饰羽毛,白鹭羽毛;羽饰
Egyptian blue 埃及蓝
Egyptian coatings 埃及棱纹布

Egyptian collar 埃及领
Egyptian colth 稀帆布;19世纪末英国丝毛柔软呢
Egyptian costume 埃及民族服饰,古埃及服饰
Egyptian cotton 埃及棉
Egyptian Cotton General Organization 埃及原棉联合会
Egyptian crepe 埃及绉(重型泡泡绉)
Egyptian design 埃及图案
Egyptian drabbly 埃及披衣
Egyptian flax 埃及亚麻
Egyptian lace 埃及花边,埃及手结细花边
Egyptian sandals 埃及凉鞋,埃及鞋
Egyptian scarf 埃及色条斜纹布
Egyptian style 埃及款式
Egyptian wig 埃及假发,古埃及假发
Egyptian wool 埃及羊毛
ehiffon 雪纺
eider 羽绒,鸭绒
eiderdown 鸭绒,凫绒;保暖长毛绒;毛毡;鸭绒制品
eiderdown blanket 双面绒毯
eiderdown cloth 长绒薄呢
eiderdown pillow 鸭绒枕头
eiderdown sack 鸭绒睡袋
eider yarn 细软毛线,细软无绒线
eight-angle star 八角星形
eighteenth century tableau 18世纪图画
eight-gore skirt 八片裙,八幅裙
eight-harness satin weave 八枚缎纹
eight heads theory 八头高理论
eight immortals 八仙
eight lock 2+2双罗纹
eight panel cycling shorts 八片骑车运动弹性短裤
eight pattern 8字形图案
eight-point cap 八角警察帽
eight-point star design 八角星花纹
eights 8号尺码(衣服、手套、鞋等)
eight single breasted jacket 镶嵌式外套
Eisenhower 军用短外套
Eisenhower jacket 艾森豪威尔夹克(美军制服上衣),野战夹克
Eisteddfed costume 威尔士服装
eis wool 艾斯双股细绒线
eis yarn 针织双股线
ekranga 印度红平布
eksuti 印度单纱棉布

E/L(export license) 出口货物许可证
elaborate design 精心设计
elaborate lace effect 精细网眼花边效应
elaborate pattern 提花花纹,织造花纹
elaborate toilet 盛服;艳妆
elacha 棉丝经条里子布
elaiche 玫瑰底白条手织绸
Elangchun ethnic costume 鄂伦春族民族服
elastane combination yarn 弹性复合纱
elastane fiber 弹性体纤维,弹性纤维
elastche 手织格子布
elastic 弹松带,弹性织物,松紧带,橡皮带,橡皮筋
elastic adjuster 松紧带调节器
elastic and mending stitch 弹性缝补线迹
elastic and tape braid feed 松紧带与编带送料装置
elastic anisotropy 弹性各向异性
elastic ankle strap 裤脚松紧带;护踝带
elastic armlet 臂环
elasticated boots 宽紧靴,松紧靴
elasticated cuff 穿松紧带的弹性袖口
elasticated ribbing 弹性针织罗纹带
elasticated top 弹力袜口
elasticated with frill 有褶皱的弹性袖头
elastication 加缝松紧带的弹性面料
elastic back waist 半橡筋腰
elastic band 松紧带,橡筋带,橡皮圈
elastic bandage 弹力绷带
elastic belt 松紧带,弹性腰带,宽紧带,松紧腰带,橡筋带,橡筋裤带
elastic belt length 松紧带长
elastic belt width 松紧带宽
elastic blind stitch 弹性暗缝线迹
elastic bottom strap 下摆(裤脚)松紧带
elastic bra 橡筋胸罩
elastic bracelet 弹性手镯
elastic braces 弹筋吊裤带,弹力背带
elastic braid 松紧带,橡筋带,橡皮筋;罗纹口;袜带
elastic braid guide 松紧带导向器,橡筋带导向器
elastic brassiere 弹性乳罩,弹性胸罩
elastic brassiere tape 胸罩宽紧带,胸罩松紧带
elastic canvas 硬衬布,重浆衬布
elastic cloth 弹性织物,松紧带,橡筋带,松紧布
elastic cord 松紧带,橡筋带,橡皮筋,松

紧绳,松紧线
elastic corduroy　弹力灯芯绒
elastic corset belt　弹性胸罩带
elastic cotton weft knit fabric　弹性棉纱纬编针织物
elastic covered yarn　弹性包芯线
elastic crepe　弹力绉
elastic cuff　松紧袖口,橡筋袖口,橡筋袖头
elastic decorative waistband　弹性装饰腰带
elastic deformation　弹性变形
elastic demand　弹性需求
elastic drawstring　松紧束带
elastic duck　厚衬里布,厚浆黑衬里布
elastic edging　弹性饰边
elastic edging stitch　弹性包边线迹,嵌边弹性包边线迹
elastic elongation　弹性伸长
elastic extenuation　弹性衰减
elastic fabric　弹性织物,松紧布
elastic fatigue　弹性疲劳
elastic fiber　弹性纤维
elastic filament　弹性长丝
elastic finish　弹性整理
elastic finish sheeting　弹性整理稀薄平布
elastic flannel　针织单面法兰绒,针织法兰绒
elastic foundation　弹力衬底布;弹力胸衣
elastic full waist　全橡筋裤腰,松紧带裤腰
elastic garter　弹力吊袜带
elastic gathering　松紧带抽褶
elastic gathering attachment　松紧带抽褶附件
elastic gore　鞋帮松紧带
elastic gum　橡胶
elastic gusset　橡筋鞋口布
elastic half boxer waistband　半橡筋腰带
elastic heel　弹力跟
elastic hem　弹性卷边
elastic hemmer loop folder　弹性卷边折边器
elastic hemming loop folder　弹性卷边折边器
elastic & hook strap　扣袢式弹性吊带
elastic in a casing　翻折穿入松紧带
elastic isotropy　弹性各向同性
elasticity　弹性,弹力
elasticity of knitted fabric　针织物弹性
elasticity stitching　弹性线缝,伸缩性线缝
elasticized　弹力织物
elasticized band　弹性带

elasticized dress and skirt　松紧式礼服裙
elasticized fabric　弹性织物
elasticized gathering　松紧绳抽褶,松紧式抽褶
elasticized low neckline　弹力低领口
elasticized neckline　弹力低领口,弹性领口
elastic shirring　松紧式碎褶
elasticized swimwear　弹力游泳衣,橡筋游泳衣
elasticized waistband　松紧腰带
elasticized waist dress　松紧式女裙装
elasticized waistline　弹性腰线,伸缩性腰线
elasticized waist trousers　弹力腰线裤
elastic jeans woven fabric　弹力牛仔布
elastic knit stitch　弹性编织线迹
elastic knitwear　弹力毛衫
elastic lace　弹性花边
elastic-leg briefs　弹性裤腿的短裤
elastic limit　弹性极限
elastic material　弹性缝料
elastic narrow fabric　弹性窄幅织物,弹性带
elastic net　弹性网眼
elastic ornamental stitch　弹性装饰线迹
elastic overlock stitch　弹性包缝线迹
elastic performance　弹性
elastic plan　弹性计划
elastic power net　弹力网眼经编织物
elastic rib　弹力罗纹,弹性带,松紧带
elastic ribbon　弹性带,松紧带,橡筋带,橡皮筋
elastic rope　松紧绳
elastic round hat　弹簧圆帽
elastic ruche tape　弹力褶裥带
elastics　弹性狭织物,松紧带织品
elastic seam　伸缩缝,弹性缝
elastic sewing machine　橡筋机
elastic sewing thread　弹性缝纫线,橡筋缝线
elastic shirring　弹性抽褶,松紧带抽褶
elastic shirring foot　弹性抽褶压脚
elastic shirring machine　弹性抽褶缝纫机
elastic shirring smocking　图案形弹性抽褶
elastic shirring smocking pattern　弹性抽褶刺绣花纹,弹性抽褶花纹,图案形弹性抽褶刺绣花纹
elastic shirring stitch　弹性抽褶线迹

elastic shoes 橡筋鞋
elastic sided boots 松紧靴，两侧松紧及踝靴
elastic side shoes 侧挡鞋,松紧鞋
elastic silk weft knit fabric 弹力真丝针织物
elastic single jersey 弹力汗布针织物
elastic smocking 弹性正面刺绣
elastic socks 弹力短袜
elastic sock top 松紧袜口,橡筋袜口
elastic stitch 弹性线迹,伸缩线迹
elastic stockings 弹力长统袜,弹力袜,医疗用弹性袜
elastic strap 松紧带,橡筋带,橡筋,橡皮筋
elastic stretch stitch 弹力伸缩线迹
elastic stretch stitching 弹性伸缩线缝
elastic supply 弹性供应
elastic support 弹性撑骨
elastic tape 松紧带,弹性带,橡筋带
elastic tape for ballet stockings 芭蕾舞袜带
elastic tape for corsets 弹性胸罩带
elastic tension binder 有弹性张力的滚边器
elastic thread 弹性线,橡筋线
elastic thread covering machine 橡筋线包覆机
elastic thread sewing-on device 弹性线缝纫装置
elastic top 松紧口,橡筋口,袜口
elastic top hosiery 紧紧口袜
elastic top socks 橡口袜
elastic triple seam 三行弹性线缝
elastic triple straight stitch 三行弹性直形线迹
elastic triple zigzag stitch 三行弹性,三行弹性曲折形线迹
elastic undershirt 弹力汗衫
elastic underwear 弹力内衣
elastic waist band 弹性腰带,松紧腰带,松紧腰头
elastic waist belt 弹性腰带,松紧腰带
elastic waist brief 松紧带三角裤,宽紧束腰三角裤
elastic waist gown 宽紧束腰长衫,松紧束腰长衫
elastic webbing 弹性织物;弹性带,松紧带;弹性罩
elastic webbing band 松紧带

elastic weft knit fabric 纬编弹性针织面料
elastic width 松紧带宽度
elastic wrinkle resistant fabric 弹性抗皱织物
elastic yarn 弹力纱,弹力丝,橡筋线
elastic zigzag seam 弹性曲折缝
elastik 薄里亚麻衬布(奥地利),薄亚麻衬布
elastin fiber 弹性蛋白纤维
elastique 急斜纹精纺呢,精纺粗条大衣呢,细毛大衣呢
elastodiene 橡皮筋
elastomer 弹性体
elastomeric fiber 弹性纤维,弹性体纤维
elastomeric-hard bicomponent fiber 纤素-硬链段双组分纤维
elastoplasticity 塑弹性
elatche 手织棉丝格子布
Elbert Hubbard tie 埃尔伯特·哈伯德领带
elbow 肘,胳膊肘儿,手肘;(衣服)袖肘部
elbow circumference 肘围
elbow cloak 披肩
elbow cuff 可翻折及肘袖口;肘翻贴边
elbow curve 小袖内撇线
elbow dart 肘省,袖衬省,肘褶
elbow flex 肘弯曲
elbow gauntlet 齐肘手套
elbow girth 肘围
elbow girth line 肘围线
elbow joint 肘关节
elbow length 肘长,至肘长
elbow length gloves 长筒手套
elbow length sleeve 中袖,及衬袖
elbow-level 肘围线
elbow line (E.L) 袖肘线,肘线
elbow pad 护肘
elbow panniers 宽大的裙撑,至肘裙撑
elbow patch 肘垫布,肘垫布,肘部贴布
elbow point 肘点
elbow seam 后袖缝,肘弯缝,弯缝
elbow seam line 后袖缝线,小袖内撇线
elbow sleeve 五分袖,中袖,半袖,肘长袖
elbow support(er) 护肘
elbow to elbow breadth 两肘间宽
elbow-to-wrist measurement 肘至腕长
elbow width 肘宽
elderberry 莲灰色
electoral cloth 双面无葛

electoral wool 美利奴细羊毛
electra 伊莱克特拉伞绸
electra cloth 伊莱克特拉伞绸
electret nonwoven 驻极体非织造布
electrical cloth cutter machine 电动裁布机
electrical colth cutting machine 电动裁布机
electrical conductivity fiber 导电纤维
electrical iron 电熨斗
electrical isolation fiber 电绝缘纤维
electrical linemen wear 修电线工工作服
electrically conductive fiber 导电纤维
electrically conductive nonwoven 导电非织造布
electrically-heatd flat bed press 电热平板压烫机
electrically-heated garment 电热服
electrically powered knife 电剪刀
electric and treadle sewing machine 电动、脚踏两用缝纫机
electric blanket 电热毯
electric blue 铁青色,电火青色,电火花蓝
electric clipper 电轧剪
electric cloth cutter 电裁衣刀,电剪,电剪刀
electric color 电光色,电气色
electric computer layout 电子计算机排料
electric conductivity 电导率
electric cord braiding machine 电动编带机
electric dress 电灯装
electric equipment 电器装置
electric eye sensor thread cleaning machine 电眼感应清线头机
electric hand iron 手提电熨斗
electric heat ctrying 电热烘燥
electric heat fabric 电热织物
electric heating blanket 电热毯,电热毛毯
electric hemming controller (EH-controller) 电动卷边控制器
electrician 电工
electric iron 电熨斗
electricity heating shoes 电热鞋
electric lighting 电气照明
electric magnetic tape cutter (EC device) 电磁吸铁切带刀装置
electric press 电烫机,电烫衣机,电压呢机
electric pressing 电熨烫
electric puff iron 弧形电熨斗
electric rezor 电剃刀
electric scissors 电动剪刀,电剪刀

electric sealing machine 高频熔缝机,电气熔缝机
electric sewing machine 电动缝纫机
electric shoes 电热鞋
electric snap fasterning machine 电动揿钮钉扣机
electric socks 电热短袜
electric soldering iron 电烙铁
electric spark 电火花
electric steam boiler 电热式蒸汽锅炉
electric stripe 电光横条织物
electric switch 电源开关
electric velvet 浅色点子丝绒
electric vest 电垫背心
electric washing machine 电动洗涤机
electrified lambskin 电光处理的小羊皮
electrify character 带电性
electrifying 毛绒光电[整理]
electrifying finish 电光整理
electrochromatic display 电子变色显示器
electrocoating 静电植绒
electrocolor method of carpet printing 电子喷印法地毯印花
electro-communication 电子信息交流
electroconductive composite fiber 导电复合纤维
electroconductive fabric 导电织物
electroconductive fiber 导电纤维
electroconductive property 导电性
electroconductive synthetic fiber 导电合成纤维
electroconductive textile 导电纺织品
electrofixer 电定形器
electroflocking 静电植绒
electrofying 电光整理
electrofying finish 电光整理(用于长绒针织物)
electromagnetic spectrum 电磁波谱
electromechanical plotter 机电式绘图仪
electromechanical thread cutter 电动机械剪线刀
electronically programmable bar tacking machine 电脑控制加固缝纫机
electronically programmable lockstitch sewing machine 电脑控制平缝机
electronically programmed bartacking machine 电子程序控制套结缝纫机
electronical programme 电子程序控制
electronic color matching device 电子配色装置

electronic commerce 电子商务
electronic control 电子控制
electronic control centre 电子控制中心
electronic controlled pattern sewing machine 电子花样缝纫机
electronic embroidery machine 电子绣花机
electronic flat machine 电子提花横机
electronic foot control 电子控制压脚
electronic gloves 电子手套
electronic jacquard 电子提花
electronic jacquard machine 电子提花针织机
electronic jacquard narrowing fabric 电子提花带织物
electronic mono-gramming machine 电子织字缝纫机
electronic multi-head embroidery machine 电子多头绣花机
electronic sewing machine 电动缝纫机,电子缝纫机
electronic shoes 电子鞋
electronic shopping 电子购物
electroplated button 电镀扣
electrostatic absorption 静电吸附
electrostatic clinging property 静电粘附性
electrostatic coating 静电植绒
electrostatic decay half-time 静电半衰时间
electrostatic flock printing 静电植绒印花
electrostatic induction fabric 静电感应织物
electrostatic nonwoven 静电法非织造布
electrostatic nonwoven fabric 静电非织造布
electrostatic printing 静电印花
electrostatic printing machine 静电印花机
electrostatic resistance 抗静电性能
electrostatic spun yarn 静电纺纱
electrum 琥珀色
ELEGANCE 《优雅》(德)
elegance 华贵,优雅,高尚,优雅装
elegant 典雅,风雅,清雅,雅致
elegant look 端丽款式,高雅款式,礼服用款式
elegantly 优雅
elegant town wear 正统高雅都市服
element 单元,要素,元素
elementary color 原色
elementary diagram 基本图形
elements of operating cost 生产费用要素
elephant bell-bottom 象腿裤脚
elephant bells 喇叭裤(35cm以上),大裤脚,钟形裤
elephant crepe 双绉,象纹绉
elephant-foot design 八角纹,象脚纹
elephant gray [grey] 象灰色
elephant-leg pants 象腿裤
elephant pants 象腿裤
elephant skin 象灰色(红灰色)
elephant sleeve 大袖窿紧口袖
elephant toweling 浮松布
eletronically programmable bar tacking machine 电脑控制加固机
elevation view 立面图
elevator dress 电梯装,电梯服装,升降机驾驶员服装,升降机装
elevator heel 男士增高鞋鞋跟
elevator shoes 埃莉维特鞋,增高鞋
elevon dolman sleeve 副翼袖
elfin hem 尖角形下摆
el hamman wool 中等摩洛哥羊毛
eliminostatic agent 静电消除剂
Elio Borhanyer 埃里奥·贝尔昂耶尔(西班牙时装设计师)
Elisabeth De Sennville 伊丽莎白·德·塞内维尔(法国时装设计师)
Elizabethan collar 伊丽莎白领(1558～1603),伊丽莎白女王领
Elizabethan costume 伊丽莎白装束
Elizabethan ruff 伊丽莎白襞领
Elizabethan style 伊丽莎白款式
Elizabeth Arden 伊丽莎白·雅顿
elk 麋鹿;麋鹿皮革,驼鹿皮革,软鞣粗牛皮,粗纹大牛面革,靴面皮革
elken 厚重棉帆布
elk leather 鹿皮革,鞋面皮革,麋鹿皮革
elkside 仿麋鹿皮革的大牛皮
ELLE 《她》[法];爱乐
Elliottine silk 埃利沃廷因针织绸
ellipse 椭圆形
ellipse-stitch 椭圆形线迹
ellipse template 椭圆样板
ellipsis mark 省略号
elliptic collar 椭圆形可卸男领
ellsworth cotton 美国埃尔斯沃思棉
elm green 榆绿色
elmwood 榆木色
elongation 伸长
elongation percentage 伸长率
Elsa Peretti 埃尔莎·佩里蒂(1940～,美国珠宝设计师)
Elsa Peretti 埃尔莎·佩里蒂(美国时装

设计师)
Elsa Schiaparelli 埃尔莎·夏帕瑞丽
(1890~1973,法国高级时装设计师)
Elunchum boots 鄂伦春靴
elysee work 贴花刺绣品
elysian 法国纬绒波纹厚呢
elysian overcoating 拷花大衣呢
Elytra cloth 伊利特拉蝉翼纱
émail [法]景泰蓝
emamel leather 漆皮,漆革
emamelled cloth 漆布
Emanuel Ungaro 伊曼纽尔·温加罗
(1933~,法国时装设计师)
embassy case 使馆包
embellished sweater 装饰羊毛衫(镶串珠、珠片及贴花等)
embellishment 装饰;用饰物装饰服装
emblaser 激光绣花机;镭射刺绣机
emblem 徽章;纹章,饰标,象征性标记;(刺绣)图案,胸袋上的刺绣
emblematic ring 象征性戒指
emboss 浮雕印花,凹凸轧花,压花,拷花
embossed 压纹
embossed backing 压花底布
embossed calico 压花棉平布,压花布
embossed carpet 拷花地毯
embossed cloth 压花布,拷花布,凹凸花纹布,浮雕布,轧纹布,轧花布
embossed composition leather 凹凸花纹皮
embossed crepe 拷花绉,轧纹泡泡纱
embossed derivative velvet 凹凸绒
embossed effect 凹凸花纹效应,拷花效应,压花效应;轧纹效果,轧花效果
embossed fabric 轧花织物,轧纹织物
embossed fake fur 拷花人造毛皮
embossed finish 浮雕轧花整理,拷花整理,凹凸轧花整理
embossed georgetee velvet 拷花乔其丝绒
embossed jeans 拷花牛仔裤
embossed label 拷花唛头
embossed leather 浮纹皮革,拷花皮革,压花革
embossed leather shoes 拷花皮鞋
embossed-like overcoating 仿拷花大衣呢
embossed overcoating 拷花大衣呢
embossed pattern 浮雕花纹,凹凸纹花纹
embossed pattern-bonded fabric 压纹粘合非织造布
embossed pattern fabric 高花纹织物
embossed plush 拷花长毛绒

embossed printed shirting 印花轧花布
embossed ribbon 拷花带,凹凸花带
embossed sweater 浮花毛衣,拷花毛衫
embossed type rug 凹凸花地毯,浮花地毯
embossed velvet 拷花棉绒,拷花丝绒
emboss finish 凹凸轧花整理
embossing 浮雕印花;凹凸轧花,拷花,压花;拷花整理,凹凸轧花整理
embossing cloth 轧纹布
embossing leather shoes 轧花革鞋
emboss print 浮雕印花
embrittlement 脆变,脆化,发脆
embroider 绣花,刺绣;绣花机,针刺绣花器;绣花师
embroidered 绣花
embroidered apron dress (童)绣花连衫围裙
embroidered back 绣花手套背
embroidered badge 绣花徽章
embroidered bag with pearls 珠绣手袋
embroidered bed sheet 绣花床单
embroidered blouse 绣花衬衫
embroidered braid 绣花带
embroidered cap 绣花帽
embroidered carpet 绣花地毯
embroidered cloth 绣衣
embroidered clothes with pearls 珠锈衣
embroidered dress 绣花服装
embroidered dress with paillettes and beads 珠绣服装
embroidered dress with pearls 珠绣服装
embroidered eyeful gloves 有孔花纹手套
embroidered fabric 绣花织物
embroidered flowers 刺绣花朵
embroidered garments 绣花服装
embroidered gauze 绣花纱罗
embroidered gloves 绣花手套
embroidered handkerchief 绣花手帕
embroidered hat 绣花帽
embroidered hose 绣花袜,吊线袜
embroidered jeans 绣花牛仔裤
embroidered lace 刺绣花边
embroidered organdy 绣花蝉翼纱,绣花玻璃纱
embroidered pajamas 绣花睡衣
embroidered patch 绣花牌(章),绣花贴牌
embroidered pattern 绣花花纹,刺绣花样,绣花图案
embroidered pillowcase 绣花枕套
embroidered Qipao with beads and paillettes

珠绣旗袍
embroidered ribbon 刺绣缎带
embroidered sampler 绣花样本
embroidered satin shoes 绣花缎鞋
embroidered scarf 绣花围巾
embroidered shawl 洗礼披巾
embroidered shoes 绣花鞋
embroidered silk pajamas 绣花丝绸睡衣裤
embroidered skirt 绣花裙
embroidered slippers 工艺鞋,绣花拖鞋
embroidered slippers with beads 珠绣拖鞋
embroidered socks 绣花袜
embroidered sweater 绣花毛衫,绣花毛衣
embroidered table cloth 绣花台布
embroidered tie 绣花领带
embroidered vest 绣花马甲
embroidered waistcoat 绣花马甲
embroiderer 绣花机
embroidering 刺绣,绣花
embroidering needle 绣花针
embroidering requisite 刺绣用具
embroider on cutted piece 绣片
embroidery 绣花,刺绣；装饰；绣品；贴花
embroidery area 绣花部位,刺绣部位
embroidery badge 绣花徽章
embroidery braid 绣花带
embroidery character 绣花性能
embroidery clipping machine 刺绣挖剪机
embroidery cloth 刺绣底布
embroidery cone 绣花头
embroidery cord 嵌绣
embroidery cotton 绣花棉线,绣花纱线
Embroidery Council of America 美国刺绣委员会
embroidery crash 刺绣亚麻布
embroidery crewel needle 刺绣针,绣花针
embroidery darning stitch 绣花补缀线迹,绣花缀纹绣,织补绣,缀纹绣
embroidery design 刺绣图案,绣花图案
embroidery design out line is uncovered 绣花不良,绣花露印
embroidery dress 绣衣
embroidery edging 刺绣滚边
embroidery effect 绣花效应
embroidery fabric 刺绣织物
embroidery floss 松捻绣花丝线,绣花丝线
embroidery foot 绣花压脚
embroidery frame 刺绣绷(架),刺绣绷子,绣花绷架

embroidery gauze 绣花纱罗织物
embroidery hoop 绣花绷架
embroidery lace 刺绣花边,绣花花边
embroidery lathe 绣花支架
embroidery linen 刺绣用亚麻布,绣花麻布
embroidery location 绣花位置
embroidery lock stitch 绣花锁式线迹,绣花锁缝线迹
embroidery machine 刺绣机,绣花机
embroidery memory 绣花存储器
embroidery needle 绣花针
embroidery on the stump 斯顿刺绣
embroidery patch 绣花徽章
embroidery pattern 刺绣花纹
embroidery plate 绣花板
embroidery plating hosiery 添纱绣花袜
embroidery plating hosiery machine 绣花添纱袜机
embroidery plating knit fabric 绣花添纱针织物
embroidery plating knitted fabric 绣花添纱针织物
embroidery repeat 刺绣机针距
embroidery ribbon 刺绣缎带
embroidery rib top machine 绣花罗纹袜口机
embroidery scissors 刺绣剪刀,绣花剪刀
embroidery sewing machine 绣花缝纫机
embroidery silk 刺绣丝线,绣花丝线
embroidery stiletto 绣花锥
embroidery stitch 刺绣线迹,绣花线迹
embroidery thread 绣花线,刺绣线;缝纫线
embroidery thread ball 绣花线球
Embroidery Trade Association 美国刺绣贸易协会
embroidery trim 刺绣饰边
embroidery warp pattern 刺绣添纱花纹
embroidery with pearl 珠绣
embroidery yarn 绣花线
emeraid 翠绿色,鲜绿色;绿宝石
emerald black 墨玉绿
emerald cut 绿宝石切割
emerald green 翠绿,鲜绿,翠玉绿,翡翠绿
emerald green sombre 深绿玉色
emergency chute 救生降落伞,急救降落伞
emergency flotation clothing 应急漂流服
emergency light 事故信号
emerge stone 诞生石(生日宝石)
emerized 磨毛
emerized effect 起绒效应

emerized fabric 仿麂皮起绒毛织物,金刚砂起绒织物,磨毛织物
emerizing （织物表面）磨毛；起绒
emerties 密织棉布
emery bag 金刚砂袋
emery board 金刚砂板
Emilio Pucci 埃米利奥·普奇(1914～,意大利时装设计师)
Emilio Schubert 埃米利奥·舒伯特(意大利时装设计师)
Emmanuelle Khanh 埃曼纽勒·康恩(1938～,法国时装设计师)
emollient 润肤剂
emotional color 感情色彩
emperor 英国棉经毛纬斜纹衬布
emperor shirt 乡村绅士红法兰绒衬衫
emphasis 强调
emphasize 强调
empire blue 青绿色,帝国绿；灰蓝色,帝国蓝
empire bodice 帝国式紧身胸衣(带有帝国风格的女用紧身胸衣,与长裙配用)
empire coat 帝国式女外套
empire cotton 帝国棉
empire-cut leotard 帝国式短紧身衣
empire dress 帝国式裙装(高腰,低领口,小袍袖,直筒窄裙)
empire fashions 拿破仑一世时期女服式样
empire gown 帝国式袍,高腰节长裙
empire house gown 帝国式高腰长裙
empire jacket 帝国式夹克
empire jupon 帝国式衬裙
empire line 帝国式高腰线；帝政式,帝国式,帝国型
empire line dresses 帝国风格
empire look 帝国风貌,帝国风格
empire ornament 帝国装饰
empire petticoat 帝国式衬裙
empire princess dress 帝国公主装
empire silhouette 帝国轮廓,帝国式服装轮廓;帝国式(法国拿破仑称帝时流行的低领口、高腰节连衣裙)
empire skirt 帝国式礼服裙,高腰裙(高腰直线条宽裙),高腰节裙,帝国裙(法兰西19世纪初)
Empire slinking skirt 帝国式紧身裙
Empire stays 帝国式高腰紧身褡
empire style 帝政风格,帝国风格(法国第一帝国时期流行的服饰风格,大领口短上衣,高腰裙)

empire styleline 帝国风格结构线
empire stylistic form 帝国款式
empire-waist evening dress 帝国式高腰节晚装
empire waistline 帝国式腰线(高腰节)
empire yellow 帝国黄
empire yoke 帝国过肩；帝国抵肩
empress cloth 女式双层细呢
empress eugenie hat 带鸵鸟羽饰的小帽
empress gauze 花纱罗,提花纱罗
empress petticoat 皇后衬裙
empty hem 空边(棉服边部缺棉花)
emulate 仿效
emulsion gloves 乳胶手套
enable 恢复正常操作
enamel 珐琅,景泰蓝
enamel blue 珐琅蓝(紫蓝色)
enamel button 珐琅扣
enameled brass 涂瓷的黄铜扣
enameling duck 涂层帆布
enamel-kid 漆皮鞋
enamel leather 漆革
enamelled cattle leather 上漆牛皮
enamelled cloth 漆布,漆皮布
enamel remover 指甲油除去液
enbifuku 燕尾服
encased seam 封边缝
encasement 包装；装箱
en cheveux ［法］昂谢弗发型
enclosed edge 包边
enclosed hem edge 包缝下摆,滚条下摆
enclosed seam 封边缝
enclosed type 封闭式
enclosed zip 封闭式拉链
en coquille 贝壳纹刺绣
encroaching satin stitch 分茎绣
end-and-end 异经
end-and-end Chambray 异经线布雷衬衫布
end-and-end cloth 异经织物
end-and-end combination twill 异经复合斜纹
end-and-end fabric 异经织物
end-and-end figuring 异经花纹
end and end hair cord 异经麻纱
end-and-end madras 异经马德拉斯狭条衬衫布,异经条纹衬衫布
end-and-end shirting 异经衬衫织物
end bar 缝端加固
end cloth 头子布
end curler 卷发杠

ended fabric 零头布
end fent 零头布,短码布,头子布
endi 印度蓖麻蚕丝绸
ending 缝线接头;两端色差,头梢色差,头尾色差
endive 菊苣色
endless fabric 环形织物,揩手巾
endless feed 循环送料
endless felt 循环毛毡
endless knot 环状结
endless tow 长丝束
endless woven belting 环形机织带;连续带
end line of collar band 底领前宽斜线
end of trousers 裤脚
end of waistband is uneven 腰头探出(前口不齐)
endorse 背书
endorsement 背书
endorser 背书人
end patch 增强衬小角
end piece 零头布
end play 轴端余隙;轴向间隙
end product 成品,最终产品
endroit 织物正面
ends breaking defect 断疵
ends per inch (E.P.I) 经密
end stitching 坯布缝头
end stopper 吊钟(拉绳末端的钟状扣)
end-to-end fabric 异经织物
end trimming 剪屑
end-use 产品用途,最终用途,消费用途
end use condition 服用条件,使用条件
end use field 服用范围,使用范围
end use product 最终产品
energy savings 节能
en-factory price 出厂价格
enfold 加衣褶
engage in garments export 经营服装出口
engagement ring 订婚戒指
engaging ring 订婚戒指
Engilish blue 英国蓝
engineered nonwoven 特制非织造布
engineered pattern 拼组图案
engineered pleat 局部褶裥
engineered print 局部印花
engineered printing 拼组印花
engineered textiles 工程纺织品
engineering and technicel personnel 工程技术人员

engineering change 工艺更改
engineering consultation 工程咨询
engineering fabric 工程[用]纺织品
engineering index 工程技术文献索引
engineering investment 工程投资
engineering psychology 工程心理学
engineering quality 工程质量
engineer's boots 工程师靴,机工靴
engineer's cap 铁路工帽
engineer's cloth 工作服布
English antique lace 英式仿古机制花边
English blue 英国蓝
English British Wools 英国羊毛
English chain 英式链条饰带
English chintz 英国光亮棉布,轧光印花棉布;18世纪麻经棉纬印花布
English coat 英式男大衣
English cottage bonnet 比比罩帽
English count 英制支数
English country dress 英国田园女服
English drape 英国男西装式样,英式夹克装,单排扣男外套
English drape suit 英国褶裥套装
English edging 英式滚边
English embroidery 英国刺绣,英国网眼刺绣
English farthingale 英式裙撑
English fell seam 英式双折边叠缝
English flannel 威尔士法兰绒
English foot 英国袜,英式平袜袜跟
English glove 英式手套
English gown 英式女大衣
English hood 山墙(软)帽,英国头巾
English hunting 英式猎装背心
English lace 英国花边
English leather 厚实五枚棉缎;厚实五枚棉麻缎
English lille 婴儿服饰花边(窄梭结花边)
English lounge 英国男夹克式样;英国式夹克装
English net 六角网眼花边
English placket 英国式门襟,英式门襟
English point 单针绣花边
English quilting 英国绗缝
English red 英国红,氧化铁红
English rib 英式罗纹组织
English rib socks 1+1罗纹短袜
English rose 英国玫瑰红色
English size 英国尺码,英国鞋号
English style 英国绅士风格

English style moustache 英国式小胡子
English system 英制
English velvet 英国丝绒
English walking jacket 英式散步服
English welt 英式光边,英国关边
English wool 英国羊毛
English wool matchings 英国并级羊毛
English work 英国白亚麻布刺绣
English wrap 英国围裹式外套
English yarn count (Ne) 英制纱线支数
en grande tenue 穿着礼服
engraved roller printing 凹面滚筒印花
enhanced background 增强背景
ENICAR 英纳格
enigma variations 城乡融情
enile 土耳其手结厚绒毛地毯
enile rug 土耳其手结厚绒毛地毯
enlarge 放大
enlarged scale 放大比例尺;放大的尺度
enlargement test 扩大试验
enlarger 放大机
enlarging a design 放大花样
enlarging machine 放样机
enlarging of size 放大尺寸[码]
enn-shell silhouette O形轮廓
enormous collar 超级大领
enredo 恩雷多裙(墨西哥)
Enrico Coveri 恩利科·科韦利(意大利时装设计师)
Enrico Massei 恩里卡·马塞(意大利时装设计师)
ensanguined color 血红色
ensemble 全套服饰配合协调的整体效果,总效果;整套服装,三件套(衣、裙、外套),配合协调的套装;服饰整体调和
ensemble costume 套装,女式套装
ensemble dressing 组合式衣着
ensemble du soir 晚会套装
ensemble fashion 组合式时装
ensemble print 组合印花
ensemble suit 三件式套装,整体套装
ensign button 徽章扣
ensign cloth 平纹旗布
ensure public security uniform 保安服
en tablier 连衣裙前片横排的缎带饰边
entangled fabric 缠结成网非织造布
entangled yarn 网络丝,交缠丝
entanglement filament yarn 交并丝
entangling filament yarn 交络丝
entari 长及踝外套

enter a bid 投标
enterprise management 企业管理
enterprise resource planning (ERP) 企业资源计划
entire height 总体高
entity 实体
entoilage 花边网地;加固用织物
en-tous-cas 晴雨两用伞,遮阳伞
entrebande 零头布(法国称谓)
entre deux 衣缝嵌线;直条拼花;衣裤滚边
entre large 法国中幅亚麻布
entre rios 阿根廷高弹硬羊毛
entretela 棉衬里布
entry measurements 开口尺寸
entry suit 火区消防服,消防服(入火区)
entwining twill 菱纹斜纹
envelope 口袋型女式内衣
envelope braid 筒状纸带
envelope chemise 无袖直筒女内衣,筒型女内衣
envelope clutch 信封式无带女提包
envelope compilation 男子信封式内衣
envelope curve 包迹,包络线
envelope dress 信封式女装
envelope handbag 信封式公文包
envelope neckline 信封式领口
envelope neck vest 信封式领圈背心
envelope pleat 信封式裙
envelope pocket 信封式口袋
envelopes 包装用帆布
envelope sleeve 三角形褶裥宽袖,封袋袖
enveloppe 打包用帆布(法国称谓)
envers 衬里;织物背面
envirolites 环保鞋
environmental biology 环境生物学
environmental button 环保纽扣
environmental engineering 环境工程
environmental factor 环境因素
environmental garment 环保服装
environmental pollution 环境污染
environmental quality 环境质量
environment containment standard 环保标准
environment control 环境控制
environment labels 环境标志
environment protection of plant 工厂环境保护
environment protection shoes 环保鞋
enzor net 六角网眼花边

enzyme washing 酵素洗（无须洗水石洗出石磨效果）
enzyme wool 生皮板毛
eolienne 风神绸
eosin 曙红
eosine 朝红,伊红
epaule drapée 有自然皱褶的肩线
epaulet,epaulette 肩饰,肩章,肩袢
epaulet and elbow patch 肩袢及肘部贴缝片
epaulet pocket 有肩章式样扣的口袋
epaulet shirt 飞行员衬衫,肩袢衬衫,有肩袢的衬衫
epaulet silk 肩章缎
epaulet sleeve 肩章袖,褡袢袖,落肩袖,肩袢式插肩袖
epaulet(te) 肩饰,肩章,肩袢
epaulette coat 肩章外套
epaulette tab 肩袢小片
epauliere 肩甲;带护肩甲防弹衣
ephebi 军用斗篷
ephod （犹太教）大祭司法衣
epicene fashion 男女通用款式
epidermis 皮肤表皮
épingle 棱纹绸,棱纹织物
épingle brocade 棱纹锦
épingle crepon 棱纹提花绉绸
épingle faconné 经棱纹提花织物
épingline 经棱纹丝毛织物
epitoga 中世纪学位袍;法国官员礼服披肩;袍状斗篷
epitome 典型,缩影
épomine hood 中世纪学位袍;法国官员礼服披肩;袍状斗篷
éponge 海绵呢;尖毛巾膨松织物花呢
éponge cloth 海绵呢,结子花呢,珠皮大衣呢
epoxy button 环氧扣
equability 均匀
equal divided method（EDM） 等分法
equalize 修版,修样
equal rhythm 匀等协调
equal value of return 资金的等值
equation line 等分线
equestrian cutfits 骑马服,骑装
equestrianne fashion ensemble 女骑手全套时装
equestrianne tights 针织女紧身衣裤
equestrian outfit 骑马服;骑[士]装
equestrienne 女骑手服饰

equestrienne costume 女骑手装
equestrienne fashion ensemble 女骑手全套时装
equestrienne tights 针织女紧身衣裤
equipage 服饰;(古时的)化妆品
equipment 设备,装置,器材
equipment appurtenances 设备附件
equipment characteristics 设备性能
equipment compatibility 设备兼容性
equipment failure 设备故障
equipment in terchangeability 设备互换性
equipment layout 设备布置;设备排列
equipment maintenance 设备维护,设备维修
equipment management 设备管理
equipment reformation 设备改造
equipment renewal 设备更新
equipment running rate 设备运转率
equipments 配备
equipment type flow process chart 设备型流程程序图
equipment upgrading 设备改造
equipment utilization 设备利用率
equity joint venture 股权式合资企业
equity theory 公平理论
eraser 橡皮擦
era silk 苎麻蚕丝
erasing knife 刮刀
erasing knife blade 刮刀片
erchless 尔琪尼丝
erea 细亚麻衬衫布
erect back 直背
erect figure 立像;腆胸体型,挺胸体型
erect narrow back high shoulder 狭背耸肩
erect pile 立绒,直立绒头
ergonomics 工效学,人类工程学
ergonomics of clothing 服装工效学(研究人体、服装、环境之间相互关系)
ergonomist 工效学家,人类工程学家
eria silk 苎麻蚕丝
Erik Montensen 埃里克·莫滕森(法国服装设计师)
eri silk 苎麻蚕丝
Erkens' worsted 厄肯氏斜纹精纺呢
ermine 白鼬毛皮;白鼬皮长袍;银貂皮
ermine cuffs 白鼬皮袖头;白貂皮袖头
Ermined 饰有貂皮的,穿着貂皮袍的
ermine fur 白鼬毛皮,白貂毛皮,扫雪毛皮,银鼠皮,扫雪毛皮
erogenous 强调暴露性感部分的服装理

论;性感的
erogenous zone　性感区
erratic stitch pattern　不规则线迹花纹,无规律花纹
error　误差
error code　错码
error controlled system　误差控制系统
error system　误差检测系统
erskine tartan　苏格兰阔格花呢
ESCADA　艾斯卡达
Escaine synthetic chamois　爱斯凯恩人造鹿皮(非织造布,商名,日本)
escalier　梯形组织花边;百叶窗带
escamis　地中海东部低级单面绒布
escapechute　救生降落伞
escape clothes　脱离都市服
escape liability　逃避责任
escape suit　救生衣
escarelle　腰袋
escargot　螺旋裙
escargot skirt　蜗牛裙(拼缝呈蜗牛壳纹样),螺旋波浪裙
escarpin　[法]薄底浅口皮鞋,舞鞋
esclavage　花式项链
escoffion　[法]爱斯考芬发饰
escot　法国衬里哔叽,厄斯科特哔叽
escurial　缠结花纹;西班牙针绣花边
E-shoes　电子鞋
Eskimoan costume　爱斯基摩服饰
Eskimoan look　爱斯基摩风貌
Eskimo cloth　爱斯基摩织物,爱斯基摩大衣呢,爱斯基摩双面大衣呢,厚毛呢
Eskimo coat　爱斯基摩外套
Eskimo fabric　爱斯基摩缎纹呢
Esmeralda cloak　防水叠省披风
espace　褶宽
espadrille　登山帆布鞋,帆布便鞋,布面平底凉鞋(后跟用带缚住在足踝上的),布面系踝平底凉鞋
espagnolette　法国粗梳毛呢
Espansione　聚氨酯非织造布(商名,日本钟纺)
esperon　法国细哔叽衬料
espiguillas　有穗饰边
Espirt　埃斯普瑞特
espresso slip　直筒型裙边衬裙
esprit of modern art & artist stylistic form　现代艺术风格和艺术家款式
esquire knot　绅士结,绅士领带结
essayage　试衣,试样

essential color　基本色
essential linen　家用亚麻布;家用亚麻布制品
essential tone　基本色调
estame　法国针织精纺呢
estamene　精仿起绒斜纹呢;法国粗毛哔叽
estampado　印花棉布
Estee Lauder　雅诗兰黛
Esterole　聚酯羽绒型絮棉(商名,日本东洋纺)
esthetic design carpet　艺术地毯
esthetic finish　美观加工
estheticism　唯美主义
esthetics　美观性,美感性,美学
estimated tare　估计皮重
estimated value　推算价格
estimation　估值
estopilla　细亚麻衣料
estrella　埃斯特里拉,埃斯特里拉丝毛绉,丝毛双绉
estrich　鸵鸟绒毛
estridge　鸵鸟绒毛
etaise　荷兰亚麻台布
etalit fabric　鬃棉交织装饰布
étamine　纱罗,仿纱罗织物
étamine a voile　毛织巴里纱
etamine glace　印花薄织物,印花丝毛薄织物
etched-out　烂花
etched-out fabric　烂花布,烂花织物,凸花布,腐蚀缕孔印花织物
etched-out georgette　烂花乔其纱
etched-out printed handkerchief　烂花[印花]手帕
etched-out velvet　烂花绒
etched-out velveteen-like fabric　烂花仿平绒
etching　烂花;蚀刻印花
etching darning foot　烂花补缀压脚
etching discharge　烂花拔染法
etching embroidery　按墨稿刺绣,按墨色图刺绣,烂花刺绣,墨色刺绣
etching silk　紧捻绣花丝线
etching stitch　墨色绣针法
etendelle　马毛织物
eternity ring　永恒戒
ethereal blue　太空蓝
Ethiopian shirtdress　埃塞俄比亚衬衫式裙
ethmoid bone　筛骨

ethnic 民族风格;民族形式
ethnical 民族风格;民族形式
ethnical culture 民族民俗文化
ethnic baroque 民族感的巴罗克调
ethnic clothes 传统民族服装,民族服装,民俗服装
ethnic costume 传统民族服装,少数民族服装
ethnic design 民族图案
ethnic embroidery 传统民族刺绣
ethnic look 民间款式,民族款式,民族风格,民族风貌
ethnic mood 民族调
ethnic pattern 民族图案
ethnic print 民族印花;民族印花布;民族印花布服装
ethno nomads look 游牧部落风貌
etiolation 褪色
etiquette 礼节;礼貌;礼仪
étoff 机织物(法国称谓)
étoff à fleurs 花卉纹锦缎
étoff d'or façonnée 织金锦
étoile [法]星状图案,星状花边图案;有光缎纹织物
Eton 伊顿型款式
Eton blue 伊顿蓝
Eton cap 伊顿帽,伊顿无檐帽
Eton clothing coat 伊顿外套
Eton coat 伊顿外套
Eton collar 宽翻领,阔翻领,伊顿(制服)领
Eton collar suit 伊顿领套装
Eton crop 伊顿短发(英国贵族学院学生发型),伊顿短发型;伊顿平头
Eton hat 伊顿礼帽,伊顿帽
Etonian uniform 英国伊顿公学男性制服
Eton jacket 伊顿夹克,伊顿式(女士,少年)短上衣,伊顿式无领短外套
Eton jacket bodice 伊顿式无领短外套,伊顿式少年用短上衣,妇女用短上衣
Eton style 伊顿型款式
Eton suit 伊顿服(19世纪,英国),英国伊顿公学男生制服
étoupe 废麻填料
étramée 法国大麻布
étrasse 绣花丝线
Etruscan 意大利农民服装
etschingo 日本最优级生丝
etschizen silk 日本最优级生丝
etshigo silk 日本最优级生丝
etui 小箱,小匣

etwee 小箱,小匣
eucalyptus 浅绿灰
Eugenie 尤金妮(法国拿破仑第三皇后,早期时装倡导者)
Eugénie collarette 尤金妮钩边领
Eugénie dress 尤金妮连衣裙
Eugenie hat 欧仁妮帽,尤金妮帽
Eugénie paletot 尤金妮宽外套
Eugénie petticoat 尤金妮衬裙
Eugénie's wigs 登山帽,打猎帽
Eulaha carpets 印度古典花纹密实地毯
Eureka cotton 尤里卡棉
Euro-American look 欧洲型美国款式
Euro-American style 欧美款式,欧洲型美国款式
Eurocoton 欧洲棉业协会
European and American style 欧美风格;欧美型
European casual 欧洲式便装
European chintz 欧洲风格印花布
European clothes 西装,西服
European committee for standardization 欧洲标准委员会
European costume 欧洲服饰
European cut 欧洲大陆式裁剪
European-cut jeans 欧洲式牛仔裤
European-cut shirt 欧洲式衬衫,西装衬衫,西服衬衫
European Disposable and Nonwovens Association 欧洲非织造布及用即弃织物协会
European Economic Community 欧洲经济共同体
European fifties 欧洲50年代装(20世纪)
European fit 欧洲型合身
European look 欧洲款式,欧洲风貌
European pattern 欧版
European press with microprocessor 欧洲型电脑熨烫机
European Union Eco-label 欧盟生态标签
Euro-Scotch style 欧陆苏格兰款式
Euro-Scotch stylistic form 欧陆苏格兰款式
Euxinet 丝毛呢
EVA interlining 乙烯·乙酸乙烯共聚体黏合衬
evaluation 估价,评估,评价
even basting 均等粗缝,均等疏缝(两面线迹长度相等)

even checks 棋盘格花纹
even-feed mechanism 上下均匀送布机构
even hemline 平裙摆
evening 晚装;去厚(制革)
evening bag 晚装手袋,晚宴手袋,晚会手袋,晚会包
evening bag with pantographic fastener 可抽带系口或收缩系口的晚装手袋
evening bag with trousse 带小袋的晚会手提包
evening boots 晚会靴,晚会用长靴
evening box 晚会盒,晚会饰包
evening cape 晚礼服斗篷,晚礼服披肩
evening clock 晚会钟形帽
evening clothes 夜礼服
evening clothing 晚礼服
evening coat 燕尾服,晚礼服,晚礼服大衣
evening dress 晚礼服,夜礼服,燕尾服
evening dress coat 男子夜礼服,晚宴外套
evening frock 男子夜礼服
evening gown (长及地的)晚礼服,夜礼服,拖地礼服,曳地晚礼服
evening handbag 晚会手袋,晚宴手袋,晚装手袋
evening hat 晚会帽
evening jacket 短礼服,晚宴服,晚会便服
evening length 离地一英寸的裙长
evening make-up 晚间化妆
evening neck 全敞领,袒肩领
evening neckline 袒胸露肩领口线
evening pajamas 裤装晚礼服
evening petticoat 及踝长的晚裙(前中线及两边侧缝处开衩)
evening pumps 晚会碰拍鞋,晚会舞鞋
evening purse 女晚用珠包,晚用珠包
evening sand 黄昏沙色
evening sandals 晚会凉鞋
evening separates 晚会单件装
evening shift 晚班
evening shirt with chitterlings 带饰边的晚装衬衫
evening shoes 晚会鞋(总称),晚宴鞋
evening skirt 晚裙(正式或半正式场合穿着)
evening slip 晚礼服衬裙
evening slippers 女子晚礼服鞋,晚宴便鞋
evening sports 晚会便装
evening suit (男)晚礼服,夜礼服
evening sweater 晚会毛衣
evening vest 礼服背心
evening wear 晚礼服,晚装
evening wrap 女宽罩衫(晚会用),晚会外套(披在晚礼服外)
even layout 平套法排料
even lock 单线不调梭缝迹;匀顺感
even sided twill 两面斜纹,双面斜纹
even toe presser foot 平趾压脚
even twill 双面斜纹,同面斜纹
even twill weave 双面斜纹,同面斜纹
EVERBRIGHT 依波
ever clean collar 保洁领,永洁领
ever crease 耐久褶裥
ever-cut finish 烂花整理
Everett(e) 艾弗雷特便鞋,爱维莱特鞋
ever fast 不褪色,永不褪色
everglaze 耐久光泽
ever green 常青,冬青绿
everlasting 不褪色;坚固斜纹呢;19世纪精纺薄花呢
everlasting cloth 永固缎纹织物,永固呢,永固织物
everlasting lace 耐用花边
ever-monte 双面针织布
ever pleat 耐久褶裥;耐久折痕
ever pleat skirt 耐久褶裥裙
ever press 永久熨烫
everyday clothes(or dress, wear) 便服,休闲装
everyday dress 便服
everyday mill 正常开工的工厂
everyday wear 便服
Ewenke ethnic costume 鄂温克族民族服
E-wool 生态防缩羊毛
E-wool treatment 生态防缩羊毛处理
exaggerated cowl 密集型垂褶领
exaggeration 夸张手法
examitum 六股丝锦缎(金银线丝花缎)
excellent 优等
excellent definition 轮廓清晰(印花)
excelsior cloth 英国色织物;英国白地色边布
Exceltech 防水透气织物
excess 余量
excess case 吃势过多,放宽过多
exchange 交流
exclusive collection 专用服装(如专用于参与评比、参赛、展示等)
exclusive sale 包销
exclusive store 专卖店
ex dock 码头交货价

execution 刺绣针迹数
exempt inspection 免检
exemption certificate 免税证明书
exercise apparel 舞蹈衣;紧身运动衣
exercise sandals 便凉鞋
exercise shorts 运动短裤
exercise suit(or wear) 训练装,训练服
ex factory 工厂交货
ex factory price 工厂交货价;出厂价格;工厂交货价格
exhaust air system 废气净化装置
exhaust dyeing 浸染
exhibit 陈列品
exhibiting and selling of products 产品展销
exhibition 展览会;展示会
exhibition and sale meeting 展销会
exhibition room 陈列室
exhibits 展品
existing model 现行型号
existing seams 原有接缝
exothermic reaction 放热反应
exotic 异国情调
exotic color 奇异颜色
exotic costume 奇异服装;奇装异服
exotic design 异国调图案
exotic style 异国情调式样
expandable bracelet 弹簧链手镯
expandable fabric 伸缩性织物
expandable waistline 有伸缩的腰围线
expanded plastics 发泡塑料
expandra 弹力劳动布
expansion 扩张,展开,张开;扩大部分;(衣服)下摆,衣裙
expansion skirt 大摆裙
expansive color 膨胀色
expectancy theory 期望理论
expended snap 外伸按扣
expensive color 奢华颜色,奢华色
experimental fashion 实验性时新款式,实验性时装
experimental plant 实验工厂
expert 高手,专家
expertise 专长;专家鉴定;专门知识
expertism 专长
expert system（ES） 专家系统
explanatory notes 凡例
exploded view 立体形象
exploitation 开发
exploratory development 探索性发展

ex point of origin 现场交货
export 出口
export agent 出口代理
export capacity 出口差
export children's wear 出口儿童服装
export commodity 出口商品
export contact 出口合同
export credit 出口信贷
export declaration 出口申报单
export duties 出口税
export duty 出口税
export goods 出口商品
export market 出口市场
export of labour services 劳务出口
export-oriented 外向型
export-oriented economy 外向型经济
export quota 出口配额
export quotation 出口报价
export textiles 出口纺织品
export trade 外销
export value 出口价
expose 袒露
exposed notch 刀眼外露
exposed seam 明线缝迹;外露缝迹
exposed zipper 明拉链,外露拉链
expose-the-breast clothes 女露胸衣,女式低胸衣
exposing 透浆(鞋)
exposure 曝光,曝露
exposure meter 曝光仪
exposure suit （紧急迫降时水中用的）抗浸御寒服,海上救生服,防护服,防水外衣
exposure time 曝光时间
express stitcher 高速纡缝机
express stripes 经面斜纹蓝白条棉工装布
ex quay 码头交货价
exquisite 服饰过于讲究的男子,花花公子
exquisite craftsmanship 技艺精湛
ex railway 铁路旁交货
ex ship 船上交货
extended facing 连挂面
extended heel 加大袜跟,扩大袜跟
extended line 延长线
extended shoulder 宽肩式,过肩袖
extended shoulder sleeve 过肩袖
extended sleeve 过肩袖
extended snap 外伸按扣
extended tab 腰夹宝剑头,裤腰宝剑头;

活动扣袢
extended waistline facing 连腰里的腰头
extended wrist 过腕袖
extended wrist sleeve 过腕袖
extended Y-heel 扩大Y形袜跟,扩展Y字形袜跟
extenders 延伸段
extensibility 伸长性,延伸性
extension 范围尺寸;伸长
extension closing 门襟,叠门
extension cuff 加长袖口,延伸袖头
extension in-seam pocket 袋口和衣片相连的内缝插袋
extension line 尺寸辅助线,引出线
extension plate 加长板
extensor 伸肌,外部
exterior image 外观形象
external form 外型
external malleolus 外踝
external malleolus point 外踝点
external packing 外包装
extra 额外,附加,特等品
extra backing 特别底布;特别加固
extractor (洗衣)脱水机
extract printing 拔染印花
extra extralarge (XXL) 超特大号
extra-fancy blue gem 艳美蓝钻
extra-fancy gem blue 格外花色蓝钻
extra filling fabric 特别纬纱织物
extra fine gauge 最细针距
extra fine yarn 特细支纱,高支纱
extra flare 附加裙摆
extra-fort 覆毛边嵌条
extra fullness 特等丰满度
extra heavy-duty 特厚,特重;特重织物
extra-heavy gabardeen 加厚华达呢
extra heavy material 特厚缝料
extra high pile fabric 超高毛圈织物
extra large (XL) 特大号
extra long shank button 特长柄纽扣,加长柄纽扣
extraneous risk 外来风险
extra pale 特淡
extra pocket 附加小袋
extra small size 特小号
extra soft elastic 超柔软松紧带
extra super carpet 优级双面提花地毯
extra tall 特大号
extra thick material 特厚缝料
extravaganza look 离奇款式

extravaganza style 离奇款式
extravehicular visor 宇宙飞船外的脸盔
extra warp fabric 特加经纱织物,特纬织物
extra white single jersey 特白汗布
extra wide cloth 超宽幅布,超阔幅布,特阔幅布
extreme dimension 极限尺寸
extreme erect figure 严重的凸胸体型
extreme low shoulder 严重高低肩
extreme shoulder low 严重高低肩
extreme stoop figure 严重的驼背体型
extreme-volume silhouetee 过分宽大廓线
extremities 四肢
extroversion 外向性
extruded fabric 挤压法非织造布
extruded foxing 挤出围条(鞋)
ex warehouse 仓库交货;仓库交货价
ex wharf 码头交货;码头交货价
eye 眼(针眼,扣眼等),孔,圈,环,索眼;眼睛,眼眶,眼圈
eye ball 眼球
eye black 黑色眼睑膏,睫毛膏
eyebrow arch 眉毛美容法
eyebrow brush 眉毛刷
eyebrow color 眉墨
eyebrow pencil 眉笔,眉毛笔,眼影笔
eyebrow shaper 眉梳
eyebrow tweezers 眉钳
eye buttonhole 圆头扣眼,圆头纽洞,圆头纽孔
eye-catching dress 引人注目的女服
eye cosmetics 眼影膏
eye dropper 滴水器(烫衣用)
eyeful gloves 有孔手套
eye-glass 单片眼镜
eye-glasses 眼镜,夹鼻眼镜
eye hole 眼眶,眼窝;穿绒圆孔眼
eye & hook 风纪扣
eye lash 睫毛
eye lash curler 睫毛夹,卷睫毛器
eyelet 纽孔,洞眼,网眼;孔眼锁缝;穿绒圆孔眼,镶边小圆孔;小孔扣眼,鞋眼;鸡眼
eyelet brassiere 有孔胸罩
eyelet button 带眼纽扣
eyelet buttonhole 鸽眼纽扣;金属鸽眼,孔眼,圆头纽孔,凤眼
eyelet buttonhole sewing machine 圆头锁眼机,凤眼机

eyelet buttonhole without taper bar 无套结圆头纽孔
eyelet buttonhole with taper bar 有套结圆头纽孔
eyelet buttonholing 锁凤眼,圆头扣眼
eyelet circular knitting machine 圆形网眼袜机
eyelet collar 鸡眼领;针孔领(别针固定领端)
eyeleteer 小孔钻;穿孔钻
eyelet embroidery 穿孔刺绣,网眼刺绣,孔眼绣
eyelet end 扣眼收针;凤眼,圆孔,扣眼圆孔,锁眼收尾
eyelet end buttonhole 锁眼收尾的纽孔
eyelet end buttonhole sewing machine 圆头纽孔收尾缝纫机
eyelet end buttonhole with fly bar 有遮盖套结的圆头纽孔
eyelet fabric 网眼织物,网眼布
eyelet fastening machine 钉鸡眼机
eyelet hole 圆头纽孔,纽孔;网眼,穿绒圆孔眼,金属鸡眼,圆头扣眼,圆头纽洞
eyeleting machine 锁圆头纽孔缝纫机;皮革扣孔机
eyelet knitted fabric 菠萝组织针织物
eyelet lace 网眼花边
eyelet machine 圆头锁纽孔缝纫机;圆头锁眼机
eyelet punch 羊眼冲头
eyelet setter 孔眼钳
eyelet shank button 带孔纽柄纽扣
eyelet stay 眼衬(鞋)

eyelet stitch 网眼绣,小孔绣;锁眼针迹
eyelet stitch buttonholing 锁凤眼纽孔,锁圆头纽孔
eyelet stitching machine 凤眼机,圆头锁眼机
eyelet tab 鞋眼衬片
eyeletting machine 锁圆头纽孔缝纫机
eyelet toe closing 缠绕法缝头
eyelet weft knitted fabric of spun silk 绢丝网眼针织物
eyelet work 网眼刺绣,穿孔刺绣,孔眼绣;打孔眼
eyelid 眼睑,眼皮
eye line 眼线
eye liner 描眼膏;眼线笔
eye looper 有眼弯针
eye make-up 眼部分妆
eye patch 眼罩,遮眼罩
eye pit 眼窝
eye reversing lever 有孔倒缝杆
eye setter 孔眼钳
eye shade 眼睑膏;眼罩,遮眼物
eye shadow 眼睑膏;眼影
eye shadow color 眼影色
eye shiner 眼睛增光剂
eyes & hooks 钩棒扣,乌蝇扣
eyeslit 眼形扣洞
eye socket 眼眶,眼窝
eye wear 眼镜类(总称)
eye winker 睫毛
ezor 伊佐腰布

F

FA 花
fabala 衣裙边饰
fabric 布料,面料,织物,布匹;结构
fabric abrasion resistance 织物耐磨性
fabric absorption 织物吸湿性
fabric/accessory inventory system 布料/辅料存货系统
fabric air permeability 织物透气性
fabric and accessory 原辅料
fabricated materials 已加工材料
fabricated part 服装裁片,服装裁片单件
fabricated textile 仿造织物
fabrication line 加工作业生产线
fabric bending resistance 织物抗弯性
fabric block garment 拼块服装,香槟衫
fabric bolt 卷板布匹
fabric breathability 织物透气性
fabric bulkiness 蓬松度
fabric care 织物保管,织物维护
fabric carrier 布带裤,布裤
fabric characteristics 织物质感
fabric clearance 缝料通道,织物松动量,织物活动范围
fabric clippings 碎布屑
fabric closeness 织物紧密度
fabric color fastness 织物染色牢度
fabric combinations 面料搭配
fabric comfort 织物穿着舒适性,织物舒适性
fabric composition 布料成分,织物组成
fabric configuration 织物结构
fabric consumption 估料,耗料,用料量
fabric costing 面料成本,织物成本
fabric count 织物经纬密度
fabric cover 布面丰满度,织物覆盖系数
fabric-covered button 包纽,布包纽扣
fabric-covered button machine 包纽扣机
fabric covered shoes 包布鞋
fabric covering factor 织物紧度
fabric crease elasticity 织物折皱弹性
fabric crease resistance 织物防皱性
fabric crease retention 织物保褶性
fabric cutting machine 裁布机
fabric cuttings 碎布
fabric defect 织物疵点,布疵

fabric density 织物密度
fabric design 面料设计,衣料设计,织物设计
fabric dimensional stability 织物尺寸稳定性
fabric distortion 织物变形
fabric draft 织物设计草图
fabric drapability 织物悬垂性
fabric draping factor 织物悬垂系数
fabric durability 织物耐用性
fabric economy 用料量
fabric edge 布边,缝料边,料边
fabric edge guide 导布边器
fabric edge sealer 封边器
fabric elasticity 织物弹性
fabric elasticity of crease 折皱弹性
fabric excess 预褶份
fabric extension 织物伸长
fabric external appearance inspection 织物外观检验
fabric-faced waistband 本料衬里的腰头
fabric fall 布料悬垂性,织物悬垂性
fabric-feed 送料,送布
fabric finish 织物整理
fabric flammability 织物可燃性
fabric flatness 织物平挺性
fabric foam laminates 织物泡沫层压制品
fabric for dress 连衣裙料
fabric for protect radiation heat 防辐射热织物
fabric for raincoat 俄罗斯防雨布
fabric for suit 套装料
fabric gloves 布手套
fabric grain 布纹理,织物纹理
fabric guard 护布,拦布
fabric handbag 布制手提包
fabric handle 织物风格,织物手感
fabric handling 织物风格
fabric heat resistance 织物耐热性
fabric heat transmission 织物导热性
fabric identification 面料识别
fabric inset waistband (连衣裙的)镶嵌布腰头
fabric inspecting table 验布台
fabric lace 布制花边

fabric layer 布层,面料层
fabric length 织物匹长
fabric light permeability 织物透光性
fabric liveliness 织物滑爽性
fabric loop 布环,布袢
fabric management system 织物面料布边处理系统
fabric material 面料,衣料
fabric melt-hole resistance 织物抗熔孔性
fabric memory 织物记忆性能
fabric mildew resistance 织物防霉性
fabric moisture transmission 织物吸湿性
fabric of judo wear 柔道服织物
fabric of medium length fiber 中长纤维织物,中长仿毛花呢
fabricoid 充皮布,漆布,仿皮布
fabric pattern 织物花纹,织物图案
fabric patterning 织物花纹设计
fabric pilling 织物起球
fabric pores 织物孔隙
fabric porosity 织物孔隙度,织物透孔性,织物紧密度
fabric property 织物性能
fabric quality 织物品质,织物质量
fabric rating standards 织物分级实物标样
fabric resilience 布料回弹性
fabric resiliency 织物回弹性
fabric roll 布卷
fabric serviceability 织物服用性能
fabric sett 织物经纬密度
fabric sewability 织物缝纫性能
fabrics for wearing 穿着用织物
fabric shears 面料剪力
fabric sheen 织物光泽
fabric shoes 布鞋
fabric shrinkage 织物缩水率,织物收缩率
fabric simulation 织物仿真
fabric-skin 隔热织物,护肤织物
fabric slitting 织物剖幅
fabric's memory 织物回复性
fabric snagging 织物钩丝
fabric softener 织物柔软剂
fabric specification 织物规格
fabric streak 织物条花疵
fabric stretch 织物弹性,织物伸长回弹性;织物伸长
fabric structure 织物结构
fabric style 织物风格
fabric surface 织物表面

fabric swatch 面料样本
fabric tab 布带
fabric texture 织物结构;织物质地;织物组织
fabric thickness 织物厚度
fabric to fabric bonding 织物与织物粘合
fabric to fabric lamination 织物与织物的层压黏合
fabric tube 管状织物,圆筒形针织物,筒状织物
fabric type 面料种类
fabric vapor permeability 织物透汽性
fabric warmth retaining 织物保暖性
fabric wash 布料洗水
fabric washability 织物耐洗性
fabric wash and use 织物免烫性
fabric wash and wear 织物洗可穿性
fabric water permeability 织物透水性
fabric water proofing property 织物防水性
fabric wearing test 织物穿着试验
fabric weight 织物重量
fabric width 织物门幅,织物幅宽,布宽,幅宽,门幅
fabrography 网印技术
face 正面,布面;表面;脸,脸面
face-bonded fabric 面黏合织物
face-bonding 面黏合
face brush 脸刷
face button 面扣
face cloth 光面女式呢,经面织物;方巾,面巾,洗脸毛巾
face cream 面膏,雪花膏
faced cloth 背面有附加纱线的织物,经面织物
faced fabric 黏合布面料
faced hem 贴边下摆
faced neckline 贴边领口
faced opening 装贴布的开襟;装贴布的开衩
faced placket 面对面开口,装贴布的开襟,装贴布的开衩
faced sleeve 装贴边袖
faced slit cuff opening 贴边袖衩
faced waistline 贴边腰线
face fabric 黏合布面料
face-finished fabric 正面整理织物
face-finished woolen 光洁毛织物
face-finishing 绒面整理;呢面整理
face flannel 洗脸毛巾

face fungus 胡须
face goods 正面整理织物;光洁毛织物
face guard 护面具,面罩
face lift 消皱膏;整容,整容术
face-lifting 整容
face mask 面罩,面具,假面具
face pack 面膜
face plate （缝纫机上的）面板,门盖,台板,护面具,屏面
face plate complete 台板组
face powder 面粉,扑面粉,铅白粉,香粉
face powder brush 粉刷
face protector 防护面罩
facer 贴边工,镶边工
face seam 毯面拼接
face shape 面形
face shield 防护面罩,面罩
face side 布面,正面
face silk 面子绸料
face stitch 面缝线迹,正面线迹
faceted briolette 三角刻面梨形(珠宝)
faceted girdle 刻小面腰缘(钻石)
faceted pear-shape 有刻面梨形(珠宝)
faceted stone 刻面翻头宝石(珠宝)
face texture 布面质地,布面结构
face-to-back shade deviation 染色正反面色差疵
face-to- back variation 染色正反面色差疵
face to face 合掌铺料
face-to-face bonded-pile carpet 双层黏合绒头地毯
face-to-face carpet 双层地毯(可割成两个割绒地毯)
face to face for directional spreading 对合铺料,和合翻身铺料法
face to face one way spreading 双幅对称铺料
face to face spreading 来回和合铺料法,面对面铺布
face towel 面巾,毛巾
face up spreading 布面向上铺布
face up well 面朝上好看(钻石)
facial 面部按摩
facial bone 面颅骨
facial expression 面部表情
facial make-up 面部化妆
facial masks 脸部蒙具
facial ointment 擦脸油膏
facial pack 润肤膏(美容)
facial play 面部表情

facial skin 脸部皮肤
facial tissue 擦面薄纸
facial towel 面湿巾
facile closing bag 轻便提包,易开合提包
facilites 工具
facility layout 设备平面布置
facing 贴条,镶边,贴边,袋垫;领面;挂面;军装上的领章;袖章
facing-casing 贴边作抽带管
facing-in-one 连贴边
facing leans out of front edge 止口反吐,里料外露
facing leather 鞋面滚边皮
facing line 挂面线,滚条线,镶边线
facing lining 挂面夹里
facing pattern 挂面纸样
facings 军装上的领章、袖章及其他装饰品,(军装上领子或袖口等处的)异色镶边,鞋帮耳护面
facing's facing 贴边的翻折边
facing silk 面子绸料,领绢,滚条绸
facing sleeve 袖口贴边,袖口花边
facing strap 贴门襟
facing tab 贴边袢
Faconnable 法松那布勒(法国时装设计师)
faconné 精巧小花纹织物,提花布,捆花布,小花纹织物,凸花纹
faconne taffeta 花塔夫绸
faconné velvet 烂花丝绒
facsimile 复制,摹写
facsimile fabric 仿真织物
facsimile printing 摹真印花
factory 工厂
factory automation 工厂自动化
factory building 厂房;厂房建筑
factory cloth 坯布
factory cotton 本地棉坯布
factory design 厂房设计
factory layout 厂房布局
factory overhead 车间管理费
factory-reeled silk 厂丝
factory sales department 工厂门市部
fad 风尚,狂热流行,流行快潮;流行一时的服装,时髦装束;一时爱好,一时风尚;一时流动商品
faddish 喜欢赶时髦的
faddist 趋附时尚的人
fade 褪色
fade away 褪色

faded denim 淡色薄斜纹布,褪色劳动布;暗淡蓝
faded look denim 退色型牛仔布;退色型劳动布
faded look type 陈旧风格,褪色风格,外观陈旧型
faded rags 褪色碎呢
faded rose 暗淡玫瑰红
faded type 褪色型
fade-in 渐显
fadeless 不褪色的
fade-out 渐隐
fade-out finishing 返旧处理,褪色处理
fade-out jeans 褪色牛仔裤
fade proof 防褪色
fading （布料、服装的）褪色
fading sample card 退色样卡
fad style 一时风行的款式
faga 丝腰带
fagara silk 樗蚕丝
fag end 散口边（布匹头尾）,绳索的散端;零头布,绳带
faggot 装饰线迹
faggot filling stitch 集束刺绣针迹
faggoting 装饰线迹接缝,折边曲折连缝;抽纱绣
faggoting effect 条子效应
faggoting machine 装饰线迹接缝机
faggoting seam 装饰线迹接缝
faggoting stitch 束心线迹,装饰线迹
faggoting trim 装饰修边
faggot stitch 束心线迹
faggot stitching 装饰性线缝
fagot 装饰线迹
fagoting 抽绣,抽纱绣;抽纱法
fagoting seam 装饰线迹接缝
fagoting stitch 连缀缝;连带编织绣;装饰线迹
fagot machine 装饰线迹接缝机
fagot stitching 装饰线缝
fagot stitching machine 装饰线迹缝纫机
fagotting machine 装饰线迹接缝机
faience 彩釉瓷（青光蓝色）
failed test sample 不合规格的样品
faille 菲尔绸,罗缎,绫纹绸,线绢,绨（丝棉交织物）
faille chinoise 全丝罗缎
faille cotton 棉纬罗缎,棉纬绨
faille crepe 罗缎绉
faille de chine 全丝罗缎

faille francaise 法国罗缎
faille marquise 里昂花缎（经摩擦轧花整理）
faille ribbon 罗缎丝带
faille taffeta 罗缎塔夫绸
failletine 薄罗缎
failletine moiré 波纹罗缎
failletine ribbon 薄罗缎丝带
faillette 19世纪末光面棱纹毛呢;圆点纹（用于薄织物）
failure 故障,失效
failure diagnosis 故障诊断
faint color 淡色
faint coloration 褪色
fair 商品展览会,商品交易会,博览会
fair average quality (FAQ) 中等品
fair color 色泽适中
fair drying 适干
fair hair 金色头发
fairing 光顺
Fair Isle (F&I) 费尔岛式,费尔岛式多色几何图案的针织品
Fair Isle sweater 费尔岛式毛衣（以多色几何图案花样设计）
fair stitching 正针法
fairway green 高尔夫球场草绿色
fair weather article 晴天服用物品
fairy costume 仙子服装
fairy tale flowers dress 仙境花女装
faja 宽腰带（南美）,南美色彩鲜艳的宽腰带
fake cheap fashion 顽世装
fake fashion 仿制时装
fake flap 假袋盖
fake fur 人造毛皮
fake fur cape 人造毛皮披肩
fake fur coat 仿皮大衣
fake fur knitted fabric 人造毛皮针织物
fake jewelry 假珠宝
fake pearl 人造珍珠
fake pocket 假袋
falaise 法国哔叽
falbala 衣裙荷叶边
falcon and uraeus headdress 鹰蛇头饰
falcon head covering 兀鹫头饰
falda 白丝绸拖裙服装
faldetta 福尔迪达外衣;齐腰彩色塔夫绸女斗篷;马耳他土著头饰
falie grijn 法利格林罗纹呢;仿驼毛棱格呢

fall 面纱;绉领,花边大翻领,领子翻下部分;(外衣的)宽下摆;大脚裤;(旧时)领带;领角衬;(织物)悬垂性;长假发,悬垂式假发;领插角片,领口片;向下飘拂的服装
fal-lal 服饰品
fallal 花饰,(服装上的)装饰品;华而不实服饰
fall clothes 秋冬服装
fall coat 秋季外衣
fall-drop repeat 间距位移(图案单位)
falling band 花边大翻领
falling band collar 带花边男式大翻领;范达克领
falling collar 宽垂领
falling ruff 不规则褶皱的无浆轮状领,垂皱领
falling tucker 盖在胸前的悬垂式抵肩
fall leaf 落叶色
fall on 叠色
fall-on print 叠色印花
fall-on printing 叠色印花
fallow 淡棕色,浅棕色
fall plate fabric 压纱织物
fall plate lace 压纱花边
falls 男裤前门襟
fall's check 仿席纹色格布
fall trousers 水手裤
fall weight 中厚春秋服面料
fall wig 垂假发
fall-winter fashion 秋冬时装
fall wool 秋毛
false bosoms 假胸
false bottom 假底
false cable 仿绞花,假绞花
false calves 男子衬腿
false color 不坚牢色,伪色,伪彩色
false-colored diamond 假彩色钻石
false crease 假折缝
false cut through 假开剪
false eyelashes 假睫毛
false face 假面具(万圣节用)
false fly 暗钮牌
false fur 仿毛皮
false gauze wool fabric 假纱罗毛织物
false gown 少女式服装;法国大革命时期女装
false hair 假发
false hanging sleeve 悬饰袖
false hem 假缝,暗缝;有贴边的下摆,假下摆
false hemp 印度麻,菽麻
false hips 假臀;垫臀衬裙
false plain 花式平素织物(外观介于平素和花式之间)
false pocket 假袋
false rump 假臀,臀部边饰
false seam 线缝不直
false selvage 假边
false sisal hemp 类西沙尔麻,类剑麻
false sleeve 假袖
false stitch 假缝;线缝不直
false tussah 柞蚕丝,中国北方野蚕丝
false twill 假斜纹织物
false twist yarn 假捻纱
false welt shoes 假沿条鞋
falsie 妇女衬胸(胸罩内的乳房衬垫)
falsies 海绵胸罩;(胸罩的)衬垫物,假乳房;假发;假胡须
falzen 毡
family color coordination 同系色调和,同族色调配
family ring 家庭戒指
family uniform 家居服
famis 法米斯绸
famous brand 驰名品牌,名牌,著名商标
famous brand and high class fashion 名牌高档时装
famous brand fashion 名牌服装
famous brand high-quality new product 名优新产品
famous product 名牌产品
fan 扇,扇子
fan-backed dress 扇形后身晚礼服
fanchon 花边装饰头巾;花边装饰女帽;发网
fanchon bonnet 头巾式无檐女帽
fanchon cap 户内薄纱小女帽
fancies 花色货品,花式织物
fanciful costume 奇异服装,奇装异服
fan collar 扇形领
fancy 花色,花式;时新织物;时兴纺织品
fancy air textured yarn 花式喷气变形丝
fancy and figured twill 花式斜纹
fancy apron 花式围裙
fancy article 花式货品,花式织物
fancy back 花色布背衣料,双面异花色织物;装饰性背部设计
fancy ball dress 化装跳舞服
fancy belt 花式饰带

fancy blanket 提花毯,花式毯
fancy braid 花色编带
fancy button 时尚扣,花色扣
fancy buttonhole 花式扣眼,花式纽孔
fancy buttonhole filling stitch 织网绣
fancy carcoat 花式卡曲衫
fancy carton 彩色纸盒
fancy chambray 花式钱布雷布
fancy check 花式格子花纹
fancy cloth 花色布,装饰布,花式织物
fancy coat 花式外套
fancy collar 花哨领,花式领,花俏领
fancy cotton braid 花式棉编带
fancy cotton cloth 花式棉布
fancy crepe 花色绉
fancy crepe silk 小花绉
fancy cuff 花式袖头
fancy cut 花式切磨
fancy cuts 花色磨翻类(珠宝)
fancy decoration stitch 花式装饰线迹
fancy design pattern 花式图案,装饰花纹,装饰花
fancy diagonals 花式斜纹
fancy diamond 花色钻石,花式钻石
fancy dimity 变化麻纱,柳条纱罗
fancy dinner suit 变化式半正式男式晚礼服,变化式女式正餐套装,变化式礼服
fancy dobby weave 花式多臂提花织物
fancy dot 花式点纹
fancy doubled yarn 花式合股线
fancy dress 花式服装,花哨服装,化装服装,化装舞会服,舞蹈服
fancy-dyed fabric 花式染色织物
fancy fabric 变化组织[针]织物;花式[针]织物
fancy flannel 花式法兰绒
fancy flap 花式袋盖;花式前襟
fancy goods 花哨商品,花式服装,花式货品,花哨服饰;花式织物;精品
fancy goods store 花哨商品店,精品屋
fancy hairdo 别致女子发式
fancy hand knitting yarn 花式绒线
fancy hose 花袜
fancy jacket (式样)奇巧夹克,花式夹克,奇特夹克
fancy knit sweater 花式编织毛衣,织花毛衫
fancy knitted fabric 花色针织物,花式组织针织物
fancy knitted sweater 织花毛衫
fancy knitting yarn 花式绒线

fancy lace 花式花边织物
fancy laying-in knitted fabric 花式衬垫针织物
fancy leather 花色皮革,花式皮革
fancy line 花辫绳
fancy lining 花色里子布,花式条纹里子布
fancy lustrine 花式闪光绸,花式袖里绸,花式斜纹袖里布
fancy mandarin button 花结扣,盘花扣,盘纽
fancy mesh fabric 花式网眼织物
fancy monk's cloth 蜂窝毛巾布
fancy net 提花网眼纱,提花绢网
fancy oatmeal 刺绣用黄褐色十字布
fancy ornamental stitch 花式饰缝线迹
fancy overcoating 花式大衣呢
fancy overcoating woolen 花式大衣呢
fancy pattern 非传统花样,奇特花样(图案)
fancy perfume 幻想型香水
fancy pique 花色凹凸织物
fancy plain 花式平纹织物
fancy plated hose 花式添纱袜
fancy plating 花色添纱
fancy pleats 花式褶
fancy plied yarn 花股线
fancy plush 花式长毛绒
fancy polyester crepe 花涤纶绉
fancy raised knitwear 花式绒毛衫
fancy rib 花罗纹,花式罗纹
fancy satin brocade 花色锦缎
fancy seam 花式缝
fancy selvage 花式布边
fancy shade 杂色
fancy shammy 花麂皮针织物
fancy sheer 花色薄绸
fancy shirt 花式礼用衬衫
fancy shirting 花色衬衫料
fancy shop 花哨商品店,精品屋
fancy socks 花袜
fancy spiral yarn 花色螺旋纱,花色加捻纱
fancy spun silk yarn 竹节绢丝
fancy stitch 花式线缝,花式线迹,装饰线迹;花式组织
fancy stitching 花式线缝
fancy stripe 花式条纹布,花色条子布
fancy style 奇特款式
fancy suit 花式套装
fancy suiting 花呢,西服料,花式套头衣料
fancy taffeta 花式塔夫绸

fancy thread 花色线,花式线
fancy top 花袜口
fancy trimming 花式饰带
fancy tuck 加饰活褶
fancy tuxedo 变化小礼服,变形半正式礼服,花式夜小礼服
fancy twills 花式斜纹
fancy twisted silk fabric 花线春
fancy venetian 色直贡
fancy vest 花式背心,异料背心,替换背心,变化背心
fancy vesting 花式背心料
fancy waistcoat 花式背心,异料背心
fancy weave 花式组织
fancy weft knit fabric 花色纬编针织物
fancy woolen 粗纺花呢,粗花呢
fancy work 刺绣品,钩编织品,钩针织物,编织品
fancy worsted 精纺花呢
fancy woven piqué 花式凹凸织物,花式席纹布
fancy wrap hose 吊线花袜
fancy yarn 花式纱线,花式丝线
fancy yarn fabric 花色线织物
fancy yarn fancy suiting 花式纱线花呢
fancy yarn knitwear 花式线羊毛衫
fancy yarn union brocade 华锦绸
fancy yarn woolen sweater 花式线羊毛衫
fancy yarn wool fabric 花式线毛织物
fan design 扇形花纹
fan dress 扇形连衣裙
fang-kong gauze 方空纱,方孔纱
fan hoop 锥形环撑衬裙
fanhua damask twill 繁花绢
fanny belt 臀部带
fanny pack 带有腰袋的彩色饰物,腰袋
fanny sweater 盖臀毛衣;法式毛线衣
fanny wrap 范尼腰带;盖臀围布(设计服装款式用)
fanny wrapper 臀部缠绕物
fanny wrap shawl 缠臀围巾
fanon (主持弥撒时)神父(或副主祭)左臂佩戴的饰带,旗布;天主教教皇穿的披肩式祭袍
fanon collar 法诺领
fanpak 平纹法兰绒
fan pleats 扇裥,扇形褶裥
fantailed hem 扇尾型后衣摆
fantail hat 扇尾帽
fantaisie 花式织物

fantasia 意大利花式织物
fantasy clothing 幻想型服装
fantasy costume 奇特装束
fantasy dress 幻想派服装
fantasy line 幻想型
fan tucks 放射性活裥
FAQ(free alongside quay;free at quay) 码头交货
farasdange 印度手织优质棉布
farash cloth 印度手织铺地板布
farasia 中东长袍,北长袍
far-away collar 荡开领,远离领,离颈领
far-away neckline 荡开领口,远离领领口
fard 东印度印花铺地布
fargrant brocade 香香锦
far-in 保守的服装
far-infrared drying 远红外[线]烘燥
far-infrared fiber 远红外纤维
far-infrared fiber and textile 远红外纤维及其纺织品
far-infrared ray 远红外线
far-infrared ray radiated fabric 远红外织物,远红外辐射织物
far-infrared sensitive fiber 远红外敏感纤维
farmer blouse 农民罩衫,女绉褶上衣,有皱褶的宽松上衣
farmer look 农民风格款式,农民时式,农民风貌
farmer's collar 农民领
farmer's satin 仿真丝缎子,棉或人丝里子布,纬面缎衬里织物,黑色棉衬布
farming fabric 农用织物
farm vijaya 罗布奇棉
farm wool 圈饲法生产的羊毛
farm work cloth shoes 农田鞋
farm work rubber boots 农田胶靴
farous 伊拉克腰布
far-out clothes 奇装异服
far-outer 反传统者
farrell prolific cotton 法雷尔棉
farthingale 裙撑,裙环;裙环裙,粗亚麻女衬裙
farthingale silhouette 裙撑型(轮廓线条)
farthingale skirt 法辛盖撑裙(16～17世纪)
farthingale sleeve 裙撑袖
fascia 带,饰带,绷带;裹布
fascinator 头帕,网眼毛披巾
fash 碎呢料

fashion 花色,花式,流行,时髦,时尚,风格;时新款式,时新式样,流行式样;时装
fashionable 时髦人物
fashionable button 带柄纽扣,时髦纽扣
fashionable clothes 时髦服装
fashionable dress 时装
fashionable dress performance 时装表演
fashionable gloves 时款手套,时尚手套
fashionable length 时髦衣长
fashionable outwear 时髦外衣
fashionable style 花色新颖,流行款式,流行风格
fashion accessories 时装配饰
fashion administrator 时装管理员
fashion advertising 时装广告
fashion adviser 时装顾问
fashion and accessories 流行服饰
fashion androgyny 男式女装
fashion announcement 时装发布
fashion apparel 流行服装
fashion appearance 时装外观
fashion art 流行服饰艺术
fashion artist 时装设计师,时装艺术大师
Fashion Avenue 第七大街(美国纽约市美国时装业中心所在地)
fashion babies 时装娃娃
fashion belt 时尚腰带
fashion book 流行服饰书籍,时装书籍
fashion boots 新款式靴子(总称),时装靴
fashion bulletin 时装公报
fashion business 服饰业,时装业,流行服饰行业
fashion button 时尚扣,时装扣,时款纽扣
fashion buyer 流行服饰采购员,时装采购人员
fashion calendar 时装发布日程表
fashion cap 时款帽,时尚帽
fashion catalogue 时装目录册
fashion centre 时装中心
fashion circle 时装界
fashion climate 流行服饰气候
fashion clinic 流行服饰诊所,流行会诊,时装讨论会
fashion clothes 时髦衣服
fashion clothing 时装
fashion cognoscente 时装鉴赏家
Fashion Collection 时装发布会,时装博览会
fashion color 流行色

fashion color area 流行色范围
fashion color cycle 流行色周期
fashion color dissemination 流行色传播
fashion color forecast 流行色预测
fashion color research organization 流行色研究机构
fashion column 时装专栏
fashion consultant 流行服饰顾问,流行服饰咨询机构,时装顾问,服饰顾问
fashion contemporarist 服饰上有时代感的人,流行款式引导人
fashion coordinator 流行款式协调者,时装调度员,时装流行导向人员
fashion copywriter 流行服饰撰稿员
Fashion Coterie 纽约小集团时装节(美国)
fashion creator 创造时尚者,时装设计师
fashion culmination 时装全盛期
fashion culture 服装文化
fashion currency 时装流行
fashion cycle 流行周期,时髦周期,时装流行周期
fashion design 时装设计
fashion designer 服饰设计师,时装设计师
fashion designer stylist 新款式设计师
fashion design master 时装设计大师
fashion director 流行服饰指导员,流行总监,时装导向者,时装调度员,时装引导员
fashion dolls 时装娃娃
fashion drawing 服装效果图,时装款式图
fashion dress 时装
fashion dynamics 时装动态
fashioned clothing 风行服装
fashioned foot 成形袜脚
fashioned hosiery 成形袜子,成形针织品
fashion editor 流行服饰编辑,时装编辑,时装杂志编辑
Fashion Editors Club (FEC) 流行服饰编辑俱乐部
fashioned knit fabric 成形针织物
fashioned outerwear 成形针织外衣
fashioned seamless hosiery 无缝成形袜
fashioned thigh 成形袜腿
fashioned V-neck 收放针V形领口
fashion element 流行要素,流行因素
fashion fabric 时髦面料,时装面料,流行面料

fashion factory 时装厂
fashion festival 时装节
fashion flash 流行快讯
FASHION FOLIO INTERNATIONAL 《流行服饰世界》(英)
fashion follower 流行追随者,时装追随者
fashion forecast 流行预测
fashion forum 时装讨论会
fashion garment 时装
fashion glasses 装饰眼镜
fashion goods 流行服饰品
Fashion Group 时装集团(1931年创立于纽约)
Fashion Group International Inc 国际时装集团公司
fashion handbag 时款手袋,时尚手袋
fashion helmet 新款式的头盔
fashion history 服饰史,服装史
fashion house 女服店,女装店,时装公司,时装商店
fashion idea 时装概念,流行意识
fashion ideal 流行观念
fashion illustration 服饰图,时装画,流行服饰画,时装画技法
fashion illustrator 流行服饰画家,时装画画家,时装图样师
fashion image 流行概念,流行印象,时装形象
fashion individualization 时装个性化
fashion industry 流行服饰产业,时装工业
fashioning 成形,收放针,收针
fashioning mark 时装标记
fashioning marks 收放针花
fashion innovator 流行开创者
Fashion Institute of Technology (F.I.T.) 纽约时装学院(美)
fashion items 流行服饰品
fashion jacquard sand crepe 时新绸
fashion jeans 流行牛仔裤,时款牛仔裤
fashion jewelry 流行首饰,时装饰品
fashion journal 时装杂志
fashion journalist 时装记者
fashion knit 针织领带绸,针织领带织物
fashion leader 流行带头人,流行款式导向者,时装领袖
fashion life 流行寿命,流行周期
fashion life cycle 时装流行周期
fashion line 成形线,收放针线迹,流行式
fashion made in refuse 再利用服装
fashion made of bamboo 竹制时装
fashion made of refuse (利用废料制的)垃圾时装
fashion magazine 流行服饰杂志,时装杂志
fashion map 时装图表
fashion mark 成形星
fashion marketing 时装销售
fashion marks 成形记号
fashion mart 时装市场
fashion merchandise director 流行总监,时装流行总监
fashion merchandising 时装销售
fashion merchant 时装经营者
fashion message 时装信息
fashion-minded children 时尚感儿童
fashion mode 流行模式,时新式样,流行
fashion model 时装模特儿
fashion monger 买卖时髦商品的商人,时装商,赶时髦的人
fashion newness 时装时新感
fashion news 流行信息
fashion obsolescence 时装衰退,时装逐渐过时阶段
fashion oriented clothing 时装
fashion oriented clothing manufacturer 时装裁缝,时装制造者
fashion paper 时装样本
fashion parade 时装表演,时装展览
fashion pattern book 服饰图形,流行服饰图刊
fashion performance 时装表演
fashion period 流行周期
fashion photograph 时装照
fashion photographer 流行服饰摄影师,时装摄影师
fashion photography 时装摄影
fashion picture 时装画
fashion plate 穿着时髦者,前卫流行者,服装穿着极时髦的人;时装图片,流行服饰画,流行服装图,时装图样;时装样片
fashion presentation 时装表演,新装发布会
fashion press 流行服饰新闻报道,流行款式报道,时装媒体,流行媒体
fashion principle 流行原则
fashion product 时尚产品
fashion promotion 服饰推销,服饰推广,流行服装促销

fashion reporter 流行服饰记者,时装记者
fashion research 流行服饰研究,流行研究
fashion retailer 时装零售商
fashion retailing 服饰零售业,流行服饰零售业
fashion rise 时装上升阶段
fashion rule 流行规律
fashion sameness 时装单调性
fashion seam 长筒袜假缝,成形缝,面缝
fashion season 时装季节
fashion sense 流行感,流行敏感度
fashion shade 流行色
fashion sheet 服饰目录单,流行服饰录单
fashion shoes 时装鞋
fashion shop 时装店
fashion show 时装表演,时装展示,时装秀,时装展示会
fashion showing 服饰展示(会);新装发表会
fashion silhouette 时装轮廓,时装轮廓线条
fashion sketch 时装效果图,效果图
fashion sketching 时装画
fashion sophistication 流行服饰精品
fashion stitch pattern 花式线迹花样
fashion story 时装报道
fashion style 流行款式
fashion stylist 时装设计师
fashion tape 装饰带
fashion technology 服装技术,时装工艺,时装工艺学
fashion technology clinic 时装工艺培训班
fashion theme 流行主题,时装趋势,流行趋势
fashion timetable 时装供销时间表
fashion trade fair 时装展销
fashion trend 流行趋势,时装潮流,时装趋势
fashion versatility 时装多面性
fashion watch 时装表
fashion week 时装节,时装周
fashion world 时装界
fashion writer 流行服饰评论家,流行服饰作家
fashion yarn 花色纱,花式毛线,流行毛线
fassement au fuseau 梭结花边
fast 牢固;不褪色
fast back 背面加固织物
fast back Marseilles 双层填芯漂白床单罩布
fast base 色基
fast color 不褪色,坚牢染料,坚牢色泽
fast color dyeing 坚牢染色
fast dyed fabric 不褪色织物
fast dyed yarn 不褪色纱线
fast-elastic deformation 急弹性变形
fastener 纽扣,揿钮,扣件,钩扣,扣合物,四合钮;拉锁,拉链;尼龙搭扣;紧固件
fastener attaching 钉纽扣
fastener cloth 搭扣布
fastener look 拉链风貌,拉链款式
fastener tape 搭扣带,拉链带
fastening 紧固件,扣合件
fastening dyed fabric 撮缬(扎缬)织物
fastening machine 钉扣缝纫机,加固缝纫机,(金属扣)钉扣机
fastening seam 保险线迹,加固线迹,加固缝
fastening stitch 保险线迹,加固线迹,叠针
fasten off 针线打结,缝牢
fasten slip stitch 贯针
fastian 棉天鹅绒;棉布灯芯绒
fastness 坚牢度;色牢度
fastness for dyeing 染色坚牢度
fastness grading 色牢度评级
fastness rating 牢度等级,色牢度等级
fastness to acid 耐酸色牢度
fastness to acid spotting 耐淡酸渍牢度;耐酸性,耐汗渍牢度
fastness to alkali 耐碱色牢度
fastness to alkali perspiration 耐碱性汗渍牢度
fastness to alkali spotting 耐淡碱渍牢度
fastness to atmospheric gases 耐大气牢度
fastness to bleaching 耐漂白牢度
fastness to bleeding 耐渗色牢度
fastness to boiling 耐沸煮色牢度
fastness to brushing 耐刷洗色牢度
fastness to chemical washing 耐化学洗涤牢度
fastness to chlorine-bleaching 耐氯漂牢度
fastness to crocking 耐摩擦脱色牢度
fastness to daylight 耐日晒牢度
fastness to dry-cleaning 耐干洗色牢度
fastness to heat 耐热牢度
fastness to hot pressing 耐热压烫牢度
fastness to hot water 耐热水牢度
fastness to ironing 耐熨烫牢度

fastness to laundering 耐机洗牢度
fastness to light 耐日光牢度
fastness to mercerizing 耐丝光牢度
fastness to milling 耐缩绒牢度,耐毡合牢度
fastness to perspiration 耐汗渍色牢度
fastness to planking 耐酸缩绒牢度
fastness to pleating 耐褶裥牢度
fastness to potting 耐沸水牢度
fastness to rain 耐雨淋牢度
fastness to sea water 耐海水牢度
fastness to soaping 耐皂洗牢度
fastness to soda boiling 耐碱煮牢度
fastness to steaming 耐汽蒸牢度
fastness to steam pleating 耐汽蒸褶裥处理牢度
fastness to sun light 光脆牢度
fastness to washing 耐洗涤牢度
fastness to water 耐水浸色牢度
fastness to water spotting 耐水渍牢度
fastness to wear 耐穿着牢度
fastness to weathering 耐气候牢度
fastnes to daylight 耐天然光牢度
fast pile velveteen 扣紧纬平绒
fast print dyestuff 坚牢印花染料
fast running 快缝
fast running vertical hook machine 直梭钩快缝缝纫机
fast selvage 光边,加固边
fast setting 快速定形
fast shade 坚牢色泽
fast to light 耐日晒牢度
fatas 金银丝饰边丝(或棉)面纱
fat colors 油溶性染料
fat figure 肥大体型,胖体型
fat form 肥胖体型
fat-free cloth 无油脂毛呢
father collar 教父领
father cut 爸爸式发型
fatigue cap 工作帽,美国武装部队军帽
fatigue clothes 军人工作服,军人劳动服,军人作业服
fatigue clothing 劳动服
fatigue dress 劳动服装,军人工作服
fatigue hat 工作帽
fatigue jeans 工作牛仔裤
fatigue look 工作服风貌,工作服款式,军工服款式
fatigue resistance 抗疲劳性
fatigue rupture 疲劳断损

fatigues 工作服,劳动服,工作裤,野外工作服;军人工作服,军人劳动服,军人作业服,军用杂役服;美国军帽(用耐磨织物制成)
fatigue style 工作服款式
fatigue sweater 法帝冈毛衣
fatigue uniform 劳动服
fatmans wear 大尺寸服装
faufil(e) [法]绗线
faulard 毛薄软绸
fault 疵点,故障,瑕疵
fault detection 疵点检测
fault diagnosis 故障诊断
fault-free cloth 无疵布,正品布
faultless cloth 无疵布,正品布
fault marking 疵点标记
fault packing 包装不良
faulty shade 错色
Fauntleroy 贵族男童装
Fauntleroy suit 梵特勒若伊套装
fausse blonde 丝绣花绢网
faussement boutonné [法]假钮,样钮
fausse valenciennes 仿瓦朗西安花边
Fausto Sarli 福斯托·萨尔利(意大利时装设计师)
Faust slippers 浮士德便鞋
fautunn 丝经毛纬棱格呢
faux camaieu [法]非一色,朦胧感配色;单色调配色
faux fur 人造毛皮
faux tweeds 仿粗花呢
faveurs 窄丝带
favors 绶带,饰带
favourite color 最喜爱的颜色
fawn 鹿毛色,浅豆灰,浅黄褐色,幼鹿棕色
fawn brown 幼鹿棕色
fawn canton 棉经毛纬斜纹防雨布
fayal lace 亚速尔群岛手编芦荟纤维花边
fayence prints 靛蓝印花棉布
fayetta 丝毛交织薄呢
fazziness 绸面起毛
fearnaught, fearnought 粗绒大衣呢;外套
fearnaught suit 粗绒大衣呢外套
fearnothing jacket 男式粗呢短上衣
fearnought 粗绒大衣呢
feasibility 可行性
feasibility study 可行性研究
feasible study report 可行性研究报告
feather 服装,服饰;羽毛;羽毛帽饰,羽

饰,羽毛饰品
feather bed 羽毛褥子
feather boa 羽毛围巾
feather bone 羽骨(妇女胸衣的撑骨);羽毛饰边绣
feather boning 羽骨撑(紧身露肩无吊带礼服的支撑物,现用锦纶丝制做)
feather brush skirt 卷羽状花边薄裙
feather calf 羽状牛皮
feather cape 羽毛披肩
feather chenille 羽毛雪尼尔线,羽毛绒绳线,羽毛纱
feather choker 羽毛首饰
feather choker plume 羽饰
feather cloth 羽毛呢
feather cord corduroy 细条灯芯绒
feather cut 卷羽发型,蓬松发型,羽毛状发型
feather edge 袜口锯齿边
feather edge braid 圈纹窄辫带
feather edging 羽毛花边
feather fan 羽毛扇
feather grey 羽灰色
feather look 羽毛款式,羽饰风格,羽饰风貌
feather pillow stuffing 羽毛枕芯
feather quilt 羽毛衣,羽毛绒衣;羽绒被
feather quilting machine 羽绒缝纫机
feather shag 长毛绒
feather smocking 羽毛缝缩褶绣
feather stitch 杨树花线迹,羽状线迹,杨树花针法;羽毛饰边绣
feather stitching 羽状线缝
feather stuffing 羽毛填料
feather style 鸟羽式发型
feather toque 羽毛豆蔻帽
feather twill 人字形斜纹,山形斜纹
feather-weight nylon coil 轻质尼龙齿链
feather-weight zipper 轻质拉链
feather yarn 羽毛纱
feature 脸型;特点,特色
feature floor 花式铺地织物
fecamp 漂白亚麻平布;本色亚麻平布
Federal Fur Products Labelling Act 联邦毛皮制品标记法
Federation International of Master Tailors (FIMT) 国际高级西装裁缝师联盟
Federico Forquet 费德里科·福尔凯(意大利时装设计师)
federitt 床用粗棉布

fedora 费杜拉帽
fedora hat 费多拉帽,美国费多拉帽,男式浅顶软呢帽
fedora lace 嵌花式针绣花边
feed (缝纫时)进给,送布
feed across regulator 横向送料调节杆;横向送料调节器
feed actuator 送布调节座,送料针距座
feed bag 圆筒平底手提袋,圆柱手提袋(皮革或帆布);饲料包
feed band 送料带
feed bar 缝纫机牙叉,送布杆
feed-cam 送布凸轮
feed connecting link 线迹密度调节杆
feed control 进布量控制
feed cut 送布切边
feed disc 送布盘
feed dog (缝纫机)送布牙
feed dog damage 狗齿痕,送布齿疵
feed dog height 送布牙高度
feed dog path 送布牙运行轨迹
feed dog without sideway vibration 无侧面振动送布牙
feed driving 送料传动
feed driving rock shaft 针距座轴
feeder 输送器,围涎(英国)
feed fork 送料叉,牙叉
feed forked connection 缝纫机牙叉,牙叉
feed forked connection slide block complete 牙叉送布滑块组件
feed forward 前馈
feed hand ratchet lever 手动送料手柄棘轮
feeding 送料,推布
feeding amount 送布量
feeding area 送布部位;送料部位
feeding cycle 送料周期
feeding device 送料装置
feeding foot 送布压脚
feeding mechanism 进布机构
feeding vest 喂食背心(胃切除病人用)
feed lifting eccentric shaft 送布凸轮轴
feed lifting mechanism 抬牙机构
feed lifting rock shaft 抬牙轴
feed lifting rock shaft crank 抬牙轴曲柄
feed lifting rock shaft crank complete 抬牙曲柄组件
feed lifting rock shaft crank roller 抬牙曲柄辊
feed motion 送布运动

feed-off arm felling 绱袖
feed-off-the-arm felling machine 逆送式曲臂[底板]平缝机
feed-off-the-arm-machine 逆送式曲臂缝纫机
feed pitch 送布间距,送料间距
feed plate 送布板
feed post 送料杆
feed ratio 送料比
feed regulator 送布调节器,送料(针距)调节器
feed regulator friction washer 送料调节器摩擦垫圈
feed regulator handle 线迹密度调节扳手
feed rock shaft 送布摆轴
feed rock shaft complete 送布摆轴组件
feed rock shaft crank 送布摆轴曲柄
feed roller 送布牙轮
feed scale 送料标尺(刻度盘,指示盘)
feed stroke 送料行程
feed-synchronized binder 同步送料滚边器
feed timing 送料定时
feed unit 送料装置
feed wheel driving lever block 送布轮传动杆滑块
feel (织物、衣料、服装的)手感
feeler gauge 测隙规,塞尺,隙片
feeling (对艺术的)感受,感觉,(艺术品的)情调,色彩感(情,觉),色感,手感
feet 呎,英尺
feetage 面积(制鞋)
fegoting seam 花式针迹接缝
feldspar 长石色
fell (衣服的)平缝;接缝;羊毛;兽皮
felled seam 对折缝
feller 缝纫机;(缝纫机的)平缝装置;合缝工,接缝工
felling 法式缝边,附边,咬口接缝,装边
felling foot 接缝压脚
felling machine 平接缝纫机,撬边机
felling marks 分匹色纬
felling stitch 折缝线迹,直针绣
fell seam 埋夹,折边叠缝,折缝
fell stitch 平接缝线迹,折缝线迹,明缲针法
felt 毛布,毛毡;毡制品
feltability 毡合性,缩绒性
felt-back elastic 毛毡底弹力松紧带,毡背松紧带
felt boots 毡靴

felt Breton 布列塔尼毡帽
felt cap 毡帽,绒帽
felt carpet 毛毡地毯
felt cleaner 毛布洗涤器,毛毯洗涤器
felt cloche 钟形毡帽
felt cushion 毡垫,毡衬
felted blanket 呢面毛毯
felted carpet 毛毡地毯,地毡
felted fabric 缩绒织物
felted flannel 双面绒棉毯
felted mattress 棉毡垫,毡垫,毡衬
felted texture 毡制品
felted wool 呢帽毡,羊毛毡
felt fabric 毡合织物,毡类,毡呢
felt floor covering 毛毡地毯,地毯
felt for shoe mattress 鞋衬毡
felt for substituting leather 代革毡
felt for wall covering 墙毡
felt goods 毡制品
felt hat 呢帽,毡帽,牛仔毡帽
felting 缩呢
felting ability 成毡性能,缩绒性
felting needle 毡合针
felting power 缩呢性,缩绒性,毡合性,毡化性
felting propensity 缩呢性,缩绒性,毡合性
felting property 缩呢性,缩绒性,毡合性,毡化性,毡缩性
felting quality 缩绒性,毡合性,缩绒质量
felt-like fabrics 毡合织物
felt mattress 衬垫毡,衬毡垫
felt pad 毡衬垫,毡垫,毡衬
felt piece 匹毡
felt proofing 防毡缩整理
felt resistance 防毡缩性
felts 毡面软底拖鞋(或便鞋)
felt shoes 毡(呢)鞋,毡靴
felt shrinkage 绒缩率,毡缩率
felt stetson 斯泰森毡帽(美国)
felt tent linen 亚麻苫布,亚麻帐篷布
felt-tip ruler 毡制粗头笔
female color 柔和色
female dual-purpose shirt 女式两用衫
female figure 女性体型
female form 妇女体型,女性体型
female pattern 凹人花纹图案
feminine 女性风格
feminine blouse 宽大女上衣,女式上衣,女式衬衫

feminine color 女性色,娇柔色
feminine dress 女性化服装
feminine ensemble 妇女套装
feminine figure 女性体型
feminine frock 女上衣
feminine hygiene 妇女卫生
feminine hygiene product 妇女卫生用品
feminine line 女性线条
feminine look 女性风貌,女性服式,女性款式,女性美型
feminine period 女性化时期
feminine sapphire 淡色蓝宝石
feminine silhouette 女性轮廓
feminine style 女性款式
femininity,feminity 女子的气质
feminization 女性化
femoralia 短裤,内裤
femur 股骨大腿
femur girth 大腿最大周长
fen 分
fencer's mask 击剑面罩
fencing 击剑服;零头布
fencing blouse 击剑衬衫;衬衫式罩衣
fencing breeches 击剑裤
fencing clothes 击剑服
fencing gloves 击剑手套
fencing jacket 击剑上衣,击剑式夹克
fencing mask 击剑护面,击剑面罩
fencing shirt 击剑衬衫,击剑衫
fencing shoes 击剑鞋
fencing suit 击剑装
fencing wear 击剑服
fencing-wear silk 击剑绸
fencing-wear silk armure 银剑绸
Fendi 芬迪(意大利皮毛公司)
fenghua twill 凤华绫
fengyi tussah crash 凤艺绸,凤艺呢
fenille bar 席纹编带
fent 零头布,短码布;坯布片;(衣衫的)领口;(裙腰上的)开口
feradje 土耳其妇女外套,土耳其贵族冬上装
Feraghan 小块波斯地毯
ferde 平纹粗棉布
Ferdinando Sarmi 费迪南多·萨尔米(美国时装设计师)
feredeza 宽松大袖斗篷,宽松大袖外套
feredje 土耳其妇女外套,土耳其贵族冬上装
fereghan 小块波斯地毯

ferlin 英国粗纺呢
ferment washing finish 酶洗整理
Fernando Sanchez 费尔南多·桑切斯(美国内衣设计师)
fern green 蕨叶绿
fern stitch 箭头针迹;蕨状绣
ferraiolo 黑圆披风(正规场合下牧师穿)
ferret (棉、毛、丝的)细带;丝绒
ferret fur 白鼬毛皮,黑足鼬毛皮,雪貂皮
ferroniere,ferronniere [法]宝石垂额的头圈
ferruginous 赤褐色
ferrule 箍,套圈
fers 固定衣服的扣环或别针
festival fuchsia 节日紫红
feston stitch 扇形边纹刺绣线迹
festoon 花彩,环状装饰物;穗边窗帘
festoon drapery 带穗花缎窗帘
festoonery 彩饰
festoon stitch 花影线迹,穗边线迹,纽孔线迹
festul 绣花红色纱罗面纱
Fettoflan 菲托夫兰绒布
feutrière 毛毡
few-o-fil polyamide hosiery 少孔丝锦纶袜
fez 非斯帽,土耳其平顶圆帽,土耳其毡帽(红色附有黑缨),圆筒形无边毡帽(地中海);摩洛哥女帽
fezel 白飘带
fiber 纤维
fiber art 纤维技术
fiber board 纤维板,样板用纸
fiber board insole 纸板内底(鞋)
fiber-bonded carpet 针刺毡合地毯,非织造地毯
fiber-bonded cloth 胶合纤维布,纤维黏合织物,非织造布
fiber-bonded fabric 胶合纤维布,纤维黏合织物,非织造布
fiber-bonded floor covering 针刺毡合地毯,非织造地毯
fiber content 纤维成分;纤维含量
fiber dyed fabric 纤维染色纱织物,色纺纱织物
fiber fill 纤维絮
fiber-filled vest 纤维填充背心
fiberglass casement fabric 玻璃丝窗帘布
fiberglass fabric 玻璃纤维织物
fiber-glass tape 玻璃纤维卷尺
fiber lace 纤维花边

fiberlock fabric 纤维锁结织物
fiberlock nonwoven 纤维锁结非织造布
fiber materials 纤维材料
fiber property 纤维特性
fiber silk 人造丝
Fiber Society 纤维学会
fiber stitch 纤维绣
fiber web sewing-knitted nonwoven 纤网型缝编法非织造布
fiber web stitch-bonded nonwoven 纤网型缝编法非织造布
fibre 纤维
fibric gloves 布手套
fibrilia 麻纤维;麻织物
fibroin 丝心蛋白
fibrous filler 纤维填料
fibrous glass 玻璃丝
fibula,fibule 搭扣;腓骨;扣衣针(古希腊、罗马的),领针,饰针
fibule 饰针;搭扣
ficelle [法]军用条纹
fichu 披肩式三角女薄围巾(18~19世纪)
fichu collar 围巾领
fichu lavalliere 法莉埃公爵夫人式披肩
fichu menteur 女式半胸围巾
fichu-pelerine 大披肩
fickle color 闪光
fickle fashion 易变款式
fictitious outline 想象线条
fidelity 重现精度
field bag 军用挂包
field boots 齐膝长筒靴,(齐膝紧裹的)长筒靴
field drying 露天干燥
field green 原野绿
field grey 军服灰(深灰色)
field pants 士兵工作裤,野外裤
field service uniform 野外勤务军服,战地服
field sports shoes 野外运动鞋
field uniform 军装,野战军服,战地服
fieltro 西班牙连帽毡斗篷
fiery red 火红色
fiesta 橘红,节日红
fiesta shirt 男式白运动衫;节日衬衫,墨西哥结婚衬衫
Fifth Avenue 第五街(纽约著名时装店集中地)
fifties(20 cen.) 50年代装(20世纪)
fifties skirt(20 cen.) 50年代裙子(筒裙)

(20世纪)
fifty-fifty 各半发型
fig 服装,衣服,盛装;穿着
Figaro jacket 费加罗外套(欧洲女式外套),费加罗夹克
fig brown 无花果棕
fighten band 束紧衣带
fig leaf 黑色丝质小围裙
figural motif[motive] 图案花型
figurative art 人体姿态艺术,造型艺术
figurative design 图案设计;具象图案,形象图案
figurative drawing 人体绘画
figurative jacquard 大提花织物
figurative language 图像语言
figurative pattern 绘画式花样,形象式花样
figurative print 绘画格调印花布
figurative thought 形象思维
figure 身材,体形(型),身段;图,图案,图形,图案花纹;数字
figure-clinging 紧身型的
figure-clinging line 紧身型
figure-clutching gown 紧身女袍
figure-control lingerie fabric 紧身内衣料
figure-control panel 紧身布片
figured 提花,提花织物
figured blister 花型胖花
figured carpet 提花地毯
figured casement 提花窗帘布
figured cloth 提花布,提花织物
figured corduroy 提花灯芯绒
figured double cloth with color effect 双面异色提花织物
figured drawing 人体素描
figured-effect 花纹效应
figure design 人体姿态设计
figured fabric 提花织物;提花布
figured felt carpet 花式毡毯
figured gauze 提花纱罗
figured habotai[habutae] 提花纺绸
figured hair cord 提花麻纱
figured hair silk 丝麻花绸
figure dictation 图形表达(服装画上)
figure dimensions 人体[三维]尺寸
figured lining 有图案的夹里
figured moire antique 花卉缎条与水浪形条子间隔的织物
figured outline 花纹轮廓线
figured pattern 提花图案

figured pique 提花凸纹织物
figured plumetes 手绘细棉布
figured plush fabric ［提花］长毛绒织物
figured poplin 提花府绸
figure drawing 人体素描
figured satin 织花绸缎，花缎
figured scale 标尺
figured silk 提花丝织物，花绸
figured skirt 花裙，图案花裙，织花布裙
figured stockings 花袜
figured stripe 提花条纹
figured tabaret 塔巴勒花绸
figured taffeta 提花塔夫绸
figured tape 花纹带
figured towel 提花毛巾；织花浴巾
figured velvet 花丝绒
figured Wilton 威尔顿提花地毯
figure face fabric 纹面织物
figure faults 体型缺陷
figure firming bodice 妇女紧身褡
figure firming fashion 妇女胸衣，妇女紧身褡，全帮
figure-hugging 贴身
figure-hugging bodices 紧贴式紧身胸衣
figure improver 小臀垫
figure outline 体型轮廓
figure's bulge 人体隆起部位
figure's contour 人体外表
figures de chinay 荷兰手工花边
figure shading 花式阴影
figure skate 花式滑冰鞋，花式溜冰鞋；花式溜冰装
figure skating shoes 花样滑冰鞋
figure's outline 人体轮廓线
figure stance 人体立姿
figure stitch 联合线迹
figure type 体型类型
figuring 图案，花纹；绣花
figuring machine 打样机
Fiji cotton 斐济长绒棉
filagree 金银丝饰品
filament 长丝，单纤维
filament core yarn 长丝芯纱
filament fabric 长丝织物
filament fancy suiting 长丝花呢
filament fancy yarn 长丝花色纱
filament yarn 长丝纱，复丝纱
filament yarn fabric 长丝织物
filatrice 丝线；法国绢丝绸
filature silk 机缫生丝，厂丝

fil de carnassiere 意大利打结式花边
fil de cren 花边外缘饰线
fil de trace 贴花花边轮廓线
filed uniform 野战军服
filed worsted 高级礼服黑呢
file feeder 锉刀状送料
filemot 枯叶色，黄褐色
file silk 横棱纹绸
files management 档案管理
filet 方眼花边网
filet brodé 绣花网眼花边
filet conté 网眼花边
filet de Bruxelles 布鲁塞尔网眼纱
filet de carnasiere 流苏花边
filet fabric 透孔织物，花边织物
filet guipure 方眼大花纹网眼花边
filet lace 方网眼花边，网眼花边
filet net 机织方网眼纱，经编方网眼纱
filet net of lace 方网眼
filet richelien 花卉纹网眼花边
filet silk 多股生丝绣花线
filght boots 第二次世界大战后飞行靴
filibeg 苏格兰短裙式童装(或女服)
filigree 金银丝饰品
filigree point 金银丝花边
filigree ring 金银丝戒指
filik 土耳其非利克山羊毛；东方山羊毛地毯
filled cloth 上浆布，增重布
filled finish 上浆整理
filled goods 低级粗纺呢；植绒呢
filled mattress sewing machine 床垫折边缝纫机
filler 填充料
filler fabric 填充物
filler fiber 填絮纤维
fillet 头带
fillet gound 方网眼帆布
filleting 本色厚亚麻带窄饰边带
fillibeg 苏格兰男短裤；短裙
fill-in 临时替代品；领带可分离的衬衫前胸
fill-in color on a sketch 设色
filling (agent) 填充料；纬纱
filling-backed fabric 纬二重织物，纬背组织织物
filling-backed serge 纬背哔叽
filling cord piqué 横向灯芯布，横向凸条布
filling-faced fabric 纬面织物
filling ikat 纬纱扎染布

filling insulating padding bed cover 衬棉床罩
filling knit 纬编针织物
filling knitting 纬编针织
filling lace 缝花边
filling lace stitch 贴花边绣
filling mass 填充料,填料
filling material 填料,填充材料
filling pile fabric 纬起绒织物
filling product 填充物,添加物
filling rep 纬棱纹织物
filling reversible 起绒毛呢,双面双色起绒织物
filling sateen 横贡,纬面棉缎
filling satin 纬面缎
filling stitch 刺绣针迹,填充刺绣针迹,集束针迹;贴线绣
filling streaks 色纬档疵
filling stretch 纬向拉伸
filling-stretch woven fabric 纬向拉伸织物
filling twill 纬面斜纹
filling-wadded double cloth 垫纬双层织物
filling yarn 纬纱
film 薄膜,面膜
film dress 电影服装
film fiber 薄膜纤维
film loop 循环影片
film screen printing 绢网印花,筛网印花
film-splitting nonwoven 膜裂法非织造布
film transplanting leather shoes 移膜革鞋
filoche 法国八枚哔叽;网状印花
filo floss 散丝线
filoselle 绣花丝线
filo silk 散丝线
filouche 平纹薄棉布
fil plat 漂白棉绣花线
fil tiré 抽线刺绣品
fimbriation 缝缘
final coat 外观涂层整理
final examination 出厂检验,最后检验
final folding machine 包装折叠机
final inspection 成衣检验,出厂检验,最后检验
finalized product 定型产品
final pattern 校样样板
final pressing 大烫,成衣整烫
final product 最终产品
final seam 终止缝,末缝
financial management 财务管理
financial year 财务年度

findings [美、复]零碎的服装附件、工具(针、线、扣、里布等)
fine 优质黄麻
fine adjustment 微调
fine belt 细皮带
fine canvas 鞋用帆布,细帆布
fine carpet 精细地毯
fine cleavage 精美不成形品(钻石)
fine cloth 细布
fine clothing 美利奴血统羊毛
fine corduroy 细条灯芯绒
fine cotton goods 细棉布
fine count 细支
fine count linen handkerchief 高支麻手帕
fine delaine wool 美利奴细毛
fine denier 细旦
fine-denier fiber 细旦纤维
fine-denier multifilament yarn 细旦多纤复丝
fine-drawing 修布
fine drill 细斜纹
fine eyelet 细网眼
fine fabric 精细织物(通常需手洗或干洗),细薄布料,细薄织物
fine fleecy sweatshirt 细绒衫
fine gathering 细褶裥
fine gauge 细针距
fine gauge knit 细针距针织物
fine goods 精品
fine grade 优等品
fine grain 细粒面(皮革)
fine hair 细绒毛
fine jewelry 高档珠宝,优质珠宝
fine knitted fleece 细绒布(针织)
fine laundering 轻洗(洗涤丝绸、毛织品)
fine linen 亚麻细布
fine medium clothing 美利奴血统羊毛
fine medium staple 美利奴血统羊毛
fine mercerized cotton thread 细丝光棉线
fine needle corduroy 细条灯芯绒
fineness 细度
fine number 细支
fine pattern 精细图案,精致图,细子花纹
fine pinwale corduroy 细条灯芯绒
fine plain 细平布
fine plain cloth 细平布
fine plane 细平布
fine pocketing 平纹口袋布
fine raised knitted fabric 细绒布,针织细绒布

fine-ribbed fabric 薄罗纹织物,细罗纹织物
finery 服饰,服装装潢,服装装饰,服装边饰,华丽的服饰,艳服,鲜艳服装;优美,优雅
finery design 服饰设计
fine silk clothes 纨(中国古代)
fine staple 美利奴血统羊毛
finette 棉哔叽料(常用于睡衣料或衬料)
fine twill 精细斜纹布
fine wool 细羊毛
fine workmanship 制作精心
fine yarn fabric 细支纱织物
finger 手指;测厚规(衣料);纤维束
finger cot 手指套,指套
Finger-Free glove 灵活指手套(商名)
finger-free gloves 立体手套
finger free stretch glove 分指手套
fingering yarn 细绒线
finger knitting 手工编织
finger machine 手套手指编织机
finger mark 指痕
fingernail 手指甲
finger press 指压
finger rug 手织地毯
fingerstall (皮革或橡皮等制成的)护指套,指端套
finger starting lever 指触起动架
finger-tip (射箭用的)指尖套,指尖
finger tip length 手指尖长的外套长度,指尖长度(上衣肩部至指尖长度)
finger-tip stitch 手套指端针迹;手套指端状线迹
finger-tip towel 小方巾,揩指尖小方巾;高档细薄小手帕
finger wave 芬格儿发型,湿烫发型,,手指卷发型,手指冷烫法
finger wave hairstyle 手指卷发型
finger weaving 手工编织
Finic costume 芬兰民俗服
finish 整理,后整理,精修,烫衣整理,修整
finish all over 竣工;全面精加工
finish depth 加工深度
finished appearance 成品外观
finished carpet 地毯成品
finished cloth width 成品织物幅宽
finished dimension 成品尺寸,熨烫后尺寸
finished edge 光边
finished fabric 成品布
finished garment dyeing 成衣染色
finished kneeband 齐膝带
finished knee breeches 齐膝短裤
finished knee breeches side vent 齐膝短裤侧开衩
finished knits 针织成品布
finished leather shoes 修面革鞋
finished market product 成品
finished measurement 成品尺寸,成品尺码
finished product 成品
finished products management 成品管理
finished room 成品间
finished side vent 齐膝短裤侧开衩
finished silk fabric 练熟绸匹,熟绸
finished-size 成品尺寸
finished width 成品幅宽
finished worsted 光洁整理精纺呢,缩绒剪毛精纺呢
finished yarn 加工纱
finisher 整烫工
finish fabric 整理织物
finish-free knitted fabric 免整理针织物
finish-free knitted goods 免整理针织物
finishing 整理,后处理,后整理
finishing allowance 精加工余量
finishing blotch 污斑
finishing board 烫台
finishing press 整烫压平机
finishing seam 成形缝,面缝
finishing spot(or stain) 污斑
finish inspection 出货验货
finish ironer 整理熨烫机
finish-ironing 整烫
finish knee breeches 齐膝短裤
finish line 成形绒,面缝线
finish line of cuff 袖口翘线,袖头止口线
finish line of front facing 挂面止口线
finish operation 整理作业
finish presser 熨烫工,整理熨烫工;整理熨烫机
finish pressing 大烫,成衣整烫,整理熨烫,整烫
finish pressing area 后整理-整烫部位
finish pressing technology 整烫工艺
finnesko 芬尼史可靴,鹿毛皮靴
finnesko boot 北极鹿皮靴,连毛鹿皮靴
Finnish Accreditation Service (FINAS) 芬兰服务认可机构
Finnish costume 芬兰民俗服饰
Finn jaguar mink 芬兰美洲虎纹貂皮
finnraccoon 加工过的浣熊毛皮

fino-fino panama	巴拿马极品帽
fins	脚蹼
fin trait	法国船帆布
fire	火彩
fire behaviour	着火性能
fire cloth	耐火布
fire control boots	消防靴
fire cracker	鞭炮色
fire dress	消防服
fire-engine red	救火车红色
fire fighter's entry suit	入火区消防服
fire fighter uniform	消防服
fire fighting garment	消防服
fire-fight suit	防火衣
fireman slicker	消防雨衣
fireman's tunic	消防服
fireman uniform	消防服
firemen exposure uniform	消防员灭火防护服
firemen's uniform	消防服
fire opal	火蛋白石
fire prevention and safety requirement	防火和安全要求
fire proof	防火性
fireproof cloth	防火布
fireproof fabric	防火织物;阻燃纤维
fireproof fiber	防火纤维
fireproof finishing	防火整理
fireproof material	防火材料
fireproof suit	防火衣
fireproof textile	阻燃纺织品,防火纺织品
fireproof uniform	消防避火服
fire resistance	耐火性,阻燃性
fire resistant coverall	防火工作服
fire resistant fabric	防火织物,阻燃织物
fire-resistant finish	防火管理,抗火整理
fire retardancy	阻燃性
fire retardancy treatment	阻燃整理
fire retardant finish	阻燃整理,阻燃处理
fire retardant flight suit	防火飞行服
fire retardant product	阻燃产品
fire retarded thread	(防火服装用)阻燃线
fire retardency	阻燃性
fir green	冷杉绿
firing inflammation	着火温度;着火点
firm	坚固
firm cord	厚实凸条布
firm feel	厚实手感,有身骨手感
firm handle	手感厚实
firmla	重绣长背心
firmness	厚实,坚实性,有身骨性
firm roll	紧卷织物
firm support bra	硬托胸罩
first backing fabric	簇绒地毯第一层底布,簇绒底布,基底布
first class products percentage of export	出口一等品率
first grade	一等品
first hand investigation	实地考察
first impression	第一印象
first marker	原始标记
first pattern	第一套样板
first quality	一等品,优质
first rate	一流,优等
firsts	一等品
first sample	初样
first sample inspection	初样检验,首件检验
fish	腰省,腰身裥,肚省
fish bone stitch	鱼骨形线迹;鱼骨绣
fish dart	肚省,长腰省,菱形褶,双向褶,鱼形褶
fisher	食鱼鼬皮
fisher fur	鼬鼠毛皮
fisherman rib	双面集圈织物
fisherman sandal	渔民凉鞋
fisherman's cloth	钓鱼装
fisherman's knit	渔夫毛衣;爱尔兰大毛衣
fisherman's lace	意大利黑白几何形棱结花边
fisherman's sandals	渔夫凉鞋
fisherman's sweater	渔夫毛衣,渔夫式厚套衫,传统手工粗线毛衣
fisherman's vest	渔夫背心
fisheye	小洞;鱼眼花纹
fisheye lens	超广角镜头,鱼眼镜头
fishing boots	钓鱼靴,防水高筒靴
fishing hat	钓鱼帽
fishing jacket	钓鱼短上衣,钓鱼夹克
fishing parka	捕鱼派克大衣(长至膝部,带帽)
fishing suit	钓鱼服
fishing vest	钓鱼背心,钓鱼马甲
fishing wear	钓鱼服
fish mouth	鱼口西装领
fish mouth collar	鱼口西装领(圆头上片西装领)
fish mouth lapel	鱼嘴领(驳领尖为圆形)
fish mouth shoes	鱼嘴鞋
fish mouth toe close	鱼嘴式缝拱头
fishnet	棉纱罗网眼布;鱼网,伪装网;花

式袜
fishnet hose 网眼连裤袜
fishnet knit shirt 圆领针织衫
fishnet poncho 透明穗饰披巾
fishnet rib fabric 网眼罗纹布
fishnet shirt 网眼保湿透湿衫
fishnet stockings 网眼长筒袜,网眼袜
fishnet underwear 网眼棉内衣
fish pearl 仿造珍珠,人造珍珠
fishscale embroidery 鱼鳞刺绣,鱼鳞绣
fishscale knitwear 鱼鳞毛衫
fishscale woolen sweater 鱼鳞羊毛衫
fishtail 鱼尾裙;鱼尾纹
fishtail dress 鱼尾式拖裙装
fishtail setting 鱼尾戒指基座
fishtail skirt 大鱼尾袖;大鱼尾裙
fishtail train 鱼尾形拖裾
fish twill 人字形斜纹,山形斜纹
fishwife costume 渔妇式宽身束腰女服
fish wife skirt 卖鱼女裙
fishy handle 滑溜手感
fissure 裂缝(钻石)
fist girth 拳围
fit (服装)合身,适宜,相称
fit and flare 紧身大摆式(上部贴身,腰围到裙摆展宽),上身合身和下身展宽的款式
fit and flare line 上贴(身)下散(开)线,上贴下散型
fit and movement style 紧身与运动组合款式
fitch 臭猫毛皮,臭鼬毛皮,鸡鼬毛皮
fitchet 前侧直衩
fitch fur 艾鼬毛皮
fit gather dress 合身褶连衣裙
fitness 合身(衣服),适应性,适用性
fitness craze 健美热潮
fitness industry 健身工业
fitness look 健身装
fit on 试穿
fit-out 配给服装,制服,全套制服
fitted block 合身样板,贴体样板
fitted bodice 紧身衣
fitted cape 合身斗篷
fitted carpeting 全室地毯
fitted coat 合身大衣
fitted midriff 上腹部育克
fitted midriff skirt 高贴腰裙
fitted princess dress 合身的开刀连衣裙
fitted set-in sleeve 窄形接袖

fitted sheet 成形床单
fitted short shorts 合身超短裤
fitted sleeve 合身袖,贴身无带女装
fitted strapless dress 贴身无带女装
fitted waist dress 合腰身的连衣裙,腰部合身的连衣裙
fitter 裁剪试样工,试穿员,试缝师
fit the dress to the figure 量体裁衣
fittige 平纹粗平布
fitting (英国服装鞋袜的)尺寸,尺码;试缝,假缝;试穿;装配;选鞋;印花对位
fitting design system 试衣设计系统;试衣系统
fitting garment 合体服装
fitting label 体形尺寸标志
fitting line 净缝线
fitting quality 合身程度
fitting room 试衣间,试衣室
fitting sweater 紧身运动衫
fitting-up 部件安装
fit well (衣服)很合身
fitzherbert hat 变形气球帽
fiume 埃及粗亚麻
five 5号尺码的衣着用品
five-eights hose 翻口中筒袜,高尔夫球袜
five-eights length 六分长
five-end satin 五枚缎纹
five-gored skirt 五片长裙
five-harness satin 五枚缎纹
five-heddle satin 五枚缎纹
five miss blister fabric 五列凸纹浮线针织物
five o'clocks 细花缎亚麻台布
five pair eyelet shoes 五眼鞋
five-pocket jeans 五袋款牛仔裤
five-point cut 马丽·匡特式发型
five precious fragments 五珍(钻石)
five-shaft satin 五枚缎纹
five-step-scale 五级灰色样卡
five thread overlock 五线锁边
five thread overlock machine 五线包缝机,五线拷克车
fivette 五页斜纹里子布
fixation 定色
fixed day shift 长日班
fixed guider contour sewing machine 固定靠模缝纫机,静止靠模缝纫机
fixed knife 定刀
fixed shift 固定班
fixed supply 固定供应

fixed-wage system 固定工资制
fixer 修车工
flabby handle 松弛手感
flabby skin 松弛皮肤
flaccidezza 松软绸
flag cloth 旗布
flagged points 织疵标记点
flagging 下垂
flaine 法国条纹被褥织物
flairs 喇叭裤
flake cloth 粗节织物
flake fabric 植绒织物
flake twist 雪花线,竹节花式线
flake yarn 雪花线,竹节花式线
flak jacket 防弹短上衣,铠装防弹短外套
flamboyant 艳丽
flamboyant color 火红色
flam-cleaning fabric 火浣布
flame 火焰,有焰燃烧;火焰绣
flame checking 抗火整理,阻燃整理
flame-cleanning fabric 火浣布
flame color 火红色
flame embroidery 锯齿形帆布刺绣,焰形刺绣
flame fiber 阻燃纤维
flame-free finish 无焰整理
flamenco boots 弗拉明科裤
flamenco dress 法兰达斯人穿的大连衣裙,西班牙吉卜赛舞蹈裙装
flamenco pants 弗拉明科裤
flame orange 火橙色
flame proof 防燃,阻燃
flameproof fabric 防火织物,阻燃织物
flameproof fiber 防火纤维,阻燃纤维
flameproof finish 防火整理
flame proofing 防燃防焰整理
flame proofing finish 阻燃整理
flame red 火焰红
flame repellency 拒燃性
flame repellent finish fabric 阻燃整理织物
flame resistance 抗燃性
flame resistant fiber 防火纤维,阻燃纤维
flame resistant finish 抗燃整理
flame retardance 阻燃性
flame retardancy 阻燃性
flame retardant fabric 阻燃织物
flame retardant fiber 阻燃纤维
flame retardant finish 阻燃整理
flame retardant finished fabric 阻燃整理织物

flame retardant property 阻燃性
flame retardant sleepwear 阻燃睡衣裤
flame retardant textile 阻燃纺织品
flame scarlet 大红色,狸红色
flame singeing 火焰烧毛
flame spread 展焰性
flame spread property 火焰蔓延性能
flame stitch 火焰缝法;火焰网眼绣;锯齿形线迹,匈牙利花边
flame work 焰形刺绣
flamingo 红橘色,火烈鸟色
flamingo dress 佛朗明哥装
flamingo pink （火烈鸟）粉红色
flaming rate 燃烧蔓延速度
flammability （织物）可燃性,易燃性
flammable fabric 易燃性织物
flammable fabric act (F.F.A) 易燃织物法
flamme 印花纱;印花毛纱织物;麻棉色织平布
flammeolum 古罗马新娘小面纱
flammete 麻棉色织平布
flammeum 罗马新娘火红长面纱
flanders lace 弗兰德斯针绣花边（一列锁眼针迹,每一线圈都用结锁住）
flanelle de chine 平纹毛织物
flanelle de rheims 法国花式线斜纹法兰绒
flange 凸缘,镶边,肩缝凸缘,冒肩;装饰流苏;耳朵皮（前身里或挂面的小块拼接布）
flange dart 肩缘褶
flanged bushing 法兰轴衬
flanged bushing pin 法兰轴衬销
flange effect 冒肩效果
flange heel 喇叭跟,凸缘鞋跟,喇叭形鞋跟
flangeing 帽形整理
flange inset 落肩嵌片,冒肩嵌片
flange line 冒肩线
flange shoulder 冒肩,凸缘肩
flange sleeve 凸缘袖
flange tuck 肩缘褶
flanging 翻口,帽形整理;镶边
flanker 系列新产品
flank side 胁腹
flannel 法兰绒,棉法兰绒;绒布;哈味呢,哈咪呢;法兰绒衣服;绒布衬里
flannelet(te) 单面棉绒布,绒布
flannelette bed sheet 拉绒床单

flannelette sheeting 被单绒布
flannelette shirt 棉绒衬衫,绒布衬衫
flannelette underwear 绒布内衣;绒布衫裤
flannelet(te) with stripe and check 条格绒布
flannelet vest 绒布背心
flannel finish 拉绒整理,仿法兰绒整理
flannel knitted fabric 法兰绒针织物
flannel-like fabric 仿法兰绒织物
flannel lining 绒布衬里
flannel pajamas 绒布睡衣裤
flannel petticoat 法兰绒衬裙
flannels 法兰绒裤;羊毛内衣
flannel shirt 法兰绒衬衫,绒布衬衫,法兰绒运动衫;法兰绒衬裙
flannel sport coat 法兰绒运动外衣
flannel twill 斜纹绒布
flannel vest 绒布背心
flano 薄法兰绒;条子法兰绒
flap 袋盖,口袋盖;帽边;前襟
flap and button down pocket 扣纽扣的有盖袋
flap and ticket pocket 盖式票袋
flap & button down pocket 扣纽袋盖口袋,扣纽扣的有盖袋
flap button-through pocket 明纽有盖袋
flap edge is uneven 袋盖不直
flap effect 袋盖效果
flap facing 袋盖贴边
flap heel sole 卷跟底(鞋)
flap interlining 袋盖衬
flap jetted pocket 盖式两滚边口袋
flap jetted side pocket 有盖贴袋
flap lining 袋盖衬里,袋盖里
flap lining lean out of edge 袋盖反吐(袋盖里子外翻)
flap out grain 袋盖丝绺不正
flap-over binding 隐嵌线袋盖
flap patch pocket 袋盖贴袋
flapper 摩登女郎;饰金属片短连衣裙
flapper dress 低腰节连衣短裙;花花少女装
flapper look 花花少女风貌,轻浮少女款式;消瘦体形;小野禽风貌
flapper pants 弗莱帕裤
flapper style 小野禽服式
flapping over 省缝
flap pocket 盖袋,盖式口袋,有盖袋,袋片
flap pressing 袋盖熨烫

flappy dress 宽松女服
flap sticking up 袋盖反翘
flap-under binding 明嵌线袋盖
flare (衣裙)喇叭形宽摆;喇叭裤,阔脚裤;长袜裤口
flare above knee 膝上喇叭造型
flare and tube style 展宽和管状组合风格
flare at knee 膝位喇叭造型
flare below knee 膝下喇叭造型
flare chute fabric 照明降落伞绸
flared clothing 宽摆外套
flared coat 宽摆外套,燕子领宽摆式上衣
flared leg 喇叭裤管
flared leg panties 喇叭形短裤
flared leg pants 喇叭裤
flared legs 喇叭裤脚;喇叭裤,展宽裤
flared leg underpants 喇叭形短裤
flared line 展宽型
flared panties 宽摆内裤,喇叭形内裤
flared pants 喇叭裤
flared petal sleeve 喇叭型花瓣袖
flared shorts 喇叭短裤,宽脚短裤,宽脚口短裤
flared skirt 宽裙,喇叭裙,宽摆裙,伞裙,斜裙,农妇裙,吉卜赛裙,波浪裙
flared skirt in stuff 宽摆呢裙
flared slip 宽摆衬裙
flared tier 喇叭层片
flared trousers 喇叭裤,宽脚裤
flare-gather skirt 碎褶喇叭裙
flare heel 展宽鞋跟
flare knitting 喇叭袜口
flare legs 喇叭裤,喇叭裤管,展宽裤
flare line 裙摆展宽线
flare panties 喇叭短裤
flare parachute 照明降落伞
flare point 裙摆展宽开始点
flares 阔脚裤,喇叭裤
flare shorts 喇叭短裤
flare skirt 喇叭裙,波浪裙,宽摆裙
flare slacks 喇叭裤
flare sleeve 喇叭袖
flare slip 宽摆衬裙
flare tiers skirt 宝塔裙,多层波浪裙,阶层式喇叭裙
flare trousers 喇叭裤
flare-tuck skirt 喇叭塔克裙
flaring at the bottom 放大下摆
flash (服装、外表的)浮华;虚饰;(军用)肩章;徽章;战士假发绸带结

flash bulb 闪光泡
flash coat 闪光外套
flashdance look 闪光舞风貌
flashdance top 闪光舞针织衫
flash lamp 闪光灯
flash satin 闪缎
flash spun nonwoven 闪纺法非织造布,纺粘法非织造布
flash temperature 闪燃温度
flash yarn 闪光线
flashy nightshirt 闪光舞式睡衣(低领,套袖)
flat 轮廓不清;平底鞋
flat abdomen 平腹
flat and button down pocket 带纽扣的有盖袋
flat and smooth 平服(加工术语)
flat-back needle 平背针
flat back type 扁平背型,平背型
flat bar knitting machine 横机,平机
flat beard-needle knitting machine 平型钩针针织机
flat bed knitting machine 横机
flat bed lockstitch machine 平板式锁式线迹缝纫机,锁式线迹平缝机
flat bed lockstitch zigzag sewing machine 曲折锁缝机,曲折锁式线迹平缝机
flat bed machine 横机,平机
flat bed one needle lockstitch sewing machine 单机针锁式线迹平缝机,单针锁缝机
flat bed press 平板热压机,平板压烫机
flat bed scanner 平板绘图仪
flat bed sewing machine 平底板缝纫机
flat bed steam press 平板蒸汽压烫机
flat bed three needle chain stitch machine 平板式三针链式线迹缝机
flat bed three needle lockstitch sewing machine 三针锁式线迹缝纫机
flat bed two needle lockstitch sewing machine 双针锁式线迹平缝机
flat blade 扁平肩胛骨
flat bottom 平底;平下摆,平裙
flat braid 平编带
flat braided elastic 平编松紧带
flat braided elastic tape 平编宽紧带
flat braid trim 饰带饰边
flat-butted seam 对头并接缝合
flat butt-end seam 对接缝头
flat buttocks 平臀

flat button 平纽扣
flat buttonhole 平纽孔,平头扣眼
flat cambric 杂色低档轧光细平布
flat canvas 轧光刺绣十字布
flat cap 低顶圆帽(英国16～17世纪),平顶帽;戴低顶圆帽者
flat chafer fabric 单纱织物
flat chenille 雪尼尔花线
flat chest 平胸
flat chest type 平胸型
flat cloche 平坦钟形帽
flat collar 平领,平翻领,袒领
flat construction 服装平面构成
flat crepe 平双绉,平纹紧捻绉,仿绉布
flat cut 平顶头
flat derriere 小臀
flat duck 双经帆布
flat elastic 扁平松紧带
flat embossing 拷花整理
flat embroidery 平针刺绣
flater seam 扁平缝
flat fabric 横机针织物,平面织物,薄织物;平幅针织物
flat face 扁脸型
flat feed dog 平板送布牙
flat feed spring 平送布弹簧
flat fell 拼缝
flat felled seam 叠接缝,平接缝,暗包缝,明包缝,外包缝,折伏缝
flat fell seam 叠接缝,平接缝,暗包缝,明包缝
flat fell seam attachment 拼缝装置,平接缝装置,折边缝附件
flat filament 扁平长丝
flat fitting collar 平摊领
flat fitting peter pan collar 铜盆领,摊在肩上的圆领式
flat foot 平脚
flat gimp 平嵌芯细辫带(锁钮孔用)
flat goods 平针织物,平幅针织物,平线织物(无绉线的绉缩)
flat heel 平跟
flat(heeled)shoes 平跟鞋
flat hem 平折摆边
flat iron 烙铁,熨斗
flat jacquard fabric 普通提花针织物
flat knit 平针针法;平针织物,平针织物;平针组织
flat knit fabric 平幅针织物
flat knit hose 全成形袜

flat knitted fabric 平针织物
flat knitting machine 针织横机；针织编织机；平型针织机
flat knit T-shirt 平针 T 恤衫
flat knot 方结,平结
flat lapped seam 平搭接缝,叠接缝
flat linking machine 平缝机
flat-lock 绷缝
flat-lock machine 绷缝机，绷缝缝纫机；平针机
flat-lock seam 绷缝，覆盖缝
flat-lock stitch 绷缝线迹,覆盖线迹,平式锁缝针迹
flat machine 针织横机
flat Milan 米兰梭结花边
flat Milanese knitting machine 平型米兰尼斯经编机
flat nose 扁鼻子
flat ounce duck 重型双经单纬帆布
flat pattern design 平面纸样设计
flat patternmaker system 平面打板系统
flat pattern making 展示图
flat pattern making system 平面纸样制作系统
flat piece transfer printing machine 平板衣片转移印花机
flat piping 扁平嵌条
flat plate press 平板压烫机
flat plate pressing machine 平板压烫机
flat pleat 平伏褶
flat pocket 平盖袋
flat point 平式针绣花边
flat point lace 霍尼顿小花纹花边,平式针绣花边
flat pressing 平铺熨烫,铺整压平
flat printing 平网印花
flat pump 平跟碰拍鞋,平跟潘普鞋
flat rib machine knitting V 形横机,双针床横机
flat roll collar 平翻领,平贴领
flat rouleau 扁形滚条
flat ruche 具有毛圈效应的经编织物
flats 平底鞋,平跟鞋,平底女鞋,低跟女鞋；双头
flat sailor 水手领
flat satin stitch 缎纹刺绣针迹,无衬垫缎纹刺绣针迹
flat screen printing 平网印花
flat seam 绷缝,拼缝,扁平缝,平缝
flat seamer 平缝机,绷缝机

flat seaming 平缝
flat seaming machine 平缝机
flat seam 3 needles 4 threads 三针四线绷缝
flat seam stitch 绷缝线迹,扁平缝线迹
flat seat figure 扁臀体型
flat set 平整定形,平整处理
flat sewing machine 平车,平缝机,平机
flat shoe lace 扁平鞋带
flat shoes 平底鞋,平跟鞋
flat shoe lace 扁平鞋带
flat-shoulder-style 平肩型
flat stitch 平针；平针线迹,平缝线迹；平面绣
flat stitched-on pocket 平贴袋
flat stitching method 搭缝法
flat stocking blank 平袜袜坯
flat stone 扁平钻石
flat straw hat 平顶草帽
flat surface 布面平整,呢面平整
flat table 平台形(珠宝)
flatten 拉皮(毛皮)
flattening 展平
flatter collar 平贴领
flattering color 讨人喜欢的色彩
flatter seam 扁平缝
flattie,flatty 平跟鞋,无跟鞋；拖鞋
flatties 平底鞋,平跟鞋,无跟鞋
flat top 平头(发式),平顶头
flat trim 平饰边
flat tummy 平腹
flat two-needle lockstitch sewing machine 平板式双针锁缝机
flatty 平跟鞋
flat type bartack 平式套结
flat underwear 平针内衣
flat underwear fabric 平针内衣织物
flat work 家用织物；整只裁剪(毛皮)
flat work ironer 大熨烫机
flat woven fabric 单层机织物
flat yarn 扁平长丝纱；仿草杆丝纱
flauk 格子花呢
flaunt 华丽服装
flavor, flavour 风韵,气息
flaw 瑕疵(钻石)
flawless 无瑕(钻石)
flawless cutting 贴身裁剪
flawless internally 内无瑕疵(钻石)
flax 亚麻,鸦麻,麻；麻布；亚麻黄(暗染黄棕色)

flax bush 新西兰亚麻
flax dacron 麻涤纶;麻的确良
flax damask 亚麻花缎
flax fabric 亚麻织物,亚麻布,麻纱布
flax-like fabric 仿麻织物
flax-like fiber 仿麻型纤维
flax linen 亚麻布
Flaxon 弗拉克森仿麻细平布(棉仿麻细布,商名,美国)
flax/polyester blended cloth 麻涤布(亚麻)
flax rug 双面亚麻地毯;双面亚麻走廊条毯
flax/wool suitings 麻毛呢
flayway lop 短而宽大的飘飞上衣
fleabag 睡袋
flea bag 睡袋
flea fur 小皮毛围巾
flea market look 跳蚤市场风貌
fleck 光斑,雀斑
fleck yarn 闪色花纱
fleco morisco 流苏花边
fleece 起绒布,厚绒头织物,绒头织物;绒衬里;长毛大衣织物,长毛大衣呢
fleece-backed 绒布里
fleeced 里子起绒;单面起绒
fleeced fabric 起绒针织布,针织绒布,柔软起毛织物
fleeced goods 单面起绒针织物
fleeced knit fabric 针织起绒布
fleeced-lined 羊皮衫
fleeced mozambique 莫桑比克格子花呢
fleeced underwear 羊毛内衣
fleece fabric 起绒织物;纤网型非织造布
fleece-faced fabric 绒面织物
fleece finish (毛毡)起绒整理
fleece gloves 绒布手套
fleece knit 起绒针织物
fleece knit fabric 起绒针织物
fleece knit goods 起绒针织物,绒面针织物
fleece-lined 骆驼绒
fleece-lined coat 驼绒衬里大衣
fleece-lined fabric 绒里面料
fleece-lined goods 厚绒布
fleece-lined knit fabric 里子起绒的针织物
fleece-lined overcoat 驼绒大衣;绒布里大衣
fleece-lined robe 羊皮袍
fleece-lined shoes 驼绒衬里鞋
fleece-lined sweater 驼绒衫

fleece-lined underwear 绒里内衣
fleece with anti-pilling 防起球绒布
fleecy backed 背面起绒的
fleecy domette 起绒布
fleecy fabric 起绒织物;纬编衬垫针织物
fleecy handle 软毛绒手感
fleecy hosiery 起绒针织物
fleecy knitting 背面起绒偏织
fleecy look 轻便温暖型款式
fleecy nonwoven textile 起绒型非织造布
fleecy sweat pants 绒裤,卫生裤
fleecy sweat shirt 棉绒衫,绒衣,棉绒T恤衫,卫生衫
fleecy sweat T-shirt 棉绒衫,绒衣,棉绒T恤衫,卫生衫
Flemish flax 佛兰德亚麻
Flemish Holland 比利时本色亚麻平布
Flemish hood 硬挺亚麻布白头巾
Flemish lace 佛兰德花边,佛兰芒花边,梭结花边
Flemish point 佛兰德针绣花边
flesh 肥胖;皮革肉面,皮里;人体表面肌肤;肉色
flesh blond 肉棕色
flesh color 肉色
flesher 剥皮工(制革);仿麂皮皮革
flesh figure 肥胖体型
fleshiness 多肉,肥胖
fleshing 生皮去肉(皮革)
fleshings (芭蕾舞演员穿)肉色紧身衣
flesh side 毛皮里层,皮肉面,肉面
flesh tights 肉色芭蕾舞紧身衣
flesh tint 肤色,肉色
fleshy arm 粗臂
fleur de jeunesse 法国泽纳斯防水防污花绸
fleur de lis 法国百合花图案;虹彩花纹图案
fleur de lys [法]虹彩花纹图案,法国百合花图案
fleur de soie 高级花丝缎
fleuret 粗纺斜纹厚呢,粗服呢,纱毛呢;花缎;法国细亚麻布
fleuret silk 绣花绸
fleur volante 针绣花边凸纹
flex abrasion 屈曲磨损
flex-fatigue resistance 抗挠曲疲劳性
flex general ruler 通用软尺
flexibility 挠性,柔曲性,弯曲性,通用性
flexibility resistance 耐折性,柔韧性
flexible container 集装包,集装袋
flexible cutting table 组合裁床

flexible glasses 可折眼镜,软性眼镜
flexible manufacturing system(FMS) 柔性制造系统
flexible reserve price 机动保留价格
flexible rule 软尺,卷尺
flexible ruler 卷尺
flexible shirt 两用衬衫
flexible stitch pattern 可变线迹花样
flexible tacker 弹性加固缝缝纫机
flexible time 弹性上班时间,灵活时间
flexible work group (FWG) 弹性工作组
fleximesh 正反线圈交替针织物
flexing abrasion resistance 耐曲磨性
flexing cycle 挠曲周期
flexing elasticity 挠曲弹性
flexiplastic 柔性塑料
flexi-stitch pattern 可变线迹花样
flexitacker 弹性加固缝纫机
flexitime 弹性上班时间
flex lift 挠曲寿命
flexor 屈肌
flexural resilience 挠曲回弹性,弯曲回弹性
flexural rigidity 挠曲刚度
flexure 挠曲,折褶
flicker 闪烁
flight attendants' uniform 空姐服
flight bag 轻便帆布包,航空旅行包
flight clothing 飞行衣
flight deck cap 宇航帽
flight jacket 飞行短上衣,飞行服,飞行夹克
flight style horsehide jacket 飞行式马皮夹克
flight uniform 飞行服,飞行制服
flimsy material 薄织物
fling skirt 便裙
flint 电石色
flip-chip dress 并块装;拼块连衣裙
flipe 折叠布
flip-flops 平底人字拖鞋,人字拖鞋
flip hairstyle 飞行式发型,外卷发型
flippy skirt 荷叶边喇叭裙,抛式裙(裙摆展宽翘起),翘摆裙(下摆翘起的展宽裙)
flips 夹趾式塑料拖鞋
flip skirt 不分前后身的裙子
flip-tie blouse 领结衬衫
flix 海狸绒;毛皮
flix courts 法国亚麻平布
float 浮针;救生衣,救生圈;(从肩部自然下垂的)伞式休闲服,伞式主妇服;跳花疵
float boots 福乐特靴,福洛特靴
floating 浮线
floatless pattern cloth 背面无浮花提花针织物,无虚线提花织物
float panel 浮动布片
float plated pattern 架空添纱图案花纹
float plating 架空添纱
floats 福乐特靴;悬浮染色织物;系带及踝筒靴(类同厚皱底和厚衬绒沙漠靴)
floccule 絮状物
floches 缝纫细丝线,细缝纫丝线
flock 毛束,棉束,植绒短绒;水手羊毛紧身衣
flockati (希腊)手工制皮底短袜
flock binder 植绒粘合剂
flock carpet 植绒地毯
flock coating 植绒涂层
flock dot 植绒花纹
flocked carpet 植绒地毯
flocked fabric 植绒织物,植绒布
flocked goods 植绒织物
flocked nonwoven fabric 植绒非织造布
flocked ribbon 植绒带
flocked sheeting 平绒
flocked suede 植绒仿麂皮
flocked twill 斜纹绒
flocket 大袖宽袍
flock finishing 植绒整理
flocking (织物上)植绒,植绒整理
flocking on nonwoven fabric 非织造静电植绒布
flock mattress 毛屑或棉屑床垫
flock print 植绒印花
flock printed handkerchief 絮状印花手帕,发泡印花手帕,植绒印花手帕
flock printed sheer 植绒印花薄绸
flock print fabric 植绒印花布
flock printing 植绒印花
flock suede 仿麂皮绒
flock yarn 竹节花线;雪花绒
flocky lace 植绒花边
Floconné 弗洛松内起绒大衣呢
flog closing 盘花纽门襟;盘纽,盘花扣
Flohpelzchen 小皮毛围巾
floor carpet underpad 地毯背衬
floor cloth 铺地织物,平纹粗厚铺地麻织物
floorcovering 地毯

flooring 帐篷铺地布
flooring felt 地毡
floor length 及地长
floor length gown 及地袍
floor level 地面线
floor mat 小地毯
floor price 最低限价
floor rug 地毯
floor space 车间面积；占地面积
flop 宽檐软帽
flop hat 垂檐帽
floppy brim hat 垂檐帽
floppy cap 垂檐帽
floppy hat （宽帽檐）下垂软帽，宽边软帽，垂檐帽，下垂帽
floppy straw hat 垂檐草帽
flor 亚麻网眼织物
floral bra 花朵胸罩
floral braid 花蕾镶饰
floral design 花卉图案，彩花式图案
floral embroidery blouse 绣花上衣
floral frog closure 盘纽，盘花扣
floral motif 花卉花纹
floral pattern 花卉图案，花卉图样
floral perfume 花香型香水
floral print 花卉图案印花，花卉印花
floral print gown 印花长衫
floral print skirt 印花布裙
floral ring stitch 环状花卉绒迹
floral sprig print 小花印花
floral unit 花型图案单元
florence 佛罗伦萨绸，衬里绸；意大利毛呢；19世纪棱纹纱罗
Florence Eiseman 佛劳伦斯·易斯曼（美国童装设计师）
Florence sandals 佛罗伦萨凉鞋
florence silk 佛罗伦萨丝里子绸
florentine 佛罗伦萨厚绸；精纺背心呢；夏季斜纹裤料
florentine drill 三上一下斜纹布
florentine embroidery 佛罗伦萨刺绣（十字布针绣，Z字针绣）
Florentine lace 佛罗伦萨花边
Florentine leather 佛罗伦萨皮革
Florentine neck 袒肩方领
florentine neckline 佛罗伦萨领口
florentine stitch 垂直线迹；佛罗伦萨针法
Florentine straw hat 佛罗伦萨草帽
Florentine work 焰形刺绣
floret 16世纪法国高级细毛呢；18世纪英国粗纺呢；有光毛花缎
floret linen 纯棉网眼织物；亚麻网眼织物
floret silk 优级绢丝
floretta 18世纪法国有光精纺呢；花缎；漂白细亚麻平布
floriation 花卉装饰艺术
Florida 佛罗里达棉；佛罗里达衬布；罗马尼亚印花衬布
Florida bowstring hemp 佛罗里达弓弦大麻
Florida cotton 佛罗里达海岛棉；上浆粗棉布；印花衬衫布
florid pink 脸红色
flor linen 网眼亚麻布，网眼布
florodora girl costume 弗洛罗多拉女服
florodor cloth 棉毛平纹薄呢
floss 丝绵；绣花丝线
floss-padded clothes 丝绵袄
floss-padded robe 丝绵袍
floss silk （绣花）丝线，绣花丝绒，绒线；丝绵
floss thread 刺绣用松软亚麻线；低捻丝线
flotation jacket 漂浮夹克
flotation vest 救生背心
flounce 衣边皱褶，荷叶边
flounce clothing 服装边饰
flounced detail 荷叶边饰
flounced dress 荷叶边喇叭裙，荷叶边连衣裙
flounced skirt 荷叶边裙，摆饰裙
flounced sleeve 摆饰袖（上部纤细，下部宽大），荷叶边袖
flounce ruffle 荷叶边，裙边
flounces 花边裙襞皱边
flouncing 荷叶边，薄纱皱褶边
flouncing lace 薄纱皱褶花边，荷叶花边，单边月牙花边
flouncy skirt 张开的裙子
flourishing thread 亚麻绣花线
flow 飘垂（衣服、头发等）
flow ateliers 做细软女装的工作间
flow back 逆流（洗衣机液流）
flow chart 程序分析图
flow-chart form of network 程序分析网络图
flow charting 控制系统图示法
flow diagram 生产流程图解
flowed-on foam backing 地毯刮浆泡沫底布

flower 花；花卉
flower and bird design 花鸟纹
flower basket hat 花篮帽
flower blue 花青
flower brooch 花饰针
flowered fabric 彩色花纹织物
flowered mink 花卉图案貂皮外套
flower garden hat 花园帽
flower garden toque 花坛式豆蔻帽
flower hat 花帽
flower hole 西服插花孔眼
flower lapel 花形翻领，花瓣式翻领
flower motif stitch 花状线迹，花卉线迹
flower pattern 花状花样
flower-petal design 花瓣花纹
flower-petal treatment 花瓣式饰巾装饰法
flower piece 花卉画
flowerpot cloche 花盆式钟形帽
flowerpot hat 倒花盆形男帽
flowerpot heel 花盆跟
flowerpot shoes 花盆鞋
flower satin brocade 金玉缎
flower skirt 花形裙
flowers patterned brocade 满花绸
flow-flow 紧身胸衣的瀑布形饰带环
flowing line 流畅线条
flow line 流水线
flow process 流水作业
flow process chart 流程程序图
flow production 流水生产
flow production line 流水生产线
flow sheet 流程表；流程图
flow through a network 网络流程
flow through hosiery boarder 流水线式袜子定形机
fluffing 抖开晾干，抖松
fluffy cut 毛茸茸发型
fluffy look 毛茸茸款式
fluffy ruffles 妇女胸衬，胸罩杯垫，奶杯垫
fluffy style 毛茸茸款式
fluffy sweater 毛茸茸毛衣
fluffy yarn 起毛线
fluid dress 连衣裙
fluid jet cutting 射流裁剪
fluid line 流畅线条；流畅型
fluid look 流动感款式
fluid style 流动感款式
fluid-tight seam 液密缝

flu mask 卫生口罩
fluorescence chromatogram 荧光色谱(图)
fluorescent bleach 荧光漂白
fluorescent brightening agent 荧光增白剂
fluorescent color 荧光色
fluorescent dye 荧光染料
fluorescent fabric 荧光织物
fluorescent lighting 荧光白色；荧光照明
fluorescent nonwoven fabric 荧光非织造布
fluorescent PVC gloves 荧光 PVC 手套
fluorescent red 荧光红
fluorescent resin pigment 荧光树脂颜料
fluorescent sand crepe 萤萤绸
fluorescent screen 荧光屏
fluorescent suiting silk 荧光呢
fluorescent tape 荧光带
fluorescent whiltening 漂白，荧光增白
fluorescent whitened wool 荧光增白羊毛
fluorescent whitening 荧光增白
fluorescent whitening fabric 漂白织物
fluorescent yellow 荧光黄
fluorescer 荧光增白剂
fluorite greed 萤石绿
fluorocarbon fiber 碳氟纤维
fluorotitanate-treated wool 钛酸氟处理毛织品
flush 经面或纬面织物的正面
flushability 可冲洗性
flushable nonwoven fabric 可冲洗非织造布
flushings 粗纺厚呢
flush line 外露线
flush setting 弗莱西基座(钻石)
flute 女装圆凹褶，褶子
fluted hem 沟褶边；螺旋饰边下摆
fluted trimming 褶裥饰带
fluting 褶裥，褶边；褶裥熨烫；条纹织物，绉纹布，绉纱，绉纹呢
fluting iron 烫褶裥熨斗
fluting machine 烫褶裥机
fluttering 波动，抖动；轻飘性
flutter jacket 飘逸夹克
flutter sleeve 悬垂袖
fly (衣、裤的)纽扣遮布，纽扣遮盖，前门襟，纽牌，裤门襟，裆
flyaway 宽大不合身的衣服
flyaway jacket 比翼夹克，女工超短全衬夹克，飘逸夹克
flyaway top 飘飞上衣(短而宽大)
flyboy glasses 飞行员护目镜
fly cap 蝴蝶帽，双翼女帽

fly closing 拉链门襟,叠门
fly closing buttoned 隐扣式叠门襟
fly edge 门襟止口
flyer's armour 防弹飞行服
flyer's helmet bag 飞行员头盔袋
flyer's uniform 飞行员制服
fly facing 门襟贴边,裤遮扣布,遮扣贴边,裤门襟,门襟
fly facing edge 门襟止口线
fly facing outside edge 门襟外口线
fly fringe 带小穗的丝边,缘饰
fly front 前襟,掩襟,暗门襟
fly front closing 暗门襟,前叠门
fly front garment 暗门襟服装
fly front opening 暗门襟开口
fly front zipper opening 暗门襟拉链开口
flying 布面拖纱疵
flying gown 18世纪宽松长袍
flying jodhpur 短马靴,焦特布尔(马)靴
flying-panel skirt 嵌色条裙,飘幅裙
flying saucer tunic 无领无纽宽松上衣;[不明]飞碟塔奈克
flying squirrel 鼯鼠皮
flying suit 飞行服
flying suit wear 飞行服
flying wear 飞行服
fly interlining 门襟衬
fly lazy-daisy stitch 辫子绣,连环绣
fly lining 门襟衬里,门襻里(子)
fly net 网眼交织物
fly notch 门襟标记
fly opening (裤)门襻,裤门襟,隐扣式开襟
fly panels skirt 飘幅裙
fly piece 门襟片
fly placket 隐扣式开襟,门襟
fly pocket 女裤袋
fly serging 拉链包缝
fly shield 裤里襟,裤子叠门,里襟
fly shield inside edge 里襟里口线
fly shield lining 裤子叠门衬里
fly shield outside edge 里襟外口线
fly stay 门襟贴布
fly stitch 比翼刺绣;比翼线迹;飞蝇缝,苍蝇针法
fly tongue 门襟暗牌;门襟里搭襻
fly zip 门襟拉链
foam attenuated fabric 发泡挤压[非织造]布
foam back 泡沫塑料衬里

foam backed fabric(or textile) 泡沫塑料衬里织物
foam backing 地毯泡沫底封;泡沫塑料衬里
foam bonded nonwoven 泡沫黏合法非织造布
foam coated fabric 泡沫塑料涂层织物
foam coating 泡沫涂层
foam diving suit 泡沫型潜水服
foamed plastic slippers 泡沫塑料拖鞋
foam fiber 泡沫塑料纤维
foam finishing 泡沫整理
foam green 泡沫绿
foaming printing 发泡印花
foam injection moulding machine 发泡射出成型机(制鞋)
foam laminated fabric goods 泡沫[塑料]层压织物
foam laminated materials 泡沫塑料层压织物
foam multicolor dyeing 泡沫多色染色
foam pad 泡沫垫,泡沫垫片;泡沫衬丁
foam plastic garment 泡沫塑料服装
foam plastics 泡沫塑料
foam plastic slippers 泡沫塑料拖鞋
foam printing 泡沫印花
foam rubber 多孔橡胶,海绵橡胶,泡沫橡胶
foam rubber padding 泡沫乳胶衬垫
foam rubber tongue 泡沫乳胶鞋舌
fob (男裤)表口袋,男裤表袋;(怀表的)短链及饰物,表链饰物,怀表短带
FOB 离岸价格
fob chain 怀表短带,怀表短链
fob pin 表袋别针,短链子别针
fob pocket (男裤上的)表袋,裤表袋,男裤表口袋
fob pocket facing 表袋贴边
fob ribbon 丝表带
fob watch 袋表,怀表
focal 男围巾,方形麻围巾
focal distance 焦距
focale 福外尔领带(古代)
focal length 焦距
fochette fabric 双层圆筒平针织物
focus 焦距
foden 福登布,蓝白条斜纹工装布
fog 发灰(摄影),雾色
fogara silk 樗蚕丝
fog green 雾绿色

foggy dew 雾露色
fogie 丝绸手帕
foil 陪衬色;陪衬物;叶形饰
foilback 箔背古(钻石)
foil button 以箔衬底的纽扣
foil dress 用即弃铝箔装
foil mask 击剑护面
foil-printing 金银箔印花
fold 折叠,打折,褶,褶子;卷边;合股(纱线)
foldability 可折叠性
fold abrasion 折边磨
fold abrasion resistance 耐折边磨
fold and press back yoke 扣烫过肩
fold and press hem 扣烫底领
fold and press reinforcement for knees 扣烫膝盖绸
fold and to press back yoke 扣烫过肩
fold and to press crotch reinforcement patch 扣烫裤底
fold and to press heel stay 扣烫贴脚条
fold and to press hem 扣烫底边
fold and to press reinforcement for knee 分烫膝盖绸
fold-back 折边
fold-back cuff 法式袖头,双袖头,翻折袖头,折边袖口,双层袖口,法式袖口
fold-back facing 连贴边
fold down 向下折
fold down casing 卷边式抽带管
fold dyeing 扎染
folded back sleeve 马蹄袖
folded body 大身折边
folded dart 折叠省
folded edge 折边
folded edge moccasin 折边软鞋
folded guide 折边导向
folded paper method 纸样映射法
folded pleat 折裥
folded selvage presser 卷边压脚
folder (缝纫机的)折边器,卷边压脚
folders 折叠式眼镜
fold facing line 翻折线
fold form 折叠形式
fold hem 折叠底边
fold hemline casing 折底边抽带机
fold here 折叠处
folding 折布
folding and cutting machine 折布裁袋两用机

folding and pressing 折烫,扣烫
folding and pressing back yoke 扣烫过肩
folding and pressing crotch reinforcement stay 折烫裤底
folding and pressing heel 折烫贴脚条
folding and pressing heel stay 扣烫贴脚条
folding and pressing hem 扣烫底边
folding and pressing knee kicker 折烫膝盖绸
folding body 折叠大身
folding cuff 双袖头
folding deformation 折叠形变
folding easel 折叠画架
folding endurance 耐折性,耐折度
folding fan 折叠扇
folding fault 折痕,折印,折疵
folding glasses 折叠式眼镜
folding hanger 折叠式衣架
folding length 折叠长度,折幅
folding lining inter stitch 衬里折边内缝线迹
folding machine 折叠机
folding make-up 定长折叠
folding plate 折布板
folding rule 折尺
folding seam 折叠缝,折边缝
folding slippers 折叠式拖鞋
folding square 折尺
folding station 折叠台
folding table 折布台
folding umbrella 折伞,折叠伞
fold inside 往里折
fold inside width 折叠宽
fold line 折叠线,折边线,翻卷线,裤子卷脚线
fold line for lapel 驳头口;(领)驳口
fold line of collar 领上口
fold line of lapel 驳口线
fold line of top sleeve 前偏袖线
fold line of under sleeve 后偏袖线
fold line of zhong shan suit pocket 老虎袋折边
fold marks 折皱印
fold-over binding 折翻式滚边
fold-over braid 活裥饰带
fold-over clutch 信封式小提包(或挂包)
fold-over collar 翻折领
fold-over flap 外折袋盖
fold packing 折叠包装
folds 衣纹

folds and pleats 衣褶
fold seam 折叠缝,折边缝
fold seam allowance 折缝头
fold three 折三褶
fold-under hem 内折底摆
fold up 向上折,折起
fold-up flap 翻折式袋盖
fold-up hat 折叠式帽子
fold-up hood 罗宋帽,折叠式风帽
fold-up raincoat 折叠式雨衣
folette 淡色三角形宽松围巾
folgorant 福格兰德交织泡泡纱
foliage design 簇叶图案
foliage green 叶绿色
foliage pattern 叶瓣花纹
foliated 剪边装饰
foliation 花叶形装饰
folk art 民间艺术
folk costume 民间服装,民俗服装
folke art 民间艺术
folklore 民俗款式,民俗服装
folklore craft-based accessories 民间手艺饰品,以民间手工艺为主题的饰品
folklore fashion 民间式时装,民俗服装款式
folklore look 民俗风格,民俗风貌
folklore print 民间印花
folklore style 民俗风格
folksy long skirt 民间特色的长裙
folkway 社会习俗
folk wear 民间服装,民俗服装
folk weave 松结构条格粗布,土布,条格粗布
follower 随动件
following of the fashion 赶时髦
follow-me-lads 女孩帽后长丝飘带
follow shot 跟镜头
follow the fashion 赶时髦
fondant prints 渗化型印花布
fond de neige 雪花纹花边底布
fond de rempli 花边地纹
fond d'or 金地花缎
fond epingle 梭结花边用六角网眼地
fond simple 简易花纹地纹
fondu printing 虹彩印花
fonfange headdress 丰当计头饰(17世纪)
Fontana sisters 丰塔纳姐妹(意大利时装设计师)
fontange [法]丰当什女帽;妇女花边高头饰;上浆打褶耸立在头上的薄纱头巾

fontanges hat 薄纱头巾;覆盖头顶的小帽
Fontangts 芳丹发型
foofaraw 华丽装饰;(衣服)褶边,华丽饰边
fool's badge 丑角挂徽
fool's cap
fool's cap, foolscap 小丑帽,滑稽帽,滑稽小丑圆锥帽,圆锥形小帽,大页纸帽,顽童纸帽,锥形帽
foot (复数为 feet)脚,(缝纫机)压脚;英尺(呎)
footage 尺码,呎长(以英尺表示的长度)
foot bag 腿筒
football face mask 橄榄球面罩
football helmet 橄榄球头盔
football jersey 橄榄球衫,足球球衣,足球衫
football knit shirt 足球衫,橄榄球衫
football pad 护腿
football shirt 橄榄球衫,足球衫
football shoes 足球鞋
football shorts 足球短裤,橄榄球短裤
football training shoes 足球训练鞋
football wear 足球装
foot-binding 缠小足(中国古代),裹脚
foot bottom 袜底
foot chain 脚链
foot channel 压脚沟槽
footcloth 马背装饰毯;地毯
foot cover 袜套
foot coverlet 盖脚毯
footed pajamas 裹袜式睡衣裤,连袜式衣裤,连袜套睡衣
footed pajamas style 连袜式睡衣裤款式
footee 儿童低筒轻便鞋,编结童鞋
foot entry 裤口,裤脚口
footgear, foot gear (总称)鞋袜
footing 饰边棉带;镶边花边
foot length 脚长
footless tights 无脚紧身裤
footlets 女套袜,脚掌套
foot lifting 抬压脚
foot lifting amount 压脚提升高度
footman's coatee 随从燕尾服
foot-mantle 骑马罩裙;骑马防尘外衣
foot muff 暖脚套,暖脚筒,腿筒
foot pad 鞋垫
foot pattern 脚型
foot pattern measuring 脚型测量
foot pedal 踏脚板

foot pressure 脚步压力
foot rule 一英尺长的尺
foot ruler 一英尺长的尺
foot sewing machine 踏板式缝纫机
foot socks 女袜套
foot splicing 袜脚加固
foot switch 脚踏开关
foot tread sewing machine 踏板式缝纫机
foot warmer 暖袜
footwear 鞋类(总称);鞋袜(统称)
footwear lining 鞋靴里衬
footwear size 鞋号
foot work 脚步动作(步法),舞蹈步法
forage cap 盔甲男童帽,哥萨克军便帽
foraging cap 军便帽,男童鸭舌帽
foraminous unitary non woven fabric 多孔单元非织造布
forbid 禁止
forbidden dyes 禁用染料
force lasting-leather shoes 套楦皮鞋
force majeure 不可抗力
forchette 手套指头狭条嵌镶,指缝档布(手套)
forchetting 狭条嵌镶(手套指头旁)
ford 时髦式样,(妇女时装中)仿高档品式样的低档品
fore-and-aft cap 前后有檐的帽
fore-and-after 猎鹿帽,前后附舌帽
fore-and-after cap 福尔摩斯前后翘猎帽
forearm 前臂,前袖
forearm seam 前臂缝
forearm sleeve seam 前袖缝
forecasting 预测
forefinger 食指,示指(食指,第二指)
forehead 额
forehead cloth 额饰布,扎头带
foreign capital 外资
foreign dress 外国服装,洋装
foreign dress woman's and girl's 妇女及少女用外国服装,女洋装
foreign exchange earning 创汇
foreign investment 国外投资
foreign trade 对外贸易
foreign trade zone 对外贸易区
forel 呢边
forelady 女领班;女工长
foreleg 前肢
fore limb 前肢
fore lock 额发
foreman 工长

forepart 前帮,前襟,前片,前幅,前围(制鞋)
forepart lining (衣服)前襟里子
fore sleeve 前袖,袖套,装饰袖
forest 墨绿色,森林色
forest boots 森林靴
forest fireproof shoes 森林防火鞋
forest green 丛林绿色(暗黄绿色),森林绿
forestieri 素色粗纺缩绒织物
fore stitch 平伏针法
forest night 夜森林色
forest shade 林荫色
forest whites 佩尼斯通粗毡呢
foresty cloth 森林警卫队制服呢
fore type 先型,原型
forfars 本色粗厚亚麻布
forget-me-not 灰蓝,潮蓝(植物色),勿忘我草色
fork (裤子的)档幅,胯幅
fork and front run 前龙门,上档弧线
forked shoe 丫巴鞋
forklore fashion 民俗服装款式
fork piece 裤衩档布,裤子衩档布
fork quantity 衩档拼插料尺寸,衩档处前后拼插料尺寸
forks 指缝档布(手套)
fork spreader 拨叉
fork to waist 裤直档,胯幅,直档
form 体型,造型,外形,形状;人体模型,模特儿;(制鞋)展开图
formability 成形性,可成形性
formal 女夜礼服,晚礼服,夜礼服
formal attire 礼服,礼仪服,正式场合服装
formal balance 形态平衡,匀称平衡
formal beauty 形式美,形态美
formal black 礼服黑
formal day wear 日礼服,昼间礼服
formaldehyde 甲醛
formaldehyde content 甲醛含量
formal dress 妇女礼服,正式礼服,礼服
formal evening wear 晚礼服,夜礼服
formal gesture 程式化姿势动作
formal gown 民族与宗教礼袍,正式礼服
formal group 正式群体
formal hat 礼帽
formalistic life style 正规生活方式
formal jeans 宴会牛仔裤,正式牛仔裤
formal rule 形式法则

formal shirt 礼服衬衫,制服衬衫,正规衬衫
formal shoes 穿正规服装时的鞋子（总称）,礼服用鞋
formal skirt 礼服裙
formal suit 礼服,正式礼服
formal utility 形式效用
formal vest 礼服背心
formal wear 正式礼服,礼服
formatic press 立体烫,立体熨烫机
formatic pressing 立体熨烫,立体熨烫机
formation ability 尺寸稳定性,形稳性
formative art 造型艺术
formback 泡沫塑料衬里
form design 产品外形设计,外形设计
formed fabric 立体成形织物;非织造布
formed plastics garments 泡沫塑料服装
former 成型机
form-filling apparel 合身型服装
form finisher 大身熨烫机,立体人形整烫机
form finishing machine 外形整理机
form fitting 紧贴合身
form fitting apparel 合身服装
form fitting sweater 紧身运动衫;紧身毛衫
forming 造型
forming and stitching knife pleat 叠顺褶
forming & stitching knife pleat 叠顺褶
formol 甲醛,福尔马林
form persuasive garment 适体性服装
form pressing 立体烫,立体熨烫
form pressing machine 立体熨烫机
form-shaping underwear 紧身内衣
form stability shirt 免烫整理衬衣
form stability suit 形稳性服装
for reference only 供参考
Forties' blazer(20 cen.) 40年代运动夹克（20世纪）
Forties look(20 cen.) 40年代风貌(20世纪)
fortifying fiber 增强纤维
Fortuny plait 福图尼褶
Fortuny pleat 福膝纳褶裥
Fortuny print 福图尼印花绸布
Forture 福琴混合股线织物
forward and backward sewing 前后缝纫,正反向缝纫,倒顺缝纫
forward baste （第二次试衣前用）长针脚疏缝
forward fashion 超前流行,超前流行款式
forward-leaning skylarker 前倾云雀帽
forward pocket 前倾型口袋,裤子斜式插袋
forward sewing 正向缝纫,顺向缝纫,向前缝纫
forward stance 前倾立姿
forward stitch 正送线迹
fossa jugularis point 颈窝点
fossys 印度棉布
Foster cotton 福字棉,福斯特棉
fota 厚实棉布（印度）
fotaloongee 印度麻丝条纹混纺绸
fote-mantle 护衣围裙
fotoz 土耳其犹太女用遮脸头巾
fottes 东南亚条格棉布,印度条格棉布
fougères 漂白家用亚麻布;粗包装帆布
foulard 薄软绸,丝斜纹绸,印花薄绸,真丝斜纹绸,桑丝绫,真丝绫,桑花绫;薄软绸围巾
foulard de chine 点子花纹薄软绸
foulard design 薄软绸图案,福拉德图案
foulardin 奥地利轧光棉袖衬布,富拉丁细平布（奥地利）
foulard poile de chevre 毛织薄呢
foulard scarf 软绸围巾
foulé 粗哔叽
foulé cashmere 开士米粗呢
foulé serge 粗哔叽
foulie 糙面粗哔叽
foul weather garments 恶劣气候防护服
foul weather gear 恶劣气候防护服,恶劣天气战士服装,坏气候防护服
foumart fur 臭貂毛皮,欧洲艾鼬毛皮
foundation 衬底;底布,粉底;基本纸型;基本内衣,紧身衣,女胸衣,上下连身整形内衣;里衬布
foundation cloth 刺绣用丝织底布
foundation cream 粉底膏霜,粉底霜
foundation dress 紧身装
foundation fabric 底布,地布
foundation garment 妇女紧身褡,妇女全套紧身内衣,妇女胸衣,妇女紧身褡;全帮;衣服衬衣;整形服装;整形内衣
foundation lotion 粉底化妆水
foundation muslin ［上胶］硬衬里纱布,硬衬里布
foundation net 上胶粗网眼纱,硬挺整理粗网眼纱
foundation pattern 单元图案,基本花纹,

基础纸样,简单图案
foundation pattern method　原型法
foundation pattern modification　基础纸样变形法
foundation plan　地脚示意图
foundation shirt　贴身衬衣
foundation silk　薄丝纱(印度),有光内衣绸
foundation stick　粉底条(美容)
foundation tape　人体模型基准线
foundation twill　衬里斜纹布,斜纹里衬布
foundation wear　妇女胸衣,妇女紧身褡
foundling bonnet　硬檐软顶小帽
fountain hair　喷水发型
four bar chain crank motion　四连杆曲柄链装置
four bar linkage mechanism　四连杆机构
fourchette　手套分指裆布,指缝裆布(手套)
four class　丁等,四级
four-cord sewing　四股缝纫
four-cord sewing thread　四股缝纫线
four-fold seam　四折缝
four-fold yarn　四股线
four-frame brussels　四经架布鲁塞尔地毯
four-frame brussels carpet　四经架布鲁塞尔地毯
four-frame carpet　四经架地毯
four-gored skirt　四幅裙,四片裙
four-gored slip　四片式连身衬裙
four-gore skirt　四幅裙,四片裙
four-gore slip　四片式连身衬裙
four-harness filling-faced twill　四页纬面斜纹薄棉布
four-hole button　四孔纽扣,四目扣,四眼扣
fouriaux　丝绸紧身衬裙
four-in-hand　活结领带
four-in-hand knot　四步活结
four-in-hand tie　德比(贝)领带,活结领带,四步活结领带
four-in-one　日本密织黑结
four-leaf twill　四页斜纹布
four needle cover stitch　四针覆盖针迹
four-needle thread　四针线
four-piece suit　四件式套装,四件套女装(衬衫、毛线背心、夹克、裙子等)
four-ply brocade　四色纬纱提花台布
four-point diamond　四尖钻石,四分钻石
four-point zigzagger　四点曲折线线迹缝纫机
four-P(4P) tone　四P绸

fourragère　[法]法国军装饰带,军服左肩穗带
fourreau　紧身连衣裙(法国名称)
fourreau dress　紧身连衣裙(法国名称)
fourreau skirt　腰线上无褶省的多片连衣裙
fourreau tunic　连体双层连衣裙
fourre-tout　[法]福托袋,杂物袋
fours　4号尺码的手套(或鞋号);灯笼裤,运动裤
four section girdle　四节腰带
four-sided stitch　四边形针法
four stitch zigzag　四点曲折线迹
fourteen　14号尺码,14号尺码服装
fourteen point　每英寸开口14次的机制花边
four thread overlock machine　四线包缝机
fourths　四等品
fourth toe　第四足趾
fourties look　20世纪40年代款式(长上衣,宽而上的翻领,粗宽的裤子翻边等)
Fowler trimmer　福乐剪线器
Fox　福克斯棉
fox　狐狸毛;狐皮,狐毛皮;狐狸棕色
fox brush　狐尾
fox cap　狐皮便帽
fox coat　狐皮大衣,狐裘
fox color　狐色
fox dress　狐皮服装
fox fur　狐皮
fox-fur robe　狐袍
fox hat　狐皮帽
foxing　后帮的后部(制鞋),围条(鞋),修补和装饰用鞋面皮
fox robe　狐皮大衣
fox tail　狐尾
FP　氧化铝纤维
frac　英式合体男晨礼服
fractional management　分级管理
fracture　破裂
fractured design　断面效果图案
fragile fabric　脆性织物,易损坏织物
fragile fiber　脆性物料
fragility　易碎性
fragment　碎块,碎片
fragmentary motif design　散点花样
fragmentation bomb chute fabric　炸弹降落伞织物
fragrance　香气,香味;香味化妆品
fragrance imageology　香味表
fragrant brocade　香香锦

fragrant fabric 香味织物,芳香织物
fragrant fiber 香味纤维,芳香纤维
fragrant smelling fabric 香味布料
fraise 法国草莓红;芙瑞丝领饰,时尚领饰
fraise à la confusion 不规则褶皱的无浆翻领
frame 画框
framed collar 镶边领子(牧场外套的毛皮领)
frame gloves 针织手套
frame heel 小方形高跟鞋
frames 眼镜框
frame tape 半漂亚麻扁带
France mode 法国模式
France twill 法国斜纹
franchise 特许加盟店;免赔率
franchised store 专卖店
Francis worsted twill 法国斜纹
Franco-Cuban heel 法国-古巴鞋跟
Franco Zeffirelli 费朗科·西菲雷妮(意大利著名戏剧服装设计师)
franella 绒布(智利,巴拉圭用语)
Frankfurt Heimtextil Trade Fair (FHTF) 法兰克福纺织交易会
Frank Olive 弗兰克·奥利弗(1929～ ,美国女帽设计师)
franneru 绒布(日本名称)
frash 弗雷雪吊袜带
fraternal ring 共济会戒指
fraternity pin 兄弟会别针
fraternity ring 兄弟会戒指
frayed cutoff shorts 毛边男短裤
fray hem 磨损衣边
fraying 散边;织物边缘磨损
frazada 拉丁美洲棉毯
Frazer tartan 格子呢
fred perry 筒状针织物结构
free alongside ship (FAS) 船边交货
free arm 筒式底板;底臂;筒式底臂
free arm machine 筒式底板缝纫机
free arm sewing 筒式底板缝纫
free arm sewing machine 筒式底板缝纫机
free at factory 工厂交货
free beauty 奔放美,自由美
free cross stitch 乱十字绣
free export 出口免税
free finger gloves 立体手套,自由式手套
free formaldehyde 游离甲醛
free form ring 自由式戒指
free-hand curved line 徒手画曲线

free-hand cutting 自由裁剪(无纸型平面裁剪)
free hand design 徒手设计
free hand drawing 手绘,徒手画
free hand embroidery 乱针绣,手工自由绣花,徒手绣花
free hand monogram 手工字母绣
free hand printing 手工描花
free hand sketch 徒手草图
free hand zigzag embroidery 徒手曲折形绣花
free hanging hem 下摆活络
free hanging lining 下摆活络的夹里
free inspection 免验
free inter-stitch 五线链式线迹,自由中间线迹
free-lance 自由职业者
free-lance costume designer 个体服装设计师;自由服装设计师
free-lance designer 自由服装设计师
free-lancer 自由设计师
free-lay carpet 拼块地毯,组合地毯
free line 自由型
free market 自由市场
free-motion work 自由缝纫;自由运动作业
free of all average (FAA) 一切海损均不赔偿
free of duty 免税
free pattern layout 自由排料
free perimeter 自由界区
free port 自由港
free rice stitch 乱针绣
freesia 小苍兰色
free size 可调尺寸
free size belt 可调节(长度)腰带
free size skirt 调腰式围裙
free sketch 写意画
free stitch 乱缎带贴绣;自由针法
free style wrestler's wear 自由式摔跤服
free-to-shrink 自由收缩
free touch 奔放(笔法)
free trade area [zone] 自由贸易区
free wig 万能假发
freezing tester 耐寒试验机
freight 班轮运费
freighter 托运人
freight forwarding agent 货运代理
freight station pier 货运站码头
freight to collect 运费到付
frelan 垂片头饰女帽

Fremish lace	佛兰芒花边
French antelope lambskin	仿羚羊皮的羔羊皮
French back	股经棉布；特经斜纹呢
French backed	纬缎背呢
French back serge	纬缎背哗叽
French back serge twill	纬缎背哗叽
French back twill	纬缎背斜纹呢
French beaver	法国海狸毛皮
French beige	米色，灰棕色
French belting	法式腰衬
French beret	法式贝雷帽
French binding	法式镶边
French blue	法国蓝（红光鲜蓝色）
French boa	皮毛或羽毛长圆筒围巾
French bottoms	法式男裤喇叭口
French bra	无乳杯法式胸罩，无胸杯法式胸罩
French braid	法式发辫
French calf	法国小牛皮
French cambric	上等亚麻薄细布
French canvas	法式帆布衬；厚重强捻纱罗女装料；平纹刺绣底布
French cashmere	法国精纺斜纹呢
French chalk	（裁缝用）滑石，划粉
French challis	加光印花薄织物
French cinch	法国腰带，束腰宽带，式肚带（有时装长吊袜带）
French clean	干洗
French cleaning	化学清洁法
French clock	法式斗篷
French collar	法式套领
French continental	法式西装型
French corsage	法国女紧身胸衣
French costume	法国民族服饰
French crepe	法式平绉
French crepe cord	法式薄绉
French crepe yarn	法国绉纱
French cuff	（法式衬衫）双袖头，法式翻边袖口，（衬衫）双袖口
French curve	曲线板，弯尺
French-cut leg	法式女泳装
French-cut panties	法国女内裤
French dart	刀背缝；曲线省，法式省（腰节上 2.5～5cm 间的省），法式褶，刀背省
French delaine	薄花呢
French diamond pique	法国菱形网纹织物
French double pique	法式双面凹凸组织（织物）
French dressmaker's hem	法式滚边下摆
French facon flannel	法国优质斜纹绒布
French faggoting	法式线迹接缝；法式装饰线迹
French fall boots	法式坠褶靴
French farthingale	法国式裙撑
French felling	法式缝边
French finish suiting	无光上浆棉平布
French flannel	法国条格法兰绒
French fold	法式折叠（两边向中间折叠并留有狭缝）
French foot	法国长筒袜；法式平袜脚
French foulé	法式毡缩织物
French front	暗门襟，法式（衬衫）前门襟
French front shirt	暗门襟男衬衫
French gathering	法式缩褶
French gauge fine inch	法式细针距
French gigot sleeve	羊腿袖
French gingham	高级轻薄方格平布
French gores	法式多片裙
French grey	浅灰色
French heel	法国（鞋）跟，法式鞋跟
French hem	法式衣边（毛边卷向里面缝牢）
French hood	法国头巾；连斗篷法式帽
French hose	男式圆形拼缝裤
French jacket	法式夹克
French kid	法国小羊皮
French knickers	法式女短衬裤
French knitting	法式针织饰带
French knot	法式线结（绕针而成的花式结）
French knot stitch	法式点结绣
French lace	法国机制花边
French lawn finish	法国细麻布式整理
French line	法国式鞋头形（方形鞋头）
French maid sleeper	法式少女睡衣（两件套，花边装饰短上衣）
French merino	绵羊毛；斜纹精纺呢，双面斜纹呢，法国美利奴呢
French navy	法国海军蓝色，无光海军深蓝色
French nightcap	法国睡帽；丝带装饰的白帽
French opening vest	法式低领背心
French panties	法式女短裤
French percale	法国高级密织薄纱
French piping foot	法式滚边压脚
French placket	法式袖口衩，半开襟袖口衩，法式袖头衩

French placket band 法式袖口衩镶边,法式袖头衩镶边
French pleat 法国装饰褶裥
French pleating 法式装饰褶裥
French pocket 滚边袋,嵌线袋,嵌袋
French policeman's cape 法式警察披风
French polonaise 法式衣裙;爱尔兰式连衣裙
French portrait buttons 法式人像纽扣
French provencal 法国乡村图案
French purse 法式钱包
French quilting 凹凸纹细布
French Revolution style 法国大革命风格
French rib 法式罗纹
French roll 法式发髻
French rose 法国红色
French ruff 法式轮状大皱领
French sailor dress 法式水兵领连衣裙
French sailor hat 法式水兵帽(海军蓝或色,棉布无檐帽,饰以海军蓝硬帽带和红毛球)
French sailor jack 法式水兵夹克
French sailor jacket 法式水兵夹克
French seam 法式线缝,来去线缝,来回缝,袋缝
French seam simulated 仿法式缝(包缝,外包缝)
French sennit 法国绳辫
French serge 法国哔叽
French shawl 法式披巾
French shoe size 法国鞋子尺码
French sinker wheel fabric 法式吊机织物
French size 法国尺码;法码
French skirt 法式后拖裙
French sleeve 法式袖,连肩袖,和服袖,超短袖,女式可卸袖,连袖(类似和服袖)
French tack 法式袢(连接面料和夹里的线袢),线袢
French terry 背圈平针织物
French Textile Machinery Manufactures Association 法国纺织机械协会
French thumb 弯曲型姆指,法式拇指;法式手套
French turnup 法式翻裤脚
French turtle neck 法式高圆套领
French twill 有光轻薄斜纹;英国右斜纹毛呢
French twist 法式卷发
French vanilla 法国香草色
French vermilion 法国红

French vest 法式背心
French welt 法式光边,法式关边;法式镶边
French whipped seam 法式锁边缝(先缝两道后锁边,用于绣花、花边或薄型织物)
French wilton carpet 威尔颜机织绒头地毯
fresco 弗雷斯科呢
fresh red 鲜红色
fresh salmon 鲜鲑肉色
fresh touch 新颖手感
Fresnel effects 弗莱斯纳尔效应
fret 回纹饰;控花刺绣;希腊饰边,卍字饰;饰网(中世纪);无边便帽
frial sales counter 试销专柜
friar brown 修士褐色
friar front shoe 僧侣鞋
friar's cloth 粗重方平棉布
frictional properties 摩擦性能
frictional resistance 摩擦阻力;耐磨性能
friction cloth 涂橡胶亚麻布
friction fabric 增阻布
friction mark 擦白
friction towel 擦背浴巾布
Friday suit 星期五套装
fried shirt 浆过的衬衫,礼服衬衫,上浆衬衫
Friend bonnet 贵格罩帽
friendship pin 友谊针
friendship ring 护身戒指,祝福戒指,友谊戒指
frieze 绒毛呢,拉绒粗呢,起绒粗呢(13世纪荷兰);厚重毛圈织物
frieze boucle 毛巾布,毛圈织物
frieze carpet 起绒地毯
frieze flannel 棉毛法兰绒
friezette 棱纹家具布
frieze velvet 毛圈丝绒
frigate cap 商船船帽,商船船员帽
frigidines 法国马尾衬
frileuse 女用斗篷
frill 饰边,荷叶边,褶边,绉边
frill collar 波褶领,褶边领,绉边衣领
frilled apron 绉边围裙
frilled elastic braid 褶边橡筋带
frill edge 饰边,荷叶边,褶边
frilled neckline 饰边领领口,褶边领口,绉边领口
frilled shoulder stap 饰边肩带
frilled sleeve 褶边式袖,绉边式袖

frill effect 褶边效应
frillery 衣褶边
frillies 百叶裙,褶边裙
frilling 褶边,织物上的褶裥
frill machine 褶裥机
frills 花边裙襞皱边
frills flouces ruffles 花边裙襞蓬松皱边
fringe （服装上的）缘饰;流苏,蓬边,毛边;前额垂发,刘海;排须;缨穗
fringe boots 流苏长靴
fringed 毛边
fringed boots 流苏装饰长靴
fringed cuff 穗饰克夫,穗饰袖口
fringed edge （布）毛边
fringed fabric 带穗织物
fringed narrow fabric 有穗狭带
fringed pocket 缘饰袋,缘饰口袋
fringed rope necklace 流苏项链
fringed scarf 流苏围巾
fringed selvage [selvedge] 毛边;加穗饰边
fringed tongue 锯齿边鞋舌
fringe foot 流苏压脚
fringe hair 前刘海发型
fringe lace 流苏花边,马克莱姆花边
fringe motion 缘饰装置,加穗装置
fringe stitch 流苏绣
fringe thread 流苏线,加穗线
fringe traction device 缘饰装置,成穗装置
fringe tuft 缨,流苏
fringing 排须,须条,须头,缘饰
fringing fancy stitch sewing machine 双行装饰线迹缝纫机
fringing machine 穗边织带机,缘饰机
friponne 女骑马服;女式三层裙
fripper 旧衣商
frippery 俗艳;艳俗低档服装
friquette 机制细网眼绣花花边
friquette lace 机制细网眼绣花花边
frisado 丝绒（西班牙名称）
frisé 高级亚麻布;毛圈粗呢;平纹结子线花呢;绒圈织物;粗棱纹棉布;素色剪毛地毯
frisé bouclé 法国毛巾布,大花纹丝花缎
frisette 妇女额前卷发
friska fabric 结子花线织物
frison 起绒粗呢
frisure 金银薄织带;金银线包芯纱
friz 卷发
frizette 妇女额前卷发;毛圈家具织物

frizz 卷发
frizzing 呢面卷结整理
frizzle 小轮状皱领
frock 罩衣,罩衫;（女）上衣;（童）外衣;连衫裙;男礼服大衣,长军大衣;工装;（水手穿的）羊毛套衫;僧袍,僧衣,道袍
frock cloth 浅灰色薄棉斜纹呢
frock clothing 礼服大衣;大礼服
frock coat 佛若克男大衣,福乐克礼服(19世纪),男礼服大衣,双排扣宽下摆男大衣;及臀合体女夹克
frocked jacket 非正式男大衣
frock great coat 双排扣室外长大衣
frocking 棉布面料（外衣用）,外衣棉织物
frock paletot 礼服大衣（男子）
frock skirt 福乐克裙
frog 搭环（腰带上挂武器或工具）;花结扣,盘花扣;腰带挂环;长纽带
frog fastening 盘纽;盘花扣
frogging 纺锤形纽扣
frog-loop buttonhole 盘花纽环形扣眼
frog pocket 裤斜插袋,骑马裤斜开口插袋,斜开口插袋（骑马裤）
frog suit 密闭防护衣
frog toggle button 盘花套绳纽孔
frog toggle button hole 盘花套绳纽孔
from pressing machine 立体熨烫机
fronce 皱褶
frong facing interlining 挂面衬
frong interlining 前身衬
front 前片,前身,襟,前胸;（衣服）正面;胸前部分,胸前物,（男人用白衬衫的）假胸;开口;下摆;假前发;鞋子前部
frontal 额带,帷子;额前盔,额前饰物;圣坛绣花挂帐
frontal bone 额骨
front and back shoulder cape 前后披肩
front and back yoke 前后覆肩
front and back yoke line 前后过肩线
front armhole 前袖笼
front armpit point 前腋点
frontayl 额饰,额饰布
front band 门襟翻边;明门襟
front bib 前兜
front blowing 前身膨胀
front bodice 前衣片,衣服前身
front body 衣服前身,前衣片
front border 门襟带
front bust area 前胸部（区）
front button 前扣

front centre line 前中心线
front closing bra 前扣式胸罩
front closings 门襟
front closing seam 下摆缝
front closure bra 前系扣胸罩
front collar 领面
front cover 前盖
front crease 裤腿前折缝,裤线,前挺缝线
front crotch 前裆,前浪,小裆
front crotch curve 前裆弧线
front crotch depth curve 上裆弧线
front crotch length 前裆长
front crotch line 前裆直线
front crotch seam 小裆缝,前裆缝
front crotch seam line 前裆内撇线,前裆直线
front crotch stay 前裤裆底,小裤底(小裆里子)
front crotch width line 小裆宽线
front crutch 小裆
front crutch stay 前裤裆底,小裤底
front cur seam 前摆缝
front curve 门襟圆角线
front curve of hood 帽前口线
front curve point line 门襟圆角线
front cut 撇门,前摆;前摆裁剪;止口圆角(前身止口末端的圆头)
front cut away 大圆角前摆
front cut line 门襟止口圆角线,撇门线
front cut point 门襟圆角点,止口圆角点
front cut silhouette 前摆轮廓线
front dart 胸省,(衣)前省,前褶,胸褶;(裤)前后省
front drape line 前造型线
front edge 门襟止口,止口
front edge is out of square 止口缩角
front edge is uneven 止口不直
front edge is upturned 止口反翘
front edge line 门襟止口线,前襟线,门襟圆角线,止口直线
front edge seam 门襟止口缝,止口缝
front edge sewer 止口缝纫机
front edges of sack-coat 男式便装短上衣的前衣边
front edge sticking up 止口反翘
front edge uneven 止口不平
frontel 额饰,额饰布
front facing (衣服)挂面,贴边
front facing cutside edge 挂面止口线
front facing figure 前身正面体形

front facing inside edge 挂面里口线
front facing interlining 挂面衬
front facing outside edge 挂面止口线
front fastening brassiere 前扣胸罩
front finish line 撇门线
front fly 门襟,门牌
front fly seam 门襟缝
front fly zipper (衣)门襟拉链;(裤)前裆拉链
front gilet 马甲,背心
front gimp 正面嵌芯细辫带
front girlet 围襟
front hem 前衣裙下摆
front hemming machine 前卷边缝纫机
front hip 前臀
front hip pocket 前臀部袋,前下插袋
front hook bra(ssiere) 前扣胸罩;前钩式胸罩
frontier branches of science 边缘学科
frontier inner lining 前夹里
frontier pants 美国西部牛仔裤,西部拓荒裤
frontier trade 边境贸易
frontings 优质亚麻平布
front inner lining 前夹里
front-in-one 一片式前片
front interlining 前身衬,前片衬
front leg 裤前片,前裤片
front length (衣)前身长
frontlet 额饰
front light 正面光
front line of hood 帽嘴线,风帽前口线
front lining 前身里子
front looper 前弯针
front looper thread cast-off 前弯针脱线
front neck 前领圈,前领
front neck depth 前领深
front neck facing 前领口贴边
front neckline 前领口线
front neck point (F.N.P.) 颈围前中心点,前颈点
front of garment 衣襟
front open dart 前身通省
front opening 前开门,(中装)大襟,对襟
front opening basic line 大襟斜线
front opening bias line 大襟斜线
front opening brassiere 胸前开襟胸罩
front opening curve 大襟弧线
front open shoes 前开口鞋
front or back shoulder cape 前后披肩

front or back yoke line 前后过肩袖
front outer 前片面料
front overlap 搭门,前搭门,衣前搭门
front overlap line 搭门线,搭门直线
front panel 前襟,前片,前幅,前身,前衣片
front part 前襟,前片,前身
front pitch line 底边翘高线,起翘横线
front placket 明门襟,半开襟
front placket opening 前半开门襟
front placket width 门襟宽
front pleat 西裤的褶裥(前腰头处),裤前褶
front pocket 前袋,前胸袋
front pocket facing 前袋口贴边
front point line 底边弧线(马甲前摆处),前身底边弧线(前身底摆夹角斜线)
front press 前身熨烫;前身熨烫机
front princess line 前公主线
front rise 前直裆,前裆,前浪,小裆
front rise curve 前裆弧线
front rise line 前裆内撇线,前直裆线,小裆线
front rise seam 前裆缝,小裆缝
front rise width line 小裆宽线
front round cut 圆角前摆
front run 前上裆弧线
front seam 前帮中缝,前袖缝线
front seam line 前袖缝直线
front shirt 明门襟男衬衫
front shirt body 前主体身片
front shoulder 前肩
front shoulder cape 前披肩
front shoulder dart 前肩省
front shoulder line 前肩线
front shoulder point 前肩点
front shoulder seam 前肩缝
front side seam 前侧缝
front skirt 前裙片
front skirt body 前主体身片
front sleeve 大袖,前袖
front sleeve drape 前袖造型
front sleeves/yoke 前过肩袖
front slit 前开衩
front square cut 方角前摆
front stiff (礼服衬衫的)硬衬胸,礼服衬衫假胸
front strap 明门襟
front strap shoes 袢带前帮鞋
front tab 前袢,前立
front tape 门襟带
front tape of trousers 小裤底

front through dart 前身通省
front torso 前躯干
front upper thigh 前大腿
front view 前视图,正面图,正视图,主视图
front waist 前腰
front waist arc 前腰弧
front waist band (裤)腰头门襟,腰间门襟,前腰头
front waist dart 前腰省
front waist length 前身至腰围长,前腰节长
front waist point (FWP) 腰围线前中点
front width 胸阔,前胸宽,门襟宽
front yoke 前过肩
front yoke line 前过肩线
front yoke sleeve 前过肩袖
front zip 前拉链
front-zipper foundation 前拉链女子整姿内衣
front-zipper jumpsuit 前开拉链连衣裤工作服;前开拉链伞兵跳伞服
front-zipper pants 前拉链裤子
front-zipping jumpsuit 前拉链连衣裤工作服;前拉链伞兵跳伞服
frost 霜色
frost appearance 霜白外观
frosted 无光泽的
frosted corduroy fabric 霜花灯芯绒
frosted yarn 霜花纱
frost grey 霜灰色
frosting 霜白疵;霜白现象;霜发,结霜发型
frosting finish 雪花整理,雪花整理
frost wash 雪花洗,霜花洗
frosty green 霜绿色
froth stain 泡斑
frothy garment 轻薄服装
frothy lace 轻薄花边
frottage 拓印法
frottle 浴巾织物
frou-frou 过分装饰(女服);沙沙声(丝绸衣服)
frou-frou dress 沙沙装
frou-frou length 沙沙长(拖地礼服等)
froufrou mantle 多层褶女斗篷
froufrou petticoat 精品衬裙
froufrou skirt 精品裙
frozen dew 冰露色
fruit's color 水果色彩
fruit stain 果汁污渍
frustration of contract 合同落空

F-test　方差比检验
fuchsia　倒挂金钟紫红色,紫红色
fuchsia pink　浅红莲色,浅莲红
fuchsia red　玫红色,梅红,深红莲色
fuchsin(e)　洋红,品红,碱性品红
fuchun batiste [habotai]　富春纺,青春纺
fucle　裤扣
fugitive color　褪色
fugitiveness　短效性,褪色性,不稳定性,易挥发性
fugitiveness to light　不耐光性,光褪色性
fugitive tinting　易褪着色
fu-gu crep　复古罗
Fuji　富士织物
Fujiette　富士纺
Fuji palace　富士派力斯
Fuji silk　富士纺,绢丝纺,平纹绢
Fukui habutai　日本最厚重的电力纺
Fukui silk　福井双绉
fukuji　日本手织棉制服布
fukurami　川端手感丰满度
fukusa　礼品盒盖绸
fuli brocade　富丽缎
fuli brocaded damask　富丽绸
fuli chin quilt cover　富丽锦被面
full　肥大,丰满;深色;盛装;缩绒,缩呢;毡合
full and clothy feel　丰满手感
full and round　丰满
full armhole　袖窿(全长);挂肩
full automatic circular hosiery machine　单程式袜机
full automatic claw type hydraulic toe-lasting machine　全自动爪式液压前帮机(制鞋)
full automatic flat knitting machine　全自动横机;全自动平型针织机
full automatic hosiery machine　全自动成型袜机
full automatic oil　全自动加油
full automatic oiler　全自动加油装置
full automatic pocket-hole sewing machine　全自动开袋机
full automatic rotary type plastic making machine　全自动轮转式塑胶鞋机
full automatic sewing　全自动缝纫
full automatic skip quilting machine　全自动跳针绗缝机
full automatic two-station type shoemaking machine　全自动双站式结鞋帮机

full-back stitch　全倒缝线迹,倒回穿刺线迹
full ball button　整球形扣,整圆形扣
full beard　络腮胡子,满脸大胡子
full bleach finish　全漂白棉布整理
full bleaching　全漂白
full blood wool　纯血统羊毛
full body　丰满织物
full body figure　成熟体型
full bosom　丰满胸部
full bosomed figure　丰满体型,丰满胸部的体型
full bottom　全覆式下装
full bottom wig　展宽假发,展式假发
full brogue　孔饰皮鞋
full brogue shoes　费欧雕花皮鞋,全雕花皮鞋
full bust　丰满胸部
full cape　全圆形披风,圆形披风
full cardigan knitted fabric　畦编针织物
full cardigan rib　全畦编
full chest　丰满胸部,满胸
full circle flare　太阳裙,全圆裙
full circle skirt　太阳裙,全圆裙
full circle sleeve　整圆摆袖
full circular skirt　太阳裙,两片喇叭裙
full color　全色,完全色
full core fiber　全芯纤维
full cover　书面细布
full-cross gauze　全绞式纱罗织物
full crotch length　全档长
full cuff　全翻边袖口,全克夫袖口,全卡夫袖头;真翻边
full cut　宽大式裁剪
full cut top　宽大式裤身;宽大式上身
full dartless sleeve　一片无省袖
full dress　燕尾服,大礼服,正式礼服,法定款式服装(指燕尾服);全装
full dress coat　男式大礼服
full dress evening coat　礼服,晚装,正式礼服,正式晚服
full dress rehearsal　着装彩排
full dress suit　燕尾服,正装套服
full dress uniform　大礼服,礼服制服
full dull　全消光;全无光
full dull filament　全消光丝
full dull polyester peach　全消光涤纶桃皮绒
full dull yarn　全无光丝
fulled fabric　缩绒织物
fulled scarf　缩绒长巾

full-electrification 全盘电气化
fuller cowl 满胸垂褶
fuller gore flare 大喇叭分片裙
fuller skirt 宽松女裙
Fullesta 聚酰胺工业缝纫线（商名，日本尤尼吉卡）
full face 黑体字;圆脸
full face finish 丰满呢面整理
full face gas mask 面部封闭式防毒面罩
full facial cover 全封闭面罩
full fashion 全成形衣服
full-fashioned hose 全成形女袜
full-fashioned hosiery (F/F hosiery) 平袜,柯登式长筒袜,全成形针织品,全成形袜
full-fashioned mark 成形星,全成形记号（全成形号）
full-fashioned outerwear 全成形外衣
full fashioned stockings 全成形长袜,全成形长筒女袜
full-fashioned straight-bar machine 全成型平型钩针机
full-fashioned sweater 全成形羊毛衫
full fashioning (F/F) 全成形
full fashion stockings 全成形长[筒女]袜
full feel 丰满手感,手感丰满
full figure 丰满体型
full-figure swim foundation 全身型泳装基础纸样
full-figure swim wear 成熟体型泳衣
full finish 双面整理
full flared skirt 全圆裙,喇叭长裙
full front 裾（中国服装）,前身
full front pleat 全前身褶裥
full fused front 全黏合衬前片
full garment knitting 全成形外衣编织
full gather smock 碎裙罩衣
full gauge 1+1 rib 四平满针罗纹;四平组织（羊毛衫）
full grain 皮革正面
full grain boots 光面皮靴;牛皮粒面靴
full grain leather 全粒面革
full grown 丰满的
full handle 丰满手感,手感丰满
full harness damask 单把吊提花缎
fulling 缩呢,缩绒,毡合
fulling fold 缩呢折痕
full-jacquard 满提花,全提花
full-jacquard fabric 满提花针织物,完全提花针织物

full knitted fabric 全成形针织品
full-legged trousers 大裤管西裤
full-length 及跟式,拖地式;拖地式服装,长达足跟式服装;与裙长等长的外套长度;（量身时）衣服总长（后颈中心到摆缘）;总长,全长
full-length fitted sleeve 全长合体袖
full-length mirror 穿衣镜,大试衣镜,全身镜
full-length skirt 拖地长裙
full length stockings 舞袜
full liability 完全责任
full lightly 适度缩绒,预缩（美国名称）
full line 丰满型,宽绰轮廓,宽松型,膨展型（轮廓）;实线
full lining 全衬里;全衬里制作;全夹
full matching 全幅匹配（指服装图案）
full mercerized finish 全丝光整理
full month shoes 满月鞋
fullness 丰满性,丰满度,宽松度;鼓起
fullness of shade 色泽丰满度
full open front 全开襟
full open front sweater 开襟卫生衫
full pants （宽松）袋形裤
full piqué 手套凸边缝
full print 饱满深色印花
full puffed sleeve 大泡袖
full range of size 尺码齐全
full regular 全成形针织品
full repeat 完全循环;完全花纹;完全组织
full roll collar 全翻领
full rough peruvian cotton 秘鲁粗绒棉
full run-down 单件直下
full scale 实物大小,实际尺寸;原尺寸;足尺的
full scale test 大规模试验
full seam 叠接缝,平接缝
full seam width 全缝宽
full shade 饱和色
full shaped hat 全成形帽;全成形男帽
full silhouette 丰满型,宽松型（轮廓）,膨展型（轮廓）
full size 原尺寸,全尺寸;全身的;无搭边排样尺寸,合理排样尺寸
full size mirror 大试衣镜
full skin 全皮
full skin fiber 全皮[层]纤维
full skirt 宽裙,宽下摆裙,喇叭长裙,全圆裙
full skirted style 宽裙风格

full sleeve 宽袖,长袖,垂袖	fully-shaped heel 全成形袜跟
full slip 长衬裙;宽松式女套裙	fully shrunk 全预缩的
full stout vest 半胸紧身式背心	fully-shrunk knit fabric 全预缩针织布
full strength orange 全浓橙色	fully trimmed overalls 可调节工装裤
full strength print 饱满深色印花	fulvous 暗黄色,黄褐色,茶色
full tee 全T恤(一块圆料构成,前后腰节抽褶,仿土耳曼式)	fulvous color 黄褐色
	fume fading 烟气褪色
full texture 紧密组织	fume-proofing finish 防烟气整理
full thread fabric 全线织物	fumishings 服饰品
full thread woven fabric 线织物	functional apparel 功能性服装
full throughly 缩绒,缩呢	functional arms span 两臂功能展开宽
full tongue shoes 整舌鞋	functional clothing 功能服装
full top 宽松上衣	functional design 功能性设计
full top grain 皮革正面	functional fabric 功能织物(具有防火、防雨、防缩、防尘、防静电等性能)
full tricot 满穿双梳栉经平织物	
full trutleneck 三重折领	functional fashion 功能时装
full turtle neck 三重折领	functional feature 功能特征
full upper shoes 满帮鞋	functional fiber 功能纤维
full value 整个明暗关系(绘图)	functional finish 功能整理(防雨、防火、防蛀、防缩、防皱等各种特种用途整理)
full view 全景,全视图	
full voile 双股巴里纱	functional property 功能性,官能性
full-volume silhouette 宽大轮廓	functional sport wear 功能性运动服
full weft warp-knitted fabric 经编全幅衬纬织物,全幅衬纬经编织物	functional textile 功能性纺织品
	function design 功能设计
full weight 毛重,全重	function finish 功能整理
full width 全幅,平幅	fundamental breach 根本性违约
full width folded 全幅平褶	fundamental color (测色)基础色,基本色,地色,基色,原色
full width weft insertion 全福衬纬	
fully aromatic polyester fiber 全芳香族聚酯纤维	fundamental measure 基本尺寸
	fundamental point 基本点
fully-articulating manikin 连接一体铜人	fundamental property 基本性能
fully automatic feather quilting machine 全自动羽毛绗缝机	fundamental tone 基调
	fun fashion 嬉戏时装
fully automation 全盘自动化	fun fur 化纤花色毛皮,人造毛皮
fully-bleached linen 全漂亚麻布	fun furs 廉价毛皮服装,人造毛皮服装;针织化纤花式裘皮
fully constructed garment 结构研究服装	
fully fashioned 全成形的	fungicide finish 抗菌整理
fully fashioned garment 全成形服装,全成形针织品	funginertness 抑霉菌性
	fungus resistance 耐霉性
fully fashioned hosiery 柯登式长袜,平袜,全成形袜子	funk fashion 乡土气息款式,朴实无华款式
	funky clothes 时髦服装,时髦衣服
fully fashioned knitwear 全成形针织物,全成形针织品	funky look 朴实无华款式
	funnel collar 筒形立领,漏斗领
fully fashioned outerwear 全成形外衣	funnel hat 漏斗帽,无檐漏斗形帽,圆筒帽
fully fashioned rib garment 全成形罗纹外衣	funnel neck 筒形立领,漏斗领
	funnel neckline 漏斗形领口
fully fashioned stockings (F/F stockings) 全成形长(筒女)袜	funnel sleeve 大口翻边袖,漏斗袖
	funnel-type folder 漏斗式折边器
fully relaxed dimension 全松弛处理后尺寸	funny bone 幽默感;肘部尺骨端
fully-shaped collar 全成形领圈	funny pants 奇异裤

fun skin 化纤花色毛皮,针织化纤花式裘皮
Fuqiang fiber 高湿高模量黏胶纤维,富强纤维,波里诺西克,虎木棉
fur 毛皮,皮子,皮草,皮毛,裘皮;绒毛;裘皮制品
fur anorak 皮猴儿
fur beetle 损害裘皮的甲虫
furbelow (衣裙)褶饰,边饰,裙饰,褶襞;俗气装饰
fur beret 毛皮贝雷帽
fur body (圆筒形)粗毡帽坯,筒式粗毡帽坯
fur brush 毛刷
fur brushing machine 毛皮滚毛机,毛皮刷毛机
fur cap 毛皮帽
fur cape 裘皮披肩
fur clean 裘皮清洁法
fur cloth 仿毛皮织物,人造毛皮,毛皮织物
fur clothes 毛皮袄,毛皮服装
fur clothing 毛皮大衣;裘;皮衣
fur cloth trousers 人造毛皮裤
fur coat 毛皮外套;裘皮大衣,毛皮大衣
fur collar 毛皮领/皮领
fur collared coat 裘皮领大衣
fur collar jacket 毛领夹克,皮领夹克
fur collar overcoat 裘皮领大衣
fur-covered button 毛皮包扣
fur cuff 毛皮袖口,毛皮袖头
fur cutting machine 毛皮裁剪机
fur design 裘皮设计,毛皮设计
fur-effect fabric 充皮织物,仿皮织物
fur fabric 仿毛皮衣料,仿皮织物
fur fabric coat 仿皮大衣
fur fabric hem 毛皮饰边
fur facing lashing machine 正面裘皮缝纫机
fur facing sewing machine 毛皮正面缝纫机,正面裘皮缝纫机
fur farming 人工饲养动物毛皮
fur-felt 绒毛毡
fur-felt hat 丝绒呢帽,绒毛毡帽
fur fiber 毛皮纤维,绒毛
fur garment 裘皮服装
fur hat 毛皮帽
fur hat body 绒毛帽坯
fur hook eye 毛皮钩眼扣
fur imitation 人造毛皮,仿毛皮织物
fur imitation plush 仿毛皮长毛绒
furisode 宽袖和服
fur jacket 裘皮短大衣
Fur Labeling Act 美国皮质商标法
fur-leather garment 裘革服装
fur-like fabric 仿毛皮织物,人造毛皮织物
fur-like fiber 仿兽毛纤维,兽毛状纤维
fur-like handle 仿毛皮手感,似毛皮手感
fur-lined coat 毛皮衬里大衣,毛皮里裘皮大衣,皮里长大衣
fur-lined gloves 毛皮里子手套
fur-lined gown 毛皮袍
fur-lined jacket 短皮袄,毛皮里短袄,皮里短袄
fur-lined robe 皮袍
fur-lined shoes 毛皮衬里鞋
fur liner 毛皮里子
fur liner wear 毛皮里服装,皮里服装
fur lining 毛皮衬里,毛皮里子,皮里
fur lining clothing 毛皮衬里大衣
fur lining coat 毛皮衬里大衣
fur lining felling machine 衬里毛皮折缝机
fur lining garment 毛皮衬里服装,毛皮衬里外衣
fur making 毛皮缝制
fur marker 毛皮划粉
fur market 毛皮市场
furnace black 炉黑
fur necklet 皮围巾,兽毛皮围脖,兽毛围巾
fur neckpiece 皮围巾
furnes flax 比利时亚麻
furnishing 服饰,服饰品(美国名称);陈设品;家具
furnishing braid 家具饰带
furnishing fabric 家具布,家具织物;装饰用织物
furnishing prints 装饰印花布
furnishings 服饰品(美国)
furnishing trimming 家具饰带
furnishing wadding 装饰衬垫
furniture cord 家具套饰带
furniture damask 家具花缎
furniture denim 斜纹家具布
furniture dressing 家具衬里,家具衬料
furniture fabric 家具[用]织物
furniture plush 家具用长毛绒,家具丝绒
furnitures 家具织物
furniture stuff 装饰织物,装潢布,家具布;家具填料
furniture twill 家具用斜纹印花布

furniture upholstery 家具装饰材料
furniture web 家具织带
fur or leather garment 裘革服装
fur overcoat 裘皮大衣
fur padding machine 毛皮衬里缝纫机
fur parka 皮猴儿
fur parka type ski jacket 皮派克滑雪衫
fur paw 动物爪皮毛饰边;动物爪毛皮拼成的皮毛衣服
fur piece 毛皮制品;裘布裤子
fur polishing and ironing machine 毛皮整烫机
furred 起毛的;毛皮的
furrier 毛皮商,皮货商;毛皮衣制作工,皮货技师
furrier knife 毛皮切刀
furrier's beetle 毛皮甲虫
furrier's craft 毛皮工艺;毛皮技艺
furrier store 皮货店
furrier's workroom 皮货加工场
furrier's workshop 毛皮工作车间
furriery 毛皮;毛皮衣制作;毛皮商
furring 贴毛皮;贴毛绒;毛皮衬里,毛皮镶边
fur robe 毛皮袍,皮袍
furrow 起皱,皱纹;沾污,染污
furrow silk 棱条绸
furry material 毛皮衣料;仿毛皮衣料
furs 毛皮围颈物;毛皮衣服,皮衣,裘皮服装
fur scarf 毛皮围脖,毛皮围巾
furscous 淡褐灰色
fur seal 海豹毛皮
fur sewing machine 毛皮缝纫机
fur sewing machine for stretching operation 弹性织物毛皮缝纫机
fur sewing thread 裘皮缝纫线
fur shading 人造毛皮润色
fur shawl 皮披肩
fur shoes 毛皮鞋
fur skin 毛皮
fur skin cape 毛皮披肩
fur skin collar 毛皮衣领
fur skin overcoat 毛皮大衣
fur skin stole 毛皮围巾
fur skin tie 毛皮领带
fur stole 裘皮披肩
fur store 皮革商店,毛皮商店
fur strap sewing machine 毛皮条缝合机
fur substitute 毛皮代用品

fur tack 毛皮扣钉
fur texture 毛皮结构
further-processing 深加工
furtive 轻快感毛皮;运动式裘皮夹克,具有轻便风格的裘皮时装
fur topcoat 皮大衣
fur trader 皮货商
fur trim 毛皮饰边,毛皮滚边
fur-trimmed cuff 毛皮袖头
fur-trimmed evening dress 裘皮装饰晚礼服
fur-trimmed evening jacket 裘皮装饰晚礼服
fur trimming 皮领;毛皮饰边,毛皮装饰;毛皮修剪整理
fur type 毛皮类型
fur-type fabric 人造毛皮
fur-type knitted fabric 人造毛皮针织物
fur worker 毛皮工人
fur working 毛皮加工
fuscous 暗褐,淡褐灰
fuscous color 暗褐色
fuse 熔(化),熔合,热熔黏合
fused belt 压熔腰带
fused collar (热熔)黏合领
fused collar fabric 热熔黏合领衬织物
fused fabric 热熔层合织物
fused interfacing 热熔衬,热熔贴面
fused joint 熔融黏结(非织造布)
fused ribbon 醋酯熔边布条
fused seam 压熔缝合
fused shirt collar 热熔黏合衬衫领
fuse fusible interlining 热熔黏合衬,热压薄膜衬
fuse interlining top collar 热黏领衬、面
fuselage cloth 飞机机身布
fuse-ring anthrone dyes 稠环蒽酮还原染料
fusibility 可熔性
fusible fabric 热熔衬
fusible interlining 黏合衬,热熔衬,热熔黏合衬
fusible non-woven interlining 无纺黏合衬
fusibles 热熔衬,黏合衬
fusible stay 热熔牵条
fusible web 双面黏合衬
fusible woven interlining 有纺黏合衬
fusiform 流线形
fusing 熔合
fusing collar in terlining 黏翻领(领面与领衬黏合)
fusing interfacings 热熔衬;热熔贴面

fusing interlining 热熔衬,黏衬
fusing machine 黏合熨烫机
fusing machine operater 压衬作业员
fusing material 熔合缝料
fusing press 黏合机,压烫熔合机
fusing press machine 黏合机,压衬机
fusion bonded nonwoven 熔融黏合法非织造布
fusion shrinkage 热熔收缩
fussies look 绒毛风貌
fusta 西班牙粗纺呢
fustanella 男式裙,希腊短裙,男短裙
fustian 粗斜纹布,法斯蒂安织物;纬起绒织物,纬起毛织物
fustian anapes 纬起毛织物
fustian cord 灯芯绒
futa 埃及光洁毛巾
futah 阿拉伯男子腰布
futahaba kanakin 日本衬衫料坯布
fu tou 幞头
futurism 未来主义风格
fuwen matelasse 浮纹花绸
Fu Zi shoes 福字鞋
fuzz 绸面起毛,微毛
fuzz hair type 绒毛型
fuzzies look 绒毛款式
fuzzing and pilling 起毛起球
fuzz resistance 抗起毛性
fuzzy edge 毛边
fuzzy pattern 轮廓模糊
fuzzy pile 绒
fuzzy style 绒毛款式
fuzzy texture 茸毛表面
fylfot 卍字饰
fylfot cross 卍字绸缎花纹图案;卍字图案

G

gaban 加邦布,法国粗布
gabarage 打包用粗布
gabardeen 华达呢,轧别丁,斜纹织物;宽大的粗布衣,工作服;犹太粗布长袍(中世纪)
gabardine 轧别丁,华达呢,斜纹织物;中世纪犹太人穿的粗布长袍;华达呢服装;(英国工人的)工作服,宽大的粗布衣
gabardine raincoat 华达呢隐扣雨衣
gabbia 头饰(意大利称谓)
gabercord 棉轧别丁,棉华达呢
gaberdine 轧别丁,华达呢;中世纪犹太人穿的粗布长袍;华达呢服装;(英国工人的)工作服
gaberum 加伯伦格子布
gable bonnet 尖拱帽檐女帽
gable hat 尖拱帽檐女帽,山墙(软)帽
gable headdress 16世纪妇女三角形头饰,山墙头饰
gable hood 16世纪英国黑丝绒头罩
Gabriell CocoChanel 加布里埃勒·夏奈尔可可(法国高级时装设计师)
Gabrielle Chanel 加布里·夏奈尔(法国时装设计师)
Gabrielle dress 加布里装;女童泡泡袖连衣袖;女式田间连衣裙
Gabrielle sleeve 泡泡袖
Gabrielle waist 紧身胸衣
Gadaz No.1 cotton 印度加达一号棉
gadget bag 化妆包
gadiapet 床上用纺织品
Gaetano-Savini Brioni 加埃塔诺·萨维尼·布廖尼(法国时装设计师)
gaffer 烫皱褶熨斗
gaffered silk 轧纹绸
gaffering 褶裥;缝褶裥
gage 定规;毛线织物的针数
gahfiya 加非阿头巾(科威特)
gaigum 印度手织格子细布
gaily-colored 色彩鲜艳的
gaily-colored pattern 色彩鲜艳花样
gainly conduct 优雅举止
Gainsborough hat 大型装饰女帽,堪斯保罗大型装饰帽,黑丝绒女帽,马尔保

罗帽
gaint panda fur 大熊猫毛皮
gaitan 保加利亚彩色窄编带,盖顿编带
gaiter 鞋罩;绑腿;有绑腿的高筒靴;高帮松紧鞋
gaiter boots 女式及踝短靴
gaiter bottoms 裤腿吊带绕过靴底的长裤
gaiter cloth 妇女绑腿呢,鞋罩呢
gaiter thigh-protector 护腿
gaiter trousers 裤腿吊带绕过靴底的长裤
gaji 盖季光亮缎(东南亚),印度光亮缎
gala 粗棉布
galabia 宽大长袍(地中海阿拉伯国家),无领对襟束带长袍
galabijeh 无领对襟束带长袍,埃及宽大长袍
galabiyah 无领对襟束带长袍,埃及宽大长袍
galage 鞋面;防雨靴;木底低腰靴
galashiels 苏格兰粗纺呢
galatea 白底蓝红条纹布,衬衫用条子棉织物,蓝白条纹布
galatea fichu 女式荷叶边围巾
galatea hat 草编童帽
gala twill 加拉斜纹
galef 粗斜纹布
galerum 伽莱伦格子(东南亚);加莱伦帽(古罗马)
galette 粗绢丝;法国绢塔夫绸
galette silk 粗绢丝
galevum 伽莱伦帽
galicha 印度绒头棉地毯
galilla 伽利拉领,带饰边的半月形领
galims 伊朗加利姆双面绒头羊毛地毯
Galitaine 加里泰因(意大利时装设计师)
galla 蕉麻
gallant 丝带装饰
gallants 现形缎带装饰缝,蝴蝶结装饰缝;缎质蝴蝶结饰物
Gallenga 加伦加(意大利时装设计师)
Galligas coins 革绑腿(19世纪);加利加斯肯裤(16~17世纪)
galligaskins 革绑腿(19世纪),加利加斯肯裤(16~17世纪),皮护腿;男式筒形裤

Gallini cotton 埃及加利尼棉
Gallipoli cotton 意大利加利波利棉
Gallo-Greek bodice 带交叉饰带的紧身上衣款式
galloon 缎带,纱带,全银花边,全银丝带,双光边饰带;金银花边;金银丝带
galloon and double 丝鞋带;丝鞋结
galloon braid 斜织细带
galloon lace 金银丝花边
galloon raschel lace 金银花边;拉舍尔缎带花边
galloon ribbon 横棱纹缎带
galloshes 防雨靴
gallowses (裤的)背带,吊带
gallow twill 加拉斜纹
galluses (裤的)背带,吊带
galoche 加罗西套鞋
galon [法]军衔饰纹(带),金线头带,丝头带
galon d'argent 银丝花边
galon de couleur [法]色线
galosh 木屐;木鞋
galosh closing 可调节夹紧式系紧件
galoshes 长筒胶鞋,防水布套鞋,高筒橡皮套鞋,高筒套鞋,硬底鞋
galoshes gums 长筒套靴
galoshs 屐,木屐,木鞋;防水套靴,加罗西套鞋
galra 印度伊斯兰教女服手织条纹绸
galuchat 坚韧的头层鲨鱼皮
galungan 皇冠状头饰
galway 红色粗厚法兰绒;爱尔兰红色厚大衣呢
galyak 加立克裘
gambade 防水绑腿布,强缚绑腿布
gambeson 棉马甲;皮马甲;软铠甲;盔甲内衣
gambeto 厚呢外套
gambettes [法]红脚鹬绑腿
gambiered Canton 拷绸,茛纱
gambiered Canton gauze 拷绸,茛纱,香云纱,拷纱,茛纱绸
gambiered canton silk 拷绸,茛绸
gambiered gauze 黑胶绸,荔枝绸,拷绸,拷纱,茛纱,香云纱
gamboge 橙黄色,橘黄,藤黄
gambo hemp 泽麻,槿麻
gambroon 平纹棉布;斜纹棉布;亚麻斜纹里子布;交织斜纹布
gamcha 手织原色条纹布

gamekeeper 巴布尔夹克
gaming purse 抽绳提包
gamin hairstyle 流浪儿发型
gamme [法]色阶;色调
gamp 伞
gamsa 仿绉背缎斜纹绸
gamut 色移
gamut of colors 色阶
gamza 仿绉背缎斜纹绸
Gandhi cap[hat] 甘地帽,印度无边白帽
gand(o)ura 衬衫式肥大服装(中近东和北非),甘杜拉装(非洲)
gand(o)urah 衬衫式肥大服装(中近东和北非),甘杜拉装(非洲)
ganga 斯里兰卡大麻
ganges 仿蛇皮革
gangetic muslin 孟加拉细棉
gangster suit 强盗式外套(20世纪30年代,宽肩宽驳头双排纽,黑色或灰色细条法兰绒)
gangway 档车弄,工作走道
ganse 带子的窄边
gant de compétition [法]比赛手套
gantlet 宽口臂套;(骑马或击剑等用的)长手套,防护手套,长宽口手套;(中世纪骑士戴的用皮革和金属片制的)铁手套
gantline 吊绳
Gantron 甘特朗亮光织物
Ganymede 凯尼米得款式(侍酒俊童款式)
Ganymede look 凯尼米德款式,凯尼米德风貌
Ganymede sandals 侍酒俊童式凉鞋;凯尼米得凉鞋
Ganymede style 希腊男童款式
gaohua jacquard brocattele 高花绸
gaohua jacquard fuji 高花绸
gaohua jacquard sand crepe 泥地高花绸
Gaoshan ethnic costume 高山族民族装
Gaoshan ethnic fancy cloth 高山族花布
Gaoshan fancy cloth 高山族花布
gap 开气口
gape 豁口
gaping 咧口(鞋)
gaping armhole 袖隆太大(与袖山不配)
gaposis 衣着不整病,因衣服太紧纽扣间出现的裂口
gara 印度手织粗棉布
Garabani Valentino 加拉班尼·瓦兰蒂

诺(意大利时装设计师)
garan finish 毛皮修剪整理
garb 服装,制服,装束,特殊服装;特种服装
garber cotton 美国加柏棉
Garbo hat 嘉宝帽(遮阳阔边软帽)
garcon look 假小子风貌(20世纪),少年款式
Garconne look 少男型,(20世纪)假小子风格
garcon style 少年发型
garde corps 宽松无带外套;宽袖防寒服
garde corps with hanging sleeves 悬饰袖防寒服
garden furnishing 户外用布(各种篷布、躺椅用帆布等)
garden green 花园绿
garden hat 宽檐下垂女帽;椭圆平顶女帽
gardening tog 园艺衣
Garibaldi blouse 红色阔罩衣;加里波第衫
Garibaldi jacket 齐腰方形짜上衣
Garibaldi shirt 红色阔罩衣;妇孺宽大衬衫,加里波第衫
Garibaldi sleeve 加里波第袖
Garibaldi suit 无领童衫;加里波第套装
gari serge 粗毛哔叽
garish 花哨
garland 花冠,花环;希腊神父头饰
garlandage 花环饰
garment 衣服,服装,外衣,衣着
garment accessories 服饰品,服装辅料;衣服配料
garment aftercare 针织外衣后整理
garment alternation 服装修改
garment assembly 整套服装
garment association 服装协会
garment bag 衣罩;放置衣服的折叠旅行袋
garment bias 斜裁滚边料
garment blank 衣片,衣坯
garment-borne-particles 服装产生的微粒尘屑
garment brand 服装品牌
Garment centre 服装中心
garment character 服装特性,服装特色
garment closure 拉链;纽扣
garment color 服装色彩
garment color combination 服装配色
garment component 服装组件
garment construction 服装结构
garment conveyer system 服装悬挂传输系统

garment conveyor 衣片传递装置
garment corporation 服装公司
garment cutting 服装裁剪
garment cutting book 服装裁剪样本
garment demand 服装需求
garment design 服装设计
garment design competition 服装设计大赛
garment detailing 服装细部
garment dip process 成衣浸渍整理(成衣免烫整理)
garment drape 服装悬垂性
garment dyeing 成衣染色,衣件染色
garment dyeing shrinkage 成衣染色缩水率
garment engineering 服装工程
garment evaluation 服装评估
garment fabric 服用织物,服装面料,衣料
garment fabrics innovation and design centre 服装面料创新和设计中心
garment factory 服装厂
garment felt 服装毡
garment finish 服装整理
garment finishing 服装整理,成衣整理
garment function 服装功能
garment hanger 衣架
garment image 服装形象
garment in Chinese style 中式服装
garment industry 成衣工业,服装工业,服装业
garment information database 服装信息数据库
garment inspection 成衣检验
garment interlining 服装衬里,服装衬料
garment knitting 外衣编织
garment laundry pressing machine 成衣洗烫机
garment length 成衣长
garment length circular knitting machine 计件圆形针织机
garment-length knitting machine 计件衣坯针织机
garment machinery 服装机械
garment making 服装制作
garment management information system (GMIS) 服装企业信息管理系统(服装MIS系统)
garment manufacture 服装制造;服装制造业,服装工业
garment manufacturer 服装制造商;服装制造业
garment manufacturing 服装制造

garment marking 服装标记
garment material for main parts 大料
garment measurement 服装测量
garment merchandising management 服装商品企划
garment molding 服装模型;服装熨烫定形
garment panel 衣片
garment parts 服装部件,衣片
garment pattern 服装纸样
garment physiology 服装生理学
garment pieces 衣片
garment press 烫衣机
garment printing technology 成衣印花技术
garment product data management system 服装产品数据管理系统(服装 PDM 系统)
garment robot 服装机器人
garments 服装,衣着
garments association 服装协会
garments basic theory 服装基础理论
garment section 裁片,衣片
garment setting 成衣定形,服装定形
garment sewing 服装缝制
garments function 服装功能
garments higher education 服装高等教育
garment shrinkage 服装收缩
garments hygiene 服装卫生学
garment size 服装尺寸,服装尺码
garment size series 服装号型系列
garment sizing 服装尺码
garments magazine 服装杂志
garments mark theory 服装符号学
garment society 服装学会
garments of plain silk gauze 素纱禅衣
garment specialist system 服装专家系统
garment specifications 服装规格
garments physiology 服装生理学
garment spray process 成衣喷洒整理(成衣免烫整理)
garments research and designing centre 服装研究设计中心
garments robot 服装机器人
garments setting 衣服成型
garments standardization 服装标准化
garment steamer 服装熨烫机;汽蒸熨烫服装用人体模型
garments trade clause 服装贸易条款
garment style 服装款式
garment sweater 外用针织套衫,纬编外用针织套衫

garment trade 服装贸易;服装业,外衣业
garment twist 洗涤后衣服的扭曲
garmenture 衣着,服装
garment vacuum packing 成衣真空包装
garment visualization 服装可视化
garment wash(ing) 成衣水洗,普通水洗;服装水洗整理
garment wear test 服装穿着试验
garment wet processing 成衣湿加工,服装湿加工
garment working sketch 服装工作图
garnache 斗篷式套头上衣;卡纳克长披风(中世纪)
garnet 石榴红;石榴石
garnet brown 石榴石棕(红棕色)
garnet red 深石榴红,石榴石红,酱红
garnet rose 石榴红
garnish 装饰,装饰物
garnishing 服装饰边
garniture 服装,服饰;饰带;修剪
garras 家用坚固棉布
garrha 印度粗平纹棉布
Garrick overcoat 加立克大衣(18 世纪)
garrison belt 弹力吊袜带;军装腰带
garrison cap (士兵)船帽,船形军帽,船形帽,军用扁平折叠布帽,无帽缘军用布帽
garrison pants 殖民地制服裤子(英国)
garter 衬衫袖带;筒袜带,吊袜带;嘉德勋章
garter band 袜口下段吊带部分
garter belt 吊袜带紧身褡;吊袜束腰带;吊袜松紧带
garter belt hose 连吊袜带袜子
garter brassiere 吊带胸罩
garter briefs 吊袜短内裤,吊袜三角裤
garter girdle 吊袜束腰,吊袜腰带,连吊袜带紧身褡
gartering 服饰吊带
garter length 半高筒短袜(长至小腿肚),袜带长的短袜
garter measure 裤脚口尺寸
garter panties 吊袜内裤,连吊袜带内裤
garter-run stop 袜口下段防脱散
garter skirt 碎褶裙
garter socks 嘉德短袜
garter stockings 连吊袜带长统袜
garter top 袜口下段
garter webbing (吊袜)松紧带
garter welt 吊袜带袜口
gas black 气黑

gas cell fabric 气囊织物
gascon coat 女骑马服
gas defence 防毒
gas fading 烟气褪色;烟薰褪色
gas iron 瓦斯熨斗
gasket 衬垫,填料
gasketing fabric 衬垫织物
gaskins 宽身裤,灯笼裤;皮绑腿
gas mask 防毒面具
gas permeability 透气性
gas-proof coveralls 防毒工作服
gas proofing 不透气
gas proofness 不透气性,气密性
gas protection equipment 防毒器材
gas protective clothing 防毒气服装
gassed georgette 充乔其纱;高支烧毛线绉布
gassed yarn 烧毛纱,烧毛纱线
gassing 烧毛
gassing frame 烧毛机
gas tightness 气密性
Gaston Berthelot 加斯东·贝特洛(法国时装设计师)
gate ball shoes 门球鞋
gate fine 细棱条绸;细棱绸
gather (衣裙)折裥,皱褶,死褶,碎褶;给(衣、裙)打褶裥,抽碎褶
gather cape 抽褶斗篷
gather control notch 抽褶定位刀眼
gathered cape 抽褶斗篷,带褶披肩,碎褶披肩
gathered dart 抽褶省
gathered dirndl 抽褶裙,碎褶裙
gathered edging 抽褶边饰,荷叶边饰
gathered frill 抽褶荷叶边
gathered neckline 束带领口
gathered pants 绉褶裤,褶裥裤
gathered petal 抽褶花瓣袖
gathered petal sleeve 收褶花瓣袖
gathered pleat 碎褶
gathered ruffle 抽裥褶边,碎裥褶边,碎褶边,抽褶边
gathered self-bound seam 抽褶自包缝
gathered shoulder 抽褶肩
gathered skirt 抽裥裙,碎褶裙,打裥裙,褶裙,收裥裙
gathered sleeve 碎褶裙,抽褶裙
gathered trousers 碎褶裤
gatherer 打裥器,打裥装置
gather flare skirt 碎褶喇叭裙
gather front open 碎褶前开襟

gathering 打裥,抽碎褶,打褶裥,缝褶裥,内褶缝,收褶
gathering attachment 打裥器,打裥装置
gathering device 打裥装置
gathering foot 碎褶压脚
gathering maximum limited dial 最大打裥限度盘
gathering tape 打裥带
gathering threads 抽褶缝线
gathering wave 褶裥波纹
gather mark 碎褶记号,碎褶符号
gather notches 抽褶刀眼
gathers 打裥裥织物;碎褶
gathers in a dart 省道中抽褶
gather skirt 碎褶裙,褶裙
gather sleeve 褶裥袖
gathers on a band 在平带上抽褶
gathers shirred skirt 碎褶裙
gather stitch 打褶线迹
gather-up curtain 上卷式幕
gatpot flower 阔花绫
Gatsby look 盖茨比款式(20世纪20年代的美国时装,法兰绒套装等)
gattar 印度丝经棉纬缎
gattered silk 轧纹绸
gatti 菱形花布(东南亚);印度菱纹棉布
gatyak 宽松及膝农夫裤
gatya trousers 戈塔亚麻男裤
gaucho 加乌乔服饰
gaucho belt 加乌乔饰带
gaucho blouse 加乌乔短罩衫(南美),加乌乔牧人短罩衫(南美)
gaucho hat 加乌乔帽,加乌乔牧人帽,牛仔帽
gaucho look 加乌乔款式,南美牧人风格,牧人风貌(南美)
gaucho pants 高乔牧人裤;(南美)牛仔裤,女宽松裤,加乌乔裤,短长裤;
gauchos 加乌乔牧人裤(裤腿塞在靴内),加乌乔裤;南美牛仔裤(裤脚肥大,踝部束紧)女式牧裤;短长裤
gaucho shirt 高乔牧人衬衫(阔领针织衫),加乌乔牧人阔领针织衬衫,牧人裙
gaucho silhouette 加乌乔轮廓,加乌乔轮廓线条,南美牧人轮廓
gaucho skirt 加乌乔牧人裙,牧人裙
gaucho style 加乌乔款式
gaucho stylistic form 加乌乔款式
gaucho suit 加乌乔裤装(大裤脚,长及小

腿肚裙裤和短皮外套)
gaud 炫丽服饰品
gaudery 俗艳廉价的服装(或服饰)
gaudivi 印花粗布(印度)
gaudy 花哨;华丽而俗气的廉价饰物
gauffer 轧花,起皱,拷花,打皱褶
gauffer calender 印纹轧压机,凹凸纹轧花机
gauffered cloth 拷花布,轧花凸纹织物,轧纹布,轧纹织物,凸凹轧花布
gauffered design 拷花花纹
gauffered fabric 拷花布,轧花织物,轧花凸纹织物,轧纹布,轧花凸纹织物
gauffered finished fabric 轧纹整理织物
gauffered finish fabric 轧纹整理织物
gauffrage effect 轧花效应
gauffré 拷花,轧花(法国名称)
gauffré ribbon 轧花带,拷花带
gauffré velvet 雕花天鹅绒
gaufré 拷花,轧花(法国名称)
gaufré crepe [轧花]泡泡绉
gaufré ribbon 轧花带,拷花带
gaufré satin 轧花绉缎
gaufré silk 轧花绸
gauge (gg) 定规,样板;隔距;毛线织物的针数,织针号
gauge doria 印度模纱织物
gauge mark 定位标记
gauging 测定;多层抽褶,缩褶(用于童装、女衫);校验
gauging batch 规测批量
gaung baung 岗邦头巾(缅)
gaunt 中世纪比利时粗纺呢
gauntlet (骑马或击剑等用)长手套,防护手套;(中世纪骑士用的皮革和金属片制的)铁手套;长宽口手套;宽口臂套,金属护手
gauntlet cuff 护手式长袖头,宽口手套式袖头,宽硬式袖头
gauntlet mitten 有前护臂的连指手套
gauntlet opening 宽口手套式的宽口
gauntlets 礼服手套
gausape, gausopam 紫色厚重呢
Gauzanda 高赞德蝉翼纱(商名,瑞士)
gauze 纱布,纱罗,(棉、丝织)薄纱,高斯织物
gauze and leno silk 纱罗类织物
gauze bandage 纱布绷带
gauze broché 花纱罗,提花纱罗
gauze cloth 薄纱布,纱罗织物
gauze curtain 纱罗窗帘
gauze curtain fabric 窗帘纱
gauze doria 印度仿纱罗织物
gauze effect 罗纹,纱罗花纹
gauze effect fabric 罗纹织物
gauze fabric 薄绢,薄纱布,纱罗织物
gauze flannel 条纹薄法兰绒
gauze gambiered 莨纱
gauze illusion 女装纱罗
gauze kerchief 纱巾
gauze leno 纱罗,透凉罗,网眼罗
gauze look 沙罗款式,纱罗风貌
gauze mask 纱布口罩
gauze nonwoven 非织造纱布
gauzes 纱
gauze silk 纱
gauze stockings 麻纱袜
gauze sylphide 缎条纱罗
gay colors 缤纷彩色
gay deceivers 假胸罩,胸衬
gay 1990s' 20世纪90年代艳丽服装
gay 1990s' swimsuit 20世纪90年代艳丽泳装(两片式;圆形横条针织裤装,长及膝的裤子和短袖及臀部的上装
gaza au fuseau 花边网眼凸棱效应,花边网眼凸边效应
gazar 挺爽绸,挺爽呢,闪光薄丝织物
gaze 纱罗,透凉罗,网眼罗
gaze barege 印花薄呢,印花毛葛
gaze brilliantine 高级丝绸有光纱罗
gaze chambery 丝纱罗
gaze damasseé 双纬花纱罗
gaze de fil 亚麻条纹纱罗
gaze de lin [法]麻纱罗
gaze de paris 巴黎纱(法国制全丝薄绸)
gaze de voilette 丝织透明薄纱罗
gaze faconnée 提花纱罗
gaze faconnée broché 绣花用纱罗
gaze faconnée raye 柳条纱罗
gaze filoche 全丝网眼纱罗
gazeline barège 19世纪末女装用半透明纱罗
gaze lissé 法国薄生丝纱罗
gazelle 瞪羚羊皮
gaze marabout 法国真丝纱罗;三色绒织物
gaze mask 纱布口罩
gaze milanaise 棉芯丝线薄纱
gaze ondée 法国波纹纱罗
gaze tour anglais 纱罗织物
gazi 印度粗棉布

gazlick 土耳其粗平布
gazzatum 亚麻纱罗,薄纱罗
gazzi 丝经棉纬手织装饰布
gear 服装,衣着,衣服;时髦行头,(年轻人)时髦服饰
gear duck 齿轮用帆布
gear stitch 齿轮状缎带贴绣
gebei lining 裙褶
gedgehog wig 刺猬假发
gee string G带(美国印第安人或歌女、舞女等当三角裤穿)
geflect 缎纹上的刺绣针迹
gefu taffeta lame 格夫绢
geisha silk 匹染纺绸;薄面纱
Gelao ethnic costume 仡佬族民族服
gelatin fiber 凝胶纤维
gelatinous fiber 凝胶纤维
gele 盖丽(非洲女戴巾或头饰),格利头饰(非洲),非洲妇女头巾或头饰
gel fiber 凝胶纤维
gel fiber filament 凝胶丝
gel filament 凝胶丝
gelled fiber 凝胶纤维,胶化纤维
gel-spun fiber 凝胶纺纤维
gel state fiber 凝胶态纤维
gel thread 凝胶丝
gel yarn 凝胶丝
gem 宝石
gem button 宝石扣
gem color 宝石色
gem cut 宝石切割;被切割成的宝石
gemel ring 嵌套的双环戒指,双环(或多环)戒指
Gemini 盖米尼(耐高温涂铝织物,商名,英国)
Gemological Institute of America (GIA) 美国宝石学院
gemologist 宝石学家
gemology 宝石学
gemstone 宝石,适合作宝石的矿石
gemstone bracelet 宝石手链镯,镶宝石手镯
gemstone pendant 宝石垂饰
gemstone ring 宝石戒指
genappe 毛棉条纹呢
Genay 盖内薄型织物
gendarme 大陆式铜扣警察夹克,法国警察服装款式
gendarme cap 及膝警察圆披肩(法国)
gendarme cape 法国警察穿的长至膝部的圆披肩

gendarme jacket 大陆式铜纽夹克(来自法国警察夹克)
General Agreement on Tariffs and Trade (GATT) 关税及贸易总协定
general arrangement drawing 总装图
general blind hemming attachment 普通暗缝[服装边缘]装置
general cleaning 普通洗涤
general embroidering 普通刺绣
general fabric 普通织物
general finish 常规整理
generalization 通用化
general layout 总平面图;总体布局
general material 普通缝料,一般缝料
general merchamdise manager(GMM) 营销总经理
general overhaul 大修[设备]
general plan 总计划,总图
general purpose machine 通用机器
general purpose pitch-based carbon fiber 通用沥青基碳纤维
general seamer 海京包缝机,普通包缝机
general town wear 一般性都市服,一般性外出服
general utility press 万能压平机
general view 概略图,总图
general viscose rayon 人造丝
general wholesale 百货批发
generic heel 通用鞋根
generic name 通称,属称;被联邦贸易委员会认定的纤维
generos crudos 本色棉布
generous marking 过松的排料
generous seam 大余量缝
genet 麝猫黄(暗绿光黄色);麝猫皮
Geneva band 日内瓦白色双饰带;日内瓦牧师领
Geneva embroidery 日内瓦刺绣
Geneva gown 黑色宽袖长法衣(日内瓦)
Geneva gown and bands 牧师黑长袍与白长带
Geneva hat 日内瓦帽,高顶宽边帽
Genghis khan style 成吉思汗款式
genghis rug 土库曼斯坦地毯
gengxin bellow silk 更新缎
geniculum 小膝
Genoa 热那亚厚花边
Genoa-back corduroy 斜纹地灯芯绒
Genoa cloak 热那亚斗篷
Genoa cord 热那亚灯芯绒,斜纹地灯

芯绒
Genoa embroidery 热那亚刺绣
Genoa lace 梭结花边,热那亚花边
Genoa plush 热那亚棉纬平绒
Genoa twill 热那亚斜纹,三页斜纹
Genoa velvet 热那亚全丝花丝绒;棉纬平绒
Genoa velveteen 热那亚棉纬平绒
Genoese embroidery 热那亚刺绣,热那亚剪花刺绣
Genoese lace 热那亚花边
genouillere 护膝甲
genou volage [法]膝盖式发型
gentian blue 龙胆蓝(浅紫蓝色)
gentian violet 龙胆紫
gentle color 柔和色
gentle curve 优美曲线
gentle cycle 缓和程序(洗衣机),洗衣机缓和程序
gentle fashion 文雅款式,优雅式样
gentlemen's 绅士服
gentlemen's carpet 光泽柔和的地毯
gentlemen's jeans 绅士牛仔裤
GENTLEMEN'S QUARTERLY 《绅士季刊》(美)
gentlemen's rubber boots 绅士靴
gentlemen's shaped line 英国式绅士风度式样
gentlemen's suit 绅士服
gentle washing program 缓和的洗涤程序
gently flared 温和型喇叭裤
gently flared pants 温和型喇叭裤
gentrice 贵族感,高贵感
gent's parka 男派克大衣
genuine bonding fiber 全黏结纤维
genuine fleecy 添纱起绒织物
genuine leather 天然皮革
genuine leather belt 真皮腰带
genuine leather buckle 真皮带扣
genuine leather handbag 真皮手袋
genuine look 纯粹款式
genuine part 正品配件
genuine ultramarine 开然群青,艳蓝色
geodesic distance 最短距离
Geoffrey Beene 杰弗里·毕恩(1927～,美国时装设计师)
geometrical drawing 几何图
geometrical pattern 几何图案
geometric bob 几何短发型
geometric cut 几何形式裁剪

geometric design 几何图案;几何纹样
geometric distribution 几何图形的分布
geometric form 几何形模型(素描用)
geometric grey scale 几何分级灰色卡(评定染色牢度)
geometric line 几何线条
geometric manipulation operation 几何花纹控制
geometric pattern 几何图案,几何纹样
geometric porosity 多孔性结构
geometric print 几何图案,几何印花
geometric properties 结构特性
geometric regularity 几何规整性
geometric repeat 几何图案花纹循环
geometric shape 几何形状
geometric spot 几何图案点子
geometry cloth 织物几何结构
geo nonwoven 土工非织造布
George 乔治靴
George Almanio 乔治·阿尔马尼奥(意大利时装设计师)
George boots 乔治靴,乔其靴
Georges Doeuillet 乔治·杜雷特(法国高级时装设计师)
George Stavropoulos 乔治·斯塔沃普洛斯(美国高级时装设计师)
georgette 乔其纱,八综棉绉
georgette brocade 织花乔其纱
georgette crepe 俄罗斯乔其纱,乔其纱,乔其绉
georgette shirt 乔其纱衬衫
georgette velvet 乔其绒,乔其丝绒
georgette with printed velvet flower 烂花乔其绒
georgetton 棉乔其纱
Georgia crepe 佐治亚绉
Georgia Institute of Technology 乔治亚理工学院(美)
Georgian crepe 佐治亚绉(链状皱纹,以小菱纹、鸟眼纹等构成)
Georgian prolific cotton 佐治亚短绒陆地棉
Georgian wool 格鲁吉亚绵羊毛
georgienne 法国娇琴绸
georgine 娇琴纱
geos 乔斯大衣呢
geranium lake 天竺葵红
geranium pink 妃色,绯红
Gerard Pipart 热拉尔·皮帕尔(1933～,法国时装设计师)
Gerber cut 格柏自动裁剪系统

Gerber cut bundles [garments]　格柏裁片
Gerber cut-pants　格柏裁片
Gerber equipment　格柏公司设备
Gerber mover　格柏吊挂线系统
gergette velvet　乔其绒
German continental　德式西装型
German costume　德式民族服
German gown　德式长袍
German helmet　德式帽盔
German hose　德式宽松开衩男裤
germania　马海毛里子缎
Germanic costume　德国民族服饰
German jacket　德式短上衣,德式夹克
German knot stitch　德式打结针法
German Patent (GP)　德国专利
German prints　德国印花粗棉布
German sailor's cap　德国水手帽
German seam　德式缝(锯齿形缝),锯齿缝
German seamer　通用绗缝机
German stitch　德国绣；德式针迹(双面针迹和斜形针迹相结合)
German wool　德国绒线
germa seamer　通用绗缝机
germ-free clothes　无菌衣
germ-free uniform　无菌工作服,无菌衣
germicide agent　杀菌剂
germ-resistant　防腐剂；防菌剂
germ-resistant fabric　抗菌织物
germuset　格缪赛绸(小亚西亚制丝棉花缎)
gerras　家用结实棉布
gertrude　童装衬裙；婴儿内衣(无袖,两肩常用纽扣)
Ge silk　绎丝
Gestalt　格式塔(流派)
Gestalt theory　格式塔理论
gesture　手势,姿态
geta　木屐,日本木屐
getaz toe　盖塔兹袜头
getee　拉吉马哈尔大麻
G.G.stylistic form　GG 款式
ghabrum　加伯伦格子棉布
ghaghara　加格拉裙(印度)
ghagi　印度丝缎
ghagra　加固利裙；加固拉裙(印度)
ghagrapet　内衣布(印度妇女用)
ghagri　加固利裙；加固拉裙(印度)
ghaher　手织纯棉条纹布
gharbasti　棉经野蚕丝纬绸
gharha　达勒布

ghatar　孟加拉湾蚕丝
ghati　手工粗平布
ghatpot　绫,真丝绫
ghatpot plain embroidered　绣花绫
ghatpot plain heavy　大绫
ghatpot plain mixed　交织绫
ghatta　扭结黄麻
ghazilieh　土耳其八码棉丝花绸
ghenting　根特亚麻平布
Ghent lace　比利时根特梭结窄花边
gherad　印度白绸
ghetee　印度漂白棉布
GH fiber　GH 阻燃纤维(商名,日本)
ghilam　南京绸
ghilian silk　伊朗丝绸
ghillie(s),gillie(s)　低帮无舌运动鞋,跟从鞋,吉利鞋
ghiordes knot　套双经地毯结
ghiordes rugs　吉奥德地毯
ghlila　及臀背心
ghosting　重像
ghost shifts　全自动车间,无人车间
ghum rug　优级波斯地毯
gi　束腰宽松服(柔道或空手道运动员穿)
giaberello　14～15 世纪意大利罩衫
giallo　铅黄
Gianfranco Ferré　詹弗兰科·费雷(1945～,意大利时装设计师)
Gianmarco Venturi　詹马科·文图里(意大利时装设计师)
Gianni Versace　吉安尼·范思哲(1946～,意大利时装设计师)
gianshan crash　千山绸
giant abutilon　大商麻,大青麻
giant collar　巨型领
giant panda fur　大熊猫毛皮
giant saddle stitch　大鞍形针迹
giant stripe　特宽条纹,特宽条纹花纹
giant zip coat　大拉链外套
gib　缝纫机夹紧轮
gibao　汗衫；(和服的)贴身衬衫
Gibson Girl　20 世纪初美国推崇的妇女服饰
Gibson girl hairstyle　吉卜森女发型(美国艺术家吉卜森设计)
Gibson girl look　吉卜森女服款式,吉布森女郎款式
Gibson girl's style　19 世纪末至 20 世纪初吉卜森女服风格
Gibson girl style　吉卜森女服款式美国艺

术家吉卜森设计
Gibson waist blouse 吉卜森女衬衫
gibus 吉巴士帽
gibus hat 可折叠黑缎男帽,夜礼帽
gift bag 礼品袋,赠品包
Gifu pongee 日本歧阜府绸
gig coat 箱形男外套;女式翻领合身长外套
gigging 起绒;起毛
Gigi dress 吉吉女礼服(长宽袖,丑角领,齐膝);吉吉少女连衣裙
Gigi look 吉吉款式
Gigi style 吉吉款式
gigot (袖隆松,袖口紧)羊腿袖
gigot sleeve 羊腿袖(袖隆松,袖口紧)
gigottes 羊腿型短裤
GI haircut 美军发型
Gil Aimbez 吉尔·艾贝兹(美国时装设计师)
gilan 伊朗丝绸
Gilbert Adrian 吉尔伯特·阿德里安(1903～1946,美国时装设计师)
gilded cattle leather 镀金牛皮
gilded skirt 金属线印花布裙
Gildo Cristian 吉度·克里斯蒂安(意大利时装设计师)
gilet (腰线以上的)短背心;女背心内衣;女式假衬衣;女式紧身马甲
gilet sport 轻便背心
gilham 南京绸
gill box 针梳机
gillies 低帮无舌运动鞋(鞋带系于踝节部的无舌鞋),吉里鞋,随从鞋
gilling oxford 19世纪浅口运动鞋
gills 男衬衫底领(俚称)
gill spun yarn 针梳纱
GI look 美国士兵款式
gilsonite 沥青色
gilt 女式假衬衣
gilt button 金纽扣
gilt coin 金币
gilt edge 金边
gilt-finished aluminum eyelet 镀铝纽孔
gilt leather 包膜金银皮,金银皮
gilt membrane 包膜金银皮,金银皮
gimian 土耳其油绒毯,姆伦精细丝绒地毯
gimmie (免费赠送的)广告鸭舌布帽
gimmie cap [hat] (免费赠送的)广告鸭舌布帽,印有商标的赠帽
gimp 嵌线;披肩;薄纱;(装饰用)嵌芯辫带;花边图案粗廓线

gimpa crepe 银波绉
gimp buttonhole 嵌线纽孔
gimp covered fur hook and eye 皮毛缠线钩眼扣
gimp covered hook & eye 缠线钩眼扣
gimped buttonhole 粗线纽孔,嵌线纽孔
gimped embroidery 衬垫刺绣,嵌芯刺绣,缀金刺绣,教堂式刺绣
gimple crepe 银波绉
gimp yarn 嵌芯花线
Gina Fratini 吉纳·弗拉蒂尼(英国时装设计师)
Gin ethnic costume 京族服饰
ginger 姜色(棕色)
gingeras 东南亚绸,印度绸
ginger brown 金色宝石;金黄绿玉
ginger snap 栗棕色
ginger spice 红棕色
gingham 格子布,柳条布,方格色织布,金汉姆织物,色织手袋布,条格平布
gingham check 彩色格子布
gingham dress 方格[条格]布女裙
gingham for handbag 色织拎包布
gingham plaid 彩色格子布
gingham print 仿色织条格印花
gingham tissue 条格薄绢
GIOIA 《欢乐》(意)
GIORDANO 左丹奴
GIORGIO ARMAN 乔治阿曼尼
Giorgio Armani 乔治·阿玛尼(1934～,意大利时装设计师)
Giorgio Armani/AX 乔治·阿玛尼
Giorgio Correggiari 乔治·科雷贾里(意大利时装设计师)
Giorgio Saint Angelo 乔尔吉欧·圣·安杰洛(美国时装设计师)
GI pants 美国士兵
gippon 基蓬衫(14世纪欧洲军用服装),紧身长袖衫
Gipsy blouse 吉卜赛罩衣
Gipsy bonnet 吉卜赛帽,妇女儿童戴的宽边帽,宽边帽
Gipsy cloak 吉卜赛外衣
gipsy cloth 高档棉法兰绒织物
Gipsy costume 吉卜赛民族服饰
Gipsy dress 吉卜赛女装
gipsy earring 大圆圈耳环,吉卜赛耳环;奴隶耳环
gipsy hat 吉卜赛帽
Gipsy longuette 吉卜赛长连衣裙

Gipsy look 吉卜赛款式
gipsy seam 蚌面鞋
gipsy stripe 吉卜赛条纹
gipsy stylistic form 吉卜赛款式
gipsy vamp 蚌面鞋
giraffe collar 折叠式立领;高领
giraffe design 长颈鹿花纹图案
giraffe print 长颈鹿印花
girandole 吉伦多垂饰
girasol 火蛋白石
giraut 色条布(印度制),印度细布,印度色条布
girdle 带子;腰带;腹带;束腹,绑肚,紧身褡
girdle band sewing machine 腰带缝纫机
girdle cloth 紧身衣料
girdle corset (女用)束腹带
girdle seam 腰带线缝
girdle stead 腰围
girdle with suspenders 有肩带的腰带或束腹
giridi dhoti 印巴手织平纹刺绣柞绸腰布
girl dress 女童装
girl frock 裙衫
girlish style 女孩风格
girl's bag 少女背包
girl's blouse 女童衫,少女短上衣
girl's clothes 女孩服装,少女装
girl's coat 少女大衣
girl scout cloth 灰绿色制服棉布,美国女童子军制服布
girl's dirndl 紧身细褶女孩裙服;紧身腰褶女孩服
girl's drawers 女童长衬裤
girl's dress 女童装
girl's figure 女孩号型;女孩体形
girl's frock (女)童裙衫,少女裙衫
girl's jumpsuit 女童吊带裤
girl's over blouse 少女罩衫
girl's size 女孩号型
girl's skiing trousers 少女滑冰裤
girl's skirt 女童裙
girl's skirt suit 少女裙装
girl's ski trousers 少女滑冰裤
girl's slip 女童套裙
girl's smock 女工作服,女便服;女式无袖宽内衣
girl's socks 少女袜
girl's style 少女款式
girl's style fly 女装钮牌
girl's suit 女童套装

girl's wear 女孩装,女童装
girl s weater 少女套装
girl's-wind-jacket 女孩防风夹克
girth (身体有关部位的)围长;横档,腰围;腰巾,束带;大小,尺寸;宽带织物;阔带织物
girth line 腰围线;腰节线
girth measurement 腰围尺寸
girth tolerance 腰围余量
girt-up gown 束腰长袍
girt-up surcoat 束腰长袍
girt-up tunic 束腰长袍;束腰罩衫
giselle 花色缎带;法国精纺薄毛纱
GI shoes 美国步兵军鞋
gispy cloth 高档棉法兰绒织物
GI style 美国士兵款式
given after 放行
GIVENCHY 纪梵希
given size 给定尺寸
givrin 棱条绉
giza cotton 埃及吉萨棉
glabella 眉间点
glacé 光泽效应
glacé cotton 蜡光饰线
glacé finish 光泽整理
glacé leather 亮面革
Glacem 聚酰胺超细纤维(商名,日本钟纺)
glacés 有光织物(棉经、马海毛纬)
glacé silk 闪光绸
glacés lusters 有光羽纱
glacé taffeta 闪光塔夫绸
glace thread 蜡光棉线,蜡光线;上光丝
glacé yarn 上光线,蜡光线
glacier 冰川色
glacis 透明感
glad clothes (口语)时髦衣服;最好的衣服;晚礼服
gladiator sandals 斗士凉鞋,格雷蒂凉鞋
glad rags 晚礼服,夜礼服,最考究的衣服
Gladstone 格莱斯顿男式双排扣短外套
Gladstone bag 格莱斯顿皮箱;铰合式手提旅行包
Gladstone collar 格莱斯顿立领
glam fashion 富有魅力的时装
GLAMOUR 《魅力》(美)
glamo(u)r 魅力;衣着入时
glamour dress 魅力装束
glamour girl 衣着入时的漂亮女郎
glannen 粗纺法兰绒

glare 显眼
glass 玻璃;镜子;眼镜
glass butt 玻璃棉毡
glass beads 玻璃小珠;南京玉
glass bra 眼镜胸罩
glass button 玻璃扣
glass-ceramic fiber 玻璃-陶瓷纤维
glass cloth 玻璃器皿揩布;玻璃纤维织物,涂玻璃粉织物;松织亚麻经条布
glass cover 眼镜盒套
glass curtain 窗纱
glass curtain cloth 窗纱织物,玻璃纤维窗帘布
glass curtain fabric 窗纱织物,玻璃纤维窗帘布
glass eraser 玻璃纤维擦图笔
glasses 眼镜
glasses pocket 眼镜袋
glass fabric 玻璃布,玻璃[纤维]织物
glass fiber 玻璃纤维
glass fiber fabric 玻璃纤维织物
glass fiber felt 玻璃纤维毡(作绝热用)
glass fiber mat 玻璃纤维垫子
glass fiber warp knitted cloth 玻璃纤维经编针织物
glass fiber woven cloth 玻璃纤维布
glass filament 玻璃[长]丝
glass glasses 玻璃片眼镜
glass-headed pin 玻璃头大头针
glassine 玻璃纸
glass silk 玻璃纱,玻璃丝
glass silk interlining 玻璃丝衬
glass suit 赛车手防火玻璃服
glass textile 玻璃纤维纺织品
glass top cloth 玻璃台板布
glass top fabric 玻璃台板布
glass top nonwoven 玻璃台板非织造布
glass toweling 玻璃器皿揩布
glass yarn 玻璃丝;玻璃纱,玻璃纤维纱
glassy finish 玻璃光泽整理
glaucous color 蓝绿,绿灰色
glaze 色泽,光泽;加光,轧光;上光;极光
glazed 高光泽
glazed calico 摩擦轧光细布
glazed chintz 摩擦轧光艳丽花布
glazed chintz finish 摩擦轧光整理
glazed color 釉彩
glazed cotton 加光棉线,上光棉线,加光线,蜡光棉线
glazed cotton sewing thread 上光棉缝线

glazed finish 高光泽整理
glazed kid 光泽整理的小山羊皮
glazed kidskin 上光小山羊皮
glazed silk 加光丝,上光丝
glazed thread 加光线,蜡光线,上光线,上光丝
glazed yarn 加光纱,上光纱
glaze rolls 轧光棉衬布
glazing 打光,光面加工;轧光,上光
glazing agent 上光剂
glazing sheet 上光板
Glcut 美国士兵短发型
gleamer 容光焕发膏
gleam green 闪光绿
glen check 格伦格子呢;小方格花纹
glengarry 斑点粗呢;苏格兰便帽;连帽大衣
glengarry cap 船形军帽
Glen Urquhart plaid 格伦厄克特格子呢,苏格兰异形格呢
Glen Urquhart checks 格伦厄克特花格纹;格伦厄克特格子呢
gliding properly 平滑性,光滑,(织物表面的)滑移性
glint cloth 亮丽布料
glissade 格利萨特里子布,丝光纬缎棉衬布,黑地细色条经条棉衬布
glitter 闪光,发光;(呢绒久磨后出现的)极光
glitter button 闪光扣
glitter effect 闪光效应
glitter hose 闪光金银丝袜子
glittering fabric 闪光织物
glittering finish 闪烁整理
glitter look 闪光风貌,闪光款式
glitters 闪烁片
glitter yarn 闪光纱
global view 全向视图
globose handle 球形伞柄
glocke 雨披式缩绒厚呢外套
gloria 环形头饰;雨伞布,细密轻薄交织伞绸
gloria cloth 丝毛交织薄绸
gloriana 低档丝经棉纬伞绸
gloria silk 丝毛交织薄绸
glorietta 棉纱细密薄布
gloriosa 棉经丝纬伞绸
gloss 光泽,光彩;使有光泽,上光
glossaret 高级毛呢;高级丝绸
gloss finish 平光整理
glossiness 光泽性,光泽度

gloss-iron 烫光
gloss number 光泽指标
Glossop printers 英国格鲁索印花布
gloss silk fabric 熟织织物(毋需后整理的色织丝织物)
glossy 有光泽的
glossy black 乌亮
glossy wool 丝光羊毛,有光羊毛(氯化羊毛)
glove 手套
glove band 手套袋
glove box 手套箱;干燥箱
glove button 手套扣
glove cuff 革柔皮;手套口布,手套克夫
glove cutting press 手套裁剪机
glove fabric 手套面料,手套织物
glove finger commencing 手套指起口
glove finger knitting machine 有指手套针织机
glove flannel 手套棉法兰绒
glove knitting machine 手套机,手套针织机
glove leather 手套皮革
glove length 手套长度
glove liner 手套衬里
glove machine 手套机
glove piqué machine 手套凹凸缝纫机
glove presser 手套压烫机
glove press machine 手套压烫机
glove prix seam machine 手套毛边外向缝缝纫机
gloves (五指分开)手套;拳击手套;棒球手套
glove seam 手套缝,手套缝合方式
glove seam type 手套缝合方式
glove silk 经编手套丝织物,制手套用经编丝织物
glove sleeve 紧身细长袖
glove's needle 手套缝针,手套专用针
glover's stitch 手套线迹
glove stitch 手套针法
glove stretcher 手套伸张器
glove string 手套带
glove suede 仿麂皮织物(制手套用)
glove toe closure 手套式缝头
glove total length 手套全长
glove with liner 夹手套,带衬里的手套
glove yarn 手套用纱线
glow 鲜艳夺目;灼热
glowing 无焰燃烧
glowing resistance 耐烧灼性;耐灼热性

glowing time 无焰延燃时间,余辉时间
glsaa wool 玻璃棉,玻璃绒
glue 胶
glued hem 胶粘式贴边
glued seam finish 粘合缝份处理
glue liner 平布上胶衬,上胶软衬布
gluteal fold(or furrow) 臀沟;臀褶
glutenin fiber 麦谷蛋白纤维
gluteus 臀肌
gnafi 露兜树叶织物
gnathion(gn) 颏端点,颌下点
goalkeeper's gloves 守门员手套
goaly 本色绸
goar 三角布,三角布条,服装上楔形花边条
goatee 山羊胡须,山羊胡子
goatee beard 山羊胡须
goat fur 山羊毛皮;山羊皮袍
goat fur robe 山羊皮袍
goat fur wear 羊毛皮服装
goat hair 山羊毛
goat hair carpet 山羊毛地毯
goat leather 山羊皮,山羊皮革
goat skin 山羊皮,山羊皮革;山羊皮袍
goatskin garments 山羊皮服装
goatskin jacket 羊皮夹克
goatskin robe 山羊皮袍
goatskin wear 羊皮服装
goat wool 山羊毛
gob cap 美国水兵帽
Gobelin 哥白林像景织物
Gobelin blue 晴青绿色,深蟹灰
Gobelin corselet 哥白林束腰,高布林束腰
Gobelin stitch 哥白林网眼绣,像景针迹
Gobelin tapestry 粗经细纬印花窗帘布;哥白林双面挂毯
gob's hat, gob hat 美国水兵帽,白帆布水兵帽,水兵帽
Godamo 哥达摩(涤黏仿麻色织物)
goddess's dress 持蛇女神服饰
goddess sleeve 女神袖
godet 倒 V 形三角布,档布,(衣裙上拼缝)三角形布片
godet hem 档布下摆,插角布下摆
godet pleat 档布褶
godet skirt 加褶裙
Godey's Lady's Book 戈戴妇女期刊(美国 19 世纪著名时装杂志)
gofer 起皱褶;起皱纹
goffer 轧花,拷花;皱褶,襞,起皱褶,起皱

纹,打皱褶;烫皱褶的熨斗
goffered cloth 波纹织物
goffered crepe 轧纹泡泡绉;轧花泡泡绉
goffered design 轧花花纹,拷花花纹
goffered ribbon 轧纹带;轧花带
goffered satin 轧纹绉缎;轧花绉缎
goffered silk 波纹纬丝;波纹绸,轧纹绸,轧花绸
goffered veil 花边头巾
gofferer 烫褶子用的熨斗
goffering 褶裥整理,打褶裥
goggles 护目镜,防护眼镜
goghair cotton 印度哥哈里棉
go-go 时髦,最新式
go-go boots 戈戈靴,中高跟长统女靴(指漆皮或光亮人造革制)
go-go skirt 歌歌裙
go-go watchband 时髦表带
going-away dress 蜜月旅行装
going-away hat 蜜月帽
going price 现行价格
golabee durreeaee 手织平纹玫瑰红绸
golbute 印花布;染色布(巴基斯坦制,做男衬衫或女衣用)
gold 金子色,金黄色,金色;浅橄榄棕
gold and silver brocade(or damask) 金银线织锦缎
gold and silver embroidery 金银线绣
gold and silver jewelries 金银首饰
gold and silver lace 金银花边,金银线花边
gold and silver thread 金银线
gold bracelet 金手镯
gold braid (制服用)金色饰边
gold brocade 金线锦缎
gold brown 金褐色
gold cloth 黄金大衣呢,两面呢,金黄大衣呢
gold coin 金币
gold color 金色
gold crocodile 金黄鳄鱼皮
gold dust cotton 砂金棉
gold earth 金黄大地色
gold embroidery 金线刺绣,盘金刺绣
golden and silver printing 金银粉印花
golden apricot 土黄,金杏色
golden bud brocade 金蕾锦
golden cream 金黄奶油色
golden fleece 浅米黄
golden glow 鲜艳金黄色
golden green 金黄绿

golden haze 金黄烟雾色
golden mist 金雾色
golden muskrat 黄金麝鼠皮
golden nugget 矿金色(金棕色)
golden oak 金橡树色
golden ochre 金黄褐色
golden olive 橄榄黄
golden poppy 金罂粟色
golden powder printing 金粉印花
goldenrod 菊科黄花色
golden section 黄金分割
golden section rule 黄金分割律
golden section ruler 黄金分割尺
golden straw 稻草黄色,稻草色
golden treasure ground brocade 金宝地
golden yellow 金黄,金黄色
gold foil 金箔
gold foil printing 金箔印花
gold in woven satin 织金缎
gold lace 金花边;金线花边;金属线辫带
gold leaf 金箔
gold necklace 金项链
gold-plated button 镀金纽扣
gold powder printing 金纷印花
gold print 金粉印花
gold ring 金戒指
gold thread 金线,镀金线
gold tissue 金丝透明绸
goldtone 起毛织物(有金线的)
goldtone effect 金点子花效果,金抢效应(毛织物)
gold tussores 金丝罗缎
gold washed 包金的
gole 头巾搭在肩部的部分
golf 高尔夫服装
golf bag 高尔夫球袋
golf cap 高尔夫帽
golf clothing 高尔夫服装
golf cloths 高尔夫呢
golf costume 高尔夫球运动衣,高尔夫服饰
golf culottes 高尔夫裙裤
golf cutfits 高尔夫装
golfer 羊毛衫,毛线衫
golf gloves 高尔夫手套
golf green 中绿,纯绿
golf hose 高尔夫袜,运动袜,高尔夫长筒花毛袜
golfing clothing 高尔夫服装
golf jacket 高尔夫夹克

golf jupon	高尔夫西裤,高尔夫裤,灯笼裤
golf knickers	灯笼裤;短裤
golf outfits	高尔夫装
golf panties	灯笼裤
golf pants	高尔夫裤,灯笼裤
golf pullover vest	(针织)高尔夫套头背心
golf shirt	高尔夫衬衫
golf shoes	高尔夫球鞋
golf skirt	高尔夫裙
golf socks	高尔夫短袜,高尔夫袜
golf stockings	高尔夫中统袜,高尔夫中筒袜
golf style	高尔夫风格;高尔夫款式
golf sweater	高尔夫毛衣,高尔夫羊毛衫
golf umbrella	高尔夫伞
golf vest	高尔夫背心
golf wear	高尔夫服装,高尔夫球服,高尔夫球衣
golgas	双面印花法兰绒
goller	女式肩巾
golon d'argent	[法]银饰带
golon d'or	[法]金饰带;金丝花边
golosh	长统套鞋,高筒橡胶套鞋
goloshe	长统套鞋,高筒橡胶套鞋
golpatheen	手织彩条绸
gombaz	意大利低级棉绒布
gombo hemp	泽麻,槿麻
gona	印度狭长地毯
gondoina	薄料女式长衣,薄罩衫
gondola pants	贡都拉裤
gondola stripe	游艇条纹
gondolier net	黑丝发网
gondolier's hat	船夫帽
gonel	长袍
gong-fu jacket	功夫衫
gong-fu pants	功夫裤
gong-fu shoes	功夫鞋
gong-fu suit	(中国)功夫服,武术服装
gong-fu uniform	功夫装
goniometer	测角计(量体),测角仪(量体形)
good class	良好级
good color	色泽良好
good color cotton	色泽良好的棉
good design mark	优秀设计标记(日本)
good fit	合身(衣服),很合身
good fitting jeans	紧身牛仔裤,合身牛仔裤
good hand	手感好
good looks	姿色
good luck knot	吉祥结
good rinsibility	易洗净性能
goods	商品,成品;货物;钻石货色;衣料(美),织物
good-size	大尺码的
good-sized	大尺码的,相当大的,宽大的
goods on approval	试用物品
goods packing	商品包装
good year welt process	缝沿条工艺(制鞋)
goodzi	东南亚粗布,印度棉布
goof proof	防滑跌(地毯用语)
goolbuti	东南亚印花粗布,印度印花粗棉布
goon	细打包麻布
goose	长柄熨斗
goose-bellied doublet	带假前片的男式紧身上衣
goose down	鹅绒
goose-down vest	羽绒背心
goose eye	菱形斜纹,鹅眼形斜纹
goose eye twill	菱形斜纹,鹅眼形斜纹
goose feather	鹅羽毛
goose gray	鹅灰色
gooses	长柄熨斗
Gopertz curve	戈泼茨曲线
Gorden Lake Clarke	戈登·卢克·克拉克(英国时装设计师)
gordian knot	装饰性方形饰结
gordon cord	棱条斜纹织物
gordon highlanders	歌登高山人粗格子呢
gore	裤裆,拼衩,拼合裆,拼块,三角布;笃裆,笃角
gore ankle boots	高腰松紧鞋
gore boots	松紧靴
gored and flared	多片裙
gored flare skirt	拼接喇叭裙
gored petticoat with dust ruffles	带防尘皱边的拼片衬裙
gored skirt	片裙,拼片裙,三角裙,多片裙
gored with godets	加三角插布的宽摆多片裙
gore-flared skirt	多片喇叭裤;喇叭口裙
gore heel	三角片袜跟
gore line	多片分割线,裙片分割线
gore panel	多片裙裙片
gore seam	三角缝,三角缝
gore shoes	松紧鞋
gore skirt	多片裙,三角裙
4-gore skirt	四片裙
8-gore skirt	八片裙
4-gore slip	四片套裙

Gore-Tex 戈尔特克斯(用聚四氟乙烯微孔膜复合的防水透气织物,商名,美国)
gore toe 三角片袜头
gorevan rug 中型波斯地毯
gorge 串口,兜领,领串口(缺嘴),领口
gorge cut seam 领口省缝
gorge dart 领孔褶,前领圈省,领口省
gorge line 领串口斜线,领圈线,(领圈)串口线
gorge line is uneven 领串口不直
gorgeous look 华丽款式
gorgeous style 豪华款式;华丽款式
gorge run 领圈弧线
gorge seam 领圈缝
gorget (中世纪妇女用于包住颈、肩部的)头巾,披巾,肩巾,项链;(盔甲的)护喉,护颈巾(中世纪),(作装饰的)衣领
gorge tape 串口牵带
gorget patch 领章
gorgonelle 荷兰平纹棉布
gorilla yarn 竹节粗花线
gorro 针织帽,绒线帽
goshenite 透绿玉
goshoo 金银丝线
goshpeck 戈什佩克羊绒织物
gospodsky 俄罗斯优质脱胶亚麻
gossamer 薄纱,薄雨衣织物,薄织物;薄雨衣,戈塞默(涤纶缝纫线)
Gossyplum 木本棉
Gossyplum barbadense 海岛棉
Gossyplum herbaceum 草本棉
Gossyplum hirsutum 陆地棉
Gossyplum Nanking 中国棉
Gossyplum religiosum 印度土产木本棉
gota 金银花边,金银丝编带
gotah 细棉布
Gothic 哥特风格;哥特式服装
Gothic costume 哥特式服装
gothic-stitch 链式针迹
go-through 穿过式花边,梭结式(花边)
go-through lace machine 梭结花边机
goton 棉花(埃及称谓)
gotra 包头巾;阿拉伯人头巾
gourgandine 软胸衣
gousset 小口袋
gouzlieh 棉条纹长袍布
government lssue helmet liner 针织军帽衬里
government silk 绌丝
governo cotton 巴西戈弗诺棉

governor 统治者帽
gove shoes 侧裆鞋
gown 罩衣,长衫,长外衣,长袍,哥翁,长外衣;非戎装(古代);女睡衣,睡袍,晨衣;妇女的正式礼服;褂
gown à la francaise 法式袍
gown à langlaise 英国裙
gown à la polonaise 波兰裙
gown à la sultane 旅行服装
gowning room 穿衣室
gown neckline 教士袍领口
gown used for buddhist monks 袈裟
goza cotton 阿富汗果札棉
grabanni 格兰班尼刺绣
grace 雅致
graceful 大方
graceful carriage 身段优美
graceful curve 优美曲线
GRACIA 《优雅》(意)
gradation (色彩)层次感,渐变,明暗法,浓淡法,晕色,云纹
gradational pattern print 渐变色彩印花
gradation of tone 色级
gradation stripe 渐变条纹(颜色或粗细),色调渐降条纹
grade 档次,等级,品级,品位,渐变层次;渐次变化;渐次调和
grade assignment 定级
graded pattern 推档样板,放缩纸样,样板推档
graded piece 放码样片
graded seam allowance 不同宽度的缝份
graded size 推档尺
grade labelling 等级标签,质量标签
grade of cotton cloth 棉布品等
grade of cotton yarn 棉纱等级
grade of goods 商品的等级
grade point 放码点,推档点
grader (纸样)放缩机;分等工,分级员,号型放码员,推板师
grade reference line 放码指示线
grade rule 放码规则,推档规则
grade rule table 放码规则表,推档规则表
grade standard 等级标准,品级标准
grading 放大缩小(裁剪样本),(纸样的)放缩,推档,放码;分等,分级
grading and classify 分等分级
grading and marker 缩放和排料设备
grading and marking equipment 放码排唛架设备

grading line 推档线
grading machine （纸样）放缩机
grading method 分级评价法,推档方法
grading pattern 分档纸样,分尺寸纸样
grading point 推档点
grading process 推档顺序
grading ruler 推档尺
grading seam 分层缝
grading system 放码系统
grading terminology 推板术语
grading tool 推档工具
gradual blending 省尖逐渐缩小（不是突然变尖）
graduated band collar 中式领
graduated chech 渐变粗细条格布
graduated checks 粗细条格子布
graduated dart 渐变组合省
graduated dress 毕业服（传统为白色）
graduated pattern 色彩渐变花纹
graduated rhythm 逐步协调
graduated ruffle neckline V形荷叶领
graduated side pleats 渐变顺褶
graduated square 分数计算尺
graduated stripe 色调渐降条纹
graduated tuck 渐变活褶
graduate gemologist 毕业宝石学家
graduation 逐变项链
graduation of curve 曲线修匀
graffiti look 粗画风貌
graffiti stylistic form 古拉夫款式
graft copolymer hollow fiber 接枝共聚物中空纤维
grafted fiber 接枝纤维
grafted hollow fiber 接枝中空纤维
grafted polyamide filament 接枝聚酰胺长丝
grafted rayon 接枝黏胶人造丝
grafted weighting of pure silk 真丝绸接枝增重
grafting 接枝;（针织衣片的）缝接
graft rayon staple 接枝黏胶短纤维,接枝人造丝短纤维
graged finish 摩擦轧光整理
grain 谷物色;粒面（皮革）;劈向（钻石加工）;纹理（织物）
grain askew 钻眼偏位
grain cross 织物横向,织物纬向
grain de ble 棱纹绒（法国名称）
grain de poule 法国哔叽
grain ditection 纹理方向

grain d'orge 法国匹染斜纹粗呢;法国漂白细亚麻布;法国八经八纬素色哔叽
graine 棱纹塔夫绸
grained composition leather 合成纹皮
grained leather 印花皮革
grainer 格林钻,格令钻(喱钻)
graing coarse cardigan 粗节纱毛衫
grain gloves 纹面皮手套
grain grossier 法国粗被套布
grain leather 光面皮革,全粒面革,纹皮
grain leather handbag 纹皮手提袋
grain leather shoes 粒面革鞋
grain-like composite fiber 木纹状复合纤维（木维截面并列多层复合纤维）
grain line（G.L.） 经向标志;经向线;丝缕,丝缕线;纹理线
grainline arrow 经向标志线
grainline indicator 裁向线
grainline placement 布纹线（丝缕）方位
grainlines 经向标志;经向线
grain marking 丝缕标号
grain sack 厚黄麻布袋
grain side 皮革的正面,粒面,纹面
grain straight 织物经向;织物纵向
grainy coarse sweater 粗节纱毛衫
grainy weave 粒状花纹
gram（g） 克
gramaskes 粗厚长袜
grammont 白色梭结花边;比利时黑绸带
grams per denier 克/旦
granada 格兰纳达薄呢,棉经毛纬破斜纹精纺黑呢
grandad shirt 爷爷衬衫
grande-assiette sleeve 礼服盖袖
grande bride 锁眼针绣六角网眼纱
granded shirt 爷爷衬衫
grande habit 男式法庭服
grande rose 漂白亚麻花缎
grande tennue 礼服,军礼服
grande toilette 大礼服,仪式服
grande venise 法国大花纹亚麻花缎桌布
grandine 加光精纺花呢
grand knit 加大针织品
grand lez 法国全毛白色军服料
grand lion 法国花式亚麻台布
grand opera dress 蓬起观戏装
grandpa look 祖父装(19世纪至20世纪初欧美民间风格装)
grandpa shirt 仿古衬衫（模仿美国西部淘金热时代的衬衫）

grandrelle cover	双层胶合防水布
grandrelle fabric	混色花线呢;19世纪女孩紧身褡棉毛混纺;英国双层防雨布,双层防水布,双层防雨布
grandrelle shirting	英国工作服彩条布
grandrelle yarn	混色花绒(异色长丝或单纱合股并捻);混色花线,混色雪花丝
grandrille shirting	花式条子衬衫料
grandrille yarn	混色花线
grand shirt	加大衬衫
grand vair	珍贵毛皮
granite cloth	花岗石纹呢
granite green	花岗石绿
granite grey	花岗石灰色
granite pattern	花岗石花纹
granite soie	(灰白色的)真丝花岗岩细纹绸
granite tag	样板用纸
granjamers	睡衣裤
grannie skirt	祖母裙
granny	老祖母式,老奶奶式,古蕾妮式
granny('s) bag	祖母包,老奶奶式手提包
granny blouse	古蕾妮式女衫(高领口,长袖,有育克的古极款式)
granny bonnet	祖母式系带帽
granny boots	古蕾妮靴(前面开槽,串带眼系紧),奶奶靴(矮腰)
granny costume	(19世纪60年代美国形成的)老奶奶服式,古蕾妮服式
granny dress	(宽松,长及足踝的)老奶奶裙,宽松连衣裙,婆婆裙;宽松装,阿婆服
granny glasses	金丝眼镜,老奶奶式眼镜
granny gown	老奶奶长袍,古蕾妮长袍,祖母袍装
granny knot	织布结
granny look	老奶奶风格,老奶奶风貌,祖母款式
granny nightgown	老奶奶式长睡衣,祖母长睡衣
granny print	古色小花印花布;祖母印花
granny's bag	古蕾妮袋
granny silhouette	老奶奶式(轮廓线条)
granny skirt	古蕾妮裙,老奶奶裙,祖母裙
granny's look	老奶奶风貌,古蕾妮风貌,仿古型款式(模仿美国西部淘金热时代的服装款式);老奶奶装
granny style	老奶奶款式,老奶奶式
granny wajst	祖母衫;紧身衣
grape	葡萄紫(深暗红紫色)
grape button	葡萄扣,葡萄纽
grape fruit	葡萄柚黄(黄色)
grape jam	葡萄果酱色
grape juice	葡萄汁色
grapevine	葡萄藤色
graph	图表;图形
graph check	方眼格子纹,细密格子纹
graph coloration technique	图形染色技术
graph coloring	图着色
graphical method	图解法
graphic design	平面造型设计;印刷美术图案
graphic designer	平面设计师
graphic geometric pattern	印刷几何图案
graphic line	绘画线条
graphic look	图案款式
graphic pattern	图案
graphic piece together	画报式拼接
graphic plotter	绘图仪
graphic print	广告衫印花,印刷美术印花
graphic style	图案款式
graphite	石墨;石墨灰色;石墨纤维
graphite crayon	石墨笔
graphite fiber	石墨纤维
graphite reinforcement fiber	石墨增强纤维
graphite type structure fiber	石墨型结构纤维
graphite whisker	石墨须晶,石墨单晶纤维
graphitized fiber	石墨化纤维
graph paper	方格纸
graph paper check	方格纸格子纹
grass	葛布;青草色
grass carpeting	草坪地毯,人造草坪
grass cloth	夏布,葛布,苎麻布
grass embroidery	草梗刺绣,麻线刺绣,花草刺绣
grass fiber	植物粗纤维
grass green	草地绿,草绿
grass lawn	麻纱织物(麻织物),细麻布,麻纱
grass linen	夏布
grass rug	棉麻经纱双面花毯,仿草地地毯
grass-seed nonwovens	草籽布
grass skirt	草裙,夏威夷草裙
grassy green	草绿色
grate	格栅
gratel	素色亚麻斜纹布
graticule	方格图,十字线
grave carpet	墓毯
grave clothes	寿衣,尸衣
gravel	砂砾色

gravity-feed iron with water tank	吊水熨斗
gravy-proofed cloth	油布,漆布
gray	同 grey
grayish	浅灰色的,微带灰色的
grazet	异色经纬交织精纺呢
grease chalk	蜡质粉片,蜡质划粉
grease fulling	生坯缩呢,含油脂缩绒
grease paint stick	化妆油彩
grease stain	油污;油渍
greasy blowing	生坯半蒸呢
greasy fabric	呢坯
greasy feel	油滑手感
greasy handle	油脂状手感
greasy-luster	润泽
greasy piece	坯呢
greasy wool	原毛
Great Britain costume	大不列颠民族服
great coat	厚外套,厚大衣,厚重长大衣（用毛皮衬里）
great coveralls	巨式连衫裤装
great gatsby look GG	款式
great hemp	博洛涅大麻
great overall	巨型连身裤装（大贴袋、大衬衫领、大后育克）
grebe	鹏鹉褐
grebe cloth	长绒棉内衣布
Grecian bend	希腊腰曲线
Grecian border	希腊回纹布
grecian braid	紧密棉带
grecian quilt	全线双面床单布
Grecian knot	希腊式发髻
Grecian-looking dress	希腊夜礼服
grecian net	小圆形图案网眼
Grecian sandals	希腊风格凉鞋
Grecian slippers	希腊式室内便鞋,低帮软便鞋（英国）
Grecian tunic	希腊式短束腰宽衣
greco-roman wrestler's wear	古典式摔跤服
Greek bag	希腊式毛料提包（方形或长方形上开口款式）
Greek belt	希腊式绑扎饰带,希腊饰带;希腊式皮带
Greek boy look	希腊男童款式,凯尼米得款式
Greek boy style	希腊男童款式
Greek check	希腊格子粗棉布
Greek costume	古希腊服饰,希腊民族服饰
Greek design	希腊式图案
Greek embroidery	希腊刺绣（贴花刺绣）
Greek fishermen's cap	希腊渔民式软帽
Greek footwear	希腊鞋饰
Greek fret	希腊回纹饰
Greek god	希腊神式[卷]发型
Greek lace	希腊花边,希腊挖花花边
Greek look	希腊款式
Greek lounging cap	希腊式居家帽
Greek ornament	希腊装饰
Greek point	希腊针绣挖花花边
Greek stripes	希腊彩条粗布
Greek style	希腊款式
Greek warrior's wear	古希腊武士装
green (G)	绿色,青色
green algae	绿藻色
green almond	青杏仁色,黄绿色
green-aproned men	穿绿裙的搬运工
green bag	绿包
green bamboo brocade	翠竹锦
green banana	香蕉绿
green belt	绿腰带,柔道绿腰带
green beret	绿色贝雷帽
green black	墨绿
green-blue-slate	带灰绿蓝
green cinnabar	朱砂绿
green clear	鲜绿
green cotton	未熟棉;绿色棉花（无污染棉花）
green crystal	绿晶色
green deep	暗绿
green design	绿色设计
green discharge	绿色拔染
greener	色光偏绿
greener-gallery costume	唯花主义服式（19世纪末英国妇女的服装款式）
greenery	葱翠色
green essence	新鲜香料绿
green eyes	绿眼睛;绿眼睛的绿色
Green fiber	绿色纤维,无污染纤维
green flash	闪电绿
green flax	生亚麻,青亚麻,嫩亚麻
green-ground prints	绿地印花布（爪哇蜡防印花布）
green house shielding	温室护罩
greening	使成绿色
greenish	略呈绿色的
greenish blue (gB)	绿光蓝色,带绿青色
greenish cast	带绿色光
greenish-grey (gG)	蟹青色
greenish-yellow (gY)	绿光黄色
green light	浅绿

green lily	绿百合花色
green linen	本色亚麻布
green mist	雾绿色
green oasis	绿洲绿,绿洲色
green off	[使]成绿色
green pale	浅绿
green plum	青梅绿
green plush	绿绒
green ramie	绿叶苎麻,未经硫磺熏白的苎麻
greens	绿制服(美军),美军绿军服
green seed cotton	绿籽陆地棉
green smalt	大青色
green tea	茶绿
Green textile	绿色纺织品,无污染纺织品
green tint	浅绿色
green yellow (GY)	黄绿色
grege	生丝色
grège yarn	丝毛混合纱线
grego	带头巾粗厚短大衣,格里高外套(希腊、近东;带风帽),希腊式粗布夹克(带风帽)
gregorian	格里高里假发发型(16～17世纪)
gregs	男式筒形开衩裤
greig	坯绸,绸坯;坯布,本色布
greige	坯绸,绸坯;坯布,本色布;生丝色
greige beige	灰褐
greige cloth	坯布,本色布
greige fabric	本色织物,原坯织物
greige goods	本色布,坯布
greige silk	生丝[厂丝];生坯绸
grenada	西印度原棉;棉经毛纬有光呢
grenade	法国格兰内德优质台布,法国格兰内德亚麻台布
grenadier boots	近卫军靴
grenadine	紧捻纱罗织物;亚麻台布花缎;领带用光亮薄绢;毛薄纱
grenadine broché	提花纱罗
grenadine cord	复捻辫绒
grenadine crepon	毛织薄纱(有方格或棱条)
grenadine motif	胖花花纹
grenadine red	石榴汁红(浅红橙色)
grenadine satin	里子用色织棉背缎
grenadine USA fabric	丝盖棉
grenal	格林内绸
grenfell cloth	格伦费尔全绒斜纹布
Greole evening dress	法国古典礼服
Gre's	格雷夫人(法国时装设计师)
gresette	法国女式呢
gress green	草绿
Gretchen neckline	格雷琴领口
Gretchen style	德式金黄长辫发型,格雷琴发型,格雷琴发型
grex	格勒克斯支数制
grey	本色,灰;坯布,本色布;灰色衣服,灰色织物;穿灰军服者;穿灰衣服者
grey allocation	坯布分批
grey area measures	灰色区域措施
grey black	黑灰
grey cloth	本色布,坯布,坯织物
grey dark	深灰色;
grey dawn	拂晓灰色
grey deep	暗灰色
grey drill	卡其坯布,厚斜纹坯布
grey dull	深灰色
grey dyeing	生坯染色
grey fabric	坯布,坯绸,坯织物
grey finder	灰色样卡罩框
grey finish	坯布整理
grey flannel	灰色法兰绒
grey flannel suit	灰色法兰绒装
grey fox	灰狐皮
grey gauze	纱布坯
grey goods	本色布,坯布,原色布
grey green	灰绿色
grey hair	花白头发
grey interlock fabric	本色棉毛布
grey inventory	坯布库仔
greyish (g)	灰味
greyish blue	灰蓝色
greyish green	浅灰绿
greyish lavender	淡灰紫色,中莲灰
greyish purple	深灰紫
greyish tone	灰色调
greyish white	灰白
grey jeans	薄斜纹坯布,细斜纹坯布
grey kit fox collar	灰狐皮领
grey knits	针织坯布
grey level	灰度级;灰色值
grey level transformation	灰度变换
grey lilac(or lilak)	淡灰紫色,浅灰丁香色;珍珠白,芡实白
grey lily	珍珠白
grey mending	坯布修补,呢坯修补
grey mercerization	坯布丝光
grey mercerized	原坯丝光的
grey mist	雾灰色
grey morn	晨灰色,银灰

grey muskrat 灰麝鼠毛皮
grey natural fabric 本色织物
grey pale 白灰色,淡灰色,灰白色,苍白色
grey poplin (drill) 府绸(卡其)坯布
grey red 灰红色
grey room 坯布间
greys 灰法兰绒裤;灰军服(美国)
grey sand 灰沙色,米色
grey scale 灰度模式;灰度等级;灰色样卡,灰色分级卡(测定色差比色用)
grey scale for assessing change in color 变色评级灰色样卡
grey scale for assessing staining of color 沾色评级灰色样卡
grey scale for change in color 褪色样卡
grey scale for color change 灰色色卡
grey scale for staining 灰色沾色色卡
grey scale of color change 色泽变化灰色分级卡
grey setting 坯布定形
grey sheeting 本白平布,细布坯,白市布,本色宽幅平布;本色宽幅斜纹布
grey shirtings 本色细平布
grey son cotton 格雷逊陆地棉
grey stock 坯布库存
grey stock room 原坯储存库
grey system 灰色系统
grey ticken 亚麻粗布(北爱尔兰制,做里子用)
grey topper 灰色圆筒帽
grey tube cloth 圆筒坯布
grey violet 浅紫色
grey washing 坯布水洗
grey weight 坯布重量
grey width 坯布幅宽
grey wolf 灰狼毛皮
grey yarn 原纱,本色纱;原丝
grey yellow 灰黄色
gridelin 深紫红色
griffe [法]缝入商标
griffin 秃鹰灰
griffin cotton 美国格里芬棉
Grilene 聚酯纤维(商名,瑞士依姆斯)
grillage 在边网眼凸棱效应
grinning effect 露底现象
grip 手提包,小提箱
grip bag 抓手提包
grip feed 夹紧送料
gripper 夹子,夹片,钩扣;免缝按扣,四合扣,四件扣,五爪纽

gripper Axminster carpet 夹片式阿克明斯特地毯
gripper closing 揿纽门襟
gripper fastener (夹钉在衣服上的)大揿纽按扣,四合扣
gripper hanger 带夹衣架
gripper-spool Axminster carpet 夹片-筒管式阿克明斯特地毯
gripper-spool carpet 夹片-筒管式机织地毯
grip sack 旅行包,手提包,小提箱
grip-snip thread cutter 夹紧剪线刀
grip sole 防滑底
grip top 弹性袜口
gris 本色棉布,棉坯布
grisaille 色织花呢;法车灰白色布;彩点效应
gris-brun 法国军服呢
grise 爵士毛皮
grisette 法国淡褐色女式呢
gris fer bleuté (蓝铁灰色)法国军服呢
grissaille 法车格里萨利布
grivelé 斑点花纹布
grizzle 灰色;花白头发;灰色假发;灰斑毛皮;灰色服装;灰白花斑色调
grizzled hair 花白头发
groffer 褶边,皱边
grog 双经布,英国双经织物
grogan 工作靴
grogram 格罗格兰姆服装;格罗格兰姆呢(丝与马海毛交织)
grogram hat 格罗格兰姆帽,罗缎帽
grogram yarn 柔软绣花丝线
Gro-lon 针织弹力丝
grommet 扣洞;扣眼,金属孔眼;金属环洞;垫圈,鸡眼;军帽顶圈;加固扣眼
grummet grommet
groove 鞋底槽
grooved inside foot 沟型管条内压脚
grooved outside foot 沟型管条外压脚
grooved pattern 凹凸花纹,波纹花纹
grooved presser foot 嵌线压脚
gros 横棱绸,法国厚重织物
gros d'Afrique 厚重横棱织物
gros d'Afrique corde 厚重棱条织物
gros d'Alger 横棱丝织物
gros de Berlin 法国柏林横棱绸
gros de Chine 横棱织物
gros de londres 伦敦横棱绸(丝或人造丝制,有宽窄相间棱条)
gros de Lyon 里昂横棱织物

gros de messine 经面罗缎
gros de Naples 厚重丝织物(意大利那不勒斯制)
gros de rome 轻薄绉绸
gros des indes 彩色横条纹绸
gros de suez 帽子衬里棱纹绸
gros de suisse 双经横棱绸
gros de Tours 提花床单布;图尔横棱绸
gros d'lspahan 伊斯法罕横棱绸
gros d'Oran 奥兰织锦缎
gros d'Orleans 奥尔良斜棱纹织物
gros drap 粗厚织物
gros forts 厚实家具用布
grosgrain 葛,罗缎,茜丽绸
grosgrain belt 罗缎腰带
grosgrain hat 罗缎帽
grosgrain ribbon 罗缎饰带,罗缎丝带
gros point [法]厚重织物花边;横棱花边,意大利棱纹针绣花边;毛圈装饰布
gros point de venise 意大利棱纹针绣花边
gross (一)罗(等于12打)
grosse 厚重的;优质羊毛
grosse chainette 法国平纹素色哗叽
grosse draperie 法国缩绒粗络毛织物
grosse grenadine 毛哗叽
grossera skin coat 也门短皮大衣
gross-heat-conductivity 总热传导
gross output 总产量
gross profit 毛利
gross weight 毛重
gross yards 总码数
gros velour 粗纬条长毛绒
Grosvenor gallery costume 唯美主义服式(19世纪末英国女装款式)
grotesque ornament 怪异装饰
grotto blue 岩洞蓝
ground 绒织物底布;地色;地纹
ground color 地色
grounded point lace 平面针绣花边,网眼花纹花边
ground end 底经,地经
ground fabric 地布,底布
grounding printing 打底印花
ground jumper 球场防寒夹克
groundnut fiber 花生蛋白质纤维
groundnut protein fiber 花生蛋白质纤维
ground pattern 地绸织,地纹
ground shade 地色,底色
ground thread 底线
ground tint 底色,浅地色;底色调,地色调

group 群体
group brand 联合商标
group dynamics 群体动力
group dynamic theory 群体动力理论
grouped check 成组格子
group float 大跳花
grouping 分组,归类
group language 行话
Group Mode Creation (GMC) 巴黎高级时装店设计师组织
group of business strategy maker 战略决策群体
group presser 组合压脚
group pressure 群体压力
group psychology 群体心理
group sampling 分层抽样
group stripe 成群条纹
group technique (GT) 成组技术
group technology 成组工艺
grouse cap 松鸡帽
grow line 连衣[缝合]线
grown-on collar 连领
grown-on garment 叠缝服装
grown-on strip 叠缝带子
grown-up clothes 成年人服装,成年人衣服
grow sleeper 可放长的儿童睡衣
growth of fabric 织物潜伸(二次水洗后),弹性织物加负荷后长度差异率,织物次级蠕变
gruardsman coat 士兵大衣
gruard's overcoat 士兵大衣
grubh hole (鞋子)虫眼
grubh sutry 丝经棉纬手织布
grumment 扣洞,扣眼;军帽顶圈
grummet 衬垫
grunge look 吉拉吉风貌,贫困款式,穷潦风貌
G string G带(美国印第安人或歌女、舞女等当三角裤穿),狭条布三角裤;作三角裤的狭布条(黄色表演穿)
G suit (=gravity suit)宇航服;抗超重飞行服
Gtarbo hat 嘉宝帽
guamero 风梨纤维
guanaco 红棕色美洲驼毛皮,栗色骆马毛皮,栗色美洲驼毛皮
guanaco wool 南美产原驼毛
guanaquito 幼栗色美洲驼毛皮
guanaquito fur 幼栗色美洲驼毛皮
Guangdong embroidery 粤绣

Guangdong patterned damask satin 花广绫
Guangdong plain satin 素广绫
Guangdong satin 素广绫
Guangdong satin brocade 花广绫
guangming brocaded velvet 光明绒
Guangzhou satin 广州缎
guanle clogue 冠乐绉
guanoguito fur 原驼毛皮
guaranello 纬起毛织物,纬起绒织物
guarantor 保证人
guard 戒指圈;运动员防护用品;刀剑护手盘;拉链底座;金银花边束带
guard boot 19世纪军警靴
guard hair 针毛(毛皮),鬃毛
Guardian Angel 阻燃衣料(商名)
guard infanta 裙撑架,裙环
guards 华丽服装饰边
guard's coat 卫兵外套(英国陆军);卫士型大衣
guard's coating 军装长大衣
guardsman clothing 卫兵外套
guardsman coat 卫兵外套(英国陆军);卫士型大衣
Guatemalan costume 危地马拉民族服饰
guayabera catalana 条纹边薄呢
Guayaberas 瓜亚贝拉衫
Guayabera shirt 瓜亚贝拉衬衫(前片四个有盖大贴袋,前后衣片有塔克,开关两用领,短袖)
guayanille 瓜阿亚尼那棉
guazhou tapestry 挂轴绸
gubba rug 山羊毛粗剪花毯
GUCCI 古奇(意大利皮具店)
Gucci loafers 古奇式平跟船鞋
gudar 厚重条子布
gud-ka-cheet 蓝地黑花棉布
Gu embroidery 顾绣
Guendje rug 根吉地毯
guepiere 法式肚带,小紧身胸衣
guerlain 娇兰
guerley 东南亚平布,印花棉布,东印度平布
Guernsey 格恩西衫
guernsey 针织紧身毛衫
Guernsey 足球运动衫;针织紧身毛衫,格恩西衫
guerrilla look 游击队军服风貌,游击队军服式
guest towel 小方巾,揩指尖小巾;高档细薄小手帕

guetres 法国绑腿
gueuse 贝格粗梭结花边;低级驼毛呢
gu-gu cap 姑姑冠,顾姑冠
guhua printed brocatine 古花缎
guibert 法国厚漂白重亚麻布
guibray 棉粗灯芯绒
guide 比尺(小型助缝工具);引导器,导引物;定距器
guide aid 辅助导向装置
guide basting 线钉,标记假缝
guide cloth 导布
guide line 标线;标准线
guide mark 标记,记号,辅助标记
guide plate 导板
guilloche 扭索形饰,绳形饰
guimp, guimpe 肩巾(修女用);镶边带,嵌芯细辫带;无袖女胸衣,高领女内衣,背心裙内穿女内衣
guimped embroidery 几姆拍刺绣,教堂式刺绣
guimpe dress 背心裙
guimp lace 绸带;凸纹花边
guimple 薄纱;披肩
guinagar 高级棉细布
guinan black lamb fur 贵南黑紫羔羊皮
Guinea 厚重蓝棉布;花棉布
Guinea cloth 英国产几内亚布
Guinea styles 几内亚式低级靛青布
guingan 金甘交织绸
guinget 法国制薄驼毛呢;大麻帆布
guinoline dyes 喹啉染料
guiote cotton 菲律宾吉奥特棉
guipon 法国毛棉平布
guipure 大花花边,凸纹花边,贴花花边,花色色纱金银线
guipure d'art 亚麻凸纹花边
guipure de binche 比利时网眼花边
guipure de bruges 凸纹梭结花边
guipure de flandres 梭结花卉花边
guipure lace 镂空花边;凸花花边
guipure lace embroidery 抽纱花边,凸花网眼绣花边
guipure renaissance 粗布刺绣
guise 穿戴,装束
guiter 阿赞加尔缎,吉特缎
gulbadan 手织经条绸
gulbandau 绸布裤料
gulbani 金线薄绡(东南亚)
guleparivane 印巴毛呢
guleron 头巾;披巾

gulf coast fading 臭氧褪色
gulf cotton 美国海湾棉
gulisant 克什米尔毛织物
gulix 白亚麻细布,漂白亚麻布
gull grey 海鸥灰
gulnagar 印度细布
gum 胶;古鲁衫(音译,低腰设计的上装)
gumbiered canton plain silk 拷绸
gumbiered canton silk 香云纱,拷绸
gum boots 橡胶鞋;高筒套靴,惠灵顿长靴,长筒套靴,橡胶高筒靴
gum club checks 复式格纹,狩猎俱乐部格纹;三色格子织物
gum drop green 橡皮糖绿(色)
gummed tape 缝补布条,黏补布条
gumpty sheeting 英国中支平纹坯布
gums [美]长筒套靴,长筒橡胶套鞋
gum sandals 橡胶底凉鞋
gum-shield 护齿
gum shoes 胶鞋,橡皮套鞋;橡胶底帆布鞋,轻便运动鞋
gum tape 松紧带
gumti 手织粗床单布
gum twill 薄软绸
gum-up 粘搭
gum yarn 宽紧线,橡筋线
gun 枪状物;喷雾器;打包细麻布
gun belt 枪带
gun boats 大尺码鞋子;一双大脚(美国)
gun brass 炮铜色
gun club check 三色格子织物,狩猎俱乐部格纹,套格拼色格子织物,复式格纹
gun flap (军用雨衣)肩部加层织物;枪垫(点缀夹克衫肩部)
gunge rug 日本坚固地毯
gun metal 蓝灰色,黑灰色;炮铜色
gun metal grey 蓝灰色;炮铜色
gunmetal mink 阿留申貂皮,铁色貂皮,蓝铁色貂皮
Gunn cotton 美国格恩棉
Gunn tartan 格恩彩色条格呢
gunny bag 黄麻布袋,黄麻袋
gunny cloth 黄麻袋布
gunny hat 麻编草帽
gunny linen 麻布衬
gunny sack 黄麻袋布
gun patch 荷枪贴片,枪托垫肩;肩部装饰布块
guota 定额

guota management 定额管理
guota sampling 定额抽样
gurlandage 花环饰
gurn 古鲁衫(长而宽松的罩衫)
gurrah 18世纪印度粗布
guru 古鲁衫;宽长外袍
gusset 护腋甲片,三角形布料,腋下插角布;三角形衬料(填补、加固、拼放用)
gusset heel 三角片袜跟
gusseting 尖角布(衬垫手套的)
gusset-insert 插角,缝纫插角
gusset pocket 接挡口袋
gusset stay 角撑条
gusset toe 三角片袜头,楔形袜头
Gustave Tassell 古斯塔夫·塔塞尔(1926～,美国时装设计师)
guttar 印度丝面棉背缎
gut thread 加固线,衬垫线
guxiang quilt cover 人丝古香缎被面
Guy Laroche 盖伊·拉罗舍(1923～,法国时装设计师)
Guyotatz cloth 罗马尼亚格约塔兹黑白呢
gwlanen 法兰绒
gym bloomers 运动灯笼裤
gym clothes 运动衣
gym-honed women 健美女性
gym look 体操款式
gymnasium costume 健身服
gymnastic shoes 体操鞋
gymnastics kit 体操服
gymnastics shoes 体操鞋
gymnastics suit 体操服
gymnastic uniform 体操服
gym outfit 运动套装,运动衫裤
gymp 花色包线
gym shirt 球衣
gym shoes 运动鞋,体育鞋,体操鞋,球鞋
gym shorts 运动短裤;运动短袜
gym slip 古姆衫,(中小学女生)体操衫
gym style 体操款式
gym suit 运动服,体操服
gym trousers 运动裤,体操裤
gymtunic 古姆衫,(中小学女生)体操衫
gym vest 运动背心
gypsy blouse 吉卜赛女衬衫,吉卜赛罩衫
gypsy bonnet 吉卜赛帽,宽边帽
gypsy cloak 吉卜赛斗篷;吉卜赛外衣
gypsy cloth 吉卜赛绒布(用于划船衣、网球衣和运动衫等)
gypsy costume 吉卡赛服装,吉卜赛民

俗服
gypsy dress 吉卜赛装（皱边,缨穗装饰,丝束腰带）
gypsy earring 吉卜赛耳环
gypsy hat （女、童）宽边帽,吉卜赛帽
gypsy longuette 吉卜赛长连衣裙
gypsy look 吉卜赛款式,吉卜赛风貌
gypsy mode 吉卜赛款式

gypsy neckline 吉卜赛领口,束带式领口
gypsy skirt 吉卜赛裙,宽摆裙
gypsy sleeve 吉卜赛袖
gypsy straw 吉卜赛草帽,饰满鲜花的小草帽
gypsy stripe 吉卜赛服装条纹
gypsy style 吉卜赛款式
gypsy vamp 对脸鞋,吉卜赛鞋面皮,蚌面鞋

H

haarlem checks 荷兰蓝色(或红色)亚麻窗帘格子布
haba 男式直角束腰外衣;粗纺毛织物
habai pure silk brocade 哈蓓缎
habai silk brocade 哈蓓缎
habberley carpet 低级棉毛交织地毯
habei silk brocade 哈蓓缎
haberdash 打扮(男子);服装设计;服装制作
haberdasher 英国缝纫用品商;美国男式服饰用品商
haberdashery 缝纫用品;[总称]男服饰用品;缝纫用品店;男子服饰用品店
haberdashery look 哈伯代舍勒风貌(仿花纹组合的女服式),男服饰品款式
habergeon 锁子铠;短鳞甲;(中世纪)无袖铠甲
haberjet 宽幅粗呢
habilatory art 服装艺术
habilimentation 穿着打扮,服装艺术;服装业
habiliments 衣服;制服;礼服;特定式样服装
habit 表示宗教级别的衣着;妇女骑装;古语服装;(某阶层尤指修道士的)特征服装
habit back placket 紧身裙后开衩
habit cloth 优质呢绒;女式骑装呢
habit d'escalier 短睡袍式晚装
habitmaker 女骑装裁缝
habit-redingote 公主款式长外衣
habit shirt 骑马服用衬衫
habotai 电力纺,纺绸,全丝薄软绸
habotai brocade 特号葛
habotai de suisse 仿电力纺(商名)
habotai mixed 交织电力纺,交织纺绸
habutae 电力纺,纺绸,全丝薄软绸
habutai 纺绸,电力纺,全丝薄软绸
habutaye 电力纺,纺绸,全丝薄软绸
hachure 影线,晕线,刻线
hacking coat 骑马用外套,夹克骑装(紧腰下摆,两侧或后面开衩)
hacking jacket 夹克骑装,骑装短上衣;骑马外套

hacking muffler 骑马用围巾
hacking outfit 骑装
hacking pocket (有盖)上衣斜袋
hacking scarf 19世纪30年代校园学生长围巾;马车驭手长围巾
hacking-style scarf 可绕颈两周系结的长方形围巾
hada 哈达
hada silk 哈达
haddat 棉印花方巾
hadjar 刺绣用手纺金线
haemostatic dressing 止血绷带,止血包扎纱布
haft sole 加固袜底
haich 爱克披巾
haick (阿拉伯)白布大罩衣,阿拉伯妇女的白色绣花面纱
haik 阿拉伯人的白罩袍;阿拉伯妇女的白色绣花面纱
hail tonic 生发剂
hainan silk 海南生丝
hainan twill faconne 海南绫
haining silk 海宁地毯生丝
haining wool 海宁地毯羊毛
hair 头发;动物毛
hair accessories 发饰;头饰
hair bag 发袋
hair-band 发箍
hair bind cord 扎头绳
hair bleach 头发漂白;头发漂白剂
hair bow 发结;发带
hair braid 发辫
hair brush 发刷
hair brushing 刷头发
hair canvas 马尾衬
hair canvas interfacing 马尾衬
hair cardiner 粗毛拔削机(毛皮)
hair care 头发保养
hair carpet 发毛地毯,兽毛地毯
hairchief (女用)三角头巾,包头巾
hair clasp 发簪,发针
hair clip 发夹
hair cloth 粗毛衬衣;黑炭衬,马尾衬,毛里衬
hair colo(u)r 发色

hair conditioner	发蜡
hair cora dimity	麻纱
haircord carpet	发毛起圈地毯
haircord muslin	细棱纹薄纱
hair cords	麻纱(棉织物)
hair cream	发乳
hair curler	卷发器[类],卷发筒
haircuts	发式,发型(男子)
hair cutting	理发
haircutting shears	剪发剪刀
hair decoration	发饰
hairdo	发式,发型(女子);做头发
hairdress	女发式;发饰
hair dresser	高级女理发师,美发师
hair dresser's gown	梳发衣,理发衣
hairdresser's tool	美发工具
hairdresser's jacket	女部美发师夹克
hair dressing	美发剂;做发
hair dryer	吹风机,干发器
hair dye	染发剂
hair embroidery	发绣
hair-ent	发网
hair felt	毛毡
hair fiber	动物毛纤维
hair grip	小发夹
hair interlining	毛鬃衬,黑炭衬
hair iron	烫发器
hairlace	发网;方网眼花边
hair lacquer	发胶
hairlessness	秃顶者
hair line	额前发际线;发型轮廓;海拉因呢(密细条纹精纺呢);细线条
hairline block	细线格子
hairline seam	发际线缝,窄的工形线迹
hairline stripe	毛发条纹,细条纹
hairline tri-color	三色经纬条纹呢
hairline	发际线;发型轮廓;细直条纹织物;密细条纹精纺呢;海拉因呢
hairliquid	洗发液
hair lotion	美容生发水;头发化妆水;洗发剂
hair mode	发式;发型
hairnet	发网,网巾,网子
hair oil	发油,头油
hairologist	头发保养专家(美国)
hair ornaments	发饰;发饰品
hair parting	头发挑缝
hair piece	饰发,小假发辫,小假发髻,小假发束;遮秃的一绺假发,男子假发,装饰小假发(女子)
hair pile carpet	粗毛绒头地毯
hairpin	夹发针,发夹,发叉;发簪
hairpin form	发夹形
hairpin lace	发夹花边,手工饰边花边
hair powder	发粉
hair pressing	火烫
hair ribbon	发饰缎带;头发丝带;发带
hair seal	粗毛海豹毛皮,海豹皮
hairside	有毛的一面
hair set	整发,整烫
hair shirt	刚毛衬衣;马尾腰布;粗毛衬衣;马尾衬衫
hair silk	加捻绢丝
hair slide	角质(或玳瑁)发夹
hair spray	喷发定形剂;喷式发胶;头发喷洒剂
hair sticks	发簪
hair stripe	精纺呢细条纹;细线条
hair stroke	毛向(裘布)
hairstyle	发式,发型
hairstyle turban	发型头巾女帽
hair styling	做发式
hairstylist	发式师;发式专家;发型师
hair thinning	削发
hair thread	发线
hairtician	发式师(美国);发型师;高级理发师
hair tonic	生发水,美发剂
hair tonique	护发素
hair treatment	头发保养
hair type	发式,发型
hair-weaving	假发植入术
hairy acrylic fiber	毛状(聚)丙烯腈系纤维
hairy appearance	外观发毛
haitien	彩色粗纬绸
hakama trousers	日本丝质长裤,日式裙裤
hakama	日本丝质长裤,日式裙裤
hakamaji	手织棱格条纹棉布
hakimono	日本木屐等凉便鞋
hakir	(印度丝经棉纬)条纹绸
hakistery	黑白地色印花(伊朗衣料印花)
Hakka suit	客家装;香港早期女性套装
hako no gohō	日本膝下长袍
halabe	蜘蛛丝
halabe silk	蜘蛛丝
half across back	半背宽
half-and-half bound	两面滚边,正反面均匀滚边
half and half sleeve	半对半袖
half apron	半截式围裙

half-around belt 半腰带
half-auto 半自动
half-back belt 半腰带(缝在上衣后腰节处);(西装背心)后祥带
half-back lining (上装或夹克衫的)后背半夹里,后身半衬里
half-back stitch 半倒针;半回针线迹;半弯穿刺线迹
half back stitch 半回针绣
half ball button 半球形扣
half-bath finishing 半浴整理
half-belt 半腰带,燕式腰带
half binder 半滚边器
half bleach 半漂白亚麻布
half bleaching 半漂白
half boiling 半脱胶,半练
half boots 半长筒靴,半筒靴,高腰皮靴
half bra 半罩式胸罩
half-breton 半布列塔尼帽
half brogue 半雕花布洛格鞋,半镂花皮鞋
half cardigan 半开襟毛衫;半畦编
half cardigan knitted fabric 半畦编针织物,玉米罗纹
half chain stitch 半链式针迹;半锁针
half chintz 半光印花布
half-circle cape 半圆斗篷
half-circle flare 半圆台裙
half-circle skirt 半圆台裙
half-circle sleeve 半圆摆袖
half circular cape 半圆形短披肩
half clothing (比短外套还短的)半长外套
half coat 开襟女短上衣,女齐腰短上衣
half combing wool 半精梳用羊毛,中级羊毛
half-compass cloak 法式半圆斗篷
half coronet 短冕状头饰
half-cross gauze 半纱罗织物;花色纱罗织物
half-cross leno 半纱罗织物;花色纱罗织物
half-cross stitch 半十字绣;锁眼针迹
half cuff 半翻袖;半卡夫袖头
half curl 半卷发器
half-cut goods 半成形针织品
half damask 半丝花缎
half degummed silk 半脱胶丝
half discharge printing 半拔印花
half dot 半圆点花纹
half dress 半正式晚装,日装

half drop 跳接法(印染图案设计)
half-drop match pattern 1/2 阶段边续纹样
half-feather stitched hem 半羽状绣缘边
half finished product 半成品
half foot 半只压脚;单趾压脚
half(full) cuff 半(全)克夫袖口
half-gauge jersey 隔针平针
half-gauge rib 隔针罗纹
half-glasses 双片式眼镜
half gown 半袍,及大腿束腰外套
half handkerchief 小三角形女头巾
half hat 饰有羽毛和花朵的半边帽
half heel 半袜跟
half hitch 半结
half hose 短筒袜;中筒袜;(男)短袜
half hose machine 短袜机
half hosiery 短筒袜
half jack boots 缘饰童靴;赛马靴
half kirtle 短外套
half lap 半搭接;半叠接
half-length figure 半身像
half(-length) sleeve 半袖
half-length sock 半截鞋衬垫
half-length supporter sole 半托底(鞋)
half-lined 半夹里
half-lined skirt 半夹裙
half-linen cloth 半麻织物
half lining 半衬里,半夹里
half made 半完工的定制服装(袖长,衣摆,腰围等为假缝;由顾客自缝);半制成品
half-mask 连帽化妆斗篷;穿连帽化妆斗篷的人;黑色小面具;半截面具;戴半截面具的人
half mercerizing 半丝光处理
half-milano rib 半米兰诺罗纹
half mitt(en) 半指手套
half-moon pocket 半圆形曲线口袋
half-moon shoulder pouch 半月形肩袋
half mourning 半丧服,半正式丧服
half pattern (仿制时的)半纸样
half persian lamb 半波斯羔毛皮
half petticoat 无上身衬裙
half pique 外翻缝;(手套手指)外翻缝缝合
half planket 半翻门襟
half pocket 半口袋;半深袋
half presser 半压脚
half resist 半防染印花

half robe 半袍,及大腿束腰外套
half roll collar 半平翻领
half-round ruche 半圆褶裥饰边
half-round splicing 袜底半圆加固
half-set net fabric 半穿经网眼织物
half shirt 男式短衫
half silk 半丝织物
half size 半码尺寸(美国胖妇女尺寸),身体矮小的妇女衣服尺寸,上身短的妇女衣服尺寸
half skirt 半截裙
half sleeve 中袖,半袖;五分袖;袖套;装饰袖
half sleeve outing shirt 半袖衬衫,中袖旅游衬衫
half sleeve shirt 半袖衬衫
half slip 斗衬裙,短衬裙,无上身衬裙
half socks 半长筒袜
half sole 鞋底前掌
half stitch 暗花绣,花边刺绣;正反锁混合线迹;针织物边部双重线圈;枕套花边的松针迹
half stitch（ing）process 正反锁混合缝程序
half stockings 短筒袜,男短袜,中筒袜
half-thicks 薄质白法兰绒
half thread fabric 半线织物
half thread poplin 半线府绸
half thread woven fabric 半线织物
half tint 中间色调,中色,半浓淡色
half-toilette 半正式晚装,日装
halftone 中间色调,半色调
halftone color 中间色
halftone effect 半色调效应
half twist 半纱罗织物,花式纱罗织物
half voile 半巴里纱
half waist 半腰围
half waistband 侧腰带(半边腰带)
half Wellington 短筒靴
half-Windsor knot 半温莎领结
half wool 棉毛交织物
hali 土耳其东方风格大地毯
halimtarakshi 鸽眼纹花缎
halina 海立拿呢,南斯拉夫粗纺格子呢绒
haling hands 搬运工手套,厚工作手套
halirug 东方风格大地毯
hallencourt 法国斜纹亚麻台布
halles crues 法国本色亚麻布
halling 帷幕;挂毯
hallstand 衣帽架

hallux 大趾,拇趾
halo （妇女戴在头后部的）小花环或锻带,小帽子等
halo bonnet 中世纪无边系带女帽
halo hat 光环帽,环檐女帽,圈环帽,晕圈帽
haloing 晕渗现象
Halston 霍尔斯顿(美国服装设计师)
halt mitt 半指手套
halter （露肩、背）女套头背心;三角背心
halter back's neckline 后背吊带领口
halter belt 吊带式饰带
halter blouse 吊带领女衫
halter bra 吊带式胸罩
halter brassiére 吊带式胸罩
halter cowl 吊带式垂褶
halter cut 吊带式裁剪
halter dress 露背吊带连裙装;背带式女装;背带式女运动装;吊带领装
halter neck 女式三角背心领,吊带领
halter neckline 套索领口,挂脖领口,吊带领口,背心式领口
halter slip 吊带式连衣衬裙
halter tank 背带式连衣泳装;背带式背心
halter top sundresses 三色背心式太阳装
halter top 三角背心式女上衣,袒肩露背胸衣
halter type bodice 三角背心式紧身胸衣
halter 颈部系带,女三角背心;（三角背心的）颈部系带
halterneck bra 吊带式胸罩
halter-top sundresses 三角背心式太阳装
halwan 白色开司米平纹呢
hamadan 哈马顿地毯
hamadanrugz 哈马顿地毯
hamamlik 东方式方浴毯
hamas 印度漂白棉布
hambel 异色斜纹地毯
hambel carpet 异色斜纹地毯
hambro line 捆扎绳
Hamburg edging 汉堡刺绣花边(商名,瑞士)
Hamburg（hat） 汉堡毡帽,高级毡帽
Hamburg hat 汉堡帽,中折帽,小礼帽,有缘中褶帽
Hamburg lace 汉堡花边
hamburgo americano 未漂白棉床单
Hamburg point 汉堡抽绣
Hamburg wool 汉堡绣花毛线
Hamburg yarn 汉堡绣花纱线

hamidieh 叙利亚丝棉织物
hamilton lace 苏格兰梭结菱纹花边
hammer-bow[design] 锤击状花纹
hammer cloth 车座套子布;马车布篷
hammered satin 压花缎,锤花缎
hammered silk 压花缎,锤花缎
hammer felt 钢琴槌毡
hammer-on snap 免缝按扣,四按合扣
hammer-on snap tape 免缝按扣带
hammersmith carpet 英国毛绒棉底地毯
hammock cloth 纱罗织物;帐篷帆布,吊床帆布
hamouli cotton 埃及哈默黑棉
hamster 仓鼠毛皮
Han costume 汉族服饰
Han nationality costume 汉族民族服
hanabishi 日本金线绸
Hanae Mori 森英惠(1925～ ,日本时装设计师)
hanaori 花织
HANATSUBAKI 《花椿》(日本)
hancaatjes 印度平纹白细布
hand 手;手感;手艺;人工的;行家,专业人员
hand accessories 手饰品
hand (auto)-open umbrella 手开(自动)伞
handbag 小旅行袋,旅行包;背囊;(女用)手提包,手袋
handball wear 手球装
hand basting 手缝针法;手工疏缝
hand blind stitch 手工暗缝线迹
hand blocked fabric 手工模板印花织物
hand block printing fabric 手工模板花布
handbook 便览,指南
hand book purse 女用窄长小包
hand brassiere 束带胸衣
hand breadth 手宽
hand buttoning 手工钉扣
hand-carry 手提
hand carpet sewing machine 手工地毯缝纫机
hand chain 手链
hand clipping 手工修剪
hand cloth 方巾,手帕
handcoverchief 方巾,手帕
hand crafted 手工蜡染
hand craft industry 手工业

hand craft look 手工制作风格,手艺风貌
hand cream 擦手香脂
hand-crochet cardigan 手工镂空开襟衫
hand crocheted shawl 手钩围巾
hand crocheted sweater (手工)钩编毛衫,镂空衫,钩针毛衫
hand cuff 喇叭形翻边硬袖
hand cutter 单件裁剪师,手工裁剪师;手推电裁刀;手动裁剪机;推剪机
hand cutting 手工裁剪,手裁
hand-done process 手工工序
hand-drawing 手工描绘,手绘
hand-drawing handkerchief 手绘手帕
hand-drawing kimono silk 手绘和服绸
hand-drawing silk crepe 手绘真丝绉
hand-drawing silk handkerchief 真丝手绘手帕
hand-drawing necktie 手绘领带
hand-drawing silk satin 手绘真丝缎
hand-drawing silk scarf 手绘丝巾
hand-drawn 抽绣品,抽花刺绣品;抽纱织物
hand-drawn embroidered handkerchief 抽绣手帕
hand-drawn silk scarf 抽绣丝巾,手绘丝巾
hand-driven flat knitting machine 手摇横机
hand-embroidered 针绣花边,手绣
hand-embroidered handkerchief 手绣手帕
hand-embroidered hat 手绣花帽
hand-embroidering 手工绣花
hand-embroidery 手绣
hand-embroidery sweater 手绣毛衫
hand evaluation 手感评定
Hand Evaluation and Standardization Committee 手感评定与标准化委员会
hand-fashioned goods 手工编织品
hand fall 花边装饰喇叭形翻边硬袖
hand fashioned 手工收放针成形
hand feed (缝制中)手推动作
hand feel 手感
[hand] feeling 手感
hand-felled 手工织缝的
hand felled binding seam 手工滚边缝
hand-filling 人工包装,手工包装
hand-felling 手缝边,手工嵌边
hand finish 手工修理
hand finished 手工整修的
hand finisher 手工整理员

hand-frame knitting machine 手摇针织机
hand girth 掌围
hand glass 手镜,手执放大镜
hand handle 手感
hand-held 手握式
hand hemming 手工卷边,手工缝边
handicraft 手工艺,手工业;手艺;[总称]手工艺品
handicraft look 手艺风貌
hand information retrieval 手工检索
handing 手感
hand iron 家用熨斗,手动熨斗,手工熨烫器
hand-iron press 手动式熨斗
hand-ironing pad 手工熨烫[衬]垫
handiwork 手工,手工制品
handkerchief 围巾;头巾;方巾;手帕,手绢
handkerchief button 手帕纽
handkerchief check 手帕格子花纹
handkerchief dress 斜拼式手帕裙;斜拼式手帕上衣;手帕式上衣(四方布制作)
handkerchief hem 手帕的锯齿边;手帕型裙摆
handkerchief hemline 垂褶手帕摆边;锯齿状裙摆线;帕角式摆边
handkerchief lawn 薄型亚麻细平布,仿亚麻细平布
handkerchief linen 薄型亚麻细平布
handkerchief point 手帕垂角型
handkerchief point skirt 手帕裙
handkerchief print 百家衣花样;手绢印花花样
handkerchief skirt 手帕裙
handkerchief skirt style 手帕裙式风格
handkerchief sleeve 手帕型方袖口
handkerchief tunic 束腰手帕裙
hand knife cloth cutting machine 手动电气裁剪机;手推裁剪机
handknit 手织,手织物
hand knitted hosiery 手工针织品
hand-knitted scarf 棒针围巾
hand knitting 手工编织,手工针织;手织;棒针编结
Hand Knitting Association [美]手工编织协会
hand knitting carpet 手工编织地毯
hand knitting knitwear 棒针衫
hand knitting machine 手摇针织机
hand knitting thread 手编毛线,手编绒线

hand knitting wool 手编毛线,手编绒线
hand knitting yarn 手工编结线;手编毛线,手编绒线
hand knitting yarn for coarse needles 棒针绒线
hand knob 手动按钮
hand knoted 栽绒地毯,手结地毯
hand knoted carpet 栽绒地毯,手结地毯
hand lace 手工花边
handle (衣料)手感
handle bars 长八字胡须
handle determination 手感测定
hand length 手长
hand lens 手执放大镜
handle operated machine 手动缝纫机
handle tester 织物风格仪
handle with care 小心轻放,小心装卸
handle 手袋把手;手感
handleba m(o)ustache 长八字胡须,翘八字胡须
handle-o-meter 织物手感测定器
handling equipment 搬运设备
handling quality 加工性能,使用质量
handling hands 搬运工手套
handling property 操纵性能,使用性能
handling 画法
handling 手感;装卸;处理
hand machine 手动缝纫机
handmade 手工制作的;手工制品;手工制造
handmade bobbin lace 手工梭结花边
handmade embroidery lace 手绣花边
handmade eyelet 手缝扣眼
handmade lace 手编花边
handmade pyjamas 手工制睡衣
handmade rug 手工地毯;手织毛毯
handmade work 手工制品
hand-me-down 穿过的旧衣服;现成的廉价服装;现成服装
hand-me-down clothes 穿过的旧衣服
hand meter 手感测定仪
hand modification 手感改良
hand needle 手缝针
hand-needlework tapestry 手工刺绣的花毯
hand-netting needle 手工结网针
handoleer 子弹带(子弹带腰饰)
hand-open umbrella 手开伞
hand over to the next shift 交班
hand painted carpet 手工描花地毯

hand-painted straw 手绘草帽
hand painting 手绘；手工印花
hand painting silk clothing 真丝手绘服装
hand pick stitch 手工刺绣针迹
handpiece 头巾
hand pinker machine 手动锯齿边缝纫机
hand-plaited seam 绳辫缝
hand point presser 翻角整平器
hand pose 手姿势
hand-pricked edge 手工缝边
hand print 手工印花
hand printed silk scarf 手工印花丝巾
hand printing 手工印花
hand properties 织物手感
handpull shoulder pads 手装肩垫
hand punch 手工打孔器
hand raising 手工起毛
hand-rolled edge 手工滚边
hand-rolled handkerchief 手工卷边手帕
hand-rolled hem 手工卷边,手挑筒形缘边
hand-run mechlin 手绣花缘的机制网眼梭结花边
hand-run tuck 手打褶裥,手工塔克
hand saddle stitch （表里交叉线迹的）手工对称线迹,手工鞍形线迹
hand sample 小样
hand screen print 手工筛网印花
hand screen printing 手工筛网印花
hand screen printing fabric 手工筛网印花织物
hand-seam pattern 手缝花样
hand sew 手缝
hand-sewed embroidery 手绣
hand sewer 手缝工,手工缝纫工
hand sewing machine 手动缝纫机,手摇缝纫机
hand[sewing] needle 手缝针
hand sewing thread 手缝缝机;手工缝线
hand sewing 手工缝纫；手工缝制
hand-sewn seam 手缝接缝
hand sewn vamp 手工缝制鞋面,人工缝制鞋面
[hand] shear 剪刀,手工裁剪刀
hand shirring 手缝抽裥
hand shoulder pads 手袋肩垫
hand sleeve 下半截袖
hand slip-stitched hem 挑针缝边
hand slitting 手工剖幅

hands off 请勿动手
handsome 俏丽
hand spray 手持喷雾器
hand spreading machine 手推式拉布机
handspun 土布；手纺纱
hand spun cloth 土布
hand spun silk yarn 手纺丝线
hand stitch 手缝线迹;手缝针法
hand-stitched appearance 手缝线迹外观
hand-stitched effect 手工线迹效率
hand stitching 手工缝纫,手工缝制
hand stitching article 挑花
hand table printing 手工台板印花
hand tailor 手工裁缝,手工缝制;特制[服装]
hand tailoring 手工缝纫,手工缝制
hand tester 手感测定仪
[hand] touch 手感
hand towel 面巾,手巾,擦手巾
hand trimmer 修线员
hand-tuned welt 翻袜口
hand tying 手工打结
hand value 基本手感值,综合手感值,织物手感值,织物风格值
hand value (HV) 织物风格值
handwarmer pocket 暖手[口]袋
handwarming slot （服装上的）暖手[口]袋
hand wash 手洗
hand washing 搓洗
Hand weavers Guild of America [美]手工织造协会
hand wheel （缝纫机头上的）手轮;上轮
hand woven 手织的
hand woven cloth 手织土布,土布
hand work 手工操作
handwork 手工,手工艺品
hand-worked buttonhole 手缝扣孔,手缝扣眼
hand-worked eyelet 手缝鸽眼孔,手缝合眼孔
handwoven cloth 土布
hand wrinkle meter 织物折皱手感仪
handy case 便盒
handy coat 方便外套
handy cutter 手提裁剪机
handy punch 手工打纽孔
hanf 大麻（德国称谓）
hang 垂挂;修整下摆;悬垂性
Hangchow 杭绉,杭绸

Hangchow silk 杭州生丝
Hangchow silk gauze 杭罗
Hangchow silk plain 杭纺
Hangchow velvet satin 杭库缎
hang crocheted sweater 手编毛衫
hang dry 悬干
hanged mark 吊袢印
hanger 挂袢,衣袢,领袢,佩带,肩饰带,衣架,挂物架,挂钩
hanger apparel 挂衣销售,挂衣展示,衣架展示
hanger braid[loop] 挂衣袢
hang la 杭罗
hanger line 衣架型
hanger loop 吊袢
hanger packing (悬)挂(包)装
hanger system 吊架系统
hang-free cardigan 对边开衫,无扣开衫
hanging assembly line 吊挂流水线
hanging calendar silk 丝织挂历
hanging dummy 吊挂人体模型
hanging loop 吊环;吊袢;领袢
hanging mosquito net 圆顶蚊帐
hanging property 悬挂性
hangings 帘布,挂布
hanging seam 挂缝,悬缝
hanging sleeve 吊袖,挂袖,假袖,悬饰袖
hanging sleeve overcoat 悬饰袖大衣
hanging snap 悬空按纽(由线袢与布料相连)
hanging strip 吊带
hanging tapestry 挂轴绸
hang pick 三角形破洞
hang shot 三角形破洞
hang style 悬垂发式
hang tag 保养说明标签;(服装上的)吊牌,挂牌;(商品)标签
Hangzhou crepe 杭绉
Hangzhou habotai 杭纺
Hangzhou leno 杭罗
Hangzhou silk 杭绸,杭纺
Hangzhou silk gauze 杭罗
Hangzhou silk umbrella 杭州绸伞
Hangzhou velvet satin 杭库缎
Hani ethnic costume 哈尼族服饰
Hani nationality's costume 哈尼族服饰
Hanizu costume 哈尼族民俗服
hank printing 绞纱印花
hankie 小手帕,手绢
hankie dress 手帕裙装

hankies 小手帕
hanks to pounds 亨司折磅
hanky 手帕,手绢
hanky dress 手帕裙装
hanneri silk 日本绢丝
hanolchade 黑白条纹毯
hanos 印度10综花缎
Hansa yellow 汉撒黄(黄色颜料)
hanseline 超短紧身男上衣
hanser 马尾衬
hanten 日本男士劳动服
hanwen lamé 汉纹缎
hanxin brocade 汉新缎
han-yeri 和服刺绣领
Hanzu costume 汉族民俗服
haol 中式及地长袍
haori 日本宽松及膝外衣,羽织外套(日)
happi 哈比外袍,(日本式)宽松外衣
happi[coat] 哈比外袍(长至臀部的连袖便服);(日常穿的)宽松外衣;日本花纹上衣
happiness crepe 快乐绉
happiness voile broché 怡悦绉
happy coat 哈比外袍(长至臀部的连袖便服);(日常穿的)宽松外衣;日本花纹上衣
happy face 衬衫或扣子上的笑脸图案
happy occasion coat 喜事服
haps 粗厚毛披巾
harabini 丝毛呢
harahche 瓦略切鞋
harakake moori 新西兰麻叶纤维
Harami 清真寺用哈腊米地毯
Harami rug 清真寺用哈腊米地毯
harbour blue 蓝绿[色]
harcha-beni guil wool 摩洛哥羊毛
hard and bast thread 麻线
hardangar embroidery 菱形抽花刺绣
hardanger 挪威针绣
hardanger bonnet 挪威四幅拼成的方帽,类似婴儿系带的挪威传统帽
hardanger cloth 窗帘织物,帷幔织物
hardanger embroidery 哈当厄刺绣,菱形抽花刺绣
hardanger lace 哈丹格几何花纹细丝花边
hardanger skaut 挪威传统头饰
hard-burned 重烧毛的织物
hard crepe 硬绉绸,硬纺黑绸
hard currency 硬货币

hardeesy 哈迪西小格粗呢
hard elastic 硬弹性
harden 最低档粗麻布
hard expert 硬专家
hard[face] finish [精纺呢绒]光面整理
hard feel 硬挺感,粗硬感
hard felt hat 高帽身帽;硬毡帽
hard fiber or leaf fiber fabric 蕉麻、剑麻织物
Hard Fibers Association 韧皮纤维协会
hard-finish 硬挺
hard-finishing 厚浆整理,硬挺整理
hard handle 粗硬手感,手感极硬
hard hat 安全帽;防护帽;圆顶窄边毛毡礼帽
hard jade 硬玉
hard knot 死结
hard lining fabric 硬衬
hard paper 硬纸
hard seam 硬缝
hard selvedge 硬边
hard-texture 致密结构
hard-to-ease fabric 不易抽细褶的面料,厚硬面料
hard-to-handle material 手感粗硬缝料
hard twist yarn 紧捻纱线,强捻纱
hard twisted fabric 强捻[丝]织物,紧捻[纱]织物
hardware 金属服装
hard-wearing 耐磨的,耐穿的
hard-wearing property 耐磨性能,经穿性能
Hardy Amies 哈代·艾米斯(1909~ ,英国时装设计师)
hare 亚麻;麻类长纤维;野兔毛皮
harebell lace 风信子花纹
hare fur 野兔毛皮
harem dress 哈伦服(对称或不对称的垂褶服)
harem knickers 哈伦半长裤,后宫灯笼裤
harem look 哈伦风貌(近东);哈伦款式
harem pajamas 后宫睡衣裤
harem pants 哈伦裤,伊斯兰后宫裤;阁裤;扎脚管宽松女长裤
harem silhouette 哈伦式轮廓线,哈伦型轮廓
harem skirt 闺阁裙;哈伦裙;(伊斯兰)后宫裙;女裙,灯笼裙;仿土耳其马裤式的裤裙
harem slippers 哈伦式便鞋,后宫便鞋

harem style 伊斯兰后宫式
harem trousers 闺阁裤,后宫裤
harem 垂褶裙;土耳其裤;近东长裙款式
haremor 伊斯兰式泡泡裙
hari seal 海豹毛皮
hari 硬挺度(日本称谓)
haris tweed 哈里斯呢
harl 亚麻;麻类长纤维
harlequin 大格子纹;精纺花呢;小丑服
harlequin checks 彩格布;方格斑纹
harlequin design 小丑服饰图案
harlequin glasses 菱形镜片的眼镜
harlequin hat 丑角帽,小丑帽
harlequin striped fabric 彩条织物
harlequin stripes 彩条织物,花色条子布
harlequin suit 丑角服
harlot 男式紧身连袜裤
Harlow look 哈洛风貌;哈洛款式(斜裁裙装,大裤脚等)
Harlow pants 哈洛裤(从臀部到裤脚呈宽大线条)
Harlow shoes 哈洛鞋
Harlow slippers 哈洛拖鞋;哈洛便鞋(露后脚跟、经缎饰边)
Harlow 哈洛鞋(块状高跟系带包鞋)
harmonious colour 和谐色
Harmonized Commodity Description and Coding System 国际统一商品分类系统
harmony 调和,和谐
harmony and unity 调和与统一
harmony of gradation of hue 色相连续调和
harmony of hue gradation 色相渐变调和
harmony of hue succession 色相连续调和
harmony of succession of hue 色相连续调和
harmony of tone analogy 色调类比调和
harn 低档粗亚麻布
harness 全套衣帽装备
harness checks 浮纹格子棉布
harness webbing 降落伞背带
harp seal 扇尾海豹毛皮
HARPER'S BAZAAR 《哈珀斯市场》(意)
Harris 海力斯,海力斯呢;海力斯织物
Harris tweed 海力斯粗花呢
harsh 粗糙,手感粗硬
harsh[feel] 手感粗糙
harsh hand 手感粗糙
harsh handle 粗糙手感,手感粗糙

harshness 手感粗糙度
hartshorn black 鹿角黑
hartsville cotton 美国哈茨威尔棉
Harvard[cloth] 牛津布；哈佛斜纹衬衫布
Harvard crimson 哈佛深红色
Harvard shirting 牛津布；哈佛斜纹衬衫布
Harvard twill 牛津布；哈佛斜纹衬衫布
harvest gold 芥末黄[色]，橄榄黄[色]
harvest pumpkin 南瓜色
Hashidah shawl 哈希达有边羊绒披巾
hasp 银色搭扣
hassock 跪垫，膝垫
hastings tapestry 刺绣挂毯
hat 冠，帽子，大檐帽，(有檐)帽；红衣主教帽
hat and cap making 制帽
hat and clothes 衣冠
hat and gown 衣冠
hat band 帽边缎带；丧帽黑带，帽圈
hat block 帽模，帽木模
hat body 帽坯
hat box 帽盒
hat brim 帽边；帽檐
hat cap 外帽内的里帽
hat case 帽盒
hatch 影线
hatcheck room 衣帽间
hatcheck stand 存帽架
hatched 叠花织物
hatched design 影线圈案
hatched surface 画剖面线的面
hatching 阴线，剖面线法
hat core 帽模
hat cover 帽套；帽罩
hat fawr 宽边夹顶帽
hat felt 帽坯，帽毡
hat-forming machine 帽坯机
hat guard 帽扣
hatienne 粗细横棱纹绸(英国名称)
hat leather 帽内皮条，帽内防汗带
hatlessness 不戴帽子
hat lining 帽衬里；帽里布
hat maker 制帽匠
hat-making machine 制帽机
hat mask 沙滩帽；钢盔帽；帽型面具
hat-net 帽网
hat peg (挂帽用)帽钉，挂帽钩
hat pin 帽饰针，女帽帽针
hat press machine 制帽水压机

hat proofing 帽檐上浆，帽子硬挺整理
hat rack 挂帽架；挂帽钩
hat ribbon 帽带
hat-shag 帽绒，男式高帽丝绒
hat shape 初制帽坯，帽型
hat skiver 帽内侧皮
hatstand (可移动的)帽架
hat steamer 蒸呢帽锅
hatter 帽商，制帽匠
hatter's felt 制帽毡
hatter's plush 帽绒，男式高帽丝绒
hatter's plush wool 制帽兔毛
hatter's silk 帽边绸，帽里绸
Hattie Carnegie 哈蒂·卡内基(1889~1956，美国时装设计师)
hatting factory 制帽厂
hatting 制帽；制帽材料
hat-tree 帽架
hat with zip pocket 带拉链小袋的帽子
haubergeon 锁子铠；短鳞甲
hauberjet 宽幅粗呢
hauberk (中世纪)领肩铠甲，无袖铠甲
hauler 承运人
hauling hands 搬运工手套，苦力手套
haunch 胯部，臀部
haunseleyns 男式超短上衣
hausa checks 英国靛蓝条格棉平布
hausse-cull 臀垫
hautae 纺绸
haute boutique [法]高级时装店；[法]高档现成服装；巴黎高级时装；高档女子时装；最新款式；高级时装设计师
haute couture 时尚女服设计师；女服时新式样；著名高级时装店；时装屋[女式]
haute couture house 高级时装店
haute couturier 时髦女服设计师；时髦女服商；时髦女服商店
haute couturiere 时髦女服女设计师
haute nouveaut'e 新式高级衣料，新颖美观高级衣料
haute-lisse 16世纪法国亚麻花缎
haut ton 弹簧裙撑
Havana[brown] 哈瓦那棕，黄棕色
Havanese embroidery 哈瓦那刺绣
havanese work 彩色丝线厚重刺绣品
havelock 哈夫洛克；军帽后遮颈布，帽后的遮阳布
havelock cap 驾车帽
havenese embroidery 彩色锁眼针迹刺绣

haversack 帆布背包;干粮袋,行军粮袋
having clothes altered 翻新服装
having the coat turned 翻新服装
haw [color] 山楂色
Hawaiian costume 夏威夷衬衫;夏威夷民族服饰
Hawaiian print 夏威夷印花花样(有大而鲜艳的热带植物)
Hawaiian shirt 夏威夷衬衫,夏威夷衫
Hawaii dress 夏威夷女装
Hawaii shirt 香港衫;夏威夷衬衫
hawking gloves 狩猎手套
Hawkins' cotton 霍金氏棉,早熟短绒美棉
hawksbill turtle red 玳瑁红
hawser-laid rope 三股九花绳
hay 枯草色
Hay's china cotton 海氏棉,密西西比晚熟中绒棉
hayk 阿拉伯裹体布
hays china cotton 美国海氏中棉
Hayti cotton 海地棉
haze blue 烟雾蓝[色]
hazel 榛子棕(浅黄褐色)
hazel brown 榛子棕
hazy 朦胧
hazy effect 朦胧效果,印染模糊效果
hazy matching 朦胧感配色
hazy view effect 朦胧效果,印染模糊效果
head 头,头部
headache band 额面带圈,头饰带,止痛头带
head and neck 头部和颈部
head and neck length 头和颈长
headband 妇女头巾,束发带,扎头带;束发圈;箍带
head circumference 头围
headcloth 头巾,包头巾
headdress 帽;头戴物,头饰,幅巾,头巾
headed ruffle 双木耳边,(上层翻下形成的)双层绉边
headend 布头印记,匹端印记
headgear 帽子,安全帽;盔;头饰
headgear and footwear 鞋帽(总称);衣着
head girth 头围
head guard 护头
head height 头高
head height line 头高线
heading 布匹头印;隔码线;毛巾开剪线;褶端
heading of terry fabric 毛巾开剪线,毛巾端线
headkerchief 头巾,包头布,四角头巾
head lamp 额灯,头灯
headlamp cord (矿工)头上戴的小型照明灯绑带,头灯带
head length 头长
head light 额灯,头灯
headlining 汽车衬里呢
head-mounted umbrella 戴在头上的伞
head ornaments 头戴物,头饰
headpiece 帽子;头巾;头盔
head rail 诺尔曼头巾,盎格鲁撒克逊头巾,有边饰的头巾,欧洲大头巾
headrest 垫头枕,头靠
headscarf 围巾;头巾
head shawl 围巾;头巾
head size (HS) 头围
head square 围巾;头巾
head tire 头饰;首饰
head-to-height proportion 头长与身长比例
head to rump 头接尾[臀](毛皮或皮革)
head veil 头纱,向颈后垂长头纱
head warp 包头布
head wear 头饰;帽子
headwear and footwear 衣着
head wide 头宽
head wrap 头部用方巾或丝带
head wrapping 围巾包头
health and sanitary regulation 卫生检疫规定
health blanket 风湿病护垫;卫生毛毯
health care shoes 健身鞋
health care textile 保健纺织品
health corset 健康胸衣
health crepe 内衣绉,卫生内衣绉
health elastic band 保健松紧带
health fashion 健身服
healthful gloves 保健手套
health giving web 健身带
health protection textile 保健纺织品
healthy complexion 健康肤色
healthy fashion 健身服
healthy glow 容光焕发
healthy sexy 健身性感装;性感健身
healthy slippers 健康拖鞋
hearing aid 助听器
heart breaker 心碎发型

heart button 鸡心扣
hearth rug 壁炉地毯
heart line 心形发型
heart neckline 心形领口
heart notch 心形豁口
heart print 心形印花
heart scarab 心脏宝石
heart-shaped brilliant 心形圆钻；鸭心钻
heart-shaped headdress 心形头饰，心形头巾
heart-shaped neck 鸡心领
heart-shaped neckline 鸡心领口
hearts jute 低档黄麻
heart yarn 芯纱
heassian 黄麻
heat bondable fabric 热黏合织物
heat bondable fiber 热黏合纤维
heat bondable filament 热黏结长丝
heat bondable polyolefin fiber 热黏结聚烯烃纤维
heat bondable synthetic conjugate fiber 热黏结合成共轭纤维
heat bonded fabric 热黏合织物
heat bonded wadding 定型棉；热熔絮片，热熔絮棉
heat bonding 热黏结
heat bonding machine 热压黏合机；压衬机
heat-calendering bonded nonwoven 热轧黏合法非织造布
heat color 暖色［彩］
heat conductivity 导热性
heat delivery surface 放热表面
heat durability 耐热性
heated motorcycle suit 电加热摩托车服
heat fading 热变色
heat from occupant 人体发热
heat-fusible scrim 热熔性稀松布
heat-generating fiber 发热纤维
heather 灰暗浅紫红色，杂色
heather effect 混色效应
heather mix carpet 混色地毯
heather mixture 混色编织绒线；混色呢
heather rose 玫瑰灰
heather silk 混色蚕丝呢
heather tweed 混色花呢
heather violet 石南紫色
heather wool 混色编结绒线；混色毛呢
heather yarn 混色纱线
heathery effect 混色效应

heathery mixture 混色毛纱
heathery tweed 混色花呢
heath wool 严寒地区绵羊毛
heating fabric 供暖纺织品
heat insulating felt 隔热毛毡
heat insulation material 隔热材料
heat insulation uniform 消防隔热服
heat-melt-adhesive interlining 热熔衬
heat melted interlining 热熔衬
heat moisture comfortable clothing 热湿舒适性服装
heat moisture comfortable clothing evaluating method 热湿舒适性服装评价方法
heat moisture property of garment 服装的热湿性能
heat producing shoes 发热鞋
heat protective gloves 防热手套，隔热手套
heat reflective finish 热反射整理
heat reserve and retention fiber 蓄热保温纤维
heat reserve and retention function 蓄热保温功能
heat-resistant fabric 耐热织物
heat-resistant fiber 耐热纤维
heat-resistant infusible fiber 耐热难熔纤维，耐热不熔纤维
heat retention 保暖性
heat retention fabric 保暖织物，保温织物
heat retentive fiber 保暖功能纤维
heat-sealable garment interlining 服装保暖衬
heat-sealable interlining 保暖衬
heat-sealed closure 热封闭
heat sealing 热密封
heat sealing machine 热焊接机；熔接缝纫机
heat-sensitivity 热敏性
heat sensitive fiber 热敏纤维
heat sensitive discolor fabrics 热敏变色织物
heat sensitive shape memory fiber 热敏形状记忆纤维
heat set fabric 热定形织物
heat set pleat 热定形褶裥
heat set stretch yarn 热定形弹性丝
heat settability 热定型性
heat setter 热定型机
heat setting 热定型；热定形

heat setting bonding	热定型黏合	heavy duchesse satin	厚女公爵缎
heat setting machine	热定型机	heavy duty	高效型;厚型[服装];牢固型
heat setting shoulder pads	热定型肩垫	heavy duty [wear]	实用服装
heat shirnkable fiber	热收缩纤维	heavy duty boot	实用靴;厚重耐用靴
heat shrinkale character	热收缩性	heavy duty economy shears	重型经济剪刀
heat shrinkage	热收缩	heavy duty feeder	厚料送布车
heat shrinkage rate	热收缩率	heavy [duty] laundering	强效洗涤
heat-storage and thermo-regulated textile	蓄热调温纺织品	heavy duty machine	重型缝纫机
		heavy [duty] material	厚重缝料
heat-storing fabric	蓄热织物	heavy duty shoes	耐磨厚皮鞋
heat-storing fiber	蓄热纤维	heavy duty slide type hook and eye	厚型搭扣
heat stretched fiber	热拉伸纤维		
heat transfer medium	导热介质	heavy duty snap	厚型按纽
heat transfer printing	热转移印花	heavy duty tack	厚型面料的加固褶
heat transmissibility	传热性	heavy duty thread	高强线
heat transport	传热	heavy duty wear	厚重型工作服
heat weld nonwoven fabric	热[熔]黏[合]非织造布	heavy elastic shirring	强弹性抽裥,强弹性抽褶
heat weld padding cloth	热熔衬布	heavy fabric	厚织物,厚重织物
heat-yellowing	加热发黄	heavy features	粗眉大眼
heave here	起吊处	heavy feeling	厚重感
heavenly pink	宇宙红	heavy fiber	粗韧皮纤维
heavier-weight fabric	较厚型织物	heavy figure	粗壮体型
heavily consolidated cloth	重缩绒织物	heavy-finish denim	重浆色织劳动布
heavily felted cloth	重缩呢织物	heavy finished plains	重浆素色棉布
heavy	深色的;粗壮的人;厚重的;沉重的	heavy finishing	增重整理,重浆整理
heavy acid wash(ing)	重酸洗	heavy flannelette	厚绒布
heavy arm	粗臂	heavy flannelette checked	厚格绒
heavy blade	凸肩胛骨	heavy fleece	厚绒
heavy bobbin lace	粗梭结花边	heavy fleecy sweatshirt	厚绒衫
heavy bosom	丰满的胸部	heavy flight clothing	冬季飞行服
heavy bust	大胸围[型]	heavy gauge	粗针距;大尺寸(机器设备)
heavy C bag	重型C字麻袋	heavy goods	打包粗麻布;厚户外用纺织品;厚重织物
heavy calf	粗壮小腿		
heavy canvas	厚帆布	heavy interfacings	厚衬布,厚内衬
heavy cardigan	粗厚开襟式毛衫	heavy knit	厚重针织物
heavy chain stitch	双层锁链绣	heavy laundering	强效洗涤
heavy checked flannelette	厚格绒;双面绒	heavy layered	重衣款式(如衬衫加衬衫,毛衣加毛衣等)
heavy cloth	厚重织物		
heavy coat	厚外套	heavy layered style	超级重衣款式
heavy color	深色,浓色;深色裆疵	heavy leaded vinyl apron	重铅乙烯基围裙
heavy costume	厚重型服装		
heavy cotton drill	厚棉斜纹布;厚卡其	heavy leather	厚皮革
heavy cotton flannel	厚重棉法兰绒	heavy line	粗实线;粗线条
heavy coverage [printing]	大面积覆盖印花	heavy lips	厚嘴唇
		heavy material	厚料
heavy cream	胖女人	heavy melton	麦尔登厚呢
heavy crepe	重绉丝织物	heavy metals content	重金属含量
heavy drill	粗斜纹;粗斜纹布	heavy muslin	粗平布
heavy drill flannel	厚斜纹绒布	heavy neck	高领

heavy outerwear 厚重外衣
heavy pongee 厚茧绸
heavy print 饱满深色印花
heavy pullover 厚套衫;厚羊毛套衫
heavy repair 大修[设备]
heavy seam 粗线缝
heavy sewn-in 厚型可缝衬
heavy shade 饱和色,深色
heavy sheer 半透明高密丝织物,半透明重绉绸
heavy shirting 厚重平布,重磅平布
heavy shoes 厚重鞋[子]
heavy sizing 重浆
heavy sole 厚重鞋底
heavy string 粗绳
heavy stroke 浓墨重彩
heavy suit 厚套装
heavy suitings 厚重衣料
heavy taxes 重税
heavy thigh 粗腿
heavy thread 粗线
heavy weave cotton cloth 厚棉布
heavy weight 重磅
heavy-weight fabric 厚重织物
heavy-weight geotextile 厚重型土工布
heavy-weight material 厚重缝料
heavy-weight texture 厚重织物
heavy-weight zipper 粗齿拉链
heavy wool 低洗净率羊毛,重脂含杂毛
heavy zipper 粗齿拉链;大齿拉链
Hebert de Givenchy 休伯特·德·吉旺希
he brocade 盒锦
hebridean cloth 海力斯粗花呢
Hecowa 海科华耐洗整理仿麻织物(商名,瑞士)
hedebo 丹麦抽花刺绣
hedebo embroidery 原野刺绣
hedge green 灰绿[色]
hedgehog wig 刺猬假发
heel 脚后跟;(鞋、袜)后跟
heel and toe 踵趾面
heel-and-toe splicing 加固袜跟袜头
heel attachment 袜跟装置
heel ball 蜡墨上光丸(擦鞋沿用);鞋跟底部
heel base 鞋跟基体
heel bolt cam 袜跟螺栓三角
heel bone 跟骨
heel breadth 足后跟宽

heel breast 鞋跟前侧,跟腹
heel breast flap 鞋跟掌口襟
heel cam 袜跟三角,补跟闸刀
heel chain 袜跟链条
heel clearing cam 袜跟退圈三角
heel comb 袜跟梭子
heel cover 鞋跟包皮
heel curve 后跟弧线
heel elevation 鞋跟高度
heel girth 足后跟围
heel gore 三角片袜跟
heel half 袜跟半圆
heel height 鞋跟高度
heel iron 袜跟重锤
heel lasting machine 制鞋用前帮机
heelless fencing shoes 无后跟的击剑鞋
heelless hose 无跟袜[子]
heelless hosiery 无跟袜[子]
heelless leg blank 长筒女袜
heelless shoes 无跟鞋
heelless socks 无[后]跟短袜
heel lift 鞋跟皮层;鞋跟托;鞋掌面
heel nail puller 跟钉拔钉器
heel piece 替换鞋跟;鞋后跟,鞋上掌
heel pitch 鞋跟斜度
heelplate （钉于鞋后跟的）金属鞋掌,鞋跟铁片
heel pocket 袋形袜跟
heel pouch 袋形袜跟
heels 有跟鞋
heel sandal 有跟凉鞋
heel seam 鞋跟缝
heel seat 鞋跟顶面;鞋跟座
heel seat machine 鞋后跟机
heel socks 鞋后跟衬垫
heel stay 裤脚贴边,贴脚边,贴脚条(毛呢裤脚口的贴条布)
heel stitch cam 袜跟成圈三角
heel tab 袋形袜跟小片
heel tap 鞋跟掌;后跟皮层
heel top 加高鞋跟的皮
heguang crepe damask 和光绉
heichima 日本棉挂毯
heide wool 德国地毯用海德粗羊毛
heifer 海虎绒;小母牛皮
height 高度;身高
heighten 颜色加深
height of reinforced heel 加固鞋跟高
height of shoulder 肩高
heim textile [德]家用(装饰)纺织品

heinsi shell button 海司贝扣
heko-obi 白绉绸柔软腰带
Helen Rose 海伦·罗斯(美国好莱坞电影服装设计师)
Helena Rubensten 鲁宾斯代恩(美国著名戏剧化妆师)
Helenic chiton 海伦式奇通衫
helenienne 海纶尼素色斜纹绸
helicon 海立康(防皱薄型棉衬衫布)
heliotrope 浅紫红色
hell 海尔哔叽
hellencourt 斜纹亚麻织物
hell serge 低级哔叽;藏青或黑色学生服
Hellstean heel 希尔斯泰因鞋跟(螺旋形鞋跟)
helm 海姆盔(中世纪)
helmet 头盔,钢盔,钢盆;防护帽,胄状帽,盔形帽
helmet and armour 盔甲
helmet bonnet 盔形软帽
helmet cap 盔形无檐帽;针织护帽
helmet liner 塑料钢盔衬帽
helmet with chin strap 下颌系带盔形帽
Helmut Lang 赫尔穆特·朗格(奥地利服装设计师)
helper 辅助工
helvetia 丝经绢纬双面印花府绸
Helvetian voile 海凡兴巴里纱(永久性爽挺整理,商名)
hem 卷边,折边,贴边,吊边;下摆,衣裾;裤口折边,缝边边缘
Hemant Sagar 海蒙·萨加(法国著名时装设计师)
hem around 摆围;下摆围
hem band 贴脚条;贴脚边
hem bottom 下摆卷边
hem cuff 袖口卷边,袖头缝边
hem depth 底边宽度,下摆卷边宽度
hem drawstring 下摆拉绳
hem-edge machine 缲边机
hem facing 底边贴边
hem front facing 前门贴边卷边,挂面下摆卷边
hem fullness 裙摆量
hem gauge 摆份定规
hem interlining 底边衬
hemline (衣、裙下摆)底边;贴边;(衣服的)底边缘,下摆线,裙摆线,底边线;裤脚口线

hemline arc 摆弧线
hemline sweep 底摆步距(裙装),裙摆量
hem lining bottom (衣服)夹里下摆卷边
hemlock 铁杉色
hem machine 卷边缝纫机
hemmed (hmd) 缝好边的;折好边的;镶好边的
hemmed bottom (无翻边)平脚口
hemmed cuff opening 折边袖衩
hemmed placket 缝边衩
hemmed seam 卷边缝
hemmed-edge opening 摆缘式开口
hemmed-edge placket 摆缘式开口
hemmer 缝纫机折边器,卷边器;缲边工
hemmer foot 卷边压脚
hemming 毛皮纸口;毛皮贴口;卷边缝,镶边缝;(缝纫中)卷边,镶边,卷边纽
hemming attachment 卷边附件
hemming band 镶边
hemming bottom 贴边,卷边,卷边折边,折缝底边(或裤脚口);卷边纽
hemming folder 卷边折边器
hemming foot 卷边压脚
hemming handkerchief 锁边手帕
hemming in zigzag stitch 卷边曲折线迹
hemming lining bottom 夹里下摆卷边
hemming machine 卷边机
hemming operation 卷边操作
hemming pants leg 裤脚口卷边
hemming pocket top 袋口卷边
hemming presser 卷边压脚
hemming presser foot 卷边压脚
hemming seam 卷边缝
hemming shirt fronts 衬衫前襟卷边
hemming sleeve 绡袖;衣袖卷边,缝袖子贴边
hemming stitch 缭边线迹,卷边线迹
hemming stitch bottom 缲底边
hemming stitch front edge 扳止口
hemming stitch hem 缝底边;折缝镶边
hemming with elastic tape 卷边缝松紧带
hem notch 底边刀眼
hemodialysis fiber 血液透析纤维
hemostatic textile 止血纺织品
hemp 大麻,汉麻,火麻,魁麻,线麻;大麻色
hemp pants leg 裤角口卷边
hemp belt 麻带(大麻)

hemp braid 麻条(制帽)
hemp carpet 大麻地毯
hemp cloth 大麻布，汉麻布；大麻织物，汉麻织物
hemp[color] 大麻色，汉麻色
hemp/cotton blended yarn 大麻棉纱
hemp/cotton blending cloth 大麻棉混纺布，汉麻棉混纺布
hemp/cotton denim 大麻棉牛仔布，汉麻棉牛仔布
hemp/cotton fabric 大麻/棉混纺织物，大麻/棉布
hemp fabric 大麻织物
hemp fiber 大麻纤维
hemp floor covering 黄麻地毯布；低级大麻地毯布
hemp hand knitting yarn and thread 大麻棒针纱线，汉麻棒针纱线
hemp line 大麻长纤维
hemp linen 亚麻和大麻交织布
hem pocket 卷口袋边
hem pocket top 袋口折边
hemp pleat 布的褶边
hemp textile 大麻纺织品；亚麻纺织品
hem pull up and stand away 裙身吊(服装缺陷)
hemp yarn 大麻纱
hem rib length 下摆罗纹长度
hem rib width 下摆罗纹宽度
hem rule 卷边尺寸
hem run 底边光顺，圆顺的底边线
hem seam 边缝，卷边缝
hem setting 下摆的制作
hem shaping 起口成形编织
hem sleeve 卷衣袖边
hemstitch 镶边缘饰；卷边线迹；花饰线迹；抽丝；线诱
hemstitch attachment 卷边器
hemstitched pillow-case 透花刺绣枕套
hemstitched serviette 透花刺绣小桌布，透花刺绣餐巾
hemstitched sheet 透花刺绣被单
hemstitched table cloth 透花刺绣台布
hemstitch embroidery 花饰刺绣；透花刺绣
hemstitcher 卷边缝纫机
hemstitch handkerchief 阔边手帕
hemstitching 卷边线缝；卷边缝纫
hemstitching satin stripes handkerchief 宽边缎条手帕

hemstitch machine 花式线迹缝纫机；卷边线迹缝纫机
hemstitch sewing machine 花式线迹缝纫机；卷边线迹缝纫机
hemstitch sheet 透花刺绣被单
hemstitch work 花饰刺绣
hem tack 可放长的裙褶
hem top 顶边卷边
hem turning elements 卷边机构
hem width 下摆宽，边宽
hem with mitered corner 折角拼缝下摆
Henan [pongee] 河南[柞丝]府绸
hench bone 腰椎骨
henequen fiber 剑麻
henequin 剑麻
henkelplush 毛圈长毛线；圈状长毛绒织物
Henley 半开襟汗衫
Henley boater 亨利蓝毡帽
Henley knit shirt 亨利针织衫；半开襟汗衫
Henley neckline 亨利领口；亨利领口线
Henley placket 亨利衫式开口
Henley shirt 半开襟汗衫；亨利针织衫
henna flower design 凤仙花纹图案
Henna pongee 内黄绢
henna 棕红色
hennaed hair 染成棕红色头发
hennaed nail 染成棕红色指甲
hennin 亨宁女帽(欧洲15世纪)；塔形垂纱帽；塔形头饰
hennup 大麻(荷兰称谓)
henrietta 19世纪纬面开士米斜纹呢，亨利塔毛葛
Henrietta 亨里塔毛葛
Henry neckline 亨利领口(前面半开襟加钮)
hepatic 猪肝色
Hepburn hair style 赫本发型
Hepburn hat 赫本帽
Hepburn sandal 赫本凉鞋
Hepburn style 赫本风格
hepepetwan 双面缎
he-ping faconné taffeta 和平绢
he-ping jacquard faille 和平葛
he-guang crepe 和光绉
Herat rug 赫拉特伊朗及阿富汗制纯毛地毯
herati pattern 赫拉蒂波斯地毯图案

herbaceous peony 芍药红,牡丹红
herbal garden 植物园色
Hercules braid 赫格利斯编带
hereke wool 土耳其粗地毯羊毛
herez 棉背密绒毯
herigot 赫利考连头巾宽外套(13～14世纪)
Heris rug 驼毛地毯,赫利斯地毯
Heriz rug 赫列兹地毯
Herkiner 赫开纳钻石
Herkiner diamond 赫开纳钻石
Hermana Marucelli 赫尔马纳·马鲁切利（意大利时装设计师）
hermani 法国薄丝毛呢
Hermès [法]赫尔墨斯服饰店
Hermes hairdo 赫尔墨斯发型
HERMES 爱马士
Hermés 爱玛仕（法国设计店）
hernani 法国丝毛呢
heron 深紫灰[色]
herringbone 海力蒙;人字纹;人字斜纹布;人字呢;海力蒙精纺呢绒;人字（鱼骨）花型;人字呢服装;针织狭带,针织物领口窄带
herringbone camlet 人字羽纱
herringbone chain 人字形链状饰物
herringbone effect 人字花纹
herringbone fabric 人字纹布料,人字斜纹布
herringbone fancy suiting 人字花呢
herringbone or cross stitch 海力蒙或人字形线迹,鲱骨或交叉线迹
herringbone pongee 小花纹绸
herringbone serge 人字哔叽
herringbone smocking 人字司马克;人字缩褶绣
herringbone stitch 人字线迹;人字形针法;三角针法
herringbone stitch fagoting 人字形拼缝绣
herringbone stitching 人字线缝
herringbone stripe 人字形条纹
herringbone tulle 人字形薄纱罗织物
herringbone tweed 人字粗花呢
herringbone twill (HBT) 人字形斜纹,山形斜纹;人字斜纹布
herringbone twill suiting silk 条影呢
herringbone twill ticking 人字斜纹床罩
herringbone woolen 人字呢
herringbone woolens 人字呢绒
hessen 荷兰粗亚麻布

hessian 黄麻平纹布,打包麻布
hessian bag 海生袋,黄麻细布袋
hessian boots 笨重高筒军靴;黑森靴
hessian cloth 黄麻麻布
hessian parachute 黄麻载重降落伞
heterochain fiber 杂链纤维
heterochromatic stimulus 异色刺激
heterocycloamide fiber 杂环（聚）酰胺纤维
heterocyclic copolymer fiber 杂环共聚物纤维
heterocyclic polyamide fiber 杂环聚酰胺纤维
hetero fiber 异质复合纤维,共轭纤维
heterofil 异质丝,共轭线
hetero-filament 异质复合丝,共轭丝
heterofil bicomponent fiber 异质比组分[复合]纤维
heterogeneous fiber 非均相纤维,非均匀性纤维（皮芯结构差异的纤维）
heterogeneous fleece 异质套毛
heterogeneous wool 异质毛
heteroprofile fiber 异形纤维
heteroshrinkage fiber 异收缩纤维
heteroyarn 异质[复合]丝
Hetex 希泰克斯烂花棉织物
heuke 及踝围裹靴,及膝围裹靴;尤克披风
heuke with hat 荷兰斗篷
heuse 骑马厚皮靴
hexads 六角配色
hexagonal filament 六角[形截面]丝
hexagon pattern 六角形纹
hezaam 阿拉伯男式羊毛长腰带
Hezhe ethnic costume 赫哲族民族服
hiapu 中国轻薄苎麻布
hibernian embroidery 爱尔兰刺绣;希伯宁刺绣
hibiscus 木槿色
hibiscus cannabinus fiber 洋麻,槿麻
hibiscus fiber 波拉纤维
hibiscus red 芙蓉红
Hibul 中空聚酯絮棉（商名,日本帝人）
Hi-carbon 碳纤维（商名,日本旭化成）
hickory cloth 条子斜纹布
hickory stripe 棕白条子斜纹布,蓝白条子斜纹布
hi-cut bottoms 高位[裁剪]下装
hi-cut briefs 高裤脚三角裤
hi-cut legs 高[位]裤脚

hide	大张皮;皮革;兽皮
hidden flow	暗伤
hidden hook	隐式搭钩
hidden in seam with reinforcement pocket	隐式嵌缝加固袋
hidden line	隐藏线
hidden pocket	内袋,隐袋
hidden seam	隐式缝
hidden-surface removal	隐藏面消除
hidden zipper	暗缝拉链,隐形拉链
hidden zipper closing	暗拉链门襟
hiding property	被覆性能,覆盖性能
high absorbed water fiber	高吸水纤维
high-absorption fiber	高吸水性纤维,高吸收性纤维
high-additional value	高附加值
high alkali glass fiber	高碱玻璃纤维
high-altitude suit	高空飞行服
high-altitude anorak	登山防风衣
high alumina fiber	高铝纤维(氧化铝含量较高的硅酸铝纤维)
high and low cowl neckline	高-低垂领口
high-arm lift	高抬手量
high-arm look	大袖窿款式
high-back cowl	高位背垂褶
high-back overalls	高后背套衫
high bleed fabric	高渗析性织物
high blending ratio	高混纺比
high boots	高筒靴,长筒靴
highbred jeans	名牌牛仔裤
high built-up collar	高耸领
high bulk acrylics	高膨体聚丙烯腈系纤维
high-bulk fiber	高膨体纤维
high-bulk hand-knitting yarn	膨体[手编]毛线
high-bulk nonwoven	高膨松非织造布
high-bulk orlon	高膨体奥纶
high-bulk staple fiber	高膨体短纤维
high-bulk textile	高膨体[纱]织物,高膨体纺织品
high-bulk yarn	高膨体纱
high bulky yarn	高膨体纱
high bust	高胸型
high button shoes	侧面鞋扣及踝靴
high cabochon	高穹顶式(珠宝)
high cam	高嵌条
high-chic seamed look	时髦线缝风貌
high-class	优等
high collar	高领
high collarbone	高锁骨
high collarbone type	高锁骨型
high-colored	色调强烈的,极鲜明的
high-colored design	色调强烈图案
high complexion	红润的面色
high contrast grandrelle yarn	高反差夹色纱
high contrast	高度对比,高反差
high count cloth	高支纱织物,精细织物
high cowl neckline	高垂领[口]
high crime fiber	高卷曲型纤维
high crime high wet modulus rayon	高卷曲高湿模量黏胶纤维
high-cut boots	高筒靴
high-cut bottom	高位[裁剪]下装
high-cut brief	高裤脚三角裤
high-cut leg	高[位]裤脚
high-cut leg bikini	高裤脚比基尼泳衣
high-cut leg opening	高位裤脚开口
high denier fiber	高旦纤维,粗旦纤维
high density	高密度
high density downproof fabrics	高密防羽绒织物
high density fabric	高密度织物
high density polyethylene interlining (HDPE interlining)	高密度聚乙烯[黏合]衬
high done	高穹顶式(珠宝)
high elasticity (HE)	高弹性
high elasticity yarn	高弹弹力丝
high energy type	高能型
higher neckline	立领口
highest bust-level width	最高胸平宽度
high extensibility fiber	高伸长性纤维
HIGH FASHION	《高级服饰》
high fashion	高档时装;时髦款式;最新流行款式;高档服饰品;高级时装
high fashion color	最新流行色
high fashion designer	高级时装设计师
high fashion fabric	流行织物
high fashion garment	高级流行服装
high fashion house	高级时装店
high-feeling fiber	高触感[性]纤维,触觉优良的纤维
high finish	高级整理
high-fired pitch fiber	高温焙烧沥青纤维
high-flux hollow fiber	高[流]通量中空纤维
high fold collar	拿破仑大衣领,翻折高领
high footing	高起点

high-frequency drying 高频烘燥
high-frequency electric sealing machine 高频熔缝机
high-frequency sewing 超声波缝制
high functional fiber 高功能[性]纤维
high-functionality fiber 高功能[性]纤维
high-function fiber 高功能[性]纤维
high gauge 细针距羊毛衫
high gauge sweater 细线毛衣
high grade 高档品
high-grade clothing 高档服装
high-grade coarse knitting wool 高粗绒线
high-grade finish 高级整理
high-grade goods 高档商品
high-grade Irish poplin 高级爱尔兰毛葛
high hat 高顶黑色大礼帽;礼帽;高帽
high head 花边装饰小帽
high heel 高跟
high heel shoes 高跟鞋
high-heeled sandal 高跟凉鞋
high-heeled shoes 高跟鞋
high heels 高跟鞋
high hip 高臀围;上臀部
high hip girth 上臀围
high hip measurement 上臀围
high hip size 上臀围
high horizontal cowl 高位垂褶
high hydroscopic fiber 高吸水纤维
high hydroscopic towel 高吸水毛巾
high impact rayon 高冲击强度人造丝
highland cape 高地斗篷;高地披肩
highland check 苏格兰彩格
highland dress 苏格兰高地服装(如男子穿的叠褶裙子等)
highlander 苏格兰方格布,苏格兰方格呢
highland suit 苏格兰高地男孩套装
highland wool 高地绵羊毛
high-leg briefs 高裤脚三角裤
high-leg rain boots 高筒雨靴
high-legged maillot 高脚泳衣
highlight 辉亮部分,高光泽,光彩照人
highlighter 轮廓色(美容)
highlighting 醒目性
high loft 高膨体,高膨松度;高膨松非织造布
high loft fabric 高蓬松度织物
high loft nonwoven 高膨松性非织造布
high-low 及踝高统靴
high-low effect 高低[花纹]效应(非织造布)

highly angled twill 急斜纹
highly combustible fabric 易燃织物
highly conductive synthetic fiber 高导电性合成纤维
highly oriented fiber 高取向纤维
highly squared shoulder 高肩
highly swellable cellulose fiber 高溶胀性纤维素纤维
high-melting fiber 高熔点纤维
high modulus fiber 高模量纤维
high modulus low shrinkage yarn 高模量低收缩丝
high modulus rayon fiber 高模量黏胶丝纤维
high modulus weave 高模量织物
high moisture-absorbent fiber 高吸湿纤维
high moisture-content fiber 高吸湿纤维
high molecular material 高分子材料
high multifilament 高复丝
high multifilament yarn 多纤维复丝
high neck 高领
high-necked halter top 高领吊带上装
high neckline 高领口
high-needle flat sewing machine 高速平缝机
highnic lace 希格尼克花边
high oriented yarn 高取向丝
high-performance 高效,高性能
high-performance antistatic fiber 高效抗静电纤维
high-performance carbon fiber 高性能碳纤维
high-performance fiber 高性能纤维
high-performance protective fabric 高性能防护织物
high-performance rayon 高性能黏胶丝
high-performance textile 高性能纺织品
high pick fabric 高纬密织物
high pile 高绒头;长毛绒
high pile coat 长毛绒大衣
high pile fabric 长毛绒织物
high pile knitted fabric 长毛绒针织物
high pile knitter 长毛绒针织机
high pile knitting machine 长毛绒针织机
high pile lining 长毛绒衬里
high platform boots 高台靴
high platform sole 高台底(鞋)
high platform sole shoes 高台底鞋
high-ply cutter 厚层裁剪机

high polymer 高聚物
high-porosity gel-state fiber 高孔隙率凝胶态纤维
high power vacuum board (熨烫用)强力抽湿台板
high-power yarn 高收缩丝,高应力丝
high pressure body press 高压大身熨烫机
high pressure body shape press 高压立体[人形]整烫机
high purity silica fiber 高纯度二氧化硅纤维
high-priced designer 高价设计师
high quality (HT-Q) 高品质
high quality fabric 优质织物
high quality finishing 高级整理
high quality material 上等料子
high quality suit 优质套装
high quality suiting 优质套头衣料;优质西服料;优质套装
high relaxed cowl 高位松垂褶
high resilience 高回弹性
high-rise 高腰款式,短直裆,深上裆
high-rise belt 高腰饰条,高腰腰带
high-rise dress 高腰裙;高腰装
high-rise pants 高腰裤
high-riser 袋状裤(臀至腰腿都宽大);深立裆的裤子
high-rise waistline 高腰线
high roll collar 高翻领
high ruffly collar 高荷叶边领
High School of Fashion Industries 时装工业高等学校
high-sensitive-leval consuming 高敏感度的消费
high shawl collar 高位青果领
high shoes 高帮鞋
high shoulder 高肩,耸肩,耸肩体型
high shoulder blade 高肩胛骨
high shoulder blade type 高肩胛骨型
high shoulder sleeve 高肩袖,耸肩袖
high shrinkage fiber 高收缩纤维,高缩率纤维
high shrinkage polyester (fiber) fabric 高收缩涤纶织物
high shrink staple 高缩率短纤维
high silica content fiber 高硅石含量纤维
high silica glass fiber 高硅氧玻璃纤维(耐温1200℃以上),高硅石玻璃纤维
high silica wool 高硅氧棉
high silk hat 高顶硬礼帽

high slit opening 高开衩
high socks 高筒袜,及膝中筒袜
high-society costomer 高层顾客
high speed chainstitch flat bed sewing machine 链式线迹高速平缝机
high speed drafting machine 高速绘图机
high speed electronic handknife splitting machine 高速电子刀带式削皮机(皮革加工)
high speed eyelet buttonhole sewing machine 高速圆头锁眼机;凤眼机
high speed flat sewing machine 高速平缝机
high speed four needle elastic ribbon loom 高速四针宽紧带机
high speed handknife splitting machine 高速刀带式削皮机(皮革加工)
high speed lock stitch machine 高速锁缝机
high speed lock stitch machine with needle feed 针送料高速锁缝机
high speed needle feed lockstitch machine 针送料高速锁缝机
high speed seamer 高速缝纫机
high speed sequential buttonhole machine 高速连续纽孔机
high speed sewing machine 高速缝纫机
high speed single-needle flat sewing machine 高速单针平缝机
high-speed tubular shape lap-seam felling machine 高速筒状卷接缝纫机
high spliced heel 加固高跟
high splicing 高跟加固
high standing collar 高立领
high strength and high modulus fiber 高强度高模量纤维
high strength aramid fiber 高强芳纶纤维
high strength fiber 高强度纤维
high strength flame-resistant fiber 高强度耐火纤维
high strength glass fiber 高强度玻璃纤维
high strength synthetic fiber 高强度合成纤维
high stress fiber 高应力纤维
high stretch fabric 高弹织物
high stretch polyamide hosiery 高弹锦纶袜
high stretch yarn 高弹丝
high style 高级女子时装;最新款式

high style fabric 高级整理织物
high-swelling viscose fiber 高溶胀黏胶纤维
high-tech 高技术
high-tech clothes 高技术服饰
high-tech crisp-appard 挺爽型高技术服装
high-tech fabric 高技术织物
high-tech fiber 高技术纤维
high-tech look 高技术风格,高技术型
high technology 高技术
high technology fabric 高科技布料
high technology fiber 高技术纤维(高性能纤维及高功能纤维等)
high technology's Indian style 高技术印第安款式
high-tech pressing machine 高科技熨烫机
high-tech textile 高技术纺织品
high-tech zone 高新技术区
high temperature fiber 耐热纤维,耐高温纤维
high temperature-pressure dyeing 高温高压染色法
high temperature protective clothing 防高温服,高温保护服
high temperature protective shoes 高温防护鞋
high temperature resistance 耐高温性
high temperature resistance fiber 耐高温纤维
high temperature-resistant organic fiber 耐高温有机纤维
high temperature setting 高温定型
high temperature shrinkage 高温收缩
high-temperature thermosetting 高温热定型
high tenacity entangled yarn 高强度网络丝,高强度交络丝
high tenacity fiber 高强力纤维
high tenacity filament yarn 高强度长丝[纱]
high tenacity high-modulus fiber 高强[度]高模[量]纤维
high tenacity industry fiber 高强度工业用纤维
high tenacity nylon filament 高强力锦纶丝
high tenacity polyvinyl alcohol 高强聚乙烯醇纤维

high tenacity rayon 高强力人造丝,强韧人造丝
high tenacity viscose fiber 高强力黏胶纤维
high texturized yarn fabric 高弹丝织物
high thermal 最高温
high-thermal resistance 耐高温性
high thread count 紧密织物
high toe shoes 高头鞋
high-tongue 中世纪高舌平跟鞋
high tongue style 高鞋舌款式
high-topped shoes 高帮鞋
high-top tennis shoes 高帮网球鞋
high touch 高敏感度
high touch fiber 触感优良的纤维,高触感[性]纤维
high tower cotton 美国高塔棉
high twist 高捻
high twist fabric 强捻丝织物,紧捻丝织物
high twist filament 强捻丝
high two 高位置两扣款式
high upper shoes 高腰鞋
high value-added product 高附加值产品
high visibility clothing 高能见度服装(学童、路工、机场人员穿的荧光保护服),信号服
high volume production 大量生产
high waist 高腰,高腰型
high waistband 高腰
high-waister 高腰女装
high waistline 高腰线
high-waist shorts 高腰短裤
high-waist skirt 高腰裙
high waist slip 高腰衬裙
high waist pants 高腰裤
high waist tapered slacks 高腰窄脚裤
high-waist tapered trousers 高腰细脚裤
high waist trousers 高腰裤
high warp tapestry 立经挂毯
high-water (裤子等)特别短的
high water absorbable polyester fiber fabric 高吸水涤纶织物
high water-absorbent fiber 高吸水[性]纤维
high water shorts 特短裤
highwayman coat 强盗大衣
highwayman collar 强盗大衣领
high wet-modulus 高湿模量
high wet-modulus fiber 高湿模量纤维

high wet-modulus rayon 高湿模量人造丝
high wet-modulus staple fiber 高湿模量短纤维
high wet-modulus viscose 高湿模量黏胶纤维
high wet-modulus viscose fiber 富强纤维
high wet-resistance viscose fiber 高耐湿性黏胶纤维,高抗湿性黏胶纤维
high wettability acrylic fiber 高吸湿腈纶
high women's wear 高级女装
hi kanakin 红色细薄棉布
hikers 旅行靴
hiking 交替拉伸放松织物;郊游服;步行服
hiking boots 徒步旅行靴,远足靴
hiking costume 旅行装,步行装
hiking shoes 旅行鞋,徒步鞋
hiking shoe with fringed tongue 带舌旅行鞋
hiking up 后翘
hiking wear 旅游服,步行服
Hilary Clark 希拉里·克拉克(美国时装设计师)
Hilda 棉经毛纬衬里布
hilda 希尔达里子呢
hilda lining 希尔达里子呢
hi-leg briefs 高裤脚三角裤
hi length 长及臀部式服装
hill dropping 撒籽种棉
Hilliard cotton 喜勒陆地棉
hillsborough hoggs 爱尔兰棕白亚麻平布
hi-lo corduroy 间隔条灯芯绒
hiloft-nonwovens 高膨松非织造布
hi-lo loop carpet 间隔条毛圈地毯
hi-lo pile 高低绒头地毯
hilow bulked yarn 高低混纺膨体纱
himalaja 仿山东绸;高级斜纹呢
himalaya 仿山东绸;高级斜纹呢
himation 大氅长方布;希马辛大长袍(古希腊);希腊披风从左肩披下缠身的长方形布
himobutton 布扣;布条扣;直扣
himro 手织真丝花缎
hinde cotton 埃及棉地中的野生棉,红褐棉
hind edge seam 暗边缝
hind quarters 后半身
hinge bracelet 铰链手镯
hinged blank foot 铰链盲压脚
hinged blinder foot 铰链暗缝压脚

hinged bracelet 铰链手镯
hinged edge cutter foot 切边活压脚
hinged foot 活压脚,铰链压脚
hinged guard 活动防护罩
hinged last 弹簧鞋楦
hinged lining foot 缝衬里活压脚
hinged-open flat seam 平摊缝
hinged-open type seaming 铰链开口状缝纫
hinged pin 铰链销
hinged (presser) foot 活压脚
hinged raising foot 活压脚
hinged taping foot 铰链滚带压脚
hinged tube foot 管状活压脚
hinged-turn piping foot 向上滚边活压脚
hinged work plate 活动工作板
hinged zipper foot 缝拉链活压脚
hinge effect 铰链效应
hinge foot 铰链压脚
hinqunhat cotton 印度浅黄棉
hinroo 印度花卉绸
hip 髋,髋部;臀围;胯部佩带(美国)
hip accent 强调臀部
hip arc 臀弧
hip arc line 臀弧线
hip bags 臀垫
hip belt 臀部腰带,臀带
hip bone 髋骨
hipbone bikini 低腰比基尼
hipbone pants 低腰裤
hip bone skirt 低腰裙(不系带,落在髋骨处),腰骨裙,挂胯裙
hipbone slacks 低腰裤
hip boot(s) (渔民,消防员等穿的)高至臀部的长筒橡胶靴,及臀靴
hip breadth 臀宽
hip button 西装后背中央褶纽扣
hip circumference 臀围
hip conscious skirt 强调臀部裙
hip depth 臀围;直裆
hip-full form 臀型,腰型
hip girdle 臀部紧身带
hip girth 臀围
hip girth line 臀围线
hip guard protector 护臀器
hip hang dress 垂臀装
hip hanger 挂臀裤,低腰裤;挂臀裙
hip hange skirt 挂臀裙,低腰裙
hip height 臀至地
hip-hugger 紧身低腰裤;短直裆裤;紧裹

臀部的
hip-hugger belt 低腰饰带
hip huggers 短直裆裤
hip hugger skirt 低腰裙,抱臀裙
hip-hugger waistline 低腰裙[裤]腰线
hip hugging skirt 强调臀部裙
hi-pile 长毛绒
hi-pile fabric 长毛绒织物
hip interest skirt 强调臀部裙
hip length 臀围长(腰围到臀的长度);臀高;臀直
hip-length 及臀式
hip length jacket 齐臀夹克
hip length semi-fitted blouse 至臀长半合体上衣
hiplets 并合式裤袜,连裤袜(商名)
hip level 臀围线
hip line (HL) 臀围线
hip measure 臀围
hipness 跟潮流
hip pad [patch] 臀垫
hippie 嬉皮士服饰
hippie bag 印第安袋,嬉皮士袋
hippie coat 嬉皮士服
hippie culture 嬉皮服饰文化
hippie look 嬉皮款式;嬉皮士风貌
hippie necklace 嬉皮珠项链
hippies 嬉皮士
hippie style 嬉皮士款式;颓废派风格
hippie stylistic form 嬉皮士款式
hip pleated skirt 襞褶裙;紧臀多褶裙
hip pocket 裤后袋,臀部口袋,臀袋;后枪袋
hip pocket button 臀袋纽
hip pocket facing 裤后袋贴布
hip pocket position (line) 后袋位置线
Hippolyte Roy 伊波利特·罗伊(1763～1829,法国时装设计师)
Hip-pop 饶舌
hippy 嬉皮士[的],臀部肥大的
hippy figure 臀部肥大体型
hippy look 颓废派风貌
hippy style 嬉皮款式,嬉皮士风格;颓废派风格
hippy vest 嬉皮士背心
hip rider 低腰裤,短直裆裤
hip-rider swimsuit 两件套低腰女子泳装
hip room 臀围松量
hip seam 裤子后中缝
hip shot 两髋一高一低的

hip size 臀围
hip slope 后臀倾角
hip-slung pants 背带裤,低腰裤
hip smocked dress 臀部司马克裙
hipsterism 超时髦,入时
hipsters 低腰裤,裤腰低及臀部的裤子
hip support 护臀
hip to ground 臀至地
hip type 臀部类型
hip up girdle 提臀紧身褡
hip width 臀围,臀宽
hip wrap scarf 臀巾
hip yoke 衬腰育克,髋部育克
hip yoke skirt 衬腰裙
hired costume 租赁服装
hi-rise girdle 高腰紧身褡
hi-riser 高沿口鞋(总称)
Hiroko Koshino 越野平子(日本时装设计师)
Hiromi Yoshida 吉田广美(日本时装设计师)
Hiromichi Nakano 中野裕通(日本时装设计师)
his and hers look 同面料男女装风格,男女同款风格
Hisashi Hosono 细野久(日本服装设计师)
Hisepare 聚丙烯非织造布(商名,日本窒素)
Hi-silkiss finished 高光泽整理
histogram 直方图
historical costume 古装,古代服装
historical jacket 时代夹克
historic style 历史性款式
historic stylistic 历史性款式
hit-and-miss effect 时现时隐效应
hitch stitch 反锁线迹,加绕线迹
Hi Tech traumas 高科技冲击
hi-top trainer 高帮训练鞋
hit-or-miss rug 无固定花纹地毯,色布条捻织地毯
hive bonnet 蜂巢式罩帽
hizaam 阿拉伯男士白色腰带
H leg H形边脚
H line H字型;H型造型;直线条形服装
HM carbon fiber 高模量碳纤维
hō 古代日本膝下长袍
hoar 灰白

hobble （20世纪初）蹒跚裙,穿下摆裙
hobble silhouette 穿摆裙式;莲步型
hobble skirt （20世纪初）蹒跚裙,莲步裙;窄摆裙;旗袍裙
hobble style 窄摆裙式
hobnail 服装上的平头钉
hobnailed shoes 平头钉鞋
hobo bag 流浪汉袋,游民包
hobo look 流浪汉风貌;流浪装
hockey boots 曲棍球靴
hockey jersey 曲棍球衫
hockey shin guard 曲棍球护胫
hockey skate 冰上曲棍球的冰鞋
hockey socks 曲棍球短袜
hockey wear 曲棍球装
hodden 平织粗呢,手织粗呢
hodden gray 白黑羊毛混合物,白灰羊毛混合物;灰色呢,苏格兰厚呢
hoddin 平织粗呢,手织粗呢
hodrunck 黄棕色平布（东非）
Hoffman press 霍夫曼式熨衣机
hogskin 猪皮,猪革;猪革制品
hoike 侧开缝披肩
ho-kwang crepe 和光绉
holanda 西班牙阔亚麻布
Holbein stitch 双平行线迹,台阶形线迹;意大利针法,霍尔班针法
Holbein work 钩线刺绣,霍尔班刺绣,罗马尼亚刺绣,台阶形花纹刺绣
holdall 旅行手提箱;旅行手提包
hold-down device 压布装置
holder 支架
hold framed sunglasses 宽边太阳镜
holder guide 导夹
hole-and-button cuff 扣扣袖口
hole and flat stitch embroidering machine 圆孔及平针针迹绣花机
hole and pad stitch 打孔线迹
hole braid 窄编带;窄枕结花边
hole profile fiber 异形中空纤维
holi 胡利粗纺毛呢（阿拉伯男服面料）
Holland 荷兰亚麻细布;洁白亚麻细布;荷兰粗支棉或亚麻重浆印花布;仿亚麻窗帘布
hollanda 细条纹亚麻布
hollandas 仿亚麻条格棉布
Holland costume 荷兰民俗服
hollande 密织亚麻床单布,平纹亚麻细布
Holland fabric 荷兰平纹棉布

Holland flax 荷兰亚麻
Holland linen 荷兰本色亚麻布
Holland shade cloth 窗帘布
hollie point lace 针绣宗教花边,宗教刺绣花边
hollie-stitch 神圣针迹（锁眼花边针迹）
hollie work 神圣针迹绣
hollow acetate fiber 中空醋酯纤维
hollow braid 空心带
hollow chest 塌胸,凹胸
hollow composite fiber 中空复合纤维
hollow cored fiber 中空纤维
hollow-core polyester fiber 中空涤纶纤维
hollow cut 棉灯芯绒
hollow cut velveteen 凸条灯芯绒
hollowed-out dress 镂空装
hollow fabric 中空织物,管状织物
hollow fiber 中空纤维
hollow figure 凹胸体型
hollow filament 空心丝,中空丝
hollow flatten fiber 中空扁平[截面]纤维
hollow flocking filament 空心植绒长丝
hollow glass fiber 中空玻璃纤维
hollow inflated rayon 中空充气人造丝
hollow instep 裤脚起翘
hollow of throat 颈窝
hollow polyester fiber 中空聚酯纤维
hollow polyester staple fiber 涤纶中空短纤维
hollow profiled fiber 中空异型纤维
hollow rayon fiber 中空黏胶纤维
hollow semipermeable fiber 中空半渗透纤维
hollow shaped article 中空制品
hollow splitting-type composite fiber 中空裂离型复合纤维
hollow synthetic fiber 中空合成纤维
hollow viscose fiber 中空黏胶纤维
hollow web 管状织物,圆筒织物
hollow yarn by emulsion process 乳液法中空丝
holly green 圣诞冬青绿
Hollywood （衣服等）花哨的
Hollywood fashion 花哨时装
Hollywood-top slip 好莱坞型连身衬裙
hologram memory fiber 全息记忆纤维
holoku 荷璐扣拖裙长袍;一种带有拖裙的长袍（夏威夷妇女在正式场合穿的）
holster pocket 枪套形口袋（常用于运动

装上)
holyberry 冬青果色
holy stitch 宗教绣,宗教针法
holy work 神圣针迹绣
home accessories 室内装饰品
home cover 室内套袜
home dress 居家服;家常服
home dressmaking 家庭裁制
home eyeballing 登门促销
home furnishing [fabric] 家具布,家具织物,装饰织物
home gown 便服
home ironing board 家用烫台
home launder 家庭洗涤
home laundering 家庭洗涤
Home Laundering Consultative Council (HLCC) 家庭洗涤顾问委员会
home laundry equipment 家用洗衣设备
home machine washing 家庭洗涤,家用洗衣机洗涤
home market 国内市场
home sewer 家庭缝纫业者
home sewing 家庭缝纫
home sewing machine 家用缝纫机
home sewing spool 家庭缝纫用线轴
homespun 彩点粗花呢,火姆司本,粗纺花呢;手织土布;手工纺织呢;手织大衣呢
homespun cloth 土布
homespun [fabric] 火姆司本,钢花呢,彩点粗花呢;手工粗纺呢
home textile 家用纺织品
home-use lock machine 家用锁式线迹缝纫机
home-use sewing machine 家用缝纫机
home wear 家居服,便服
homienchow 平纹花绢
homing device 倒缝装置
homochromatic 同色的
homo-fiber 均质纤维,单组分纤维
homofil 单组分纤维,单组分线
homogeneity 均匀性
homogeneous filament 均质丝
homopolyester fiber 均聚酯纤维
honal kladi 横条纹毛毯
honan pongee 河南榨丝府绸
honan 河南榨丝府绸
HONEY 《蜂蜜》(英)
honey [yellow] 蜜黄[色]
honeycomb 蜂窝纹;蜂窝纹织物

honeycomb canvas 小方格刺绣底布;蜂窝纹帆布
honeycomb effect 蜂窝效应
honeycomb fabric 蜂巢织物,蜂窝状三维织物
honeycomb knit 蜂窝针织物
honeycomb pique 蜂窝凸纹衣
honeycomb quilt 蜂窝纹床单布
honeycomb reseau 菱形网眼花边地布
honeycomb rib fabric 蜂巢网眼针织物
honeycomb smocking 蜂窝司马克绣,蜂窝状褶饰
honeycomb stitch 蜂窝状线迹,蜂窝针迹,蜂窝绣
honeycomb suitings 毛蜂巢呢
honey dew 密露橙色(浅橙色)
honey gold 黄棕色
honey mustard 芥末色
Hong Bang tailor 红帮裁缝
Hong Bang tailor's shop 红帮店堂
Hongchow 杭绸,杭绉
Hongchow silk plain 杭纺
hong kong 丝经亚麻纬印花平布
Hong Kong Fashion week 香港时装节
Hong Kong shirt 香港衫
Hong Kong shoes 香港鞋
hongroise 纯色法国哔叽,素色法国哔叽
honiton applique 小花纹花边
honiton appliqué lace 霍尼顿缝饰花边
honiton braid 椭圆花纹细编带
Honiton lace 霍尼顿小花纹花边,小花纹梭结花边
hood 兜帽,头巾,风兜;(以颜色标志职位高低的)披肩布;锥形制帽毡;后垂布
hood cape with liripipe stump 有长飘带的学位服披肩
hood collar 兜帽领
hood drawstring 兜帽系带,风帽拉绳,头巾拉绳
hooded Anorak 带风帽的夹克衫,爱诺瑞克风雨夹克
hooded circular cloak with ball fringe 带风帽的绒线流苏斗篷
hooded clothing 带风帽的外套
hooded coat 连兜帽外套,有头巾外套;风雪衣
hooded fleecy sweat jacket 带帽绒布夹克

hooded heel 连帮跟
hooded jack 兜帽夹克,风帽上衣
Hooded jacket 兜帽夹克,风帽上衣
hooded neckline 风帽围巾形领口,头巾兜帽形领口
hooded parka 连帽式派克大衣
hooded raincoat 连兜帽的雨衣
hooded robe 连帽长袍
hooded sweatshirt 连帽式厚运动衫
hooded trainer 带风帽运动衣
hood fabric 顶篷布,帐篷布,罩盖织物
hood front line 帽嘴线
hood height 风帽高
hood jacket 风帽夹克;带风帽的上衣
hoodlum fashion 流氓款式
hoodlum style 流氓款式
hood neckline 帽下口线
hoohoo 平纹格棉布
hood under line 帽下口线
hood width 风帽宽
hood with liripipe 带飘带的头巾、兜帽
hook 钩,挂钩
hook and bar 钩棒扣;裤扣;裙扣;钩扣,钩袢,裙钩、裤钩
hook and closing 钩眼钩合
hook and eye 风钩;风纪扣;钩扣;领钩
hook and eye buckle 对钩环扣
hook and eye closing 钩眼扣合
hook and eye fastening 钩眼扣合
hook and eye tacker machine 领钩和平纽孔缝纫机,套圈锁纽孔机
hook and eye tape 钩扣带
hook and loop fastener 压合带,尼龙搭扣带;魔术带;子母带
hook and loops （尼龙）搭扣
hook and loop tape 搭扣带
hook body 钩梭体
hook buckle 钩形带扣
hook closure 套圈锁眼,钩眼锁紧
hooked fur over coat 皮猴儿
hooked rug 钩编地毯
hooked stitch 钩针针法
hooked vamp shoes 勾背鞋
hookflex 钩袢
hook handle 钩形柄
hook-loop fastener 搭扣
hook needle 钩针
hook point 缝纫机梭尖
hook reel 旋梭绕线轴
hook tape （装订在衣领上的）衣钩条

hook type buckle 弹簧环钮
hook vent 明单衩;明后衩;钩形后身衩
hoop 戒指,耳环;撑裙圈,裙箍;加撑裙,圈环裙,裙环式衬裙
hoop bagging 粗黄麻制酒花袋布
hoop bracelet 串环手镯
hoop earring 大圈耳环;吉卜赛耳环,奴隶耳环
hoop for darning and embroidering 绣花棚架,织补棚架
hooping Holland 亚麻布
hoop petticoat 裙环式衬裙
hoops 裙撑,有裙环的裙
hoopskirt 箍环式,圈环式,加撑裙
hoover apron 护士服;护士围裙
hop fiber 啤酒花韧皮纤维
hopi acala cotton 美国高强度陆地棉
hop pocketing 酒花袋布(粗黄麻制)
hopsack 方平织物,席纹呢,板司呢
hopsacking 粗纺厚呢;低档粗棉布;席纹粗黄麻袋布
hopsack tweed 粗纺板司花呢
horizon blue 淡天蓝色
horizontal abduction 水平拉伸(肌肉);水平收缩(肌肉)
horizontal balance line（HBL） 水平基准线
horizontal buttonhole 横向扣眼
horizontal chromaticity plan 等色彩度平面
horizontal collar 水平领(衬衫领开成水平)
horizontal corded interlock fabric 横棱棉毛布
horizontal cylinder machine 横式圆筒底板缝纫机
horizontal cylinder sewing machine 横式圆筒底板缝纫机
horizontal dart 水平省
horizontal division 水平分割
horizontal edge trimmer 水平式修边器
horizontal flow theory 泛流理论(一种时装款式流行理论);横向流行理论(时装)
horizontal knit 横条针织物
horizontal leno 横罗
horizontal line 横条,环纹;水平线
horizontally rotating shuttle 水平旋梭
horizontal measurements 水平尺寸

horizontal missing 横向拉长线圈,横向延展线圈
horizontal mode 水平模式
horizontal neckline 一字[形]领[口],船形领口
horizontal plane 水平面
horizontal pointed twill 横山形斜纹
horizontal repeat 横向循环
horizontal ripple 横波纹织物
horizontal satin stripe towel 缎档毛巾
horizontal side pocket 侧缝横袋
horizontal slit (牛仔裤前面)水平袋口
horizontal streak 横条疵
horizontal stripe 横条纹,横格条纹;横条子布;色纬
horizontal style line 横向款式线
horizontal tuck 横向褶裥织物;水平线襞,横襞
horn button 角制纽扣(如羊角扣)
horned cap 修女大白帽
horned head-dress 双角形头饰
horn raglan 一片插肩袖
horn shaped pocket 兽角形袋
horny handle 角质[手]柄
horse blanket 马鞍被,马鞍褥
horse cloth 马鞍毯(正面黄麻反面羊毛,且双面斜纹)
horse cover 马鞍被,马鞍褥
horsehair (HH) 马毛,马鬃;马毛带;马尾衬,人造合成马尾衬
horsehair braid 马毛织带
horsehair brush 马鬃刷
horsehair cloth 马尾衬
horsehair hem 夹毛衬缘边
horsehair interlining 马尾衬
horsehair stiffening cloth 马毛硬衬布
horse-hide 马皮,马革
horse hoof collar 马蹄领
horse-hoof neckline 马蹄形领口
horse-hoof sleeve 马蹄袖
horse leather 马皮,马革
horse leather shoes 马皮鞋
horse-shoe collar 马蹄领,马蹄形领
horse-shoe collar life preserver 马蹄领救生衣
horse-shoe cuff 马蹄袖,马蹄袖头
horse-shoe jumper 马蹄式背心裙(前后很低的马蹄领口,肩带与裙身同料)
horse-shoe neck 马蹄领
horse-shoe neckline U 形领口,马蹄形领口
horse's hoof shoes 马蹄鞋
horsetail liner 马尾衬
horticultural textile 园艺用纺织品
hose 袜子,袜类;短袜;长筒袜;(旧时)男紧身裤(从腰至脚);胡斯裤(中世纪)
hose-bottom pants 窄脚裤
hose-bottom trousers 窄管裤
hose clip 袜夹
hose-examining machine 验袜机
hosen 胡斯裤(中世纪);袜类,护腿;袜子,齐膝短裤15世纪紧身裤
hose setting machine 袜子定型机
hose top 袜口
hose-tops 无跟长袜
Hoshimoto 桥本纪子(日本时装设计师)
hosier 袜商,内衣经销商
hosiery 袜子;针织袜类;针织品;针织生产;制袜业
hosiery board 袜子定形板
hosiery boarder 袜子定形机
hosiery boarding machine 袜子定形机
hosiery closed-toe 头封闭袜
hosiery concave 袜凹
hosiery fabric [袜类]针织物
hosiery for climb 登山袜,山地袜
hosiery form 烫袜板
hosiery frame 织袜机
hosiery heel 袜跟
hosiery industry 针织工业
hosiery inspecting apparatus 验袜器
hosiery knitted by double cylinder knitter 双针筒袜
hosiery knitter 织袜机
hosiery machine 袜机
hosiery made of few-o-fil polyamide faber 锦纶少孔丝丝袜
hosiery mill 针织工厂,织袜厂
hosiery package machine 袜子装袋机
hosiery seamer 缝袜机
hosiery shape 袜子定形板
hosiery tester 量袜器
hosiery toe closing 袜头封口
hosiery unboarded 非定形袜
hosiery whole length 袜子总长
hosiery with blended yarn 混纺线袜
hosiery with color cast 多彩袜
hosiery with conjugated yarn 复合丝袜
hosiery with polyamide few-o-fil yarn 锦纶少孔丝丝袜

hosiery with polyamide tight twist filament 锦纶紧捻丝袜
hosiery with sharp toe 尖足袜
hosiery with triblended yarn 三合一交织袜
hosiery yarn 针织纱[线],针织[袜类]用丝
hospital garment 医用服装
hospital gauze 医用纱布
hospital uniform 医务服
host coat 主人外套
hostess 家居服
hostess coat 女服务员装;主妇家居长袍
hostess culottes 及跟睡衣(大脚裤)
hostess dress 主妇服装
hostess gown 女服务员装;主妇长袍;主妇袍装
hostess pyjama 家居睡衣裤
hostess robe 主妇长袍;主妇袍装
host fabric 基底织物,基质织物
host machine 主机
hot air sealing tape 热风胶贴带
hot air seam sealing machine 热风胶贴机,热风机
hot air shrinkage 干热收缩
hot calendering bonding 热轧黏合
hot carpet 电热毯
hot clip 电热发夹
hot color 暖色,热色(彩)
hot comfort of clothing 服装热舒适性
hot compress bag 热敷袋
hot curler 电热卷发器
hot glueing 热胶合
hot iron 高温熨烫
hot item 热销的服饰
hot mask 热面具
hot pants 热裤,女式超短裤,冬季短裤
hot pants with bib 连护胸的热裤
hot peg 挂帽钩
hot pink 鲜艳的粉红色
hot rack 挂帽架
hot skirt 超短裙
hot stamping machine 烫金机(制鞋)
hot top 热衫
hot water extraction 现场清洗地毯
hot water post-boarding 热水后定形
hot water resistance 耐热水性
hot water washing 热水洗涤
hotaori 保多织
hotel stuff uniform 酒店制服
hothead press 热压头压烫机
hot-house thermal insulation cloth 温室保温布
hot-melt adhesive interlining 热熔衬
hot-melt adhesive padding cloth 热熔黏合衬布
hot-melt interlining 热熔衬
hotoz 土耳其犹太妇女遮脸巾
hot-press fusible interlining 热压热熔衬
hot-set 热定型;热定形
hounds' ears 犬耳袖口,大翻边袖口
hounds' tooth check 犬牙格花纹,千鸟格花纹
hound-tooth 犬牙格
hound-tooth check 犬牙格花纹
hounscot 轻薄精仿呢
houppelande 奥布兰袍;男式宽松大袍(14~15世纪)
houppeland with dagges 带饰边的奥布兰袍
hourglass dress 沙漏装
hourglass heel 沙漏型鞋跟
hourglass line 沙漏线形,沙漏型
hourglass-shaped skirt 蜂腰形,滴漏形;沙漏形裙
hourglass silhouette 沙漏型轮廓
hourglass style 沙漏型款式
houri coat 土耳其外套
houseaux [法]皮绑腿
houseboy pants 男服务员裤(七分长,腰间有同料带结),工作裤
house-cap 宿舍帽
housecoat (女用)宽大家便服,家居服;主妇服;妇女宽松式外衣;工作罩裙
housedress 家居服,女便服;家庭主妇装
house fabric 家用织物
house flannel 家用法兰绒
housefrock (妇女)家居裙衫,家居服;女便服;家常罩衣
housegown 长浴衣
household detergent 家用洗涤剂
household dye 家用染料
household linen 家用亚麻布
household lock stitch sewing machine 家用拷边机,家用锁式缝纫机
household sewing machine 家用缝纫机
household stitcher 家用缝纫机
household textile 家用纺织品
household washing machine 家用洗衣机

household zigzag lock stitch sewing machine　家用曲折锁边缝纫机
house jacket　家用夹克,女便装上衣
house jeans　厂家牛仔裤
house mannequin　时装店模特
House of Legroux　莱格诺斯帽店(法国女帽店)
House of Paquin　帕坎高级时装店(法国高级时装店)
House of Premet　普雷校高级时装店(法国高级时装店)
House of Redfern　瑞德芬高级时装店(英国高级时装店)
house shoes　拖鞋;(室内)便鞋
house slippers　拖鞋;便鞋
house textile　家居纺织品
house wear　家居服
housewife('s)　针线盒
housewife's cloth　中等亚麻平布
house wives　针线盒,针线匣
housings　服饰
housse　[法]马衣;短盖袖套头外衣
howell　粗平布
howling bags　彩色花纹男裤
hsiang-yun-sha　香云纱
Hsining wool　西宁羊毛
H style　H型款式
HT-4 fiber　HT-4纤维
hu fu　胡服
hua mosi　花摩丝
hua-yao crepe　花瑶(绉)
huacaya　一种羊驼毛
huachun batiste　华春纺
huachun habotai　华春纺
huadahua damask　花大华绸
huajin brocade　华锦绸
hua-le damask satin　花累缎
huandi huckaback　环涤绸
huangsi　首期收获的蚕丝
huanian jacquard mock crepe　丁香绫;花黏绫
huanian mock crepe　花黏绫
huarache　瓦略切鞋
huaraches　墨西哥农民木拖鞋;条带鞋帮拖鞋
huarong brocaded damask　花绒绸
huaru jacquard crepon　花如呢
huaxin mixed brocade　华新纺
huaxin polyester batiste　华新纺
hua-yao crepe　花瑶

huazhu polyester marquisette　华珠纱
Hubbard blouse　荷叶边宽松抽腰衬衫
Hubert de Givenchy　休伯特·德·纪梵希(1927～ ,法国时装设计师)
huccatoon　低档棉色布
huck towel　小花纹毛巾
huck　浮松布
huckaback　浮松布
huckaback embroidery　浮松布刺绣
huckaback stitch　浮松针法;浮松绣;织补线迹
huckle　髋部
huddersfield　精仿呢
Hudson bay clothing　哈得孙湾外套
Hudson bay coat　哈得孙湾外套,水兵外套(厚毛料,对排扣)
Hudson's bay blanket　哈得孙拉毛粗毛毯
Hudson seal　哈得孙仿海豹毛皮
hue　色彩;色相;色调
hue analogy　色相类比
hue circle　色相环
hue contrast　色相对比
hue dependency　色相依赖性
hue difference　色相差异
hue-difference angle　色差角
hue gradation　色相渐变
hue of color　色相
huepilli　无袖衬衫
hue succession　色相连续
huge button　大型纽扣,特大纽扣
huge furry hat　超大毛皮帽,夸张毛皮帽
hugger belt　臀带
hug-me-tight　(女)紧身短马甲;(女)针织背心或内衣;合身背心;V字领针织背心
Hugo Boss　老板
Huguenot lace　胡格诺花边
Hui ethnic costume　回族民族服,回族服饰
huipil　无袖宽松直身上衣
huipil grande　墨西哥印第安妇女多褶头饰
huipil[-li]　胡披儿(中美)
huke　胡克斗篷外衣(11世纪),男式松垂外套
hula　夏威夷草裙
hula dress　草裙舞服(夏威夷),夏威夷草裙舞服
hula skirt　夏威夷草裙;呼啦圈舞裙
human anatomy　人体解剖,人体解剖学

human behavior	人体行为
human body	人体
human body functional fiber	人体功能纤维
human body measuring system	人体测量系统
human body modeling technology	人体建模技术
human body proportion	人体比例
human circumstance	人类环境
human ecology	人类生态学
human engineering	人体工程学,人体工效学
humane standard	人道标准
human factors engineering	人类因素工程学
human figure	人体,人体体型
human hair (HuH)	人发
human perception analysis (HPA)	人体出汗分析
human perception value	人体舒适感觉值
human proportion	人体比例
human resources	人力资源
human scale	人为标度
human science	人类科学
human skeleton	人体骨骼
human torso	人体躯干
humeral veil	(天主教教士的)长方形丝披肩
humerus	肱骨
humhum	印度棉布,素色粗棉布
humidity	湿度
humidity-sensitive fiber	湿度敏感纤维
humpback	驼背者
humped figure	驼背体型,有隆肉体型
humphrey cotton	美国汉弗里棉
Hunan embroidery	湘绣
hunchback	驼背,弓背,驼背者
hunchbacked type	驼背型
hundred flowers brocade	百花锦
HUNDRED IDEES	《一百种花样》(法)
hundred pleater	厨师帽
hung	薄粗大衣呢
Hungarian	18世纪轧光花呢
Hungarian cord	匈牙利丝绸滚绳饰边
Hungarian costume	匈牙利民族服饰
Hungarian dress	匈牙利传统服装
Hungarian embroidery	匈牙利针绣(平针或缎纹针迹)
Hungarian hat	匈牙利帽
Hungarian hemp	匈牙利大麻
Hungarian lace	匈牙利花边(梭结花边)
Hungarian point	仿弗罗伦萨式刺绣
Hungarian stitch	匈牙利针法
Hungarian suit	匈牙利套装;双排扣束腰少男装
Hungary blue	匈牙利蓝色,钴蓝色
hungback	薄粗大衣呢
hunge	男式松垂外套
hungry	厚薄横档(织疵)
Hunnicutt cotton	19世纪早熟陆地棉
hunt cap	狩猎帽
hunt coat	猎狐外套
hunt derby	狩猎硬防护帽
hunt dress	猎装
hunter breeches	狩猎裤
hunter cloth	猎装布;猎装呢(印度)
hunter clothing	猎装
hunter jacket	猎装上衣
hunter's green	猎装绿,深暗绿色
hunter's pink	粉红色猎装布
hunting bag	狩猎袋,狩猎挂肩袋
hunting beret	狩猎贝雷帽
hunting boots	狩猎靴
hunting breeches	狩猎马裤
hunting calf	反面牛皮
hunting cap	鸭舌帽;猎帽(红色或荧光色)
hunting check	狩猎格子花纹
hunting cloth	猎装布;猎装呢(印度)
hunting clothing	猎装外套
hunting coat	猎装短外套
hunting costume	打猎装
hunting culture	捕猎文化
hunting design	地毯的狩猎图案
hunting dress	狩猎装(双排扣燕尾服骑装)
hunting jacket	打猎衣,猎装夹克(粗毛料单排扣,肩有垫布)
hunt(ing) look	英国狩猎款式
hunting morning	狩猎晨礼服
hunting necktie	狩猎宽领带
hunting outfits	猎装
hunting pants	狩猎裤
hunting pink	深红色;深红色猎装布;(猎狐者穿的)红猎装
hunting plaid	猎装方格呢
hunting scene	地毯的狩猎图案
hunting shirt	红色狩猎毛料衬衫
hunting stock	狩猎领巾

hunting style 英国狩猎款式
hunting suit 猎装
hunting tartans 猎装格子呢
hunting vest 猎装背心
hunting watch 猎用表,双盖表
hunting wear 猎装
hunt look 猎装款式;猎装风貌
huntsman's hat 猎人帽
hunt style (英国)狩猎款式
hurden 最低级麻布
huseau 骑马长靴
hush cloth 吸音绒布
HUSH PUPPIS 暇步士
Husking cloth 劳动手套布
husky build 高大体型
husky sizes 高大男童尺寸
hussar boots 轻骑兵中筒皮靴
hussar breeches 骑兵马裤
hussar funic 轻骑兵束腰外衣
hussar jacket 轻骑兵女短外套
hussar 女式毛边短外套
Hussein Chalayan 胡森·查拉扬
hussy 针线盒
hwachow 平纹上浆绸
hwa mien chow 纯棉府绸
hwasienchow 柔软绫纹绸
hwayong 中国丝绒
hwayutwan 毛葛,毛府绸
hyacinth 紫蓝色,风信子色,紫丁香色
hyacinth blue 风信子蓝(深紫蓝色)
hyacinth red 风信子红(暗红橙色)
hyacinth violet 风信子紫(紫蓝色),紫罗兰(色)
hybrid cotton 杂交棉
hybrid organic-inorganic fiber 有机-无机混合纤维,有机-无机杂化纤维
hybrid silk 真丝包覆丝,复合真丝(以锦纶为芯包覆真丝的包芯丝)
hybrid style 混合款式
hybrid 混纺织物
hybrid-type fabric 混纤型复合丝织物
hyderabad carpet 花纹丝绒毯
Hyderabad Gaorani cotton 印度海德拉巴棉
hydrangea blue 绣球花蓝
hydraulic-autobalance sole attaching machine 液压自动平衡压鞋底机
hydraulic cloth cutter 液压裁剪机
hydraulic cutting press 油压裁剪机
hydraulic die cutting machine 液压裁断机
hydraulic press 液压冲压机
hydraulic pulling over toe cement lasting machine 液压上胶前帮机(制鞋)
hydraulic sole attaching machine 液压压底机(制鞋)
hydraulic swing arm cutting press 摇臂式液压裁断机(制鞋)
hydraulic top fuse machine 油压式衣领压衬机
hydraulic travelling head cutting machine 龙门式液压裁断机(制鞋)
hydroentangled 水刺法非织造布
hydroentangled fabric 水刺成网非织造布
hydroentangled nonwoven 水刺成网非织造布
hydro-extracting cage 脱水机
Hydrofil 吸湿性聚酰胺纤维(商名,美国)
hydrogel fiber 水凝胶纤维
Hydron 海昌蓝(商名)
Hydron blue 海昌蓝(商名)
hydrophilic fiber 亲水性纤维
hydrophilic hollow fiber 亲水性中空纤维
hydrophilicity 亲水性
hydrophilic property 亲水性
hydrophility 亲水性
hydrophobic fiber 疏水性纤维
hydrophobic property 疏水性
hydrophobic synthetic fiber 疏水性合成纤维
hydrophobicity 疏水性
hydroscopic fiber 吸湿性纤维
hydroscopic property 吸湿性
hydroscopic synthetic fiber 吸湿性合成纤维
hydroscopicity 吸湿性
hydrosetting 湿[热]定型
hygenic towelette 湿纸巾
hygiene fiber 卫生用品纤维
hygienic-absorbent products 吸湿性卫生用品
hygienic belt 卫生带
hygienic fabric 保健织物
hygienic finishing (织物的)卫生整理
hygienic performance 卫生性能
hygienic standard 卫生标准
hygienic textile 保健纺织品,卫生[用]纺织品
hygral expansion 湿膨胀

hygrometer 湿度表,湿度计
hygroscopic behavior 吸湿性
hygroscopic clothing 吸湿性服装
hygroscopic finish 吸湿整理
hygroscopic property 吸湿性能
hygroscopic synthetic fiber 吸湿性合成纤维
hygroscopicity 吸湿性

hymo 马海毛衬布,马海毛垫布
hymonette 毛衬布
Hypalon coating 海帕伦橡胶涂层
hyperpeach 超桃皮绒
hypin 缅甸本色粗布
hypnotic cloth 催眠布
hypotenuse 斜边

I

Ian Thornas 伊恩·托马斯(英国时装设计师)
ice and snow sports shoes 冰雪运动鞋
iceberg green 冰山绿
iceberry green 湖灰,浅青灰
ice blue 冰蓝(淡绿光蓝)
ice cap 冰帽
ice-cream cones hat 冰淇淋蚕卷帽
ice dyeing 冰染
ice flow 冰流色
ice-hockey set 冰球鞋(镶上冰刀)
ice-hockey shoes 冰球鞋
ice-hockey skate 冰球鞋(镶上冰刀)
icelandic costume 冰岛服饰,冰岛民族服饰
icelandic sweater 冰岛毛衣,斯堪的纳维亚毛衣;冰岛衫
iceland wool 冰岛羊毛
ice skating boots 滑冰鞋
ice-snow blue 冰雪蓝
ice washing 霜花洗,加漂白剂的石磨洗
ice wool 艾斯双股细毛线
ichella 智利印第安妇女流苏长披肩
icicle 冰柱色
icon 圣像
icy morn 冰封黎明色
ida canvas 本色软亚麻布(刺绣或女衣用)
idea (设计)构想;想象;观念;意念;意见
ideal arm 标准臂
ideal back type 标准背型
ideal figure 标准体型,理想体型
ideal hip type 标准臀型
ideal image 理想形象
idealized drawing 示意图
ideal shoulder type 标准肩型
identification 鉴别
identification bracelet (I.D. bracelet, IB)身分手镯,鉴别镯
identification markings 确认标记
identification of product 产品标识
identification of textile fiber 纺织纤维鉴别
identifying and bundling 分片
idiom 格调,特色

idiom mittens (由一根穿过外衣袖子的细绳连着,以防失落的)儿童连线手套
idiosyncrasy 特有风格,特质
idle running 空转
idle time 空载时间;停机时间,停台时间
idle unit 闲置设备
idria lace 伊德里阿花边(粗棱结花边,几何花纹,意大利,南斯拉夫)
ie sina 手织长绒毛布
ie taua 树叶纤维布
igloo mitt 爱斯基摩手套;伊格卢运动手套
ignitability 着火性
ignition point 着火点,着火温度
ignition resistance 抗点燃性
ignition retardance 阻燃性
ignition temperature 燃着温度
iguitos cotton 秘鲁短绒棉
ih-ram 麦加进香朝服;伊兰白棉布衣;伊兰粗呢
ikat 纱线扎染布,纱线蜡染;印经丝绸
ikat design 伊卡图案,伊卡扎染图案
ilaicha 手织条纹绸
ilicha (印度)手织棉/丝织物
iliospinale anterius 髂骨上棘点
ilium 髂骨
ill-fitting 不合身
ill-fitting garment 不合身服装
ill management 管理不善
illuminance 照度
illuminated discharge printing 着色拔染印花
illumi yarn 光亮丝(单丝较粗、较亮的人造丝)
illusion 错觉,假象;真丝薄网眼纱,透明面纱
illusion blue 面纱蓝色
illusion of actor 化装面纱
illusion pattern 错视图案
illusion setting 细纽式装座(钻石)
illustration 插图,图解,效果图
illustrator 插图师
ilocano cloth 伊洛干诺条格布
image 形象
image exhibition 形象展示

image fiber　导像纤维,映像纤维
image map　风格设计想像图
image pattern　形象纸样
image printer　图象打印机
image processing　图象处理
image quality　影像质量
imaginary line　参考线,假想线
imagination　想象;想象力
imaginative vents　装饰开衩
imago　伊梅戈羊皮化效应印花布
imbabura cotton　秘鲁因巴布拉棉
imbalance　不平衡
imbecile sleeve　落肩灯笼袖
imberline stripe　因伯莱因条纹布
imbricate shape　鳞片花纹
imbrocado　金银线滚边绸,金银丝绸,因布罗卡多绸
imdienne　轻薄印花条子布;法国小花纹棉布
imirat　印度平纹棉布,印度平布
imirillus myzinym　中世纪轻薄丝织物
imitate　模仿
imitated furs printing　仿兽皮印花
imitation　模仿,仿造
imitation alligator　人造鳄鱼皮
imitation antique carpet　仿古地毯
imitation astrakhan　仿(俄罗斯)羔羊皮织物
imitation backed fabric　充加背织物
imitation bark button　仿树皮扣
imitation batik prints　仿蜡防花布
imitation beetle finish　仿搖光整理
imitation beggar apparel　乞丐服
imitation bone button　仿骨扣
imitation border　假罗纹边,假罗口,假罗纹
imitation coconut shell button　仿椰壳扣
imitation corded seam　仿嵌条缝
imitation cuff　假袖口,假翻袖,活袖口
imitation diamond　人造钻石(合成钻石)
imitation embossed overcoating　仿拷花大衣呢
imitation embroidered fabric　仿绣织物
imitation fabric　仿生织物
imitation fashion marks　仿点状纹
imitation fingernail　人造指甲
imitation flower　人造花
imitation French piping　仿法式滚边
imitation French seam　仿法式缝
imitation fur　仿毛皮,仿裘皮,人造毛皮

imitation fur blanket　仿裘皮毛毯
imitation fur fabric　仿毛皮织物,人造毛皮
imitation gauze　仿纱罗织物,多孔织物
imitation gingham　仿条格平布
imitation gold yarn　充金线
imitation haircloth　充马尾衬,仿马尾衬,人造马尾衬
imitation horn button　仿角制扣
imitation horse hair　人造马毛
imitation horse hair fabric　充马毛织物,仿马毛织物;仿马尾衬,人造马尾衬
imitation horse hair lining　人造马尾衬
imitation jewelry　人造首饰;人造珠宝
imitation lace　仿手工花边
imitation lace pattern　假网眼花纹
imitation lambskin　仿羔皮
imitation leather　人造(皮)革,合成革,仿皮革;人造革涂层织物
imitation leather belt　仿革裤带,仿皮裤带,人造革裤带
imitation leather buckle　仿皮带扣
imitation leather button　仿皮扣
imitation leather handbag　充皮手袋,仿革手提包,人造革手提包
imitation leather sole　仿皮底(鞋)
imitation leather wear　人造革服装
imitation leno　充纱罗织物,仿纱罗织物,多孔织物
imitation linen　仿亚麻布
imitation linen finish　仿麻整理
imitation linen finish with gelatine solution　明胶溶液仿麻整理
imitation moccasin shoes　仿莫卡辛鞋,仿软帮鞋,仿烧卖鞋
imitation moleskin　厚毛绒斜纹棉布
imitation natural wool　仿原色羊毛
imitation organdie　仿奥甘迪,仿蝉翼纱(合成纤维薄织物经合成树脂整理而成)
imitation otter　仿水獭皮
imitation peach fabric　仿桃皮绒织物
imitation pearl　仿珍珠,人造珍珠
imitation pearl button　仿珍珠扣
imitation pocket　假袋,假口袋
imitation rabbit hair　充兔毛皮织物,仿兔毛皮织物
imitation raglan　仿连袖大衣
imitation rep(p)　仿棱纹平布
imitation safety-stitch seam　仿安全线

迹缝
imitation satin stripe printed handkerchief
仿缎条印花手帕
imitation sealskin　仿海豹绒
imitation seam mark　仿收放针花
imitation sheep skin　仿羊皮
imitation sheep skin finishing　仿羊皮整理
imitation sheepskin　仿老羊皮
imitation shell button　仿贝壳扣
imitation silk　仿丝,仿真丝
imitation silk fabric　仿丝织物
imitation silk finishing　仿丝绸整理
imitation silk floss　仿丝绵
imitation silk floss filler　仿丝绵
imitation silk technology　仿丝技术
imitation silver yarn　充银线,仿银线
imitation spot knop　仿点纹集圈
imitation stone button　仿石扣
imitation suède　仿麂皮;人造麂皮,人造羔皮
imitation tweed　充钢花呢
imitation velvet　仿天鹅绒
imitation wax print　仿蜡防印花
imitation wax printing　仿蜡防印花
imitation wood button　仿木扣
imitation wool　仿毛,仿羊皮,人造毛
imitation wool fabric　仿毛织物
imitation wool finishing　仿毛整理
imitation wool gabardeen　仿毛华达呢
imitation wool processing　仿毛加工
imitation wool serge　仿毛哔叽
imitation wool suiting　仿毛西服料
imitation wool technology　仿毛技术
imitation-worsted cloth　半精梳毛织物
imitative design　仿像设计
imitative kabe crepe　涤玉绉
imitative satin stripe printed handkerchief 仿缎[条]印花手帕
imizillus　中世纪薄绸
immaculate process　耐久褶裥整理
immediate elastic deformation　即时弹性变形;即时弱性变形
immediate payment　即期付款
immediate recovery　即时回复
immediate shipment　立即装运
immune from taxation　免税
impact resistance　抗冲击性
impact strength　冲击强力
impact type insole stapling machine　冲击式中底鞋钉钉机,冲击式钉中底针机
impala　黑斑羚色
impatiens pink　凤仙[花]红色
imperceptive perspiration　无感出汗
imperfect adhesion　贴布不良;贴布浮起(台板印疵)
imperfect color step　花纹分段色变(印花花纹)
imperfect dyeing　染色不良
imperfect figure　特殊体型
imperfection　疵病,疵点
imperfect selvage　不良布边
imperfect tension　张力不当
imperial　帝髦;厚绒布;意大利金银线织棉;金绒绸;比利时彩色条格布;特大号,花式棉布
imperial blue　景泰蓝
imperial coating　防水精纺细呢
imperiale　法国精纺哔叽
imperial ottoman　粗横棱柔软织物
imperial pongee　厚塔夫型茧绸
imperial purple　暗紫色
imperial red　紫红色(带红色的深紫色)
imperial robe　朝袍,衮服,龙袍(中国古代皇帝着装)
imperials　厚绒布
imperial sateen　厚棉绒工装布
imperial serge　防水细哔叽,精纺斜纹布
imperial shirting　英国漂白细平布
imperial sizing　英制尺寸
imperial skirt　特大加撑衬裙
imperial tape　厚棉带
imperial valley cotton　皇谷棉
imperial velvet　条子丝绒(丝绒条和比它窄一半的棱条构成条纹)
impermeability　不透[气或水]性,不渗透性;防水性
impervious backing　[地毯]防水衬底
impervious finish　涂层整理
imperviousness　不渗透性
impervious water proofing　不透气性防水
implements　工具;服装
import　进口
import agent　进口代理人
import and export license system　进出口许可证制
import customs clearance　进口报关
import declaration　进口报单
Imported wool　进口羊毛

import goods 进口商品
import license 进口许口证
import quota 进口配额制;进口限额
import quota system 进口配额制
import restraints 进口限制
Importer-Retailer Textile Advisory Committee 纺织品进口及销售咨询委员会
imports and exports trade 进出口贸易
import tax 进口税
import technology and equipment 引进技术与设备
import textile 进口纺织品
import value 进口值
impregnated fabric 浸轧织物,浸渍织物
impregnated nonwoven fabric 浸渍法非织造布
impregnating bonded nonwovens 浸渍黏合法非织造布
impression 压痕,印痕,压印;印象
impressionism 印象派;印象主义
impressionistic print 印象派印花
impression mark 印痕
imprime marble 大理石纹花样
imprimé marblé [法]大理石花样
imprint 服装压痕
improper finish 修剪疵
improper milling 缩呢不良
improper tension 张力不当
improve 改进
improved prolific cotton 美国改良丰产棉
improved upland cotton 美国改良种陆地棉
improved wool 改良毛
improved zanza 改进盛泽真丝纺绸
improvement trade 加工贸易
improving business management 改善经营管理
impulse merchandise 冲动商品(触引起消费者购买欲的商品)
in 向内携裆
in-a-dart sleeve placket 简易式袖衩
in-and-outer 内外衬衫,轻便衬衫
in-and-out seaming 内外缝纫
in between curtain 结构窗帘
in between neckline 中间领圈
in board 内装式
in-boots look 塞入靴子款式
in-boots style 塞入靴子款式
incandescent light 白炽光
incarnadine 肉色,血红;肉色的,淡红色的

inca traditional costume 南美印加传统服饰
inch 英寸;身高(复);身材
inchering 量体裁衣
inching 微动
inching motion 寸动装置
inch ruler 英制尺
inch size 英制尺码
inch tape 英制软尺
incidence angle 入射角
incidental looping 偶发浮线
incised design 划花
incle 带子
inclined portion 线圈的圈柱
incombustibility 不燃性
incombustible fabric 不燃性织物
income tax 所得税
incompatibility 不配伍性,不相容性
incongruity 不相称;不协调
incons picuous seam type 暗缝型
incontinence pad 失禁衬垫
incontinence product 成人失禁用制品
incontinence textile 失禁用纺织品
incontinent package 失禁用包
incorporated into incorporation 垫入
incorporated rayon staple 黏合共混短纤维,碱化黏胶纤维
incorrect marking 标志不清,标志错误
increasing skirt hem 增加裙下摆
incroyable bow 女式花边蝴蝶结
incroyable coat 因可瑞博上衣;女式宽翻领长后摆外套
incroyable cravat 因可瑞博领结
incroyables cut 奇特男子发型
indanthrene cloth 阴丹士林蓝布
indent 订单,缺口,凹口
indentation 凹痕,压痕
indentation look 犬牙款式
indentation style 犬牙款式
indented tail bottom 曲形圆下摆
independent legal person status 独立法人地位
independent pattern 单独纹样
inderkins 德国窄幅麻布
indestructible flat chiffon 丝巴里纱
indestructible voile 丝巴里纱
index 索引
index finger 食指
index plate 刻度板
index point 指标

indexed colour 索引颜色
India chintz 印度大花轧光家具布
India costume 印度民族服饰
India cotton 印度提花轧光厚棉布
India cuttanee satin 印度麻棉交织缎，丝麻交织缎
India design 印度图案
India embroidery 印度刺绣
India hemp 印度大麻，印度麻，菽麻
India lake 印度胭脂红
India lawn 印度上等细布
India linon 印度高级漂白细布
India Madras 印度马德拉斯织物
India muslin 印度手织细棉布，细软棉平布
Indian 印第安服饰
Indiana 印第安纳高级细布
Indian abutilon 印度苘麻
Indiana cloth 印第安纳高级细布
indian appliqué 金银丝嵌花刺绣
Indian bag 印第安袋，嬉皮士袋
Indian beads 印第安念珠
Indian belt 印第安腰带
Indian blanket 印第安花纹手织毛毯
Indian bonnet 印第安帽
Indian broadtail 印第安大尾羔羊毛皮
Indian carpet 印度地毯
Indian ciciclia 印度西西克里亚锦
Indian costume 印第安民族服饰，印度民俗服，印度民族服装
Indian cotton 印度棉
Indian Cotton Mills' Federation 印度棉纺织厂联合会
Indian crepe 印度绸
Indian cut 金银丝浮纹织物
Indian design 印度式图案
Indian dhurrie 印度手工纱棉毯
Indian dimity 印度棱条格薄细布
Indian embroidery 印度刺绣（链式针迹，衲缝贴花等针绣）
Indian gown 18世纪丝长袍，印度及踝长袍
Indian head 印第安人头商标平纹棉布；印第安人头棉布；印第安人头像标志
Indian head cotton 印第安人头商标（平纹）棉布；印第安人头像标志
Indian headwork 印第安珠绣
Indian hemp 印第安大麻
Indian ink 墨汁

Indian jewelry 印第安首饰
Indian J Fiber & Text Res（Indian Journal of Fiber & Textile Research）《印度纤维和纺织学报》（期刊）
Indian jute Industries' Research Association 印度黄麻工业研究协会
Indian kurta 印度柯泰衫
Indian lace 印度梭结花边，印度丝花边，包芯花边
Indian lake 印度胭脂红
Indian lamb 羔羊皮，印度羊羔皮
Indian lawn 印度细平布
Indian linen 印度亚麻平布；印度仿亚麻精梳棉布
Indian look 印第安风格，印第安风貌，印度安款式
Indian madras 印度条子细布，马德拉斯大彩格平布
Indian mallow（hemp） 苘麻，青麻
Indian moccasin 印第安鹿皮软鞋；印第安无后跟软鞋
Indian motif 印第安式印花花纹
Indian mull 印度高支无浆细纺布
Indian muslin 印度薄细布，印度平纹细布
Indian necktie 印度男式平纹细布领巾
Indian nightgown 印度睡袍
Indian okra 印度槿麻，印度洋麻
Indian pants 印第安裤
Indian prints 带有印度图案特色的印花棉布，印度式花纹印花棉布；印第安手工印花布
Indian purple 印度紫
Indian red 印第安红（红棕色）
Indian rug 印度地毯
Indian sandals 印度凉鞋
Indian shawl 开司米披巾；东方色彩毛披巾
Indian shirting 印度细平布（英国制，对印度出口）
Indian silk 印度手织薄绸
Indian stylistic 印第安人款式
Indian suit 印第安套装
Indian sweater 印第安厚毛衣（北美）
Indian tan 印度棕黄色
Indian teal 印度水鸭色
Indian turban 印度缠头布
Indian war bonnet 印第安羽毛头饰
Indian wedding blouse 印度婚礼女上衣
Indian yellow 印第安黄（老黄色）

India silk 印度手织薄绸
India tape 印度窄编带(英国制,细而挺)
indicate plate 标牌
indication mark 指标标志
indigo 靛蓝,靛青;军服呢
indigo [blue] 靛蓝,靛青,深紫蓝色;靛蓝底色花布
indigo blue printed fabric 靛蓝花布;蓝印花布
indigo carmine 靛胭脂蓝
indigo copper 铜蓝
indigo denim 靛蓝牛仔布
indigo dye 靛蓝染料
indigo print 蓝白花布
indigo printed fabric 靛蓝印花布
indigo printing 靛蓝印花
indigo prints 靛蓝印花布,靛蓝印花织物,蓝花土布
indigo pure 化学靛蓝
indigo red 靛红
indigo serge 藏青哔叽
indigotine 纯靛蓝
indigotine fabric 毛蓝布,毛青布
indigo warp dyeing 靛蓝经纱染色
indigo white 靛白(靛蓝还原后的隐色体)
indirect 间接
indirect investment 间接投资
in dishabille 穿着便衣
indispensable 必备手提袋,丝绸(丝绒)抽口小手袋
individual design 单独花型
individual economy 个体经济
individual fashion 个人化服饰
individual garment 个性服装
individualistic manner 独特的[个人]风格
individual liability 个人责任
individual package 单件包装
individual taste 个人爱好
indonesian costume 印尼民俗服,印尼民族服
indoor apparel 室内服装
indoor casual 室内便装
indoor dress 室内服装,室内衣
indoor furnishings 室内装饰[织物]
indoor-outdoor boots 室内外兼用靴
indoor-outdoor carpet 室内外两用地毯
indoor shoes 家居鞋,室内鞋
indoor slippers 室内拖鞋
indoor sportswear 室内休闲运动服

indoor wear 家居服,室内服装
indo-Portuguese embroidery 具有葡萄牙风格的印度刺绣;具有印度风格的葡萄牙刺绣
indorse 背书
inducing colour 诱导色,感应色
industrial accident 工伤事故
industrial and commercial sales tax 工商税
industrial art 工艺;工艺美术
industrial clothing 工业制服
industrial corporation 工业公司
industrial cost system 工业成本制度
industrial design 工业产品设计
industrial design of garment 工业性服装设计
industrial drying tumbler 工业干衣机
industrial duty 工业税
industrial engineering 工业工程学
industrial enterprise 工业企业
industrial fabric 产业用纺织品
Industrial Fabrics Association International [美]国际产业用纺织品协会
industrial fiber 产业用纤维
industrial flame-barrier fabric 产业用防火织物
industrial flat bed lock-stitch sewing machine 工业平缝机
industrial form 工业用人台(胸架)
industrial garment 工业制服
industrial garment design 工业性服装设计
industrial insurance 劳动保护
industrial knit 产业用针织物
industrial know-how 工业生产专门技术
industrial laundering 工业化洗涤;工业洗衣
industrial laundry 工业化洗涤;工业洗衣
industrial mask 工业用面罩
industrial net output value 工业净产值
industrial pattern 工业用纸样
industrial plush 工业用长毛绒
industrial pollution 工业污染
industrial property license 工业产权许可证
industrial property right 工业产权
industrial protective clothing 工业防护服
industrial rainsuit 工业雨衣
industrial safety 安全生产
industrial-scale 大规模,工业规模

industrial service 工业性作业
industrial sewing 工业缝纫
industrial sewing machine 工业缝纫机
industrial size 工业化尺寸
industrial stitcher 工业缝纫机
industrial textile 工业用纺织品
industrial thread 工业用线
industrial use sewing machine 工业用缝纫机
industrial washing machine 工业洗衣机
industrial whizzer 工业脱水机
industrial zipper 塑钢拉链；树脂拉链；工业用大尺寸拉链
industry 工业
industry and commerce 工商界
industry standard 工业标准
ineffibles 男裤
inelastic deformation 非弹性变形
inequi-front vest 异襟背心
inexpressibles 男裤
infantado 西班牙美利奴羊毛
infanta style 西班牙公主服式风格
infant bib 婴儿涎布
infantees 襁褓；束襁褓带子；包布
infant sack 婴儿上衣
infant's bonnet 婴儿软帽
infant's cap 婴儿软帽
infant's coat 婴儿外套；婴儿无檐帽
infant shoes 婴儿鞋
infant's layette 婴儿全套衣物用品
infant's off the face 婴儿鸭舌帽
infant's outfit 宝宝装；婴儿套装
infant's suit 婴儿套装；幼儿服装，幼儿衣着
infant's wear 婴儿服
in fashion 现在正流行
inferior garments 低档服装
inferior [goods] 次品，低档商品，劣质商品
inferior quality 劣品
inferiority feeling 自卑感
inflammability 易燃性，可燃性，燃烧性
inflammable fabric 易燃织物
inflatable bra 充气胸罩
inflatable dunnage 充气衬垫
inflatable fabric 可充气织物
inflatable life jacket 充气救生衣
inflatable sac 可充气的囊，可膨胀的囊
inflated garment 充气服
informal daytime clothing 便服，日常服

informal dress 便服，日常服
informal habit 非正式骑马装
informal military dress 军便装
informal pumps 日常皮鞋，非正式碰拍鞋
informal suit 便服，便装，日常服
informal wear 便服，日常服装；准礼服，半正式礼服
information 情报；信息；资料
Information Council in Fabric Flammability 织物耐燃性情报理事会
information data 情报资料
information label 产品说明标签
information network 情报网络
information retrieval 情报检索，资料检索
information technology 信息技术
informative labeling 质量说明标签
infrared dryer 红外线干燥机
infrared heating 红外线加热
infrared microscopy 红外显微术
infrared protective visor 防红外线目镜
infrastructure facilities 基础设施
infula 英富勒垂饰，主教冠带，僧帽垂饰带
infusibility 不熔性
in general 流行一时
in general wear （服装，衣着）流行一时
Ingo Jones 琼斯（英国著名戏剧服装设计师）
ingrain carpet 双面提花地毯
ingrain dye 显性染料
ingrain hose 原纱染色袜
ingrain jute carpeting 混色黄麻地毯
inguinal circumference 下肢根围，大腿根围
ingyi 印其上衣（缅甸克耶族）
inharmony 不协调
inherent quality 内在质量
in-house use 自用
initation leather man-made leather 人造革，仿皮革
initial brooch 首字母胸针
initialed belt 织字腰带，织字裤带
initial handkerchief 绣字手帕
initial modulus 初始模量
initial product 起始产品
initial ring 图章戒指
initial trial 初试
2 in 1 jacket 二合一夹克；两用夹克

injecting rubber sole cloth shoes 注胶布鞋
injection machine 塑胶注射机(制鞋)
injection molded system 注塑成形法(制鞋)
injection molding cloth shoes 注塑布鞋
injection molding plastic shoes 注塑塑料鞋
injection molding shoes 注塑鞋,注压鞋
injection-moulded shoes 注压鞋,注塑鞋
injira cotton 哥伦比亚棉
ink blue 墨水蓝,蟹青[色]
inkiness 墨黑
inking 打墨印;修补着色
ink jet printing 喷墨印花
inkle 亚麻带子;带子
ink printed handkerchief 石印手帕
inlaid 嵌花
inlaid appliqué 金银丝嵌花刺绣
inlaid bent trimmers 嵌花弯剪刀
inlaidnet 嵌花网眼
inlaid straight trimmers 嵌花直剪刀
inlaid trimmers 嵌花剪
inlaid yarn 镶嵌纱;镶嵌线
inlay 嵌花;镶嵌;镶嵌料;镶嵌物
inlay decoration 镶嵌装饰
inlay design 镶嵌图案
inlaying 镶
inlay material 衬料
inlay pattern 衬纬花纹,镶嵌花纹,镶嵌花样
inlay patten design 镶嵌花纹图案
in-leg 内衬,下裆
inlet 镶嵌;镶嵌物;床品用粗棉布
inlet material 被套料
in livery 穿制服
inner and outer dressing 两用衫
inner beauty 内在美
inner belt 内腰带,裙裤腰带,内侧松紧带
inner bind 隐缝
inner box 内箱;内盒
inner construction 内衬结构;内结构
inner construction fabric 内衬织物
inner curve 内凹线
inner foxing 内围条(鞋)
inner heightened shoes 内增高鞋
inner hook 内钩
inner lapel 小襟
inner line of facing 贴边里口线
inner line of front facing 挂面里口线
inner-lining 夹里

inner malleolus 内踝
inner packing 内包装,销售包装
inner pocket 衣里口袋
inner quality 内在质量
inner retaining ring 内扣环
inner sleeve 内套筒,内袖,袖瘪肚
inner sole 里底,鞋垫
inner-vest jacket 内背心式短夹克
innerwear 贴身时装,内衣
inner workings 服装内部的缝制
innocent look 纯洁型
innocuousness 无害性
innovation 创新,革新,改革
innovation of style 款式翻新
innovative fashion 创新的流行款式
innovative modeling 创新造型
innovator 革新者
inodorous leather 无气味皮革
inorfibre 无机纤维
inorfil 无机纤维
inorganic fiber 无机纤维
inorganic oxide fiber 无机氧化物纤维
inorganic ultrafine fibrer 无机超细纤维
in or out blouse 两用下摆女衫
in or outer shirt 两用下摆衬衫
in-process inspection 半成品检验,工序检验
inquiry 询盘
in-regular jacquard knitted fabric 不规则提花针织物
inseam (衣、袖)内缝;(裤)内长;下裆线;下裆缝;股下线;手套内接缝;
inseam and outseam (裤)内、外缝;(裤)内、外长
inseam buttonhole 剪接线扣眼,门襟缝道处的纽孔
inseam leans to front 前袖缝外翻
inseam length 内缝长度
inseam line 内缝线,前袖缝线;裤内缝线,下裆线
inseam opening 缝道开口
inseam pocket 接缝处插袋,缝道插袋,内缝接袋(袋布与衣片连在一起)
inseam pocket reinforcement 缝道插袋袋口加固
inseam run 下裆缝线;袖缝线
inseam sleeve placket 缝道袖衩
inseam trouser pocket 缝道裤插袋
insect bar 条格蚊帐布
insect proof finish 防蛀整理

insect repellent 防蛀整理
insect repellent finish 防虫整理
insect resistance 防虫性；抗蛀性
insect resistance finish 防虫整理
insect screen 驱虫窗纱
insensible perspiration 气态汗（潜热）
insert （缝纫）镶；嵌；补；镶块；镶嵌料；镶嵌物
insert cord 嵌绳带
inserted back belt 后嵌腰带
inserted pocket 插袋，挖袋
inserted seam 夹入缝，嵌缝
insert elastic 嵌松紧带
insert facing 裤口贴边
insert fur 镶毛皮
insert gloves 衬里手套，镶边手套
inserting （衣服上）嵌、饰、绣饰；（缝纫中）镶嵌；补；绱；镶
inserting cord 嵌绳带
inserting elastic 嵌松紧带，装松紧带
inserting elastic in hems 卷边装松紧带
inserting fur 镶毛皮
inserting label 装标签
inserting lace 镶花边
inserting layout 排料叉套法；排料镶套法
inserting leather 嵌饰皮革
inserting of fur 毛皮镶边
inserting of leather 嵌饰皮革
inserting patch 嵌补片
inserting pocket 插袋，挖袋
inserting sleeve 绱袖，装袖
inserting slide fastener 装拉链
inserting stay strap 绱衬带
inserting tape 装牵带
inserting tape between binding and body material 缝料嵌带滚边
inserting waistline 嵌接腰带
inserting zipper 装拉链
insertion 嵌入；嵌饰，嵌饰花边，衣缝嵌线，窄花边
insertion lace 夹花边，饰边，花边细带，镶边花边，绣饰花边
insertion raschel lace 拉舍尔嵌花花边
insertion stitch 连缀缝，连带编织绣
insert lace 镶花边
insert leather 皮条镶边；嵌饰皮革
insert patch 补（布）片
insert pleating machine 嵌镶折裥机
insert pocket 插袋；斜袋；挖袋
inscrt repellent 防蛀整理

insert retaining ring 插片护圈
insert sleeve 装袖；绱袖
insert strip 嵌条带
insert type diaper 插入式尿布
insert waistline 嵌接腰线
in-service use 实际使用
inset 嵌料，镶料；嵌片；镶边
inset neckline 有嵌绣开衩的圆领口
inset pocket 镶嵌口袋，挖袋
inset sleeve 装袖
insetting 绱
insetting layout 镶嵌法
insetting sleeve 绱袖，装袖
inset vest 镶嵌背心
in-set waistband 镶嵌的裙腰（连衣裙）
inset yoke 内嵌式育克
inside belt 内裤腰布，内裙腰布，腰衬布，腰衬带
inside bound pocket 嵌线里袋
inside breast pocket 里胸袋，女内胸口袋
inside button 内扣
inside chest pocket 胸口内袋
inside collar 领里
inside collar band 领里口
inside corner 内角
inside edge of front facing 挂面里口线
inside facing 内贴边
inside fold 内折边
inside-garment microclimate measuring system 衣内微气候测量系统
inside hemming 贴边，下摆贴边
inside hemming facing 内贴边（缝纫）
inside-leg 下落裆，下裆缝
inside-leg length 内腿长
inside-leg seam 下裆缝，下落裆
inside length 股下；裤内长，下裆长
inside line of collar 翻领上口线
inside line of right fly 裤里襟里口线
inside lining 衬头
inside-out clothes 反穿的衣服（将缝合线露在外面）
inside-out look 反穿款式
inside-outside 内衣外衣化
inside-outside effect 内外效应
inside-out style 反穿款式
inside-out-stocking 双面组织袜子
inside packing 内外装
inside pocket 服装内袋，里袋，暗袋
inside seam 暗缝，内侧缝，下裆缝
inside seam line 内缝线；下裆线

inside selling 内销
inside selvage 布边以内
inside shop 自产自销服装店，内制服装店（服装制作全程都在店属厂或定点厂内完成）
inside sleeve 内袖，袖瘪肚，袖下片，小袖片，衬袖
inside sleeve length 内袖长
inside sleeve seam 袖腋下缝
inside slit 内裂缝
inside split （织物）内裂缝
inside tab （西服）里垂片
inside tape 裙腰衬带，腰衬带，腰衬布
inside trouser leg 内侧裤长
inside waistband 藏在裙子或裤子里的腰头，翻向（裙、裤）内侧的腰头
inside width 净宽
inside zip 里面拉链，内侧拉链
insignia 徽章；奖章带
insignia blue 勋章蓝
insignia on cap 帽徽
insolation 日光暴晒
insole 鞋垫；鞋内底
insole leather 鞋垫革
insole rib 鞋内底卡口
insolubility 不溶性
inspect 检查，检验
inspected by oneself 自行检查
inspect exported clothes 检验出口服装
inspect imported fabrics 检验进口布料
inspecting certification 检验证
inspection 检验，商检，验货
inspection at random 抽查，随取验货
inspection certificate 检验证书
inspection certificate of origin 产地检验证书
inspection certificate of quality 品质检验证书
inspection certificate of quantity 数量检验证书
inspection certificate of value 价值检验证书
inspection certificate of weight 重量检验书
inspection certificate on damaged cargo 验残检验证书
inspection in process 中间验货
inspection level 检验标准
inspection mark 检验标记
inspection of end products 成品检验

inspection of goods 货物检验
inspection of origin 产地检验
inspection of packing 包装检验
inspection of preparation for delivery 交货预检
inspection of quality 质量检验
inspection report 检验报告
inspection routine 检验程序
inspection stand 检验台
inspector 检验员
inspiration （设计）灵感
instability 不稳定性
installation 装配
installation dimension 安装尺寸
instant sunglasses 防碎太阳眼镜
instant threading 快速穿线
instep 脚背，袜背，伏面（鞋面的覆盖足背部分），鞋面
instep half 袜背半圆
instep leather 鞋面革
instep length 及脚背长
instep seam 鞋面缝
Institute of Clothing and Physiology in Hohnstein 霍恩施泰因服装和生理研究所
Institute of Textile Technology 纺织工艺研究所
institutional furnishing fabric 公用设施纺织品
institutional laundering 规定洗涤法
In-Store 店中店
instructed person 熟练工
instruction book 参考书，指导书
instruction manual 使用说明[书]
instrumental analysis method 仪器分析方法
instrumental [color] matching 仪器配色
instrumental error 仪器误差
instrumental evaluation 仪器评定
instrumental mannequin 仪器式人体模型
instrumental measurement 仪器测定
instrumentation 仪表化
insulated 防寒服
insulated ability 保暖性能；绝缘能力
insulated boots 保温靴；绝热靴；防水保暖靴
insulated gloves 御寒手套
insulated jacket 防寒夹克
insulated underpants 御寒衬裤
insulated wear 绝缘服

insulated working boots 防寒工作靴
insulating blanket 保温套,保温层
insulating cap 绝缘帽
insulating capacity 隔热能力;绝缘能力
insulating felt 绝缘毡
insulation leather shoes 绝缘皮鞋
insulation rubbe boots 绝缘胶靴
insulation uniform 阻燃服
insurance 保险
insurance policy 保险单
intact garment (仿制时)保持原样的服装
intagliatela 凸纹绣花花边
intangible assets 无形资产
intarsia 几何图案羊毛衫;嵌花;嵌花编织
intarsia collar (针织)嵌花领
intarsia fabric 嵌花织物
intarsia flat knitting machine 嵌花横机
intarsia knit fabric 无虚线提花针织物,嵌花针织物
intarsia knit sweater 虚线提花毛衫
intarsia pattern 镶嵌花纹,嵌花花纹
intarsia sweater 嵌花毛衣,嵌花毛衫
integral design 整体设计
integral garments 全套服装;整体服装
integrated bib vest 组合式护胸背心
integrated finishing 多功能整理
integrated quality control 综合质量控制,整体质量控制
integrated sewing device 综合缝纫装置
integrated sewing station 综合缝纫台
integrated sewing unit 综合缝纫装置
integration 一体化
integrity 均匀完整性,完整性
Inteiro cotton 巴西因特罗棉
Int'l Kids Fashion Show 国际童装展(美国)
intellectual property 知识产权
intelligent sewing system 智能缝纫系统
intelligent shoes 智能鞋
intelligent textile 智能纺织品
intense color 浓烈色彩
intensity 强度;强烈度
intentation 缺口
interactive marker 交互式排料
interactive textile 功能性纺织品,互感纺织品
interbank rate 同业买卖汇率;银行间汇率
intercalated fiber 镶嵌纤维,掺杂纤维

(碳纤维中含有其他化学元素或化合物)
interchangeability (可)互换性
interchangeable bag 可互换式手提袋
interchangeable double cloth 表里交换双层织物
Inter.color 国际流行色;国际流行色协会
interconstruction 复合材料结构
interfaced facing 服装加衬挂面
interfaced hem 加衬下摆
interfaced stiffening 内衬式硬衬
interfacing 连接;内衬,衬布,衬头
interfacing cloth 衬料,衬布
interfacing line 界面线,连接线,面际线
interfacing materials 黏衬材料
interference color 干涉色,干扰色
interference-color chart 干涉色图谱
interference fringe 干涉条纹
interference pattern 干涉图
interflow system 交替流水作业线
interior corner 内角
interior craft 室内工艺品,室内装饰品
interior decoration 室内装饰
interior decorative nonwoven 室内装饰非织造布
interior fabric 室内装饰织物
interior lace 室内装饰花边
interior textile 室内装饰纺织品
interlace 交织花纹
interlaced twill 芦席斜纹
interlaced yarn [喷气]交络纱,网络丝,交络丝
interlacing 编织,交织,连锁成环
interlacing hosiery 交织袜
interlacing net hosiery 交织网眼服
interlacing ornament 交织纹样
interlacing point 交织点,组织点,交络点
interlacing stitch knitted fabric 绞缠组织针织物
interlayer 隔层,夹层,间层
interleaving paper 衬纸
interline 在(衣服)内层装衬里
interlining (服装)内层衬布;衬;黑炭衬;马尾衬,毛里衬,衣服衬里;夹层;中间衬料
interlining below waistline 下腰节衬
interlining bleached cloth 漂白布衬
interlining canvas 粗布衬
interlining [cloth] 垫布,衬布

interlining domett 盖肩衬
interlining fabric 垫布,衬布
interlining flannel 绒夹里布,彩点绒衬布
interlining for front part 胸衬
interlining for shoe and hat 鞋帽衬
interlining grey cloth 本白布衬;法西衬
interlining hair cloth 马鬃衬,马尾衬
interlining-lining 衬里和夹里
interlining material 衬料
interlining pattern 衬纸样
interlining under the waist line 下节衬,下腰节衬
interlining woolen 驼绒
interlock 联锁;双罗纹;棉花布;双罗纹组织
interlock chainstitch 双联线迹,联锁线迹,多线链式线迹,链式绷缝线迹
interlock compound fabric 双罗纹式复合针织物
interlock fabric 棉毛布,双罗纹针织物
interlock flat knitting machine 双罗纹双反面袜机,双面横机
interlocking buckle 对抱式带扣
interlocking fasten 对抱式带扣
interlocking floral[design] 缠枝花纹
interlocking lace stitch 连锁式花边绣
interlocking tapestry 双面组织挂毯
interlocking twill 芦席斜纹
interlock knit 双罗纹编织品
interlock machine 联锁缝纫机,棉毛机,双罗纹针织机
interlock pants 棉毛裤
interlock pile fabric 双罗纹绒头针织物
interlock rib 棉毛布
interlock safety stitch 多线安全链式线迹,联锁安全线迹
interlock seam 锁缝,联锁缝,多线链式缝
interlock sewing machine 多线链式线迹缝纫机,联锁缝纫机
interlock singlet 棉毛衫
interlock sports pants 棉毛运动裤
interlock sports suit 棉毛运动衫裤
interlock sports sweater 棉毛运动衫
interlock stitch 多线绷缝线迹,多线链式线迹,联锁线迹;锁缝线迹,销缝线迹双罗纹组织
interlock stitch machine 多线链式线迹缝纫机,联锁缝纫机
interlock stitch sewing machine 绷缝缝纫机

interlock stitch with top cover 绷缝,覆盖锁缝线迹
interlock stitch without top cover 底绷锁缝线迹
interlock trousers 棉毛裤
interlock T-shirt 棉毛短袖圆领衫,棉毛衫
interloop 互连,连接,连锁
interlooper 内弯针
interlooping 互连成环
intermediary substance 中间产品
intermediate color 中间色,过渡色
intermediate finish 中间整烫
intermediate garment 中层衣
intermediate goods 中间产品
intermediate hue 中间色
intermediate inspection 中间检查
intermediate-jacquard fabric 小花型提花织物
intermediate lining material 中间衬料
intermediate point 中间点
intermediate press 中间压平机
intermediates 中间产品
intermeshing 相互串套
intermingling yarn 交络丝,交缠丝
intermitted stripe 断续条纹
intermittent pattern 间隔性图案
intermittent process 间歇性工艺过程
intermittent production 间歇生产
intermittent ruffler 间歇打裥器
intermittent spot effect 间隔点纹效应
intermittent stripe 断续条纹
internal lining 内衬
internal malleolus 内踝
internal packing 内外装
internal point 内线点
internal standard 内部标准
international 国际仲裁
International Article Numbering Association (IANA) 国际物品编码协会
International Association for Testing Materials (IATM) 国际材料试验协会
International Association for Textile Care Labeling 国际纺织品保养标签协会
International Association of Clothing Designers (IACD) 国际服装设计师协会
international (China) fashion 国际(中国)时装
international clearing 国际结算

International Clothing Machine Fair　国际服装机械交易会
INTERNATIONAL COLOR AUTHORITY 《国际流行色》(意)
International Colour Authority (ICA)　国际色彩局
International Colour System (ICS)　国际羊毛局采用的配色程序
International Commission for Colour in Fashion and Textiles (INTERCOLOUR)　国际时装纺织品流行色委员会
international company　国际公司
international cooperation　国际合作
International Cotton Advisory Committee　国际棉业咨询委员会
International Disposables Exhibition (IDS)　国际用即弃制品展览会
International Disposables Exhibition and Association　国际用即弃产品展览及协会
international double taxation　国际双重征税
International Drycleaning Research system (IDRC)　国际干洗研究委员会
International Dyer　《国际染色家》(月刊)
international economic cooperation　国际经济合作
International Fabric Association　国际织物协会
International Fabricare Institute　国际纺织品洗涤研究所
International Fair　国际博览会
international fashion　国际化时装
International Fashion Council (IFC)　国际流行[服装]评议委员会
International Federation of Association of Fextile Chemists and Colorists　国际纺织化学家及染色家协会联合会
International Federation of Cotton and Allied Textile Industries　国际棉花与棉纺织工业联合会
International Fur Trade Federation (IFTF)　国际皮草贸易联合会
international gray scale　国际灰色样卡
International Hosiery Exhibition (IHE)　国际袜品展览会
International Institute for Cotton　国际棉业协会
International Laboratory Accreditation Conference (ILAC)　国际实验室认可合作组织
International Ladies Garment Workers Union　国际女装生产者协会
International Lady's Garment Worker's Union (ILGWU)　国际女装工会
International Laundry Association (ILA)　国际洗涤协会
International Law　国际法
International Law Association　国际法协会
International Linen and Hemp Confederation　国际亚麻和大麻协会
International Linen Promotion Commi-ssion　国际亚麻促进会
international market　国际市场
international market competition　国际市场竞争
International Mohair Association　国际马海毛协会
International Nonwoven and Disposables Association (INDA)　国际非织造织物和用可弃物品协会
International Nonwovens and Disposables Association　国际非织造布和用即弃制品协会
International Nonwovens Conference & Exhibition　国际非织造布会议及展览会
International Old Lacers　国际带子制作者及收藏者协会
International Organization for Standarditzaion (ISO)　国际标准化组织
International Rayon and Synthetic Fibers Committee　国际再生纤维和合成纤维委员会
International Sericultural Commission　国际蚕丝业委员会
International Silk Association　国际蚕丝协会
International Society of Industrial Fabric Manufacturers　国际产业用纺织品生产者协会
International standard　《国际标准》
international support　对外援助
international technical support　对外技术援助
international technology exchange　国际技术交流
international technology trade　国际技术贸易

International Textile and Apparel Association 国际纺织服装协会
International Textile and Garment Worker's Federation（I.T.G.W.F）［德］国际纺织和服装工会
International Textile Care Labelling Code（I.T.C.L.C）国际纺织洗整标志
International Textile Club 国际纺织联合会
International Textile Institute 国际纺织学会
International Textile Machinery Exhibition（ITME）国际纺织机械展览会
International Textile Manufacturers Federation 国际纺织生产者联盟
International Textile Manufactures Association（ITMA）国际纺织业协会
International Textile Manufactures Federation（ITMF）国际纺织业协会
INTERNATIONAL TEXTILES 《世界纺织》（荷）
international trade 国际贸易
International Trade Fair 国际贸易展览会
International Trade Organization（ITO）国际贸易组织
international unit system 国际单位制
International Wool Secretariat 国际羊毛局（IWS），国际羊毛咨询处
International Wool Secretariat Award 国际羊毛局奖
International Wool Secretariat Fashion Studio（IWSFS）国际羊毛局服装款式创作室
International Wool Secretariat Technical Center（IWSTC）国际羊毛局技术中心
International Wool Textile Organization 国际羊毛纺织协会
international wool trade 国际羊毛贸易
internipple breadth 两乳头宽，胸点宽
inter packing 内外装
interpersonal skills 人际关系技巧
inter-ply shifting 布层间移动
intersected twill 飞斜纹，间断斜纹
Inter Selection 国际时装批发展销会（法国）
inter shirring stitch 内抽褶线迹
intersia pattern hosiery machine 嵌花袜机
interspersion 点缀，散置

inter-stitch 联缝线迹
inter-stitch sewing 联缝线迹缝纫
Interstoff 国际纺织服装展（德国）
Interstoff world 英特斯多福国际时装节（德国）
intertexture 交织；交织织物
interting slide fastener 装拉链
interweave 交织
interweaved habotai 交织纺绸
interwind fiber 捻合纤维，缠结纤维
INTERWOOLLABS 国际羊毛工业试验室联合会
interwoven bag 交织麻袋
interwoven fabric 交织布；交织织物；毛粗纺混纺（交织）织物
inter woven linen 亚麻交织布
in the gray 坯布，绸坯
in the grease 呢坯
intimacies 亲密装（对衬衣、内衣等进行外衣化设计而成）
intimate apparel 贴身内衣，个性服装
intralacing 自连锁成环
intralooping 自连成环
in-trend 顺应趋势
intricate design 复杂花样
intriguing style 引起好奇的款式，有迷惑力的款式
intrinsic humidity 内在湿度
intuition 直感，直觉
inturned welt 双层袜口，袜子翻口
INUOVI 伊娜薇
invecauld 印福柯德
inventory 存货；存货单
invermark 印福马克
inverness 长披风；可卸式无袖长披风；披肩外套；羊毛格子长披风；紧身圆领披肩
inverness［cape］紧身圆领披肩；可卸式无袖长披风，披肩外套
inverness clock 可卸式无袖长披风
inverness clothing 羽袖大衣（袖似披肩）
inverness coat 可卸式无袖长披风，披风外套，羽袖大衣，披风大衣
Inverness suit 因弗内斯套装
inverse hemming 反卷边
inverse slopped sole 反披头底（鞋）
inverted box pleat 倒对裥；倒箱式褶裥
inverted（box）pleated pocket 阴（明）裥袋
inverted frame 金属框架不外露手袋
inverted leg-o'-mutton sleeve 女式倒置半

褾袖
inverted plait 倒褶边；倒褶裥
inverted pleat 阴裥,阴褶,暗裥,倒褶裥
inverted pleated pocket 暗裥袋,阴裥袋
inverted pleats 暗裥
inverted pleat skirt 倒褶裙；阴褶裙,内翻褶裥裙
inverted skirt 倒裥裙,阴裥裙
inverted triangle 倒三角形
inverted vent 倒置骑马缝叠襞(无剪口的叠襞骑马缝),阴衩,暗衩
investiture 装饰物,装饰；服装
investment 投资
investment clothes 值得花钱的服装,值得投资的衣服(高质量、高价格服装)
investment clothing 投资性服装(高质量、高价格服装)
investment dressing 投资性服装(高质量、高价格服装)
investment jeans 高价牛仔裤
invisea 印维锡棉人造丝交织物
invisible chromatogram 不可见色谱
invisible depletion 无形磨损
invisible fleecy fabric 添纱衬垫针织物
invisible green 深绿,墨绿
invisible hem 暗卷边
invisible hook 暗钩
invisible mending 精细织补
invisible plaid 暗纹方格呢
invisible seam 暗缝
invisible shoes 隐形鞋
invisible stitch 暗缝线迹
invisible thread 暗线
invisible tucks 暗褶,暗裥
invisible zipper 暗缝拉链；隐形拉链
in vogue 正在流行
invoice 发票
inward collection (I/C) 进口托收
inward corner 凹角
inward curve 内凹线
inward mitered corner 内折45°缝角
in-work dress 上班装
inwrought 织物外加装饰花纹
iolite 堇青石
Ionic chiton 艾奥尼克式奇通衫(古希腊)
Iowa State University of Science and Technology 衣阿华州立科技大学(美)
ipspection lot 检验批号
Iraki cotton 伊拉克短绒棉
Iranian carpet 波斯地毯,伊朗地毯

Iranian costume 伊朗民族服饰
Iranian woman's 伊朗女式蹬脚裤
Iranic costume 伊朗民俗服
Iraqian costume 伊拉克民族服饰
iraser 红外激射,红外激光；红外激光器
Ire 尼日利亚裙
Irene 艾琳(美国时装设计师)
Irene Castle bob 艾琳·卡斯尔短发
Irene Galitzine 伊雷内·加利特津(意大利时装设计师)
Irene Gilbert 艾琳·吉尔伯特(爱尔兰时装设计师)
Irene twill 艾琳斜纹里子呢
iridescent 闪色衣料；荧光色
iridescent damask satin 闪光花线缎
iridescent fabric 闪光织物；闪色织物
iridescent hosiery 闪色袜
iridescent luster 荧光光泽,虹彩光泽
iridescent pointille taffeta 直隐绸
iridescent polyamide gauze 闪光锦丝纱
iridescent rayon denim 有光真丝牛仔布
iridescent satin brocade 朝霞缎,闪光缎
iridescent stripe tissue crepe 闪碧绡
iridescent style 不对称型
iridescent suiting silk 闪光呢
iridescent velvet 闪金立绒
iris 彩虹色
irisated prints 彩虹[印]花布
iris blue 鸢尾蓝(浅蓝色)
irised green 鸢尾绿
irised prints 彩虹[印]花布
iris green 鸢尾绿
Irish beetle sinish 爱尔兰捶布整理
Irish cambric 爱尔兰漂白全亚麻平布,爱尔兰漂白全亚麻纱
Irish cape 爱尔兰披肩
Irish cloth 中世纪爱尔兰毛衬布
Irish cochef lace 爱尔兰手工钩编花边
Irish crochet 爱尔兰手工钩编花边
Irish donegal 多呢盖尔粗花呢
Irish duck 爱尔兰纯亚麻帆布
Irishes 北爱尔兰漂白平布
Irish eye diaper 爱尔兰菱形斜纹织物；菱形花纹亚麻布
Irish finish 爱尔兰仿亚麻整理
Irish flax 爱尔兰亚麻
Irish frieze 爱尔兰厚呢
Irish green 爱尔兰绿
Irish knit 爱尔兰羊毛衫(绳编花纹)

Irish lace 爱尔兰农村花边,爱尔兰花边	熨烫定形联合机
Irish lawn 爱尔兰上等细麻布	ironing board 熨烫工作台,熨衣板,烫凳
Irish linen 爱尔兰亚麻布,爱尔兰漂白纯亚麻布	ironing board cover 熨烫板罩布
	ironing cylinder 熨烫滚筒
Irish point 爱尔兰针绣花边	ironing fastness 耐熨烫牢度,熨烫牢度
Irish point crochef lace 爱尔兰手工钩编花边	ironing felt 熨烫毡
	ironing head 熨斗头
Irish point lace 爱尔兰针绣花边	ironing machine 熨烫机
Irish poplin 爱尔兰无葛;横条丝无绸	ironing mat 熨斗垫,熨烫垫
Irish stitch 爱尔兰线迹,爱尔兰绣	ironing material 熨原辅料
Irish trimming 爱尔兰机织饰带	ironing padding 熨烫垫布
Irish tweed 爱尔兰手纺毛纱粗花呢,爱尔兰(粗)花呢	ironing resistance 耐熨烫性
	ironing sensibility 熨烫敏感性
Irish wool 爱尔兰绵羊毛	ironing shoe pad 熨鞋垫
Irish work 爱尔兰白色刺绣,爱尔兰刺绣	ironing shrinkage 熨烫收缩率
Irish(rose)point 爱尔兰玫瑰花纹花边	ironing sponge 熨烫海绵垫(烫布的衬垫)
iris leaf 鸢尾叶色	ironing station 熨烫工作站
iris orchid 鸢尾蓝花色	ironing table 熨烫台板,熨衣台
iris pattern 鸢尾花形	ironing technique 熨烫技术
iris prints 彩虹[印]花布	ironing temperature 熨烫温度
irisated prints 彩虹[印]花布	ironing test 耐熨烫牢度试验
irlanda (古巴)条子亚麻细布,条子棉细布	ironing to melt 烫熔
	iron melting hole test 织物熔孔试验
Iro 尼日利亚裙	iron off 烫掉
iro momen 仿南京土布	iron-on 用熨烫法粘附于织物表面
iron 烙铁;熨斗;熨烫,烫平;铁色	iron-on cotton 有胶棉衬
ironability 可熨烫性,耐熨烫性	iron-on interfacing 热熔衬,胶衬
ironability finish 耐熨烫性整理	iron-on motif 熨压成形图案
iron black 铁黑	iron-on patch 热熔补钉,热熔补片
iron blue 铁蓝(带蓝灰色),深蓝	iron-ons 黏合衬,垫熔衬
iron board 烫衣板	iron out 熨干(衣服)
iron brown 铁棕	iron out wrinkles 熨平皱襞
iron buff 铁黄(淡棕花色)	iron oxide black 铁黑
iron damp 湿烫	iron oxide red 铁红
iron dry 干熨,熨干	iron oxide yellow 铁黄
ironed fold 烫痕	iron piece 铁云(鞋)
ironer 熨烫工,烫衣服的人;熨烫器	iron press 熨烫
iron even 烫散	iron pressure 熨烫压力
iron fabric 烫料,硬挺整理加光织物,增加摩擦扎光棉布	ironproofing 耐熨烫的;耐熨烫整理
	iron rest 熨斗垫,熨斗盘,熨斗架
iron fabrics 熨烫衣料	iron setting 熨烫定形
iron fold 烫痕	iron-shine (熨烫不良产生的)亮光,熨烫亮光,亮光疵
iron-free 免烫	
iron-free finish 免烫整理	iron shirt 熨烫衬衫
iron gray 铁灰(带绿光深灰色),铁灰色	iron shoes 熨斗套,熨斗鞋(放置熨斗用)
iron hat 礼帽;熨盔,铁帽	iron stand 铁凳(烫衣肩等部位的工具);熨斗垫,熨斗盘,熨斗架
ironed fold 烫痕	
ironing 熨烫;整烫;熨平;(总称)烫过的(要烫的)衣服	iron stiffly 烫煞
	iron temperature 熨烫温度
ironing and preboarding machine [袜子]	iron tester 熨烫[收缩]试验仪

iron-toe-end 熨斗头形鞋头
iron top collar for preshrinking 热缩领面
iron well 熨斗架
iron with water container 吊水熨斗
irradiated wool 辐射处理羊毛
irregular 不匀
irregular curve 变化弯尺,自由曲线尺
irregular cutting pile 剪绒不正
irregular hem line 不规则下摆线
irregular pattern 稀密不匀
irregular piece-length 不定匹长
irregular pile bar 长短绒档
irregular pleat 变化褶
irregular pocket mouth 袋口不方正
irregulars 带残疵的降价服装,轻微疵品（不列一等品）
irregular seam 线缝不匀,线缝调整不良
irregular selvage 布边不齐
irregular skirt 不规则裙（下摆不在同一水平线上）
irregular stitch 乱纹花;乱针迹疵,线圈不匀,线迹不匀
irregular stitch setting 线圈不匀,线迹不匀
irregular stripe 不规则条纹,随意条纹
irregular width 门幅不齐
irreversible 单面织物
irreversible fabric 单面织物
irreversible flannelette 单面绒布
isabel 灰黄色;伊莎贝尔细薄精梳毛织物
Isabella sleeve 伊莎贝拉袖
Isabelle 灰黄色;八经六纬服用织物
Isao Kaneko 金子功（日本时装设计师）
Isfahan carpet 伊斯法罕马海毛毯
Isfahan yarn 伊斯法罕马海毛绒线
Ishan cotton 西非伊尚棉
isitebe 南非韧皮纤维垫
island green 岛屿绿,海岛绿
island-in-the-sea matrix fibril type bicomponent fiber 天星状双组分复合纤维
island print 岛屿印花料
isle of wight lace 威特岛花边（英国式梭结花边和针绣机制网眼花边）
ismaili （东非）红边彩条棉布,条子棉布
isochromatic 等色的;等色线
isochromatic curve 等色曲线
isochromatic line 等色线
isochromatic stimulus 等色刺激
isochromatic triangle 等色线三角形
isochromes 等纯色量

isochrome series 等色系列
isolated color 孤立色
isolated pattern 大花纹
isolated raised effect 大花纹起绒效应
isolated sample 单件样品
isolation garment 绝热服
isomeric colors 同质同谱色,同谱同色,无条件等色
isomeric match 无条件配色
isotemperature line 等色温线
isotints 白色量,等白系列
Isotoner slippers 伊泽托纳鞋（芭蕾舞鞋形,由轻、软可洗的弹力布制成）
isotones 黑色等量系列,等黑系列
isotonic gloves 等压手套
isotropic carbon fiber 各向同性碳纤维
isotropic fiber 各向同性纤维
isovalent color 等价色
Ispahan rug 伊朗伊斯法罕马海毛绒地毯
Ispahan yarn 伊斯法罕马海毛绒线
isree 印度手织细棉布
Issey Miyake 三宅一生(1938～,日本时装设计师)
Issuing bank 开证银行
istaberk 印度伊斯塔伯缎
Italian 意大利缎纹布
Italian belt 意大利式腰带
Italian boy cut 意大利男孩发型
Italian casual 意大利式便装
Italian cloth 意大利纬面缎纹布;意大利棉毛呢;黑色直贡呢
Italian collar 意大利式领
Italian color 意大利色（红、绿、白对比色）
Italian continental 意大利的欧洲款式西装
Italian continental suit 意大利欧式西装
Italian cords 意大利条纹棉布
Italian corsage 意大利低领口紧身衣
Italian costume 意大利民族服饰
Italian cross stitch 双边十字绣
Italian cut 意大利式发型
Italian cut jeans 意大利裁法的牛仔裤
Italian cut shoes 意大利鞋型
Italian cutwork 意大利挖花织物
Italian farthingale 意大利轮形裙环
Italian ferret 意大利衣边用丝带
Italian hemp 意大利大麻
Italian hemstitching 意大利抽绣,意大利缘边绣

Italian hose 意大利威尼斯男裤
Italian insertion stitch 意大利拼缝绣
Italian insertion buttonhole stitch 意大利嵌饰锁眼绣
Italian iron （圆筒形）意大利熨斗（可烫绉花边等的）
Italian lace 意大利式花边
Italian line 意大利型
Italian lining 意大利里子布，意大利里子呢
Italian mixed wool cloth 毛缎子；意大利毛呢
Italian mode 意大利式；意大利式西装
Italian neck 意大利式领
Italian neckline 意大利领口线
Italian nightgown 意大利式晚袍；爱尔兰式连衣裙
Italian quilting 意大利式绗缝
Italian roll 意大利卷型领
Italian silk 意大利蚕丝
Italian stitch 意大利针法，双平针线迹
Italian straw 意大利麦杆色
Italian stripe 意大利条纹
Italian wool 意大利美利奴杂交羊毛
itarsi 印度麻，菽麻
itching feeling ［苎麻］刺痒感
item （服装的）种类；项目，品目
item number 品号
item of business 业务项目
itemized price 分类价
ITO 国际贸易组织
ito nishike 金丝花缎
Itsuko Nōkashima 中岛伊津子（日本时装设计师）
iverted pleat placket 暗褶衩
ivory 象牙；象牙色；象牙白；乳白（釉色）；象牙黄
ivory black 象牙黑
ivory brooch 象牙饰针
ivory button 象牙扣

ivory cream 象牙奶油色
ivory deep 白淡黄色
ivory look-alikes 假象牙
ivory necklace 象牙项链
ivory nut button 象牙椰子核纽扣；橡树果纽扣
ivory rose 象牙玫瑰
ivory white 象牙白，乳白（釉色），牙白
ivory yellow 象牙黄（乳白色）
ivy 常春藤图案
Ivy blazer 常春藤运动夹克
ivy cap 常春藤帽
ivy fold 艾维装饰法，胸袋中手帕装饰法
ivy［green］ 常春藤绿（暗橄榄绿色）
Ivy League look 常春藤［联合会］风貌；常青藤联合会款式（20世纪50年代美国以及60年代日本流行的自然肩、窄翻领、三个纽扣的直筒形西装，裤子细长）
Ivy League model 常春藤［联合会］款式
Ivy League pants 常青藤［联合会］裤
Ivy League shirt 常春藤［联合会］衬衫
Ivy League style 常春藤［联合会］款式
Ivy League suit 常春藤［联合会］西装
Ivy look 常春藤［联合会］风貌；常春藤［联合会］型
ivy shirt 常青藤衬衫，藤纹衬衫
ivy slacks 常青藤风格的长裤，藤纹裤，艾维裤
Ivy stripes 常春藤条纹
ivy style 轻装，常青藤式轻装
ivy sweater 藤纹毛衣
Iws fastness specifications 国际羊毛局色牢度标准
izar （伊斯兰教的）伊札尔（女穆斯林穿的可遮蔽全身的棉布外衣，男穆斯林朝觐服的围腰白布）；白色棉面纱（阿拉伯）
iznak 巴勒斯坦民俗服中的粗重黄金项链
Izzue 伊祖

J

jabot 女服胸襟部的褶裥;皱纹花边;男服领圈前部的花边领饰;垂胸领饰;波状胸饰
jabot blouse （后扣式）垂胸领饰衬衫
jabot collar 胸饰领
jabot frill 衬衣绉边
jabot ruffle[frill] 衬衣皱边
jacaranda 蓝花楹色
jacconets 细薄布;细棉薄纱
jacerma 波斯尼亚无袖合身女夹克
Jacgues Doucet 雅克·杜塞（法国高级时装设计师）
Jacgues Fath 雅克·法思（法国高级时装设计师）
Jacinth(e) 红锆色;淡橙色,淡橘红色;橘红色宝石
jack 无袖皮军衣（中世纪步兵护身）;（口语）夹克
jackal 豺毛皮
Jack & Betty fashion 杰克与贝蒂服饰
jackboots 过膝长筒靴
jacket (jkt) 夹克衫;短上衣;短外衣;套毯
jacket button 夹克纽扣,上衣纽扣
jacket cape 夹克式披肩,外套型披风
jacket dress 两用夹克套装;上衣连裙装
jacket earring 片柱式耳环
jacket felt 套毯
jacket hem 夹克下摆
jacketing 做短上衣的衣料
jacket of single layer 单层夹克
jacket shirt 夹克式衬衫
jacket shorts 夹克短裤套装
jacket side panel 夹克衫侧面嵌条
jacket suit 短夹克套装
jacket vest 夹克背心
Jackie Rogers 杰基·罗杰斯（美国服装设计师）
jacking finish 卷布式轧光整理
Jackson Pollock Style 杰克逊·波洛克风格
jack suit 短夹克套装
jack towel 环挂揩手巾
Jacobean embroidery 詹姆斯刺绣
jacobean work 英国东方风格绒线刺绣
Jacobite 贾科拜特苏格兰格子呢
jaconas 细薄布;细棉薄纱
jaconet 棉织细薄纱;单面轧光色布;细白布
jaconnette 棉织细薄纱;单面轧光色布
jaconnot 棉织细薄纱;单面轧光色布
jacqmar 贾克玛细薄毛织物
jacquard 提花,花式;纹织;提花织物;提花机
jacquard à jour 提花纱罗
jacquard bag 提花[手提]布包
jacquard batiste 花富纺
jacquard bed sheet 提花床单
jacquard bed spread 提花线毯;提花床罩
jacquard bed spread fringed 排须提花线毯
jacquard bed-cover 提花床罩
jacquard blanket 提花毯,提花绒毯
jacquard check 提花格纹
jacquard chine 花格锦缎,印经锦缎
jacquard circular knitting machine 圆形提花针织机
jacquard cloth 提花织物;提花针织物
jacquard cotton-linen handkerchief 提花[亚]麻棉手帕
jacquard crash 提花麻纺绸
jacquard crash matelasse 提花仿麻绸
jacquard crepe 花绉绸,花绉
jacquard curtain 提花窗帘
jacquard cylinder 花枕头,提花机花筒,花辊筒
jacquard Derby 提花罗纹
jacquard double knit 双面提花针织物
jacquard doupion crepe 花双宫绉
jacquard drill 麻棉厚斜纹布
jacquard ecossais 英国格子提花织物
jacquard embroidering 提花绣花
jacquard embroidering half-hose machine 提花绣花短袜机,提花吊线袜机
jacquard embroidery machine 提花绣花机
jacquard embroidery plating hosiery machine 提花绣花袜机
jacquard fabric 提花织物,贾卡提花织物,大花纹织物

jacquard fake fur knitted fabric 提花人造毛皮
jacquard flannel 提花法兰绒
jacquard flano 提花条子法兰绒
jacquard flat knitting machine 提花横[编针]机
jacquard floats 提花
jacquard gauze 花纱；提花纱罗织物
jacquard glace 闪光提花织物
jacquard gloves 提花手套
jacquard grisaille 黑白提花织物
jacquard hair sweater 提花毛衫
jacquard handkerchief 提花手帕
jacquard hard satin 花累缎
jacquard hose 提花袜
jacquard hosiery machine 提花袜机
jacquard jersey 提花针织物
jacquard knit 提花针织品
jacquard knit fabric 提花针织物，提花针织布
jacquard knit jacket 提花针织夹克衫
jacquard knit rib 提花罗纹针织物
jacquard knitted fabric 提花针织物
jacquard knitted terry 提花[针织]毛巾布
jacquard knitting machine 提花针织机
jacquard knitwear 提花毛衫
jacquard label 提花商标带
jacquard lace 提花花边
jacquard lace fabric 提花花边织物；贾卡花边织物
jacquard lady's dress 提花女衣呢
jacquard lady's dress worsted 提花女衣呢
jacquard lance 提花浮纹织物
jacquard loom 提花织布机
jacquard mock crepe 黏胶丝提花四维呢，人造丝花四维呢
jacquard mock leno 似纱绸
jacquard multicoloured handkerchief 多色提花手帕
jacquard ou crepe 花偶绉
jacquard pattern 大花纹织物；提花图案，大提花织纹
jacquard pattern socks 提花短袜
jacquard pile blanket 提花绒毯
jacquard plush knitted fabric 针织提花长毛绒
jacquard poult 葛
jacquard purl fabric 花式双反面针织物
jacquard raschel machine 拉舍尔提花编机
jacquard raye 色经条子提花织物
jacquard rayon poult 织绣绸；绒花绸
jacquard ribbon 提花带
jacquard rib fabric 提花罗纹布，提花罗纹针织物
jacquard scarf 提花围巾
jacquard sheet 大提花被单布；大提花家具布；提花床单
jacquard silk quiet covering 真丝被面
jacquard stripe 提花条纹
jacquard sweater 提花运动衫；提花毛衫；提花毛线衫
jacquard table cloth 提花台布
jacquard T/C moke leno curtain fabric 涤棉透空大提花窗纱
jacquard terry bed cover 提花毛圈床单
jacquard terry cloth 提花毛巾布
jacquard terry weft knitted fabric 提花毛圈针织物
jacquard tie silk 提花领带绸
jacquard towel 提花毛巾
jacquard transfer knitter 提花移圈针织机
jacquard-tuck combined knit(ted) fabric 提花集圈复合针织物
jacquard tuck lace 提花集圈网眼
jacquard tuck rib fabric 提花集圈罗纹织物
jacquard twill 花绫
jacquard umbrella silk 提花伞绸
jacquard upholstery brocade 提花家具装饰织物
jacquard upholstery matelasse 马特拉塞提花家具布
jacquard velour(s) 提花天鹅绒，仿提花天鹅绒
jacquard warp knitted fabric 提花经编织物
jacquard warp knitter 提花经编机
Jacquard warp-knitting machine 贾卡提花经编机
jacquard weave 提花织物
jacquard weft-knit fabric 纬编提花针织物
jacquard weft knitted fabric 纬编提花针织物
jacquard woolen sweater 提花毛衫
jacquard woven denim 提花斜纹布
jacquard woven towel 提花毛巾

Jacqueline de Ribes 雅克琳·德里贝斯
（法国时装设计师）
Jacqueline Jacobson 雅克琳·雅各布森
（法国时装设计师）
Jacques Delahaye 雅克·德拉海（法国时装设计师）
Jacques Doucet 雅克·杜塞（1860～1932,法国时装设计师）
Jacques Esterel 雅克·埃斯特雷尔（法国时装设计师）
Jacques Fath 雅克·法恩（1912～1954,法国时装设计师）
Jacques Griffe 雅克·格里夫（法国时装设计师）
Jacques Heim 雅克·埃姆（1899～1967,法国时装设计师）
Jacques Kaplan 雅克·卡普兰（美国时装设计师）
Jacques Tiffeau 雅克·蒂福（美国时装设计师）
jacquette cape 夹克式披肩；外套型披风
jac-shirt 衬衫式无里平下摆短夹克衫
jade 翡翠；玉；玉制品；绿玉色，浅绿色
jade adornment 玉佩
jade belt（yudai） 玉带（中国戏装）
jade bracelet 玉手镯
jade button 玉石扣
jade clothes sewn with gold thread 金缕玉衣
jade gray 灰粉绿
jade green 翠绿色，绿玉色
jade grey 灰粉绿，玉灰
jade hair pin 玉簪
jadeite 硬玉
jadeite button 翡翠扣
jade jewelry 玉首饰
jade lime 果绿色，翡翠橙色
jade pendant 玉佩
jade ring 玉戒指；玉环
jade sheen 玉光色
jade uniforms 柔道装
jade white 玉石白
Jaeger 雅格尔纯毛料；雅格尔纯毛料服装；雅格尔纯羊毛衫
Jaeger underclothes 雅格尔卫生衫裤
Jaffa Orange 雅法橙色
jaffer 英国闪光色织棉平布
jag 织品上的V形凹口
Jaganath(s) 坚固平纹棉布，鞋靴衬里棉布

jager 杰格钻石
jagged line 凹凸不平线条
jagging 剪边装饰
jago 家用亚麻平布
jags 剪边装饰
jague 中世纪夹克
jaiam carpet 印度手织印花毯
jaidar 印度网纹细平布
jainamaz 棉织拜毯
jak ［日］背心
jaki 印度手织棉细布
jamadane 手织色经平布
Jamaica cotton 牙买加棉
Jamaicas 牙买加裤
Jamaica shorts 牙买加短裤；男用田径短裤；裤长为臀围到膝盖一半的短裤
jamawar 印度贾马瓦尔开司米织物；印度贾马瓦尔粗纺毛披巾
jambeau[greave] 胫甲
jambiéres ［法］绑腿，裹腿护腿；绑腰带
jamdanee 印度高级棉平布
jamdani 浮纬花纹布
James Deen style 詹姆斯款式（以T恤衫、牛仔裤、夹克相组合）
James Galanos 詹姆斯·加兰诺斯（1929～，美国时装设计师）
James Laver 詹姆斯·雷费（美国服装史学家）
jamkalam rug 印度棉拜毯
jamkalam 印度棉拜毯
jammed construction 弹力织物的伸张性触；最密织物
jammed fabric 极限密度织物
jams （口语,pajamas之缩略）睡衣裤；（冲浪运动用）束带泳裤，（腰间束带,长及膝部的）睡裤式游泳裤
janamaz 棉织拜毯
janapan 印度麻，菽麻
Jane Blanchot 简·布朗绍（法国女帽设计师）
Jane Burns 简·伯恩斯（美国服装设计师）
Jane Cattlin 简·卡特琳（英国服装设计师）
Jane Derby 简·德比（美国服装设计师）
Jane Regny 简·瑞格尼（美国首位运动装设计师）
janeo 祭祀用线
Janet Colombier 雅内·科隆比耶（法国时装设计师）

jangipuri 印度低档黄麻
jannequin 土耳其粗棉布
Janseniste panniers 简森特裙撑
janus cloth 双面异色呢
janus cord 经棱纹呢
jaob's ladder 断线脱散
Japan Bunka College 日本文化服装学院
Japan Chemical Fibers Association 日本化学纤维协会
Japan curlies 日本废丝，日本丝吐
Japanese costume 日本民族服饰
Japanese cotton sock 日式棉袜
Japanese crepe 日本绉布
Japanese darning stitch 日本式缀纹绣
Japanese design 日本式图案
Japanese dress 和服
Japanese embroidery 日本刺绣；日本刺绣色线或金银丝缎纹针迹
Japanese fashion 日本风格时装
Japanese hemp 日本大麻
Japanese Industrial Specification (JIS) 日本工业标准；日本工业规格
Japanese Industrial Standard 日本工业标准；日本工业规格
Japanese look 日本款式，日本风貌
Japanese marten 日本貂鼠皮
Japanese mink 日本貂皮
Japanese native cloth 日本土布
Japanese parasol 日本阳伞
Japanese pongee 日本茧绸
Japanese prodigal style 日本浪人款式
Japanese prodigal stylistic form 日本浪人款式
Japaneseque 日本式的，日本风格的
Japaneseque design 日本风格设计
Japaneseque fashion 日本风格时装
Japanese rug 日本绒毯
Japanese sandals 日本式凉鞋
Japanese satin 日本丝缎
Japanese silk 日本丝；日本丝麻缎
Japanese snow shoes 日本雪地鞋
Japanese spun silk 日本绢丝
Japanese Standard Association (JSA) 日本标准协会
Japanese style 日本服饰；日本款式
Japanese traditional 日本传统装
Japanese velvet 日本丝绒
Japanese wrapper 日式居家长袍
Japan Fashion Colour Association (JAFCA) 日本流行色协会

Japan Men's Fashion Unity (JMFU) 日本男式时装协会
japanned 涂漆织物，漆布
japanned cloth 涂漆织物
Japan silk 日本生丝；日本丝麻缎
Japan Silk and Synthetic Textiles Exporter's Association 日本丝和合纤纺织品出口商协会
Japan Textile Colour Design Centre 日本织品色彩设计中心
Japan Textile News 《日本纺织新闻》(期刊)
Japan uniform 日本制服
Japan velvet 日本丝绒
Japan Wool Products Inspection Institute Foundation 日本毛制品检验协会
japara 印度贾帕拉丝绣薄纱
japer 火油钻
japergonsi 日本金丝边薄纱
Jap fox 乌苏里江浣熊毛皮
japilline 双面缎纹呢
japonais 日本薄纺绸
japonette 印花棉绉布
jappe 细纺平布；平底牛津鞋
Jap raccoon 乌苏里江浣熊毛皮
japrak 杂色的土麦拿地毯
jaquar 美洲豹毛皮
jaquenolle 印度条子薄细布
jardiniere 树叶形花纹；水果形花纹；花朵形花纹
jardiniere velvet 丝绒花缎(缎地，丝绒花，毛圈有割、有不割，且高低不等)
jardiniere 花、果、叶花纹
jari(l)a 布罗奇棉
jarit 贾里特蜡防印花布
jarre 粗兔毛
jarretelle [法]吊袜带
jasey 贾西假发，毛线假发
jasmin(e) 淡黄色，茉莉黄，素馨色
jasmine gloves 茉莉香手套
jasmine green 茉莉绿
jaspe carpet 碧玉花纹地毯
jaspé check 贾斯佩格子
jaspe cloth 贾斯佩绸
jaspelline 贾斯佩林薄大衣呢
jaspe mouline yarn 彩色桔子花式丝绒
jasper 墨绿色；芝麻绸；芝麻呢
jasper cloth 贾斯珀布
jasper flannelette 芝麻绒
Jasp(er) opal 碧玉蛋白石

jasper red 碧玉红色,碧玉红的
jasper velvet 贾斯珀绒;经向彩条丝绒
jaspopal 碧玉蛋白石
jass Modern 爵士现代主义
jastaucorps 裘斯特可外衣(17～18 世纪)
Java 爪哇红地色条布
Java cantala 坎塔拉剑麻
Java canvas 刺绣用黄褐色十字布;爪哇十字布
Java cotton 木棉
Java lizard 爪哇蜥蜴皮
Javanese Batik Prints 爪哇蜡防印花布(一种以绿地为主的深色花布)
java print 瓜哇蜡防印花
Java stripes 英国爪哇红蓝条子布
Java supers 爪哇平布
javelin shoes 投标枪鞋
jaw 颚,颌
Jawaharlal coat 贾瓦哈拉尔外套
jawline 下巴轮廓,下巴外形
jaws 颚毛
jay 中蓝色
jazerant 鳞片式铠甲
jazim 手织粗地毯
jazz 颜色过鲜,花俏,不和谐
jazz garter 爵士弹性宽吊袜带
Jazz oxford 平底牛津鞋,爵士牛津鞋(鞋底用透明塑料制成)
Jazz style 爵士风格
Jazz suit 爵士套装(第一次世界大战后流行)
J-cut 栽绒地毯的不匀割绒
jeam pocket 斜口贴袋
jean 粗斜纹棉布;牛仔布;劳动布;三页细斜纹布
jeanage 适合穿牛仔裤的年龄
jeanager 适合穿牛仔裤的年龄的人
jean back 斜纹底棉平绒
jean back velvet 斜纹地棉平绒
Jean Baptisle Isabey 阿比(法国著名戏剧服装设计师)
Jean Barthet 让·巴塞(法国时装设计师)
Jean Cacheral 让·卡舍莱尔(法国时装设计师)
jean character 牛仔品味
Jean-Charles de Castelbajac 让-夏尔·德卡斯特巴加(法国服装设计师)
Jean-Claude de Luca 让-克洛德·德吕卡(法国时装设计师)

Jean Colonna 让·科洛纳(法国时装设计师)
Jean Dessès 让·德塞(1904～1970,法国时装设计师)
jean dress 牛仔裙装
jeanette 三页斜纹呢;细斜纹布
jeaning 牛仔裤化
jeaning look 牛仔裤朴素款式
jean jacket 粗斜纹棉夹克
jean jumpsuit 牛仔连身裤,牛仔式连身裤
Jean Louis 金·路易斯(美国著名戏剧服装设计师)
Jean-Louis Scherrer 让-路易·谢莱尔(1936～ ,法国时装设计师)
Jean-Marc Sinan 让-马克·西纳(法国时装设计师)
Jean Muir 让·缪尔(1933～ ,英国时装设计师)
Jeanne Lafanrie 让娜·拉方丽(法国时装设计师)
Jeanne Lanvin 让娜·朗万(1867～1946,法国时装设计师)
jeannet 棉经毛纬斜纹呢,斜纹呢
jean pants 紧身裤,紧身工作裤,牛仔裤,斜插袋,斜纹棉布裤,坚固呢裤
Jean Patou 让·帕图(1887～1936,法国时装设计师)
Jean paul Gaultier 让·保罗·戈尔捷(法国时装设计师)
jean pocket 斜口贴袋,牛仔裤裤袋,斜插袋
Jean Pomeréde 让·波墨雷德(法国时装设计师)
Jean-Remy Daumas 让-雷米·多马(法国时装设计师)
jeans 牛仔裤;斜纹棉布裤;紧身工作裤;紧身裤;坚固呢裤
jean satin 缎纹布,厚斜纹布
jeans button 牛仔纽扣;金属牛仔扣;工字纽
Jean Schlumberger 让·施卢姆贝格尔(法国时装设计师)
jeans clothing 牛仔衣裤
jeans hat 牛仔帽
JEANS INTERW [德]《牛仔流行》
jeans item 牛仔装类
jeans look 牛仔风格,牛仔风貌
jeans-pantaloons 牛仔裤
jean stripes 条子斜纹

jeans styling　牛仔风格
jeans twill　牛仔绫
jeanswear　牛仔装(总称);牛仔裤
JEANSWEST　真维斯
jedim　缝合地毯
Jeff Sayre　杰夫·塞尔(法国服装设计师)
jellaba　带风帽斗篷(阿拉伯男子穿);杰莱伯罩衣
jellies　软底模压足球鞋
jellybag　睡帽
jelly bean sandal　豆形软底透明凉鞋
jemmy　男式射击服
jemmy boot　男式轻便骑马靴
jenappe yarn　烧毛纱
jenfez　土耳其横棱纹绸
jenkins cotton　美国詹金斯棉
jennets　棉粗哔叽
Jenny Lind dress　19世纪中期林德裙
Jenny(Mme.Sacerdote)　詹尼(萨司铎特夫人)(法国高级时装设计师)
Jenny Sacerdote　燕妮·萨塞多特(法国时装设计师)
jerga　墨西哥粗纺格子呢
jerkin　女用背心;(旧时)男紧身无袖短上衣;短上衣,合身毛背心,杰金背心,紧身皮袄,贴身男背心
jerkin suit　紧身背心套装,无袖上衣背心,杰金女套装,无袖上衣套装
jesselmere wool　印度纬纺用低档羊毛
jersey　乔赛;平针织物;女用紧身内衣(卫生衣);紧身运动套衫;运动衫;紧身毛衣
jersey cloth　全毛或毛交织斜纹织物;有伸缩性的绒毛布;匹头针织布;仿针织物;特里科经编针织物;匹头针织布(统称)
jersey cloth gloves　绒布手套,针织绒布手套
jersey costume　紧身长针织衫
jersey crepe　仿平针绉绸
jersey dress　针织紧身外衣,针织服装
jersey flannel　针织法兰绒
jersey foundation　女子针织紧身内衣
jersey-jacquard fabric　乔赛提花针织物,小提花针织物
jersey knit　单面针织物
jersey knitgoods　乔赛针织物
jersey machine　乔赛机,单面圆机,单面针织机

jersey piece goods　匹头针织布(统称)
jersey shirt　针织衫
jersey singlet　汗布文化衫
jersey sportswear　夏奈尔式运动装
jersey stitch　平针
jersey sweater　紧身毛衫
jersey tweed　素色粗呢(手感柔软)
jersey velour　平针丝绒
jersey wear　针织坯布,针织物;弹力针织物
jersey wool　优级羊毛;优级细毛纱
jester costume　滑稽丑角服,小丑服,小丑装,丑角服饰
jester dress　小女型
jester's cap　弄臣帽;小丑帽
jester's costume　丑角服饰
jester suit　中世纪弄臣套装
jesuit cloth　宗教服装用黑色粗厚强捻平纹毛织物
jesuit lace　爱尔兰钩编花边
jet　煤玉黑,煤玉色,黑玉色
jet-black　乌亮,乌黑,深黑
jet-black hair　乌黑头发
jet-black suit　乌黑套装
jet clip　固扣夹,包装用胶夹
jet dyeing　喷射染色
jetee　拉吉马哈尔大麻
jet ink printing　喷墨印花
jet printing　喷墨印花,喷射印花
jet spray printing　喷墨印花
jet spun yarn　喷气纺纱
jetted pocket　双滚边袋,双嵌线袋
jetted with flap pocket　有袋盖的双嵌线袋
jetted with re-inforcement pocket　加圆双嵌线袋
jetted with zip pocket　拉链式袋口的双嵌线袋
jet-tex yarn　喷气变形丝
jetting　嵌线
jetty　乌黑发亮,黑玉色
jet with tab pocket　有带袢的双嵌线袋
jeu look　游戏款式
jewel　宝石
jewel artist　珠宝艺术家
jewel case　首饰盒
jeweled buckles　装饰扣环
jeweled button　装饰纽,装饰纽扣
jeweled collar　宝石领
jeweled diadem headband　珠冠式束发带

jeweled metal girdle 镶宝石金属腰带
jeweled pantyhose 莱茵石透明连裤袜
jeweler 珠宝匠;珠宝商
jewelled backs 镶珠背片
jewelled buckles 镶宝石;装饰扣环
jewelled button 装饰纽扣
jewelled sandals 镶宝石凉鞋
jeweller's velvet 装饰锦盒用短丝绒
Jewel(le)ry 珠宝,宝石
Jewel(le)ry neckline 宝石领口,小圆领口
jewel neck(line) 珠宝领口,小圆领口,珠宝高圆领
jewel ring 钻石戒指;宝石戒指
jewelry 首饰;珠宝;宝石蓝
jewelry blue 宝蓝色
jewel-tone 珠宝色调
jew's mallow 长果种黄麻
jeypore print 手工模板印花方巾
jeypore rug 印度贾波地毯
jhapan 印度丝绣花卉纹细薄布
jhapara 印度贾帕拉丝绣薄纱
jharan 印度手织粗梳棉布
jhibamdik 印度粗棉纱罗
jhilmeel 印度薄丝纱
jhouta 黄麻
jhuganat 印度漂白有光粗棉布
jhuna 印度稀松细布
jiafeng fine plain 嘉丰细布
jianchun satin striped voile 建春绡
jianchun silk gauze 建春绡
jian-dao silk 间道
jibba(h) 长布袍(穆斯林男子用)
jiffy dress 速缝自制衣服;粗针速编毛衣商标
jiffy-knit sweater 粗针膨体毛衣,速编毛衣
jiffy threading 瞬间穿线
jigger 卷染机
jigger [coat] 宽松女短上衣
jigger button 双排扣服装翻领下的隐藏扣
jiju brocaded damask 集聚绸
jilda 聚丙烯腈长丝(商名,日本)
Ji-li shoes 吉利袜
Jilis silk 辑里丝
Jimmy look 吉米款式(短外套加裤的组合);吉米风貌
jimper skirt 背心裙
jimping 锯齿切裁;英国衣边剪花
jinbo damask 金波缎

jindiao brocatelle 金雕缎
jinduan satin striped lame 金缎绉
jinduan ticking 锦缎枕面
jinfeng satin brocade 锦凤缎
jing embroidery fabric 京绣品
jinge mock crepe 金格呢
Jing ethnic costume 京族民族服
jinghua jacquard mock crepe 晶花呢
jinghua jacquard suiting silk 晶花呢
jinghua kimono crepe 精华绉
jingi brocade 锦益缎
jinglun sheer 晶纶绡
Jingpo ethnic costume 景颇族民族服,景颇族服饰
Jingpozu costume 景颇族民俗服
jingyin crepon brocade 晶银绉
jingzi tussah oxford 井字格
Jingzu costume 京族民俗服
jinhe moss crepe 锦合绉
jinhui lame 金辉绸
Jinichi Abe 阿部寻一(日本时装设计师)
jinle brocade 锦乐缎
jinlei brocade 金蕾锦
jinlian 金莲鞋
jinling brocade 锦玲缎
jinling broche 金陵锦
jinnah cap 真纳帽
Jino ethnic costume 基诺族民族服,基诺族服饰
jinxing faille 金星葛
jinxin upholstery silk 锦新装饰绸
jinxing matelasse [faille] 金星葛
jinxiu brocaded velvet 锦绣绒
jinxiu twill damask 锦绣绸
jinxuan brocatelle 金炫绸
jin-yi brocade 锦益缎
jinyi jacquard sand crepe 锦艺绸
jinying velvet 锦莹绒
jinyu brocade 锦裕缎
jinyu leno brocade 锦玉纱
jinyun cloque 锦云绸
jinyu satin brocade 金玉缎
jinzhuang velvet ecrase 锦装绒
jinzi tussah oxford 井字格柞绢绸
jipijapa 巴拿马草帽
Jipins rug 杰平斯帷幕呢
jircaza 色花提花布(印度);平纹花卉棉布
jit dyeing 卷染
jitney bag (美国盛硬币的)小钱包;小手

提包
jitterbug 宽松牛仔裤,(美国)摇滚爵士牛仔裤
jiu-xia crepe satin brocade 九霞缎
jixiang design 吉祥图案
jixuan brocatelle 金炫绸
ji-yi crepe damask 集益绸
ji-yun burnt-out sheer 集云绡
Joan 女式系带室内帽
jobber 外发加工者;织物销售商
jobbing-on 套口
jobbler 中间商
job cap 职业帽
job dispatcher 作业调度程序
job-dyeing 衣件零料染色
job hat 职业帽
job library 作业库
job lot 零码商店
job process card 工作卡片
job production 单件生产
job queue 作业排队
job responsibility 岗位责任制
job responsibility system 岗位责任制
jobs 零头布
job's tears 印度念珠簇
job simplification 作业单纯化
job step 加工步骤;作业段
job study 工作研究;作业研究
job suit 职业服,职业装
job training time 作业培训时间
job wage system 岗位工资制
jocelyn mantle 女式无袖及膝双层斗篷
jock 男式运动紧身短裤
Jockey 乔其裤,男式三角裤;赛马骑师装
jockey blouse 骑士罩衫
jockey boots 骑士靴;缘饰童靴;赛马靴
jockey cap 长鸭嘴帽;骑手帽;赛马帽
jockey cloth 英国骑师毛葛
jockey coat 骑师外套
jockey pants 骑师裤
jockey satin 鞋面花缎
jockey shirt 骑装女衬衫
jockey shorts 运动短裤,骑士短裤
jockey silks 骑师服;骑士绸衣
jockey stripe 骑士条纹
jockey waistcoat 骑士直筒背心
jockey's stripe 骑士条纹
jockstrap (男运动员用弹性织物制的)下体护身
Jodd's USC system 乔特USC测色系统

jodhpur 短马靴(复数);马裤
jodhpur boots 短马靴
jodhpur breeches 马裤
jodhpur pants 马裤
jodhpurs 小脚管马裤;短马靴;骑马裤
jodhpur shoes 短马靴
jodhpurs-style jeans 马裤式牛仔裤
jods 马裤
jog seam 变宽缝
jogangara 镶嵌花样
jogger's jacket 慢跑夹克
jogging 慢跑服
jogging [running] suit 跑步服
jogging outfits 慢跑服,跑步套装
jogging pants 慢跑长[运动]裤
jogging set 运动套装
jogging shoes 跑步鞋,漫步鞋,慢跑[运动]鞋
jogging shorts 竞走短裤;夏天短跑裤
jogging style 慢跑服款式
jogging suit 慢跑服;运动套装;跑步装
John Anthony 约翰·安东尼(1938~,美国时装设计师)
John Bates 约翰·贝茨(英国时装设计师)
John Cavanagh 约翰·卡瓦诺(1914~,英国时装设计师)
John Galliano 约翰·加里亚诺
John Isle cotton 约翰·海岛棉
johnney collar 竖领,小型女衬衫领
johnnie 短袖开背上衣(住院病人用)
johnny 病员罩衫(无领短袖后开);短袖开背上衣(住院病人用)
Johnny collar 约翰尼领(女衬衫小方角领)
johnny johnnie (病号)短袖无领罩衫
John Tsiattalou 约翰·恰塔路(美国时装设计师)
John Weitz 约翰·韦茨(1923~,美国时装设计师)
Joid 充填用中空聚酯纤维(商名,日本东泽纺)
join 缝合;连接;接合线
joined width dirndl skirt 多幅抽褶裙
joining 缝合,连接,(皮革,毛皮)拼接
joining band 缝合带子
joining bias strips 接斜条
joining blouse and skirt at waist 缝合上身和裙子腰部
joining center back seam 合背缝

joining collar interlining 拼领衬
joining crotch (seam) 合前后裆缝
joining cuff and sleeve 绱袖头
joining elastic 缝合松紧带,缝合橡筋带
joining facing pieces 缝合两片贴边
joining front edges 合止口
joining hood seam 合帽缝
joining inside seam 合裙缝;合下裆缝
joining joke to shirt 缝衬衫覆肩
joining-join up （把相配的皮)连在一起
joining knitting 针织缝合
joining mark 搭头痕
joining or securing seam 搭头缝或保险缝
joining overlap strips 叠接带子缝合
joining panels 缝合衣片
joining parts of unequal contour 不等轮廓材料的缝合
joining right and left yoke 缝合左右育克
joining seam 缝合缝;接合缝
joining seamline 接缝线
joining shoulder 缝肩
joining shoulder seam 合肩缝
joining side seam 合侧缝,合摆缝;合裙缝
joining sleeve bottom 合袖头
joining sleeve seam 合袖缝
joining top collar 合领面
joining under collar and top collar 合领子
joining waistband and lining 合腰头
joining waistband to skirt 接裙腰头
joining with overlapping strip 叠接带子缝合
joining yarn 缝合线
joining yoke 绱过肩
join in trousers 裤接缝
join marks 花纹拼缝
join piece 联匹
join shoulder 缝肩
join skirt 连接裙
joint 断匹疵;接点;接缝;接头
Joint Advisory Council for the Carpet Industry 地毯工业联合咨询委员会
joint cloth gunny bag 接腰麻袋
jointed hemmer 组合卷边器
joint filler 填缝科
joint inspection 共同检验
joint mark 花版接头
joint meeting 技术交流会;技术座谈会
joint operation 联营
joint panel 相叠衣片
joint production 合作生产
joint products 联合产品
joint skint 连接裙,接缝裙
joint venture 合资经营;合资企业
joint venture enterprise 合资企业
joint wear 拼接服装
join up 毛皮拼接
joinville 男式宽领巾
joker 腰牌;挂卡
Jole Veneziani 若·韦内齐亚尼(意大利时装设计师)
Jones Herlong cotton 琼斯赫朗棉
Jones improved cotton 琼斯改良棉
Jones long staple cotton 琼斯长绒棉
Jones number 1 cotton 琼斯1号棉
jonny collar 小竖领;紧立领
Joria wool 印度乔里阿绵羊毛
jornade 宽袖骑马短外套
joseph 18世纪女用骑马外套
Joseph's coat 包袱式样布;多色外衣
Josephine costume 18世纪末、19世纪初法国约瑟芬服式,皇后服式
josephine knot 装饰结
josephine trcot 手工网眼钩编织物
Josette 乔瑟特斜纹棉布
joshaghan rug 伊朗农村织毯
jour 花边网眼刺绣
jour deux place 双列沙眼纱罗
jour trois place 三列纱眼纱罗
journade 宽袖骑马短外套
Journal of the Textile Institute ［英]《纺织学会会志》(期刊)
Journal of the Textile Machinery Society of Japan 《日本纺织机械学会志》
journey man 技工
journey suit 旅行套装
Jouy Print 法国式小花卉印花绸布
Jove Poplin 乔弗府绸
Jowers cotton 美国乔沃斯棉
jowl 颚骨;垂肉;二下巴(下颌垂肉);下颚
J-shoe 丁字形鞋
juan-gong doupioni taffeta 绢宫绸
juangong douppioni pongee 绢宫绸
juban 棉汗衫;绸汗衫
jubbah 久巴长衣;中东长袖及踝宽松外衣;长布袍(穆斯林男子用)
jubbulpore carpet 印度朱巴耳珀地毯
jubbulpore hemp 优质印度大麻
jubbulpur carpet 印度朱巴耳珀地毯
judge 裁定

judgement 鉴定,评价
judge's robe 法衣,法官服
judo belt 柔道腰带,宽腰带
judo clothes 柔道服
judo coat 柔道服
judo-gi 柔道服
judo jacket 柔道外套
judo suit 柔道装
judo uniform 柔道装
judo vest 柔道家上衣
judo wear 柔道服
jufti 地毯结子
Ju-fu 具服(中国古代的官服)
jugan nathi 印度粗棉布
Jugoslavian embroidery 南斯拉夫刺绣
Jugoslavian look 南斯拉夫风貌
jugular hollow 颈窝
juicy 富有色彩的,绚丽的
juive tunic 大V领及臀吸腰式外形
jujnabe 中亚手结地毯
jujube red 枣红
Juki 重机(日本缝纫机公司)
Jules Francois Crahay 朱尔·弗朗索瓦·克拉耶(法国时装设计师)
Juliet cap 朱丽叶帽(华丽的珠宝无边帽)
Juliet dress 朱丽叶裙
Juliet gown 朱丽叶长袍(尼龙长袍,高腰,可有花边或起皱下摆,小泡袍袖)
Juliets 朱丽叶鞋(戏剧)
Juliet sleeve 朱丽叶袖(顶端泡袍状,以下贴体的短袖型)
Juliet slippers 朱丽叶便鞋
jumbo cotton 美国琼博棉
jumbo needle 粗针
jumbo package 特大卷装
jumbo sewing machine 重型缝纫机
jumbo zipper 粗牙拉链
jump 外套,贾普外套;贾普胸衣;长及大腿士兵外套
jump boots 跳伞靴
jump coat 贾普外套(长及大腿的便装)
jumper 无袖连衣裙;水手短上衣;宽松工作夹克;针织女套衫;连帽皮外衣;连衫裤童装
jumper-blouse 水手装
jumper coat 男式套装上衣;夹克;背心式外套
jumper dress 背心装;背心裙装
jumper frock 罩衫
jumper pajamas 连身睡衣裤

jumpers 童嬉服,连衫裤童装;背心装
jumper shift 衣领衣身呈对比色罩衫
jumper skirt 马甲裙;无袖连衣裙;背心裙;女学生裙
jumper suit 女式罩衫;女短外衣套装;连衫裤工作装;女紧身连衫裤;跳伞装
jumper underpress 下热式熨烫机
jumpiness 跳跃式
jumping boots 长统皮靴
jumps 紧身胸衣无袖套领罩衫
jump shorts 连衫短裤装
jump stitch 间隔缝线迹,跳缝线迹
jump suit 连衫裤工作服;跳伞服;女紧身连衫裤便服;连体服
jumpsuit pajamas 连衫睡衣裤(前中线开襟)
Jun Saito 斋藤纯(日本时装设计师)
juncture 接缝,接合[点]
jungle 丛林印花纹样;丛林战斗靴
jungle boots 丛林[战斗]靴
jungle cloth 防风棉布
jungle fashion 森林探险家款式
jungle fighter boots 棕色防水皮靴
jungle green 丛林绿(深墨绿色);深草绿;墨绿色衣服
jungle look 丛林风貌
jungle print 丛林印花
jungle stylistic form 丛林款式
juni-hitoye 日本皇后加冕服饰
junior 少女型;美国瘦小的女服尺寸
junior bag 小钱包(盛硬币)
junior bodice 少女上衣
junior figure 少年体型
junior measurement chart 少女服装尺寸表
junior miss 少女;少女服尺寸;苗条妇女衣服尺寸
junior miss skirt 少女尺寸的短裙
junior petite size 少女小号尺寸
juniors' 少女式(毛衣)
junior size 少年尺寸(约15~18岁);瘦小的女服尺寸
junior style 少年款式
junior suit 少年装
junk jewel(le)ry 廉价首饰品,假珠宝
Junko Koshino 小筱顺子(日本时装设计师)
Junko Shimada 岛田顺子(日本时装设计师)
junoesque (女子体型)高大丰满的;(女

子)仪态大方的;华贵美丽的
junoesque figure 高大丰满体型(女子)
junoesque manner 仪态大方(女子)
jupe 法国女裙;苏格兰女上衣;女紧身马甲
jupe-culotte 裙裤
jupe fronce 皱褶裙
jupel 14世纪欧洲铠甲罩衫
jupe pantaloons 裙式马裤
jupe petale 花瓣裙
jupe plisse 襞裙
jupe plissée 细襞裙
jupon 14世纪欧洲铠甲罩衫;衬裙;女骑马服;法国毛棉平布;裤子;西裤;紧身袄(穿在铠甲内)
jupon style 连衣衬裙风格
jushuang batiste faconne 聚爽绸
jusi 久西条格细布
jussara 杰希卡
just and sisal handbag 麻质手提包
just appreciable fading 始感褪色
justaucorp 男上装;紧身上衣
justaucorps 裘斯特可外衣(17~18世纪),紧身衣
juste-au-corps 裘斯特可外衣(17~18世纪),紧身衣
justicoat 紧身衣
just in time (JIT) 准时生产
just in time (JIT) system 及时生产供应系统

Just in time production 即时生产
just knee 及膝长
just mark 勉强级
just size 正确尺寸
just waist mark 恰当腰线
jute 黄麻,络麻,草麻,火麻,夹头麻,绿麻,台湾麻,幼麻
jute and sisal handbag 麻质手提袋
jute backing 黄麻底背
jute bag 麻袋
jute bagging 黄麻包装布,黄麻袋布
jute canvas 粗黄麻袋布,黄麻帆布
jute carpet 黄麻地毯
jute fabric 黄麻织物
jute rug 黄麻地毯
jute rug backing 黄麻地毯底布
jute sacking 黄麻袋
jute scrim 平纹网眼黄麻布
Jute Technological Research Laboratories 黄麻工艺研究所
jute thread 黄麻线
jute yarn 黄麻纱
jutting cap 耸袖山
jutting collar 细嵌线领
jutting derriere 突出臀部的裙子
juvenile clothing 儿童衣着,儿童衣料
juxtapose 并列循环图案
juxtaposed repeat pattern 并列循环图案
jzsper flannelette 芝麻绒

K

kaaba karaman rug 卡巴卡腊曼地毯
kaai finish 卡艾整理(棉毛交织物无张力丝光整理)
kaasha cloth 卡沙法兰绒
kabaya 印度尼西亚女上衣,卡巴雅长袖女上衣
kabaya jacket 卡巴雅外套
kabe crepe 碧绉,单绉,素碧绉,印度绸;日本金丝双绉
Kabel stitch [德]绞花链状绣,钢绳链状绣
kabe-twist yarn 璧绉绒
kabistan rug 卡比斯顿地毯
kabuki dress 卡博基服(日本卡博基剧院演员的围裹式服装)
kabuki robe 和服式短袍
kabuki sleeve 和服袖
kabuki song 歌舞伎鞋
kabul carpet 喀布尔地毯
kabul pashm wool 喀布尔羊毛
kabul rug 喀布尔地毯
kabyle 卡比尔毛披巾
kachoji 日本蚊帐纱,网眼纱
kaddar shawl 卡达披肩
kaddhar 克什米尔凯德哈尔平绒织物
kadiva 丝绒
kadu 树叶制睡垫
kadungas 美国卡邓加斯印花细平布
kaffir sheet 卡菲尔粗斜纹布
kaffiyeh 凯菲手织印染布;头巾(沙漠地带阿拉伯人戴)
kaffiyeh-agal 阿拉伯方形头巾
kaftan 土耳其长袍;阿拉伯男上衣;有腰带的长袖袍
kaga 日本中等丝绸;日本加贺绸
kagayuzen 丝绸花卉图案
kahing silk 嘉兴生丝
kahnami 卡纳米棉
kahnami cotton 卡纳米棉
kaifeng taffeta faconne 开封汧绸
kailies 印度男式披肩布
kaimakani 土耳其薄纱布
kain kapala 凯恩卡帕拉莎笼
kain pandjang 凯恩潘詹莎笼(印尼)

kains 美国凯恩斯色纱格子布
kairens 优质土耳其毯
kairens rug 优质土耳其毯
kairuan 北非羊毛地毯
kaiui 日本甲斐绸
kakeda 日本细生丝
kakaya jute 孟加拉黄麻
kaki cotton 棕乳色埃及棉
kako obi 日本男式正装和服腰带
kala 卡拉帽;伊朗绵羊皮男帽
kalabatun 细金属包芯线;贴金线棉布
kalamal 土耳其斯坦白地条子布
kalameit 易着色黄麻
kalamkari 印度手织手绘棉布
kala patadar 印度手织条纹绸
kalasiris (古埃及)紧身长袍,卡拉西利丝裙
kaleidoscope 万花筒图案
kaleidoscope pattern 万花筒花纹
kalemkar 印花棉布
kalgan lamb 库尔勒羔羊毛
kalgan wool 张家口羊毛(地毯羊毛)
kalgpira 古希腊妇女罩脸面纱
kalhapur kolhapure 印度夹带式皮凉鞋
kali 伊朗卡里地毯
kalicha rug 印度卡林地毯
kalimtarakshi (印度)小花丝花缎,印度小花纹锦缎
kalin rug 印度卡林地毯
kalings silk 中国缝线用生丝
kall 网帽
kalmal 白地条子布(土耳其斯坦)
kalmkar 克什米尔卡尔姆卡尔平绒织物
kalmuc 卡尔穆克粗呢地毯
kalmuck 卡尔穆克粗大衣呢;卡尔梅克粗毛大衣;双面异色厚绒布;伊朗低级平布
kalmuck priest's collar 卡尔穆克牧师领
kalmuck rug 卡尔穆克粗呢地毯
kalpak 黑毡帽
kalpatadar 丝棉交织条子缎(印度)
kalso 丹麦厚底凉鞋;丹麦镂空凉鞋
kalyptra [希]古希蜡面纱
kamaeri (西装或外套的)平领
kamarchin 卡马钦衫(波斯)

kambayams 印度格子棉腰布
kamdani 细棉布
kamelaukion [希]卡美劳金帽
kamerijk 荷兰细薄布
kamik 卡米克靴(爱斯基摩人穿),开密克靴
kamiks 爱斯基摩高筒皮靴
kamis 伊斯兰教男长宽衬衫;卡米衫(阿拉伯国家)
kamishimo 日式宽肩武士服
kamizool 女式短袖衬衣
kampskatcha slipper 尖头高面低跟女鞋
kamptulicon 棉麻涂层地毯
kanakin 日本细棉布
kanazawa silk 色织平纹绸
kandahar carpet 坎达哈地毯,印度色块地毯
kandinsky style 康定斯基风格图案
kandys 紧身袖系带服装;坎弟斯(古代波斯)
Kanebo 嘉娜宝
kanee roomal 开士米方披布
kangara 印花棉方巾(沙特阿拉伯)
kangaroo 袋鼠型毛皮;袋鼠革
kangaroo dress 袋鼠式裙装
kangaroo [leather] 袋鼠革
kangaroo look 袋鼠款式
kangaroo pocket 袋鼠袋(前身大贴袋);袋鼠式口袋;特大口袋
kangaroo pocket skirt 袋鼠式口袋裙
Kangaroo's hair 袋鼠毛皮
kangaroo shirt 袋鼠衬衫(有大贴袋);前中部带圆孔的孕妇裙
kangaroo style 袋鼠款式
kangfu brocade 康福缎
kangle jacquard mock crepe 康乐呢
kaniki 卡尼基平纹布
kaniki marduf 卡尼基马杜夫棉斜纹布
kankura 苎麻(孟加拉湾称谓)
kannikar 肯尼喀羊毛披肩;仿肯尼喀山羊绒花披肩,肯尼喀山羊绒花披肩
kano cloth 美国卡诺坯布
kanoko 日本鹿斑绸;头饰薄纺
kanova 聚酰胺原液着色长丝(商名,日本钟纺)
Kansai Yamamoto 山本宽斋(日本时装设计师)
kanvi cotton 印度堪维低级棉
KANZ 凯茨
kanzu 康祖长袍(非洲)

kaoala 澳洲树袋熊毛皮
Kaoliang red 高粱红
kapa 树皮布;南斯拉夫黑色盒形帽;夏威夷手织布
kapar 稀格子布;孟加拉湾棉织物
kapata 印度手织女裤用织物
kapok 木棉
kapok life jacket 木棉救生衣
kapok pillow 木棉枕头
KAPPA 背靠背
kappel 波兰无边便鞋
kappn 葛布
kapron 卡普纶
kapta 卡普泰上衣;宽大套头衫
kapuzen-kleio 带头巾女装
karabagh rug 卡腊巴羊毛地毯
karaca sweater 卡拉卡毛衫(土耳其);镶土耳其饰边的双翻领套头毛衫
karachi cotton 卡拉奇棉
karadagh rug 卡腊达地毯
karaja rug 卡腊达地毯
karakul cloth 卡拉库尔呢;羊毛厚绒大衣呢
karakul fabric 卡拉库尔呢;羊毛厚绒大衣呢
karakul lamb 中亚羊毛皮
karakul 卡拉库尔羔皮
kara man 土耳其手织卡拉曼地毯
karamanian kilim rug 土库曼斯坦卡巴卡腊曼地毯
karamushi fiber 日本苎麻纤维
karankas 金银丝花丝缎(印度)
karat 克拉
karate costume 空手道服装
karate gi 空手道服
karate jacket 空手道短夹克
karate pajamas 空手道服式睡衣(两件套)
karate suit 空手道装
karcher 方围巾,方头巾
Kare(e)ba 短袖衬衫(牙买加)
kareya 卡里亚粗平布
karimganji 印度浅色黄麻
Karl Lagerfeld 卡尔·拉格菲尔德(1939~,法国时装设计师)
karma 披肩;围巾;格子棉布
karmanian Khilim 卡曼尼基林地毯
karnak cotton 埃及卡纳克棉
karrildock 荷兰卡里尔多克粗亚麻篷帆布

karsey 混色斜纹呢
kart 幼驼毛布
kasabeh 细网眼羊绒织物
kasan 卡桑粗呢;毯用绒毛
kasawari 印花平布(印度)
kasaya 法衣,袈裟
kas cotton 苏丹卡斯棉
kasha 卡沙棉法兰绒;卡沙细哔叽;卡沙细呢
kasha cloth 卡沙法兰绒
kashan rug 卡香波斯地毯
kasheda 手织刺绣品
kashgar 喀什羊毛;喀什羊毛呢;喀什地毯
kashgar cloth 喀什驼绒呢;喀什长绒呢
kashgar rug 喀什地毯
kashgar weave effect 喀什长绒呢花纹
kashgar wool 喀什细白羊毛
kashmir 开士米,山羊绒,紫羊绒,羊绒,开士米织物,羊绒织物;精纺毛纱针织物;松软斜纹棉织物
kashmir coat wool 低质印度羊毛
kashmir rug 克什米尔地毯
kashmir shawl 开士米披巾,克什米尔开士米披巾
kashmir worsteds 克什米尔手工精纺呢
kasida 手织刺绣品
kaskkai 克尔曼地毯
Kasper 卡斯珀(美国时装设计师)
kasuri 扎染纱手工织物
kasvin rug 波斯仿古图案地毯
kata aya 日本卡塔阿亚斜纹衬衫布
katari cloth 手织丝棉交织色织物
Kate Greenaway bonnet 格林纳维罩帽,凯特·格林纳维软帽
Kate Greenaway costume 格林纳威童装,凯特·格林纳维装束
Kate Greenaway inspired dress 格林纳维装
Kate Greenaway look 格林纳维风貌
katey 手织柞蚕丝织物
katha silk 印度绣花丝绒
kathee 篷帐粗布
Katherine Hamnet 凯瑟琳·哈姆涅特
katifet 天鹅绒
kat-no 朝鲜伞形防雨帽
Katsushige Muraoka 村冈胜重(日本时装设计师)
kattun 德国印花棉布,德国棉平布
kauagoeorimono 川越织物

kaunakes clothes 康纳克斯服
kaunakes 古代美索百达米亚柯纳克裙,柯纳克裙(古代西亚)
kaveze 土耳其无边锥形高帽
Kawabata Evaluation System 川端织物手感的评价体系
Kawabata Evaluation System-F handle tester (KES-F handle tester) 川端织物风格仪
Kawabata's fabric handle instrument 川端织物手感风格仪
kawachimomen 河内木棉
kawamata 日本低档毛织品
kawo fiber 木棉纤维
kawzaw 印度绒头地毯
kawzaw rug 印度绒头地毯
kaya 日本蚊帐布
kayadi wool 伊拉克地毯用毛
kayhamu 克什米尔凯哈穆低级呢
kayseri rug 土耳其凯西里地毯
Kazak ethnic costume 哈萨克民族服饰
kazak rug 哈萨克长毛绒地毯
kazue Ito 伊藤和枝(日本服装设计师)
Kazuko Hayashi 林和子(日本服装设计师)
kebaja 开巴亚上衣(印巴)
kebaya 开巴亚上衣(马来)
keche carpet 土耳其山羊毛地毯
keche rug 土耳其山羊毛地毯
kechi carpet 土耳其山羊毛地毯
kechi rug 土耳其山羊毛地毯
kedis 土耳其重型棉布
keds 开兹鞋
keel 红印色
keel [mark] 码印,匹印
keep sample 存档样;留底样
keeper 搭环;拎包,小提包
kef(f)iyeh 凯菲手织全棉或丝经棉纬印染布;凯菲头巾;阿拉伯头巾
kefieh 凯菲头巾;凯菲手织全棉或丝经棉纬印染布(阿拉伯);阿拉伯头巾
Keiko Suzuki 铃木庆子(日本时装设计师)
keister 拎包,小提包
keith cotton 美国早熟棉
keityomoyo 庆长花样(日本),日本庆长时代和服花样
kekchi cotton 危地马拉凯奇棉
kelat wool 伊朗·巴基斯坦俾路支地毯毛

kelim 双面无绒头地毯
kelim embroidery 彩色几何图案刺绣,凯利姆刺绣
kelim rug 双面绒头地毯
keli satin brocade 克利缎
kelly 凯利绿,鲜黄绿色
kelly bag 凯莉手袋,马鞍型手袋;凯利提包;名演员梯形提包
kelly cotton 美国凯利晚熟棉
kelly green 果绿,凯利绿(深黄绿色),鲜黄绿色
kelly satin brocade 凯利缎
kelly satin plain 素金玉缎
kelp 海藻褐
kelpie process 克尔派毛织物整理
kelt 苏格兰克尔特呢
keluodine 缎纹卡其,克罗丁
kemban 缠身布(远东地区)
kemea 印度花塔夫绸
kemes 英国女式内衣
kemise 英国女式内衣
kempy tweed 抢毛粗呢
kempy yarn 粗抢毛纱
kemse 英国女式内衣
Ken Scott 肯·斯科特(意大利时装设计师)
kenaf 熟红麻,槿麻,印度络麻,野麻,洋麻,南方型洋麻
kenari 印度金银线花边
kendal green 英国肯德尔绿色粗呢
kender 罗布麻,红野麻
kender fabric 罗布麻织物
kender knit fabric 罗布麻针织物
kender textile 罗布麻纺织品
keneri 剑领(西装礼服领)
kennel hood 16世纪妇女头罩
kennet 威尔斯肯尼茨粗呢
Kenneth Jay Lane 肯尼斯·杰伊·莱恩(美国珠宝设计师)
Kensho Abe 安部由章(日本时装设计师)
kensingion quilt 肯辛顿被单
Kensington Fashion Fair 肯辛顿时装博览会(英国)
kensington stitch 肯辛顿针迹
kente 肯特服(美国)
kent collar 上下盘领
kente cloth 加纳花布;加纳丝绸
kenting 爱尔兰肯廷亚麻平纹里子布
kentucky jeans 棉毛牛仔布;肯塔基棉毛呢

kentucky rough prime hemp 肯塔基优质大麻
kentucky single dressed hemp 肯塔基大麻
kenyan tobe 肯尼亚印花裹裙
kenzo sleeve 贤三袖(日本时装设计师高田贤三设计)
Kenzo Takada 高田贤三(日本服装设计师)
kepi 法式士兵帽,法国军帽
kep stitch 绣挂毯用十字针迹
kept 储存
keratin gene-transferred cotton 角蛋白转基因棉
kerbas 伊朗粗棉布
kercher(e) 长方形头巾,方形头巾
kerchief 方头巾,三角头巾,围巾
kerchief neckline 方围巾领凹
kerf 划痕
keri-ne-chardani 手织花卉彩缎
keri velo 手织花卉彩缎
kerkezi ethnic costume 柯尔克孜族民族服
kerman 波斯手织克尔曼地毯
kerman rug 波斯手织克尔曼地毯
kerman shah rug 波斯手织克尔曼地毯
kermel fiber 聚酰亚胺结构耐高温合成纤维
kermer 克尔默纯丝披巾(埃及)
kermes 虫红,虫胭脂
kermes[scarlet] 胭脂红
kermiss 英国低级棉衣料
kerry cloak 爱尔兰黑外衣
kersey 克尔赛密绒厚呢;克尔赛呢衣服;克尔赛手织粗呢;棉毛绒(粗绒布)
kerseymere 克尔赛梅尔短绒大衣呢
kerseymere twill 克尔赛梅尔斜纹
kerseynet(t)e 棉毛交织呢;双层棉毛交织呢;阿尔帕卡呢;羊驼毛花呢
kerseys 克尔赛呢裤
kesis 条子布;色布(色泽鲜艳,印度、巴基斯坦)
kestos bra 凯斯托斯胸罩
keswa el kbira 希伯来长裙装束
kethoneth 希伯来及腓束腰外衣
kett satin 欧洲大陆经面缎
ketteki-hō 日式长袍
kevenhuller hat 男式三角帽
kevlar ballistics vest 凯夫拉防弹背心
kevlar composite 凯夫拉纤维复合材料
key 关键码;键;主调

keyback 用可弃手巾	khaki light 卡其灰
keyboard 电键码;键盘	khaki look 卡其色款式;陆军款式;卡其风貌
keyboard entry 键盘输入	
keyboard neckline 键盘形领口	khaki pants 卡其裤
key border 键匙式拷边,地毯键盘式饰边	khakis 卡其服装;卡其裤
key button hole 圆头纽孔,锁眼纽孔	khaki shorts 卡其布短裤
key case 钥匙袋(小贴袋)	khaki style 陆军款式
key chain 钥匙袋	khalak 锥形头饰上的罩纱
key color 基本色	khalamkar 印度手工画卡兰卡花布
key currency 关键货币	khalat 土耳其女用深色长裹巾
keyhole 钥匙孔领口	khali 针绣毡布
keyhole bra 钥匙孔胸罩	khalin pashmina 克什米尔手织卡林羊绒织物
keyhole button hole 圆头纽孔,锁眼纽孔	
keyhole collar 钥匙孔式领	khalkhal 波斯脚镯
keyhole neck 凹口圆领,钥匙孔式领圈	kham 土耳其低档平纹棉布
keyhole neckline 锁孔圆领口,钥匙孔圆领口	khan 印度手织女裤织物
	khandala 肯达拉平纹织物
keyhole opening 钥匙孔开口	khandeis cotton (印度)肯地斯棉
key-in 键盘输入	Khandesh Roseum (印度)肯地斯混种棉
key item 关键商品	khanga 肯加布(东非);英国服用色棉布
keymo finish 凯莫整理(用硫酸溶液处理的毛织物防缩整理,商名,英国)	khange 冈吉衣
	khansu 肯聚长袍(东非)
key number 索引号	khapedi ma phul 印度手织色织花缎
key operation 关键作业	khari 印度粗棉布
key pattern 万字花纹,卍字饰,卍形花样;主样板	khariasri 手织条纹柞蚕丝绸
	kharwa 红色平纹粗棉布
key pattern piece 主样板衣片	khas 达卡细棉布
key pocket 钥匙(小贴)袋	khasa 印度细棉布
key products 基础[关键]产品	khasida 手织刺绣品
key resource 主要货源	khat 埃及亚麻头巾
key ring 钥匙圈	khatchli bokhara 布哈拉地毯
keyword 关键字	khathdar 克什米尔优质开士米披巾
kezutaka katoh 加藤和孝(日本时装设计师)	khatherast 克什米尔中级开士米呢
	kheetee 东南亚轧光布;印度轧光印花棉布
khab 印度金银丝锦缎;印度金考布锦	
khabbikutah 波斯短绒头栽绒地毯	kheikeli [俄]驱魔包
khada 印度卡迪手织粗平布	khelim rug 基里姆地毯;花毯
khaddar 印度卡迪手织粗平布	khemir 埃及丝披巾
khadie 手织带毛边粗平布	kherche [俄]串珠刺绣
khahua 印度匹染粗厚布	khersek rug 波斯长毛绒厚地毯
khaiki 日本甲斐绸	khes 印度粗棉斜纹布
khaki 黄褐色;卡其黄;卡其色,土黄色;暗棕黄色织物;卡其布;卡其布服装(尤指军装)	khesi 印度宽幅粗斜纹布
	khikois scarves 英国希科伊斯色织领巾
	khilim rug 双面无线头地毯
khaki cloth 卡其布	khio wool 伊朗高级地毯毛
khaki cotton 浅紫红棉花	khirs 托钵僧衣
khaki drill 卡其	khiva rug 精细羊毛小地毯
khaki drills 卡其布	kho 和服式及膝束腰长衫
khaki gray 卡其灰色	khodar 印度粗棉布
khaki grey 卡其灰色	khodia 印度手织粗棉布

khoi wool　伊朗高级地毯毛
khokti　科克提布(印度)
khoktibanga　印度科克提本加棉
khombal　印度粗毛毯
khoodbauf　库德鲍绸
khorassan　喀拉逊羊毛;喀拉逊地毯
khorasson knot　科腊桑手工地毯结
khorji　手织霍吉锦缎
khoseb　埃及棉平布
khotan　和田地毯
khotan rug　和田地毯
khudurangi　东非粗棉布
khulchack wool　库尔恰克山羊毛
khum　土耳其染色粗平布;新式伊朗地毯
khurkeh　巴勒斯坦喇叭袖长裙
khurta　克塔装,无领宽松长衬衫(印度)
kian pakkian　竹条布
kibachijoo　[日]合丝织
kibr　开伯外衣(阿拉伯)
kichorkay　印度棉布
kick　钱包;裤袋(俚);小瓣子[织]疵
kickback　倒转
kicker　鞣革机;婴儿鞋
kicker boot　足球长筒靴
kick pleat　倒裥(裙底部);助行裥;跨步裥
kick pleat skirt　暗褶裙;倒褶裥紧身裙;助行褶裙
kicks　裤子;鞋子(俚)
kick tape　裹脚布
kid　山羊皮革;小山羊毛皮
kid caracul　卡拉库尔山羊皮
kid cloth　羊绒厚呢
kidde net　弹力网眼经编织物
kidder carpet　基德地毯
kidderminster　基德明斯特地毯;双面粗呢
kiddy print　儿童喜爱的印花图案
kidfinished cambric　细薄棉布
kid fur　小山羊皮
kid gloves　小山羊皮手套,羊皮手套
kid karakul　卡拉库尔山羊皮
kid leather　小山羊皮;上光山羊皮
kid mohair　马海羔羊毛,安哥拉羔羊毛
kid mohair motte　马海羔羊毛
kidney belt　护腰带
kidney cotton　基德尼棉
kidney pad　护腰垫
kids　小山羊皮制品(如手套、皮鞋等)
kid's button　童装纽扣

kid seam　凸边缝
kids fur　小山羊毛皮
kid shoes　小山羊皮皮鞋,羊皮鞋
kidskin　小羚羊皮;小山羊皮;幼兽皮
kid's size　小童尺码
kid's sweater　童毛衫
kid's wear　儿童服装;童装
kidungas　基东加斯印花布
kienchow　中国手织凸纹绸
kier mark　煮练斑渍
kier stain　煮练斑渍
kietch cotton　美国基思棉
kiginu　生绢
Kiki Byrne　凯凯-伯恩(英国时装设计师)
kiki　提提裙(南太平洋)
kikkosha　蚊帐用珠罗纱
kikoi　基科伊色边条子厚棉布;南非手织棉布,肯尼亚手织棉布
kikois　基科伊色边条子厚棉布;南非手织棉布,肯尼亚手织棉布
kikoy　基科伊色边条子厚棉布;南非手织棉布,肯尼亚手织棉布
kilc pleat　苏格兰褶
kilim carpet　基里姆地毯;花毯
kilim rug　基里姆地毯;花毯
kilmarnock bonnet　宽顶方格子呢帽
kilmarnock　苏格兰粗哔叽;苏格兰混色毛毯
kiloyes　基科伊色边条子厚棉布
kilt　卷起;使有直褶;苏格兰短裙;男装裙;苏格兰褶裥短裙式童装(或女服);(苏格兰式男用)叠褶短裙;(儿童穿)苏格兰短裙
kilt dress　苏格兰连衣裙
kilted cloth　褶裥织物
kilt fold skirt　苏格兰百褶裙(男)
kiltie　穿褶裙的人;花样鞋舌;花样长舌鞋;穿褶裙的苏格兰高地士兵
kiltie dress　苏格兰裙装
kiltie flats　苏格兰鞋舌平跟鞋
kiltie look　苏格兰高地服装风貌(款式)
kiltie oxford　苏格兰鞋舌牛津鞋
kiltie shoes　苏格兰鞋
kiltie tongue　披巾式鞋舌;苏格兰鞋舌
kilting　打褶
kilt pin　安全别针
kilt pleat　苏格兰褶,顺风褶;助行褶
kilt shoes　苏格兰鞋
kilt skirt　(苏格兰式男用)褶叠短裙;苏格兰短裙

kilt suit 苏格兰套装
kilty 穿褶裙的人;穿褶裙的苏格兰兵;花样鞋舌;花样长舌鞋
kilty tongue 运动鞋镶边皮鞋
kim 印度金银丝锦缎;印度金考布锦
kimcha 中国花缎
Kim Dong Soon 金东顺(韩国服饰设计师)
Kimijima Ichiro 君岛一郎(日本时装设计师)
kimkhab 金考布锦
kimono 和服;和服式女晨衣;各种连袖服装
kimono brocade 和服花绸
kimono crepe 和服绸;和服绉
kimono design 连身袖设计
kimono dress 连袖服装;连袖裙
kimono flannel 和服法兰绒
kimono for judo 柔道外衣
kimono jacket 和服夹克
kimono lapel 和服领;和尚领
kimono overcoat 连袖大衣
kimono shoulder 无肩缝(用于女装);和服型轮廓
kimono silk 和服绸(有中国、日本传统图案的轻薄丝织物);素和服绸
kimono sleeve 和服袖;连袖
kimono sleeve with gusset 加裆布和服袖
kimono sleeve without gusset 无裆布和服袖
kimono slip-on 和服式外套
kimono style 和服式
kimono wrap 和服式披肩
kimono yoke 和服式过肩
kinaesthesia 运动感觉
kinanthropometric study 动态人体测量学
kinanthropometry 动态人体测量法
kinari 波斯走廊地毯
kincob 印度金银丝锦缎;印度金考布锦
kinder 罗布麻,红野麻,夹竹桃麻,茶叶花,茶梨子
kindergarten children's garment 幼儿园童装
kindergarten cloth 美国童装织条子布
kindergarten uniform[wear] 幼儿园服
kindestan rug 金迪斯顿地毯
kindred fabric 同类织物
kinds of cotton cloth 棉布种类
kinds of goods 货品

kinematics 运动学
kinesiology (人体)运动学
kinesthetic sense 运动感觉
kinetic art 活动服饰
kinetic fashion 动感流行款式,运动感的流行服装款式,活动服饰
kinetic friction 动摩擦
kinetic friction coefficient 动摩擦系数
kinetic silhouette 动感型
king cotton 美国金字棉
kingfisher 翠鸟色
king improved cotton 美国改良金字棉
king package 特大卷装
king's blue 品蓝,钴蓝(青光蓝色);蓝色矿物颜料
King's clothing[coat] 英国军服
king's color 英军国旗图案
king's yellow 雌黄
king's gold 雌黄
king's green 巴黎绿(黄光绿色)
king size (X) 特长号;特大号;特大尺码
King's own 亲王
king tiro satin 金雕缎
kingston blazer 金斯顿运动夹克
kinik wool 土耳其羊毛
kinik 土耳其羊毛
kinisol 菲律宾蕉麻,马尼拉麻
kinji shusu 日本嵌金线缎
kinkale 东南亚金银线花缎,印度金银丝花缎
kink band 扭结带
kinkhab 印度金考布锦,金银丝锦缎
kink hair 卷缩头发
kink yarn 起圈花线
kinky hair 卷缩头发
kinky boots 长筒女靴(长及膝或股),黑皮革长筒女靴
kinky clothes 奇装异服,奇异服装
kinky hair 卷缩头发
kinky look 奇装款式,反常款式
kinky style 奇装款式,反常款式
kinlochewe 肯诺其威
kinran 日本金阑缎
kinscord 花式沟纹织物
Kintail 肯德尔
kintted corduroy fabric 针织灯芯条
kintted escape chute 针织救生管道
kintted sofa-cover 针织沙发布
kiotonan 中国花缎
kip 幼仔毛皮;羔皮革;小牛皮革

kip calf-skin 犊皮
kipper 特宽领带
kipskin 幼仔毛皮；小牛皮革，幼兽革
Kirgiz ethnic costume 柯尔克孜族服饰，柯尔克孜族民族服
Kirgiz nationality's costume 柯尔克孜族服饰
Kirgizzu costume 柯尔克孜族民俗服
kirkagatsch cotton 小亚西亚柯卡加奇棉
kirkcaldy stripe 克卡尔迪条子布
kirman rug 伊朗克曼地毯
kirman wool 伊朗克曼羊毛
kirman-levehr rug 伊朗克曼地毯
kirmanshah rug 伊朗克曼地毯
kirriemuir 斜纹亚麻刺绣底布
kir-shehr carpet 土耳其基尔谢地毯
kirtle 柯特尔服装（中世纪）；男短外衣（中世纪）；外披斗篷（中世纪）；女长袍（中世纪）；带风帽长披风；女长衫；女裙；外套内的女工长袍；衬袍；骑马女装；短外套；男式及膝束腰长衫
kis-ghiordes rug 吉奥德地毯
kismess 印度印花棉布
kismiss 东南亚平布
kiss curl 卷鬓
kissable neckline style 开脖领款式
kissable zipper attachment 缝拉链附件
kissable zipper foot 缝拉链压脚
kisscurl 垂于额前（或耳前、后颈）的一绺鬓发
kissing strings 头巾式女帽的系带
kiss-me not 勿吻我帽
kiss-me quick 戴在脑勺上的小女帽（19世纪后半叶）；"快吻我"帽
kiss-me quick hat 戴在脑勺上的小女帽（19世纪后半叶）；"快吻我"帽
kissua 方格色织布，格子布
kissuto 东非棉印花布
kit 服装（英）；特殊场合服装；小动物的毛皮；整套零件
kitay 棉经丝纬色织布
kit-bag 行囊，长型帆布用具袋
kit-cat canvas 画像帆布
kitchen cloth 厨房用布
kitchen dress 厨房工作服
kitchen linen 厨房用亚麻布
kitchen towel 揩布
kit fox 北美灰狐皮，小狐毛皮，狐毛皮
kite print 风筝印花
kite sleeve 风筝袖

kitsch 通俗型；装饰过分的纹样
kitsch fashion 庸俗服饰
kitty fur 小猫皮
kityang 手织苎麻平布
kiv yar brocade 九霞缎
ki-wata 原棉（日本称谓）
kiwi green 几维绿，猕猴桃绿
klaft 尼密赛头巾布（古埃及）
kleanka 粗硬麻布
K legs K型裤脚
K-line K字型
klomp 荷兰木鞋
klompen 荷兰木鞋；克龙潘木屐
knapsack 背包，军用背包，旅行帆布背包，青春背包，质地坚实的书包
knap surface 绒面
knatherast 中级开士米呢（克什米尔）
kneading action 揉搓作用
knee 膝部；膝盖；中裆
knee band 齐膝带
knee-baring style 露膝款式
kneebend 屈膝
knee bent measurements 弯膝尺寸
knee boots 高筒靴，膝靴
knee brake 膝刹车
knee breeches 鞋口有纽扣的及膝男靴；齐膝短裤
knee buckles 男短裤金属搭扣
knee cap 膝盖骨，膝甲；运动员用护膝
kneecap supporter 护膝，护膝支撑
knee circumference 膝围，膝围线
knee circumference line 膝围线
knee control 压脚抬杆
knee dress 齐膝装
knee flex 膝弯曲
knee-fringe 裤口有丝带边饰的开襟长裤
knee girth 膝围
knee height 膝高
knee-hi 齐腰长袜；齐腰裤子
knee-high 齐膝，齐膝长袜
knee-high boots 高筒靴，女长筒靴
knee-high hose 齐膝袜
knee-high length 及膝高，膝盖高度
knee-high riding boots 长筒骑马靴
knee-highs 及膝高；齐膝中统袜
knee-high socks 长筒袜，及膝袜，齐膝袜
knee-high stovepipe boots 烟囱靴
knee-hi hose 及膝袜
knee-hi socks 齐膝袜
knee hole 容膝空隙

knee-hoverers 及膝款式
knee joing 膝关节
knee kicker 护膝布;跪毯;膝盖绸
knee-length 齐膝长;齐膝衣服
knee-length Bermuda shorts 齐膝百慕大短裤
knee-length boots 齐膝长筒靴
knee-length hose 齐膝中筒袜
knee-length hosiery 中筒女袜
knee length lining 膝长衬里
knee length shorts 及膝短裤
knee-length skirt 齐膝裙
knee-length socks 长筒袜;齐膝短袜
knee-length stockings 齐膝中筒女袜
knee-length turnout coat 齐膝消防服
knee level 膝围线
knee lever 膝杆
knee lift 膝抬压脚(装置)
knee line 中裆线,中裆围线;膝线
knee lining 膝盖绸
knee measure 膝围
knee over hose (过膝)长筒袜,舞袜
knee pad 膝垫;保健套,护膝;髌
kneepad felt 护膝毡
knee pan 膝盖骨
knee pants 齐膝短裤
knee patch 裤膝加固布片,膝垫
knee piece 膝甲;膝铠
knee protecting felt 护膝毡
knee reinforcement 膝部加固
knee escape skirt 露膝裙
knee sheath 刀鞘袜
knee shield 护膝
knee slacks 刀鞘袜;齐膝短裤;齐膝中筒袜
knee splicing 膝部加固
knee-string 收紧裤子膝下的抽带
knee stockings 中筒袜
knee strap 过膝束带,护膝扣带
knee-to match 匹配的裤子
knee-to-knee breadth 两膝宽
knee trousers 中长裤
knee warmer 暖膝套
knee width 膝部宽;膝盖围
kneipp linen 克内盖普亚麻布
knicker band 踢脚条
knickerbockers 运动裤;宽松裤;少男裤;灯笼裤;女用扎口短衬裤(英国);扎口女短裤
knickerbocker tweed 运动服装用异色结子粗呢
knickerbocker yarn 彩点花式毛纱,彩点线
knicker breeches 猎装裤;骑马裤
knickers (膝下扎起)灯笼裤;尼卡裤;女用扎口短衬裤;男短裤;女用短衬裤
knickers to match (与衣或裙)匹配的裤子
knicker suit 灯笼裤套装;尼卡套装(同料上衣、背心、裤三件套)
knicker yarn 彩点花式毛纱
knickknack 小饰物,小衣饰
knife 裁刀;刀片;切刀
knife bar 刀杆;刀架
knife cut bar 刀杆
knife cutting bar 刀杆
knife cutting width 刀切宽度
knife driving bellerank 切刀驱动曲柄
knife driving connection 剪线刀连杆
knife plaits 狭褶裥
knife pleat 狭褶裥,窄褶裥;剑褶裥
knife pleat skirt 剑褶裙,刀褶裙
knife-sensing unit 刀口传感装置
knife switch-out level 切刀旋出杆
knife thread holder 剪线刀钩;剪线钩
knife tucking 机械打裥
knightly girdle 臀部金属饰带
knight of the needle 裁缝
knight's armour 骑士盔甲
knit 针织,编织,编结,成圈;针织物
knit-and-tuck rib cloth 畦编织物;集圈织物;集圈罗纹织物
knit-and-welf cloth 编织-浮线针织物
knit article 针织品
knit back 修布针(针织)
knit banding 针织物滚镶边
knit binding 针织物缝合;针织物滚边
knit braid 针织绳索
knit clip-front blouse 夹扣胸襟针织上衣
knit coat 针织短外衣
knit cotton slip 针织棉套裙
knit cuff 针织克夫,罗纹克夫,罗纹松紧袖口,针织袖口
knit-de-knit fabric 拆编纱织物,假编纱织物
knit-de-knit[textured]yarn 假编变形丝
knit dress 针织服装
knit ensemble 针织套装
knit fabric 针织物
knit fabric construction 针织物结构

knit fabric finishing 针织物整理
knit fabric yield 每磅针织物码数
knit float work 编织-浮线色织物;针织-浮线针织物
knit garment 针织服装
knit gingham 针织条格纹布
knit gloves 编织手套
knitgoods 针织品
knit imitation center plait folder 针织物中央限位褶裥器
knit jacket 针织[毛线]夹克
knit mark 罗纹号
knit material 针织物
[knit-miss] bourrblet fabric 花式双罗纹针织物
knit pajamas 针织睡衣
knit panel 直条针织物
knit paper fabric 纸线针织物
knit pattern 针织花纹
knit pleat 助行褶
knit polo 针织马球衫,针织波罗衫
knit sewing 针织物缝纫
knit shirt 针织衫;针织衬衫
knit shirt category T恤衫,T恤衬衫
knit shirt collar 针织衬衫领
knit skipper 针织滑雪衫
knit stit 针织套装
knit stitch 针织成圈
knit stitch machine 缝编机
knitted 编,织
knitted apparel 针织服装
knitted artificial blood vessel 针织人造血管
knitted astrakhan 仿羔皮针织物
knitted backing 针织底布
knitted bed cover 针织床罩
knitted belt 针织线带
knitted beret 针织软帽
knitted blouse 编织女上衣;针织女衬衫;针织女罩衫
knitted bootee 针织婴儿鞋
knitted braid 针织饰带
knitted cape 针织披肩
knitted cardigans 开襟羊毛衫
knitted carpet 针织地毯
knitted children's wear 针织童装
knitted cloth fulling 羊毛针织物缩绒
knitted clothing 针织短外套
knitted coat 针织大衣;针织短外套;针织外衣

knitted collar 针织衣领
knitted colored stripe ribbon 针织彩条带,织彩条带
knitted cord 针织编带;针织滚条
knitted corduroy fabric 针织灯芯绒
knitted crepe georgette 针织乔其纱
knitted cuff 针织袖口,针织克夫
knitted curtain 针织窗帘
knitted design 针织图案
knitted dress 针织服装
knitted edge 针织边(防脱散)
knitted elastic bandage 针织弹力绷带
knitted elastic webbing 针织弹性带
knitted fabric 针织物;针织坯布
knitted fabric for food industry 食品工业用针织品
knitted fabric for military 军用针织物
knitted fabric material 针织布料,针织面料
knitted fabric of blended yarn 混纺纱针织物
knitted fake fur 人造毛皮针织物
knitted fashion 针织时装
knitted fire hose 针织消防水龙带
knitted fishing net 针织渔网
knitted flat ruche 无圈效应经编织物
knitted fleece 针织起绒布,起绒针织物
knitted fleece shirt 针织卫生衫
knitted fleece shirt and trousers 针织绒衫裤
knitted footwear 针织袜类,针织鞋制品
knitted garment 针织服装
knitted gloves 针织手套
knitted goods 针织品
knitted ground fabric 针织底布
knitted hat 编织帽,针织帽
knitted helmet 针织防护帽
knitted highpile 针织长毛绒,针织长毛绒织物
knitted high pile fabric 针织长毛绒
knitted-in seam (女长袜)假缝
knitted interlining 针织衬
knitted iron-on 针织黏合衬
knitted jacket 针织短大衣;针织外套;针织夹克
knitted jacquard fake fur 提花人造毛皮针织物
knitted jogging suit 针织运动套装
knitted knickerbockers 针织灯笼裤
knitted knickers (美国)针织男孩短裤;

（英国）针织女内衣,针织女扎口短衬裤;针织灯笼裤
knitted lace　针织花边
knitted lace hose　针织网眼袜
knitted lace pattern　针织网眼花纹
knitted laid-in fabric　衬垫针织物;衬纬针织物
knitted leno fabric　纱罗针织物
knitted lifesaving hose　针织救生管道
knitted linings　针织衬里布
knitted loop cloth　添沙线圈针织物;毛圈针织物;毛圈针织布;针织毛巾织物
knitted material　针织料;针织物
knitted necktie　针织领带
knitted net lace　棉(腈纶)爱丽纱;网眼针织花边
knitted net work　网眼针织物
knitted-on protion　加编部分
knitted outerwear　针织外衣
knitted outwear　针织外衣
knitted overtop　下摆罩外面的针织衫,针织套头衫
knitted packing bag　针织包装袋
knitted pajamas　针织睡衣裤
knitted piece goods　针织面料
knitted pile fabric　无圈针织物;针织长毛绒织物
knitted plush　针织长毛绒;针织长毛绒织物
knitted pyjamas　针织睡衣
knitted rib　针织带,针织牵条
knitted ribbon　针织丝带,针织缘子
knitteds　针织物
knitted scarf　针织围巾
knitted seal goods　针织海虎绒;针织海豹绒织物
knitted selvage　针织布边
knitted selvedge　针织边(防脱散),针织布边
knitted shirt　针织衬衫;针织裙
knitted sofa cover　针织沙发布
knitted sports wear　针织运动装
knitted stitch　弹力纱针织物
knitted stockings　针织长筒袜
knitted stretch　弹力针织物
knitted stretch fabric　针织弹力织物
knitted substrate　针织底布
knitted suit　针织套装
knitted sweater　毛线衫,针织绒衫
knitted table cover　针织台布

knitted terry　毛圈针织物;针织毛巾布
Knitted Textile Association　[美]针织品协会
Knitted Textile Dyers' Federation　针织染色业联合会
knitted tie　针织领带
knitted tie fabric　针织领带布
knitted tropical suitings　针织凡立丁
knitted trousers　毛线裤;针织裤
knitted underpants　针织内裤
knitted undershirt　针织内衣
knitted underwear　针织内衣
knitted upholster fabric　家用装饰针织物
knitted velour　针织丝绒;针织天鹅绒
knitted vest　针织背心;毛线背心
knitted waistband　罗纹下摆
knitted waistcoat　针织背心,针织西服背心
knitted wall bangings　针织墙布
knitted wear　针织服装
knitted welt　针织边(防脱散),针织罗口,针织饰边
knitted wool trousers　毛线裤
knitted wool-like flannel　仿毛法兰绒针织物
knitted wool-like gabardine　针织仿毛华达呢
knitted wool-like serge　针织仿毛哔叽
knitted wristlet　针织护腕
knitted yardgoods　针织匹头
knitter　针织机;编织机
knit thermals　针织保暖内衣裤
knit tie　针织领带
knitting　针(编)织;针(编)织品;针(编)织法
knitting and braiding machine　针织和编织机
knitting bag　针织包
knitting clothing of braid　桑蚕丝带编织服装
knitting computer　针织计算机
knitting cotton　棉针织物
knitting density　编织密度
knitting embroidery　编绣
knitting eyelet　网眼针织物;针织网眼纱;菠萝组织针织物
knitting finish　编织收口
Knitting Guild of America　美国针织协会
knitting heedle　织针
Knitting International　《国际针织》(月刊)

knitting line 针织横列
knitting machine 针织机
knitting needle 针织机针;棒针;毛衣针
knitting-on 起编
knitting pile carpet 针织绒线地毯
knitting pin 绒线针,手编棒针,针织棒针
knitting pin sweater 棒针毛衫
knitting stitch 编织绣;针织圈距
knitting stitch type 针织针法
knitting tool 针织工具
knitting wear 针织服装
knitting width 编幅;针织坯布布幅;编织宽度;针织坯布布幅
knitting wool 毛线;绒线;针织绒线
knitting wool socks 毛线袜;绒线袜
knitting yarn 编织纱;针织纱[线];针织用纱
knit top 针织上衣
knit transfer top 套口罗口
knit tubing 圆形纬编坯布;筒装针织物;圆形针织布
knit underwear 针织内衣
knit up 织补
knit vest 针织内衣;针织背心;针织衬衣
knitwear [总称]针织品;编结物;毛衫;针织服装,针织品
knitwear binder 针织物滚边器
knitwear press 针织品熨烫机
knitwear set 针织套装
knitwear setter 针织服装定形机
knit-weaving fabric 针织-机织物,织编织物
knit-weaving machine 织编机
knit-woven fabric 针织-机织物,织编织物
knit wool cap 针织羊毛帽
knit wool insweater[stockings] 用毛线织成毛衫[毛线袜]
knob 按纽;袋纽;旋纽;帽顶子
knob style 夹趾男拖鞋
knob toe 结纽式鞋头
knob yarn 疙瘩花线,结子花线
knock-down 装配生产
knock knees 外翻膝
knock knees figure X型腿身体型
knock-off 翻制设计;脱套;(时装样本等的)翻印本;下班时间;下班
knockoff 廉价仿制品,廉价复印品;时装设计复制品
knockoff accessories 翻制设计的服饰品

knockoff apparel 翻制设计的服装
knockoff designer 翻制设计师
knockoff manufacturer 翻制品制造商
knockout jacket 驱虫衣
knop 地毯绒结;领结;帽结;肩章,肩饰
knop cloth 无圈织物
knop design 集圈花纹
knop fabric 结子布
knop knitting yarn 结子绒线
knop tweed 结子粗呢
knop work 集圈针织物;胖花针织物
knop yarn 疙瘩花线,结子花线;异色结子粗纺毛纱
knot 结;结纽;肩饰;领结;帽结;结头;(装饰)花结;蚊子结
knotched one piece shawl collar 缺口单面青果领
knotch lapel 缺角下领片,缺角驳头
knot dyeing 扎染
knotless warp-knitted net 无结经编网
knot line 结纹
knot stitch 结粒线迹,结粒绣
knot tail 接头尾
knotted blanket stitch 结式毛毯锁边绣
knotted buttonhole stitch 结式扣眼线迹,结式锁眼绣
knotted cable chain stitch 结式花纹链状绣
knotted carpet 手工地毯
knotted fabric 网罗织物
knotted feather stitch 打结杨树花针迹,结式羽状绣
knotted fringe 结式缘饰
knotted frotte 结头粗毛圈花线织物
knotted girdle 打结腰带
knotted insertion stitch 结式拼缝绣
knotted lace 打结[式]花边,扣结花边
knotted net 打结网
knotted pile 手结栽绒;地毯结
knotted pile carpet 打结绒头地毯
knotted rug 手结栽绒地毯
knotted seam stitch 结粒拼缝绣
knotted stitch 结粒绣
knotted stitch fagoting 结粒拼缝绣
knotted wig 打结假发
knotted work 打结[式]花边
knotter 打结器
knotting 打结编网
knotting device 打结器
knotting stitch 打结线迹

knot tying machine 打结机
knotwork 编织物,打结产品
knot yarn 疙瘩花线;结子花线;异色结子粗纺毛纱
know-how 技术诀窍,技术秘密,专有技术
know-how textile 特殊用途织物,专用织物
knuckle yarn 疙瘩花线;结子花线
knurling 滚花
koala 树袋熊毛皮
kobai 棱纹织物
kodrung 东非粗棉布
koffo 蕉麻
kohaba 窄幅织物
ko hemp 葛麻,葛藤韧皮纤维
kohl (阿拉伯)化妆墨,(阿拉伯妇女用)眼圈粉
koja 四页斜纹布
kojah mink 变异水貂毛皮
kokaya silk 日本绸坯
koko 针织麻袋
kokochnik 俄罗斯已婚妇女头饰
kokoshnik 俄罗斯珍珠镶边女帽
kokoshnik hat 俄罗斯珍珠镶边女帽
kokti 科克提布
kokura 手织横棱棉布
kokuraori [日]小仓织
Kolbe [德]考佩发型
kolhapuri sandals 库尔哈帕凉鞋;印度夹带式皮凉鞋
kolinski 貂皮,西伯利亚貂皮,红貂毛皮
kolinsky 貂皮,西伯利亚貂皮,红貂毛皮
kolinsky skins 元皮
kolobium 库罗宾上衣(古罗马)
kolpos 由于束腰、系带生成的褶
kompon 中国亚麻平布
kona bukoroji 面粉袋布
konieh carpet 土耳其科尼埃地毯
kontush 波兰贵族悬饰袖长袍;德国女袍
kooletah 爱斯斯摩库尔塔外套
koolin 仿麻人造丝绸(商名)
koomach 库默克布
koordistan 科迪斯顿波斯地毯
koork 伊朗白山羊绒;驼毛
kooroon 库龙平纹呢
kootnee 库特尼花布
kopia 考比阿帽(印尼)
koplon fiber 科普纶纤维
korako 菽麻纤维

korako flax 新西兰亚麻
Korean boots 韩式宽趾卷头皮靴
Korean child's jacket 韩式儿童夹克
Korean costume 朝鲜民族服饰,韩国民俗服
Korean dress 韩式服装
Korean hat 朝鲜高顶宽边系带帽
Korean hat and band 韩国帽子及发带
Korean overcoat 韩式缎面外衣
Korean rubber shoes 朝鲜族雨鞋
korotes 粗印花棉布
kose 高丝
koshi 硬挺度(日本称谓)
koshimaki 棉布长衬裙;灰绉绸裙
koshta 黄麻(孟加拉湾称谓)
K'ossu 缂丝(中国传统工艺美术织品)
kossu fabric 缂丝(刻丝,克丝)
kota 平纹棉布
kotan 考顿胸衣(印尼)
kothornos 古悲剧舞台用鞋
kotzen 奥地利库曾毛毯
koujong 斜纹柔软呢
koulah rug 土耳其库拉拜毯
kraerisal yarn 圈圈花式线
kraft board 牛皮纸板
kraft muslin 薄型粗纺烛芯纱盘花簇绒织物
kraft (paper) 牛皮纸
kraft sheeting 灯芯纱盘花簇绒床单
kredemnon 克莱迪浓面纱
krefeld velvet 棉底丝圈丝绒
krepis 克利匹斯凉鞋(4世纪,古希腊)
krimmer 克里默羔皮;仿克里默羔皮织物
Krizia 克利兹阿时装店
Krizia/Krikzia Poi 科利亚/科利亚波伊
kron 优级俄罗斯亚麻
K'sa 摩洛哥男式长披肩
ktex (kilotex) 千特,千号
kuba 长细绒东方毛毯
kuba rug 长细绒东方地毯
kuban wool 俄罗斯库班羊毛
kubistan rug 卡比斯顿地毯
kub-kob 土耳其硬底鞋
Kublai Khan clothes 忽必烈汗装(立领,宽袖,纹织缎子)
kudistan rug 克迪斯顿粗毛地毯
kufiyah 库菲亚头巾(伊拉克)
kuhlchack 银灰色的羊绒
kui design 夔纹

kuitou headdress 盔头(中国戏曲)
kulah 阿訇帽(伊斯兰教)
kulah rug 长绒大地毯
kulijah 真丝或毛皮衬里骆驼毛外套
kulkan 库尔坎披巾(伊朗)
kulkan shawl 库尔坎披巾(伊朗)
kulyahi rug 波斯毛毯;库耳耶希毛毯
kumees 手织达卡细平布
kummerband 印度腰带;绶带;徽带;宽腰带
kumptas cotton 印度孔塔斯棉
kumya 库亚衬衫;摩洛哥衬衫
kunawey 库纳威绸
kung-fu pants 功夫裤,中国拳裤
kung-fu shoes (中国)功夫鞋
kung-fu uniform 功夫装
kunstseide 黏胶丝(德国称谓)
kurbelstickerei 机制绣花
kurdish leather boot 科迪士皮靴
kurdistan rug 克迪斯顿粗毛地毯;克迪斯顿细毛地毯
kurk wool 库凯山羊毛

kurkee 库基粗厚毛毯
kuron 库朗橡胶弹性织物
kurrachee cotton 印度库拉奇棉
kurta 短袖衬衫;无袖衬衫;克塔装(印度)
kurtah [shirt] 柯泰衫(印度),印度男式无领套头衬衫
kurtosis 夹峰值;(峰态)峰度
kusak 土耳其男子用围腰布
kutaree (印度)腰布花边(丝或棉制)
kutars 条子纱罗(印度平纹地纱罗条子)
kuttamromee 手织丝棉布
Kuwaiti costume 科威特民族服饰
kuzuzukajima [日]葛塚缩
kwangtung gauze plain gambiered 莨纱
kwangtung silk plain gambiered 莨绸
kylin design 麒麟纹样
kyooyuzen [日]京友禅
kyrle cloth 凯尔卷毛呢(经向羊毛纱与马海毛纱相间排列)
kyudayan longyi 方形花纹织物,菱形花纹织物

L

lab 实验室
lab clothing 实验室工作外套
lab coat 实验室工作服
Lab colour Lab 模式
lab dip 色样
label 标签;商标牌(带);标记;标号;唛头;徽章牌
label applying machine 缝商标机
label cloth 标签布;商标带
labeled length 标定长度
labeled size 标码
label hole 标签洞眼
labelling machine 钉唛头机;上标签机;贴商标机
labelling unit 贴商标装置
label making machine 贴标签机;打商标机
label ribbon 商标带
labels 学院帽上的飘带
labels cutting and folding machine 剪折商标机
label sewing machine 缝标签机
label silk 商标绸;商标缎
label size 标签尺寸
label tacker 缝标签机
label tape 机织商标带
labial commissure 口角(口唇联合)
labor and materials 工料
laboratory 实验室
laboratory accreditation 实验室认可
laboratory approval system 实验室认可体系
laboratory coat 实验室工作服
laboratory garment 实验室工作服
laboratory sample 实验室试验样品
laboratory test comparisons 实验室间对比检验
laboratory test sample 实验室试验样品
labor guota 劳动定额
labor-saving 节省劳力
labour cloth 牛仔布,坚固呢,劳动布,劳动呢,劳动卡,中联呢
labour concentrated type 劳动密集型
labour condition 劳动条件
labour contract 劳动合同

labour force 劳动力
labour insurance 劳动保险
labour intensive enterprise 劳动密集型企业
labour intensive industry 劳动密集型产业
labour intensive machinery 劳动密集型机械
labour intensive process 劳动密集加工
labour management 劳动管理
labour organization 劳动组织
labour productivity 劳动生产率
labour protection 劳动保护
labour protective coveralls 劳保(工作)服
labour raincoat 劳工雨衣;工作雨衣
labour saving sewing equipment 省工缝纫设备
labour strength 劳动强度
labour system 劳动制度
labret 唇饰
lab-to-bulk reproducibility 小样大样重演性
laburnham 莱本汉呢
laburnum 莱伯南斜纹薄呢
lace 花边,蕾丝;网眼织物;编带;饰带;鞋带;系带;有图案网眼织品
lace and tiny tuck trim blouse 网眼细褶裥上衣
lace appligue 花边镶饰
lace attaching folder 装花边器
lace blouse 花边女衫
lace bonnet 花边圆帽
lace boots 系带靴
lace brassiere 花边胸罩
lace clock 网眼绣花花纹
lace clock design 网眼绣花花纹
lace cloth 累丝纱,细纱罗织物;网眼织物
lace collar 花边领
lace-collar blouse 花边领上衣
lace-collared sweater 系带领毛衫;飘带领针织衫
lace cover 花边毛毯;花边垫
lace-covered button 花边包纽
lace curtain 花边纱,花边窗帘
lace d'amour 编结织物;亚麻台布

laced applique 花边镶饰
laced belt 穿绳紧身腰带
laced boots 系带靴
laced closing 系带门襟,穿线系紧的门襟
laced closure 系带门襟,穿线系紧的门襟
laced corset 穿绳紧身衣
laced foundation 穿线收紧的内衣
laced girdle 束带紧身褡
lace double knit 透孔双面针织物
laced sample 花边样本
laced tie closure 系带扣
laced-up 系带;系绳;系紧绳带
lace edging 饰边;花边镶边
lace effect 花边效果;花边效应,网眼效应
lace elastic 花边弹力松紧带
lace fabric 花边织物;纱罗织物,网眼织物,多孔织物
lace filling stitch 填花边孔绣
lace fond 网眼底布
lace fontange 高框架饰带女帽;花边女帽
lace foundation 花边束腰胸衣
lace garter band 花边吊袜带
lace gloves 花边手套;网眼手套
lace ground 花边底布
lace hair sweater 孔花毛衫
lace hankie 花边手帕
lace hose 网眼袜
lace insertion 窄小的直花边;镶嵌花边
lace inspection 花边检验
lace jacket 花边料夹克
lace knitwear 孔花毛衫
lace lozenges 菱形蕾丝花纹
lace machine 花边编结机
lace making 花边编织;手工编织花边;饰带制造;鞋带制造[工艺]
lace marquisette bow 玛基赛特花边蝴蝶结(束发装饰品)
lace net 花边网
lace pattern 网眼花纹
lace pillow 编花边用垫子
lacer 鞋带
laceration 划破
lacerna [拉]莱塞纳披风,罩在宽外袍外的短披风(古罗马)
lace ruffle 花边褶边;花边荷叶边
lace runner 滚花滚子;滚花花边

lace seam 花边缝迹
lace shoes 系带鞋
lace silk 花边丝线
lace spring 网地贴花花边
lace stitch 花边网眼组织;纱罗组织
lace stockings 网眼袜
lace stripe 纱罗条纹;提花充纱罗条子布
lace stripe cloth 提花充纱罗条子布
lace stripes 纱罗条纹织物
lacet 菜山装饰带,防缩带;带子;辫子
lace table cloth 网眼花边台布
lace tape 花边带
lace-to-toe sneakers 布拉佳型帆布便鞋(束鞋子至鞋尖)
lace-trimmed bra 饰边胸罩
lace-trimmed loose-fitting housecoat 饰边宽松长袍
lace-trimmed loose-hanging gown 饰边宽松长袍
lace-trimmed slip 镶花边套裙
lace-trimming 花饰边;花边装饰
lacets bleu 法国人字或山形斜纹裤料
lacets tape 编结带
lacets work 带子编织物;辫子编织物
lace-up 系带子的;系带靴
lace-up belt 系结腰带,打结腰带
lace-up boots 长筒系带皮靴;系带靴
lace-up corset-type vest 系带式束身背心
lace-up front 系带前开口;鞋带式开口
lace ups 系带靴
lace warp fabric 经编花边织物
lace warp-knitting machine 花边经编机
lace woolen sweater 孔花毛衫
lacework 花边;带子编结品;网眼针织物;辫子编结品
lace yarn 花边纱线;花边用线
lacey neck 花边领
lace york frock 花边覆肩外长衣
lachak 棉斜纹布
lachdan 粗毡毯
lachka 印度拷花绸(丝经,银线纬)
lachorias 印度棉织物
lacing 花边装饰;(衣服)镶边;衣服系带;鞋带
lacing attachment 花边编织附件
lacing cord 编结绳
lacing front 系带前开门
lacing thread 扣线;扎绞线
laciniate 有穗[边]的

lacis 方网眼花边
lacking levleness 染色不匀
lack-lustre 无光泽
lack of fullness at chest (衣服)塌胸
lacoste eyelet knitted fabric 拉科斯网眼布
Lacoste knit shirt 拉科斯针织衫
lacovries 印度棉织物
Lacquer 暗漆红
lacquer cloth 树脂涂层织物;涂膜布
lacquer coated fabric 涂膜织物;上漆织物
lacquer finish 表面涂膜整理;上漆整理
lacquer printing 漆印印花
lacquer prints 漆印花布(漆料印花)
lacquer red 漆红
lacquered satin 涂膜缎
lacrimal bone 泪骨
lacs 结实的细绳
lacs d'amour 法国织花亚麻台布
lacto button 合成树脂纽扣
lacy 花边状的;网状的;有花边的
lacy briefs 花边短内裤
lacy half slip 花边衬裙
lacy knit 网状针织物;花边针织物
lacy neck 花边领
lacy neckline 花边领口
lacy stockings 网状女长筒袜;花边长筒袜
ladam cotton 印度拉大姆棉
ladder 抽丝,梯脱(线圈脱散);梯状物
ladder bar 防脱散横列
ladder boots 梯状靴
ladder braid 梯形编带;格子网眼编带
ladder fiber 梯形聚合物纤维
ladder hem stitch 梯形抽绣
laddering run 线圈纵向脱散
ladderless stockings 防抽丝袜
ladder-net mesh underwear 梯形网状内衣
ladder polymer fiber 双股形聚合物纤维,梯形聚合物纤维
ladderproof 防脱散
ladderproof fabric 防脱散织物
ladderproof stockings 防脱散袜子,不脱散袜子
ladder resistant band 袜口防脱散边
ladder stitch 梯缝;梯纹刺绣;双辫梯形绣;脱散线圈;脱散线迹
ladder stitcher 梯缝缝纫机

ladder stitch machine 梯缝缝纫机
ladder tape 百叶窗带
ladder wed 百叶窗带
ladderwork 挖花绣织物
ladha charkana bufta 手织棉丝交织平纹布
ladies' 女式
ladies' apron 女家用围裙
ladies' boots 女式长筒靴
ladies' cap 女士便帽
ladies' cloth (粗纺)轻薄女式呢;优质法兰绒
ladies' cloth and dress worsted (薄型精纺)女衣呢
ladies' coat 女式外套(大衣)
ladies' coating 女式呢
ladies' crosets 胸罩
ladies' dress gloves (女士)晚装手套;晚宴手套
ladies' hat 女士宽檐帽
ladies' hairdresser 女部理发师
ladies' hairstyles 妇女发型
ladies' knee-high stockings 及膝女长袜
ladies' lingerie 女内衣裤
ladies' long coat 女大衣
ladies' nightwear 女睡衣
ladies' one-piece ski suit 妇式滑雪连身服
ladies' pantsuits 女式紧身裤套装
ladies' pullover shirt 女套衫
ladies' sandals 女式凉鞋
ladies' shorts 女式短裤
ladies' side button 女士侧系扣高帮松紧鞋
ladies' slacks 女裤
ladies' sweater 女式毛衣
ladies' wear 妇女衣着;女服;女装
ladies' wear department 女服部
Ladik rug 土耳其拉迪克地毯
Ladine 18世纪精纺花式轧光呢
lado cotton 西非拉多棉
ladoga carpet 花卉边花地毯
ladylike fashion 像贵妇人的时装(款式)
lady sanitary napkin 妇女卫生巾
lady's cloth 薄型女式呢
lady's dress worsted 女衣呢
lady's hat 女帽
lady's skirt 凤尾裙
lady's suit 女套装;女西装
lady's textured stockings 女用网眼长袜

lady's wing type sanitary towels　妇女翼状卫生巾
laer-cut label　激光裁切唛头
laffis　绢丝薄绸,绢丝纺
lagais　优质轧光棉布
lagais du roi　特优质轧光棉布
la garconne　[法]华丽年代
laggard　滞后消费者
lagging　滞后性
lagging material　防护材料
lagoon　礁瑚蓝
lagos cotton　非洲拉各斯棉
laguary cotton　哥伦比亚拉瓜伊拉棉
la guyra　印度拉加伊腊棉
la guyra cotton　印度拉加伊腊棉
laharia　印度扎染布
Lahore　拉舍尔羊绒花呢
Lahore chaddar shawl　拉舍尔仿羊绒披巾
lahore cloth　拉舍尔羊绒花呢
Lahu ethnic costume　拉祜族服饰
laid embroidery　衬垫刺绣;嵌芯刺绣;教堂刺绣
laid fabric　无纺织物;无纬织物
laid-in　衬垫;衬纬
laid-in effect　衬垫效应
laid-in fabric　衬垫针织物;短程衬纬针织物
laid-in selvage　折边
laid-in stitch　衬纬针法;衬垫针法
laid-in weft-knitted fabric　纬编衬纬针织物
laid laffis　绡丝薄绸;绢丝纺
laid-on　贴放
laid pile　排绒整理织物
laid stitch　长针迹;席编式贴线锈
laid-up　拆卸修理
laid work　贴线刺绣品
laine　羊毛(法国称谓)
laine de carmenie　波斯山羊毛
laine d'été　法国黏胶羊毛混纺织物
laine de terneaux　法国美利奴羊毛
laine elastique　轻薄绉呢;轻薄棱纹呢
laingah　手织平纹细棉布
laisot　亚麻平布;优质低支亚麻帆布
laisse-tout-faire　家居大围裙
laizes　小花网地花边
lake[red]　胭脂红[色]
lake black　漆黑色
lake blue　湖泊蓝(浅暗蓝色)
lake green　湖绿[色],摩洛哥绿[色]
lake-munia　色织花绸(印度制,花卉纹,涡形纹边花)
Lakoda　勒科德海豹皮(阿拉斯加海豹,琥珀色,光滑毛皮)
laky　胭脂红色的
Lalio cotton　印度拉黑奥棉
lama　金银皮,金属薄条;毛衬布
lama barchent　异纬双面绒布
lama's cap　拉玛帽(有帽耳、饰带、织锦帽冠、棉布帽檐的女帽)
lamb　羔羊皮,小羊皮,(仿)羔羊皮
lamba　马达加斯加兰巴布;婆罗洲叶纤维织物;羔羊毛彩色披巾
lamba mena　裹尸布
lamba shawl　马达加斯加兰巴披巾
lamballe　[法]朗巴尔围巾
lambert cotton　苏丹兰伯特棉
lamb fur　羔羊皮,小羊皮
lamboy　盔甲裙(15～16世纪)
lambrequin　装饰性挂帘;盔饰盖布
lambsdown　驼绒;棉背毛绒面针织物,驼绒针织物
lambsdown knit fabric　驼绒针织物
lambsdown sweater　驼绒衫
lambskin　羔羊皮,小绵羊皮;小绵羊革,仿羔羊皮织物;纬面缎纹绒布
lambskin cloth　纬面缎纹绒布;仿羔羊皮织物
lamb suede　羊皮
lamb's wool　羔羊毛;羔羊毛色
lamb's wool knitwear　羔羊毛衫,羊仔毛衫
lamb's wool knitwear sweater　羊仔毛衫
lamb's wool padding　羔羊毛衬垫
lamb's wool sweater　羔羊毛衫,羊仔毛衫
lamb wool　羔羊皮
lame　[法]金属线;金银线织物
lamé　金银锦缎(金银丝交织)
lamé cloth　金银锦缎
lamé damask　金银锦花缎
lamé lace　金银丝花边
lamé stockings　金属线长袜
lame thread　金银线
lamé wool cloth　金银毛花呢
lame yarn　金银线,金银丝花式线
laminate　层压材料
laminated board nonwoven　层压非织造布
laminated board nonwoven fabric　层压定型非织造布
laminated bonded fabric　层压黏合织物

laminated cloth 多层黏合布,胶合布
laminated composition fabric 层压复合纺织材料
laminated denim 层压(涂层)牛仔布
laminated fabric 层压织物,叠层织物,胶合织物,黏合织物
laminated finishing 层压整理,叠层整理
laminated heel 堆跟
laminated jersey 层压针织品
laminated knit 叠压针织物,胶合针织物
laminated nonwoven fabric 叠层非织造布
laminated press felt 层压毡毯
laminated sole shoes 层压底鞋
laminated synthetic leather 层压革
laminated tarpaulin 层压帆布
laminated water-proof fabric 叠层防水织物
laminates 层压织物;多层黏合布;胶合布
laminating 层压,叠制;复贴整理;胶合;涂层
laminating finishing 叠层整理
lamination 织物胶合工艺
lamination machine 贴合机
laminator 层压机
laminette 金银线织物;金属扁丝
lammy 防寒水手短上衣;连帽厚呢大衣(部队官兵穿);衬有垫料的短上衣
lamont 拉蒙特条格呢
Lamoose Artificial Leather 拉莫斯人造麂皮(用聚氨酯浸渍工艺制,商名,日本)
lamot 蕉麻,马尼拉麻
lampas 彩花细锦缎;印度印花绸;装饰花缎
lampas du japon 法国经棱纹丝花缎
lampasette 彩色斜纹绸
lampe 毛纱罗
lamp black 灯烟色
lamphouse 灯箱
lampshade bonnet 罩形圆帽
lampshade hat 罩形帽,灯罩帽
lampshade straw 罩形草帽
lamp sleeve 灯泡袖
lampwick 管状织物,灯芯
lampwicking 管状织物,灯芯
lamsa 伊朗优质印花棉布
lana 羊毛;羽绒(拉丁语)
Lancashire flannel 兰开夏色织白边法兰绒
Lancaster cloth 兰开斯特油布
lance 小纬浮纹

lance stitch 短直线纹刺绣针迹
Lancetti 兰塞蒂(意大利时装设计师)
LANCOME 兰蔻
land bridge transport 大陆桥运输
landing 到岸;卸货
landlady shoes 伍尔沃思鞋
landmark 测量定位点
land otter fur 獭皮
landrine 波浪翻边中筒马靴
LANDUN 蓝顿
laneé 浮纬纹织物;斑点花纹织物
laneé embroidery 镶边刺绣
laneé jacquard 浮纹织物
Lanella 拉纳拉棉毛法兰绒
lanfet 高级大麻织物
lange 毛织物
langère 粗纺毛床单布
langet 服装皮带;荷兰梭结粗花边;头盔羽饰
langhua crepe 浪花绉
lang-kiln red 郎窑红
Langtry hood 伦奇兜帽
languette 裙用舌状饰物
langutis 低档印度腰布
lanilla 硬挺整理精纺哔叽
lanital 人造羊毛(酪蛋白纤维,商名,比利时)
Lankart cotton 美国兰卡特棉
lannov velvet 法国丝绒
lansdown 兰斯唐毛葛
lansdwowne 兰斯唐毛葛
lanshan 褴衫(中国古代的一种短袍)
lansquenet pantaloon 雇佣军服式长裤
lanstinized 改性麻纤维,改性韧皮纤维
lantern jaw 瘦长下巴;突出下巴
lantern shape bag 灯笼包
lantern skirt 灯笼裙
lantern sleeve 灯笼袖
lantern[type]sleeve 灯笼袖
LANVIN 兰文
lanyard 双色编织绳
laos 劳斯绉
lap (衣服)下摆;接头,搭接;盖住两膝和大腿的衣裙(人坐着时);腰膝间大腿部(人坐着时);裙兜,衣兜;衣服叠门
lap allowance 叠门量
La paz 莱巴珍珠
lapchi [俄]树皮鞋
lap dress 膝衣
lapel 驳领,驳头,翻领,卷边;下领片;

折边
lapel bias （西装上衣）驳领斜度
lapel buttonhole 驳头眼；下领片纽孔
lapel cardigan 有领开襟毛衫
lapel collar 驳角领；驳头领
lapel dart 驳头省
lapeled cardigan 有领开襟毛衫
lapel edge appears loose[tight] 驳口外口松[紧]
lapel edge curve 驳头止口弧线
lapel edge line 驳头止口弧线
lapel edge to appears loose[tight] 驳头外口松[紧]
lapeled vest 有领背心
lapel facing 驳领面；驳头贴边
lapel interlining 驳头（细布）衬
lapelless jacket 无领夹克
lapel line 驳头线
lapel microphone 西装翻领或口袋上的微型麦克风
lapel notch V形翻领；西装领的V形部分；驳头剪口
lapel outer edge too loose 驳头外口松
lapel outer edge too tight 驳头外口紧
lapel pad 驳领垫
lapel padding 翻领衬垫
lapel peak 驳领尖，驳头尖，翻领尖角
lapel pin 下领片饰针
lapel point 驳领缺口
lapel press 驳领熨烫机
lapel roll 卷领，翻领
lapel roll line 驳口线，驳头线
lapel roll line is uneven 驳口不直
lapel roll padding automat 自动翻驳领机，自动翻领衬垫机
lapel run 驳头外圆线
lapels 翻驳领
lapel seam 驳领口缝
lapel shaper 驳领造型
lapel step 驳领宽，驳头宽
lapel style line 驳头止口弧线
lapel vest 有领背心
lapel width 驳领宽，驳头宽，折边宽度
lapful 满兜，满膝
lapidary 宝石工，玉石工；宝石商；宝石专家
lapin cotelé [法]棱状兔毛皮
lapin 兔毛皮；兔毛皮镶边
laping up 叠布，铺幅落料
lapis 宝石绿，雀绿，靛蓝绿，雀蓝

lapis lazuli 青金石，天青石
lapis lazuli[blue] 群青色，天青石蓝（蓝色），琉璃蓝；
lap joint 叠接
lapland boot 拉普兰靴
la pliant 弹性裙撑
lapnis 菲律宾苎麻纤维
laportea gigas 仰光大麻
lapot 桦树皮系带粗制鞋
lap over 印花图样交搭，花样重叠
lappa 披巾（利比里），印度真丝锦缎
lappas 英国拉珀斯色条棉布
Lapp costume 拉普民族服饰
lapped 骑缝
lapped band cuff 相叠的袖克夫
lapped dart 叠合省（剪开后叠合缝省）
lapped flat seam 叠平缝
lapped joint 卷接缝，折缝，搭接缝
lapped seam 搭接缝，折边搭边缝，骑缝，压缉缝，叠缝
lapped seam attachment 叠层装置，叠缝装置
lapped sections 叠条
lapped sleeve closing 叠合袖口
lapped stitches 叠缝
lapped zipper 暗拉链
lappet dotted swiss 浮纹点子薄细布
lappet 垂襞，垂片；翻领；垂饰
lappet fabric 浮纹织物
lappets 垂饰；浮纹织物
lappet thread 刺绣线；浮纹经线
lappet yashimagh 伊斯兰教妇女用浮纹面纱
lapping 搭接；印花毛衬布
lapping cloth 燕呢用[棉]衬布
lapping seam 搭接缝；坐倒缝；骑缝
lap pocket 有盖口袋；直盖口袋
lapponica 方格花呢披肩
lapp trousers 拉普兰裤
lap robe 旅行毛毯
lap seam 压缉缝，搭接缝，骑缝
lap seamer 搭接缝纫机；叠接缝纫机
lap seam felling 搭接折缝
lap seam felling industrial sewing machine 搭接折缝工业缝纫机
lap seam felling operation 搭接折缝作业
lap seam folder 搭接折缝卷边器
lap seaming 搭接缝纫，下摆缝合
lap seaming machine 搭接缝纫机
lap-shoulder style 叠肩式

lap strap 绑腰安全带	lariat necklace 系结长项链
large（L） 大号；大号服装	lariat tie with aiglet points 金属饰结绳状领带
large abdomen 凸腹(体型)	
large arm 大臂	Laristan 拉利斯顿地毯
large back 宽背	Laristan rug 拉利斯顿地毯
large batch 大布卷	lark 泥灰[色]，云雀灰黄色
large-brimmed cloche 宽边钟形帽	lark spur 飞燕草色
large bust 大胸	larme 珠形花边花纹
large buttocks 大臀部	larné stockings 金属线长袜
large check bed sheet 大方格床单	larrigan 北美长筒麂皮靴，油皮靴
large collar 大领	larrigan leather 油鞣革
large cone 大筒装	laser cloth inspector 激光验布器
large contrast 大色差	laser-cut label 激光裁切唛头
large-coverage design 覆盖面大的[印花]图案	laser cutter 激光裁床；激光裁剪刀
	laser cutting 激光裁剪
large cup 大乳罩窝；大胸罩杯	laser embroidery machine 激光绣花机
large-cut shirt 宽大衬衫	laser fabric inspecting machine 激光验布机
large front 大襟	
large front clothes 大襟衫(旧时)	laser fabric inspection system 激光验布系统
large godet skirt 大叶形裙	
large head pin 巨帽大头针；图钉	laser plotter 激光绘图机
large hip 大臀部	laser processing 激光加工
large-hole sew-through button 大眼平缝扣	laser-scan display 激光扫描显示
	lashed pile velveteen 扣紧纬平绒
large lapel 大襟	lassis 轻磅废丝绸
large order 大批定货	lasso line 套索款式
large package 大卷装	last 鞋型；鞋楦
large-part stacker 大部件堆叠器	lastex 橡皮线
large pattern 大型图案	lastex yarn 包纱橡筋线
large pattern brocade 大锦	lastics 弹塑性体
large repeat 大单元花样	last-in first-out storage 后进先出栈
large repeat construction 大花纹循环	lasting 厚实斜纹织物(制鞋用)，强捻厚斜纹织物；坚固斜纹呢
large repeat pattern 大循环花纹	
large roll 大卷装	lasting boot 深色开司米面靴
large-scale cooperation 大协作	lasting crease 永久性烫缝
large-scale design 大花纹	lasting pincers 绷楦钳(制鞋)
large-scale pattern 大花型	last pattern 楦型(鞋)
large seat trousers 大臀围裤子	last pattern design 楦型设计
large size（L） 大尺寸；大号	lastrine 黑色塔夫绸，无色塔夫绸
large sliding caliper 杆状计(量体形用)	latch buckle 带扣，闩锁扣，拴锁扣
large spun rayon square 人造丝大方巾	latch closure 环扣
large taper cane 大锥角锥形筒子	latch elastic hemmer 锁闩弹性卷边器
large thighs 粗大腿	latchet 鞋带
large touch 大胆	latch hemmer 锁闩卷边器
large twill 急斜纹	latch hook 钩编器
large unfolded wing（da la chi） 大拉翅（满族妇女头饰）	latch needle 舌针
	latch tacker 线辫收尾器
large upper arm 大上臂	latch tool 钩编器
large waist 粗腰，大腰围	late majority 随流消费者
laria cotton 印度拉里亚棉	late-day clothes 晚昼服，午后服

late-day dress 晚昼裙装，午后裙装
latent heat 潜热
latent-type defects 潜在型疵点
lateral clamps 横向夹头
lateral layout 横向排料
lateral malleolus 外踝
lateral neck root point 颈侧根点
lateral stability 横向稳定性
laterite 土红
latest fashion colour 最新流行色
latex 胶乳；树脂(如聚氯乙烯)乳剂；天然橡胶松紧纱芯
latex disposable gloves 一次性乳胶手套
latexing 胶乳整理
latex yarn 伸缩线，橡筋线
latigo leather 矾鞣革
latissimus dorsi 背阔肌
lattice 装饰性透孔织物
lattice basket stitch 格子绣花线迹
lattice braid 格子网眼编带，格子花编带
lattice pattern 网格图案；格子花纹
lattices 小方格
lattice seam 格子缝
lattice umbrella 小方格伞
launch 推出新产品
launch planning 新产品计划
launder 洗衣，洗涤，浆洗；(洗后)熨烫，洗烫
launderability 可洗性，耐洗性能
launderable cloth 可洗衣服
laundered appearance 洗涤后外观
launderette 自动洗衣店，快速洗衣店，自助洗衣店
laundering 洗烫，熨烫衣服
laundering and dyeing shop 洗染店
laundering durability 耐洗涤性
laundering test 洗涤试验
Launder-o-Meter 耐洗牢度试验仪
launderproof 耐洗性能
laundress 洗烫衣物女工，洗衣妇
laundrette 自动洗衣店
laundromat 自动洗衣店
laundry 洗衣；洗衣房，洗衣店，干洗店
laundry bag 洗衣袋
laundry bill 洗衣凭单
laundry blue 绀青，洗涤蓝
laundry brush 洗衣刷
laundry cycle 洗涤周期
laundry duck 洗衣滚筒包覆用帆布
laundryman 洗衣男工；取送衣物的人；洗衣店货车司机
laundry mark 洗衣店标签
laundry nets 洗衣用网袋
laundry net(ting) 洗衣用网袋；染色用网袋
laundryproof 耐洗性能
laundry-proof property 耐洗涤性
laundry resistance 耐洗[烫]性
laundry shrinkage 洗涤收缩率
laundry soap 洗衣皂
laundry trade sheeting 耐洗被单布；重磅棉布，厚棉布
laundry tray 洗衣槽
laundry tub 洗衣槽
laundry van 洗衣店货车
laundry washing 机洗
laundry woman 洗衣女工
Laura Ashley look 劳拉·阿什利风貌
Laura Ashley type 劳拉·阿什利型
Laura Ashley 劳拉·阿什利(美国时装设计师)
Laura Biagiotti 劳拉·比亚焦蒂(意大利时装设计师)
laurel crown 月桂冠
laurel green 黄绿色，浅绿色，月桂绿
laurel oak 桂叶栎嫩叶色
laurel pink 月桂粉红色(淡红色)
laurel wreath 桂冠
laval 亚麻布
lava lava 英国重浆印花布；花腰布；短裙；玻利尼西亚花腰布；玻利尼西亚短裙；印花布短围裙(萨摩亚及其他太平洋岛屿)
lavalier(e) 垂饰式项链
lavalliére [法]莱弗立尔项链(系有垂饰物)
lavena 拉维纳原色薄呢
lavender 浅蓝莲，熏花草紫(淡紫色)；漂白印花细亚麻布
lavender blue 淡紫蓝色
lavender frost 霜紫色
lavender grey 紫灰色
lavender luster 淡紫色蓝
lavender water 熏衣草香水
lavent linen 大麻织物
laventine 里子薄绸，全丝袖里绸
laver 洗衣盆
Laver mode 雷弗模式
law of intermediary colours 中间色律
law of value 价值规律

lawful rights and interests 合法权益
LAWLANDEE 派克兰蒂
lawn 上等细布;细麻布;细竹布;细纺
lawn fabric 人造草坪织物
lawn finish 上等细布整理
lawn sleeves 上等细布袖
lawn-party dress 户外午后会客装
lawn-tennis apron 围裙式细布女网眼服
lawn-tennis costume 19世纪80年代欧洲草地网球服
lawyer's robe 法衣;法官服;律师服
laxey 浅圆钻石外观
lay 铺放
layed lip 层叠式口红
layer 纱层
layer cut 分层式发型
layered look 多层式款式;多层风貌;假叠层款式
layered necker 双层套领
layered neckpiece 双层领圈
layered skirt 多层裙
layered sleeve 多层式袖
layering 多层穿法;重叠穿法
layette 初生婴儿全套用品;新生婴儿全套衣物;婴儿服;婴儿全套用品
lay figure 服装人体模型
lay flat to dry 放平干燥
laying[and top stitched]seam 坐缉缝
laying edge seam 坐倒缝
laying-in 衬垫;衬纬;衬经
laying-in circular weft knitting machine 衬经衬纬圆纬机
laying-in elastic 衬垫橡筋线
laying-in knitted fabric 衬垫针织物
laying-in machine 摊布机
laying-in thread 衬垫纱
laying-in weft knitting machine 衬垫纬编机
laying-up 叠布;叠幅落料;铺幅落料
laying-up machine 叠布机
lay marker 划料;排料;铺放排列图
lay off 下料
layout 布局;布置图;排列图;裁片排版图;排料;(穿着)搭配;设计;草图;放样,划线,布置,规划
layout arrangement 缝纫排料
layout chart 缝纫排料图,缝纫平面布置图
layout to cut defeets 借疵
lay planning 缝纫排料;排码克;排唛架;

铺幅编排
lay-stay flat 地毯铺平性
lay time 装卸时间
lay-up process 排料过程
lazarine 波浪翻边中筒靴
lazear jacket 便服
lazies 拉齐斯手工花边,点纹花边
lazy-daisy stitch 平式花瓣线迹,皱菊线迹,套扣针法
Lchiro Kimijima 君岛一郎(日本时装设计师)
le dernier mode 最新服饰潮流
Lea carpet 李氏地毯
leach resistant antimicrobial fabric 耐沥滤的抗微生物织物
lead chromate treated cotton 铬酸铅处理棉布
lead disc weights 铅盘挂锤
lead 铅色;先导
leaden 铅色,暗灰色
leaden gray 铅灰色
leader 导布
leader line 引线,指示线
lead grey 铅灰色
lead impregnated clothing 铅浸渍服
leading article (为招揽生意的)特别廉价商品
leading strings 儿童裙的窄长肩带
lead pellet weights 铅丸挂锤
lead powder 铅粉
lead time 设计与生产相隔时间
lead white 铅白
leaf design 卷叶纹
leaf edge 领边
leaf fiber 剑麻;叶纤维
leaf green 叶绿(黄绿色);绿色颜色
leaf hat 斗笠
leaf rain-cloak 蓑衣
leaf stitch 叶形绣;叶形针法
league table 名次表
leakage 漏出;渗出
leaking sewing 漏针
Leamington Axminster 利明顿阿克明斯特小地毯
lean body look 苗条款式
lean body style 苗条款式
lean dress 纤细女装
lean handle [涂层]干燥手感
lease trade 租赁贸易
leasing 租船运输;租凭

least thigh circumference 大腿最小围
leather 皮子;熟皮;羊皮;皮革;皮革制品;皮箱;皮短裤;皮绑腿
leather[top] 皮上衣
leather apron 皮围裙
leather armo(u)r 皮甲
leather article 皮件
leather article thread 皮件线
leather badge 真皮徽章
leather bag 皮包
leather belt 皮带;皮裤带
leather belt hook 皮带钩
leatherboard 人造皮革板;再生革板
leather boots 皮靴
leather brown 皮革棕(棕色)
leather button 皮革纽扣;皮扣
leather-buttoned corduroy jacket 皮纽灯芯绒外套
leather bonnet 圆皮帽
leather cap 革皮帽,皮便帽,皮帽
leather cap with fur flaps 毛皮翻边的皮便帽
leather cloth 皮革布,人造布;人造(皮)革;漆(皮)布;棉毛麦尔登
leather clothes 皮衣;皮革服装
leather clothing 皮大衣
leather clothing[coat]inserted with fur 镶饰毛皮的皮大衣
leather coat 皮革上装;皮大衣;皮革大衣
leather color 皮革色
leather craft 皮革制品(总称);皮革制作工艺
leather cutting board 切皮板
leather cutting shears 皮革切片
leather dressing 皮革修整
leather dye 毛皮染料
leatheret 人造革,仿皮革
leatheret gloves 人造皮革手套;仿皮手套
leatheret jacket 人造皮夹克
leatheret sandals 人造革凉鞋
leatheret(te) 仿皮革,人造革
leatherette gloves 人造革手套
leatherette jacket 人造皮革夹克
leatherette sandals 人造革凉鞋
leather fabric 皮革织物;人造革;漆布;棉毛麦尔登
leather facing 皮衬片
leather fiber 皮革纤维
leather finishing 皮革整理

leather for garments 服装革
leather for gloves 手套革
leather for shoe cover 鞋面革;鞋里革
leather fringe 皮革流苏
leather garment 皮革服装
leather gloves 皮手套
leather goods 皮革制品
leather handbag 皮手袋;皮手提包;真皮手袋
leather hat 皮帽
leather hollow 有孔皮革
leather hook 皮带钩
leatherine 人造皮,人造皮革
leathering 镶皮靴
leather jacket 皮夹克
leather jerkin 男式皮外套
leather jewelry 皮首饰
leather label by hot-stamping 压印皮牌
leather label 皮牌
leather lining 皮革衬料
leather-like 皮革状的
leather-like coating 仿皮涂层
leather-like fabric 仿皮面料,仿皮织物
leather-like material 人造皮革;革状材料
leather look 光亮风貌,光亮外观
leather lookalike 仿皮革
leather mesh 皮编鞋
leather money pouch 皮钱袋
leather nailing machine 钉皮机
leather necktie 皮领带
leather nub 包皮纽扣;皮革包扣
leatheroid 人造革,纸布,纸皮
leather overcoat 皮外衣
leather packing 皮垫衬
leather palm gloves 皮掌手套
leather palmed wool gloves 皮掌绒手套
leather patch 皮革贴饰;皮牌
leather point needle 缝皮车针
leather printing 皮革印花
leather product 皮革产品
leathers 骑马皮裹腿;皮马裤
leather sandals 皮凉鞋
leather-sewing needle 缝皮料的针
leather shears 皮革剪刀
leather shoes 皮鞋
leather shoes style rain shoes 皮鞋式雨靴
leather shorts 皮短裤
leather side 皮面,皮革面
leather silk 日本皮革绸

leather skirt 皮裙
leather skiving machine 削皮[革]机
leather sleeve 皮袖子
leather slippers 皮拖鞋
leather socks 皮短袜
leather sole 皮(鞋)底
leather sole and heel construction 皮底皮跟结构
leather sole and wooden heel construction 皮底木跟结构
leather sport clothing 皮质运动衣
leather sport jacket 运动皮夹克
leather strap 皮套环
leather strips 皮条
leather substitute 皮革代用品
leather suspender 皮吊裤带
leather tassel 皮制饰坠,皮制穗饰(流苏)
leather throngs 皮制人字凉鞋
leather top 皮上衣
leather trimming 皮革饰边,皮革装饰
leather trimming machine 削皮[革]机
leather trousers 皮裤
leather trunk 皮箱
leather vest 皮背心
leather ware 皮革制品
leather ware sewing machine 皮革制品缝纫机
leather wear 皮革服装,真皮服装
leather worker 皮革[制作]工人
leather work 皮革制品;皮革制作
leather working 皮鞋加工,皮革制作
leather work(ing) gloves 皮革劳动手套
leather wove 布皮
leathery look 似皮革外观
L'EAU 依泉
leaven 潜移默化的影响;色彩;气味
leaver lace 列韦斯精细花边
Lecien Lelong 勒西安·勒隆(法国时装设计师)
leclercq dupire yarn 棉芯毛纱花线
Lecoanet Hemant 勒科安涅·埃蒙(法国时装设计师)
Lectrolite fabrics 累克屈罗赖特多功能织物(商名)
leda cloth 毛天鹅绒
lederhosen 吊带花饰皮裤;围裙式皮短裤
LEE 李
Lee shin woo 李信雨(韩国服饰设计师)
Lee Young hee 李阳姬(韩国服饰设计师)

leek green 韭葱绿(暗黄绿色)
leeway 余量
left bank look 左岸风貌(美国)
left elevation 左视图
left feed top stitch folder 左送料面缝折边器
left fly (衣)门襟;(裤)门袢;左门襟
left fly edge (衣)门襟止口;(裤)门袢止口
left fly edge line 左门襟止口线
left fly inside line 裤门襟里口线
left fly interlining 裤门襟衬
left fly lining (衣)门襟里布;(裤)门袢里布;左遮扣盖贴边衬里
left fly outside line 裤门襟外口线
left folder 左向折边器
left forepart 左前襟
left front 左前胸;左前身;左前片
left front opening 左大襟;左衽
left front opening edge 左盖右的门襟
left-handed sewer 左手缝纫机
left-hand machine 左手缝纫机
left-hand needle 左面针
left-hand presser 左压脚
left-hand twill 左斜纹
left-hand twist S捻,左手捻,顺手捻
left-off (衣服等)穿旧的,不用的
leftover (bit and pieces) 边角料
leftover material 余料
left-over-right 男式门襟
left pocket T-shirt 左胸袋T恤衫
left side 反面;左边,左侧
left sidebuttoning front 左大襟;左衽
left side panel 左嵌条
left sleeve 左袖
left topside 左腰头
left-to-right twill 右向斜纹
left twill 左斜纹
left twist S捻,左手捻,顺手捻
left view 左视图
leg 花边连线;线圈一侧;假腿,腿,下肢;裤管;裤脚;袜筒,靴筒
legal operation[dealing] 合法经营
legal person 法人
legal representative 法定代理人
legal unit of measuremement 法定计量单位
leg armour 腿甲;护腿甲
leg art 大腿艺术
leg binding 裤脚管滚边
leg boots 长筒靴,深靴

leg counter 袜筒线圈横列计数器
leg forming press 裤管成形熨烫机
leg gather 腿部褶裥
legger press 裤管压烫机
leggin 裹腿；开裆裤，儿童护腿套裤
legging hose 西伯利亚庆典用鹿皮靴
leggings 裹腿，绷腿，腿罩；护脚；儿童护腿套裤，开裆裤，儿童防寒紧身裤；袜筒
leg guard 护腿
leggy 双腿修长匀称的，裸露大腿的
leggy figure 双腿修长匀称的体形
leggy look 露大脚款式
leggy style 露大脚款式
lleg harness 护腿甲
leg hem 裤脚管滚边
leg hole 裤脚口；裤口
eghorn 草帽黄（淡橘黄色）；意大利麦秆草帽
leghorn hat 意大利草帽，意大利女帽，意大利麦秆辫草帽
legionary cap （旧时）军团帽
legion blue 军团蓝
legionnaire's cop 士兵帽
leglet 手镯，脚镯，脚饰带
legline 裤腿线
legline style 腿线造型
leg knit 袜子
leg length 袜筒长，袜子长度
LEGO 乐高
leg-of-mutton 羊腿形袖子
leg-of-mutton sleeve 羊腿袖
leg opening 裤脚口；脚口宽；裤口；腿部开口
leg opening is uneven 脚口不齐
leg opening rib 裤口罗纹
leg opening width 裤口宽
legs 裤脚
leg's ankle 脚踵
leg seam 裤管缝；袜筒缝
leg shape 脚形
leg type 腿型
leg warmer 演员针织护腿；保暖腿套，（保暖）袜套
leg wear 腿部着物；袜子；护腿
leg width 裤管宽；裤中裆
leg wrappings 绑腿
leg yarn 袜筒纱；袜筒丝
lei 夏威夷花环
Leicester Polytechnic College 莱斯特理工学院（英）

Leicester wool 莱斯特有光长羊毛
leiderhosen pants 莱德豪森短裤（面料用仿麂皮，装有背带）
Leipzig yellow 贡黄，铬黄，莱比锡黄
leisure 丝绒边，绸缎边；休闲，闲暇
leisure and sport wear 休闲与运动服装
leisure bra 家常胸罩
leisure clothes 休假服
leisuer clothing 休闲服
leisure coat 休闲外套
leisure goods 休闲用品
leisuer jacket 休闲夹克衫
leisuer shirt 休闲衬衫
leisuer shoes 轻便鞋，休闲鞋
leisuer suit 便装，休闲套装
leisure sport wear 休闲运动装
leisure style 休闲风格
leisure suit 便服套装，休闲套装
leisure wear 便服，休闲装，假日服装
Leitz lustremeter 莱兹光泽计
lemon 柠檬黄
lemon drop 柠檬[水果]糖色
lemon[yellow] 淡黄色，柠檬黄
lemon yellow pale 浅柠檬黄
lemonade 柠檬水色
lenet linnel 薄麻绸
L'ENFANCE ET LA MODE 《儿童装苑》（法国）
lenghas 印度条格或纯色土布
length （身体、服装的）长度；一段布；服装下摆长度
length ease 长度放松量
lengthen （衣服）放长
lengthening arm (extension bar) 伸长臂（接长杆）
length excluding fault portion 净长
length from shoulder-neck point to showlder point 肩项宽
length grain (L.G.) 直丝绺；经纱方向，经向
length line 袖长线；衣长线；裙长线；裤长线；上平线
length of back 背长
length of clothes 身长；衣长
length of curved parts 人体体表实长
length of cut 匹长
length of effective pilk 有效绒头长度
length of foot 脚长
length of front 前身长
length of repeat 花纹循环长度；印花天地

尺寸
length of sack 上衣长度
length of top 袜口长
length over all (LOA) 全长,总长
3/4 length sleeve 七分袖
lengths of material 件头衣料
length-to-width ratio 长宽比
lengthways grain 经向丝绺
lengthwise (针织物)纵向;直向
lengthwise fold 排料时直向对折,纵向折叠
lengthwise grain 经纱方向;直丝绺
lengthwise grainline 直向布纹
lengthwise setting 纵向定位
lengthwise streaking 纵向条痕
lengthwise stripe pattern 纵条花纹
lengupa cotlon 哥伦比亚海岛棉
Lenin cap 列宁帽
Lenin clothing 列宁装上衣
Lenin coat 列宁装;列宁装上衣
Lennon specs 伦农太阳眼睛
leno 罗,纱罗,纱罗织物,(棉型)珠罗纱,通花布
leno and gauze 纱罗;纱罗织物
leno brocade 花纱罗,纱;提花纱罗织物
leno cellular 圆筒形纱罗织物
leno cloth (棉型)纱罗织物;罗;通花布
leno cotton 印花棉布
leno edge 纱罗布边
leno fabric 纱罗织物
leno-fastening 纱罗边
leno handkerchief 纱罗手帕
leno jacquard handkerchief 提花纱罗手帕
leno lady's dress 纱罗女衣呢
leno lady's dress worsted 纱罗女衣呢
leno-like cloth 罗布
leno-marquisette 薄纱罗
leno selvage 纱罗布边,绞边
leno selvedge 纱罗布边,绞边
leno silk 罗
leno stripe 纱罗条纹
lens 透镜
lens barrel 镜筒
lensu 手织菱格方布巾
lens wipes 擦镜布
Leon Baskt 利昂·巴斯克(俄国著名戏剧服装设计师)
Leonard 莱那多(法国著名戏剧发型设计师)

Leonardoda Vinci 利那多·达·维茨(意大利著名戏剧服装设计师)
Leontine 莱昂廷斜纹绸
leopard 豹皮;豹皮外衣
leopard cat 斑点野猫皮
leopard fur 豹[毛]皮
leopard pattern 豹皮花纹
leopard print 豹皮印花,仿动物毛皮印花
leopard skin 豹[毛]皮
leop yarn overcoating 花圈线大衣呢
leotard (杂技、舞蹈演员穿的)低领紧身连裤袜;(运动员或舞蹈演员穿的)紧身衣,高领长袖紧身衣;(妇女穿的)紧身下装,连裤袜;芭蕾紧身衣,长袖紧身连衣裙,低领口紧身衫裤,短裤紧身连衣裤,针织紧身连衣裤,女子体操服
leotard shirt 高领长袖紧身衣
leper's protective shoes 麻疯防护鞋
lepsha leggings 莱沙裤
Leroy 利洛(法国著名戏剧服装设计师)
lesenphants 丽婴房
less ease 吃势不足,放宽不足
lesser panda fur 小熊猫毛皮,小猫熊毛皮
less-expensive garment 廉价服装
lesso 浅色印花布
less over-lap 前叠合处很浅的双排扣型
let 让,放
let down 服装放长
letered sweater 字母毛衣
leting out 服装改宽
let out (衣服)放大,放宽,放出(吊边);抽刀(皮革)
let out round waist (衣、裤)放腰
let-out sections 伸长片(毛皮)
let-out skin 抽刀的皮,展长毛皮
let-out strips 伸长片(毛皮)
letter buckle 字母带扣
letter cardigan 绣字开襟衫,绣字毛衫;文字贴布刺绣毛衣;字母卡迪根毛衫
lettered buckle 有字母的带扣
lettered jacket 字母夹克
lettered silk 字纹花绸
lettering 文字图案
lettering stencil 镂空字母型板
letter of credit 信用证
letter sweater 字母毛衣
lettice 白色毛皮;灰色毛皮
lettice bonnet 貂皮帽
lettice cap 貂皮帽

lettice ruff 环形绉领
letting down （服装）改长
letting out （服装）改宽
lettuce edge 莴苣叶滚边；无褶荷叶边；波浪边
lettuce edging 曼柔缘边；莴苣叶形滚边；小皱边
lettuce green 莴苣绿（黄绿色）
leuco acid dyeing process 隐色酸染色法
leuco dyeing process 隐色体染色法
leucorosaniline 隐色品红
levant cotton 利凡特棉
levantine 利凡廷；利凡廷衬里布，利凡廷里子绸；斜纹薄绸；19世纪软丝绒
levant leather 利凡特皮革（摩洛哥）
levator 提肌
level 程度；水平；水平面
level cut carpet 平绒地毯
levelling 片匀（制革）
levelling acid dye 匀染性酸性染料
levelling machine 匀革机
level loop pile floor covering 平齐绒圈地毯
level pile 平齐绒头织物
level pile thickness 平齐绒头厚度
level shade 均匀色泽
lever 杆
levers' lace 机制提花列韦斯花边
levers' lace fabric 列韦斯提花花边织物
leviathan canvas 绒线刺绣十字布
leviathan stitch 双十字绣，路轨线迹
leviathan wool 刺绣毛线
leviathan yarn 刺绣多股毛纱；羊毛刺绣缝纫线
levigate 磨光
levis 牛仔裤，深蓝铜铆钉的包腿牛仔裤
Levi's（Levis） 李维斯；李维斯牛仔裤（商名）
Levi's 501 直线型李维斯暗纽牛仔裤（裤襟处用铝暗纽）
Levi's 502 直线型李维斯拉链牛仔裤
Levite gown 莱文蒂袍
Levi shade 深蓝牛仔裤色泽
Lewis stitch 刘易斯线迹；刘易斯缝纫
Lewis style 刘易斯型
Lewis tweed 刘易斯粗呢
lēzard ［法］蜥蜴皮革

lezarde 法国家具用带
LFI 《她》（意大利）
Lhenga 印度女短圆裙
Lhoba ethnic costume 珞巴族服饰
liabilitor 责任人
liang-di gauze 亮地纱
liang-sha striped pongee 凉纱绸
liang-yan leno brocade 凉艳纱
lian-jin leno brocade 帘锦罗
liao ghatpot 缭绫
liao ghatpot 缭绫（中国古代织物）
liao-feng boucle 辽凤绸
Liaoning pongee 辽宁柞丝绸，辽丝纺；柞丝平纹绸
Liaoning tissue gauze 辽宁柞丝绸
Liaoning tussah silk 辽宁柞丝绸
Lias 突出胸部的金属丝框
libas 埃及膝男裤
libau 利博粗亚麻
libelle 利贝儿发束
liberation shoes 解放鞋
liberty blue 解放蓝
liberty bodice 自由紧身背心（旧时儿童穿的厚棉布内衣）
liberty cap 自由帽（无檐锥形软帽，18世纪法国），古希腊自由帽，法国圆锥帽，自由帽（帽檐狭窄），弗里吉亚无檐帽
liberty cloth 手工模板印花布；英国优质自由布；利伯蒂印花布
liberty knot stitches 乱结粒绣
liberty line 自由型
liberty prints 利伯蒂印花布，手工模板印花布，自由式印花花样
liberty satin 自由缎；光亮缎带
liberty stripe 棉条子斜纹劳动布
Libform 列韦福姆锦纶织物
library buckram 书面帆布，书面硬布
libret 埃及粗亚麻布
licence 许可证
licensed trademarked fiber 注册商标纤维
lichen 青苔色
licorice 甘草色
lid 帽子
Lido collar 燕子领［欧洲］
lie down shape （垫肩式）大翻领
liegn 扣子尺码
lienceillo 棉坯布
lienzo 南美本色平布
Li ethnic costume 黎族民族服饰

life 活体模型;有效期
life belt 保险带,安全带;救生圈
life buoy 救生带;救生圈;救生衣
life cycle 生命期,流行期,循环周期
life cycle assessment 生命周期评价
life device for the iron 熨斗提吊设备
life duration 耐用期,使用寿命
life idea 生活意识
life jacket 救生衣
life mask 面模,含碳过滤器的口罩
life model （真人）模特儿
life preservers 救生器
lifes 人体模特儿
life-saving fabric 救生[器具]用织物
life-saving jewelry 救生手饰
life-saving vest 救生背心
life-size 等身像;真人大小
lifesome 充满活力的,生气勃勃的
life style 生活方式
life style trends 生活方式趋势
life-support system 生命线（宇航员提供氧气路线）,维持生命系统
life time 使用期限;使用寿命
life vest 救生背心,救生衣
lift 鞋后跟皮
lift curl 莱夫脱针型发卷
lifter 压脚扳手
lifting arm 举起臂
lifting lock motion 升降定位机构
ligament 韧带,系带
ligature 系结物
lige crepe 丽格绉
limiting oxygen index [极限]氧指数
light 光,光线;光泽;浅色
light accent 亮部重点
light adaptation 明视适应性
light ageing 光老化
light and shade contrast 明暗对照,明暗反衬
light and shade 明暗
light azalea 淡杜鹃色
light bean green 浅豆绿
light blue green 淡蓝绿色
light blue 淡蓝色,浅蓝色;浅青色
light bluish green 豆青色
light brown 淡褐色
light casque 轻便头盔
light cerulean 浅青色
light chartreuse 淡黄绿色,淡荨麻酒色
light chestnut 淡栗色

light clear blue 淡青蓝,浅湖绿[色]
light clear series 明澄色系
light clothing 轻便服装;轻装;轻薄服装
light color 明色,浅色
light-colored clothes 浅色服装
light-colored shirt 浅色衬衫
light color effect 浅色效应
light contour 轻笔勾勒轮廓
light crack 光致裂缝
light cutter 带灯光的裁剪机
light dancing shoes 轻便跳舞鞋
light duty 轻型
light effect finish 光效应整理
light elm green 浅榆树绿
lighten 变亮,发亮;色彩调淡
lighten end product 轻质产品
lightness 明度
lightening change 闪电式换装（模特等）
lightening stitch 闪电形线迹,伸缩线迹
lighter pendant 轻型垂饰
light fabric 薄织物,轻薄织物
light fastness 耐光性,耐光牢度
light fastness rating 耐光色牢度等级
light filling bar 细纬档
light fleece 薄绒
light fleecy sweatshirts 薄绒衫
light gold 浅金色
light gray 浅灰色
light grayish green 灰豆绿色
light grayish(ly) 明灰色
light green 浅绿色,品绿色
light greenish blue 淡青色,粉青色
light grey 浅灰色
light ground 明亮背景
lightgun 光笔
lighting plot 舞台照明分布图
light ironing 微烫,轻烫（不需上浆、喷水,用力小）
light jack boot 有边饰的软皮靴
light jeans 轻薄牛仔裤
light knitted fabric 薄针织物
light leather goods 轻皮衣服
light leather material 薄型皮革料
light leather stitching 薄型皮革缝纫
light lilac 浅丁香色
light lime green 浅灰绿[色]
light line 细线
lightly printed pattern 浅[色]印花图案
light mahogany 浅赤褐色

light material	薄缝料
light milled cloth	轻质缩绒毛织物
light mohogany	浅红木色
light multi	纯彩
lightness	亮度,明度;浅淡色;轻巧感
lightness basic tone	明度基调
lightness contrast	明度对比
lightness difference	明度差
lightness index	明度指数
lightning express cotton	美国湿地棉
light olive	浅橄榄色
light olive green	淡橄榄绿
light orange	淡橙黄色,浅橘黄色
light overcoat	夹大衣
lightpen	光笔
light pink	浅粉色
light place	薄段织疵,细纬裆
light plot	舞台照明分布图
light polyurethane weft knit	薄型氨纶纬编针织物
light power	光度
light print	浅色印花
light purple	淡紫色
light rain boots	轻便雨靴
light raincoat	轻便雨衣
light red	淡红色
light-reflectant finish	织物光反射整理
light resistance	耐光性
light rose	浅玫瑰红色
light rose violet	浅玫瑰紫色
light rubber beach shoes	微孔胶拖鞋
light rubber shoes	胶便鞋
light sapphire	浅蓝色
light scarlet	大红色
light sensitive color printing	光敏印花
light sensitive discolor fabric	光敏变色织物
light sensitive fiber	光敏纤维
light sew-in	薄型可缝衬
light shade	浅色
light source	光源
light source color	光源色
light spotted	一级白度棉
light stone wash(ing)	轻石磨洗
light suit	轻薄套装,夏日套装
light sulfur	浅硫黄色
light sulphur	浅硫黄色
light tan	驼色
light tangerine	浅橘红色
light taupe	浅灰褐色
light to medium heavy material	轻薄至中厚缝料
light tone	浅色调
light trimmers	轻质剪刀
light value	亮的层次
light weight	轻量
lightweight denim	轻身牛仔布;薄牛仔布
lightweight fabric	薄料,薄型织物,轻织物
lightweight fabric sewing	薄料缝纫
lightweight fancy suiting	薄花呢
lightweight interfacing	轻薄衬布;薄内衬
lightweight jacket	轻薄夹克衫
lightweight lace	轻质花边
lightweight material	轻薄缝料
lightweight nylon canvas	轻质锦纶帆布
lightweight plush	轻薄长毛绒
lightweight poncho	轻便斗篷
lightweight sewing machine	轻型缝纫机
lightweight zipper	细牙拉链
light yellow	鹅黄色,浅黄色
light yellowish grey	淡黄光灰
light zipper	细齿拉链;小齿拉链
ligia	利吉奥围巾(秘鲁)
ligne	莱尼(号,用于纽扣)
lignette	低级薄哔叽
lignin-based carbon fiber	木质素系碳纤维
ligno-cellulose	木质素纤维
ligong lame	立公缎
liguette	衬衫式外衣(男用)
liguid ammonia finish fabric	液氨整理织物
liguid ammonia mercerizing	液氨丝光
liguid marking pen	液体划线笔
lihaf	手纺纱棉地毯
liihographic printing	平版印花;石印印花
lilac	淡紫,丁香紫,淡雪青,紫藤色
lilac gray	淡紫灰色
lilac grey	淡紫灰色
lilac rose	丁香红
lilac snow	淡紫色
lilac white	淡紫白
lilies	缠足,三寸金莲(旧时中国)
lille arras lace	利累阿腊里花边,梭结网眼花边
Lille lace	里尔花边
lille lace	六角网眼地棱结花边
lille tapestry	法国里尔挂毯

Lillian costume 莉莲·拉塞尔装束;莉莲女服
Lillian Russell costume 莉莲·拉塞尔装束
Lillian Russell silhouette 莉莲·拉塞尔型轮廓,莉莉安·罗茜型轮廓
Lilly Daché 莉莉·达歇(1904～,美国帽子设计师)
Lilly Pulitzer 利利·普利策(1932～,美国时装设计师)
lily 百合花;洁白的
lily Benjamin 白色男工作服
lily[color] 百合色
lily-footed look 三寸金莲风貌(旧时中国)
lily green 百合花绿,淡黄绿色
lily prince's dress 百合花王子服饰
lily sleeve 百合花袖,小喇叭袖
lily white 纯白的,纯洁的
lily-yarn 针织线绳
lima bean 利马豆色
Lima cotton 秘鲁利马棉
limbric 英国林姆布里克高级细平布
limbus 色差极大的明显边缘
lime 酸橙绿色
limeade 酸橙汽水色
lime blossom shade 椴绿(浅黄绿色)
lime cream 酸橙奶油色
lime green 酸橙绿(暗黄绿色)
lime juice 酸橙汁色
lime light 石灰光色
lime peel 酸橙皮色
lime punch 酸橙混合饮料色
limerick lace 爱尔兰挖花针绣花边,网地针绣花边,利默里克花边
limerick gloves 羊羔皮手套,利默里克手套
lime stone 石灰石色
lime yellow 酸橙黄色
limit 限额
limit[ed]centre plait folder 中间限位折裥器
limited wash fastness 有限洗涤牢度
limiting-oxygen-index (LOI) 极限[耗]氧指数,限氧指数
limiting quality level 质量合格标准,质量合格限度
limit size 极限尺寸
limoges 褡套条子布;法国大麻粗袋布;青宝蓝

limousine 利莫森精纺缩绒呢;精纺山形斜纹呢;女式及地披风
limp cloth binding 软布面
limp collar 软领
limp fabric 处理不当的过软织物
limpness 柔软度
lincere 亚麻细布
linceul 法国床单布
Lincoln green 林肯绿;林肯绿呢
Lincoln lamb 林肯羔羊皮
Lincoln wool 林肯羊毛(林肯绵羊毛)
lincord 粗线纽孔
Lind star dress 林德星夜礼服
Lindbergh jacket 林白夹克衫(早期朴素的飞行员夹克)
Linden cord 林登绳
linden green 椴绿(浅黄绿色)
linden yellow 椴黄
linde star dress 林德星夜礼服
lindiana 毛经丝纬绉布
Lindner look 林德纳款式
Lindner style 林德纳款式
lindsay 林赛彩色格子呢
line (设计、制图的)线条;线形;给(衣服)装衬里;划线于;线带,绳,袢;流水线;品种系列;纹路;行业;型
line a garment with fur 用毛皮衬里
LINEA INTIMA 《内衣款式》(意大利)
LINEA ITALIANA 《意大利服饰》《意大利》
LINEA ITALIANA UOMO 《意大利男子服式》(意大利)
lineaments 面部轮廓;面貌
line and colour 线与色
linear art 线条艺术
linear compositon 线条构图
linear correlation 线性相关
linear definition 勾线,线条勾出轮廓
linear density 线密度
linear optimization 线性优化
linear pattern 线型图案
linear programming (LP) 线性规划
linear regression 线性回归
linear yard 码长,(统幅)每码
lineation 斑纹,画线,皱纹(皮肤表面)
lined 衬里的,有里子的
lined atlas 起绒经缎织物
lined cloth 双层毛织物(正、反面不同组织或色泽)
lined clothes 夹衣,夹大衣

lined dress 夹衣
lined garment 夹衣
lined hood with lappets 有垂带的亚麻头巾
lined jacket 夹衣;夹袄
lined leather 布衬里皮革
lined legging 夹绑腿
lined long-gown 夹长衫,夹长袍
line drawing 铅笔素描;线描;线条描
line drawing(in traditional ink and brush style) 白描,素描,线条描,线描;绘线图型[计算机]
lined robe 夹长袍
line dry 悬挂凉干
lined satin 三梭栉经缎织物
lined short gown 短袄
lined skirt 全夹裙
lined sleeve 夹里袖
lined tricot 三梭栉起绒经编织物
lined trousers 夹裤(有衬里)
lined undercoat 夹内衣
lined undervest 夹贴心内衣;夹贴心背心
lined vest 夹背心
lined warp knitting 起绒经编织物
lined work 菱形光纹织物
line effect 线状花纹
line fiber 亚麻纤维
line-for-line copy 完全仿造,依样仿制
line from line 依样仿制
line from shoulder to waist girth 前腰节线
line graph 线图
line layout 生产线布置
line linen yarn 长纤亚麻纱
lineman's boots 养路工皮靴
linen 白色衬衣裤;白色内衣裤;亚麻,亚麻布(或纱、线);亚麻织物;亚麻制品;仿亚麻制品;亚麻色
linena 充亚麻平纹棉布
linen and worsted dress unions 麻经毛纬条子呢(亚麻)
linen basket 脏衣物筐
linen batiste 童装用漂白亚麻平布
linen blouse 亚麻女衬衫
linen button 布扣,亚麻布包扣
linen cambric 平纹亚麻织物,亚麻平布
linen canvas 亚麻帆布
linen cap 亚麻布便帽,亚麻布帽
linen checks 亚麻条格布
linen[cloth] 亚麻布,亚麻织物

linen/cotton blended fabric 麻棉布(亚麻)
linen/cotton blended yarn 麻棉纱(亚麻)
linen/cotton handkerchief 棉麻手帕
linen/cotton shirt 麻棉衬衫(亚麻)
linen crash 亚麻粗布
linen crepe 亚麻绉布
linen damask 亚麻锦缎;亚麻花布
linen duck 亚麻帆布
linen duster 女式亚麻布宽松外衣;亚麻风衣
linene 仿亚麻平纹棉布
linen embroidery 亚麻布刺绣
linenette 仿亚麻织物
linen fabric 亚麻织物;亚麻布
linen finish 仿亚麻整理
linen frieze 充马海毛粗花呢;亚麻卷毛厚绒织物
linen goods 亚麻布料;亚麻制品
linen handbag 亚麻手袋
linen handkerchief 亚麻手帕,亚麻手帕布;麻纱手帕布
linen hands 亚麻的手感
linen industry 亚麻工业
linen-like 仿麻的,仿麻型
linen-like fabric 仿麻织物
linen-like finish 仿麻整理,仿亚麻整理
linen-like polyester crepe 涤纶仿麻绉
linen-textured 仿亚麻布的
linen-textured rayon fabric 仿(亚)麻黏胶纤维织物
linen lining 亚麻布衬里
linen look 麻型,麻型织物外观
linen mattress 俄罗斯亚麻褥垫布
linen mesh 亚麻网眼布
linen out pattern 花样超边
linen oyster 半漂或浅色刺绣亚麻布
linen plains 亚麻平布
linen pot 亚麻联匹
linen roughs 帆布型亚麻(衬里)布
linen rug 走廊地毯;亚麻地毯布
linen scrim 剧院纱罗
linen sheeting 亚麻被单布
linen silk 里子绸;蜡纱美丽绸;蜡线羽纱
linen straw 整理过的亚麻编织物
linen taffeta 里子塔夫绸
linen tape 亚麻带
linen tape ruler 皮卷尺
linen textile 麻纺织品
linen thread 亚麻线

linen touch 亚麻布触感
Linen Trade Association ［美］亚麻贸易协会
linen tussore 条格亚麻色织布
linen type cloth 仿麻布,仿亚麻织物
linen type polyester filament cloth 合纤长丝仿麻布
linen underwear 亚麻内衣
linen velvet 丝绒衬里
linen yarn 亚麻纱
line of beauty 线条美,曲线美,造型美
line of collar point 领尖点线
line of front cut point 门襟圆角点线
line of sewer 缝纫机系列
line panty stocking 有缝连裤袜
line production 流水作业法
liner 班轮;衬里;夹里;托布;胆料;衬里针织手套;胶合麻袋;描唇笔;眼线笔;眼线膏;眼线刷
liner band 衬带
liner band sewing machine 衬带缝纫机
line ring 线戒
liner leather 皮衬里
liner look 条纹款式
liners 衬里针织手套;胶合麻袋
liner shipping 班轮运输
liner term 班轮条款
liner waybill 班轮运单
lines 概况,设计思路;地毯纵向结头行数
line softener 眼线刷
line stitch 线状刺绣针迹
line supervisor 流水线监督员
line system 流水线系统
linet 法国未漂亚麻衬布
line tape 卷尺
linette finish 仿麻整理
line-up 使平直
line work 菱形花纹织物
linge 法国亚麻布
lingerie 女内衣;女衬衫,女式贴身衣裤,女睡衣;亚麻制品;内衣用亚麻布
lingerie and corsetry 妇女套装
lingerie crepe 平绉
lingerie dress 内衣式礼服
lingerie elastic 内衣松紧带
lingerie gown 女式睡袍
lingerie hat 法国刺绣帽,内衣帽
lingerie hem 花边
lingerie look （女）内衣款式;衬裙式;内衣外穿风貌

lingerie making up 内衣裁缝
lingerie ribbon 内衣缎带,内衣窄带
lingerie satin 女式内衣丝缎
lingerie scissors 薄布剪刀
lingerie seam 女式衬衣缝
lingerie set 女内衣缘饰
lingerie shears 薄布剪刀
lingerie stitch 女式衬衣线迹
lingerie strap guard 按扣牵带,按扣线绊
lingerie style 内衣款式
lingerie tape 内衣系带
lingette 阴影条子缎（商名）
Linguette 法兰绒,轻薄哔叽
lingyan damask satin 灵岩缎
linin facing stay 护条皮
lininess 纵条疵点
lining （服装的）里料;衬里;里子;装衬里;夹里;衬料
Lining 李宁
lining back 后身夹里,后夹里
lining cambric 衬里细薄布;衬里麻纱
lining cloth 衬布;里子布
lining drawing 衬里拉伸
lining duck 棉帆布衬里
lining fabric 衬里织物;黑炭衬;里料;衬里,里子
lining facing 鞋里贴边
lining facing stay 护条皮
lining felling 缝衬里,绷衬里
lining felling machine 缝衬里机
lining felt 衬毡,隔垫毡衬,毛毡衬里
lining flannelette 绒布衬里
lining front 前夹里,前身夹里
lining full 全里,全衬
lining fur 毛皮衬里
lining guarter 肘部衬（肘衬）
lining half 半里,半衬
lining hem 夹里下摆
lining lambsdown 驼绒衬里
lining mohair cloth 马海呢衬里
lining pants 有夹里的裤子;夹裤
lining pattern 夹里［的］纸样
lining plate 衬板
lining pocket 衬里袋
lining quarter 肘部衬;胸部衬
lining satin 里子缎
lining sewing machine 衬里缝纫机
lining silk 蜡线美丽绸;蜡线羽纱;里子绸;纱纬美丽绸
lining sleeve 袖夹里;有夹里的袖子

lining stay 衬里牵条
lining stretcher gauge 衬里伸缩规
lining taffeta 里子塔夫绸
lining thickness 衬里厚度
lining too loose 衬里布反吐
lining too tight 衬里布松份不足
lining velvet 丝绒衬里
lining-faced waistband 腰里,腰夹里,腰头夹里
lining-interlining 夹里及衬
linings 衬里布,里子布
link 连接;链环;链节;纽带;鸳鸯扣,(衬衫的)袖口链扣
link and link 双反面针织
link bracelet 链状手镯
link button 连接型双扣
link cuff 系扣袖口,系扣袖头
linked 用金属片、毛皮、塑料等缝合的服装
linked dress 将金属、毛皮、塑料等几何图形衣料缝合而成的服装
linker 套口机;缝袜头机
linking 缝合,套口,缝合
linking bar 套口架
linking course 套口横列
linking jig 套口架
linking machine 缝袜头机;针织缝盘机;合缝机;套口机;捆边机
linking seam 套口缝,连圈缝合
linking strand 地毯底布连接线
link mark 连接线符号,延展连接线符号
link motion 连杆运动
link powdering stitch 环点绣
links （衬衫袖口上的)扣钮;袖口纽
links and links hosiery machine 双反面袜机
links and links jacquard hosiery with three colors 三色凹凸提花袜
links and links jacquard hosiery with two colors 双色凹凸提花袜
links and links knitting 双反面编织
links and links machine 双针筒圆形针织机;平板针织横机
links and links socks 凹凸提花短袜
links and links stitch 双反面线圈
link[shirt] 袖头链扣衬衫
links-jacquard 凹凸提花
links jacquard embroidery hosiery machine 绣花双反面袜机
links jacquard hosiery machine 提花双反面袜机
links-links 双反面
links-links circular knitting machine 双针筒圆形针织机
links-links design 双反面花纹,双反面图案
links-links fabric 凹凸针织物,双反面针织物;凹凸织物,双反面织物
links-links flat bar knitting machine 平板机,双反面横机
links-links knitting machine 双反面针织机
links stitch 链状线圈,反面线圈
link type thread take-up lever 连杆式挑线杆
linlang damassin 琳琅缎
linli satin brocade 琳丽缎
linnea 非洲市场印花棉布
linneas 印度棉布
linnet 仿麻织物
linnet finishing （织物)仿麻整理
linnette 防绒布,仿麻织物
lino 丝织面纱;丝织薄纱罗
linoleum 漆布;油毡
linomple 法国细麻布
linon a jour 亚麻纱罗织物
linon 法国仿亚麻薄纱;上等细布
linsay 亚麻经羊毛纬交织物
linsel 法国毛麻交织物
linsey 林赛布;亚麻经羊毛纬交织物;含棉毛织物
linsey-woolsey 麻经毛纬交织物,毛麻交织呢,棉经毛纬交织物
linshang 双宫竹节绸
linsle[thread]stockings 莱尔长筒袜
lint ball 起球疵
lint 皮棉,棉绒;(织物上磨落的)纤维屑,软麻布;软麻布色
linters 棉短绒
lint-free uniform 无尘服
lintheamina 亚麻床单
lint index 衣指
linting 织物上磨脱纤维屑
Linton tweed 林顿粗花呢
lint pick-up 起毛起球
lintress 丝织物
lintrius 亚麻床单
Linum usitatissimum 亚麻
lion 法国平纹亚麻布;法国小花纹亚麻布;雄狮色

lion ornament 狮子纹样
lion pattern 狮子纹样
liotardo stripe 利奥塔多条子布
lip 唇,嘴唇;端,边,缘
lipbrush 唇膏刷;唇笔
lip cream 油质唇膏
lip gloss 珠光唇膏
lip layered 层叠式口红
lipophilicity 亲油性
lippish white linen 大麻布
lip pomade 油质唇膏
lip rouge 唇膏
lip shiner 闪光唇膏
lipstick 唇膏;口红;唇红(玫红色)
lipstick color 唇膏颜色
lipsticking 搽口红
lipstick pencil 口红笔
lipstick red 口红色
liquette 男用外衬衫(宽敞,前领口低);衬衫式外衣
liquid chemise 流水般女衫
liquid chromatogram 液相色谱(图)
liquid-cooled suit 液冷服
liquid crystal cloth 液晶布
liquid crystal polymer fiber 液晶聚合物纤维
liquid disperse dyes 液状分散染料
liquid face powder 水白粉
liquid reactive dyes 液状活性染料
liripipe 法国帽垂布;尾状长飘带(旧学士帽用);披风(旧法官、教士用);披巾
liripiplum 尖头头巾;法国帽垂布
liripoop 法国帽垂布;尾状长飘带(旧学士帽用);披风(旧法官、教士用);披巾
Liropol 利罗波尔缝编机
Liropol sewing-knitting machine 利罗波尔缝编机
Liroslor knitting machine 利罗斯洛尔针织机
lisardes 埃及粗亚麻布;波斯棉布;印度棉布
lisere 滚边狭带;帕巾狭边;棱纹花缎;窄边,窄带
liseurse 女用室内衣服;休闲便衣
lisiere 布边
lisieux 乡村亚麻布
lisle fabric 高级丝光针织品
lisle hosiery 棉线袜
lisle lace 透明细白花边;莱尔花边

lisle stockings 莱尔长筒袜;丝光线袜
lisle thread 莱尔线
lisle thread stockings 莱尔线袜
lisle top 莱尔(高级丝光线)袜口
lisle yarn 莱尔高级丝光针织线
Lison Bonfiles 利松·邦菲尔(法国时装设计师)
lissé 极薄全丝绉
list 一览表;目录,缘边,布边;饰带;(头发、胡须的)分缝
listados 英国利斯塔多色棉布
listas 英国利斯塔多色棉布
list carpet 碎布地毯
listed piece 色花匹头
listing 缘边,布边;饰带,狭布条
liston 狭丝带
listónes 饰边;饰带;纽扣;窄丝带
list slippers 布边制的拖鞋
list work 衣边贴花,衣边贴饰
Lisu ethnic costume 傈僳族服饰
liteau 横向红蓝条亚麻桌布;织物边部的质量标记
litham 阿拉伯遮脸巾,阿拉伯妇女面纱
litharge 铅黄
lithee 丝塔夫绸
lithographic printed handkerchief 石印手帕
lithograph printing 石印印花
lithopone 锌钡白
Lithuania costume 立陶宛民族服
liti 构树皮纤维织物
litmus blue 石蕊蓝
Litrico 利特雷科(意大利时装设计师)
litt 印染织物
little black dress (L.B.D) 朴素的黑色服装;基型黑色裙装;黑色小衣裙
little blouse 小型女衬衫
LITTLE BOBDO 巴布豆
little-boy legline "小孩"装裤腿线
little boy's and girl's clothes 小儿服
little boy shorts 男童短裤,翻贴边短裤
little brannon 布兰农棉
little cap 无边小帽
little dress 小型连衣裙
little finger (fifth finger) 小[手]指;小指(第五指)
little girl collar 女式小圆角领
little girl look 幼女风貌;幼女款式
little hennin 小塔形垂纱帽
little iron 微烫,稍烫(织物具有免烫的一

般性能)
Little Lord Fauntleroy suit 方特罗贵族小男孩套装;小公子套装
little maid dress 娇小少女装
little nothing dress 简便装
little slit legline 小开叉裤腿结构线
little suit 小型套装
little toe 小趾
little tricorn riding hat 小三角骑士帽
Littleway shoe construction 钉鞋内底
Little Women dress 小妇人装;带蝴蝶结翻领抽褶童裙
liturgical headdress 主教冠,白色锥形僧帽
liuse chiemyong 中国红丝绒
liuxiang crepe jacquard 留香绉
liuxiang jacquard crepe 留香绉
live line garment 带电操作服
liveliness 鲜明感
live model 真人模特儿
live modelling 真人模特儿
livening treatment 丝鸣整理;增艳处理
liver[brown] 肝棕色(红褐色)
liver chestnut 深栗色
liver color 赤褐色;猪肝色
liveried 穿成套衣服的;穿制服的;穿号衣的
livery (侍从,司机)特殊制服;带金银边饰的服务员制服;行会会员制服;号衣;装束(诗)
livery cape 仆人披肩
livery cloth 制服料;制服呢
Livery jacket 利佛略夹克(汽车比赛车手穿)
livery tweed 制服呢;制服料
livery uniform 号衣,制服
livery vest 仆人马甲
livid 青黑色;青灰色
lividity 铅灰;铅色;青黑,青灰;土色
living model (真人)模特儿
living skirt 活力裙
Livio di Simone 利维奥·迪·西莫内(意大利时装设计师)
liyi crepe jacquard 丽谊绉
liyi jacquard crepe 丽谊绉
lizard 军草绿,橄榄绿;粒状皮纹皮带;蜥蜴皮;蜥蜴色
lizard belt 蜥蜴皮带
lizard handbag 蜥蜴皮手提包
lizard leather 蜥蜴皮;蜥蜴革

lizard skin 蜥蜴皮
Liz Claiborne 利兹·克莱本(1929～,美国时装设计师)
llama 美洲驼;美洲驼毛
llama coating 美洲驼绒大衣呢
llama croise 细绒精纺呢
llama lace 黑精纺毛纱梭结花边
llama shirting 高级棉驼毛混纺衬衫布;高级棉驼毛混纺衬衫织物;睡衣织物
llanos 闪光绸(白色棉经,紫或黑色丝或马海毛纬)
load balance 负荷平衡
loadable nonwoven fabric 承重非织造布
load-elongation curve 负荷伸长曲线
loading 装货
loading charge 装货费
loading period 负戴周期
loafer 懒汉鞋;平底便鞋,乐福便鞋;平跟船鞋;拖鞋;洛弗夹克
loafer shoes 懒汉鞋;乐福鞋
loafer with tassles 挪威无带低跟便鞋
lobby carpet 走廊地毯
lobster bisque 龙虾浓汤色
lobster red 龙虾红
local chintz finish 局部擦光整理
local color 独特色;风味;乡土色彩;专属色;自然色
local defect 局部性疵点
local deluster 局部消光
local effector 局部效应
local embossed finish 局部轧纹
local expert 土专家
local flavour 风味
local flocking 局部植绒
local flocking finish 局部植绒
local industry 地方工业
local raised finish 局部起毛
local sanding 局部磨毛
local stretching 局部拉伸
localization 国产化
locate-and-sew procedure 定位与缝制过程
Lochmore 诺其谟
lock (缝纫中)锁缝;锁缝机;发饰;鬈发
lock chain stitch 双线链式线迹
lock-corner 锁角
locket 刀鞘台部;(装有照片或贵重金属纪念品的)项链;牢记项链;(挂在项链下的)保藏纪念品的贵重金属小盒
lock hemming handkerchief 锁边手帕

lock knitting 锁缝编织
lock knot 织布结
lock machine 锁缝机
lock necklace 锁式项链
locknit fabric 经平绒针织物
lock pendant 锁式项链
lockram 亚麻织品;低级亚麻织物
lockstitch 双线锁缝;二重缝;锁式线迹;锁针法;锁扣眼
locksitch bar tacker 锁式线迹套结加固缝纽机
lockstitch bar tacking machine 锁缝打结机;锁式线迹套结加固缝纽机
lockstitch buttonholer 平缝锁眼机,锁眼机,锁纽孔机
lockstitch buttonhole machine 锁扣眼机
lockstitcher 锁式线迹缝纽机
lockstitch hemmer 锁式线迹卷边器
lockstitch(ing) 锁式线缝
lock stitching 连锁缝,锁式针迹
lock stitching button hole 锁扣眼
lock stitching machine 平缝机;锁缝机
lockstitch machine 锁缝机;锁式线迹缝纽机
lock stitch 1 needle 2 threads 单针双线锁式线迹
lock stitch 2 needles 3 threads 双针3线锁式缝迹
lockstitch on buttonhole 锁扣眼
lockstitch pattern tacker 锁式线迹花样加固机
lockstitch seam 锁式线迹缝,包缝
lockstitch seamer 锁式线迹缝纽机
lockstitch sewing machine 平缝缝纽机;锁式线迹缝纽机
lockstitch zigzag 锁式曲折形线迹,Z字形曲折线迹
lockstitch zigzag sewing machine 锁式人字形缝纫机
loco suit 拘束衣(囚犯穿);囚犯拘禁衣
locrenan 法国大麻篷帆布
locstitch pile fabric 双面绒头(毛圈)织物
loden 深灰绿色;洛登缩绒厚呢;洛登缩绒厚呢做的衣服
loden cloth 防水缩绒厚呢;洛登缩绒厚呢
loden coat 洛登(缩绒厚呢)外套;草绿防水布大衣
loden green 深灰绿色;洛登缩绒厚呢绿色
loden hat 草绿防水布帽;洛登帽

loden jeans 洛登牛仔裤
loden skirt 草绿防水布裙
lodeve 法国毛织物
LOEWE 罗意威
L'OFFICIELDES TEXTILES 《纺织》(法国)
loft 膨松;弹性(指毛织物)
loft drying 室内风干,室内晾干
loftiness 松软丰满手感,膨松感
loft wool 优级羊毛,富弹性毛
lofty fabric 膨松织物
lofty hand 弹性感,膨松手感
lofty handle 弹性感,膨松手感
log 裁缝工作时间表[管理]
loganberry 罗甘莓色
log card 记录卡
logger's 伐木工人短夹克
logger's boots 伐木靴
logo 边印;(衣、帽上面的)标识;标记
logo button 标识扣(扣面有特定标记)
logo jeans 标识牛仔裤
logwood 苏木(天然染料)
logwood black 苏木黑
lohi 克什米尔手工纺织洛希厚呢
loi 罗伊细呢
loincloth (热带地区原始民族作为衣服的)缠腰布;遮盖布;遮盖带
loin[cloth]skirt 腰布裙
loita 粗棉布
Loktuft 烯烃纤维无纺地毯
lollipops 棒糖色
Lombardi pattern warp machine 朗伯迪花式圆形针织机
Lome convention 洛美协定
lona 墨西哥洛纳帆布
London binding 伦敦色线棉带
London boy 伦敦男孩发型
London cord 伦敦坚实粗条布
London Fashion week 伦敦时装周(英国)
London Fog 伦敦雾雨衣
London International Collection 伦敦国际服装展(英国)
London Interseason 伦敦时装展(英国)
London Interseason show 伦敦成衣展(英国)
London look 伦敦风貌,伦敦型,保守的优雅男士风格
London Prét 伦敦成衣展(英国)
London shrunk 伦敦预缩整理
London smoke 伦敦烟灰色,暗灰色

London Textile Testing House 伦敦纺织试验所
London Wholesale Millinery Manufacturers' Association(L.W.M.M.A) ［英］伦敦女帽和妇女头饰成批制造商协会，伦敦女帽商协会
londres 伦敦女士呢；法国伦敦细呢；缩绒呢
londrin 隆德林缩绒薄呢
Lone star cotton 得克萨斯棉
loneta 南美棉帆布
long（L）（服装）长尺寸；长裤
long and short stitch 叉针绣；长短针刺绣线迹；长短线迹
long and short tacking 长短粗缝
long armed 长臂羽状绣
long armed cross stitch 长臂十字线
long armed（feather）stitch 长臂（羽状）绣
long arm sewing machine 长臂缝纫机
long-bellied doublet 带假前片的男士紧身上衣
long bell sleeve 钟形长袖
long blouse 长罩衫
long boots 长筒靴
long cape 长斗篷
long carriage machine 长三角座滑架横机，长三角座滑架平机
long chain synthetic polymer 长链合成纤维
long chain zipper 加长拉链
long cloth 上等棉布；漂白平纹细棉布
long clothes 婴儿衣服（初生时）
long coat 长大衣；长毛套
long core hemmer 长芯子卷边器
long corset 长紧身褡
long cut 长发型，长发式
long darner 长织补针
long dash line 长划线；长虚线
long diagonal basting 长斜针假缝
long diagonal tacking 长对角粗缝
long drawers 宽松长袜裤
long dress 女长衣
long ells 粗斜纹呢，粗纺呢
long ends 红色轧光粗斜纹呢，拖纱印疵
longer-than-shoulder length 过肩长
longest 特长号（西装裤尺寸）
longevity shoes 寿鞋
long feed serging machine 长送料包缝机
long figure 高体型

long finger 中指（第三指）
long fitted cuff 紧身长袖口；合体长袖头
long flax 长亚麻
long focus lens 长焦距镜头
long fur coat 长皮大衣
long gaunt woman 身材细长的瘦女子
long gilet 背心式长外套
long gloves 长手套
long-gown 大褂；长衫
long hair 长发型，长发式
long haired fabric 长绒面料
long haired rabbit 长毛兔皮
long haired rug 长毛［地］毯
long hair finish ［毡帽坯］长毛修剪整理
long half slip 长半裙，无上身的长衬裙
long handles 连体内衣
longhee 印度腰布；缅甸腰布；头巾织物
long hemp 长大麻
long hood 带长丝带的女头巾
longies 长衬裤；长内裤；冬季长裤（男孩穿）
LONGINES 浪琴
longitudinal 纵向缝
longitudinal clamps 纵向夹头
longitudinal elasticity 纵向弹性
longitudinal repeat 纵向花纹循环
longitudinal seam 纵向缝
longitudinal shear mark 直剪印
longitudinal shrinking 纵向收缩
longitudinal slot 纵向窄缝
longitudinal view 纵视图
long jacket suit 长上衣套装
long-johns 长衬裤；长内衣裤
Long John trunks 朗·约翰式男泳裤（至膝盖上针织裤，鲜红、鲜绿等与黑色条子间隔，系腰带）
long jumping shoes 跳远鞋
long knee breeches 法式紧身中裤
long lantern sleeve 长灯笼袖
long-lasting 耐久
long lean look 瘦长风貌
long lean style 瘦长款式
long-leg cross stitch 长脚十字缝
long-legged 腿长的
long legged figure 腿长体型
long legged panties 长腿紧身短裤
long leg panties 及腿紧身短裤
long-leg panty girdle 束腰长裤，及大腿紧身束腹
long length 全长，长号长度

long length sleeve　长袖
long-limbed figure　四肢细长体型
long line　长线形
long-line bra　长形胸罩
long-line bra-slip　长线形胸罩衬裙
long-line brassiere　长形胸罩
long-line sweater　长线形毛衣
long-long dress　超长装；特长服装
long man　高个男子
long nap fabric　长绒面料,长绒织物
long neck　长颈
long neck figure　长颈体型
long necklace　长项圈；长项链
long nose pliers　尖嘴钳
long or giant saddle stitch　长或大鞍形线迹；马鞍形线迹
longotte　朗戈蒂平布
longs　长裤
long panel　长嵌条
long pants　长内裤(英国男子)；紧身长衬裤；长裤
long pile fabric　长绒织物
long pile shag　长毛绒粗呢
long piled fustian　长毛纬绒织物
long point collar　长尖角领
long pointed collar　长尖角领
long poll　蓬乱绒头长毛绒织物
long prongs　服装长尖头
long puffed sleeve　抽裥长袖
long robe　教士服；法官服
long ruler　长尺
long-run dumping　长期性倾销
long saddle stitch　长鞍形[刺绣]针迹
long scarf　长围巾
long seam　长缝；长条成形缝
long seamer　长缝机
long seam stitching　长缝线缝
long seam unit　自动长缝[缝纫]机
long skirt　长裙
long sleeve　长袖；水袖(中国戏装)
long sleeve blouse　长袖衫；长袖上衣
long sleeve cotton shirt　长袖棉衬衫
long sleeved body　长袖紧身衣
long sleeved shirt　长袖衬衫
long sleeve qipao　长袖旗袍
long sleeve jacket　长袖女夹克衫
long sleeve jumper　长袖套衫
long sleeve knit shirt　长袖针织衬衣
long sleeve length　长袖长[度]
long sleeve long gown　长袖长衫

long sleeve mandarin dress　长袖旗袍
long sleeve robe　长袖袍
long sleeve shirt　长袖衬衫
long sleeve vest　长袖内衬衣；长袖汗衫
long socks　长短袜
long staple cotton　长绒棉
long stick（LS）　长码尺
long stitch　长线迹；长针脚；缎纹刺绣针迹；疏迹
long-staple cotton　长绒棉
Long-Staple Uplands　美国长绒陆地棉
long stocks　高筒男袜
long straight-hanging seam　直挂长缝
long teazeling　纵向拉绒,纵向起绒
long-term agreement　棉制品长期合作协定
long-term plan　长远规划
long thread end　长绒头,长线头
long togs　(海员)上岸穿的衣服
long torso　长躯体,低腰[节]线,低腰型；合身低腰裙
long torso line　长身线形,长身型
long torso silhouette　长身躯轮廓,低腰型轮廓
long torso skirt　合身低腰裙
long-travel bottom feed　长行程下送料
long triangular marking　长三角形开省记号,(纸样上表示开省的)长三角形记号
long trousers　长裤
long T-shirt　长T恤衫
long tunic　长塔奈克；拖地长袍
longuette　迷地装,中长裙装,过膝长裙装,中庸装
longuis　朗久伊斯棉格子布；印度格子棉塔夫绸
long vegetable fiber　麻纤维
long velour　经向条纹长毛绒
long vest　长背心,长马甲
long waist　低腰身
long-waisted dress　低腰裙
long waisted figure　低腰体型
long wool　长羊毛,精梳毛,有光长羊毛
longyi　缅甸腰布；头巾用织物；腰布用织物
lono　细白蕉麻
Lons star cotton　得克萨斯棉
loodiana khes　手织棉布
look　风貌,风格；服式,款式；时式；式样；脸色；模样；外表
looking glass　穿衣镜

look mask 防阳光面罩
look of clothes 服装总主题；服装总外观效果
look of costume 服装风貌
look-over 验布
looks 容貌
loom 织布机
loo mask 妇女半截面罩（化装舞会），遮住脸上部的面罩
loom finished linen 原纱漂白亚麻布
loom goods 坯布
loom-state fabric 坯布；熟织织物（毋需后整理的色织丝织物）
loom state printing 坯布印花
loong(h)ees 印度男式披肩布
loong(h)ie 印度棉格子布
loongies 印度男式披肩布
loongyes 印度男式披肩布
loons 喇叭裤
loop 线环，线套；(布)环带；带袢
loop and button[closing] 结环纽眼门襟
Loop and loop 连袜裤（商名）
loop bikini 线环比基尼
loop bonding 缝编
loop brabant 列纬斯机棉制花边
loop buttonhole 环扣纽眼；纽袢孔；纽环扣
loop carpat 毛圈地毯
loop closing 环扣门襟
loop closure 环扣门襟
loop cloth 毛圈织物；毛巾布
loop column 地毯绒圈纵行
loop course 套口横列
loop cutting machine 割绒机，剪绒机
loop distortion 线圈歪斜，线圈变形
looped 结环纽眼；有手缝针迹的
looped braid stitch 圈编针迹，圈形编织绣
looped brocade 绒圈锦
looped buttonhole 纽带扣
looped carpet 毛圈地毯
looped dress 带裙撑的双层抽褶双层裙
looped edge ribbon 起圈花边饰带
looped fabric 毛圈织物，毛巾织物
looped fancy yarn 线圈花色线
looped filling 纬起圈织物
looped filling pile fabric 纬起圈织物
looped fringe 结环流苏饰边
loop edge 毛圈边
looped hosiery 手工缝制针织品

looped lace 起圈花边
looped plush 毛圈长毛绒
looped thread 环圈线，起圈花线
looped yarn 毛圈线，结子线
looper 套口机；缝袜头机；弯针
looper thread 缝合底线，圈结线
looper thread guide 弯针导线器
loop fabric 毛巾织物
loop fastening 绳圈式衣扣
loop filling 纬起圈织物
loop filling pile fabric 纬起圈织物
loop formation 成圈，毛圈形成，成圈过程
loop gingham 毛圈条格平布
loop guider 导缝器
loop hosiery 缝制成形针织品
looping 缝袜头；浮底线；浮线环疵；环缝；套口；连圈缝合
looping machine 套口机；缝袜头机
looping operation 缝袜头作业；套口作业
looping seam 套口缝；缝头缝；环缝
loop knitting yarn 圈圈绒线
loop lace 起圈花边
loop length 线圈长度，线环长度
loopless toe 无缝袜头
loop-pile carpet 毛圈式地毯，起圈地毯
loop-pile fabric 毛圈绒头织物
loop-pile tufted carpet 毛圈簇绒地毯，起圈簇绒地毯
loop-pile tufting machine 起圈绒头簇绒机
loop plush 毛圈式长毛绒
loop plush effect 毛圈效应
loop ply yarn 小辫子花线，线圈花线；卷缩纱线，卷曲股线，结子纱线
loop-raised fabric 拉绒织物；毛圈织物；经编起绒织物
loop row 地毯圈绒横列
loop ruche 毛圈褶裥饰边
loops 环结；裤耳
loop selection 提花编织
loop selvedge 毛圈边
loop sole 毛圈袜底
loops on the top 毛（服装缺陷）
loop-state printing 坯布印花
loop stitch 套口线迹；锁缝线迹；饰缝线迹；环形装饰线迹；环圈花式缝
loop stripe 带子
loop tape 裤袢条
Loop the Loop 连裤袜，连袜裤

loop thread	起圈花线
loop tie	线环领带;环绳带
loop towel	毛巾,面巾
loop turner	翻带器,线环转动器
loop tweed	毛圈粗花呢
loop twist	起圈花线
loop type needling machine	毛圈型针刺机
loop type yarn	起圈花式纱线
loop velvet	毛圈丝绒,起圈天鹅绒
loop wheel machine	台车
loop yarn	卷毛纱;毛圈线,结子线,环圈线,圈形线
loop yarn fabric	起圈花线织物
loop yarn overcoating	圈圈绒大衣呢
loopy edge	毛圈边
loopy yarn	多圈纱线;起圈变形丝
loose	衣服太肥
loose-back	松背织物
loose-bodied	宽大服装
loose bodice	宽松大身
loose breeches	宽长水手裤(膝部宽松灯笼裤)
loose clothing	外套
loose diamond	散粒钻石
loose ease	宽松式服装
loose ends	松线
loose fashion	宽松时装
loose feeling	宽松感
loose fit	宽松合身[的衣服];松尺寸;松身式
loose fit boots	宽筒靴
loose fitting	宽松的
loose-fitting apparel	宽松的服装
loose-fitting design	宽松式服装设计
loose-fitting garment	宽松服装
loose-fitting hood	宽松风兜
loose-fitting housecoat	宽松长袍
loose-fitting pants	宽松裤
loose floating yarn	松散浮纱疵
loose folds	松驰衣纹
loose goods	散货(钻石)
loose grain	松面(鞋)
loose grain leather	松面革
loose hook	定位钩;活动钩
loose installation	(地毯)覆盖铺设
loose-laid tiles	非固定组合地毯
loose lapel	驳头外口松
loose measure	宽松尺寸;松身尺寸
loose neckline	领离脖子
loose pile	松纬起绒,浮纬起绒
loose pleat cuff	宽松褶皱袖头及袖口
loose selvage	松边
loose shirt front	衬衫的胸襟;假衬衫;女式无袖胸衣
loose silhouette	宽松式,宽松型轮廓
loose sleeve	宽松袖,大袖
loose stitch	线迹松弛,浮线线迹;稀松织物
loose-stock	散装货;散纤维;稀松织物
loose-stock dyeing	散纤维染色
loose structure cloth	松结构呢绒
loose structure fabric	松结构织物
loose structure ladies cloth	松结构女士呢
loose style	宽松式
loose texture	质地稀松
loose trousers	宽松裤
loose weave	松织
loose woven goods	稀松织物
Lord Byron shirt	拜伦[公爵]衬衫
Lord Fauntleroy suit	美国男孩装束
lord trousers	西装裤
L'OREAL	欧莱雅
lorette	卢勒特混纺呢
lorgnette	长柄眼镜;长柄眼镜式望远镜(观剧用)
lorgnon	单片眼镜,夹鼻眼镜
loricas	(古罗马)皮胸甲,铠甲;兜甲,护身硬壳
lorna forte	重磅棉帆布
lorraine	洛兰花呢
lorum	罗拉姆外衣(古罗马)
lose fit boots	宽筒靴
loss of color	褪色
lot	批;套;组
lot acceptance inspection	成批验收检验
lotanza	白亚麻布(古巴)
lot building	配批
lot by lot sampling	逐批抽样
lot card	批号卡
lot inspection	全批检验
lotion	化妆水;洗剂,洗液
lot number	批号;批数
lotorine	丝毛牧师服织物
lot product	批量产品
lot sample	批样,批量试样
lot size	批量,批量大小
lot to lot colour variation	批间色差
lot to lot shading problem	批间色差问题
lotus design	莲花纹

lotus red 莲红
lotus-flower shape 莲花式
lotus-root-like fiber 天星型复合纤维,藕截面型纤维
loud 俗艳的,过分花哨,[色彩]刺目的,耀眼的;刺耳的
loud colour 过分花俏色彩;俗艳色彩
loudish wear 俗艳服装;衣服有点俗艳的
loud socks 醒目袜;刺眼的袜子
Louis XIII collar 范达克领(17世纪流行的男式大翻领,常缀有花边);带花边男式大翻领
Louis Dell'Olio 路易斯·黛尔·奥里佛(美国服装设计师)
Louis heel 路易斯鞋跟
Louis Feraud 路易·费罗(法国时装设计师)
Louis heel shoes 路易斯跟鞋
Louis philippe costume 路易斯·菲利浦式女装
Louis quinze lace 路易窄花边
Louis treize braid 窄亚麻编带
LOUIS VNITTON 路易威登
Louis Vuitton 路易·维通(法国时装设计师)
Louis 路易斯(美国时装设计师)
Louise Boulanger 路易丝·布朗热(法国时装设计师)
Louis(XV) heel 路易(十五)鞋跟
louisiana cotton 美国路易斯安那棉
louisine 卢伊辛绉;卢伊辛绸
Loulan shoes 楼兰靴
lou mask 遮住脸上部的面具
lounge coat 休闲西服
lounge dress 随意连衣裙
lounge jacket 休闲夹克;普通夹克;西装上衣
lounge pajamas 随意睡衣裤
lounger 家常衣着;休闲便服
lounge shirt 随意衬衫装
lounge slippers 室内连袜便鞋
lounge style 便式服装
1ounge suit 家居套装;日常西装,男式常服;男普通西装
lounge suit jacket 双排纽男子上装
lounge wear 家居服;家常便服;休闲服;双排纽西装便服
lounging cap 绅士居家帽
lounging gown 家常长袍
lounging jacket 男士居家夹克;男西装;西装工作服
lounging pajamas (长至踝关节两件套或上下连装的)裙裤睡衣;便袍
lounging robe 休闲长袍,室内长袍
loup [法]半截面具;女用黑色天鹅绒面罩;上质(绒质)轻女面罩
loupe clean 明净(钻石)
louped 已检视宝石
lousiness 毛丝;织物起毛
louver sunglasses 百叶窗式太阳眼镜
lovat 灰绿杂色,蓝绿杂色,棕绿杂色
love 真丝绉
love beads "爱情与和平"的彩色念珠;爱珠
love bracelet 爱心手镯,心心手镯
love knot 装饰蝴蝶结
lovelocks 爱心发型;娇发
lover linen 仿爱尔兰漂白纯亚麻布
love ribbon 缎条纱带;孝服缎带
lover's knot 情人结
lover's watch 情侣表
low and high hip figure 高低臀体型
low back 低背;下背部
low back bra 低背胸罩
low-back cowl 低位背垂褶
low back dress 低背女装
low-backed corset 低背胸衣
low back overalls 低背女装裤
low blouson line 低摆型(夹克)
low boots 短[腰]靴
low bust 低胸[型]
low cape 微黄色级(钻石)
low collar 低领,低敞领
low color 淡色
low contrast 弱反差
low-count fabric 粗支稀疏织物
low-count plain weave fancy suiting 粗支平纹花呢
low-count yarn 粗支纱
low-coverage design 低覆盖面印花图案
low cowl 低位垂褶
low cowl neckline 低垂领
low cross-bred wool 粗支杂交羊毛
low-cut (领口、鞋帮)开得低的;低位裁剪;低位后片
low-cut back 低开背
low-cut bra 低开胸罩
low-cut dress 低领连衣裙
low-cut evening gown 低领晚礼服;低领夜礼服

low-cut front 低领口前片
low-cut neck 大开领口
low-cut neckline 大开领口,袒胸露背式领圈
low-cut waistline 低裁腰布线
low density polyethylene interlining (LDPE interlining) 低密度聚乙烯[黏合]衬
low down cardigan 深露型针织外套
low dress 低领口连衣裙,低领口女外衣;袒胸装
low duty 轻型
low elastic 低弹
low elastic fabric 低弹织物
low elastic yarn 低弹丝
low embroidery 一般缎纹刺绣;平针刺绣
low end 低档的,低级织物
low end mill 低级纺织品工厂,低档产品工厂
low end woolen 低级粗纺毛织物
lower abdomen 下腹,下腹部
lower and lift motion 上下运动
lower base cloth 双层起毛织物的里层
lower buck 下烫衣板
lower bust girth 下胸围
lower bustline 下胸围线,乳根围线
lower drop-feed 下送布;下送料
lower end of waistband 腰头里襟端
lower eyelid 下眼睑
lower feed 下送布;下送料
lower garment 下装;下身衣着
lower heel 低跟
lower hem 服装下摆
lower-hem slit opening 下摆开衩
lowering hook 底开口拉钩
lower knee girth 下膝围
lower leg 小腿
lower leg length 小腿长
lower limb bone 下肢骨
lower limbs 下肢
lower lip 下唇
lower palpebra 下眼睑
lower pocket 下口袋
lower sleeve edge 底袖边
lower-slung 低腰
lower stocks 羊毛男袜;真丝男袜
lower tension 底线张力
lower textile 织物的反面;里层织物
lower texture 里层织物
lower thread break 底线断头,断底线
lower thread 缝纽底线

lower threading 穿底线
lower torso 下躯干;下身,下体
lowest price limit 有限价,最低售价
low extension fiber 低伸长纤维
low-fired pitch fiber 弱火处理的沥青纤维,低温沥青纤维
low flammability 不易燃性,难燃性
low flash product 低燃点产品
low-formaldehyde resin finishing 低甲醛树脂整理
low-grade 下品低级,低级的
low-grade clothing 低档服装
low heel 平跟,低鞋跟
low-heel boudoir slippers 低跟闺房便鞋
low-heeled sandals 平跟凉鞋
low-heeled shoes 平跟鞋
low high 中长袜
1ow[high]collar 低[高]领
low hip 低臀
low jaw 下颌
low key[tone] 暗基调,暗色调
low level loop pile carpet 短绒地毯
low margin store 薄利多销商店
low/medium grades 中低档
low melting point fiber 低熔点纤维
low modulus elastic fiber 低模量弹性纤维
low neck 大袒胸领
low-necked 露背低领的
low-necked mode 露背低领式
low neckline 低领口线,低领口;大袒胸式服装
low notched collar 低领豁口领
low notched jacket 低驳领夹克
low-notch lapel 低领嘴西装领
low 2 piece knotched collar 低开式两片缺角领
low pile 低绒圈;低绒头
low-pilling variant 低起毛无起球改性纤维
low power stretch 低[功能]弹性,低载荷弹性织物
low relief 浅浮雕
low-rise 短裆紧身的,裤子浅上裆
low rise briefs 浅上裆三角裤
low round neckline 低圆领口
low serge 低级哔叽;藏青或黑色学生服
low shoes 低口鞋;浅口鞋;半筒鞋,半筒靴;短靴
low shoulder 低肩,垂肩,塌肩;溜肩;溜

肩体型
low shoulder figure 溜肩体型
low slung 低腰抱臀裤
low slung belt 束在腰下的带
low soiling finish 防污整理,抗污整理
low strength high extension type 低强高伸型
low stretch fabric 低弹织物
low stretch polyester sewing thread 涤纶低弹丝缝纫线
low stretch texturized polyester facy suiting 低弹涤纶长丝花呢
low stretch yarn 低弹变形丝(假捻后经定形的变形丝);低弹丝
low support 贴肉衣着;贴身衣着
low technology 低技术
low temperature resistance 耐低温性;耐寒性
low texturized yarn fabric 低弹丝织物
low thermal 最低温
low thread count fabric 组织稀疏织物
low topline oxford shoes 耳式浅口鞋
low topline shoes 浅口鞋
low topline tongue shoes 舌式浅口鞋
low torso 胴长装(细长躯体装);细长躯体
low turtle neck 低领口;下垂式高圆领
low turtle neckline 低领口线
low twill 缓斜纹
low twist 低捻;弱捻
low two 纽位很低的西式上装
low upper shoes 矮腰鞋
low V neckline 低V字领口
low waist 低腰身
low-waisted bells 低腰喇叭裤
low-waisted skirt 低腰裙
low waistline 低腰[节]线
low-worsted 粗纺
loynes 洛尼斯[浅色印花]薄呢
lozenge 菱形;菱形纹
lozenge design 菱纹
lozenge front 菱纹装饰前片的女紧身衣
lozenge motive 菱形花纹,菱形花纹图案
lozenge shape 菱形花纹;花边网眼
lozenge-shaped 菱形的
lozenge twill 菱形斜纹,鹅眼形斜纹
L-peak lapel suit L形尖驳领西装,尖驳领西装
L rating 耐光色牢度等级
L-revers collar L形翻领,L形西装领

L-square 直角尺,L形角尺
L-shape tacker L形加固机
L-shaped fastening seam L形定位缝
L-shaped lapel 西装L形翻领
L-shaped rule L形尺
L twill bag L形斜纹麻袋
Lu embroidery fabric 鲁绣品
Lu embroidery (Shandong embroidery) 鲁绣
Luana 罗阿拿织物
Luau pants 夏威夷宴会裤(印有夏威夷图案,长至小腿肚的男式裤)
lubricating oil 润滑油
lubrication period 加油周期
lubricious resistance 滑脱低抗力
lucas shawl 犹太妇女用卢卡斯披巾
lucca cloth 金银丝织物
luchage 阿朗松花边的加光整理
Lucian Foncel 卢西安·丰塞尔(法国时装设计师)
Luciano Papini 卢恰诺·帕比尼(意大利时装设计师)
Luciano Soprani 卢恰诺·索普拉尼(意大利时装设计师)
Lucien Lelong 卢西恩·勒隆(1889～1958,法国高级时装设计师)
Lucile (Lady Duff Goryon) 露西尔(杜夫·戈登女士)(1863～1935,英国时装设计师)
lucite button 人造荧光树脂扣
lucite embossed button 有机压花扣
ludgas 印度妇女披身布
ludhiana khes 手织棉布
Luft 聚酯中空复合短纤维(商名,日本钟[浦]纺)
luftspitze 烂花花边(德国称谓)
lug 孔眼;突缘;针钩
luggage 红褐色;行装
luggage band 行李带
luggage cloth 提包布,行李布
luggage handle 行李箱提手
Luis Estevez 卢斯·埃斯(美国时装设计师)
luisine 英国优质丝光布(用埃及棉制)
lukchoo 中国蓝丝棉布
lule 东方厚地毯
lule rug 东方厚地毯
LuLu Cheung 张路路
lumbar region 腰部
lumber clothing[coat] 伐木外套

lumber coat　伐木外套
lumberdine　黑薄纱
lumber jack　伐木工短夹克衫;短夹克衫
lumber jacket　伐木工短夹克衫;短夹克衫
lumber jack shirt　伐木工衬衫(大格子厚毛料制)
lumbermen's over　伐木靴
lumbermen's overshoes　伐木靴
lumbermen's socks　罗口羊毛袜
lumber shirt　伐木衬衫(填棉格绒衬衫)
lumen staple fiber　中空短纤维
Lumiart　聚酯异形截面短纤维(商名,日本东洋纺)
Lumicell　黏胶短纤维(商名,日本兴人)
luminal art　彩光艺术,灯光艺术
luminescence　发光
luminescent dye　荧光染料
luminescent effect　发光效应
luminescent fiber　发光纤维
luminescent printed handkerchief　发光印花手帕
lumineux　亮光绸,闪光绸;闪光丝带
lumineux ribbon　有光丝带
luminist　灯光艺术家
luminist art　彩光艺术,灯光艺术
luminosity　亮度,明度,明亮,发光度;发光体
luminous cloth　发光布
luminous color　夜光色
luminous fiber　发光纤维
luminous printing　发光印花
luminous thread　夜光线
luminous watch　夜光表
Lumiyarn　聚酯金银丝(商名,日本东丽)
lump　凹凸不平;坯布,双联匹坯布;双幅布;长匹,超长匹
lumps　长匹,块,团;坯布,双幅布
lunardi hat　气球帽
lunar fashions　登月服,登月装束
lunar look　登山风貌;登月服装款式
lunch box[bag]　午餐提包
lunchbox　餐盒形提袋;饭盒肩包
luncheon suit　午餐轻松套装(上下不同衣料轻松套装)
lune　月牙形,半月形
luneburg flax　德国优质亚麻
lunette　月牙图案
Luneville lace　伦内维耳花边(机制网眼地加棱结花卉纹)

lungee　头巾(印度,缅甸);腰布
Lungee　印度棉腰布;印度棉格子布;印度丝棉绣花布
lungi　头巾(印度,缅甸);印度棉腰布;印度棉格子布;印度丝棉绣花布
lun taya　缅甸手织伦塔耶绸
Luoba ethnic costume　珞巴族民族服
Luoba nationality's costume　珞巴族服饰
luobo flax　罗布麻
luo-chi　罗绮
L'UOMO VOGUE　《男子时装》(意大利)
lupine　羽扇豆蓝(淡蓝色)
lupis　优质马尼拉麻;优质蕉麻
Lurex　卢勒克斯(金属丝线)
lurex　卢勒克斯金银丝;卢勒克斯金银丝织物
lurid　苍白的,惨白的;死灰的,青灰的
lurirug　伊朗鲁瑞部落地毯
luristan rug　伊朗鲁瑞部落地毯
lushan leno　庐山纱
Lushanchow　河南鲁山柞丝绸
luster　光泽,光彩;(棉经毛纬)有光呢;亮光
Lusterella　丝光醋酯变形丝
lusterer　光亮织物
luster cloth　有光呢;有光织物
luster effect　光泽效应
luster fabric　棉毛有光呢
luster finish　上光整理
luster finished rug　上光整理地毯
lustering　上光整理;加光丝带;光亮绸
lustering gauzy silk　光亮薄绸
lusterless　无光泽
luster lining　亮光里子绸;亮光里子呢;羽纱;有光羽纱
lusterne　丝光斜纹里子布
lusterness　光泽度
luster orlenans　有光棉毛交织物
luster pile fabric　光亮绒头织物
luster raye　有光提花绸
luster rug　有光地毯
luster umbrella　绢伞
luster wool　有光羊毛
luster wool yarn　有光毛纱
luster worsted　精纺有光毛纱
luster yarn　蜡线;有光色线;有光线
lustestring　加光丝带
lustre　[法]贴疵
lustre fancy suitings　珠光花呢
lustre finish　上[柔]光整理

lustreless 无光泽
lustre lining 羽纱
lustrene 丝光斜纹里子布
lustreux 匹染棱纹绸
lustrine 全丝光亮塔夫绸；有光织物；有光斜纹袖里子棉布
lustring 光亮绸；上光丝带；上［柔］光整理
lustring gauze silk 光亮薄绸
lustro-silk 闪光丝（旧称铜氨人造丝）
lustrous 光泽；深橘黄色的
lustrous color 鲜艳色
lustrous colors 鲜艳色彩
lustrous fiber 有光纤维
lustrous filament 有光长丝
lustrous finish 有光整理，光泽整理
lustrous furniture fabric 有光家具布
lustrous rayon 有光人造丝
lustrous rayon fabric 有光人造丝织物
lustrous thread 有光线
luteous color 深橘黄色
lutescent 带黄色的
lutescent light 带黄色光
lutestring 光亮绸；细棱纹绸；黑丝绸；有光丝带（系眼镜）
lutherine 丝毛牧师服织物
lux, lx 勒［克斯］
luxeuil 凸花花边
luxor 罗纹丝缎
luxor satin 卢克索缎
luxuriance 奢华风格
luxurious hand 丰满手感
luxury 华贵,豪华；豪华装；奢侈品
luxury fabric 复杂花纹织物
luxury goods 奢侈品
LVMH 法国时尚用品集团
lyard 灰白色；银灰色；带灰白色条纹的
lyart 银白色；灰白色；带灰白色条纹的
Lycee Recital look 高中女学生独唱会款式
lyceenne look 高中女学生装；高中女学生风格
lyceenne style 高中女学生风格
Lycra 莱卡（聚氨酯弹性纤维,商名,美国杜邦）；莱卡弹性织物；莱卡纤维
lycra-based fashion 莱卡时装
lycra denim 弹力牛仔布
lycra fashion 莱卡时装
lycra hoses 弹力袜；莱卡袜
lycra jeans 弹力牛仔裤
lycra rib 弹性罗纹
Lycra seam 莱卡缝
Lycra sleeve 莱卡袖
lycra socks 弹力袜；莱卡袜
lycra stretch knit 莱卡弹力针织物
lyme regis lace 梭结针锈花边
Lynel 聚氨酯纤维（商名,意大利）
lynx 山猫毛皮
lynx cat 山猫毛皮
lynx cat fur 小山猫毛皮
lynx cat skin 小山猫毛皮
lynx[fur] 山猫［毛］皮；猞猁毛皮
lynx (1eather) 山猫毛皮
lynx skin 山猫皮；猞猁皮
Lyocell 天丝；莱塞尔纤维
Lyonnesse pongee 里昂府绸
Lyons blue 里昂蓝
Lyons gold thread 里昂金线（丝或棉芯外包镀金铜丝）
Lyons[lace] 里昂花边（网眼地,精细梭结花边）
Lyons satin 里昂缎；丝背棉缎
Lyons silk 里昂丝绸
lyons thread 铜芯金线（编带及镶饰用）
Lyons Velvet 里昂丝绒
lyre 高级粗纺呢
lyric couture 抒情型时装
lyric line 抒情型

M

maakbar 马克巴,马克巴钻石(低劣钻石)
maarad cotton 比马棕色变种棉,马拉德棉
maaypoosten 服装用丝绸
mabel [twill] 马贝尔斜纹里子呢
mabogany color 红木色,赤褐色
mabroum 松散结构的平纹棉布
MaC 雨衣
macabre 小花纹丝毛织物
Macallister cotton 美国麦卡利斯特棉
macana 默卡纳彩格平纹棉布
macaroni cravat 花花公子领结
macaroni fiber 中空纤维
macaroni ragon yarn 中空人造丝
macaroni [style] 花花公子式样
macaroni suit 花花公子套装,马卡路尼套装(18世纪70年代)
macaroni yarn 中空丝
macaroon 窄管状滚边
macaw green 鹦鹉绿
macbride cotton 中等长度陆地棉
maccall cotton 晚熟陆地棉
Macclesfield tie silk 麦克莱斯菲尔德领带绸
maceio 粗梭结花边;巴西柔韧棉
macfarlane 带风帽的大方格厚呢大衣,披风大衣
Mach nozzle washing machine 马赫喷嘴水洗机(商名,荷兰)
machanical flocking 机械植绒
machine 机器;机械
machine alignment 机器校准
machine ancillary time 机器辅助时间
machine arm (缝纫机)机头臂
machine backstitching 机缝回针线缝
machine bar tack 机缝加固封口结
machine base 底座
machine basting 机制粗缝,机制假缝
machine bed 机板
machine blindstitch 机缝暗缝线迹
machine blind stitched hem 机缝暗缲下摆
machine bobbin 缝纫机梭芯
machine buff 机割皮革
machine buttonhole 机器开纽孔
machine buttonhole stitch 机缝纽孔线迹
machine cabinet 机柜
machine calender 拷花轧压机,浮雕轧压机,凹凸轧花机
machine chain tack 机缝袢
machine clipping 机械修剪
machine configuration 机器配置
machine cotton 缝纫用棉线
machine cover 机罩;机套
machined buttonhole 机制纽孔
machine dimensions 机器尺寸
machine downtime 机器停工时间
machined-worked eyelee 机缝鸽眼孔
machine embroidered dot 机绣小圆点
machine embroidered handkerchief 机绣手帕
machine embroidered hat 机绣花帽
machine embroidered skirt 机绣围裙
machine embroidery 机绣
machine exploitation 设备利用率
machine fell seam 机缝包缝
machine finishing 机械整理
machine for covering button with cloth 包扣机
machine for covering button with fabric 包扣机
machine for feet of socks 柯登式袜脚机
machine for invisible backing 衬垫织物纬编圆机
machine for twisted fringes 捻缨穗机
machine-foundation 地脚
machine frame 机架
machine-fused buttonhole 胶熔纽孔
machine-hooked rug 机织钩针地毯
machine hour 台时工作量;台小时
machine head 车头,机头
machine idle time 机器闲置时间
machine installment 机器安装
machine knife cut 机器裁剪刀
machine knitting 机器编织;机器针织
machine knitting needle 机[器]编[织]针
machine-knotted pile carpet 机结绒头地毯
machine lace 机制花边
machine-made carpet 机制地毯

machine-made lace 机制花边
machine maintenance 机器保养
machine model 机型
machine neatening 机器缝光边
machine net 机制网
machine oil 机油
machine operator 机器操作工人
machine overedge stitch 机缝锁边线迹，机缝包边线迹
machine padding stitch 绗缝线迹
machine parts 机件(零件、配件)
machine pedestal 机架
machine pinking of cloth 机器锯齿裁切
machine-pleated fabric 机器压褶织物
machine print 机器印花
machine printing 机器印花
machine problem 机器故障
machine processor 机器洗涤机
machine products 机械产品
machine pulley 缝纫机上轮
machine rest (缝纫机)机头撑
machine retrieve 机器检索
machine-rolled hem 机缝卷边
machinery 机器；机械
machinery maintenance 机械维修[维护]
machine set 机座
machine sewing 机缝，机器缝纫
machine sewing of grey 坯布机器缝头
machine sewing silk 缝纫机丝线
machine sewing thread 缝纫机用线，机缝线
machine silk 机缝丝线
machine size 机器尺寸
machine specification 机器规格，机种
machine stand 机台；机座
machine stitch 机缝线迹，压边针迹；装饰线迹；缝纫机切边
machine-stitch back centre seam 合背缝
machine-stitch chest interlining 缉胸衬
machine-stitch collars together 合领子
machine-stitch dart 缉省道
machine stitched 机缝的
machine-stitched buttonhole 机缝纽孔
machine-stitched hem 机缝缘边
machine-stitch french dart 合刀背缝
machine-stitch front edge 合止口
machine stitching 机缝，机缝线迹
machine-stitching belt loop 合裤带袢
machine-stitching front and back rise seam 合前后裆缝
machine-stitching inside seam 合下裆缝
machine-stitching side seam 合裙缝
machine-stitching sleeve tab 合袖袢
machine-stitching waistband 合腰带
machine-stitching waist pleat 缉裤腰裥
machine-stitch inside seam 合下裆缝
machine stitch seam 机缝线缝
machine-stitch shoulder seam 合肩缝
machine-stitch side seam 合摆缝
machine-stitch sleeve seam 合袖缝
machine-stitch waistband 合腰带
machine-stitch waist pleat 缉裤腰裥
machine table board 机板
machine thread 缝纫机用线
machine time 机器运转(或作业)时间
machine tool 机床
machine twist 缝纫机用线；三股丝线
machine type 机型,机种
machine wash 机洗
machine wash and dry 机器洗涤与干燥
machine wash cold 冷水机洗
machine washability 可机洗性,耐机洗性
machine washable 可机洗的
machine washable finished wool 机洗整理羊毛制品,经机可洗整理的羊毛制品
machine washable silk 机洗丝绸
machine washable wool 可机洗羊毛制品,机可洗羊毛制品
machine washing 机洗
machine washing severity 机器洗涤考验
machine-worked buttonhole 机缝纽孔
machine-worked eyelet 机缝鸽眼孔
machinist 机械师；机工；缝纫工
macho look 男子汉风貌
macintosh 防水胶布；防水胶布雨衣；防水外套；马金托什防水布；马金托什外套；防水胶布雨衣
macintosh cloth 橡胶防雨布
maciver cotton 马西韦棉
mack 防水胶布；防水胶布雨衣；防水外套；马金托什防水布；马金托什雨衣；轻薄防水织物；麦基诺厚呢；麦基诺厚呢外套
mackay stitcher 内线缝边机制鞋
mackinac 厚重防寒织物,马金瑙毛织物
mackinaw 厚重防寒织物,麦基诺[方格拉毛]厚呢,再生粗纺呢
Mackinaw blanket 马金瑙厚毛毯
Mackinaw cloth 马金瑙毛织物

Mackinaw coat 麦基诺厚呢短大衣；麦基诺厚呢外套
Mackinaw coating 马金瑙双面大衣呢
Mackinaw flannel 马金瑙全毛法兰绒
Mackinaw jacket 马金瑙夹克（长至臀部），麦基诺厚呢夹克[及臀部]
mackintosh 马金托什防水布（外套，雨衣）；轻薄防水胶布；轻薄防水织物；防水胶布；防水胶布雨衣；麦基诺厚呢；麦基诺厚呢外套
mackintosh [blanket] [印花机]防水橡胶衬布；防水胶布
mackintosh cloth 橡胶防雨布
mackintosh rubber blanket [印花机]防水橡胶衬布；防水胶布
mackluk 慕克拉克靴
Mac Nab harris 马克纳海力斯粗呢
Maco cotton 埃及马科棉
Maco foot 马科袜，黑色毛袜；自然色棉袜底；异色袜
Maco percale 马科密织薄细布
Ma-coual 中式男宽袖短上衣；中式女真丝外套
macrabine 埃及半漂白粗亚麻织物
macrame belt 流苏花边腰带；马克莱姆腰带
macrame choker 马克莱姆高领；马克莱姆项链
macrame fringes 马克拉莫流苏
macrame knot [编成流苏或花边等饰物的]装饰结，马克莱姆结
macrame lace 流苏花边；马克莱姆花边
macrame stitch 流苏花边绣，马克莱姆绣
macrame 流苏花边
Macr Bohan 马克·博昂（法国时装设计师）
macro quality 宏观质量
Madagascar 马达加斯加海水蓝；手织椰纤维织物
Madagascar lace 马达加斯加花边
madam （土耳其）白细布
Madame Cheruit 谢吕夫人（法国时装设计师）
madapolams 马德波勒姆细布；上浆平纹棉布
ma-da-qui brocade 马打球锦
madar 棉混纺法兰绒
Mad Carpentier 麦德·卡彭特（法国服装时装店）
maddeira embroidery 白亚麻布绣花

madder 茜草（染料）；鲜红色；茜草印花小花纹布
madder brown 鲜红棕
madder red 鲜红色，茜草红，紫红
madder red deep 暗红色
made in China 中国制造
Madeira embroidery [work] 马德拉刺绣；白亚麻布绣
Madeira lace 马德拉[白亚麻布上刺绣的]花边
Madeleine Chernit 玛德琳·谢吕特（法国时装设计师）
Madeleine de Rauche 玛德琳·德劳克（法国时装设计师）
Madeleine Vionnet 玛德琳·维奥内（1876～1975,法国时装设计师）
made to measure 定制服；度身打板
made to measure firm 定制服装店
made to measure garment 定制服装，定做服装
made to order 定制的；定制服装
made to order clothing 定制的服装
made-up goods 制成品
made your order 定制的服装
Madonna 阿尔帕卡毛花呢
Madonna bracelet 麦唐纳手镯
madrapa 印度粗平布
madras 马德拉斯手帕；马德拉斯纵条衬衫布
Madras check 马德拉斯格纹织物
Madras cotton 印度马德拉斯棉
Madras gauze 马德拉斯花纱罗
Madras gingham 马德拉斯条格布，棉条格薄细布
Madras goods 印度马德拉斯棉布
Madras handkerchief 马德拉斯手织棉布；马德拉斯包头布；马德拉斯手帕；（在后整理时）纬纱由经纱浸染着色的织物
Madras hemp 印度麻,菽麻
Madras lace 马德拉斯梭结黑白花边
Madras muslin 马德拉斯窗帘纱
Madras pearl 马德拉斯珍珠
Madras plaid 马德拉斯格子
Madras rug 马德拉斯地毯；19世纪印度地毯
Madras shirtings 马德拉斯高级细平布
Madrassee silk 马德拉斯生丝
Madras stripe 马德拉斯条纹布
Madras work 马德拉斯刺绣

Mae West （飞行员用）海上救生衣,海上救生背心
Mae West silhouette 梅·卫斯特型轮廓线条,美腰型轮廓线条
mafors 麦福头纱,窄长面纱,修女面纱;女式斗逢
magazine 仓库,货栈
magazine type weft-insertion 复式多头衬纬
magenta 紫红,洋红,品红,牡丹红
magenta greys 品红布头重浆坯布
magenta haze 淡品红
magenta purple 洋红紫
Maggy Rouff 玛吉·鲁夫(1897～1971,法国时装设计师)
magic 神秘美
magic chain stitch 交错链状绣,魔法链状绣
magic lantern 幻灯机
magic tape 魔术贴;尼龙搭扣;尼龙搭扣带;尼龙粘扣带
magic wear 魔术装束
MAGLIERIA ITALIANA 《意大利针织品》(意大利)
Magnamite 腈纶基碳纤维
magnet 吸缝针用磁铁;磁体
magnetic cure shoes 磁疗鞋
magnetic fiber 磁性纤维(磁保健纤维)
magnet line U型磁铁型
magnetotherapy fabric 磁疗织物;磁性布
magnet-shaped neckline 磁铁状凹领口
magnette 亚麻平布
magnification 放大,放大倍数,放大率
magnificence 豪华,华丽
magnificent costume 服饰华丽
magnificent military stylistic form 华丽军队款式
magnifying glass 放大镜
magnifying power 放大率
magnolia 浅桃红色,浅紫色
magoya silk 日本细双绉
magpie 黑白花纹
magrabine 埃及手织粗亚麻平布
magruder cotton 马格鲁德棉
magua 橙色外套(中国清朝);马褂(中国清朝)
Magyar blouse 马扎尔外套;马扎尔上衣（袖子与前后身整块裁剪而成）
Magyar dress 马扎尔服装;匈牙利民俗服
Magyar sleeve 匈牙利式袖,马扎尔袖(匈牙利民族服装中与前后身整块裁剪的连袖)
mah 亚麻
mahals 伊朗长绒地毯
maharaja coat 大君外套(印度)
Maharaja turban 马哈拉贾头巾式女帽
mahilda 印度手织羊绒呢
mahioli helmet 夏威夷波形帽
mahlda 印度山羊绒呢
mahmudi 手织细布;印花布
mahogany 柳桉木棕色
mahogany colour 赤褐色,红木色
mahoitres 14～15世纪法国袋形袖
mahout 埃及粗毛织物;法国低级毛织物;轻质缩绒织物
mahuva cotton 印度马胡瓦棉
maiden pearl 处女珠
maid's suit 妈祖装,白衫黑裤套装
mail 铠甲,锁子甲
mailbag belt 邮包袋
mailbag duck 邮袋帆布
mail cloth 蜂窝纹绸;刺绣用蜂巢纹绸
mail coat 锁子铠甲
mailisi 麦里司
maille 花边针织物网眼;针织面纱网眼
mailles de bas 法国梅勒德巴哔叽
maillot ［法］紧身衣;紧身连衣裙装;（衣连裤）女泳衣;舞蹈紧身衣;运动员紧身衣
maillot bloomer 灯笼裤式泳衣
maillot nageur ［法］优质泳装,运动员泳衣
mail net 三角网眼织物
mail pouch 邮递袋,邮件袋
mail transfer 信汇
maimal （印度）最优质细棉布
mailman's jacket 邮递员夹克
Mainbocher 梅因博歌(1890～1976,美国高级时装设计师)
main(de toilette) ［法］浴用手巾手套
maine hunting boot 梅因猎靴
maines lace 丝网眼纱
main-fu 冕服(中国古代的皇帝服装)
main interlining 主衬;硬衬织物
main label 主唛;主商标
mainliner 曼利纳绉
main part 主体
main shaft 主轴
main stream 主流;主要倾向
maintainability 可维修性

maintenance	日常维修
maintenance facilities	维修设备
maison	高级时装店
maize	玉米黄
majolica blue	花饰陶器蓝
major brand	主要牌号，名牌
major defect	大疵点
major fashion	大众化时装，多数派时装
major overhaul	大检修；大平车
major repair	大修理
major products	主要产品
majorette boots	半高白靴(军乐队女指挥穿)
mak	亚麻
makaloa	夏威夷垫子
Makat	法国马卡特斜纹毛织物
makatlik	土耳其东方走廊地毯
makaty	波兰马卡提手工纺织呢
make	做(衣)；缝制；制作；制造
make a marker	服装排料
make an opening for pen on flap	做(胸袋)插笔口
make belt loop	缝裤带袢
maker brand	制造者的商标
make Chinese frog	盘花扣，盘花纽
make cloth	织布
make collar	制领
make coat and skirt	缝制衣裙
make collar band tab	做领舌
make cuff	缝袖口边，缝制袖头
make down	改小(衣服)
make flap	做装盖
make french tack at cuff	叠卷裤脚
make hat	制帽
make into	制成
make lay	裁剪排版；制唛架
make longer	改长些
make marker	裁剪排版；制唛架
make of	制造
make over	改制，翻新
make patch pocket	做贴袋
make pleats	缝制褶裥
make pocket flap	做袋盖
make pocket welt	缝袋牙
make ready time	生产准备时间
make replacements and technical innovations	更新改造
makers-up	服装工人；成衣匠；裁缝；制品装配工
make sample	做样品；打样
make shoes	制鞋
make shorter	改短些
make shoulder pads	制肩垫
make tailor's tack	打线钉
make-through	单件操作(一件衣服全由一人缝制)，单甩
make tuck	打褶
make up	包装；缝制；化妆品；化妆；打扮
make-up artist	化妆师
make-up box	化妆盒
make-up cape	化妆用罩衣
make-up color	化妆色彩
make-up gown	化妆工作服
make-up products	美容化妆品
make-up puff	化妆棉
make-up smock	化妆工作服
make-up table	化妆台
make waistband	缝制腰带(腰头)
makhi	红地印花面纱
makhmal	印度马克马尔绒
Makhtul [silk]	印度马克图尔丝
making	做；制作；缝制
making [a] marker	排料
making an opening for pen on the flap	在袋片上做插笔口
making a range of samples	系列取样
making belt loops	合串带袢
making box or inverted pleats	缝制和合裥或阴裥
making box pleats	缝制和合裥
making Chinese frog	做盘花纽
making clothes	做衣服
making collar	制领
making collar tab	做领袢
making cuff	缝制袖头
making French tack	拉线襻；拉线袢
making French tack at cuff	叠卷脚
making full	打绉裥
making hood	做风帽
making inverted pleats	缝制阴裥
making left and right fly	缝制裤门袢；里襟
making longer	改长
making of original sample	打初样
making of photo sample	打宣传样
making operation	制鞋操作，制鞋底操作
making out	制图划线
making oversized	改肥
making patch pocket	做贴袋
making pen opening	做插笔口

making piped openings 缝制开口滚边
making pocket 做袋
making pocket flap 做袋盖
makings 所需材料
making shorter （衣服）改短
making shoulder pad 缝制肩垫,做垫肩
making straps 缝制线带
making straps or loops 缝制线带或线环
making tailor's tack 打线钉
making [the] lay （裁剪间）排料
making-up 包装标志；缝制；裁剪；生头；织物包装
making-up defect 缝制疵病
making-up room 包装间
making-up stitch 缝合线迹
making-up technique 缝制技术
making-up technology 缝制工艺
making-up yard lapped 折叠成包
making waistband 缝制腰带
making welt pocket 做袋爿
makkan jin 微起绒厚帆布
mako cotton 埃及马科棉
mako jumel 埃及马科棉
malabar 马拉巴印花布；东印度印花手帕料
malabar carpet 印度马拉巴地毯
malachite 孔雀石
malachite green 灰松绿,孔雀石绿（鲜艳黄绿色）,品绿
malaki cotton 埃及吉萨棉
malar [bone] 颊骨,颧骨
malass 叙利亚马拉斯纱罗
Malaysian costume 马来西亚民族服饰
malborough 法国小花纹毛哔叽；英国小花纹毛哔叽
malconformation 不完善形态,畸形结构
maldahi cloth 丝绸锦缎
male figure 男子体型
malefique 比利时重磅精纺斜纹呢
maleform 男子体型
male pattern 凸纹图案
malformation 变形,畸形[体]
mal garan 中亚地毯
malgaran rug 中亚地毯
mali 缝编织物
Mali bracelet 马里手镯（皮制手工艺手镯）
malida 刺绣用山羊绒薄呢
Mali fabric 缝编非织造布（商名,德国）
Malifil [machine] 马利菲尔缝编机

Malifil stitch-bonding machine 马利菲尔缝编机
Maliknit machine 全部用经纱成圈的缝编机（商名,德国）
Malimo [fabric] 马利莫缝编织物
Malimo machine 马利莫缝编机
Malimo sewing-knitting machine 马利莫缝编机
Malimo thread layer sewing knitting machine 马利莫线层缝编机
Malimo-pol 马利莫-波尔缝编机
Malimo-pol fabric 马利莫波尔缝编织物
maline(s) 马林丝纱罗；平纹精纺呢
malines lace 薄纱；绢网；丝网眼纱
Malipol fabric 马利波尔缝编织物
Malipol machine 马利波尔缝边机
Malipol sewing-knitting machine 马利波尔缝编机
Mali technique 马利缝编技术
Malivlies machine 马利弗里斯缝编机
Malivlies sewing-knitting machine 马利弗里斯缝编机
Maliwatt machine 经纱-纤网型非织造布缝编机（商名,德国）,马利瓦特缝编机
Maliwatt sewing-knitting machine 马利瓦特缝编机
malla gussa cotton 土库曼斯坦棕色棉
mallard blue 野鸭蓝
mallard green 鸭蓝绿,野鸭绿,绿头鸭绿
mali fabric 马利织物
mallius cotton 马利斯棉
mallow [purple] 锦葵红（浅紫红色）
Malmal 马尔木刺绣
malo （夏威夷）马楼腰带,夏威夷羽毛网纱；夏威夷男式缠腰布；构树韧皮纤维
maltese 机制棉粗花边
Maltese embroidery 流苏绣品,马尔他刺绣
Maltese lace 马耳他几何形梭结花边；机制棉粗花边
Malvi cotton 印度马尔瓦棉
Malwa cotton 印度马尔瓦棉
mama coat 妈妈外套
mamaki 落尾木纤维平纹织物；落尾木属韧皮纤维
mambo style 曼波风格
mameluke sleeve 奴隶袖（17世纪）
mamilla 乳头[点]

Mamma cotton 美国马默思棉
mammal 平纹细布
mamma shoes 妈妈鞋(老年妇女鞋)
mamoudi 印花布；密织细布，手织细平布；细软微黄色亚麻
management 管理；经营
management expenses 管理费
management information system 管理信息系统
management regulation 管理制度
management system 管理体制
managing staff 管理人员
manbag 男式背包
manch aileron 鱼鳍形短袖
manche balloon 气球袖，灯笼袖
manche lampion 灯泡袖
mancheron [sleeve] 曼丘洛装饰袖(从肘部至肩部的上半袖)，假袖；短罩袖
Manchester brown 曼彻斯特棉(碱性棉)
Manchester cotton 曼彻斯特棉布
Manchester goods 棉布类
Manchester velvet 曼彻斯特棉天鹅绒
Manchester yellow 曼彻斯特黄，马休黄
manchette 套袖，克夫；女式午后装的袖口荷叶边
manchu crepe 色条纹绉
Manchurian dogbane 中国东北罗布麻
Manchurian ermine 中国貂皮
mandarin 柑色；法国棉经丝纬布；中式紧身马褂
mandarin blue 深蓝色
mandarin clothing 中式大衣；中式对襟马褂(外套)
mandarin coat 女式对襟绣花外套；女式织锦长外套；妇女晚礼服外套；橙色外套(中国清朝)，对襟马褂(中国清朝)，中式紧身马褂；清朝官吏外套
mandarin collar 中式领；长衫领；高领；马褂领；清朝官服领；旗袍领
mandarin dress 对襟绣花服装；旗袍；中式服装
mandarin duck flannel 鸳鸯色法兰绒
mandarine 棉经丝纬布
mandarine jacket 中式夹克；马褂；对襟绣花上衣
mandarine neckline 中式领口线
mandarin-neck shirt 中式斜领口式衬衫
mandarine orange 柑橘色
mandarine red 朱红[色]，橙红[色]
mandarine sleepcoat 立领睡袍，中式睡袍

mandarine sleeve 中式袖，满服袖；大袖口喇叭袖
mandarine style gown 中国官服式长衫
mandarine style pajamas 中式睡衣
mandarine work top 中式工装上衣
mandarin jacket 对襟绣花上衣，马褂，中式上装
mandarin neckline 中式领口
mandarin orange 柑橘色
mandarin red 橙红色
mandarin robe 旗袍
mandarin shoes 旗鞋(中国满族)
mandarin sleeve 满服袖，中式袖
mandarin style gown 中国官服式长衫
mand atkins cotton 尤里卡棉
mandatory arbitration 强制仲裁
mandatory imspection 法定检验
mandatory standard 强制性标准
mandici collar 女服高硬领
mandil guzrati 旁遮普金丝头巾纱；印度金银丝薄绸
Mandile 曼代尔头巾，曼代尔丝毛黑头巾
mandilion 曼德林宽松夹克
mandrenague 菲律宾棉经棕纬布
Mandrian dress 蒙德里安式直身裙
mandrin jacket 橙色外套(中国清朝)；马褂
mandyas 主教紫色斗篷；修士黑色短斗篷
mandypyta cotton 曼地塔棉
Mandyu cotton 巴拉圭曼地乌棉
mane 鬃毛
Man ethnic costume 满族服饰
maneuverability 灵活性
maneuvering curve 可操纵的曲线
manga 墨西哥套头外套
manganese bronze 锰棕色
manganese brown 锰棕色
mangle 干洗轧液机
mangle crease 轧皱印
mangled hessian 轧光打包麻布，轧光麻袋布
mango 芒果色
Mangrol cotton 印度曼格罗棉
man-hour 工时
man-hour quota 工时定额
manhua brocade 满花绸
manicure 修指甲；修指甲师
Mani-Hose (20世纪70年代)曼尼男式连裤袜商标，男裤袜

manihot silk 木薯蚕丝
manikin 人体模特,胸架;女模特儿,时装模特儿;矮人
manila 蕉麻纤维
manila board 白纸板
manila cloth 蕉布,蕉麻布,马尼拉麻布
manila [hemp] 马尼拉麻,蕉麻,菲律宾麻
manila maguey 坎搭拉剑麻
manilion 宽大短上衣;无袖外衣
manilla 金属轮
maniple (神父左臂上佩带的)弥撒带
manipulated 毛棉混纺呢
manipulated cloth 混纺呢
manipulation 操作法;移省道
manipulator 操纵装置;机械手
mankin 时装模特儿
man-machine interactive method 人机交互方法;人机交互方式
man-made fabric 化纤布
man-made fiber 化学纤维,人造纤维;化纤
man-made fiber blanket 化纤毯
man-made fiber knitting yarn 化纤绒线
Man-made Fiber producers Association [美]化学纤维生产者协会
man-made fur 人造毛皮
man-made leather 人造皮革,合成皮革,仿革
man-made mineral fiber 人造矿物纤维
man-made protein fiber 人造蛋白纤维
man-made spider fiber 仿蜘蛛丝纤维,人造蜘蛛丝纤维
man-made vitreous fiber 透明化学纤维,人造透明纤维
man-milliner 制作女帽(或头饰)的男工匠;男性女帽[或头饰]商人
mannequin 服装人体模型;女模特儿,(表演)时装模特儿;胸架;矮人
manner 派头(气派)
mannerism 服饰矫揉造作;个人惯用的表现手法,个人惯用的格调;癖性;习性
manner of packing 包装风格
mannish 男子气的;女装男服风格的
mannish collar 男式领
mannish jacket 男式短上衣,男性化外套
mannish look 仿男式服装风貌,(女装具有的)男服风貌;男子气款式;男性款式
man of parts 多面手
man's 男衬衫

man's fasten 男式系扣
man's figure 男子体型
man's fly front with French bearer 有法式腰带的男式暗门襟
man's hat 男帽
man-shi-lu flannelette 玛什鲁布
mansion maker 公寓服装商
man's ring 男式戒指
man's sandal 丁字凉鞋;男式凉鞋
man's three-piece suit 男式三件套装
man's trousers 男裤,绅士裤
man's two-piece suit 男式两件套装
man's upper outer garment 男上装
manta 女披肩(西班牙,美洲),女式薄方巾;男式方格围巾;毛毯;未漂细棉布,棉床单坯布,廉价棉布
man-tailored 男式女用的;按照男子服式裁制的
man-tailored blouse 男式女衬衣
man-tailored collar 男用西服领
man-tailored jacket 男式女[用]夹克
man-tailored jeans 男式女[用]牛仔裤
man-tailored suit 男式女[用]套装;男式女西装
manta suit 墨西哥本色细布男式两件套
manteau 曼特斗篷,披风,披肩;女式开襟长外衣;芒多裙(17世纪)
manteau domino 多米诺外套(化妆舞会用)
manteau veston [法]西装大衣
mantee 蔓蒂外套(18世纪,露三角胸衣和衬裙的女式外套)
manteel 披风;小斗篷;短外套;曼提儿披肩(18世纪)
mantelet 中世纪毛边短披肩;小斗篷;披风;短外套;短氅
mantel grijn 凸纹驼毛呢
mantelle 精纺毛织物
mantelletta 无袖法衣(红衣主教)
mantellum 曼塔路外套
mantelot 女式长披肩
mantilla 薄头罩;黑丝披巾;晚礼服斗篷;(西班牙、墨西哥)连披肩薄头罩,女用薄纱短披风
mantilla lace 黑丝披巾花边
mantle 披风,斗篷,披肩;无视罩衫;斗篷状长外套
mantle cloth 斗篷织物
mantle/core bicomponent fiber 皮芯型复合纤维;皮芯型双组分纤维

mantle fiber 灯罩纤维
mantlet 小斗篷;披巾;短披肩
mantling 徽章彩饰;斗篷料
manto 女黑披肩
manton de manilla 真丝绉刺绣大方巾
mantua (17～18世纪)女用长外衣,披风,斗篷;平纹丝织物
manual 手工的;说明书,手册
manual control (M/C) 手控,人工控制
manual embroidery pattern 人工绣花花样,手绣花样
manual felt carpet 手工毡毯
manual handling 人工操作
manual oiling 手动加油,手加油
manual operation 人工操作,手工操作
manual pattern sewing 手工花样缝纫
manual program 手编程序
manual punch 手动穿孔器
manual sampling 手工取样
manual snap fastening machine 手动揿钮钉扣机
manual utility press 手动万能熨烫机
manufacture 制造
manufactured fiber 人造纤维,化学纤维
manufacturer 制造商
manufacture's name 制造厂名称
manufacture's part No. 制造厂件号
manufacture's serial No. 制造厂批号
manufacture's source code 制造厂货源代号
manufacturing flow chart 工艺流程图;制造流程表
manufacturing process 制造过程;制造程序
manufacturing schedule 生产日程
manufacturing to buyer's samples 来样定制
manufactures 制品
manuscript 底稿
manve wine 紫酱
Manx tweed (英国)曼克思粗呢
many-way print 多向印花
Mao collar (中国毛式)立领,毛式领,中山服式衣领
Mao jacket 毛式上衣,中山装
Mao [suit] 毛式服装;人民装
maolao su chiyong 中国红绒纱
Maonan ethnic costume 毛南族服饰,毛难族民族服
mao-qing cloth 毛青布

mao-qing fabric 毛青布
map 映象
map cloth 地图布
maple 灰黄色
maple sugar 槭糖棕(黄棕色)
map mounts 裱地图布
mapping 映射技术
mar 火星牌粗格子呢
marabou jacket 鹳毛夹克
marabou 单丝经缎(织物);薄绉绸,薄绉丝带;羽毛边饰;鹳羽;马拉布生丝织物
marabout 单丝经缎;薄绉绸,薄绉丝带;羽毛边饰;鹳羽
maracaibo lace 抽线花边
marama 罗马尼亚薄纱头巾
maramato 阿拉伯金钱花缎
marana 弹性绉呢
maranham cotton 马兰汉棉
marathon shoes 马拉松鞋
marathon socks 马拉松袜
marble cloth 云石纹呢;云石纹书面布
marble print 大理石纹印花
marble silk 云石纹绸
marble wash 云纹石洗
marbre 云石纹色织布;云石纹丝毛呢
marbrinus 精纺云石纹刺绣底布
Marc Bohan 马克·博昂(1926～,法国时装设计师)
Marcasiano 马卡西诺(美国时装设计师)
Marcel 马赛尔(法国著名戏剧发型设计师)
Marcel Boussac 马赛尔·布萨(法国时装设计师)
marceline 马瑟林绸(丝织薄丝绸);双经单纬亮绸;单经多纬横棱绸;衬里绸,羽纱
marcella (英国)凹凸纹细布
Marcel Lassance 马塞尔·拉桑瑟(法国时装设计师)
Marcel Maronqiu 马塞尔·马龙丘(法国时装设计师)
Marcel Rochas 马赛尔·罗莎(1902～1955,法国时装设计师)
marcel [wave] 马塞尔卷发,大波浪发型
marche 法国挂毯
marchey 东印度条格平布
marduff 东非斜纹重帆布
marengo 皮板毛黑地白点呢
maretz 真丝绉
Margaret 玛格瑞特(意大利时装设计师)

margherita 意大利机制网眼纱
marginal seam 边缝
marginal sewing operation 缝纫机缘操作
marginal stitching operation 缝纫边缘操作
margin cutting 边幅修切
margin to seam 缝头
Marglass 玻璃纤维(商名,英国)
Margot lace 玛戈特花边;丝网地棉线花边
marguerite 意大利机制网眼纱
Maria Antonelli 玛丽亚·安东内利(意大利时装设计师)
Mariano Fortuny 马里阿诺·福图尼(1871~1949,意大利时装设计师)
Marie Alphonsine 马里耶·阿方西娜(法国时装设计师)
Marie Antoinette costume 玛丽·安托瓦内特装束
Marie Antoinette sleeve 玛丽·安托瓦内特袖子
Marie Antoinette style 玛丽·安托瓦内特款式
Marie-Antoinette stylistic form 玛丽·安托瓦内特款式
MARIE CLAIRE 《玛丽·克莱尔》(法国)
Marie galante cotton 马利亚加朗特棉
marigold(yellow) 金盏花黄,橙色
Mariko Hohga 甲贺真理子(日本时装设计师)
marine 海蓝色,藏青
marine bill of lading 海运提单
marine blue 海蓝色
marine cap 航海帽
marine chronometer 航海天文钟
marine cotton 叙利亚棉布
marine fiber 海草纤维
marine green 海水绿
marine look 海军风貌,海军风格;海员款式,海滨款式
marine stripes 海军条子布
marine style 海军风格
mariniere look 水兵服型服装风格,水手型
mariniere 水兵服
Marino Faliero sleeve 玛丽诺·法里埃罗袖
Mario Valentino 马里奥·瓦伦蒂诺(意大利时装设计师)
mariposa 条纹缎
Mariuccia Mandelli 马鲁西·曼黛利(意大利时装设计师)
mark 标记,标志,记号;型号;污渍;着色沾污;标记于,标明;(裁片上)打号;包装标志,商标;唛
mark a breath 打开销路
mark button and buttonhole posi-tion 标扣位和扣眼位
mark button position 点扣位
mark buttonhole position 划扣眼位,画扣眼位
marked price 标价
markeen 棉经毛纬厚呢
marker 标记;裁片纸版(在表层布上)划样,划样板;排版师;画皮;唛架;打印工;排料,排料图,排料长度;码克,码克长度
marker copier 码克复制机,排料图复制机
marker cutter 排板师;排版师
marker duplication 码克复制,排料图复制
marker fabric spread (计算机)排料,衣料展开方式
marker fallout 排料落料
marker filp 排料图翻动(计算机)
marker lay 排料
marker laying 排唛架;排纸版
marker lay making 按纸版划样;唛架制作
marker length 排料图长度
marker maker 码克师,排料员,排版师,码克师
marker making 描样;码克排版,排码克;排唛架;码克师
marker paper 排板纸
marker pen 记号笔
marker planner 排版师;排版,排料,排码克;排唛架
marker plotter 码克排版;码克绘印机,唛架绘印机
marker plotting 唛架制作
marker system 排料系统
marker width 排料图宽度
marker yardage 码克长度,排料图长度
market 市场,行情
marketable commodities 对路商品
marketable garment 畅销服装

marketable goods 畅销货,畅销织物
market bleach 普通漂白;普通漂白布
market forecast 市场预测
market one's own products 自产自销
market risk 市场风险
market sample 推销样品
market segmentation 市场细分(化)
market share 市场占有率
market supply & demand 市场需求
market survey 市场调查
market transform 市场转型
mark fashion 运动衣符号款式
marketing centre 交易中心
marketing ideas 营销理念
marketing internationalization 营销国际化
marketing positioning strategy 市场定位
marketing research 市场调研
marking 画样,描样,划码克;标记,打号;打号机;鸟羽斑纹;鸟羽斑点;兽皮斑纹
marking back 反面搭色
marking buttonhole position 划扣眼位
marking button position 定扣位;定眼位
marking chalk 划粉(片)
marking cotton 绣花打样用棉线
marking device 绘标用具
marking fabric 商标标记织物
marking fluid[ink] 印墨痕
marking graphite 裁衣用划线粉
marking-in 划样,划码克
marking ink 标记墨水
marking line 裁剪制图线
marking loss 划线损耗
marking machine 打号机,打印机,刷唛头机,鞋面画记号机
marking motion 打印装置
marking notch 对刀口
marking off 搭色疵点
marking of push button 按钮标记
marking out 划线;标记
marking paper (裁剪)纸样,贴纸
marking pencil 裁剪划线笔
marking pin 记号钉
marking quilting line 划绗缝线
marking ruler 放码尺
marking smooth 划顺
marking stitch 十字形刺绣针迹;十字缝,印记十字绣
marking stitch foot 底缝压脚
marking symbol 裁剪制图符号

marking system inspection 记分制检验法
marking tool 标记工具
mark number 唛头号数
mark of conformity 合格标志
mark & pattern adhesive 商标和花样胶黏剂
mark pin 商标钉
mark pocket opening line 标记开袋口处
marks 唛头,标记
Marks&Spencer 马莎
mark stitch 记号缝迹,线钉
mark stitching 十字形刺绣针迹;十字绣
mark-up artist 化妆师
mark-up cape 化妆用披风
mark-up cream 化妆雪花膏
mark-up man 化妆师;化装师
mark-up table 化妆台
mark-up woman 化妆师;化装师
mark yarn 仿合股双色纱
marl 仿斑点纱;泥土色
Marlborough hat 马尔伯勒帽
marl effect yarn 斑点效果纱,仿斑点纱
marli 贴花用六角网眼纱;薄亚麻布
marlotte (16 世纪)马洛特短斗篷式女外套
marly 贴花用六角网眼纱;薄亚麻布
marl yarn 夹色纱线,双色纱线;斑点花式纱
marmato 金银丝花缎
marmot 旱獭毛皮;土拨鼠毛皮
marmot fur 黄狼皮
marmotte (法)旱獭毛皮;驯旱獭女郎包头巾
marocain 波纹织物
marocain crepe 马罗坎绫纹绉
marocs 单面起毛毛哔叽
maroon 暗棕灰色;赤褐色;褐紫红色,栗色;深豆沙色;枣红色;紫酱色
marquise 卵形;卵形宝石戒指;卵形金刚石戒指;马眼形(珠宝);丝经棉纬缎,提花家具缎,侯爵夫人缎;女式三角帽
marquise finish 缎纹整理;棉贡缎加光整理
marquise gem cut 卵形宝石
marquisette 薄纱罗;方孔经编织物
marquisette curtain 薄纱罗窗帘
marquisette curtain net 薄纱罗窗帘纱
marquisette lace 薄纱罗花边
marquisette lace bow 薄纱罗花边蝴蝶结
marramas 金丝挂毯

marron 栗色;紫酱色
marrow bone 骸骨
marrow edge 曼柔缘边
Marry Jane 玛丽·简(美国时装设计师)
marrymuff 廉价粗呢
Marseilles 马塞提花床单布
Marseilles quilt 马塞提花床单布
Marseilles spread 马塞提花床单布
marsella 马塞拉漂白厚亚麻布
marsella linen 马塞拉纯亚麻布
mars red 樱红[色]
marston 马斯顿棉
mart 商业中心,市场
marten 貂毛皮,貂皮
marten coat 貂皮大衣
marten dress 貂皮服装
marten fur 貂皮
Martha Washington costume (19世纪90年代)玛莎·华威顿服
martial attire 戎装
Martin cotton 马丁棉
Martine Sil Bon 玛蒂·西尔·邦(法国时装设计师)
martingale 半腰带,燕尾腰带
martinique abutilon 苘麻
Martin's Human measuring kit 马丁计测器
marveilleux 玛维卢兹缎
marvella 有光长毛绒
marvello 有光长毛绒
Mary Jane shoes 玛丽·简童鞋,低跟钝头搭袢皮鞋;(20世纪80年代)包鞋式高于脚背有扣带的低跟、无跟或坡跟的玛丽·简鞋
MARY KAY 玫琳凯
Mary McFadden 玛丽·麦克法登(1936～,美国时装设计师)
Mary Quant 玛丽·匡特(1934～,英国时装设计师)
Mary Stuart cap 玛丽帽,玛丽·斯图亚特帽(苏格兰女王玛丽·斯图亚特使用的一种边缘上用金针别薄纱,周围装饰花边的心形帽)
Masahisa Shimura 老村雅久(日本时装设计师)
masalia 默萨利厄斜纹波纹布
Masayuki Abo 英保优之(日本时装设计师)
mascades 马斯卡德头巾绸
mascara 睫毛油,染睫毛膏;染眉毛膏

mascaret 精纺提花缎纹花呢
mascle (13世纪)菱形金属片穿成的盔甲,鞭形金属片盔甲
masculine feminine 男式女装
masculine figure 男子体型
masculine garment 男性化服装
masculine look 男式女装款式
masculine pantaloon 男式风格的裤子
masculine style 男式,男式女装款式
masculine 男性风格
mashaju 印度白色棉花缎
mashlah 玛诗拉长衫
mashru 棉背丝绸;丝光棉布;丝盖棉条子缎
masi 麻丝
masi cloth 马西布
mask 面具;面罩;口罩;护面罩;面膜;环形大墨镜;掩蔽
maskati 东非小花纹棉布
mask hat 面具帽
masking 伪装
masking tape 绘画遮蔽胶带
masloff 马斯洛夫厚呢
masquerade 化装舞会服饰
masquerade [costume] 化装舞会服
mass 色块;块面
Massachusetts Textile and Apparel Council 马萨诸塞州纺织服装委员会
massage 按摩
massage health care shoes 按摩保健鞋
massage shoes 按摩鞋
mass apparel 大众化服装
mass chromatogram 质量色谱图
mass coloration 原液着色
mass consumption 大规模消费
mass customization 批量定制
mass fashion 大众化时装,大众流行;大批量生产时装
massicot 铅黄
massiru 东南亚平纹薄绸
massive jaw 宽大的下巴
Masslinn 马斯林无纺毛巾
mass merchant 超级商场
massotherapy fabric 按摩织物
massotherapy pants 按摩裤
mass per unit area 单位面积质量
mass-produced garment 大批量生产的服装
mass production 大量生产;批量生产
mass sale store 批发店

mass shell button 马司贝扣
mass tone 主色,浓色
master cam 主凸轮
master grade 推挡基样
master pattern 母板,基本样板,原始图案
master's gown and hood 硕士服
master's hood 硕士悬垂披肩
mastic 乳香黄
mastic cloth 宽条刺绣底布
mastodon cotton 密西西比短纤陆地棉
Mastomo Yamaji 山地正伦(日本服装设计师)
masulipatam 印度裁绒地毯
mat 垫;麻袋;闷光织物,密结花纹;席子;消光的,无光的;毡化
mata 平纹棉床单布
matabie 金银花纹织物
mat braid 服装镶边;毛织饰边带
matador hat 斗牛士帽
matador pants 斗牛士裤子
matador shirt 斗牛士衬衫
matador 斗牛领巾;西班牙斗牛士服装
matador's jacket 斗牛士夹克
mat-border cloth 席垫镶边布,席垫滚边布,席垫镶边带
match 搭配;相配;配色打样;配色;[地毯]拼接配合
match belt 配衬腰带
matchbox seaming 沿裥缝切线
match coat 印第安披肩大衣
matched check 对格;配称的格子织物
matched design 阴阳图案,鸳鸯条格图案
matched pairs 配对染料
matching 配批;匹配;相衬;配色;搭配
matching and stitching sleeve lining seam 叠袖里缝
matching button 配色纽扣
matching carpat 配色地毯
matching colour 拼色,相配色
matching cutter 电脑对格裁料系统,对格裁床
matching flap facing 拼袋盖里
matching garment 配套服装
matching jacket 比赛服
matching jewelry 相配的首饰
matching of color 照样配色
matching pants 配色裤子,配套裤子;配合件
matching pillow sack 成型枕套

matching plaid 对格子
matching rule 对条对格规则
matching seam 配色线缝
matching short-sleeve sweater 竞赛短袖毛衣
matching sleeve 配袖
matching stitching 配色线缝
match maker foot 与格子织物相配的压脚
match mate 配套,相配(服饰品)
match to shade 配色
mate 相配
matelassé 马特拉斯,提花凸纹双层织物
matelasse organdy 马特拉塞泡泡纱
matelasse silk-like fabric 浮纹花绸
matelasse wool cloth 马特拉赛呢
matelot 水手式短上衣
matelot sweater 水手装套衫
material 原料;材料;织物;布料;斜纹布,轻量劳动布
material balance 物料平衡
material civilization 物质文明
material coordination 组合不同材料创造新感觉的技巧
material drifting 斜向进料;缝料斜边
material end 缝料端
material feeding 物料运输
material flow 原料周转
material interest 材料趣味
materiality 质感
material mismatch 衣料失配设计
material pucker 缝料起皱器
material puckering 缝料起皱
material reguirement planning (MRP) 物料需求计划
material release 缝料卸下
material science 材料学
materials for sewing 缝纫用料
material spread 铺缝料
material storing 物资储备
material testing 缝料试验
material thickness 缝料厚度
material to kill diseases 杀菌布料
material transportation 物料运输
material weight 缝料重量
maternity blouse 孕妇衫,孕妇罩衫
maternity brassiere 孕妇胸罩
maternity clothes 孕妇服
maternity dress 孕妇服
maternity garter 孕妇吊袜带

maternity girdle 孕妇帮肚,孕妇用束腰带
maternity napkin 产垫
maternity pants 孕妇裤
maternity robe 孕妇装,产妇服,宽松服装
maternity shoes 孕妇鞋
maternity skirt 孕妇裙
maternity slip 孕妇连身衬裙
maternity style 孕妇款式
maternity top 孕妇上装
maternity wear 孕妇装
mate's receipt 大副收据
mate threads 搭配花式线
mat finish 消光整理;制鞋乌光整饰
math maker foot 缝对格织物压脚
Mathandar shawl 马森德开司米披巾
mathematical model 数学模型
Mathio cotton 印度马思奥棉
matiere 材料质感
matinee 棉布连帽外套;女式晨衣;喝茶服
matinee coat 婴儿短外衣;婴儿短外套
matinée [法]马丁尼长项链
matinée-length necklace 长串珠链,日间社交用项链
Matisse style 马蒂斯风格
mat jersey 消光乔赛
matka 印度绢绸
mat kid 消光羔皮
matow 广东双宫丝
matrimonio 漂白床罩
matrix fiber 基质型[复合]纤维
matrix-fibril bicomponent fiber 基质-原纤型双组分纤维
mat-rush 包装用草席
mat silver 闷光银色
matta cotton 巴西马塔棉
matt coating 芝麻薄花呢
matte 暗淡色调,无光泽
matted material 草垫制品;毡制品
matte 暗淡色调;无光整理
matte finish 消光处理
matte jersey 无光乔赛,无光针织物
mattern notches 刀眼剪
matte style 无光(亚光)型
matt fabric 无光织物
matt fiber 无光纤维
matt finish 无光整理,消光整理
matt finishing 消光整理,无光整理

matt gold 黯金色,黯金色的
Matthews cotton 美国马休斯棉
matting 消光
matting cloth 十字编织布
matting cord 四经四纬席纹布
matting oxford 牛津席纹衬衫布
matting power 缩呢性;缩绒性;毡合性
Mattique 聚酯超细纤维(商名,美国杜邦)
Mattis cotton 美国马蒂斯棉
matton 印度开士米披巾花纹
matt prints 无光印花布
mattress 褥,垫子
mattress duck 工业用单纬帆布
mattress ticking 床垫布
mattress twine 褥垫线
matts 席纹布
matt shirting 席纹衬衫布
matt umbrella 田字格伞
mature fiber 成熟纤维
mature figure 成人体型;成熟体型
mat worsted 垫席型粗纺毛织物
maual knotted carpet 手工栽绒地毯
maucilli 棉细布
maud 柳条灰呢;莫德披肩;(苏格兰牧人穿的)灰格子呢披巾或披衣
maud fabric 莫德呢,棉经毛纬呢
Maud Roser 莫德·罗塞尔(法国女帽设计师,时装设计师)
mausari 稀格子蚊帐布
mausoorie 稀格子布
mauve 紫红色
mauve diamond 淡紫钻石
mauveglow 鲜红紫色
mauveine 紫红色
mauve morn 晨紫色
mauve orchid 浅紫蓝色
mauve wine 紫酱[色],酒红[色]
mawata 日本丝绵
mawsie 苏格兰羊毛衫
Max & Co 麦克斯·库
Maxey cotton 美国马克塞棉
maxi 迷喜长度;超大的,超长的;迷装;长女服(长裙,长大衣等)
maxi cape 及踝披风,特长披风
maxi-chemise 特大衬衫;过膝衬衫装
maxi clothing 迷喜外套;长外套
maxicoat 特长式大衣;及踝长袍;迷喜大衣;加长外套
maxidress 迷喜女装;加长女服

maxi length 超长;及踝长
maxilength 拖地裙装,拖地装
maxi-length trench coat 超长风雨衣
maxi line 超长型;超大型
maxi look 超长风貌,迷喜风貌
maximal color 最全色
maximizer bra 增大型胸罩
maximizer brassiere 增大型胸罩
maximum calf girth height 最大腓围高
maximum head breadth 头最大宽
maximum head length 头最大长
maximum lift 最大抬高量
maximum number of stitches per minute 每分钟最大针迹数
maximum pattern area 最大花型范围
maximum stitch width 最大线迹宽度
maximum thigh girth 最大股围
maximum working width 最大工作宽度
maximum zigzag width of throw 最大曲折线迹宽度
maxine taffeta 马克辛塔夫绸
maxi-order 大订单
maxis 超大
maxi scarf 特长围巾
maxi seam pitch 最大线缝节距
maxi-shorts 长短裤
maxi silhouette 超长轮廓;超大型;及踝型轮廓
maxi-skirt 长大衣;超长裙;及踝长裙
maxi-sweater 特大号毛衣
Max Mara 麦克斯·玛瑞
max-speed 最大缝速;大速度
maxi [style] 及踝时装式样
MAXSTUDIO 麦克斯
maybasch silk 日本生丝
MAYBELLINE 美宝莲
mayenne 法国精细漂白亚麻布
Mayo dress fabric 梅奥斜纹织物
Mayo twill 梅奥斜纹
mazachar 印度银波花缎
mazamet [wool] 皮板毛;马扎默呢,法国麦尔登呢
mazarine 深蓝色(服装)
mazarine [blue] 深蓝色,深紫蓝色;深蓝色服装
Mazu suit 马祖装
Mckey process 制鞋透缝工序
mcnair 235 cotton 美国东南地区陆地棉
M-cut collar M形驳口大衣领
meadow green 草地绿,草绿

mealy 小浓斑疵,有小浓斑的
mean clothes 破烂的衣服
meander 回纹波形饰,雷纹
meander motif 回纹图案;回纹花纹
mean deviation 均差
mean line 中线
mean-square deviation 均方差
mean-square regression 均方回归
mean-square value 均方值
mear crepe 马尔绉
measure 度量,测量;量身;量具;尺寸,大小
measure cloth 量布
measure cutter 按体缝制的裁剪工
Measure for winding point 卷尺
measure frame (圆筒形针织物用)拉幅框
measurement 量身,量体;度量,测量;尺寸,围度,尺度
measurement chart 尺寸表(服装裁剪)
measurement level 围度位置
measurement leveler 服装围度位置
Measurement Standard Committee (MSC) 人体测量标准委员会
measurement taking 尺寸测量
measurements 尺寸,大小;妇女三围;刻度
measures 尺码
measure sb. for new suit 量身做衣
measure tape 带尺;卷尺;皮尺
measuring amplitude 测量范围,测量幅度
measuring at your home 到府量身
measuring chart 测量表
measuring cylinder 量筒
measuring devices 量度用具
measuring recording chart 人体测量记录表
measuring stick 短尺
measuring tape 带尺;卷尺;皮尺
measuring technique 测量技术
measuring time 测时
measuring tool 量度用具
mebane triumph cotton 得克萨斯棉
Mecca rug 麦加地毯
mechanical bonding 机械粘合
mechanical characteristics 机械特性
mechanical drawing random process 机械牵伸杂乱法生产非织造布
mechanical efficiency 机械效率
mechanical fabric 工业织物

mechanical finishing 机械整理
mechanical grader 机械放缩仪
mechanical installation 机械安装
mechanically-bonded fabric 机械结合法非织造布
mechanical move time 机作时间
mechanical property 力学性能,机械性能
mechanical shrinkage 机械防缩
mechanical stretch [织物]弹性整理
mechanical watch 机械手表
mechanic's pocket 工装大贴袋
mechano-electrical integration 机电一体化
mechano-electronic integration 机电一体化
meche colore [法]异色发绺发型
Mechlin embroidery 梅希林花边
Mechlin lace 梅希林花边,细网眼梭结花边
Mechlin machine 梅希林网眼花边机
Mecklenburg thread 梅克伦伯爾亚麻线
Mecklenburg 梅克伦伯提花缎纹呢
mecroaching satin stitch 分茎绣
medal 勋章
medal ribbon 奖章带
medallion 花边装饰纹;团花;饰孔革;圆形或椭圆形浮雕[颈饰]
medallion carpet 团花花毯
medallion design 团花图案
medallion lace 图案花边,花卉花边
medallion necklace 圆垂饰重链项链
medallion shoes (鞋头有小孔的)装饰鞋,小孔装饰鞋
medallion toe 美达利翁鞋头,孔饰鞋头,孔隙鞋
meddle lightness 中明度
medias 针织物;针织服装
medial longitudinal arch 足弓
medial malleolus 踝关节,内踝
medias shirt 针织卫生衣
medias wear 针织服装
medic 医生白衬衫式外套;医生白外套立领
medic alert bracelet 纯银刻字手镯,药性显性手镯
medical and hygienic textile 医药卫生用纺织品
medical bandage 医用绷带
medical clothing 医用服装
medical fiber 医用纤维

medical fiber fabric 医用纤维织物
medical gauze 医疗纱布
medical huck 缩绒厚呢
medical mop 外科手术用拭布
medical pad 药用鞋垫
medical synthetic fiber 医用合成纤维
medical textile 卫生[用]纺织品,医用纺织品
medical uniform 医务服
medical wear 医务服
medical-protective clothing 防药物服
medicated cotton 药棉,脱脂棉
medicated finishing 疗效整理(在织物上添加治疗药物的加工技术)
medicated cotton wool 药棉,脱脂棉
medici 荷叶边花边,荷叶边状花边扇形花边
Medic collar 本·凯撒立领(医用白外套立领)
Medic shirt 本·凯撒衬衫式外套,医用白衬衫式外套(立领,侧襟)
Medici collar (18~19 世纪)美第奇扇形大立领,女式滚边小立领
Medici lace (16 世纪)美第奇花边,梅迪契花边
medicinal cotton 药棉
medicinal fiber 药物纤维,治病防病纤维
medicis 法国梭结花边
medieval cloak 中世纪连帽斗篷
medieval embroidery 中世纪(式)刺绣(平针粗花纹用锁边针迹固定线脚);中世纪刺绣挂毯
medieval hunting hat 中世纪猎帽
medievalism 中世纪精神
medieval overshoe 中世纪鞋套
medieval style 中世纪风格
meditate on the past 怀古
meditation shirt 冥想衫(宽松,开启式衣袖,套头束腰衬衣)
medium (M) 中等尺寸;中厚
medium and pinwale corduroy 中细条灯芯绒
medium ball point 中等球形针尖
medium blue 中蓝色
medium closing 中领开口
medium cloth 中档上浆棉坯布;条纹薄绒呢;中厚毛织物
medium color 中色
medium counts 中支数
medium count yarn 中支纱

medium cross-bred wool 中支交配种羊毛
medium draft 中绵,中切绵
medium fabric 中厚织物
medium fine 中细
medium fine wool 半细羊毛
medium finish 中度上浆整理
medium-firm knit fabric 中密度针织织物
medium fit 贴身式;贴身衣着
medium focal length lens 中焦距镜头
medium foot 中压脚
medium grade coarse hand knitting yarn 全毛中级粗毛线
medium gray 中灰[色]
medium green 中绿色
medium grey 中灰色
medium-heavy fabric 中厚面料
medium-heavy material 中厚料子,中厚缝料
medium heel (shoes) 中跟鞋
medium-high heel 中高跟
medium interfacings 中厚衬布
medium jacquard fabric 中花型提花织物
medium length fiber 中长纤维
medium light fabric 中轻织物
medium line 中间线,中线
medium overcoat 中大衣
medium pattern printings 中型花纹印花
medium pink 中桃红
medium plain cloth 中平布
medium point 中等针尖
mediums 上浆中档坯布
medium set point 中等规格针尖
medium sew-in 中型可缝衬
medium shade midfibre fancy suitings 中色中长花呢,中间色中长花呢
medium shade wool-like fancy suiting 中复色中长花呢
medium shade 中等色泽,中色
medium size 中号
medium & small size enterprises 中小型企业
medium spread collar 半展开领,中等八字领
medium staple 中纤维,中长纤维,中绒
medium staple cotton 细绒棉
medium staple fiber 中长纤维
medium stripes 提花缎纹条子布
medium sweep length 中等拖曳长
medium thick material 中厚缝料

medium throw machine 中等摆幅缝纫机
medium-to-long neck 中长颈
medium tone 中间色调
medium twist 中等捻度
medium wale corduroy 中条灯芯绒
medium weight 中等重量衣料
medium weight fabric 中厚织物
medium weight fancy suitings 中厚花呢
medium weight interfacing 中等克重的里衬,中厚里衬
medium weight material 中厚缝料
medium [weight] zipper 中牙拉链
medium wool 中级羊毛;半细毛;中等长度羊毛
medium yarn 中支纱
medium yarn fabric 中支纱织物
medium yellow 中黄色
medium yellowish gray 中黄光灰
medium yellowish grey 中黄光灰
medium zipper 中牙拉链
medufehdi 高级手织棉布
medullary ray design 木纹图案
medullated fiber 有髓纤维,毛髓纤维
meen pow 广东地区棉织物
meet 衣服合身
megastore 大型卖场
megila 印度黄麻织物
megnificent military style 华丽军服款式
meherjun 波斯粗地毯羊毛
mehramat 克什米尔梅拉麦平绒面纱
meili lining twill 美丽绸,美丽绫,高级里子绸
meisen 日本铭仙绸
mekla 裙用粗棉布
mekli dhorka 手织红蓝条黄麻布
melamine fiber 蜜胺纤维
melamine-formaldehyde fiber 蜜胺纤维
melange 混色哔叽;混色毛纱;手工梭结丝花边
melange cloth 混色布,混色织物
melange luster 混色有光织物
melange printings 毛条印花
melange serge 混色哔叽
melange suiting 混色薄开士米毛织物
melange yarn 混色纱
melaye 叙利亚梅拉伊绸
Melbes 聚酰胺中空细旦丝(商名,日本帝人)
Melbourne wool 墨尔本羊毛
meldable fiber [可]熔合纤维

melded fabric 双组分熔纱织物;熔焊型非织造布
melded-fiber structure product 热熔黏合非织造布
meldeds 熔焊型非织造布,热熔黏合非织造布
meld-fiber bonded fabric 熔合纤维[黏结非织造]布
mele silk 混色丝线
Meles rug 小亚西亚梅累斯地毯
melimeli 漂白棉平布
melis 法国大麻船帆布
mellofehdi cloth 高级手织棉布
mellow colour 悦目的颜色
mellow fashion 成熟时装
mellow finish 柔软,柔软整理
mellow finishing 柔软整理
mellow green 柔和绿
mellow hand 柔软手感
mellowing 柔和;柔软处理;揉布
mellow yellow 柔和土黄,芽黄[色]
melon (法)瓜帽;浅红橙色,瓜瓤红;隆起的大肚子
melon-based orange 瓜瓤橙
melon belly 隆起的大肚子;大腹便便的人
melon hat 瓜形帽
melon line 南瓜型
melon(of balloon)sleeve 气球袖
melon silhouette 南瓜轮廓,南瓜型
melon sleeve 气球袖,羊腿袖,瓜形袖,三角袖
melt resistance 抗熔性
meltable fiber [可]熔合纤维
melt-blown fabric 纺黏布,熔喷法非织造布
melt-blown fiber 熔喷纤维
melt-blown microfiber 熔喷微细纤维
melt-blown nonwoven 熔喷非织造布
melt-bonded webs 热熔黏合非织造布
melt-colored fiber 熔体着色纤维
melt-dyed fiber 熔体染色纤维
melten-bonded nonwoven 热熔黏合非织造布
melting bonding fiber 熔体黏结纤维
melting-hole behaviour 熔孔性
melton 麦尔登(粗纺呢绒),麦尔登呢,重缩重起毛呢绒
meltonette 麦尔登薄呢
melton finish 麦尔登整理;呢面整理

melt-spun fiber 熔纺纤维
melt-spun metal fiber 熔纺金属纤维
Melty 聚酯低熔点絮棉(商名,日本尤尼吉卡)
melusine 制帽用梅留辛茸毛毡呢
membership group 实属群体
membrane collar interlining 领角薄膜衬
memmeru 绒布
memo 备忘录
memory color 记忆色
memory crease 耐久褶裥
memory-delay device 记忆延时机构
memory shirt 记忆衬衫,保形免烫衬衫
memory stitcher 电子缝纫机
Memphis cotton 美国产孟菲斯棉
Menba ethnic costume 门巴族民族服
men chijimi 棉绉布
mend 修补;缝补;织补;打补丁
mend clothes 补衣
mende 法国滑细里子哔叽
mender 缝补者
mender crock 织补好的破洞
mending 缝补[用]纱线;织补;缝补;修补
mending bagging 补包用黄麻布
mending cotton 缝补棉线
mending hem 缝补卷边
mending mark 修补疵,织补疵,修痕
mending needle 织补针
mending stitch 缝补线迹
mending tape 背胶牵带
mending tears 修补破洞
mending thread 缝补线
mend shoes 修鞋
mend stitch 织补线迹,缝补线迹
mend woolens 织补毛料衣服
menhofu 棉制帆布
Menin lace 梅宁花边
menkwa[cotton] 日本原棉
Menni shirt 曼妮衬衣,驼色针织内衣;纽扣门襟针织衬衫
Menouffieh cotton 埃及梅努弗棉
men's and women's apparels 男女服装
men's backpack 男式背包
men's Bermuda 男式百慕大短裤
men's bib and brace overalls 男工作背带裤
men's boilersuits 男工装连衣裤
men's boots 男式靴
men's briefs 男三角裤
men's cap 男式便帽

men's clothes 男装
men's clothing 男装
men's clothing design 男装设计
men's cotton suitings 男线呢
men's dress materials 男服料
men's dress socks 绅士袜
Men's Fashion Association（MFA）［美］男式时装协会
men's figure 男子体型
men's formal attire 男礼服
men's hat 男帽,男式帽子
men's high leg boots 男式长筒靴
men's lingerie 男内衣裤
men's look blouse 男式女衬衫,女衬衫,连腰带短罩衫
men's nightwear 男睡衣
men's pullover 男式羊毛套衫
men's ring 男式戒指;男用戒指;男子戒指
men's shirt 男衬衫
men's shirt-body 男式汗衫
men's shirt-sleeve 男式长袖衬衫
men's shoes 男鞋
men's short-sleeved pullover 男式短袖羊毛套衫
men's short-sleeved sweater 男式短袖厚运动衫,男式短袖毛线衣
men's shusu 绉面缎纹
men's sizes 男子尺寸,男子服装尺寸
men's slacks 男式宽裤
men's slacks-casual 男式便裤
men's suit 男套装;男西装
men's suit jacket 男式夹克
men's suit slacks 男式套裤
men's suitings 男式服装面料,男子衣料
men's sweater 男毛衫
men's three-piece suit 男式三件套装
men's trousers 绅士裤;男裤
men's two-piece suit 男式两件套装
men's underclothes 男内衣
men's undershorts 男式短衬裤
men's underwear 男内衣
men's underwear and shorts 男内衣裤
men's V neck rib knit pullover 男式V领针织罗纹套衫
men's waistbanding 男裤腰头
menswear 男服（总称）,男服料
MEN'S WEAR 《男子服装》(美国)
men's wear 男服,男装;男子衣着
men's wear department 男装部

men's wear design 男装设计
men's wear look 男服风貌,男服女穿风貌
men's wear Retailer's Association（MRA）男式服装零售商协会(美国)
men's wear store 男式服装店
men-tailored collar 男式西装领
mental 斗篷;外套
mental outlook 精神面貌
men wear store 男式服装店
Menzies tartan 门齐斯格子呢
Mephistopheles suit 靡菲斯特套装
Meraline Raye 梅拉林赖条子薄呢
meraline 窄条呢
mercantile paper 商业票据
mercenary dress （15～16世纪)雇佣兵装
mercer 布商;绸缎商
mercerization 丝光
mercerization of knitting goods 针织物丝光
mercerization style 丝光风格
mercerize 丝光处理
mercerized 丝光,丝光缝纫线
mercerized and bleached canvas 丝光漂白帆布
mercerized and printed bed sheet 丝光印花床单
mercerized bed sheet 丝光床单
mercerized cotton 丝光棉纱;丝光棉布
mercerized cotton cloth 丝光布
mercerized cotton thread 丝光棉线(2～6股,缝纫刺绣用）
mercerized fabric 丝光织物
mercerized finish 丝光整理
mercerized khaki 丝光卡其
mercerized sewing thread 丝光缝纫线
mercerized sheet 丝光床单
mercerized single jersey 丝光汗布
mercerized stockings 丝光袜
mercerized striped shirt 丝光柳条衬衫
mercerized stripes 丝光纱条子棉布
mercerized thread 丝光线
mercerized towel 丝光毛巾
mercerized wool 丝光羊毛
mercerized wool knitwear 丝光羊毛衫
mercerized yarn 丝光纱
mercerizing 丝光处理
mercery 绸布商店;绸布业;绸布;丝织品;天鹅绒织物
merchant tailor （制作定制服装)裁缝

mercilline 梅雪林绸
mercury 水银色
merge & acquisition tactics (M&A tactics) 并购策略
merino comeback 美利奴归宗羊毛
merino crepe 美利奴闪光丝光绉
merino fiber 美利奴再生毛
merino hosiery 棉毛混纺针织物
merino quality terms 美利奴毛品质名称
merino tulle 法国美利奴网眼纱
merino wool 美利奴羊毛
merino yarn 美利奴精纺毛纱;美利奴针织毛纱;美利奴毛棉混纺纱
merino 美利奴绵羊毛;美利奴毛织物,美利奴斜纹呢;美利奴头巾;棉毛混纺针织物;长弹毛织物;精细斜纹呢;混纺针织毛纱;菲律宾窄幅棉布
merinyl 羊毛锦纶混纺纱
merletto 意大利梭结花边
merletto a maglie 意大利网眼花边
merletto a retine ricamate 意大利绣花网眼地花边
merletto a tombola 意大利梭结花边
merletto biondo 原色丝花边
merletto prombini 意大利梭花边,意大利花边
mermaid dress 美人鱼装
mermaid lace 威尼斯精细针绣花边
mermaid line 美人鱼型
mermaid [line] skirt 美人鱼式长裙
mermaid sheath 紧身细长晚礼服
mermaid silhouette 美人鱼轮廓
Merovingian period costume 墨洛温王朝服饰
merrin wool 美利奴皮板毛
Merry Widow 玛莉寡妇胸衣,坡莉内衣(与胸罩连为一体,"玛莉寡妇"源于1905年著名歌剧名称)
Merry Widow hat 玛莉寡妇宽边帽,风流寡妇帽
merve 奇异高级里子缎
merveilleux 奇异高级里子缎
merv wool 默符粗羊毛
mesankooria 纯白野蚕丝
mesched rug 梅什ohm地毯
meseri 农村头巾布
meseritsky 俄罗斯厚呢
mesh 网眼;网眼布;网织品;网状物
mesh bag 网袋,网兜;网眼袋布
mesh belt 网孔饰带,网状腰带

mesh blouse 网眼衬衫
mesh bracelet 精致的金属链节手镯
mesh cap 网眼帽
mesh cloth 网眼布
mesh decorative cloth 网眼装饰布
mesh fabric 网眼物;网眼布
mesh gauge 网眼扣针
mesh hat 网眼帽
mesh hose 网眼袜
meshi 丝;丝绸,绸缎;丝织品
mesh knit 网状编结构
mesh pin 网眼扣针
mesh shirt 网眼衬衫
mesh shoes 网眼鞋
mesh socks 网眼短袜
mesh stockings 网眼长筒袜
meshta 洋麻纤维,槿麻纤维
mesh undershirt 网眼汗衫
mesh vest 网眼背心
meshwork 网眼织物
mesot 洋麻,槿麻
messaline 梅萨林丝缎,细软缎
Messaline finish 梅萨林整理(摩擦轧光整理,使织物有光泽和柔软的手感)
mess jacket 白色的晚餐礼服;餐厅侍者短上衣;军官晚餐服
mess kit 英国军官晚礼服
mess uniform 军晚礼服
mest 土耳其袜
mest(h)a 熟红麻,槿麻,印度络麻,野麻,洋麻
mestiza dress 菲律宾妇女民族服装;女混血儿服(19世纪末菲律宾上流社会的混血女性喜欢的着装)
mestiza wool 南美洲交配种羊毛
mestizo dress 混血儿女服
meta-aramid fiber 间位芳族聚酰胺纤维
metabolic rate 能量代谢率
metabolism 新陈代谢
metacarpal bone 掌骨
metachrome dyeing process 同浴铬媒染法
metaculture 超级文化
metal and whale bone corset 紧身垫撑胸衣
metal badge 金属徽章
metal bag 金属包
metal belt 金属腰带
metal bridge 金属桥式连接的(拉链)
metal button 金属纽扣

metal cane 金属手杖
metal-chelated fiber 金属螯合纤维
metal clip 金属夹
metal cloth 金属线织物
metal-coated fabric 金属涂层织物
metal-coated fiber 镀金属纤维,金属镀覆纤维
metal color 金属色
metal complex dye 金属络合染料
metal-containing carbon fiber 含金属碳纤维
metal cuff links 金属袖口纽
metal fastener 金属拉链
metal fiber (MTF) 金属纤维,金属丝
metal filament 金属长丝
metal filled chenille 金属芯绳绒线
metal filled glass filament 金属芯玻璃丝
metal handle 金属柄
metal hanger 金属衣架
metal hangtag 金属吊牌
Metalian 聚酰胺导电纤维
metalized fabric 金属涂层织物
metal label 金属商标牌
metal lace 金银线花边
metallic asbestos yarn 金属石棉线
metal(1ic) belt 金属腰带
metallic braid 金属编织带
metallic buckle 金属带扣
metallic button 金属扣
metallic chain belt 金属链式腰带
metallic cloth 金属线织物
metallic cuff link 金属袖扣
metallic embroidery 金属刺绣,金属绒刺绣
metallic fabric 金属线织物,金属涂层织物
metallic fiber 金属纤维,金属丝
metallic filament 金属丝
metallic gauze 金属纱罗(印花)
metallic glare 耀眼金属光
metallic gray 金属灰(浅红灰色)
metallic grey 金属灰(浅红灰色)
metallic hand 金属样手感
metallic jacket 金属衣
metallic jersey 金属线乔赛
metallic knit 金银线针织物;嵌金属丝针织物
metal(lic) lace 金属线花边
metallic look 金属款式
metallic plaid 金属格子花纹
metallic prints 金属粉印花布
metallic satin 金属光泽缎织物
metallic screen 金属网
metallic staple 金属短纤维
metallic style 金属光泽款式
metallic thread 金属线;金银线
metallic tone 金属色调
metallic transfer 金属烫贴
metallic vest (击剑)金属背心
metallic weighting [丝织物的]增重整理
metallic wire braid 金属丝编织带
metallic wire fabric 金属丝织物
metallic woven cloth 金属线织物
metallic woven fabric 金属丝织物
metallic yarn 金银丝,金银线,金属线
metalline 仿金银线织物
metallized cloth 金属化织物
metallized fabric 金属涂层织物
metallized fiber 金属涂层纤维,镀金属纤维
metallized glass fiber 镀金属玻璃纤维
metallized polyester filament 镀金属聚酯长丝
metallized thread 金银丝,金银线
metallized yarn 含金属纱线;金银线;[含]金属丝,镀金属丝
metal mesh bag 金属网包
metal ornament 金属装饰品
metalot sweater 水手装套衫
metal panniers 金属裙撑
metal plate (饰于服装上的)金属牌
metal press button 揿纽
metal prong 拉链齿尖
metal ring 金属环
metal rivet 金属铆钉
metal rod 金属人体架
metal shank 金属扣柄
metal shank button 金属单柄纽扣
metal slip-on [shoes] 有金属装饰的套穿鞋
metal snap 金属四合纽
metal stick 金属手杖
metal thread 金属丝,金属线
metal-toothed chain 金属齿链
metal-toothed chain zipper 金属齿拉链
metal walking stick 金属手杖
metal yarn 金属纱线
metal zipper 金属拉链
metameric color 同色异谱色
metameric color match 条件配色

metamerism 位变异构现象	mezzo transfer 金石烫贴,金属烫贴
metamorphose 变形,变样	mhabrum 土耳其棉料纹布
metastable fiber 亚稳态纤维	Miao brocade 苗锦
meter 公尺;米;仪表;仪器	miaochun crepe damask 描春绉
meter cell 米数计	Miao ethnic costume 苗族服饰
meter counter 米尺计长表	Miao(nationality)brocade 苗锦
meter rule 公制尺;米尺	micado 米卡多薄型塔夫绸
meter ruler 米尺;公制尺	Michael Jackson Jacket 迈克尔·杰克逊夹克(红色皮夹克)
Meters/bonwe 美特斯邦威	
method 方法	Michael Jackson look 迈克尔·杰克逊造型
method of Boston matrix analysis 波士顿矩阵	
	Michael Kors 米歇尔·科斯(美国时装设计师)
methyl blue 甲基蓝	
methylated cotton 甲基化棉	Michel Goma 米歇尔·戈马(法国时装设计师)
methyst orchid 青莲	
meticulous cutting 精致裁剪	Michel Klein 米歇尔·克莱因(法国时装设计师)
metre 米;公尺	
metre counter 公制测长器	Michel Schreiber 米歇尔·施赖伯(法国时装设计师)
metre tape measure 米制带尺	
metric chromaticity 米制色度	Mickey 米奇妙
metric conversion chart 尺寸对照表	Mickey Mouse 米老鼠(图案)
metric conversion table 公制换算表	Micrell 聚酯超细长丝(商名,意大利)
metric counts 公制支数	micro 超短的;特短超短裙;迷哥装;超迷你装;露股装
metric lightness 米制亮度;米制明度	
metric number 公制支数	micro accordion plait 细风箱褶
metric size 公制尺寸	microbe 微生物
[metric] tape measure 卷尺	microcapillarity 微毛细管作用
metro mill 红粉英姿	microcapsule dye 微胶囊染料
metropolitan look 大都市风貌	microcapsule fabric 微胶囊织物
metz cord 梅兹罗纹呢	microcapsule printing 微胶囊印花
meuliquin 紫红布;细亚麻布	microclimate 小气候,微气候(衣服与皮肤间的环境温湿度)
mexicaine 法国花边塔夫绸	
Mexican costume 墨西哥民族服饰	microclimate under clothing 服装内气候
Mexican cotton 墨西哥棉	micro-cord 细凸纹
Mexican drawn thread work 墨西哥抽绣	microcrater fiber [表面]微细凹凸纤维
Mexican drawnwork 墨西哥抽绣	micro-crimpy filament 微卷曲长丝
Mexican embroidery 墨西哥刺绣	microdenier fiber 微细[旦]纤维
Mexican look 墨西哥风貌,墨西哥款式	microdenier yarn 微细[旦]长丝
Mexican-look shoes 墨西哥型鞋类	micro dress 超超短连裙装(长及上腿部分),特短超短裙装
Mexicans 杂色平布;杂色斜纹布;英国出口的坯布	
	microfiber 超细纤维
Mexican sisal 墨西哥西沙尔麻	microfibre 超细纤维
Mexican stylistic form 墨西哥款式	microfibre fabric 超细纤维织物;桃皮绒
Mexican wedding boots 墨西哥婚礼靴	microfibre jacket 桃皮绒夹克
Mexican wedding shirt 墨西哥婚礼男衬衫(白色,绣花)	microfil 微细长丝
	microfilament 微细长丝
Mexican wool 墨西哥粗羊毛	Microfine 聚酰胺细长丝(商名,美国杜邦)
meyer linen 大麻布	
Meyers Taxas 马克塞棉	microfine denier fiber 超细纤维
mezzo punto 法国粗梭结花边	microfine fiber 超细纤维(0.1~1.0分特)

Microlene 聚丙烯细短纤维(商名,意大利)
micro length 超短的;超短迷你型长度,超短型长度,迷哥型长度
Microlith 玻璃纤维(商名,德国)
Microloft 聚酯细[旦]纤维絮棉(商名,美国杜邦)
Micro mattique 聚酯超细纤维(商名,美国杜邦)
micro-mesh 微型小网眼
micrometallic transfer 微型金属烫贴
micro-meteorological evaluation 小气候评定
micro-mini (M.M.) 超迷你裙;超短的;迷哥裙
micro-mini length (M.M. length) 超超短型长度,超迷你型长度
micro-mini skirt (M.M. skirt) 特短超短裙,微型超短裙,超迷你短裙
Micronesse 聚酯细[旦]纤维(商名,美国赫斯特—塞拉尼斯)
Micropake 辐射屏蔽性聚丙烯纤维(商名,英国考陶尔)
micro peach fabric 微细桃皮绒织物
micropile 超细绒[毛](俗称桃皮绒)
micropile fabric 超细绒织物,桃皮绒织物
micropore 微孔
microporous coated fabric 微孔胶[防雨]织物,微孔涂层织物
microporous fiber 微孔纤维
microporous hollow synthetic fiber 微孔[性]中空合成纤维
microporous synthetic fiber 微孔型合成纤维
microporous touch fabric 微粉状手感织物
microprocessor controlled sewing machine 微控缝纫机,微机控制缝纫机
micro quartz fiber 微石英纤维
microscope 显微镜
microskirt 超短裙,迷哥裙,超迷你裙,露股裙
Microspun 聚酯细纤维(商名,美国纤维工业)
microstaple fiber 微细短纤维
Microstar 聚酯/聚酰胺6机械裂离型超细复合纤维(商名,日本帝人)
micro stretching finish 微量伸缩整理
Microsupplex 聚酯细长丝(商名,美国杜邦)

microwave dyeing 微波染色
microwave radiation protective coverall 防微波辐射工作服
microwave radiation shield textile 微波辐射屏蔽纺织品
mid finger 中指(第三指)
mid heeled sling 中跟吊鞋
mid hip girth 中臀围
mid length short pants 半短裤
mid length shorts 中短裤
mid life 中年
mid night blue 午夜蓝
midani 土耳其手织米达尼绸
mid-ankle 中踝节部
mid-armhole 袖窿弧线中点,袖窿中点
mid-armhole back 袖窿后中点
mid-armhole dart 中袖胸省
mid-armhole front 袖窿前中点
mid-back cowl 中位背垂褶
mid-calf 腿肚(膝踝间);中腓裙
mid-calf boots 及小腿肚高靴,及腓长靴
mid-calf length 及腿肚长
mid-calf pants 长至腿肚的裤子,腓中裤
mid-calf skirt 腓中裙,及腿肚的裙子
mid-cut 中针距
mid-depth cowl 中位垂褶
middie 水手领宽上衣
middle 中间;腰部,身体中部
middle age 中年
middle-age spread 中年发福
Middle Ages style 中世纪风格
middle and old aged people's wear 中老年服装
middle back 中背部
middle ball point 中球针尖
middle chest circumference 胸中围
middle finger (third finger) 中指(第三指)
middle grade 中档货
middle grade coarse kintting wool 中粗绒线
middle grade coarse knitting yarn 中粗绒线
middle heel 中跟;中跟鞋
middle-heeled shoes 中跟鞋
middle/high grade 中高档(等级)
middle hip 中臀围
middle hip line (MHL) 中臀围线
middle hip measure 中臀围

middle last forming pattern machine　腰帮定型打钉孔(制鞋)
middle light colour　中明色
middle lightness　中明度
middle/low grade　中低档(等级)
middle of the road　中庸之道
middle price　中等价格
middles　中等货
Middlesex blue flannel　米德尔塞克斯靛蓝色法兰绒
middle shade　中间色
middle size (M)　尺码中号,中尺寸
middle sole　夹层底(制鞋)
middle tint　中间色,过渡色
middle tone　中间色调
middle twill　中等斜纹,漂白厚棉经面斜纹
middress　中空装
middy　水手领上衣(罩衫)
middy blouse　水兵式女衫;水手领女罩衫;水手衫
middy braid　水手服饰带;水手服织带;斜纹饰边
middy collar　水手领;迷蒂领(用于女生制服及上衣与连衫裙)
middy dress　水手领女服;水手领童装
middy ensemble　水手女衫
middy jacket　水兵服式上衣,水兵服,水手夹克
middy look　水兵风貌,水兵风格;水手风格;水手领款式
middy skirt　中长裙
middy-style sweater　水手式套衫
middy suit　水兵领三件套童装
middy sweater　水手领毛衫
middy-top pajamas　套头式[无扣]两件套睡衣
middy twill　斜纹布(泛称),细薄料斜纹布
midfiber　色织凡立丁;中长纤维
midfiber fabric　中长化纤织物
midfiber yarn-dyed crepe　色织高绉中长织物
midfiber yarn-dyed valitin　化纤凡立丁;色织中长凡立丁
midfiber yarn-dyed whipcord　化纤马裤呢;色织中长马裤呢
midfibre bulked overcoating　中长膨体大衣呢
midfibre fabric　中长化纤织物;中长布

midfibre fancy suitings　中长花呢
midfibre gabardeen　中长华达呢
midfibre serge　中长哔叽
midfibre yarn-dyed cavalry twill　色织中长马裤呢
midfibre yarn-dyed fancy suitings　色织中长花呢
midfibre yarn-dyed herringbone　色织中长海力蒙
midfibre yarn-dyed herris　色织中长海力斯
midfibre yarn-dyed medium weight fancy suiting　色织中长厚花呢
midfibre yarn-dyed modelon　色织凉爽呢,中长凉爽呢
midfibre yarn-dyed polyester/viscose hopsack　色织中长板司呢
midfibre yarn-dyed semifinish serge　色织中长啥味呢
midfibre yarn-dyed tropical suitings　色织中长薄型花呢
midfibre yarn-dyed valitin　色织中长凡立丁
midfibre yarn-dyed whipcord　中长马裤呢
mid-grade clothing　中档服装
mid-hip　中臀处
midi　迷地装;迷地裙;中长裙,中长裙式样;中庸装;中等长度
midi cape　迷地式披风(长及小腿的中长披风)
midi clothing　迷地外套;中长外套
midi coat　中长外套,迷地外套
midi dress　迷地裙装,中长裙装
midi length　长及小腿的长度,及腓长度
midi look　迷地风貌;迷你风貌;迷地款式
midinette　[法]时装店女店员,女裁缝;法国巴黎高级时装店裁缝女工
midi silhouette　迷地型轮廓线条
midi skirt　中长裙,迷地裙;中庸裙,芭蕾舞裙
midi slacks　及腓长裤
midi style　中长裙式样
midium yellow　中黄
mid-knee　中膝
mid-knee level　膝围线
mid-knee skirt　中膝裙(裙长在膝中部)
midleg　腿的中部
mid-length fiber　中长[型]纤维
mid-length short pants　中短裤,半短裤
mid-length viscose staple fiber　黏胶中长

纤维
mid-length wool type chemical fiber fabric 中长化纤仿毛织物
mid-line 中线
mid-neck 颈弧线中点;领中
mid-neck dart 中领省
midnight black 深黑色
midnight black [blue] 深黑[蓝]色
midnight blue 午夜蓝(深暗蓝色)
midnight [color] 午夜月色
midnight navy 黑蓝色
Midori Matsumoto 松木翠(日本时装设计师)
midriff 膈;女露腰上衣;蜜多夫装;中腹部,女服身体中部;嵌腰片
midriff [area] 中腹区;中腹部;腰腹区
midriff back 后嵌腰
midriff bathing suit 露腹女泳装
midriff costume 露腰装
midriff dress 嵌腰装(腰部有宽带状剪接布)的连衣裙;系肚带[的]连衣裙
midriff front 前嵌腰
midriff jacket 露腰上衣,嵌腰夹克
midriff line 横膈线
midriff shell 嵌腰片档布
midriff shirt 露腰衬衫
midriff style 嵌腰式风格
midriff top 露腰女上衣
mid-shoulder 肩线中部;肩中点
mid-shoulder dart 中肩省
midsize 中号;中型的,中号的
mid-sole 夹层鞋底
mid-sole overlock machine 包中底机(制鞋)
mid-thigh 大腿中部
mid-thigh-length (裤、裙等到)大腿中部长度
mid-thigh-length overblouse 至大腿中部的女长罩衫
mid-wale corduroy 中条灯芯绒
midway socks 短袜
midway thigh girth 中间大腿围
midwife's gown 助产士服装
mien 风采,风度
mienchow 四川中级蚕生丝;中国丝绸
mi-fils 法国米菲尔细平布
mi-florence 米佛罗伦萨里子薄绸
mignardise 嵌带编织,衬细窄条带的钩编品
mignonette 圆纬针织物;粗毛针织品;法国细丝花边;黄绿色,灰绿色,木犀草绿
mignonette lace 梭结六角网眼,梭结细网眼花边,娇小花边
mignonette net 机制网眼织物,机制网眼纱
migot 西班牙羊毛
migrating cationic dye 迁移性阳离子染料
migrating dye 泳移性染料
Miguel Cruz 米格尔·古鲁兹(意大利时装设计师)
mihaba kanakin 日本衬衫用棉布
mikado 英国米卡多塔夫绸
milady 时髦女人
milaine 棉毛薄花呢
Milan braid 米兰马海毛编带
Milan collar (意大利)米兰领,米兰式领
Milan hat 米兰细草帽
Milan lace 米兰梭结花边,机制花边,网地图案花边
Milan point 米兰金银丝梭结花边
Milan straw 米兰细麦杆(制帽辫)
milanaise 米伦内窄编带
milanaise yarn 棉芯丝线
Milanese 米兰尼斯花边(菲律宾);米兰尼斯经编织物;低档棉布
Milanese cord 米兰尼斯凸纹织物
Milanese flat warp-stitch knitting machine 米兰尼斯平型经编机
Milanese hose 米兰尼斯经编袜
Milanese knit 机制经编针织物
Milanese knitting 米兰尼斯针织物
Milanese lace 米兰尼斯花边,米兰梭结花边
Milanese loom 米兰尼斯经编机
Milanese machine 米兰尼斯经编机
Milanese rib 米兰诺罗纹
Milanese warp-knitting machine 米兰尼斯经编机
milano collar 米兰式领
milano hair sweater 空气层毛衫
milano knitwear 空气层毛衫
milano rib 米兰诺罗纹
milano rib fabric 罗纹空气层织物;米兰诺罗纹针织物;四平空转织物
milano rib knit 罗纹空气层针织物;米兰诺罗纹针织物
milano rib modified 变化米兰诺罗纹
milano woolen sweater 空气层毛衫
Mila Schoen 米拉·肖恩(意大利时装设

计师)
milassa 土耳其金色条纹毯
mildew 霉,发霉;丝绸灰疵
mildew cond rot 霉烂
mildew proofing 防霉处理
mildew proofing finish 防霉整理
mildew resistance 防霉性
mildew stain 霉斑
mildewy mark 霉斑
mild finishing 柔软整理;轻度整理
mildly warm water 温水
mild scouring 轻洗
mile-fleur print 小花印花花样
Milenaise 米伦奈斯平布
milfa 美尔发衫(科威特)
military 军装风格
military and defense textiles 军事国防用纺织品
military back 挺胸背
military bag 军用袋
military bedford cord 军用马裤呢,军用棱纹呢
military boots 军靴
military braid 军服辫带,军服饰边,粗平饰带
military cap 军帽
military cloth 军服呢
military clothing 军用服装
military coat 军大衣
military collar 军服领
military cord 经向灯芯布
military duck 军服帆布
military equipage 军装
military fabric 军用织物
military fashion 军服式
military folding hat 军用折叠帽
military garment 军装
military hell 军靴跟,直鞋跟
military jacket 军装式夹克
military look 军装风格,军装风貌,军装装式
military m(o)ustache 军人髭
military olive 军用橄榄绿
military police rubber boots 武警胶靴
military poncho cloth 军用雨衣
military protective clothing 军队防护服
military rank tape 军衔带
military shirt 军用衬衫
military shoes 解放鞋,军鞋
military specification 军用服装规范,军用规范
military standard (MIL-Std) 军用标准(美国)
military stock 军用领带
military style stormcoat 军服式风雪大衣
military styling 军装线条,军装款式
military trench coat 军用战壕大衣
military type apparel 军用型服装
military uniform 军服,军装
Milium 美丽姆金属涂层织物
milk casein fiber 乳酪蛋白纤维
milkiness 乳白色,浊白色
milkmaid hat 挤奶女工帽
milkmaid shirt 牛奶姑娘裙
milkman's pocket 送奶人的大口袋
milk pack 乳酪润肤膏
milk white 乳白,乳白色
milky blue 乳白蓝色
milky coating 乳白涂层
milky diamond 乳色钻石
milky green 乳白绿色
milky hat 布制毡帽,柔软帽
milky tone 乳白色调
milky viscose 乳白黏胶丝
milky-white 乳白
mill 纺织工厂;缩绒,缩呢
milled 缩呢织物
milled asbestos 精加工石棉
milled cloth 缩呢织物,缩绒织物
milled finish(ing) 缩绒整理
milled goods 毛毡,缩绒织物
milled scarf 缩绒长巾
milled serge 缩呢哔叽
milled tweed 缩绒粗呢
milled worsted 缩绒精纺毛织物
milled worsted fabric 轻微起绒整理的精纺[仿]毛织物
Mille-feuille dress 蜜露芙颐尔衣,多层薄纱服装
Millefiore 小碎花图案
mille-fleur prints 千朵花印花花样,小花印花花样
millefleurs [法]百化香粉;挂毯花卉纹小型对称花卉纹
mill ends 零绸;零头布
mille point 英国小花斜纹呢
Mille raye 经向细条纹绸;窄里白条密织薄纱;窄条纹丝棉绸
mill finish 出厂整理
mill finish worsted 薄绒精纺毛织物

mill flower stitch	闪光点缀绣
mill rigs	缩呢折痕
mill run	等外品(纱线,织物等),次货
mill store	工厂直销店
millennium fashion and accessories	千禧年服饰
Millerainised fabric	米勒林织物
millerayes	虹彩绸,虹衫绸,千条绸
millet color	米色
mill-finished fabric	色织布
mill-finished ticking	色织被单布,色织被套,色织床罩布
millimetre	毫米
milliner	女帽及妇女头饰的设计、制作、整修和销售者;女帽的制作者或销售者
millinery	女帽商;女帽制造业,整修业或销售业;女帽;妇女头饰
millinery blocker	女帽模工
millinery felt	女帽毡
millinery hair	透明女帽辫(瑞士等国)
millinery ribbon	女帽带,女帽缎带
millinery work	女帽及妇女头饰制作
milline's needle	疏缝针
milling	缩呢;缩绒
milling contraction	缩绒收缩
milling in the length	纵向缩绒,纵向缩呢
milling power	缩绒性,缩呢性,毡合性
milling property of wool	羊毛的缩绒性
milling rate	缩绒率
milling scrimp	缩呢折痕
milling shrinkage	缩绒收缩
milliskin knit (4-way stretch knit)	超弹针织物
millitex	毫特[克斯]
millstone ruff	轮状襞领,磨盘式绉领
Milpa	聚酯仿毛型复合变形丝(商名,日本帝人)
miltons	米尔登猎装呢
mimic colouration	拟色
mimollet	过膝裙
mimollet skirt	米莫莱中长裙
mimosa	含羞草黄
mina cloth	米纳交织厚呢
Mina Ricci	米纳·里奇(意大利时装设计师)
minaret silhouette	伊斯兰教寺院轮廓
minaret tunic	伊斯兰教寺院风格的束腰外套
minas gerais cotton	米纳斯杰拉斯棉
minaudiere	化妆盒,晚礼包
minchew silk	中国生丝
mindo cotton	明杜棉
mined worsted	缩绒精纺毛织物
mine-laying chute fabric	布雷降落伞织物
mineral blue	深蓝色,石青
mineral cotton	矿棉,石棉
mineral dye	矿物染料
mineral fiber	矿物纤维
mineral green	石绿
mineral grey	矿石灰
mineral khaki	矿物草黄;矿物卡其黄
mineral tannage	矿物鞣皮
mineral wool	矿物棉,石纤维
mineral wool fiber	[炉]渣绒纤维,岩石纤维,矿棉纤维
mineral yellow	石黄,夕金色
miner's cap	矿工帽
miner's hat	矿工帽
mine's proof shoes	防雷鞋
miner's protective rubber boots	工矿胶靴
miner's suit	矿工服
miner's wear	矿工服
minette look	美尼特款式
minette style	美尼特款式
minever	白毛皮;生兽皮
mine worker's garment	矿工服
Ming blue	藏青,明瓷蓝
mingchow	杭州锦缎
Ming green	明瓷绿(艳绿色)
mingguang brocade	明光花绸
mingguang brocaded damask	明光花绸
minghua jacquard poult	明华葛
mingled yarn	混色纱
Ming yellow	明瓷黄
mingyue burnt-out nylon sheer	明月绡
mingyue satin	明月缎
minhua jacquard poult	明华葛
mini	迷你长度,超短型长度;迷你装,超短裙,迷你裙
mini and micro garment	超迷你服装
miniature	小型
miniature camera	小型照相机
miniature check	细格子纹
miniature electric cutter	迷你裁剪机;微型电剪
miniature flower	纤细花纹
miniature marking	极小花样
miniature pattern	极小连续花样,微型花样,纤细花纹,纤细花样
miniature size scale	缩样尺

miniature skirt 纤细短裙
miniature top-hat 袖珍礼帽
miniature topper 小型高顶礼帽
mini-bag 迷你包
minibikini 超短分体式女游泳衣
miniboiler type steam 微型蒸汽炉熨斗
miniboiler with iron 带微型蒸汽炉的熨斗
mini bustle 超小裙撑
mini button 迷你纽扣
minic 仿制品
Minicare 易维护织物;洗可穿织物
mini-check 小格花纹
minicoat 超短上衣,超短外套;迷你大衣,超短大衣
mini corset waistline 束腹腰线
minicoveralls 迷你背带连衫裤装
mini-crini 迷你衬裙
mini cutter 迷你裁剪机,微型电剪
minidress 超短衬裙;超短裙套衫;超短连衣裙
mini fusing press 迷你(简易型)黏合机
mini-fall(wig) 超短卷假发,短悬垂式假发
minification 缩小
minifil(ament) yarn 粗旦单纤复丝
minigrain 混色丝
mini handbag 迷你小包
mini-jacquard 小型提花
mini-jupe 迷你裙
mini kerchief 超短围巾
minikilt 超短(苏格兰)褶裙
mini knitting suit 超短针织套装
mini length 超短长度,迷你长度
mini look 超短款式,迷你风貌
minimal avt 最简单派艺术(抽象派艺术的一个变种);最简式派艺术
minimal brassiere 迷你胸罩,肤色胸罩
minimal fashion 黑白服饰
minimalism 极少主义,最低限度主义(主张服装设计力求质朴,与超摩登倾向相反)简约主义,最简单派艺术(抽象派艺术的一个变种)
minimalist 简约主义
minimal top 最短上衣
mini maxi skirt (M.M. skirt) 踝上 5cm 长裙
mini-mini-skirt 超超短裙,超迷你裙
minimizer bra 减小型胸罩
minimizer brassiere 缩小型胸罩

minimum care 洗可穿织物,易维护织物
minimum-care finish 不需保养式整理(服装免烫等整理)
[minimum] harden 最低级亚麻布(用于工作服)
minimum leg girth 最小脚腕围
minimum lift 最小抬高度
minimum specifications 最低规格,最低标准
minimum standard 最低等级
minimum wearing tolerance 最小放松量
minimus 小[手]指;小脚指
mini-page 迷你侍童式发型
mini-pants 超短裤;热裤
mini-petti 超短衬裙
mini-pettipants 半长内裤
mini riveting machine 小型铆钉机(制皮革品)
mini sack 迷你上衣;迷你袋装(造型像袋子的女装)
mini-shorts 超短裤
mini-skirt 迷你裙,超短裙
mini-slip 超短衬衣,迷你衬裙
ministerial standard 部标准
mini-stock 最低存货
mini-style 迷你风格,超小风格
minisuit 短西服上衣
mini top 超短上衣,迷你衫
minium 猩红色,朱红色
miniver 白(鼬)毛皮,生兽皮
mink 貂皮,水貂毛皮;貂皮外衣;深褐色
mink cap 貂皮便帽,无檐貂皮帽
mink clothing 貂皮大衣;貂裘
mink coat 貂皮大衣,水貂皮大衣
mink collar jacket 貂皮领夹克
mink dress 貂皮服装
mink finish 貂皮整理
mink fur 貂皮
mink hat 貂皮帽
mink imitation 人造貂皮;貂皮仿制品
mink jacket 貂皮短大衣
mink skin 水貂皮
mink tail 貂尾皮
minlang brocade 明朗缎
Minnie Mouse Shoes 米老鼠鞋
mino 日本稻草雨衣
Minoans costume 米诺斯服饰
minor defect 小疵点
minor discrepancy 小瑕疵
minor fashion 少数派时装(仅由少数人

接受的款式)
minor maintenance　小修
minor overhaul　检修,小平东
minpow　进口棉织品
minsmere cap　明斯米尔便帽(英国)
minstrel cloth　唱诗绉
mint　薄荷色,薄荷绿
mint green　薄荷绿,浅翠绿
mint leaf　薄荷叶色
mintalik drill　手织粗斜纹棉布
minter cotton　美国明特陆地棉
minto　明突
minus ion fiber　负离子纤维(负碱性离子)
minutiae　小花纹
Minyon　三醋酯变形丝(商名,日本三菱人造丝)
mir [saraband] rug　优质萨拉班德地毯
mirabll　美丽宝
miracourt lace　梭结花边;枝状花纹嵌花花边
Miranda pumps　米兰达浅口无带鞋(高厚的宽跟)
mirecourt　贴花梭结花边
mireshka　俄罗斯挖花抽绣品
mirganji　印度硬质黄麻
Miro style　米罗风格
mirror　镜
mirrored color　镜映色
mirrored piece　镜像衣片,对称衣片
mirror effect yarn　镜面效应花线
mirror finish　光亮整理
mirror image　对称形象,镜面形象,完全对称
mirror lens　镜面透镜
mirror-like satin　镜面缎
mirror line　对折线
mirror piece　对称衣片
mirror repeat　镜面对称循环
mirror set out　镜面对称排列
mirror symmetry　镜面对称
mirror velvet　镜面丝绒
mirror work　镜片刺绣
mirzapur rug　植绒地毯
misalign　对花不准
misaligned collar　鸳鸯领
misaligned feeding　送布歪斜,送料不准
misaligned notch　刀口不齐
miscellaneous hair felt　杂毛毡
miscellaneous wool　强力粗羊毛,杂羊毛

mischievous look　淘气鬼款式
mischievous style　淘气鬼款式
misfit　不合身,不匹配,对花不准
misfit of knitting pattern　花纹不齐
misfit of printing pattern　对花不准,花纹走样
misfitting　对花不准
mismatch　不匹配;对花不准
mismatched checks　对格不准
mismatched stripes　对条不准
mispick　缺纬,百脚疵
misplaced collar　领子不正
mispunching pattern　错花,提花错花
misrating　错评(级)
misregister　印花错位,印花对花不准,套版不正(即染疵病)
Miss Dior　迪奥小姐
missed end　缺经
missed stitch　漏缝,漏针,跳针
misses　淑女装
misses'　少女尺码,妇女尺码
misses' body measurement　少女(或妇女)人体测量
misses' figure　少女体型
misses' sizes　少女(或妇女)尺码
miss filling　缺纬
misshape　不合形,形状不对
missinet　英国彩色调
missing　漏针,漏圈,不编织
missing button　错用纽扣
missing height　不编织高度,浮线高度
missing needle　不编织针,漏针
missing pattern　缺花疵
missing stitch　漏缝
mission cloth　粗厚方平织物
mission net　大网眼粗窗帘布
Mississippi [delta 51] cotton　美国密西西比棉
miss-match　失谐;失谐美(现今时装设计中采用的奇特的不和谐搭配)
Missoni　米索尼(家族)(意大利时装设计师)
Missoni Rosita & Ottavio　米索尼[夫妇]罗茜达和沃特维奥(意大利时装设计师)
Miss petite size　娇小姑娘尺码
miss stitch　空针;跳花
miss stitch effect　浮线效应
mis-stitching　错缝
miss tuck　花针疵

missy figure 小姑娘体型
misted yellow 雾黄色
mist [grey] 淡红灰色,雾灰色
mist green 雾绿色
mistletoe 豆绿[色]
mistral 米斯特拉尔结子呢
Mistress of the Robes （女王或王后的）服装侍从女官长
mist spray damping 喷雾给湿
misty blue 雾蓝色,影青
misty grey 雾灰
misty lilac 雾丁香色
Mit Afifi cotton （埃及）米阿菲菲棉
mitaines [法]独指手套,露指手套,单袂手套
mitcheline quilt 毛巾图案;提花双层床单布
mite-proof fabric 防螨织物
miter 斜接缝;斜接面;斜拼接,45°折角拼缝;主教礼冠,大祭司冠,古希腊妇女束发带;犹布神职人员高帽
miter collar 领角拼接领,斜角领
mitered continuous strip 长条贴边折角处拼缝法
mitered corners 45°缝角,相等角
mitered double banding 折角拼缝的双层滚镶边
mitered hem 折角拼缝的下摆
mitered trim 折角拼缝的贴边
mitering 拼缝成直角的斜拼接;折角
miter joint 斜接,[地毯]斜面对接
miter line 缝纫斜接线
mitkal 米特卡尔窄幅粗棉布
mitorse [silk] 米托斯松捻绣花丝线
mitra 主教礼冠
mitre 45°折角拼缝;斜接缝,斜接面,斜拼接;主教礼冠;大祭司冠（古代犹太教）;束发带（古希腊）
mitre collar 斜角领;斜角式拼接领
mitre joint 斜接
Mitsuhiro Matsuda 松田光弘（日本时装设计师）
mitt 女用露指手套;独指手套;连指手套;拳击手套;棒球手套;劳动手套;防护手套
mitten 独指手套;漏指手套;拳击手套;连指手套;女用露指手套
mitten-sleeve gown 婴儿睡衣;有连接手套袖的服装
mitts 连指手套,露指长手套,拳击练习手套

MiuMiu 缪缪
miwen upholstery silk 密纹装饰绸
mix 男女混用
mix and match 混合与配合（色彩,花样和衣料等）
mix and match look 混合再搭配的款式,混合再搭配风貌
mixed brocade 天孙锦
mixed cambric 混纺细布
mixed checks 英国交织条格布
mixed chromatogram 混合色谱(图)
mixed color effect 混色效应
mixed colors 混合色
mixed cotton rayon satin 人造丝背缎
mixed crepe 交织绉绸
mixed damask 交织锦缎
mixed fabric 混纺织物;交织物
mixed fancy suitings 三合一花呢
mixed filament yarn 混纤丝
mixed filling 错纬
mixed gabardine 混纺华达呢
mixed habotai 交织电力纺
mixed jacquard 花饰绸
mixed jacquard bed cover 交织大提花床罩
mixed knitted fabric 交织针织物
mixed lace 针绣梭结联合花边
mixed multi-layer material 混纺多层缝料
mixed muslin 混纺市布
mixed net hosiery 交织网眼袜
mixed paj 交织洋纺
mixed pique 凸花绸
mixed polymer fiber 混抽纤维,聚合物混纺纤维
mixed poplin 抽丝绵绸;交织毛葛
mixed satin 软缎（交织缎类丝织物）
mixed satin brocade 交织花软缎;交织绒面缎
mixed satin crepe 花软绉
mixed satin plain 交织素软缎;交织苏贝缎
mixed serge 混纺哔叽
mixed shameuse 留香绉
mixed-shrinkage yarn 异收缩性复合丝
mixed silk piece goods 交织绸段
mixed soochow brocade 交织古香缎
mixed spinning fiber 混纺纤维
mixed spun silk yarn 混纺绢丝
mixed spun tussah yarn 混纺柞蚕绢丝

mixed suit 混合套装
mixed suit style 组合式套装款式,混合色套装款式
mixed suzhou brocade 交织古香缎
mixed tapestry satin 交织织锦缎
mixed-weft brocatelle 特纬缎
mixed twill 交织绫;复合斜纹,混合斜纹
mixed worsted suitings 混纺花呢
mixed woven linen 亚麻交织布
mixed yarn 混纺纱;混色纱;混纤丝
mixi 超短与特长的组合[款式];迷你裙和迷喜裙的组合款式
mixing ratio 纤维混合比
mixture 交织;混纺纱,混色纱,混色织物,交织织物,混织物
mixture crepe 交织绉布;交织绉绸
mixture fabric 混纺织物;交织物
mixture serge 混纺哔叽,雪花哔叽,混色毛哔叽
mixture yarn 混纺纱;混色纱
mix-woven 交织物
mizpah necklace 米士巴[情人]项链
Mizuno 美津浓
mizz 中东软皮脚罩
MNG 芒果
moat collar 离颈立领,宽式立领
mobcap (18世纪室内戴的)头巾式室内女帽,睡帽
mo-beel 动态耳环,摇动耳环
mobel linen 莫贝尔亚麻条格布
mob hat 婴儿太阳帽
Mobile cotton 美国莫比尔棉
mobile earring 动态耳环,晃动耳环,摇耳环
mobile textile 汽车用纺织品
Mobile Thermo 自动保温型服装(智能型服装,商名,日本)
mobility 可动性
MOBILON tape 日本无比耐松紧带
moc 软帮鞋,烧卖鞋
moca-loafer 乐福鞋
mocca 珠网线迹(刺绣)
moccador 摩卡多绸
moccasin 莫卡辛鞋,莫卡辛式高筒靴;平底鹿皮鞋,软帮鞋,软拖鞋,烧卖鞋
moccasin-foed shoes 莫卡辛鞋
moccasin stitch 软鞋缝迹
moccasin stitch machine 软鞋缝纫机
moccasin toe 莫卡辛鞋头
moccasin shoes 莫卡辛鞋

moccasin-type shoes 莫卡辛型鞋
mocha 摩卡咖啡色,带灰色的深咖啡色;细亚绒绵羊手套革
mockador 手帕;儿童围嘴
mocha leather 摩卡手套革
mock buttonhole 驳头眼;假扣眼
mock casing 抽带管(利用缝头或狭贴边制成)
mock chenille yarn 充(仿)雪尼尔花线
mock crepe 充绉布,仿绉布;充绉绸,仿绉绸;平绉
mock cuff 假克夫,虚翻边
mock dog skin 仿狗皮
mock dyed fabric 假染织物
mock dyeing 纱线假染定形
mock-Egyptian cotton 仿埃及棉;仿埃及棉纱;纺埃及棉织物
mocker 衣服,服装
mocket[er] 手帕,涎布,餐巾
mock-fagotting stitch 假束心线迹;假装饰线迹
mock fashion 充收放针,肩部缀袖开口
mock fashioning mark 充收放针花
mock fly 假门襟,女裤假门襟
mock French seam 仿法式缝
mock gauze 仿纱罗织物
mock grandrelle yarn 仿双股双色纱的单纱,混色纱
mock gussetting 无缝圆袜假收针花
mock lace 假网眼织物
mock leather 人造革
mock leno 充纱罗,仿纱罗
mock leno for summer clothing 夏衣纱
mock narrowing 假收针
mock open front 假开襟
mock-pearl collar 仿珍珠领,假珠领
mock plain 花色平素织物,假平素织物
mock-pleated fabric 仿褶裥效应织物,褶裥效应经编织物
mock plied yarn 充股线
mock pocket 假口袋
mock printings 仿印花
mock quilting 充凹凸布,仿凹凸布
mock rib 假罗纹
mock romaine[crepe] 充罗马绉,仿罗马绉
mock run and fell 假包口接缝
mock safety stitch 仿(假)安全缝线迹
mock satin 充花缎,仿花缎
mock seam 长筒袜的假缝,假缝

mock-seam hose 长筒假缝圆袜
mock seaming 假缝
mock seaming machine 假缝缝纫机
mock-seam stockings 假缝长筒袜
mock selvage chain 假缝辫
mock shirt cuff 短袖外翻边
mock split 假添纱花纹编织
mock turtle[neck] collar 单层半高领,单层翻领
mock turtleneck[sweater] 单层翻领毛衣;双层针织罗纹领毛衣
mock twins 假叠层式套衫
mock-twist[yarn] 仿合股双色线,仿双股双色纱
mock two piece 仿套装裙
mock-up 样品,模型(与实物同样大小);试衣模式(裁剪戏装使用)
mock vent 假开衩
mock voile 充巴里纱,仿巴里纱
mock worsted yarn 半精梳毛纱
mock-wrap hose 充吊线袜,仿吊线袜
mocmain 木棉
moco cotton 巴西莫科棉
mod 新潮;摩登风格(1958年前后英国)
modacrylic(MAC) 变性聚丙烯腈纤维;改性腈纶
modacrylic fiber 改性聚丙烯腈纤维(含35%～85%丙烯腈的共聚纤维)
modacrylics 改性聚丙烯腈纤维(含35%～85%丙烯腈的共聚纤维)
modacrylic self-extinguishing fiber 改性聚丙烯腈自熄纤维
modal 莫代尔纤维素纤维
modal fiber 高湿模量黏胶纤维,莫代尔纤维(高强度和高湿模量黏胶纤维的属名,国际代码 CMD)
MODA SPORTS 《摩登运动服》(意大利)
mod boots 摩登靴(20世纪60年代中期)
mod cap 摩登帽
mode [服饰等的]式样;风尚,风气;阿拉莫德绸;浅灰色,浅褐色
mode fashion 时髦式样
MODE INTERNATIONAL 《摩登世界》(法国)
mode 风格;风尚;服式,款式(妇女衣帽等),样式,模式;流派;流行;型
model 型,型号;样式,款式;样板;时装模特;人体假人(人形模型);模型
model agency 模特介绍所
model arm form 连臂人台,连臂人体模型

model B hosiery machine B字圆袜机
model cap 样帽
model case 样板,典型事例
model coat 展样大衣
model corridor facility 地毯试验用模拟走廊设备
model dress 展样女服
model fiber 样品纤维,模型纤维
model form 人台,衣架;人体模型
model girl 模特儿女郎,女时装模特
model hat 样品帽,示样帽
model[indicate] plate 型号板;型号牌
modeling 当[时装]模特;造型;模特职业
modeling line 造型线;成型线
modeliste [法]服装造型师
model K hosiery machine K字袜机
modeller 制鞋面和鞋帮的裁剪工人
modelling 立体感;造型;模特职业;模特技术
modelling beauty 造型美
model(l)ing collar 领型
modelling line 造型线
model(l)ing of trousers 裤型
model(l)ing skirt 裙型
modelliste 服装造型师,独创设计制样师
model machine 样机
model number 型号
model of human body 人体模型
model of morphology 形态模型
modelon 毛涤纶
modelon fancy suitings 凉爽呢
model plate 型板
model prototype 标样;样品;模型
model school 模特培训学校
model weft-warp insertion circular knitting machine 衬经衬纬针织圆机
modena 深紫色;意大利棉毛薄织物;绢、棉、毛混纺织物
moderate 适度的,中等的
moderate color 中间色
moderate cost 价格公道
moderate price 公道价格,中等价位
moderate tone 适度色调
moderate-stretch knit 适度拉伸针织物
modern 现代派的人
modern art 现代艺术
modern colour 摩登色彩
MODERN DRESS & DRESSMAK-ING 《现代服装》(中国)

modern dress　现代服装
moderne　摩登呢
modern fashion　摩登式样;时下流行
modern fur industry　现代皮草工业,现代毛皮工业
modernism　现代主义
modernity　现代性;时尚,新式
modernist fashion　黑白对比花型流行式
modernized equipment　现代化设备
modern look　现代装,现代风格;摩登风貌
modern pump　摩登无带女鞋
modern ready-to-wear　现代成衣
modern style　现代风格
modern-style brooches　新式饰针
modern-style diamond ring　新式钻戒
Modern Textiles　《现代纺织品》(美,月刊)
mode shade　流行色泽,时髦色泽
modeste　女式三层裙
modesty　端庄
modesty-bit　(穿在低领女服里的)遮胸小背心
modesty-piece　(穿在低领女服里的)遮胸小背心
modesty pants　温裤
modesty vest　(穿在低领女服里的)遮胸小背心
modifiable factor　可变因素
modification　变性,变质,改性
modification of synthetic fiber　合成纤维改性
modified　变性的,改性的
modified acetate [fiber]　改性醋酯纤维
modified acryl(ic) fiber　改性聚丙烯腈纤维
modified acrylics　改性聚丙烯腈纤维
modified back stitch　星止针法
modified cellulose staple　改性纤维素人造短纤维
modified cellulosic fiber　改性纤维素纤维
modified continuous filament　变形长丝,改性长丝(旧称)
modified cotton　改性棉[纤维]
modified cross-section fiber　异型截面纤维
modified fiber　变性纤维,改性纤维
modified high wet modulus rayon　改性高湿模量人造纤维
modified milano rib fabric　变化罗纹空气层针织物
modified nylon　改性尼龙,改性锦纶
modified polyamine fiber　改性聚酰胺纤维
modified polyester fiber　改性聚酯纤维
modified rayon　改性人造丝
modified [rayon] staple fiber　改性人造短纤维
modified treatment　变性处理
modified triacetate fiber　改性三醋酯纤维
modified viscose fiber　改性黏胶纤维
modified viscose staple　改性黏胶短纤维
modified wool　改性羊毛
modified wool fiber　改性羊毛纤维
modified yarn　变形丝,改性纱
modify　修正
moding　造型
modish　时髦的,流行的
modiste　女裁缝,时髦女服裁缝;女帽商;女装店
mod(s) look　摩登风貌,新潮风貌;摩登款式,新潮款式,新式样;现代装
mod(s) style　时髦风格,现代风格
modular carpet　组合地毯,拼块地毯
modular length　定形长度
modular manufacturing　模块加工
modular manufacturing system　模块式生产系统
modular programming　模块化程序设计
modular quick response sewing (MQRS)　模块式快速反应缝制
modulus　模数
moff　阿塞拜疆莫夫绸
mog[g]　毛衣;皮上衣
mogador　彩色领带绸
mogador wool　摩洛哥褐色羊毛
mogi　皮上衣
moguang brocade　摩光缎
mo-guang rayon brocade　摩光缎
mohair　马海毛,安哥拉山羊毛;马海毛织物;马海毛混纺织物;马海毛服装
mohair beaver plush　马海毛海狸绒(呢),马海毛长毛绒
mohair braid　马海毛编带;俄罗斯编带
mohair brilliantine　马海毛有光呢
mohair cloth　马海毛呢(衬里用)
mohair cloth linen　马海呢衬里
mohair coney seal　黑色马海毛海狸绒
Mohair Council of America　美国马海毛协会

mohair fancy suiting 马海毛花呢
mohair fleece 马海毛大衣呢；银枪大衣呢；银戗大衣呢
mohair floor rug 马海毛地毯
mohair hat 安哥拉山羊毛帽；马海毛帽
mohair hat with feathers 带翎马海毛帽
mohair knitwear 马海毛衫
mohair-like acrylic knitting yarn 仿马海毛腈纶膨体针织绒线
mohair lustres 马海毛低级有光呢
mohair pile fabric 马海毛绒头织物
mohair plush 马海毛长毛绒
mohair rug 马海毛割绒地毯
mohair serge 马海毛哔叽
mohair sicilian 马海毛有光薄呢
mohair sweater 马海毛衫，马海绒衫
mohair tropical 马海毛平纹薄花呢
mohair velvet 马海毛丝绒，马海毛立绒
mohair yarn 马海毛纱
Mohawk hairstyle 莫霍克发型
Mohegan haircut 莫希干式发型
mohlaine 莫兰绉呢
moho cotton 非洲莫霍棉
mohor bandi 金丝花缎
moire 云纹绸；波纹绸；波纹布；波纹；云纹型织物；云纹马海呢
moiré anglaise 波纹轧光条影绸
moiré antique 波纹轧光条影丝织物；云纹绸；波纹绸
moiré à pois 小缎点波纹绸
moiré à rétours 对称波纹织物
moiré bayadere 波纹横条绸
moiré calendering 波纹整理
moiré cloths 云纹布，波纹布
moiré cffect 莫尔效应，云纹效果，波纹效应
moiré effects fashion 波纹效应时装
moiré fabric 波纹织物，云纹织物
moiré figure 刮花波纹
moiré finish 波纹轧光整理，云纹整理
moiré francaise 法国条状波纹织物
moiré-free pattern 无云纹花型
moiré imperial 较模糊的波纹效果；波纹罗缎
moiré ineraillable 波纹条影绸
moiréing 云纹整理，轧纹整理
moiré lining 波纹衬里布
moiré lisse 无波纹轧光织物
moiré metallique 霜花波纹绸
moiré mille fleurs 印花生丝毛葛

moiré miroir 镜光波纹绸
moiré nacre 珠光波纹绸
moiré oceam 浪花波纹绸
moiré pattern 云纹花样，波纹图形，波纹花样(毛皮)；波纹莫尔图像
moiré poplin 波纹毛葛
moiré racré 珠光波纹绸
moiré ribbon 波纹带
moiré ronde 年轮波纹绸，木纹波纹绸
moiré scintillant 特亮波纹缎
moiré silk 云纹绸
moiré soleil 波纹光亮绸
moiré supreme 优质波纹缎
moiré tabisée 波纹织物
moiré taffeta 波纹塔夫绸
moirette 波纹薄线呢，粗纬波纹绸，粗纬波纹呢
moiré velours 波纹线
MOISELLE 慕诗
moist 潮湿
moist colors 水彩颜料
moist heat setting 湿热定型
moisture 吸湿性
moisture absorbing synthetic fiber 吸湿性合成纤维
moisture absorption 吸湿
moisture content 含水率
moisture-laden air 含湿空气
moisture liberation 放湿
moisture permeability 透湿性
moisture permeable waterproof fabric 透湿防水织物
moisture permeable waterproof finish 防水透湿整理
moistureproof paper 防潮纸
moisture regain 回潮率
moisture resistance 透湿阻抗
moisture sensitive discolor fabric 温敏变色织物
moisture sensitive fiber 湿[度]敏[感]纤维
moisture temperature 温湿度
moisture vapour transmission (MVT) 透湿率，透水气性，透湿性
moist rouge 乳脂红
mojo 黄槿韧皮纤维
moka 摩卡咖啡色(带灰深咖啡色)，灰深咖啡色；细亚绒绵羊手套革
mokador 手帕，涎布，餐巾
mokho cotton 西非莫库棉

mola 美洲彩花装饰布
molaine 仿印花毛洋纱；棉毛交织品；英国棉马海毛交织品
molano 莫拉诺长毛绒
mold 霉，霉菌，模具，模型，模子；冲模；样板
moldable fabric 可模塑变形织物，造型织物
mold center 模压主鞋跟
molded boots 模压滑雪靴
molded dress 模压裙装，热定形裙装
molded edge 模压边缘
molded fabric 模压织物，模塑织物
molded felt 木型模制毡帽
molded garment 模压外衣，浇铸外衣
molded pile product 模压绒头织物
molded sole 模压鞋底
molding 压模件；压制件；装饰线条；制模
mold inhibitor 霉菌抑制剂
mold-making 造型
mole 鼹鼠毛皮；痣；深灰色
mole brown 鼹鼠棕
mole grey 鼹鼠灰
mole(skin) 鼹鼠毛皮
moleskin 鼹鼠皮，仿鼹鼠皮；鼹鼠皮呢；厚毛头斜纹棉布
moleskin cloth 摩尔斯根呢；鼹鼠皮呢
moleskin fabric 仿鼹鼠皮织物
moleskins 厚毛头斜纹棉布服；厚毛头斜纹棉布裤；鼹鼠皮呢
moli cloth 摩力呢
molik 摩立克（仿毛色织布）
Moline 重磅丝塔夫绸
molinos 莫利诺中支纱平纹棉织物；墨西哥棉
Molinos cotton 墨西哥莫利诺棉
molleton 莫利通双面绒；法国麦尔登呢；双面绒布
Mollie Parnis 莫利·帕尼斯(1905～1992，美国时装设计师)
mollielast 软弹性体
mollitan 莫利通双面绒；法国麦尔登呢；双面绒布
Mollofehdi 高级手织棉布
moltem metal dyeing process 液态金属染色法
moltese stitch 垂落针绣
molybdenum oxide fiber 氧化钼纤维
momen-mon 日本棉纺织品（商名）
momentum 气势

momie cloth 马米呢；马米绉；花岗石纹织物
momie crepe 绉纹棉布，花岗石纹棉布
momie towel(1)ing 粗厚毛巾布[料]；花岗石纹毛巾布
momie wool mixed cloth 马米呢
momohiki 防寒衬裤，御寒衬裤；单片女泳装
mompe 裙裤
mompei 裙裤（日本妇女劳动时穿）
monastic 僧袍
monastic robe 僧袍
monastic silhouette 僧袍型轮廓
Monba ethnic costume 门巴族服装
mon-chirimen 日本花绉绸
Mondrian dress 蒙德里安式直身裙
Mondrian look 蒙德里安风貌，蒙德里安服装风格，抽象派服式
Mondrian style 蒙德里安风格，蒙德里安款式
money belt 钱包腰带，铁夹腰带（有拉链式开口暗袋）
Money Bush cotton 美国钱丛棉
money mitt 藏钱针织手套
money pouch 钱袋，钱囊
Mongolian boots 蒙靴
Mongol[ian] ethnic costume 蒙古族服饰
Mongolian jacket 蒙古夹克
Mongolian lamb 蒙古羔皮
Mongolian look 成吉思汗款式，蒙古款式
Mongolian style 蒙古款式
Mongolian wool 蒙古羊毛
Monika Tilley 莫尼卡·蒂蕾（美国泳装设计师）
monitor panel 监控面板
monitor program 监督程序
monitor routine 监督程序
monitor screen 监控屏
monitor system 监控系统
monitored control system 监控系统
monitoring device 监控装置
monitoring standard 监测标准
monitoring unit 监控装置
monk 方平组织厚布
monk cap 僧帽
monk dress 僧侣装，修道士装，僧袍式连裙装
monkey 长毛猴毛皮（非洲），猴皮；隆肉
monkey cap 猴帽，圆顶无檐小帽
monkey clothes （美国俚语）礼服；制服；

男子晚礼服;军礼服
monkey jacket 紧身短上衣(妇女、孩童用);水手紧身上衣;住院病人长睡衣
monkey suit 军服;制服;礼服;男子晚礼服
monkey wrench 活络板手
monk front shoes 僧侣束带鞋,修道士束带鞋
monk habit 僧袍
monk hat 僧帽,圆顶无檐帽
monk kini 超短女三角游泳裤,单片女泳装,男子超短裤
monk's 僧侣服
monk's belt 僧侣腰带
monk's cape 僧侣披肩,修道士披肩
monk's cloth 粗厚方平织物;席纹粗呢
monk's cowl 僧侣兜帽
monk's dress 僧袍式连裙装
monk's habit 道袍;僧袍;修道士服装(会衣、会服)
monk shoes 修道士鞋,僧[侣]鞋
monk shoe strap 僧侣鞋带
monk's obi 僧侣宽腰带
monk's robe 僧侣袍
monk's seam 平接缝,绷缝,搭接缝,双折边叠缝,法式缝
monk's shoes 僧侣鞋,孟克鞋,半高腰鞋
monk strap shoes 僧侣束带鞋,修道士束带鞋,孟克鞋
monk's woolen cloth 席纹粗呢
mono hand 单项手感
monobloc machine 整件缝纫机
monochromatic color 单色,一色
monochromatic harmony 单色协调
monochromatic light 单色光
monochromatism 单色性
monocle 单片眼镜
monofiber yarn 纯纺纱,单一纤维纱
monofilament 单丝
monofilament suture 单丝手术缝线
monofilament thread 单丝缝线
monofilament yarn 单丝;单丝长纤纱
monoflex 莫诺弗莱克斯弹性松紧织物
monogram 标记用交织字母,刺绣字母,服装师姓之首字母
monogram embroidery 交织字母绣花
monogramming 缝绣交织字母
monogramming and embroidery 缝绣交织字母与绣花
monokini 超短露胸比基尼;超短女三角泳裤;(男子)超短裤;无胸罩女泳装;男式三角泳裤;无上装泳衣;一件式比基尼
monopanty 连袜裤
monosex 无性别装
monosex fashion 超越性别流行服饰,无性别流行服装,无性别装
monotone 单色调,黑白调
monotone design 单色效果花纹,单色效果图案
monotone look 同色调套装款式;同色而深浅不同的服式
monotone print frock 单色印花长外衣
monotone style 同色套装款式
monotone tweed 单色调粗呢
monotonous design 单色调设计
monotony 单调
monotop welt 单袜口
Monsanto modacrylic self-extinguishing fiber 孟山都改性聚丙烯腈自熄纤维
monster peak stetson 怪顶牛仔帽
monster shoes 怪形鞋(粗陋外形,鳞茎式鞋尖和巨型跟)
Montagnac [cloth] 蒙塔纳克卷毛呢
Montago mery beret 蒙哥马利贝雷帽
MONTAGUT 蒙特娇
montana coating 经纬异色女用外套;经纬异色女大衣呢
Montana 蒙塔纳(法国时装设计师)
montbeliard 厚床单布;厚被套布
Monte Fiber 蒙特纤维公司(意)
Montecristi fino-fino 超级蒙特克里斯蒂帽
Montecristi supertino 特级超细蒙特克里斯蒂帽
Monteith 芒蒂斯大氅(英国)
Monteiths 芒蒂斯手帕(色底白点花,英国制)
montera (西)黑色斗牛士帽,布帽
montero cap (西)猎人帽,(帽檐可翻下盖住耳朵的)圆猎帽
montero 有帽沿[可翻下护耳的]圆猎帽
Montespan sleeve 蒙铁斯潘袖
Montevideo merinos 蒙特维迪奥美利奴短羊毛
Montevideo wool 蒙特维迪奥羊毛
Montgomery beret 蒙哥马利帽
MONTHLY STYLE MAGAZINE 《流行通信》(日本)
montre-bracelet 表手镯(法)

montserrat [cotton] 印度孟特塞腊特棉
montuno 巴拿马白粗布大袖绣花男衬衫
monzome shusu 条纹丝缎
moo moo 夏威夷穆穆衫,宽松式姆姆装
moocha [南非土著人作腰布的]短围裙
mood 风格,情调,基调,心情
mood ring 情绪戒指
mooltan 手工棉地毯
Mooltan cotton 印度莫尔坦棉
moon beam 月光色
moon blue 月光蓝
moon bow 月虹色,银色
Moon cotton 美国月棉
Moonga silk 印度蒙加丝
moon glow skin colour 月光肤色
moon light 浅米灰
moonlit nauve 月光紫色
moon mist 月雾色
moon pocket 弯月形口袋
moon rock 月岩色
moonstone 月长石
moor 鼓花缎
Moore College of Art Philadelphia 费城莫尔艺术学院
Moorish lace 摩洛哥窄花边
mop towel 揩布
mopen 摹本缎,贡缎
moppet 布制玩偶
moquette 家具用长毛绒;绒头织物,割绒织物;布鲁塞尔地毯;羊毛天鹅绒
moquette carpet 机织割绒地毯
moquette type fabric 绒头织物,割绒织物;布鲁塞尔地毯;羊毛天鹅绒
moquette yarn 割绒用纱
moqui blanket 美国穆快毛毯
moqui cotton 美国穆快棉
mora 摩勒衫(马拿马)
moravian 八股缝纫线
moravian blouse 摩拉维亚宽松女上衣
mordant dye 媒染染料
mordant print 媒染印花
morea 希腊原棉;棉织物;莫里阿缎
moreen 横凸条波纹织物
moreen stripes 波纹条子布
morees 英国细布
more of less clause 溢短装条款
morenos 本色亚麻布
mores 风俗
morfil 比利时莫菲尔厚呢
morganite 红绿玉

morine 波纹织物
morion 高顶轻头盔(16～17世纪)
morisco 流苏花边
morlaix 捷克粗亚麻布
Morman Norell 莫曼·诺雷尔(美国时装设计师)
morning clothing 晨礼服;早礼服;昼间礼服
morning coat 晨礼服,男礼服,晨燕尾服,常燕尾服
morning crepe 晨绉
morning cut 晨礼服裤脚线
morning dress 男式晨礼服;女式家常服
morning frock coat 男合体晨礼服;臀合体女夹克
morning glory 牵牛花红
morning glory skirt 牵牛花裙,喇叭片裙
morning gown 晨袍,晨衣
morning mist 晨雾色,淡湖蓝[色],清水蓝[色]
morning robe 居家装
morning shift 早班
Moro hemp 菲律宾莫罗大麻
moro 马尼拉麻
morocain crepe 莫洛干绉
Moroccan bag 摩洛哥皮革工具手提包;满地缉线酒色皮革手提包
Moroccan fez 土耳其帽
Moroccan lace 摩洛哥窄花边
Moroccan leather 仿搓纹革,仿摩洛哥革,仿植揉山羊搓纹革;摩洛哥革,植揉山羊搓纹革,搓纹革
Moroccan slippers 摩洛哥拖鞋
Morocco(leather) 麻洛哥革,植鞣山羊搓文革,搓文革;仿摩洛哥革,仿植鞣山羊搓纹革,仿搓纹革
morphologically deviating wool 变态羊毛
morpho-structured fabric 仿[大闪]蝶翅结构织物(由异收缩扁平截面聚酯混纤丝织成,通过光的反射和干涉,商名,日本可乐丽)
morpho-yarn 仿真[形态结构]丝
morral (西)莫罗袋
Morris carpet 马立斯机织厚绒头地毯
morris rug 莫里斯机织印花地毯
morse 嵌宝金(或银)扣子;镶宝石的法衣祥扣
mortarboard [cap] 学士帽,方顶帽
mosaic 镶嵌式,拼成的
mosaic blouse 镶拼衫

mosaic blue　马赛克蓝
mosaic canvas　绣花粗十字布
mosaic effect　马赛克效果,镶嵌效果
mosaic fashion　镶嵌流行式
mosaic hat　镶嵌帽
mosaic heel　马赛克跟
mosaic lace　镶嵌花边
mosaic-like composite fiber　镶嵌状复合纤维
mosaic print　镶嵌式花型印花
mosaic rug　黏合绒头地毯
mosaic stitch　马赛克绣,镶嵌缝迹
mosaic tape　马赛克黏结带
mosaic tilling　马赛克填绣
mosaic wool work　拼花绒头地毯,拼花绒线编织品
mosambique　方格仿毛织物;棉经毛纬条格织物;莫桑比克毛纱罗
moschettos　配靴合体长裤
Moschino　莫斯基诺(意大利时装设计师)
moscovite　丝经棉纬凸纹绸
Moscow　莫斯科针绣花边;莫斯科大衣呢
Moscow beaver　莫斯科海狸绒
Moscow canvas　绣花十字布
moshi　苎麻(朝鲜语)
moskowa canvas　本色绣花十字布
mosolin　平纹细布
mosquito bar　细眼蚊帐布
mosquito boots　防蚊靴
mosquito-net　蚊帐[纱];网眼纱
mosquito net fabric　蚊帐布
mosquito netting　蚊帐[纱];网眼纱
mosquito repellent clothes　(防蚊叮咬的)驱蚊衣
mosquito resistance　防蚊性
mosquito resistant fabric　蚊帐布
mosquito resistant finishing　防蚊整理,驱蚊整理
moss　苔藓绿,秋香绿,黄绿色
moss agate　苔纹玛瑙
moss bege　法国苔绒
moss cord　螺旋形线,苔绒多股线
moss crepe　人造棉纱绉,苔绒绉
mosscrepe cloth　苔绒绉
mosscrepe fabric　苔绒绉织物
mosscrepe yarn　苔绒绉线
moss effect　苔绒效果(轻度起绒)
moss effect finishing　苔绒整理
mosser　苔绒,苔绒织物

mosses　中国手工大绞丝
moss finish　苔绒式整理(呢绒轻度起绒)
moss gray　沼泽灰
moss green　苔藓绿,秋香绿,黄绿色
moss grey　沼泽灰
mossing　拉绒,拉毛
mossing finish　苔绒式整理(呢绒轻度起绒)
moss microfibre fabric　苔(藓)纹桃皮绒
mossoul stitch　人字形线迹
mossoul wool　阿瓦西绵羊毛
moss pile　短绒头
moss stitch　桂花针法;桂花针组织
moss yarn　刺绣用起毛纱,[起]茸毛纱
mossy cotton　短绒棉
mossy crepe　苔绒绉
most favored nation　最惠国
most formal wear　正式礼服
mosul rug　伊拉克几何图案地毯
mota　印度厚棉布
motercycle jacket　摩托夹克,摩托装
moth　蚊虫,蠹虫;蛾色
moth ball　樟脑丸,卫生球
moth-eaten　虫蛀[的],蠹蛀[的]
Mother Hubbard　女式宽松长罩衣,哈伯德主妇装
Mother Hubbard costume　哈伯德大妈装束
Mother Hubbard dress　赫伯德主妇装
mother-of-pearl　珍珠母;珍珠母纽扣;珠母层
mother-of-pearl button　珍珠母扣;螺钿扣
mother-of-pearl shell button　螺钿纽[mother] pearl button　贝壳纽扣
motheye　蛀眼,蛀孔
mothicide　杀蛀虫剂
moth-infested goods　虫蛀品,虫蛀织物
mothproof　防蛀[的]
mothproofer　防蛀剂
mothproof finish fabric　防蛀整理织物
mothproof finished fabric　防蛀整理织物
mothproofing　防蛀;防蛀处理
mothproofing agent　防蛀剂
mothproofing finish　防蛀整理
mothproof knitting yarn　防蛀绒线
mothproof knitwear　防蛀羊毛衫
mothproof quality　防蛀性能
mothproof treatment　防蛀处理
moth repellency　防蛀性
moth repellent　防蛀剂;防蛀织物

moth resistance 防蛀虫性
moth resistance insect proof 防蛀
moth resistance finish 防蛀整理
moth test cloth 蛀虫测试布
motia 手纺纱平纹棉布
motichur 破斜纹边织物
motif 花边,蕾丝;花纹图案,(图案)基本花纹,主题花纹,基本色彩;(设计)主题
motif art déco 直线条,简朴为主的花纹,装饰派艺术主题
motif from ecology 生态学主题
motif lace 花卉花边,图案花边
motif of design 设计主题
motif of garments design 服装设计主题
motif pattern 花纹图案,基本花纹图案
motif printing 主题花纹图案印花
motifs cachemire 开士米花纹
motion diagram 运动图,作业图
motion management 巡视管理
motion mark (织物的)厚薄段;(毛织物的)稀密路;(丝织物的)稀密档
motion study 动作测定;动作分析
motique 流动精品服装店
motivation dress 主题礼服,专门礼服
motive 花纹图案,基本花纹图案,主题花纹,基本色彩
motley 杂色布;杂色呢;杂色布衣服;杂色五角装;杂色丑角服,小丑彩衣
motley woolen 杂色呢
motor 电动机
motorcycle boots 摩托靴
motorcycle bustier 摩托车紧身胸衣
motorcycle jacket 摩托夹克;摩托装
motor-driven scissors 电动剪刀
motor-driven sewing machine 电动缝纫机
motoric function 肌肉运动功能
motoring cap 摩托帽
motoring veil 驾车面纱
Motril cotton 西班牙莫特里尔棉
mottle 杂色;斑驳;斑纹,斑点;斑点纱线
mottle color 斑点色
mottled appearance 斑纹外观
mottled carpet 斑纹地毯
mottled color 斑点色
mottled effect 斑点效应,色点效应
mottled silk 斑点丝线,异色合股丝线
mottled wilton 威尔顿斑纹地毯
mottled yarn 斑点纱线,异色合股花式纱线
mottles 斑点纱线织物;黑白斑起毛灯芯绒
mottle yarn 斑点花式纱线
mottling 色点染色,斑点染色
mouchoir 手帕,手绢
mouflon 摩弗伦野羊皮(南欧)
moujik 女用宽披肩
moujik pants 俄罗斯农人裤
moulage design 立体造型设计
mould 霉菌;模具,模型;冲模
mould centre 模压主跟
moulded boots 模压滑雪靴
moulded dress 模压裙装
moulded fabric 模压织物
moulded garment 模压衣服
moulded shoes 模压鞋
moulded sole 模压[鞋]底
moulded sole shoes 模压底鞋
moulding cutting 模塑裁剪
moulding press 模压机
moulding technology 模压工艺;模压技术
moulding 压模件
mould inhibitor 抑霉剂
mould-on 模压
mould pressing 模压
moulds 冲模
mould-soled shoe 模压底鞋
mould stain 霉斑
moule 法国轻软大衣呢;条格隐纹织物
mouline twist 多股花式线,混色纱
moulinee 多色股线
mouliner yarn 杂色花色线,杂色花式线
mound 隆墩
mountain [climbing] boots 登山靴
[mountain] climbing boots 登山靴
mountain cloth 石棉
mountain cork 石棉
mountaineering boots 登山靴
mountaineering jacket 登山外套
mountaineering rope 登山绳索
mountaineering suit 登山服
mountaineering wear 登山服
mountain flax 石棉
mountain hosiery 山地袜;上山袜
mountain leather 石棉
mountain panther fur 雪豹毛皮,美洲狮毛皮
mountain shoes 登山鞋,登山靴
mountain skiing shoes 高山滑雪鞋
mountain tartary 鞑靼地毯毛
mountain wood 石棉

mounted goods　镶毕钻石,镶正的钻石
mounted stone　镶毕钻石,镶正的钻石
mountie cap　(西)猎人帽
Mountie's hat　加拿大皇家骑警帽(宽边尖顶帽);西班牙圆猎帽
mounting　镶;装座(首饰),座架
Mountmellick embroidery　蒙特梅利克刺绣,爱尔兰针绣
mount stitch　山形缎带绣
mourat wool　苏格兰棕色细羊毛
mourn dress　丧服,葬礼服
mourning　丧服(做丧事穿),丧用面纱
mourning [apparel]　丧服,葬礼服
mourning armlet　服丧缠臂黑纱
mourning badge　黑纱
mourning band　黑纱,丧章
mourning bonnet　丧服圆帽
mourning color　丧服色
mourning costume　丧服,葬礼服,孝服
mourning crepe　丧服绉
mourning dress　丧服,葬礼服
mourning stuff　丧服料
mourning tartan　苏格兰丧服
mourning veil　女式葬礼黑色面纱,黑色披纱
mouse　灰褐,鼠灰
mouse color　灰褐色,鼠灰色
mouse costume　小鼠装束
mouse-dun　略带褐色的深灰色
mouse gray　灰褐色,鼠灰色
mouse grey　灰褐色,鼠灰色
mousers　毛瑟式女皮裤靴(齐腰,琼琦靴,富有光泽皮质)
mouse skin　鼠皮呢,起绒织物
mouse skin fabric　鼠皮布
mousquet [rug]　土耳其细密地毯
mousquetaire coat　火枪队式大衣;穆斯可特上衣
mousquetaire collar　火枪队式领;穆斯可特领;中翻领
mousquetaire costume　穆斯可特服饰;17~18世纪法国皇家近卫兵的火枪服饰
mousquetaire cuff　火枪队克夫,喇叭式宽克夫(袖口),翻克夫袖
mousquetaire gloves　翻筒手套,火枪手式手套,穆斯可特手套,易戴式晚宴长手套
mousquetaire hat　火枪手帽,宽檐帽,穆斯可特帽
mousquetaire mantle　火枪手披风
mousquetaire sleeve　火枪手式袖,穆斯可特袖,近卫骑兵袖,翻折宽大袖口
mousquetaire style　火枪手风格
mousseline　透明平纹薄织物;真丝薄绸
mousseline de laine　薄精纺花呢
mousseline de soie　薄雪纺绸,薄丝纱
mousseline matte　无光平纹薄丝织物
mousseline satin　薄丝绸
mousselinette　粗平布
moustache　髭,八字须
moustachio　八字须
moustiquaire　透明薄纱罗
moutan muslin　英国浮花薄纱
mouth　口;手袋袋口;鼠标器
mouth corners　嘴角
mouth piece　口罩;拳击护齿
mouth-piece charge　口罩棉填料
mouth-piece tape　口罩带
mouth veil　口罩面纱;(阿拉伯妇女)裹头长面纱
mouton　染色绵羊毛皮
mouton collar flight jacket　羊皮领飞行员式夹克
mouton lamb　羊毛皮(经处理),羊皮
mouton processed lamb　染色绵羊毛皮
movable clothes rack　活动挂衣架
movable presser foot　可拆压脚,活动压脚
movable sewing box　活动(缝纫)箱
movable shear blade　活动剪刀片
mova silk　绢丝
movement allowance　衣服宽份,宽余量
movement folds　动态衣纹
movement sketch　动态素描
movement studies　动态练习
movement theory　移情说
moving cutter　活络剪刀
moving guider contour sewing machine　活动靠模缝纫机
moving trapper　活络夹紧剪刀
mowlds　男式衬裤,内裤
moyle　密儿拖鞋
mozambique　莫桑比克毛纱罗;莫桑比克条格薄呢
mozzetta　天主教用短斗篷
Mr. John　约翰先生(美国女帽设计师)
M. silbon　M.西尔邦(法国时装设计师)
M. TSUBOMI　子苞米
muaf　穗饰扎发带

muckinder 儿童小手帕
muckluck （爱斯基摩人穿的）海豹[或鹿]皮靴；海豹皮鞋；软皮底毛绒袜；（底部缝有软皮的）羊毛袜
mud cracking 龟裂效应；起霜花
muddy color 土色，浊色
muddy complexion 灰暗肤色
mud pearl 蓝珍珠；混珠
muff 袖窿；揣手儿；暖袖，暖手筒，（女用）皮手筒；皮手笼
muff bag 手筒皮包，手笼皮包（手笼与皮包两用）
muffettee 毛线围巾；[手套的]腕套；罗纹翻袖口；老人用的羊毛针织或皮手套
muff handbag 皮手笼提包
muffin hat 松饼型男帽
muffin cap 松饼帽
muffler 长围巾（12英寸宽，格子或一色、毛、丝或人丝制），围巾；白色丝巾，男式配晚礼服的丝巾；厚手套，拳击手套，无指手套；手筒
muffler cummer-band 围巾式腰带
muff pocket 暖手[口]袋
mufti 便服，便衣，休闲装；（官员、军官等的）便服
Muga silk 印度蒙加丝
mugginess 闷热感，湿闷性
Mughlai cotton 印度穆格莱棉
muka 新西兰大麻
mukharech cotton 伊拉克穆哈里什棉
mukhayyar 马海毛
mukluk （爱斯基摩人用）穆克拉克靴，海豹毛皮长靴；慕克拉克套鞋；皮长靴
mukluk slipper 慕克拉克套鞋；暖脚套鞋
Mulam ethnic costume 仫佬族服饰
mulan damask satin 目澜缎
Mulao ethnic costume 仫佬族民族服
mulatto 黄褐色[的]
mulatto color 黄褐色
mulberry 桑果紫红色，桑椹色，深紫红
mulberry bark fiber 桑皮纤维
mulberry paper clothes 桑纸服装
mulberry silk 桑蚕丝，家蚕丝
mulberry silk ball 桑蚕绵球
mulboos khas 马尔布斯高级细布
mule 室内连袜便鞋（女式）；室内拖鞋（女式）；拖鞋式女鞋；无踵女式拖鞋
muleedah-pushmina 印度绒面女式呢
mule hide 骡皮
mules 无后跟的拖鞋，室内拖鞋

mule [sandals] 密儿拖鞋
muleta cape 西班牙迷蒂披肩；西班牙中长披肩
mule twist 走锭纺棉纱
mule yarn 走锭纱
mull 麦尔纱；漂白细软薄布；重浆帽里布；漂白细布；丝或人丝薄绸
mull-chiffon 瑞士雪纺薄布
mulle 麦尔纱；漂白细软薄布；丝薄绸；重浆帽里布
mullet 纹章的星形图案
mullmull 细软薄布，薄纱，丝质薄绸
mull muslin 白细布；细薄衣料
mulmul 细软薄布，薄纱，丝质薄绸
mulmulkhas 最优质达卡细布
mulquinerie 亚麻织物坯布；帆布坯布
multan 木尔坦棉；手结厚重棉地毯
Multan cotton 木尔坦棉
Multan rug 木尔坦地毯
multihead embroidery machine 多头绣花机
multi bar fabric 多梳栉经编针织物
multiaxial fabric 多轴向织物，多向织物
multiaxial raschel warp knitting machine 多轴向拉舍尔经编机
multiaxial stitch-bonding machine 多轴向缝编机
multiaxial warp knit fabric 多轴向经编织物
multiaxial warp knitted fabric 多轴向经编织物
multiaxial warp knitting fabric 多轴向经编织物
multibar fabric 多梳栉经编织物
multibar raschel lace machine 多梳栉拉舍尔花边机
multibar raschel machine 多梳栉拉舍尔经编机
multibar tricot machine 多梳栉特里科经编机
multibar warp knitting machine 多梳栉经编机
multibar warp-knitted fabric 多梳栉经编针织物
multiborder printed handkerchief 多边形印花手帕
multibrand strategy 多品牌策略
multi collar 复合领
multicolor cloth 多色织物
multicolor(ed) damask 彩色花缎

multicolored jacquard 多色提花
multicolored stone stud 彩石花鞋扣
multicolor effect 多色效应
multicolor folded yarn 多色股线
multicolor Nanking brocade 彩库缎,彩库锦
multicolor pattern 彩色图案
multicolor plaid shirt 彩色长方格衬衫
multicolor printings 多色印花
multicolor warp 多色经纱
multicolour 多种色彩;多彩,彩色
multicolour check 多色格子花纹
multicolour cloth 多色织物
multicoloured damask 彩色花缎
multicoloured stripes 多色条子
multicoloured thread 多色股线
multicoloured T-shirt 色条T恤衫
multicolour embroidery work 彩绣
multicolour fancy yarn 杂色花式线
multicolour matching 多色彩搭配
multicolour pattern 彩色图案
multicolour plaid shirt 彩色长方格衬衫
multicolour printings 多色印花
multicolour stripe 多色条纹
multicomponent fiber 多组分[复合]纤维
multicomponent polymer fiber 多组分共聚物纤维,共聚物纤维
multicomponent textile 多组分纺织品
multiconstituent fiber 多成分[复合]纤维
multi coordination 多种成分服装组合
multi-cord 复丝,多纤长丝
multi-cord composite yarn 多芯型复合纤维(海岛型复合纤维)
multiculture 网格多元(服装设计思想)
multidimensional cloth 多维织物
multidimensional fabric 多向立体织物
multidirectional feed system 多向送料系统
multi-direction type 各向同性型(非织造布)
multi dull taffeta 无光塔夫绸
multi-fiber 复型纤维
multifiber adjacent fabric 多种纤维贴衬织物
Multi Fiber Arrangement (MFA) 多边纤维协定,纺织品多边协定
multi-fiber blends 多纤维混纺织物
multi-fiber fabric 多纤维织物
multifibre arrangement 多种纤维协议
multi-fibrous yarn-like strand 立构形成纤丛纱线
multifil [ament] 复丝
multifilament bundle 丝束
multifilament yarn 复丝,复丝纱
multifilamentary composite wire 复丝状双金属线,双金属复丝
multifilamentary tow 复丝[丝]束
multifilamentary wire 金属复丝
multifilaments mixed brocaded damask 集聚绸
multiflora cotton 美国多花棉
multi-folded yarn 多股线
multifunctional clothing 多用途服装
multifunctional fabric 多功能织物
multifunctional finishing 多功能整理
multifunction embroidery machine 多功能绣花机
multifunction fabric 多功能织物
multi-grader 尺码放缩仪
multi-hand 综合性手感
multi-head automatic embroidery machine 多头自动绣花缝纫机,多头电子刺绣机
multi-head embroidery machine 多头绣花机
multi-head flat knitting machine 多系统平型针织机
multi-height carpet 多层绒头地毯
multi-industry 多种经营
multi-items of products 多品种
multi-item-small-lot 多品种小批量
multi-job operation 多重作业操作
multi-kroine 多色效应
multilaycr air-cooling underwear system 多层空气冷却内衣系统
multilateral agreement 多边协议
multilateral contracts 多边合同
multilateral trade 多边贸易
multilateral trade negotiation 多边贸易谈判
multilayer brcomponent filament 多层型双组分长丝
multilayer conjugate fiber 多层型共轭纤维,多层型复合纤维
multi-layered fabric 多层织物
multi-layered look 多层风貌,多层装,多层型款式
multi-layered skirt 多层短裙
multi-layered style 多层型款式
multilayer fabric 多层织物

multilayer fiber 多组分纤维；多层纤维
multilayer material 多层缝料
multilayer woven fabric 多层机织物
multilevel pile 多层绒头，不平齐绒头
multilobal 多叶形(截面)的，异形截面
multilobal cross section 多叶形截面
multi-lobed filament 多叶型长丝，异形截面长丝
multi-microporous fabric 多微孔织物
multinational corporation 跨国公司
multinational enterprise 多国公司
multi-needle 多针
multi-needle flat lock machine 多针绷缝机
multi-needle flat sewing machine 多针平缝机
multi-needle lockstitch machine 多针锁缝机
multi-needle sewing machine 多针缝纫机
multi-needling zones inclined both sided needling machine 多针刺区斜向对刺针刺机
multioriented fiber 多次取向纤维
multipattern 复合花样
multipiece bonded lining fabric 多段黏合衬织物
multi-pleat 群褶
multiple brands 多种商标
multiple cloth 多层织物；多层组织
multiple color 多色
multiple colors 多重色
multiple composite finished fabric 多层复合整理织物
multiple coordination 多种调和法(制作套装)
multiple cord foot 多线嵌线压脚
multiple correlation 复相关
multiple darts 多重省
multiple decorative stitch 复合装饰线迹
multiple direction print 多方向印花
multiple fabric 多层织物
multiple fancy yarn 组合花色线(由多股不同花色线组成)，组合花式线
multiple-fiber material 多种纤维材料
multiple-fiber producer 多品种纤维生产厂商
multiple-head embroidery machine 多头绣花机
multiple-layer fabric 多层织物
multiple-needle machine 多针缝纫机

multiple-needle tucking 多针打裥
multiple-ply thread 多股线
multiple-ply yarn 多股线
multiple-row seam 多行缝
multiple sampling 多次抽样
multiple sheath-core type 多层皮芯型
multiple side-by-side composite fiber 多层并列型复合型纤维
multiple small-lot production 多品种小批量生产
multiple stitching mechanism 复合缝纫机械
multiple stitch zigzag 复合曲折形线迹
multiple unit knitting machine 多节针织机
multiple wound yarn 并合纱，并绕纱
multiplex yarn 复合长丝
multiplicity 多重性
multiplied tone 多层次色调
multiplied yarn 多股线
multiplies 叠布层
multiply cloth 多层织物
multiply cutting 多层裁剪
multiply stack 多层堆置
multiporous acrylic fiber 多孔性[聚]丙烯腈纤维
multiprints 多色印花
multiprocess yarn 多次变形纱
multiprogramming 多道程序设计
multipurpose design 多用途花样设计
multipurpose electronic sewing machine 多功能电子缝纫机
multipurpose finish 多用途整理
multipurpose finishing 多防整理
multipurpose foot 多功能压脚
multipurpose gauge 万能量规
multipurpose outfit 多功能外套
multipurpose sewing machine 多功能缝纫机
multipurpose underpressing unit 多功能中间熨烫机
multipurpose use of wastes 废物综合利用
multi-range sewing needle 多量程缝纫机针
multirow stitching 多行线缝
multi-seaming 多道缝迹
multi-season coordination 多季组合
multi-section cotton machine 多节柯登机
multi-segment composite fiber 多股裂离异形复合纤维(裂离型复合纤维)

multi-size pattern 多码板样
multi-sleeve 多袖装;组合袖;多袖
multi-spiral fiber 多重卷曲纤维,三维卷曲纤维
multi-stage outsole 多段底(鞋)
multi-step zigzag stitch 多重曲折线迹
multi-stitch zigzag 多针曲折线,宽 Z 形线
multi-stitch zigzag stitch 多针工形线迹,多针曲折形线迹,多针 Z 形线迹
multi-strand pearl necklace 多层式珍珠项链
multi-strand thread 多股线
multi-stretch stitch 多伸缩性线迹,复合伸缩形线迹
multi-stripe 多种条纹组合花样
multi-suture heel Y形接缝袜跟
multi-tape 多条织带
multi-thread chain stitch 多线链式线迹
multi-thread chain stitch one sided ornamental stitch 单面绷链式线迹
multi-type 多品种
multi-variance 多元方差
multi-variance regression 多元方差回归
multi-variate analysis 多元分析
multi-variate normal distribution 多元正态分布
multiwall bag 多层袋
mummy cloth 马来绉
mundeel 军官头巾
municipal pollution 城市污染
Munsell chroma 芒塞尔彩度
Munsell color space model 芒塞尔色立体
Munsell color system 芒塞尔表色制
Munsell color tree 芒塞尔色树
Munsell hue 芒塞尔色调
Munsell notation system 芒塞尔表色系统
Munsell photometer 芒塞尔光度仪
Munsell renovation system 芒塞尔修正表色系,芒塞尔修正表色制
Munsell solid 芒塞尔色立体
Munsell value 芒塞尔值
Munsell value scale 芒塞尔明度标尺
muntjac fur 麂皮
murga 虎尾兰麻叶纤维
Murgala 印度莫加拉彩色织锦
murgavi 印度虎尾兰麻叶纤维
murrey 桑果紫红色;桑椹色
murrey color 黑紫色

murva 印度大麻
musahri 印度蚊帐布
musa textilis 麻焦
muscle 肌肉,健美;袖臂;上袖
muscle knit [shirt] 健美紧身针织衫
muscle shirt 健美紧身针织衫,健美无袖圆领衫
muscle sleeve 健美袖,马甲袖
muscle structure 肌肉结构
muscle tissue 肌肉组织
muscular arm 肌肉发达的臂
muscular shoulder type 强健肩型
musculature 肌
musette bag 肩背皮包,背包;小行囊(美国士兵与徒步旅行者用)
mush 小胡子,髭
mushaddah 手织棉粗平布
mushajjar 印度印花缎
mushka 印度穆什卡绸
mush rat fur 小麝鼠毛皮
mushroo 印度穆什鲁缎
mushroom (女用)蘑菇形扁帽;浅褐棕色,蘑菇色
mushroom cloche 蘑菇式钟形帽
mushroom cut 蘑菇发型
mushroom fastener 尼龙搭扣
mushroom gown 蘑菇装
mushroom hair style 蘑菇式发型
mushroom hat 蘑菇帽,蘑菇形草帽,蘑菇式扁帽,覃形帽
mushroom ironing press 蘑菇式压烫机
mushroom pleat 极细的热定型褶裥
mushroom press 蘑菇式熨烫机
mushroom sleeve 蘑菇袖
mushroom-type 蘑菇式
mushroom weft knit fabric 针织蘑菇布
mushru 印度穆什鲁缎
mushy wool 枯羊毛
music cord 英国全棉平纹纹线
musical sandal 响声鞋,音响凉鞋
Musiel Grateau 穆西耶尔·格拉托(意大利时装设计师)
musin finish 细布整理,薄纱整理
musin fitting shell 细布试身样
musin pattern 样服,白坯布试衣
musin yarn 中高支棉纱
musjoor 穆斯朱尔纱罗
musk 麝香
musk deer 麝毛皮,香獐毛皮
musk deer fur 麝毛皮

muskmelon 甜瓜色
muskobads 伊朗默斯科贝地毯
musk ox 麝牛毛皮
musk-ox fur 麝牛毛皮
musk-ox wool 麝牛绒毛
muskrat 麝鼠毛皮
muslin 平纹细布;市布,中平布,白平布,龙头市布,五福市布;薄纱织物,麦斯林纱,麦斯林呢,麦斯林;细薄平纹毛织物
muslin bed sheeting 床单细布
muslin delaine 细薄平纹毛织物
muslinet 英国厚麦斯林布,厚条纹布
muslinette 厚条纹布
muslin fitting 试衣模式(裁剪戏装使用)
muslin fitting shell 细布试身样
muslin mock-up 试衣模式(裁剪戏装使用)
muslin model tryout 试衣模式(裁剪戏装使用)
muslin pattern 细布样板
musquash 麝鼠毛皮
musquash cap 麝鼠皮便帽,麝鼠皮帽
muss (织物)捏皱,弄皱
mussel silk 贝须丝
mussiness 布面轻微折皱
mussing (织物)捏皱
mussing resistance 防捏皱性
muss-resistant finish 防捏皱整理
mustache 髭
mustang 熟褐色
mustard 暗黄色,芥末黄,芥子酱色
mustard yellow 芥末黄,浅暗黄色
mustard brown 芥末棕,黄棕色
mustard gold 稻草黄[色]

Mustela erminea 白鼬皮;扫雪皮
mutant look 突变装
mutation 变异(毛皮)
mutation mink 变异貂皮
mutation mink fur 变异水貂毛皮
mutch 亚麻布女帽(或童帽)(苏格兰)
muted box plaid frock 暗方格外长衣
muted gray 柔和的灰色
muted grey 优雅柔和的灰色
muted shade 柔和色调
muted stripe 朦胧条纹,晕色条纹
muted stripe frock 暗条纹外长衣
muti-head flat knitting machine 多节平型针织机,多幅柯登机
mutiple stitching mechanism 复合缝纫机构
mutka 手织穆特卡粗绸,粗质腰带绸
mutton chops 羊排络腮胡子;圆形络腮胡子
mutton cloth 较松平针织物
mutual correlation 互相关
mutual inspection 互检
muumuu 夏威夷宽松及踝衫,夏威夷穆穆衫,姆姆装
Muzzy's button 马吉扣
Mygrene de Premonvile 迈丽斯·德·普雷蒙维尔(法国服装设计师)
myoto 精细日本羊毛地毯
mysore 迈索尔染色棉布;低档套结地毯
mysore silk 印度家蚕丝制迈索尔花绸
mysore wool 印度迈索尔羊毛
mystery 行业
mystical style 神秘风络
myzinum 薄绸

N

nabo[fiber] 菲律宾乌檀纤维
nacarat 鲜艳橘红色;橘红色亚麻布;肉色细布
NACCB 英国国家认证团体认可委员会
nacre pigment printing 珠光涂料印花
nacre printing 珠光印花,渗化印花
nacre velvet 珠光丝绒(绒头与地组织异色闪光)
nacré 珍珠母;珠光;闪光效应
nacreous finish 珠光整理
Nadam cotton 印度纳丹短绒棉
nae 细密网眼纱织物(夏威夷)
naga-juban 和服式及地内衣
nagapore 印度轻软绸
nagli pashmina 仿开士米呢,印巴仿山羊绒呢
nago nodzi 黑白横条毛毯
Nagoya silk 名古屋双绉
nail 钉皮;纳尔鞋;指甲
nail bleach 指甲漂白剂
nail brush 指甲刷
nail clipper 指甲钳,指甲轧剪
nail cream 指甲膏
nailed 爪钉
nail enamel 指甲油
nail file 指甲锉
nailhead 爪钉;点纹细呢,点子花纹细呢
nailing board 钉皮板
nailing pincher 钉钉机(加工裘皮、皮革用)
Nailon 聚酰胺66长丝(商名,意大利蒙特纤维公司)
nail polish 指甲磨光剂,指甲油,趾甲油
nail process 钉皮过程,钉皮工艺
nail revealed 露钉[鞋]
nail sandals 分趾拖鞋
nail scissors 修甲小剪刀
nail shoes 钉鞋
nain 现代波斯地毯
nainoo 色经平纹棉布;小花纹白棉布
nain rug 伊朗森纳结手工地毯
nainsook 南苏克布,全棉薄平布,平纹细薄棉织物,奈恩苏克细布

nainsook checks 奈恩苏克格子布,平纹棉格子布
nainsook finish 奈恩苏克整理,不上浆柔软整理
nainu 色经平纹棉布;小花纹白棉布
naive chemise 天真贴身内衣
naive design 天真图案
naka 二上二下斜纹布
naked foot sandals 裸脚凉鞋
naked sweater 紧身毛线衣,紧身羊毛衫
naked wool 全毛薄呢
nakhai bicllidi 墨西哥红蓝白黑纬向色条毛毯
nakli daryai 印度手织细布
namad carpet 纳马德嵌花毡毯
namazlik[rug] 土耳其制祈祷用地毯
nambali 印度南巴利印花绸
namdas 西藏南姆达司毡化呢
name 店名标签;服装内标签
nameboard 招牌,标签
name brand 铭牌;名牌产品;名牌商标
name cloth 标记带
named place 交货地点
name label 标签名,名称标签
name merchandise 名牌商品
name of commodity (goods) 商品名称,货名
name of article 品名
name order list 名次表
name plate 铭牌,商标
name selvedge 织物字边
name tape 标有姓名的布条(衣服等上)
namgali 印度手织绸
namihaba kanakin 本色棉衬衫布
namitka 俄罗斯手工纺制的面纱
nammad carpet 纳玛毛毡地毯
namud carpet 纳玛毛毡地毯
namunah 手织条纹布
nana haircut 光后脑发式
nanako 日本南库绸,纳纳库绸
nanas sabrong 坎塔拉剑麻
nancy 抽线刺绣
nanduti 南美南杜里花边
nanduty 南美南杜里花边
Nanjing fabric 南京布,紫花布

Nanjing silk　宁绸
Nanjing silk twill　素宁绸
Nanjing velvet　建绒
Nanjing yellow　南京黄色(浅灰黄色)
Nanjin twill　素宁绸
nanka　南京绿地黑条纹斗篷绸
nankeen　手纺纱织土布;南京浅棕黄色土布;芝麻布;细亚麻网;紫花布;南京棉布(本色棉布),土[棉]布;本色棉布裤
Nankeens　土布裤,紫花布长裤
Nankeen twill　斜纹色布
Nankeen yellow　南京黄(浅灰黄色)
Nankin　南京棉布(本色棉布),土[棉]布;本色棉布裤;宁绸;南京丝绒
Nankin button　南京丝吐,南京废丝
Nankin cotton　南京棉,中国棉
nankinet　低级土布
Nanking　南京丝绒花边
Nanking cotton　南京棉,中国棉
Nanking silk　南京宁绸
Nanking silk brocade　南京花绸,库锦
Nanking silk plain　(南京)素宁绸
Nanking velvet　建绒
Nanking yellow　南京黄(浅灰黄色)
Nannex　阻燃聚酯纤维(商名,日本可乐丽)
nano fiber　纳米纤维(超微细纤维)
nanometre fabric　纳米面料
nanometer fiber　纳米纤维(超微细纤维)
nanometer material　纳米面料
nanometer technique　纳米技术
nanoprevent uniform　纳米防污服
Nanshan pongee　烟台丝制南山府绸
Nanshanssu silk　东北柞蚕丝制南山素绸
nansi　由中国进口的原棉
nansu　平纹细软棉织物
Nan-yang　南阳府绸;南阳柞绸
Nanyang pongee　中国南阳二六柞绸
Naopoeon high coat　拿破仑翻折高领上衣
nap　起绒,拉绒;绒毛;拉毛
Napa leather　纳帕皮(美国)
napa leather　手套皮
nap bar　毛档疵,撬档疵(丝绸)
nap cloth　绒面织物,起绒织物
nape　后颈,颈背;领背
nape necklace　颈背项圈
napery　亚麻布;抹布;餐巾
nape to centre waist　后长;背长
napery hem　手帕缝边;手帕卷边;台布缝边

nap fabric　绒类织物,起绒织物
nap finish　起绒整理,搓绒整理
nap[finished]overcoating　珠皮[涂层]呢
naphthalene　卫生球,樟脑丸
naphtholated articles　纳夫妥染色织物,纳夫妥印花织物
naphtholated goods　纳夫妥染色织物;纳夫妥印花织物
naphtholate printing　色酚印花
naphthol dye　纳夫妥染料;色酚染料
naphthol printing　纳夫妥印花
napier　纳皮尔双面大衣呢;纳皮尔黄麻地毯
napier mat　纳皮尔黄麻地毯布
napkin　餐巾,小毛巾;苏格兰围巾;头巾;手帕;尿布;卫生巾
napkin-cap　男式家居帽
napkin cloth　餐巾布
naples hemp　意大利大麻
napless　(呢绒上)没有绒毛的;磨破了的
napless finish　光洁整理
Naples Yellow　那不勒斯黄(略带红色)
Napoleon　拿破仑服饰;拿破仑靴
Napoleon boots　拿破仑靴(19世纪)
Napoleon coat　拿破仑外套
Napoleon collar　拿破仑领(高立领宽驳头),波拿巴领
Napoleonic collar　拿破仑领,波拿巴领
Napoleon mode　拿破仑服饰
napolitaine　窄编带;纳波利坦全毛法兰绒
napolitaine cord　纳波利坦起棱织物
nap overcoat　拉绒大衣,卷绒大衣
nap overcoating　珠皮呢
nappa　纳帕革;纳帕手套羊皮
nappa bra　羊皮乳罩
nap pattern　毛结花纹
nappe　台布;抹布;餐巾;法国家用亚麻布
napped cloth　绒布,起绒织物
napped cotton fabric　棉绒布
napped danmsse chine　印经拉绒缎
napped fabric　拉绒织物,拉毛织物,起绒织物,绒布
napped-finish goods　拉绒织物,拉毛织物,起绒织物
napped finishing　拉毛整理
napped grain　绒面
napped hosiery　拉毛袜
napped jersey　起绒乔赛;起绒针织物
napped seam　搭缝

napped suede 起绒仿麂皮,拉绒仿麂皮
napped tricot 起绒特利考
napperon 小台布
nappiness 毛羽;发毛程度
napping 拉毛磨绒,拉绒,拉毛,起绒;搓呢
napping cotton 起绒性能好的棉花
napping finish 起绒整理,搓绒整理
napping in the reverse direction 反向起绒
napping printing 起绒印花,发泡印花
napping property 起绒性能,拉绒性能
napping resist 抗拉绒[印花]
nappy 起绒的,起毛的,拉绒的;尿布
naps 珠皮呢
Nara Women's University 奈良女子大学(日)
Narainganji [jute] 孟加拉纳拉因甘吉黄麻
narcissus 水仙花色
narma cotton 那玛美种陆地棉
Narman embroidery 诺曼[绒线]刺绣
narrow and broad ribs 宽窄罗纹条
narrow and natural shoulder 狭肩,窄肩
narrow belt 细窄腰带
narrow braid 窄幅饰带;花边
narrow carpet 机织窄幅地毯
narrow chemical lace 窄幅烂花花边
narrow cloth 窄幅布,窄幅织物
narrow cording foot 窄幅线压脚
narrow crepe 窄幅绉绸
narrow cross point 细交叉形针尖
narrow dart 狭省,窄省
narrow decorative stitch 装饰窄缝线迹
narrow dress 纤细女装
narrow duck 窄幅帆布
narrowed piece 收针衣片
narrow elastic fabric 窄幅弹力织物
narrow eyelet buttonhole without taper bar 无套结窄圆头纽孔
narrow fabric heading 有穗夹带,有穗窄带
Narrow Fabric Institute 窄幅织物协会
narrow fabric 带织物,狭幅织物,窄幅织物;带
narrow fabric skirt 流苏,缘饰
narrow goods 窄幅织物;带
narrow hemming 狭卷边,窄卷边
narrow hemming foot 狭卷边压脚
narrow hip 窄臀
narrowing 改狭,改窄

narrowing flat knitting machine 收针型横机
narrowing mark 收针花
narrow lace 狭式花边,窄幅花边
narrow lapel 小翻领,窄驳领;窄驳头
narrow list kersey 克瑟窄幅低级粗呢
narrow look 瘦长款式
narrow notch lapel 小驳领;小方领
narrow pants 细裤
narrow pointed collar 窄尖领,小尖角领
narrow rib 狭罗纹
narrow sheeting 窄幅床单布
narrow shoulder 窄肩
narrow skirt 窄裙,一步裙
narrow straight buttonhole without taper bar 无套结窄直形纽孔
narrow straight buttonhole with short wide taper bar 短宽套结直形纽孔
narrow style 瘦长款式
narrow type inside foot 小内压脚
narrow type outside foot 小外压脚
narrow wale 窄棱条,窄条灯芯绒
narrow ware 带;窄幅织物
narrow width 布幅不足
narrow woven goods 带;窄幅织物
narrow wrist 窄腕
narumi-shibori 扎染绉
nasal bone 鼻骨
nasolabial fold 鼻唇沟(鼻唇褶)
nasturtium red 旱金莲红(暗红橙色)
nasturtium yellow 旱金莲黄(橙黄色)
Natasha look 娜塔莎款式
Natasha style 娜塔莎款式
natelassé wool cloth 马特拉赛呢
National Academy of Needlearts (NAN) [美]全国针绣学会
National Aeronautic and Space Administration look (美)宇宙服风貌,宇宙服款式
National Association of Decorative Fabric Distributors (美)全国装饰织物经销商协会
National Association of Fashion and Accessory Designers (NAFAD) (美)全国服装服饰设计师协会
National Association of Textile Supervisors (NATS) (美)全国纺织专家协会
National Association of Glove Manufacturers (NAGM) (美)全国手套制造商协会
National Association of Hosiery Manufac-

turers(NAHM) （美）全国袜类生产者协会
National Association of Importers and Exporters of Hides and Skin(NAIEHS) （美）全国皮革进出口商协会
National Association of Men's Sportwear Buyers (NAMSB) （美）全国男式运动服装采购商协会
National Association of Testing Authority (NATA) 澳大利亚检验管理协会
National Association of Textile & Apparel Wholesalers (NATAW) （美）全国纺织品及服装批发商协会
national blue 美国蓝
national brand 全国性商标,制造业者商标
National Bureau of Standards （美）国家标准局
national costume 民族服装,民族服饰
National Cotton Council of America （美）全国棉花委员会
National Cotton Ginner's Association （美）全国棉花轧花协会
National Council for Textile Education （美）全国纺织教育委员会
national flag design 国旗图案
National Institute of Drycleaners （美）全国干洗商协会
nationality 民族风格,民族性
National Joint Committee for the Carpet Industry （英）全国地毯工业联合委员会
National Knitted Outwear Association （美）全国针织外衣厂商协会
National Knitwear and Sportwear Association （美）全国针织品和运动衣协会
National Knitwear Manufacturers Association （美）全国针织品生产者协会
National Needlework Association （美）全国针绣协会
National Outerwear and Sportwear Association (NOSA) （美）全国外衣和运动衣协会
national private brand 全国性个人商标
National Retail Federation （美）全国零售联盟
National Retail Merchants Association ［美］全国零售商协会
National Standard 国家标准
national style 民族风格
national technology 民间工艺

National Textile Center （美）全国纺织中心
National Textile Processors Guild （美）全国纺织加工协会
national traditional fabric 民族传统织物
national treatment 国民待遇条款
native 国产的,本地产的
native brocade 小锦(宋锦品种之一)
native cloth 英制出口非洲的染色棉布
native cotton 本地棉
native pattern 原始图案
native products fashion show 国货时装表演
natives 土丝(未复摇土缲丝)
native silk 桑蚕土丝,土(缲生)丝,土丝(未复摇土缲丝)
native silk piece goods 本土绸
native silk rereeled 白摇经(复摇土缲丝)
native stripes 手织非洲土布
native wool 国产羊毛;美国东部绵羊毛;土种羊毛;未改良羊毛
natte 色织席纹绸
nattier ［blue］ 深蓝色
natty 漂亮的;整洁的
natty accessories 整洁服饰
natural 本色,天然色,土黄色,浅黄褐色;自然蓬松发式;原色毛纱
natural abundance 自然丰度
natural bleaching 天然漂白
natural blended fabric 天然纤维(为主的)混纺织物
natural body 正常人体
natural bra 自然胸罩
natural brassiere 自然型胸罩
natural bristle 天然鬃毛
natural color 自然色,天然色,本色
natural(ly) colored cotton 天然彩色棉
natural color uplands cotton 天然彩色细绒棉
natural color system (NCS) 自然［颜］色系统
natural comfort jeans 舒适型牛仔裤
natural convolution （棉纤维的）天然扭曲
natural cotton rope 原色棉绳
natural crimp 天然卷曲,自然皱缩
natural dress 自然装束
natural dye 天然染料
natural dyestuff 天然染料
natural elasticity 自然弹性

natural fiber 天然纤维
natural fiber color 天然纤维色
natural fiber material 天然纤维材料
natural function jeans 机能型牛仔裤
natural fur 天然毛皮
natural grain 天然粒面
natural gray yarn 天然本色纱
natural hairstyle 自然式发型
natural hairtype 自然蓬松发式
naturalia 原色薄呢
naturalistic print 自然写实花样印花
natural length 自然长度,普通长度
natural life-supporting eco-system 自然生物生态系统
natural light 自然光
natural lighting 日光照明,自然采光
natural line 自然线形
natural look 天然外观,自然外观,自然风格,自然风貌
natural-looking 天然的外观,自然的风貌
naturally colored cotton fabric 天然彩色棉织物
naturally pigmented wool 天然有色羊毛
natural material 天然材料
natural material button 天然原料扣
natural mineral pigment 天然矿物颜料
natural mink 天然貂皮
natural motif 自然图案主题
natural neckline 自然型领口
natural pastel 自然蜡笔色
natural pattern 自然写实花样
natural protein fiber 天然蛋白质纤维
natural rubber suit 天然橡胶服
natural selvedge 织物自然布边
natural shade 天然色,本色
natural shoulder 正常肩型,自然肩型,不垫肩式
natural silhouette 自然型轮廓,自然型轮廓线条
natural silk 蚕丝,真丝,天然丝;真丝绸
natural silk-like 仿真丝[的]
natural straw 天然编织草
natural taste 天然[纤维]风格
natural tinted fabric 天然本色织物
natural tone 自然色调
natural tone buckle 云花带扣,云花扣
natural twist 天然转曲
natural waistline 自然腰节线
natural wave 天然卷发
nature 性质,性格,特性;自然,自然界,自然态;真人裸体模特儿
nature colour 自然色,本色
nature fabric 本色织物
nature inspection 性能检查
naturell 内丢里尔内衣细平布
nature pattern (回归)自然花样,自然写实花样
nature stylized 自然风格
naught duck 零号帆布
Nauker 诺克尔绿色斜纹布,乌兹别克斯坦绿色斜纹布
Nautica 啄木鸟
nautical blouse 水手领罩衫;女兵式女衫
nautical clothing 航海服装
nautical design 航海风格图案
nautical look 船员款式,海员风貌;海员服造型,海员服风貌
Navajo blanket 美国纳瓦霍手织花纹厚毛毯
Navajo prints 美国纳瓦霍印花花纹,印第安式印花纹
Navajo rug 美国纳瓦霍手织花纹厚毛毯
Navajo Wool 美国纳瓦霍羊毛
naval 海军帽
naval cap 海军帽
naval lace 海军服饰带,军服金饰带
navette 梭子形(珠宝)
Navsari cotton 印度纳夫萨里棉
navy [blue] 藏蓝,藏青,海军蓝,深蓝
navy canvas 海军亚麻帆布
navy cloth 海军[制服]呢,细制服呢
navy coating 海军呢,细制服呢
navy look 海军款式
navy pea coat 海军水手外套
navy serge 海军哔叽
navy stripe 海军条纹
navy striped single jersey 海军条汗布
navy style 海军款式
navy suiting 海军呢
navy twill(ed) flannel 藏青斜纹厚法兰绒
Nawar 印度手工织制棉带
Naxi ethnic costume 纳西族民族服饰
naxin 典礼时首领披的毛毯
nazca embroideries 秘鲁古代绣品
ndargua cotton 西非那古棉
neadend 布头
neal seal 充海豹皮
Neapolitan hat 那不勒斯锥形高帽
near-bias edge 近斜丝绺的边

near-infrared 近红外[的]
near-infrared radiation 近红外线辐射
near-infrared ray 近红外线
near-seal 充海豹皮
near-ultraviolet ray 近紫外线
near-white 近白色
near-gem 近似宝石级
near-silk 丝光衬里棉布
neatened seam 项缝,领缝,修毛缝,整洁缝
neatening 修整,修补,光整;防松散
neatening neck facing 修整领贴边
neat fashion 清爽时装,整洁雅致的时装
neat finish 光边整理
neatly dress 衣着整洁
neat silk 加捻丝线
neat's leather 牛皮[革]
neat writing 字迹工整
necanee 印度尼卡尼条子平布
necessaire box 男子随身用品盒
neck 领,领口;颈,颈围;领围,领圈
neck apart 领圈宽
neck area 领部;颈区
neck around 领窝,领围
neck badge 领章
neckband (装饰用)领圈;领巾;衬衫领;领口;立领;领圈带
neckband shirt 饰狭窄领带的无领衬衫
neck base 颈根,蝴蝶结,领花
neck base girth 颈根围
neck base line 领口;领圈线;领根围线
neck breadth 颈宽
neck chief 围巾,领巾
neckcloth 颈巾,领巾;领饰;(旧式)领结
neck dart 领省,颈省
neck depth 领深
neck depth line 领口深浅,领深斜线,领深,前领深线
neck drop 领深
necked button 颈扣
necked yarn 竹节纱
necker (领和前胸一部分连在一起的)套领
neckerchief 围巾;领巾;颈巾;妇女颈饰
neckerchief slide 领巾环
necker dress 套领装(领和胸部连在一起的针织套领的连衣裙)
necker layered 假叠层领圈,圆高领
neck facing 领面
neck fashioning 领圈成形

neck flap 护颈
neck fleece 领带
neck formation 细颈形成,肩颈形成
neck girth 颈围;领弯
neck guard (盔甲的)颈甲,护颈
neck handkerchief 围巾,三角头巾,领巾
neck height 领高
neck hole 领圈,领口,领孔
neck hook 领钩,风化扣
neck hook and eye 风纪扣;领钩
neck-in 卷内边,缩幅
necking phenomena 颈缩现象
necking zone 颈缩区
neckkerchief 领巾
necklace 项链,项圈,项饰
necklace clasp 项链卡子
necklace pendant 项链垂饰
necklet 皮围巾;装饰围领;小项圈,项链,项饰
neck level 颈围线
neckline 领口,开领,领线,颈窝线;领圈
neckline-and-placket band 连领门襟滚镶边
neckline balance 领圈线对位
neckline curve 领圈弯尺
neckline dart 领口省;领孔褶
neckline drape 领圈垂褶
neckline drat 领口省(领口颏)
neckline edge 装领线
neckline gap 领圈太大
neckline opening 领围
neckline placket 明门襟,半开襟,半开襟领口
neckline shape 领口形状,领圈形状
neckline sort 领口种类
neckline styling 领口式样
neck opening 领圈,领口,领窝;领脚长
neck piece 领圈,领座贴边;颈饰,毛皮围巾;颈甲(盔甲)
neckpiece 领饰,领圈,围脖,皮围巾
neck point 颈肩基点,顶顶,颈点,领点
neck point height 领尖高,领尖长
neck point to breast point 肩颈点至乳峰长
neck rib 领口罗纹
neck rib height 罗纹领高
neck rib width 领口罗纹宽
neck ring 环圈项链
neck ruche 女式褶裥饰边围巾
neck ruff(ling) 褶裥领

neck run 领圈弧线
neck scarf 围巾,领巾,颈巾;妇女颈部服饰;领带;领饰
neck seam 领缝
neck seaming 领缝,项缝
neck seam line 领缝线,颈缝线
neck shoulder point 领肩点
neck size 颈围
neck stock 宽大硬领带(18～19世纪)
neck tab 领(口)袢
neck tap 领(口)袢
neck temperature 肩颈温度
necktie 斗牛领巾;领带;领结
necktie clip 领带夹,领带别针
necktie fabric 俄罗斯领带绸
necktie holder 领带杆,领带夹
necktie in the first button 端纽领带
necktie interlining 领带衬
necktie knitter 领带编织机
necktie lining 领带衬绸
necktie material 领带绸
necktie pin 领针
necktie silk 领带绸
necktie with zipper 拉链领带
neck to bust 颈至胸
neck to hem button 从领至下摆的纽扣
neck-to-hem closing 从颈至下摆的门襟
neck to knee hollow 颈至膝弯长
neck to waist 颈至后腰
neck to waist(anterior) 颈至腰长(前身)
neck to waist front centre 颈至腰部前中心
neck to wrist length 颈至腕长
neck-type zipper 领部拉链
neck vent 颈部开衩
neck waist 背长
neck waist length 后长;背长
neck-waist line 背长线,颈至腰长度
neckwear (总称)颈部服饰
neck wear 颈饰;围颈用品
neck width 领宽,领内宽
neck width line 背长线;领宽线
neck with dickie neckline 虚襟领衫
neck yoke 项圈;领座圈
nectarine 蜜桃色
nect pattern 净板;净样
need hierarchy theory 需求层次论
needle 针;缝衣针;编织针
needle and thread 针线
needle back 针后

needlebar (缝纫机)针天心;针杆;针把
needle bar connecting link adjusting screw 针杆连杆调节螺丝
needle bar crank 针杆曲柄
needle bar frame 缝纫机针杆座
needle bat 针柄
needle bight 机针弯曲
needle blade 针刃;针身,针杆
needle board 针毯烫垫
needle bonded fabric 针刺非织造布
needle book (书形)插针垫,插针布
needle breakage 断针
needle cage 针盒
needled carpet 针刺地毯
needle chew 针疵
needle clamp 针夹;针轧头
needle cooling 缝针冷却
needle cords 优质灯芯绒;细条纹光面呢
needle count 成品编织密度;针号
needle craft 刺绣技巧;缝纫;刺绣
needle cut 扎断纱;针疵
needle cutting 针损
needle cycle 针迹循环
needle damage 针损
needle density 针刺密度
needle detector 验针器
needle(d) fabric 针刺织物,针刺非织造布
needle(d) felt 针刺毡
needled floor cover 针刺地毯
needle dip 针迹
needled mat 针刺毡
needled nonwoven 针刺[法]非织造布
needled paper-making felt 针刺造纸毛毯
needle driving bar 针杆
needled top felt 针刺上毯
needled weftless felt 针刺无纬毛毯
needled wet felt 针刺造纸湿毯
needled wool sweater 针织羊毛衫
needled woven felt cloth 针刺毡合织物
needle eye 针眼
needle feed 针送料
needle feeding mechanism 机针送料机构
needle feed sewing machine 针送料缝纫机
needle felt 针刺毡
needle-felted carpet 针刺毡毯
needle felt loom 针刺制毡机
needle fabric 针刺织物
needle foot 缝纫机压脚

needle for sewing machine 缝纫机针
needle front 针前
needle-full 穿在针上一次所用的线
needle gauge 针幅
needle groove 针槽
needle guide 导针器
needle heating 针发热
needle holder 针夹
needle hole 针眼,针孔
needle impingement 机针冲刺,针刺
needleizing 易缝纫处理
needle lace 针绣花边
needle line 条花疵
1-needle lock stitch 单针锁边缝迹
2-needle lock stitch sewing machine 双针锁缝机;双针缝线迹缝纫机
needle loom 导纬针织带机;针刺机
needle loom carpeting 针刺机制地毯
needle loom for ceramic fiber 陶瓷纤维针刺机
needle loom for glass fiber 玻璃纤维针刺机
needle loom selvage 针织边;针刺窄织物布边
needle loop 针编弧
needle loop interlacing 针编弧串套
needle loop transfer 针编弧移圈
needle lubricator 缝针润滑器
needle machine 单针刺绣机
needle-made lace 针绣花边
needle of size 14 14号针
needle-out structure 抽针结构
2-needle overedger 双针包边机
needle penetration 针刺,针刺穿刺
needle pitch 针距
needle plate (缝纫机)针板
needle plating socks 闪色花袜
needlepoint 针尖;针绣花边,刺绣品
needle-pointed tracer 针状轮点线器
needlepoint embroidery 十字布或网眼布刺绣,斜针绣法,点刺绣法
needlepoint fabric 细结子女衣呢,细珠皮女衣呢
needlepoint lace 针绣花边,钩针花边
needlepoint stitch 针绣针法
needle position 缝针定位[线]
needle positioner 缝针定位器
needle position misalignment 缝针定位线偏离
needle position width 缝针定位幅度

needle-punched carpet 针刺地毯
needle-punched fabric 针刺织物
needle-punched felt 针刺毡,针刺毛毡,针刺呢
needle-punched filter felt 针刺过滤毯
needle-punched nonwoven 针刺法非织造布
needle-punched synthetic leather 针刺合成革
needle-punched web 针刺纤[维]网
needle punching 针刺法
needle-punching nonwoven 针刺法非织造布
needle-punching shoulder pads 针刺肩垫
needle rod 针杆
needle run lace 溜针花边
needle scarf 针槽
needle sidewise movement 针侧向运动
needle size 针号
needle spacing 针距
needle stroke 针刺冲程;针刺动程
needle tapestry 机绣仿挂毯织物
needle tapestry work 针缝织锦绣;针绣挂毯
needle thread 针线,上线,面线
2-needle 4-thread 扎骨机
needle thread demand 面线需要量
needle threader 引线器
needle thread eyelet 穿线器孔
3-needle 4-thread flat seam 三针四线绷缝
2-needle 3-thread interlock stitch 双针三线绷缝线迹
3-needle 5-thread interlock stitch 三针五线绷缝线迹
3-needle 6-thread interlock stitch 三针六线绷缝线迹
needle thread knife 针线切刀
2-needle 3-thread lock stitch 双针三线锁式缝迹
needle thread loop 针线环
needle thread monitor 面线监控器
3-needle 4-thread multi-chain stitch 三针四线多链式线迹
needle thread nipper 针线夹
2-needle 4-thread overedger 双针四线包边机
1-needle 3-thread overedge stitch 一针三线包边线迹
needle thread pull-off 针线拉钩

2-needle 5-thread safety stitch 双针五线安全线迹
needle thread supply 面线供给量
needle thread tension 针线张力
needle thread tension spring 夹线簧
needle tip 针尖
needle toe 夹鞋头,尖头
needle-toe shoes 尖头鞋
needle trade[s] 服装工业,成衣业
needle-woman 女裁缝;女缝纫工
needlework 女红,裁缝;针线活;刺绣活;缝纫业
needlework button 包纽
needlework tapestry 刺绣壁画,刺绣挂毯
needling and stitch-bonding technique 刺编技术
needling density 针刺密度
needling depth 针刺深度
needling intensity 针刺密度
needling interlining 针刺棉(保暖絮片)
needling machine 针刺机
needling machine with down and up stroke 上下针刺机
needling stroke 针刺冲程;针刺动程
needling width 针刺幅宽
Nega-stat 皮芯型复合抗静电防污纤维(芯为碳纤维,皮为聚酯纤维,商名,美国杜邦)
negative air pressure 负气压
negative and positive effect 阴阳效应,阴阳效果
negative contrast 负对比
negative heel shoes 负跟鞋
negative making 制负片
negative mark 否定号
negative prints 花纹留白
negative requirement 否定需要
negative thermal fabric 消极式保温织物
neghelli 尼盖利菽麻粗布
négligé 女式长睡衣;便服
negligee 宽松便服;休闲装;女式长睡袍,女长睡衣;不整齐的穿戴
negligee costume 便服,家居服
negligence 风格粗松
negotiating bank 议付银行
negotiation 议付
negotiation credit 议付信用证
negrepellis 粗纺长毛绒;黑色缩绒织物
negritude 黑色
negro cloth 大麻粗布

negro cotton 西非棉
Negroid skin 黑人皮肤
Nehru 尼赫鲁上装
needle thread supply 面线供给量
Nehru cap （印度）尼赫鲁帽
Nehru clothing [jacket] 尼赫鲁外衣
Nehru coat 尼赫鲁上装
Nehru collar 尼赫鲁领（立领）
Nehru hat （印度）尼赫鲁帽
Nehru jacket 尼赫鲁上装（立领,印度男子的日常服）
Nehru neckline 立领领口
Nehru style 尼赫鲁风格;尼赫鲁款式
Nehru suit 尼赫鲁式服装;尼赫鲁套装
neige 白色效应
Neigelli 印度尼盖莉菽麻粗布
neigeuse 斑点纹粗呢
Neighborhood Cleaners Association 社区洗衣协会
neighborhood style 地区风格
Nelson coat 尼尔逊外套
nemed 嵌花地毯
Nemes head covering 尼姆巾
nemesit 尼密赛头布(古埃及)
nemuriana 睡扣眼;睡扣孔(不开鸽眼孔)
nenneko banten 日本式背幼儿用外套
neoclassic style 新古典主义风格(18~19世纪)
neoclassical look 新古典主义风格(18~19世纪)
neoclassicism 新古典主义
neo-dadaism 新达达主义
neo-folklore 新民间款式
neo-hippie look 新嬉皮士风貌
neo-impressionism 新印象派;点彩派
neo-modernism 新现代主义型,超摩登型
neon color 霓虹灯色
Neopolitan hat 那不勒斯锥形高帽
neo-pop 新波普艺术
neo-povertism 新贫民风格式样
neoprene 氯丁橡胶
neoprene coated fabric 氯丁橡胶涂层织物
neoprene raincoat 塑胶雨衣
neoprene rain jacket 塑胶短雨衣
neoprene suit 氯丁橡胶服
neo-psychedelic art 新幻觉艺术
neorealism 新写实派;新现实主义
neorealist 新现实主义者
neo-romantic 新浪漫主义
neo-romanticism 新浪漫主义

neo-traditionalism 新传统派	nether stocks (16世纪)超膝袜子
neo-transparence 新透明款式	nether yarments 裤[子]
nep 棉结,白星;毛粒;麻粒	net kint fabric 针织网眼织物
nep and moits 棉结杂质	net knitted fabric 针织网眼布
nep cloth 棉结纱织物	net knotting machine 结网机
nephila silk 蜘蛛丝	net lace 薄纱花边,丝网花边,网眼花边
nephrite 软玉	net length 净长
neplune green 浅粉绿[色]	net leno 网眼纱罗
nep pattern 毛结花纹,集圈花纹	net-like desigh 网纹
nep potential 拉毛性能,起毛潜力	net-like effect 网眼效应
nepprene raincoat 塑胶雨衣	net look 简洁款式
neppy fabric 多粒结织物	net mass 净重
neppy web 多粒结纤维网	net pattern 净样;格子花纹
Neptune green 海洋绿,浅粉绿	net roller 网带
nep tweed 粒结花式粗呢	net sales 产品销售净额
nep yarn 结子纱;粒结花式纱线	net silk 加捻生丝纱,生丝加捻丝线
nep yarn coating 结子线衣料;粒结花线大衣料	net socks 网眼袜
	net stitch 网眼针法
nequen cloth 内昆布,中美洲产龙舌兰纤维织物	net stockings 网眼长筒袜
	net suke 坠子
neri silk 蚕衣丝,蚕衣	net sweater 网状毛衣
nerve 回缩性;复原性	netting 结网,绳网;网状织物
nessel garn 德国荨麻纱	netting needle 结网针
nessel tuch 德国荨麻织物	netting silk 编结丝线
nessu 奈恩苏克布	netting thread 结网线;编网线
nested layout 嵌套式样板排料	nettle cloth 荨麻织物;棉漆布;苎麻织物
nested pieces 嵌套式样片	nettlestuff 海军吊床用粗麻绳
net 网眼织物,网状物,绣花边地网	net weight 净重
net aid design (NAD) 网络辅助设计	network 网络;网眼织物
net baby doll 新洋娃娃款式(小领,紧袖,装饰斜截荷叶边)	network fiber 网状结构纤维,网络结构纤维
net band 网巾(中国明代男子束发用)	network method 统筹方法
net binding 网包边	network yarn 发泡丝
net-bound seam finish 包网缝份处理	network yarn fabric 网络丝织物
net canvas 刺绣十字布	net yards 无疵布匹净长
netch 领豁口	net yarn 结网纱
netcha 尼枪外套(爱斯基摩)	neuilly 仿哥白林双面挂毯
net curtain fabric 网眼窗帘布	neutral 非彩色,中立色
net draft (无放缩的缝纫)净样图	neutral acid dyes 中性浴染色酸性染料
net effect 网眼效应	neutral color 不鲜明色;与灰色相协调的颜色;中和色;中间色
neteldoek 苎麻织物	
net embroidery 网眼刺绣,网眼纱作底布	neutral dye 中性染料
net eyelet 网形鸡眼	neutral gray 中性灰
net fabric 花边网眼纱,网状织物	neutralization 中和
net gauze 龟甲纱,六角网眼纱	neutral pack 中性包装
net ground 绣花网眼地布	neutral packing 中性包装,无商标无厂名的包装
nether garments 裤子	
nether integuments 男裤	neutral point 非彩色点(色彩学用语);中和点,中性点
nether lip 下唇	
nether stockings 超膝袜子(16世纪)	neutrals 中性色(冷暖色之间的色泽)

neutral step wedge 灰色分级比色楔样
neutral tint 灰色,青灰色,暗淡色,不鲜明色
neutral tone 中间色调
neutron-shielding fiber 防中子辐射纤维
never press 免烫,无需熨烫
never shrink 防缩整理
never shrink finish 预缩加工
Never-press 免烫整理(商名)
new adult 新成年人(指20岁左右的男性)
New Age 新时代
new age punk 新时代朋克服饰
new arrival 新款时装
new baby doll 新洋娃娃
new baggy jeans 新袋形牛仔裤
NEW BALANCE 纽巴伦
new black 新黑色
new blue 新蓝色
new breed 新型发式
new button down collar 新型扣领尖领(用纽扣固定宽领)
new Chinese style 新中装
new city image 新都市形象
new classic 新传统款式
new clips 裁剪下脚碎布,碎呢;新碎布;新碎呢
new clothes 新装
New Cotton 免烫新棉布(商名)
new cuttings 裁剪下脚碎布,碎呢;新碎布;新碎呢
new establishment 新体制派
new face 新脸(时装模特儿)
new family 新家庭派(指70年代以来选择新的生活方式的青年,服装市场用语)
new feather 新鸟羽发型
new feel fiber 新触觉纤维,新感觉纤维
new formal 新正式服装
new frontier fiber 新型尖端纤维,新领域纤维
new generation synthetic fiber 新一代合成纤维,新合纤
new ivy look 新常春藤风貌
new ivy stylistic form 新常青藤款式
new jacquard poult 新纹绨
new layered 新叠层式款式
new layered style 新叠层式款式
new life fashion 新生活服饰
new life style 新生活方式

New Look 迪奥新风貌(1947),新风貌
new look 新面貌;新式样,新款式,新潮流
new material look 新材料款式
new material style 新材料款式
new material technology 新材料技术
New Mexico wool 新墨西哥羊毛
new military style 新军装款式
new mini 新超短裙
new model 新款式
new op art 新光效应绘画艺术
New Orleans cotton 美国新奥尔良棉
new packing material 新包装材料
new product 新产品
new product development 新产品开发
new production development 新产品开发
new salary style 新工薪族款式
new sample making 打新样
new sexy 新性感
new structural hair sweater 新结构毛衫
new structural knitwear 新结构毛衫
new structural woolen sweater 新结构毛衫
new stuff dress 新材料装
new style (n.s.) 新型,新式,时髦式样
new sweathearts [neckline] 新情人领口线;新爱心领口线
new technique 新技术
new technology 新工艺,新技术
new tent line 新篷帐型(20世纪70年代末)
new thirties 刚30岁的青年
new trad 新传统装,新的美国传统型服装款式
new traditional 新传统装;新的美国传统型服装款式
new trend 新趋势
new T-shirt 新T恤
new type cutting method 新式裁剪方法
new type sewing method 新式缝制方法
new wave 新潮流,新浪潮,新趋势
new white shirt 织花白衬衫
new wool 新羊毛
New York Cotton Exchange 纽约棉花交易所
New York Fabric Show 纽约面料展,纽约织物展
New York Premier Collections 纽约先导展(美国)
New York Prét 纽约成衣展(美国)

New York Textile Group　纽约纺织集团,纽约纺织联合会
New zealand flax　新西兰麻
New zealand hemp　新西兰麻
New zealand twill　新西兰黄麻斜纹粗袋布
New Zealand wool　新西兰羊毛
New Zealand Wool Board　新西兰羊毛管理局
newar　手织棉带
newborn baby　初生婴儿
Newbury coat　纽贝里[单排纽]外衣
New-Edwardian look　新爱德华风貌
New-Edwardian style　新爱德华风格
new-fitted silhouette　新型合身轮廓
newmarket[coat]　女式长大衣,紧身外衣
new-military look　新军装款式
newness　花绉绸
newness retention　保持永新性能
Newport slipon　纽波特外衣(室内用宽袖对襟女外衣)
Newport-pagnel lace　梭结花边
news conference　新闻发布会
newsbag　报刊包
newsboy cap　报童帽,报童鸭舌帽(帽身平顶呈蓬松状)
newsy　新颖别致的女式服装
newsy dress　新颖别致女服
Nextel　铝硼硅氧化物纤维(可耐热至1300℃,商名,美国明尼苏达矿业)
next-to-skin　贴身的
next-to-skin finish　贴肤整理(用于提高织物贴肤穿着舒适性)
next-to-skin wear　贴身服装,贴身衣着
next to the skin　贴身
neyge　装饰边
nguine cotton　西非泛红棉
niagara　尼亚加拉绉布
niantic foot　双缝袜脚,成形袜脚,收放针袜脚
niantic heel　无缝袜跟片
nicalon　碳化硅纤维(商名,日本碳公司)
nicanie　英国尼卡尼条子垫褥布
nice curler　有刷卷发器
niced　围巾;裹胸布
nick　刻痕
nickel　镍灰色(红光浅灰色)
nickel-coated carbon fiber　镀镍碳纤维
nickel green　镍绿色,浅暗绿色
nickerbockers　灯笼裤,女用扎口短裤

Nicole Groult　尼科·格鲁特(法国高级时装设计师)
nid d'abeille　蜂窝结构图案
nien　中国北方绢丝
nifty clothes　时髦服装
Nigel Preston　奈杰尔·普雷斯顿(英国时装设计师)
Nigeria Allen　尼日利亚艾伦棉
Nigerian costume　尼日利亚民族服饰
Nigerian cotton　尼日利亚棉
nigger[brown]　鼻烟色(深棕色)
nigger head　织物起球
niggerhead　疙瘩花纹;卷曲花纹织物;织物起球疵点
niggerhead curl　混纺珠皮呢
night attire　睡衣
night blue　夜蓝
night cap　睡帽
night chemise　贴身睡衣
night cloak　旅行斗篷;夜用斗篷
night clothes　睡衣,睡袍,夜间家常服
night cream　夜霜
night drawers　睡裤
nightdress　(女、童)睡衣;睡衣,睡袍
night emulsion　晚乳液
night full dress　晚礼服,夜礼服,燕尾服
nightgown　(女、童)睡衣;睡袍;妇女长睡衣;夜间室内衣;舒适女装;抵肩
night-gown top　领肩
nightie　女睡衣,小睡衣,睡袍
night joggers　慢跑运动员夜用服装,夜行服
night rail　女式睡衣,女式宽松晨袍
night rayl[e]　女式睡衣,妇式晨衣
night robe　睡衣,睡袍
nightshade　颠茄色
night shift　衬衣式睡袍;夜班
nightshirt　长及脚踝的衬衣式睡衣;男长睡衣;男睡衣
night suit　睡服;睡衣;晚礼服
nighttime wear　晚间家常服,晚装
night vision　夜视觉
night wear　(总称)睡衣;晚服;夜间家常服;睡衣
nighty　(女、童)睡衣
nighty-suit　睡衣
nigrescence　黑色(皮肤、头发、眼睛等)
nihau　夏威夷垫子
Nike　耐克
nikerie cotton　南美内可瑞棉

Nikita Godart 尼基塔·戈达尔（法国服装设计师）
NIKKO 日高
Nile bule 尼罗河蓝,淡绿蓝色
Nile green 尼罗河绿,浅青绿色
nilla 尼拉丝麻交织绸
nilsaria 尼尔萨丽蓝点纹条格布
nimbus gray 雨云灰色
nimbus grey 雨云灰色
Nimes 尼米斯呢,匹染精纺平纹呢
Nina Ricci 尼娜·丽西（1883~1970,法国时装设计师）
nine 9号尺码的衣服
nine heads theory 九头高理论
NINE RICCI 莲娜丽姿
nine-stap-scale 九档色卡
NINETEEN 《十九岁》（法）
nineteen 19号尺码
nine-tenths [length] 九分长
ninety 90号尺码
ninety days cotton 美国90日棉
Ningbo tailor 红帮裁缝
Ningbo tailor's shop 红帮店堂
Ninghai cord 中国宁海双丝绸
NINI WEST 玖熙
Nino Amalfi 尼诺·阿马尔菲（法国时设计师）
ninon 尼龙绸;薄绸;英国薄绸
ninon volie 化纤仿巴里纱
niobium carbonitride fiber 碳氮化铌纤维
nip creases 小折边
nip dyeing 面轧染色
nip tuck 节粒裥饰
nipis 剑麻平纹织物
nipped edge 起伏波状边
nipped-in 腰部很紧的,紧身的,贴身的
nipped-in waistline 紧箍的腰线,扣紧腰线
nipped-up edge 封边
nipped waist 紧腰;掐腰
nipple 乳头[点]
nipple area 乳头区
nipple breadth 乳间宽
nipple breath 乳[间]宽
nipple line 乳头线
nippers 夹鼻眼镜;镊子;钳子
nippongee 仿柞丝府绸
Nippon Kaiji Kentei Kyoka (NKKK) 日本海事鉴定协会
Nishang silk 霓裳绸

nishijin [brocades] 京都织锦缎
nishiki 日本金银丝锦缎
nite cap 有网睡帽
Nitis rug 尼里斯地毯,波斯羊毛地毯
Nitivylon 聚乙烯醇缩醛工业用丝（商名,日本）
nitrile alloy fiber 腈合金型纤维（聚丙烯腈系混抽纤维）
nitrile-butadiene rubber(NBR) 丁腈橡胶
nitrile rubber 丁腈橡胶
nitrile synthetic fiber 腈系合成纤维
nitrocellulose rayon 硝酸纤维素人造丝,纤维素硝酸酯人造丝
nitrocellulose silk 硝酸纤维素人造丝,纤维素硝酸酯人造丝
nitro-cotton 硝化棉（用棉短绒制的硝酸纤维素）
nitrogen fading 氮气褪色
nitro-hydrocellulose 硝基水解纤维
nitro-rayon 硝酸[法]人造丝
nitrosilk 硝酸[人造]丝
nitroso dye 亚硝基染料
NIVEA 妮维雅
Nivion 聚酰胺纤维（商名,意大利埃尼）
no back lining 前夹后单（服装）
Noba cotton 苏丹努巴棉
no bleach 勿用漂白剂;无漂白
noble look 高贵风貌,高贵款式
noble metal 贵金属
noble wool 英式精梳用羊毛
Noboru Yamafuji 山藤升（日本服装设计师）
no-bounce support 防震护托
no-bra bra 无胸罩胸衣（穿着舒适,不强调胸部造型）
no-bra look 袒胸式,无胸罩款式
no-bra style 无胸罩款式
Nobuo Nakamura 中村武夫（日本服装设计师）
Nobuyuki Ota 太田伸之（日本服饰界评论家）
no contrast 无色差
no-core braid 无芯编带
no-crease wool 免烫毛织物
no-crush finish 无折绉整理,抗皱整理
nocturne satin 夜景缎
no-cut sew 无裁缝服装（20世纪70年代后期）
no-dart garment 无省服装
node 节;节点;节瘤花线衣服;英国结子

线织物
node cloth 节瘤花线织物；英国结子线织物
no dress 请穿便服（邀请帖之客套语）
Noele 抗静电聚丙烯腈短纤维（商名，日本）
no-element fabric 非织造布
no fixing 衬里下摆不缝合
no-form 无固定形服装
nogs 大麻
noh costume 日式丝缎演出服
no heel 无跟鞋
no-heel shoes 无跟鞋
noil 精梳短毛；精梳落毛；落棉；落麻；落毛
noil cloth 绵绸丝织物，绵绸；䌷丝织物，䌷丝绸
noil poplin 绵绸，绵绸丝织物；䌷丝织物；䌷丝绸
noil silk 䌷丝（绸），绵绸
noil silk yarn 䌷丝
noil silk yarn fabric 䌷丝纺织物
noil stripes 䌷丝条子布；精纺短毛纱条子呢
noil yarn 短纤维纱
noily wool 含短纤维低级羊毛
noir de vigne 葡萄黑
no-iron 免烫；免烫织物
no-iron cotton 免烫棉布
no-iron finish 免烫整理
no-ironing 免烫
no-ironing suit 免烫服
no-iron shirt 免烫衬衣
noise criteria 噪声标准
noise jewelry 金色金属饰物
noise pollution 噪声污染
noise standard 噪声标准
noise-free performance 低噪声性能；低噪声作业
noise-free process 低噪声作业
noisette 榛子褐
noisy 色彩过分艳丽；服装过分鲜艳
no lining 无里衬
nomad carpet 诺曼德地毯，波斯羊毛地毯
nomad fashion 游牧民族款式
nomad look 牧民款式,游牧民族风貌
Nomex 诺梅克斯（芳族聚酰胺耐高温纤维，美国）
Nomex nylon 诺梅克斯（芳族聚酰胺耐高温纤维，美国）
nomenclature 名称,命名,术语
Nomex endless blanket 诺梅克斯环状毯（耐高温）
nominal count 名义支数,公称支数
nominal number 名义支数,公称支数
nominal count of hand knitting yarn 绒线名义支数
nominal denier 公称旦数
nominal machine width 机器公称宽度
nominal size 公称尺寸
nominal tex 公称号数
nominal width 公称幅度,名义门幅
nomud carpet 嵌花毡毯
non battue 法国低级亚麻帆布
non button coat 无扣女外衣
non cling 防静电加工
non color 无彩色（非彩色）
non color jacquard 非彩色提花
no necktie 不打领带
non iron 免熨烫
non jacquard fabric 非提花织物
non jeans 不像牛仔裤的牛仔裤,似牛仔裤[的牛仔裤]
non perfect register 对花不准
non protecting duty 非保护关税
non regular requirement 不规则需求
non suit 变化套装,搭配成套的服装,轻便套装
nonagon 九边形；九角形
non-apparel fabric 非服用织物
non-assertive colour 不确定色
Nonbalon 阻燃性黏胶纤维（商名,日本兴人）
non-bra bra 隐形胸罩；无胸罩感的胸罩
non-bulk theory 无凸出部位理论
nonbulky sweater 紧身卫生衫,紧身毛衫
Nonbur 阻燃聚丙烯腈纤维（商名,日本旭化成）
non-cellulosic fiber 非纤维素纤维
non-cellulosic synthetic fiber 非纤维素合成纤维
nonchlorine bleach 非氯漂白
nonchlorine retentive finish 耐氯整理
noncircular fiber 异形纤维,非圆形截面纤维
noncircular filament 非圆截面丝
noncoated fabric 非涂层织物
non-cohesion silk 无抱合生丝
non-collar shirt 无领衬衫

noncolor 非彩色
non-color jacquard 素色提花，组织[结构]提花
non-combustibility 不燃性
non-combustible fabric 不燃性织物，非燃性织物
non-conductive fiber 不导电纤维，绝缘纤维
non-conforming item 不合格产品
non-conformity item 不合格品
non-contact 非接触式
non-contact measurement 非接触测量法
non-convertible collar 单用领
non-corrosive fiber 不腐蚀纤维
non-crease 抗皱
non-crease fabric 抗皱织物
non-crease rayon 抗皱黏胶丝绸，抗皱人丝绸
non-creasing finish 抗皱整理
non-crushable linen 抗皱亚麻布
non-cuff shirt 无袖头衬衫
non-cylindrical filament 非圆形[截面]丝
non-defection article 合格品
non-deformed fiber 未形变纤维
non-directional fabric 无方向性面料；无方向性织物
non-directional fused interlining 无方向性热熔衬
non-directioned cloth pattern 织物无方向图案
non-discoloring 不脱色，不变色
non-dude 不讲究打扮，服饰上不讲究
nondurables 不耐用物品
non-dyeing fiber 无染力纤维
nonelastic fabric 无弹性织物
nonelastic webbing 无弹性带子
nonelastic woven tape 无弹性织带
nonfading 不褪色
nonfelting property 不缩绒性
non-fire-retardant fabric 无阻燃剂织物（纤维本身有阻燃作用）
Nonfla 阻燃性聚丙烯腈短纤维（商名，日本东邦人造丝公司）
non-flame fiber 不燃性纤维
non-flame property 不燃性，阻燃性
non-flammability 不燃性
non-flammable fabric 难燃织物
non-flammable textile 阻燃纺织品
non-fogging glasses 防雾眼镜
non-formaldehyde DP finish 无甲醛耐久压烫整理
non-formaldehyde finishing 无甲醛整理
non-formaldehyde resin finishing 无甲醛树脂整理
non-glitter effect 无极光效应
non-homogeneity 不均匀性
non-hydroscopic property 不吸湿性
non-ignitability 不着火性
non-industrial clothing design 非工业性服装设计
non-inflammability 不燃性
noniron finish 免烫整理
nonironing 免烫
nonironing fabric 免烫织物
non-irritating 无刺激性
non-jacquard fabric 针织变换组织织物，非提花针织物
non-jeans 仿牛仔裤的裤子
non-keratinous fiber 非角蛋白质纤维
non-knit fabric 非针织织物
non-language exchange system 非语言交流系统
non-lock 未锁边
non-lock slider 无锁拉链头
non-luster wool 无光羊毛
non-marker economy 非市场经济
nonmelting cross-linked fiber 不熔融交联纤维（如酚醛纤维）
non-mercerized cotton fabric 本光布
nonmetal material 非金属材料
non-metameric match 无条件等色配色
non-metric information 非定量情报
nonneedle knitted fabric 无针针织布
nonobjectivism 艺术的抽象主义，抽象艺术
non-open end spinning 非自由端仿纱
non-oriented fiber 未取向纤维
non-pareilles 仿驼毛呢
non-perfect register 对花不准
non-petroleum synthetic fiber 非石油型合成纤维
non-pile floor covering 无绒头地毯
non-pile knitting carpet 无绒头针织地毯
non-pilling finish 不起球整理
non-polar fiber 非极性纤维
non-polluting 无污染
non-press cloth 免烫衣服
nonproductive personal 非生产人员
nonproductivity 非生产性

non-random cross-linked polymer fiber 非无规交联聚合物纤维(梯形聚合物纤维)
non-ravel 袜口不脱散
non-ravelling protion 不脱散部段
non-ravel top 防脱散袜口
non-reciprocal treatment 非互换待遇
non-repeating design 不重复光纹
non-reversible fabric 异面织物,有正反面织物
non-round filament 非圆形[截面]长丝
nonrun 不脱散,不抽丝
nonrun finish 防脱散整理,防抽丝整理
nonrun hosiery 防脱散针织物;防脱散袜子
nonrun stockings 防脱散长袜
nonrun top 防脱散袜口
non-separating zipper 不分离拉链
non-set yarn [假捻]未定形丝,弹力丝
non-sew covered snap 非缝式包布按扣
non-sewn selvage 漏缝[布]边
non-sew snap fastener 非缝式按扣
non-shattering glasses 不碎眼镜
non-shift finish 防滑移整理
non-shrink 抗缩,防缩
non-shrinkable wool 防缩羊毛,氯化防缩羊毛
non-shrinkage 防缩性
non-shrink treatment 防缩处理
non-sinkable suit 救生衣
non-sizing yarn 交络丝,免浆丝
non-skid latex back 地毯防滑衬垫
non-skinned fiber 无皮层纤维,全芯纤维
non-skip blindstitch 连续暗缝线迹,无空针暗缝线迹,无跳缝暗缝线迹
non-skip stitch 连续线迹,无空针线迹,无跳缝线迹
non-sleeve blouse 无袖女衫
non-slip driving shoes 防滑驾驶鞋
non-slip finish(ing) 防滑整理,抗滑整理
nonsoil retentive softener 抗污柔软剂
non-splitting conjuagate fiber 非裂离型复合纤维
non-staining property 不着色性
nonstandard 非标准;非标准零件
nonstandard body conformation 非标准体型
non-stationary 非静止状态
non-stitch stitch 连续线迹,不空针线迹

nonstore retailing 无店铺零售
non-stretch 无伸缩
non-stretchability 无拉伸性
non-stretchable 不能伸缩的,无弹性的
non-stretch braided rope 无伸缩性编织绳索
non-stretch bulked yarn 非伸缩性[膨化]变形纱
non-stretch fabric 无弹性织物
non-stretch pants 无弹性裤子
non-stretchy 非伸缩性
non-stretch yarn 非弹力丝,非伸缩性丝
non-suit 变化套装,搭配成套的服装
non-symmetrical design 不对称花纹
nontariff barrier 非关税壁垒
non-thermal underwear 隔热内衣
non-thermoplastic filament yarn 非热塑性长丝
non-thermoplastic polymer fiber 非热塑性聚合物纤维
non-thermoplastic textured yarn 非热塑性变形丝
non-torque textured crimping 无扭矩变形长丝,无捻回变形丝
non-torque type coil like filament 无扭矩螺旋状卷曲
non-torque yarn 无捻回弹力丝,无捻回变形丝
non-toxic cotton 低酚棉,无毒棉
non-toxicity 无毒性
nontraditional suit 现代西装,现代服装
nonutility 不实用的(衣服等)
nonwax chalk 非蜡划粉笔
nonwoven 非织造的
nonwoven adhesive interlining 无纺黏合衬
nonwoven adhesive web 非织造黏合纤维网,非织造黏合织物
nonwoven bag for medical use 医用非织造布袋
nonwoven bonded wrapping 非织造黏合包装材料
nonwoven carpet 非织造地毯
nonwoven composite material 非织造布复合材料
nonwoven durables 耐久性非织造布
nonwoven fabric 非织造布,非织造织物;无纺织物,无纺布
nonwoven felt 非织造[布]毡
nonwoven filling material 非织造填充

材料
nonwoven flannelette 非织造黏合绒,非织造绒
nonwoven flocked cloth 非织造植绒布
nonwoven floor covering 非织造地毯
nonwoven quilting 绗缝被
Nonwoven Industry 《非织造织物工业》(月刊)
nonwoven interlining 非织造黏合衬,无纺布衬
nonwoven ironing felt 非织造熨烫毡
nonwoven-like film fiber 非织造薄膜纤维(用压纹技术制造的一种膜裂纤维)
nonwoven mat 非织造垫;非织造席
nonwoven medical padding for shoes 非织造药用鞋垫
nonwoven powder puff 粉拍纸
nonwoven quilt 绗缝被
nonwoven quilting 绗缝被子
nonwoven radiation-proof fabric 非织造防辐射织物
nonwoven roll goods 非织造布卷材
nonwoven rug 非织造地毯
nonwovens 非织造布,无纺织物
nonwoven sanitary shorts 非织造卫生短裤,旅游短裤
nonwoven scrim 非织造纱布
nonwovens cushion 非织造衬垫
nonwoven surgical gown and hat 非织造手术衣帽
nonwovens for automobile industry 汽车工业用非织造布
nonwovens for book spine 书脊用非织造布
nonwovens for meridian tire 子午线轮胎用非织造布
nonwovens for packing bag 包装袋用非织造布
nonwovens industry 非织造布工业
nonwovens machinery 非织造布机械
nonwovens mask 非织造口罩
nonwovens textile 非织造纺织品
nonwovens web formation 非织造布成网
non-yellowing 不泛黄
non-yellowing softener 不泛黄柔软剂
nop yarn 结子花线,异色结子粗纺毛纱
Nordic design 北欧式图案
Nordic Environmental Label 北欧环境标签(从纤维原料到纺织最终产品的生态标志)

Nordic sweater 北欧毛衫
no requirement (NR) 无要求
Norfolk cotton 美国诺福克棉
Norfolk Down wool 英国诺福克塘种羊毛
Norfolk jacket 诺福克外套(有腰带而前后有褶,单排扣,齐臀长)
Norfolk suit 诺福克套装,上衣背部打褶套装
Norfolk suiting 英国诺福克套装
Norihisa Ohta 太田记久(日本时装设计师)
Noriko Kazuki 香月娜里子(日本时装设计师)
norm 标准,规格,准则;定额
Norma Kamali 诺尔玛·卡迈利(1945～,美国时装设计师)
normal 正常;本色毛棉针织品
normal color vision 正常色觉
normal condition 标准状态,常规条件
normal figure 正常体型
normal fitted 正装合体衬衫
normal knitting 正常编织
normal length 标准长度
normal lens 标准镜头
normalization 统一化
normal nylon 锦纶66
normal pressure and temperature 标准压力与温度,常温常压
normal production program 正常生产程序
normal running 正常运转
normal seam 普通缝
normal sleeve 普通袖
normal socks 短筒袜,男短袜
normal straight stitch 普通直线迹
normal temperature and pressure 常温常压
normal trichromation 正常色觉
normal twill 正则斜纹,正规斜纹
[normal type] one-way fasteners 单向拉链
normal waistline 中腰点,自然腰线(普通腰线)
normandie 列韦斯织机生产的棉纱花边
Normandy lace 法国诺曼底花边
Normandy val 诺曼底谷机制暗花细花边
Norman embroidery 诺曼刺绣
Norman Hartnell 诺曼·哈特内尔(1901～1979,英国时装设计师)
Norman Norell 诺曼·诺雷尔(1900～

1972,美国时装设计师)
normothermia 正常体温
Norse blue 挪威蓝
North American Free Trade Agreement 北美自由贸易协定
North Carolina Textile Manufacturers Association 北卡罗来纳纺织生产者协会
Northampton[shire] lace 北安普敦群细网眼梭结花边
northern dozens 约克群密绒厚呢
northern jute 印孟产乌坦亚黄麻
northern muskrat 北美麝鼠皮;棕麝皮
Northerns cotton 印度北方棉
northern soul 大喇叭裤
Northern Stal [cotton] 美国北极星棉
northwester 海员长雨衣;海员风雨帽
northwest frontier province cotton 巴基斯坦西北边区棉
north wool 莱斯特边区绵羊毛
Norwegian costume 挪威民族服饰
Norwegian pattern 北欧情调的几何花纹,挪威花样
Norwegian yarn 挪威羔羊毛针织纱
nor'wester 海员长雨衣;海员风雨帽
Norwich crepe 诺里奇绉;丝棉交织绉绸;丝经毛纬细呢
Norwich grograine 诺曼奇细呢
Norwich shawl 诺里奇丝披巾
no-seam 无缝
no-seam garment 无缝服装
nose ornament 鼻饰
nose piece 护鼻
nose rag 手帕
nose ring 鼻环
nose veil 盖鼻面纱,(遮眼鼻)短面纱
no-sew snap 免缝按扣;四合扣;四件扣
no-sew snap setter 免缝钉扣器
no-sew snap tape 免缝按扣带
no-shape garment 无省无结构线的简单服装
no-side-seam bodysuit 无侧缝紧身衣
nosing 地毯梯级凸边
no-sleeve dress 无袖长裙,无袖长衣
no-sleeve turtle 高领无袖毛衣
nostalgia 怀旧,重新流行
nostalgic fashion 怀古服式,怀旧款式;怀古服装
nostalgic feeling 怀旧情思
nostalgic look 怀旧风貌,思乡风貌,思乡款式

nostalgic yearning 怀旧情思
no-swell finish 抗溶胀整理
no tailor 无裁缝服装(20 世纪 70 年代后期)
no-tariff barrier 非关税障碍
notary office 公证处
notary public 法定公证人
notation 符号,标志;符号表示法,标志表示法;编织示意图,意匠图
notation paper 意匠图,意匠纸,花样设计纸
notch U形剪口,V形剪口,凹形剪口;选择器标记;刀眼,刀位,打刀眼;领口,领嘴,领豁口;衩口;剪口;刀口;切口
notch collar 衩口领;缺嘴领;翻领;豁口领;刻槽领,U形领,豁角西装领
notch collar pajamas 翻领睡衣
notched 凹凸襟;U形或V形剪口
notched collar 刻槽领,西装领
notched lapel V形翻领;平驳领;平驳头
notched lapel collar V形领,缺角西装领
notched roll collar 圆缺嘴领,平驳头圆西装领
notched rule 刀眼尺
notched shawl collar 缺角新月领,圆缺嘴领
notched tie collar 豁口领带领
notcher 牙口剪,记号剪
notches U形剪口,V形剪口
notching 打刀口;打刀眼
notching centerline 作中心线刀眼
notching corner 作转角刀眼
notching hemline 作底边线刀眼
notching inward curve 作内弧刀眼
notching machine 齿形刀片裁剪机
notching seam allowance 作缝份刀眼
notching shoulder tip 作肩端点刀眼
notching underarm seam allowance 抬裉缝剪口
notching waist 作腰节刀眼
notch lapel 平驳领,菱领,V形翻领,菱领,缺嘴领,平驳领;平驳头,豁口驳头
notch lapel collar 豁口西装领
notch mark 领嘴符号;刀口符号,刀口标记
notch marker 牙口剪,记号剪
notch placement 刀眼位置
notch point 衣片间缝合处刀口点
notch position 领嘴线

notch position line 领嘴线
notch shawl collar 豁口青果领;豁口新月领
notch symbol 刀眼标记
note 票据;色彩配置;速写草图;构图特色
notebook 笔记本
notecase 皮夹子,钱夹,钱包
no tension pucker 无张力起绉器
notes on talks 会谈纪要
not fast 毛边
not for sale 非卖品
no-throw silk thread 无捻多股丝线
no-throw silk yarn 无捻多股丝绒
notion counter 针线等用品的杂货柜
notion counter store 针线等用品的杂货店
notions 服装配件,服装附件(针、线、纽扣等),个人衣物(针线等小件用品);小件佩带物;小件日用品(美国)
notion trade 女服业;缝纫用品业
not piece knitted garment 非织片针织服装
Nottingham lace 诺丁汉花边,V形图案花边
Nottingham lace curtain machine 诺丁汉装饰花边机
not transferable 不准转让
Notts Wool 诺兹有光长羊毛
no tuck 裤前无裥型款式
Notwegian costume 挪威民俗服
nougat 杏仁糖色
nouka 诺卡地毯用毛
no-um netting 帐篷用超细网
no-understructure 无里层的[服装]结构
no-understructure construction 无里层的服装结构
nourish cream 滋养霜
nourishing cream 滋养霜
nouveau riche look 暴发户款式
nouveauté 饰边,滚边;编带,饰带;新花式织物
Nouvel chignon [法]最新发髻
nouvelle collection 时装展销,新款式服装展销
Novas 耐氯漂聚氨酯弹性纤维(商名,日本钟纺)
novel effect 花式效应
novel in style 款式新颖
novel jacquard snad crepe 意新绸

novel textile desigh 新奇编织花纹,新颖的织物设计
novelty 花式纱;新颖的织物设计;新产品
novelty button 花式扣
novelty gloves 拼花手套
novelty pattern 新颖花样
novelty stripe 花式条纹
novelty suitings 花式衣料
novelty sweater 花式运动衫,花式毛线衫
novelty thumb 手套拼花拇指
novelty weave 花式组织
novelty yarn 花色纱,花式纱
novoloid 含线型酚醛树脂不少于85%的聚酚醛纤维(属名)
novo paulista cotton 诺沃波利塔棉
now 非常时髦的
no-waistline-seam garment 无腰线服装
no wale cord 平面绒
no way print 无方向印花
now-you-don't[-see-it] 不直接看到[衬衣裤]
now-you-see-it 直接看到(衬衣裤)
noyales 法国漂白亚麻布;法国大麻船帆布
N-type of fiber 抗缩纤维
Nu ethnic costume 怒族民族服饰
nuage (织物的)斑点花纹,暗影效应,云层效应,迷雾效应
nuance 色彩细微差异
Nuba cotton 苏矾努巴短绒棉
Nubari cotton 埃及阿菲菲改良棉
nubby wool fabric 结子花呢;海绵呢
nubby yarn 结子花式线
nubia 女工羊毛头巾,女式羊毛头巾
nubs 结子花线
nub tweed 结子粗花呢
nubuck 牛皮里绒靴;软不克牛皮里绒靴
nubuk 努伯克革(有麂皮感的软皮)
nubuk boots 牛皮绒面靴;努伯克革靴
nub yarn 疙瘩花线,结子花线
nucha 项;项背面
nuche 项;颈背面
nude 橙红色,肉色;裸体;肉色袜子;肉色针织物
nude brassiere 迷你胸罩,肤色胸罩
nude drawing 人体绘画
nude dress 肉色女服
nude figure 裸体人像
nude knit fabric 肉色针织物
nude leg look 露腿款式

nude look 露腿风貌,裸露风貌;裸露式,透明式;裸体款式
nude model 裸体模特儿
nude size 赤脚尺码
nude stockings 肉色袜子
nude tan 皮肤色
nude toe 露趾
nudy style 裸体款式
nue 深浅色丝绒
nugatory agreement 无效的协议
Nugger cotton 印度纳格棉
nugget［gold］ 矿金色,金棕色,柑橘橙［色］
nugget ring 嵌天然金块的宝石戒指
nugget［toe］ 露趾
nuisanceless technology 无公害工艺
number 打号
numbered shirt 足球衫
numbering （裁片）打号;编号
numbering machine 打号机;编号
number of stitch 针(脚)数
number shirt 数字印花圆领衫,胸前印有数字的圆领衫
number tape 号码带
numdah rug 印度绒绣毛毡地毯
numeral 数字
numeral［knit］shirt 足球衫,数字衫
numerical control automatic sewing unit 数控自动缝纫设备
numerically controlled technology 数控技术
numnah rug 印度绒绣毛毡地毯
numud 诺曼德地毯
nun habit 尼姑服,修女服
nunny bag 海豹皮背包
nun's cloth 缪斯薄呢,修女黑色薄呢
nun's cotton 刺绣用纯白棉纱
nun's habit 尼姑服,修女服
nun's thread 细白亚麻线
nun's veiling 缪斯薄呢;修女头巾;修女面纱
nun's work 刺绣(旧);修女花边;修女刺绣
nun tuck 横褶(裙子);缩弧塔克
nuque 颈背
Nureyev shirt 纽列也夫式衬衫（低圆领口、滚边、褶裥克夫、长袖）
Nurma［cotton］ 印度布罗奇棉
nurse apron 护士围裙;护士工作服

nurse cloth 护士布
nurse dress 护士服
nurse gingham 护士蓝白条子平布
nursery cloth 托儿所用布
nurse's cap 护士帽
nurse's cape 护士披风
nurse's cloth 护士服布
nurse's gingham 护士蓝白条子平布
nurse shoes 护士鞋
nurse's uniform 护士(制)服
nursing bra 护理胸罩
nursing brassiere 护理胸罩
nursing corset 孕妇胸衣,哺乳内衣,保育胸衣
nursing gown 护理服
nursing wear 护士服
nut button 果壳纽扣
nutmeg 豆蔻灰
nutmeg green 豆蔻绿
nutria 河鼠毛皮,海狸毛皮;淡棕色
nuts button 坚果纽扣
Nyesta fabric 尼丝塔织物
Nylfrance 聚酰胺66纤维(商名,法国罗纳一普朗克)
nylon 聚酰胺纤维,锦纶,耐纶,尼龙;尼龙制品
nylon bag 尼龙袋
nylon belt 锦纶带,尼龙带
nylon boning 尼龙衬带
nylon bra band 锦纶胸罩带
nylon bristle filament 锦纶鬃丝
nylon buckle 尼龙带扣
nylon button 尼龙扣
nylon cap 尼龙帽
nylon chiffon 锦纶雪纺,锦纶薄绸,尼龙雪纺
nylon coated fabric 锦纶涂层织物,锦纶涂覆织物,尼龙皮,尼龙涂层织物
nylon coating jumper 锦纶软缎短夹克,尼龙软缎短夹克
nylon coil 尼龙齿链
nylon coil zipper 尼龙环扣拉链
nylon cord holder 尼龙线团芯
nylon corded velvet 锦纶棱条丝绒
nylon cords 起口锦纶线
nylon/cotton trousers 锦棉交织裤
nylon/cotton twill 锦纶棉绫,尼棉绫
nylon-covered rubber thread 锦纶橡胶线
nylon crepe gown 锦纶绉纱长衫,尼龙绉纱长衫

nylon crepe satin 锦纶绉缎,尼龙绉缎
nylon crinkle fabric 尼龙绉布,尼龙绉绸
nylon crushed velvet 锦纶压花丝绒
nylon curtain grenadine 锦纶窗帘纱,尼龙窗帘纱
nylon fabric 尼龙织物;尼龙布
nylon fabric with PU coating 尼龙 PU 涂层布
nylon fastener 尼龙拉链
nylon fastener tape 魔术贴;尼龙搭扣;尼龙搭扣带
nylon fiber 锦纶短纤维
nylon fil(ament) 锦纶长丝,尼龙长丝
nylon filet gloves 锦纶网眼手套,尼龙网眼手套
nylon flock-adhesive cloth 尼龙植绒布,信美绒
nylon gauze 闪光尼丝纱,闪光锦丝纱
nylon gingham 格子纺
nylon gloves 尼龙手套
nylon handbag 锦纶手袋,尼龙手袋
nylon hose 锦纶袜,尼龙袜
nylonic acrylic fiber 锦纶丙烯腈系纤维
nylon interweave 尼龙交织布
nylonized polyester fiber 锦纶化聚酯纤维
nylon jersey 锦纶运动衫
nylon jogging suit 尼龙运动套装
nylon knitted fabric 锦纶针织布,尼龙针织布
nylon lace 锦纶花边,尼龙花边
nylon lining satin 尼龙里子缎
nylon lustrine 锦纶丝绫,尼丝绫
nylon lycra 尼龙莱卡,尼龙弹力布
nylon magic tape 锦纶(或尼龙)搭扣带,锦纶褡链,锦纶带扣
nylon mesh 锦纶(尼龙)网眼花边
nylon mesh cap 尼龙网眼帽
nylon meshes 尼龙网眼纱
nylon-mesh screen 锦纶网筛
nylon microfibre fabric 尼龙挑皮绒
nylon military-ballistics vest 尼龙军用防护背心
nylon mixed pique 锦纶交织凸花缎
nylon-modified phenolic resin fiber 锦纶改性的酚醛树脂纤维
nylon monofilament 锦纶单丝
nylon moss microfibre fabric 尼龙苔纹挑皮绒
nylon mousse 锦纶弹力丝
nylon multifilament sewing thread 锦纶复丝缝纫线
nylon oxford 尼龙牛津布
nylon palace 锦纶派力斯,锦纶纺,尼龙纺,有光纺,尼龙绸,尼丝纺
nylon panties 尼龙三角短裤
nylon pants 尼龙裤
nylon petticoat 锦纶衬裙,尼龙衬裙
nylon plume 尼龙羽毛
nylon plush 锦纶长毛绒
nylon/polyester gingham 涤锦绡,尼涤锦
nylon puckered fabric 锦纶绉纹织物
nylon/PVC coated fabric 尼龙/PVC 涂层布
nylon/PVC coated jacket 尼龙/PVC 涂层夹克
nylon/PVC jacket 尼龙/PVC 填棉夹克
nylon/PVC padded jacket 尼龙/PVC 间棉夹克
nylon raincoat 锦纶雨衣,尼龙雨衣
nylon rayon abbatre 欢欣缎,锦纶(或耐纶、尼龙)人造丝凸纹缎
nylon-reinforced belt 锦纶增强[橡胶]带,锦纶带
nylon rib 尼龙罗纹
nylon rip-stop 尼龙格子布
nylon rope 锦纶绳
nylon rope trick 锦纶绳现象
nylons 尼龙长袜
nylon seersucker taffeta 锦纶绉条纹薄塔夫绸,锦纶塔夫泡泡纱
nylon serge 尼龙哔叽
nylon sheer 锦纶绢,尼绡,素丝绡
nylon sheer trim slip 锦纶(尼龙)丝边套裙
nylon shioze 锦纶纺,耐纶纺,尼丝纺
nylon shorts 尼龙短裤
nylon slipover parka 锦纶(尼龙)套头派克短外套;锦纶(尼龙)套头派克大衣
nylon socks 锦纶袜,尼龙袜
Nylon Solar-α 以碳化锆微粒为芯的聚酰胺纤维(能吸收太阳能,商名)
nylon sole 尼龙鞋底
nylon spandex lace 尼龙弹力花边
nylon-spandex bicomponent fiber 锦纶-聚氨基甲酸乙酯双组分纤维(弹性纤维)
nylon/spun rayon union taffeta 锦黏塔夫绸,尼新纺
nylon staple 锦纶短纤维
nylon staple fiber 尼龙短纤
nylon stockings 尼龙袜

nylon stopper 尼龙绳索扣
nylon straw 锦纶草秆状单丝
nylon stretch sewing thread 锦纶弹力缝纫线
nylon stretch socks 弹力尼龙袜
nylon stretch tights 锦纶(尼龙)紧身弹力连袜裤
nylon stretch vest 弹力锦纶背心
nylon stretch yarn 弹力锦纶丝
nylon sweater 锦纶衫,尼龙衫
nylon tactel 尼龙塔克特
nylon taffeta 锦纶塔夫绸,尼龙塔夫绸,尼龙绸;锦纶纺,尼龙纺
nylon taffeta with PVC(PU)coating PVC(PU)涂层尼龙布(风雨衣、羽绒衣面料)
nylon tape fastener 尼龙搭扣,尼龙子母扣
nylon taslan 尼龙塔斯纶
nylon taslan shorts 尼龙塔斯纶短裤
nylon thread 锦纶线,尼龙缝线,尼龙线
nylon tooth-brush 锦纶牙刷
nylon top 尼龙毛条
nylon tow 尼龙丝束
nylon tow sewing thread 锦纶复丝缝纫线
nylon tricot 锦纶经编织物
nylon tricot bathing suit 锦纶(尼龙)经编泳衣
nylon tricot shirt 锦纶(尼龙)经编衬衫
nylon trilobal 尼龙闪光布料
nylon trilobal suit 尼龙运动套装
nylon tulle 锦纶经编薄纱,锦纶绢网
nylon twill 尼龙斜纹布;尼丝绫,锦纶丝绫
nylon tyre cord 锦纶轮胎帘子线
nylon tyre yarn 锦纶轮胎[用]丝
nylon umbrella 锦纶伞,尼龙伞
nylon umbrella silk 尼龙伞绸
nylon velvet 锦纶丝绒,尼龙丝绒
nylon watch-guard band 锦纶手表带
nylon/wool union mock leno 锦纶羊毛交织呢
nylon/wool union suiting silk 锦纶羊毛交织呢
nylon yarn 锦纶纱,尼龙纱;锦纶丝,尼龙丝
nylon zipper 尼龙拉链
nylon zipper parka 尼龙拉链派克大衣
nymphet 美貌少女,早熟少女,性感少女
nysilk 超声无光处理法
nytril fiber 奈特里尔纤维,(聚偏氰乙烯纤维)

O

oak 青叶色
oak brown 橡树棕
oakleaf braid 橡树叶辫带,橡叶图案帽圈织带(用于英国传统的官员制服)
oak silk 柞蚕丝
oak tag 厚样板纸,样板用纸
oak-tag pattern 厚纸样板
oakwood 栎树棕(棕色)
oasis ［沙漠］绿洲绿(浅暗黄绿色),浅秋香色
oasis [green] 沙漠绿洲绿(浅暗黄绿色)
oatcake linen 麦饼纹亚麻布
oatmeal cloth 燕麦纹织物
oatmeal effect 燕麦花样
oatmeal 点纹布,粗纬毛巾布;燕麦纹样
oats cotton 美国燕麦棉
oban 美国奥斑厚花呢
obdurability 坚硬,韧
obese 过分肥胖的
obese figure 过分肥胖体型
obi（belt） （日本）欧比宽腰带,和服腰带,阔腰带
obiji 日本和服腰带绸
obi sash （日本）欧比宽腰带,和服腰带,阔腰带
obi-styled sash （日本）欧比宽腰带,和服腰带
object colour 物体色
object-colour stimulus 物体色刺激
objective evaluation 客观评定,仪器评定
objective measurement 客观测定,仪器测定
object lens 物镜
object printings 实物图形印花
object teaching 实物教学,直观教学
oblique design 斜纹花型设计,斜纹图案设计;斜纹图案
oblique ironing 熨烫歪斜,熨烫不正(疵点)
oblique line 斜线
oblique method 斜面法测试织物刚度
oblique neckline 单肩斜领口,斜肩领口,斜领口
oblique pocket 斜袋;斜插袋
oblique toe-end 斜头(制鞋)

oblique weft 斜纬
oblong 椭圆形
oblong ball 椭圆形线球
oblong collar 长方领,长椭圆形领(比翼领)
oblong cross stitch 长方形十字绣
oblong neckline 长方形领口,椭圆形领口;椭圆形领口线,长方形领口线
observation check 外部检验
observation error 观察误差
obsolescence 过时,时装衰退期
obstinate stain 经久不褪的污渍
ocal 双宫茧;双宫绸
occasional dress 特定场合礼服
occasional dressing 应时衣着
occasional wear 应时服装
occiput 头枕部
occupational clothing 职业装,职业服
ocean 海蓝色,绿蓝色
ocean bill of lading 海运提单
ocean blue 海蓝色,绿蓝色,海洋蓝色
ocean carriage 远洋运输
ocean carrier 海运承运人
ocean gray 浅灰色,海灰色
ocean green 海洋绿(淡黄绿色)
oceanic colour 海洋色
ocean wave 海浪绿
ocelot 豹猫毛皮,美洲豹猫皮
ocelot coat 美洲豹猫皮大衣
ocelot fur 豹猫毛皮
ocelot pattern 黄褐斑纹,豹猫斑点花纹
ocepa （直筒绣边的）安迪式衬衣裙
Ochanomizu University 茶水女子大学(日本)
ocher 黄褐,浅暗橘黄色
ocher red 红光藏蓝
ochre 浅暗橘黄色
ochre red 红光藏蓝
ochroid 深黄赭色的
ochroid colour 深黄赭色
o'coat 外套,大衣
o'coat collar 大花领
octagon tie 八角形领巾
octagonal cabochon 八角穹顶式(珠宝)
octagonal cap 八角帽

octagonal crosscut 八角形相交磨翻(珠宝)
octagonal flat table 八角平台形(珠宝)
octagonal hat 八角帽
octagonal stepcut 八角形台阶磨翻(珠宝)
octapulum 绸布
odd bits of cloth 布头
odd jacket (套装中的)单零夹克；不配套夹克，单件夹克，替换夹克
odd lot 零码商店
odd material 零料
odd piece 短码，短码布
odd scraps 碎布
odd sizes 服装特殊尺寸
odd vest (套装中)单零背心；单肩背心，替换背心，异质背心
odd waistcoat 单肩背心，替换背心，异质背心
odd-come-short 破旧衣服；碎布片，零头布
odhani 奥德哈尼平布披巾(印度)
odjaklik 土耳其东方式壁炉地毯
odor 气味
odor adsorbing finish 吸臭整理
odorant 香料
odor destroying insoles 除臭鞋垫
odour 气味
odour destroying insole 除臭鞋垫
odour-proof hosiery 防臭袜
odour resistant 防臭[味]
oegge 鞋边棱，装饰边
oeil de perdrix 八经四纬哔叽；花边结子地
oeillet 穿带子用洞眼，穿带子用开口
Oeko-Tex certification 生态纺织品标准证书
Oeko-Tex label 生态纺织品标签
Oeko-Tex standard 生态纺织品标准
Oeko-Tex Standard 100 生态纺织品标准100(国际上认可的纺织品生态标准)
offal 碎皮
offbeat colour 自由配色
offbeat fashion 不落俗套的款式
off-black 黑色不黑；黑色偏离
off-body[line] 宽松式
off-body look 宽松款式，宽松风貌，离身款式
off-body silhouette 离身轮廓
off-body style 离身风格
off-centered circle 偏心圆

off centre 偏离中心
off-centre front 前搭扣不正
off-centre opening 偏襟
off-clip 脱边
off collar 一字领
off colour 不合色样,染色走样,色差
off colour diamond 杂色钻石
off colour material 不易着色的材料
off-colour wool 色污毛,尿污毛,变色羊毛
off design 非传流设计
off duty 下班款式
off-duty wear 下班后穿着,休闲装
offer 报价,报盘
off-gauge 不合规格,非标准
off-grade 等外品(纱线、织物等),副次品；不合格的,等外的
off-grain printed fabric 丝缕不直的印花面料
off-grain stitching (大针距)拨开缝制；丝缕紧偏
off-grain 丝缕歪斜；丝缕歪斜
office automation 办公室自动化
Office of Textile and Apparel ［美］纺织服装管理处
office uniform 办公服
office wear 办公服装
officer collar 女式军官制服立领；中式立领；官领
officer's cape 海军蓝交领中长披风(美国海军军官制服)；军官披风,官职外套,军官外套,卫兵大衣
officer wear 办公服装
official acceptance 正式验收
official ceremonial costume 正规礼仪服装
official clothing (guanyi) 官衣(中国戏装)
official dress 官服；朝服
official hue 公定色相
official's boots 官靴
official shade 公定色调
official uniform 礼服
official wear 正式场合穿服装的总称,正式服装,公开场合服装(办公服,工作衣,半礼服等)
official weight 公定重量
off-key 不协调
off neck 露颈领,一字领
off neckline 宽开领口,离脖领口,露颈领

口,一字领口
off-needle position 偏离基线
off olive 浅黄绿色,非纯粹橄榄色
off parts 分离部件
off pattern 花形不符
off-patterned cloth 错花织物,漏花织物
off-pressing 最终烫平;最终压烫
off-price 廉价
off-price apparel 价格优惠的服装
off-price store 降价商店
off registration 脱版
off-season 淡季
off set 抵消
off-set twill weave 变化网形斜纹
off-shade 色差,染色走样
off-shore production 国外生产产品;境外加工
off-shore textile 海外纺织品
off-shoulder 露肩式,离肩型
off-shoulder bra(ssiere) 露肩型胸罩(无肩带或肩带在外侧)
off-shoulder design 露肩设计
off-shoulder neckline 露肩式领口,大一字领口
off-size 成形不良,尺寸[码]不符
off specification 不合规格的
off specification goods 不合规格产品
off spool 缝纫线脱轴,脱线轴
off-square 近似平衡织物;经纬缩率差
off-square fabric 不平直织物,不方正织物
off-square selt 织物的非均衡密度
off standard 不合标准;等外级;等外品;等外的
off-the-arm machine 递送式曲臂[底板]缝纫机
off-the-face 不遮住脸的(发式);无檐女帽;窄边上翻的(女帽)
off-the-face hat 卷边帽
off-the-neck 开领领口
off-the-peg 成衣,现成服装;购买成衣
off the peg dress 成衣
off-the-rack 成衣,现成服装;购买成衣
off-the-shoulder 离肩型,露肩式;露肩的,裸肩的
off-the-shoulder de'colletage 低胸露肩服
off-the-shoulder dress 露肩装,露肩衣,袒胸露肩的衣服
off-the-shoulder leotard 露肩式短紧身衣
off-the-shoulder neckline 露肩领口

off-time 淡季的
off-tone 色光不一致(染色疵点)
off-tone fading 全褪色
off turtle neck 宽高圆套领;宽松双翻领;离脖高套领
off-white 非纯白色;本白;灰白,;米白色,黄白色
off-white silk shirt 米色丝衬衫
off-winding 退绕
Ohio merino 美国俄亥俄美利奴毛
Ohio merino wool 美国俄亥俄美利奴毛
ohrna 方形薄布
oil 油;油布雨衣;油布衣裤
oil belt duck 重型防水帆布
oil blue 油蓝色
oilcan (缝纫机用)油壶;加油器
oil check window (缝纫机头)加油窗
oil cloth 油布,漆布
oil colours 油溶性染料
oiled cloth umbrella 油布伞
oiled cotton raincoat 油布雨衣
oiled paper umbrella 油纸伞
oiled pick 油纬
oil(ed) silk 油绸
oiled slicker [raincoat] 油布雨衣
oiled sweater 原毛毛衣,含脂毛衣
oiled wool 原毛,含脂毛
oil feed component 给油部件
oil field boots 油田鞋
oil flax 亚麻,胡麻
oil hole 油眼
oiler 防水油布外套;加油器
oilers 防水油布衣裤
oil leather shoes 油皮鞋
oillet 穿带洞眼,穿带开口
oil looking fabric 油光布
oil proofing 防油[整理]
oil-painting linen 亚麻油画布
oil red 油红
oil repellent finish 拒油整理
oil repellent finish fabric 拒油整理织物
oil repellent finished fabric 拒油整理织物
oil repellent finishing 拒油整理
oil resist finish 防油整理
oil resistance leather shoes 耐油皮鞋
oils 油布雨衣
oil shoes 油鞋
oil silk (防水)油绸
oilskin 防水油布,油布雨衣,防水套装

oilskin coat 油布雨衣
oil spot 油渍,油污
oil stain 油渍,油污
oil tannage 油鞣革
oil tanned leather 油鞣革
oil-tight 不透油
oil-treated fabric 涂油织物
oily coating 含油涂层
oily hand 油滑手感
oily stain 油迹
oiselle hemp 槿麻,洋麻
Okinawa jyofu 日本香蕉茎纤维布
okra 美国奥克拉棉;秋葵纤维
okro 美国奥克拉棉;秋葵纤维
oksalon 奥克萨纶
olanes 古巴印花棉布
OLAY 玉兰油
old bess cotton 西印度群岛老贝斯低级棉
old-age comfort 老年人穿着舒适性
old clothes man 旧衣商
old face 老式花边
old fashion 陈旧式样,旧式
old fashioned 过时的
old gold 古金色,浅黄至浅橄榄棕色
oldham 英国粗呢
old heliotrope 浅紫色,古紫色
oldie's fashion 怀旧式时装
old lace 老式花边,梭结花边,针绣花边
old liberated area look 老区风貌
old mauve 暗紫红色
old moss green 老苔绿(橄榄色)
old olive 橄榄棕色
old rose 老玫瑰红(浅暗红色)
old school tie 昔时小学生领带,校友领带
old wash 洗旧[状]整理
olecramon 肘头点
olefine fiber [聚]烯烃类合成纤维(含85%以上乙烯、丙烯或其他烯类)
Oleg cassini 奥莱格·卡西尼(美国时装设计师)
oleopneumatic 油压后踵按磨机(制)鞋
oleopneumatic heel seat machine 油压后踵按磨机(制鞋机器)
O-line O字形;O字形服装造型
olivauto cloth 草绿色细呢
olive 茶青色,黄绿色,橄榄色,黄褐色,橄榄绿
olive black 橄榄黑,墨橄榄色

olive brown 橄榄棕,黄褐色;橄榄形扣
olive drab 灰橄榄色,浅橄榄棕
olive drab (O.D.) 草黄[色],草绿[色],灰橄榄色;(美国陆军军服用的)草黄色呢
olive drabs 草黄色军服(美国)
olive eyeleting machine 长型鸡眼扣孔机
olive gray 橄榄灰
olive green 橄榄绿,草地绿,草绿
olive grey 橄榄灰色
olive yellow 橄榄黄
Olivier Lapidus 奥立维·拉比杜斯(法国高级女装设计师)
Olivier Montague 奥利维·蒙塔古(法国著名时装设计师)
olivine 橄榄石
Ollyet 17世纪挪威、英国毛织物
olone 法国原色麻篷帆布
Olympian blue 奥林匹亚蓝,中天蓝色
o-maku 日本锦缎
omber 由深到浅的色调
ombre (法)深至浅色调;深浅色条纹布;深浅条纹织物
ombré check 彩虹方格花纹,深浅色格纹
ombré dyeing 虹彩染色,深浅色染色
ombré effect 彩虹效应,雕刻云纹效应
ombreé stripe 深至浅条纹
ombré fabric 深浅条纹织物
ombrelle (法)小阳伞
ombré moire renaissance 深浅色波纹织物
ombré printings 虹彩印花
ombré rayé 虹彩与地色交错条纹
ombré silk 月华色丝
ombré stripe 彩虹条纹
OMEGA 欧米伽
omission examination 免验
omnibus design 多用途花样设计
omni-dry clothing 速干衣裤
OMO (on my own) 靠我自己
omoplate 肩胛骨
omphalion 脐点
O'more College of Design 奥莫尔设计学院(美)
on consignment 代销
ondé 异色棉毛交织呢;波纹整理织物
ondée 波纹纱;粗细股合股线
ondé yarn 波形纱线
ondine 重棱纹罗缎织物(有波曲状效

应),粗线罗缎;光亮软呢
ondulé 经向波纹织物,棱纹布;长毛绒
on duty 上班款式
one-and-one ribbed goods 1+1罗纹针织品
one-and-one top 1+1罗口,袜口罗纹
one bath dyeing process 一浴法染色
one-button cuff 单扣袖头,单纽窄袖口
one button frock coat 单纽男大衣,单纽男礼服大衣
one-button single breast 单排单纽的
one-button single-breasted jacket 单襟式单扣上衣
one-button style 一颗纽款式
one-button suit 单扣西装
one by one rib 1×1罗纹
one-by-one rib knitted fabric 1+1罗纹针织物,细罗纹针织物
one collar 单片领
one-colour coordination 同色相配色
one-colour material 单色缝料
one continuous line 一条龙
one-dart skirt 单省裙
one-dart skirt foundation 单省道裙片原型
one dollar blouse 一美元女衫
one-face fabric 单面织物
one-head embroidery machine 单头绣花机
one-hour dress 速成装(一小时制成)
one-hundred denier crepe 百旦绉
one-hundred pleater 厨师帽
one-hundred-pleat skirt 百褶裙
one line production 单线生产,线流水作业
one-man show 个人作品展
one needle lockstitch machine 单针锁式线迹缝纫机
one-part mold 单粒包纽模,个人作品展
one-piece (服装)上下一体,一件式的
one-piece back 无背缝
one piece combination 连衫裤
one-pice dress 连衣裙
one-piece armhole cowl 整片式袖窿垂褶
one-piece back (服装)无背缝
one-piece bathing suit 连体露背泳装,连衣裤泳装
one-piece bodysuit 连身裤
one-piece clothes 一体服装
one-piece collar 一片领,上盘领,单片领
one-piece combination 连衫裤
one-piece corset 紧身胸衣裙
one-piece cuff 一片式克夫,一片式袖口
one-piece diamond-shape gusset 单片菱形(腋下)插角布
one-piece dress 连衣裙,连衫裙
one-piece dress suit 连衣裙套装
one-piece facing 连裁挂面
one-piece fire-fighting suit 密闭式消防服
one-piece fitted sheath dress 一片式合体连衣直裙
one-piece flap 一片裁的袋盖
one-piece garment 深衣,衣裤相连的服装
one-piece jersey suit 上下连身针织运动衫,整件针织运动衫(童装)
one-piece jumpsuit (衣裤相连)无腰线连裤装
one-piece lapel 一片式驳领
one-piece look 一片式风貌,一片式款式,连衣裙款式,形似连衣裙的两件套款式
one-piece molded construction 制衣整片模压结构
one-piece playsuit 一件式娱乐服
one-piece pocket 一片袋,片式袋片
one-piece raglan sleeve 连肩袖
one-piece silhouette 一片式服装轮廓线
one-piece skirt 一片裙,单片裙,单页裙,整裙
one-piece sleeve 整片袖,大裁袖,一片袖
one-piece sleeve placket 一片式袖衩
one-piece style 一片式连装款式
one-piece suit 连身工作服,连身套装
one-piece vamp shoes 整帮鞋
one-ply yarn 单股纱
one-point blouse 一点饰女衫
one-point look 单独标记款式,一点饰款式
one-point mark 单独标记
one-point shirt 星标饰女衫,有单独标记的女衫(在衣服局部或某点装饰标记)
one-point style 一点饰款式
one-process automatic hosiery machine 单程式自动织袜机
one-process garment knitting 整体编织
one repeat 完全组织
one-seam skirt 一片裙
one set corset 女用束腹套衣,一套紧身衣
one-shoulder 不对称单肩式
one-shoulder dress 单肩连裙裤
one-shoulder gown 单肩长袍,单肩袍装

one-shoulder neckline 斜肩领口, 单肩式领口
one-shoulder style 单肩款式
one-sided 单面的
one-sided covering stitch 单面缝线迹
one-sided drill 单面卡其
one-sided fabric 单面织物
one-sided fullness 单向增摆量
one-sided satin handkerchief 单面缎纹手帕
one-sided spiral transport system 单面螺旋状运行织物
one-sided terry 单面毛圈织物
one-side fancy suiting 单面花呢
one-side printings 单面印花
one-size bra 单码胸罩, 弹力胸罩
one-size-fits-all [model] 单一尺寸服装, 均码服装
one-suiter 男用小衣箱
one-third tone 三分之一色调
one-thread blindstitch hemming machine 单线暗缝卷边机
one-thread overseaming stitch 单线包缝线迹
one-thread roll hemming machine 单线卷边机
one way spreading 单层单向铺料法
one/two-head embroidery machine 单头/双头绣花机
one-time use 一次性使用
one-wash finishing 一次水洗[牛仔裤]
one-way cloth pattern 织物单向图案
one-way design 单向花样, 单向图案
one-way fastener 单向拉链
one-way layout 一顺排, 衣片一顺向排料
one-way pile 单向绒毛
one-way placement 一顺排, 衣片一顺向排料
one-way pleat 顺风褶, 顺褶裥, 顺裥, 单向褶
one-way print 单向印花
one-way spreading 单向铺料法, 一顺铺
one-way strip chromatogram 单向条色谱
on-grain 沿经向(或纬向); 无纬斜效果
onion bagging 网眼袋布
onion cloth 网眼袋布
onion green 葱绿, 葱青
onion skin paper 葱皮纸
onix 米黄色
on lay 服装镶嵌物

on-line color continuity monitor 在线色连续监控系统
on-line color metric metering system 在线测色系统
on-line console 在线控制台, 联机控制台
on-line control 在线监控, 在线控制, 直接控制
on-line detection 在线检测
on-line operation 联机操作, 在线操作
on-line quality monitoring system 在线质量监控系统
on-line test 在线测试
on-line to textile literature 纺织文献联机检索
on-location dyeing 地毯定位喷染
only coat fiber 全皮纤维
only-core fiber 全芯纤维
on shade 近似色
on stream 在运转中; 在操作中
onteora rug 美国手织地毯
on-the-dot registration 准确对花
on the grain 顺着织物纹理
on-tone 色光一致
on-top skirt 上层裙(用于欧洲民族风格的多层裙设计)
on trial 试用
onyx 缟玛瑙, 石华
onyx colour 缟玛瑙色
oomph 热情; 性感; 魅力
Oomra cotton 印度乌姆拉棉
Oomrawuttee cotton 印度奥姆拉伍蒂棉
oormuck 欧默克驼毛呢; 羊绒呢
oo-sze 广东生丝
ooze 绒面小牛皮; 茸毛; 皮革鞣液
ooze calf 绒面小牛皮, 植鞣[小牛]绒面革
ooze leather 仿麂皮织物; 植鞣[小牛]绒面革
Op Art 光效应艺术, 视幻艺术; 视幻艺术服装风格;
op-art 光效应艺术, 欧普艺术纹样, 现代造型艺术, 光学艺术图案, 简易草图纹样
Op art print 光效应绘画艺术印花
opacity 不发亮, 不透光, 不透明性
opague material 不透明材料
opal 乳白, 乳色, 白宝石色; 蛋白石, 水蛋子石; 欧派尔细平布
opal blue 乳蓝色
opalescence 乳[白]光, 乳[白]色

opalescent veil 乳白色面纱
opal finish 烂花整理,烂花印花
opal finishing 仿烂花整理;仿烂花印花;消光整理;消光印花
opal finished georgette crepe 烂花乔其,烂花绉
opal gray 乳白灰
opaline 高级细软白布
opaline green 蛋白石绿(浅灰绿色)
opal printings 消光白色印花
opals 轧光细棉布
opaque 不透明的,无光泽的;不透明织物
opaque color 覆盖色,不透明色
opaque hose 不透明袜子
opaqueness 不透明度;不透明性
opaque pantyhose 不透明连裤袜
opelon 聚氨酯弹性纤维(商名)
open 铺开
open air drying 晾干,风干
open-back shoes 露后跟鞋
open band twist 反手捻
open band yarn 反手纱
open basket stitch 粗孔方平式贴线绣
open bidding 公开招标
open-bottom all-in-one 妇女胸衣
open-bottom garment 裙摆口服装(指裙类部分整姿内衣等)
open bust bra 露乳型胸罩
open-button 紧身胸衣
open case 敞式旅行包;无盖箱
open-chain stitch 方型链状绣;开式链状绣
open circuit 开放式呼吸装置(潜水运动用)
open clock 袜子的网眼花纹
open collar 敞开领,开领,开门领
open-crown hat 无顶女帽,开顶帽
open-cut look 开放式发型
open dart 开褶
open design 现成花样
open double spreading 单向平摊铺料(用于起毛、起绒或有图案方向的铺料)
opened economic region 经济开发区
opened-end zipper (尾部)分开拉链
open-ended casing 有开口的串带管
open-ended zip 开式拉链
open-end fastener 开口式拉链
open-end one-way fastener 开口式单向拉链
open-end slide fastener 开口式拉链

open-end yarn 气流[纺]纱,自由端[纺]纱,转杯[纺]纱
open-end zipper 开口式单向拉链,开口式拉链
opener 开布机;开证人
open fabric 松结构织物,组织稀松织物
open face 稀疏织物;[织物]稀路疵
open fastener 敞开型拉链
open fold 平幅折叠
open free area 织物透孔面积率
open front 门襟,开前襟
open front collar 前开门领
open front neck(line) 前开领口,前开领圈线
open front skirt 前开口裙子
open girdle 及腿长的束腹
open gown 无袖罩袍
open-heel shoes 后露鞋
open hem 敞口缝边
opening (衣服的)开襟;门档,开门;端口;(裁片上的)剪口
opening end (服装)开口止点
opening net 开口网眼
opening seam 开缝
open knit 网眼针织物
open lace 网眼花边
open lap seam 开口缝
open lazy-daisy stitch 跨线针迹,开孔平式花瓣针迹,套针
open-line 开架服装
open-meshed fabric 网眼织物
open mesh structure 网眼结构
open-necked shirt 敞领衫,开领衫,翻领衫
open neckline 开叉领口
openness 放开度,放开角度(以裤腿角度表示腿部姿势,表示服装质量)
open nonwoven fabric 疏松非织造布
open out 拉开
open place 松档;稀弄
open porosity 表面多孔性
open robe 露出里层的双层裙
open seam 开式缝,分开缝,手推缝,套口缝合
open-seat pants 童开档裤
open-set mark 松档
open setting 稀经稀纬,低密度织物
open-shank sandals 露趾系带凉鞋
open shank(shoes) 侧空鞋,侧露鞋
open shirt 敞领衬衫,敞领衫,开襟衬衫

open shoes 透空鞋(总称)	opera glasses 观剧眼镜,看戏望远镜
open sleeve 开放式袖,开启式袖	opera gloves 观剧或社交用的长手套
open square neck 开式方领口	opera hat 观剧帽,夜礼帽,折叠式大礼帽
open square neckline 开式方领口	opera-hood 妇女头巾(观剧或晚宴用)
open stitch 透气针法	opera hose 长筒舞袜
open stripes ribbon 透孔条纹带	opera length 加长长筒女袜
open-structured knit 稀松针(织物)	opera length gloves 观剧或社交用长手套
open texture 稀松组织结构	opera [length] nacklace 晚礼服项链,看戏长项链
open-texture fabric 透孔织物,网眼织物	opera mantle 晚礼服斗篷
open [the] seam 分缝	opera pumps 晚宴鞋,歌舞鞋,观剧鞋,欧琶拉潘普鞋
open toecap shoes 开包头式鞋	opera slippers 侧空式男鞋;看戏鞋,观剧鞋;男用室内拖鞋,浅口便鞋
open-toed pumps 露趾浅口无带鞋,无尖舞鞋,前露舞鞋	operate time 动作时间
open-toed shoes 前露鞋,头空鞋(露趾鞋)	operating 经营
open-toed slides 前空拖鞋	operating button 操纵按钮
open-toed slingback 前后空皮鞋	operating characteristic 运转特性
open-toe hinged foot 开趾铰链压脚	operating condition 操作条件
open-toe mule 露趾无踵带拖鞋,前空密儿拖鞋	operating cycle 操作周期,运转周期
open-toe shoes 头空鞋,露趾鞋	operating days 运转天数
open-top circular knitting machine 多三角圆形针织机	operating dress 手术服
open-top jeans button 通心工字纽	operating gown 工作外套;外科手术工作服
open-top sinker type machine 多三角圆形针织机	operating instruction 操作说明
open type 开口式,敞开式	operating line 操作线,工作线
open waist shoes 腰窝透空鞋	operating method 操作法
open weave knits 网眼针织物	operating parameter 运行参数
open-welt seam 衣片正面褶状翼片缝合方法,活页接缝	operating personnel 操作人员,挡车工
open width 平幅,原幅	operating program 操作程序
open-wing collar 敞领,平翼领,开襟衫领	operating sequence 操作程序
openwork 透孔织物,网眼织物,透孔制品	operating skills 操作技术
openwork embroidery 透孔刺绣	operating specifications 操作规程
openwork gloves 通花手套;网眼手套	operating system 操作系统
openwork hose 网眼袜	operating table 操作台,工作台
openwork jacquard 网眼提花	operating test 操作试验
openwork panel 网眼花边镶边	operation 操作,工序,运转,作业
openwork rib fabric 网眼罗纹织物	operation handbook 操作手册
openwork stripe 网眼条子花纹	operation instruction 使用说明[书]
openwoven back 地毯的抽绞地布	operation manual 操作手册
open zipper bottom 开口拉链下摆	operation method 工作法
opera bag 观剧包,戏剧包	operation process chart 作业程序图
opera cape 短斗篷,观剧披风,观剧披肩;披肩,男子中长披风,晚礼服斗篷	operation quality 运行质量
	operation schedule 生产作业计划
opera cloak 女士(观剧或晚宴时穿着的)晚礼服斗篷	operation sheet 工艺卡
	operation study 操作研究
opera drawer 社交或晚宴时穿在里面的长衬裤	operation wear 作业服
	operative temperature 操作温度
opera flannel 浅色全毛法兰绒	operative workload 挡车工工作量;挡车工劳动强度

opera top 袒胸背心
operator 操作员,挡车工,技工
operator's stand 操作台
opera wrap 观剧披风
Oporto wool 波尔图粗羊毛
opossum 负鼠毛皮,负鼠皮
opponent colors 补色
opportunity cost 机会成本
opposite colour 相反色,对立色
optical art 视幻艺术纹样;光效应艺术,现代造型艺术
optical art design 视幻艺术图案
optical brightening agent(OBA) 荧光增白剂
optical character recognition(OCR) 光学字符识别
optical dye 荧光染料
optical fiber 光学纤维,光导纤维
optical illusion 光幻视,幻觉,错视
optical print 光学印花
optical spectrum 光谱
optical whitening 荧光增白
optic fiber 光导纤维
optics functional fiber 光学功能纤维
optimal value 最优值
optimization 优化,最优化
optimization technique 最优技术
optimized process 最佳工艺过程
optimum efficiency 最佳效率
optimum estimation 最优估计
optimum lot size 最优批量
optimum performance 最佳性能
optimum production lot size 最优生产批量
optimum value 最佳值
optional clause 选择性条款
optional coordination 任意组合
optional equipment 备用设备,附加设备,选用设备
optional print 光学印花
option look 任选搭配型,自选款式
opulent beard 浓须
opulent look 富豪款式,豪华服装款式
opulent style 富豪款式
opus anglicum 盎格鲁刺绣;英国刺绣工艺
opus araneum (手工制的)网眼梭结亚麻花边
opus byssinum 海丝纤维
opus filatorium 网眼花边
opus pulvinarium 柏林毛线刺绣;椅垫刺绣
Or filé 丝芯金线
orange 橙色,橘黄色;黄色颜料
orange blossom 白色香橙花(英美婚礼常用)
orange brown 橙褐色,橘棕色
orange chestnut 橙
orange clear 浅橙黄色
orange cloth 橙黄色布
orange colour 橘黄色
orange diamond 橙色钻石
orange garment 橙黄色衣服
orange light 淡橙黄色
orange ocher 黄赭色
orange ochre 黄赭色
orange pale 淡白橙色
orange peel 橘皮色
orange-peel defect 橘皮状起皱疵点
orange-peel effect 橘皮纹效应
orange popsicle 橘子棒冰色
orange red 橙红色
orange-red cast 橙红色光
orange stick 橙木棒(用于修剪指甲)
orange system 橙色类
orange[-wood] stick 橙树棒
orcein 苔红素,地衣红(红棕色植物染料)
orchid 淡紫,兰花紫
orchid bloom 兰花色
orchid mist 灰紫,朦胧兰花色
orchid pink 兰花粉红[色]
orchil 苔色素
order 订单;订货(下单子),定购,定制;秩序
order a suit 定制一套衣服
order B/L 指示提单
order by sample 凭样订货
order clothes 定做服装
ordered benzoheterocycle-imide copolymer fiber 有序苯并杂环-酰亚胺共聚纤维
ordered linear network polymer fiber 有规线形网络聚合物纤维
ordered oxadiazole-imide copolymer fiber 有序噁二唑-酰亚胺共聚纤维
ordered polybenzoxazole-amide fiber 有序聚苯并噁唑酰胺纤维
order form 定货单
ordering 定货
ordering cycle 订货周期

ordering meeting 订货会
orderly growth 有秩序地增长
order made ［服装的］定制
ordermade clothes 订做服装
order number 订单号
order of uniform 定做制服
order processing 订单处理
order sheet 订货单
ordinary 不宜精梳的羊毛
ordinary bleach 常规漂白
ordinary bright knitting yarn 普通有光针织丝，普通有光针织纱
ordinary clothes 室内便装
ordinary dress 日常服，便服
ordinary drop feed 普通下送料
ordinary fabric 素织物
ordinary feed 正常送料
ordinary fiber 常规纤维，普通纤维
ordinary loop knot 平结，筒子结
ordinary maintenance 日常维护
ordinary net 经编网眼织物
ordinary quality 中等品
ordinary sewing 普通缝纫
ordinary twill 正规斜纹，正则斜纹
ordinary uniform 日常制服
ordinary welt 一般袜口
ordinary wool 粗纺用绵羊毛
ore button 矿石扣
orenburg shawl 奥伦堡纬编披巾
organ felt 风琴用毡，琴毡
organburg 粗袋布
organdie 奥甘迪（蝉翼衫），蝉翼纱；玻璃纱
organdie applique look 蝉翼纱贴花款式
organdie applique style 蝉翼纱贴花款式
organdie finish 奥甘迪（蝉翼纱）整理
organdy 蝉翼纱，奥甘迪；玻璃纱
organdy finish 奥甘迪整理，蝉翼纱整理（用浆料、树脂或化学品对细薄平纹织物进行的挺爽整理）
organdy lawn 奥甘迪蝉翼纱
organic base fiber 有机纤维
organic fiber 有机纤维
organic heat-resistant fiber 有机耐热性纤维
organic-inorganic hybrid fiber 有机-无机杂化纤维
organic optical fiber 有机光学纤维，有机光导纤维
organic style 天然款式

organic unity 有机的统一
organization 组织
organoleptic evaluation 感官检验；感官评定
organza 透明硬纱
organzari 硬挺整理的丝织物
orhna 印度长头巾
orient 珍珠光泽，珠光色
oriental and western clothing 东西方风格［结合的］服装
oriental blue 东方蓝（暗蓝色）
oriental carpet 手织东方地毯
oriental cloth 东方布
oriental costume 东方女式服装
oriental crepe 广绫；东方绉，重磅双绉
oriental design 东方图案
oriental dress 东方服装；东方女服
oriental embroidery 东方刺绣
oriental fashion 东方流行服饰
oriental fiber 拉伸纤维，取向纤维
oriental knot 东方结
oriental lace 东方式花边
oriental look 东方（指地中海以东地区）款式，东方风貌，东方式样
oriental pattern 东方花样
oriental pearl 东方珍珠
oriental pendant 东方垂饰
oriental red 正红，大红色
oriental reproduction 东方地毯复制品
oriental reproduction rug 机制仿手织东方地毯
oriental rug 手织东方地毯
oriental rug knot 东方地毯结
oriental satin 东方缎
oriental silk 东方绸
oriental stitch 东方绣，东方针迹（长列平行刺绣针迹中央有斜纹短针）
orientation 定向性
original 独创服饰，独创服饰品；原型原样（设计师设计不加修改的原型）
original brand 独创商标，专有商标
original certificate 原始凭证
original colour 原色
original design 图案原稿
original drawing 原图
original film 黑白稿，黑白片
original form 原型人台
original inspection 初始检验
original pattern 花样原稿，基本图，原纸样

original pattern line 原型线
original piece 原布
original sample 原样
original shade 原色度
O-ring 单环扣;环形扣
orion blue 猎户星蓝
oripeau 法国奥利博金丝带
orissa cotton 印度奥里萨棉
Orlane 幽兰
orle 毛皮边缘
orleans 奥尔良花哔叽;棉经毛纬密织布
orleans cloth 奥尔良棉毛呢
orleans cotton 奥尔良棉
orleans linings 奥尔良棉毛里子呢
orleantine 奥尔良廷花哔叽
orlon 聚丙烯腈纤维(商名,美国杜邦)
ormuk 奥穆克驼绒细呢
ornaburg 粗袋布
ornament 点缀,装饰;装饰纹样,点缀品,饰品,饰物;级样
ornamental belt 装饰带,装饰腰带
ornamental belt buckle 装饰性皮带扣
ornamental braid stitch 辫状装饰性线迹
ornamental button 装饰纽;装饰纽扣
ornamental buttonhole edge 纽孔边饰
ornamental design 花饰
ornamental edge finishing 装饰性修边
ornamental fabric 俄罗斯服饰布
ornamental [pattern] seam 装饰缝
ornamental seam 装饰缝
ornamental seam configuration 装饰缝图样
ornamental sewing 装饰缝纫
ornamental sole rim seam 鞋底边装饰缝
ornamental stitch 花式线缝,装饰线缝
ornamental stitching 装饰缝纫;装饰线缝
ornamental stitch type 装饰线迹型
ornamental thread 装饰线
ornamental thread eyelet 装饰线纽孔
ornamental thread guide 装饰线导架
ornamentation 装饰,装饰品
ornamentation linen 亚麻装饰布
ornament neckline with shirring 用抽褶装饰领口
ornament stitch of jeans pocket 牛仔裤袋上的装饰线迹
ornate metal closure 装饰性金属扣件
Oroyqenzu ethnic costume 鄂伦春族服饰
orphrey 奥费利刺绣;奥费利绣带;金(或银)线刺绣饰边(法衣前襟上),精致刺绣

orraye 奥赖双面绣花缎
orrelet 女士遮耳头巾
orrice 金银丝带子,装饰编带
orris (装饰用)金银线编带;金银丝刺绣,花边
orsa lace 本色亚麻粗花边
orseill 苔色素
orsey silk 加捻生丝
ortho mixture 不同纤维交织物
orthoblend 交织织物
orthodox look 正统款式
orthodox style 正统款式
orthodox textile fiber 常规纺织纤维,传统纺织纤维
orthogonal woven fabric 正交机织物
orthopaedic figure 畸形体型;体型矫正
orthopaedic strapping 外科矫正畸形用带
orthopedic shoes 矫形鞋,整形鞋
ortigao 巴西苎麻韧皮纤维
ortigues 法国奥尔蒂格粗帆布
Orvieto lace 奥尔维也托花边
Osaka City University 大阪市立大学(日本)
Oscar de la Renta 奥斯卡·德拉伦塔(1932~,美国时装设计师)
oscillating arm 拐臂
oscillating rock shaft 缝纫机摆轴
oscillating shaft 缝纫机下轴
oscillating shuttle 缝纫机摆梭
oscillating traverse motion 往复摆动运动
oscillating washer 摆动式洗涤机
oscillogram 波形图
osnaburg 粗平布,低支纱色格棉布,柳条或色格粗棉布,废纺棉织物
osprey 帽饰鸟羽
ossan 苏格兰高级羊毛长筒袜
ossein silk 骨胶原人造丝,生胶质人造丝
Ossie Clark 奥希·克拉克(英国时装设计师)
ostrich down 鸵鸟绒毛
ostrich feathers 鸵鸟羽毛
ostrich(leather) 鸵鸟皮(革)
ostrich leather shoes 鸵鸟革皮鞋
ostrich plumes 鸵毛花纹
ostrich yarn 卷曲绒毛线
Ostwald 奥斯瓦特(色彩学著名教授)
Ostwald colour space model 斯特瓦尔德色立体
Ostwald colour system 奥斯瓦特[表色

系]制
O style　O形[服装款式];O字形款式
OTC　贸易合作组织
Ottavio Missoni　奥塔维奥·米索尼(意大利时装设计师)
otter　水獭皮;沃特绵羊
otter imitation　仿水獭皮,人造水獭皮
ottoman　粗横棱纹织物,粗直棱纹织物
ottoman cord　粗直棱纹织物
ottoman plush　19世纪棱纹地丝绒
ottoman rib　粗横棱纹织物,粗直棱纹织物
ottoman silk　四维呢
ottoman wool and mixed cloth　奥特曼呢
O type knot　O形结
ou crepe　偶绉
ouate　棉絮
ouch　宝石胸针(或饰针、扣子),饰物;珠宝
oulemari　树皮纤维
oumejima　青梅缟
ounce duck　盎司帆布
ounce thread　细亚麻线
oursine　法国奥尔辛粗呢
oushak rug　奥沙克松软绒头地毯
out　向外推档
outbacks　原野款式
out basting　外假缝;锯齿形纽孔锁缝
outdoor accessories　户外饰物配件
outdoor apparel　户外服装
outdoor clothes　室外服装
outdoor coat　室外外套
outdoor dress　14～15世纪的户外裙;14～15世纪的室外装
outdoor exposure　露天暴露,室外暴露
outdoor exposure test　室外露置试验
outdoor fabric　户外用织物
outdoor furnishings　户外用布,野外用布
outdoor look　室外款式
outdoor ourtain　室外帷幔
outdoor resistance　耐室外气候老化性
outdoor sportswear　户外休闲运动服
outdoor style　室外款式
outdoor textile　室外[用]纺织品
outdoor weathering　室外气候老化,室外风化
outer apparel　外衣服装
outer belt　外用腰带
outer carton　外箱
outer circle　外圆

outer clothing　外套,外衣,轻便大衣
outer coat　外套,大衣,轻薄大衣,女式宽大衣
outer collar look　外翻领款式,外翻领型(衬衫领翻出外衣)
outer defect　外观疵点
outer dimension　外形尺寸
outer fabric　面料
outer foxing　外围条(鞋)
outer garment　外衣,外套
[outer] garment knitting　外衣编织
outer garments　外衣类
outer [inner] carton　外[内]纸箱
outer jacket　外衣;罩衣
outer malleolus　外踝
outer man　男子的外表装束,男子外表
outer packing　外包装,外箱
outer-shirt　春秋衫
outer sleeve　外袖[片]
outer vest　户外背心
outerwear　(总称)外衣;外套;户外装
outer woman　女子外表;女子装束
out-fashioned　裁剪成形
outfit　服装;全套衣服;套装;专用套装
outgoing quality limit　出厂品质下限
outh　饰针
outing cloth　户外运动服织物;条子呢;素色呢;软绒布
outing costume　室外服装,外出服
outing dress　(欧洲)外出女装
outing flannel　软绒布,轧绒布
outing shirt　旅游衫
out knee　膝内翻
outlandish costume　奇装异服;奇异服装
Outlast fiber　调温纤维(腈纶型,商名)
outlet　工厂直销商场;男式长裤放缝头
outlets　打折店
outlet seam　缝份;缝头,毛边,毛头
outline　略图,轮廓,外形;(服装的)轮廓线;(制图)实线
outlined buttocks　紧臀纹
outline design　外形设计
outline drawing　草图,略图,轮廓图,外形图
outline embroidering　勾边绣
outline embroidery　轮廓刺绣
outline mark　裁片净样号,净样符号
outline of pattern　花纹轮廓
outline pocket　明袋,贴袋,明贴袋
outline quilting　随花绗缝

outline seam 轮廓缝,外型缝
outline smocking 轮廓缩褶绣,缩褶状轮廓
outline stitch 轮廓线迹,凸凹花纹针法
outline stretch stitch 外廓伸缩线迹
outline work 轮廓刺绣
outmoded machine 老机器
out of colour sample 与色样不符,不合色样
out of date fashion 过时服装
out of door fabric 户外用织物
out of fashion 不时兴,过时,不时髦
out of grainline 偏丝绺
out of livery 穿便服,穿便衣
out of mode 不流行;过时
out of order 不正常
out of phase 失真
out of register 不对齐,对版不准,对花不准
out of round diamond 不圆钻石
out of roundness 不圆度
out of season 不合时令,过时
out of service time 停机时间;停台时间
out of set 安装不良
out of shape (服装)走样,变形
out of style 不时兴
out of uniform 穿便服
out pleat patch pocket 压片袋,压片贴袋
out pocket 明袋,贴袋,明贴袋
output 产量
outseam 明缝;外接缝;外露缝;裤栋缝;裤侧缝;手套外接缝
outseam length 外缝长;裤长
outseam on trousers 裤中缝,裤子外缝,裤中外缝线
outseam run 裤子外缝线
outseam trousers 明缝裤
outshot 等外品;废品
outside 外观,外表
outside breast pocket 外胸袋
outside collar 表领;领面(里)
outside dimension 外形尺寸
outside edge of collar 翻领外口线
outside face 外表面
outside fold 外折边
outside foot 外压脚
outside in 服装翻里
outside leg length 腿外侧长(自腰际线沿臀部至地面距离)
outside length 裤长

outside line of collar 翻领外口线
outside line of collar band 底领下口线,领座下口线
outside line of left fly 裤门襟外口线
outside line of right fly 裤里襟外口线
outside packing 外包装
outside pocket 外口袋
outside seam 外侧缝
outside shop 外加工店,外加工服装厂
outside sleeve 外侧袖,外袖[片],大袖片
outside sleeve length 外袖长;外臂长
outside sole 鞋外底,鞋前底
outside structure 外型结构
outside welting foot 贴边缝外压脚
outside wrapping 外包装
outsize 裁剪毛头;特别高大的人;超常尺码;非标准尺寸制品;特大尺码的服装;特大号;特大号服装;特大袜筒女袜
outsize clothes 特大号衣服;超大码服装
outsole 皮鞋(皮靴)跟;鞋外底,大底,鞋前底
outsole with heel 连跟底(鞋)
outstanding loan 未偿还的贷款
out-to-inside colour difference 内外色差
out-to-out 总尺寸,总长度,总宽度,全长,全宽
outturn 产量;产品质量
outturn sample 到货样品
outward beauty 胴体美,人体曲线美
outward collection (O/C)出口托收
outward corner 凸角
outward curve 外凸[的]弧线
outward-curved back 外弯背型
outward documentary bills 出口押汇
outward processing 外加工
out wear 外衣;穿破,穿旧
out wear machine 外衣针织机
ouvrage 法国刺绣
ouvré 刺绣图案;多臂小花纹织物
oval cabochon 橄榄穹顶式(珠宝)
oval face shape 椭圆脸型
oval flat table 橄榄平台形(珠宝)
oval hexagonal crosscut 橄榄六角形相交磨翻(珠宝)
oval line 椭圆形
oval neck U字领,椭圆领
oval neckline U字形深领口,椭圆形领口,蛋型领口
oval silhouette 椭圆形轮廓,蛋形轮廓
oval silk 松捻纬丝线(刺绣)

oval toe （女式）椭圆形鞋头
oval trunk hose 男式椭圆形拼缝裤
oven （整烫用）烘箱
overall 罩衣,宽大罩衫(家里穿);工作服,防护服,背带工装裤;紧身军裤;总体
overall adjuster and buckle 工装裤调节袢及纽扣
overall buckle 工装裤环扣;工作裤环扣;滑带扣环
overall cloth 俄罗斯劳动布
overall culottes with self belt 背带裙裤
overall design 满地花纹图案设计
overall dimension 总尺寸,外形尺寸
overall drawing 总图
overall efficiency 总效率
overall flocking 满地植绒
overall height 全身长
overall jeans 背带牛仔工装裤,牛仔套裤,牛仔围兜套裤,木工牛仔裤
overall length 总体长度
overall plan 总体规划
overall project 总体设计
overalls for medical personnel 医务工作服;医护服
overall size 总尺寸,外廓尺寸
overall twill 工作服用斜纹织物
overall view 全貌图,全面图
over and over stitch 倒回线迹,倒回针
overarm length 外臂长
overarm seam 外袖缝
overarm sleeve length 全袖长
overbleach 过漂,漂白过度
overblouse 女式长罩衫,女罩衫
overblouse swimsuit 两件式女子泳装
overblown 腰围过大
overblue 荧光钻石
overbodice （穿在女式连衣裙外的）小背心外套
overboots 高腰套靴,胶皮套鞋
over briefs 高腰三角裤
overcast 锁边;拷边;包缝
overcast armhole 锁袖窿边
overcast bottom 锁缝下摆
overcast edge 锁缝边,包缝边
overcasting （片料）锁边;包缝;拷边
overcasting machine 锁边机;包缝机
overcasting stitch 包缝线迹,包缝线迹,锁边线迹
overcast running stitch 绕线平针绣

overcast seam 锁边缝,包边缝
overcast shell edge 包边缝贝形缘边(锁边缝贝形缘边)
overcast stitch 锁缝线迹,包缝线迹,环形针法,绕线
overchain stitch 链式线迹
overcheck 套格子花纹(织物),相重格子
overcheck sport coat 大方格运动外衣
over closed shedding 下开口
overclothes 外套,外衣,罩衣
overcoat 大衣,男大衣,外套,大衣
overcoat button 大衣纽扣,有柄纽扣
overcoating 厚大衣呢;外套料,大衣料
overcoating finish 表面涂层整理
overcoat suiting 大衣呢
over collar 领面;外层领
over collar style 外领款式
over-construction 结构过紧,织物组织过密
over-curly hair 深度卷发
over dampening 给湿过分
overdecoration 装饰过度
over-depreciation 超提折旧
overdrape 厚窗帘
overdress 薄外衣,外套;大胆奇异的装束
over-dyeing 多重染色
overedge 拷克,锁缝,包缝,锁(拷)边
overedge chain stitch 包缝链式线迹
overedge foot 包缝压脚
overedge hem 包缝边,拷克边
overedge intermittent shirring 包缝间歇打褶
overedge machine 扎骨机
overedge machine with 2-needle 4-thread 二针四线包缝机,二针四线拷边机,扎骨机
overedger 包缝机,包边机
overedge seam 包(边)缝,锁边缝
overedge sewing 包边缝纫,锁边缝纫
overedge sewing machine 锁边机,包缝机
overedge stitch 锁(拷)边线迹,包边线迹,包缝线迹
overedge stitch 1-needle 3-threads 一针三线包边线迹
overedge stitch seam 锁边线迹缝
overedge stretch stitch 伸缩性包边线迹
overedge width 包边幅度
overedging 包边缝纫,包缝
overedging machine 拷边机,包边机

overedging machine with 2-needle 4-thread 二针四线拷边机,扎骨机
overedging operation 包边缝纫
overedging seam 包边缝
overedging width 包边幅度
over-exposure 曝光过度
overfeed fabric 经编超喂织物
overfeed ratio 超喂比
overfeed tenter 超喂拉幅机
overflow dyeing 溢流染色
overgaiter 呢鞋罩
overgarment 大衣,罩袍;外衣类总称
overgassed 烧毛过度
overgown 外袍,外长袍;无袖罩袍,长罩袍
overgreen 深绿色
overgrown 过长,太宽大
overgrown wool 过长毛
overhand 锁缝,平式缝接
overhanding 锁缝,平式缝接
overhanding seam finish 平式缝接处理
overhanding stitch 锁缝线缝,边缝针迹
overhand knitting 平式缝接编织
overhand knot 反手结
overhandling stitch 边缝线迹,接缝线迹
overhand seam 搭缝
overhand sewing 卷边缝合,卷边缝纫
overhand stitch 对接缝线迹,绞针线迹
overhand tuck 平式缝接细褶,绕边缝裥饰
overhang 空出余量
overhanging cutting knife 悬垂切刀
overhanging edge 松边
overhaul 大修[设备],修配
overhaul period 检修期
overhead 管理费
overhead conveyor 悬挂式输送器;悬挂式输送线
overhead measurements 头围尺寸
overhead-plating weft knit fabric 架空添纱编针织物
overhead seam 搭缝,搭接缝
overhead sewing 卷边缝合
overheating 加热过度
over-jacket 罩衣;外褂
over-jupon 罩裤,外裤;罩衣
over knee 过膝
over knee boots 过膝长筒靴
over knee length (服装)超膝长度,过膝盖长

over knee socks 过膝长袜,长筒袜
over knee stockings 过膝长袜,长筒袜,舞袜
overlaid seam 搭缝,搭接缝
overlap (门襟)交搭,重叠;袖衩搭边
overlap belt 有搭头的腰带
overlap dart 叠合缝的省道,压缝的省道
overlap extension closing 外伸式腰头,探出腰头
overlapped seam 折叠缝,搭接缝
overlapped working 工序重叠
overlapping 重叠,交叠,搭接
overlapping design 重叠印花花纹,搭交印花花纹
overlapping mark 裁片交叉重叠号
overlapping peaks 重叠峰
overlapping seam 压缝,搭缝
overlap seam 搭缝;搭接缝,压缝
overlap stitching 接针线缝
overlap stitching method 搭缝法
overlay 装饰鞋面层(制鞋)
overlay mat 表面薄毡
overlay supervisor 重叠管理程序
overlay tape 较粗的包层带
overlock 包缝,锁(拷边)
overlocked seam 包缝
overlocker 包缝机
overlock hem 拷克边
overlocking 包缝,锁(拷边)
overlock machine 包缝机,锁边机,拷边机,拷克车
overlock seam 包缝;锁缝
overlock seaming machine 包缝机
overlock seam width 包缝宽度
overlock sewing 锁边缝纫;包边缝纫
overlock sewing machine 包缝机,锁边机,拷克车
overlock stitch 包缝线迹
overlock stitch sewing machine 包缝缝纫机
overlock with five-thread 五线包缝
overlock with three-thread 三线包缝,锁三线
over looper 上弯针
over mature fiber 过成熟纤维
over-mercerization 过度丝光
overnight bag 短途旅行小提箱;小旅行包
overnight case 过夜旅行箱
overnight clothes 供一夜或短期过夜用替

换的衣服
overnight case 过夜旅行箱
overnit fabric 单面浮线针织物
overpants 外裤,套裤,罩裤
overplaid 套格子花纹,相重格子
overpress 烫过头,烫黄
overpressing 烫过头,烫黄
overprinted resist 罩印防染
overprinting 叠印,套印,罩印
overprint over printings [地色上]罩印,[印花上]套印,[织纹上]盖印
overpull 套衫,套头衫
over-raising 起绒过度
over relax 过分松弛
over-robe 罩袍,袍褂
over sack 短大衣,短外套
over seamer 包缝机
overseam(ing) 包缝,绷缝
overseaming machine 包缝机
overseas cap 海外帽,军用船形帽
overseas market 国外市场,海外市场
over seat length (服装)超臀长度
oversetting 过度定形
oversew 对缝
overshirt 罩衣,罩衫,外套式衬衫
overshoes 套鞋,鞋套,鞋罩,罩靴
oversho(o)t 浮纬花纹
overshot coverlet 几何图案床单
over-shoulder 横宽肩型
over-shoulder bag 肩背包
over-shoulder measure 肩宽尺寸
over shrinking 过缩
oversize (OS) 加大尺寸,特大尺寸,特大型,尺寸过大;特大型尺码(服装)
oversized 超量上浆的,重浆的,特大的
oversized cardigan 大型卡迪根
oversized collar 特大型领
oversized garment 特大号服装
oversized jumpsuit 特大式连身裤装
oversized knit shirt 特大型T恤衫
oversized look 大款式,特大款式;超大风格,特大型风貌
oversized shirt 超大号衬衫
oversized style 特大款式
oversized T shirt 特大型T恤衫
oversized top 特大套衫
oversize line 特大型
oversize silhouette 特大型轮廓
oversize top 特大套衫
overskirt 半裙;上套裙;外裙;短罩裙;两页裙;套裙
over-slashed bar 重浆档
oversleeve 袖套,袖筒,罩袖,套袖
over soft finishing 超柔软整理
over square 套格子花纹
overstitch (装饰用)明线迹;覆盖线绣;面缝线迹
overstitching and serging stitch 包缝和包边线迹
overstitching seam 锁边缝,包边缝
overstitch machine 包缝机
overstrain 张力过大
over stripe 套条子花纹,格子花纹,相重条纹
oversuit 外套装,防护罩衣
over sweater 运动服外套
over-the-calf stockings 中筒女袜
over-the-counter trading 现货交易
over-the-knee look 过膝风貌
over (the) knee socks 盖膝长筒袜
over-the-knee style 过膝风格
over the knee [length] stockings 过膝长筒女袜
over the seat length 长至臀下,及臀衣长,盖住臀部的衣长
overtight package 过紧包装
overtop 套头衫
overtuft 超量簇绒
overtuft fabric 超长簇绒织物
overturned collar 翻领
over vulcanization 过硫(鞋)
overtwisted caterpillar 强捻毛虫花线
overweight 超重
overwidth 布幅过宽
over-wraping 搭接;外包装
owl glasses 枭型太阳眼镜(特大、粗宽镜框)
ownbrand 自有商标(指零售商不用制造商的商标)
owning stripe 遮日条纹(鲜艳粗条纹)
ox leather 公牛皮,牤牛皮
ox leather shoes 牛皮鞋
ox tendon sole 牛筋底(鞋)
ox tendon sole leather shoes 牛筋底鞋
Oxalon 奥克萨纶
oxazine dyes 嗯嗪染料
oxblood [red] 牛血红(棕红色)
Oxford 牛津布;深灰色混纺针织纱;灰色合股线毛呢;条格竹布;牛津鞋,浅帮鞋
Oxford bags 牛津袋裤子,牛津裤,(宽松)

袋形裤
Oxford bag trousers　袋型裤
Oxford blue　牛津蓝,深蓝色,深紫色
Oxford bottom　宽松直裤脚,牛津裤裤脚
Oxford cap　牛津帽,学位帽,学士帽
Oxford chambray　青年布,牛津钱布雷织物(平纹,色经白纬)
Oxford check　牛津格
Oxford cloth　牛津纺,牛津布
Oxford coatee　单排扣粗呢男夹克
Oxford down　牛津塘种羊毛
Oxford fabric　牛津纺,牛津布
Oxford gray　牛津灰,深灰色;牛津灰色呢
Oxford gray mixture flannel　牛津灰法兰绒
Oxford grey　牛津灰,深灰色;牛津灰色呢
Oxford lenger　牛津革
Oxford mixture　牛津灰;牛津灰色呢
Oxford ocher　深色黄,深土黄
Oxford shirt　牛津衫,细条纹衬衫
Oxford shirting　牛津布,牛津纺;牛津衬衫料
Oxford shoes　牛津鞋,浅口便鞋,耳式靴
Oxford tie　牛津领带
Oxford trousers　牛津裤
Oxford-grey mixture flannels　牛津灰法

兰绒
oxhide　大牛皮;大张牛皮革
oxidation dye　氧化染料
oxidation strip　氧化剥色
oxide fiber　氧化物纤维
oxidized colour　氧化染料
oxidized fiber　氧化纤维
oxidized wool　氧化羊毛
oxo wool　替代羊毛的亚麻
Oxonian [jacket]　牛津装
oxter　腋窝(苏格兰、爱尔兰)
oxygen consumption　耗氧量
oxygen depletion　缺氧
oxygen fading　氧气退色
oxygen index method　含氧指数法
oxygen mask　氧气面具,氧气面罩(高空飞行员、潜水员用)
oxygen umbilical cord　脐带式供氧管
oyah lace　(土耳)奥亚赫花边;土耳其针绣花边,土耳其钩编粗花边
oyster gray　牡蛎灰(淡灰绿色),浅灰色
oyster linen　奶油色刺绣亚麻布
oyster white　牡蛎白
ozier　博布棉
ozone fading　臭氧褪色
ozu aya　日本棉斜纹布

P

pabbled 满地皱纹,满地小卵石纹
pabnapar 印度小花纹细布
paboudj 摩洛哥尖头平跟便鞋
pac 派克靴;高筒鹿皮靴;(寒冷天气穿的)防水系带高筒鞋;(寒冷天气穿在皮靴或套鞋内的)无跟系带羊皮(或毡)鞋
pac boots 派克靴,高筒鹿皮靴
pachas 马海毛凸条呢
pachras 孟加拉湾彩条粗布
pacific nettle 太平洋荨麻
pacific 深蓝绿
pack 包(纤维包装单位),捆;单元,部件,组件;组装;包装;包装材料;背包;发乳,香脂,护肤霜;捆扎;打包,装箱;软皮鞋,鹿皮鞋(穿在靴内)
packable hat 可压式帽
packable straw 可压式草帽
packable style 可压式
package 包装,卷装
package bag 包装袋
package bale 件(包装单位)
package density 包装密度,卷装密度
package design 包装设计
packaged production 小批生产
package dress 包装服(可叠成很小尺寸且不皱的轻巧服装)
package introduction 成套引进
package linen 亚麻包装布
packages in damaged condition 包装破损
package size 包装型号(大小),包装大小,卷装大小
package suit 组合套装(多套服装互换配套)
package transfer 成套转让
package without former for sewer 缝纫机无管卷装
packaging 包装,装箱;内包装;包装风格;装配,组装;包装术;包装物;包装业
packaging bag 包装袋,打包袋
pack band 行礼带
pack belt 打包带
packboard 帆布背包
pack cloth 包装布;打包布
pack clothes after ironing 熨烫后包装衣服

pack duck 打包帆布
pack dyeing 筒装染色
packed 包装
packed in bag 袋包装
packed in box 盒包装
packed in canvas 帆布包装
packed in carton 纸板箱包装
packed in case 箱包装
packed in cloth 布包装
packed in fancy carton 彩色纸盒包装
packed in gunny sack 粗麻袋包装
packed in jute bag 麻袋包装
packed in kraft paper 牛皮纸包装
packed in mat 席子包装
packed in paper 纸包装
packed in paper box 纸盒包装
packed in plastic bag 塑料袋包装
packed in woven polypropylene bag 聚丙烯编织袋包装
packer 包装工,打包工;打包机
packet 小包,小盒,小袋,小捆
pack goods in box 货物装箱
packing 包装;装箱;打包;包装法;包装标志;外包装
packing bag 包装袋,打包袋;包装袋用布
packing bag fabric 包装袋用布,打包袋布
packing box 包装箱;包装盒;装货箱
packing braiding machine 包装袋编织机
packing case 包装盒;包装箱;装货箱
packing charge 包装费
packing clause 包装条款
packing cloth 包装布,打包[粗]布,打包麻布
packing cord 包装绳;打包绳
packing cost 包装成本
packing damage 包装损耗,包装损伤
packing design 包装设计;包装图案
packing dimension 打包尺寸,包装尺寸
packing felt 毡垫,毡衬
packing inspection 包装检验
packing linen 亚麻包装布
packing list 包装单,装箱单,花色码单
packing list and weight memo 装箱单和重量单

packing machine 打包机,包装机
packing material 包装材料
packing measurement 包装尺寸
packing needle 包装针,缝包针,缝包用大号针,打包用针
packing of exports 出口商品包装
packing pad 包装衬垫
packing paper 包装纸
packing press 包装机,打包机
packing room 包装间
packing rope 捆扎绳
packing rubber belt conveyor 包装工段橡胶带传送机
packing scale 包装秤
packing sheet 包装布;包装单(纸)
packing sleeve 衬垫套子
packing specification 包装规格
packing textile 包装用纺织品
packing thread 捆扎麻线
packing ties 打包铁皮
packing tool 打包工具,包装工具
packing twine 打包绳,包装麻线
pack rope 打包绳
pack sack 旅行背包
pack sheet 高级打包麻布,打包布,包装布
pack thread 包装缝线,捆扎线,打包绳,包装缝线
paco 羊驼毛;阿尔帕卡织物
Paco Rabanne 帕科·拉巴纳(1934～,法国时装设计师)
Paco Rabanne helmet 帕拉·拉巴纳头盔
paco wool 羊驼毛
pacputan 纬纱用粗毛
pad 垫,肩垫,衬垫;垫塞;衬垫料;装衬垫;浸染,轧染;波纹丝表带
padaya 帕达亚腰布(斯里兰卡),漂白棉腰带
padded armo(u)r 棉盔甲
padded-back lining 黑背印花里子布
padded bra(ssiere) 加垫型胸罩,镶垫乳罩,有垫胸罩
padded clothes 衬垫服装
padded doublet (peasecod) 棉紧身上衣(豆荚式)
padded girdle 衬垫腰褡
padded gloves 加垫手套
padded hat 防震填充帽,有填充料的帽子
padded hem 垫衬边缘

padded jacket 登山防风衣,东方式夹袄,填棉夹克,充棉夹克
padded satin stitch 包芯缎绣
padded shirt 填棉衬衫
padded shoulder 垫肩
padded stem stitch 垫梗包针法,立体包梗针法
padded vest 棉背心
padder brassiere 加垫(海绵等)乳罩
padding (衣服用)衬垫,垫塞,毛里衬;垫料;垫肩;袢丁;箱形垫;衬里织物;缝衬头
padding cloth 垫布,衬布;衬垫织物;棉絮;绣花垫底棉线
padding cotton 绣花垫底棉线
padding rolled lapel 翻领衬里
paddings 西装麻衬布
padding shoulder 垫肩肩线
padding stitch Z字疏缝线迹,工字疏缝线迹,人字疏缝线迹,衬里线迹,扎缚线迹
padding thread 衬垫线,加固线,填充线
paddock coat 男用紧腰骑马外衣
paddock [coating] 帕多克防水细呢(英国制,一般为棕色)
paddock model 小牧场夹克型(单排双纽扣夹克)
paddy field design dress 水田衣
paddy field hosiery 水田袜
pad fitting seam 试穿用衬垫缝,装垫缝
pad hood with down 将羽绒填入风帽
pad-ink 打印色
padisway 意大利帕迭斯威棱纹绸
pad joining 缝合衬里,上衬垫
padlette 帕德勒特凸花刺绣;贴花花样
padou 窄丝带
padre hat 神父帽
pad roll dyeing process 轧卷染色法
pads 波纹细丝带;眼镜丝带;衬垫
pad shoulder with sponge lumps 衬海绵垫肩
pad stitch 衬垫线迹,扎缚线迹;扎针,衲针(针法),扎缚针法
pad stitch for lapel 扎驳头
pad stitching 衬垫线缝,扎缚线缝;扎针,衲针(针法)
pad-stitching lapel 衲驳头,扎驳头
padua say 精纺斜纹呢
padua silk 意大利厚棱纹绸

padua soy 意大利厚棱纹绸;棱纹绸服装
paejama 印度裤
pa'ejamas 穆斯林男裤,印度裤
paenulla 披奴拉披风(古罗马)
pagari 印度手织棉布
page (boy) bob 侍童短发
pageboy hairstyle 侍童式发型;妇女发梢向内卷曲的齐肩发型
pageboy jacket 侍童服,侍童夹克
pageboy style 侍童短发,侍童式发型,发梢向内卷曲的齐肩发型
pagehop jacket 侍童夹克;侍童服
pagliaccio 意大利喜剧的丑角衬衣
pagne 丝经蕉纬织物;非洲棉腰带
pagnes sor 高档棉布
pagnon 高级粗纺呢
pagoda[blue] 塔蓝(暗绿光蓝色),蟹青色
pagoda dress 塔式女装
pagoda hat 宝塔帽
pagoda line 宝塔型
pagoda sleeve 宝塔袖,佛塔袖
pagoda toque 宝塔豆蔻帽,塔形女帽
pagri 印度手织棉布;印度头巾
paguin 帕坎(法国时装设计师)
PA handle 聚酰胺织物手感
pahom 披巾(泰国男女作上衣用,色彩鲜艳)
pahone 披巾(泰国男女作上衣用,色彩鲜艳)
pahpoon 闪光棉布
pah-poosh 波斯高跟丝绒拖鞋
paiangposh 床单;被单;床罩
pail 洗衣桶
paile 中世纪一种皇室织物
paillette 亮片,闪光珠片;珠片装饰;闪光织品
paillette de soie 亮晶绸,珠片绸
paillette noir 有光细斜纹内衣绸
paillette pants 珠光裤
paillette satin 闪光缎,珠片缎
pailotisha 白色真丝纱罗
paina limpa 木棉
painsi 印度粗双经布
paint 化妆品,油彩,颜料
paint box 颜料箱
paint brush 画笔
painted cloth 17世纪英国油漆格言帆布
painted design 花布图案,手绘花样
painted fabric 手绘花布,手描花布

painter coat 油漆工罩衫,画家工作罩衫
PA interlining (polyamide interlining) 聚酰胺(黏合)衬
painter pants 画家工作裤;油漆工裤
painter's 画家裤
painter's canvas 油画帆布
painter's cap 画家帽,油漆工帽
painter's coat 画家外套
painter smock 画家罩衫,油漆工罩衫
painter's pants 画家裤,油漆工裤
painter style 画家风格
paintex 服饰画艺
painting 绘画,着色
painting art 绘画艺术
painting dress 绘画装(印有肖像画或风景画的服装)
painting table 画桌
paint printed handkerchief 涂料印花手帕
paint wool 带涂漆标志的羊毛
paira pratti cotton 印度泛红棉
pair blouse 配对女衫
pair buckle 字母带扣;对扣
paired comparison 成双对比
paired rings 双套环,双套环戒指
paired style 成双服装款式
pair fabric 配对织物
pairing 配对,配袜
Pairs 巴黎
pair style 成双款式,配对款式
pair style helmet 对式头盔
paishet 亚麻
paisley 佩斯利涡旋花纹;佩斯利涡旋纹花呢;佩斯利涡旋纹花呢制品(如披巾、领带等)
paisley motif 佩斯利图案
paisley pattern 佩斯利花纹,佩斯利涡旋花纹,火腿花纹;佩斯利图案
paisley shawl 佩斯利细呢披巾,佩斯利细毛披巾
paissan 粗黄麻纤维
paisseau 法国粗毛哔叽
paisu 中国东北产白桑蚕丝
paita 秘鲁培塔棉
paitusu 中国江西产白生丝
paj 洋纺全丝平纹薄织物,洋纺,小纺
pajam 印度帕贾姆棉布
pajama 休闲喇叭裤(女)
pajama bottoms 睡衣裤
pajama check 格子睡衣布,睡衣格子
pajama cloth 睡衣布

pajama coat 睡衣
pajama flannel 睡衣(用)法兰绒
pajama jacket 短睡衣
pajama look 睡衣款式
pajama pants 睡衣式长裤,宽松裤
pajamas 睡衣,睡衣裤,宽松裤
pajama set 成套睡衣,精做成套睡衣
pajama stripes 条子睡衣布,睡衣条纹
pajama style 睡衣款式
pajama trousers 睡衣,宽松裤
pajamas trousers 宽松裤,睡衣
pajamas with drawing waistband 束带睡衣
pajamas with waistband 束带睡衣裤
pajama webbing 宽型松紧带
pajama with drawing waistband 束带睡衣
pajunette 妇女睡衣
pakama 格子花绸
pakamas 帕卡玛围巾布,英国棉格子布
pakea 夏威夷平纹细席子
pakele 波罗的海毛纺细带
pakelite 波罗的海毛纺细带
Pakistani costume 巴基斯坦民族服饰
Pakistani vest 巴基斯坦式合体背心(镶金边的长青果领,合体型)
pakistan wool 阿富汗羊毛
Pakki 中国量布单位
palace 派力司,派力司织物,派力司花呢,精纺呢绒
palace brocade 宫锦,花纺,提花全丝派力司
palace crepe 派力司绉
palace mesh 宫纱
palace plain 全丝派力司
palace satin 贡缎;库缎
palace shoes 宫廷鞋
palai 泰国潘农布(重浆印花)
palain taffeta 塔夫绸
palambangs 棉头布,棉腰布
palampore 印度轧光印花布
palanche 法国麻经毛纬衬里布
palate bone 颚骨
palatinate 淡紫色;淡紫色运动上衣(作为荣誉标志)
palatine 毛巾披肩;薄纱小围巾,毛皮领巾
palazzo pajamas 宽松女式套装(半正式场合穿),松身装
palazzo pants 大裤腿女裤,阔脚裤,宫殿女裤,帕拉柔裤,松身裤

palazzos 阔脚裤(女子),松身裤
pale 苍白,浅色;灰白的,苍白的;中世纪一种皇室织物
pale agua green 淡水绿色
pale azalea 浅杜鹃花色
pale blue 淡蓝,淡蓝色
pale cerulean 淡青
pale colour 淡雅色
pale coloured shirt 淡色衬衫
pale fawn 淡黄褐色,淡糙米色
pale green 葱绿,淡绿,粉黄绿
pale grey 白灰
pale khaki 浅土黄色
pale lalic 浅莲灰
pale lavender 淡紫色
pale lilac 浅莲灰
pale lime yellow 淡酸橙色
pale look 苍白款式
pale lotus 浅灰莲色
pale mauve 浅豆沙,浅豆沙色
pale olive green 浅橄榄绿
pale orange red 浅红橙色
pale pink 淡粉红色
pale pinkish gray 藕色
pale purple 淡紫,青莲,浅青莲,紫丁香色
pale red 浅红
pale red purple 浅灰莲色
pale shade 淡色,浅色
pale shell pink 淡血牙色
pale star 淡星色
pale stinian embroidery 阿拉伯刺绣
pale strength prints 淡色印花
pale tone 淡色调,浅色调,淡雅色
pale yellow 苍黄,浅黄
palen pore 印度轧光印花布
Palestine head dress 巴勒斯坦头巾
paletot (男)礼服大衣;(男、女)宽外套,女式紧身上衣(19世纪);男上衣;双排扣宽摆男上衣
paletot jacket 沛乐杜外套
paletot sac 直筒大衣
palette 调色板
palette colour 调色板色彩
palette sweater 多色彩毛衣
palghat mat 草席
Palicat 帕利卡手帕(小亚细亚),土耳其手帕
palitana cotton 印度帕里塔纳棉
palium 帕留姆(拜占庭时期套头衫)

pall （教皇、主教的）披肩；教会用家具布；天鹅绒盖布；外衣细呢
palla （古罗马）妇女宽大罩袍,帕拉色裹衣（古罗马）
pallas 帕拉斯色织平纹棉围巾；帕拉斯毛皮绒
pallas fur 法国帕拉斯毛皮绒
pallet 调色板，鞋样模板，盔甲腋窝金属图片；腋窝甲
pallium （古希腊、古罗马）男子大披肩；（天主教教皇或大主教的）白色羊毛披肩带；袈裟；帕利姆披巾；中世纪金银丝绸布
pallor 灰白,苍白
palm 手掌,掌心；（手套等）掌部；掌尺；（手缝粗厚材料时顶针用的）掌皮
palm-bark rain cape 蓑衣
palm beach 胖比司,夏服呢,胖比司呢
palmbeige 胖比司
palm circumference 掌围
palmering 柔化加光整理；柔软加光整理
palmette 法国毛围巾；扇状叶；棕榈树叶纹
palmette design 地毯云纹图案
palmetto 菜棕绿（暗黄绿色）
palm fan 蒲扇
palm fiber 棕；棕榈科纤维统称
palm girth 掌围
palm-leaf cap 葵帽
palm-leaf fan 葵扇
palm-leaf sandal 古埃及棕榈叶凉鞋
palm ornament 棕榈装饰
palm size 掌围
palm wares 棕编织品
palmyra[fiber] 扇叶树头榈叶纤维
palo borracho fiber 南美木棉籽纤维
palomino 淡黄褐色
Palomino mink 巴洛米诺貂毛皮
palsano 高级帕尔瑟诺大麻（商名）
paltock 佩尔托克外衣（14~15世纪）
paludament 皇帝大氅；战袍（古罗马）
paludamentum 帕鲁达曼特披风（古罗马）
palungao 印度洋麻,槿麻
pam 羊绒,山羊绒,开司米（克什米尔用语）
pambah 棉花
pamdani 达卡细平布
pam hair 山羊绒
pamna hazara 印度细布

pampas grass 浅绿色,蒲苇草色
pampootie 莫卡辛风格鞋
pamut 手工编织和绣花毛线
panache 彩色花纹效应；羽饰
panache style 羽饰发型
panama 巴拿马（仿毛呢）；巴拿马薄呢,花呢；匹染平纹呢；方平粗布；手编草帽辫；巴拿马草帽
panama canvas 巴拿马帆布
panama cloth 巴拿马,花呢,巴拿马布
panama hat 巴拿马草帽,巴拿马帽（麦秆制遮阳帽）
panama hat palm 巴拿马帽棕
panama suiting 巴拿马西服呢
panama zephyr 巴拿马提花彩色条布
PAN-based carbon fiber 聚丙烯腈基碳纤维
pancake 化妆用的湿粉饼,水粉饼,皮质鞋垫
pancake bag 鼓状提袋,铃鼓包,烤饼状提包（形状如烤饼,或鼓的提包）
pancake beret 法式贝雷帽,盘型贝雷帽
pancake panama 巴拿马饼式帽,薄饼式巴拿马帽
pancela 西服背心
panchang knot 盘长结
pan check umbrella 细格子伞
panchpat 孟加拉湾染色绸
panchromatic film 全色片
panda fur 大熊猫毛皮
pane 长方格（棋盘格等）
panecia 法国金银线绣花丝绒
paned 用杂色小布片拼成的
paned hose 接缝裤（用绸条或丝绒条拼做）
panel 布块,裁好的衣片,女服嵌料,嵌条或饰条,拼块（皮革,裘皮）,镶片
panel coat 后背双饰缝女大衣
panel design 嵌花花纹,镶板花纹；片花图案,派内尔图案
panel dress 垂幅装,饰缝装（前后有饰缝的服装）,镶嵌装,派内尔连衣裙（缀有长方块饰布）
paneling 宽纵条花纹,阔纵条花纹
panel jacket 分割式外套
panel line 剪接线
panel line seam 衣片嵌条缝
panel pantdress 垂幅式连裤裙装
panel pleat 片饰褶
panel print 派内尔印花,画框型印花花

样,镶嵌图案
panel prints 画框型印花花样,镶嵌图案
3(4,5,6,7) **panels cap** 三(四、五、六、七)瓣帽
panel seam 镶嵌缝
panel seamer 女服嵌料缝纫机
panel skirt (裙上缀缝布块的)掩块裙,幅片裙,饰缝裙
panel stripe 粗条编织花纹,派内尔条纹
panel styleline 多开身结构线,结构分割线
panel warp 绣花添纱
panels cap 瓣帽
paneva 俄罗斯传统衬衫
Panex 氰基碳纤维,碳纤维(商名,美国)
pang 中国服用丝绸
pangalo cotton 埃及潘加洛棉
pangfil 南京丝绸
pangnio cotton 埃及褐棉
panier 垫臀衬裙,鲸骨圈,裙撑
panier de poche [法]鱼篮式提包
panjam 低级生丝
panne 平绒
panne effect 平绒效果
panne satin 高级厚重平缎
panne skirt 平绒裙
panne velvet 平绒
pan[n]ier 裙撑,鲸骨圈;垫臀衬裙,鼓裙
pannier bag 挂包
pannier crinoline 裙撑用衬裙,裙撑用硬衬布
pannier double 双拼裙撑
panniered overskirt 裙撑罩裙
panniers à bourelet 底边有裙环的衬裙
pannier silhouette 驮篮型
pannier skirt 撑裙式大裙,帕尼撑裙(18世纪)
panning 泰国潘农布(重浆印花)
panno combrido 东南亚平布,印度平布
pannonia leather 仿皮漆布
pano cru 厚棉床单布
pano ferro 帕诺弗罗粗亚麻织物
panoply 礼服;全副甲胄;全副盔甲
panossare 红条棉腰布
panox 聚丙烯腈预氧化纤维(商名,英国 RK 碳纤维公司)
panrigcs 印度花卉纹绸
pans 内裤
pan stick 棒状油质香粉

pansy 三色堇紫(深紫色)
panta court 短裤
panta-court shorts 白色束腰带紧身羊毛短裤
pantagraph 缩放仪
pantalets (19世纪)露在裙外的宽大女长裤;灯笼裤;宽松女裤
pantalettes (19世纪)露在裙外的宽大女长裤;灯笼裤
pantalon 长裤,裤子
pantalon collant 紧身裤
pantalon dress 连裤裙装
pantalon masculin 男式风格的裤子
pantalon mince 细长裤子
pantalon suit 长裤套装
pantaloon [法]裤
pantalo(o)n de grandpere [法]祖父裤
pantalo(o)n de maharajah [法]君王裤
pantalo(o)n Indian [法]印度风格裤
pantalo(o)n masculin [法]男式风味裤
pantalo(o)n mince [法]细窄裤
pantaloons 马裤,长裤,窄裤;(英国查理二世时期)宽大灯笼裤;美国中裤
pantaloon suit 长裤套装
pantaloon training 运动长裤
pant coat 驾车外套;连裤套衫;轻便型大衣
pant draft 裤装制图
pantdress 穿在相配短裤外的女服;连裤裙;裙裤套装
panted lace 打褶花边
pantee 童裤;女裤,女运动短裤
pant fit 裤子合身性
pant garment 裤装
pantgown 全长连装裤(正式场合)
panther skin 美洲豹皮
pantie 裤衩裆布
pantie belt 束腹女衬裤
pantie briefs 紧身短衬裤,紧身三角裤
pantie-corselette 妇女紧身褡,连胸衣紧身短裤
pantie garter 内裤吊袜带
pantie girdle 束腹健美裤,束腹短裤,紧身女衬裤,绑裤
pantie hose 连裤袜,连袜裤,紧身裤
pantie-less style 无胯款式
panties (女、童)紧身短裤;短内(衬)裤;女裤;女式运动短裤
pantie-shorts 紧身短衬裤
pantihose 紧身衣裤,连裤袜,连袜裤

panting 裤料
panting machine 打褶机
panti-slip 连短裤衬裙;背心热裤装
pantistockings 连裤袜,连袜裤,紧身衣裤
panti-tights 卫生衣裤
pant-jumper 无袖连身裤装
pant leg 裤腿
pantliner 紧身束腹内裤,贴身长衬裤
Pantofile 潘特富尔鞋
pantofle 室内用柔软便鞋;拖鞋
pantograph 缩放仪
pantone 全色调
pantopat 缩放花纹试样机
pantoufle 室内用柔软便鞋
pant proportion 裤子比例
pantry jacket 餐厅男服务员短上衣
pants 裤子,长裤,(宽松的)便裤;女式运动短裤;(女、童)紧身短衬裤;[英]男短衬裤;紧身长衬裤
pants boots 踝高靴;及踝配裤短靴;连袜鞋紧身裤
pants[coat]backpart 裤[衣]后片
pants dress 连裤裙装,裤裙女装,百慕大装
pants ensemble 裤子组合装
pants foundation 束腹连裤胸衣
pants girdle 裤腰带;紧身女衬裤;绑腹健美裤;束腹短裤;绑带
pant shoes 厚底鞋;与喇叭裤匹配的皮鞋;松糕鞋
pants hose 女用袜裤
pants inseam 下档长
pant skirt 大脚管裤裙;裙裤
pants length 裤长
pants lines 腹带,收腹内裤
pants look 以裤子为中心的款式;注重裤子的款式
pants loons 马裤
pants seat 裤臀围
pants shoes 配喇叭裤的鞋
pants shop 裤子专门店
pants size 裤子尺寸
pants skirt 裙裤
pants socks 裤用袜
pant-suit 长裤套装(上衣与裤子配套的妇女服装)
pants waist 裤腰围
pants waist flat 裤腰围
panty 短内裤,女短裤,童短裤
panty boots 连袜鞋紧身裤

panty dress 搭裤连裙装
panty foundation 束腰连裤胸衣
panty girdle 绑腹健美裤,短衬裤紧身褡,紧身女衬裤,内裤吊袜带,束腹短裤
panty hose 连裤袜,连袜裤,紧身裤
pantyhose blank 连袜裤袜坯,连裤袜坯
pantyhose for fit 适体连衣袜
panty set 连身裤
panty-shape diapers 纸尿裤
panty skirt 打褶短裤裙
panty slimmer 紧身内裤
pantyslip 小女孩短衬裙
panty stockings 连袜裤,紧身衣裤
panty waist 连裤童装,幼儿衣裤
Panung 潘农(腰)布(泰国)
pan yarn 马毛纱,人造马毛纱
PAN yarn 聚丙烯腈纱
pao 袍
paoningkuopan 中国四川最优级生丝
pao velvet 帕昂天鹅绒
papache hat 俄罗斯黑羊皮帽子
papal cross 有三条横档十字架
papalina 意大利毛葛
papeline 丝经绢纬绸
papal yellow 菩提树色
paper 纸;文件;票据
paper bag 纸袋
paperbag waistline 抽绳式腰线
paperboard 纸板
paper brush 织物砂纸起绒整理
paper cambric 窄幅细纺,窄幅细薄布
paper chromatogram 纸色谱(图)
paper cloth 纸布,涂纸浆棉布或麻布
paper cone 宝塔纸管
paper cord 纸纱,纸绳
paper-cut 剪纸艺术
paper cutter 裁纸刀
paper-cutting scissors 裁纸剪刀
paper doll dress 纸娃娃蓬裙式连衣裙(紧身,收腰)
paper doll look 纸娃娃款式,纸娃娃风貌
paper doll style 纸娃娃款式,纸娃娃风貌
paper dress (一次性)纸衣服,纸服装,用即弃服装
paper fabric 纸质织物,纸背复合织物,

用即弃织物
paper fan 纸扇
paper felt 造纸用毡,造纸毛毯
paper felt duck 造纸帆布
paper from synthetic fiber 合成纤维[制成的]纸
paper guide 纸做的小样板
paper hangtag 纸吊牌
paper hat 纸帽
paperiness 纸感,纸样感(织物风格)
paper knife 裁纸刀
paper knit 纸编针织物
paper knitting 纸线编织
paper lining 衬纸
paper machine felt 造纸毛毯
paper-made-dummy 胶纸人体模型(用胶纸和针织紧身衣做成)
paper makers' felt 造纸毛毯
paper making canvas 造纸帆布,造纸毛毯
paper making felt 造纸毛毯
paper muslin 轧光硬挺里子细布,轧光硬挺细布
paper notation 方格纸意匠图
paper padding 纸衬
paper panties 纸裤
paper pattern 样板图,服装纸样
paper press 电压机,纸板压呢机
paper pressing 纸板压呢
paper printing 纸印花,热转移纸印花
paper programmed tape 程序控制纸带
paper roll 图纸卷
paper sculpture 纸雕
paper shears 剪纸剪刀
paper shirt 纸衬衫(一次性穿用)
paper stand 图纸架
paper stencil printings 纸板印花
paper string 纸绳
paper taffeta 硬挺塔夫绸
paper twine 纸绳,纸线
paper weight 镇纸
paperworn diamond 纸磨钻石
papery 棉布上浆整理后的光滑性
paper yarn 纸纱,纱芯纸线
paper-yarn fabric 纸线织物
papery-finish 仿纸光滑整理
papery handle 纸状手感
papillion 女式元宝螺帽
papilion rep 横棱织物,精纺平纹织物
papillon taffeta 点纹或花卉纹塔夫绸,蝶塔夫绸
papoon 珀蓬布(泰国民族服饰),泰国色织潘农布
papoula de st Francis 洋麻纤维,槿麻纤维
pappreserve printings 仿蜡防印花
paprica 辣椒红(橙红色)
paprika 辣椒红(橙红色)
papyrus ornament 纸莎草装饰(埃及)
Paquin 帕坎(法国时装设计师)
para-aramid fiber 对位芳香聚酰胺纤维
paraboloid 椭圆抛物面
parachute 降落伞
parachute bag 跳伞包;桶状提袋
parachute button 伞式纽扣
parachute cord 降落伞绳
parachute dress 伞式连衣裙
parachute fabric 降落伞绸,降落伞织物
parachute hat 气球帽,降落伞帽
parachute jumping shoes 跳伞鞋
parachute pants 降落伞裤,伞兵裤(脚口到腿部装拉链)
parachute silk 降落伞绸
parachute skirt 降落伞式裙,伞裙
parachute sleeve 降落伞袖
parachute suspension line 降落伞绳
parachutist pantaloons 降落伞裤
parachutist suit 跳伞运动员装
para cotton 巴西帕拉棉
parade 模型时装展览(英国);丝头绳
parade armour 流行铠装
para-dichlorobenzene 对二氯苯杀虫剂
paradise green 苑绿色
paradise pink 伊甸园玫瑰色
paraffin coating 蜡涂层
paraffin duck 石蜡防水帆布
paraffin duck pants 石蜡处理防水帆布裤
parafil 平行纺纱
paraflame 聚酯阻燃纤维(商名,日本钟纺)
paraguay 花边
paraguay lace 巴拉圭花边
parahyba cotton 巴西帕拉伊巴棉
paraiba cotton 巴西帕拉伊巴棉
parallax 倾斜线;视差
parallel dart 平行省
parallel French dart 胁下平行省
parallel line grating 织物密度板
parallel padstitch 平行纳针线迹

parallel raised seam 平行凸起缝
parallel rule 平行尺
parallel-laid bonded fabric 平行铺置黏结织物
paramatta 棉毛呢,毛葛
parament 祭衣(基督教);领饰,袖口装饰;贴边
parangon 丝绸
paranitraniline red 巴拉红(橙红)
Paraquay lace 巴拉圭花边
parared 巴拉红(橙红),毛巾红
paras 手织斜纹棉布
parasisol 仿亚麻整理的西沙尔麻织物
parasoil 聚丙烯腈抗污纤维(商名,日本)
parasol 遮阳伞(女用)
parasol skirt 阳伞多片裙,阳伞裙
parate shirt 海盗衫
paratroop boots 伞兵靴
paratrooper jeans 伞兵裤
paratrooper suit 伞兵套装
parche 醋酯/黏胶混纤丝(商名,日本三菱人造丝公司)
parchment 羊皮纸色(浅棕色)
parchment cotton 充(仿)羊皮纸细平布
parchment dressed lambskin 羊皮纸型羔羊皮
parchment lace 金银丝梭结花边,凸花边,羊皮纸花边
parchment (yellow) 羊皮纸黄
pardah 印度面纱
pardessus [法]男大衣;男外套
pardessus chemise 衬衫式外套
pardia kufr 珀丽地亚绣花肩巾;绣边细棉布,印度珀丽地亚织花肩巾
parement 装饰织锦
pareo 帕瑞欧裙
pareu 泊勒印花腰布;(波利尼西亚群岛)印花裙布,印花腰布;长方形大花棉布
parew 波利尼西亚群岛印花裙布;印花腰布
parfleche 生皮革,水牛革;生皮革制品,水牛革制品
parguet weave 镶嵌织纹
parhdar 印度素色或金边棉布
paridia kafar 印度珀丽地亚绣花(织花)肩巾;花边细棉布
parietal bone 顶骨
paripasha 印度萨条子软绸
paripurz 毛圈绸,开司米毛圈呢
Paris binding 巴黎硬挺斜纹布

Paris blue 巴黎蓝(深蓝色)
Paris Collection 巴黎时装博览会;巴黎时装发布会
Paris cord 巴黎细横棱绸
Paris doll 模特儿(服装店用),裁缝用女服模型人
Paris embroidery 巴黎刺绣(用细白线在棱纹地上绣缎纹针迹)
Paris green 巴黎绿(黄光绿色),砂绿
Parisian blue 巴黎蓝,天蓝
Parisian silk 巴黎丝
Parisisienne 法国小花纹丝织物;巴黎黑色薄花呢
Paris mode 巴黎服式,巴黎时装
Paris point 环缝,手套背面的饰缝
Paris stitch 巴黎线迹,环缝线迹
Paris white 巴黎白
Paris yellow 巴黎黄(黄色),铬黄
paritanewha fiber 新西兰麻叶纤维
paritaus 多彩绸;开司米彩呢
parka 带帽风雪衣,风雪大衣;派克大衣(带风帽)
parka coat 派克大衣,风雪大衣
parka hood 结带头巾,派克头巾;派克服兜帽
parkal 高密薄纱
parkal(i)a 手织印花布
parka-type ski jacket 派克式滑雪衫
Park Hangchi 朴恒冶(韩国服饰设计师)
Park Younsoon 朴润洙(韩国服饰设计师)
parma 中度紫色,深紫色;印度手织方格棉平布
parm-narm 克什米尔山羊毛高档细呢
parquet weave 镶嵌织纹
parramatta 棉毛呢,毛葛
parrot green 鹦鹉绿(黄绿色)
parsley green 欧芹绿(橄榄绿色)
parsnip 防风草色
parson's hat 牧师帽
part (衣服、机器的)零件,部件;衣片;头发分缝;头路
parterre 丝绒;薄花缎;地毯
parthenos 法国巴丹诺丝线
parti 多色,彩点花纹效应
partial assembly drawing 部件装配图
partial casing 局部抽带管
partial delivery 局部交货
partial interfacing 部分加衬,局部衬
partial knitting 部分针编织

partial lengthwise fold 局部纵向折叠
partial lining 部分夹里,半夹里
partially acetylated cotton 部分乙酰化棉,PA棉
partially aromatic polyamide fiber 部分芳基聚酰胺纤维
partially carboxymethylated cotton 部分羧甲基化棉
partially drawn yarn 部分拉伸丝,预拉伸丝,低拉伸丝
partially etherified cotton 局部醚化棉,低醚化棉
partially oriented yarn 部分取向丝,预取向丝,低取向丝
partial matching 部分匹配(服装图案)
partial pattern cancellation 局部花纹消除
partial pleats 局部褶饰
partial pressure suit 部分高压服,高空代偿服
partial roll collar 半翻领
partial shrink 部分收缩
partial weft warp knitted fabric 局部衬纬经编织物
partical lengthwise fold 局部纵向折叠
particle measurement computer 颗粒影像测量计算机
particolor 杂色
parti-colored 多样色的,杂色的;彩点花纹效应
parti-colored yarn 杂色纱,杂色丝
particolor hose 杂色胡斯裤
particulary fine fiber 特细纤维(直径为16～23μm的纤维)
parting 头发分缝;头路;衣片
partition chromatography 分配色谱法
partitioned view 分块视图(计算机)
partition panel fabric 隔离[用]织物
partlet 打褶绣花紧身衫;无袖上胸衣(16世纪)
partly bald head 半秃顶
part of mesh 部分网饰
part pressing 部件熨烫
partridge cloth 斑点灯芯绒
partridge color 鹧鸪色
partridge cord 斑点灯芯绒
part row knitting 部分横列编织
part sectioned view 局部剖面图
parts list 零件单,配件目录
parts sewing process 零件缝纫流程

part-time stylist 业余设计师
part width 部分幅宽
part wool blanket 混纺毛毯
party 宴会服饰
party blouse 宴会短上衣
party clothes 社交服,交际服
party color 杂色
party-colored 多样色的,杂色的;彩点花纹效应
party dress 宴会(礼)服,社交服,晚会服
party pajamas (适合于晚餐或舞会的)派对裤,睡衣裤
party pants (适合于晚餐或舞会的)派对裤,睡衣裤
party shoes 舞会鞋
party skirt 宴会裙
parure 全套首饰
paruthi cotton 印度派鲁锡棉
parwalla 孟加拉湾彩边棉布
pash(i)m 羊绒,山羊绒,开士米(克什米尔用语)
pashmina 羊绒,山羊绒,开士米(克什米尔用语)
pashmina alwan 阿尔万羊绒白哗叽
pashmina shawl 羊绒披巾
pashmina tweed 克什米尔羊绒粗呢
pashum 开士米
paslel shade 浅淡优美色彩
paso 帕索色织棉布;缅甸帕索色织绸
pasodis 床单
pass due notes 过期票据
passe 绣花绷子,刺绣架子
passement 羊皮纸刺绣花纹;花边
passement a l'aiguille 手工针刺花边
passement au friseau 棱结花边
passementerie 金银花边;绳饰;金银线镶边(穗边);边饰;珠饰镶边
passe montagne [法]山地防寒帽
passenger side bag 乘客座侧安全气袋
passepil 滚边线条
passepoil 丝织边带,毛织边带
pass-fail level 合格与不合格标准
passing 刺绣用金银花线
passing braid 刺绣用金银花线;细辫带;金银花线狭辫带
pass-fail criterion 合格与不合格标准
pass-fail judgment 合格与不合格判定
pass-through buckle 腰带夹扣
passtpoil 毛织边带
past due note 过期票据

paste 玻璃宝石,仿宝石玻璃(人造宝石);浆糊;粘贴;浆、膏、胶	patching with tab pocket 有袢带的贴袋
paste board 纸板,硬纸板	patch leather handbag 拼缝皮手袋
paste dot 浆点法(黏合衬热熔胶)	patch-o-matic darner 半自动织补机
paste grain 丰满纹理整理(皮手套)	patch pattern 补丁图案
paste knife 浆刀	patch pocket 贴袋,贴式口袋
pastel 粉笔画;浅色调;粉色调;色彩柔和;菘蓝染料	patch pocket skirt 贴袋裙
	patch pocket with flap 有盖贴袋
pastel blue 淡青蓝	patchquilt design 补缀图案
pastel color 柔和浅淡的色彩	patch-reinforcement 加固垫布
pastel green 淡绿色	patch sportswear 补缀运动服
pastel green blue 淡绿蓝	patch stay 袋牵布
pastel lavender 淡紫色	patch test 贴布实验
pastel lilac 淡紫色,淡丁香花色	patch up 衲
pastel mink 淡色貂毛皮	patch with flap pocket 有袋盖的贴袋
pastel peach 浅桃色,肉色	patch with pleat pocket 有褶裥的贴袋
pastel pink 淡红色	patch with tab pocket 有袢带的贴袋
pastel shade 浅色,柔和色调	patch work (将布片)拼缝;补花
pastel solid pants 浅色裤子	patchwork dress 拼缝裙装,杂色拼布裙装
pastel tone 淡色调,浅色,柔和色调	
pastel yellow 淡黄色	patchwork jeans 补缀牛仔裤,贴布牛仔裤
paste piece 附属片,小片	
paste point 浆点法(黏合衬热熔胶)	patchwork knitting 嵌花块编织
pasties 黏附式胸罩,(脱衣舞女戴的)乳头罩;乳饰	patchwork pattern 接缝图案;接缝花样
	patchwork prints 拼缝图案印花,补缀型印花花样;贴布印花;拼缝印花;拼布绗缝
pastille 点子花纹,织物上的大圆点纹	
pastoral look 牧歌型,田园型款式	patchwork quilt 拼布绗缝
pastoral print 田园画印花	patchwork sweater 补缀毛衫
pastoral staff 主教权杖	patchwork variation look 贴式变化款式
pastourelle 法国牧羊女哔叽	patchwork 接缝品,拼缝物,贴布
pastron 护胸革(击剑)	patella 膝盖骨
pat 山羊毛;中国蚕丝;黄麻;刺绣用土丝线	patella centre 膝盖骨中点
	patent 专利
pata 泰国潘农布(重浆印花)	patent Axminster 雪尼尔地毯
patadeones 莎龙;莎龙布料;围裙	patent Axminster carpet 阿克明斯特雪尼尔地毯
Patagium 翼膜饰	
patagonian wool 巴塔哥尼亚长绵羊毛	patent-back carpet 背面涂胶地毯,特制底地毯
patalo 派多拉绸(有扎染的或手工木版印边纹的花纹)	
	patent beaver 防水海狸呢
patch 补丁,补片,补缀布片,贴片;(加固用)垫布;臂章、饰章、眼罩;饰牌、徽章;(17~18世纪)饰颜片	patent belt 漆皮革
	patent cloth 蜡布
	patent cord 毛或棉经毛纬的长毛绒织物
patch and flap pocket 有袋盖的贴袋	patent cordage 机制绳
patch bone 插片	patent croc bag 漆皮鳄鱼皮包
patch coat pocket 大衣明兜	patentes 漂白床单布
patcher 缝补工,缝补者	patent flannel 防缩薄法兰绒
patches 贴花	patent leather (黑亮)漆皮,漆面革
patching 修补	patent leather handbag 漆皮手提袋
patching and mending stitch 修补线迹	patent-leather look 漆皮风貌
patching from inside 从里面补	patent leather look hair 漆皮风貌发型

patent leather shoes 漆皮鞋
patent satin quilt 提花双层床单布
patent shoes 漆皮鞋
patent specification 专利说明书
patent twist 紧捻纱平布
patent velvet 灯芯绒,棉条绒,趟绒
patent-selvedge hessian 中边黄麻布
paternoster 大珠(天主教)
path robe 女浴衣
patha 格子棉潘农布
patient's gown 病号服,住院服
patina 铜绿,绿锈;古色;铜绿色
patina green 铜锈绿(浅黄光绿色)
patio carpet 游廊地毯
patio dress 抽象图案的直筒连裙装;室外就餐裙
patio pants 一般传统女裤子(内院裤)
patka 手织山羊绒呢
patlet 女式低领内衣
patna 轧光印花棉布;几何图案绒头地毯
patola 派多拉绸(有扎染的或手工木版印边纹的花纹)
patoli 丝绳;金属编带
patora 佩多拉(合成皮革商标)
Patrick De Barentzen 帕特里克·德巴伦特增(意大利时装设计师)
patroller 验布工
patrolling cycle 巡回周期
patrolling line 巡回路线
patrolling operation 巡回操作
patronier 纸样师
patte boutonne 有纽扣前襟
pattee 革制绑腿,螺旋型绑腿;(在湿地或泥地上行走时穿的)木套鞋,木底鞋;防水低帮套靴
patten 高齿木履,木底鞋
patten golosh 防水套靴
patter coat 花型涂布,印花涂布
pattermaker 样板师
pattern 样品;样布;式样;花样,提花式样,花纹组织;(衣服的)纸型;纸样;(裁剪)样板;图案;图样
pattern acquisition 花纹探测
pattern adjustment 图案搭配
pattern-aided design (PAD) 纸样辅助设计
pattern analysis 图案分析,花样分析
pattern area 花型范围,花纹面积
pattern arrangement 纹样排列
pattern autogeneration 纸样自动生成

pattern blanket 配色包袱样
pattern bonding 花纹黏合法
pattern book 样本集;纸样本;纸样书
pattern calculation 花纹参数计算
pattern chart 纸样图表
pattern combination 花纹组合,花样组合
pattern company 样板公司,纸样公司
pattern construction 纸样结构;纸样制作
pattern coordinate 花样搭配
pattern copy 图样复制
pattern correction 纸样修正;纸样修改
pattern creation 纸样创作
pattern cutter 纸样师,样板师,打样师
pattern cutter's shears 样板剪刀
pattern cutting 打板;纸样裁剪
pattern cutting and making up 裁剪与缝制
pattern cutting illustration 裁剪图
pattern definition 花纹符号表示,花样符号表示法
pattern depth 花型高度
pattern design 纸样设计,花纹设计,图案设计;花纹图案;排纸样
pattern design system (PDS) 样板(纸样)设计系统,服装结构设计系统;开头样
pattern design technology (PDT) 纸样设计工艺
pattern developer 纸样设计者
pattern development 纸样变化
pattern development process 花纹显像过程
pattern development system (PDS) 纸样开发系统
pattern direction 花型方向
pattern draft 花纹组织图
pattern drafting 原型制图,纸样制图
pattern draping 纸样立体裁剪
pattern drum 提花滚筒
pattern duplicating machine 纸样(图样)复印机
patterned 满地花纹
patterned carpet 花纹地毯
patterned effect 花纹效果,叠绕效应
patterned envelope 纸样套
patterned fabric 花纹织物
patterned grosgrain 特号葛
patterned hose 满地花纹袜子
patterned satin brocade 修花缎
patterned shirt[blouse] 花样衬衫
patterned silk twill 桑花绫

patterned stockings 花纹长袜,花袜
patterned stripe 花样条纹
patterned tack 花式打结
patterned tacking machine 花式打结机
patterned tull lace 花饰丝网花边
patterned web 花纹纤维网
pattern effect 花纹效果
pattern envelope 纸样套
patterner 打板师
pattern fit allowance 对花公差
pattern fitting 对花
pattern flocking 图案植绒
pattern folds 式样褶
pattern generation 纸样生成
pattern generation system 花纹编织系统,样板生成系统
pattern goods 花洋布
pattern grade 纸样放码(推档);纸样放缩
pattern grader 号型放缩员,样板放缩员,纸样推档员;图样缩放仪,样板放缩仪
pattern grading 样板缩放,样板推挡,纸样放码
pattern grading machine 纸样放缩机
pattern grid 花纹格子图,意匠图
pattern grid paper 意匠纸
pattern hook 样板挂钩
pattern identification 样板注脚
pattern identification panel 花型识别样板
pattern industry standard 纸样工业标准
pattern information 花纹信息,纸样线条构成,纸样资料
patterning 形成花纹,组成图案
pattern interlock 提花双罗纹织物
pattern knit 花色针织法
pattern lack 花型不全
pattern landmark 服装样板位置记号
pattern layout 花纹排列,排版
pattern layout by one closed side 一头齐的缝纫排料
pattern layout by two closed sides 两头齐的缝纫排料
pattern limitation 花型范围
pattern line 服装样板划线
pattern magazine 样板杂志
pattern maker 打样师,纸板师,样板师,制纸样者,服装设计师
pattern making 打板,服装纸样制作
pattern making process 打板过程
pattern making scissors[shears] 纸样剪刀

pattern making shears 样板剪刀
pattern making skill 纸样制作技巧
pattern making symbol 制板标记
pattern making system 纸样制作系统
pattern making term 制板术语
pattern making tool 打板工具
pattern manipulating process 纸样处理过程
pattern manipulation 花纹变换,纸样处理
pattern marker 纸样排板
pattern marker maker 样板排板师
pattern marking 纸样标记
pattern marking system 描样系统
pattern master 打样专用曲直线尺
pattern measurement 服装样板尺寸,纸样尺寸
pattern misfit 对花不准
pattern mix 混合图案
pattern modification 纸样修正;纸样修改
pattern motif 花型花纹
pattern needle loom 花纹针刺机
pattern notation 设计意匠图
pattern notches U形轧剪,刀眼剪,打眼刀
pattern on pattern 上下衣花样相配,双重花样,重叠花样
pattern operation 图形操作(计算机)
pattern outline 纸样轮廓线
pattern output 花纹输出
pattern paper 花样设计纸,样板纸,意匠图纸,纸型,纸样
pattern perfection 纸样修正;纸样修改
pattern perforator 纸样打孔器
pattern piece 衣片样板,纸样裁片
pattern piece layout 试排用料
pattern placement 服装纸样的排放
pattern planning 服装纸样设计
pattern plate 样板
pattern plot 纸样标识
pattern potential 花型变换可能性
pattern preparation 样板准备(计算机)
pattern preparation system 花纹准备系统
pattern production 纸样制作
pattern punch 样板打孔机
pattern range designing 花纹排列设计
pattern reader 花纹读出器
pattern recognition 图形识别
pattern recording unit 花纹记录装置
pattern repeat 花型循环,花样循环,图案

循环

pattern room 样板房
patterns 木套鞋(木底鞋)
pattern sample 花样样品,款式样,纸样样品
pattern scanner 花纹扫描器
pattern schema 花型;花型图
pattern scissors 纸样剪
pattern section 纸样部件
pattern set 成套纸样,整套纸样
pattern settings 花纹配置
pattern shade difference 花纹色泽深浅不符
pattern shape 纸样形状
pattern sheet 花纹设计板;服装纸样本;样本集;衣服纸样
pattern sign 纸样上的标记
pattern size 样板尺寸;纸样尺寸
pattern sketch (服装)式样设计图;(布料)花纹图样;花纹设计图;图案写生
pattern skins 提花裤袜
pattern slide 提花底脚片
pattern's mathematical model 样板数学模型,纸样数学模型
pattern specifiation 纸样规格,纸样说明
pattern stability test 模型稳定性试验(人造革模制品热稳定性试验法之一)
pattern stage 制样板过程
pattern stitch 花式线迹,花纹线迹
pattern stitching 花纹线缝
pattern stockings 花式长袜,装饰长袜
pattern symbol 样板上的记号;纸样上的符号
pattern tacker 花纹假缝机
pattern thread cropping machine 剪花机
pattern thread 浮纹线,提花线
pattern tracer 描花纹机
pattern transposition 花纹位移
pattern unit 花纹单元,图案单元
pattern walewise transpasition 花纹纵移
pattern warper 花式整经机
pattern work 色织织物,提花织物,提花织品
patti 帕图织物(印度);山羊发毛;山羊发毛粗呢
pattia 旁遮普彩色棉带
patti coat 衬裙
pattina shoes 派迪纳合成革鞋
patto 帕图织物(印度);山羊发毛;山羊发毛粗呢

pattoo 帕图织物(印度);山羊发毛;山羊发毛粗呢
patto woollen cloth 巴托粗呢
pattu 帕图织物(印度);山羊发毛;山羊发毛粗呢
patu 帕图织物(印度);山羊发毛;山羊发毛粗呢
patu knudrang 克什米尔绣花驼毛呢
pau 人造马毛
paukas 手织粗平纹印花棉布
Paul Poiret 保罗·普瓦雷(1880～1944,法国时装设计师)
Paul Smith 保罗·斯密斯(英国服饰设计师)
Paul's rearrangement 保罗重排作用
pauldron 肩甲
paulette 波莱特(美国时装设计师)
paulin 篷帆布,防雨帆布
Pauline Trigere 波林·特丽吉尔(1912～,美国时装设计师)
paultock 男短夹克
paunch 大肚子
paunch figure 大肚子体型
paunch mat 绳垫,绳席
pauni 漂白细棉布
pauper look 贫穷型风貌,乞丐服款式
paux 垂襞,垂片;翻领;翻袖
paw 翻领;脚部毛皮;爪
Paw crosses 小块皮
payback period 回收期
payback time 回收期
pay by instalments 分期付款
pay date 结算日
pay duty 纳税
payee 收款人
payer 付款人
paymaster 美国帕马斯特棉
paymastor cotton 美国帕马斯特棉
payment 付款
payment for goods 贷款
payment in full 全额付清
payta cotton 秘鲁帕塔棉
pay tax 纳税
pazy 聚酯仿短纤纱长丝(商名,日本帝人)
PBT fiber (polybutylene terephthalate fiber) 聚对苯二甲酸丁二酯纤维(弹性涤纶纤维)
pca coat 海员扣领短上衣,水手外套
PDS fiber 聚对二氧杂环己酮纤维

peach 桃色(浅红橙色),桃红,浅藕红
peach basket hat 美人草帽
peach beige 浅藕红,血牙红
peach blow 紫粉红色
peach blossom 桃红色(粉红色)
peach blossom alizarin red 山茶红,茜红
peach blow 紫粉红色
peach brushed 桃皮起绒,桃皮刷绒
peach bud 桃蕾色
peach button 桃形扣
peach cobbler 桃汁色
peach fabric 磨毛布;细绒面织物
peach-face micropowder fabric 桃皮绒微粉感织物
peach-face touch 桃皮绒手感
peach fuzz 桃皮(绒)色
peach fuzz type 桃皮绒毛型
peach nectar 桃汁色
peach pearl 桃红珍珠色
peach red 桃红色(粉红色),黄光绯色
peachskin 桃皮绒;桃皮绒手感,桃皮绒效果;桃皮绒织物,磨毛布
peachskin effect 桃皮感,桃皮效应,微细绒毛表面效应
peachskin fabric 桃皮织物,桃皮绒织物
peachskin finish 桃皮绒加工
peachskin finished fabric 桃皮型织物
peach-skin jacket 桃皮绒夹克
peachskin-like fabric 仿桃皮绒针织物,短茸毛织物
peachskin touch 桃皮绒织物手感
peach velvet 桃皮绒
pea clothing 水手粗呢上衣
pea coat 海员扣领短上衣,水手粗呢上衣,水手外套
peacock blue 孔雀蓝(暗绿光蓝)
peacockery 花花公子浮华服饰
peacock feather 花翎
peacock feather fabric 孔雀裘
peacock green 孔雀绿(黄光绿色)
peacock revolution 孔雀革命(男装),孔雀革新(指促进男用时装色彩丰富化的改革用语)
peacock's eye 孔雀眼花纹
pead cap 女鸭舌帽
pea green 豆绿,豆青,青豆绿色,黄绿色
pea jacket 海员厚呢上装,领航外套,水手短上衣,双排纽厚呢上衣
peak (衣着上的)突出部分;帽檐
peak capacity 最大生产能力

peaked 有帽檐的
peaked cap 无舌尖顶帽,鸭舌帽,有帽檐帽
peaked collar 尖形领,尖领
peak(ed) lapel 尖驳头,尖领,尖角翻领,戗驳领,戗驳头,剑领
peaked lapel collar 尖角式西装领
peaked mink cap 有帽檐貂皮帽,遮檐貂皮便帽
peaked shawl collar V形丝瓜领
peaked shoes 喙形尖头鞋,细尖头鞋
peak lapel 戗驳头,尖领,剑领
peak season 旺季
peal embroidery 珠球刺绣,圈纹刺绣
peality 女用披肩
peanut fiber 花生[蛋白质]纤维
PE apron PE塑胶围裙
pearce cotton 皮尔斯棉
pear design 梨形图案
peare 缎带装饰环
pearl 圈纹边饰,牙边;珍珠,念珠;珍珠花;珠光色
pearl bandings 珍珠带
pearl(beaded) bracelet 珍珠手镯;珠饰手镯
pearl blue 浅蓝灰色,珍珠蓝
pearl blush 珍珠红
pearl braid 波状花线编带
pearl brooch 珍珠饰针
pearl button 珠母扣,珠光扣,贝壳纽扣,有机玻璃扣,波状扣,珍珠扣
pearl cotton 珠光线,珍珠棉线,绣花棉线
pearl diadem 珍珠王冠
pearl dog collar 珠饰项圈
pearl dress 有珍珠饰带的裙装
pearl edge 珠球边,圈纹边,防脱散横列
pearl embroidery 圈纹刺绣,珠球刺绣,珠绣
pearl essence 珍珠粉,珠光粉
pearlescent 珠光般的
pearlescent color 珠光色
pearl fabric 双反面针织物
pearl gray 珍珠灰(淡黄灰色或浅蓝灰色)
pearl grey 珍珠灰
pearlies 珍珠纽扣,贝壳纽扣,珠母扣,珠母纽;有贝壳纽扣的服装
pearlin 圈纹狭花边,丝线花边,亚麻花边
pearlina yarn 针织用丝毛合胶强捻线
pearling 丝线花边;亚麻花边;圈纹狭花

边;棉或麻牙边细布
pearl lame 珠光金银丝织物
pearl-like coating 珠光涂层
pearl luster printings 珠光印花
pearl lustre wool/polyester 珠光毛涤纶
pearl lustre wool/polyester fancy suitings 珠光毛涤花呢
pearl knitting machine 平板横机,双针筒圆形针织机,双反面针织机
pearl luster printings 珠印花
pearl mink 珍珠貂毛皮,珠光貂毛皮
pearl necklace 珍珠项链;珍珠项圈
pearl necklace model 珠项链模型
pearl net 珠状网眼
pearl pigment 珠光颜料
pearl pin bar 牙边花边
pearl pin 珍珠针
pearl powder 珍珠粉,珠光粉
pearl printings 珠光[效应]印花
pearl rack 波纹畦编
pearls 珍珠项链
pearls and jade 珠翠
pearls-embroidered clothes 珠绣衣
pearl sequins 珍珠亮片
pearl shape 珍珠形体
pear-shaped busk 梨形胸衣
pear-shaped figure 梨形体型
pear-shape gem cut 梨形宝石
pearl silk 伊朗细生丝
pearl snap 珠光四合纽
pearl stitch 曲折锁边缝线迹,三角线迹,双反面线圈,珠状线迹
pearl string 珠子丝
pearl string model 珠[项]链模型
pearl tuck 串珠饰缝褶
pearl white 锌钡白,珍珠白
pearly 用珍珠装饰的;珍珠纽扣
pearl yarn 珍珠纱,珠球色纱
pearly boucle yarn 球圈线
pearly button 珍珠纽扣
pearly yarn 高级双股绣花棉线
pear white 珍珠白
peasant blouse 农妇衫;农家女衫;村姑衫
peasant cloth 农民布,厚实细布仿欧洲土布
peasant coat 农民长大衣;止口、底边、克夫等用毛皮饰边的长大衣
peasant collar 农民领
peasant costume 农民服装;农民装束
peasant darwers 中世纪欧美农妇长衬裤,村姑长内裤
peasant dress 村姑装,农妇装
peasant fashions 乡村风格服饰,农民节日装
peasant lace 粗平纹花边
peasant look 农妇风貌;农民款式;欧美农妇服装款式,农妇服装款式
peasant neckline 农妇式领口;农民式领口;欧美农民式领口,束带式领口
peasant shirt 农民衫
peasant skirt 农妇裙欧美农民穿的碎褶裥裙,村姑裙,宽摆裙
peasant skirt opening 村姑裙式开口(一边为光边处理,一边为贴边式扣合份)
peasant skirt placket 村姑裙式开口(一边为光边处理,一边为贴边式扣合份)
peasant sleeve 村姑长袖,农妇长袖
peasant-style dress 村姑型服装,农妇型服装(根据欧美农民服装风格设计的服装,以立领、灯笼袖、碎褶裙和劳动布工装裤为特征)
pease cloth 比斯呢(间隔型提花条格色织布)
peat 泥炭色
peat fiber 泥炭纤维
pea tulle 平织网眼纱
peat wadding 泥炭填絮
peat yarn 泥炭纤维纱线
peau 起绒织物,仿皮绒织物(用粗纺毛纱、人造丝或蚕丝织制,常采用金刚砂起绒工艺加工)
peau d'ange 天使缎
peau de chamois 麂皮绉
peau de cygne 法国天鹅皮绸
peau de diable 法国印花棉布
peau de gant 白丝花缎
peau de mouton 普德蒙顿仿羊皮女大衣呢
peau de peche 桃皮绒,仿丝绒薄绒布
peau de poule 法国仿鸡皮绒
peau de prepe 光滑软绉
peau de rayon 织闪绸
peau de singe 仿猴皮绒
peau de soie 仿新娘礼服用有光泽和细棱条的厚重织物
peau de souris 鼠皮绸(柔软丝织物)
peau de suede 格子花呢;仿麂皮织物
peau d'ours 熊皮绒
pebble 皱纹,卵石色,小卵石纹,碎石

花纹
pebble cheviot 绉面大衣呢
pebble crepe 苔绒绉
pebbled finish 碎石纹整理
pebbled grain 多卵石纹粒面,小卵石纹理
pebble effect 泡泡效应
pebble finish 碎石纹整理
pebble leather 碎石粒纹革
PEB fiber 聚对苯甲酸乙氧酯纤维,荣辉纤维
pecan brown 山核桃棕色
pecari 美洲野猪皮,西貒皮
pecary 美洲野猪皮,西貒皮
pecary leather 美洲野猪革
PE/C cambric 涤棉细布
peccary 美洲野猪皮,西貒皮
peccary leather 美洲野猪革
PE/c combric 涤棉细布
pechiyong 白长毛绒
PE/c khaki 涤棉卡其
PE/C sewing thread 涤棉缝纫线
pectoral 胸饰,胸铠,护胸垫;胸肌
pectoral cross 主教胸前十字架
pectoralis major 胸大肌
pedal （缝纫机的）踏板;踏脚
pedal collar 圆驳领（踏板形领）
pedal control backtack 踏板控制回针
pedal pushers 及腓女运动裤,骑车中长裤,过膝女裤
peddler's wool 劣等羊毛
pedestal mat 浴室草垫,盥洗室席垫
pedicure 美足术;修脚;修脚师
pediment headdress 英国兜帽,三角形兜帽
PE disposable raincoat PE 一次性雨衣
peds 脚掌套
peduru 手编草垫
peejays 男人睡衣裤
peek-a-boo blouse 雕绣薄质女衫,半透明薄质女衫
peek-a-boo earring 小宝石穿孔耳环
peek-a-boo fashion 半透明装
peek-a-boo panties 薄质褶边短裤
peek-a-boo waist 雕绣薄质女衫,半透明薄质女衫
PEEK fiber 聚醚醚酮纤维
peel bond strength 剥离强度
peeler 剥皮工（制革）
peeler cotton 皮勒陆地棉

peeling 剥皮整理,仿真丝剥皮整理
peep-toe sandals 露趾凉鞋
peep-toe trainer 露趾训练鞋
peep-toe wedge with sling back 踝部吊带露趾坡跟凉鞋
peerless cotton 美国皮勒斯棉
peg （晒衣）夹子;（木制）假腿;裤子;装假腿的人;陪衬小花;上宽下窄
pega moid 防水布
peg board knitting 纹板编织
pegged boots 木钉靴
pegged leggers 木钉裤（轮廓似楔子,强调裤腰部分）
pegged pants （上宽下窄的）锥形裤,陀螺裤,木钉裤（上部大,下部逐渐变小,轮廓像楔子形的裤子）
pegged shoes 木钉鞋
pegged skirt （上宽下窄）陀螺裙,楔形裙,锥形裙
pegged skirt silhouette 锥形裙廓体
pegged trousers （上宽下窄）木钉裤,楔形裤
peggy collar 蓓姬领（前身领角呈深裂贝纹形状）
pegleg 木制假腿;装假腿的人;锥形裤（打腰褶,小脚口,20 世纪 80 年代初流行）
pegleg pants 陀螺裤,木钉裤（上部大,下部逐渐变小,轮廓像楔子形的裤子）
pegs 裤子
peg skirt 上宽下窄裙,陀螺裙
peg-top 陀螺裙
peg-top pants 萝卜裤,陀螺裤,木钉裤（上部大,下部逐渐变小,轮廓像楔子形的裤子）
peg tops 上宽下窄裤[裙]子;陀螺裤
peg-top silhouette 陀螺型轮廓
peg-top skirt 陀螺裙
peg-top trousers 陀螺裤,木钉裤（上部大,下部逐渐变小,轮廓像楔子形的裤子）
peg yarn 袜筒用纱线
PEI fiber 聚醚酰亚胺纤维
peignes fleur 花梳子
peignoir 女晨衣,女式宽大便衣,妇女化妆衣;梳妆披巾,理发披巾
peignoir set 睡衣-长袍配套装
peiguang brocaded damask 蓓光绸
peihua brocatelle 蓓花绸
peilice 皮衬里长外衣

peit 毛皮,生皮,皮货;毛皮围脖
pekan 鱼貂毛皮
pekin 北京条纹绸,北京宽条子绸经条纹布,壁挂及装饰布
pekin blue 北京蓝
pekin crepe 北京条子双绉
pekinese stitch 北京缝针法
Peking 北京条纹绸
pekin gauze 绒条纱罗织物
Peking Opera 京剧
Peking style 京派风格
pekin nouveauté 彩棱纹条子绸
pekin satin brocade 北京花缎
pekin satin plain 北京素缎
pekin stripe 等宽纵条纹织物
pekin velours 棉丝交织绒头条子绒布
pekin victoria 经条纹布
pel silk 松捻股丝
pelade 法国皮板毛
pelang 丝绸
pelerine [法]倍利暖女士披肩,细长女用披肩
pelerine design 菠萝花纹
pelerine stitch 菠萝组织
pelerine work 菠萝组织针织品
pelican 塘鹅色
pelisse 佩利瑟大衣,饰长外衣,皮衬里长外衣,皮长外衣,女式阔领饰皮轻便外衣;毛皮披风,毛皮衬里披风,女用皮制长外套,童大衣
pelisse cloth 斜纹粗呢
pella 大氅,斗篷
pellestrina lace 葡萄叶纹梭结花边
pelleton 帽用山羊毛
pellmell wearing 混乱穿着
pelo 松捻股丝
pelone 黑色毛绒
pelotage 秘鲁低级骆马毛
pelote 山羊毛,毛束;毛线球
pelt 毛皮;皮袄,皮货(总称);毛皮围脖
pelt-belt 可折皮带型轻夹克
pelt muffle 毛皮围脖
peltry (总称)毛皮;皮货,生皮
pelt wool 皮板毛,短皮板毛
peluche 法国长毛绒
peluche argent 法国银线丝绒
peluche double 双面长毛绒
peluche double face 丝轻棉纬长毛绒
peluche duvet 法国天鹅绒
peluche épinglée 法国毛圈长毛绒

peluche ombre 月华丝绒
peluche velvet 金丝绒
pelure d'oignon 硬挺整理径面缎
pelvic girdle 盆骨带
pelvic proteetion 骨盆保护器
pelvis 骨盆
penalty 罚金
pen-and-ink drawing 钢笔画
pen-and-ink sketch 钢笔速写
pen and pencil 美工笔
penang 印度手织平纹厚棉布
pencil 铅笔,眉笔,唇笔;画法,笔调,笔法
pencil attachment 铅笔插腿
pencil blue 靛蓝地色布;深紫蓝色
pencil craser 铅笔橡皮
pencil drawing 铅笔画
pencil dress (直筒形衣身的)铅笔装,直筒装,直线条服装
pencil eraser 铅笔橡皮
pencil lead 铅芯
pencil pants 笔杆裤
pencil pocket 笔袋
pencil shoes 铅笔鞋
pencil silhouette 笔杆形轮廓
pencil sketch 铅笔速写,铅笔素描
pencil skirt 笔杆裙,铅笔裙
pencil slim skirt 铅笔式女裙
pencil slim top 铅笔式上衣
pencil stripe 铅笔线条纹,细点条纹
pencil stripes 细点条纹织物,铅笔线条纹织物
pencil-thin moustache 八字小胡子
pencil-type eraser 笔式橡皮
pend (苏格兰)垂饰,挂件
pendant 垂饰,垂饰物,挂件,有垂饰的项链,坠子
pendant earring 垂挂式耳环
pendant necklace 垂饰项链
pendeloque [法]坠子;鸡心形垂饰;鸡心形挂件
pendent 垂饰物,挂件
pen drawing 钢笔画
pendulum clock 摆钟
penelope 十字布,绣花底布;无袖针织夹克
penelope canvas 刺绣用粗十字布
penetrating printings 渗透印花
penetrating property 渗透性
penguin 企鹅灰
penguin gray 企鹅灰

penguin shell button 企鹅贝扣
penguin suit 太空服,宇航服,企鹅服
pengy tropical suiting 废丝混纺薄型织物
peniche 佩尼契大花黑白梭结花边
penistone 佩尼斯通粗毡呢
penne 平绒
Penng loafers 便士乐福鞋
penning seam 手工锁缝(用于丝绒或针织物边缘)
penny loafer 便士浅底鞋
pennystone 佩尼斯通粗毡呢
pen painting 钢笔画
pen picture 钢笔画
pen shan ssu silk 本山素绸
pentachlorphenol content (PCP content) 五氯苯酚含量
pentad 五色相配色
pentalobal 五角异形丝
peony 牡丹红(暗红色)
peony brocade 牡丹绸
people hood 民族性;民族意识
PEPCO 小猪班纳
pepita 牧人黑白格子布;双色格子花纹
peplos 古希腊披袍(派帕洛斯);女式长外衣(古希腊);女式大披肩(古希腊);披巾式女外套
peplum 裙腰剪接片;腰褶,腰部装饰褶襞;褶襞短裙;装饰短裙;波形褶襞
peplum blouse 腰褶罩衫
peplum dress 腰褶裙装
peplum jacket 宽摆束腰夹克
peplum suit 细腰狭裙,有腰裙的套装,褶襞短裙套装
peplum top 腰褶上装
peplus 女式长外衣(古希腊);女式大披肩(古希腊);佩柏勒斯衫(古希腊)
pepper and salt 芝麻呢,芝麻布;夹花条纹布,夹花条纹呢;芝麻点纹
pepper corn 胡椒籽色
pepperell drill 尔优级斜纹坯布
peppermint green 薄荷绿
PEPSI 百事
Per Spook 佩·斯波克(法国时装设计师)
perajes 佩拉头巾(危地马拉)
percal(e) 高级密织薄纱(匹染或印花)
percale bed sheeting 密织床单布
percale stripes 条纹衬衫布
percaline 丝光薄纱,丝光高级细薄纱;上光软衬里布

percan 巴拉坎厚呢
percé 网孔花纹
perceived color difference 观感色差
percentage of A-class goods 正品率
percentage of false cut through 假开剪率
percentage of wear and tear 服装折旧率
perception 知觉
perception fatigue 感觉疲劳
perceptive perspiration 有感出汗
perceptual knowledge 感性知识
perch 验布台
perched quilt 小菱形花纹双层布
percher 验布工
perches 法国中等亚麻布
perching 验布
perching of gray good 坯布检查
perchlorethylene 四氯乙烯(除油渍剂,干洗剂),全氯乙烯
perchlorinated polyvinyl chloride fiber 氯化聚氯乙烯纤维
perchlorovinyl fiber 过氯乙烯纤维
percussian 敲击(美容)
percussion cap 大帽
perfect fiber 理想纤维,完善纤维
perfect figure 标准体型
perfect register 对花准确
perfect stance 标准立姿
perfect stitch 正确线迹,完整线迹
perfect stone 完美石
perfect yellow 纯黄
perfecto dress 雪茄裙装
perfluorcarbon fiber 全氟化碳纤维,碳氟纤维
perforate 打眼
perforated card 打洞纸板
perforated fabric 网眼织物,仿纱罗织物
perforated leather 小孔皮革,有孔皮革
perforated marker 漏空样板,镂空画样
perforated weave 透孔织物
perforater 打眼器
perforating (在裁片上)打眼
perforation 挖孔
perforation machine 打眼机
performance 性能
performance art 表演艺术
performance characteristics 表现特性,工作特性,效能特性
performance figure 质量指标
performance in wear 穿着性能
performance index (PI) 性能指标,性能

指数
performance life 使用寿命
performance number 性能参数
performance property 织物性能,服用性能
performance specification 织物性能要求
perfume ring 香味戒指
perfume 香料;香水
perfumed bath powder 浴后香粉
perfumed fabric 芳香织物
perfumed fiber 香味纤维,芳香纤维
perfumed finishing 香味整理
perfumed printing 香味印花
perfumery 香料类
peridot 橄榄石
perils of the sea 海上风险
perini fiber 洋麻,槿麻
period dress 时代服
periodical maintenance 定期保养
periodic inspection 定期检验,周期检查
periodic repair 中修(设备)
peripheral equipment 外围设备
peripheral product 外围产品
periscope hat 潜望镜式帽
periwig 贝瑞假发;带假发;假发(男子)
periwinkle[blue] 长春花蓝(浅紫光蓝色)
perkan 厚呢,厚毛织物
perky bow 警长领带,西部领带,细绳领带,鞋带型领带
perky knot 快乐领结
perky suit 时髦套装
perle 珍珠绒整理;珍珠呢,珠皮呢,栗鼠呢
perle cotton 丝光刺绣棉线
perle flannel finishing 珍珠绒整理
perlin 苏格兰花边
perlines 丝光细布
per-loft 聚酰胺袜用弹力丝(商名,日本帝人)
perlon 贝纶(聚酰胺6纤维,商名,德国拜耳)
perm 电烫,电烫发
permachem treatment 珀马琴[防菌]处理(纺丝液中添加防菌药品,纺制耐洗的防菌纤维)
permanent antistatic fiber 持久性抗静电纤维,永久性抗静电纤维
permanent antistatic treatment 耐久性抗静电处理
permanent blue 耐洗蓝

permanent bright finish (PB finish) 耐久光泽整理
permanent care labeling 耐久性保养说明标签
permanent crease 永久折痕,永久褶痕
permanent creasing 永久褶痕
permanent cuff 固定袖头
permanent deformation 永久形变
permanent dye 长效染发剂
permanent elongation 永久伸长
permanent embossed finish 耐久性拷花整理;耐久性轧花整理
permanent finsh 耐久整理,永久性整理
permanent fire-retardant rayon 永久性阻燃黏胶丝
permanent flame-proof finish 永久性阻燃整理
permanent flame retardant (PFR) 耐久性阻燃剂
permanent growth 永久性形变,次级蠕变,第二次潜伸
permanent heat setting 永久热定形
permanent luster 永久性光泽
permanent lustered finish 耐久性上光整理
permanently crimped fiber 永久[性]卷曲纤维
permanent pleat 持久性褶裥,耐久褶裥,永久褶裥
permanent pleating 耐久褶裥加工
permanent pleating machine 耐久褶裥机,永久褶裥机
permanent pleat set 耐久[性]褶裥定形,永久[性]褶裥定形
permanent press (P.P.) 耐久压烫,耐久定形;PP整理;PP整理(的)衣料
permanent press fabric 耐久性定形织物,耐久性压烫织物
permanent press finish (P.P. finish) 耐久定形整理,耐久压烫整理
permanent press(ing) 耐久熨烫;定型熨烫
permanents 不退色薄棉布;轧光细棉布
permanent set 永久定形,永久定型
permanent setting of worsted fabric 精梳毛织品的永久定形
permanent shinkage 永久性收缩
permanent sizing 耐洗上浆整理
permanent starchless finish 耐久性无淀粉整理

permanent stretch 永久伸长,永久变形
permanent wave 烫出的波浪型发型(电烫或化学烫);烫发;永久性卷发
permanent waviness 永久性卷曲,永久性波纹
permanis cotton 巴西珀马尼斯棉
perma-press finish 耐久压烫整理
permeability 透气性,通透性;透水性;渗透性;渗透率
permeability index 透湿指标
permeable fabric 透气性织物
permeable to light 可透光(的)
permeable to vapor 可透蒸汽(的)
permo 珀莫纺呢
pernam cotton 秘鲁伯南棉
pernambuco cotton 巴西伯南布哥棉
peroneus longus 腓骨肌
perpendicularity 垂直度
perpendicular line 正交线
perpetuelle 珀佩蒂尤尔细哗叽
perprogrammed pattern 预编程序花样
perraje 佩拉头巾(危地马拉),危地马拉围巾
perroguet 提花绉绸;大麻船帆布
perroguet crepe 提花丝质绉布
perrotine printings 模板印花
Perry Ellis 派利·埃利斯(1940~1986,美国时装设计师)
persain 波斯绸
perse 暗蓝色,暗紫色,蓝色,浅蓝色,蓝灰色;手工模板印花上光棉布;多色珠饰镶边;蓝色记号线
persening 防水麻织物(黄麻或亚麻制)
perse rug 东方地毯
perse serge 蓝色毛哗叽
perse silk 金丝花锻
Persia silk rug 波斯地毯
Persian (a) 波斯绸
Persian blue 波斯蓝
Persian carpet 波斯地毯
Persian coat 波斯外套
Persian cord 波斯双经棱纹细呢
Persian costume 波斯民族服饰
Persian design 波斯图案
Persian embroidery 波斯刺绣;波斯羔羊皮
Persian jewel 波斯宝石色
Persian knot 波斯结
Persian lamb 波斯羔羊皮,波斯羔羊
Persian lamb skin 波斯羔羊皮

Persian lawn 波斯细薄亚麻平布
Persian morocco 棕色羊皮
Persian orange 波斯橙色
Persian ornament 波斯装饰
Persian prins 波斯印花布
Persian red 波斯红,天然红色染料,红色颜料
Persian rug 东方地毯,波斯地毯
Persian shawl 波斯羊绒披巾
Persian slippers 波斯拖鞋
Persian stitch 波斯线迹
Persian turban 波斯包头巾帽
Persian wool 波斯羊毛
persimmon 柿棕(棕色),柿棕红
persimmon red 柿红(红橘色)
persistance pleat 持久性褶裥
persistent quality 持久性,恒定的质量
personal appearance 面容,仪容
personal color 肤色
personal-fit pattern 个体适体纸样
personal flotation devices (PFDS) 个体漂浮装置
personal hygiene 个人卫生
personality 个性,个性化;个性时装;人格
personalization 个性化
personal predilection 个人爱好
personal preference 个人爱好,个人偏爱
personal proprietor ship 独资经营
personal psychology 个体心里
personal style 个人风格
personal taste 个人爱好,个人鉴赏力,个人品味
personal washing agent 皮肤洗涤剂
personhood 个性
perspective depth 透视深度
perspective grid 透视网络
perspective projection 透视投影
perspective scale 透视比例,远近比例
perspective sense 远近感
perspective view 透视图
perspiration 出汗,发汗;汗液
perspiration absorptive finish 吸汗整理
perspiration fastness 耐汗渍牢度
perspiration proof 耐汗渍,防汗渍
perspiration-resistant 耐汗渍的
perspiration spot 汗斑
perspiration-sunlight fastness 汗渍与日光综合牢度
perspiration test 耐汗渍(色)牢度试验
perspirometer 耐汗渍(色)牢度试验

perte 法国佩太本色大麻帆布
peru hat 秘鲁帽
peruke 男子假发,佩鲁基假发
peruquier 假发师
peru sweater 秘鲁毛衫
Peruvian 花卉厚织物
Peruvian cloth 秘鲁刺绣亚麻台布
Peruvian costume 秘鲁民族服饰
Peruvian cotton 秘鲁棉
Peruvian embroidery 秘鲁刺绣
Peruvian hat 秘鲁帽
Peruvian Pima 秘鲁比马棉
Peruvian Pima cotton 秘鲁比马棉
Peruvian Sea tsland cotton 秘鲁产海岛棉
Peruvian wool 秘鲁绵羊毛
peruvienne 双面异色缎,花卉重磅织物
perverted image 反像
perwitsky 鼬皮
PES interlining 聚酯(粘合)衬
pesticide 杀虫剂
pesticide protective clothing 防杀虫剂服装
pesticide-protection garment 防农药服
peta 印度二等绞丝
petal collar 花瓣领
petal dress 花瓣装
petal hem 波浪缘边,波浪下摆
petal-like conjugate fiber 花瓣状(截面)复合纤维
petal-like cross section 花瓣状横截面
petal neck 花瓣领
petal neckline 花瓣领口
petal pants 花瓣短裤
petal pink 玫瑰花瓣色
petal pocket 花瓣形贴袋
petal pusher 女式中长裤(骑自行车时穿用)
petal-scalloped sleeve 花瓣-扇贝形袖
petal skirt 花瓣裙
petal sleeve 花瓣(短)袖,兰花瓣袖
petal stitch 花瓣针法
petal style 花瓣式;花瓣式饰巾;克莫德头巾;塞入式饰巾
petal treatment 花瓣式饰巾装饰法;克莫德头巾;塞入式饰巾
petanella 潘德尼拉呢(泥炭纤维与毛混纺织物)
petasos 彼泰索丝帽;佩塔索斯帽(古希腊);阔边低顶帽(古希腊);锥形阔边帽(古希腊);宽边帽(古希腊戏剧)

petenioche 低级绢丝
Peterkin cotton 彼得金棉
Peterkin limb cluster cotton 彼得金多铃棉
Peterkin mew cluster cotton 彼得金多铃棉
petenlair 18世纪披腾莱尔女士上衣
peter pan collar 铜盆领,小飞达领,彼得潘领,(女装、童装)小圆领
Peter Pan costume 彼得潘装束
Peter Pan hat 彼得潘帽
petersburg flax 彼得堡亚麻
petersham 彼得沙姆织物;厚重珠皮大衣呢;彼德沙姆硬衬,连腰裙的腰衬;佩特香呢;罗纹丝带
petersham cloth 佩特香织物
petersham coat 彼得沙姆大衣
petersham costume 彼得沙姆服饰(1790～1851)
petersham elastic 彼得沙姆松紧带
petersham mode 彼得沙姆服饰(1790～1851)
petersham ribbon 彼得沙姆棱条丝带(女用束发带,帽子边饰)
petersham tape 彼得沙姆衬带
Peter Thompson dress 彼得汤姆逊装
petex 佩特克斯(以聚酯纤维为原料的纺丝黏合法制成的非织造布,商名)
petin 法国佩廷呢,羊毛与驼毛混纺织物
petinet 花边,花边网;纱罗织物
petit 法国轻薄或低档织物
petit col plat 小平领
petit col volante 皱边小领子
petit drap 法国平纹薄呢
petite 服装中等偏小号,服装最小号
petite brides 法国小六角形网地针绣花边
petite draperie 耐潮呢绒,抗毡缩毛织物
petite figure 娇小体型
petite size 女装小号尺寸
petite-téte [法]娇小发型
petit grain 法国小粒纹厚丝塔夫绸
petit-gris [法]灰鼠毛皮
petit gulf 小海湾陆地棉
petit motif 法国小花纹花边
petit paletot 宽短外套
petit point 精细斜缝刺绣,针绣挂毯,小花边编织法
petit point stitch 小斜针迹
petit poussin 梭结花边花形,窄花边

petits carreaux 法国小方格毛哔叽
petits pois 小点纹,小斑点
petit toile 法国优质条格亚麻布,小麻布
petit veiours 法国轻薄棉天鹅绒
petrissage 揉摩动作
petrol 深绿
petrochemical fiber 石油化学纤维
petroleum pitch fiber 石油沥青纤维
pettibockers 及踝真丝紧身衬裤
petticoat 内裙,衬裙;苏格兰男子短裙;女装(古)
petticoat breeches 裙裤
petticoat dress 衬裙式女装
petticoat lace 衬裙花边
petticoat look 衬裙款式,衬裙型
petticoat net 衬裙网眼布
petticoat trousers 裙裤
petticulotte 衬裙裤,短裙裤
pettipants 半长女衬裤,女半长内(衬)裤
pettiskirt 衬裙
pettiskirt briefs 连裙紧身裤
petti-slip 衬裙,短衬裙,无上身衬裙
petty size 小号(尺寸)
petunia 深紫红色
peussian velvet 德国纬平绒
pewter (grey) 锡镴灰色
peyia 彩边绸
pH balance fiber pH值平衡纤维
pH Value 酸碱度值
phalange 指骨,趾骨
phalangeal joint 指节点
phalian silk 阿富汗丝绸
Phal-wadi 印度法瓦迪扎染棉布
phantom 人体模型
phantom green 幻想录
phantom line 假想线
pharaoh double crown 法老双重王冠
pharas 印度法拉斯棉布
phase 动物皮毛颜色的变异
phase change fiber 相变纤维,蓄热纤维
phase measurement profilometry 白光相位测量技术
phase-separated fiber 相分离纤维
pheasant 雉鸟色
pheasant's eye 鸟眼花纹
phenix crown 凤冠
phenix-tail skirt 凤尾裙
phenol-formaldehyde fiber 酚醛纤维,苯酚甲醛纤维
phenolic fiber 酚醛纤维

phenolic resin fiber 酚醛树脂纤维
philabeg 短裙;苏格兰短裙;英格兰短裙式童装(或女服)
Philadelphia College of Textile & Science 费城纺织综合大学(美)
philanising 仿毛棉布[硝酸]处理法
philene 酚醛阻燃纤维(商名,法国)
philibeg 苏格兰短裙;英格兰短裙式童装(或女服)
philibit black hemp 菽麻
Philippe Venet 菲力浦·韦内(法国时装设计师)
philippine costume 菲律宾民俗服,菲律宾民族服饰
Philippine embroidery 菲律宾(手工)刺绣
philippine maguey 坎塔拉剑麻
philips POISE standard 菲利普 POISE 标准
philtrum 人中
phiri 二级山羊绒
phloem fiber 韧皮纤维
phlox 福禄考花色
phlox pink 福禄考红
phloxine 竹桃红
phoenix coronet 凤冠
phoenix crown 凤冠
phoenix design 凤凰纹样
phoenix-tail 凤尾裙
phoenix-tail fabric 凤尾布,凤尾纱色织布
phoenix-tail skirt 凤尾裙
phony face 假面具
phool kary 印度刺绣条纹棉细布
phoras 印度薄棉布
phormium 新西兰麻叶纤维
phormium fiber 新西兰麻叶纤维
phoson 亚麻帆布
phosphonic reactive dye 膦酸活性染料
phosphonomethylated cotton 膦羧基甲基化棉
phosphorylated cotton 磷酸化棉
phosphoryl type cation ion exchange fiber 磷酰基型阳离子交换纤维
photee cotton 印度福替棉
photocell control 电眼控制
photochemical printing 感光印花
photochemical textile printings 感光[织物]印花
photochromatic printings 光敏变色印花

photochrome 彩色照片
photochromic 光致变色的
photochromic dye 光致变色染料
photochromic fiber 光致变色纤维
photochromic spectacles 变色眼镜
photochromism 光致变色[现象]
photoconductive fiber 光导纤维
photoconductivity 光电导性
photocopy method 复印法
photodegradation of fiber 纤维的感光降解[作用]
photoelectric inspection of cloth 光电验布
photo fading 光褪色
photoflash photography 闪光摄影术
photographer's stylist 时装摄影设计师
photographer's vest 摄影师背心
photographic color selection 照相选色
photographic printings 感光印花,照相印花
photographic rating standard 分级样照
photographic stylist 摄影(采样)设计师
photographic weaving 丝织像景,照相织造
photography 摄像法;摄影;摄影术;照相
photo luminescence 光致发光
photo marker 摄影排版
photomeasure 照相量体机
photo order 照相定制西服(不同尺量,不用试样)
photo printing 照相印花
photo rapid dyestuff 感光显色染料
photo sample 目录样,照相样,宣传样
photosensitive discolor microcapsule 光敏变色微胶囊
photosensitive dye 光敏染料
photoshop 图像处理软件
photostability 光稳定性,耐光性
photostatic copy 直接影印件
photo stylist 摄影造型师
phototendering 光脆损
phototropic dye 光变性染料
photo yellowing 光致泛黄
Phrygian bonnet 弗里吉亚罩帽(狭窄帽檐,帽身成圆锥形,在法国作为自由的象征);自由帽
Phrygian cap 弗里吉亚无檐锥形帽,垂尖圆锥帽,费里吉亚帽
Phrygian mitre 费里吉亚法冠,弗里吉亚(无檐)帽,自由帽
Phrygian needlework 弗里吉亚(金银线古代风格)刺绣
phthalocyanine dyes 酞菁染料
phulkari 东印度花卉图案刺绣,花卉图案刺绣织物和服装织物
phulshuta cotton 孟加拉弗尔舒塔白色短绒棉
phul silk 中国出口孟加拉的生丝
phulwar 印度丝线起花疏松棉布
phum 山羊绒,开司米
phylactery 装饰穗
physical actions 形体动作
physical and chemical test 理化实验,理化试验
physical-based modeling 物理建模
physical beauty 人体美,形体美
physical-chemical properties 理化性能
physical development 身体的发育
physical dimension 外形尺寸
physical education uniform 体育课制服,练服
physical error 人为误差
physical imperfection 物理疵点
physical jerks 体操;健身操
physical modification 物理变性
physical power 体力
physical production 产品产量,实际产量
physical protection 人体防护
physical shape 人体形态
physical standard 实物标样
physical strength 体力
physical style of acting 形体表演风格
physical test methed 物理检验方法
physico-chemical modifications of fiber 纤维的物理化学改性
physico-chemical surface modification 物理化学表面改性
physiological mechanism 生理机理
physiological psychology 生理心理学
physiotherapy health fabric 理疗保健织物
physique 男子体格;男子体形
phytotron 人工气候室
PIAGET 伯爵
piano shawl 钢琴披巾,西班牙丝质穗饰绣花披巾
pianta bag 晚宴手袋,小型宴会袋
piara 秘鲁皮亚拉棉
piara cotton 秘鲁皮业拉棉
picarde ratine 法国细毛圈棉布
picasso style 毕加索风格图案;毕卡索

风格
picatta and fagoting finish 皮科特装饰光边
picatta and fagoting stitching 皮科特装饰线缝
piccadills 齿形边饰
piceme 派赛姆
pichina 棕色斜纹呢
pick 袖珍梳子
pick-and-pick 芝麻点花纹；芝麻呢
pick-and-pick brocade 胸衣花缎，胸衣锦缎
pick-and-pick design 芝麻点纹
pick density 纬密
pickelnaube 德国军帽，德国军盔
picking out size 采寸，量体
picking pincers 镊子，修布钳，修呢钳
pick(ing) stitch 落穗针法
pickle 腌泡（皮革）
pickle (color) 咸菜色
picklock 西西里亚绵羊毛
pick-on-pick moiré 匹克匹克云纹绸
pick or pin point saddle stitch 细点鞍形线迹，捻针线迹
pick out threads 地毯绒头脱落，编织松散
pick point saddle stitch 鞍形针迹，捻针线迹
picks per inch 每英吋纬密
pick stitch 落穗针迹，细线绣花线迹
pickup felt 引纸毛毯
pick yarn 纬纱
picot 锯齿边，牙边；饰边小圈；毛圈边
picot braid 锯齿编带
picot edge 毛圈边，锯齿边
picot edging 毛圈缘边加工
picot elastic braid 毛圈边弹性平编带
picot festoons bar 扣眼锁缝线
picot hem 狗牙边起口，锯齿边起口，月牙边
pictorial art 绘画艺术
picot seam 毛圈缝合
picot selvedge 圈形边缘
picot stitch 锯齿形线迹，毛圈缘边绣
picotta finish 皮科特光边
picotte 18～19世纪低档薄驼毛呢
picot trim 牙边饰
picture 图，画
picture album 画册
picture book 画册，图册

picture collar 象形领
picture flame heel 影纹袜跟
picture hat 大型装饰女帽；马尔保罗帽；堪斯保罗帽；肖像帽，车轮帽，宽边花式女帽，羽饰宽边女帽
picture in line 线条画
picture knitting 图案编织
picture magazine 画报；有插图的杂志
picture plane 画面
picture velvet 像景丝绒，印经丝绒
picot welt 锯齿边袜口
picot yarn 毛圈花线，无圈花式丝
picot zigzag stitch 人字形线迹，锯齿形线迹
piebald 杂色的，有花斑的
piebald design 花斑花纹
piebald pattern 花斑图案
piece 块；片；件；布匹；部分；布段；颈巾，领巾；修理；拼合
piece-alignment 衣片对准
piece broken 零头布商（古时）
piece counting 计划
piece dressing 单件组合装束
piece dyed 匹染织物
piece dyed cloth 匹染布，匹染织物
piece dyed fabric 匹染织物
piece dyeing 单件染色，匹染，后染
piece dyeing cloth 匹染布
piece dyeing fabric 匹染织物
piece end 短码布，零头布，头子布
piece glass 织物分析镜
piece goods 按匹出售的织物；(总称)布匹，匹头
piece goods dyeing 匹染
piece knitted garments 织片针织服装
piece length (布)匹长
piece mix 混合套装，单件衣组合
piece number 件号
piece orientation 衣片［样板］定向
piece perimeter 衣片轮廓线(计算机)；衣片周界(计算机)
2-piece raglan sleeve 两片式插肩袖
3-piece raglan sleeve 三片式插肩袖
pieces of old cloth or rags pasted togather to make cloth shoes 袼褙
piece sewing machine 缝布机，环缝机
piece size 裁片尺寸
2-piece sleeve 两片式袖
3-piece split sleeve 三片式袖
piece to piece variation 匹间差异

piece together	拼合；拼接
piece together flange	拼接耳朵皮（前身里或挂面的小块拼接布），拼接镶边
piece together gore to sleeve	拼袖角
piece together mark	拼接号，拼接记号
piece together under collar	拼领里
piece up	修补
piece verification	衣片检验
piece verify	衣片检验
piece with patch up	拼匹
piece work	单件生产，计件工作
piece worker	计件工
pie chart	圆饼图表，圆圈图表
piecing	缝料拼接
piecing fabric	接头布
piecing-up	接头
piecing-up work	流苏花边，编结物
pied	斑纹的，杂色花纹的
pied-de-poule（check）	鸡脚花格
pied hose	异色紧长筒袜
pie frill	混染皱边
pielles cabrados	薄毛裤料
pielles negros	薄毛裤料
pieman rod	摇杆
pie-pan hat	饼锅形帽
pierced[-ear] earring	穿耳耳环
pierced earring	穿孔式耳环
pierced fabric	纱罗织物
pierced-look earring	充穿孔耳环
piercing	穿孔，刺破
pier glass	大穿衣镜
pier mirror	大穿衣镜
pierre a cotton	石棉
Pierre Balmain	皮尔·巴尔曼（1914~1982,法国时装设计师）
Pierre Berge	比尔·贝热（法国时装设计师）
Pierre Cardin	皮尔·卡丹（法国服装设计师）
Pierre Dalby	皮尔·达尔比（法国时装设计师）
pierre value	印度皮尔维卢棉
pierrette costume	女丑角装束
pierrot	丑角面具，意大利喜剧的丑角衬衣，紧身低领上衣
pierrot collar	丑角领，小丑领，波褶领
pierrot collar choker	丑角领圈
pierrot costume	丑角装束
pierrot cuffs	丑角袖头
pierrot look	丑角服式,丑角款式,丑角装
pierrot pants	丑角裤子,小丑裤
pierrot ruff	丑角襞领
pie-shape（pi-shape）	π形
piezosensitive discolor fabric	压敏变色织物
pig hair yarn	衬头用猪鬃线（传统缝制工艺中使用）
pig hide	猪皮
pig leather	猪革,猪皮
pig leather shoes	猪皮鞋
pig nappa garments	光面猪皮服装
pig nappy	猪光面皮
pig's bristle	猪鬃
pig skin	猪皮,猪革
pig skin interlining leather	猪里皮
pig split	猪二层革［皮］
pig split garments	猪二层皮服装
pig split gloves	猪〔二层〕皮手套
pig split vest	二层猪皮背心
pig suede	猪绒面革［皮］
pig suede garments	猪绒面皮服装
pig suede gloves	猪绒面手套
pig suede jacket	猪皮绒面夹克
pig suede（split）vest	猪绒面（二层）皮背心
pigeon	鸽灰（紫光灰色）
pigeon blood	鸽血红（暗红色）
pigeon breast	鸡胸
pigeon breasted type	鸡胸型
pigeon chest	鸡胸
pigeon-chested figure	鸡胸体型
pigeon's blood	鸽血红色,深红
pigeon's neck	鸽颈蓝（浅蓝光灰色）
pigeon's wing	鸽翼假发
piggy	儿童脚趾
pigmenet before extrusion	纺前着色的
pigment	色素,颜料,涂料
pigment apparel printing	涂料织物印花
pigment coloration	涂料着色
pigment dye	涂料,色素染料
pigment dyed canvas	涂料染色帆布
pigment dyeing	颜料染色,涂料染色
pigmented fiber	着色纤维,消光纤维
pigmented polyester	［纺丝前］着色聚酯
pigmented rayon	着色人造丝,消光人造丝
pigmented wool	（天然）有色羊毛
pigmented yarn	纺［前］染［色］丝,无光纤纱,无光丝
pigment fiber	纺丝前着色纤维

pigment fiber yarn 无光化纤纱,着色化纤纱
pigment melanin 黑色素
pigment padding 浸轧染色
pigment pad dyeing 悬浮体轧染,涂料轧染
pigment paste 涂料浆
pigment print 涂料印花
pigment printing 涂料印花法;颜料印花法
pigment rayon 无光人造丝,着色人造丝
pigment taffeta 涂料塔夫绸,消光塔夫绸
pignas 彩色棉手帕
pigskin 猪皮,猪革
pigskin garments 猪皮服装
pigskin interlining leather 猪里皮
pigskin jacket 猪皮夹克
pigskin piqué [法]花式凹凸织物,猪皮棱纹布
pigskin shoes 猪皮鞋
pigskin upper leather 猪面革,猪面皮
pigskin wear 猪皮服装
pigtail 辫子,小辫子,发辫
pigtailing 辫结
pigtails 发辫式发型
pigtail wig 猪尾假发,辫子式假发,带辫假发
pije 荷兰佩伊厚粗呢
pik 皮克(织物长度单位)
pike gray 蓝灰色
pike grey 蓝灰
pike staff 金属头手杖
piked shoes 细尖头鞋,喙形尖头鞋,翘头鞋
pikeman's armour 枪兵铠甲(16世纪)
pilch 三角色布(尿布);婴儿外衣
pile 毛茸;起绒;绒毛,绒面;绒头织物,割绒织物,起绒织物;针织长毛绒
pile blanket 绒毯
pile carpet 绒头地毯
pile coated carper 栽绒地毯
pile coated fabric 绒面织物
pile coating 植绒
pile containing product 绒头织物
pile cotton blanket 棉绒毯
pile crush 绒头压扁
pile dishcloth 洗碗绒毛巾布
pile dressing 绒头排直整理
Piledus 派爱立帽
pile fabric 毛绒织物,起绒织物,绒头织物;绒布
pile fabric fastener 尼龙搭扣
pile face 毛面,绒面
pile face brocade 绒面缎
pile fastening 绒头固结
pile figure 绒头花样,绒头纹样
pile floor covering 毛绒地毯
pilegrim 皮格里披肩
pile ground faconné taffeta 绒地绢
pile height 绒头高度
pile hosiery machine 毛巾袜机,毛圈袜机
pile jacket 绒毛短外衣
pile knit fabric 毛绒针织物
pile knitted fabric with brushed pattern 刷花绒针织物
pile lay 绒头倾斜
pile length 绒头长度
pile lined flight jacket 飞行式衬绒夹克
pile lined horsehide surcoat 衬绒马皮外套
pile liner 衬绒(布)
pile lining 毛绒衬里
pile loop 毛圈,绒圈
pile loop velvet 绒圈锦
pile napping 拉绒,起绒
pileolus 派爱立帽(天主教),天主教牧师无边帽,罗马教皇无边帽
pile-on-pile 高低绒头织物,叠花丝绒
pile-on properties 染深性能
pile overcoating 立绒大衣呢
pile pressure 绒头耐压力
pile resilience 绒头回弹性
pile resistance 绒头抗压力
pile retention 绒头保持性
pile reversal 倒绒,绒头倒状
pile ruche 绒毛褶裥饰边
pile rug 绒头地毯
pile setting 起绒
pile shearer 剪绒头机
pile shearing machine 剪绒机
pile slippers 绒毛拖鞋
pile socks 毛巾袜,毛圈短袜
pile stitch 长毛绒组织
pile stitch bonded machine 毛圈型缝编机
pile stitch bonded nonwoven 毛圈型缝编法非织造布
pile stitch-bonding with substrate 底布-毛圈型缝编
pile synthetic leather lining 合成绒面里子革
pile thickness 绒头厚度,绒头高度

pile tricot 长毛绒针织物
pileus 无边毡便帽(古罗马希腊),无檐帽
pile yarn 起绒纱线,绒毛纱线
pilgrim 朝圣披肩(妇女用在罩帽后的打褶的装饰披肩)
pilgrimage pouch 进香袋;朝圣袋
pilgrim apparel 朝圣服
pilgrim[collar] 朝圣领(延伸到肩部的大圆领,胸前的领尾有两个长夹角)
pill 球粒疵
pillar stitch machine 嵌线缝纫机
pillbox 无边[檐]平顶女帽,筒状女帽,圆盒帽
pillbox hat 筒式女帽,圆盒帽,药盒帽
pilled-in selvage 紧边疵
pilleolus 天主教无边帽,罗马教皇无边帽
pilleus 无边毡便帽
pilling 起球痴;织物表面起球
pilling of knitted fabric 针织物起毛起球
pilling propensity 起毛起球倾向
pilling resistance 抗起毛起球性
pilling resistant finish 抗起球整理
pillion 鞍褥,后鞍
pillow 凸纹棉布;凸纹亚麻布,麻纱;拳击手套;枕头
pillow bag 大方包,枕头包
pillow bar 花边饰边网底,牙边网底
pillowcase 枕套
pillowcase linen 漂白平纹细亚麻布
pillow cases 枕套织物
pillow-cotton 棉枕套布
pillow cord 枕头边绷带
pillow cover 枕套
pillow for health 保健枕头
pillow fustian 细棱条灯芯绒,狭(窄)棱条灯芯线
pillow lace 梭结花边;枕结花边
pillow linen 亚麻枕套布
pillow sack 枕套,成形枕套
pillow sham 绣花枕套,靠枕套;座垫套
pillow-slip dress 枕套装(像枕套一样直线条感的平坦长服)
pillow towel 枕巾
pillow tubing 圆筒形枕套布
pill-pront textile 易起球纺织品
pill resistance 抗起球性能
pill-resistant fiber 抗起毛起球纤维
pill-resistant finish 抗起毛起球整理
pillsworth 漂白棉布,仿亚麻棉布
pill wear off 毛球脱落

pilly 成团,起球
pilos 皮路斯毡帽(希腊),皮洛斯帽,圆锥顶帽
pilot cloth 军服呢,海员厚绒呢,水手呢
pilot coat 飞行员大衣,领航员大衣,飞行服式外套
pilot jacket 飞行员外套,领航员外套;海员厚夹克
pilot lamp 指示灯
pilot model 试选样品
pilot run 试运转
pilot shirt 飞行员衬衫(有肩章,大口袋),肩袢衬衫
pilot's suit 女飞行员两件套装
pilot test 中间试验
pilou 棉法兰绒
pima cotton 比马棉
pimean wool 松捻针织绒线
pimiento 椒红色,鲜红色
pimento red 椒红色,鲜红色
pimsole ducks 防雨平布
pin (用针)别住,钉住;大头针,饰针,扣(衣)针;发夹;别针徽章,别针像章
pina cloth 菠萝叶纤维;菠萝叶纤维织物(菲律宾及西印度)
pina fiber 菠萝叶纤维
pinafore (小孩用)围涎布,围嘴,围涎,反穿衫;围裙,围裙装,饭单,饭单裙
pinafore and guimpe 围裙装及高领内衣
pinafore dress 围裙服;围裙装(像围裙般后开口,前身一般有实用大口袋,裙身宽敞);无袖无领服
pinafore heel 童鞋跟,围涎式扁平跟
pinafore jumper 围兜式背带裙,胸兜背心裙
pinafore pleats 裥裙
pinafore swimsuit 带护胸的泳装
Pinara cotton 秘鲁皮纳拉棉
pinarette 短围涎(小儿);小围裙,短裙;短学生裙;无领无袖连衣短裙
pinatusa 菲律宾皮纳尤瑟大麻色布
pin-back shirt 背扣衬衫
pin basting 针别式假缝,大头针假缝
pin bone lace 细梭结花边
pinbyu 印度手织粗棉布
pin case 针盒
pin center setscrew 定位销固定螺钉
pincette 小镊子,小钳子
pinchback clothing 背部紧身外套(半腰带,打褶)

pinch back coat 皱背外套,皱背夹克,后扎腰外套
pinchbeck 仿制品,廉价货,冒牌货,赝品
pin check 细格子,小方格;细格子纹
pin check umbrella 细格子伞
pincher 钉钉机(加工裘皮、皮革用);钉皮机
pinchers 钳子;(美国)鞋
pinch front 前凹帽(前上端凹进帽子的总称)
pinchina 低级斜纹呢,平纹厚呢
pinch pleats V形裥,浅形裥,浅褶裥
pinch pleat tape 浅形褶带
pinch waist 小腰身
pin collar 饰针领,针孔领
pin curl 发夹式卷发,针卷发
pin cushion 插针包,针垫;针插
pin-cushion hat 针垫帽
pin-cushion jewelry 针插式贴饰
pin dot 小水珠花纹,圆点花纹,针夹点子花纹
pin down 倒针
pineapple bobbin 菠萝筒子,双锥形筒子
pineapple cloth 菠萝纤维织物;手帕亚麻布;上浆全丝薄纱
pineapple cone 菠萝筒子,双锥形筒子
pineapple fiber 菠萝纤维
pineapple hemp 菠萝麻
pineapple knit fabric 菠萝针织物
pineapple lace 菠萝花纹花边
pineapple leaf fiber 菠萝叶纤维
pineapple ornament 菠萝装饰
pineapple package 双锥形筒子,菠萝筒子
pineapple pattern 菠萝花纹
pine bark 松树皮色
pinecone 松果色
pinecone ornament 松果装饰
pinefore 防脏围裙
pine green 松树绿
pine leaf stretch stitch 包边松针弹性线迹,曲折线迹
pine marten 褐貂皮
pine marten fur 松貂毛皮
pine needle 松针绿(暗绿色)
pine needle wool 松叶纤维
pine ridge scout hat 松条童子罩帽
pine silk 人造丝纺绸
pine tree 松树呢
pine wool 松叶纤维
pin-fit 用大头针固定

pin fitting 缝纫前试样,大头针试样
pinhead 别针尖,别针头;针头状花纹,细圆点花纹;针头状花纹织物,细圆点花纹织物,细圆点花纹服装;头小的人
pinhead check 针尖格子纹;针尖格子呢
pinhead staple 针尖状纤维
pin(head)stripe 针点条纹
pinhead weave 针头纹织物
pinhole 针眼;针孔;针洞疵
pinhole collar 别针扣领,针孔领,鸡眼领,饰针领
pin-in bra(ssiere) 针别式胸罩(两侧用搭钩或别针与服装固定,中间由橡筋连接的胸罩)
pink 粉红,粉红色;粉红色衣料;红色猎狐上衣(英国)
pink carnation 淡红色,肉色
pink coat 猎狐外套
pink diamond 粉红钻石
pinked and double-stitched seam finish 双次缝锯齿剪边缝份处理
pinked and stitched seam finish 压线锯齿剪边缝份处理
pinked edge 荷叶边疵,锯齿边疵;锯齿边
pinked flat hem 锯齿缘边
pinked hem 锯齿下摆
pinked seam 锯齿边接缝
pinked seam finish 锯齿剪边缝份处理
pinked shoes 细尖头鞋
pinkers 狗牙剪刀,花边切刀,锯齿剪刀
pinkers pinking scissors 齿边布样剪刀
pink hemming 狗牙边
pinkie 小(手)指
pinking 剪锯齿边,锯齿裁切;衣边剪花;打饰孔
pinking cutter 齿边布样切裁器
pinking iron 齿边布样切裁器
pinking machine 扎驳头机,轧锯齿边机,锯齿边布样切裁机
pinking scissors 齿边布样剪刀,花齿剪,锯齿剪
pinking seam 锯齿缝
pinking shears 齿边布样剪刀,花齿剪,锯齿剪
pinkish red 品红
pink lavender 浅雪青,浅粉莲
pink lipstick 粉红色唇膏
pink lock slider 拉链钩锁拉头
pink lotus 浅粉莲色
pink mist 薄雾玫瑰色

pink out　装饰,打饰孔
pink red　桃红
pinks　旧时美陆军军官淡褐色裤子
pink tint　淡粉红色
pinky taupe　淡红鼬鼠皮色
pinky white　粉红白
pin lock slider　拉链钩锁拉头
pin mark　布边针孔疵,大头针作记号
pin marking　大头针标志,针列记号
pinnacle stitch　小尖塔形线迹,尖顶形线迹
pinna marina　贝足丝,海丝
pinna silk　贝足丝,海丝
pinna wool　贝足丝,海丝
pinned-up hair　掠起式发型
pinner　庇娜头巾
pinners　垂片头饰(17～18世纪),女帽式头巾,围涎
pinni　天然红色棉土布
pinning　用大头针固定
pinning blanket　婴儿裹衣,新行儿全套用品
pinning machine　订商标机
pinny　围裙,饭单,饭单裙,围诞布;针尖白点疵
pinokpok　菲律宾蕉麻织物
Pino Lancetti　皮诺・兰切蒂(意大利时装设计师)
pin pad　针垫
pin pendant　别针式垂饰
pinpoint collar　尖角领
pin pointed stripe　针点条
pinpoint eyelet　细点子网眼
pin point mest　小网眼
pinpoint saddle-stitch　细点鞍形线迹
pinpoint stitch　细点针绣线迹
pin rib　细棱条薄细布,经向细棱纹效应
pin seal　海豹纹理;海豹皮,海豹革;海狗皮
pinsonic process　超声波热熔黏合工艺
pinsonic quilting　超声波衲缝法
pin spanner　内六角扳手,销子扳手
pinspot　细点花纹
pin stitch　手绣包缝
pin stripe　细条纹,细条子;细条纹织物,细条纹服装
pin stripe duck　超细白条子帆布
pin stripe effect　细条纹效应
pin stripe fancy suitings　牙签条花呢
pin stripe shirting　细条纹衬衫料

pin tack　大头针假缝,细裥假缝
pintade　18世纪法国模板印花装饰布
pintado　16世纪印度低级模板印花布,17～18世纪印度轧光印花平布
pin top collar to under collar together　覆领president
pintor　聚丙烯腈抗起毛起球纤维(商名,日本东邦人造丝公司)
pin tuck　针纹褶饰;针形缝褶;细褶,细绉;狭裥;折褶固缝;细小突出缝褶,细缝褶,细条缝褶;双罗纹集圈组织
pintucked seam　细褶缝
pintucked shirt　细褶衬衫
pintuck foot　细褶压脚
pin-tucking　折褶固缝
pin-tucking design　折褶固缝花纹
pin-tucking machine　折褶固缝缝纫机
pin-tucking method　折褶固缝法
pin-tucking width　折褶固缝宽度
pin tuck seam　折褶固缝线缝
pin tucks pocket　针纹褶饰口袋
pinwale　灯芯绒细条纹,细棱条
pinwale corduroy　细棱条灯芯绒,细条灯芯绒
pinwale corduroy jacket　细条灯芯绒夹克
pinwale pique　细条凹凸织物
pin wheel hosiery boarder　针轮式袜子定形机
pinwheel skirt　转轮焰火裙
pin work　别布造型,细小凸出饰纹
pioneer costume　拓荒者服装(美国18～19世纪流行)
pioneer product　先驱产品
pip　金属肩章,军衔星
pi-pa button　琵琶扣
pipe　滚边,为(衣服)滚边;凸缘,拷边
pipe buttonhole　滚边扣眼
piped　皮革折曲时起皱的;滚边的,镶边的
piped buttonhole　滚边纽孔
piped edge　(衣服)滚边,镶边
piped edge sewing machine　滚边缝纫机
piped neckline　嵌条领口
piped placket　滚边袖衩
piped pocket　滚边袋,嵌线袋,无盖暗袋
piped seam　嵌条缝,嵌线缝,细滚边缝合;直筒裤
piped tuck　嵌条褶
pipe edge　镶滚(衣、布)边
pipe inside line of facing　滚挂面

pipe orgn folds 管形有硬衬折叠,圆形折叠
pipe orgn pleats 管风琴裙,皱裙
pipe pocket mouth 滚袋口
piper (缝纫机)滚边(拷边)装置;镶边器;包缝机,拷边机
pipe-shaped hollow fiber 管形中空纤维
pipestem pants 瘦腿裤,直筒裤
pipestems 瘦腿裤
pipestem silhouette 直筒形轮廓
pipestem trousers 瘦腿裤
piping 滚边;镶边;包缝,拷边;缝饰带;窄幅毛丝织物
piping and brief rounded collar blouse 滚边小圆角领上衣
piping around collar 滚边领
piping around pajamas 滚边睡衣
piping binder 滚边器
piping buttonhole 滚扣眼
piping cord 包边绳,滚边线绳
piping cuff 滚边袖口,滚边袖头
piping edge felling 滚边的边缘折缝
piping folder 折边滚边器
piping foot 滚边压脚
piping front facing 滚挂面
piping hole 滚边洞眼
piping insertion 滚边嵌线
piping machine 滚边机
piping pocket 滚边袋,嵌线袋,嵌袋
piping pocket mouth 滚袋口
piping ruffler 滚边打裥边,滚边荷叶边
piping seam 滚边缝
piping strip 嵌线皮
piping trim 滚边,嵌线,捆条
piping (trimmed) pocket 嵌线袋
pique 凹凸毛织物,凹凸组织,灯芯布;法国仿纱罗织物
pique anglais 素色毛哔叽
pique braid 凸边带
pique crepe 凹凸绉;棉绉布;棉绉纱
pique damas 法国八综素色斜纹绸
pique dimity 凸纹巴里纱
pique embroidery 白色凸条刺绣,棱条刺绣
pique polo-shirt 珠底马球衫
pique seam 凹凸接缝,内包缝
pique stripe 灯芯绒条纹,凸花条纹
pique voile 凸纹巴里纱
piranshahi siah 染色平纹棉布
pirate boots 海盗靴
pirate costume 海盗装束
pirate hat 海盗帽
pirate pants 海盗裤(细长,长至足踝)
pirate shirt 海盗衫
pirates look 海盗型服装款式
pirate style 海盗(型)款式
piratical felt 海盗式毡帽
piri 螺旋卷绕的金银线
pirle finish 珀尔整理(呢绒的防雨、防缩、防污整理,商名,英国)
pirl thread 螺旋形金银线
pisa stitch 梯形线迹
pi-shape π形
pis (pocket) 臀部后袋;臀部口袋;后枪袋
pistache 淡黄绿色
pistachio green 淡黄绿色
piste 滑雪夹克
pistol pocket 臀部后袋;臀部口袋;后枪袋
piston 活动芯杆
pitambar 金边手织布
pitch 沥青色
pitch-based carbon fiber 沥青基[质]碳纤维
pitch-black 乌黑
pitch-dark 深黑,漆黑
pitch fiber 沥青纤维
pitch-pin 对花小钉
pitch time 节拍
pitgaveny 皮加芙呢
pith 遮阳盔
pith hat 软木帽,木髓盔帽,皂角木髓遮阳帽
pith helmet 软木帽,木髓盔帽,皂角木髓遮阳帽
pittman cotton 美国皮特曼棉
pitman (rod) 缝纫机摇杆
pivot 钻石轴尖
pivotal-transfer technique 旋转式移动技术
pivoted lever 摆臂
pivoted-transfer technique 纸样旋转技法
pivoting 旋转法
pivot pin 大平头别针
pivot point 支[枢]点,转动中心点,中心点,基准点
pivot point technology 移转定点技术
pivot sleeve 楔形袖
pivot technique 旋转法;旋转技术

pixie 皮克茜发型
pixie cap 尖顶帽,顽童式毛线绒球帽
pixie cape 夹兜帽斗篷,连帽斗篷
pixie cut 顽童式女子短发型,皮克茜发型
pixie hairstyle 顽童式女短发型
pixie hat 尖顶帽,顽童式毛线绒球帽,妖精帽
pixie hood 尖兜帽
pixy cut 顽童式女子短发型
pixy 皮克茜发型
pizzazz 时髦派头,大胆创新的品质
pizzo 意大利原色丝花边
place 地点
place in selling 畅销排名
placement 形体定位
place of tuck 打褶处
place satin 贡缎
placing material (裁剪前)排料
placket 衬裙(古);开襟,(裙腰)开口;衩口,裙衩;(男衬衫)袖衩;女裙口袋;内裙,围裙
placket band 半开襟滚镶边,衩镶边
placket extension band 衩叠门
placket facing 袖衩条
placket folder 衩折边器
placket folding machine 袖衩折叠机
placket front 明门襟;前半开襟;袖衩
placket front pullover 半开襟衫
placket fusing press 门襟黏合机
placket hole (裙腰)开口
placketing 女裙开衩
placket lapel 襟贴
placket length 门襟长
placket-line 袖衩线
placket machine 门襟机
placket neckline 半开襟领口,开衩领口
placket overlap 大衩边
placket seam 门襟缝
placket shirt 半开襟套衫
placket tongue 襟舌
placket undershirt 扣子衫
placket undershirt with roll collar 翻领扣子衫
placket width 门襟宽
placket with gathers 袖褶开衩
plaga 猎网
plague 襟上饰物,饰板,名誉奖章
plagula 床围布

plahta 乌克兰女裹裙
plaid 格子花纹;格子花布;苏格兰格子花呢;格子花呢服装
plaid-back 背面格子双层大衣料
plaid blouse 长方格上衣
plaid cross stitch 普通十字绣
plaid flannel 格子法兰绒
plaid gingham 方格色织布
plaiding 密绒厚呢
plaid pattern 苏格兰格子花纹
plaid placement 对格部位
plaid repeat 花格循环
plaid stitch 缎带格子绣
plaid stripe interlock fabric 抽条方格棉毛布
plaid stripe knitter 条格花纹针织机
plain 简朴;正针;普通的,平的,素的;平纹的,平针的;平纹布,素色布
plain all-over flocking 单色植绒
plain and purl knitted fabric 正反面针织物
plain back 平纹地,背面平纹哗叽
plain back velvet 平纹地经绒
plain back velveteen 平纹地纬绒
plain bengaline 素绨
plain blanket 素毯,素色羊毛毯
plain bottom 平下摆,平裤脚,平裙;平脚口
plain box silk 板绫
plain braid 平编带,平辫带
plain carpet 素色地毯
plain circular knitting machine 单面圆纬机,单面圆形针织机,平纹圆形针织机
plain cloth 平布;平纹布;平纹织物;素色布
plain clothes 便衣,便装,便服,常服,休闲装
plain coating 素呢
plain collar 普通(衬衫)领
plain color 素色,单色,纯色
plain colored bed sheet 素色床单
plain colored pile blanket 素色绒毯
plain colored satin brocade 素软缎
plain colored scarf 素色围巾
plain colored table cloth 素色台布
plain colored towel 素色毛巾
plain crepe 平绉
plain crepe satin 素绉缎,普通十字绣
plain cuff 普通袖头
plain cutting 平面剪裁

plaindin 苏格兰毛哔叽
plain dowble warp bag 双经平纹麻袋
plain drop 光滑水滴形(珠宝)
plain dyed 素,单色;素色的,一色的,单色的
plain dyed bed sheet 素色床单
plain dyed cloth 素色织物
plain dyed fabic 单色织物
plain dyed satin stripes handkerchief 素色缎条手帕
plain dyed table cloth 素色台布
plain dyed towel 素色毛巾
plain dyed yarn 单色纱线
plain edge 平边
plain edge machine 平边(缝纫)机
plain embossed design 素色轧花
plain eyelet fabric 单面纱罗提花织物,单面网眼花纹
plain fabric 平纹织物,光面织物,素织织物
plain face 光面
plain faced neckline 普通贴边领口
plain facial 普通脸部处理
plain fake fur 素色人造毛皮
plain fancy suitings 素花呢
plain felt 平面毯
plain finish 本色整理,不丝光整理
plain flannel 素色法兰绒
plain flat knitter 针织横机
plain foot bottom 平针袜底
plain front 暗门襟
plain furnishing fabric 平纹装饰织物
plain gauze 平纱罗,素纱罗
plain goods 平纹坯布,平纹织物
plain habutai 纺,平织电力纺
plain hem 平折(包缝)缘边
plain interlining woolen 素驼绒
plain jersey 针织平布,平针织物
(plain) kabe crepe 素碧绉
plain knit 平针针织物;平针织物
plain-knit gloves 平针手套
plain-knitted fabric 平针织物
plain knitwear 平针毛衫
plain knot 平易领结,普通领带结
plain ladies cloth 平素女式呢
plain lucter 棉经粗马海毛纬织物
plain mesh 平纹网眼,素网眼
plain muslin 素色细布
plain neckline 平领口,颈根领口
plain net 平织网

plain net lace 无花纹网状花边
plain net lace machine 编带机
plain nylon sheer 素丝绡
plain overcoating 平厚大衣呢
plain overlap seam 平叠缝,简单搭接缝
plain pear-shape 光滑梨形(珠宝)
plain petal sleeve 无褶花瓣袖
plain plaiting 简单添纱
plain plush 素色毛圈织物,素色平纹长毛绒,素色长毛绒
plain poly-bag 包装手帕用白胶袋
plain poplin 素府绸
plain pump 普通碰拍鞋,普通潘普鞋
plain rep 平纹简单楞条子织物
plain rib body machine 平罗纹大身圆形针织机
plain rib fabric 平色罗纹针织物
plain ruffle 普通褶边
plains 平布,平纹布;素色布,重浆素色棉布,平纹坯布
plain safety hemmer 普通安全卷边器
plain satin 平缎,素缎,手织素累缎,厚重轧光缎
plain seam 平缝,坐倒缝
plain selvage 平纹布边
Plains Cotton Growers (美)得克萨斯州棉花生产协会
plain sewing 普通缝纫
plain sewing machine 平缝机;平机,平车
plain sewn closure 普通缝纫门襟
plain shade 素色,单色,一色
plain shade links and links hosiery 素色凹凸花袜
plain shade rib hosiery 素色抽条袜
plain shampoo 普通香波
plain shirting 平纹衬衫织物
plain side 技术正面,工艺正面
plain silk 绢(纺)
plain skirt 普通裙
plain smocking 普通图案形衣褶;普通正面刺绣线迹
plain sole 平针袜底
plain sphere 光滑球面形(珠宝)
plain spun silk 平纹绢丝纺
plain stitch 普通线迹,平缝线迹,平针针迹,平针线圈
plain stitched bottom 平切下摆,平切裤脚
plain stitch fabric 平针针织物
plain stitch sewing 平缝线迹缝纫,普通

线迹缝纫
plain stitch tack 平针粗缝
plain taffeta 素塔夫绸[绢]
plain toe 平鞋头；普通鞋头，没有装饰的鞋头
plain toe cap 普通鞋头，没有装饰的鞋头
plain toe cap shoes 素头鞋
plain toe shoes 平头鞋
plain top gloves 平口手套
plain top hosiery 平口袜
plain towel 全色毛巾，素色毛巾
plain trousers 普通西裤
plain tubular knit fabric 圆筒形平针织物
plain two-piece shirt 简单两片裙
plain velvet 平绒，平丝绒，素丝绒
plain weave 平纹组织
plain weave fabric 平纹织物
plain web fabric 素色网眼织物
plain web machine 多三角圆形针织机
plain weft-knitted fabric 纬平针织物
plain wilton 平纹威尔顿地毯
plain wool 素呢；无卷曲羊毛
plain woven 平纹织物
plain woven label 平面织唛，平纹织唛
plain woven ribbon 平织绸带，平织缎带，塔夫绸带
plain wpholstery fabric 平纹家具装饰织物
plain yarn 素色纱；单色纱
plain zigzag stitch 普通曲折形线迹
plait 褶；裥；褶裥；编织的绳子；打辫编成；辫状物
plait braid 辫形编带
plait down 折布，折叠
plaited dress 褶裥连裙装
plaited fabric 单纱花边织物，花边织物
plaited face 褶裥花边，金银丝花边，编织带
plaited heel 添纱袜跟，加固袜跟
plaited hosiery 添纱针织物
plaited inlay knit fabric 添纱衬垫针织物
plaited insertion stitch 互编嵌饰针法
plaited lace 褶裥花边
plaited patch pocket 有褶贴袋
plaited purl knit fabric 添纱双反面针织物
plaited seam 辫状缝，褶边缝
plaited skirt 褶裥裙；褶裙
plaited sole 添纱袜底；编织鞋底
plaited stitch 辫状线迹，人字形线迹，席纹刺绣针迹
plaited yarn 包芯线，嵌芯花线
plaiter 码布机
plait form 折叠形式
plaiting 码布；帽坯毡合；针织添纱
plaiting length 折幅长度
plaiting machine 码布机
plaiting seal 折叠式封口
plaiting stitch 辫状线迹，人字形线迹
plaits 发辫
plait securing sewing machine 折褶缝纫机
plait trim 辫状饰边
plan 计划，设计
planar cutting 平面裁剪
planchette 法式垫衬紧身衣
plane and square toe shoes 铲头鞋
plane coating 薄膜涂层
plane cutting 平面裁剪
plane figure 展示图；平面图
plan layout 工厂布置图
plan sketch 平面草图，设计草图
plane spread figure 平面展开图
plane toe shoes 偏平头鞋
plane view 平面图
plangi 绞缬，帕伦杰染色，扎染
planimeter 侧面仪
planking 毡合，缩绒
planned economy 计划经济
planned framework 谋划框架
planned management system 计划管理体制
planning board 经济排版，台板
planning commodity 计划商品
planning selling 计划销售
planning target 计划指标
plantain-leaf design 蕉叶纹
plantar arch 足底弧线
plantation 茶绿
planted design 花纹移植
planted fabric 花纹织物
planter's hat 宽边手编农场主帽，种植者帽
planter style 农场主风格
plant fiber 植物纤维
plant floor 现场
plant location 工址，厂址
plants 第一次摘的棉
plant seed button 植物种子扣
plant size 工厂规模

plaque 襟上饰物
plaquette 门襟
plasma gas cutter 等离子体裁剪口
plasma treated spun silk 等离子体处理绢丝(改进绢丝性能)
plasma treated wool 等离子体处理羊毛(改进羊毛性能)
plastelast 弹性塑料,塑弹性物;塑弹体
plaster bandage 石膏绷带
plaster cast 石膏绷带,石膏模型
plaster color 石膏色
plaster figure 石膏像
plaster mould 石膏像
plastic 塑胶片眼镜
plastic apron 塑胶围裙,塑料围裙
plastic arts 塑造艺术;造型艺术,雕塑艺术
plastic bag 塑料袋;胶袋
plastic belt 塑胶裤带,塑胶腰带,塑胶带
plastic bib apron 塑胶围胸裙
plastic boning 塑料衬带
plastic braided bag 塑料编织袋
plastic button 塑胶扣,塑料扣
plastic clasp 塑料扣子,塑料钩子
plastic cloth 塑胶布,塑料布
plastic clothes-pin 塑胶别针
plastic-coated fiber 涂塑[料]纤维
plastic-coated glass fabric 涂塑[料]玻璃纤维织物
plastic-coated gloves 浸塑手套
plastic-coated metal fiber 涂塑[料]金属纤维
plastic-elasticity 塑弹性
plastic eraser 塑胶橡皮
plastic extruded zipper 塑胶挤压拉链
plastic fastener 塑胶拉链
plastic film 塑料薄膜
plastic foam 多孔塑料,泡沫塑料,发泡塑料
plastic form 立体模型
plastic frame crinoline 塑料裙撑
plastic garment bag 塑胶衣袋
plastic garments 塑料服装
plastic handbag 塑胶手提袋
plastic handle 塑胶柄
plastic hanger 塑料衣架
plastic hangtag 塑料吊牌
plastic high boots 塑料长靴
plasticised fabric 塑[性]化织物
plasticity 可塑性

plasticized fiber 增塑纤维
plastic jacket 塑胶夹克
plastic leather 塑胶皮;塑料革
plastic lining 塑料衬里
plastic nylon 尼龙塑料
plastic operation 整形手术
plastic overall 塑胶外套
plastic pants 塑料防水裤
plastic peg 塑料衣夹
plastic pocket 塑料口袋
plastic raincape 塑胶雨披
plastic raincoat 塑料雨衣
plastic raincoat with hood 塑胶连帽雨衣
plastic ruler 塑料尺;塑料卷尺
plastics 塑料;塑料制品;整形
plastic sandals 塑料凉鞋
plastic shield 塑料罩
plastic shirt clip 衬衫胶夹
plastic shoes 塑料鞋
plastic slide fastener 塑料拉链
plastic slippers 塑料拖鞋
plastic snap 塑料四合钮
plastic sole 塑胶鞋底
plastic strain 塑性变形,塑料应变
plastic suspender 塑胶吊裤带
plastic TPO (plastic time; place; occasion) 万能穿衣法,一套衣服可以适合任何时间、场所、场合的穿用方法
plastic wedgies 坡跟塑料(女)鞋
plastic zip fastener 塑料拉链
plastic zipper 塑胶拉链
plastoelasticity 塑弹性
plastoelastic deformation 塑弹形变
plastron 兜状领;护衣;击剑护胸革;女胸衣;上浆衬胸;胸饰;(男用)带有部分前胸的衬领;皮护胸;硬衬胸
plate (金属)牌子;板;毛皮块;铠甲;铠甲金属片
plate armor 金属盔甲,棱鳞甲
plateau (plat.) 平顶女帽
plate cover 水布,烫布
plated 鸳鸯布;拼色,拼匹
plated block in toe 添纱加固袜头
plated cloth 鸳鸯织物
plated hat 呢底毛巾帽
plated hosiery 添纱针织物,棉毛布
plated knit 鸳鸯针织物
plated knit goods 添纱针织物,棉毛布
plated knit intarsia 添纱镶接
plated package 并包包袋,拼件成包

plated piece 拼匹
plated purl fabric 添纱双反面织物
plated sole 加固袜底
plated work 添纱针织物；添纱组织
plated yarn 嵌芯花线，编结线，涂料线
plate press felts 平压榨毛毯
plate printing 平版印花
plate singeing 铜板烧毛
platfond knot 藻井结
plate singeing machine 平板烧毛机，铜板烧毛机
platform (plat.) 木屐式坡形鞋跟；木屐式坡形高跟鞋
platform balance 台秤
platform boots 厚底靴
platform pumps 平底浅口鞋，砖形跟浅口鞋
platform sandals 厚底(坡跟)凉鞋，平底高跟凉鞋
platform shoes 厚底鞋，高台鞋，木屐式坡形高跟鞋，松糕鞋，月台式皮鞋
platform skiing shoes 跳台滑雪鞋
platform sole 高台底，松糕鞋底，砖形鞋底(鞋头至跟连在一起)
platform with sling back 踝部系带的厚底鞋
plati 辫状物
platied rope 编织绳
platilla 普莱蒂拉漂白亚麻织物，平纹细亚麻布
platina color 白金色
platina fox 白金狐皮
plati shirt front 衬衣前门直口褶
plati trim 辫状饰边
plating 添纱，印纹，熨平(皮革)
plating knitted fabric 添纱针织物
plating stitch 添纱组织
platinum 白光；白金色
platinum alloy 铂合金
platinum blond 淡金黄色；淡金黄色头发的[女]人
platinum silver fox 银灰色狐皮
platno 保加利亚普拉托夫丝毛呢，丝毛混纺钢花呢
platove 保加利亚普拉托夫丝毛呢，丝毛混纺钢花呢
plat shoes 高台鞋；木屐式坡形高跟鞋
platter collar 平贴领(大圆角，中等大小圆领，平贴在肩上)
platter hat 宽檐女帽

platt lace 机制平花边
platyrrhine 阔鼻人
Plauen lace 普劳恩花边(法国)，普劳恩烂花花边
plauen machine 花边机
PLAYBOY 花花公子
play clothes 轻便装，游乐装
play dress 剧装
play set 针织童装
play shoes 轻便鞋，游戏鞋
play shorts 游戏短裤
playsuit (女、童)运动衫裤；游戏套装；轻便服；游乐装，游玩装
playwear 轻便装，游乐装，运动服
plaza taupe 集市灰褐色
plchette 装饰吊袋，小口袋
plchonne double cloth 英国提花丝毛呢
pleasing handle 柔软的手感
pleat (衣服、裤、裙的)褶；褶裥；活褶；线带，绳，裢；打裥
pleatability 成褶性，褶裥率
pleat abrasion 褶裥磨损
pleat backing 裥底
pleat depth 裥深
pleat direction 搭裥方向
pleated back 背肩，背裥
pleated bell sleeve 钟式百裥袖
pleated bosom 襞胸，胸前褶片
pleated bosom blouse [shirt] 胸褶衬衫
pleated collar 有褶宽领，褶饰领
pleated cowl 折裥式垂褶
pleated cowl-draped pants 垂褶裤
pleated crinoline wheel 折叠式裙撑
pleated cuff gloves 折口手套
pleated dress 打褶连衣裙，褶裥连衫裙装
pleated effect 褶裥效应
pleated fabric 褶裥织物；褶裥布
pleated frill 折叠花边
pleat(ed) form 褶裥形式
pleated ruffle neckline 褶皱荷叶领，渐层褶饰V字领
pleated nylon lace 褶裥尼龙花边
pleated overalls 工装裤
pleated pants 褶裥裤
pleated patch pocket 明裥贴袋
pleated pocket 褶裥袋；打褶袋；褶饰口袋
pleated set 褶裥处理
pleated sheet structure 褶裥的片状结构
pleated shorts 打褶裥短裤
pleated shorts with turn-ups 打褶翻边

短裤
pleated skirt 对褶裙;百褶裙;褶裥裙
pleated sleeve 褶袖
pleated tennis skirt 褶裥网球裙
pleated tricot 褶裥经编织物
pleated trousers 腰围前打褶的西裤,有褶裤
pleated warp knit fabric 褶裥经编针织布
pleated warp knitted fabric 褶裥经编针织物
pleated-gore skirt 拼片褶裥裙
pleater 打褶裥者;打褶装置
pleat-facing 褶裥贴布
pleat finish 打褶裥加工
pleat fold 打褶折边
pleat-gore skirt 片褶裙
pleat-held by stitching 褶裥缝纫,缝住一段的褶裥
pleat-held after washing 洗后褶裥保持
pleating 打褶;打裥;褶裥处理
pleating attachment 打褶附件
pleating machine 褶裥机,打裥机
pleating pattern 织物打裥用模纸
pleat inside width 折叠宽,褶里深
pleatless trousers 无褶裤,合身无褶裤,平脚口裤
pleat line 裥位线
pleat making machine 打褶(裥)机
pleat operation 打裥作业
pleat pattern 织物打裥用模纸
pleat pocket 褶饰口袋
pleat position(line) 褶裥线;裥位线;打褶线
pleat (position) line 裥位线
pleat press 压褶机
pleat [pressing] machine 打褶机
pleat retention 褶裥保持性
pleats 蔽膝
pleat seam 褶裥缝
pleat seaming 褶裥缝纫
pleat skirt 褶裥裙;褶裙
pleat spacing 裥距
pleats retentivity 褶裥保持率
pleat stay 褶牵条
pleat top 褶山
pleat trousers 卷脚口裤
pleat under 褶裥窝边;褶里
pleat underlay 裥叠量
plentiful 丰盈,丰腴
pleochroism 多向色性

pleuche 金丝绒
pleuche velvet 金丝绒
plexiglas(s) 有机玻璃(镜片用)
plexiglass button 有机玻璃扣;珠光扣
pli religieuse 横襞,横褶,水平线褶
pliable feather 细羽毛
pliable hand 柔软手感
pliancy 柔韧性,易弯性
plicose 普利科斯防雨绸(商名,美国)
plied yarn 股线,合股线
plied yarn duck 线帆布
plied yarn of different nature of strand 不同纱的股线
pliers 钳子,老虎钳
plies 叠布层
plimsoll duck 鞋用帆布;橡胶底帆布鞋
plimsolls 胶底帆布轻便鞋,橡胶底帆布鞋
plissé 泡泡纱绉条,泡泡纱效应
plissé crepe 泡泡纱;细绉布
plissé fabric 泡绉织物
plissé finishing 绉缩整理
plissé printing 泡泡纱印花
plissé satin 人造丝绉缎
plissé skirt 细褶裙
ploc 牛毛,废粗毛
ploc fiber 牛毛,废粗毛
plodan 格子粗呢
plommet 丝毛或丝麻织物
plonese frock 男式礼服
plonkete 蓝色;粗纺毛呢
p-looper 单链缝套口机
plotter 绘图机
plotting scale 绘图刻度尺,制图比例尺
plough bone 犁骨
plow boots 家耕靴,农耕靴
plucking 拔鬃毛(毛皮),去掉外层粗毛的拔毛工艺;剥棉
plug 高顶礼帽
plugged oxford 栓帮牛津鞋
plug hat 高顶礼帽,高礼帽
plug oxford 低帮系带牛津鞋
pluie 闪光呢,金银丝织物
plum 梅红(暗红青莲色);紫红色;青紫色
plumage 羽毛束,羽毛填料;漂亮精致的衣服;羽衣
plumb 织物加重
plumber's bag 管工用袋,铅管工用袋
plumbers' ticking 人字斜纹床品织物

plumb line 铅锤线,铅垂线,准绳
plum-blossom design 梅花纹
plum color 梅花色,茄色,深紫色
plume 驼鸟长羽毛;羽毛帽饰;羽饰;羽毛
plumed hat 羽饰帽
plumes 羽毛束,羽毛填料;漂亮精致的衣服;羽衣
plumetic de coton [法]包花绣
plumetis [法]包花绣;薄花呢,清地小绒花织物;小花薄洋纱;羽状绣花针迹
plumette 毛呢,丝毛呢
plumety 透明印花亚麻纱;薄细布,点子薄细布
plume velvet 缎条丝绒
plum green 梅青色,梅子青
plumose fiber 籽壳纤维
plump figure 丰满身段
plumpness 丰满,胖
plum purple 梅红(暗红青莲色),青紫,紫红色
plumule helmet 羽饰盔
plum wine 茄色
plumy helmet 羽饰盔
plunge back 低V形后背
plunge brassiere 深露式胸罩
plunge front 低V形前胸
plunge neckline 女服深开式领口
plunging 深入型,深露式款式
plunging brassiere 深露式胸罩
plunging collar 插缝领
plunging neckline 深尖领,深开式领口,深露式领口
plunging style 深露式款式
plunket azures 蓝色粗纺呢
pluralism 多元化
plus fours 高尔夫球裤,短灯笼裤
plush 长毛绒;(差役穿的)毛线裤;长毛绒[织物]
plush astrakhan 仿羔皮长毛绒
plush astrakhan cloth 仿羔皮长毛绒织物;仿羔皮长毛绒
plush carpet 剪绒地毯
plush cloth 长毛绒织物,毛圈织物
plush clothing 长毛绒大衣
plush coat 长毛绒大衣
plush cockle back 长毛绒背弓
plush collar 长毛绒衣领
plush dress material 长毛绒服装料
plushed 仿长毛绒的

plushes 长毛绒裤
plushette 低级长毛绒织物
plush fabric 长毛绒大衣织物,长毛绒物,毛圈织物,毛绒织物
plush finish (地毯)长毛绒整理
plush for clothing 服装用长毛绒
plush hosiery machine 毛巾袜机
plush leather 长毛绒毛皮
plushlike 仿长毛绒的
plushlike fabric 类似长毛绒织物
plush lining 毛面棉里经编织物
plush loop 长毛绒圈,毛圈
plush overcoating 长毛大衣呢
plush sole 弹性袜底,毛圈袜底,绒袜底
plush stitch 仿土耳其地毯针迹
plush topcoat 长毛绒大衣,绒大衣
plush trousers 长毛绒裤,绒裤
plush velour 长毛绒织物,密绒长毛绒
plush velveteen 长毛棉绒,纬向天鹅绒
plushy 具有长毛绒外观的
plush yarn 毛圈纱,毛绒纱
plus look 添加款式
plus one 三股线精纺呢
plus one dress 添一件式装束
plus one fashion 增加一件的款式(在已流行的款式上增加一件),加一时装
plus size garment 加大尺寸服装
ply 纱线股数;织物层数;股线;折叠,绞合
2-ply 两股
ply alignment 缝料铺层边对齐
ply allowance time 缝料铺放时间
pl yarn (parallel yarn) 包缠纱,花式纱
ply fabric 多层织物
plyprism thread 特彩线
ply twist 合股捻
ply weave 多层组织
plywood 夹板;胶合板
ply yarn 股线,合股纱
ply yarn drill 线卡(其)
ply yarn fabric 全线织物,线织物
pneumatic 女子体型匀称的,胸部丰满的;充气的
pneumatic bar tacker 气动套结机
pneumatic cellular fiber 含气空腔纤维(具有细胞状结构的纤维);泡沫纤维
pneumatic collar press machine 气压翻烫领机

pneumatic control 气动控制
pneumatic figure 胸部丰满体型；匀称体型
pneumaticity 含气性
pneumatic knife switching unit 气动刀具开关装置
pneumatic needle bar control 气动针杆控制
pneumatic pocket creasing cachine 自动风压折袋机
pneumatic sewing head 气动控制缝纫机头
pneumatic snap fixing machine 气动揿钮钉扣机
pneumatic tacker 气动钉钉机(加工皮革用)
pneumatic utility press 万能风压熨烫机，万能整烫机
pocahontas costume 波卡洪塔斯装束(美国印地安妇女)
pocahontas dress 美国印第安裙装
poche bateau ［法］船形袋
poche fendue 剪口长缝口袋
poche gousset （裤腰内侧或背心的）小口袋
poche jeans ［法］牛仔裤斜袋
pochette 小口袋；小腰袋；小钱袋；小手绢；装饰小背带
pochet welting machine 开袋机
pocho jeans 牛仔裤式的斜口袋
pochon ［法］大背袋
pocka 女紧身短上衣
pocket 衣(裤、裙)袋，口袋，兜；钱包，钱袋，腰包；双层织物的中空部分
pocketable coat 袋装外套
pocket and bows 衣袋蝴蝶结
pocket backing 袋里布，口袋里
pocket bag 袋兜
pocket bearer 吊袋祥；口袋祥，口袋挂祥
pocket blank 袋坯
pocket book 女用手提包；皮夹(子)(美)
pocket cascade 褶饰袋
pocketchief 胸袋饰边；(西装)胸袋饰巾，胸袋手帕
pocket cloth 口袋布
pocket creaser 折(口)袋机
pocket creasing machine 折袋机
pocket drill 斜纹衣袋布
pocket entry 袋口
pocket facing 袋口牵布；口袋面；袋口

贴边
pocket flap 口袋盖；袋盖
pocket flap folding machine 折袋盖机
pocket flap forming press 袋盖压烫机
pocket flat creasing machine 折袋盖机
pocket folding press 口袋折烫机
pocket handkerchief 口袋手帕，[西装]胸袋手帕
pocket heel 袋形袜跟
pocket hem 口袋折边
pocket hemmer 口袋卷边器，折袋口器
pocket-hole （衣服）袋口
pocket-hole sewing machine 开袋机
pocketing （口）袋布，衣袋用布
pocket jetting 袋口
pocket line 开袋线，袋口线
pocket linen 口袋布
pocket lining 袋里布，袋衬，袋布
pocket magazine 圆筒形纱布
pocket marker 口袋划线器
pocket mouth 袋口
pocket-mouth interlining 袋口衬
pocket [mouth] stay 垫袋口布
pocket opening 袋口
pocket or flap out grain 袋或袋盖丝绺不正
pocket out grain 袋布丝绺不正
pocket patching 贴袋
pocket piece 袋布，口袋布
pocket piping 衣袋装饰滚边
pocket placement 口袋部位，口袋装法
pocket position 袋位
pocket position line 袋位线
pocket pouch 袋囊
pocket reinforcement patch 袋角衬；袋角加固布
pocket rivet （牛仔裤）袋口撞钉
pockets 袋；圆筒形织物
pocket setter 钉袋机，绱袋器
pocket siphonia 长风雨衣
pocket size 口袋尺寸
pocket square 胸袋方巾，胸袋饰巾，装饰方形手帕
pocket stay 垫袋布，垫料
pocket style 口袋式样
pocket top bands 袋口边
pocket trim height 袋边高
pocket T-shirt 左胸袋T恤衫(浅圆摆)
pocket watch 挂表，怀表
pocket weave 中空双层织物

pocket welt 袋嵌条;口袋开线;袋牙边
pocket welting machine 袋口踏边机,开袋机
pocket width 袋宽
pod green 豆荚绿(鲜黄绿色)
poepoes 坎塔拉剑麻
poerce-ersistance shoes 防刺穿鞋
poesy ring 纪念金戒指,诗句金戒指
poet's collar 诗人领(无领衬、不上浆的柔软方角领,因18世纪末19世纪初英国诗人喜欢而得名)
poet's shirt 诗人衬衫(长袖,长尖领)
pogonotomy 剃须
pogonotrophy 蓄须
poh 蓖麻蚕丝
poil trainant 挂经织物
poinsettia 一品红
point 点;尖;具体地点;手工针缝花边(统称);装饰用蝶领结;尖色头系带(古);手套背面的三条装饰缝;花边及刺绣的各种针迹及线缝仿针绣挂毯;(设计)焦点,要点
point à brides 连接窄条的针绣花边
point à carreaux 法国方格花梭结花边
point à l'aiguille 法国针绣花边
point à la minute 十字形和星形针迹
point à la turgue 土耳其针绣花边
point and creaser 翻角熨平器
point-and-tie 尖角蝶带
point and tube turner 制衣翻转钳
point appliqué 贴花花边,嵌花花边
Point Arabian 阿拉伯窗帘花边
point à réseau 法国针绣花边
point bisette 白亚麻梭结花边
point blanket 起毛厚毛毯
point-bonded-staple fabric (PBS fabric) 点黏合热熔非织造布
point bonding 点黏结非织造工艺
point-bonding-staple 点黏合纤维
point brodé 凸纹梭结花边
point chaudieu 花边的链状线迹
point coating 点子涂层,点涂布法
point colbert 凸纹花卉针绣花边
point conté 绣花网地花边
point coralline 珊瑚花纹花边
point coupé 法国挖花花边
point crochet 钩边花边
point d'Alencon 手工丝花边
point d'Angleterre [法]昂格莱特雷梭结花边

point d'angleterre 布鲁塞尔针绣花边
point d'anvers 安特卫普花边
point d'arabe 大花型梭结花边
point d'argentan 阿根坦针绣花边
point d'arras 法国阿拉斯低级梭结花边
point de bailleul 拜约耳梭结花边
point da brabancon 链式扣眼线迹
point de boutonnière 锁眼针迹针绣花边
point de brabant 布鲁塞尔花边,比利时针绣花边
point de bruges 白色凸纹枕套带
point de bruxelles 布鲁塞尔针绣花边
point de campane 梭结花边条
point de champ 网眼地花边
point de chant 六角网眼花边
point de cone 花边用锥形图案
point de cordova 三线平行打籽花边线迹
point de crete 几何形花纹的梭结花边
point de dieppe 梭结针绣花边
point de feston 扣眼线迹
point de filet 网眼格花边
point de flandres 佛兰德梭结花边
point de France [法]法国花边
point de france 针绣凸纹花边,花边
point de galles 亚麻钩结花边
point de gauze 针绣花边地布
point de gaze [法]薄纱花边;马鬃点纹花边
point de gerbe 花边用扣眼线迹
point de gibecière 花边用链状线迹
point de gobelin 戈贝林凸纹线迹
point de grecque 花边用四角八角交替网眼地
point de hongrie 嵌花花边带;彩色塔夫绸
point de la reine 皇后绉
point de lille 六角网眼梭结花边
point de malines 花边用曲折扣眼线迹,梅希林花边
point de maxli 露地梭结花边
point de mechlin 梅希林花边
point de Milan 米兰大花花边
point de moscow 俄罗斯式针绣花边
point de neige 蛛网钩编;雪点花边;点纹网眼织物
point de paris [法]巴黎花边;低级棉机制花边六角网眼花边
point de paris ground 梭结花边用网眼地
point de pyramide 花边用锥形图案
point de raccroc 细结缝迹

point de ragusa	腊纠萨花边
point de régence	厚边梭结花边
point de repassé	平纹组织线迹
point de reprise	针绣花边用角形线迹
point de riz	[法]萨克森刺绣；米粒形花纹针织机
point de rose	玫瑰纹网地花边
point de Sedan	[法]色当花边
point de sedan	卵形纹针绣花边
point de sorrernto	针绣花边用松圈地
point d'espapne	西班牙花边
point d'esprit	薄绢网，花珠罗纱
point d'etoile	方网眼地针绣花边
point de toile	平纹组织线迹
point de tricot	钩编织物的大方眼
point de tulle	梭结花边细网地；白亚麻梭结花边
point d'eu	粗瓦朗西安花边
point de valenciennes	瓦朗西安花边，方眼针迹
point de velin	打底纹针绣花边
point de venise	针绣凸纹花边
point de venise a' réseau	亚麻方眼花边
point-devise	穿着合身
point d'hongrie	人字形图案刺绣品
point d'lrelamde	爱尔兰机制粗花边
point double	六角网眼花边
point duchesse	白色凸纹枕套花边
pointed brush	细锋毛笔
pointed collar	尖领，尖角领
pointed cuff	尖角形克夫，尖角袖头，尖角形袖口
pointed end bow	尖角蝴蝶结领带，两端成三角形的领结
pointed end necktie	尖头领带
pointed end tie	尖端领带，尖角领带
pointed finger nail	尖指甲
pointed fox	白毛红狐毛皮
pointed heel	宝塔形加固袜跟；袜子高跟
pointed hemline	手帕型尖底摆
pointed hood	尖顶风帽
pointed khaki drill	人字卡其
pointed lace	北意大利大花花边；意大利梭结花边
pointed marker	点状标记
pointed pocket	尖角底(的)袋
pointed shoes	尖头鞋
pointed sleeve	尖袖
pointed toe	尖鞋头
pointed toe-end	尖鞋头
pointed top godet	尖裆片
pointed topline shoes	尖口鞋
pointed twill	人字(形)斜纹，山形斜纹
pointed twill knitted fabric	人字形斜纹布
pointelle	罗纹网眼布
pointelle embossed effect	小点子凹凸花纹
point end necktie	尖头领带
pointer and creaser	翻角熨平器
point evantail	扇形花纹针绣花边
point faisceau	人字纹针绣花边
point gaze	[法]纱罗花边；针绣花边
point ground	梭结花边的网眼地
point guipure	针绣花边
point half tones effect	云纹效应
point heel	高跟
pointilistic effect	点画效应，点子花纹效应
pointille	[法]点画法；小点子花纹
pointillism	新印象派，点彩派，点画派；点描法花样，点描画法
pointillisme	泥点纹样，点描法花样
pointilliste knits	夹花针织服装
pointillistic effect	点子花纹效应，点画效应
pointilte	小点子花纹
pointing	手套绣饰工艺
point jesuit	爱尔兰仿威尼斯钩编花边
point kant	网地瓶形花纹梭结花边
point lace	针绣花边
point lache	针绣花边的三角形扣眼线迹
point lapel collar	尖角式西装领
point luxeuil	针绣花边
point make-up	重点化妆法
point net	机制针绣花边
point noué	针绣花边用结子扣眼线迹
point of sales (POS)	零售点，销售点
point ondulé	流苏花边的双棱纹
point paper	意匠纸
point paper design[draft]	意匠图
point paper draft	意匠图
point pecheur	黑白几何形梭结花边
point plat	针绣扁平花纹花边
point plat appliqué	网地枕套花边
point presser	服装小型熨烫台，小烫板
point puy	梭结花边
point saracene	仿土耳其挂毯
point seaming	套口，连圈；缝针头；边缝缝合
point serré	针绣花边的曲折线迹

point shoes 硬芭蕾舞鞋,足尖舞鞋
point [splicing] heel 宝塔形加固袜跟
point tiellage 针绣花边交叉线迹
point tiré 抽绣品
point to point seaming 套口缝合,双圈缝合
point toe embroidery shoes 勾尖绣花鞋
point toe shoes 尖头鞋
point tresse 发绣梭结花边
point turc 针绣花边线迹
point turner 尖角翻特器
point turque 仿抽华线迹
point type 针夹类型
point width 胸点间距;乳间宽
poiret twill 普瓦勒特轧别丁
poisee 花卉缎
poisoning 中毒
poitrine 女性丰满的胸脯
poke (朝前撑起的)阔边女帽;口袋;女帽朝前撑起的宽边
poke bonnet 波克软帽,宽前沿女帽,波克罩帽前倾系带宽边女帽
poke collar 波克领,礼服衬衫领(古典晚宴服衬衫所用的极高领型)
poke high 波克软帽
poke out 服装起翘
poke out pocket 乞食袋
poker chip button 薄片形纽扣
poke sleeve 袋形袖
pokey [bag] 收袋口的小袋
pokrovatz carpet 手织山羊毛毯
poky 衣着褴褛的,不整洁的
polar bear fur 北极熊毛皮
polar bearskin plush 仿白熊皮长毛绒
polar boots 观赛靴
polar fleece 摇粒绒
polar fleece cap (涤纶)绒布帽
polarian 仿羔皮织物
polarization glasses 偏振眼镜
polarizing microscope 偏振光显微镜
polaroid glasses 偏光片眼镜
pole boots 极地靴
polecat 黄鼠狼毛皮
polemite 荷兰波勒迈特羽纱
pole tent 支柱结构帐篷
poleyn 护膝铠甲,膝甲
police badge 警察徽章
police boots 警靴
police jacket 警服
police officer's vest 警察(用)背心

policeman look 警察款式
policeman uniform 警服,警官服
policeman's cap 警帽,警察帽
policemen's uniform 警服,警官服
polinosic 虎木棉
polish 擦亮剂;加光,磨光
Polish carpet 波兰地毯
polished composition leather 光滑合成革,光滑皮
polished cotton 光亮棉织物,加光棉织物
polished cotton sewing thread 打光棉线
polished girdle 抛光腰带
polished leather shoes 光面革鞋
polished thread 打光缝线,加光线,上光纱线
polished twine 上光纱线
polished yarn 上光纱线
polish greatcoat 男皮饰紧身长大衣
polishing 打光,抛光,上光[整理]
polishing cloth 抛光用布,抛光用呢
polishing felt 抛光用毡,擦光用毡
polishing machine for imitation fur fabric 人造毛皮烫光机,人造毛皮整理设备
polishing process (环保)深度处理
polish jacket 及腰女夹克
polish mantle 连披风及膝女斗篷
polish remover 除光液
Polish rugs 波兰地毯
Polish (shoes) 波兰鞋
polisse cloth 斜纹粗呢
polka 紧身短夹克,针织紧身衫,紧身针织女上衣,女紧身短上衣
polka dot 色地白圆点花布图案;圆点花纹衣料
polka dot frock 大圆点外长衣
polka dot muslin 圆点花细平布
polka dots 圆点花样
polka dot trim pajamas 大圆点镶边睡衣
polka dot umbrella 圆点花伞
polka gauze 织绣点子纱罗
polka jacket 紧身短夹克,针织紧身衫,紧身针织女上衣
pollera 秘鲁手织多层绣边女裙,拉丁美洲裙子
pollock cotton 美国波洛克棉
pollution treatment 三废治理,污染处理
polo belt 马球护腰带
polo boots 马球靴
polo cap 马球帽
polo cloth 马球牌双面厚绒呢(商标)

polo clothing[coat] 马球外套
polo coat 菠萝呢大衣,厚绒呢轻便大衣,马球外套;英国外套(第一次世界大战时,英国陆海军士官穿着)
polo collar 马球领,男式可翻立的白色硬领
polo dot 印花大圆点
polo jmas 圆领长袖套衫长裤;圆领套衫与长裤组成的睡衣
polo look 马球款式
polonaise (18世纪流行的)波兰式围裙;花式连衣裙;四股装饰花线
polonaise lining 交织里子绸
polonaise style 波兰式
polonaise style coat with bramdenburg fastenings 有勃兰登堡系带或钩扣的波兰式大衣
poloneck 波罗领,马球领,半前开襟小翻领,围脖圆领
poloneck jumper 马球领套衫
poloneck(line) 马球衫领;圆高领
poloneck sweater 高圆翻领套衫
polo placket 马球衫式开口
polo pyjamas 圆领长袖套衫长裤;圆领套衫与长裤组成的睡衣
polo shirt 马球衬衫(开领,短袖);翻领衫,扣子衫
polo shirt suit 开领马球衬衫两件套
polo style 马球衫式
polo sweater 开领马球运动衫,马球毛衣
polooutsole 坡式鞋跟
polo-shirt dress 加长T恤衫(盖肩袖,系腰饰带);开领短袖衬衫式连衣裙,T恤连裙装
polrchrome 彩色
poltalloch 波塔罗其
polyacrylonitrile staple fiber spun yarn 腈纶纱
polyacrylonitrile yarn 聚丙烯腈纱
polyamide 锦纶,尼龙
polyamide checked silk 锦丝格子绸
polyamide/cotton plush velvet 人造毛皮
polyamide/cotton twill 锦棉绫
polyamide curtain gauze 锦纶窗帘纱
polyamide/dull rayon mixed matelassé 松花绸
polyamide fiber 聚酰胺纤维;锦纶
polyamide filament-fabric for raincoat and jacket 俄罗斯锦纶雨衣绸
polyamide filament hosiery 锦纶长丝袜
polyamide filament scarf 锦纶丝围巾

polyamide filament sewing thread 锦纶缝纫线
polyamide filament tape 锦丝带
polyamide filament thread 锦纶丝线
polyamide gauze 闪光尼丝纱,闪光锦丝纱
polyamide ground etched-out velvet 锦地绒
polyamide imide fiber 聚酰胺-酰亚胺纤维
polyamide interlining 聚酰胺黏合衬
polyamide mixed fencing-wear silk 银剑绸
polyamide monofilament 聚酰胺单丝
polyamide monofil hosiery 锦纶单丝袜
polyamide multifilament 锦纶复丝
polyamide multifilament hosiery 锦纶复丝袜
polyamide palace 锦丝纺,尼丝纺
polyamide parachute silk 锦丝跳伞绸
polyamide/polyester gingham 锦涤绡
polyamide profiled filament hosiery 锦纶异形丝袜
polyamide/rayon mixed twill damask 锦绣绸
polyamide/rayon union brocade 彩金缎
polyamide profiled filament hosiery 锦纶异形丝袜
polyamide socks 尼龙袜;锦纶袜
polyamide spun-bonded nonwovens 锦纶纺黏布
polyamide staple fiber 锦纶短纤维
polyamide staple fiber spun yarn 锦纶纱
polyamide 6 stretch yarn 聚酰胺6弹力丝
polyamide/textured polyester huckaback 环涤绸
polyamide thread 锦纶线,尼龙线
polyamide transparent sewing thread 锦纶透明缝纫线,聚酰胺透明缝纫线
polyamide treatment 聚酰胺树脂处理
polyamide tufted carpet 聚酰胺簇绒地毯
polyamide twill 锦丝绫
polyamide yarn 锦纶线
polyaniline fiber 聚苯胺纤维
polybag (服装包装用)胶袋,聚乙烯袋
polybenzimidazole fiber 聚苯并咪唑纤维,PBI纤维(美国研制,耐高温纤维)
poly bis-benzimidazo-benzophenanthroline fiber 聚双苯并咪唑-苯并菲绕啉纤维(耐高温纤维,商名,美国)
polybutylene terephthalate fiber 聚对苯二

甲酸丁二酯纤维
polycaprolactone fiber 聚己内酯纤维
polychloroprene 氯丁橡胶
polychromatic dying 多色淋液印花
polychromatic printing 多彩喷射印花
polychrome 多色彩,多彩饰,彩饰法;色彩装饰;多色画法
polychrome printing 多色印花(印花滚筒由多个花纹色拼合并制成)
polycot[blend] fabric 涤棉混纺布
polycotton yarn 涤棉混纺纱(有时亦指合成纤维与棉的混纺纱)
polyester 聚酯纤维,涤纶
polyester/acrylic blended yarn 涤腈纱
polyester/acrylic fancy suiting 涤腈花呢
polyester/acrylic mixed fancy suitings 涤腈花呢
polyester/acrylic serge 涤腈哔叽
polyesteramide fiber 聚酰胺酯纤维
polyester and cotton blend 涤棉混纺织物
polyester and cotton blended thread 涤棉混纺线,涤棉缝纫线
polyester-based elastic yarn 聚酯基弹性丝
polyester bleached sewing thread on cone 涤纶漂白宝塔线
polyester blended tussah square 混纺呢
polyester blended tussah tweed 混纺呢
polyester brocade 涤纶花缎,涤纶锦缎
polyester brocade satin 涤纶花缎
polyester button 聚酯纽扣(塑料扣);仿壳纽;涤纶扣
polyester capillaries 聚酯丝
polyester celtic 海南绫
polyester-cored sewing thread 聚酯芯缝线
polyester colored sewing thread on cone 涤纶染色宝塔线
polyester-core thread 涤纶芯线
polyester/cotton(PIC) 涤纶混纺织物
polyester/cotton batiste 涤棉绸
polyester/cotton blended yarn 涤棉纱
polyester/cotton blend fabric 聚酯[纤维]-棉混纺织物
polyester/cotton denim 聚酯[纤维]-棉粗斜布,涤棉牛仔布
polyester/cotton disperse dye 涤/棉分散染料
polyester/cotton fabric 涤棉混纺织物,涤棉织物,涤棉布
polyester/cotton fine fabric 涤棉细布

polyester/cotton khaki 涤棉卡其,涤卡
polyester/cotton liangsam 凉爽绸
polyester/cotton liang-shuang 凉爽绸
polyester/cotton mixed 涤棉绸
polyester/cotton plain cloth 涤棉平布
polyester/cotton plains 涤棉平布
polyester/cotton/polyamide blended yarn 涤棉锦纱
polyester/cotton shirt 涤棉衬衫
polyester/cotton striped 绦纤绸
polyester/cotton twill 涤爽绸
polyester/cotton twill damask 涤棉花绸
polyester/cotton union dye 涤/棉通用染料
polyester/cotton vat dyes 涤/棉还原染料
polyester/cotton yarn 涤棉纱
Polyester Council of America 美国聚酯纤维协会
polyester crepe 涤丝绉,涤纶绉
polyester curtain tissue 涤纶窗帘绡
polyester double-faced fleece 涤纶双面起绒布
polyester elastic fiber 聚酯[型]弹性纤维
polyester elastic tape 聚酯松紧带
polyester embroidery thread 涤纶绣花线
polyester fabric 涤纶织物;涤纶布
polyester fancy suiting 涤纶花呢
polyester fiber 聚酯纤维
polyester fiber-containing textile 聚酯纤维混纺织物
polyester filament 涤纶长丝
polyester filament and staple combined yarn 聚酯长丝和短纤维的复合纱
polyester filament/blend yarn union poult 似纹葛
polyester filament sewing thread 涤纶长丝(束丝)缝纫线
polyester filament thread 涤纶丝线
polyester-filling fancy tussah 星月绸
polyester film lumiyarn 聚酯薄膜金银丝
polyester film tape 聚酯薄膜带
polyester gabardine 涤纶华达呢
polyester gauze 涤纶纱(织物)
polyester georgette 涤纶乔其纱
polyester georgette blouse 柔姿纱衬衫
polyester georgette printed 涤纶印花乔其纱
polyester georgette sand crepe 涤乔绉
polyester gingham 涤格纺

polyester habotai 涤纶绸,涤纶纺,涤丝纺
polyester interlaced brilobal set yarn 有光三叶涤纶低弹网络丝
polyester interlining 聚酯黏合衬
polyester jacquard cloth 涤纶提花布料
polyester jacquard linen 涤纶提花里子绸
polyester jacquard woven tie 涤纶提花领带
polyester kabe crepe 涤纶碧绉
polyester knitted tie 涤纶针织领带
polyester/lamé gauze 银蝶纱
polyester linen type filament fabric 涤纶长丝仿麻布
polyester lining satin 涤纶里子缎
polyester mesh 聚酯[纤维]网
polyester meshes 涤纶网眼布
polyester metallized fiber 涤纶金属化纤维
polyester metallized thread 镀金属聚酯线,涤纶金银线
polyester microfibre pants 涤纶桃皮绒裤
polyester mock leno 涤丝透凉绸
polyester mohair 涤纶马海毛
polyester monofilament 聚酯单丝,涤纶单丝
polyester multifilament 涤纶复丝
polyester necktie 涤纶领带
polyester/nylon mixed brocade 涤锦交织提花绸
polyester/nylon silk 涤锦绸
polyester/nylon suiting 涤锦丝交织提花绸
polyester/nylon peach 涤锦复合桃皮绒
polyester oearlized button 聚酯仿珠纽扣
polyester over cotton doubled-faced [weft] knit fabric 涤盖棉双面[纬编]针织物
polyester over cotton pliated (single) jersey 涤盖棉单面针织物
polyester oxford 涤纤绸
polyester padding 喷胶棉;涤纶棉
polyester patterned satin brocade 涤霞缎
polyester peach skin fabric 涤纶桃皮绒
polyester peach skin pants 涤纶桃皮绒裤
polyester pearlized button 聚酯仿珠纽扣
polyester polar fleece 涤纶双面绒布
polyester polar fleece printed 涤纶双面印花绒布
polyester pongee 涤丝绸[纺],涤纶塔夫绢,涤纶仿柞丝绸
polyester poplin 涤府绸

polyester-polyamide alloy fiber 聚酯-聚酰胺合金型纤维
polyester/polynosic blended yarn 涤富纱
polyester/polynosic mixed broken twill 形格绸
polyester printing button 聚酯印字扣;耐热印字纽
polyester/PVC coated fabric 涤纶PVC涂层布
polyester(nylon)/PVC coated jacket 涤纶(尼龙)PVC涂层夹克
polyester/ramie blended fabric 涤麻(麻涤)混纺布
polyester/ramie blended palace 涤麻派力司
polyester/ramie blending cloth 涤麻混纺织物
polyester/ramie blending fancy suiting 涤麻混纺花呢
polyester/ramie cloth 涤麻布
polyester/ramie fancy suitings 涤麻花呢
polyester/ramie tropical 涤麻薄花呢
polyester/rayon burnt-out sheer 太空绡
polyester rope 聚酯[纤维]绳,涤纶绳索
polyester sand crepe 特纶绉,涤乔绉
polyester satin 涤纶缎
polyester satin brocade 涤纶花缎
polyester serge 涤纶哔叽
polyester sewing yarn 聚酯缝纫线
polyester silk scarf 涤纶丝巾
polyester silk-like satin brocade 涤美缎
polyester silk-like yarn 涤纶仿真丝
polyester single jersey 涤纶汗布
polyester spun thread 涤纶短纤维缝纫线
polyester staple fiber 涤纶短纤,涤纶棉
polyester staple fiber spun yarn 涤纶纱
polyester striped batiste 涤丝直条绸
polyester suitings 涤纶西服料
polyester taffata 涤纶塔夫绸
polyester textured fabric 涤纶变形织物
polyester textile 聚酯纤维纺织品
polyester textured fabric 涤纶织物;涤纶布
polyester thread 涤纶线,涤纶缝纫线
polyester tie by hand-printed 涤纶手工印花领带
polyester tissue faille 涤纶薄罗缎
polyester top 涤纶毛条
polyester tow 涤纶丝束
polyester tricot tracksuit 涤纶经编料运动

套装
polyester trousers 涤纶裤
polyester twill 涤丝绫
polyester twill faconne 涤丝绫
polyester type elastic fiber 聚酯型弹性纤维
polyester union gingham 华格纺
polyester/viscose blended fabric 涤黏布,快巴的确良
polyester/viscose blended yarn 涤黏纱
polyester/viscose fabric 涤黏织物;涤黏布,快巴的确凉
polyester/viscose fancy suitings 快巴(涤黏花呢)
polyester/viscose Harris 涤黏海力斯
polyester/viscose low elastic linen-like fabric 涤黏低弹仿麻织物
polyester/viscose mixed nacre window holland 涤黏丝交织珠光窗帘布
polyester/viscose plain cloth 涤黏平布
polyester/viscose serge 涤/黏哔叽
polyester/viscose valitin 涤/黏凡立丁
polyester/viscose wool-like fancy suiting for women 涤黏仿毛女式呢
polyester/viscose worsted flannel 涤黏啥味呢
polyester voile 全涤绡
polyester wadding 涤纶絮料
polyester/wool fancy suitings 涤毛花呢
polyester/wool/ramie tropical 涤毛麻薄花呢
polyester yarn 涤纶丝;涤纶纱
polyester zipper 涤纶拉链,聚酯拉链
polyether/ester elastomeric yarn 聚醚-酯弹性丝
polyether fiber 聚醚纤维
polyethylene fiber 聚乙烯纤维
polyethylene oxybenzoate fiber 聚对苯甲酸乙氧酯纤维,荣辉纤维
polyethylene rope 乙纶绳索
polyethylene terphthalate fiber 聚对苯二甲酸丁二酯纤维
polyethylene yarn 乙纶纱
polyfilament yarn 复丝纱线
polyfilled clothing[coat] 喷胶棉大衣;涤纶棉大衣
polyformaldehyde fiber 聚甲醛纤维
polyglycollide fiber 聚乙交酯纤维
polygon 多边形
polygraph 多重图
polyhedron 多面体
poly linen 涤纶仿麻织物
polymer 聚合物
polymer extruded nonwoven 聚合物挤压法非织造布
polymeric compound 聚合物
polymeric dye 聚合染料
polymerization 聚合作用
polymer-laid nonwoven 聚合物直接成网非织造布
polynesian inspiration 腰蓑衣,印花衣膝衣
polynesium inspiration 玻里尼西亚灵感
polynondiurea fiber 聚壬二脲纤维
polynosic fabric 富纤织物
polynosic fiber 波里诺西克纤维(高湿模量黏胶纤维),富纤(富强纤维),虎木棉
polynosic fine poplin 富纤细布
polynosic plain cloth 富纤平布
polynosic poplin 富强纤维府绸,富纤府绸
polynosic rayon 波里诺西克纤维(高湿模量黏胶纤维),富纤(富强纤维),虎木棉
polynosic yarn-dyed lawn 富纤细纺
polyolefin[e] fiber 聚烯烃纤维
polyolefine monofilament 聚烯烃单丝
polyoxadiazole fiber 聚噁二唑纤维
polyoxamide fiber 聚乙二酰胺纤维,聚草酰胺纤维
polypeptide fiber 多肽纤维
polyphenol-aldehyde fiber 聚酚醛纤维
polypivalolactone fiber 聚叔戊内酯纤维
polypropylene and cotton plated knit fabric 丙盖棉针织物
polypropylene/cotton blended yarn 丙/棉纱
polypropylene fibre 丙纶,聚丙烯纤维
polypropylene filament 聚丙烯长丝,丙纶长丝
polypropylene over cotton double jersey 丙盖棉针织物
polypropylene pile 丙纶绒
polypropylene sewing thread 丙纶缝纫线,聚丙烯缝纫线
polypropylene spun-bondde nonwovens 聚丙烯纺黏布
polypropylene staple fiber 丙纶短纤
polypropylene staple fiber spun yarn 丙纶纱
polypropylene tape 聚丙烯带

polypropylene top 丙纶毛条
polypropylene tow 丙纶丝束
polypyromellitimide filament 聚苯均四酰亚胺长丝
polyreflets 法国压花长毛绒
polyruethane leather 聚氨基甲酸酯合成革
polystyrene fiber 聚苯乙烯纤维
polysulfonamide fabric 聚砜酰胺纤维织物
polysulfone fiber 聚砜纤维
polysulphonamide fiber 芳香族聚砜酰胺纤维
polytene 聚乙烯纤维
polyterephthaloyl oxalamidrazone fiber 聚对苯二甲酰-草酰-双脒腙纤维(耐高温抗燃纤维,德国)
polytetrafluoroethylene fiber 聚四氟乙烯纤维
polytetramethylene naphthalate fiber 聚萘二甲酸丁二[醇]酯纤维
polythene fiber 聚乙烯纤维
polyurea fiber 聚脲纤维
polyurethane 氨纶
polyurethane coates fabric 聚氨基甲酸酯涂层织物
polyurethane elastic fabric 氨纶弹力织物
polyurethane [elastic] fiber 聚氨基甲酸酯[弹性]纤维,氨纶(中国商名)
polyurethane fiber hosiery 氨纶丝袜
polyurethane finishing 聚氨酯整理
polyurethane foam 聚氨酯泡沫塑料
polyurethane leather PU革,聚氨基甲酸酯合成革
polyvinyl alcohol-based carbon fiber 聚乙烯醇基碳纤维
polyvinyl alcohol/cotton blended yarn 维/棉纱
polyvinyl alcohol fiber 聚乙烯醇[系]纤维;维纶
polyvinyl alcohol filament yarn 聚乙烯醇长丝
polyvinyl alcohol staple fiber spun yarn 维纶纱
polyvinyl alcohol vinyl chloride fiber 维氯纶
polyvinyl chloride 聚氯乙烯
polyvinyl chloride fibre 氯纶
polyvinyl chloride leather PVC合成革
polyvinyl chloride sole PVC鞋跟
polyvinyl chloride staple fiber PVC短纤维;氯纶棉

polyvinyl fabric 聚乙烯织物
polyvinyl film 聚乙烯薄膜
polyvinylidene chloride lining 聚偏氯乙烯衬
pomade 发膏,头油
pomander 香袋,香盒
pomegranate 石榴红(暗红色);石榴纹样
pomegranate red 石榴红(暗红色)
pommel slicker 骑马用雨衣
pomona green 嫩绿,带黄绿
pompadour 小花卉纹;带绿的蓝色;女式紧身胸衣,蓬巴杜纵漩涡发型;头饰;帽饰;臀垫
pompadour duchesse 花卉条纹缎
pompadour gros de tour 高级罗缎,杜尔横棱绸;高级细棱纹布
pompadour hairstyle 后掠式发型
pompadour polonaise 蓬巴杜高级罗缎绣花裙,波兰裙
pompadour serge 英国小花卉纹哔叽
pompadour skirt 蓬巴杜裙
pompadour taffeta 法国蓬巴杜条子花塔夫绸
pompeian ornament 庞贝装饰
pompeian red 庞贝红(浅棕红色)
pompeian velvet 色织丝绒
pompon (衣帽装饰)绒球,丝球,(军帽的)毛球
pompon bobble 装饰绒球
pompon socks 绒球短袜
ponceau 罂粟红,朱红
poncho 棕色条边穗粗纺呢;南美穗饰披巾,雨披,斗篷装
poncho cape 有穗饰的斗篷,披巾雨披
poncho cloth 军用防雨披;防雨厚毛毯;朋卡织物
poncho dress 披肩式连衣裙
poncho skirt 雨披式裙,庞裘裙
poncho sleeve 穿头披巾袖
poncho(-style) coat 雨披式大衣,披肩式大衣,庞裘式大衣
poncho trimmed 镶边穗饰
ponderability 有重量感
POND'S 旁氏
pongee 山东府绸,茧绸,棉织仿茧绸,野蚕棉绸,柞绸
pongee imperial 厚重光亮丝府绸
pongee print 棉印花细布
pongee silk 柞丝绸,茧绸
pongee spun silk 桑绢纺绸

ponson velvet 庞森丝绒
pont 绒毛浮起,浮线
Ponte (di) Roma 双罗纹空气层组织,罗马组织,变化罗纹空气层组织
Ponte di Roma fabric 棉、毛空转织物；蓬托地-罗马组织针织物；双罗纹空气层织物
pontiac 深灰针织防雨呢
pontificalia 主教全套装饰
pony boots 小马皮靴
pony cloth 仿马皮长毛绒(棉地,绒头为马海毛),仿幼马毛皮织物；小马皮
pony fur 小马毛皮
pony skin 仿马皮长毛绒(棉地,绒头为马海毛),仿幼马毛皮织物；小马皮
pony skin boots 小马皮靴
ponytail 马尾发型
ponytail interlining 马尾衬
ponytails 马尾辫
poodle 卷曲绒头织物
poodle cloth 卷曲绒头织物,仿长卷毛狗皮织物
poodle cut 卷曲式发型,鬈毛狗发型
poodle hair 狗毛
poodle skirt 长卷毛狗图案圆裙
poodle woolen 卷毛呢
pool blue 池水蓝
pool green 潭绿色
poole cloth 英国普尔光面呢
pooled fabric 混纺织物
poor boy 罗纹紧身运动衫
poor boy shirt 顽童装
poor boy sweater 罗纹紧身运动衫,报童紧身便服
poor-country style 田园朴素风格
poor feeling 手感不良
poor fitting garment 不合身服装
poor lasting 鞋帮不正
poor line 不光滑线,线条不良
poor look 破旧型款式；破旧装,贫穷装
poorly formed stitch 劣态线迹
poor man's cotton 美国佃农棉
poor mark 印花不良
poormitton 克什米尔花卉纹开士米披巾
poor posture 有缺陷的体型
poor quality product 低档织物
poor registration 脱版
poor section 皮质差的部段
poor selvedge 布边不良
poor sewing workmanship 缝工不良

poor shade 浅色绸
poor shoulder slope 溜肩,斜肩
poor smoothness 平滑性不良
poor taste 粗俗,低级趣味
poor workmanship 做工粗劣
pop 通俗的,流行的；用撤纽扣住(英国)
pop art 通俗艺术,波普艺术,流行艺术；波普艺术服装风格
pop art make-up 波普式化妆
pop beads 中型塑料珠项链
pop clothes 流行服装,时兴服装
popcorn 玉米花形状,结子花线效果
popcorn fiber 玉米花状纤维,爆玉米状纤维
popcorn stitch 玉米花图案针法,玉米花针法,玉米花状线迹
poplar flower 杨树花
poplin 毛葛,府绸
poplin broche 花毛葛,花府绸,织花府绸
poplin coat 风雨衣
poplin(e) 毛葛,府绸
poplin grosgrain 葛
poplin jacket 毛葛短外衣
poplin lama 英国拉玛毛府绸
poplintte 纱府绸
popliteal space 腘窝
pop look 波普风貌；波普型款式,通俗型款式
popover 家事服
popper 子母扣,按扣；撳纽,按纽；有孔小珠
poppits 中型塑料珠项链
pop plus humor 波普加幽默
poppy red 芙蓉红,罂粟红(硃红色)
pop socks 波普袜
pop style 波普型款式
pop stylistic form 波普款式
popular brocaded rayon poplin 大众绸[呢]
popular button 流行纽扣
popular button size 流行纽扣尺寸
popular fashion 大众时装,流行时装,普通流行时装
popularization and application 推广应用
popular pattern 流行服装款式
popular-priced market 大众价格市场
popular slip 背带式女式衬衣,背带式普通衬裙
popular suit 背带式套装
popular theme 热门主题

popular ware 热门货
popular [woollen] cloth 大众呢
populin 府绸
popver 家事服
Popy Moreni 普碧·莫雷妮(法国时装设计师)
poral 波拉呢
porcelain 瓷色
porcelain accessories 瓷器服饰品
porcelain blue 瓷蓝色
porcelain green 瓷绿色
porcelain lace 瓷质花边
porcelain lace process 蓝白仿瓷印花布
porcupine poattern 菠萝花纹
pore 汗孔,毛孔,细孔
pore volume 孔体积
porgee 手织平纹绸
pork-pie hat 凹顶帽,馅饼式男帽
poro neck 围脖圆领
poromer 透气性合成革
poromeric 多孔聚合物,通气性强的塑料,透湿人造革,透气性人造革
poromeric shoe material 通气性鞋料
porosity 多孔性,孔隙率,透气性
porous antitoxic clothing 透气防毒服
porous cloth 多孔织物
porous fabric 多孔织物
porous fiber 多孔纤维
porous hollow fiber 多孔中空纤维
porous skin(ned) fiber 多孔皮层纤维
porraceous 韭菜绿
porraceous color 韭菜绿色
port 港口
portable bag closer 手提封包机
portable life support system 手提式生命维持系统
portable machine 轻便型缝纫机,手提式缝纫机
portable moisture meter 便携式测湿仪
portable sewing machine 轻便型缝纫机,手提式缝纫机
port cabello cotton 委内瑞拉卡贝略港棉
port canons 靴花边
port congestion surcharge 港口拥挤附加费
port duty 港口税
porter hat 搬运工帽
porter's knot 搬运工用的垫肩
portfolio 公文(皮)包
portfolio clutch 纸夹形手握袋

portiere 门帘,帷幔;厚装饰织物
portly 肥胖的人,大个儿;特大号服装
portmanteau 旅行皮箱,衣箱
port of delivery 交货港
port of import 进口港
porto Rico cotton 波多黎各棉
port of shipment 装运港
port philip wool 澳洲菲利普港羊毛
portrait 前领垂皱的宽大折领
portrait collar 肖像领
portrait hat 羽饰宽边女帽,宽边花式女帽;肖像帽
portrait lapel 肖像西装领(大开口翻折领)
portrait tapestry 丝织人像,肖像缀锦
PORTS 宝姿
PORTS MEN 宝姿
port surcharge 港口附加费
port to port 港到港
Portuguese band stitch 葡萄牙环箍针法;葡萄牙缘饰针法
Portuguese border stitch 葡萄牙环箍针法;葡萄牙缘饰针法
Portuguese costume 葡萄牙民族服饰
Portuguese knot stitch 葡萄牙包梗针法;葡萄牙细包针
Portuguese knotted stem stitch 葡萄牙包梗针法;葡萄牙细包针
portuguese stem stitch 葡萄牙包梗针法;葡萄牙细包针
port wine 葡萄牙酒红(深紫红色)
posahuanco 手织彩色棉布
pose 身体呈现的样子
poshet 小背袋
positioning 定位
positioning collar stay 领角薄膜定位
positioning marking 裁剪定位描样,排版描样
position pressing (衣服)部位熨烫
positions and parts for garments 服装部位
positive contrast 正对比
positive correlation 正相关
positive-ionic cotton fabric 阳离子棉织物
positive pressure [inflated] garment 正压力[充气]服
positive print 正像复印品
positive stripe 阳条子图案
positive thermal fabrics 积极式保温织物
post 四合扣纽桩,柱,杆

post and telecommunication uniform 邮电服
post and telecom uniform 邮电服
post-bed 柱式底板高缝台
post-bed basting machine 柱式底板假缝机
post-bed lockstitch sewing machine 柱式锁缝缝纫机
post-bed sleeve attaching machine 柱式底板上袖机
post-boarder 后定形机
postboarding 后定形(整理时的热定形工艺)
post box in patch pocket 贴袋内的方形小贴袋
postboy waistcoat[vest] (旧时)邮递员背心外套
post-cure 延迟焙烘(成衣后的焙烘)
post-cured permanent press 后定型加工(焙烘)
post earring 后侧耳环
posted price 标价
posteen 阿富汗羊皮大衣
poster 海报,招贴
poster cloth 广告招贴布
poster color 广告色
poster dress 广告衫
poster rug 彩边风景地毯
posterior arm length 全后臂长
posterior armpit point (PAP) 后腋窝点
posterior chest width 背宽,后胸围宽度
posterior full length 身长,身高
posterior hip arc 后臀弧
posterior interarmpit breadth 腋窝后宽
posterior neck length 颈后长
posteriors 后身,身体后部
posterior shoulder width 后肩宽,背肩宽
posterior waist length 后腰长(度)
postilion boot with gambade 带绑腿布的骑士靴
postilion coat 御者外套
postilion waistcoat 马车夫背心
postil[l]ion 高顶窄卷边骑马女帽;骑士式女短上衣
postil[l]ion hat 骑士帽
postin 巴基斯坦普司丁拉绒毛呢,起绒粗呢
post-inflation finish 后蓬松整理
post-modern 超便装;超摩登型款式;超现代

post modernism 后现代主义
postolui [俄]喀尔巴阡鞋
post pocket 开贴袋(贴袋上有开袋)
posts 杯型乳饰;胸罩杯
post-sale service 售后服务
post & stud 四合扣
post-treatment 后处理
post-type sewing machine 柱式底板缝纫机
posture 模特儿摆的姿势;身段;体态;仪态;姿色
posture aid corset 健美紧身衣,整形紧身衣
post wage system 岗位工资制
posy 诗铭(戒指)
posy green 花束绿
posy ring 纪念金戒指,诗句金戒指
pot hat 圆顶高帽,常礼帽,烟囱管帽
pot lace 网地瓶形花纹梭结花边
pot linen 多匹缝结的亚麻布
potassium titanate fiber 钛酸钾纤维
potbellied figure 凸腹体型
potbelly 凸肚,大肚子;大腹便便的人
potential shrinkage 潜在收缩
potential value 潜值
poterior armpit point 后腋窝点
potholder vest 采用锅垫图案的钩编背心
potten kant 网地瓶形花纹梭结花边
poturi 保加利亚男白哔叽马裤
pouch 袋,口袋,小袋,衣兜;皮弹药袋;小钱包;邮袋;眼袋
pouch bag 囊形袋;束腰圆形袋;小钱袋
pouch belt 装有腰袋的皮带
pouch heel 袋形袜跟,袜子的梯形高跟
pouch heel socks 袋形袜跟短袜
pouching quantity 鼓起量
pouch pocket 囊袋形口袋,钱包形口袋
pouch pocket skirt 荷包袋裙
pouf 向外膨起部分(衣服打褶裥后);高鼓并卷起的发型;高髻发型(18世纪);波夫头饰
pouff(e) 高髻发型;蓬松部分;向外膨起部分(衣服打褶裥后);高鼓并卷起的发型;波夫头饰
poulaine 普廉尖鞋,细尖头鞋
poulangy 法国素色麻毛厚呢
poulet sleeve 鸡腿袖
poult de la reine 皇后绸,花岗石纹绸布
poult [de soie] 真丝横棱绸(质地较厚重),波纹绸,绉绸

poultry feather 家禽羽毛
pounce 用细砂纸磨光毡帽
pound goods 论重量出售的纺织品
pounding （制鞋）敲钉
pounding block 压板
pouritache 厚饰带,粗饰带
pourpoint 男式棉夹衣,棉紧身衣（14～15世纪）
poussin [lace] 法国精细窄花边
Poussin lace 小鸡花边
pover(r)a 概念派艺术
povertism 乞丐式时装
povertism fashion 贫困主义服式；乞丐式时装
poverty fashion 穷汉装（由T恤衫与牛仔裤组合）
povilion 尖顶大帐篷
powder 香粉,香松
powder blue 粉蓝色,粉末蓝
powder-bonded nonwoven 粉末黏合非织造布
powder-bonded nonwoven fabric 粉末黏合非织造布
powder box 香粉盒
powder compact 香粉盒
powdered chalk 记号粉
powdering 满幅点子花纹,满幅夹子花纹
powdering mantle 梳妆斗篷,化妆衣
powder method 粉印法（码克复制）
powder off 降低面部化妆色调（使柔和）
powder point fusible interlining 粉点热熔黏合衬
powder puff 粉扑；化妆师；化妆艺术家
powder puff nonwoven 粉拍纸,粉拍用非织造布
powder-room 化妆室；女洗手间
powder rose dust 藕灰
powder silk 火药袋[用绢丝]绸
powel davis 粗帆布
power dressing 弹力装
power fabric 弹性针织物,弹性布
power fiber 能量[转化]纤维（吸收太阳能或人体热能后能辐射远红外射线的纤维）
power-knit 强弹力针织物
power loom habotai 电力纺,纺绸
power net 弹性针织物；弹力网,弹力网经编织物
power net fabric 弹性针织物
power net girdle 弹性腰带

power net machine 弹力网眼经编机
power & public relationship 2P策略
power stretch 高弹性伸缩,强弹力
power stretch fabric 强弹性布料
power supply 电源
power woven habutai 电力纺
poyas 保加利亚红色饰带
poynet 下臂袖,下半袖
PP tape 聚丙烯带
practical apparel 实用服装
practical art 工艺美术；实用美术
practical dress 实用服装
practical property 实用性能
Prada 普拉达
praid 大格子纹
prairie dress 草原装；大草原连衣裙（直立领口,肩裥袖,多片裙,荷叶边）
prairie look 大草原开拓者风貌,大草原开拓者款式（高领口,长袖,两片长裙,印花细布）
prairie skirt 大草原喇叭裙（腰部抽褶,裙边有1～2道荷叶边,平布或印花细布）
prairie style 大草原开拓者款式；美国女拓荒者风格
prairie sunset 草原晚霞色
pram jacket 童年夹克；婴儿短外衣
pram set 婴儿全套衣物用品
pram suit 童车装；婴儿装
prat 臀部口袋,裤子后袋
pratkick 臀部后袋；臀部口袋；后枪袋
prayer bones 双膝,膝盖
prayer rug 跪拜毯
prayer seam 普通接缝,平行接缝
prayer veil 祈祷者（小三角花边）面纱
PRD-14 fiber 聚酰亚胺纤维（商名,美国）
preshape 预先成型
preassemble 预装配
preboarding 预定形
prebonding bulky nonwovens 预黏合膨体非织造布
precieuse 条子平纹绸
preciosities 贵重物品
precious metal 贵金属
precious miter 嵌宝红衣主教冠
precious stone 宝石
precise pattern 精确花样
precision 精密度,精确度；精确性
preclosed elastic 预封松紧带
preconditioning 预调湿；预调温湿度

precreping 预压花,绉布压花
precreping calender 预压花机
precuffed trousers 翻边裤
precured permanent press 预定形处理
precutting attachment （服装）预裁附件
predatory dumping 掠夺性倾销
predilection 偏爱;偏好
predominant fiber 混纺中主要纤维
preen 胸针,饰针,别针
prefab 预制品
prefade 预褪色
preference 趣味
preferential duties 特惠关税
preferential system 特惠制度
prefixing 预先固定缝料
prefixing armhole 预先固定袖窿
preformed bra cup 预成形的胸罩窝
pregnancy 孕妇胸衣,保育胸衣,哺乳内衣
pregnant figure 怀孕体型
prein 压平布面（将绒毛压平）
pre-inspection 预查
preky bow 团长领带
prelas 防水棉或亚麻织物（用油和蜡处理过的）
prelate 法国涂柏油大麻帆布,涂沥青大麻织物
preliminary drawing 初稿
preliminary finish 前处理,预整理
preliminary test 预试
preloaded fabric 预浸渍织物
prema cotton 陆地棉
premanent embossed finish 耐久性拷花整理
pre-markering 预排法
premature infant 早产婴儿
pre-mercerization 预丝光
premetalized acid dye 金属络合酸性染料
premier 服装发表会的第一天（首场）;首场（发布会）
Premier Collections 先导展（英国）
premier diamond 总理钻石
premiere vision 第一视觉
premise 场所
premium grade 优等品
pre-neatened edge 预修整边
pre-operation work 日常生产准备,生产作业准备
preoxidized acrylic(s) 预氧化聚丙烯腈纤维

preoxidized fiber 预氧化纤维
preparation for delivery 交货准备
preparatory finish 前处理,预整理
preparatory machinery for knitting and braiding 针织和编织用准备设备
preparatory time 准确时间
prepared-for-print 待印花的半制品织物
prepared material 经过前期处理的织物
preppie look 预科生款式;预科生装
preppy look 预科生款式;预科生装
pre-production 试生产
pre-production inspection 试生产检查
pre-programmed pattern 预编程序花样
pre-programmed stitch 预编程序线迹,编程线迹
pre-Raphaelite look 拉斐尔前派;拉斐尔前派服式
pre-Raphaelite stylistic form 拉斐尔前派款式
pre-relaxing treatment 预松弛处理
pre-sale testing 出售前检查
presbyopic glasses 花镜
preschooler's wear 小童服,学龄前儿童服装
prescouring 预煮练,预洗涤
presensitizing 织物树脂定形
presentation 提示
present fashion 时兴;时款
preserves 防风镜,防护用品,太阳眼镜（墨镜）
preset control 程序控制
presetting 预定形
presetting machine 预定形机
pre-shape 预先成形
preshaping tape 预先定型的牵条
preshrink 预缩
preshrunk brushed denim 防缩磨毛牛仔布
preshrunk finish fabric 防缩整理织物
preshrinking 预缩
preshrinking fabric 预缩织物
preshrinking machine 预缩机
president 双层棉毛交织厚呢缎,提花家具绸,再生毛起绒呢
president braid 斜纹滚条
presize 预定的孕妇尺寸[尺码]
presoak 事先浸泡
prespotting 干洗前预去污处理
press 烫平,压烫,熨烫,熨烫机,压平机,打包机;压辊脱水,模压

pressability 熨烫性
press ball 布馒头熨斗垫
(press) block 熨板,熨烫模板
press board (熨烫用)烫板;烫台;压板;熨烫工作台;手工印花台板;电压纸板
press bonded fabric 压黏布
press bonding fabric 压黏布
press button 按纽,揿纽,四合扣
press camera 新闻摄影机
press chest interlining 烫胸衬
press cloth 水布,烫水布,熨烫垫,熨烫覆布
press clothes 熨平衣服
press collar interlining 压领衬
press collar point 热压领角定型,压领角
press crease 烫皱痕
press crease retention 熨烫皱痕持久性
press dish 烫压板
pressed crease retention 熨烫褶裥持久性
pressed felt 压制毡
pressed foot 压脚
pressed heel 压跟
pressed heel sole 压跟底(鞋)
pressed-in crease 熨烫折痕,熨烫皱痕
pressed-in crease retention 熨烫折痕耐久性
pressed kersey 薄型白色克尔赛呢
pressed pile 瘪绒,倒绒
pressed pleat 定型裥
pressed plush 拷花长毛绒,压花长毛绒
presser 熨烫工;打包工;压脚;压料铁,压铁,压具
presser bar 压脚杆
presser bar lifter 手抬压脚
presser bar pressure 压杆压力
presser bar spring bracket 压杆弹簧导架
presser foot 压脚
presser foot arm 压脚臂
presser foot height 压脚高度
(presser) foot lifter 压脚扳手,压脚提升器
presser foot symbol 压脚符号
presser foot throw 压脚升程
presser garment 压力服
presser lifting amount 压脚提升高度
presser pad 烫衣机毡制衬垫
presser regulator screw 压脚调节器
presser sewing 压紧缝纫
press fastener 揿纽
press felt 熨烫垫,熨烫垫布

press finishing [针织品]热板压烫定形
pressfit 压烫合身(服装)
press gloss 压光,轧光
press-in crease retention 熨烫折痕耐久性
pressing 冲压;烫衣;熨烫;压呢
pressing and setting machine 连续压烫机,压烫定形机
pressing bench (熨烫用)烫凳
pressing bun 熨烫馒头
pressing by conventional iron 火焰熨烫
pressing chest interlining 烫胸衬
pressing cloth 水布,熨烫衬布,熨烫垫布
pressing collar point 压领角
pressing cushion 熨烫垫
pressing dart 压烫省道
pressing dart open 烫省缝
pressing dummy [成衣]蒸烫机
pressing equipment 熨烫设备,熨烫器
pressing figured velvet 压花丝绒
pressing finish 熨烫整理
pressing gadget 服装打包机
pressing interlining 烫衬,压衬
pressing iron 熨斗
pressing lever 紧压杆
pressing lining seam 坐烫里子缝
pressing machine 熨烫机,压烫机,整烫机
pressing mark 光印,极光;亮光疵
pressing mat 熨烫垫
pressing mitt 手套式烫垫
pressing open 分烫,烫开
pressing open collar seam 分烫绱领缝
pressing open dart 烫开省缝
pressing open gorge line seam 分烫领串口
pressing open seam 熨平线缝,烫开线缝;分缝
pressing open shoulder seam 分烫肩缝
pressing open side seam 分烫侧缝
pressing open sleeve seam 分烫袖缝
pressing pad 熨烫垫
pressing plate 烫衣板,压呢板
pressing princess line 烫刀背缝
pressing quality is not good 熨烫不良
pressing room 整烫间
pressing seam open 熨平线缝;劈缝
pressing shrinkage 压烫收缩
pressing sponge 熨烫垫
pressing stand with an exhauster 抽气烫台
pressing technique 熨烫技术
pressing template 熨烫模板

press iron 烙铁,熨斗
press line 烫出的折痕
press machine 大型熨衣机
pressman shirt 记者衬衫(英国)
press mitt 手套式烫垫
press-off 脱套,拷针
press-off design 一隔一的并圈花纹
press-off detector 脱套自停装置
press open 烫开
press open collar seam 分烫领缝
press open dart 分烫省缝
press open french dart 分烫刀背缝
press open gorge line seam 分烫领串口
press open lining seam 分烫里子缝
press open seam 烫平线缝,烫开线缝
press open shoulder seam 分烫肩缝
press open side seam 分烫摆缝
press open sleeve seam 分烫袖缝
press release (服饰)新闻稿
press retention 压烫保持性
press room 整烫室
press seam 劈缝,分缝(两片缝合后,再将缝分开熨平)
press seam open 烫平线缝,烫开线缝
press seam opening 烫开缝
press sheet 水布,熨烫垫布
press skirt 熨平裙子
press stand (裁剪、缝制用)工作台板
press steaming 熨烫汽蒸
press stud 揿扣,大白扣,揿扣
press stud fastener 揿扣,揿纽,揿纽
press stud in four piece set 四件扣
press tenter 加压拉幅机
press trousers crease 烫裤挺缝线
pressure cover 压力服
pressure mark 压斑;压痕
pressure regulating thumb screw 压脚(调节)螺丝
pressure roller 压辊
pressure sensitive adhesive 感压接贴剂,压敏胶黏剂
pressure suit (高空飞行用)增压服
pressurized helmet 增压头盔
pressurized suit 增压服
prestige 优越性
prestige price 超高价位
prestige store 名店
prêt-à-couture 高级定做服装,高级时装成衣
prêt-à-porter 高级女装成衣

prêt-à-porter designer 成衣设计师
preteens' dress 少年(少女)服
pretest 预先试验
pretintailles [法]剪花饰
pretreatment 前处理,预处理
pretreatment processes 预处理工程
pretty 俏丽
prettyism 过分讲究修饰;矫揉造作
pretty polly type 老式连袜裤
prevailing 流行;现行价格
prevailing price 牌价
prevailing style 流行式样
preventive maintenance 预防性保养;预防性维修
preventive measure 预防措施
pre-walkers 婴儿软底鞋
prewash polyester poplin 涤裤水洗布,预洗涤纶府绸
prewelt process 预缝沿条工序(制鞋)
prewetting 预湿(处理)
prexillas crudas 本色或半漂废麻帆布
price boom 大幅度上涨
price competitiveness 价格竞争性
price list 价格表,价目表
price relation 比价
pricked pattern 针刺花纹
pricking stitch 拱针
pricking wheel 点线器
prickle [对人体皮肤的]针刺感(纤维测试项目);刺痛感
prick seam 星止缝(将料边重叠以拱针针法缝合,主要用于手套)
prick stitch 短回针(用于厚料;拱针(针法)
pride of Georgia cotton 美国佐治亚大铃棉
priest frock 道袍;道衣
priest king's dress 祭司国王服饰
priestly 英国精纺呢
priest robe 僧侣袍
priest's hat 祭司帽
priest shoes 僧(侣)鞋
prima 高湿模量黏胶纤维(商名,荷兰凯米拉)
primaloft 仿羽绒纤维(商名,美国)
primary additive colors 加法(三)原色
primary backing [地毯]主要衬底
primary backing fabric 簇绒毯的基底布
primary block 原始样板
primary color 基色,原色(红、黄、蓝三

色）
primary effect　先人效应
primary group　小群体
primary hand value（HV）　基本风格值（织物）
primary heater value　织物基本风格值
primary market　一级市场(指服装、服饰的原材料制造商)
primary material　原料
primary product　初级产品
primary textile plant　纺织染整厂,纺织品前加工工厂
prime　最佳采集毛皮期;粗纺用毛,两侧或肩部毛;优级亚麻经纱
prime cost　直接成本,主要成本
prime pelt　优质毛皮
primi costume　普米族服饰
primitive clothing　原始服装
primitive colors　三原色
primitive design　朴质图案,原始图案
primitive print　经典原始印花
primrose　报春花色,淡黄色
primrose pink　报春花红
primrose yellow　樱草黄(黄色)
primuline yellow　樱草灵黄(黄色)
Prince Albert [coat]　昼间的正式礼服,双排扣式男大衣,男礼服大衣
Prince of Wales check　威尔士大方格纹
Prince of Wales plaid　威尔士大方格纹,威尔士亲王格子
Prince of Wales shoes　威尔士王子(系带无舌)鞋
princess　棉毛里子呢;公主装;紧身连衫裙
princess blue　公主蓝
princess chemise　公主式连身衬衣裙
princess clothing[coat]　公主线外套(束腰大摆)
princess coat　公主式大衣,紧身连衣裙式大衣
princess dress　公主装;紧身连衣裙(上身紧,无腰节线,裙子逐渐展开)
princesse　单丝经缎
princesse cashmere　开司米单面棉绒
princesse gown　紧身连衣裙礼袍
princesse robe　紧身连衣裙礼袍
princesse silhouette　公主线式轮廓
princess gown　紧身连衣裙礼袍
Princess Irene Galitzine　艾伦尼·加里吉恩公主(意大利服装设计师)

princess kimono sleeve　公主线和服袖
princess lace　公主花边,优质仿凸纹花边,刺绣网眼花边
princess line　公主式;公主线
princess line bikini top　公主线比基尼上衣
princess line coat　公主线式大衣
princess line dress　公主线紧腰连衣裙
princess line overdress　公主线外衣
princess maillot　公主线泳衣
princess petticoat　公主线衬裙,紧身连衣裙
princess robe　紧身连衣裙礼袍
princess seam　高背缝
princess silhouette　公主线式轮廓
princess skirt　公主式裙
princess slip　公主线衬裙,紧身连衣裙
princess stock　公主式蝴蝶结高宽领
princess style　公主款式,收腰式
princess style effect　收腰式效果,吸腰式效果
princess style gown　公主式长袍,吸腰式服装
princess style house coat　吸腰式家居服
princess style jacket　吸腰式外套
princess styleline　公主结构线
princess twill lining　阿尔帕卡里子布
princess waistline　公主线,女装上的双条竖线,公主式腰线
Princeton orange　普林斯顿橙
principal　委托人
principal color　主色
principal component　主成分
principal part　主体
principal to principal　买卖关系
principle of elasticity　弹性原理
principle of exception　例外原则
principle of hierarchy　能级原理
principle of impetus　动力原理
principle of system　系统原理
principles of color　色彩原理
prink [up]　装饰,打扮漂亮
print　印花;印花布;印花布服装
print-bonded nonwoven　印花黏合[法]非织造布
print-bonding　印花黏合
print back gray　印花衬布,印花底布
print base fabric　印花地布
print bonded fabric　印花黏合[法]非织造布

print bump gray 印花衬布,印花底布
print cloth 印花坯布,印花布
print coat 印花涂层
print design 刷花
print dress 印花布女装,印花服装
printed apron 印花围裙
printed bed sheet 印花床单
printed blouse 印花衬衫
printed blue nankeen 毛蓝印花布
printed border handkerchief 印花花边手帕
printed calico 印花[平]布;印花织物
printed canvas 印花帆布
printed carpet 印花地毯,色绒纱机织地毯
printed casement 印花窗帘布
printed cloth 印花布
printed corduroy 印花(染色)灯芯绒
printed cotton linen 全棉印花里布
printed crepe fabric 印花绉纹呢
printed cut pile carpet 印花割绒地毯
printed denim 印花牛仔布
printed drill 印花卡其
printed (dyed) polyester georgette 印花涤纶乔其纱,柔姿纱
printed dyed velveteen 染色平绒
printed fabric 印花织物
printed fake fur 印花人造毛皮
printed flannel handkerchief 绒布印花手帕
printed flannelet(te) 印花绒布
printed flannels 印花法兰绒
printed foxing 涂色围条(鞋)
printed fur 印花毛皮
printed georgette velvet 印花乔其绒
printed georgette velvet with flower 烂印乔其绒
printed gloves 印花手套
printed goods 花布,印花布
printed grain 皮革印刷粒纹,皮革的印花粒纹
printed grandrelle yarn 仿合股双色纱
printed grosgrain 印花葛
printed ground 印花底色
printed handkerchief 印花手帕
printed hosiery 印花袜
printed interlock bed spread 提花印花线毯
printed interlock fabric 印花棉毛布
printed jacquard towel 提花印花毛巾

printed jeans 印花牛仔裤
printed knit fabric 印花针织物
printed knits 印花针织布
printed knitwear 印花毛衫
printed label 印唛,印花商标带
printed label tape 印花商标带,印刷商标带
printed lady's dress 印花女衣呢
printed lady's dress cloth 印花女衣呢
printed lady's dress worsted 印花女衣呢
printed mittens 印花独指手套
printed moguette 印花绒头织物
printed muslin 印花细布
printed nylonpalace 印花尼丝纺
(printed) nylon tape (印花)尼龙塔斯纶
printed panne velvet 光辉绒
printed pattern 销售式纸样,印刷式纸样;印花图案
printed pattern pajamas 印花睡衣
printed pique 印花灯芯布
printed rain boots 印花雨靴
printed register 印花对花
printed sarong 印花裙布
printed scarf 印花围巾
printed sheet 印花床单
printed sheeting 印花细布
printed shirt 印花衬衫,花衬衫
printed shirting schreinered 电光花布
printed shoes 印花鞋
printed silk 印花绸
printed single jersey 印花汗布
printed stitch-bonded fabric 缝编印花织物
printed tape 印带,印字带;(装饰)吊带
printed T/C fabric 印花涤棉布
printed ticking 印花被套布
printed towel 印花毛巾
printed twill 印花斜纹布
printed umbrella 花伞
printed velveteen 印花平绒
printed voile 印花巴里纱
printed warp matelasse 印经凸花绸
printed wool sweater 印花羊毛衫
printed yarn 印花纱线
printed yoke style pajamas 印花覆肩式睡衣
printer 打印机;印花工
printers 印花平纹坯布
printer's blanket 印花衬毯

printer's felt 印花衬毯
printextile 印花织物
print felt 印刷毡
print fold 印花折皱
print gingham 印花格子布
print industrial silk 印刷绸
printing 印花,印花工艺
printing and dyeing 印染
printing and dyeing machine 印染机
printing apron 转移印花压毯
printing blanket 印花衬布
printing block 手工印花木模
printing board 印花台板
printing bonding 印花黏合法
printing by the all-in method 一相法印花
printing by two-stage-method 两相法印花
printing design （在裁片上）印花,刷花;印花设计图案;印花设计
printing design technique 印花图案设计技术
printing fabric 印花衬布
printing felt 印刷呢
printing ground 印花坯布
printing lot 印花起定量
printing on cut piece 裁片印花
printing on print 叠色印花
printing pattern design 印花图案
printing process 印花工艺
print marking machine （缝纫机用）标印机
print on 直接印花法
print on cutted piece 裁片印花
print on print 重叠式印花
print raising gingham （仿色织）印格绒布
print round neck shirt 印花圆领衫（印花T恤衫）
prints 俄罗斯印花布
print skirt 印花布裙
print stain 印渍
print work 摹花刺绣,印花刺绣,印花作业
print worsted ladies outerwear cloth 俄罗斯印花女式外衣呢
print yarn stripes 印经条子织物
prior limitation 预定限额
priority product 重点产品
prismatic colours 光谱色彩,棱镜色彩
prismatic shank button 棱柱柄纽扣
prism pink 浅粉红
prisoner wear 囚服
prisoner's wear 囚服

prison garb 囚服
private brand (PB) 店属商标,个人商标,私有商标
private enterprise 私营企业
private firm 私营企业
private label 个人标签,商店标签,自有品牌
private labeling 自有品牌
private parts 阴部
private pocket 暗袋
private wear 个人便服,私人便服
privotal point 旋转点
prix seam 手套毛边外向缝,露出缝份
Peking blue 北京蓝
probability 概率
probability analysis 概率分析
probability curve 概率曲线
probability distribution 概率分布
probability error 概率误差
probability factor 概率因子
probability ratio test 概率比检验法
probability sampling 概率抽样
probability statistics 概率统计
probation 验证,检验,鉴定
probing 探测,测试
problem material 难加工缝料
procedural specifications 作业程序说明书
procedure 步骤,程序,工序,工业规程;过程;方法,措施
procedure flow shart 工作程序示意图
process 工序
processability 加工性能
process analysis 生产过程分析
process balance 工序平衡
process capability index 工序能力指数
process chart 工艺流程图
process design 工艺设计
processed filament yarn 变形长丝
process equipment 工艺设备;工艺装备
process flow 工艺流程
process flow diagram 工艺流程图
process formation 流程构成
process in advance 布料缝前处理
processing 加工,作业
processing according to buyer's sample 来样加工
processing according to customer's sample 来样加工
processing according to investor's sample 来样加工

processing behavio(u)r 加工性能
processing cost 加工成本
processing design 工艺设计
processing fur steamer 毛皮蒸汽清理机
processing quality 加工质量,工艺质量
processing line 工艺流程线
processing performance 工艺性能
processing property 工艺性能,加工性能
processing raw materials on client's demands 来料加工
processing time 加工时间
processing with customer's materials 来料加工
process layout 工序布置,流程布置
process management 工艺管理
process mix 生产过程组合
process optimization 过程最佳化
process printing 彩色套印
process program 工艺方案
process quality 工序质量
process record 工作记录
process route 工艺路线
process-scale chromatography 工业色谱法
process specification 工艺规程
process velocity 加工速度
procinyl dye 普施尼染料(活性分散染料,染聚酰胺纤维用,商名,英国)
producer 厂商,制造商
producer clored fiber 原液着色纤维,纺前着色纤维
producer color polyester 原液染色聚酯,原液着色聚酯
producer's dyed fiber 原液着色纤维
product 产品;(设计)作品;创作;作品
product appearance 产品外观
product buyback 产品返销
product catalog(ue) 产品目录
product category 产品类目,产品目录
product classification system 产品分类系统
product competitiveness 产品竞争性
product cost 产品成本
product deep-processing strategy 产品深加工策略
product defect 产品缺陷
product design 产品设计
product development 产品开发
product diversification 产品多样化
product engineering 生产设计

product feature 产品特性
product image 产品形象
product in process 在制品
product inspection for delivery 产品发货检验
production 产品;生产;制造,制作
production according to sample 来样生产
production activity control(PAC) 车间作业管理
production capacity 生产能力
production card 作业卡
production control 生产管理,生产控制
production cooperation 合作生产
production cost 生产成本
production cutter 叠层裁剪师
production cycle 生产周期
production days 生产日数
production department 生产车间
production dispatching 生产调度
production drawing 产品设计
production efficiency 生产效率
production equipment 生产设备
production expenses 生产费用
production factor 生产要素
production fit 生产适应性
production flow 生产流程
production flow chart 生产流程图
production forecasting 生产预测
production group 生产班组
production in bulk 大量生产
production in lots 成批生产
production inventory 生产储备
production level 生产水平
production line 生产线,流水作业线,装配线
production load 生产负荷
production lot 生产批量
production management 生产管理
production manager 生产经理
production norm 生产定额
production order 生产订货单
production order quantity 生产订货量
production order sheet 生产指定书
production-oriented management 生产型管理
production (output) per machine hour 台时产量
production pattern 生产用纸样,生产用样板
production pattern maker 生产纸样师

production personnel 生产人员
production plan 生产计划
production process 生产程序；生产过程
Production Quality Supervising Inspection Center 国家级产品质量监督检验中心
Production Quality Supervising Inspection Laboratory 产品质量监督检验机构实验室
production quantity 生产量
production quota 生产定额，生产指标
production record 生产记录
production sample 封样，生产样品
production schedule control 生产进度控制
production scheme 生产流程图
production sketch 产品草样
production specialization 生产专业化
production standard 产品规格
production system 生产系统
production target 生产指标
production team 生产班组
production time 有效工作时间
production variety 生产品种
production volume 生产量
product item change 品种翻改
productive capacity 生产能力
productive efficiency 生产效率
productive pattern 生产用纸样，生产用样板
productivity 生产[能]力，生产量，生产率
productivity per machine-hour 台时单产
product liability 产品责任
product life cycle 产品寿命周期，产品周期
product line 产品线
product management 产品管理
product mix 产品组合
Product-O-Rail system（P-O-R） 吊挂系统
product overstock 产品积压
product performance 产品特性
product plan 生产计划
product planning 产品规划
product positioning 产品定位
product pressing 成品熨烫
product price 出厂价格
product, price, plate & promotion 4P策略
product property 产品性能
product putting up 成品折叠

product quality 产品质量
product quality inspection 产品质量检验
product routine test 产品倒行试验
products buyback 产品回购
products catalogue 产品样本
product segmentation 产品细分化
product series 产品系列
products folding machine （包装）成品折叠机
product shop 产品车间
products inspection 产品检验
products of quality 优质产品
profession 行业
product standard 产品标准
product strategy 产品策略，产品战略
product structure 产品结构
product style 产品风格
product suitability to selling & purchasing 产品适销对路
product upgrading 产品升级换代
product variety 品种
product Youtine test 产品例行测试，产品例行试验
professional apparel designer 专业服装设计师
professional care 纺织品专业保养
professional clothing 工作服
professional design 设计合理
professional garments 职业装，职业服
professionally dry-clean only 只供专业干洗
professional overalls 专用工装裤
professional show 服装专业展示会
professional sports shoes 专业运动鞋
professional standard 专业标准
professional techniques 专业技巧
professional training 专业训练
professional waistbanding 专用腰带衬
proficiency certificate 专业证书
proficiency testing 能力对比试验
profile 分布图，轮廓，外形；异形丝；皱痕高度（干洗）
profiled fancy suitings 异形纤维毛花呢
profiled fiber 异形[截面]纤维
profiled fiber fabric 异形纤维织物
profiled filament 异形[截面]长丝
profiled film fiber 异形[截面]裂膜纤维，压丝毫裂膜纤维
profile-extruded yarn 异形[截面]孔挤出丝，异形丝

profile fiber 异形[截面]纤维
profile hat 半面帽,侧面帽,侧影帽
profile seam 轮廓缝,外形缝
profile silhouette 侧影型轮廓
profile stitcher 仿形缝纫机
profile stitching 仿形线迹,轮廓线迹
profile stitching machine 仿形缝纫机
profit 利润
profitability 经济合理性;可赢利性
profit and loss 盈亏
profiteer look 暴发户服饰
profit-related wages 效益工资
proglass 聚丙烯涂层玻璃纤维(商名,美国)
program 程序,方案,计划
program card 程序穿孔卡片
program checkout 程序检查
program computer [卡片]程序计算机
program-controlled sequential computer 程序控制时序计算机
program-controlled top feed 程序控制上送布
program control system 程序控制系统
program design 程序设计
program flow chart 程序流程图
program for the development of science and technology 科技发展规划
programmable 程序可控的
programmable controller 可编程序控制
programmbale digital microprocessor 可编程序数显式微处理机
programmable logic controller 可编程逻辑控制器
programmable memory 程序可控存储器
programmable [sequence] controller 可编程序控制器
programmable sleeve placket machine 程控绱袖机
programme 程序
programmed automatic cutting 程序自动裁剪
programmed check 程序检验
programmed control 程序控制系统
programmed function keyboard 程序操作键盘
programmed sewing 程序缝纫
programming and archiving system 归档系统
programming control shrink 编程控制缩布

programming controller 程序控制器,自动程序调节器
programming system electronic straight sewer 电子程序系统直形线迹缝纫机
progress chart 工作进展图表
progressive bundle system 递进式捆扎装置
progressive bundle unit production system 渐进式扎束裁片生产方式
progressive bundle unit system 逆进式捆扎单元袋置(系统)
progressive consumer 前行型消费者
progressive fashion 超前流行款式;进步流行
progressive proofs 套色打样
progressive tax 累进税
prohibitive import 禁止进口
pro jeans 正统牛仔裤
project 投影
project department system 规划部制
project evaluation and review technique (PERT) 项目评审技术
project file system 跟单系统
prolecting gown 防护服
project negotiation 项目洽谈
prolo look 工装式;劳动者风貌;劳动者服式;劳动者款式
prolon 再生蛋白质纤维(属名)
promenade 妇女便装;法国毛织带
promenade beret 散步用软帽
promenade costume 散步便裙
promenade dress 散步便裙
promenade skirt 钢圈裙撑式散步衬裙
promilan 聚酰胺66长丝(商名,日本东丽)
prominent bone 后颈处突骨
prominent bust 高胸
prominent hip 翘臀,凸臀
prominent seat 凸臀
prominent-calves 畸形小腿
promise to pay 承诺付款
promissory note 本票;期票
promix 半合成蛋白质纤维(属名)
promix fiber 普罗米克斯纤维(乙烯系单体与蛋白质共聚物纤维)
prompt payment 立即付款
prompt shipment 即期装运
proneness to snagging 易钩丝性
prong 分规尖脚;环圈纽面
prong-and-eyelet buckle 用扣针和洞连接

的皮带扣
pronged ring 按纽的固定圈
prong snap 五爪按扣;五爪纽
pronounced local colour 浓厚地方色彩
pronounced staining 明湿色渍
proof 论证,凭证
proofed breathable cloth 防水透气布料
proofed breathable fabric 防水透汽织物
proofed cloth 防水布
proofed hat 硬挺帽
proofing 防水处理,防护处理,帽檐上浆
proofing sheetings 标准坯布,无疵坯布
proofing twills 防雨卡坯布
proof of delivery 交付证明
proof of export 出口证明
prop 服装等道具
proper color 固有色
proper properties; processing properties and product properties (3D properties) 固有性能、加工性能和产品性能
properties of colour 色的特性
property 服装等道具;特性,性能
propitious omen design 龙凤呈祥(图案)
proportion 比例;均衡;调和;相称
proportional compasses 比例(两脚)规
proportional distribution cutting 比例分配裁剪法
proportional divider 比例分规
proportional grading 比例放码
proportionality 比例性,均衡性,相称性
proportional measurement chart 比例尺寸表
proportionate figure 匀称体型
proportionate measurements 比例尺寸
proportioned 尺寸齐全的服饰
proportioned fit 合格尺寸
proportioned hose 弹性袜(适应不同脚长的袜子)
proportion illusion 比例错觉
proportion of human body 人体比例
proprietary brand 特殊品种
proprietary design 专利设计
proprietary name 专利商标名
proprietary term 专利商标名
pro rata distribution 按比例分摊
prospective purchaser 潜在购买者
prospector's shirt 勘探者衬衫(贴身、轻质、御寒衬衫);针织贴身运动装
pro-sport wear 纯运动服装
prosthetics 修复术(整容;装假肢)

protanopia 红绿色盲
protease 蛋白酶
protecting hosiery 针织护膝
protectire mask 防毒面具
protective apparel 防护服装
protective athletic clothing 运动用保护服
protective clothing 防护服,防护衣;防毒衣;宇航服;劳动保护服
protective coating 保护膜;保护涂层
protective colour 保护色
protective colouration 保护色
protective covering 保护层
protective earwear 防噪声耳套
protective fabric 防护织物
protective finish 保护性整理
protective foil 保护箔
protective footwear 劳保鞋
protective garment 防护服
protective gear 防护服
protective glasses 防护眼镜
protective gloves 劳保手套
protective gown 防护衣
protective hat 防护帽
protective helmet 护盔
protective hood 防护兜帽
protective mark 保护标志
protective mask 防护面具
protective mitten 防护手套(骑马或击剑用),劳保手套
protective mitts 防护手套(骑马或击剑用)
protective plates 裆
protective ribbon 防护用带
protective rubber shoes 劳保胶鞋
protective tariff 保护关税
protective technology garment 防护技术服,工业生产防护服
protective textile 纺织保护用品
protective theory 保护说
protective trousers 防护裤
protective under-clothing 护身内衣
protective wearing fiber 防护服用纤维
protector 防护用品
Protect Wear 防护服
protein [base] fiber 蛋白质纤维
protein fiber 蛋白质纤维
protein regenerated fiber 再生蛋白质纤维
protest 据付证书
protex 再生蛋白质纤维(属名)

protocol 协议,草案,规程
proto fiber 原[生]纤维
prototype 原型;样板,样机,标准;原型裁剪法
prototype cutting 原型裁剪法
prototype design 原型设计,标准设计
prototype garment 原型服装
protractor 分度规
protruding abdomen 凸肚,凸腹
protruding tummy 小腹外凸
Provencale look 普罗旺斯款式
Provencale prints 普罗旺斯印花
Provencale stylistic form 普罗旺斯款式
provincial blue 乡土蓝色
provocative panties 女紧身妖冶短裤
proximity suit 近火区消防服
prune 梅干红(暗红青莲色),深茄色,深紫红色
prunella (昔日法官、牧师等袍料用)普鲁涅拉毛葛;(做女鞋面料用的)普鲁涅拉厚呢
prunellas 用普鲁涅拉呢作鞋面的鞋;女用普鲁涅拉斜纹薄呢
prunelle 法国普伦尼尔毛葛,普伦尼尔斜纹呢
prunelle batarde 法国普伦尼尔巴塔素哔叽
prune(-purple) 深紫红色,绀青色
prussian binding 普鲁士漆条布
prussian blue 普鲁士蓝(深蓝色)
prussian collar 普鲁士圆形立领,上大下小的竖领;普鲁士直领
prussian shawl 普鲁士披巾
prussion blue 普鲁士蓝,深蓝色
prutik 色线抽绣(俄罗斯);色线挖花绣
prylanit 阻燃纤维(由聚丙烯腈纤维经过环化、氧化、脱氢制成,商名,德国)
pschent 古埃及双重王冠
pseudo-colors 伪彩色
pseudo leather 仿皮革,人造革,合成革
pseudo-phase-change fiber 伪相变纤维
pseudophototropic effect 假光致变色效应
psoas 腰肌
psychedelic 幻觉派情调;幻觉性(色彩)
psychedelic print 迷幻亮色印花
psyche knot 后髻发式
psycho analysis 精神分析
psychographics 消费心态,消费心理学
psychological complementary 心理的补色
psychological distance 心理距离

psychology 心理学,行为科学
pu 补,补子,官补(中国古代服饰)
pua fiber 普阿麻
pua hemp 普阿麻
pu and leather badge spu 皮和真皮徽章
PU badge PU 皮徽章
pubic bone 耻骨
pubis 耻骨
public appearance 社交形象
public ownership 公有制
public relations 公共关系
Pucci helmet 普奇型头盔(由塑料玻璃制成,面部开孔)
Pucci shirt 普奇女衬衫(方形下摆,西式豁口翻领的女式长衫)
Pucci-type 普奇型
puce 暗红色,紫褐色,深紫色
puched felt 针刺毡
pucker 皱褶;皱纹;起绉;缩拢;起皱器;折叠
pucker cloth 绉纹布
pucker crinkle 起皱
puckered 起皱效应,泡泡效应
puckered cloth 泡泡纱,波纹布
puckered fabric 绉纹织物;绉纹布
puckered fabric bed-cover 绉布床罩
puckered pants 皱纹裤
puckered selvage 褶边疵
pucker-free 无皱缩
pucker-free felling 无皱缩折缝
pucker-free marginal seam 无皱缩边缝
pucker-free seam 无皱缩缝
pucker-free stitching 无皱缩线缝
puckering 抽皱;缝纫皱纹;起皱;折叠;缩皱;皱折
puckering lapel 驳头起皱
puckering on shoulder 绺肩
puckering pattern 泡泡绉;褶裥款式
puckering seam 皱缝
puckers at quilting 绗棉起绉
puckers at shoulders 裂肩(小肩起皱)
puckers at underarm seam 抬裉缝起绉
pucker up round neckline 领口皱缩
puck-free 无皱缩
puck-free stitching 无锁缝线缝,无皱缩线缝
puck-free stitching 无皱缩线缝
PU coating PU 涂层
PU color coating PU 彩色涂层
pudding cap 布丁帽

pudding cloth	烹调用棉布,布丁布
pudding face	大胖脸
pudenda	阴部
puebla velveteen	普布拉棉绒
puerto cabello cotton	委内瑞拉卡贝略港棉
puerto rico cotton	波多黎各棉
puer trad	纯粹传统款式
puff	粉扑;衣服的蓬松部分;尖角布;垫片,整片,皱褶,胖皱;被子,鸭绒被;蓬松的,松软的
puff ball	褶裥泡泡裙
puffbox	粉扑盒
puffed cap-sleeve	短泡泡袖
puffed selvedge	散边
puffed shoulder	蓬松肩线
puffed sleeve	泡泡袖,灯笼袖,胖袖
puffed style	常春藤式装饰袋巾,蓬松型装饰袋巾;胸袋中手帕装饰法
puffiness	蓬松性,松软性
puffing	(织物的)蓬松装饰
puff printing	凸纹印花
puff quilting	泡式绗缝法
puff shoulder line	蓬松肩线
puff shoulder pad	泡泡袖垫肩
puffy shoulder line	膨松肩线
pug	发髻
puggaree	[印度人用的]薄头巾,[帽子后的]遮阳布
puggree	[印度人用的]薄头巾,[帽子后的]遮阳布
pugliese	意大利低级棉
pug(nose)	狮子鼻
pugree	[印度人用的]薄头巾,[帽子后的]遮阳布
puke	优质粗纺呢
PU label	皮牌(PU皮牌)
pulan	木棉籽纤维
PU leather	PU革[皮],聚氨酯合成革
pull	把手;后蓬式服装,套衫;套头
pullayne	细尖头鞋
pull-back	后蓬式
pullback skirt	后蓬裙
pullback style	后倾发型
pull blouse	宽大套衫
pull blouson	套衫宽上衣;宽松罩衫
pulled put together	(以针织服装为中心的)新服装组合技巧
pulled thread work	抽绣
pulled together	(以针织服装为中心的)新服装组合技巧
pulled waste	碎呢再生毛
pulled work	扯纱绣;抽绣
pull en V	法国V领套衫
puller	制革工;楦鞋工
puller feed	拉轮送料;牵拉送布器
pulling	牵紧
pulling at inside seam	服装吊脚
pulling at outseam or inseam	吊脚
pulling at outside seam	服装吊脚
pulling out	抽丝
pull kimono	和服袖夹克套衫,和服袖的套头夹克
pull loose	拉回
Pullman case	普尔曼式大手提箱
Pullman slippers	普尔曼折叠式便鞋
pullnot cotton	美国普尔诺特棉
pull off	脱下
pullom	木棉籽纤维
pull on	穿上(套头的)
pull-on	套穿的;套衫,套穿衣;套穿式;套头;橡筋腰裤子
pull-on beret	套戴贝雷帽
pull-on blouse	套头女衫
pull-on girdle	套装紧身褡,松紧束腹
pull-on gloves	易戴式手套
pull-on pants	松紧带女长裤
pull-ons	易戴式手套
pull-on shorts	松紧带短裤
pull-on skirt	松紧带裙子,松紧腰裙子
pull-on sport bra(ssiere)	套头式运动胸罩
pull-on sweater	套衫,套头毛衣
pull-out hook	拉线钩
pullover	套头毛衣;套衫,无领无扣衫;套头式
pullover blouse	套头女衫
pullover coat	套头外套
pullover dress	套头连衣裙
pullayne seam	套衫接缝
pullover shirt	套头毛衣;套头衬衫
pullover stitch	套衫线迹
pullover suit	套头套装
pullover sweater	套头毛衣;套头毛衣
pullover top	套衫领
pullover vest	套头背心
pullover with zipper	拉链套衫
pullover woolly	粗毛线套衫
pull ring	拉链拉环

pull-strap 提带
pull tab 滑块,拉链的拉头
pull-through collar 穿带领
pull-through string 串带
pull-under 毛线背心
pulnott cotton 美国普尔诺特棉
pulp board 纸板
pulp felt 浆粕毛毯
pulp-like short fiber 浆粕状短纤维
pulu 氆氇(呢)
pulu cloth 氆氇呢
Puma 彪马;美洲狮毛皮
pumbi 柔软绢丝
Pumi ethnic costume 普米族民族服
Pumi nationality's costume 普米族服饰
pump 潘普鞋,包鞋,女式无带浅口鞋,无带轻软舞鞋,橡胶底浅口帆布网球鞋
pump compass 小圈圆规
pump iron 泵式熨斗
pumpkin 南瓜橙(橙色)
pumps 潘普鞋;无带轻软舞鞋;浅口无带皮鞋;橡胶底浅口网球鞋;圆头鞋
pump shoes 轻便舞鞋
punasa Pratti cotton 印度普那沙普拉提棉
punch 冲模,冲头;打孔,打眼;冲印;冲印器;打孔机;
punch card 冲孔卡,穿孔卡
punch collar staycutting marks 冲领角薄膜
punched felt rug 针刺植绒毡毯
punch(ed) work 抽绣
puncher 冲头;打孔工;打孔器
punch hole 孔位记号
punching （裁片上）锥眼;冲压
punching collar interlining 冲领衬,冲上下领衬
punching collar stay 冲领角薄膜,冲领角衬
punching machine 打眼机
punch machine 冲孔机
punch mark 打孔标记
punchwork 扯纱绣;抽绣
punch work linen 亚麻抽绣布
punch work stitch 抽绣针法
puncture mark 打洞标记
puncture resistance 抗穿刺性
pungee 山东府绸,茧绸,仿茧绸
punjab-American cotton 旁遮普美种陆地棉

punjab cotton 旁遮普产美种棉
punjab Deshi cotton 印度旁遮普棉
punjab pants 旁遮普裤(印度)
punjab silk 旁遮普生丝;旁遮普绸
punjab wool 旁遮普羊毛
punjam 低级生丝;马德拉斯坚固棉布
punk 朋克;朋克前卫风格
punk culture 朋克服饰文化
punk fashion 朋克式样;朋克款式;朋克摇滚装
punk hairstyle 朋克发型
punk look 朋克风格,朋克型款式
punk style 朋克式样,朋克款式
punta [arenas] wool 蓬塔绵羊毛
puntilla 窄花边
punto 线迹;针绣花边
punto a festone 意大利锁眼针迹
punto a fogliame 针绣凸纹花边
punto a gioie 串珠针绣花边
punto a groppo 流苏花边
punto a maglia quadrata 方网眼地花边
punto a piombini 梭结花边
punto aquila 亚麻梭结花边
punto a tuli 棉网眼纱
punto a vermicelli 螺旋纹梭结花边
punto calabrese 亚麻地挖花绣品
punto cardinale 高级针绣花边
punto catena [意]绞花链状绣,钢绳链状绣
punto ciprioto 稀疏针迹;金银线花边
punto d'arcato 锯齿边抽绣花边
punto d'avorio 意大利牙雕针绣花边
punto de Aquja 西班牙古哈针绣花边
punto de bobine 梭结花边
punto de cantella 扣眼线迹针绣花边
punto dei nobili 高级针绣花边
punto de malla 网眼织物
punto de media 袜用网眼织物
punto de napoli 粗圆网眼地梭结花边
punto de neve 雪花花边
punto de Ragusa [法]拉古萨花边
punto de spagna 西班牙花边
punto de tul 网眼织物
punto d'hungaria 帆布刺绣
punto di burano 无网地针绣花边
punto di genoa 无网地针绣花边
punto di ragusa 螺旋纹梭结花边
punto di rapallo 圈边梭结花边
punto di rose 玫瑰花纹六角网眼花边
punto disfatton 抽绣花边

punto di zante	抽绣花边
punto fiamengo	雕绣花边,抽绣花边
punto gaetano	抽绣窄花边
punto gotico	罗马花边
punto greco	雕绣花边,抽绣花边
punto in Aria	针绣挖花花边
punto in stuora	梭结网地刺绣花边
punto ingarseato	仿纱罗花边针迹
punto lace	西班牙花边,意大利花边
punto moresco	流苏花边
punto reale	挖花花边
punto ricamato a maglia	粗线亚麻梭结花边
punto rilevato	花边和刺绣中突出的线迹
punto rinascente	帆布刺绣
punto sopra punto	花边中的凸起线迹
punto spagnuolo	绸绣花边,抽绣花边
punto surana	东方花型抽绣花边
punto tagliato	挖花花边
punto tagliato a fogliame	针绣凸纹花边
punto tirato	抽绣花边
punto trapunto	梭结网地刺绣花边
pupil	瞳孔
pupil's hosiery	学生袜
pupil's wear	小学生服
puppy teeth	犬牙花纹
purbbi	最优级生丝
purchase	购买
purchase confirmation	购货确认书
purda(h)	[印地]印度面纱
purdah	帏幔,帘子;印度高级面纱;印度蓝白条棉布
pure aluminum silicate fiber	高纯硅酸铝纤维
pure bamboo fiber	竹原纤维
pure bamboo fiber fabric	竹原纤维织物
pure bamboo fiber/silk blended fabric	竹原纤维/丝混纺织物
pure bamboo fiber/wool blended fabric	竹原纤维/毛混纺织物
pure Berlin blue	纯柏林蓝
pure color	纯色
pure cotton handkerchief	纯棉手帕
pure cotton super density fabric	纯棉超高密织物
pure cotton yarn	纯棉纱
pure dye silk	纯丝;真丝染色绸
pure dye(d) silk	不增重的染色丝绸,真丝染色绸
pureepuz	印度普里普茨毛圈绒地毯
pure fabric	纯纺织物
pure finish	清水整理
pure gauze	真纱罗
pure gold	纯金
pure gold thread	真金线,细银丝外镀金线
pure indigo	海昌蓝,纯靛蓝
pure mild soap	纯肥皂,纯皂液
pure purple	纯紫色
pure ramie cloth	纯苎麻布
pure ramie fabric	纯蓝麻布,纯苎麻布
pure ramie plains	纯(苎)麻平布
pure ramie sheer	爽丽纱
pure ramie white sheeting	纯麻漂白细布
pure raw fabric	纯纺织物
pure red	大红,大红色
pure silk	纯丝;纯丝绸,真丝绸
pure silk brocade	真丝花缎,真丝锦缎
pure silk embroidered clothing	真丝绣衣
pure silk goods	真丝织物,纯丝织物
pure silk material	真丝料子
pure silk necktie	真丝领带
pure silk quilt cover	真丝被面
pure silk tie	真丝领带
pure stainless steel fiber	纯不锈钢纤维
pure starch finish	纯淀粉上浆整理
pure trad	纯粹传统款式
pure white	纯白,纯白色
pure white and translucent	洁白透明
pure white snow	纯白雪色
pure womenswear	纯女装博览会(英国)
pure wool	全毛;纯毛
pure wool fabric	纯毛织物
pure wool fancy suiting	全毛花呢
pure wool gabardeen	全毛华达呢
pure wool gabardine	全毛华达呢
pure wool knitting yarn	纯毛毛线
pure wool mark	纯羊毛标志
pure wool melton	全毛麦尔登
pure wool overcoating	全毛大衣呢
pure wool plush	全毛长毛绒
pure wool unitting yarn	纯毛绒线
pure wool yarn	纯毛纱
pure yarn	纯纺纱线
pure yarn fabric	纯纺织物
purfle	衣边,镶边,刺绣镶边,刺绣花边的边纹;饰带
purified cotton	消毒棉,吸水棉
purified linter	精制棉短绒
purifying finish	防臭整理

purism 纯粹主义
puritan 珀里坦里子呢,棉经毛纬上光布
puritan collar 清教徒式领,白色方角披肩平领
Puritan costume 清教徒服式
puritan hat 清教徒帽(圆锥形宽大帽檐的黑色高呢帽)
puritanical collar 小斗篷领(清教徒领)
Puritanism 清教徒主义
purity 风格纯正
purl 流苏,金银绉边,花边,珠边,边饰;刺绣;双反面针织物
purl bartacking 花边套结,花边加固
purl circular knitting machine 圆形双反面袜机
purl edge 流苏边,穗边,荷叶边
purl edge stitch 荷叶边线迹
purl edging 绣边
purl fabric 双反面针织物
purl flat knitting machine 平板横机,平型双反面袜机
purl knit 反针;双反面针织物
purl knit cardigan 双反面开襟绒线衫
purl knitting machine 平板横机,双针筒圆形针织机
purl loop 反面线圈
purl side 技术反面,工艺反面
purl stitch 荷叶边缝线迹,曲折锁边缝线迹,三角线迹;双反面线圈组织
purl stitching 曲折扣眼锁边缝纫,曲折锁边线缝
purl stitch machine 平板机,双针筒圆形机
purl stitch pattern 双反面组织提花
purple 大红袍(红衣主教);紫色,紫红色;紫色布;紫衣,紫袍
purple (P) 紫色
purple band 紫色镶边
purple black 墨绿红
purple blue (PB) 青紫
purple border 紫色镶边
purple bronze 紫铜色
purple cloak 紫色大氅
purple cloud 紫云色
purple deep 暗绛红色
purple drab 雪青
purple dusk 暗紫色
purple gray 淡紫灰色
purple grey 淡紫灰色
purple haze 紫雾色

purple heather 石南紫
purple light 浅绛红色
purple orchid 紫蓝花色
purple robe 紫袍
purple sage 灰紫色
purplish 略呈紫色的,微紫色
purplish blue 藏蓝,士林蓝,紫光蓝,紫蓝
purplish red (PR) 枣红,紫光红色
purply 略呈紫色的,微紫色
purpose made 定做的,特制的
purse 女用手提包(美),女用小包,小钱袋,腰包
purse belt 悬挂钱包的皮腰带,钱包带
purser collar 巴莎领
purse silk 粗软光滑的刺绣丝线
purse twists 黄色仿金丝线
push button 电纽;按纽
push-button boots 按扣靴
push-button rubber boots 按纽雨靴
purshed velvet 网形花丝绒
pushing prints 透印双面花布
pushmina 羊绒,山羊绒
pushmina hair 羊绒,山羊绒,开士米
pushmina shawl 羊绒披巾
push out 抹大(行语)
pushum 山羊绒
push-up bra [ssiere] 抬高型胸罩
push-up cups 抬高型乳杯
push-up sleeve 推高袖,上推袖,宽紧袖
PU sole shoes PU底鞋
pussy willow 柔柳绸(一种软薄绸)
pussy willow grey 褪色柳条灰
pussy willow taffeta 柔柳绸(一种软薄绸)
PU(polyurethane) synthetic leather 聚氨酯合成革
put 印度普特栽绒羊毛地毯
putang 窄幅粗棉布
put clothes into carton 装箱
put clothes into polybag 入胶袋
putia 标明地位等级的绣品
put in storage 入仓,入库
put into different categories 分门别类
put kraft paper into carton 垫防潮纸
put off 脱下
put on 穿上,戴上
puttee 裹腿,护腿,绑腿
puttie 绑腿,裹腿,护腿
put together 西装款式组合新技巧
put together look 重新组合衣着的风貌,

新组合款式,个性化组合搭配风貌
puttoo　帕图粗羊绒呢
putty　油灰色(淡灰褐色)
pu-tu　幞头(中国古代男子头饰)
PU two-tone coating　PU双色涂层
PU wax coating　PU蜡光涂层
puy lace　法国佩伊梭结花边
puyuenchow　山东府绸,茧绸
puyuenchowkin　方形茧绸
PVA fiber　聚乙烯醇纤维
PVAL　聚乙烯醇系纤维国际代码
PVC　聚氯乙烯
PVC artificial leather　聚氯乙烯人造革
PVC baby pants　聚氯乙烯塑胶婴儿尿裤
PVC belt　PVC皮腰带
PVC coating　PVC涂层
PVC fusible interlining　聚氯乙烯(PVC)热熔衬
PVC jacket　聚氯乙烯仿皮夹克
PVC label　皮牌(PVC皮牌)
PVC leather　PVC革[皮],聚氯乙烯合成革
PVC leather boots　PVC皮靴
PVC leather handbag　PVC皮手袋
PVC long clothing　PVC长外套
PVC long coat　PVC长外套
PVC/polyester fabric　(做雨衣用)单面胶布
PVC/polyester raincoat　单面胶雨衣
PVC/polyester/PVC fabric　(做雨衣用)双面胶布
PVC/polyester/PVC raincoat　双面胶雨衣
PVC poncho　PVC雨披
PVC raincape　聚氯乙烯塑胶雨披
PVC raincoat　PVC雨衣,聚氯乙烯塑胶雨衣
PVC shell coat　聚乙烯塑胶里衬雨衣
PVC sole　聚氯乙烯鞋底
PVC sole shoes　PVC底鞋
PVC sponge leather　PVC软皮,聚氯乙烯软皮,聚氯乙烯涂料海绵布
PV fabric　聚乙烯织物

PV film　聚乙烯薄膜
PV mixed fancy suitings　涤黏薄花呢
PV two-tone coating　PV双色涂层
PWA　丝披肩巾(缅甸)
pwa　丝手帕
pygal　臀部
pyjama(s)　宽松裤(印度和巴基斯坦的伊斯兰教徒);宽大服装(印度);女子休闲喇叭裤;睡衣裤;沙滩装
pyjama bottoms　睡裤
pyjama check　格子睡衣布
pyjama clothing　睡衣
pyjama flannel　睡衣用法兰绒
pyjama jacket　夹克式睡衣,睡衣上装
pyjamas　睡衣裤;宽松裤
pyjama stripes　宽条子睡衣布
pyjama top　睡衫;睡衣上装
pyjama trousers　睡裤
pyramidal　聚酯三角形截面含陶瓷微粒短纤维(商名,日本东洋纺)
pyramid checks　宝塔式格纹
pyramid coat　金字塔形大衣;窄肩宽摆帐篷形女外套
pyramid heel　宝塔形高袜跟;倒金字塔跟
pyramid shape　金字塔型
pyramid silhouette　金字塔形轮廓
pyramid splicing　宝塔形加固袜跟
pyramid topper　金字塔形短大衣
pyren　聚丙烯长丝(商名,日本三菱人造丝公司)
pyrenean woll　法国地毯用长羊毛
pyrograph　烙花,烫花
pyrography　烫(烙)花品;烫(烙)花术;烫(烙)花图形
pyromex　聚丙烯腈预氧化纤维(商名,日本东邦)
pyrope [garnet]　红榴石
python　蟒蛇皮
python robe　蟒袍
python skin　大蟒皮
python [skin] robe　蟒蛇皮长袍

Q

qauze bed cover 沙罗床罩
QD zipper 快速组装拉链
Qiana 奎阿纳纤维；脂环族聚酰胺纤维（商名，美国杜邦）
Qiang ethnic costume 羌族民族服饰
qian-shan tussah cloth 千山绸
QIAODAN 乔丹
qiaojin lame 俏金缎
qiapan 袷袢
qi-bao knot 七宝结
Qing dynasty design 清代纹样
Qing dynasty ornament 清代纹样
qing-chun tissue georgette 青春纱
qing-kuai sand crepe 轻快绸
qing-kuai satin 轻快缎
qing-yun burnt-out checked sheer 青云绡
Qipao 旗袍
Qipao petticoat 旗袍衬裙
Qipao style skirt 旗袍裙
qi silk 绮（中国古代丝织物）
qiu-ming twill lame 秋鸣绸
qiviut 麝牛绒，麝牛毛，北极金羊毛
Qixie(high heel) 花盆底鞋
Qixie(low heel) 旗鞋（平底鞋）
QR zipper 快速拆卸拉链
quacha ［西］贵客带
quadrate 方骨；方肌
quadrature-lagging 后移，滞后90°相位差
quadriceps 四头肌
quadric surface 二次曲面
Quadriga cloth 夸德里加高密薄织物
quadrille 法国格子布；法国小格子花纹
quadrille paper 方格纸
quadrille taffeta 格子塔夫绸
quadrochromatic designs 四原色图案
quadrochromatic designs printings 四原色图案印花
quadrochromatic printings 四色印花
quadruplet 四件套
quadruple cloth 四层织物，四层组织
quadtone 四色调
quail 鹌鹑色，茄质
quail-pipe boot 男式软皮高筒靴
quaintise 挖剪装饰
Quaker bonnet 教友派罩帽；奎克罩帽；

贵格罩帽；教友会女帽
Quaker cap 室内系带女帽
Quaker collar 贵格会领（教友会领）
quaker hat 奎克帽（教徒戴的低帽身卷边帽）；教友会男帽；贵格帽；三角帽，教友派大卷边男帽
qualification 合格证明
qualification inspection 合格鉴定检验
qualification test 质量鉴定试验，合格试验
qualified class （产品检验的）合格级
qualified product 合格品
qualified products list (QPL) 合格产品目录
qualin 卡林地毯
qualities 气质
quality 质地，质量，品质，优质；品级，等级；参数；(颜色的)鲜艳度；色饱和度
quality appraisal 质量鉴定
quality assurance 质量保证，品质保证
quality authentication 质量认证制度
quality binding 地毯边带
quality brand 名牌
quality certificate 品质证明，质量证书
quality certification center 质量认证中心
quality circle 质量环
quality class 品质等级，质量等级
quality classification 品质分级
quality control (QC) 质量管理，质量检验，质量控制，品质控制
quality control chart 质量管理图[表]，质量控制图
quality controller 质量检验员
quality controlling on line 在线质量控制
quality control system 质量管理系统
quality cost 质量成本
quality cotton 优质棉
quality evaluation 质量鉴定，质量评价，品质鉴定
quality fabric 优质织物
quality factor Q值，品质因素
quality first 质量第一
quality improvement 品质改善
quality index (Q.I) 质量指标，质量指数
quality inspection 质量检验

quality label	品质标记,品质标志
quality level	质量等级,质量标准
quality management	质量管理
quality mark (Q mark)	品质标记,品质符号
quality monitor	质量监控器
quality number	品号;羊毛品质支数
quality of aspect	外观质量
quality of enterprise	企业素质
quality of fit	配合等级
quality of inspect	外观质量
quality product	优质产品,正品
quality sample	品质样品
quality specification	质量说明书,质量标准
quality standard	质量标准
quality supervision	质量监督
quality symbol	品质标签,品质标记,品质符号
quality system	质量体系
Quality System Accreditation Rule (QSAR)	质量体系国际承认制度
quality system certificate	质量体系认证证书
quality tolerance	品质公差
quality wool	优质毛
quality workmanship	做工精湛
quanta	数量
quantitative regulation of imports	进口限额
quantity	数量,程度,大小
quantity of order	订货量
quantity ratio	配料比
quarry	石矿色
quarter	后帮(制鞋)
quarter blanket	马背小毯
quarter brogues	鞋帮雕花布洛格鞋
quarter-circle skirt	1/4 圆裙,小喇叭裙
quarter-circle sleeve	1/4 圆袖(袖子展开图为 1/4 圆)
quarter corner	后帮耳(制鞋)
quarter diamond	英国精纺呢
quarter-drop match pattern	1/4 阶段连续纹样
quartered cap	男童帽,四片式扁平圆顶帽
quarter goods	夸特织物(以夸特计算宽度的织物)
quartering sleeve	四片袖
quarter-lined	短夹里的,短夹里
quarter lining	胸部衬,肘部衬(肘衬);肩里,后帮里
quarter mark band	1/4 标记带
quarternary hues	由几种原色混合成的颜色,以其中两种原色或一种合成色为主
quarter reinforcement at the top of the toe	加固袜头
quarters	臀腰部
quarter sleeve	1/4 袖,短袖
quarter socks	踝下短袜
quarter thicks	粗纺呢,薄型白色克尔赛呢
quartz (QZ)	石英,水晶
quartz clock	石英电子钟
quartz fabric	石英布
quartz fiber	石英纤维
quartz optical fiber	石英光导纤维
quartz pink	石英玫瑰色
quartz watch	石英表
quashgai rug	波斯羊毛地毯
quasi-movable state show	准动态展示
quasi-shawl	类似围巾的物品
quatre	棉絮,填料,填絮,衬垫
quatre fils	四股线帆布
quebracho	坚木,制鞣革用坚木鞣质;坚木色
quebradinho	巴西木本棉
quechquemet	[西]墨西哥女披肩
Queen Anne Satin	安妮皇后缎(运动服用),运动服用丝缎
queen cord	棱纹棉裤料
Queen cotton	美国皇后棉
queen headdress	王后头饰
queen's cloth	漂白细衬衫布
queen silk	英国皇后绸,皇后绸
queen size	大号(仅次于特大号)
Queensland cotton	澳大利亚昆士兰棉
Queensland hemp	澳大利亚昆士兰大麻
Queensland wool	澳大利亚昆士兰羊毛
queen's mourning	英国单纱白条纹黑呢
queen stitch	回形刺绣线迹
Queenter	聚酰胺荧光单丝(商名,日本)
quenkas	经条手织绸
quercitron	黑栎黄;栎皮粉(黄色植物染料)
quernsey	绒线衫,紧身羊毛衫
quernstone	磨石
queue	发辫,辫子
quichi cotton	凯奇棉
quick	紧绷绷的穿着;紧身(美俚)

quick-access memory	快速存取存储器
quick assets	现金资产
quick change manufacturing	快速产品更新
quick check	快速试验
quick delivery	快交货
quick disassembly zipper	快速组装拉链
quick-drying	快干
quick feel	紧抱感
quick ironing spray	速烫喷雾
quick processing	快速工艺
quick release snap	快速打开搭扣
quick release zipper	快速拆卸拉链
quick response control system	快速反应控制系统,高灵敏度控制系统
quick response sewing system	快速反应缝制系统
quick response system (QRS)	快速反应系统
quick response	快速反应(市场经营)
quick sketching	速写
quick slipstitch	快速缲针线迹
quick style change	快速产品更新
Quickwash Plus technology	快速测定技术
quiet	不华美,素淡;额发
quiet clothes	朴素衣服
quiet color	素净色
quiet comfort	静止舒适性
quiet shade	素净色
quiet suit	闲适套装
quilited jacket	填棉绗绣夹克,填棉刺绣夹克
quility brand	名牌
quilled	折叠的,褶裥的
quill embroidery	北美部落缀绣品,毛饰刺绣
quill feather	水鸟毛
quilling	网眼纱褶裥边饰
quills	粗羽毛
quill work	印第安豪猪毛服装服饰
quilot	蕉麻,马尼拉麻
quilot fiber	蕉麻,马尼拉麻
quilot overcoating	银枪大衣呢,马海毛大衣呢
quilt	被子,被褥;床单,被单;床垫;绗缝被;绗缝
quilt bag	被套
quilt bed sheet	床单
quilt cover[ing]	被面
quilted and embroidered jacket	充棉刺绣夹克
quilted car clothing	填棉短外套,太空服
quilted car coat	绗缝驾车外套
quilted cotton bag	棉绗缝包
quilted cotton vest	绗缝棉背心(马甲)
quilted design	绗缝图案
quilted effect	衲缝效果,绗缝效果
quilted fabric	绗缝织物,绗缝夹层织物
quilted filling	绗缝填料
quilted jacket	充(填)棉刺绣夹克,填棉绗缝夹克
quilted knitted fabric	针织绗缝布
quilted leather	绗缝革
quilted lumber jack shirt	绗缝棉衬衫
quilted pants	绗缝裤
quilted quide bar	绗缝导向规
quilted robe	绗缝棉长袍
quilted seam	绗缝线迹
quilted skirt	衲缝裙子,绗缝裙,填棉裙
quilted stitch	绗缝线迹
quilted stockings	棉袜
quilted suit	风雪装(充填料缝合装);绗缝充填装
quilted vest	衲缝背心
quilted waistcoat	绗缝背心,夹层充填料背心
quilted zipper foot	绗缝拉链压脚
quilter	(缝纫机)绗缝附件,衲缝机,绗缝机
quilter fabric	绗缝夹层织物
quilter zipper foot	绗缝拉链压脚
quilt filling	被褥填料
quilting	衲缝,绗缝;绗针;绗针法;绗棉,绗缝料,绗缝衣物,被褥料;管状褶裥;填棉刺绣
quilting attachment	绗缝附件
quilting cloth	绗缝布
quilting coat	绗缝棉大衣
quilting cotton	棉胎,填絮
quilting foot	绗缝压脚
quilting goods	绗缝物,绗缝制品
quilting lace	绗缝花边,褶裥边饰网眼花边
quilting line	绗缝线
quilting machine	绗缝机
quilting needle	绗缝针
quilting padded coat	紧身绗缝棉短上衣
quilting seam	绗缝
quilting stitch	绗针法

quilt(ing) wadding 填絮, 棉胎
quilting twine 绗缝线
quilting with bound edges 滚条贴边
quilting with wadding 带絮片的绗缝
quilt padding 被套
quilt stitching 绗缝线缝; 绗缝线脚
quilt wadding 被胎, 絮被; 填絮
quilt warp 缝经线
quincunx 梅花式; 梅花形; 五点形
quinette 法国仿驼毛呢
quinite 西班牙丝毛混纺织物
quintain 细亚麻布
quintes 法国优级亚麻布
quintet 五重线
quintise 挖剪装饰
quint uplet 五件套服装
quinze-seize 真丝塔夫绸

quirk 袜子边花花纹; 手套指间斜角镶条
quirk thumb 法国手套, 弯曲形拇指
quissionet 裙环用臀垫
quiver 箭袋
quizzing glass 带柄单片眼镜
qum rug 优质波斯地毯
quoif(e) 头饰; 发型; 修女帽, 考福帽, 白色软布无边帽; 头巾
quota 配额, 定额, 限额
quota for garments 服装配额
quota management 配额管理
quota restriction 配额限制
quotation 报价; 报价单; 行情
quotation sheet 报价单
quotidian look 日常款式
quviut 麝牛毛, 麝牛绒(从麝牛下腹部取得)
Q Value 优值

R

R(red) 红色;红衣;红衣料
[R] (registered trademark) 注册商标
rabanna 马达加斯加手织粗织物
Rabat 拉巴领（牧师礼拜袍服上系带的翻领）;（教士服上由衣领垂下的）黑色胸带;衬衫前胸
rabatine 拉巴丁领
rabato 拉巴托领,(16～17 世纪有宽饰边的)大竖领,硬领（常竖颈后,也可翻折覆盖双肩）,轮状皱领的支撑物
Rabat rug 摩洛哥拉巴特地毯
rabattue 法国拉巴蒂轻薄亚麻布
rabbi 衬衫前胸,黑色垂带
rabbit (fur) 兔皮;兔毛
rabbit hair 兔毛
rabbit hair cloth 兔毛呢
rabbit hair fiber 兔毛纤维
rabbit hair knitwear 兔毛衫
rabbit hair overcoating 兔毛大衣呢
rabbit overcoating 兔毛女大衣呢
rabbit's ear bow 缚结饰
rabbit skin 兔毛皮
rabbit wear 兔毛皮服装
rabbit yarn 兔毛纱
ra(c)coon 浣熊,浣熊毛皮
raccoon coat （剪过毛的）浣熊毛皮大衣
raccoon dog fur 貉毛皮,貉绒,浣熊毛皮,浣熊绒
raccoon [fur] 浣熊毛皮
raccroc stitch 布鲁塞尔花边用的细洁缝迹
race cloth 赛马服,赛马披布
racer-back tank T字形后身的背心
raceway fabric 几何形针织物,交叉集圈针织物
racing 赛马场风衣,赛马用风衣;赛马用服饰
racing gloves 竞赛手套,赛马手套,易动手套
racing jacket 赛车夹克（防风、防雨,洗可穿,拉链门襟,抽绳下摆,双股尼龙,条纹）
racing neckline 运动领口
racing shoes （同 running shoes） 田径鞋,钉鞋

racing suit 比赛服,竞赛泳装,竞赛套装上下相连
rack 搁物架,支架,机架,挂ች架,衣架,线轴架;仔兔子皮;腊克(经编密度单位)
rack drying 架上晾干
racked and tucked pattern 扳花集圈花纹
racked full gauge 1+1 rib （羊毛衫）四平扳花
racked half-milano rib stitch （羊毛衫）三平扳花
racked pattern 波纹图案;波纹花边
racked stitch 波纹线迹;波纹组织;扳花组织
racket 球拍形雪地鞋,网球拍靶
rack stitch 波纹线迹
Racop warp knitting machine 拉克普经编机
racquet [ball] wear 壁球装
radames 拉丹绸
radar figure of management 经营雷达图
raddle cheeks 浓抹双颊
radia 法国里昂拉迪阿蚕丝薄绸
radial composite fiber 放射型复合纤维
radiale 桡骨点
radial mark 经向号,径向号
radial point 辐射中心点
radial seam 径向接缝
radian 弧度
radiance 深粉红色
radiant 容光焕发;深粉红色
radiation heat resistant fabric 防辐射热织物,防热辐射织物
radiation-proof glass fiber 防辐射玻璃纤维
radiation-proof overall 防辐射工作服
radiation-proof suit 防辐射服
radiation protecting cloth 防辐射布,防射织物
radiation protection textile 防辐射纺织品
radiation resistant fiber 抗辐射纤维,耐辐射纤维
radiation resistant finish 防辐射整理
radiation resist textile 防辐射纺织品
radio 光亮绸（法国西服料）
radioactive fabric 放射性织物

radiole point 桡骨点
radiological finish 辐射整理
radiopaque fiber 防辐射纤维
radio punch work 团花抽绣
radio-ulna 桡尺骨
radium 镭绸
radium silk 加捻电力纺
radium taffeta 镭锭塔夫绸
radius 桡骨；半径
radiux chiffon 拉迪克斯雪纺
radizelli 针绣挖花花边
RADO 雷达
Radnor cloth 拉德纳丝光棉纱提花装饰织物
radsimir 拉西米尔丧服绸
radya cotton 印度拉德耶棉
radzimir 拉西米尔丧服绸；绫（女服用，丝的密织物）
raffia lace 酒椰叶纤维花边
rag 碎布,碎呢,碎料；破旧衣服
ragbag 装破布袋
rag business 服装业（俚语）；外衣业
rag carpet 碎呢地毯
rag fair 旧衣市场
rag game 服装业（俚语）
ragged 衣衫褴褛的
ragged clothes 乞丐装；百衲衣
ragged edge 碎布,不整齐的布边
ragged look 褴褛风貌,褴褛风格
ragged selvedge 破边
ragg wool 毛针织品
raglan 棱纹棉天鹅绒；包肩,插肩,连肩,斜肩；套袖大衣；连肩（插肩,斜肩）袖大衣；连肩（袖）,插肩
raglan and yoke 连前后育克的套肩大衣,育克式插肩袖
raglan back sleeve 后连肩袖
raglan cape 连肩型披风,连肩袖披肩
raglan clothing[coat] 连肩袖外套
raglan-coat 连肩袖外套,套袖外套,插肩袖大衣
raglancord 棱纹棉天鹅绒
raglan Dolman sleeve 插肩德尔曼袖,套肩蝙蝠袖
raglan front sleeve 前连肩袖
raglan overcoat 插肩大衣,套袖大衣
raglan shoulder 插肩
raglan sleeve 包肩袖,插肩袖,连肩袖,斜肩袖,套肩
raglan sleeve coat 套袖大衣,插肩袖大衣

raglan sleeve fashioning 斜袖全成形
raglan sleeve outline 前后连肩线
raglan sleeve shirt 连袖衬衫
raglan slope line 连肩袖袖窿斜线
raglan smock 插肩袖罩衫,连肩袖罩衫,套袖工作服
raglan three piece sleeved overcoat 连肩三片袖外套
raglan with dart sleeve 有袖裆的连肩袖,有肩省的插肩袖
raglan with seam sleeve 有分割线的插肩袖
raglan with shoulder dart 有肩缝的连肩袖
rag look 破旧款式
ragman 收买烂布者,破布商
rag rug 碎呢地毯
rag trade 服装业（口语）
rag trader 服装商,服装零售商（口语）
ragusa guipure 腊纠萨镂空花边（南斯拉夫制）；罗马雕绣；拉古萨雕绣
ragusa lace 克罗地亚花边,拉古萨斜绣花边
ragyogo 匈牙利传统新娘头饰
rahben Knitting machine 集圈针织横机
rah rah suit 啦啦队套装,学生套装
rahri 拉里粗哔叽；拉里粗毛毯
railroad canvas 重浆整理十字布
railroad stripe 黑白条纹,铁路条纹
railroad trousers 金线条裤
railway bill of lading 铁路提单
railway consignment note 铁路托运单
railway head sewing machine 路轨式缝纫机
railwaymen's uniform 铁路制服
railway pocket 旅行钱袋
railway repp 火车绒,火车座垫绒头织物
railway sacks 粗麻袋
railway stitch 钩编针迹,链式针迹
raiment （书面语）衣服；服装；古装
rain bonnet 软雨帽,折叠雨帽
rain boots （半高筒）雨靴
rainbow 彩虹色
rainbow colors 彩虹色,七色光谱
rainbow effect 色混效应,虹彩效应
rainbow shell button 彩虹贝壳扣
rainbow stripe 彩虹条纹
rainbow yarn 彩虹线,虹彩花线
raincap 雨帽
raincape 斗篷,雨衣,雨披
raincloak 斗篷雨衣；(带帽)防雨斗篷
raincloth 防雨布

raincoat 雨衣,风雨衣	raised knitted fabric 起绒针织物
raincoat yarn 雨衣纱	raised knitting yarn 拉毛绒线
raindress 雨装,雨衣	raised left side 反面拉绒
rain drop repellency 防雨性能	raised line 凸条
rain drop test 雨淋试验	raised loop 起毛线圈
rain-dust coat 防雨防灰罩衣	raised neck 竖立式领;(针织衫的)高圆套领
raingear 雨衣	
rainhat 雨帽	raised neckline 抬高的领口线(相对于正常的领口线)
rain jacket 防雨夹克;雨衣	
rain or shine coat 风雨衣	raised overcoating cloth 起毛大衣呢
rain poncho 雨披	raised pattern 浮雕花纹,凸纹
rainproof 雨衣,雨披;防雨的,防水的	raised pile 立绒,丝绒
rainproofing 防雨整理	raised pile overcoating 立绒大衣呢
rainproofness 防雨性,防雨效能	raised point 凸纹针绣花边
rainproof suit 防雨西装	raised printing effect 立体印花效应
rainproof type 防雨型	raised satin stitch 衬垫缎纹针迹
rain shoes 雨鞋	raised scarf 拉绒围巾
rain slicker 大而宽的油布雨衣	raised seam(同 seam run-offs) 凸缝,线缝隆起
rain spots 雨渍	
rain stains 雨渍	raised sheet 拉绒床单
rainsuit 雨衣,雨衣套装	raised stripe 凸凹条纹;起绒凸条,起毛凸条
rainsuit for motorcycle rider 摩托雨衣(骑摩托穿)	
	raised style printings 凸版印花,凸纹[方式]印花
rain test 防雨试验	
raintight 防雨,防水	raised style 起绒织物
rain trousers 雨裤	raised thread 浮丝
rainwear (总称)雨衣,雨披;防雨服装	raised under side 背面起毛织物
rain wear fabric 雨衣布	raised velvet 凸花丝绒
rainy daisy 雨天短裙	raised weft knitted fabric 纬编起绒针织物
raion 人造[纤维素]纤维,人造丝(意大利名称)	raised work 小枝花纹凸边
	raised woven fabric 起绒织物
raised back 单面绒布;起绒夹里,毛绒夹里	raisin(brown) 葡萄干色
	raising (布料)起绒(工艺),拉绒,抓绒
raised bed sheet 拉绒床单	raising against the hair 反[毛]向起绒
raised blanket 拉绒毛毯	raising against the nap 反向起绒
raised brown 葡萄干色	raising band 拉绒损伤疵
raised buttonhole 凸纽孔	raising confusion 绒头凌乱
raised checks 浮纹格子布	raising defect 刮绒伤,起毛疵
raised crossover rid 十字凸花纹	raising density 绒毛密度,起绒密度
raised effect 凸纹效应	raising gingham 格绒布
raised embroidery 凸花刺绣	raising plain 低级平布,平纹绒布
raised fabric 经编绒布;绒类织物,起绒织物	raising satin 拉绒缎
	raising streak 起绒条纹
raised finish 拉绒整理,起绒整理	raising teasel machine 刺果起毛机
raised fishbone stitch 凸纹鱼骨绣	raising with the hair 顺[毛]向起绒
raised flat bed sewing machine 突起式平底板缝纫机	Rajah 柞丝绸
	rajah collar(同 Nehru collar) 印度拉惹立绒,尼赫鲁立领,罗闍领
raised gray fabric 起毛织物坯布	
raised jacguard design 凸纹提花花纹	rajah jacket 拉惹上衣,仿尼赫鲁外套,罗闍上衣
raised knit 起绒针织物	

rajah neckline 立领式领圈(立领的变化领圈)
rajah pants (印度)邦主短裤
rajah style (印度)邦主服式
rajah suit 邦主套服
rajeta 粗色织布
Rajputana cotton 印度拉普塔纳棉
RAKAM 《拉凯姆》杂志(意)
rake 钉鞋
ralchak 银灰色山羊绒
Ralph Lauren 拉尔夫·劳伦(1939~,美国时装设计师)
Ralph Lauren/Polo 拉尔夫·劳伦/马球
ramage 花卉图案;绉纹花毛葛;丝绉
rambler rose stitch 攀缘蔷薇针迹
Rambouillet wool 兰布莱羊毛
ramch mink 牧场貂皮
ramee 苎麻
Ramese cotton 美国拉梅斯棉
ramie 麻,苎麻,白苎,绿苎,线麻,紫麻
ramie/acrylic blended yarn (苎)麻/腈纱
ramie and linen blending cloth 苎亚麻格呢
ramie cloth 苎麻布;苎麻织物
ramie/cotton blended fabric 麻棉混纺(布)
ramie/cotton blending textile 麻棉混纺纺织品
ramie/cotton denim 麻棉牛仔布
ramie/cotton garments 麻棉服装
ramie/cotton grey plain 麻棉平布坯
ramie/cotton interweave 麻棉交织布
ramie/cotton interwoven fabric 麻交布
ramie/cotton jeans 麻棉牛仔裤
ramie/cotton knitwear 麻棉针织品
ramie/cotton mixed plains 麻棉交织布
ramie/cotton pants 麻棉裤
ramie/cotton shirt 麻棉衬衫
ramie/cotton shorts 麻棉短裤
ramie/cotton sweater 麻棉混纺衫
ramie/cotton yarn sweater 麻棉混纺线衫
ramie fabric 苎麻织物;苎麻布
ramie fiber 苎麻纤维
ramie heald twine 麻通丝线
ramie hosiery 苎麻交织袜
ramie knitwear 苎麻毛衫,苎麻针织品
ramie lining 苎麻衬布
ramie plains 苎麻平布
ramie/polyester blended yarn (苎)麻/涤纱
ramie/polyester(cotton)blended cloth 麻/涤(棉)混纺织物;麻/涤(棉)布
ramie sewing thread 苎麻缝纫线
ramie shirting 麻衬衫料;苎麻衬衫料,苎麻细平布
ramie/silk blended yarn (苎)麻/丝纱
ramie single jersey 苎麻汗布
ramie thread 苎麻线
ramie tow 苎麻条
ramie twill 苎麻斜纹布
ramie/viscose blended yarn (苎)麻/黏纱
ramie/viscose cloth 麻/黏混纺布
ramie white sheeting 苎麻漂白细布
ramie/wool blended yarn (苎)麻/毛纱
ramie yarn 苎麻纱
ramie yarn dyed cloth 苎麻色织布
Ramillie(wig) 拉密里假发
Ramona 拉莫纳平纹彩色棉衣料
rampart 壁垒式线迹
rampur chudder 兰普尔手织披肩呢
rams wool 公羊毛
ranch clothing 牧场外套
ranch coat 牧场大衣,牧场外套,牧人外套
ranched mink 养殖貂皮
rancher(o) coat 牧人外套,牧场外套,牧场大衣
ranch jack 牧场夹克
ranch mink 人工饲养貂皮
ranch wear 牧工服
rand 鞋后跟垫皮,(后)掌条
random access device 随机存取设备
random access memory 随机存取记忆《计算机》
random copolyamide fiber 无规共聚酰胺纤维
random dyeing [纱线]多色间隔染色,多色局部染色,扎染
random fluctuation 随机波动,偶发波动《贸易》
randomized conjugate fiber 无规[共轭]复合纤维
randomized conjugate yarn 无规[共轭]复合长丝
randomized pile 随机起绒,无规起绒
random laid print-bonded nonworven 无规网印花黏合非织造布
random linking 针织料拼缝
random looping 针织料拼缝
random-loop yarn 不规则毛圈花线
random pleats 随意褶

random sample 随机样本
random-scan device 随机扫描装置《计算机》
random-scan graphical display 随机扫描图形显示器《计算机》
random scatter 漫射
random shear 不规则剪毛地毯
random slub 不规则粗节花线
random stitch embroidery 乱针绣
random stripe 不规则条纹
random yarn 无规彩色丝；断续色芯花线
rang 蓝格薄细布
range 范围；色阶，色程；同类羊毛；同类织物
ranger BB 53 cotton 美国西南地区陆地棉
range of product 产品品种，产品范围
range of shade 色泽分布范围
Ranger's hat 兰杰帽
range wool 美国西部羊毛
rangoon cotton 仰光棉
rangoon hemp 仰光苘麻
rantering 暗缝；织补
rapid-access 快速[存取]存储器
rapid dyeing 快速染色
rapid fast dye 快色素染料
rapid iron 快烫，稍烫
rapidogen dye 快胺素染料
rapid press 快速熨烫，快速熨烫机
rapid-sequence camera 快速连续摄影机
raploch 原色粗呢（苏格兰）
rapolin 瑞士帽饰带
rapture rose 迷人玫瑰红
ra-ra 拉拉裙
rare-faction （空气）稀薄状态
rark （苏格兰）衬衣
ras 素色毛哔叽，平纹素色起毛哔叽
raschel 拉舍尔经编织物
raschel blanket 拉舍尔毛毯
raschel carpet 拉舍尔经编地毯
raschel cord 拉舍尔经向凸条织物
raschel crepe fabric 拉舍尔起绉织物
raschel-crochet fabric 拉舍尔钩编织物
raschel-crochet machine 拉舍尔钩编机
raschel curtain net machine 拉舍尔窗纱经编机
raschel double needle bar fabric 拉舍尔双面经编织物
raschel double rib fabric 拉舍尔双罗纹织物

raschel fabric 拉舍尔经编织物
raschel knit 拉舍尔针织物
raschel lace 拉舍尔（经编）花边；拉舍尔网眼织物；拉舍尔饰带
raschel loom 拉舍尔经编机
raschel machine 拉舍尔经编机
raschel plush 拉舍尔经编长毛绒
raschel power net machine 拉舍尔弹力网眼经编机
raschel single needle bar fabric 拉舍尔单面经编织物
raschel warp loom 拉舍尔经编机
raschel warp-knitting machine 拉舍尔经编机
raschel weft insertion 拉舍尔经编机衬纬
ras de cypre ［法］西波尔横棱绸
ras de florence 佛罗伦萨斜纹花呢
ras de maroc 摩洛哥薄哔叽
ras de perse 轻起毛制服呢
ras de saint cyr 丝经哔叽
rasete 纬面缎纹
rasha 醋酯经黏胶纬织物
rashuina wool 拉雪纳白色山羊绒
rasi 低级净棉
rasi cotton 印度拉希棉
raso 经面缎纹
raspberry 紫莓色，木莓色（暗红），紫红色，紫绛色
raspberry red 木莓红
raspberry rose 树莓红
raster 光栅《计算机》
raster display 光栅显示《计算机》
raster plotter 光栅绘图机《计算机》
ratal 纳税额
ratas 金线织物
ratcatcher 狩猎装，打猎便装
ratchet buckle 弹簧环扣
ratchet spring and hook 箱包用弹簧卡扣
rateen 纯毛哔叽
rate fixing 制订定额《管理》
rate of taxation 税率
ratine 勒丁，平纹结子花呢，珠皮大衣呢，机制粗棉纱花呢
ratine drape 珠皮呢
ratine effect 花线结子效应
ratine fries 结子花呢
ratine lace 机制粗纱花边
ratine yarn 结子线，花圈线
ratio 比率；（服装按花色、尺码）装箱配比
ratio-delay study 窝工比率研究《管理》

rational dress 骑车套装(英国妇女骑车时所穿的宽短裤及其相配的服装)
rationalization of operation 操作合理化《管理》
ratite 有平胸的
rat-tail (镶边用)鼠尾带
rat-tail braid 鼠尾形丝带
rat-tail comb 鼠尾梳
rat-tail cord 机织空心带
rattan 浅橘黄,浅橙,藤条色
rattan handbag 藤编手提袋
rattanhandle 藤柄
rattan hat 藤帽
rattan helmet 藤帽
rattan yellow 藤黄
ratteen 粗毛呢,平纹结子花呢,珠皮大衣呢
ratti coating 拉蒂厚斜纹呢
rattine lace 起圈花边
rattinet 法国薄珠皮呢
ratton 浅橘黄,糙米黄
rattvik lace 打褶花边
raumois 法国本色粗亚麻布
rave(l)ling (裁剪后织物裁边纱线脱出的)散开的纱,拆散的纱
ravelling of knitted fabric 针织物脱散
ravel-proof double chainstitch 防散脱双链线迹
ravel-proof piping 防散脱滚边
raven locks 乌黑的头发
ravensduck 棉帆布;18世纪俄国亚麻厚帆布,英国重磅亚麻帆布
rawa 拉瓦真丝面纱;旁遮善真丝面纱
raw cotton 原棉
raw edge 毛边,坯边
raw edge exposed 露毛边
raw edge hemming 毛边缝边,毛边卷边
raw edge leans out of seam 毛露,毛头
raw edge out 毛露
raw fabric 本色织物
raw fiber 纤维原料
raw flax 原亚麻,生亚麻
raw fur 生毛皮
raw hair 原毛
raw hair and underfleece of yak 牦牛原绒(西藏牛绒、马尾牛绒)
raw hemp 原大麻,生大麻
raw hide 生皮
raw jute 原黄麻,生黄
rawkiness 条边,条花疵,条痕
raw mamie 原苎麻,生苎麻

raw material 原料
raw muslin 原色细布
raw raising 煮练坯布拉绒,坯布拉绒
raw seam 毛缝,粗缝
raw sienna 生赭石色,浓黄土色,鲜黄棕色
raw silk 生坯丝绸;生丝;厂丝
raw silk fabric 生丝丝绸,生坯丝绸
raw stock 原料,未加工纤维
raw thread milling 原坯缩绒,原坯缩呢
raw thread raising 原坯起绒
raw umber 棕土棕,红棕色,赭色
raw white (RW) 原色,本白色
raw wool 原毛
raw wool of goat 山羊毛
raw wool of sheep 绵羊毛
raxa 厚毛织物
ray 线条呢;射线,光线
rayadillos 蓝白条纹棉布
Raycott 强力黏胶纤维(商名,日本东邦人造丝公司)
raye 法国细条子花纹
raye romain 色条纹丝带,罗马色条带子(色彩鲜艳)
rayleigh 凸花花边中的不规则横条
Raymahal hemp 印度雷马哈尔大麻
raynes 细亚麻布
rayon 黏纤;人造丝织物;人造丝(Rn),人造纤维
rayon/acetate shot twill 黏闪绫
rayon/acetate(union)shot twill 黏闪绫
rayon alpaca 人造丝阿尔帕卡绸
rayon and cotton lace 丝纱交织花边
rayon back cutting georgette crepe 人造丝提花乔其
rayon back satin crepe 缎背绉,人造丝缎背绉
rayon band 人造丝带,黏胶扁丝,人造扁带
rayon baronet satin 人造丝男爵缎
rayon-based carbon fiber 人造丝基碳纤维
rayon-based composite 人造丝基复合纤维
rayon-based fiber 人造丝基纤维
rayon beihua 蓓花绸
rayon bengaline 人造丝罗缎
rayon blouse 人造丝上衣
rayon brocade 细花绸,提花人丝绸,人丝花绸;人丝织锦缎,摩光缎

rayon canton crepe 人造丝重双绉
rayon cashmere 人造丝开司米绸
rayon challis 人造丝仿毛呢
rayon chiffon 人造丝薄绸,人造丝雪纺
rayon cloth 人造棉布,人造棉织物
rayon/cotton blended yarn 人棉和棉混纺纱
rayon/cotton twill 棉纬绫
rayon/cotton union faille 丝罗葛;一号绨
rayon/cotton union poplin 素毛葛
rayon/cotton union quilt cover 线绨被面
rayon/cotton union upholstery 装饰绸
rayon crash 黏纤粗布
rayon crepe 人造丝绉
rayon crepe de chine 黏胶丝双绉,人造丝双绉
rayon crepe georgette 人造丝乔其
rayon crepe jacquard 茜灵绉,人造丝绉
rayon crepe pajamas 人造丝绉睡衣
rayon crepe petticoat 人造丝绉衬裙
rayon cut staple 人造切段纤维,人造短纤维
rayon elastic suspender 人造丝宽紧吊裤带
rayon embroidered georgette crepe 人造丝绣花乔其纱
rayon fabric 人造丝织物,人造丝罗缎
rayon fiber 人造纤维素纤维
rayon fibre 黏胶纤维
rayon filament fabric 人丝织物
rayon filament [yarn] 人造丝
rayon filling foulard 蚕黏绫
rayon flannel 人造丝斜纹绒
rayon flannel twill 绒面绫
rayon flat crep 人造丝平绉
rayon flock 人造丝短绒(用于植绒)
rayon French crepe 人造丝法国绉
rayon georgette 黏胶丝乔其,人造丝乔其纱
rayon georgette skirt 人造丝乔其纱裙
rayon habotai 人造丝纺绸,人造丝电力纺
rayon habutai 人造丝电力纺
rayon horsehair 人造丝鬃毛
rayon-HP 高性能人造纤维
rayon jacquard crepe 茜灵绉
rayon jersey 人造丝经编乔赛
rayon linen satin 软缎里子,人造丝里子缎
rayon linen twill 美丽绸

rayon lining 羽纱,夹里绸,里子绸,棉纱绫,纱背绫;人丝羽纱,人丝里子纺,黏胶丝羽纱
rayon lining satin 人造丝里子缎
rayon lining silk 美丽绸,人造丝美丽绸,人造丝羽缎,人造丝斜羽绸
rayon lining twill 美丽绸
rayon lining twill staple filling 棉纬美丽绸
rayon lint 黏纤短绒,人造棉绒
rayon luster lining 人造丝羽纱
rayon marquisette 人造丝薄纱罗
rayon materlasse 人造丝凸花绸缎
rayon mignonette 人造丝细结网眼织物
rayon moire 人造丝波纹塔夫绸
rayon monofil 人造丝纤维单丝,人造纤维单丝
rayonné 人造丝,人造[纤维素]纤维
rayon net 人造丝绢网,人造丝网眼纱
rayon/nylon union twill 交织绫
rayon organdy 人造丝奥甘迪,人造丝蝉翼纱
rayon palace 有光纺,人造丝有光纺
rayon palace checked 彩格纺
rayon patterned crepe 丁香绫,花黏绫
rayon patterned poplin 黏胶丝采芝绫,人造丝采芝绫
rayon petite check frock 人造丝小格子长外衣
rayon plain satin 人造丝素缎,人造丝平缎
rayon pointille taffeta 麦尔纱
rayon poplin 人造丝府绸
rayon rhythm crepe 人造丝泡泡绉
rayon ribbon 人造丝饰带
rayon rough crepe 人造丝粗绉,人造丝鸡皮绉
rayon satin 玻璃缎,黏胶丝软缎,人造丝软缎
rayon satin bathing suit 人造丝缎浴衣;人造丝泳衣
rayon satin brassiere 人造丝缎胸罩
rayon satin plain 人造丝素缎
rayon satin stripe pajamas 人造丝条纹缎睡衣
rayon seam binding 人造丝缝口滚条
rayon semi-shameuse 人造丝采芝绫,黏胶丝采芝绫
rayon serge 人造丝棉哔叽
rayon shioze 人造丝无光纺

rayon shirt 人造棉衬衫
rayon shirt printed 人造棉印花衬衫
rayon shot satin 人造丝闪光缎
rayon silk 人造丝
rayon slub yarn 竹节人造丝
rayon soochow brocade 人造丝古香缎,人造丝风景古香缎
rayon staple [fiber] 人造短纤维,黏胶短纤维
rayon staple muslin 人造丝棉麦斯林纱
rayon stockings 人丝袜
rayon strip 黏胶条带,人造扁丝,人造丝条带
rayon striped batiste 缎条青年纺
rayon striped brocade 人造丝条子提花绸
rayon striped gauze 人造丝条花纱
rayon striped lustrine 人造丝条子花绡
rayon suiting 人造丝西服料
rayon taffeta 人造丝塔夫绸
rayon taffeta faconne 领夹纺
rayon taffeta pointille 麦浪纺
rayon tapestry satin 人造丝织锦缎
rayon/tinsel mixed tapestry satin 金银人造丝织锦缎
rayon transparent velvet 人造丝透明立绒
rayon tricolette 人造丝轻质网眼针织物
rayon trimming 人造丝装饰带
rayon triple sheer 人造丝半透明薄绸
rayon trousers 人棉裤
rayon tulle 人造丝绢网,人造丝网眼纱
rayon twill 人棉斜纹绸,人造丝斜纹绸
rayon twill damask 人造丝花绸
rayon underwear 人造丝内衣
rayon velvet 雪维绒,人造丝绒
rayon velveteen 人造丝平绒
rayon velvet suit 雪维绒套装,人造丝绒套装
rayon voile 轻丝绡
rayon yarn 黏纤纱,人棉纱,人丝纱
rayon yarn embroidery thread 人造丝绣花线
raypour 印度生丝
ray-stitch 辐射型绣花线迹,双边抽丝线迹
ray tracing 射线跟踪《计算机》
rayure bayadeur 法国棉经纬横条纹织物
raz 紧贴剪毛单色(本色)织物,法国平滑短绒织物,素色毛哔叽,平纹素色起毛哔叽
razor-edge 鲜明轮廓线

razor knife 美工刀
RB (regular bright) 普通有光
RD (regular dull) 普通消光
reach-me-down (同 hand-me-down) 成衣,现成廉价服装,旧衣服
reactive dye 活性染料,反应染料
reactive dyeing 活性染料染色
reactive dye printed handkerchief 活性染料印花手帕
reactive fiber 反应性纤维,活性纤维
reader (注明售价等的)标签
ready made 现成服装,成衣;现成物品
ready made clothes [goods] 成衣
ready-mades 成衣
ready made shop 成衣店
ready-to-wear(RTW, R-T-W) 现成服装
ready-to-wear clothes 成衣,现成服装
ready-to-wear (clothing) 成衣,现成服装
real Axminster carpet 阿克明斯特手织绒头地毯
real dress 实用装,货真价实的服装
real lace (不包括针织和钩编的)手制花边
real net weight 实际净重法
real silk 真丝绸;真丝
real silk fabric 真丝绸,真丝织物
real silk-like polyester fabric 涤纶仿真丝绸
real silk necktie 真丝领带
real suit 实用套装
realia 实观教具,示教实物
realism 现实主义,写实派
reality 现实,逼真
realization 实感,现实,现实化
real-time animation 实时动画《计算机》
reamy yarn 单双花线
reanimalising 丝绸加重处理
rear drape line 后片(部)造型线
rear elevation 后视图
rear end 后部,臀部
rear panel 后片
rear pleat 后背褶裥
reary yarn 单双花线
reb tie 棱纹领带
rebashka 俄罗斯衫
rebate 折扣
rebato 女式白色硬领
rebayn 蓝地金花绸
reboso (西班牙、墨西哥等国妇女用的)头肩大披巾,长围巾

rebound elasticity	回弹性
rebound resilience	回弹性
rebozo	哥伦比亚缩绒呢裙料；墨西哥披肩；西班牙女式头巾
rebs	棱条织物
recaded color	后退色
recall	检索《计算机》
recamo	凸花刺绣
receding colour	远感色，后退色
receipt	收据
recel	西班牙雷赛条子挂毯
recess shearing	凹式剪花
recherché satin	人造丝经毛纬绉缎
reciprocal knitting	往复编织
reciprocating knitting	往复式编织
reciprocating machine	往复式织袜机
reclaimed fiber	回用纤维
reclaimed textile fiber	回用纤维
reclaimed wool	再生毛
reclining twill	缓斜纹
reclothe	使重新穿衣，使重新披上
reconstituted fiber	回纺纤维；再生纤维
reconstituted protein fiber	再生蛋白质纤维
recordonner	镶边
recouvées	法国本色厚亚麻布
recovered fiber	再生纤维；回收纤维
recovered wool	回毛，再生毛
recovery reflective fabric	回复反射织物
recreational jacket	娱乐夹克（在进行高尔夫球、网球等娱乐活动时所穿）
recruit fashion	应聘装
rectangle	矩形，长方形
rectangle handbag	长方形手提包
rectangle line	矩形线型
rectangular bartack	矩形套结
rectangular line	矩型（款式），长方型（款式）；长方形轮廓线
rectangular neckline	矩形领口
rectangular tacking machine	长方形针缝缝接机
recurability	再压烫性
recurrent neutral	反复出现的非彩色
recurrent short seam	往复短缝
recurring fashions	再现时装
recycled fiber	回用纤维，回收纤维
recycled jeans	更生牛仔裤（几条旧裤拆散拼合而成）
recycled manufactured fiber	再生化学纤维
recycled wool	再生毛
red（R）	红色；红衣；红衣料
Red Baron helmet	红男爵飞行员帽盔，雷德·巴伦飞行员帽盔
red belt	（柔道）红腰带
red box calf	红珠皮（小牛皮）
red brown	红棕
red cap	红帽子
red clay	红黏土色
red cloth	（制鞋）红粉带，红布，纳夫妥红布，旗红布
red coconada cotton	科科拿大棉
red collar tab	红领章
red cross gingham	蓝白条色织布，红十字会条子布
red cross nures's cap	红十字会护士帽
red crown	上埃及红王冠
redder	[色光]偏红
red diamond	红色钻石
reddish	带红色的，微红的，淡红的
reddish black	天青
reddish blue	品蓝
reddish gray	红光灰
reddish orange（rO）	红光橙色，橘红
reddish ultramarine	红光群青色
reddish yellow（rY）	红光黄色
red earth	红地球；土红，锈红
redesigning	修改设计，重新设计
red exide	褐色
red fox	赤狐毛皮，红狐皮
red fringe hat	红缨帽
red hat	红衣主教帽
redingote	（双排扣）女骑装式外套；长大衣
redingote	细腰型外套；开襟女式轻外套；（18世纪）双排纽男式骑装长外套；骑装式礼服；前襟镶有对照色三角布片的女装
red jute	红麻
red lamp	红灯，危险信号
red leaf cotton	红叶棉
red mahogany	红木色，赤褐色
red marble	红大理石色
redness	红色
red ocher	红光藏蓝
red orange	柿红，橘红
red peruvian cotton	红秘鲁棉
red purple（RP）	红紫
red ribbon	红绶带；红绶勋章
red riding hood	骑马头巾，兜状连颈帽

red sable 红貂毛皮,亚洲貂皮
red sandalwood 红木色,檀香木色
red satin skirt 红缎裙
red scarf 红领巾
red selvedge melton 红边麦尔登
red shortness 热脆性
red silk cotton 印度红丝棉
red spot cotton 红斑棉
red tape 法官红带
reduced heel 收针袜跟,锥形袜跟
reduced pitch clip 小布夹
reduced scale 缩尺,缩小比例尺
reduced Y-heel 缩小Y字形袜跟
reduce the price 减价
reducing of size 缩小尺寸[码]
reduction dye 还原染料
redundant labour force 剩余劳动力《管理》
red violet 红紫,红紫色
red wood 红木棕,红棕色
red wool 澳洲含红土羊毛
Reebok 锐步
reed green 芦苇绿
reed yellow 芦苇黄
reefer 女式紧身双排纽上衣;双排纽水手上衣;长围巾
reefer clothing 双排扣水手外套
reefer coat 水手外套
reefer collar 帆形叠领,瑞福领,蟹钳领
reefer fabric 夹克用厚呢织物
reefer jacket 女式紧身双排纽上衣
reefer sack double 双排纽西服上装
Reefer suit 瑞福装
reefing jacket (男式)双排纽厚呢短夹克,双排扣水手上衣
reef knot 平结
reel 卷,线轴;伞骨状衣服干燥架
reel daryai 印度女裤布料
reeled pongee 纺绸
reel of thread 木蕊线
reeve needle 穿针
refajo 危地马拉女裙
refashion 翻新,改制,重作,改变形式
referee's wear 裁判服
reference chromatogram 参比色谱(图)
reference line 基准线,参考线
reference plane 基准平面,参考平面
reference width 基准门幅,参照门幅
reference yardage 参考用布量
refin 优选毛

refina wool 西班牙优质美利奴绵羊毛
refine 粗纺绒面呢;精练,精制,提纯
refined and bleached single jersey 精漂汗布
refined waste silk 精干品,精绵
refine look 精炼款式
refinishing 重新整理,复修
refit 重新装修,改装
reflection 反射色,影像;反射,反映
reflective brilliance (作显示用的)反射光辉
reflective clothes 反光服
reflective fabric 反光织物
reflective patch 反光牌(章)
reflective PVC/nylon vest (缝有荧光条带的)单胶安全背心
reflective safety vest 反光安全背心
reflective tape (作显示用的)反射带
reflective trimmed clothing 反射镶边服装
refleuret 优质美利奴绵羊毛
reflex blue 深蓝色
reflexion 反射色,影像
reflex mirror 反光镜.
reflex pattern V形花纹,对称花纹
refolding 再折叠
refoot 换袜底
reform 服装翻新
reformer 经面斜纹厚绒布
refractory fiber 耐高温纤维,耐火纤维
refractory fibrous material 耐火纤维材料
refractory oxide fiber 耐高温氧化物纤维
refreshing feeling 爽快感
rega-bootha 小花卉纹
regain percentage 回潮率
regain standard 标准回潮率
regal attire 王服
regalia (表明官阶的)礼服;(共济会等团体成员的)制服;华服;盛装;徽章;王权标记
regal robe 推检服
regate 水手式领带
regatta 蓝白条粗纺呢;条格棉布,里格特条子布
regatta blazer 划船夹克,赛船运动夹克
regatta stripe 里格特粗条纹
regatta stripes 里格特等宽条子花呢,英国蓝白条纹粗纺花呢
regeance 法国丝经黏胶纬花绸
regeance diagonal 法国丝经黏胶纬斜纹花绸
regence 上光整理领带绸

Regency coat 摄政期双排纽大衣
Regency collar 摄政领(略小于拿破仑领)
Regency costume 摄政服
regency point (19世纪)针绣网眼花边
regency stripe (织物上的)等宽彩色条纹
regenerated animal fiber 再生动物纤维
regenerated bamboo fiber 再生竹纤维
regenerated cellulose fiber 再生纤维素纤维
regenerated cellulose hollow fiber 再生纤维素中空纤维
regenerated cellulose rayon 再生纤维素人造丝
regenerated fiber 再生纤维
regenerated filament 再生丝
regenerated protein fiber 再生蛋白质纤维
regenerated silk yarn 再生绢丝
regenerated wool 再生毛
Regent pump 摄政碰拍鞋
Regent(style) 摄政发型
regimental 灰紫蓝色
regimentals (军队)团的制服,军装
regimental stripe 英国军服色彩条纹样,英国团队条纹,(用于围巾及帽带上)阔条彩色条纹
regina 里律纳细斜纹布
Regina Kravitz 里贾纳·克拉维茨(美国时装设计师)
Regina Schrecker 雷吉纳·施雷克(意大利时装设计师)
regina twill 英国细经粗纬粗精梳斜纹布
region motif 地域主题
register mark 对花记号,十字规矩线
register prints 双面印花
registered design 注册图案
registered feed (织物)对齐给料,织物对齐送进
registered number 服装厂注册编号
registered trademark 注册商标
registered trade name 注册商标
registration 对花;记录
reglan 套袖大衣,插肩袖大衣,插肩,包肩,连肩,斜肩;棱纹棉,天鹅绒
regny 法国雷格尼细亚麻布
regrettas 菲律宾雷格利太窄幅条纹布
regular 一般尺寸,中号;普通体形服装
regular braid 正则编带
regular bright(RB) 普通有光
regular chain 正规连销店《销售》

regular collar 普通(衬衫)领
regular corner 回复角位
regular cut 普通式裁剪
regular dull(RD) 普通消光
regular fiber 正规纤维
regular figure 正常体型
regular finish 一般整理,基本整理
regular folding 规整折叠,正常折叠
regular form 正常体型
regular hip 正常臀形
regular knitting welt 短袜袜口
regular notch lapel 方角驳领,普通西装领
regular pattern 基础纸样
regular point collar 普通尖角领,普通衬衫领
regular rayon 普通人造丝
regular saddle stitch 正规鞍形(刺绣)线迹,普通双针缝线迹,捻针线迹
regular satin 正则缎纹
regular shank button 普通有柄纽扣
regular shoulder 正常肩,平肩
regular size 常规尺寸[码],一般尺寸
regular-sole 正规鞋底
regular staple 普通[化学]短纤维
regular turned up front foot 普通翻襟压脚
regular twill 正则斜纹
regular twist Z捻,反手捻
regular way 顺向
regular zipper 普通拉链,封尾型拉链
regulation cap 制服帽
regulation zipper(同 conventional zipper) 普通拉链,封尾型拉链
rehani 彩边红绸布
reheaarsal shorts 排练短裤
Rei Kawakubo 川久保龄(1942～,日本时装设计师)
Reiko Hirako 平子札子(日本时装设计师)
reindeer hair 驯鹿毛
reinforced button 加固扣
reinforced buttonhole 加固纽孔
reinforced cloth fitting seam 加固布接缝
reinforced crotch 加固裤裆
reinforced design 加固花纹
reinforced hosiery 加固袜
reinforced inverted pleat 加固暗褶裥
reinforced knee 加固膝盖部
reinforced mock safety stitch 装饰安全线迹

reinforced piping 加固滚边
reinforced seam[ing] 加固缝纫
reinforced seams 加固缝,来回边
reinforced seat 加固臀部
reinforced selvage 加固边,袜跟加固边
reinforced stitch 加固线迹
reinforced tape 加固带
reinforced tape attachment 加固带附件
reinforced toe 加固袜头
reinforced turned overedge seam 加固翻边包缝
reinforced twill 加强斜纹
reinforced two-needle overstitching operation 双针加固包边操作
reinforcement fabric 增强用织物
reinforcement fiber 增强纤维
reinforcement for knees (裤)膝盖绸
reinforcement patch 袋牵布,加固布
reinforcement patch for pocket 袋角衬,袋角加固布
reinforcement seam 加固缝
reinforcing button 加固纽扣
reinforcing cord 加固线
reinforcing crotch 加固裆
reinforcing end of seam 缝端加固
reinforcing for knees 贴膝绸
reinforcing gusset 加固三角布料
reinforcing hosiery 加固袜
reinforcing operation 加固操作
reinforcing patch 加固布
reinforcing seam 加固缝
reinforcing sewing 加固缝纫
reinforcing strip 加固带
reinforcing tape 加固带
reinport trade 复进口
rein pressed pleat 柔和褶裥
reja 印度雷加窄条子斜纹布
reject 次品,废品,等外品
rejected product 不合格的产品
rejected tow 等外短亚麻
rejects 等外品,等外毛,下脚麻
Relanit single cylinder knitting machine 雷莱尼特单面针织机
related color 相关色,亲缘色
related shade 相似色调的色泽,深浅相似的色调
relative cover 相对覆盖度
relative density 相对密度
relative firmness (织物)相对紧度
relative humidity 相对湿度

relative metabolic rate (R.M.R.) 能量代谢率
relaxation dimensional change 回缩尺寸变化
relaxation shrinkage 回缩
relaxed clothes 便服
relaxed fabric 松弛织物,弛豫织物
relaxed state of knitted fabric 针织物松弛状态,针织物弛豫状态
relaxed suiting 体闲西装
relaxing conveyor washer 履带松式水洗机
relaxing heat-setter 松式热定形机
relax jet dryer 松式喷射烘燥机
relaxor 卷发矫直剂(美国)
released tuck 钉(缝)住部分的活裥,一头钉(缝)住,一头放松的裥,或中间钉(缝)住,两头放松的裥
release without payment of duty 免税放行
relief 凸纹,浮雕花纹
relief embossing 轧花整理
relief fabric 凸纹[浮浅]针织物
relief knitwear 浮纹毛衫
relief printing 凸纹(辊筒)印花
relief sculpture button 浮雕扣
relief-type bartack 松散式加固,凸纹套结,安全套结
religious medal 宗教垂饰项链
reline 换(衣服)的衬里
relining 重新粘衬
rellay dooray 格纹粗棉布
relustering 再上光整理
remaining thread 余线
remal 卢亩面纱,卢亩头巾
rematch 再配色,重配色
Rembrandt hat 伦勃朗帽
rembrandt rib 伦布兰特直罗纹
remeasure the length of fabric 复尺寸(复查布料长度),复米
remé yarn 金银薄膜线
reminiscence 怀旧
remittance 汇付
remitter 汇款人
remnant 短码布,零头布,边角剩料
remnant skirt 碎布拼裙
remodel 改造
remote manual control 手工遥控
remould 改型
removable 脱卸式的(衣服);可移动的;可拆装的

removable cape 可拆卸披风,可脱卸披肩
removable cuff 开衩袖口
removable table sewing machine 活动台板缝纫机
removal table sewing machine 活动台板缝纫机
remove buttons 除去纽扣洗涤(洗衣机用语)
remove lining 除去衬里洗涤(洗衣机用语)
remove promptly 立即取出(洗衣机用语)
removing bastings 拆除疏缝线,拆除疏缝针迹
renaissance 文艺复兴(服饰);文艺复兴时期风格
Renaissance cloth 法国再生毛呢
Renaissance embroidery 文艺复兴时期动态设计的透孔刺绣风格;透孔刺绣品
Renaissance lace 小花纹细花边,文艺复兴花边
Renato Balestra 雷纳托·巴莱斯特拉(意大利时装设计师)
re-needling 换针
renewable 可翻新的
renewable finish [服装]翻新整理
renewal 更新,[服装]翻新
renewed collar 换领
renew period 更新周期
renforce 平纹棉布
rengue 菠萝叶纤维细布
Reno 聚酰胺袜用弹力丝(商名,日本帝人)
renovate (衣服)翻新,更新
rent (衣服等的)裂缝,绽线处
rentering seam 暗缝,修补接缝
rentraire, rentraiture 修补
rep worsted cloth 棱纹平布呢
rep 横棱纹布,棱纹平布
repacking 重新包装,改装
repair 修理,回修;织补,修补,维修
repair boots and shoes 修鞋,补鞋
repairing needle 修补用针
repair sewing machine 修缝纫机
repair-sole 修补过的鞋底
repeat 完全花型,完全纹样;重复编织,图案重复;(花纹)完全组织,组织循环
repeatability 重演性,再现性,反复性
repeated pattern 连续纹样
repeated seam 重复缝
repeat of color effect 配色花纹循环

repeat of design 纹样循环,意匠图的循环
repeat pattern 完全纹样,完全花型
repeat precision 对花精确度
repeat size 花样循环尺寸
repellants 防水呢,防雨呢
repellent treatment 织物防护处理,织物防水处理
repetition 反复
repetition of weave 组织循环
repidazol dyes 快磺素染料
repin 重新别大头针
replace color 替换颜色
replacement parts 备件,备换件
replacement rinse 冷热洗涤
repousse lace 凸纹花边
repp (同 rep) 棱纹平布
repp gobelin 棉毛棱纹呢,印经棱纹棉布
repp effect 纬向棱纹效应
rep poplin 横棱府绸,丝光府绸
repp quilt 双面粗棱纹提花织物
reproducibility 重现性
reproduction of marker 复制唛架
reproduction of sample 复样
reps 棱纹平布
reps alternatifs 交错棱纹织物
rep sarcenet 天鹅绒,粒纹绸
rep stitch 绣花十字线迹,起棱线迹
rep stripe 棱纹条纹
rep tie 棱纹领带
reptile leather 爬行动物类皮革
reptile necktie 蛇皮领带
reptile shoes 爬行动物类皮鞋
reptile(skin) 爬行类动物皮革
republican robe 民国男装长衫
requet 法国漂白亚麻平布
requirements in technology 工艺要求
rerebrace 上臂护甲
resale shop 旧货店
resample 重新取样,再取样
research of dress and decoration 服饰研究
reseau 花边
reseau rosace 针绣花边的网眼地纹
reseda(green) 浅绿色,木犀草绿(灰光绿色)
reserve printing 防浆印花,防染印花
reshmi jari-kinar 金边丝绸
Resht work 拉什特拼缝布
resident buying office 居民购买服务处《销售》

resident execute 常驻(内存)管理程序《计算机》
residual shrinkage 剩余收缩,残余收缩
residual thread 余线
resilience, resiliency 回弹,回弹性能,弹性变形,回挺性
resilient fabric 弹性织物
resilient seam 弹性缝
resin anticrease finish 树脂抗皱整理
resination 树脂整理
resin-bonded pigment printings 树脂固着涂料印花
resin button 树脂扣
resin collar interlining 树脂领衬
resin finishing 树脂整理
resin finish blanket 树脂衬布
resin interlining 树脂衬
resin-modified fabric 树脂整理织物
resirator 纱布口罩,军用防尘口罩;防毒面具,呼吸器
resist dyed yarn 防染纱
resist printing 防染印花
resistance to abrasion 抗磨,耐磨损性
resistance to ageing 耐老化性
resistance to bending 抗弯曲性
resistance to bleading 抗渗色性,抗渗化性
resistance to creasing 抗皱性,防皱性
resistance to creep 抗蠕变性
resistance to crushing 抗皱性,防皱性
resistance to discolouration 防褪色性
resistance to flame 防燃性
resistance to heat flow 热阻
resistance to insects 防蛀性
resistance to laundering 耐洗涤性
resistance to light 耐光性
resistance to mildew 防霉性
resistance to moth 抗蛀性
resistance to mothproofing 抗蛀性
resistance to perspiration 耐汗渍性
resistance to pilling 抗起毛起球性
resistance to plucking 植绒抗拔性
resistance to pressing 耐熨烫性
resistance to ripping 抗撕裂性
resistance to scuffing 抗擦毛性
resistance to shrinkage 抗缩性,防缩性
resistance to slippage 抗缝线滑脱性,防滑移性
resistance to snagging 抗钩丝性
resistance to soiling 抗沾污性,防污性
resistance to staining 抗沾污性,防污性
resistance to sunlight 耐日光性,耐晒性
resistance to tearing 耐撕裂性
resistance to ultraviolet radiation 抗紫外线辐射性
resistance to water penetration 抗水渗透性
resistance to water spotting 耐水滴色牢度
resistance to wear 耐磨性,耐穿性
resistance to weathering 耐气候老化性,耐气候牢度,耐风蚀性
resistance to wetting 抗着湿性
resistance to wrinkling 防皱性,抗皱性
resistance to yarn slippage 防脱缝能力
resistance to yellowing 耐泛黄性
resolving drawing (衣服部件结构)分解图
resort fashion 休闲装式,休假服式
resort hat 度假地帽
resort set 郊游服
resort wear 旅游服,休假服
resource 货源《销售》
respirator 口罩,纱布口罩,军用防尘口罩,防毒面具
respite 延期付款
respond 反馈
restalgic fashion 怀旧款式,怀旧时装,复古款式
restaurant dressing 宾馆人员穿着
resting comfort 静止舒适性
restoration 恢复,复原;修补,修理
restraint (画法)严谨
restricted sudsing detergent 低泡洗涤剂
restyle (根据新款式)重新设计
restyling 翻新,改造
retail 零售
retaille 零头碎呢
retail marketer 零售营销者《销售》
retail P·O·S system 零售系统
retail price 零售价《销售》
retail stores 零售店
retentivity 保持性
retexture (织物、服装等的)质地整理
rethreading 重新穿线
reticella 针绣挖花花边
reticella a fuselli 意大利针绣挖花花边
reticella lace 针绣挖花花边,雕绣花边,抽绣花边
reticence (艺术风格)严谨
reticle (旧时女用)收口网格包,手提

网兜
reticulated fabric 网状织物
reticule 抽口式袋,手提网兜,(旧时女用)收口网格包
retinal art 网膜艺术
retouch 润色,润饰
retrace 再描摹,重新描绘
retractable collar turning and pressing machine 自动翻领烫领机
retractable needle protector (缝纫机)可拆卸的坏针自停装置
retracting spring 复位弹簧
retrieval strap opening (常用于拉链后面的)里襟
retro fashion 怀旧款式,怀旧时装,复古款式
retro look 怀旧风貌,怀旧款式
retro reflection 回复反射光
retro-reflective coated fabric 反光涂层织物
retro-reflective finishing (织物)反光整理
retro-reflective pigment printings 反光涂料印花
retro-spective fashion 怀旧款式,怀旧式时装
retro-spective style 怀旧款式
retro-spectus 回顾性评论;简要回顾
return 退货,退回的商品
returned work 顾客退货《销售》
return line 回线,回管,折线
return on investment 资金回收,投资收益《管理》
returns to vendor 商店退货《销售》
revamp 修补;翻新;翻新物,修补过的东西;给(鞋、靴)换新面
revealing blouse 袒露的女衫
Revena 雷维纳绢丝色织天鹅绒
rever collar 翻领
revers 驳领,翻领,翻袖,翻边
reverse (织物)背面,反面;起绒粗绒呢(法国制);反向,回程;相反条纹型领带
reverse bar 回针杆
reverse blend fabric 倒混织物,天然纤维为主(的)混纺织物
reverse calf 反面牛皮
reverse coat 两面穿大衣
reverse colouring 倒色,互换色
reverse crepe 反绉
reversed calf 反面皮
reverse device 倒后缝装置

reversed feed 倒送料,倒缝,回针
reversed jeanettes 斜纹里子细布
reverse down-turn hemmer foot 翻转卷边压脚
reverse edge guide hemming foot 反向导边卷边压脚
reverse feed 倒缝,反向送料,倒回针
reverse jacquard 双面反色提花针织物
reverse knit 反面作正面的针织物;针织物反面起花
reverse loop 反面线圈
reverse plating knitted fabric 变换添纱针织物
reverse pleat 内襞(裤子腰围前的内倒襞),反向褶裥
reverse preference 反向优惠
reverse quiche 后卷鬓
reverse satin 缎背织物,纬面缎纹织物
reverse sewing 倒缝
reverse side [织物]反面;倒缝边
reverse side loop (针织物)反面线圈
reverse stitch 倒缝线迹
reverse stitching 倒缝
reverse stitch lever 倒缝手柄
reverse toe 反缝头
reverse twill weave 人字形斜纹,山形斜纹,破斜纹
reverse twist S捻,顺手捻
reverse twist suiting 经纱异捻服料
reverse twist worsted 隐条精纺花呢
reverse way 反向
reverse welt 翻口袜口
reversible 双面织物;双面卡其;双面式服装,正反两用大衣,晴雨两用大衣
reversible bonded fabric 双面黏合织物
reversible cloth 双面织物
reversible clothing[coat] 双面大衣,两面穿外套
reversible coat 双面用上衣,双面用外套
reversible color printings 变色印花
reversible damask 八枚花缎
reversible drop feed 倒顺送料
reversible fabric 正反两用织物,双面织物
reversible feed 倒顺送料,可逆向送料
reversible flannelet(te) 双面绒布
reversible gabardeen 双面华达呢
reversible garment 正反面针织服装
reversible herringbone twill 双面人字斜纹

reversible homespun 双面粗花呢
reversible imperial 双面粗棉布,双面八枚棉缎
reversible jacket 两面穿夹克
reversible jumper 两面(穿)无袖外衣
reversible khaki 双面卡其
reversible linings 反面白地黑花亚麻衬里布
reversible raincoat 晴雨大衣;风雨衣
reversibles 双面织物;两面穿衣服;(晴雨)两用衣服
reversible sash 双面带
reversible satin 双面八枚棉缎纹,双面缎
reversible stitch 倒缝线迹
reversible style 可反穿式服装,两面穿服装
reversible surcoat with zipper front 两面穿前胸拉链外衣
reversible top-stitching 双面线缝
reversible tweed 双面粗花呢
reversible twill 双面斜纹
reversible vest 两面穿背心
reversible warp pile 双面经天鹅绒
reversible warp pile structures 双面经起绒织物
reversible wool sweater 两面穿毛衣
reversible zip (per) 双面拉链
reversing motion 反向运动,换向运动
reviathan stitch 双十字绣,星状绣
revival look 再兴风貌;再兴款式
revival style 复古型
revive 衣服翻新者
REVLON 露华浓
revolutionary romanticism 革命浪漫主义,革命浪漫派
revolutions per minute 每分钟转数
revolutions per second 每秒钟转数
reweaving 织补
rework 返工
reworks 回修品,返工品
Rex Harrison 哈里森裙
Rex Harrison hat 雷克斯·哈里森帽,窄边粗花呢男盆帽
reyerse stitch lever 倒缝手柄
Reykjavik sweater (同 Icelandic sweater) 雷克雅卫克毛衣,冰岛毛衣,冰岛衫,斯堪的纳维亚毛衣
reza 印度粗制手纺纱印花棉布
RGB colour (Red Green Blue colour) 红绿蓝颜色《计算机》

RGB colour model (Red Green Blue colour model) RGB色彩模型《计算机》
rhadames 丝或丝棉交织光亮缎,丝或丝棉交织斜纹绸
rhadzimir 拉西米尔丧服绸英国丧服绸(同 radzmir);绫(女服用,丝的密织物),八页破斜纹黑绸
rhadzimir surah 棱纹绸
rhinestone 仿制金刚钻,人造钻石
Rhinestone 莱茵石
Rhinestone banding (饰品)莱茵石带
rhinestone buckle 钻石带扣
rhinoceros hide 犀牛皮
rhododendron 杜鹃花色
rhodophane 法国罗多芬透明织物
rhomb net work check effect knitted fabric 菠萝丁,菱形网格针织物
rhombic design 菱纹
rhubarb 大黄色,黄褐色
rhumba [dress] 伦巴舞装
rhumba panties (幼女的)伦巴内裤(臀部有些皱纹横条)
rhumba sleeve 伦巴舞衣袖(细小横皱纹的筒状袖)
rhythm 裙子规律性摆动;(光线、色彩配置的)匀称,调和,和谐;节奏,韵律
rhythm crepe 人造丝泡泡绉
rhythmic production 节奏生产《管理》
rhytidectony 整容术
riabaul 印度窄幅粗棉布
ria velvet (人丝)利亚绒
ria velvet plain 利亚平纹
rib 边缘;凸条;罗纹;肋骨
riband (同 ribbon) (装饰用)缎带,饰带,丝带
rib argyle hosiery top 菱形花纹罗口
rib at leg opening 裤口罗纹
rib-bait 肘翻贴边
rib band 罗纹带
rib banding 针织罗纹镶边
ribbed waist 罗纹背心,罗纹紧身胸衣,儿童罗纹内衣
ribbed band 罗纹带
ribbed banding(同 ribbing) 针织罗纹镶边(用于领口、袖口、腰部等)
ribbed banding [inset] (连衣裙的)罗纹腰,针织罗纹镶边
ribbed banding inset waistband (连衣裙中)镶嵌的针织罗纹腰头
ribbed border 罗纹边

ribbed bottom 罗纹边
ribbed cloth 起棱条织物
ribbed cotton elastic braid 棉罗纹松紧带
ribbed cuff 罗口,罗纹袖口
ribbed elastic band 罗纹松紧带
ribbed fabric 凸纹织物,棱纹织物
ribbed half-hose 罗纹短筒袜
ribbed hose 罗纹袜
ribbed hosiery 罗纹袜,罗纹针织物
ribbed plush 条子长毛绒,凸条长毛绒
ribbed seamless hose 罗纹无缝圆袜
ribbed socks 罗纹短袜
ribbed sweater 罗纹带运动衫
ribbed tank 罗纹背心
ribbed top hosiery machine 罗口直下圆袜机
ribbed top 罗纹袜口
ribbed trim 罗纹饰边
ribbed twill 凸条斜纹
ribbed undershirt 凸条内衣,棱纹内衣
ribbed velvet 棱条丝绒,灯芯绒
ribbed velveteen 灯芯绒,棱条丝绒
ribbed waist 罗纹背心,罗纹紧身胸衣,儿童罗纹内衣
ribbed welts 罗纹关边
ribbing 针织罗纹镶边(同 ribbed banding);罗纹,凸条
ribbings 针织罗纹带(有弹性和装饰性)
ribbon 带,丝带,缎带,饰带
ribbon band 饰带
ribbon binding(stitcher) 镶边机
ribbon border 镶边,饰带镶配
ribbon bow 假发蝴蝶结,细带蝴蝶结
ribbon cravat 缎带领带
ribbon cutter 切带机
ribbon cutting machine 切捆条机
ribbon effect 缎带效应
ribbon end 罗纹边
ribbon-faced waistline 缎带贴边腰线
ribbon folder 卷带器
ribbon for insignia 奖章带
ribbon insertion 缎带式镶嵌
ribbon lace 缎带花边,饰带花边
ribbon loop garter 缎带环饰吊袜带
ribbon straw braid 扁平草杆状丝编带
ribbon straw fabric 仿麻织物
ribbon tie 缎带领带
ribbon waist band 缎带裙(裤)腰
rib border 罗纹边
rib bottom 罗纹下摆
rib cage 胸廓,胸腔

rib circular knitting machine 圆形罗纹机
rib crepe 罗纹绉布,棱纹绉绸
rib cuff 罗口,罗纹袖口
rib double face 双面罗纹
rib end 罗纹边
ribetillo 绒带,丝带
rib fabric 罗纹针织物,凸条织物,棱纹织物,罗纹布
rib-hip padding 肋骨,臀部垫物
rib jacquard 罗纹提花织物
rib jacquard fabric 罗纹提花针织物
rib knit 罗纹针织品;罗纹组织
rib knit cuff 罗纹袖口
rib knit cuff machine 罗纹袖口机
rib knit neckline 罗纹(编)领圈,罗纹(编)领口
ribless corduroy 灯芯布,丰满绒(棉布)
rib knit panel 罗纹针织嵌布
rib knit pile fabric 罗纹毛圈针织物
rib knitted pullover (针织)罗纹套衫
rib knitting 罗纹针织品
rib knitting hosiery machine 罗纹组织袜机
rib knit T-shirt 抽条T恤衫,罗纹T恤衫
rib knit underwear 凸纹针织内衣,罗纹针织内衣
rib knit welt 罗纹边
rib length 罗口长,罗纹长(度)
rib machine 罗纹机
rib mark 罗纹符号
rib neckband 罗纹领圈
rib neckline 罗纹领口
rib silk 桑绢罗
rib stitch 罗纹组织,罗纹线圈
rib stripe 罗纹条纹
rib structure 罗纹
rib tickler swimsuit 两件式露腰泳装
rib top 罗口,袜口罗纹,袜腰
rib top aslant 罗口套歪斜
rib top gloves 罗口手套
rib top hosiery 罗口袜
rib top machine 罗纹袜口机
rib trim 罗纹饰边
rib velvet 灯芯绒,棱条丝绒,棉条绒,趟绒
rib waist 罗纹背心,罗纹紧身胸衣,儿童罗纹内衣
rica 里卡头巾
rice and gaudy 浓艳
rice braid 花式辫带,米形织带

rice cloth 米纹布
rice-looking stitch 米状线迹
rice net 硬衬布,粗棉线网(作帽坯用)
rice stitch 米状线迹,米字绣
rice weave 花斜纹,米纹组织
rich assortment 花色齐全
rich-clad 穿着华丽
rich colour 浓色,富丽色泽
rich elasticity 富有弹性
rich fabric 华丽织物
rich gradation 层次丰富;多层次
rich in colours 色彩丰富,重彩,浓色
rich material (词尾)表示:"以……为主的材料"
rich peasant look 富农风貌(19世纪60年代)
rich purple 浓紫色
richelieu [法]黎塞留靴
Richelieu embroidery 黎塞留刺绣
Richelieu guipure 黎塞留锁眼针绣,黎塞留镂空花边
Richelieu rib socks 2+1罗纹短袜
rich-looking stitch 美观线迹
rickrack 荷叶边,波形花边带,Z字形花边,山形装饰
rickrack braid 荷叶边,波形花边带,Z字形花边,锯齿形织带
rickrack stitch 波形线迹,荷叶线迹
rickrack stretch stitch 波形弹性线迹,荷叶伸缩线迹
rickrack trim 荷叶饰边,荷叶边滚带
ric-rac braid (rickrack braid) 锯齿纹细棉带
ric-rac stitch 荷叶线迹,波形线迹
riddle 土红
ride cords 经棱斜纹马裤料
rider jacket 骑士夹克
rider's jacket 骑士夹克
riders look 骑士型款式,骑士风格
ridge 脊梁;(尤指背部的)肌肉脊状粗隆
ridge design 波纹花纹,凹凸花纹,沟形花纹
ridge teeth 尖齿形
ridgy cloth 起拱织物,凹凸不平的布,有织疵的布
ridicule 女式小提袋
riding boot ribbon 马靴带
riding boots 骑马(长筒)靴
riding bowler 带面纱女用马术圆顶高帽
riding breeches (同 show breeches) 马裤,灯笼短裤
riding cap 骑马帽,骑士帽
riding clothing [coat] 骑马外套
riding coat 骑士外套,骑装上衣
riding costume 骑装
riding gloves 骑士手套
riding habit 女式骑装,骑马装(马裤与夹克装相组合)
riding hat 骑马帽
riding hood 骑马连颈帽,骑马头巾
riding jacket 骑马外套,骑士夹克,马褂
riding skirt 骑马裙
riding suit 骑(士)装,骑马套装
riding trousers 长马裤
riding wear 骑马服
rifle green 草绿色,军绿色
rig 服装,束装;织物对折;(口语)华美奇特的装束
rigged cloth 折幅织物,对折织物
rigging 服装(口语)
right-and-left effect 从右到左的效果
right angle 直角
right angle bias binder 双面光直角斜线滚边机
right angle cross over stitch 直角交叉线迹
right angle mark 直角符号
right-angle triangle 直角三角尺
right elevation 右视图
right facing (裤)里襟里子
right fly 里襟,右门襟,裤底襟
right fly interlining (裤)里襟衬
right forepart 右前襟
right front 右前片,右前胸,里门襟
right front opening edge 右盖左门襟
right-handed sewer 右手缝纫机
right-hand flyfacing lining 右门襟贴边衬里
right hand machine 右手机
right hand twill 右向斜纹
right hand twist Z捻,反手捻
right hinged raising foot (靠右边)活压脚
right line 直线
right side 正面(RS);织物正面
right-side facing 正面贴边
right-side forepart 右前襟
right-side hem 正面下摆
right-side marking 正面标记
right-side panel 右嵌边,右嵌条
right-side topside 右腰头

right sleeve 右袖
right topside 右腰头
right twill 右斜纹
rigid braid 硬腰带
rigid fabric 硬挺织物
rigid pad 刚性垫衬
rigid presser foot 固定压脚
rig marks 折痕疵
rigolette (女用)羊毛头巾,莱考立脱头巾
rigorism (艺术上)严谨风格
rig out 一套服装
rim (衣、帽)边
rim cloth 包边布,纽孔布
rimless glasses 无框架眼镜
rimple 皱纹;起皱
rinceau [法]叶饰
rindy 牛仔裤(宽裤腰,袋状脚口,宽吊边)
ring 戒指,指环;环形饰物;(军服袖上的)金环;环纹疵;条痕
ring belt 链形饰带(同 chain belt),链带
ring bracelet 手镯环(同 bracelet ring)
ring buckle 环形扣,双环系带
ring button 有外环的纽扣
ring collar 环领
ring finger 无名指
ring gem stones 戒指宝石
ring heel 环形袜跟
ringing machine 滚领机
ringless hose 无环纹丝袜
ringlet (毛发的)一小卷;(下垂的)长鬈发;古埃及头环,发卷
ring knife cutter 环形剪刀
ring scarf (将长领巾的尖角缝合在一起的)环形领巾
ring-shaped elastic 环形松紧带
ring stitch 环状绣
ring yarn 起圈环线
rinsability 可漂洗性,可洗净性
rinsing 淋洗,水洗,漂洗,漂清
riot 色彩丰富
rip knitting fabric 罗纹针织物
rip-off selvage 剥边,沿边
ripped to fashionable shreds jeans 时尚破烂型牛仔裤
ripper 拆缝刀;拆线器
ripping knife 拆缝小刀
ripple 波纹布,涟漪布
ripple and tuck combined knitted fabric 波纹-集圈复合针织物

ripple cloth 波纹粗呢,有光长绒呢,波纹织物;赖普勒(波纹)织物,立波尔织物
rippled collar 波纹领
ripple fabric 波纹织物
ripple knitted fabric 板花针织物,波纹针织物
ripple (pony) cloth 波纹织物,有光长绒呢
ripple puckers (织物上的)波形小皱纹
ripple skirt 波纹裙
ripple sole (橡胶制的)锯齿形防滑鞋底
ripple washing machine 波动平洗机
ripplette 丝光棉条格绉布,里普莱特绉织物
rippling 超波纹
ripstop 锦纶纱加固织物
rip-stop thread 防破裂缝口线
rise (裤)立裆,直裆,裆;时装上升阶段
rise curve 直裆弧线
rise fit 直裆合身性
rise length 直裆长
rise line (straight) 裤(直)裆线
Ritsuko Shirahama 白浜利司子(日本服装设计师)
rittaisaidan [日]立体裁剪
ritzy 时新的,最新式的
rivel 皱纹,褶裥,褶襞
river blue 铁青色
river transportation 河运
rivet (饰于衣服上的)撞钉,爪钉,铆钉
rivet and burr 角钉
rivet for jeans 牛仔裤撞钉
rivet holder 铆钉托
riveting machine 铆钉机(制皮革品)
rivetting tool 铆钉工具
riviere 钻石项链
rivière (de diamants) [法]钻石项链
R.M.R. (relative metabolic rate) 能量代谢率
Rn (rayon) 人造丝
RO (reddish orange) 红光橙色
roach 额前鬈发
road safety garments 道路安全服装(如反光背心)
road show 巡回展示
roan 红白间色,红棕色;绵羊皮仿摩洛哥革
roan rouge 褐红色
robe 罩袍,长袍;晨衣,浴衣,睡袍;法衣,圣衣;(标志职位、级别等的)袍服,礼

服;印花斜纹布
robe ailee　[法]展翅裙(长裙摆卷在手上,举手时象展翅)
robe à lànglaise　[法]珑格葛袍服
robe ascenseur　[法]电梯装,升降机装(用腰带使上身衣服膨松下垂)
robebain de soleil　[法]日光浴装,太阳装
robe biblique　[法]圣经时代装(丝绸长袍)
robe cachecoeur　[法]隐心型装(门、里襟交叉,前襟类似日本服)
robe chemister　[法]衬衫式装;衬衫裳
robe corolle　[法]花开装,花开形装
robe culotte　[法]裤裙装
robe dapres-midi　妇女午后礼服
robe-de-chambre　睡袍,晨衣,浴衣;(法国化装时穿的)外衣
robe decolletee　女袒胸露肩夜礼服
robe du soir　[法]妇女晚礼服
robe gag　[法](可供多样穿着的)组合式创新服装
robe gitane　[法]吉普赛女装
robe housse　[法]裹身连裙装(宽敞型),宽敞型连裙装
robe montante　[法]妇女普通礼服
robe oiseau　[法]鸟形装
robe papillon　[法]蝶翅装,披肩装
robe-paysage　[法]风景花样装
robe polo　波罗装(马球领装)
robe pull　套头夹克,夹克套衫(长至臀围)
Robert Nelissan　罗贝尔·内列森(法国时装设计师)
Roberto Capucci　罗伯托·卡普西(意大利时装设计师)
Robert Piguet　罗贝尔·皮盖(法国时装设计师)
robes　袍
Robespierre collar　罗伯斯比领
Robin Hood costume　罗宾汉戏装大领口短上衣
Robin Hood hat　罗宾汉帽(帽身高而尖,帽缘后反折而前面向下)
Robin Kahn　罗宾·卡恩(美国珠宝设计师)
robin's egg blue　知更鸟蛋壳蓝,绿蓝色,蓝绿色
Rocco Barocco　罗科·巴罗科(意大利时装设计师)
rochelles shirting　亚麻衬衫布

rochet　白色法衣,白色线衣
rock climb wear　攀岩服
rock crystal　无色水晶
rocker　弯刀冰鞋;美国海、陆军等军士V形臂章上的弧形条纹;(制革用的)吊鞣液
Rocker look　摇滚乐风貌;洛克型,摇滚乐服式
Rockford sock　罗克福特工作短袜
Rocknroll　摇滚乐
rocknroll look(同 rock and roll look,rockn-roller look)　摇滚风貌,摇滚风格;摇滚装
rocknroll stylistic form　摇滚款式
rock pants　贴脚细窄中长裤,摇滚裤(细窄贴脚裤)
rockshaft　(缝纫机)摇臂轴
rock skirt　摇滚裙(高腰大摆裙)
rock stitch　摆动线迹
rococo　洛可可式;洛可可艺术
rococo art　洛可可艺术
rococo embroidery　洛可可丝带绣,中国丝带绣
rococo flowers　洛可可花朵
rococo ornament　洛可可式装饰
rococo print　洛可可式印花
rococo style　洛可可风格
rodeo suit　竞技表演套装
Rod Measures　杆状计测器
rodondos　漂白亚麻布
roe deer　獐皮
roe hair　獐毛
rogue ends　(经编织物的)条花疵
rogue red　胭脂红
Roica　聚氨酯弹性纤维(商名,日本旭化成)
roji　布罗奇棉
role theory　角色理论
ROLEX　劳力士
roll　(一)卷,(一)匹;(衣服)翻边;中档亚麻布;妇女弹力紧身胸衣
roll buckle　旋筒带扣
roll calender　滚筒轧光机
roll collar　大翻领,翻领
roll down　松口短袜
roll-down hem　朝下卷边
roll-down hemming　朝下卷边
roll-down overedger　朝下包缝机
roll lapel　卷领,驳头
rolled collar(同 turn-down collar)　翻领,

卷领
rolled down hemming 朝下折边
rolled dressmaker collar 女装翻领
rolled-edge handkerchief 滚边手帕
rolled-edge machine 滚边缝纫机
rolled end 松紧条痕
rolled hem 卷边下摆
rolled hemming 卷边,卷边下摆
rolled latex 橡筋芯线
rolled seam 卷缝
rolled seam finish 卷边缝处理
rolled selvage 卷边,翻边
roller 卷边帽
roller clutch (缝纫机)滚柱式单向超越离合器
roller finish 辊筒轧纹整理
roller foot 滚柱压脚,双边固定轮压脚
roller iron 滚轮熨斗,滚筒熨斗
roller printing 滚筒印花
roller skate 旱冰鞋,四轮溜冰鞋,轱辘鞋
roller towel 环状擦手巾
rollette 罗兰特细亚麻布
roller washing machine 导辊式水洗机
roll forthingale 管形裙
roll-fitting collar 竖翻领
roll hem 卷边
roll holdall 网罩式圆形手提袋
rolling (制鞋)打光
rolling hose 男式长筒袜
rolling line 拨折线
rolling list 荷叶边疵
rolling plan 滚动计划
rolling ring 多环式戒指
rolling selvedge 卷边疵
rolling stockings 男式长筒袜
rollio 滚条,滚边
roll-leg pants 卷脚裤
roll line 驳品线,绗线,折领线,翻折线
roll neck 翻领,翻领套衫,翻领服
roll neck shirt (可翻领)高领衫;樽领衫(领口高而紧)
roll-on (女子)滚展式弹力紧身褡
roll-over collar 翻领
roll point collar 不下翻的燕子领
roll press 滚筒熨烫,滚筒轧压
roll printing 滚筒印花
roll puff curl 鬈发
roll sleeve 卷折袖
roll stitch 绕边缝
roll style 卷式发型

roll-up belt 卷袖袢,卷袖饰带
roll-up [cuff] 卷口袖,翻边裙口,翻克夫
roll-up sleeve 上卷袖;卷折袖;翻克夫袖口
roll welt 单吃线罗纹关边,英式关边
Rolph Lauren 拉尔夫·劳伦(美国时装设计师)
roly-poly 矮胖子,胖孩子
Romaian hem stitch 罗马尼亚抽绣
Romaian stitch 组合线迹,罗马尼亚针迹
romain 法国罗曼缎
romaine[crepe] 罗曼绉
romal 18世纪手帕料,印度洛美塔夫绸
romal handkerchief 洛美格子麻纱手帕
Roman collar 罗马领,教士领
Roman cut work 拉古萨雕绣,罗马雕绣
Romanesque 罗马风格,罗马式服装
Romanian couching stitch 罗马尼亚贴线绣
Romanian embroidery 钩线刺绣,霍尔班刺绣,罗马尼亚刺绣
Roman(ic) lace 古罗马花边
Roman lace 罗马雕绣花边,罗马抽绣花边
Roman leggings 罗马短裤
Roman pearl 罗马珍珠
Roman point 罗马雕绣花边,罗马抽绣花边
Roman sandal 罗马凉鞋
Roman sepia 墨棕色
Roman stitch 罗马针迹(一列长的平行针迹,用短的平针迹交于其中央)
Roman stripes 罗马条纹,罗马横条缎
romantic 浪漫派的,浪漫主义的
romantic charm 神韵
Romantic color 浪漫色
romanticism 浪漫主义
romantic look 浪漫风貌,浪漫风格
romantic style 浪漫主义风格
romantist 浪漫主义者
Romeldale sheep 美国罗梅尔达尔绵羊毛
Romeo 罗密欧式拖鞋,男式室内便鞋,罗密欧式便鞋
Romeo Gigli 罗密欧·吉利(意大利时装设计师)
Romeo slipper 罗密欧式拖鞋,男式室内便鞋,罗密欧式便鞋
Romes toga 宽外袍
rommany stripe 吉卜赛条纹
Romney Marsh wool 英国罗姆内沼地长

羊毛
rompers （儿童）连裤外衣（或背心）；连裤装，背心连装裤，田鸡装；连衫短睡衣
romper suit 田鸡服
romps 儿童宽松的连裤外衣
Ronald Amey 罗纳德·埃米（美国时装设计师）
Ronald Paterson 罗纳德·佩特森（英国时装设计师）
rondolette 法国龙多莱亚麻布，法国低档绢
rone 环状针迹，蛛网针迹
rone stitch 环状针迹，蛛网针迹
rongdi taffeta faconne 绒地绢
rongeant 烂花绣品，烂花图案
ronghua jacquard poult 绒花绸
rongmian flannel twill 绒面绫
rongmian satin brocade 绒面缎
Rongomy 马达加斯加褐色原棉布
rongye striped crepe grenadine 榕椰绸
rong he 绒褐（古代）
Ronoaks cotton 美国罗诺克斯棉
roofing felt 油毛毡
room shoes 室内便鞋
room socks 室内短袜，便袜
room temperature 室温
roomy line 宽松型；宽松型服装，舒展宽松的轮廓线
roomy pocket 大型贴袋
roons hat 龙斯帽
ropa 高领女装
rope 绳，索，超长项链
rope belt 粗绳腰带
rope dyeing 绳状染色
rope form 绳状织物
rope necklace 双套项链；超长项链
rope sandal 草鞋，绳编凉鞋
ropes and tapes 绳带
rope silk 弱捻多股刺绣丝线
rope stitch 绳纹线迹；绳纹绣
rope-stitch cording seam 绳纹线迹嵌线缝
rope washer 绳状洗涤机
rope washing machine 绳洗机，绳状水洗机
roquelaure （18～19世纪）罗克洛尔服
rosaline 意大利罗萨赖因花边
rosary 念珠（天主教用），珍珠
rose 玫瑰红，玫瑰色；玫瑰花饰，玫瑰形琢型的宝石；玫瑰香水
rose amethyst 水晶紫
rose beige 灰褐色

roseberry 杂色横条布
Rose Bertin 罗丝·贝尔廷（法国女帽设计师）
rose bloom 粉红
rosebud pR-80 cotton 玫瑰蕾棉
rose carmine 玫瑰紫红
rose cloud 淡红云色
rose color 玫瑰红，玫瑰色，淡红色
rosecran 法国亚麻平布
rose de nymphe 米色，象牙白，蛋壳色
Rose Descat 罗丝·德卡（法国女帽设计师）
rose design 玫瑰图案
rose dust 豆灰
rosehube 瑞士施维茨女式花边帽
rose light 浅玫瑰色
rose madder 玫瑰红
rose mauve 玫瑰紫红，紫红
rose mist 雾玫瑰色
rosenhube （瑞士）网眼帽子
rosen pale 淡玫瑰色
Rosen tape 罗森式条带
rose ornament 玫瑰装饰
rose pale 淡玫瑰色
rose pink 玫瑰粉红（淡粉红色）
rose point lace 玫瑰花纹网地花边
rose red 玫瑰红
roses 红润的面色
rose smoke 玫瑰烟雾色
rose tan 淡红棕色
rosette 玫瑰花饰品；豹斑
rosette button 玫瑰花形扣
rosette chain stitch 玫瑰链状绣
Rosette closed toe 罗斯特缝袜头
rosette design 玫瑰花结图案
rosette toe closing 扭结法缝头
Roseum cotton 印度罗森棉
Rose Valois(Mme Fernand Cleuet) 罗斯·瓦洛斯（法国女帽设计师）
rose violet 玫瑰紫，红莲色
rose water 玫瑰香水；玫瑰香水色
rose window 中世纪镂花鞋，玫瑰窗鞋
rose wine 豆沙色
rosewood 浅红棕色，黄檀棕
rosh anara 罗纱纳拉绸
rosiness 玫瑰色，（浅）玫瑰红，红润
Rosita & Ottavio Missoni 米索尼夫妇（意大利时装设计师）
rosole 换鞋底
rossete 玫瑰花形花纹

ROSSINI 罗西尼
rostano 西班牙金银丝织锦
rostoff 罗斯托夫绵羊毛
rosy beige 带粉红色的淡咖啡色
rotary body finisher 滚动式衣身熨烫机
rotary body finishing press 滚动式衣身熨烫机
rotary braiding machine 圆形编织机
rotary cutter 转盘裁刀
rotary hook （缝纫机）旋梭,旋转钩
rotary ironing calendar 旋转熨压机
rotary knife 圆刀式电剪
rotary knife cloth cutter 回转式裁衣刀
rotary knife cutter 旋转式圆刀电剪
rotary screen printings 圆网印花
rotary sewing machine 回转式缝纫机
rotatable jeans button 旋转工字纽
rotatable reception table 回转式承布台
rotating shuttle （缝纫机）旋梭
rot-fastness 耐腐牢度,抗腐烂性
rotonde 仓当特披风,圆亭披风（指一种形状与圆顶建筑物相似的披风）
rotorspun yarn 转杯纺纱
rot proofing 防腐处理
rot resistance 抗腐性
rouche 穿线或穿带抽褶,褶裥饰边
roucou 胭脂树红,浅橙红
roudao sportwear fabric 柔道运动服织物
rouenneries 法国鲁昂花布
rouge 口红,胭脂;胭脂色;镶边用细棉布
rouge red 胭脂红
rough 布面毛糙
rough clothes 粗布衣
rough crepe 粗绉面绸,鸡皮绉,象纹绉
rough dry 未熨衣服;晒干不烫,不加熨烫的干燥
roughing 起毛
roughness 粗糙度
rough Peruvian[cotton] 粗秘鲁棉
rough point （衬衫）圆领端
Rough Rider shirt 骑马衬衫,骑兵式衬衫
roughs 帆布型亚麻坯布
rough selvedge 毛边
rough sketch 速写
rough terrain suit 越野服
rough-textured silk 粗加工丝（绢丝原料）
rough wear 粗布服装
rough wool fabric 粗服呢
rouleau （女服）滚边,滚条,滚条盘花
rouleau fastenings 盘花纽
rouleau loops 纽袢
rouleau neck 滚边领
rouleau necktie 圆筒形领带
rouleau tie 细饰带,滚条饰带,筒形领带
roulette ［法］点线轮盘,点线器,擂盘
roulle(a)u 装饰用哗叽
round back 驼背
round bertha 圆形披肩领
round bottom 圆下摆
round braid 圆编带
round brocade knot 团锦结
round buckle 圆形带扣
round collar 圆领
round collar pyramid coat 角锥形圆领大衣
round corner （短手套腕部的）圆形收口;圆领角
round dress 拖地长裙
rounded back 驼背,圆背
rounded flap pocket 有袋盖的圆角口袋
rounded front 西装上衣的圆形前摆
rounded pocket 圆底衣袋,圆角袋
round elastic 圆松紧线,圆橡筋线
round ends 圆领角
round[er] ends 圆领角
round eye 圆头纽孔
round face shape 圆脸型
round floral pattern 团花图案
round goods 圆编针织物,圆筒形针织坯布
round head nailing machine 圆头牙钉机（制鞋）
round heel 圆袜跟
round hip 全臀围,臀围
round jacket 圆形夹克
round knife 圆形裁刀,圆刀
round knife cutter 圆刀裁剪机
round knife machine 旋转式圆刀机
round knot 筒子结,一把结（俗称）
round legs （未烫缝）圆裤脚
round lower hem 圆摆
round neck 圆领
round neckline 圆领口
round neckline slip 圆领套裙
round neckline slip on 圆领套裙,圆领连衣裙
round neckline sweater 圆领毛衣
round neck shirt 圆领衫
round neck striped top 圆领间条衫
round neck sweater 圆领衫

roundness 圆形
round net 圆形网眼花边
round(plain)toe shoes 圆(平)头鞋
round pocket(同 circle pocket) 圆形贴袋
round point needle 圆锥形针尖缝针
round point with triangular tip 三角尖圆针
round seam 包边缝
round shoulder 弓肩
round specs 圆眼镜
round stitch 圆形线迹
round stocks 长筒袜
round the middle 腰围
round thread linen 刺绣用亚麻布,抽花用亚麻布
round thumb (针织手套)镶缝拇指
round toe 圆鞋头,圆头鞋
round toe-end 圆鞋头
round-toe leather shoes 圆头皮鞋
round top 圆领端;圆线头灯芯绒
round-twill 曲线棱缎纹织物
rousted 法国劳斯泰特粗呢
roustet 法国低档粗呢
routine inspection 常规检验,定期维修
routine library 程序库《计算机》
routine maintenance 例行维修,日常维修
routine quality control 常规质量管理《管理》
routing 制造路线设计
routing flexibility 工艺路线的灵活性
rouzet 法国低档粗呢
Rovan 黏胶短纤维(商名,法国罗纳-普朗克)
rove yarn 粗梳纱
Rowden cotton 美国罗登棉
rowdy (织物)条痕,条花疵
3-row teeth feed dog (缝纫机)三排齿送布牙
roxana 充双绉,仿双绉
roxano 棱纹绉布,棱纹绉绸
royal 双经领带绸;深艳的,鲜亮的
royal armure 窄幅绉绸
royal blue 品蓝,红光蓝,深宝蓝色,深绿蓝色
royal cashmere 英国高级薄呢
royal costume 衮服
royale 八综丝绸;纬向断棱纹绸
royalette 英国五综棉毛断纹呢
royal light 蓝紫
royal lilac 浓艳丁香色

royal piquett 色经斜纹绸
royal purple 深蓝紫色,深紫红色
royal red 深红色
royal rib 双经直棱平布
royal satin 皇家缎
royal twills 高级斜纹绸
royalty 提成费;专利权税
royal yellow 雌黄
roybon 法国洛埃蓬仿开司来薄呢
Roy Gonzales 罗伊·冈萨雷(法国时装设计师)
Roy Halston Trowick 罗伊·哈尔斯顿·特罗威克(美国时装设计师)
rozano 直棱纹绉织物
RP (red purple) 红紫
RS (right side) 正面
RTW (R-T-W, ready-to-wear) 现成服装,成衣
ruana 条纹棉披巾,斗牛士小方披巾
rub (织物)擦伤痕
rub down seam (皮鞋后跟的)防磨缝
rub marks (呢面)擦痕
rubakha 俄式宽松女罩衫
rubam envers 法国缎背绒带
ruban velour 丝绒带
rubashka 俄式上衣,俄式罩衫,罗巴斯卡衫
rubber 橡胶,橡皮,橡胶制品;橡皮擦;橡胶套鞋
rubber baby pants 婴儿橡胶紧身短裤
rubber-backed fabric 涂橡胶织物
rubber band 胶带;橡筋带
rubber belt 橡筋带,松紧带
rubber belt conveyer 橡胶带传送机
rubber boots 胶靴,高筒雨靴,橡胶高筒套鞋
rubber bust pad 橡胶胸垫
rubber button 橡胶扣
rubber cloth 防水布,雨衣布,涂(橡)胶布,胶布,防水布
rubber clothing 雨衣,橡胶服装
rubber-coated fabric 橡胶涂层织物,涂橡胶织物
rubber-coated textiles 橡胶涂层织物,涂橡胶织物
rubber coating 橡胶涂层
rubber core yarn 橡胶芯线
rubbered raincoat 涂胶雨衣
rubber fabric 涂橡胶织物
rubber fiber 橡胶纤维

rubber filament 橡胶长丝
rubber footwear 橡筋鞋(防水防雨)
rubber heel 橡胶跟
rubber gasket 橡皮衬垫
rubber gloves 胶皮手套,橡皮手套
rubberized fabric 橡胶涂层织物,涂橡胶织物,橡胶布
rubberized shoes 雨鞋,橡胶鞋
rubber lining of knitted fabric 套鞋夹里
rubber mask 胶乳面具
rubber-proofed sheeting 橡胶涂层平布,橡胶涂层被单布
rubber raincoat 橡胶雨衣
rubbers (橡胶)套靴;浅口橡胶套鞋(美国);橡胶底帆布运动鞋(英国)
rubber shoes 套鞋;胶鞋
rubber silk 涂胶绸
rubbers linings of knitted fabric 套鞋夹里
rubber sole 胶(鞋)底
rubber sole auto hydraulic press 橡胶鞋底自动液压机
rubber-soled suede shoes 橡胶底羊皮鞋
rubber thong sandal 橡皮带凉鞋
rubber thread 橡筋线
rubber thread count 橡筋线号数
rubber wear 雨天外套,橡胶服装
rubbing fastness 耐摩擦色牢度
rubbing machine (制革)轧平机
rubboard 洗衣板,搓板
rubicund 血色,红润色
rubine 宝石色(暗红色),红玉[色]
ruby 红宝石;红宝石色,红玉色,深红色
Ruby Keeler shoes 鲁比·基勒鞋,露比·凯勒鞋(低跟,脚背系缎带蝶结)
ruby red 红玉色,宝石红
ruby spinel 红宝尖晶石《饰品》
ruche 褶裥饰边
ruched fabric 褶裥织物
ruche thread 褶裥线
ruching 褶裥饰边
ruchle 褶裥,小皱
ruchsack 背囊,帆布背包
ruck 皱,褶
ruckle 小皱,褶裥
rucksack 帆布背包,背囊,双肩背包,旅行背包
ruddiness 红色
rude drawing 草图
Rudi Gernreich 鲁迪·简雷齐(1922～1985,美国时装设计师)

ruff (16～17世纪流行的高而硬的)轮状皱领
ruffle 褶边;荷叶边,皱褶;皱纹
ruffled bow-tie blouse 荷叶边蝴蝶结领上衣
ruffled collar 荷叶边领,波褶领,皱褶领
ruffled edging 荷叶边缘饰
ruffled skirt 褶边裙
ruffle finish 耐久皱裥整理
ruffle lace 皱纹花边
ruffle lapel 荷叶领
ruffler 打皱褶装置,打褶器
ruffles and fur below 女服饰物
ruffle set in seam 在缝头内缅荷叶领
ruffle shirt 褶边礼服衬衫;褶边裙,皱褶裙
ruffle trimmed print pajamas 荷叶边印花睡衣
ruffle wool 双股松捻粗绒线
ruffling 皱褶花边带,荷叶边;打褶,起皱;内褶缝
ruffling blade 打裥片
ruffling machine 打裥机
ruffling stitch 打裥线迹
rufous 红褐色的
rug blanket 毛毯披巾
rugby 橄榄球色织条子布
rugby shirt 橄榄球衫
rugby shorts 橄榄球短裤
rugged brushwork 粗犷笔法
rug gown 毯料长袍,毛糙织物
rug wool 六股粗绒线
ruled paper 意匠纸
ruler 尺,直尺,划线板
ruler for pattern grading 纸样放码尺
ruler plate 划线板
ruler pocket 尺袋
ruler set 套装尺
ruling pen 直线笔,鸭嘴笔
rullion 皮凉鞋
rumal 卢亩面纱;卢亩头巾
Rumanian stitch 罗马尼亚针迹
rumba costume 伦巴舞服装
rumba dress 伦巴舞衣
rumba panties 伦巴内裤
rumba sleeve 伦巴袖
rump 臀部
rump piece 臀甲
rumple 皱纹,褶裥
run 抽丝,(针织物)纵向脱散
run and fell seam 包口接缝

run-down stocking 无跟长袜
run-length encoding 行程编码《计算机》
runnability 运行性能
running 渗色;渗化;运行;跑合
running apparel 赛跑服
running characteristic 运转特性,操作特性
running down 倒毛
running factor 运转率
running hemming(同 blind hemming) 暗卷边,暗缝缲边,暗缲针
running hemming stitch 暗卷缝线迹,暗缝缲边线迹;暗缲针
running-in 试车,校**
running knot 活结,滑结
running marks 经向长皱痕,绳状擦伤痕,绳状色条痕
running pants 赛跑短裤
running repair 巡回小修,日常修理
running shirt 运动背心;无袖低领运动衫;无袖低领汗衫;背心内衣
running shoes (同 racing shoe) 田径鞋,钉鞋
running shoes with 4 spikes 四钉跑鞋
running shoes without spike 无钉跑鞋
running stitch 初缝线迹,绗缝线迹;撩针线迹,撩针(针法)
running stock 周转库存《管理》
running test 试探性试验
running trunks (没有开口或边开口)运动短裤
running up 顺毛
running yard 码长(织物长1码)
run number 批号,批数
run-of-the-mill 等外品,次品,未经检验的纱布
run proof 防脱散,防抽丝
run resist 防脱散,防抽丝
run resistant hosiery 防脱散袜子
runstitch 初缝线迹,撩针,串缝刺绣针迹
runstitcher 初缝缝纫机,撩缝缝纫机
runstitching 初缝线缝,撩缝
runstitching waist band edge 平缝腰带边
run stop 防脱散
run up seam 拔长缝迹
runway (时装表演用)栈桥式舞台,时装表演T形舞台

run work 网眼绣花
Russ costume 俄罗斯族民俗服
russel lace 雪撬花边
russet (brown) 黄褐色
russet sheepskin 褐色羊皮
Russian blouse 俄罗斯衫,俄式女衫
Russian blue 蓝灰色,浅蓝色
Russian boots (有宽翻边的)俄罗斯长靴
Russian braid 俄罗斯织带
Russian chain stitch 俄罗斯链状绣
Russian collar(同 cossack collar) 俄罗斯领,萨克领
Russian cord 俄罗斯凸条布
Russian costume 俄罗斯民族服饰
Russian embroidery 俄罗斯刺绣
Russian fur hat 俄罗斯皮帽
Russian leather 俄罗斯小牛皮
Russian muskrat 俄罗斯麝鼠皮
Russian pony 俄罗斯矮种马皮
Russian rat 俄罗斯灰鼠皮
Russian sable 俄罗斯黑貂毛皮
Russian shirt-dress (同 zhivago dress) 俄罗斯衬衫式裙装(偏襟,饰边高领口,抽褶袖)
russienne 鲁津斜纹绸
rust 褐色,铁锈色,锈迹
rustic 乡村式;质朴的
rustic brown 铁锈棕(红棕色)
rustic fabric 粗面织物
rusticity 乡村风味
rustic look 粗俗款式
rustic simplicity 淳朴
rustling(同 scroop) 丝鸣,(女服的)窸窣声
rustling finish(同 scroop finish) 丝鸣整理
rust-proof 防锈
rust red 铁锈红
rust spot 黄锈渍
rust stains 锈斑
rusty 铁锈色的,赭色的
rutile 金红石
ruxin gauze broche 如心纱
ruyby stripe 橄榄球条纹
ruyi crepe 如意绉
ruzzai 棉布绗缝
RW (raw white) 本白(色)
RY (reddish yellow) 红光黄色

S

S (small, small-size) 小尺寸，小号；(standard deviation) 标准差；(strong)(色彩) 浓、深、强烈；S 字形，S 形物，S 曲线

SA (Seventh Avenue)（美国纽约市）第七（条）街（美国服装业中心所在地）

saaij 斜纹精纺呢
sabat on 鸭嘴式盔甲套鞋
saber gloves 佩剑手套
sable 紫貂，黑貂；紫貂皮，黑貂皮，貂皮；深褐色的，赭石红（咖啡偏红）；黑貂皮围巾，黑貂皮外衣（或领）；丧服
sable coat 黑貂皮短大衣
sable fur 黑貂毛皮
sable hair 黑貂毛
sable mouille 湿沙色，带绿色的灰色
sables 貂皮短大衣；貂皮围巾；丧服
sablier line 沙漏计时器型（上半身为合身线条，裙子展宽），沙漏线条
sabot （欧洲农民穿）木鞋；木底平鞋；荷式沙宝，木屐；（凉鞋的）鞋襻；有袢鞋
sabot-strap shoes 袜带鞋，沙宝鞋
sabre gloves 佩剑手套
sabre mask 击剑面罩
sabretache 马刀挂套
sabrina neckline 萨布里纳（式）领口（用绳带系结支撑肩部的一字领）
sabrina pants 短紧身裤，萨布里纳裤
sabrina shoes 萨布里纳鞋，萨布丽娜鞋
sabrina work 萨布里纳花片贴花刺绣，萨布丽娜刺绣
sac 袋形外衣
saccharilla mull 萨卡里拉面纱头巾用漂白薄棉细布
sac de plage ［法］海滩提袋
sac dress 16～18 世纪的宽身女袍
sac en bandouliere ［法］斜挂肩袋
sacfilet （装泳衣用的）网袋
sachet 香袋，香料袋，香粉
sack 袋形外衣，布袋装，妇女、儿童用的宽松外衣；幼儿针织小外套；睡袋，袋，粗布袋，麻袋，硬纸袋，塑料袋，登山帆布背包，背囊，箱；（附着于衣肩的）丝绸褶裥长拖纱
sack-back 背部两个大箱形褶的宽松女裙

sack bag 包
sack closing machine 封袋机
sackcloth 麻衣，忏悔服；粗平袋布
sack clothing[coat] 男便装上衣；婴儿针织上衣
sack coat 袋形外套，婴儿针织上衣，男便装短上衣
sack crepe 绉缎
sack dress 袋式服，(20 世纪 50 年代流行的)袋式直筒女装；松身衣裙
sack drill 斜纹袋布
sacket 小袋，小皮夹子
sackgue （妇女、儿童穿的）宽大短夹克；(18～19 世纪)宽身女袍；（从宽身女袍肩上往后披下的）拖地丝裙裾
sacking 素色法兰绒；麻袋布，粗平袋布
sack jacket(同 **stroller jacket**) 散步外套（半正式男礼服，套装式外套），休闲随意夹克
sack needle 缝袋针
sack silhouette 布袋型
sack suit 袋形套装，普通西装
sacoche ［法］鞍囊；包
sac polochon 枕形袋
sacque 宽身女袍；拖地丝裙裾(18～19 世纪)；（妇女、儿童穿的）宽大短夹克；婴幼儿短上衣；袋，箱
sacred tree 圣树
sac wrist 松紧带手套腕口
sad 素淡的，暗淡的
sada 萨达开士米披巾（印度）；萨达棉平布（孟加拉）
sad browns 暗棕色
sad color 暗淡色
sad colored 颜色暗淡的，深暗颜色的
saddenin 色泽暗淡处理
saddle back waistband 鞍状腰带
saddle bag （自行车等鞍座后）挂包，鞍形口袋；鞍囊；鞍装；工具包
saddle bag pocket 手风琴式口袋
saddle coat 骑马雨衣
saddle jacket 骑马外套
saddle leather 鞣制真牛皮，马鞍皮革
saddle nose 鞍状鼻，塌鼻

saddle oxford 鞍饰牛津鞋,鞍形牛津鞋(鞋帮中部似马鞍形)
saddle oxford 鞍形牛津鞋
saddle pants 骑马裤
saddle raglan sleeve 鞍形插肩袖
saddle seam 鞍型缝
saddle shoes 鞍背鞋;(同 saddle oxford)鞍饰牛津鞋,鞍形牛津鞋
saddle shoulder 插肩,鞍形肩
saddle shoulder sleeve 插肩袖,鞍形肩袖
saddle sleeve 鞍形袖
saddle stitch 萨德尔针法,裤脊线迹,鞍形刺绣线迹(装饰压线缝)
saddle stitched seam 鞍形线迹
saddle stitching 鞍形线缝,鞍形线迹
sad drown 暗棕色
Sade 孟加拉萨达棉平布;旁遮普地区低档山羊绒披巾
sadha 孟加拉萨达棉平布;旁遮普地区低档山羊绒披巾
sadhabafi 素色平纹棉布
sadha chadder 白色萨达山羊绒披巾
sadhie 莎丽布
sadin 萨丁布
sad iron 大熨斗
sadowa 萨多瓦花式起绒呢
sadra 印度宗教衬衫
Sadtler standard spectra 萨特勒标准光谱
safa 金边真丝大围巾
safaline 聚丙烯腈长丝(商名,日本)
safari 瑟法里式;猎装;淡土黄色;遮阳盔
safari bag (旅行,狩猎)手提包
safari belt 狩猎腰带,瑟法里腰带
safari blouse 猎装短外套
safari boots 狩猎靴,瑟法里靴,旅游靴
safari clothes 瑟法里狩猎装
safari coat 狩猎外套,旅游外套,瑟法里外套
safari dress 瑟法里装,狩猎装
safari hat 瑟法里帽,狩猎远征帽;旅游帽
safari hat 旅游帽,狩猎远征帽
safari jacket 狩猎外衣,旅行夹克
safari look 狩猎型款式,瑟法里风貌
safari pocket 瑟法里口袋,猎装口袋
safari set 猎装
safari shirt 瑟法里衫,狩猎衫
safari shorts 瑟法里短裤,旅行短裤
safari suit 瑟法里猎装,狩猎装,旅行装
safeguard 骑马装外裙
safeguard glasses 防护眼镜

safe ironing temperature 安全熨烫温度
safekeeping 保养
safe operation 安全操作
safety 安全服饰
safety and protective textile 安全防护用纺织品
safety belt 救生带,安全带
safety bike boots with steel toe 钢头安全骑车靴
safety boots 劳保靴
safety check 安全检查
safety device 安全装置
safety fastener 安全拉链
safety garment 救生衣
safety glasses(同 goggles) (不碎玻璃的)护目镜,风镜
safety goods mark (SG-mark) 安全制品标志
safety harness 安全背带
safety hat 安全帽
safety helmet 安全帽
safety lighting 安全照明
safety overedge stitch 安全包缝线迹
safety overlock stitch 安全包缝线迹
safety pin(同 kilt pin) 安全别针
safety production 安全生产
safety protective coverall 安全防护服
safety shield 塑料罩
safety shoes 劳保鞋,安全鞋,防护鞋
safety standard 安全标准
safety stitch 安全线迹
safety stitch machine 安全线迹缝纫机
safety stitch machine with 2-needle, 5 thread 二针五线安全线迹缝纫机
safety stitch(ed) seam 安全线迹缝
safety stitcher 安全线迹缝纫机
safety stitching 安全线迹
safety switch 紧急开关,安全开关
safety textile 安全[用]纺织品,防护[用]纺织品
safe working load 安全工作负荷《管理》
saffian 着色羊皮
saffraan [荷]番红花;藏红花
saffron 番红花色,藏红色,橘黄色
saffron yellow 藏红花黄(金黄色),橘黄色
sag (衣服)下垂度,下垂
SAGA (Scandinavian Mink Breeders Association) 北欧四国(丹麦、挪威、瑞典、冰岛)的水貂生产者协会的商标
SAGA Design Collection 萨茄设计展示会

Sagar cotton 印度萨加尔棉
sagathy 细精梳斜纹毛织物
sage 鼠尾草绿(色),灰绿色
sage gray 鼠尾草灰色
sage green 灰绿色,鼠尾草绿
sagging 倾斜,下垂
sagittal suture 箭形接缝
sagittate sahade 箭形接缝
sag-no-mor 精梳毛乔赛的不下垂处理(商名)
sagum 古罗马军大衣,毛毯斗篷,罩衫,宽大袍子
Sahara shade 撒哈拉沙漠色
sahare 红边黄白条纹织物
sahries 斜纹棉布
saicai satin brocade 赛彩缎
saidan 裁剪;裁断
said [cotton] 叙利亚赛德棉
saie 细纺毛呢,细哔叽
sailcap 水手帽
sailcloth (作衣料用)帆布
sailing wear 航海服
sailmakers 用于帆布和厚皮革的针
sailor 水兵衬衫,水手衫;扁平硬边草帽
sailor beret 水手贝雷帽,水手软帽
sailor blouse 水手上衣,水手式女衫
sailor bottoms 水手裤
sailor cap 海员帽;水手贝雷帽
sailor collar 水手领,海军领
sailor dress 水手领童装,水手领女服
sailor hat 水兵帽,(童)水手帽,扁平的硬边草帽
sailor jupon (裤脚肥大的)水手裤
sailor knot 水手结
sailor look 水手款式,水手风貌
sailor pants 水手裤
sailor scarf 水手方巾,水手围巾
sailor shorts 水手短裤
sailor's jacket 水手式外套
sailor's knot 水手式领结
sailor skirt 水手裙
sailor suit (儿童或妇女穿的)海军装;水兵服,水手服
sailor's striped shirt 海魂衫
sailor's striped shirtlet and trousers 海魂衫
sailor tie 水手领带
sail yarn 鞋底缝线,帆布缝线
Saint Andrew's stitch 圣安德罗针迹(四针缎纹针迹构成十字架形)

Saint Gall lace 圣加尔花边
Saint Georges 圣乔治本色亚麻布
saint jago 棉织品(塞拉利昂用语)
saint jean 法国本色粗亚麻布
Saint Laurent helmet 圣洛朗帽盔
Saint Louis cotton 美国圣路易斯棉
Saint maur 法国丝绒丝毛圣莫尔哔叽
Saint nicholas 圣尼古拉斯哔叽
Saint Rambert 圣朗贝尔本色亚麻布
Saint Vincent cotton 圣文森特棉
sakalia [cotton] 印度萨加利亚棉
sakalio cotton 瓦加德棉,萨卡里奥棉
sakellarides [cotton] 埃及萨克拉里德斯棉
sakker 泡泡纱,绉条纹薄织物,条格绉布
saksette 菽麻,印度麻
sakusan pongee 柞蚕丝府绸
sakusan silk 柞蚕丝
salad color 色拉色
salampore 萨兰波水彩格布
salang 萨兰呢(喜马拉雅山区产坚牢粗纺毛织物)
salaoag 蕉麻,马尼拉麻
Salar ethnic costume 撒拉族服饰
Salar nationality's costume 撒拉族服饰
salari 巴基斯坦手织条格织物
Salarzu costume 撒拉族民俗服
salatiska 俄罗斯双峰驼绒织物
salem cotton 萨勒姆棉
salempore 萨兰波水彩格布
salems cotton 柬埔塞棉
salendang 印度东部色绉织棉布
sales 销售
sales as per origin 凭产地买卖
sales by description 凭说明书买卖
sales by grade 凭等级买卖
sales by sample 凭样买卖
sales by standard 凭标准买卖
salesman sample 展销样《销售》
salestte hemp 印度麻,菽麻
saling 萨麻呢
salisbury [white] 萨利斯布里白色法兰绒
salla 莎丽花格布
sallam 手纺纱上等平纹棉布
sallet 15世纪有护颈的轻头盔
sallo 萨罗红色平纹或斜纹棉布
sallow 灰黄色,土色,菜色(指人的皮肤色);鲑鱼红;橘红
sally jess bag 萨利·吉斯手提包
sally Victor 萨莉·维克托(美国时装设

计师）
salmon［color］ 肉色,橙红色
salmon pink 橘红色,橙红色,浅橙色,鲑鱼肉粉红色
salon apron 花围裙,花围身
salon dresses 沙龙时装
salona cotton 东欧产萨罗那棉
salonique cotton 希腊马其顿地区棉
salon selling 沙龙式销售（通过举办沙龙聚会向顾客进行销售）《销售》
salopette jeans 坚固呢工装裤
salopette［pants］ 沙罗佩套裤,背带工装裤,胸垫布套裤,猎人罩裤,滑雪服
salopette shorts 工装短裤
salopette skirt 工装裙
salor jupon 水兵裤
sal soda 洗衣用苏打,洗衣用碳酸纳
salt and pepper 芝麻呢
salt sacking 粗糙织物（粗糙似盐袋布）
Salvation Army bonnet 黑麦杆窄帽檐软帽,救世军帽
Salvatore Ferragamo 萨尔托雷·费拉加莫（意大利的鞋子设计师）
salwar 波斯女裤
SAM（seam abrasion machine） 线缝磨损机
samardine 法国哔叽
samba chit 旁遮普地区产绿地印花棉布
sambhal 印度手织平纹红边棉莎丽
Sam Browne belt （英国宽幅）武装皮腰带,山姆·布朗佩带
samcloth 绣花样本
samfoo 中国女式紧腰套装,衫裤（旧时中国妇女套装,出自粤语）
samfu(同 samfoo) 中国女式紧腰套装,衫裤（出自粤语）
samilis 厚重华丽丝锦缎,金银丝花缎
samis 厚重华丽丝锦缎,金银丝花缎
samita 厚重华丽丝锦缎,金银丝花缎
samite 厚重华丽丝锦缎,金银丝花缎
samiton 麻经丝纬花缎
samittum 厚重华丽丝锦缎,金银丝花缎
sammal 平纹粗毛布
sammeron 优质亚麻布
sammet 厚重华丽丝锦缎,金银丝花缎
sample 样品,样本
sample blanket 大样
sample card 样品卡
sample checking 样品确认
sample collection 包袱样

sample copy 样本
sample cutter 试样切取器,切样器
sample garment 样衣
sample hand 样品制作员
sample machinist 制样衣师
sample maker 样品制作员
sample making 制版
sample making for approval 打确认样
sample of buttons 纽扣样本
sample of fabrics 布料样本
sample order 订样品
sample picking machine 布样剪齿机
sampler 绣花样本,样品检查员
sample room 设计室,打样间,样品间
sample sewing 样品缝制
sample size 样品尺码（尺寸）；样板尺寸（在数字化仪输入的尺寸）《计算机》
sampling 抽样；做样
sampling inspection 抽样检验《管理》
sampling system 打样系统
sampot 柬埔寨彩色围腰绸带
samsum 土耳其萨姆松布
samut 金银丝花缎
samy 金银丝花缎
sanas 印度漂白或蓝色薄棉布
sanat 印度森纳特低档手织印花平布
sand 沙灰色（浅黄灰色）,沙土色,浅棕色
sandal 凉鞋,拖鞋；浅口套鞋；鞋袢；（古希腊）带子鞋；绸衬布,（桑达尔）条纹塔夫绸
sandale mordoree ［法］金褐色凉鞋
sandale plate ［法］平底凉鞋
sandales à lanieres ［法］带状凉鞋
sandal foot 不加固袜底
sandal foot hose 袜底不加固的连裤袜,不加固袜头的袜子
sandal foot pantyhose 穿凉鞋用裤袜
sandal foot splicing 袜底加固
sandals 屣,凉拖鞋
sandal shoes 凉鞋
sandal wood fan 檀香扇
sand beige 淡黄色,干砂色
sand-colored uniform 沙滩式制服
sand crepe 纱面绉
sanded cloth 磨毛布料,磨毛织物
sanded finish fabric 磨绒整理织物
sanded tease finished fabric 磨绒整理织物
sanded washer finished fabric 砂洗整理

织物
sand grain button 沙砾扣
sandiness 浅茶色,沙色,(头发等的)浅黄灰色,浅棕色
sanding (布料)磨毛;磨绒;砂洗
sand shoes 沙滩软底鞋,胶底帆布鞋
sand stitch 打粒线迹,砂粒状线迹
sandstone 浅灰棕色,砂岩棕
sandstorm 风沙色
sand wash 砂洗
sand-washed corduroy 砂洗灯芯绒
sand-washed silk 砂洗绸
sand-washed silk blouse 砂洗丝绸衬衫
sand-washed silk dress 砂洗丝绸服装
sand-washed silk shirt [blouse] 砂洗丝绸衬衫
sand-washed silk wear 砂洗丝绸服装
sandwash finish 砂洗整理
sand washing 砂洗
sand weave 醋酯沙面绉
sandwich fabric 衬垫织物(针织);袜子夹底;叠层织物,夹心织物
sandwich pocket 内贴袋,夹层袋
sandwich ruffler 层状打裥器
sandwich seam 夹层接缝
sandwich-board jumper 广告牌式无袖连衣裙,前后两片背带裙
sandwiched fabric 衬垫织物(针织);袜子夹底;叠层织物,夹心织物
sandwich-type lamination 三明治织物,夹心织物
sandy 淡茶色
sandy crepe 沙绉,苔茸绉
sand yellow 砂黄
sanforized collar 防缩领
sanforized compressive shrinkage 桑福防缩整理
sanforized fabric 防缩(棉)织物
sanforize roughness 预缩布面粗糙
sanforizing 防缩处理
sangales 漂白薄亚麻布
sangati 桑加蒂薄细布
sangbo crepe damask 桑波缎,桑波绉
sangbo satin 桑波缎
sanghua sheer twill 桑花绫
sangi 桑吉缎
sangle [法]扁带
sanglier 硬挺粗呢,桑格利尔粗呢
sangria 血红色
sanguine 血红色

sanitary band 妇女卫生巾,月经带
sanitary belt 妇女卫生巾
sanitary fiber 抗细菌纤维
sanitary finish 卫生整理
sanitary knickers (女用)月经裤,生理裤
sanitary napkin 卫生棉,卫生巾
sanitary napkins and diapers 卫生巾,尿布
sanitary shorts 卫生短裤
sanitary tampon 卫生棉塞
sanitary towel 妇女卫生巾,月经带,卫生纸巾
sanitary towel fabric 卫生巾
sanitary wear 卫生服
sanitary wool 原毛
sanna 印度漂白或蓝色棉布
San Remo hat 圣雷莫帽
sans-couture [法]无裁缝服装(19世纪70年代后期出现的时装风格,几乎不经裁缝)
sans culottes 直裁裙裤
sans douture 无制作(采用无衬,元里,不缭缝)
sans envers 双面织物
sansfabric 非织造布
sans-forme [法]无固定形服装
sanski 丝、绸缎(日本用语)
Santa Claus suit 圣诞老人套装
santipur 孟加拉手织绣花棉细布
sap green 暗绿色
sapoit 萨波伊特棉围巾(中部格子花纹,分剪处有边纹)
sapphire 蓝宝石;宝石蓝,天蓝色,深紫蓝色,青玉色
sapphire mink 蓝(毛)水貂;蓝水貂毛皮,蓝宝石貂皮
saracenet 里子薄绸
Saracenic ornament 撒拉逊人装饰
saradi 印度回教徒无袖背心
sarafan 高腰无袖刺绣连衣裙,俄罗斯农妇莎拉凡装
sarape 毡斗篷,无毯披巾
sarasa 红白色相间的印花布
sarashi cariko 漂白平纹衬衫布
sarashi kanakin 漂白衬衫棉细布(日本用语)
saratoga (女用)旅行大皮箱
sarau 工作服,防护罩衫
sarcenet 里子薄绸,平纹薄丝带,有光里子布,素纺

sarciatus 13世纪英国粗毛织物
sarcilis 13世纪英国粗毛织物
sardasi 印度产金银丝刺绣丝绒
sardinian 英国八页斜纹重磅珠皮大衣呢
sardinian sac 19世纪单排扣宽松男大衣
Sardonyx 缠丝玛瑙;深红色
saree(同 sari, sarrie) 莎丽织物,印度莎丽;莎丽布;(印度)莎丽装
sargasso green 褐绿色,绿褐色
sarge 粗哔叽
sargette 窄幅丝绸,轻薄哔叽
sargia 中世纪意大利丝无斜纹织物
sargues 法国粗纺毛麻哔叽
sari 莎丽织物,印度莎丽;莎丽布;(印度)莎丽装
sari borders 莎丽花边
sari dress 莎丽装
sarihan 柞丝绸
sark 女式无袖衬衣,苏格兰衬衣
sarong 莎笼,莎笼裙料;围裹式长筒裙,(马来群岛)围裙;莎笼式女服
sarong apron 短围裙(有胸兜的围裙)
sarong dress 莎笼装,围裙装
sarong kapala (远东男用)方包布头
sarong pants 莎笼短裤
sarong skirt 围裹式长筒裙,莎笼裙,褶皱裙
sarong swimsuit 莎笼式泳装,前围裹式泳装
sarpu 印度北部粗毛呢
sarrau 工作罩衫
sarraux 法国蓝格亚麻帆布
sarrie 莎丽织物,印度莎丽;莎丽装;莎丽布
sarrouel 伊斯兰裤(灯笼裤),萨洛埃尔裤
sarsanet 里子薄衫,平纹薄丝带,有光里子布
sarsenet 里子薄衫,平纹薄丝带,有光里子布
sarsenet ribbon 精细平纹缎带
sartor 裁缝,成衣师,补衣师
sartorial 男士服装的;缝纫的,裁缝的
sash (女、童)腰带;肩带;彩带;(军装)饰带
sash belt 腰带、饰带
sash blouse 女饰带上衣,饰腰带上衣,饰腰带罩衫,饰腰式女衫
sash closure 系腰带的门襟
sashed 系腰带的,有腰带的
sash marmar aal 漂白细软布

sassard coat 莎莎外套(有肩章、肩垫、宽西装领、束腰带)
sassoon cut 萨松发型
sassoon hairstyle 萨松式发型,短直发型
sassy look 俊俏款式
satalian 英国废纺粗支衬衫里布
satara 有光缩绒棱纹呢
satarra twill 八综花式斜纹
satchel (皮或帆布)书包,小背包,小提包
satchel bag 书包式小提包
satchel handbag 书包式手提袋
satchel pocket 贴袋
SATCHI 沙驰
sateen 纬面缎纹、纬缎、棉缎、仿缎里布、横贡(缎,呢)
sateen-back crepe 缎背绉
sateen drill 缎纹卡其
sateen shirtings 英国色织五枚经缎裙布,经缎裙子布,缎条衬衫布
sateen tick 装饰用条纹厚棉缎
satellites 英国印花棉布
sati-drap 缎纹交织呢
satin 丝织缎纹织物,缎,缎纹;缎子衣服;轧光
satinage 轧光整理
satin à la reine 重丝缎,皇后缎
satin alcyonne 表里双色缎
satin amazone 俄罗斯毛呢
satin and sateen cloth 缎纹织物
satin-back 缎背,缎背织物
satin-back coating 缎背花呢
satin-back crepe 缎背绉
satin-backed rib 缎背(经)重平织物
satin-back gabardine 缎背华达呢
satin berber 有光缎纹呢,缎纹呢
satin bonjean 精纺缎纹呢
satin border (手帕的)缎边
satin brocade 库缎,花库缎,织锦缎,摹本缎,软缎
satin brocade quilt covering 软缎被面
satin cafard 丝毛交织缎
satin cashmere 缎光细呢
satin charmeuse 轻薄光面软缎,查光斯缎
satin check 缎格布,缎格绸,格子缎
satin-checked voile 缎格巴里纱
satin cloth 有光缎纹细呢
satin crepe 绉背缎;水洗皇后
satin cuttance 丝麻交织缎(印度),丝棉交织缎
satin damask 亚麻花缎,厚花缎

satin de bruges （18世纪荷兰、比利时）丝毛交织缎
satin de chine 消光厚缎,法国紧密缎
satin de Chine 中国缎
satin delhi 德里精纺缎纹呢
satin de hollande 缎纹呢,薄开士米呢,法国精纺装饰缎
satin de laine 精纺缎纹呢;薄呢
satin de Lyon 里昂缎
satin dorure de nankin 金色线满地花织锦缎,金宝地,织金缎
satin double face 双面缎
satin drill 斜纹缎,泰西缎,贡缎,直贡
satin duchess(e) 高密缎,全丝硬缎
satine 法国毛缎(作女服用);仿缎里子布
satin ecossais(e) 埃科赛全丝条格缎
satin embroidered cheongsam 软缎绣花旗袍
satin embroidered piece 软缎绣片
satinet 衬里布(棉经毛纬或棉经棉纬);仿毛条格布(棉经毛纬);缎纹绒里棉布;羽绸(伞绸);充经缎;全丝薄缎
satinette 全丝薄缎;棉毛缎;羽绸
satin fabric 缎纹织物
satin face 缎面(双层织物)
satin faced silk crepe 绉缎
satin facon(n)e 花缎
satin faille crepe 横罗绉缎
satin feutre 绒背缎
satin figaro 18~19世纪法国16枚经缎,菲加罗缎
satin finish 缎光整理,轧光整理
satin finish leather 无光整理的皮革
satin foulard 薄亮软缎
satin francais 法兰西缎纹呢,精梳缎纹呢
sating 仿真丝缎纹织物(经轧光上蜡)
satin gaufre 拷花缎
satin georgette 缎纹乔其纱
satin georgette crepe 缎纹乔其纱
satin grec 丝织衬缎
satin grenadine 交织薄缎
satin hermine 19世纪法国仿貂缎带。银鼠缎带
satin imperial 印花缎,纬面缎纹绒布
satinisco 低级里子缎
satinize 缎光整理;(棉织物)耐久性光滑整理
satin jean 光亮牛仔布,光洁厚斜纹布
satin leather 缎光革

satin levantine 法国利凡廷鞋面缎
satin liberty 自由缎带,自由缎
satin like (同 satiny) 仿真丝缎纹织物
satin lissé 印花经面斜纹棉布;印花棉缎；利西棉缎
satin look 缎子外观,缎子风貌
satin luster 缎纹光泽
satin luxor 柔软棱纹缎
satin lyonnais 全丝有光里子缎
satin marabout 女帽用细薄单丝经缎
satin merveilleux 美妙缎
satin messaline 薄缎,梅萨林薄丝缎
satin moss 精纺细呢(平纹,稀疏)
satin mousseline 薄缎
satin net 缎子网眼布
satin onde 单丝缎
satin oriental 东方缎
satin ottoman 厚棱纹缎
satin pancross 缎背印花绸
satin panne 高级厚重平缎
satin plain 素缎,软缎
satin plain mixed 交织素缎
satin (plain) woven label 缎(平)面织唛
satin raye 缎条
satin regence 纬向细纹缎
satin rhadamas ［法国］拉丹缎
satin rib 棱纹缎
satin ribbon 缎带,缎纹丝带
satin royal 皇家缎
satins 缎
satin sans envers 双面缎带,双面缎
satin serge 缎面哔叽
satin shoes 缎面鞋
satin silk 缎,缎类织物
satin soleil 光亮丝缎
satin stitch 缎纹形线迹,缎状绣,密针线迹,手绣线迹
satin stitch dart 打底刺绣
satin stitch embroidery 缎纹针迹刺绣
satin stitching 缎纹线迹
satin stitch seam 缎纹线迹缝
satin stripe 缎条,缎子条纹
satin stripe canvas 缎条刺绣帆布
satin-striped crash 缎条疙瘩绸
satin-striped fabric 缎条织物
satin-striped puckered taffeta 缎条泡泡塔夫绸
satin-striped volie 缎条巴里纱
satin stripe gown 缎条长衫
satin stripe handkerchief 缎条手帕

satin stripes	交织条子缎		服店集中的街道)风貌,沙比路街风格
satin stripes fabric	缎条织物	savina	人造麂皮
satin stripe slip	条纹缎套裙	sawable	可锯品
satin sultan	光亮缎,苏丹缎	sawed stone	锯钻
satin surah	缎光整理细斜纹薄软绸	sawn	印度萨翁棉布
satin taffeta	塔夫缎	saw-toothed cut	锯齿裁切
satin tape	缎带	saw-toothed hem	锯齿边
satin tick	缎纹装饰棉布	sawtoothed seam	锯齿缝
satin tokko	托科缎	saxe blue	灰光浅蓝
satin tops	高级纬面缎	saxon blue	萨克森蓝(淡青色)
satin tulle	色织缎	saxon camblet	萨克森呢
satin turc	土耳其缎,经棉纬缎,法国蚕丝精纺毛纬光亮缎,精仿羊毛缎纹织物	saxon embroidery	萨克森刺绣(长针迹加金属或丝线的斜针迹)
		saxony	萨克森法兰绒;白绒布(加拿大);美利奴花呢;光毛呢(大衣呢)
satin velvet	缎地天鹅绒		
satin victoria	19世纪英国有光条纹女士呢	saxony cord	萨克森棱纹呢
		saxony flannel	萨克森法兰绒
satin vigoureaux	混色缎纹呢	saxony gauze	萨克森羊毛薄型织物
satin wear	缎子衣服	saxony lace	萨克森烂花花边
satin weave	缎纹	saya	菲律宾齐膝短裙,赛亚裙
satin weave fabric	缎纹组织织物	say(e)	细哗叽
satiny(同 satin-like)	仿真丝缎纹织物	Sayette	塞耶特呢
satin zephyr	法国棉经毛纬缎纹织物	sayla	塞拉棉围巾布
sat longyi	印度条子绸腰布	S.B.(Single breast)	单排纽扣
satlor collar	领巾领	s-band silhouette	S带形轮廓,S带型
saturated color	纯色,饱和色	SC(Shopping center)	购物中心《销售》
saturation	色品度;纯度,纯色性;饱和度;浓度	S.C.(sequence control)	程序控制
		scabbard	(剑)鞘,枪套
saturation chroma	彩度	scadinavian pattern	斯堪的纳维亚图案
saturation difference	纯色差	scadinavian sweater	斯堪的纳维亚毛衣,冰岛毛衣
saturation intensity	颜色强度		
saturnia	衬经衬纬纬编机	scale	比例;缩尺;比例尺;等级;样卡;鳞纹
saucer brim	茶碟帽		
saujoo gool goshen	俄罗斯索乔花卉丝缎	scale armour	鳞状盔甲
saulgan shi	印度索甘希平布	scale down	按比例缩小
sauna suit	桑拿训练带,训练带,训练装(同 exercise suit)	scale drawing	(按比例的)缩尺图
		scale model	比例模型,缩尺模型
sausage bag	香肠状手提包	scalene(muscle)	斜角肌
sausage curl	香肠状鬈发	scale ornament	鳞形装饰
sautoir	[法]长项链,宝石金链;马蹬吊带;三角围巾	scalet	猩红
		scale up	按比例扩大
sauvagagi	印度索瓦加几本色或漂白棉布	scaling down	(按)比例缩小
sav	细纺毛呢,细哗叽	scallop	荷叶边,扇形边,月牙边
savage fashion	粗犷款式	scallop buttonhole stitch	扇形纽眼绣
savage look	粗犷款式	scalloped center slit	荷叶边中开衩(衬裙)
savage touch	粗犷发型		
save-all	围裙,围涎,工作服,防护服;罩衫服	scalloped cloak	荷叶边披风
		scalloped collar	海扇领,荷叶领
saved list cloths	白边色织物	scalloped collar shoestring tie blouse	海扇领鞋带结上衣,荷叶领鞋带结上衣
Savile Row look	沙比路街(伦敦第一流西		

scalloped cuff 扇形袖头
scalloped edge 荷叶边
scallop(ed) edge machine 月牙边锁边机
sallop gauge 弧形定规
scalloped handkerchief 月牙边手帕
scalloped hem 扇形边缘
scalloped neckline 荷叶形领口,扇贝形领口
scalloped pocket 荷叶边口袋
scalloped sack 荷叶边宽松外衣
scalloped selvage 扇形边,荷叶边
scalloped trimming 月牙边带子
scalloped tucks 月牙塔克,月牙缝褶
scalloped walking slit 荷叶边侧开衩(衬裙)
scallop-finish edging 扇形边饰
scalloping 扇形饰物
scalloping scissor 贝纹剪刀
scalloping shears 贝纹剪刀
scallop pump 扇形口碰拍鞋
scallop seam 荷叶边缝,扇形边缝
scallop stitch 扇形刺绣针迹,扇形边线迹,荷叶边线迹
scallop zigzag stitch 人字缝,人字线迹
scalp 头皮
scalp lock 剃光头顶上的一绺头发
scan conversion 扫描变换《计算机》
Scandinavian pattern 斯堪的纳维亚花型(多用于针织衫);斯堪的纳维亚图案
Scandinavian style 斯堪的纳维亚风格
Scandinavian sweater (滑雪用)斯堪的纳维亚羊毛衫;冰岛毛衣(用防水羊毛线手工编织的厚型毛衣)
scan-line algorism 扫描线算法《计算机》
scanties (口语)女短内衬裤
scanty 超短紧身女衬裤
Scapa 斯卡帕(法国时装设计师)
Scapula 肩胛骨
Scapulaire 圣牌
Scapular 肩衣(天主教修道士的无袖全服),(天主教徒披的)肩布;无袖外衣,无袖工作服
scapular medal (天主教)肩衣徽章
scapulary 修士服(附兜头帽),无袖法衣
Scarab 圣甲虫雕像
scarab bracelet 圣甲虫雕饰手镯
scarab pendant (古埃及)金龟子垂饰
scarap pocket 扇形口袋
scarf 头巾,领巾,披巾,围巾,腰巾,领带;绶带;肩章

scarf belf 围巾腰带,围巾饰带
scarf cap 方巾帽
scarf cape 围巾式披肩
scarf choker 紧顶链式围巾
scarf collar 方巾领,围巾领
scarf for western-style clothes 西装围巾
scarf hat 飘带软布女帽
scarf mask 露眼头巾面具
scarf neckline 围巾形领口
scarf pin 领带夹针,围巾针,肩巾别针
scarf print 围巾印花
scarf ring 围巾扣(环),领带扣环,领带夹
scarf slide 领巾环
scarf tie 围巾领带
scarf tie collar 围巾式领带领
scarlet 深红色,绯红色,腥红色;红衣,红色制服;鲜红色布
scarlet corns 虫胭脂,虫红
scarlet hat 红衣主教帽子
scarlet red 鲜红,猩红
scarpetti 登山麻底鞋
scattered motif design 散乱花纹,散乱图案
scatter pin 小饰针(通常以两或三枚为一组别在妇女衣服上)
scavilones 男衬裤,男内裤
scenic design 风景图案
scenic print 风景印花,风景印花花样
scenograph 透视图
scenography 透视法,透视画法
scent 香味,气味
scent bag 香囊,香袋
schappe 绢丝织物
schappe voile 绢丝薄绸
schaube 女式灰色羊毛长外套,战士长外套;(德国)黑色无袖长袍
schedule 生产进度表《管理》
schedule time 预定时间
schema 图解,略图
schematic drawing 示意图
scheme of color 着色法
Schiaparelli 斯基亚帕雷利(意大利时装设计师)
Schiffli embroidery 飞梭刺绣,席弗里刺绣
Schiffli embroidery machine 飞梭刺绣机,席弗里刺绣机
Schiffli Lace and Embroidery Manufacturers Association 美国席弗里编带与刺绣商协会

Schiffli machine 飞梭刺绣机,席弗里刺绣机
schizzo 速写画,草图
Schleswing lace 石勒苏盖格花边
schlump 衣衫邋遢的人
schmalband 花边带,窄幅饰带
schmatte 破旧衣服;旧布,破布
scholastic attire 校服
school-age's wear 少年服装;学生装
school badg(e) 校徽
school bag 布书包,书包
schoolboy belt 男生腰带
schoolboy scarf 校服围巾,男生方巾
schoolboy's suit 男生装
school cap 学生帽
school girl look 女生风貌;女生款式
school jumper 女生背心裙
school ring 校名戒指,学生戒指
school shoes 校服鞋,校鞋
school sweater 嵌字毛衣,校园毛衣,字母毛衣
school uniform 学生装,校服
school uniform cloth 学生[校服]呢
school wear 学生装
schreiner calendar 缎光整理,电光整理
schreiner finish 缎光整理,电光整理
schreiner finished fabric 电光整理织物
schreinering 电光工艺,缎光工艺
Schrrer 雪柔(巴黎高级时装店)
schusspol machine 舒斯波尔缝编机
schusspol sewing-knitting machine 舒斯波尔缝编机
science fabric 化学纤维织物
science fashion 科研流行式样,科研时兴产品(体现新材料,新工艺,新功能等产品风格)
science of art 艺术学
science of beauty 美学
science of color 色彩学
scientific management 科学管理
sci-fi look SF 款式,幻想小说款式
scissoring 裁剪;剪下的布条
scissors 剪刀
scissors tuck 剪刀型活褶
scob 跳花疵
scollop 月牙边,荷叶边,扇形边
scolloped collar shoestring tie blouse 海扇领鞋带结上衣,荷叶领鞋带结上衣
sconece 头盔
scoop-back U 形后身

scooped hip 垂臀
scooped neck 汤匙领
scooped neckline 汤匙领口,勺形领
scooped [prominent] hip 垂[翘]臀
scoop neck 椭圆领,勺形领
scoop neckline 勺形领口,椭圆形领口,汤匙领口
scoop neckline blouse 椭圆领口女衫
scoop neckline tank 勺型领口连衣式泳装
scoop pocket 弧形袋
scoop stitch 包缝线迹
scoop up the thread 钩住线
scooter shorts 与迷蒂裙组合的短裙裤
scooter skirt 与迷蒂裙组合的短裙裤
scope 视界,眼界
scorching 烫黄,烫焦
scorch resistance 抗烫焦性
Scotch 苏格兰纺织品,苏格兰粗呢
Scotch brogue shoes 苏格兰雕花皮鞋
Scotch cap 苏格兰帽,苏格兰式无檐帽
Scotch cashmere 苏格兰开司米呢(斜纹)
Scotch check silk 苏格兰(金丝)格子绸
Scotch cloth 细麻布,细布
Scotch grain (鞋革上)碎石压花粒面,粗纹粒面
Scotch grain leather 苏格兰皮革
Scotch heel 英式平袜袜跟,苏格兰式袜跟
scotchlite 回归反射织物(商名,美国 3M)
Scotch pattern 苏格兰格子花呢
Scotch satin 法国条纹段
Scotch suit 苏格兰套装
Scotch-tape 黏胶带,透明胶带
Scotch tartan 苏格兰格子呢
Scotch tweed 苏格兰粗花呢
Scotch tweeds 一套苏格兰粗花呢服装
scots 斯科兹哔叽
Scottie 无帽檐苏格兰便帽
Scottish cambric 苏格兰仿平纹亚麻织物
Scottish cashmere 羊绒斜纹呢
Scottish checks 苏格兰格纹呢
Scottish finish 短毛绒粗呢
Scottish Gingham 苏格兰细格呢(商名)
Scottish highlander costume 苏格兰高地传统男装
Scottish lace 苏格兰梭结花边
Scottish mixture 苏格兰混纺粗呢
Scottish plaid 苏格兰格子呢
Scottish satin 法国条格绸
Scottish tartan 苏格兰格子呢

Scottish tweed 苏格兰粗花呢
scourability 洗涤性能
scoured fabric 煮练织物
scoured skin wool 洗净皮板毛
scoured wool 洗净毛
scoured yarn 煮练纱线
scouring 精练
scouring and bleaching 练漂
scout cap 童子军帽
scout suit 童子军套服
SCR（subjective comfort rating） 主观舒适度
scrabbing board 洗衣板
scrambled eggs 宇航帽
scrambled merchandising 跨行业销售《销售》
scrambled pattern 不规则的花样,混杂的花样
scrap 废料
scrap costs 废料成本
scraper 刮刀
scraper mark （印花布）刮浆印疵,擦痕
scrap-heap 废料堆
scrap leather 皮革碎料
scrap of cloth 衣服小片
scrappage 报废率
scrap pocket 扇形袋(有盖袋)
scratch felt 仿驼绒织物
scratchiness 刺痒感,搔痒感
scratch mits 婴儿护手套
scratch wig （只盖住部分头顶的）半头式假发
scratchy ［服装］刺痒感(纤维性能评价项目)
screen dot 网眼点纹
screening effect 屏蔽效应
screen prints 筛网印花
screen printings 筛网印花,绢网印花
screen-print top 丝网图案印花针织上衣
screw-back earring 拧住式耳环（不用穿耳孔）
screwdriver （螺丝）起子
screw fancy suiting 罗丝呢
screw neck shirt 螺旋领衬衫
screw-on earring 旋合固定的耳环
screw（type）earring 螺旋式耳环
screw wrench 活络扳手
scriber 划线针,划线器
scrim cloth 稀布,粗支纱稀平布
scrim fabric 稀布,粗支纱稀平布

scrimp 折皱,折纹；（旅行者、朝圣者）小袋,小提包,小背包
scrip 小背包,小提包,旅行者小袋
scripper 斯克利帕靴
scriptliner 勾线笔
scroll 涡卷形装饰
scroll work 云纹花样
scroop（同 rustling） 丝鸣
scroop finish（同 rustling finish） 丝鸣整理
scroop neck 汤匙领
scroopy handle 丝鸣感,绢鸣感
scrubbed wool 拉毛织物,刷净皮板毛
scrubbing agent 洗涤剂
scrubbing board 洗衣板
scrubbing brush 洗衣刷
scrub suit 手术服
scruff 后颈,领背,颈背皮
scrunched-up sweater 腰部以上松密的针织运动衫
scuba diving 潜水服
scuba diving wet suit 带呼吸器的深水潜水服
scuba mask 潜水面罩
scuff 家用平底拖鞋,斯卡夫拖鞋,斯克夫拖鞋,露趾平跟凉鞋
scuffing 擦痕
scuffs 平底拖鞋；丝克夫拖鞋
scuff slippers 平底拖鞋
sculpted heel 雕塑型鞋跟
sculptural fashion 雕塑风
sculptured effect 浮雕效应,泡泡纱花纹效应
sculptured heel 雕塑形鞋跟
sculptured knitwear 浮雕毛衫
sculptured pattern 凹凸花纹,浮雕花纹
sculptured pile fabric 浮雕绒头织物
sculptured velvet 凹凸绒
sculptured wool sweater 浮雕羊毛衫
scumble 柔和色调,暗淡柔和
scumming 罩色,沾色
scutcheon （装饰性的）锁孔铜盖；铭牌
scuttle bonnet 煤斗式帽
scuturm 膝盖骨
scye 袖窿；袖孔
scye depth 腋深
scye line 袖窿线
scye width 袖窿宽
SD（service dress） 军便服
S.d.（sport dress） 运动服

sea bag 水手袋
sea beaver 海獭
sea blue 海蓝(绿光蓝色)
sea boots 高筒防水靴,高筒橡皮靴
seacrest 海浪峰绿(淡绿色)
sea dog fur 海豹毛皮
sea foam 海面泡沫色
seafoam green 海沫绿(浅艳绿色)
seaforth 西福斯
sea gear 海洋服,海滨服
seagrass handbag 海草手提袋
sea green 海绿(黄绿色),淡蓝绿色
sea-island composite fiber 海岛型复合纤维
Sea Island cotton 海岛棉
sea leather 海产皮草
sea legs 男子紧身沙滩裤
sea lion 海驴毛皮
sea mist 海雾色
sea otter [fur] 海獭毛皮
sea pink 海红色
sea side costume 海滨女服
sea side wear 海滨服
sea spray 深灰绿
seal 海豹毛皮;海豹皮制品;人造海豹皮;火漆色(红棕色),暗褐色
seal boa 海豹皮围巾
seal brown 海豹帽;暗褐色
seal carton by sealing tape 封箱
seal coat 海豹皮大衣
sealed edge 封边
sealed pattern (军队服装的)标准型(英国)
sealed samples 封口标样
sealed zipper 密封拉链
sealette 仿海豹皮绒布
seal fur 海豹皮
sealine 西兰皮(澳洲兔毛皮),假豹皮
sealion skin 海狮毛皮
seal plush 海豹绒
seal rabbit 仿制用兔毛皮
seal ring 图章戒指,封印戒指
sealskin 海豹皮;海豹绒,海虎绒,海豹皮制品;海豹皮服装
sealskin cloth 海豹绒,海虎绒,仿海豹皮织物
sealskin coat 海豹皮上衣
sealskin fabric knitting machine 仿海豹皮针织机
seam 缝;线缝;缝口,缝合;接缝;缝纫;缝型;针线活
seamability 可缝性
seam abrasion machine (SAM) 线缝磨损机
seam abrasion resistance 缝纫耐磨性
seam accuracy 缝纫精度
seam allowance 缝头,缝份,缝合允许量,缝接允差
seam allowance short 缝份不足
seam amount 缝份量
seaman cap 水兵帽
seam assemble 缝料组件,缝纫组件
seam bar mark 缝形条痕
seam bartacking system 加固线缝装置
seam basting 疏缝,临时接缝
seam beading 镶珠缝
seam binding 滚边;滚条,滚边材料
seam board 托垫
seam breakage 缝纫破裂,断线缝
seam breakage strength 缝纫破裂强力
seam cockling 接缝起皱
seam configuration 缝型轮廓
seam contour 缝型外形
seam contour data 缝型外形数据
seam corners and angles 缝型弯角
seam counter timer 缝头计数定时器
seam covering 覆盖缝
seam covering machine 覆盖缝缝纫机
seam cracking 缝线开裂
seam crossing 接缝处
seam damage 缝纫损伤,缝迹破损
seam dart 合身短缝,省缝
seam depth 接缝深度
seam detector 线缝探测器
seam direction 接缝方向
seam distance 接缝长度,缝距
seam distortion 标准线歪斜
seam edge 缝边
seamed hose 有缝袜,全成形有缝袜
seamed stockings 拼缝袜
seamed toe 拼缝袜头
seamed waistline 剪接腰缝
seam efficiency 缝口强度与材料强度之比,缝合效率
seam elasticity 缝线弹性
seam elongation 线迹弹伸性
seam elongation shortage 线迹弹伸不足
seam end 缝端,缝止
seamer 缝纫机;缝纫工
seamer eye 缝头检验装置

seam finish(同 edge finish) 缝份(缘边)加工处理
seam finishing 缝型精加工
seam finish(ing) type 接缝形式
seam flammability 接缝耐磨性
seam folder 搭缝折叠器
seam folding direction 倒缝份的方向
seamfree 无缝的
seamfree pantyhose 无缝连裤袜
seamfree stockings 无缝长袜
seam gauge 导边器
seam grinning 线缝裂开
seam guide line 缝宽标志线
seam heading 两次缝间距,头份
seaming 缝纫;缝合;缝接
seaming armhole 绱袖
seaming bow 缝迹歪斜
seaming defect 缝迹疵点
seaming front edge 缝止口
seaming lace 窄网眼花边
seaming machine 缝纫机
seaming necktie 缝领带
seaming operation 缝纫操作
seaming penetration 穿透缝
seaming position 缝纫定位
seaming security 缝接牢度
seaming stitch 缝纫线迹
seaming stitch density 缝纫密度
seaming stitch extension 缝迹延伸性
seaming toe fault 缝头疵点
seaming tool 缝纫工具
seam interruption 线缝遗漏
seam joint 缝合,缝接
seam jumper 线缝探测器
seam jumping 遇缝自动跳跃,跳缝
seam length 接缝长度
seamless 圆形编织,无缝的
seamless bra[ssiere] cup 无缝胸罩窝
seamless cloth 无接缝织物
seamless-cup bra[ssiere] 无缝杯形罩杯
seamless gloves 针织手套,无缝手套
seamless hose 无缝圆袜,无缝管状织物
seamless hose machine 无缝圆袜机
seamless hose raschel machine 拉舍尔无缝袜机
seamless hosiery 无缝圆袜,无缝管状织物
seamless hosiery machine 无缝圆袜机
seamless knit 圆机针织物
seamless opera pump 无缝欧瑟拉潘普鞋(观剧或晚宴时穿着)
seamless repeat 无拼缝花型循环
seamless rib machine 无缝罗纹圆袜机
seamless shoes 无缝鞋
seamless stockings 无缝长筒袜
seamless socks 无缝袜
seamless (top) shoes 无缝鞋
seamless toe pouch 无缝袜头
seamless tubing 无缝管状织物
seamless wear technology 无缝衣着技术
seam let-out 线缝松弛,尾份
seam line 沿缝,缝路,接缝线
seam line pocket 摆缝袋,沿缝袋
seam margin 缝型边缘
seam mark 缝头压痕,缝头色疵
seam open 缝型开口
seam opening 接缝线开口;裂缝;开缝;缝型;缝式
seam pattern 缝样,缝式
seam-penetration 穿透缝
seam performance 缝迹性能
seam piping 滚边缝
seam placket 接缝线开口,用缝头开衩
seam plait 缝合打裥
seam pocket 裁片剪接处口袋,摆缝袋,侧缝插袋
seam puckering 线缝缩拢,线缝起皱,综合起拱
seam puckers 线缝皱缩
seam quality (服装)缝制质量
seam raveling 线缝脱散
seam raw edge exposed 接缝外露
seam reinforcing 接缝加固
seam resistance 缝合强力
seam ripper 拆线器
seam roll 烫袖垫,袖馒头
seam run 缝迹光顺
seam run-offs(同 raised seam) 凸缝,线缝隆起
seams 接缝缝头,缝纫毛头
seam security 缝口牢度,缝合牢度
seam sensing system 接缝传感系统
seam shrinkage 缝缩
seam slippage 缝口脱开,脱缝
seam slippage test 缝合牢度试验
seam smoothness 衣缝的平整度
seam spacing 缝型间距,缝迹间距
seam specifications 缝型规范
seamster 裁缝
seamstering 女裁缝的工作

seam stick　托垫
seam stitch　缝型线迹
seam strength　缝纫强力,缝合强力,缝边断裂强度
seam strength shortage　缝迹强度不足
seamstress　女裁缝师,缝衣妇,做针线活的妇女;缝纫应力
seam tape　缝合斜布条,滚边条
seam-tear resistance　抗接缝撕裂性
seam tensile strength　缝纫拉伸强度
seam thickness　接缝厚度
seam toe　拼缝袜头
seam tracking device　接缝追踪装置
seam turner　缝纫机假缝压脚
seam twist　接缝卷曲
seam type　缝纫线迹类型,缝型,缝式
seam undulation　线缝波浪形
seam welding　热熔缝接
seam width　接缝宽度,缝份宽
seam width gauge　接缝宽度仪
seam width range　接缝宽度范围
seamy side　服装反面,里子,夹里
sear cloth　蜡布
seasonable change　季节变化
seasonality　季节性
seasonal shade　流行色
season color　流行色,应时色,季节色
seasonless fashion　无季节时装
seasonly-held show　季节性时装表演
season's wear　季节性服装
seat　(人或裤的)臀部;上裆;裤裆;臀围
seat belt webbing　窄编织带
seat circumference　臀围
seat girth　坐围,臀围
seat level　臀围线
seat line　臀围线
seat measure　臀围
seat piece　裤腰坐份
seat reinforcement　臀围加固
seat seam　后裤裆缝,裤后缝,臀部接缝
seat suit　运动衫裤,运动套装
sea-water fastness　耐海水色牢度
sea-water resistance　耐海水性
seaweed green　海藻绿,浅灰绿色
se-back heel　后置跟
sebastopol　经色条斜纹呢
sebenia　摩洛哥条子绸
secco cocnie　深地暗彩条布
secondary block　修正样板
secondary color　间色,二次色;混合色,合成色
secondary shades　混合色,调和色,次色
secondary skin　第二皮肤(指服装)
secondary tint　柔和毛,柔和的颜色
secondbag　第二提包,小型包(只装化妆品等随身物件,宴会时携带,可装在手提包内)
second cut　重剪毛
second fitting　第二次试穿
secondhand　旧货
secondhand clothes　旧衣,二手服装
secondhand goods　旧货
secondhand saleman　旧货商
secondhand shop　旧货店《销售》
seconds　二级品,二等品,次货,等外品
second skin　第二层皮肤(指服装对人身的关系)
second skin look　第二皮肤款式
seco silk　塞科绵绸
secrete　三层裙的最里层衬裙
secret finish　保密整理(织物用新专利整理)
secret-print　隐蔽印花
sectional cutting table　组合裁床
sectional view　剖视图,断面图
sectional warper　分段整形机
section mark　分条痕
section shape　裁片形状
sector compasses　两脚规
secure stitching　安全缝纫
securing seam　保险缝
seda　蚕丝,丝绸(西班牙称谓)
sedge green　蓑衣草绿色
seduction　魅力
seductive look　诱惑款式
seeded fabric　低级棉织物
seed effect　满地点子花纹
seed embroidery　种子刺绣,细点子花纹刺绣,籽粒刺绣
seeding stitch　播种针法
seedling　幼苗绿
seedpearl　小粒珍珠,芥子珠;(作饰物镶嵌用的)小珠;米珠色
seed smocking　种子缩褶绣
seed stitch　小点刺绣线迹
seed voile　疙瘩花纹巴里纱
seed yarn　粗节花式线
seehand muslin　西亨特细布
seeing glass　镜子
seer　泡泡纱,条格绉布,皱条纹薄织物

seerband 印度缠头纱
seerhandconat 印度薄细布
seerhand muslin 印度薄细布
seersucker 泡泡纱,条格绉布,皱条纹薄织物,弹力绉
seersucker bed-cover 泡泡纱床罩
seersucker gingham 格子泡泡纱
seersucker knitted fabric 针织泡泡布
seersucker printings 泡泡纱印花
see-saw motion 上下(往复)运动
see-through 透视装,透明装;(泡沫衬里)底可见黏合针织物
see-through blouse 透明女衫
see-through dress 透明裙装
see-through fabric 纱罗织物,透光薄纱织物
see-through fashion 透明款式,透视装式样
see-through look 透空风貌,透视风貌,透明款式
see-through pants 透明裤
see-through shirt 透明衬衫
segmentae 拼合装饰
segmental body padding 身体局部填充料
segmentation 细分化
segovie 法国起绒毛织物
segovienne 14～18世纪斜纹法兰绒
seide 蚕丝,丝绸
seietterie 塞耶特尼
selected bidding 选择性招标
selection figure of leader style 领导风格选择图
selective colour 选择性颜色
selendang [印尼]萨伦丹披肩
self-adhesive tag 魔术贴,尼龙搭扣
self-belt 同料腰带,同种布料腰带
self-binder weaves 自身结接的多层织物
self-bonded nonwoven 自黏合非织造布
self-bound seam 漏落缝
self-bound seam with sink stitch 漏落缝针迹
self-casting 同料翻折串带管
self-color 原色,天然色,本色,单色
self-color backing 单色衬里,本色衬里
self colored 同色
self-contained pressing unit 整装的压烫装置
self-control 自控
self-covered belt 同料腰带(用服装同料制成的腰带)

self-covered button 同料纽扣,自包扣
self-edge 织边,布边
self-edge seam 织边缝,布边缝
self-edge tape 布边滚带
self-enclosed seam 来去缝
self-extinguishing character 自熄性
self-fabric 同料(滚边、纽扣等用料与衣料相同)
self-fabric tubing 同料串带管
self-facing 连裁过面,连裁,连挂面
self-fasten belt with D-rings D型环扣腰带
self-figure 本色织花,织花
self-fold 连贴边,连门襟
self-gripping fastener [自锁]尼龙搭扣,自搭扣,魔术贴
self-gripping nylon tape fastener [自锁]尼龙搭扣
selfham 旅行圆斗篷
self-inflammability 自燃性
self-inspection 自检《管理》
self-ironing 免烫整理
selfix sock 橡口短袜
self-lined fabric 自编衬里缝编织物
self-lining 同料夹里
self-locking slider (SL slider) 自锁拉链头
self-lubricating resin 自润滑树脂
self material 同料,主料
self-neatened finish 自身整洁加工,包缝加工
self-neatening 包缝
selfnylon tape fasteners 尼龙搭扣
self-piping 同料边饰,同料嵌线
self-programming contour seamer 自动程序控制外形缝纫机
self-sash 用服装同料制成的扎结饰带
self-selvedge 织边,布边
self-service discount department store (SS-DDS) 无人售货廉价店《销售》
self-shade 本色,单色,一色
self-shank button 单柄纽扣
self-shine polish 液体鞋油
self-smoothing 免烫
self-smoothing cellulose fabric 免烫纤维素织物
self-smoothing fabric 免烫织物
self-stick sewing tape 自黏性助缝胶带
self-stripe 同色条纹,本色条,织花
self-tone 单色调

self-tone embossd 素色凸花
self-trimming 同料边饰
self-twisting yarn out of phase 相差自捻纱
self-twist spun yarn 自捻纺纱
self-twist twisted yarn 加捻自捻纱
self twist yarn(ST yarn) 自捻纱
self-welt pocket 一字嵌线袋,单嵌线袋
seliesia 坚牢轻软的亚麻布(或斜纹布)
seller 出卖人
seller's sample 卖方样
selling sample 推销样
seloso 塞洛瑟裙装(非洲妇女裙装)
selvage 布边,织边,边缘
selvage curling of knitted fabric 针织物卷边性
selvaged 有织边的
selvage edge 布边
selvage guide (缝纫机)导布器
selvage hem 布卷边;布折边
selvage list 织边,布边
selvage rand 布边,边饰
selvage-seaming machine 包缝机,拷边机,缝边机
selvage to selvage 布边连布边,布边接布边
selvage waistband 布边作的腰头
selvedge 布边,织边
selvedge crease 皱边疵
selvedge curling 卷边
selvedge defect 边疵
selvedge gumming 布边上胶
selvedge hem 折边
selvedge legend 边印
selvedge machine 包缝机
selvedge mark 布边折痕疵;边记,边字
selvedge rand 布边,边饰
selvedge reinforcing 布边加固
selvedge-seaming 包缝,缝边
selvedge-seaming machine 包缝机
selvedge tear 破边
selvedge thread 织边线
selvedge to selvedge 布边接布边
selvedge trimmer 切边器,修边装置
selvedge trimmers and splitting device 切边分割装置
selvedge-trimming device 切边装置
selvedge-trimming machine 剪边机,修边机
selvedge trim remover 除边装置

selvedge turndown 翻边疵
selyem 蚕丝,丝绸(匈牙利称谓)
sematic color (动物的)警戒色
semi-auto snap fastening machine 半自动钉揿纽机
semi-automatic flat knitting machine 半自动横机
semi-automatic hose machine 半自动织袜机
semi-automatic oil 半自动加油
semi-automatic pressing machine 半自动熨烫机
semi-axes 色差对比卡
semi-baggy pants 半袋形裤
semi-bal collar 关驳领
semi-better price 较好价格《销售》
semi-brogue 半布洛格鞋
semichrome (奥斯瓦尔德制)纯(彩)色
semi circle 半圆,半圆弧,半圆形;半圆规
semi-circular skirt(i) 半圆裙,两片喇叭裙
semi-clover collar 西装圆上领,苜蓿叶形西装领上领,半苜蓿领
semi-clover lapel 苜蓿叶形西装领下领,西装圆下领
semi-combed yarn 半精梳纱
semi-concealed zip 半隐形拉链
semi cope 短披肩,短上衣
semi cut-away collar 八字领
semi-cord 棱纹棉天鹅绒
semi-custom made 半定做的(服装)
semi-cut-away collar 八字领
semi-dress 便礼服,简式礼服
semi-dress shirt 便礼服衬衫(介于白衬衫与运动衫之间)
semi-dull 半无光
semi-durable finishing 半耐久性加工
semi-durable goods 一般耐久性织物
semi elliptical tacker 半椭圆形加固机
semi-employed 半就业
semienyoung 黑天鹅绒
semi-evening-dress 简式晚礼服
semi-fashioned hose 半成形袜
semi-fashioned hosiery 仿全成形袜子
semi-finish 半光面整理
semi-finished denim 白坯牛仔布
semi-finished good 半制品,半成品
semi-fit 半紧身式的,半紧身式服装
semi-fitted 半紧身式的
semi-fitted blouse 较合体上衣

semi-fitted dress 半紧身连裙装
semi-fitted jacket 半紧身夹克衫
semi-flared skirt 半圆裙,小喇叭裙
semi-fold binder 半折叠滚边器
semi-formal 半正式礼服
semi-formal suit 半正式礼服套装
semi-formal wear 半正式礼服
semi-gloss 半(有)光,近有光的
semi-gloss finish 半光整理
semi-[half] slip 短衬裙
semi-hard 半硬(色彩谐调)
semi-industrial machine 半工业用缝纫机
semi-made 半成品
semi make-through 近乎一半由一人缝合完成
semi-manufactures 半成品
semi-matt 半无光
semi-mounted sleeve 加角连袖
semi-notched lapel 半刻领,半菱形领,半V字形领
semi-open design 半网眼花纹
semi-order (服装)看样定货,半定货(根据样品选择设计款式和衣料,量体,不经试穿的定做方式)
semi-peaked collar 仿戗驳领,仿夹角西装领
semi-peaked lapel 半戗驳领,半夹角翻领,半尖领,半菱领,半刻领
semi-permanent flameretardent (SPER) 半耐久性阻燃
semi-permeable hollow fibre 半渗透性的中空纤维
semi-plain 花式平素
semi-plant 中间工厂
semi-product 半制品
semi-production 半成品
semi-pumps 透孔鞋,仿碰拍鞋,透空包鞋
semi-raglan 半插肩袖式
semi-raglan clothing 半连肩袖外套
semi-raglan coat 半插肩外套,半套袖大衣
semi-raglan sleeve 半斜包肩袖,半连肩袖,仿连肩袖
semi-rep 棱纹布
semi-service stocking 中厚型长丝袜
semi-sheer 半透明薄绸
semi-sheer style 半透明服装风格
semi-short hairstyle 半短发型
semi-sleeve 短袖
semi-sleeve shirt 短袖衬衫
semi-slip 短衬裙
semi-soft collar 半软硬领
semi soutien collar coat 半二重领外套
semi-staggered repeat 半交叉型图案循环
semi-spread collar 半展开领
semi-square neckline lace trim gown 半方形花边领口长衫
semi-staple 半大路货织物,半畅销织物
semi-step-in 有开口的绑肚
semi-stiff collar 半硬领
semi-support apparel 半支持型服装
semi-synthetic fiber 半合成纤维
semi-thread fabric 半线织物
semi-tight skirt 半紧身裙
semitone 中间色调
semi-transfer 半自动化
semitransparent 半透明
semitransparent blouse 半透明女衫
semi-up hairstyle 中髻发型
semi-V neckline 半V形领口
semi-voile 半巴里纱
semi-widening 半放针
semi-Windsor knot 半温莎领带结
semi-works (试制新产品或试行新工艺的)小规模工厂
semi-worsted fabric 半精纺毛织物
semi-worsted fancy suiting 半精纺花呢
semi-worsted yarn 半精梳纱
semmit (苏格兰)贴身内衣
sempiterne 森皮特纯毛哔叽
sempstress 女服装工,女裁缝
sendal 里子薄绸,森丹绸,森丹薄袍
sender 送进装置
senorita 西班牙女工齐腰短外套
sensation of color 色觉
sense 感官;感觉;意识
sense impression 感性知觉,感性印象
sense of beauty 美感,审美感
sense perception 感性知觉
sense test 感官检验
senshaw 平纹棉布
sensibility 感受力,敏感性
sensibility of form 形式感
sensible heat 显热
sensible outfit 实用耐穿的服装
sensible shoes 实用鞋
sensitive product 敏感性产品
sensitized fabric 敏化织物
Sensua 赛绍弹力织物(商名)
sensual color 肉感色,性感色

sensual curves 性感曲线
sensual look 肉感,性感;性感装
sensuous 给人以美的享受的,激发美感的
sensuous look 美感款式
senu 赛奴绣花细布,赛奴绣花白网眼布
separable fastening fabric 尼龙搭扣
separable zipper 分享式拉链,开尾型拉链
separate collar 驳领的领片,可分式领
separate dress (女)套装,组合套装(上下衣用料不同的套装)
separate facing 分离贴边
separate fastening fabric 尼龙搭扣
separate flap 与袋口分离的袋盖
separate pants 分开开内裤
separate plate 分离板,隔板
separate pocket 袋布与面料分开的口袋
separate-pocket seam 袋布与面料分开的插袋缝
separates 自由组合套装;不配套的服装,单件衣着(指妇女不配套穿着的衣、裙等);上下装;运动休闲装
separate skirt 分开穿着的裙子,女裤裙
separates style 组合款式(指上、下衣自由组合的服装)
separate stand 分离领座
separate style 上下衣自由组合的款式,组合式服装款式
separate underpants 不配套的衬裤(内裤)
separate-welt pocket 嵌线袋
separate zipper 分离式拉链,开尾型拉链
separating-end-zipper 分头拉链
separating zipper 分离式拉链,开尾型拉链
separation lace 抽线花边,分离花边
separatrix (表示分数,日期的)斜线分隔符号(即"/")
sepia 棕黑色;乌贼墨棕,乌贼墨色;黑褐色;深棕色
septain 法国塞普坦装饰带
sequin (服装)闪光装饰片,金属小圆片,珠片
sequin applique and trimming 珠片贴绣
sequined fabric 缀有金属圆片的面料
sequin lace 西昆花边,钩编彩色花边
sequins(同 toreador suit) 斗牛士装
sequins and beading 钉珠
sequoia 红杉色
serafin 白地印花毛织物

serape 瑟拉佩披肩(流苏饰边,墨西哥人用),毛毯披巾,毡斗篷(中美地区)
serbattes 金边细平布
serf-restraint agreement 自限协定
serf-twist yarn in phase 同相自捻纱
serge (毛)哔叽;赛鲁,锁(毛边)
serge back at yoke 过肩包缝
serge bottom 底边包缝
serge canvas 小方块纹哔叽
serge cloth 哔叽呢
serge d'aumale 法国毛哔叽,奥玛尔子呢
serge de berry 19世纪英国厚重毛哔叽
serge de Rome 罗马薄哔叽
serge de soie 丝哔叽
serge double cloth 英国双层丝哔叽
serged seam 粗缝,接缝,包缝
serge front facing 前贴边包缝
Serge Lepage 塞尔热·勒帕热(法国时装设计师)
serge lining 哔叽衬里布
serge moire 波纹丝哔叽
serger 锁边工人,包边缝纫机
sergette 窄幅丝绸,轻薄哔叽织物
serge twill 本色斜纹布
sergical sutures 医疗缝线
sergine 日本丝哔叽
serging 锁边,包缝,粗缝,包边
serging machine 包边缝纫机,锁边缝纫机
serging operation 包边缝纫
serging pine leaf stretch stitch 松叶状包边的弹性线迹
serging stitch 包边线迹
series design 系列设计
series garments 系列服装
series of apparel size 服装号型系列
serigraph 绢网印花
serpentine 人字形斜纹,锯齿形斜纹
serpentine belt 波形腰带
serpentine braid 弯曲编带
serpentine crepe 蛇纹绉
serpentine seam 蛇形缝
serpentine skirt 螺旋形花色裙,蛇裙
serpentine stitch 蛇形线迹
serpentine twill 锯齿形斜纹,人字形斜纹
serpent ring 盘蛇戒指
serrated edge 锯齿边
serrated edge pattern cutting machine 锯齿边样本剪切机

serrated knife-edge shears 齿形刃剪刀,齿形刀边剪刀
serré 紧密织物
sers 棉毛(运动)衫裤
seru 日本薄哔叽
serul 土耳其式宽松长裤
serviceability 服用性,耐用性,实用性
serviceable life 使用期
serviceableness 穿着性能,耐用性,实用性
service cap 平顶有檐的美国军帽
service cap cover 军帽罩
service dress (SD) 军便装
service durability 实用耐久性
service hat 美国陆军便帽
service leather 两层皮
service lift 运货电梯
service mark 服务标志《销售》
service needle 备用机针
service parts 备用零件
service performance 服用性能
service serge 军服哔叽
service stockings 耐穿袜,厚袜子
service stripe 军龄袖条(军服左袖上斜条,陆军每条3年,海军4年),工龄袖条(指铁路制服衣袖上的横条)
service test 穿着试验,服用试验,试穿,试用
service uniform 制服,军装
service weight hosiery 耐穿袜,耐磨袜
service weight stocking 耐穿袜,耐磨袜
servoactor 随动件
servo gear 伺服机构,助力机构
sesame 芝麻色,浅橄榄灰,玉灰
set (一)套,组;定形;姿势,身材;(衣服穿在身上的)样子;安装,装置;缝、绱(in);(衣服)合身;织物经纬密度
set-back heel 后置跟
set check 对称格子
set hood 绱兜帽
set-in 另外缝上的,装带
set-in back yoke 绱过肩
set-in belt 嵌入型腰带,固定腰带
set-in collar 绱领子
set in hood 绱风帽
set-in pocket 挖袋;插口袋;斜袋
set-in shoulder pad 装垫肩
set-in sleeve 普通袖,圆袖;装袖
set-in sleeve tab 绱袖袢
set-in sleeve with center 中缝圆袖

set-in sleeve with center seam 中缝圆袖
set-in thumb 固定式拇指;装拇指手套(剪接拇指部分的手套,主要用于配礼服)
set-in waistline 嵌接腰线
set lotion 头发定型液
set mark (织物)密路
set off 陪衬物,饰物;衬托,点缀,装饰
set off color 陪衬色
set-on 装领(与连领区分)
set-on pocket 贴袋
set out 服装,装束;(商品等的)陈列,展览
set pocket 缝口袋
set screw 固定螺钉,定位螺钉
set shade zone 调整阴阳色调《计算机》
set sleeve 装袖,绱袖
set square 三角板,三角尺
setting 镶嵌;镶嵌式样;定型,定形;安装,固定
setting belt 装腰带
setting collar 装领
setting comb 定型梳
setting cuff 装袖口边,装克夫
setting interlinng 贴衬
setting of gray fabric 坯布定形
setting pad 固定衬垫
setting pocket 缝口袋
setting sleeve 装袖,绱袖
setting zipper 装拉链
settlement 结算
set to shape 定型
set-under heel 下置跟(跟后及两侧内凹)
setup (身体的)姿势,姿态;体格;体制《管理》
set-up suit 组合式套装
set width 设置门幅宽
set work 放样(常指绒线的绣花)
set yarn 假捻定型变形丝,低弹变形丝,定形丝
set yarn fabric 低弹[丝]织物
seven 7号尺码的衣服
seven and a half heads theory 七个半头高理论
seven-eights 八分长
seven-eights hose 高尔夫球袜;中筒运动袜
seven-eights length 膝上长度
seven-eights pants 九分裤(长至小腿肚下)

seven-eights sleeve length 八分袖长
seven heads theory 七个头高理论
sevens 7号尺码的手套(或鞋等)
seven-star-frog-shaped shawl 蛙形七星图案披肩
seventeen 17码的衣服、鞋(或袜等)
Seventh Avenue (S.A.) 第七大街(美国纽约时装业中心)
seventh cervical vertebra 第七颗脊椎
severe milling 重缩呢,重缩绒,重缩毡
severity 纯洁;朴素;严谨;加工深度
seville lace 塞维利亚镶边花边
sew 缝,缝合,缝拢,缝入,缝制,缝补
sewability 可缝性
sewability of fabric 织物可缝性
sewable 可缝纫的
sew-back button 长脚扣,无眼纽扣
sew button loop 钉扣袢
sew down to here 缝止处
sewed toe 缝合袜头
sewer 缝纫工具,缝纫机;缝衣匠,裁缝
sew front and back rise togerther 合前后裆缝
sew hook and eye 钉领钩袢;装钉领钩
sew-in 织补,缝补;缝合
sewing 缝头,缝纫;缝纫业;缝纫法;缝制品
sewing accessories 成衣辅料
sewing accuracy 缝纫精度
sewing action 缝纫动作
sewing aids 缝纫辅助用具
sewing and embroidery scissors 缝纫、绣花剪刀
sewing area 缝纫作业面,缝纫部位
sewing attachment 助缝装置
sewing basket 针线篮
sewing bobbin 缝纫用线团,缝纫机梭心
sewing book 服装裁剪样本
sewing bow 缉线偏离,弓形缉线
sewing box 针线盒
sewing button 钉纽扣
sewing button loop 钉纽袢
sewing capacity 缝纫能力
sewing characteristic 缝纫特性
sewing circle 缝纫周期;妇女缝纫小组
sewing condition 缝纫条件
sewing cotton 缝纫棉线
sewing cuff to sleeve 缩袖口
sewing cycle 缝纫周期
sewing damage 缝纫损伤

sewing dart 缝省
sewing directton 缝纫方向
sewing disk 缝盘
sewing edge guide 缝边导轨
sewing efficiency 缝纫效率
sewing element 缝纫零件
sewing elongation 缝纫线(缝纫时)伸长
sewing equipment 缝纫设备
sewing facing to front 缝前片(衣片)
sewing finish 提高缝纫性整理
sewing forward and reverse 倒顺缝纫
sewing front and back rise together 缝合前后裆
sewing function 缝纫性能,缝纫功能
sewing gauge 缝纫标尺;缝距
sewing hook and eye 钉领钩扣
sewing inspection 缝制检验
sewing kit 针线包
sewing-knitting method 缝编法
sewing-knitting technique 缝编技术
sewing length 缝纫长度
sewing light （缝纫机上的)缝纫照明灯
sewing machine (S.M.) 缝纫机
sewing machine attachments 缝纫机附件
sewing machine for household purposes 家用缝纫机
sewing machine needle 缝纫机针,缝纫针
sewing machine oil 缝纫机油
sewing machine operator 车缝工,缝纫工
sewing machine part 缝纫机零件
sewing machine stitch 机缝线迹
sewing machine with edge trimming mechanism 切边缝纫机
sewing mark 缝头色疵,缝头漏色(在缝头处,印浆渗出污染坯布)
sewing material feeder 缝料输送器
sewing mechanism 缝纫机构,切边缝纫机
sewing method 缝纫方法
sewing model 缝纫机型号
sewing necessities 缝纫必需品
sewing needle 缝纫针
sewing notch 缝纫对位刀眼
sewing on label 钉商标
sewing on belt loop sewing machine 钉裤带环缝纫机
sewing on snaps 钉按扣
sewing operation 缝纫操作
sewing out 缝纫卷边,将毛边折入
sewing paper 裁剪纸样
sewing part 缝纫部件

sewing performance 缝纫性能
sewing pitch 缝纫节距
sewing placket to sleeve 绱袖衩条
sewing plate 缝(制)台
sewing position 缝纫位置
sewing procedure 缝制步骤,缝纫步骤
sewing process 缝纫过程
sewing programme 缝纫程序
sewing reliability 缝纫可靠性
sewing room 缝纫车间
sewing sample 缝纫样本,缝样
sewing scissors 缝纫剪刀
sewing shop 缝纫店
sewing silk 缝纫丝线
sewing skill 缝制技能
sewing space 缝纫间距
sewing speed 缝纫速度
sewing station 缝纫工位
sewing station system 缝纫工位系统
sewing table (缝纫)台板,机板
sewing tag 挂牌
sewing technician 缝纫技师,缝纫技术人员
sewing technological requirements 缝纫工艺要求
sewing technology 缝纫技术
sewing test 缝纫试验
sewing thickness 缝纫厚度
sewing thread 缝纫线,绣花线
sewing thread for leather garment 裘皮缝纫线
sewing thread yarn count 缝纫线特数
sewing time 缝纫时间
sewing tool 缝纫工具
sewing trouble 缝纫故障
sewing twist 缝纫丝线
sewing type 缝纫方法,缝纫类型
sewing-up 缝合,缝头
sewing versatility 缝纫多面性,缝纫多功能性
sewing workshop 缝纫车间,缝纫工厂
sew-in label 缝上的标签
sew-knit fabric 缝编织物
sew-knitting machine 缝编机
sewlight 缝纫灯
sewn-on pocket 贴袋
sewn-on strip 缝带
sewn positioning 缝后定位
sewn product 缝制品
sewn products industry 缝制工业

sewn seam 已缝好的接缝
sewn selvedge 缝合式布边
sewn welt 缝制贴边
sew on 缝制;缝上;钉上
sew on button 缝线纽扣
sew one-way pleat 叠顺裥
sew on label 绱商标
sew-on pocket 贴袋
sew-pink machine 锯齿边缝纫机
sew placket sleeve 绱袖衩条
sew pocket 贴袋
sew pocket welt 绱口袋嵌线
sew-positioning 缝后定位
sew-through button 平缝扣,有眼纽扣
sew together bodice and its lining 合大身里,面
sew together front and back seam 合前后裆缝
sew together hood and its lining 合帽面里
sew together sleeve and its lining 叠袖里缝
sew together waist band and its lining 合腰头
sew to here 缝至此
sew twice 缝两道线
sexangle 六角形
sex appeal 性感
sexless corner 男女兼用装饰用品售货处
sextyping 按性别分类
sexuality 性征,性别
sexy clothes 透明或暴露式的极富性感的服装
sexy costume 性感服装
sexy fashion 性感时装
sexy look 性感型,性感装,性感款式
seyong 黑丝绒,蓝丝绒
s/f 打样(配色样本)
SFH standard fading hours 标准褪色时间,标准暴晒时间
S-finishing S-整理,表面皂化整理(改进织物手感,减少静电)
SF look (衣)起翘横线
SG mark safety goods mark 安全制品标志
S&H 百分百
shaatnez 毛麻混纺织物
shabby clothes 褴褛衣服
shabby style 褴褛风格
shabnam 印度达卡细薄布
shabri 白色羊绒,银灰羊绒

shade 色泽,色光,色调,色度,明暗程度;(保护眼睛的)遮阳帽檐
shade bar 色档疵
shade card 配色样卡
shade change 色纬档(织疵)
shaded check 渐变色方格花纹
shaded cloth 色泽深浅不匀的织物
shaded design 蜡防印花花纹
shaded effect 阴影效应,阴影花纹
shade deviation 色泽差异
shaded filling 色纬档,色纬影
shaded parts 色泽不匀的部位
shade-dried 阴干
shaded spruce 暗云彩色
shaded stripe 阴影条纹
shaded twill 阴影斜纹
shade duplication 色调重现性
shade lines 阴影线
shade marking 色泽标记
shade matching 色泽匹配
shade number 色号
shade official 公定色泽
shade pitching 打色样,拼色
shade range 色谱
shades 太阳眼镜,墨镜
shade satin 云纹缎纹,深浅色缎纹
shade sorting 色差分类
shade standard 色样
shade ticket 色泽标签
shade variations 色差
shadiness 影条疵,晕影疵
shading 调整色光;染色差异,(倒顺毛)色光差异;经编织物线圈歪斜(疵点)
shading buttonhole stitch 深浅锁眼绣
shading effect 花纹深浅效应,云纹效应
shading in tailoring 缝制时配色
shading model 浓淡效果图
shading-off 色光过头
shading property 遮蔽性
shadings 隐条织物,鸳鸯条子织物,阴阳条子织物
shadow 阴影,隐纹;八字胡子
shadow-applique stitch 阴影嵌线线迹
shadow-boxing wear 太极拳服
shadow braid 六角网眼棱结窄花边
shadow check 阴阳格子花纹
shadow checks fabric 隐格织物,阴阳格子织物,鸳鸯格子织物
shadow clock 闪色花袜,隐纹花袜
shadow cloth 影纹织物,印经织物

shadow effect 阴阳花纹,阴影效应
shadow embroidery 暗花绣花,影绣
shadow fabric 印经织物,影纹织物
shadow green 荫绿色,秋香绿
shadow lace 暗花细花边
shadow-mask colour display 节日罩式彩色显示器《计算机》
shadow organdy 涂料印花薄纱
shadow-panel slip 带防透幅片衬裙
shadow-proof skirt 防透衬裙
shadow rep 深浅色棱纹布
shadow silk 闪光绸
shadow stitch 暗花针法,松套花边的松针迹,山形针迹
shadow stripe 阴影条纹;隐条织物
shadow stripe crepe 月华绉
shadow striped crepe 月华绉
shadow striped fabric 隐条织物
shadow striped taffeta 影条纺
shadow stripes 鸳鸯条子织物,阴阳条子织物,纵向影条纹
shadow stripes fabric 隐条织物
shadow stripe voile 影条巴里纱
shadow tuck 影纹皱褶
shadow voile 影纹巴里纱
shadow warp cretonne 印经影纹布
shadow work 暗花刺绣,影绣
shadow work embroidery 泡花平绣
shafts (人的)大腿(美国俚语)
shaft sewing machine 通轴缝纫机
shag 蓬乱发式;粗毛,长绒;长绒粗呢;长绒地毯
shag hairstyle 蓬松式发型
shaggy fabric 起毛织物
shaggy look 起毛款式
shaggy thread 起毛线
shagreen 鲨革,绿皮,仿皮布
Shaili 夏丽斜纹披巾(用安哥拉山羊毛手织)
shairi 莎里羊绒细呢
shaiwar (南亚女子穿)宽松裤
shaiwarkameez(e) 宽松女套装
shakeout 摊平
shaker 套领衫,厚运动衫
shaker flannel 双面法兰绒
shaker knit 粗平针织物
shakespeare collar 莎士比亚领
shake sweater 厚运动衫
shako(cap) 圆筒高帽,高顶硬军帽,高冠毛皮无檐帽

shal 查达棉布;手织格纹内衣绸
shalamar tweed 谢拉马斜纹粗花呢
shale 披巾,法式披肩
shalee 薄型披巾布
shalli 手织披巾斜纹织物
shalli-phiri 手织山羊绒围巾
shalloon 斜纹里子薄呢,二上二下斜纹
shalwar 灯笼裤,(土耳其女子)宽松裤(脚踝处收紧的裙裤)
sham 男式短裤;摆设用床罩,摆设用枕套
Shamask 沙玛斯克(美国时装设计师)
sham button hole 花式纽孔
shameuse 留香绉,留香缎
sham hanging sleeve 悬式长袖
shamiya 保加利亚红、白、绿色头巾
sham jewellery 仿造的珠宝
shammy 麂皮,麂皮的
shammy fabric 仿麂皮针织物
shammy finish 仿麂皮整理
shammy gloves 仿麂皮手套
shammy leather 麂皮革,羚羊柔皮
shamoy 麂皮,羚羊皮
sham plush 假长毛绒
sham pocket 装饰口袋,假口袋
shampoo 洗发剂
shampooing 皂洗,水洗
shamrock 亚麻手帕
shamrock lawn 纱姆罗克高级细布
shamyanas 丝绣华盖毛织物
shanbin lame 闪滨缎
shanbi tissue crepe 闪碧绡
shangai 优质柞丝绸
Shanghai Nonwovens Conference and Exhibition 上海国际非织造布研讨会暨展览会
Shanghai shoes 绣花鞋,上海鞋
Shanghai Tang 上海滩
shanguang sand crepe 闪光呢
shanguang satin brocade 闪光缎
shank 胫,小腿;腿部;胫骨;(纽扣)柄;箍;(有脚纽扣背面的)扣腿;扣腿上的孔环;(使纽扣不紧贴织物的)线绕小梗;纽扣袢;鞋底的脚弓部分,鞋轴,弓形垫;长袜袜筒
shank button 有脚纽扣,有柄纽扣,杆式纽扣
shank button attachment 有柄纽扣附件
shank button clamp 有柄纽扣夹
shank buttonholer 有柄纽扣锁孔器

shank button spacer 有柄纽扣衬垫
shank button with stay button 有衬有柄纽扣
shank button without stay button 无衬有柄纽扣
shank foot 有柄压脚
shanking 钉扣;打结
shank wrap 缠纽扣把
Shantung 手织草杆织物;山东绸(柞蚕丝绸),仿山东绸
Shantung pongee 山东府绸
Shantung silk 山东绸,山东丝
Shanzhi habotai 盛纺
shaoxing gauze 越罗
Shaoxing habutai 绍纺
shaoxing leno 越纱
Shap cotton 缅甸夏普棉
shape 款式;样式;形状,(女子的)体形,身段;戏装,戏装衬垫;无檐帽,帽模
shape cloth 精纺全毛呢
shaped apron 合身围裙
shaped casting 定型的串带管
shaped cloth 精纺全毛呢
shaped coat 合身大衣
shaped cuff 成形夹克,成形克夫
shaped fiber 异形纤维
shaped leg 形成袜筒
shaped line 成型线;造型线;英国式绅士风度式样
shaped look 紧身款式(20世纪60年代后半期流行于英国的男服款式),成形款式(西装基本风格,沿身体线条而形成清晰轮廓)
shaped memory 耐久压烫保形整理
shaped narrowed collar 收针成形领
shaped opening 曲襟
shaped pocket 弧形袋
shaped sash 腰带,头巾,造型饰带
shaped seam 成形缝
shaped shawl collar 成形青果领
shaped underwear 成形针织内衣
shaped waist 造型腰线
shaped waist band 曲线形腰带
shape-faced 棉背真丝天鹅绒
shape-faced velvet 棉背绵纺丝绒
shapeless figure 不匀称体型
shapely 样子好的,匀称的,形状美观的
shape memory 耐久压烫保形整理
shape memory cloth 形状记忆织物
shape memory crease 形状记忆折痕

shape memory suit 形状记忆服(免烫整理)
shape pants 紧身短裤
shaper 合体服装,形体服;成形装置
shape retention 保形性
shape-retentive finish 保形整理
shape round 圆顺
shape-set 定形,定型
shape stability 形稳性,形状稳定性
shape up 紧身;成形;体形优美;顺利发展
shape-up jeans 紧身牛仔裤
shaping 花样,花式,成形
shaping device 成形装置
shaping machine (毡帽)成形机
shaping pad 造型垫,样垫
shaping seam 成型缝,定型缝
shaping side seam 成形缝,侧缝线
shapka ［俄］裘皮(或羊皮)无檐帽
shappe 绢丝
shaps (美国牛仔穿的)皮护腿套裤
sharak 沙拉克密织缎条细布(阿拉伯)
sharbati 沙巴蒂细布(东南亚)
shark 深莲灰
sharkskin 鲨皮布,鲛皮革,鲨鱼皮革;雪克斯金细呢,雪克斯金雨衣呢
sharkskin fabric 仿鲨皮织物,鲨皮布
shark swimsuit 仿鲨革泳装
sharong satin feutue 沙绒缎
sharp corner seam 尖角缝,夹角缝
sharp cut 轮廓清晰的
sharp green 耀眼绿
sharply indented seam 夹角加固缝,尖角加固缝(在尖角略缩进处加缝一道针迹)
sharp mark 轮廓清晰(印花边缘无渗色)
sharpness 清晰度
sharp-pointed needle 尖头缝针,夹头缝针
sharp-pointed pin 夹头大头针
sharp-pointed trimming scissors 尖头修剪刀
sharps 缝衣长针
sharps needle 手缝针
sharp toe 尖头鞋,尖鞋头
sharp toe leather shoes 尖头皮鞋
sharskin 鲨皮
shash 阿拉伯优质细布;中东披肩;中东棉衬衫
shashi mock crepe 纱士呢
shashi suiting silk 纱直呢

shash mantahi 莎什山羊绒方巾
shatnez 毛麻混纺织物
shattering 破碎发型
shatweh 巴基斯坦女锥形帽
shaub 印度肖布丝棉交织绸
shave 刮脸;修面
shave coat 男子膝长浴衣,男盥洗袍,修面用外套,男子便装
shaver 理发师;电动剃刀
shaving lotion 修面香液
shaving soap 剃须皂
shawl 披肩,围巾,披巾
shawl collar 围巾领,青果领,丝瓜领,斜阔领,新月领,香蕉领
shawl-collared cardigan 青果领开襟绒线衫,青果领羊毛衫
shawl fabric 披巾织物
shawl pattern (仿东方披肩的)杂色图案
shawl roll collar 披肩卷边领
shawl tongue 披巾式鞋舌,苏格兰鞋舌
shawl waistcoat 围巾制作的马甲,青果领男式马甲
She ethnic costume 畲族服饰,畲族民族服
sheaf-filling stitch 集束刺绣针迹,束状贴线绣
sheaf stitch 束状绣
sheaker 轻便运动鞋
shear 剪刀
sheared beaver 剪毛海狸毛皮
sheared towel 割绒毛巾
shearing 剪毛,剪绒;绵羔皮,一岁羊毛皮;(刚剪毛后制革的)绵羊革;羊毛衬里(美国)
shearing coat 剪毛大衣
shearing jacket 羊皮外套
shears 剪刀
sheath 紧身衣,紧身袍,紧身衬裙,紧身连衣裙;鞘形,细长直筒形;外皮,外罩;护套屏板
sheath-core composite fiber 皮芯型复合纤维
sheath corset 紧身胸衣,紧身腹带
sheath dress 鞘形连裙装,紧身连衣裙,紧身女装
sheath silhouette 细长形轮廓,鞘形轮廓
sheath skirt 鞘形裙,紧身裙,细长裙
sheath sleeve 细筒袖
sheath swimsuit 女式鞘形紧身泳装
sheathy (女装)紧身的

shed-all 防雨织物(商名,用聚合蜡分散液处理)
sheen 光泽;有光泽的纺织品,华丽的服装
sheen gab 有光轧别丁,有光华达呢
sheen gabardine 有光轧别丁,有光华达呢
sheep fur 羊皮,绵羊皮
sheep hide 羊皮
sheep leather 绵羊皮,羊皮
sheep lined horsehide surcoat 羊皮衬里马皮外套
sheep lined leather vest 羊皮衬里皮背心
sheepskin 绵羊毛(皮),羊皮纸,羊皮革
sheepskin coat 羊皮袄
sheepskin effect 仿羊皮效应
sheepskin garments 绵羊皮服装
sheepskin jacket(同 shearling jacket) 羊皮外套
sheep wool 羊毛,绵羊毛
sheer 透明薄纱;透明薄织物及制品;绡
sheer band 薄纱头巾,(伊斯兰教包头用)薄细纱带
sheer crepe 薄绉,透明绉
sheer fabric 薄纱,透明薄面料
sheer garbardine 稀薄华达呢(一般为丝织),稀薄轧别丁
sheer hose 透明袜
sheer lilac 透明丁香色
sheer linen 细薄亚麻布
sheer material 极薄的材料,透明材料
sheer muslin 薄纱(麦斯林织物)
sheer nylon blousette 锦纶(尼龙)薄纱小上衣
sheer pantyhose 透明连裤袜
sheer pink 浅粉红色
sheer silk 绡,绡类织物
sheer silk satin 薄缎
sheer silk stockings 透明丝袜
sheer stockings 透明长裤
sheer stretch hosiery with polyamide 薄型锦纶强力丝袜
sheer sucker (棉织物)绉布,薄绉,绉纱
sheer taffeta 稀薄塔夫绸
sheer-topped nightgown 上衣透明的晚装
sheer wool 精纺羊毛薄呢
sheet 被单;纸张;成幅的薄片
sheeting 细布,阔幅平布;被单布
sheet-shooting jacket 双向飞碟射击夹克
shelga 希尔加粗花呢

shell 雪尔衫,贝壳衫,无袖后钮套头衫,(女式)宽松外穿背心;贝壳形,外表,表面;贝壳鞋,雪尔鞋;米白色
shell belt 弹药带
shell blouse 紧身无袖套衫
shell button 贝扣,贝壳纽
shell cloth 英国炮弹哗叽
shell edge 贝壳形边
shell edging stitch 贝壳形边线迹
shell fabric (做服装的)面料,保暖夹衣面料,面层织物
shell foot 贝壳式加固袜底
shell form stitch 贝壳形装饰线迹
shell garment 紧身服装
shell grey 贝壳灰,淡黄灰色
shell hem 荷叶边,贝壳式衣边
shell hemmer 贝壳形卷边压脚,贝壳式衣边卷边器
shell jack 热带筒礼服
shell jacket 军用紧身夹克,男式紧身夹克
shell-less washer 无内壳式洗衣机
shell-lined coat 毛皮领毛皮袖骑马大衣
shell lining 半袖里,半夹里
shell pink 贝壳红,血牙红,贝壳粉色,带淡黄的粉红色
shells 女式浅口无带皮鞋
shell stitch 贝壳形线迹
shell-stitched edge 贝壳形缉线边
shell-stitch fabric 贝壳形经编织物
shell suit 贝壳装
shell sweater 贝壳形毛衣,无袖套头女毛衫
shell tuck 贝壳形皱裥,荷叶边皱裥
shell tucks 荷叶边,贝壳式衣边
shell-type edge seamer [缝制]贝壳形边缘缝纫机
Shem-eez(chemise)(同 sack dress) 袋式直筒女装,布袋装
shepherd check 黑白小格子花纹呢,对比色格子呢;黑白小格子花纹
shepherd's check 牧人黑色格子布
shepherd's plaid 格子花纹织物,牧人黑色格子布
shepherd's smock 牧人服
sherbet tone 果汁冻色调
sheriff tie 警长领带;团长领带;西部领带;细绳领带;鞋带型领带;州长领带,郡长领带,窄领带
Sherlock 舍洛克帽

sherry 雪利酒色
sherryvallies 骑马用厚布护胫,骑马用厚布外衣
sherwani （印度男子穿的）高领长外套
shesh 六股亚麻纱织物,埃及谢什布
shetland 设得兰毛织物；雪特兰呢（苏格兰呢）；雪特兰织物
shetland clog 设得兰木屐
shetland falls 设得兰毛网眼披巾
shetland knitwear 雪特兰毛衫,设得兰羊毛衫
shetland lace 设得兰花边
shetland point lace 设得兰针绣花边
shetland shawl 设得兰羊毛针织披巾
shetland sweater 设得兰毛衣
shetland veil 设得兰羊毛网眼披巾
shetland wool （用于针织物的）设得兰羊毛
shetland wool sweater 设得兰毛衫
Shezu costume 畲族民俗服
shibori 日本提花绸；[手工]扎染
shichun jacquard sand crepe 时春绸
shi-di gauze 实地纱
shield 铠装,腋下垫布,吸汗垫布
shiffon velvet 薄天鹅绒
shift （方言）衬衫；女用衫；内衣；女式无袖衬衣；无腰线直筒裙；换衣；烫衬衣；班次
shiftable cutting mechanism 活动裁剪机构
shift coordination 可自由替换组合着装法
shift dress 可变化连衣裙；舍弗特服装,女衬衣式服装；无腰线直筒连衣裙；（旅游）替换衣服
shifting （织物）经纬滑移
shifting sand 流沙色
shift-jumper 无袖舍费特背心装
shift rate 替换率
shift silhouette 舍弗特衬衣式,舍弗特型轮廓
shignon style 发髻发型
shikargah 狩猎图案锦缎
shike habutai 双宫绸,粗丝纺绸
shikeginu 日本双宫塔夫绸
shikii 厚丝织物（似山东绸）
shima momen 彩条棉布
shimmey 无腰身宽女服,衬衫式女服,女式无袖宽内衣
shimmy 女式衬衣

shim slacks 细脚裤
shin 胫部,胫骨
shin bone 胫骨
shine 光泽,光彩；擦皮鞋
shiners （人丝织物的）极光（疵）
shingle 屋面板发型,欣格型短发型,（女子）屋盖式短发发型
shingle cut 欣格型短发
shingle hair 欣格型短发
shingle hairstyle 欣格短发型,（妇女）男型短发
shin guard 护胫,护胫套
shining 极光
shinny fabric 有光泽的面料
shin protector 护胫器,护胫物
shiny 发亮的、擦亮的,有光泽的、磨光的,磨损的
shiny material 发光衣料,发光衣料
shiozawatsumugi 盐泽䌷
shioze 绢纺双绉；加重电力纺,有光纺,无光纺
shioze color striped 彩条纺
shioze habotai 加重电力纺
shioze silk 绢丝双绉
ship 船
ship freight 水运费
shipment 装运
shipper 托运人
shipping 装运
shipping mark 箱唛,装运标记；唛头；运输标志
shipping order 装货单《贸易》
shipping package 运输包装
shipping quality 离岸品质
shipping room （工厂等的）发货仓库
shipping sample 装船样品《贸易》
shipping space 舱位《贸易》
ship tire 帆船头饰
shir(r) 宽紧线,橡皮线；抽褶
shirlan extra 霉菌抑制剂
Shirley Instituty 锡莱研究所
shirn 山羊毛混纺厚呢
shiroji-sha 白纱罗
shiro momen 平纹棉布,白色手织布
shirred-back slip 后背打褶套裙
shirred banding 抽褶腰头
shirred bodice blouse （抽胸襟褶的）紧身上衣
shirred fabric 宽紧织物,松紧织物
shirred patch pocket 袋口抽褶的贴袋

shirred sleeve 平行皱缝式袖
shirred top 抽褶上身
shirrer blade 打褶器
shirring 抽褶,缩缝,多层收皱,平行皱缝
shirring attachment 多层收皱附件
shirring blade 多层收褶打褶器
shirring blouson 缩缝装饰短夹克（总称）
shirring edge guide 多层收褶导边器
shirring elastic 松紧带,橡筋带
shirring feet 褶裥压脚
shirring pocket 抽褶口袋
shirring smocking （装饰用的）抽褶刺绣
shirring stitch[ing] 抽褶线缝
shirring waist 腰部收褶,松紧腰
shirt 男衬衫,(仿男式)女衬衫,贴身衣,汗衫,内衣,(男用)衬衫式睡衣;标志衫;制服;穿标志衫的人
shirt band 衬衫袖衬布,衬衫领衬布,衬布
shirt blazer 装衬衫领的短上衣
shirt blouse 男式女衬衣,衬衫式罩衣
shirt blouse sleeve 女衬衫袖
shirt button 衬衫纽扣
shirt buttonhole 衬衫纽扣,平眼纽孔
shirt cardboard 衬衫纸板
shirtcoat 衬衫式外套
shirt collar 衬衫领;小方领
shirt collar stay 衬衫领插角片
shirt cuff 衬衫袖口
shirt-cuff edge 双压线缘边
shirt dress 衬衣式连袖裙
shirt factory 衬衫厂
shirt folding table 叠衫台;折衫台
shirt front 胸衣;衬衫的假前胸,衬衫的硬前胸
shirt hem 衬衫下摆
shirt holder 衬衫衣架
shirting 本色细平布,衬衫料子
shirting chambray 细平布,钱布雷布
shirting flannel 衬衫用法兰绒
shirting poplin 衬衫用府绸
shirting waist 连腰带短罩衫,男式女衬衫
shirting-weight fabric 细平布
shirt-jac 衬衫式夹克、外套
shirt jacket 衬衫式夹克
shirt jumper 衬衫式短上衣,衬衫式夹克
shirt knife blade 衬衣裁切口
shirt length 衬衫长
shirt maker 做衬衫的工人;衬衫厂商

shirt maker seam 衬衫缝
shirt of famous brand 名牌衬衫
shirt on shirt 双重穿衬衫方法
shirt on shirt on shirt 三重穿衬衫法
shirt opening 半开襟
shirt paperboard 衬衫纸板
shirt pin 衬衫别针
shirt placket 半开襟
shirt pleat 衬衫褶
shirt pocket 小袋,衬衫袋
shirt's collar turning and pointing machine 翻领角机
shirt seam 衬衫缝
shirt-shaped jacket 宽摆外套,松下摆外套
shirt sleeve 衬衫袖
shirt sleeve opening 衬衫袖口衩
shirt sleeve placket 衬衫袖口衩
shirt square collar （衬衫）小方领
shirt studs 衬衫夹
shirt style 衬衫款式
shirt suit 衬衫套装
shirt tail 衬衫前部圆下摆,衬衫后背下摆,衬衫式下摆
shirttail hem 衬衫圆下摆
shirtwaist 衬衫式连衣裙,衬衫裙,仿男式女衬衫;衬衫腰部
shirtwaist blouse(同 tailored blouse) 衬衫式女衫
shirtwaist dress 衬衫腰线连衣裙;仿男式（束腰）衬衫裙,仿男式女衬衫
shirtwaister 衬衫式连衣裙,女式衬衫裙
shirtwaist silhouette 衬衫式连裙装轮廓
shirtwaist sleeve 女衬衫袖
shirtwaist style blouse 束腰衬衫式上衣
shirtwaist style frock 束腰衬衫式长外套
shirt with ruffle in front 前胸饰褶衬衫
shirt yoke 衬衫覆肩
SHISEIDO 资生堂
shitagi 室内男和服
shita-juban 日本女短衬衣
shiti 小花卉花布
shitkicker 靴子
Shive （布料等上的）线头
shixin jaqouard sand crepe 时新绸
shizhou tapestry 诗轴绸
shober 伏尔加粗亚麻服
shock 肖克袜
shocking colour 惊骇色,特艳色
shocking pink 鲜艳粉红色

shoddy 赝品
shoddy dropper (澳大利亚、新西兰)叫卖廉价(或伪劣)服装的小贩
shoe(s) 鞋,履;鞋面;穿着时髦的人(美俚)
shoe awl 鞋锥,鞋钻子
shoe bag 鞋袋
shoe bell 鞋铃
shoe black 擦鞋匠
shoe-blacking 黑鞋油
shoe block 鞋楦
shoe brush 鞋刷
shoe buckle 鞋扣环,鞋扣袢
shoe button 鞋纽扣
shoe canvas 鞋面帆布
shoe (cap) stretcher 鞋(帽)撑
shoe clip (装饰用)鞋夹
shoe cloth 鞋面呢
shoe counter 主跟
shoe cover 鞋面
shoe cream 鞋油
shoe cushion 鞋垫
shoe designer 鞋子设计师
shoe duck 鞋面帆布
shoe flap 鞋盖,鞋舌
shoe fold (织物)两端相向折叠法(整匹长度折成12～16层)
shoe hammer 鞋锤
shoe-head's base cloth 鞋头布
shoe heel 鞋跟
shoe horn 鞋拔
shoe interlining 鞋垫
shoe keeper 鞋楦,鞋撑子
shoe lace 鞋带
shoelace tie 鞋带型领带;警长领带;团长领带;西部领带;细绳领带
shoe last 鞋楦
shoe leather 皮鞋(总称);制鞋皮革
shoeless 不穿鞋的,没有鞋子的
shoe lift 鞋拔
shoe lining 鞋里,鞋衬布
shoe maker 鞋匠,补鞋工人,制鞋工;鞋店营业员
shoe maker knife 鞋匠刀
shoemaking 制鞋业
shoe making machine 制鞋机
shoe manufacturing industry 制鞋工业
shoe material 鞋材
shoe pac (冬天穿厚袜外的)缚带保暖防水长筒袜;(北美印第安人)长筒鹿皮鞋

shoepack 肖派克鞋
shoe pad 鞋衬垫;鞋垫
shoe pak(同 shoe pac) 缚带保暖防水长筒袜;长筒鹿皮鞋
shoe pattern 鞋样
shoe polish 鞋油
shoe powder 鞋粉
shoe punch pliers 鞋跟钳
shoe quarter 鞋后帮
shoe rack 鞋架
shoe repair(ing) machine 补鞋机
shoe repairs 修鞋(补衣)
shoe repair sewing machine 补鞋机
shoe rest 鞋砧
shoe satin 鞋面缎
shoe shop 鞋店
shoe sides 鞋帮
shoes material 鞋材
shoeshine 擦鞋;擦鞋者;鞋油
shoeshine boy 擦鞋男童
shoe-stockings (新式)连袜鞋
shoe store 鞋店
shoe strap 鞋搭扣带
shoe stretcher 鞋撑
shoe string 鞋带
shoestring strap 双扣袢,鞋扣状袢
shoestring tie 鞋带型领带
shoe stud 鞋扣
shoe tab 鞋拉袢
shoe tack 鞋钉
shoe tap 鞋掌;鞋跟铁片
shoe tape 鞋带
shoe throat 鞋口
shoe throat flattening machine 鞋口整平机
shoe throat strengthened tape 鞋口加强带
shoe tie 宽鞋带
shoe toe 鞋头
shoe tongue 鞋舌
shoe-top silk 鞋面绸
shoe-top velvet 鞋面丝绒
shoe tree 鞋楦
shoe upper 鞋面
shoe upper leather sewing machine 皮革鞋面缝纫机
shoe upper sewing machine 缝鞋面机
shoe velvet 鞋面丝绒
shola 合萌芯防晒帽
shooda 苏达斜纹薄呢
shooting boots 猎人靴

shooting clothing 狩猎上衣
shooting coat 狩猎上装,射击服
shooting gloves 射击手套
shooting jacket 射击夹克
shooting shoes 射击运动鞋
shooting suit 射击服;狩猎套装
shooting vest 狩猎坎肩
shooting wear 狩猎装
shop 店铺
shop apron 工作围裙,工作饭单
shop-assistant 店员
shop boy 男店员
shop-keeper 零售商
shop parts 非标准零件
shoppe 专柜
shopping bag 购物袋
shopping mall 商业大厦
shop romal (印度)色织棉手帕,罗美杂色手帕
shops in shop 店中店,(百货公司内的)专营店《销售》
shoptalk 行话,行业术语
shop walker 巡店员
shop window 商店橱窗《销售》
shorn lamb 剪毛羔羊皮
shorn velvet 剪绒
short 矮个子的衣服尺码;空头;苎麻短纤维
shortalls 有护胸的短童裤
short and long stitch 长短缎纹线迹,锁眼线迹,叉针绣
short and slight figure 矮小体型
short and small figure 矮小体型
short and stout figure 矮胖体型
short and thick neck figure 短粗颈轮廓
short back and sides 盖式发型
short-back sailor hat 窄后檐帽
short bartack 短套结
short bartack machine 短套结机
short boots 短筒靴
short boxy jacket 女式箱型短上衣
short boxy line 短箱型,短箱式
short cape 短披肩,短斗篷
short clothes 童装
short clothing 短外套;短大衣
short coat 短外套,(婴儿脱离襁褓后所穿的)婴幼装
short coat of single layer 单层短衫
short colonial 殖民地式肥大短裤
short crotch 兜裆

short cut mark 省略符号
short dash line 虚线,点线
short diagonal basting 短斜针假缝
short doublet 紧身短上衣
short drawers 女用短衬裤
shorten 改短
short figure 矮体型,矮小体型
short gloves 短款手套
short-gown 短衫,短(大)褂
short hair 短发型
short-haired fur fabric 短毛仿毛皮织物
short head 圆头型的人
short hood 女士打褶薄头巾
short hoses 短袜
shortie (同 shorty)矮子,矮小身材;超短服装,短裤,短手套
shortie coat 女式短大衣
shortie gown 迷你睡衣
shortie nightdress 迷你睡衣,短睡衣
shortie nightgown 短睡衣,迷你睡衣
shortie pajamas 短睡衣裤
shorties 特短服装
shortie socks 及踝短袜
short jacket 短夹克
short lapped placket 裙子短开衩
short legged 腿短的
short length 短码,短外套长度
short-life garment 短期使用的衣服(医用无菌的非织造布工作服)
short line bra(ssiere) 短胸罩
short-nap fabric 呢料,短绒面料
short-narrow zigzag stitch 短窄的 Z 形针迹
short neck 短颈
short pageboy style 侍童超短发型
short paletot 女式及臀方形夹克
short panties 女用短衬裤,女用短汗裤,三角裤
short pants 短裤,三角裤
short pants look 短裤款式
short piece 短码,缺尺
short-pile fabric 短绒织物
short placket 短裙衩
short-point button-down collar 纽扣短尖领
short point collar 短尖领
short pointed collar 短尖角领
short prongs 短尖头(饰品)
short puffed sleeve 短灯笼袖
short riding 短马裤

short robe　短袍
short rounded collar　短圆领
short-run　小批量的;短期的
short-run fashion　短期流行的时装
short runs　小批量
shorts　短裤,球裤,[美]男用短衬裤;(婴儿)短装,小件衣服(尤指内衣);短匹
short seam　短缝
short seat　兜裆
short shank button　短柄纽扣
short shorts　超短裤,短短裤
short short-skirt　超短裙
short skirt　短裙
short sleeve　短袖,四分之一袖
short sleeved jacket　短袖夹克,短袖短外衣
short sleeved shirt　短袖衬衫
short-sleeved upper garment　半臂
short sleeve length　短袖长
short sleeve pullover　无纽短袖套衫
short sleeve robe　短袖长袍
short sleeve shirt　短袖衬衫
short sleeve vest　短袖背心,短袖内衬衣
short socks　及踝短袜
short square collar　小方领(短领)
short stick　短尺码
short stitch　短线迹
short suit　短裤运动套装
short, thick neck figure　短粗颈体型
short torso　短躯干(身材)
short trapezium skirt　梯形短裙
short under arm　袖下不足
short-visor cap　短舌帽
short wadded doublet　短紧身棉上衣
short waist　短腰节,高腰身
short waisted figure　高腰体型
shorts with adjustable waist　腰头可调节的短裤
short with built-in briefs　男式双层短裤
short wool　粗梳毛
shorty　超短服装;短裤;短手套;矮子
short yards　短码布
shorty gloves　短手套
shorty jacket　衬衣夹克
shorty panty　小短裤
shot　闪光效应;闪光条痕疵
shot cloth　闪光绸
shot color　闪光
shot damage　伤洞,割绒伤洞
shot gabardine　闪光华达呢
shot lisle　细麻纱袜子;细麻纱手套
shot proof　防弹的
shot-proof vest　防弹背心
shot silk　闪光绸
shot taffeta　闪光塔夫绸
shotte butadar　金色边细布
shot velvet　纬天鹅绒
shot weave rayon　人造丝闪光绸
shoulder　肩,肩胛,肩型;肩宽;(衣服)肩部;针肩;前肩革;提花机横针凸部;垫塞物
shoulder area　肩区,肩部
shoulder bag　挂肩提包,肩背包
shoulder belt　肩饰带,军用肩带,子弹带
shoulder blade　肩胛,肩胛骨
shoulder blade level　背幅线,背宽线
shoulder bone　肩胛骨
shoulder button　肩纽扣
shoulder button closing　装纽肩门襟
shoulder cape　覆肩短披风,短披肩
shoulder closing　肩襟
shoulder collar　披肩领
shoulder cut　肩省,通天省缝
shoulder cut seam　肩省缝,上肩缝
shoulder dart　肩褶,肩省
shoulder drape　肩挂
shoulder duster　垂肩型长耳环
shoulder emblem　臂章
shoulderette　两端呈袖状的女围巾,鲍莱罗围巾
shoulder flash　袖上部肩章
shoulder handbag　肩挎式手袋
shoulder height　肩高
shoulder height line　肩高线
shoulder highly squared　高肩
shoulder interlining　肩衬
shoulder knot　肩垫;肩饰花结;(17～18世纪时髦男子佩戴的)肩饰;(官员制服或仆从号衣上的)肩章
shoulder length　肩宽,肩阔;(头发等)长至肩部的
shoulder-length hair　披肩式长发
shoulder line　落肩线,肩宽线,小肩线
shoulder loop　肩袢,肩斜
shoulder marks　肩部标志,肩章,肩部符号
shoulder neck intersection　肩领相交部位
shoulder off press　肩部完成压烫机
shoulder outer section　肩袖相交部位
shoulder/overarm seam　连袖的肩袖缝线

shoulder pad 垫塞物,垫肩,裃丁,箱形垫;肩垫	shovel hat ［英］(教士)上翻宽边帽;铲形宽边帽
shoulder pad basting machine 垫肩假缝机	show 服装表演;展览,展览会;展品
shoulder padding 肩衬,垫肩衬,运动衣护肩	show bill 海报;招贴,广告
	show breeches(同 riding breeches) 马裤
shoulder pad pocket 肩垫套	show-business clothing 影视服装
shoulder pad width 肩垫宽	show card 广告单,样品卡,展品说明卡
shoulder patch 肩章	show case 陈列橱
shoulder peak 肩峰	show coat 半正式观马戏外套(背开衩,腰部贴身,窄驳头,三扣门襟,两侧倒裥,后中线长切缝)
shoulder piece 肩衬	
shoulder point (S.P.) 肩端点,肩袖点	
shoulder point setting 肩端点固定处,肩袖点固定处	shower 淋浴帽,沐浴拖鞋
	shower cap 淋浴帽
shoulder protector 护肩	showerproof 防雨衣服,雨衣
shoulder-puff sleeve 泡肩长袖	showerproof cloth 防雨布
shoulder run 肩线	showerproof coat 雨衣,防雨衣
shoulders 上背部	shower-proofing 防雨整理
shoulder seam 肩部缝,肩缝	shower repellency 防雨性能
shoulder seam linking 拼肩缝	shower-slipper 淋浴拖鞋
shoulder seam opening press 肩缝熨开机	shower-washer 淋式水洗机
shoulder seaming machine 肩部缝缝纫机	showing 时装展览;展示会,展览会;展示,陈列
shoulder shape 肩形	
shoulder-shape pad 肩形垫	showing display 展览
shoulders highly squared 高肩	showing horn 脱鞋器,鞋拔
shoulder shirt 背心式衬衣,露肩衬衣	show-off clothes 展示服
shoulder/sleeve seam 肩袖缝	show piece 陈列品,样品,展品
shoulder slipover 带肩套领衫	show room 展览室,样品陈列室
shoulder slope 肩斜(度)	showroom model 试装模特
shoulder slop line 肩斜线	show shop 展销商店《销售》
shoulder slope drop 肩斜落差	show-side 正面
shoulder slope line 肩斜线	show window 商店橱窗《销售》
shoulder socket 肩插口	show window girl 橱窗服装模特
shoulder socket bone 肩窝骨	showy dress 过分艳丽的女服
shoulder strap 肩徽,肩章,肩带,肩饰;(西裤的)吊裤带;肩吊装	shrag stitch 施拉格针法
	shred 碎布条,零头布,破布
shoulder strap bag 背带包	shrimp 藕红,小虾色
shoulder strap effect 肩带效果	shrimp pink 暗红色,深粉红色
shoulder strap handbag 挂肩提包	shrink 缩水,缩水性;使(布料)收(皱)缩;收(皱)缩;矫整衣料;(下胸围至腰节底边为罗纹边的)什赖因克短毛衣;归(整烫工艺之一)
shoulder tab 肩章,肩袢	
shoulder-tip 肩端	
shoulder tip to tip 总肩宽,背肩宽	
shoulder-to-elbow measurement 肩肘长	shrinkage 收缩,皱缩,缩水;收缩量;缩率;收缩程度
shoulder type 肩型	
shoulder vest 肩带背心	shrinkage control finish 防缩整理
shoulder width 肩宽	shrinkage of knitted fabric 针织物缩水率
shoulder width line 肩宽线,肩头外倾线(过肩腋斜线)	shrinkage preventing finishing 防缩整理织物
shoulder wing 肩翼	shrinkage restorability 收缩回复性
shoulder yoke 肩覆势,过肩,抵肩	shrink and stretch back piece 归拔后(衣)片
shou silk 绶	

shrink and stretch sleeve inseam 归拔偏袖
shrink and stretch top collar 归拔领面
shrink and stretch under collar 归拔领里
shrink and to stretch back piece 归拔后背
shrink and to stretch sleeve inseam 归拔偏袖
shrink and to stretch top collar 归拔领面
shrink and to stretch under collar 归拔领里
shrink finished fabric 预缩整理织物
shrink in 归缩
shrinking （将布料）缩水，收缩
shrinking machine 缩缝机；抽褶机
shrinking potential 收缩潜力，潜在缩率
shrinking power 收缩潜力，潜在缩率
shrink leather 绉纹皮革
shrink mark 归缩号
shrink-proof 防缩；不缩水
shrink-proof finish 防缩整理
shrink-proof lining 防缩衬里
shrink-proof wool knitwear 防缩羊毛衫
shrink-resistant finish 防缩整理
shrink sleeve cap 收袖山
shrink sweater 无袖短毛衣
shriveiled 压瘪，压皱
shroud 寿衣，尸衣
shrug 女式带袖短披肩；女式短套领衫，单扣女式短上衣；耸肩
shrunk band 束带
shrunken leather 皱纹革
shrunk finish 预缩整理
shrunk finish fabric 预缩整理织物
shrug sweater 短袖露肩毛衣，披肩式毛衣
shrunk yards （布匹）缩后码长
shtreimel 黑貂皮饰边帽
shuba 长皮大衣
shu brocade 蜀缎，蜀锦
shuchin 花罗缎
shuddar 单面素色半漂棉平布；平纹色织棉布；披肩
shuguang brocade 曙光绉
Shui ethnic costume 水族服饰
shuijing brocaded velvet 水晶绒
shuixi crepe 水洗绉
Shuizu costume 水族民俗服
shuja khani 棉经丝纬织物
shuka 半漂横条布
shuka gamti 半漂横条布
shuka ulayiti 半漂横条布
shule crepe 舒乐绉
shu leno 蜀纱
Shu Uemura 植春秀
shunwen crepe(suzette) 顺纹绉
shusu 缎，日本丝缎
shusu habutai 日本全丝薄缎
shutter cap 遮阳帽
shuttle （缝纫机）梭子，摆梭；梭芯
shuttle bobbin （缝纫机）梭心
shuttle bobbin case 梭心套
shuttle body 摆梭，梭身
shuttle cage （缝纫机）梭壳
shuttle carrier （缝纫机的）摆梭梭床
shuttle case （缝纫机）梭心套
shuttle cylinder 梭心柱
shuttle driver 缝纫机弹簧摆梭托
shuttle embroidery machine 飞梭刺绣机
shuttle hook 摆梭
shuttleless loom 无梭织机
shuttle line 梭子型
shuttle race （缝纫机）梭床
shuttle race back （缝纫机的）梭床圈
shuttle thread （缝纫）底线，梭线
siage ［日本］服装整烫成型
siam 深红色
siamose［fabric］ 色织条格布，西阿莫斯布
siauni 开士米，条纹边薄呢
Siberian anorak 西伯利亚皮夹克
Siberian hooded coat 西伯利亚连帽外套
Siberian man's 西伯利亚毛皮男装
Siberian ruby 西伯利亚红宝石《饰品》
Siberian smock 西伯利亚男式罩衣
Sibirienne 西伯利亚厚重毛呢
Sichuan embroidery 蜀绣
Sicilian 西西里织物
Sicilian embroidery 西西里贴花刺绣
Sicilian 西西里灯芯毛葛，夏季用棉经马海毛纬平纹补布
sicilienne 平纹薄绸；丝经棉毛纬横纹布
side （人体的）两胁，侧边
side and back pocket 侧袋和后袋
side and sleeve seam 侧面与袖子接缝
side back panel 后小侧片
sideband 单面彩面带
sideband checks 宽边小格子布
sidebands 带形边纹装饰织物
side body 男上衣两侧，侧衣片
side body panel 胁片

side burns 短腮须;连鬓胡子
side-by-side composite fiber 并列型复合纤维
side-by-side stitch 并列线迹
side-center-side color difference 左中右色差
side comb (固定女子头发用的)头侧梳
side cover 侧盖
side cover gasket 侧盖衬垫
side dart 横省,边省,胁褶
side-draped skirt 侧垂褶裙,边垂饰裙
side drapes 边褶;胁褶
side fell 侧折缝
side front panel 前小侧片
side front seam 前侧缝
side gore 鞋的侧裆布
side gore boots 侧裆靴
side gores 侧裆靴,松紧短靴
side gore shoes 侧面有松紧带的鞋,侧裆鞋
sidehand 侧章(装饰布条)
sidekick (裤两边)插袋
side-laced 侧带鞋
side-laced oxford 饰边牛津鞋(低跟,白色,常作护士鞋)
side-laced shoes 侧面系带牛津鞋
side laster 腰窝棚楦机
side leather 半张革
side length 裙长,裤长,侧缝长度
side line 胁线,侧体线
side mark 侧唛
side neck point (S.N.P.) 侧颈点
side opening 大襟,偏襟;胁开口
side-open shoes 侧空鞋,侧露鞋
side panel 侧嵌条;侧嵌边
side panel seam 侧面嵌料线缝
side parting 偏分头
side pleat 顺褶,单褶,同边褶
side pleat(ed) skirt 边褶裙,侧褶裙
side pleater 侧面打裥机
side pocket 侧袋,插袋,侧缝直袋,(上衣)横袋,(裤)斜直袋
side pocket facing 插袋贴边,侧袋贴边
side pocket position line 侧袋位线
side press 侧身熨烫机
side proportion 侧面比例
side seal 侧封
side seam 边缝,侧缝,腰缝,摆缝,胁缝,腋缝;外长
side seam dart 侧缝省

side seaming 侧缝缝纫
side seam length 侧缝长
side seam line 摆缝线,侧缝线
side seam open press 侧缝熨开机
side seam pocket 侧缝插袋
side seam shoes 侧缝鞋
side slippage (缝料)侧向滑移
side slit 边衩
sidesplit (衣服)衩;叉口;边衩
side stitch 暗缝,边缘线迹
side strap buckle 体侧扣,边扣
side strap shoes 侧袢带鞋
side swirls 双涡发型
side tied suit 侧结装
side-to-centre shading 边中部色差
side-to-side shading 左右色差
side vent 旁摆衩,胁开衩,双开衩
side view 侧视图,侧面图;侧面形状
sideway collar 侧开领,偏侧领
side whiskers 络腮胡子,连鬓胡子,鬓脚
side zipper pants 侧拉链裤子
sidpat 印度帆布
sidshillat 印度锡希拉印花亚麻布
sienna (brown) 赭色,浓黄土色
sienna point 西纳手工绣花网眼花边
sigapur checks 色织条格绸
sight 见(票);视力,视觉
sighting 标色,着色
signal clothing 荧光信号服装,信号服
signature 设计师姓名标记
signature bag 名仕提包,设计师标志提包,签字提包
signature scarf 名仕饰巾;签字饰巾
signet ring 封印戒指,图章戒指,印章戒指
sign fashion 象征记号服装
sika skin [日本]梅花鹿毛皮
sikunsi [日本](梅兰竹菊)四君子花样
silai khata 旁遮普密条纹细布
silara 旁遮普彩色横条细布
silesia 西里西亚里子布,西里西亚亚麻布(德国制);色织毛呢;德国低档轻薄亚麻斜纹布
silesian lace 西里西亚花边
silesian lawn 细麻布,西里西亚细布
silesian lining 西里西亚斜纹里子布(德国)
silesie 色织毛呢
silhouette 轮廓,体型;型;式;剪影,线条;法国棉经麻色纬平布

silhouette line 轮廓线,外形线
silhouette pattern 轮廓样板
silicious 西里歇斯山羊毛织物
siliconized fabric 硅化处理织物
silistrienne 丝毛厚呢
silk 丝,蚕丝;丝绸,绸缎;丝绸服装,绸袍
silk/acrylic blended fabric 丝腈混纺织物
silkaline 仿丝薄棉呢
silk and satin 华丽衣着,绫罗绸缎
silk baratbea 巴拉西厄领带绸,素锦
silk batiste 透明薄绸
silk beaver 海狸丝绸,海狸丝绒
silk blends 夹丝混纺织物
silk break yarn 断丝线
silk/bright rayon union satin brocade 花软缎
silk broadcloth 真丝绸
silk brocade 真丝花缎
silk brocade velvet 提花丝绒
silk camlet 丝羽缎
silk canvas 英国棉芯丝包缠纱绣花底布
silk cashmere 三页斜纹绸,开司米绸
silk cord 丝罗缎
silk costume 丝绸服装
silk cotton 木棉;木丝棉;丝光木棉
silk cotton goods 绨,丝棉交织物
silk cotton mixed fabric 桑棉织物
silk covered yarns 桑丝复合丝
silk covering cotton knitting fabric 丝盖棉针织物
silk crepe 真丝绉
silk crepe damask 桑花绉
silk crepe satin 真丝绉缎
silk croisé 全丝斜纹绸
silk curtain grenadine 窗帘纱
silk damascene 达马西恩缎
silk doup(p)ioni 双宫斜纹绸,熟织双宫绸
silk/dou(p)ioni twill 双宫斜纹绸
silk/dyed rayon brocade damask 金银龙缎
silkeen 仿绸细棉布
silk effect 丝绸效果
silk embroidery shawl 丝绸绣花披巾
silk embroidery thread 蚕丝绣花线
silken ease 雍容华贵
silken garment 丝绸服装
silkette 丝棉交织里子绸,缎纹光亮棉布
silk fabric 丝织物,绸缎
silk faille 罗缎锦

silk faille crepe 金丝罗缎绉
silk fan 绢扇
silk floss 绣花丝线,丝绵
silk floss pillow 丝绵枕头
silk floss vest 丝绵背心
silk floss wadded jacket 丝绵袄
silk floss wadded leggings 丝绵套裤
silk floss wadded robe 丝绵长袍
silk floss wadded short-gown 丝绵短袄
silk floss wadded trousers 丝绵裤
silk floss wadded vest 丝绵背心
silk flower 绢花
silk for garments 服装绒
silk garments 丝绸服装
silk gauze 建春绸
silk gauze fan 罗扇
silk ghatpot 真丝绫
silk gloria 丝毛交织伞绸
silk gloves 丝绸手套
silk goffer 冠乐绉
silk goods 丝绸,丝制品(含交织丝织物)
silk gown 绸袍
silk hat 缎面礼帽,高筒礼帽,大礼帽,丝质高礼帽
silk hosiery 真丝袜
silk illusion 真丝薄网眼纱
silkiness 丝光性,类真丝性
silking 手套背面装饰线迹
silk jacket 丝绸夹克
silk jersey blouse and pants 丝绸针织套装,丝针织衫裤
silk knit 丝针织物
silk knitted article 真丝针织品
silk knitwear 丝针织品
silk lapel 丝质翻领
silk-leather 绢制人造革
silk-like aesthetic 丝样美感
silk-like effect 仿丝绸效果
silk-like fabric jacquard 纬长丝提花仿绸织物
silk-like feel 丝状手感
silk-like fiber 仿丝型纤维
silk-like handle 丝绸感
silk-like knitted fabric 仿绸针织物
silk-like nonwoven fabric 仿丝绸非织造布
silk-like sheen 丝状光泽
silk mechlin 圆眼地精细梭结丝花边
silk mixture 丝毛细呢
silk mixtured cloth 丝毛细呢

silk mohair tropical 丝马海薄花呢
silk mull 软薄细绸
silk muslin 全丝薄纱
silk nankeen 英国真丝缎条棉布
silk necktie 真丝领带
silk noil pants 绵绸裤;䌷丝绸裤
silk noil pongee 绵绸;䌷丝绸
silk noil skirt 绵绸裙
silk noil twill 桑䌷绫
silk noil yarn 䌷丝,桑蚕䌷丝
silk of life 热内斯花绸,生活绸,防水防污花绸
silkolene 仿丝薄棉布
silk organdie 蝉翼纱
silk organdy 素纱
silk piece goods 绸缎
silk pin 细大头针(用于真丝的大头针)
silk point 手套背饰缝
silk polar 丝波罗呢
silk pongee 山东府绸,茧绸
silk/ramie mixed fabric 鱼冻布
silk/ramie mixed tapestry satin 桑麻交织锦缎
silk/ramie union batiste 丝麻交织绸
silk/rayon cushion covering 织锦靠垫
silk/rayon mixed tapestry 桑黏交织锦缎
silk/rayon twill damask 采芝绫
silk/rayon union brocade 珍领缎
silk/rayon union chin pillow silk 织锦靠垫
silk/rayon union satin brocade 花软缎
silk/rayon union twill 桑黏绫
silk rep 横纹绸
silk ribbon 丝带
silk rough crepe 全丝粗绉
silk rug 丝地毯
silks 彩色绸制的赛马服
silks and satins 华丽衣着;绸缎
silk satin 真丝缎
silk scarf 丝绸围巾;丝巾
silk screen printing 丝网印花
silk scroop 丝鸣,绢鸣
silk seal 仿海豹丝绒
silk seersucker 泡泡绉
silk serge 丝哔叽
silk sewing thread 蚕丝缝纫线
silk shag 厚丝绒织物,拉绒丝织物
silk shirt 丝绸(乔其纱)衬衫
silk shorts 丝绸短裤

silk single jersey 真丝汗布
silk skin 柔软光洁的皮肤;丝绸裙
silk socks 丝袜
silk spandex lace 丝弹性花边
silk spandex pants 真丝弹力裤
silk spandex swimwear 真丝弹力泳装
silk/spun tussah(union)pongee 桑柞绢纺
silks sateen 熟纺缎子,熟缎
silk stockings 长筒丝袜,丝袜
silk strap 丝带
silk stuff 丝织品
silk suspender skirt 吊带丝裙
silk/sussah pongee 桑柞纺
silktone and goldtone 夹金银丝厚呢
silk taffeta 蚕丝塔夫绸
silk tartan 格子绸
silk textile 丝纺织品
silk thread 丝线
silk top 绸面
silk trousers 丝绸裤
silk/tussah satin 桑柞缎
silk/tussah pongee 桑柞纺
silk/thssah union pongee 桑柞纺
silk/tussah union satin 桑柞缎
silk/tussah satin 桑柞缎
silk twill 真丝斜纹绸,真丝绫,真丝绫
silk twill brocade 花斜纹绸
silk twill mixed 锦桑绫
silk type 丝型(指化纤)
silk umbrella 绸伞
silk underwear 真丝内衣
silk velvet 丝天鹅绒,丝绒
silk voile 真丝绡
silk wadding 丝衬垫;丝绵
silk wadding-imitation aprayed nonwovens 纺丝绵非织造布
silk warp flannel 丝经毛纬法兰绒
silk wear 丝绸服装
silk/wool blended fabric 丝毛混纺织物
silk/wool cloth 丝毛呢
silk/wool fancy suiting 丝毛花呢
silk/wool-like fabric 仿丝/仿毛织物
silk/wool poplin 丝毛绸
silk/wool union poult 绡罗缎
silk/wool worsted flannel 丝毛啥咪
silk yarn 绢丝,丝线,针织丝线,绣花丝线
silk yarn-dyed denim 纯丝牛仔布
silky aspect 真丝感
silky blouse 丝绸女衫

silky fabric 仿真丝织物	silvery lustre 银光
silky finish 仿绸整理	silvery raccoon 银浣熊,银浣熊毛皮
silky handle 真丝般手感	simar(同 cymar)(17～18世纪)宽松女袍,宽松女上衣
silky material 丝质材料	
silky touch 真丝感[觉],丝绸感	similar color 同种色,同类色
sillon silk 棱条绸	similar color scheme 同类色配色
silp blind hemming stitch 暗缝撬边线迹,暗缝缲边线迹	simili binding 丝光滚边条
	simlah 希伯来外套
silper 西尔珀斜纹绸	simlah with tsitsith over kethoneth 希伯来大带流苏装饰的披肩
siluo faille 丝罗葛	
silver 银色,银灰色,银白色;银色奖章	Simonetta 西莫内塔(意大利时装设计师)
silver blue 银蓝灰色	
silver bracelet 银手镯	simpatico 辛派蒂可漂白平布
silver brocade 银线织银缎	simple color 原色
silver brushed 哑银色	simple form bill of lading 简式提单
silver cloth 银纱晚装绸,毛与马利筋纤维混纺呢,银光呢	simple hem stitch 简单边针法
	simple knot stitch 简单结粒绣
silver cloud 银光云色	simple miter 主教冠
silver coating (面料)涂银;银涂层	simple shape 简单条纹
silver cord 窄棱纹棉平绒	simple sole construction 单底结构
silver fern 银蕨色	Simplex fabric 辛普勒克斯经编织物
silver fox 银狐毛皮	Simplex [knitting] machine 辛普勒克斯经编织机
silver gray hair 银灰色头发	
silver grays 军用衬衫布	Simplex warp-knitting machine 辛普勒克斯缝边机
silver green 银绿色	
silver grey 银灰色,银鼠色	simplicity 简洁
silver-grey hair 银灰色头发	simplicity knot stitch 单线结粒绣
silver greys 军用衬衫布	simplification 单纯化
silver hair 银发	simply cosy 质朴闲逸(服装设计思想)
silver high-pile fabric 长毛绒	simulated buttonhole 假扣洞
silver irise 银贝color	simulated decorative hem stitch 仿形装饰折边线迹
silverised cloth (制鞋)银粉带	
silver lace 银线花边	simulated French seam 仿法式缝
silver lining 云朵银边色	simulated fur 仿裘皮
silver mink 银貂毛皮;浅水泥灰	simulated fur fabric 人造毛皮织物,仿皮织物
silver necklace 银项链	
silver pantyhose 银色连裤袜	simulated hand-knit 仿手工编织
silver peony 银光牡丹色	simulated leather 人造革,仿革织物
silver pink 银光粉红色,米红	simulated linen finish 仿麻整理
silver polished 抛光银色	simulated pearl 人造珍珠
silver ring 银戒指	simulated pleat 虚饰褶
silver sage 银灰绿色	simulated pocket 假袋
silver streak 银痕法	simulation of the consumer's Apparel purchse 消费者衣物购买[动机]模拟
silver thread 银丝	
silver tissue 银丝薄纱	simultaneous and combined motion 复合动作
silvertone 银钱夹ельной厚呢	
silvertone and goldtone 夹金银丝厚呢	simultaneous motion 同时动作
silvertone vercoating 银钱大衣呢	simultaneous sewing 同时缝纫
silver white 银白色	sina may 菲律宾条纹麻布
silver wing 银翼色,渣灰色	Sinclair 苏格兰格子花呢

sinew 肌键,筋
sinews 肌肉;体力
singapatti 东非印花棉头巾
Singapore supers 新加坡色织条纹棉平布
singed and mercerized single jersey 烧毛丝光汗布
singed yarn 烧毛纱线
singers machine 日本胜家缝纫机
single 单式;单排纽扣
single account 单边账户
single and double darts 单双省缝
single and double piped pocket 单双滚边袋
single atlas 单梳栉经缎针织物
single bar fabric 单梳栉织物
single bar tricot 单梳栉经编织物
single bed flat knitting machine 单针床平机,单针床横机
single blister jacquard knitted faric 单胖提花针织物
single border 裙边花纹
single border lace 单面边纹花边
single bottom 平脚口(无卷边)
single boxcoat 单排扣箱形外套
single breadth 单幅
single-breast (S.B.) 单排纽扣,单襟
single-breasted (上衣等)单排扣的
single-breasted closing 单排纽门襟
single-breasted coat 单排纽外套,单排纽大衣
single-breasted [coat] frock 单排纽男礼服
single-breasted jacket 单襟式上衣
single-breasted lapel 单襟驳头
single-breasted overcoat 单排扣大衣
single-breasted sack 单排扣西装
single-breasted suit 单排扣西服(套装)
single-breasted vest 单排纽背心,单襟背心
single brush stroke 单线勾勒
single buttoned 单边纽扣
single canvas 绣花用的十字布
single chain stitch 单环缝,单链线迹
single chain stitch looping machine 单链缝套口机
single chain stitch machine 单链式缝纫机
single check 单式格纹
single cloth 单层布
single collar 单片领,单(折)领
single color effect 单色花纹

single cover 特加单纬提花织物
single cuff 单层袖口,单袖头
single cut 短式短发型
single cylinder circular hosiery machine 单针筒圆袜机
single daisy stitch 单代西绣
single damask 五枚花缎
single dart 单称身省缝
single drill 纱卡其,纱斜纹
single edge fusion 单边融合
single edge razor blade 单刃裁刀
single fabric 单层织物,单面针织物,单组经纬织物
single face 单面
single faced flannelet(te) 单面绒布
single faced terry cloth 单面毛巾布
single face gabardine 单面华达呢
single face twill 单面斜纹
single fagot stitch 单针束心线迹,单针装饰线迹
single fancy knitted fabric 单面花色针织物
single fancy suiting 单面大衣花呢
single feather stitch 单羽毛饰边绣
single-feed machine 单系统袜机
single fiber adjacent fabric 单种纤维贴织物
single filament 单丝
single filling [flat] duck 双经单纬针织物
single filling knit 单面纬编针织物
single flap patch pocket 连盖式贴口袋
single-fold 单折
single footage 单幅
single gabardine 纱华达呢
single-inverted pleats 单暗裥,阴裥
single jacquard 单面提花
single jersey 单面针织物,汗布,单面乔赛
single jersey for basque shirt 海军条汗布
single khaki drill 纱卡其
single knit 正针;单面针织
single knit fabric 单面针织物
single knitted atlas fabric 单面经编缎纹织物,单面经缎织物
single knot 普通领带结,平易领结
single laid-in weft-knitted fabric 单面纬编衬纬针织物
single layer collar 单层领
single layer dress 单衣
single layer fabric 单层织物
single layer garment 单层服装

single lift heel 单层鞋跟
single line 单线线缝
single line drawing 单线绘图
single line twill 单棱线斜纹
single loops 单纽袢式
single-needle 单针
single needle bar Raschel machine 单针床拉舍尔经编机
single needle bobbinless machine 单针无梭芯缝纫机
single needle chainstitch 单针链式线迹
single needle differential feed over seaming machine 单针差动送料包缝机
single needle embroidery frame 单针刺绣机
single needle flat sewing machine 单针平缝机
single needle lockstitch 单针锁式线迹
single needle lockstitch machine 单针锁式线迹缝纫机
single needle lockstitch straight sewing machine 单针直线锁式线迹缝纫机
single needle machine 单针机
single needle seam 单针缝纫
single-needle sewing machine 单针缝纫机
single picot 单锯齿边
single picot stitch 单锯齿形线迹
single-piped 单滚边的
single-piped flap pocket 带盖单滚边口袋
single-piped pocket 单滚边袋,单嵌线袋
single-piping pocket 单滚边袋,单嵌线袋
single pique interlock fabric 凹凸纹网眼棉毛布
single pleat 单褶
single pleat trousers 单褶裤
single plush 单面长毛绒针织物
single ply 单层
single-pointed dart 钉形省,直省
single poplin 纱府绸,薄府绸
single-purl buttonhole stitch 纽眼单锁线迹
single raincoat 单排纽雨衣
single rib 3+1罗纹
single rib top 单罗纹口
single sampling plan 一次抽样检验
single satin ribbon 单面缎带
single seam 单行线缝
single-section machine 单节全成形横机
single serge 纱哔叽
single shape 单条纹

single shoulder belt 单式肩带
single side curved displacement 单面弯斜纬疵
single sided twill 单面斜纹织物
single-spring peg 单弹簧梭芯
single stage operation 一步连续操作
single step work clamping system 一步法工件夹紧系统
single stitch 单线缝迹,直线缝迹
single stitching 单线缝;单线缝纫;单线缝迹
single stripe 单条纹
singlet 男汗衫,文化衫,背心
single tack 单边固缝,单加固缝
single tariff 单式税则
single-thickness sash 单腰带,单层饰带
single thread chainstitch 单线链式线迹
single thread linking machine 单线套口机
single thread lockstitch 单线锁式线迹
single thread sewing machine 单线缝纫机
single transit document 单程过境单据
single tricot 单梳栉经编织物
single turn cord facing hemmer 单翻嵌线贴边卷边器
single twilled satin 纱直贡
single twilled merino 简易斜纹美利奴毛织物
single twist 中央扭扎抽绣;单纱捻度
single unit machine 单针床针织机,单程式针织机,单节针织机
single vent 西装的单开衩
single warp tricot 单面经编织物
single welt 单贴边
single welt pocket (同 welt pocket, stand pocket) 立式口袋,单贴边袋,一字嵌袋,挖袋,开缝口袋
single welt pocket with flap 有盖单贴边袋
single width 单幅,单幅宽
single woven 单层织物,单组经纬织物
single woven fabric (同 single yarns fabric) 纱织物
single yarn 单纱,单丝
single yarn drill 纱卡(其)
single yarns fabric 纱织物
Sing machine 胜家缝纫机(日本制)
singosirae 制衬
sinkarah [印尼]辛卡腊棉布
sinker drop socks 纵条纹短袜
sinker loop 沉环,沉降弧

sinker plating socks 闪色添纱花袜
sinker top circular knitting machine 多三角圆形针织机
sinker wheel machine 吊机
sink stitch 绗缝,衲缝
sink stitched finish 漏落针法
sink top machine 带有沉降片的圆纬机
siren look 美人鱼款式,美人鱼风貌;海妖风貌;海妖款式(服装)
siren suit 连衣裤,上罩衣,连身工作服
sirisaf 旁遮普西里赛夫细布
Sir Norman Hartnell 诺曼·哈特尼尔爵士(英国服装设计师)
siro spun 赛络纺
siro-spun yarn 赛络纺纱
sirsake 泡泡纱
sisal 剑麻,西沙尔麻
sisal hemp 剑麻,西沙尔麻
sisal wall paper 剑麻贴墙纸
sisha jacquard mock leno 似纱绸
sisi 斐济妇女串珠腰带
SISILEY 希思黎
sister print 姐妹型印花图案(以大小不同的同一花型重复构成)
sister's thread 刺绣用精白棉线
sistresay 西斯德里舍锦缎
sit (衣服等的)合身;摆好姿势
sit and flare line 上贴下散线形
site mark 定(标记)点;定标记号
siter print 姐妹型印花图案(以大小大同的同一花型重复构成)
sittara 印度锡泰拉本色棉布
siwash sweat[er] 印第安厚毛衣
siwei cloth 四维呢
six 6号尺码的衣服
six-color circle[wheel] 六色相轮
six-cord thread 六股线
sixes 6号尺码的手套(或鞋等)
six footer 六英尺长的领巾
six-gore skirt 六幅裙,六片裙
six-inch ruler 六英寸尺
six panel skirt 六片裙
six-quarter goods 六夸特幅宽的织物
six-stitch 六点线迹
six-stitch zigzag 六点曲折线迹
sixteen 16号尺码的衣服
sixty 60号尺码的衣服
si-zambi 西赞比棉平纹布
size 大小,尺寸;号型,尺码
size assortment 尺码搭配,尺寸配比

size break 型号点《计算机》
size chart 尺寸表
size designation of clothes 服装号型
size dimension 尺寸大小
sizing 量身,上浆,上浆糊
sizing finish 上浆整理
size form 尺寸模型
size grading 尺码放缩,尺码推档
size label 尺码标志,尺码唛
size range 尺码档次,全档尺码
size ribbon 帽子尺码带,帽里缎带
size specification 尺码表,尺寸规格
size stain 浆斑
size stamp 尺码印戳
size step 型号跨度《计算机》
size stick (鞋匠用的)量脚尺
size tab 尺码标签
size tariff 尺码表
sjadra 印度斯遮德拉本色粗布
sjappolen 印花细棉布
skate 冰鞋,四轮滑冰鞋,旱冰鞋,冰刀
skate cap 滑雪帽
skating boots 冰鞋,冰刀,四轮溜冰鞋,旱冰鞋
skating costume 溜冰服装
skating dress 滑冰服装,长袖紧身短连衣喇叭裙
skating outfits 溜冰衣装
skating shoes 冰鞋,冰刀,溜冰靴
skating skirt 长袖紧身短连衣喇叭裙,滑冰裙
skating wear 溜冰服,溜冰衣
skaut 头巾(挪威妇女用)
skeeler 旱冰鞋,三轮滑冰鞋
skeet-shooting jacket 射击用夹克(有皮垫肩)
skein dyed silks 色织绸缎
skeletal coding 程序纲要,程序轮廓编码《计算机》
skeleton 骨骼,轮廓
skeleton back 单夹后间衬里
skeleton edge 织物锁边,小边
skeleton sketch 轮廓草图,构架图
skeleton skirt 笼式衬裙
skeleton suit 肋形装
skeleton suit with ruff 绉领肋形装
skelning thread for art trimming putpose 工艺装饰线
skent 斯坎特短裙(古埃及男子用)
sketal system 骨骼系统

sketch	草图,示意图,速写,写生画,素描
sketch draft	速写,草图
sketch drawing	简图,草图,略图
sketcher	绘图员,描图员;描画师
sketch from nature	写生
sketch of seam	线缝图形
sketch scheme	示意图
skew (distortion)	纬斜,歪斜
skew weft	纬斜疵
ski bag	滑雪袋,滑雪包
ski band	滑雪带
ski blouson	滑雪服
ski boots	滑雪鞋,滑雪靴
ski cap	滑雪帽
ski cloth	滑雪织物
ski clothes	滑雪衣
ski costume	滑雪服饰
skidded seam	滑缝
skidlid	防护帽,防撞头盔
skidproof	防滑的
ski gloves	滑雪手套
skiing cap	滑雪帽
skiing goggles	滑雪风镜
skiing jacket	滑雪上衣
ski jacket	滑雪夹克;滑雪装
ski jupon	滑雪裤
ski knit	滑雪用针织物
SKIKO	精工
skill training	技术培训《管理》
ski mask	滑雪帽,防冻针织面罩
skim coating fabric	涂橡胶织物
skimmer	平顶宽边草帽,勺形帽;圆领无袖紧身女服;女式低跟浅口无带皮鞋
skimmer wool	皮板毛
skimp	特短衣服,特窄衣服,超短迷你裙
skimp marking	过紧排料,节约排料
skimp skirt	超短裙
skimp sleeve cap	袖山不足
skimpy hair	削薄发型
skimpy sleeve cap	袖山不足
skin	皮肤;毛皮,整张皮革,兽皮,皮张;外衣
skin color	肤色
skin-contact clothing	接触皮肤的衣服(内衣)
skin-contact property	触肤性能
skin-crosslinked fabric	表层交联织物
skin-digested wool	皮板毛,皮块毛
skin dress	紧身衣,紧身装,贴身装
skin food	滋肤乳液
skin hair	超短发型
skin hurting property	皮肤伤害性
skin irritation	刺激皮肤
skin jewel(1e)ry	贴肤人造宝石
skin lesion	皮肤损伤
skin-like fabric	仿皮肤织物(具有类似皮肤的透气输湿功能)
skin lotion	润肤剂
skin mesh	网眼皮鞋
skin model	人造皮肤模型
skinner	兽皮加工者,皮革商,皮货商
skinners satin	光亮厚缎
skinny	细腰衣服;消瘦的
skinny look	紧身款式,TT款式紧身型服装样式
skinny pants	紧身裤
skinny rib	紧身高级弹力衫
skinny rib sweater	紧身罗纹汗衫
skinny sweater	紧身毛衣
skin on skin	毛皮砖状拼接法
skin pants	紧身裤,皮裤
skin-shorts	紧身短裤
skin-side comtort	皮肤舒适感
skin socks	爱斯基摩兽皮短袜
skin temperature	皮肤温度
skintight	紧身内衣裤;紧身衣
skintight fit	紧身
skintight pants	紧身裤
skintight shorts	紧身短裤
skin tone	肤色调
skintight tricot	针织紧身芭蕾舞服
skinwear	紧身服装
skin whitener	皮肤增白霜
skin wool	皮板毛
skip	(缝纫)跳针
ski pajamas	两件套针织睡衣长裤
ski pantalets	滑冰裤
ski pants	滑雪裤
skip bundle sampling	逐包抽样《管理》
skip dent	夏装细布
skip lot sampling	逐批抽样《管理》
skipped stitch	(缝线)跳针
skipped thread	跳纱疵
skipped twill weave	飞斜纹组织
skipper	有领羊毛衫和V领羊毛衫重叠穿用,斯基珀(毛衫)叠穿款式
skipper blue	机长服蓝
skipper suit	机长服
skipping	(缝纫中的)跳针;跳缝
skipping stitch	跳针线迹;跳针法

skip-proof sole 防滑鞋底
skip quilting 跳跃绗缝
skips 跳花、跳纱
skip stitch 跳针,跳缝(性能)
skip twill 飞斜纹
ski rompers 填棉童连衣裤
skirt 女裙,裙子;裾,下摆
50's skirt 20世纪50年代裙(直线型款式)
skirt belt 裙带,附有裙腰剪接片的腰带
skirt binding 裙滚边
skirt braid 裙边编带
skirt brief 兜肚
skirt buckle 裙带扣
skirt circumference 裙围
skirt crease 裙子皱
skirted bathing suit 裙式泳装
skirted crease 裙子皱
skirted outfit 裙式套装
skirted suit 连裙套装
skirted vest 接腰背心
skirt flare is uneven 裙浪不匀
skirt gauge 量裙尺
skirt hang 裙子悬垂性
skirt hanger 挂裙架
skirt hem 裙下摆,裙边
skirt hem interlining 裙边衬
skirt hemline 裙下摆线
skirt hem line rides up 裙摆起吊
skirt hem width 裙边宽
skirting 裙料
skirt lace 裙边镶饰花边
skirt length 裙长
skirt lining 裙衬里,裙里
skirt lining ribbon 裙衬带
skirt marker 裙摆记号器;裁裙片样板
skirt on skirt 裙外裙
skirt-pants 裙裤,裙裤式
skirt plait 裙褶,裙裥
skirt pleat 裙褶,裙裥
skirt-pocket 裙口袋,裙侧袋
skirt press 裙压烫机
skirt seam 裙子线缝
skirt side seam 裙子侧面缝
skirt style seam 裙型分割缝,裙子侧面缝
skirt suit with knitted collar and sleeves 针织领、袖裙套装
skirt waist 裙腰围
skirt with laces 镶花边裙
skirt zipper 裙拉链

ski shirt 滑雪衣
ski socks 滑雪袜
ski-style pajamas 滑雪式睡衣裤
ski suit 滑雪装
ski suiting 滑雪服装面料
ski sweater (针织提花)滑雪毛衣
ski tricot 滑雪针织物
ski under wear 滑雪内衣
skive 片过的皮革,经过片边的鞋帮件;削匀(大原料皮);修平
skiver 粗纹羊皮,绵羊粒面剖层皮;片皮革工;(鞋厂的)片边工,削匀工,修平工;片边机,修平机
ski vest 滑雪背心
skiving (制革)片边
skivvies 男式内衣(指汗衫和短裤等的总称)
skivvy (男)圆领汗衫;内衣裤,短袖内衣,海军领内衣
skivvy shirt (男)圆领汗衫
ski wear 滑雪服
sko clothes 滑雪衣
skoncho 方毯披巾,流苏,斯康裘披风(整块条格花呢,中央斜开16英寸长缝,亦可作裙子)
skull 颅骨,头骨;颅罩
skull cap 无檐便帽,瓜皮帽;头顶(骨),天灵盖
skull guard (建筑工人戴的)防护帽
skunk 臭鼬毛皮
sky blue 天蓝色,蔚蓝色,淡蓝色
sky-clearing blue 霁青
sky diver's jumpsuit 跳伞服
skyey 天蓝色,淡蓝色,蔚蓝色
sky gray 天灰(淡灰蓝毛)
sky grey 天灰色(淡灰蓝色)
sky-star type composite fiber 天星型复合纤维
skyteen 深色条天蓝色底经缎织物
slack cloth 疏松布
slack course work 套口横列织物,稀眼横列织物
slackened needle 面线松弛量,残线量
slack filling 纬弓疵,松纬
slack jeans 牛仔式便裤
slack length hoses 大众袜,低筒袜
slack list 松边疵
slack mercerization 缩碱,松式[碱]丝光
slackness 不硬挺,无身骨
slack pick 松纬织疵

slack porosity 膨松性,疏松性,多孔性	slashed trunk hose 男式开衩长裤
slacks 裤,便裤,宽松长裤	slashed virago 有切口的藕节袖
slack seam 宽松缝,松散缝	slashing 剪开法;(衣服上的)开衩;(装饰性)长嵌缝;(异色衬里暴露的)切口
slack season in business 营业淡季《销售》	
slack seat 后裆下垂(裤臀部下沉)	slashing method 剪开法
slack selvedge 松边,起伏不平布边	slash line 剪开线
slacks length 女长裤	slash neck 一字形领口,一字领
slack socks 无翻口短袜	slash neckline 一字形领口,一字领
slacks suit 裤装	slash opening 嵌缝式开口
slack suit 宽松套装,便装	slash placket 嵌缝式开口
slaned pocket 斜袋	slash pocket 挖袋,斜袋,插袋,切缝口袋,长缝袋
slant 斜线,斜面;歪斜	
slant-cut bottom 斜下摆	slat bonnet 板条[支撑帽檐的]罩帽,丝蕾特罩帽
slanted arm sewing machine 斜机头缝纫机	
	slate 暗蓝灰[色],石板色,石板灰
slanted hemming stitch 明缲针,明缲缝迹	slate black 石板黑,灰黑色
slanted inset pocket 斜嵌袋	slate blue 石板蓝(暗灰蓝色)
slanted pocket 斜袋	slate color 石板色,深蓝灰色
slanted trouser pocket 斜裤袋	slate green 石板绿(灰绿色)
slanted welt pocket 斜嵌线袋	slate grey 石板灰(青灰色)
slanting front 偏襟	slave bracelet (同 bracelet ring) 奴隶镯子,手镯环
slanting gobelin stitch 缎纹哥白林刺绣线迹	
	slave earring 黑奴耳饰,奴隶耳环
slanting line 斜线	slazy knitting 布面不清,线圈不匀
slanting padding stitch 衬垫斜缝线迹	sleaze 蹩脚货,低劣物品《销售》
slanting seam 斜缝;缭缝	sleazy holland 三页斜纹里子布
slanting shoulder 斜肩型;垂肩	sleek (男装用)棉里布;口袋布;轧光斜纹棉布
slanting stitch 对角线,斜缝	
slant overedge stitch 斜向包边线迹	sleeked dowlas 捷克低级亚麻平纹布
slant over lock stitch 斜包缝线迹	sleek line 柔美型
slant pocket 斜袋	sleek look 闪光款式
slant saddle stitch 斜包缝线迹	sleek style 光滑发型
slant side pocket 侧缝斜袋	sleep bonnet (压发)睡帽
slant stitch 对角线,斜缝;明缲针法	sleep bra 睡眠用胸罩
slant tacking 斜针假缝	sleep clothing 睡衣外套;女长睡衣
slanty cuts 斜下摆	sleepcoat (长及过膝并束腰带的)男睡衣,门襟有纽扣的睡衣;有腰带的长外套
slash (衣服上的)长缝,切缝,开衩,切口;(皮革加工)砑光	
	sleeper (宽敞的长袍状)睡衣;滞销货《销售》
slash and spread 切缝和展开	
slash and spread method 切缝与展开方法	sleepers (一直包到脚的)儿童睡衣,婴儿睡袋
slashed and stitched effect 开衩及缝合效果	
	sleeping bag 睡袋
slashed breeches 开衩马裤	sleeping bonnet 睡帽
slashed doublet 切缝紧身上衣,开衩紧身上衣	sleeping bra 睡眠用胸罩
	sleeping coat 外套型睡袍
slashed leather boot 16世纪开衩皮靴	sleeping jacket 短睡衣
slashed neckline 一字形领口,长裁领口	sleeping stock 滞销库存《管理》
slashed-puff sleeve 切缝灯笼袖,西班牙袖	sleeping suit 儿童睡衣,童寝装,睡衣裤
slashed sleeve 开衩袖,切缝袖	sleeping wear 睡衣,睡服

sleep mask 黑色弹力睡眠用脸罩
sleepsafe 阻燃处理绒布(商名)
sleep set 两件套睡衣
sleep shirt(同 night shirt) 衬衣式睡袍
sleep shorts 男式橡筋睡裤
sleepwear 睡袍,睡衣;(总称)睡衣裤
sleeve 袖子,袖套;袖山线,袖山弧线
sleeve à la folle 落肩灯笼袖
sleeve action 袖子活动量
sleeve and shoulder seam 袖子和肩部线缝
sleeve attaching machine 缩袖机
sleeve back 袖子偏后
sleeve backward 袖子偏后
sleeve band 袖袢
sleeve block 袖子样板
sleeve board 烫袖板,压袖板,袖子烫板,烫马,袖型烫板,烫凳
sleeve bottom 袖口
sleeve bracelet 上臂袖手镯
sleeve button 袖口饰纽,袖扣
sleeve cap 袖山(袖上端联袖笼部位),袖头
sleeve cap height 袖山高
sleeve cap line 袖山线,袖山弧线
sleeve cap support 袖山撑条(垫),袖山支撑物
sleeve cap underlining 袖山衬(防袖山凹陷,支撑袖山)
sleeve centre line 袖中线
sleeve centre seam 袖中缝
sleeve clips 袖夹
sleeve crown 袖冠
sleeve-cuff 袖口,袖头
sleeve cushion 灯笼袖的袖衬垫
sleeved slip 连袖衬裙(兼作睡袍)
sleeve end 袖头,袖端
sleeve end line 袖口缝线
sleeve facing 大袖衩
sleeve facing folder 袖衩叠缝器
sleeve fitting 袖的假缝
sleeve forward 袖子偏前
sleeve girth line 袖长线,袖围线
sleeve [hanging] backward 袖子偏后
sleeve [hanging] forward 袖子偏前
sleeve head 袖山头,袖口
sleeve head rolls 袖头烫垫
sleeve hem 袖口边
sleeve hem width 袖口边宽
sleeve inseam 袖内缝,前袖缝

sleeve inseam swingout 前袖缝外翻
sleeve joining seam 拼角袖缝
sleeve leans to back 袖子偏后
sleeve leans to front 袖子偏前
sleeve length 袖长
sleeve length line 袖长线
sleeveless 无袖
sleeveless blouse 无袖上衣
sleeveless bodice 无袖紧腰衣,无袖式衣肩
sleeveless jacket 西装背心,坎肩
sleeveless over-dress 无袖罩衣,无袖外套
sleeveless pullover 套背心,无袖套衫
sleeveless robe 无袖长袍
sleeveless slim-fit dress 无袖紧身连衣裙
sleeveless slipover 无袖套领衫
sleeveless style 无袖款式
sleeveless sweater 无袖毛衣,毛线背心
sleeveless V-neck top V领无袖衫
sleeveless woolen sweater 毛背心
sleevelet 袖套
sleeve lining 袖里
sleeve lining inserting machine 缝袖衬里机
sleeve lining silk 袖衬里绸
sleeve link 袖纽,袖口链扣
sleeve modeling 袖型
sleeve muscle line 袖围线
sleeve opening 袖口,袖头
sleeve opening line 袖口线
sleeve overlap 袖衩搭边
sleeve pitch mark 袖子定位记号
sleeve placket 袖衩,袖袢,琵琶袖
sleeve placket binder 做琵琶袖衩器
sleeve-placket machine 缝袖衩机
sleeve placket stay 袖衩条
sleeve placket tape 袖衩条
sleeve placket top 袖衩顶端
sleeve point 袖山
sleeve press 衣袖熨烫机
sleeve press pad 烫袖垫
sleeve protector 护袖
sleeve roll 烫袖垫
sleeve seam 袖缝
sleeve seam opening press 袖缝烫开机
sleeve set 小袖子,袖套
sleeve setting line 缩袖线
sleeve setting machine 上袖机,缩袖机
sleeve shape 袖子形态,袖型
sleeve silhouette 袖子轮廓

sleeve silk matelasé 袖背绸
sleeve slip 有袖衬裙
sleeve slit 袖(开)衩
sleeve silt line 袖衩线
sleeve snap closing 按扣袖口,按纽袖头
sleeve string 袖带
sleeve style 袖型
sleeve styling 袖子款式
sleeve tab 袖袢,袖片
sleeve top 袖山点,袖山头
sleeve type 袖型
sleeve type take-up 滑杆式挑线
sleeve under facing 小袖衩
sleeve vent 袖衩
sleeve width 袖宽
sleeve wigan 袖子里料
sleeve zipper closing 袖口拉链
sleeving 管状织物
sleeving machine 缡袖机
sleevings 男式宽大马裤
sleigh curve 雪橇型弯尺
slendang 英国棉披巾,棉围巾
slender body 细瘦型身材
slender build 细瘦型身材
slender figure 苗条身材
slender form 苗条体型
slenderize 使身材显得苗条
slender line 苗条型
slender look 纤细款式
slender shoes 锥形鞋
slender silhouette 细长型,苗条型轮廓
slender skirt 细腰身裙子
slender thigh 细瘦大腿
slender welt pocket 细嵌线袋
slendong 棉披巾,棉围巾(英国)
slesia 轻薄亚麻斜纹饰布
sleve type take-up 滑杆式挑线
slick chick 衣着鲜艳的少妇(美俚)
slicker 长而宽的油布雨衣,雨衣
slicker fabric 雨衣织物,防水织物
slick hand 光滑手感
slide 便鞋,凉鞋;发针
slide body 拉链头
slide fastener 拉链
slide fastener placket 拉链式开口,拉链口袋
slide gauge 卡尺,滑尺
slide holder 拉链把手
slide loop 滑带扣环
slide plate (缝纫机)滑板
slide pull 拉链拉头
slide pull-tab 拉链头拉袢
slide ruler 计算尺
slide sandal 室内凉鞋
slide tab 拉链滑块,拉链头子
sliding caliper 游标卡尺
sliding clasp fastener 嵌合式拉链
sliding duty 滑动关税
sliding fastener 拉链
sliding gauge 曲面仪(量身体曲面用)
sliding pad 裤腿滑叠衬垫
slight hip 瘦臀
slight touching 搭色疵
slim 细长的,苗条的,纤细的
slim and ease skirt 合身裙
slim arm 细臂
slim-hipped figure 细臀型体型
slim-hipped type 细臀型
slim hook 细衣钩,暗扣
slim jeans 细长型弹力牛仔裤
slim jim 细长裤子,女式中长裤
slim limbed pants 细腿紧身裤
slim line 苗条线型,紧身线型
slim-line silhouette 细长型轮廓
slim long line body 细长身段
slim look 苗条风格,苗条款式
slimmer 女套衫
slimming line 苗条线条,苗条型
slims 细腿紧身裤
slim silhouette 苗条型轮廓
slim skirt 窄下摆裙子,旗袍裙,细长裙
slim slacks 纤细裤脚,细脚口裤
slim slip 窄型衬裙
slim style 细长款式,苗条款式
slimsy 稀薄织物;结构不良织物
slim waist 细腰
sling 印度细山羊毛织物;背带;三角巾
sling back 吊跟鞋(鞋后帮呈带状的)露跟女鞋
sling back court shoe 后袢式浅口球鞋
sling back pump 后袢式露跟浅口舞鞋
sling back shoes 后袢式浅口鞋
sling back thong sandal 夹指露跟凉鞋
sling bag 长带挂包
sling cape 吊腕式披风
sling neckline(同 one shoulder neckline) 单肩式领,单肩领口
sling pump 露跟浅口皮鞋,后袢带碰拍鞋
slink 轻飘合身的连衣裙

slinky 紧身的,苗条的,线条优美的
slip 活络里子;套裙,(有背带的)女式长衬裙,女无袖衬裙;童外衣,儿童围兜;男游泳裤;枕套
slip-baste 定针,疏缝,假缝
slip basting 定针,疏缝,隐式假缝,藏针式假缝
slip-blouse(同 blouse slip) 连衫长衬裙
slip buttonhole 绷扣眼
slip dress 衬裙式连裙装
slip gown V形领裁长睡袍
slip hemming 暗缝绷边
slip hemming stitch 暗缝绷边线迹
slip knot 滑结,活结
slip noose 滑结套,活结套
slip-on 套头式;套头衫;套裙;无扣手套;松紧鞋,无带便鞋,无扣便鞋
slip-on blouse(同 pullover blouse) 套头女衫
slip-on casual 懒汉鞋
slip-on garter 吊袜带,套入式吊袜具
slip-on gloves 无扣手套,宽口半长手套
slip ons(同 pull ons) 易戴式手套
slip-on shoes 易套鞋,松紧鞋,无带便鞋
slip or blind hemming stitch 暗缝撬边线迹
slipover 套头式连衫裙,套衫,套领运动衣
slipover blouse 套头式罩衫,套衫
slipover dress 套头式连衫裙
slipover pajamas 套头式睡衣
slippage-free stitching 无滑移缝纫
slipped stitch 跳针,脱线缝
slipper 拖鞋,浅口便鞋,轻便舞鞋,滑屐
slipper carpet 提花经起绒织物
slipper satin 鞋面花缎
slippers 拖鞋;便鞋
slippers for health 健康拖鞋
slipper socks 室内皮底短袜,暖袜,连鞋袜;连袜便鞋
slippery fabric 易滑脱织物
slippery hand 手感光滑
slips 低级棉绒布;水手裤,妇女衬裙,女式长胸衣
slip sheet 衬纸
slips height 旗袍衩高
slip shemming stitch 暗缝绷边线迹
slip sleeve 开衩袖
slip slop 沙滩便鞋
slip sole 鞋底嵌料

slipstick 计算尺
slip-stick behavior 滑黏特性
slip stitch 短而松的暗缝线迹;漏针;挑针;撩针法(绷针)
slip stitch armhole 绷袖窿
slip stitch buttonhole 绷扣眼
slip stitch button loop 绷扣袢
slip stitch collar to bodice 绷领下口
slip stitched seam 隐针缝,绷针缝
slip stitch facing 绷暗门襟
slip stitch gorge line 绷领串口
slip stitching knee kicker 绷膝盖绸
slip stitch patch pocket 绷口袋
slip stitch reinforcement for knee 绷(裤)膝盖绸
slip stitch shoulder seam 绷肩缝
slip stitch sleeve slit 绷袖衩
slit 裂缝,窄缝;裂口;槽;服装开衩,衩口
slit armhole 开衩袖窿
slit collar 开衩领,开缝领,狭缝领
slit dart 细长的褶子
slit-eyed 有细长眼睛的
slit fabric 斜裁料子,滚边料子,窄幅料
slit height (旗袍)衩高
slithery 滑溜溜的
slit neckline 一字领口,开缝领口,开衩领口,切口式领口
slit opening 隙状开口
slit placket 隙状开口
slit pocket 切开式口袋,开缝口袋,挖袋
slit ribbon 裁制带子
slit skirt 开衩裙
slit sleeve 长缝开衩袖,开衩袖
slit tape 狭条料子(从阔幅裁剪而成)
sliver 断头;裂纹;薄片、裂片;丝条,纱条
sliver high-pile fabric 梳条长毛绒针织物
sliver knit fabric 背面起绒针织物
sliver lay-in circular knitting machine 毛条喂入式圆形针织机
Sloane Ranger 英国上层社会中拘泥于传统而又衣着时髦的年轻女子
slob 褴褛
sloea(e) 古罗马凉鞋
sloop 设置服装单元
slop 外衣,宽大罩衣,工作服;廉价成衣
slope 斜度,坡度
sloper 服装样板,尺寸样板,服装原型(美国);服装原理
slope shoulder 斜肩,溜肩
slop(ing) shoulder 塌肩,斜肩,溜肩;斜

肩型
sloppy chic 邋遢装
sloppy hat 宽大宽边软帽,软毡帽
sloppy joe （女子穿的）宽松套衫;不修边幅的男人
sloppy trousers 肥大的裤子
slops 宽松裤,宽腿短裤;现成低档衣服
slop seller 廉价现成衣服销售商《销售》
slops finished with fringed sashes 束腰镶边宽松裤
slop shop 低档服装商店,现成低档服装商店
slop work 廉价成衣;低档成衣制造业
slot buttonhole 细缝纽孔
slot neckline 窄长缝形领口,狭长缝领口
slot pocket 投币孔式口袋,狭口袋,窄缝式口袋
slot seam 嵌条缝,嵌档缝,双暗缝,双闷缝,褶饰缝
slot-thread 芯线
slot width 窄缝宽度
slouch 及膝针织睡袍;帽檐,耷拉帽边;休闲短袜
slouch gown 短睡袍服
slouch hat 宽边毡帽,宽软边呢帽,垂边女帽
slouch hat trimmed with flowers 饰花垂边帽
slouch socks 三色袜口的锦纶短袜
slouchy look 懒散风貌,懒散款式,邋遢款式
sloven （衣着,外表）邋遢的人;不修边幅的人
slow burning fabric 慢燃织物,耐燃织物
SL slider （self locking slider） 自锁拉链头
slub 粗节(疵点),花式线粗节
slubbed 竹节布
slubbed fabric 竹节布,竹节花式线织物
slub bourette 疙瘩绸,麻纺绸
slub repp 棱纹竹节纬平布
slub suit 粗[纺]呢绒服装
slubsuitlit 粗(纺)呢绒服装
slub ya-jiang bourette 疙瘩鸭江绸
slub yarn 竹节花式纱线
slumber cap 压发帽,睡帽
slumber net （睡觉时用）发网,网式发帽
slumber wear 睡衣类(商店用语)
slun 蹩脚货,(摊子上出售的)廉价商品《鞘售》

S.M.（sewing machine） 缝纫机
small 小号服装;腰背部;个子矮小的人
small back 窄背
small bag 荷包
small batch 小批量
small bust 小胸
small checks 小格子纹
small check tweed 小方格呢
small clothes （18世纪）紧身齐膝短裤,小件衣着用品
small cup 扁平胸部
small falls 男裤前门襟
small hip 小臀,小臀围
small jacket 小外套
small lapel 上翻领,小驳头
small lapel suit 窄驳领西装
small load launder 轻量洗涤
small of knee 下膝线
small of the back 小腰
small parts profile machine 小缝件仿形缝纫机
small pattern brocade 小锦
small pattern fabric 小花纹织物
small round shoulder 小圆肩
smalls 小件衣着用品(如内衣、手帕等);（18世纪）紧身齐膝短裤;小钻
small scale 小尺寸
small-scale print 小花印花
small scale test 小规模试验
small scissors 小剪刀
small serrated tracing wheel 细齿轮点线器
small size(s) 小号,小尺寸
small sleeve 小袖,内袖,袖下片
small sliding caliper 滑动计(量体用)
small stitch 细针迹,密缝
small waist 细腰,小腰围
small wares 小商品,带类织品
smalt 深蓝色,大青色
smaragdine 翡翠绿的,鲜绿色的
smart 漂亮的,时髦的,潇洒的
smart case 方便盒
smart fiber 智能纤维
smartness 呢面平挺
smarts 聪明,智慧,能干;学时髦的人,装聪明的人
smart set 最时髦人士(总称)
smart shape 时髦款式
smash 破产,经营失败《销售》
smashable hat 折帽,折叠式帽

smear 油渍
smearing 刮浆
smear(ing) paste 刮浆
smemo sewing thread 形状记忆缝纫机
smile badge 微笑徽章
smock （画家、医生等的）抽袖罩衫,罩衣,工作服;（欧洲农民穿的）长罩衣,长劳动衣;（士兵穿的）伪装服;古代宽腰身女服,女式无袖宽内衣;打褶裥,司马克装饰性缩缝褶
smock coat 罩衫
smock dress 罩袍式连裙装,罩衫裙
smock frock 长劳动服,长罩衣,（欧洲农民穿）工作服
smock helmet 救火防毒面具
smocking 刺绣装饰;伸缩绣缝,缩褶绣,正面刺绣针迹;打揽;装饰性缩缝;装饰性抽褶刺绣
smocking machine 缩褶刺绣机
smocking pattern 缩褶刺绣花样
smocking stitch(ing) 正面刺绣线迹,缩褶刺绣装饰线迹
smock overall 女罩衫
smock suit 罩衫套装（宽松）
smock thread 装饰线
smock top 长袖罩衫
smoke 烟灰色,烟青色
smoke blue 烟青色,青灰色
smoke goggles 防烟雾护目镜
smoke gray 烟(灰)色
smoker 略式晚礼服
smok(e)y grey 烟(灰)色
smoking jacket 吸烟衫,吸烟装外套,半正式晚礼服外套
smoking suit 女休闲服
smoke ring 烟圈形管状围巾,晚间便围巾
smoked pearl 灰色珍珠;珍珠层
smoked topaz 烟黄色水晶
smoko 工间休息时间《管理》
smoky grape 淡紫藤色
smoky grey 烟灰色
smoky quartz 烟水晶
smoky tone 薄雾色调
smoldering 发烟阴燃
smooth 烫平,熨平
smooth burning 稳定燃烧
smooth drying 免烫
smooth drying fabric 免烫织物
smoothed curve 光滑曲线

smooth-edged tracing wheel 圆盘轮点线器
smoothing iron 熨斗,烙铁
smooth knitting fabric 双罗纹针织物
smooth leather 光面革,光滑皮革
smooth line 光滑线,圆顺线
smooth machine 双罗纹圆编机
smoothness 光滑性,滑爽性,滑糯度
smooth-touch 手感光滑
smooth type synthetic leather 光滑型合成革
smudge （揩擦后留下的）污迹
smyrna cross stitch 士麦那十字绣
smyrna stitch 士麦那绣
snag 钩丝疵;擦毛;戳破处,钩破处,抽丝处
snag-free finish 防钩丝整理
snagging 抽皱
snail button 蜗牛状纽扣
snail edge 毛边
snake boot 防蛇靴;高帮时装靴
snake bracelet 蛇形手镯
snake chain belt 蛇形链条腰带
snake hip 瘦削的臀部
snake skin 蛇皮,蛇皮革
snake skin handbag 蛇皮手提袋
snap 揿纽,按扣,钩扣,四合钮,子母钮
snap attacher 钉（四合）扣机
snap-brim 可以翻起的帽檐
snap-brim hat 折檐帽,翻檐帽,软宽檐礼帽
snap button 揿纽,四合扣,子母扣
snap button hammer-on snap 四合扣
snap clamp 揿纽夹
snap closing 按扣门襟
snap fastener 帽钉;揿纽,按扣,子母扣,四合扣
snap fastener clamp 揿纽夹
snap fastener crotch pants 开裆裤
snap fastener tape 按扣带,揿纽带
snap fastening [attaching] machine 钉四合扣机
snap front vest 按扣背心
snap front waist 装揿纽腰头
snap handle 揿纽柄
snap link 弹簧扣
snap-on collar （用按扣连接的）脱卸领
snap-on foot 可折压脚
snap panties 揿纽短裤,婴儿开裆裤
snap pants 揿纽短裤,婴儿开裆裤

snapped shoes 按扣鞋,揿纽鞋
snappiness 时髦,漂亮
snappy look 俊俏型;俊俏风貌
snap ring 开口环,扣环
snap tape 揿纽带,按扣带
snarl 扭结
snarly yarn 辫子线
sneaker 网球鞋,旅游鞋,胶底帆布鞋;运动袜
sneaker middle 中年轻便装(由牛仔裤、软底鞋组合而成)
sneaker socks 运动鞋短袜
snip (剪金属片的)平头剪刀;裁缝;剪下小片;切口;剪,剪断,剪去
snip cloth 剪布
snip off thread residue 剪线头
snip paper 剪纸
snip slip 下摆长度可调节的衬裙,裁摆式连身衬裙
snip-bottom slip 裁摆式连身衬裙
snip-grip thread cutter 夹紧剪线刀
snippers 手剪
snippet 衣片
snippings 裁剪碎料
snood 发网,发套,束发带,头带,网状帽,发套式帽子
snood with straw 草编束发网
SNOOPY 史努比
snoprag 手帕(俚语)
snorkel coat 通气管式外套
snorkel jacket 通气管式外套(其防风兜帽似潜艇的通气管,门襟拉链可到面颊)
snot-rag 手帕
snow ball 起球
snow boots 雪地靴,橡皮靴,棉靴
snow cloth 冬季织物,冬季运动服装用料
snow coat 风雪大衣,防雪大衣
snowflake 雪花呢
snowflake beige 夹丝雪花呢
snowflake cloth 雪花呢
snowflake corduroy 雪花灯芯绒
snowflake interlock fabric 雪花棉毛布
snowflake lake 雪花呢
snowflake overcoating 雪花大衣呢
snowflake square 雪花纹
snow jacket 防雪衣
snow leopard 雪豹;雪豹毛皮
snowmobile boots 雪地机动车靴
snowmobile gloves 雪地机动车手套

snowmobile mitt 雪地机动车露指手套
snowmobile suit 雪地机动车装
snows 白发
snow shirt 冬季保暖;服装
snow shoes 雪地鞋
snow suit 防寒服,雪季套装,儿童风雪服
snow washing 雪洗
snow wear 滑雪服装
snow-white 雪白
snow white costume 白雪公主装束
snowy 雪白钻石
snowy white 雪白
SNP (side neck point) 颈(根外)侧点
snub-nosed flats 扁平翘头平底鞋
snuff color 黄褐色,鼻烟色
snug 紧的;钩丝
snug fit 舒适合身,紧贴合身
snug fit gloves 紧贴舒适手套
snug fitting underwear 贴身保暖衣
snuggies 女式中长保暖内裤,针织保暖衬裤
Snuggies 斯纳吉内裤
snuggle bag 婴儿包
snugness of fit 穿着舒适性
soakers (婴儿尿布的)吸水垫
soap chalk (划线用)皂粉笔
soap-fast 耐皂洗的
soaping fastness 皂洗牢度,耐皂性
soap-mark 皂渍,灰点疵
soap-off 皂洗
soap powder 洗衣粉
soap soda scouring 皂碱洗涤
soapy handle 滑腻手感,似皂手感
soatts boots 裹腿靴
soccer boots 英式足球鞋
social-psychological assessment 社会心理评价《销售》
social psychology 社会心理学《销售》
social shoes 社交鞋(比正式礼服更华丽而高雅的鞋子总称)
social wear 社交服
socioeconomics of fashion 服饰社会经济学
sock 短袜;轻软鞋
sock cover 袜套
sock lining 鞋(衬)垫
sock loading 套袜
sock panel 短袜袜筒
sock press 短袜压烫机,短袜定形机
sock stripping 脱袜装置

sock suspender 吊袜带
sock toe 袜头
sock treat 短袜处理机
socket 钮座；插座
socket bone 臼骨
socking bag 麻袋
socklet 套袜,翻口短袜,薄形足底袜
sock-of look 突出袜子的款式
sock-o-look 突出袜子的款式
socks 短袜；护套,鞋垫
sock-suspender 吊袜带
socque （法）高木底鞋；演员短靴
socquette 短袜
sodolin 亚麻大麻混合织物
soesjes 色条纹包头布,沙埃司吉条子包头布（印度）
sof 毛织花缎（小亚细亚）,沙夫闪光薄呢
sofit 沙菲特仿花纱罗棉布
sofrina 索夫里那（高级人工人造革）
soft 呢帽；软领；柔软,柔和,模糊；柔性
soft and stiff filler 软硬棉
soft and supple 软细腻
soft and tight 柔软紧贴
soft-back elastic 柔和弹性的橡筋
softball protector 垒球保护器
softball shoes 垒球鞋
softball wear 垒球装
soft cap 软便帽
soft cloth 软布
soft collar 软领
soft color 柔和色,嫩色
soft cowl 无褶垂领
soft crepe 绢丝绉
soft cup 柔性胸罩窝,高弹胸罩窝
soft cup bra[ssiere] 高弹胸罩
soft cup nursing bra[ssiere] 高弹护理胸罩
soft cup style 高弹乳罩窝式
soft cut 软性发型
soft dress 柔型连衫裙
soft enamel 软漆皮
softener wash 轻度水洗；柔洗
softening 柔顺；柔软整理,软化
softening treatment 柔软处理
softer look 柔和款式
soft-expert 软专家
soft feel 柔软手感
soft felt hat 软毡帽
soft-filled sheeting 松捻纬纱绒布
soft finish 柔软整理

soft finished denim 经柔软整理的粗斜纹布
soft finish（ed）fabric 柔软整理织物
soft fold 自然而柔软的折叠
soft garment 柔型服装
soft goods 软商品；（服装,服饰等）非耐用品；纺织品
soft green 柔绿
soft grey 柔灰色
soft hand 柔软手感
soft handle 手感柔软
soft hat 柔软中折帽,呢帽,软毡帽
softing feeling 蓬松度
soft jeans 软身牛仔裤
soft knot 软结
soft leather 软皮
soft leather shoes 软皮皮鞋
soft-leather upper 软皮鞋面
soft-legged boots 软皮靴
softline bra[ssiere] 缩褶少女胸罩
soft look 柔和风貌；柔和款式
soft make up 柔和化妆法
softness 柔和感,柔软度
soft peach 淡桃红色
soft pencil 软芯铅笔
soft pleat 不定型褶,活裥,软褶
soft red 粉红
soft ruler 软尺
softs 软商品；（服装等）非耐用品；纺织品
soft seam 软缝
soft shoes 软底鞋（尤指跳踢踏舞时穿的软底鞋）
soft shoulder 软垫肩
soft shoulder look 柔肩款式；柔肩风貌
soft shoulder model 柔和肩线型
soft silhouette 柔软轮廓线条
soft sole [shoes] 软底,软底鞋
soft soled shoes 软底鞋
soft standing collar 软立领
soft suit 柔性套装,突出女性美的套装
soft tailoring 简做（的）缝制
soft tone 柔和色调
soft twist yarn 弱捻纱
software 软件,软设备
soft wares 软商品,（服装等）非耐用品,纺织品
soft yellow 淡黄色
sofu 日本本色平纹棉布
soie batiste 全丝细薄绸
soie d'inde 印度手织薄绸

soie ondée ［法］波纹丝线
soie platte 刺绣丝线
soil 污迹
soilability 易沾污性
soil affinity 吸污性
soilage 肮脏
soil away 织物防污处理
soil-hiding quality 藏灰性,吸灰性
soil-out 织物防污处理
soil release (SR) 易去污,去污
soil-release fabric 防污织物;防污布
soil release finish[ing] 易去污整理
soil removal 去污
soil repeilency 拒污性
soil resistance 抗污性
soil resistance fiber 抗污纤维
soil resistant finish (S. R. finish) 防污整理
soiree 晚礼服,晚会服
soisette 高级仿茧绸棉布
sola 合萌芯防晒帽
solail velvet 有光平绒
solar absorbing and retaining fabric 太阳能吸收和储存织物
solar optical property 日照性
solarisation 黑白反转花纹
solarization 日晒;负感作用,反转作用
solar-o 沙拉乌
sola topee 防晒帽,合萌芯防晒帽
sola topi 防晒帽,合萌芯防晒帽
soldano 索尔达诺(意大利时装设计师)
sole 脚底;鞋底;袜底;脚掌
solea 罗马木底凉鞋
sole beel reinforced nailin machine 大底鞋跟加固打钉机
sole button hole 剪接线扣眼
sole heel reinforced nailing machine 大底鞋跟加固打钉机
soleil 索列尔;斜纹呢,棱条缎,加强缎纹织物,加光织物
sole-in-sole 双底袜
sole-in-sole splicing 袜子双底加固
soleil velvet 有光平绒
sole leather 鞋底革,鞋底皮
sole length 袜底长
sole (of shoe) 鞋底
soles lace 特纳里夫花边
sole splicing 袜底加固
sole thread 鞋底线
solferino 鲜紫红色,品红

solid business 稳固的行业《贸易》
solid color 单色,素色
solid colored cloth 素色织物;单色织物
solid colored single jersey 素色汗布
solid design 独花(手帕用语)
solid foot 整体压脚
solidity （织物)厚实感
solid line 实线
solid modelling 实体造型
solid packing 包装牢固
solid relvet 密实丝绒
solid shade (同 plain shade) 单色,素色,一色
solid suit 单色套装
solid velvet 密实丝绒
solid woven 紧密的织物;密实编织,多层交织
solid zipper foot 固定拉链压脚
solisooty 印度手钏弱捻薄细布
solitaire ［法］索立台儿缎带;单石戒指
solitaire ring 独粒宝石戒指
solo 稀织棉布
solo shoulder 单肩款式
sologesses 印度索洛吉斯优质细布
solora 夏季薄毛料
sols 巴西索尔蛛网花边
soluable cloth 水溶布
solubilized vat violet 可溶性还原紫
solubilized sulfur dye 可溶硫化染料
solubili vat dye 暂溶性还原染料
solution concept 交换穿着观念,分解组合的穿着观念
solution-dyed fiber 原液染色纤维
solvent dyeing 溶剂染色
solvent spotting 溶剂污迹
somatology 身体学
somatometry 身体计测
somber hue 暗淡色
somberro 墨西哥宽边帽
somberro Cordobes hat (同 gaucho hat) 加乌乔牧人帽
sombrero grey 暗灰色
sombrero [hat] 墨西哥宽边帽
sombrer tone 灰暗色调
some old shit (SOS) 老一套,老套头(俚语)
some packs of cut parts 几捆(裁)片料
some waist suppression 小窄腰
sommiere 法国桑米尔哔叽
soneri 松纳利嵌金线条纹织物(印度)

Song (dynastry) bricade 宋锦
songket 金属丝纬花纹织物
Sonia Knapp 索尼亚·纳普(法国时装设计师)
Sonia Rykiei 索尼亚·丽基埃尔(法国时装设计师)
sontag 毛线短上衣；毛线披风，编织披巾；桑塔格披巾
soochow brocade 古香缎
sooner 衣衫褴褛的人；便宜货，劣质货
soosee 手织平纹绸
soot black 炭黑色
soot brown 烟炱棕，烟煤褐色
soots romal 苏洛马条子披巾布
sophisticated clothing 高级服装
sophisticated design 复杂的花样设计
sophisticated product 深加工产品
sophisticated style 不落俗套的风格(款式)
sophisticated technique 尖端技术
sophisticated traditional 精致的传统款式
sorcrote 印度索克洛平布
Sorelle Fontana(Fontana sisters) 索雷尔·佛恩塔纳(意大利高级时装店)
sorptive clothing 吸附性外衣，吸着性外套
sorptivity 吸着性
sorquenie 女式紧身束腰外衣
sorrel 赤褐，栗色，红棕色
sort 种类，类别；品种，品级；品性；姿态；派头
sorter 分档者，分类者；分皮者；分类机；分类装置
sortie 无扣合具短手套
sorting 分束；分皮；挑选，分级，分类
sorting out fabrics 衣料分类
sorting penistone 佩尼斯顿再生毛粗呢
sort out fabrics 分织物门幅；分幅宽
SOS (some old shit) 老一套，老套头(俚语)
soucanie 女式紧身束腰外衣
soufflé 绉织物上的大凸花纹
soulierà la poulaine ［法］波兰靴
sound occlusion suit 吸声服
sound-proof wall furnishing fabric 消声贴墙布
soup and fish 男式晚礼服
source of fund 资金来源
sousee 印度苏西绸
s(o)uslik 花金鼠皮

soutache 镶边饰带，辫带状镶条
soutache braid 辫带状饰带
soutache braiding machine 编带机
soutache-edged spencer 滚边短大衣
soutane (天主教)祭司法衣,黑色法衣
South West African Karakul-lamb (Swakara) 西南非洲产的卡拉库尔改良种羔羊的注册商标
southern colonel tie 窄领带，鞋带型领带，团长领带，警长领带，西部领带
southern tie 南方绑带鞋
southwester (同 sou'wester) 海员防水帽，风雨天防水帽
soutien collar 撑领(14～17世纪流行)
soutien-gorge 胸罩
souvenir look 纪念品款式(旅游地出售的服饰)；留念型服装
souvenir scarf 留念用大方丝巾(印有风景名胜画作旅游留念用)
sou'wester 海上油布长雨衣；海员防水帽
sox 儿童短袜
soxong flannel 萨克森法兰绒
soyan cloth 尼日利亚南部索扬绸
SP (sales promotion) 推销《销售》
SP (shoulder point) 肩端点,肩袖点
spa blue 矿泉蓝
space 宇宙型款式；宇航风貌；太空装
space age look 宇宙时代面貌
space blanket 太空绝缘披风，太空棉披肩
space clothes 太空服
space dye 间隔染色
space dyed knitting yarn 彩虹绒线
space dyed silk 间隔染色绸
space dyed yarn 间隔染色纱，多色线，花色纱，彩虹纱
spaced braid 露孔棉饰带
spaced pattern 飘飞花样(小图案,大部空白),留白图案
spaced tucks 间条塔克,间条缝裥
space dyeing 间隔染色
space hat 宇宙帽
space helmet 宇宙飞行帽,宇航员头盔
space helmet visor 宇航员头盔遮阳
space look 宇航风貌；宇航型款式；太空装
space out 飘飞花样(小图案,大部空白),留白图案
space perception 空间感
spacer 衬垫；衬套；空气层；(确定纽扣,

扣眼位置)间隔器
spacer textile 衬垫纺织物
space shoes 太空鞋
space shuttle uniform 航天服
space stitching 间隔线缝
space suit 太空服,航天服,宇航服
space wadding 太空棉,宇航服填絮
spacial concept 空间概念
spacie 斯佩西吸汗快干针织物,斯佩西针织布
spacious-pin wale corduroy 粗细条灯芯绒
spacious waled corduroy 粗条灯芯绒
spade beard 铲形胡须,长方形胡须,上端修圆而下端呈尖形的胡须
spade bone 肩胛骨
spade sole 铲形鞋底
spadiceous 浅褐色,鲜褐色,栗色
spaghetti belt 细腰带
spaghetti sash 细实心饰带,实习套管式圆腰带
spaghetti strap 女服细肩带,面条带
spaghetti tie 意大利细条式领带
spaier 直向开衩,垂直开衩
spair 男裤门襟
Spalding 斯伯丁
span 指距(以手指度量,通常为 23 厘米或 9 英寸);跨度;变化范围
spandex 斯潘德克斯(聚氨基甲酸酯弹性纤维商品名);氨纶;氨纶织物
spandex core spun 斯潘德克斯包芯弹力布
spandex knit fabric 氨纶弹力针织布
spandex silk knit fabric 丝弹力针织布
spandex stretch fabric 氨纶弹力织物
spandex wndergarment 氨纶[弹性]内衣
spangle 发光饰片,亮片,珠片
spaniel's-ear collar(同 dog's earcollar) 垂耳领
Spanish black work 西班牙黑刺绣
Spanish blonde[lace] 西班牙丝花边
Spanish boots 西班牙靴
Spanish cape 西班牙披风
Spanish clothing 西班牙外套
Spanish coat 西班牙式外套(有垂片的大领子)
Spanish costume 西班牙民俗服,西班牙民族服饰
Spanish embroidery 西班牙刺绣
Spanish flounce 西班牙荷叶边裙
Spanish grain 西班牙皱纹皮革

Spanish green 西班牙绿(浅暗绿色)
Spanish guipure 西班牙厚花边
Spanish heel 女鞋后直跟,柱式后跟,西班牙跟
Spanish hose 西班牙紧身裤
Spanish jacket 西班牙式夹克
Spanish lace 西班牙花边
Spanish linen 西班牙亚麻布
Spanish point 西班牙金银丝刺绣
Spanish shawl(同 piano shawl) 西班牙穗饰绣花丝披巾
Spanish sleeve 西班牙袖,灯笼式切缝袖
Spanish stitch 西班牙十字刺绣
Spanish stripes 条纹边薄绒呢
spanker(同 clapper, pounding block) 压板
span width 净宽
sparable 无头小鞋钉;防滑板
spare 备用品
spare attire 休闲服装
spare button 备用纽扣
spare collar 备用领
spare cuff 备用袖口
spare jacket 备用上衣,替换夹克
spare needle 备用针
spare parts (机器等的)备用零件
spare parts list 备件表
spares 备件
spare-set 瘦削的
spare stitch 节约针法
spare stylistic form 节约款式
spare trousers 不成套西裤,替换裤
spare tyre (人体腰腹部)多余脂肪组织,肉肚
spark 小宝石,金刚钻(宝石等的)闪光;闪光点
sparkle 小宝石
sparkled overcoating 闪光大衣呢
sparkle prints 闪烁印花布
sparklet 妇女衣裙上的小饰片,亮晶晶的小物件
sparkling colour 灿烂色
sparkling fabric 闪光织物
sparkling grape 晶莹葡萄色
sparrow 雀灰(浅灰褐色)
spar stitch 回纹缝针法
Spartacus sandal 斯巴达克凉鞋
sparterie 三三席纹硬帆布
spats 鞋套,靴套,鞋罩
spats boots 裹腿靴

spat(t) 短绑腿
spatter dash 短绑腿
spatter dashes 裤腿套,防泥水护腿
spearmint 松石绿,粉绿色
spear point 手套短V形装饰缝
special inspection 专检《管理》
special body shape press 特种人形整烫机
special export 专门出口
special figure 特殊体型
special finish 特种整理
special form 特异体型
special garments 特种服装
special hosiery 特种袜,特种功能袜
special import 专门进口
specialist shop 专门商店《销售》
specialist sports shop 专业运动商店
speciality fabric 特种织物
speciality standard 专业标准《管理》
speciality store 专营商店,特制品商店《销售》
specialization 专业化
specialized corporation 专业公司《销售》
specialized standard 专业标准《管理》
specializer 专用设备
special loom 特种织布机
special measurement clothes 特体服装
special measurement size 特体尺码
special notes 专门用语,术语
special play clothes 专门游乐服
special purpose feet 特种压脚
specials 特大品
special sale tax 特种销售税
special sewing equipment 专用缝纫设备
special sewing machine 特殊缝纫机,专用缝纫机,花式缝纫机
special-shaped button 异形纽扣
special-shaped shoulder pads 特型肩垫
special shot 特写镜头
special sports shoes 专用运动鞋
special stitch 特殊线迹,专用线迹
special store 专卖店
special trade 专门贸易
specialty 专业,特殊行业;专长,精制
specialty fibre 特种纤维
specialty hair (fibre) 特种动物毛
specialty shop 特色商店,专卖店《销售》
specialty thread 特种用线
special wool 特种绵羊毛
special working uniform 特种工作服

species 种类
specification 规格;说明书
specification of goods 商品规格
specifications 规格,规范;明细表;(产品等的)说明书
specification seam 规格接缝
specific figure (type) 特殊体型
specificity 特性,专一性
specified quality 合格质量
specified stone 特定钻石
specimen 样品,标本;试销品
speck 斑点
speckle 小斑点,色斑,斑纹,亮斑
specky dyeing 色斑疵
specs 眼镜;说明书;工作设计书;规格,明细单
spectacles 眼镜,护目镜
spectacles for long sight 远视眼镜
spectacles for nearsighted person 近视眼镜
spectacles for the farsighted 远视眼镜
spectator clothes 观赛运动装
spectator pumps 观赛浅口鞋,观赛碰拍鞋,船形中高跟鞋
spectator shoes 观赛鞋
spectator-sports syle 观赛轻便服风格
spectator's sport wear 观赛轻便运动服,观看运动表演服
spectator's wear 观赛服,参观服
spectral colors 谱色
spectrum 光谱,色谱
spectrum blue 光谱蓝
specular reflection 镜面反射
spedal pigment printing 特种涂料印花
speed demon 工作快手《管理》
spell 魅力
spellbound 漫舞艳裳
spelle work 棱结花边
spencer 斯宾塞外套,合身短外套(常镶以毛皮边饰),(长至腰围线的)短夹克,短上衣
spencer cloak 斯宾塞女式斗篷
spencer croise 双排扣短上衣
spencer jacket 斯宾塞外套,短外套式夹克
spencer suit 斯宾塞套装(短上衣与裙组合)
spendex yarn 氨纶纱
SPER (semi-permanent flame-re-tardent) 半耐久性阻燃

spherical button 球形纽扣
spherical compasses 曲线形两脚规
spherical leno 网眼纱罗
spi (stitches per inch) 每英寸针数
spice colour 香料色,肉桂色,胡椒色
spice finish 香味整理
spider leno 网眼纱罗
spiders 蜘蛛布;高级纱罗;英国妇孺衣料
spider stitch 蛛网形针迹
spider work 粗线亚麻梭结花边;蛛网刺绣
spike 防滑鞋钉,细高跟
spike coat 燕尾服
spike heel （女子皮鞋）细高跟,钉状跟
spikes 钉鞋;细高跟女鞋
spike shoes （赛跑用）钉鞋
spike tail 燕尾式外套
spiking （裁剪时）钉大钉
spin 手工饰线针绣
spin dryer 离心脱水机
spinach green 菠菜绿
spinal column 脊骨,脊柱
spindle legs 细长腿;腿细长的人
spindle line 纺锤型
spindle-shaped button 梭形纽扣,纺锤形纽扣
spindly legs 细长腿
spindly shanks 细长腿,腿细长的人
spindrying 离心脱水
spine 手工花边饰线针绣;脊骨,脊柱
spine bone 脊骨,脊柱
spinel 尖晶石
spinning 绞干;纺纱
spiral 螺旋(形)
spiral core yarn 螺旋包芯线
spiral drape 螺旋裥
spiral hosiery 螺旋纹花袜
spiral hosiery machine 螺旋花纹圆袜机
spirality （针织物）横列歪斜
spiral line 螺旋型
spiral pattern knitted fabric 纬编回纹布
spiral seaming 螺旋缝合
spiral skirt 螺旋裙,漩涡裙
spiral sleeve 螺旋袖
spiral socks 螺旋纹短袜
spiral stitching 螺旋线缝
spiral wire bustle 铁丝螺旋型裙撑
spiral yarn 螺旋线
spirit gum （演员用）粘假发胶水
spit curls 斯皮特卷发,假鬓卷发

spitalfields 英国锦缎,满地小花领带织物
splash （动物皮毛的）色斑
splash check 零散格子,溅洒格子
splash effect 泼墨效果
splash print 溅洒印花纹
splash umbrella 散点花伞
splatter-dash 高帮鞋套
splayfoot 外翻足,八字脚
splaykneed 膝外翻的,膝部转向外侧的
splay legged 弓形腿的,膝内翻的
splay toed 足趾张开的
splendid attire 盛装
splendone 希腊妇女头带
spleuchan （苏格兰）烟袋,钱袋
splice 加固袜头,袜跟,袜底,袜口;接匹
spliced heel 加固袜底
spliced heel-and-toe hosiery 加固袜头袜跟
spliced hem 楔口拼接下摆
spliced sole 加固袜跟
spliced toe 加固袜头
spliced top 加固袜口
splicing 袜子局部加固
splicing mark 接匹记号
splicing pocket flap linen 拼袋盖里
splicing thread 增强线
spline 曲线尺,活动曲线规
split 多幅织物;裂缝;层;短衩;衩口;（皮革）剖层;（横部的一层）兽皮
split and tied sleeve 裂缝蝴蝶结袖
split at back vent 背衩豁（背衩搭叠过少）
split at front edge 止口豁
split at pocket mouth 袋口裂
split fabric 分片多层织物
split falls 男裤前开口
split foot 加固袜底,双层袜底,光夹底
split foot hose 夹色底长袜
split foot hosiery machine 夹底袜机
split foot knitter 光夹底织袜机
split foot socks 光夹底短袜,双层短袜,加固底短袜
split hemline 开襟摆
split hemmer 开口卷边器
split leather 底层皮;二层皮
split maker 分割排板图《计算机》
split mandarin collar 中式翻领
split piece 切割样片《计算机》
split raglan 单边连肩袖,分开连肩袖,半连肩袖

split raglan coat 套袖大衣,分割插肩式大衣
split raglan (sleeve) 分开连肩袖,两片插肩袖,前圆后连袖
split skirt 开衩裙,分缝裙
split slab 底层皮,榔皮
split sleeve 两片插肩袖,仿连肩袖
split-sleeved coat 衬衫袖外衣,半插肩外衣
split-sleeved overcoat 连肩袖式大衣
split sole 光夹底,双层楺底
split stitch 双重线迹,复合线迹,链式线迹
split suit 分开套装(衣、裤、背心用不同衣料)
splitting (皮革加工)剖层
splitting ability 撕裂性能
splitting resistance 抗剥离力
split yarn 裂膜丝
split zipper 开口式拉链;密尾拉链
splotch 斑点,污点,污渍
splotchy 污渍斑斑的
S.P.M. (stitch per minute) 每分钟线迹数
spnyrion 内髁点
spodumene 鋰辉石
spoke stitch 辐射缝迹;分割针法
spoke stitching 双边拉丝线迹,双边花饰线迹
sponge 泡沫材料,海绵(状物);人造海绵;(熨烫前)用湿海绵揩拭;[织物裁剪前]缩水
sponge cloth 松软棉布,海绵布,烫衣润湿布,挡车布;粗纱罗
sponge effect 点描效果,云纹效果
sponge-insoled shoe 海绵底球鞋
sponge pillow 海绵枕头
sponge plastics 多孔泡沫塑料
sponger 成品整理检验员(美国)
sponge rubber 橡胶海绵,海绵橡胶,多孔橡胶,泡沫橡胶
sponge rubber slipper 海绵拖鞋
sponge shoulder pad 海绵垫肩
sponge silk 海绵绸
sponge weave 海绵组织
sponginess 手感松软
sponging (呢绒)润湿预缩
sponging damping 润湿预缩
spongy 海绵状的
spongy filler 蓬松棉;散棉

spool 线团,木纱团,塑料线团
spool cotton 木纱团,线团
spool heel 线轴跟
spool of thread 木纱团线
spool pin (缝纫机)线架
spoon busk 匙状胸衣垫
spoon ring 纯银汤匙戒指
sporadic dumping 偶然性倾销
sporran [bag] (苏格兰高地人挂在腰带上或系在短裙前的)毛皮袋
sport bag 运动袋
sport blouse 运动罩衫
sport boots 运动鞋,运动靴
sport bra(ssiers) 运动胸罩
sport cap 运动帽;便帽
sport clothing 运动型外套;两用衫;替换上衣
sport coat(同 sports coat) 两用衫;运动上衣,轻便型大衣
sport collar 运动衫领
sportex 司邦特克斯,斯保克斯织物,运动呢
sport glasses(同 opera glasses) (用于比赛、打猎的)双筒望远镜,双筒狩猎望远镜
sport gloves 运动手套
sportif 适合运动时穿的
sporting wear 运动衫,轻便衣
sportive elegance 轻快而优雅
sportive feminine 女轻便装
sportive longuette 轻快长裙
sportive look 轻便款式,运动装式,轻便袋
sport jacket 运动衣;粗花呢夹克,轻便夹克,轻便服
sports 运动调
sports and recreation textile 运动及娱乐用纺织品
sport satin 运动缎
sports bag 运动背包
sports cap 运动帽
sports casual look 运动轻便装,轻便装,轻便款式,运动款式
sports clothes 运动装,便装
sports coat 运动上衣,轻便型大衣
sports comfort 运动舒适性能
sports denim(同 faded denim) 淡色薄斜纹布,褪色劳动布
sports dress (s.d.) 运动服
sport set 运动套装,休闲套装,便装组件

sports gloves 户外运动手套
sports hat 运动帽
sports head band 运动束发带,运动头饰带
sport shirt 运动衫,休闲衬衫,便衬衫(男子作便服穿,下摆留在裤外),不系领带的衬衫
sport shirt sleeve 运动衫袖,便衫袖
sport shoes 运动鞋
sport shoe(s) sewing machine 运动鞋缝纫机
sport singlet and trousers 运动衫裤
sports jacket 运动衣;粗花呢夹克,轻便夹克,轻便服
sports look 运动服型,运动款式,运动风貌;体育竞技服
sport socks 运动短袜
sports outfit 运动服装
sports shirt 运动衫,便衬衫(男子作便服穿,下摆留在裤外,不系领带的衬衫
sports shoes 运动鞋,跑鞋
sports shoe sewing machine 运动鞋缝纫机
sports shorts 运动短裤;球裤
sports socks 运动袜
sports suit 运动衫款式的套装,专业运动套装,轻快感套装
sports sweater 运动服装
sports sweater and trousers 运动衫裤
sports tank 运动背心
sportster 运动衣
sport suit(同 sports suit) (专业性)运动套装,轻便套装;运动衫款式的套装
sports wardrobe 运动服装
sport sweater 运动服装
sports wear 运动服装,便装,轻便风格单件服装组合的运动装;穿运动服装的场合
sportswear fabric 运动衣料
sportswear for agonistic 竞技运动服装
sportswear for aquatics 水上运动
sportswear for archery 射箭服
sportswear for athletics 田径服
sportswear for ball Games 球类运动服
sportswear for equestrian 马术服
sportswear for rowing 赛艇服
sportswear for sailing 帆船服
sportswear for water polo 水球服
sportswear for wrestling 摔跤服
sportswear in acrylic 腈纶运动服

sportsy 适合体育运动时穿的;(设计、裁剪等方面)像运动服装的
sport textiles 运动用纺织品
sport vest 运动衫,运动背心
sport wear 运动服;两用衫
sporty 花哨的,轻便的,非正式场合穿的
sporty look 轻快款式
sporty style 运动衫风格,轻便款式
sporty suit 花哨的服装
spot 污迹;(图案的)点子,斑点,斑渍,[色谱]斑
spot check 现场检查,抽样检查,抽样调查
spot goods 现货
spot lace 点纹花边
spotless finish 防污渍整理
spot lifter 去斑渍剂
spot marking 污渍标记
spot muslin 点子花纹麦斯林,点子花纹薄细布
spot proof 防污渍(的)
spot removal 除污迹
spot remover 去污剂
spot-resistance 抗污性
spots 现货《贸易》;现货坯布,点子花纹棉布
spot stitch 点纹钩编线迹
spotted cat 斑猫皮
spotted handkerchief 圆点图案手帕
spotted lace 点子网眼花边
spotted lamb 斑点羊羔皮
spotted material 带点料子
spotted shunk 斑点臭鼬毛皮
spotted stone 斑钻
spot test 点滴试验法
spot welding machine 点衬定位机
spotty dyeing 染斑
sprain 扭伤
sprang 网眼织物
spray 喷雾器
spray bonding interlining 喷胶棉,喷胶絮棉,喷浆絮棉
spray colour 喷染法
spray dyeing 喷溅染色
spray earring 小枝耳饰
sprayer 喷水壶,喷雾器,水枪;喷雾者
spray(green) 浪花绿
spray gun 喷枪
spraying-bonded wadding 喷胶棉,喷胶絮棉,喷浆絮棉

spray method （码克复制）喷印法
spraynet 喷式发胶
spray printing 喷雾印花
spray starch 喷雾上浆，喷浆
spreadability 覆盖性
spread collar 宽展衬衫领，方领，八字领，温莎领
spread defect 分散性疵点
spreader 拉布工；裁剪铺布机
spread fabric 铺布
spreading （裁剪前）铺布，铺料；平布；拉布
spreading caliper 触角计（量体用）
spreading fabric （裁剪前）铺料
spreading finish 上浆整理
spreading machine 拉布机；铺料机
spreading operation 铺料操作
spread(ing) out 扩开型双排纽夹克，扩开型双排纽上衣
spreading table 展幅台，扩幅台
sprig 网地贴花花边；枝叶花纹，枝状装饰花纹
sprig print 小枝印花
spring and autumn clothes [coat] 春秋衫
spring and autumn women's wear collection 春秋女装系列
spring and hook type buckle 弹簧环扣
spring beautiful jacquard sand crepe 春美绸
spring bracelet 串珠柔韧手镯
spring bud 春叶芽绿
spring clothing 春季服装，夹大衣，风衣
spring coat 春季外套，风衣，夹大衣
spring crocus 春藏红花色
spring fabric 春季织物
spring green 嫩绿
spring handle 弹性手感
spring heel 弹跳低平鞋跟，弹跳跟
spring jacquard sand crepe 时春绸
spring needle 全成型平型钩针机
spring or autumn suit 春秋套装
spring overcoat 夹大衣
spring shirt 短袖大翻领衬衫，短袖敞领衬衫
spring sleeve 春袖
spring wardrobe 春装
spring water 春水色
spring wear 春装
spring weave 弹性衣料
spring weighted 弹簧加压的

springy handle 弹性手感
sprinkler 喷水烫衣工，喷水器
sprit at lower part of skirt 裙裾豁开
S.P.T. (Standard Pitch Time) 平均节拍《管理》
spruce 云杉绿（浅暗绿色）
spruce yellow 云杉黄
spun cassava pongee 木薯绢纺绸
spun chemical fiber yarn 化学纤维短丝纱
spun crepe 绢纺绉
spun crepe de chine 绢丝经双绉
spun dyed 原液染色法
spun-dyed fiber 原液染色纤维；色纺纤维
spun fabric 短纤维织物
spun fancy silk 绢纺疙瘩绸
spun filling foulard 绢纬绫
spun-filling pongee 四季料
spun glass 玻璃纤维，玻璃丝
spun-like fabric 仿短纤纱型织物，仿纱型织物
spun-like yarn 仿短纤纱
spun linen 最优级手织亚麻布
spun polyester thread 纺涤纶线；PP线
spun raw silk 分纤丝
spun rayon 人造棉纱；黏胶短维纱；人造棉织物
spun rayon fabric 人造棉布；人造棉织物；黏纤织物
spun rayon yarn 人造棉纱，黏胶短纤维纱
spun replacement fabric 仿纱型变形丝与短纤纱交织物
spun replacement yarn 仿纱型长丝，仿短纤纱长丝
spun silk 绢丝，绢丝纺绸
spun silk blended or interwoven fabric 绢丝混纺（交织）织物
spun silk cotton mixed 绢棉交织绸
spun silk crepe 纺建呢
spun silk fabric 绢丝织物；绢丝绸
spun silk filling taffeta 绢纬塔夫绸
spun silk gingham 绢格纺
spun silk habotai 绢丝纺
spun silk hosiery 绢丝袜
spun silk piece goods 绢丝纺[绸]，绢丝织物
spun silk pongee 绢纺，绢丝纺[绸]
spun silk sewing thread 绢丝缝线
spun silk taffeta 绢纬塔夫绸，绢丝纬塔

夫绸
spun silk yarn 绢丝,桑蚕绢丝
spun silk yarn of tussah 柞蚕绢丝
spun silver 包芯银线
spun tussah cloque 凹凸纺,柞绢凹凸绉
spun tussah herringbon suiting silk 柞绢人字呢
spun tussah pongee 柞绢纺
spun tussah yarn 柞蚕绢丝
spun viscose 黏胶短纤维纱
spun yarn 精纺用纱,细纱,短纤维纱;松捻大麻绳索
spun yarn fabric 短纤纱织物
spurs 刺马靴
spurt-ink printer 喷墨式打印机《计算机》
squab 矮胖子
squab mose 阔鼻
squam 防水油布帽
square 直角尺,矩尺;正方形;平方
square and bow 方形领带
square angular shoulder 方(角)肩
square armhole 方形袖口
square bertha 方形披肩领
square bow tie 俱乐部蝶领结;平结领带
square buckle 方形带扣
square built 体型宽阔的
square cap 四角帽;方形帽
square chain stitch 方形链式缝迹,方型链状绣
square cloth 平衡织物
square collar 方角领,方领
square college 学士方帽
square college cap 学士方帽
square construction 平衡织物
square corner 方角腕口
square cut bottom 平下摆,平裙
square cut collar 方领
square-dance dress 方块舞衣
square-dance skirt 方块舞裙
square dancing dress 方块舞衣
square(d) wig 方形假发型
square-edge finishing stitch 方形面缝光边线迹
square-end necktie 方头领带
square-end tie 方形领带
square face shape 方脸形
square-front 平下摆,方形前摆
square-front shoulader pad 方形垫肩
square heel 方后跟
square hemmed bottom 平下摆

square hoop 长方形裙撑
square jaw 方颚型
square knot 平结
square measure 直角尺
square meter [metre] 平方米
square mould 四角模(制模)
square muffler 方围巾
square neck 方领女服,方翻领
square neckline 方形领口
square net 有光机织花
square out 成直角
square out and down 作直角
square paper 方格线
square pocket 方角袋
square repeat 方形连续
square ruler 裁缝用角尺
square scale 正方形,四方形,方格,方块;直角尺,丁字尺
square scarf 方围巾
square shaped cap 四角帽
square shoulder 平肩,阔肩,方肩
square shoulder figure 平肩体型
square skirt 细褶裙子,细襞裙
square sleeve 方袖隆,方形直袖
square stole 方形披肩
square tie 方形领带
square toe 方头鞋,方鞋头
square toe leather shoes 方头皮鞋
square yoke 方形育克,方形过肩
square-toe-end 鞋子方头
squaring 成方形
squash 南瓜发型
squash blossom (hairstyle) 耳边做圈的女发型
squash hat 可折叠软毡帽,宽边软帽
squash necklace 大型华丽项圈
squash patent 压纹漆皮
squaw 印第安女人风格服装
squaw bag 嬉皮士袋;美洲印第安包
squaw blouse (美洲印第安女人)重绣女衫,服装风格
squaw boots 印第安女靴,丝蔻女靴
squaw bootie 印第安短筒靴,丝蔻短筒靴
squaw dress 方块舞裙装(细褶长裙,绣花长袖紧身衣)
squaw skirt 方块舞裙
squib 商品标签《管理》
squiggle 花体字
squinny 斜视眼
squint 眯眼;斜眼

squirrel 松鼠毛皮,灰鼠毛皮
squirrel fur 松鼠毛皮,灰鼠毛皮
squirrel grey 松鼠灰
squirrel skin 灰鼠皮
SR(soil release) 易去污,去污
sravel 阿尔及利亚斯拉弗尔原色呢
SR finish(soil-resistant finish) 拒污整理
SSDDS(self-service discount de-partment store) 无人售货廉价(百货)商店《销售》
ssernall trousers 埃及宽裆锥形裤
S-shaped silhouette S形轮廓
ssitar 西斯塔上衣,斜纹短上衣
S style S型
st(stitch) 线迹,针迹,针脚,缝,缉；一针；一件衣服；一块布
stability 稳定性
stability against laundering 洗烫稳定性
stabilized azoic dyes 稳定不溶性偶氮染料
stabilized Dacron sailcloth 聚酯帆布
stabilized finish 定型整理,稳定整理(指织物的防缩防皱整理)
stabilized type (非织造布品种的)稳定型
stable fabric (尺寸)稳定织物
stab stitch 与织物成直角穿过的线迹
stacked (女子)体态丰满匀称的
stack(ed) heel 叠层鞋跟,堆积式鞋跟
stacker 叠料器
stadium 观赛外套；运动场靴
stadium boot 观赛靴
stadium clothing (观赛穿)运动场外套
stadium coat (露天看台穿的)御寒外套,竞技场外套
stadium jumper 棒球夹克,竞技场夹克,运动短外衣
staff 厚重麻布
staffel stitch 施达菲尔针法
stage costume 舞台服装,戏装
stagen 印度尼西亚女装带状布
stagged seam 接线双轨
staggered stitch 交错线迹,错开线迹,线迹不直
staggering stitch 错开线迹,交错线迹,线迹不直
staghorn button 鹿角纽扣
stagnation 萧条
stain 污渍,污点,污迹；色斑
stain cleaning agent 去污剂
stain resistance 防沾污性

stain resistant finish 防污渍整理
stained cloth 褪色布,油污布；油画挂毯
stained glass effect 彩色玻璃效果
stainer 皮革染色工
stainless steel hanging rack 不锈钢挂衣架
stainless steel ruler 不锈钢直尺
stairing wig 阶梯式假发
stale grey 鼠灰
stamatte 全毛色纱织物
stamboul 廉价粗纺毛呢
stambouline 土耳其官袍
stam collar 双层衣领
stamer coat 小生皮旅行大衣
stamin (手感粗糙的)精纺毛织物
stammel 斯坦默尔粗毛呢；鲜红色
stamped linen 印花刺绣亚麻布
stamped plush 拷花长毛绒,压花长毛绒
stamped velvet 拷花丝绒
stamp tax〔duty〕 印花税
stance 站立姿势；姿态
stand 站直姿势；领座；领脚高；人体模型；衣架；位置
stand and fall collar 立翻领
standard 标准,规格；规范；准则；水平；最低规格,最低等级
standard abradant fabric 标准磨损织物
standard adjacent fabric 标准贴衬织物
standard allowance 标准留量
standard-as-apple-pie seam 苹果派式标准线缝
standard bill of lading 标准提单
standard block pattern 基准型样板,原型
standard color 标准色
standard cost 标准成本
standard depth 标准深度；标准颜色深度卡
standard(equilibrium) moisture regain 标准回潮率
standard faded stripes 标准褪色样条
standard fading hours(SFH) 标准褪色时间
standard figure〔form〕 标准体型
standard foot 标准压脚
standard garment 标准服装
standard goods 正品
standard grade 标准级
standard grey scale (染色牢度检验用)标准灰色分级样卡
standardization 标准化
standardization of packing 包装标准化

standardized letter digital key board 标准字母数字键盘《计算机》
standardizing body 标准体型
standard leno 半纱罗织物
standard light source 标准光源
standard (moisture) regain 标准回潮率
standard of fading 褪色标准
standard of safety 安全标准
standard pattern 标准式样,标样
standard-piece rate 正品率;正布率
Standard Pitch Time (S.P.T.) 平均节拍《管理》
standard presser foot 标准压脚
standard sample 标(准)样(子)
standard sample photo 标准样照
standard sewing technique 标准缝制技术
standard shoulder 标准型肩,标准肩型
standard size 标准尺寸,标准尺码,标准型
standard-size pattern 标准尺寸纸样
standard skirt 标准型短裙,标准裙
standard temperature and humidity 标准温湿度
standard tolerance 标准公差
standard zipper 标准型拉链,普通拉链
standaway 不贴住身体的,撑开的
stand-away collar 离颈领,(不贴颈)宽式立领,直离立领
stand-away neckline 不贴颈领口,离颈领圈,直离领领口
stand camera 支架摄影机
stand collar 中式领,立领,学生服领
stand collar jacket 立领夹克衫
stand-fall collar 二重领
stand-fall collar coat 二重领外套
stand height 领座高
standing band 17世纪护面高领
standing collar 主领,竖领
standing height 身长,身高
standing length 总体高,身高
standing model 做模特儿
standing order 长期订单
standing whisk 扫帚花边立领,女式花边大翻领
stand interlining 领座衬
stand off collar 宽式立领,离颈领
stand off neckline 离颈领口
stand offish collar 离颈领
stand out collar 前翻领
stand pocket (同 welt pocket) 一字嵌袋,单嵌线袋,单贴边袋,立式口袋,有盖挖袋,开缝口袋
stands (士兵的)全副装备
stand-up collar 立领,竖领
stand-up collar blouse 竖领上衣
stand-up curl 耸起卷发,竖发型卷发
stand-up frill 直立绉边
stand-up neckline 直立领口
stand wash 经洗,耐洗
stap shoes 搭扣鞋
stape goods 大路货
staphylococcus aureus 金黄色酿脓葡球菌
staple U形钉;主要产品,主要商品;天然原料,原材料;贸易中心,主要市场;切断纤维;纤维(平均)长度;(人造)短纤维;订书钉
stapled seam U形线缝
staple fabric 大路织物(销路历久不衰)《销售》
staple fiber muslin 人造棉细布
staple goods 大路产品;大路货《销售》
staple linen 大众化亚麻织物
stapler 订书机;(皮革加工用)骑马钉
staple rayon 人造短纤维;黏胶短纤维,人造棉;喷胶棉;人造毛
staple rayon fabric 人棉布
staples 大路货,主要产品
star braid 星带
star burst "星状"起皱,星放射状起皱
starch collar 浆领,硬领
star check 星形格子花纹
starched bosom 上浆衬胸,硬衬胸
starched collar 上浆领
starched fabric 上浆织物
starched interlining 上浆衬布
starched mull 里子粗布
starched thread 加光线,上光线
starcher 男式上浆领巾;给衣服上浆的人
starch marks 浆斑
star design 星辰图案
star filling stitch 星形贴线绣
star jeans 明星牛仔裤
star light blue 星光蓝
star look 影星款式
star-moon mixed tussah 星月绸,涤弹丝柞蚕丝交织绸
star ornament 星形装饰
star ruby 星光红宝石
star sapphire 星光蓝宝石

star shaped collar shirt 星形领衬衫
star stitch 双十字绣,星形线迹,星状刺绣线迹
start from scratch 从零开始《计算机》
starting point (制图)基点
starting torque 起动力矩
startup 农村靴
stash 髭,小胡子
state run 国营的《管理》
static beauty 静态美
static behaviour 静电性
static characteristics 静态特征
static cling 静电导致织物贴服
static electricity 静电
static flocking 静电植绒
static friction 静摩擦
static inhibitor 抗静电剂
staticky 贴体性,粘身性(静电缘故)
stating 暗缝针迹
statuary art 雕塑艺术
statue of Liberty Visor 自由神头饰带
stature 身高;身材
status jeans 名家牛仔裤,署名牛仔裤,设计师牛仔裤(有著名设计师署名的商标)
statute gallo(o)n 卫生衫裤用的棉纽带,丝纽带
statutory tax rate 法定税率
stay 撑条,拉条,滚边窄带,加固布条(块);(领尖)插角片;紧身胸衣;女用束腹带
stay binding 胸罩带,紧身带
stay button 支纽
stayed pleat 死裥,死褶
stayed seam 压条缝
stay fit jeans 永帅牛仔裤(合身而永不走样)
staying 推边固缝
staying stitch 绳状线迹,起梗线迹
staylace 妇女束腹带,胸衣带,紧身褡带
stay maker 女用胸衣制造商
stay-on slipper 系带拖鞋
stays 紧身褡,紧身胸衣,紧身背心,紧身马甲
stay stitch 定位线迹,折边固定线迹
stay stitching 稳定缝纫;稳定线迹,折边固定线缝
stay tape 滚边窄带,定位带,接缝窄带,胸衬带,牵条,过桥布
stay-up hose 紧口长筒袜

steady dial linking machine (连续传动的)缝袜头机
steam and dry iron 蒸汽电熨斗
steam-baste 蒸汽熨烫(粘合)定形
steam-basting 蒸汽熨烫定形
steam box 蒸汽箱
steam (electric) press 蒸汽(电)熨烫
steamer 蒸汽机;汽蒸器;蒸汽美容法
steamer trunk 旅行衣箱
steam extractor 抽蒸汽器
steam iron 蒸汽熨斗
steam leather softening machine 蒸汽皮面软化机(制革)
steam machine for processing fur 毛皮蒸汽清理机
steam moulded shoulder pad 蒸汽定型肩垫
steam permeability 透汽性
steam-press 蒸汽压烫
steam pressing 蒸汽熨烫
steam pressing stand 蒸汽压烫机,蒸汽烫台
steam puffer 蒸汽喷射烫衣机
steam setting 蒸汽定型
steam setting machine 蒸汽定形机
steam shrink 蒸汽预缩
steam spray gun 蒸汽喷枪
steam style 汽蒸显色印花法
steam table (整烫)蒸汽台
steam twills 粗支斜纹里子布
steek (缝纫或编织的)一针,针脚
steel (撑开紧身腰围或衬裙的)松紧钢丝
steel band knitting machine 钢带提花圆形针织机
steel blue 钢青色,深蓝色
steel cap (鞋子)钢包头
steel (cloth) tape 钢(布)卷尺
steel gray 青灰色,铁灰色
steel grey 蓝灰,青灰
steel helmet 钢盔
steel maker's 炼钢工人衬衫
steel nail file 钢指甲锉
steel tape 钢卷尺
steel tape measure [ruler] 钢卷尺
steel-toe boots 钢头靴
steelworker's suit [wear] 炼钢服
steep-crowned (帽子等)有高尖顶的
steeple crown 尖塔帽
steeple headdress 尖塔形垂纱帽

steeple hennin 塔形垂纱帽
steep twill 急斜纹
steer hide 小牛毛皮
steering 控制,操纵,调整,转向
steer leather 菜牛革,阉公牛革
steils 飞梭刺绣;往复线缝
steinkirk 斯坦克松结长领巾
Stelos point 修补用针
stem fiber 韧皮纤维
stem stitch 绳状线迹,起梗线迹,包梗绣
stem-pipe trousers 直筒裤,小裤脚裤
stencil 唛头板;(镂空)模板,硬质衣片样板《计算机》
stencil marker 漏画纸版
stencil printing 镂空版印花;模版印花
step 步态,步姿,步伐
step collar 梯形领,直角西装领
step diagonal 阶梯形斜纹织物
step drive 寸动运转
stephane 环状头饰,月外形头饰,王冠,史蒂芬头饰
stephanie lace 斯特凡内手工花边
Stephen Burrows 斯蒂芬·伯罗斯(美国时装设计师)
Stephen Jones 斯蒂芬·琼斯(英国时装设计师)
Stephen Sprouse 斯蒂芬·斯普劳斯(美国时装设计师)
step-in 易穿式;易套鞋,船鞋,大舌船鞋;女式内衣,易穿式服饰
step-in chemise 女式无袖胸衣(下套式)
step-in dress 易套连衣裙
step-ins 女式窄裆喇叭衬裤,女内裤;女式内衣;船鞋
step-in shoes 易穿式鞋,易套鞋
stepped joint 搭接;搭接缝纫
stepped sole 梯形袜底
stepped sole splicing 袜底梯形加固
steps 船鞋;女内裤
3 steps 三段式
4 steps 四段式
steps-in blouse 连裤女衫
steps stitch 阶梯线迹
2-step zigzag 二步曲折缝
stereo 立体照相机;立体摄影术
stereo boarding press 立体熨烫
stereo camera 立体照相机
stereo color 色立体
stereograph 立体像片
stereographer 立体摄影师

stereology 体视学,立体测量学
stereometry 立体测量学,立体几何学
stereopsis 立体视觉
sterile gown 消毒白罩衣
sterli lization 杀菌(作用),灭菌(作用)
sterling blue 纯蓝色
sterling price 以英磅计算的价格
stern 臀,臀部
sternum 胸骨
stertop 高帮鞋
stetson 美国斯泰森毡帽,(帽缘宽而低垂的)软呢帽,牛仔帽
Steve Steinberg 史蒂夫·斯坦伯格(美国时装设计师)
St. George Cross(同 crusader's cross) 马耳他十字架挂饰项链
stick 手杖
stick collar 棒形领
sticker 背面有黏胶的标签;胶贴;贴纸
stickerei 牙边编带
stickiness 黏性,黏着性
stick-on label 粘贴商标
stick-pin 领针,(领带用)装饰别针
stick-shaped button 棒形扣
stickum 润发脂,头油;黏合剂,胶水
stick umbrella 手杖伞
stick-up collar 男式衬衫硬领
sticky feeling 黏附感,不舒服感
stiff 坚挺
stiff and crispy handle 挺爽手感
stiff collar 硬领
stiff cover 仿皮面,假皮面
stiffening 硬挺整理
stiffening cloth 硬衬布,硬衬织物
stiff hat 圆顶硬礼帽,常礼帽
stiff-legged boots 硬筒靴
stiffness 硬挺性,硬挺度
stiff silk 绢绸
stiff standing collar 硬立领
stilbene dye 二苯乙烯染料
stiletto 穿孔锥
stiletto heel 超瘦型鞋跟,特细高跟,丝蒂蕾朵跟,特细高跟女皮鞋
still 静止摄影
still air 静止空气,死空气
still air layer 静止空气层
still hat 常礼帽,圆顶硬礼帽
stiltto 打眼锥
stilyaga (追求西方青年服饰和趣味的)俄罗斯时髦青年

stingy 窄边帽子(美国)
stink prevention hosiery 防臭袜
stint 定量,限额;定额工作;工作期限
stippled effect 点画效果,点描效果,云纹效果
stirrup pants 踏脚裤,马镫裤
stitch 缝纫;缝合;缝缉;缝缀;刺绣;锁孔绣;锁眼绣;线迹;针脚;针法;装饰线
stitchability 可缝性,可编织性
stitchable corner angle 可缝的尖角角度
stitch back centre seam 合背缝
stitch belt loop 缝裤带袢
stitch bite 横针距
stitch-bonded artificial fur 缝编仿毛皮
stitch-bonded artificial goat fur 缝编仿山羊皮
stitch-bonded artificial velvet 缝编仿丝绒织物
stitch-bonded bandage 缝编绷带布
stitch-bonded bath wear fully elastic 缝编弹力浴衣
stitch-bonded bedcover 缝编床罩布
stitch-bonded camping cloth 缝编帐篷布
stitch-bonded children's trouserings 缝编儿童裤料
stitch-bonded cloth 缝编布
stitch-bonded covering lamination 缝编铺盖层压材料
stitch-bonded fabric 缝编布;缝编织物
stitch-bonded fiber-web fabric 纤网型缝编织物
stitch-bonded fully elastic bath wear 缝编弹力浴衣
stitch-bonded goat fur 缝编仿山羊皮
stitch-bonded nonwoven 缝编法非织造布
stitch-bonded packing fabric 缝编包装材料
stitch-bonded pile fabric 毛圈型缝编布
stitch-bonded shirting 缝编衬衫料
stitch-bonded shoe liner 缝编鞋内材料
stitch-bonded thread layer fabric 衬线缝编织物,线层缝编织物
stitch-bonded toweling 缝编毛巾布
stitch-bonded velvet fabric 缝编仿丝绒织物
stitch-bonded velour 缝编绒
stitch-bonding machine 缝编机
stitch chain 线辫
stitch chest interlining 缉胸衬
stitch closure 缝袋口

stitch cloth sole 纳布鞋底
stitch collars together 合领子
stitch condensation 线迹加密
stitch condenser 密缝装置,密针缝装置
stitch condensing 线迹加密
stitch course 线迹轨迹
stitch dart 缉省缝
stitch dart in interlining 缉衬省
stitch defect 线迹缺陷
stitch density 线迹密度
stitch down piping lip 滚边
stitch down process (制鞋)压条工序
stitch down shoes 镶边装饰缝鞋,压条镶边鞋子
stitched-check 珀皮塔格子花纹
stitched double cloth 双重接结织物
stitched elastic swing hemmer 缝松紧带的摆动卷边器
stitched fabric 缝编织物
stitch edge 针缝边
stitched leather shoe 上线皮鞋
stitched pique 凹凸条纹布
stitched piping 压条
stitched pleat 压线褶,绣缝褶裥
stitch(ed) stripe 针法条纹
stitched tape 缝边带
stitch elasticity 线迹弹性
stitcher 缝纫机;缝工
stitchery 缝纫,编结;刺绣;缝纫品;编结品;刺绣品
stitches per inch (spi) 每英寸针数
stitch finishing 线迹修整
stitch foot interlacing 袜脚连接
stitch form 线迹形式
stitch French dart 合刀背缝
stitch front edge 合止口
stitch gauge 针迹密度
stitch indicator plate (缝纫机)针距指示牌
stitching 缝纫,缝合;线缝;刺绣;饰缝;锁扣眼
stitching accuracy 缝纫精度
stitching area 缝纫部位
stitching data 缝纫数据
stitching down piping lip 滚边
stitching ends of pocket mouth 封袋口
stitching guide 反面车缝标记(车缝反面结构线标记)
stitching horse 压脚
stitching interlining 缉衬

stitching interlining dart 缉衬省
stitching jig 缝纫模板
stitching machine 套口机,合缝机,缝袜头机
stitching margin 缝合边宽
stitching mark 缝纫针迹
stitching needle 缝针
stitching of heavy weight fabric 厚料缝纫
stitching operation 缝纫操作
stitching parameter 缝纫参数
stitching pocket lining 缉袋布
stitching quality 缝纫质量
stitching quality is no good 缝制不良
stitching seam width (两行线缝的)间宽
stitching speed 缝纫速度;缝速
stitching tensile strength 缝口强度
stitching type 针迹类型
stitching width 缝纫宽度
stitch inside seam 合下裆缝
stitch knit goods 缝编织物
stitch-knitting machine 缝编机
stitch length 针距;针脚长度
stitch length measuring system 线圈长度测定器
stitch-length mechanism 线迹长度调节机构,针距调节机构
stitch length regulate 线迹长度调节
stitchless jointing 无缝连接
stitch line 线迹线
stitch looping (缝纫机的)线迹松紧
stitch mark 针迹,针脚
stitch marking apparatus 线迹加标记装置
stitchology 缝纫长度
stitch-out 绗缝
stitch pattern 线迹花样
stitch performation 线迹性能
stitch per inch 每英寸针迹数
stitch per minute (S.P.M.) 每分钟线迹数
stitch pinking machine 锯齿边布样切裁缝纫机
stitch plate (缝纫机的)针板;缉裥
stitch plate opening 针板孔
stitch presser foot 线迹压脚
stitch pucker 线迹皱缩
stitch range 线迹范围
stitch regulating dial (缝纫机)针距旋钮,针脚调节盘
stitch regulating plate 针距调节板

stitch regulator 线迹(密度)调节器,针距调节器
stitch reinforced nonwovens 线缝加固非织造布
stitch scale 线迹刻度尺
stitch seam is uneven 双轨接线(接线不准)
stitch seam leans out the line 缉线上下炕(线缉得或上或下)
stitch shoulder seam 合肩缝
stitch side seam 合衣摆缝;合裤侧缝;合裙缝
stitch size regulator 针迹密度调节器
stitch skip 跳缝性能
stitch skipping 跳针
stitch sleeve seam 合袖缝
stitch space 线迹间距
stitch spacing 针距
stitch tear resistance 线缝抗裂强度,针脚抗裂强度
stitch thread 缝纫线
stitch type 线迹形式
stitch type category 线迹类别
stitch up 缝补
stitch waist band 缝合腰带
stitch width 线迹宽度,(两行缝线间的)横针距
stitch width regulate 线迹宽度调节
stitch-width regulator 线迹宽度调节器
stitch work 刺绣,缝纫刺绣,装饰缝
stitch yarn 绣花线
stoat 白鼬毛皮
stoating (缝合织物的)暗缝针迹
stock 存货,库存;(产品的)原料;备料;(18世纪男用)宽大硬领圈;(牧师用的)绸领带;史脱克领巾(系于领子内的宽带型饰巾)
stock collar 飘带领(狭带中央固缝在颈后,两端在颈前扎结下垂)
stock dyeing 散纤维染色
stock goods 库存商品;存货
stockinet 松紧织物;弹力织物,平针织物
stockinet goods 弹力针织物
stockinette 松紧织物,弹力织物;平针织物
stockinette stitch 隔行正反针编结法
stocking 长袜,长筒女袜
stocking-back cloth 密面稀背织物
stocking band 宽紧袜带,袜带
stocking blank 长筒袜坯

stocking board 袜子定形板
stocking bodice 针织弹性筒状紧身胸衣
stocking boots 长袜型长筒靴
stocking cap 针织帽,圆锥形绒线帽;袜型帽
stocking foot 袜脚
stocking form 袜子样板
stocking frame 织袜机
stocking knitter 织袜工,袜机
stocking machine 袜机
stocking mask 尼龙面罩
stocking mender 补袜工
stocking purse 管状小钱包
stocking setting machine 长筒袜定形机
stocking shoes 连袜鞋
stocking size 袜子尺寸
stocking size chart 袜子尺寸表
stocking-suspender 吊袜扣,吊袜带
stocking toe 袜头
stocking turning device 翻袜装置
stocking welt 袜口,袜口边
stocking with run-down heel 无跟袜
stocking yarn 针织纱
stock in process 在制品
stock neckline 斯托克领口
stock pattern 存档样板
stocks press machine 烫袜机
stock-tie blouse 扎结领衬衫,斯托克领衬衫
stock-tie shirt 扎结领衬衫,斯托克领衬衫
stogie 粗重长靴
stogy 粗重长靴
stola （古罗马）斯多拉女衫,女式宽松长外套
stole 女式长围巾,女披肩;教士长绸带,圣带;长巾;(古)长袍
stole coat 披肩式外套
stole collar 长披巾领
stole neckline 披巾形领口
stole pin 披巾别针
stomach 肚;腹部
stomach control 束腹
stomach dart 腹褶;肚省
stomacher 抱肚,围肚;兜肚,兜包;三角女胸衣(古),15~16世纪宫廷穿戴的V形胸饰片
stomach line 腹围线
stomach pin 披巾别针
stone 宝石

stone blue 灰蓝色,石青色
stone button 石扣
stone color 石头青,石青色,褐灰色,暗青灰色
stone grey 石头灰,青灰色
stone wash （牛仔服）石磨洗
stone-finishing real silk 砂洗桑蚕丝绸
stone-marten 扫雪貂,宝石貂皮,岩貂毛皮(亚洲与非洲产的一种貂毛皮),扫雪貂皮
stone wash and bleach 石磨加漂洗
stone wash chemicals 石磨水洗化学品
stone washed 石洗,石磨水洗褪色
stone-wash (blue) jeans 石磨水洗蓝牛仔裤
stone-washed indigo 石磨蓝
stone-washed jeans 石磨水洗牛仔裤
stone washing 石磨洗
stone washing and bleaching 石磨加漂洗
stone-washing machine 石磨洗水机
stone-wash jeans 石磨牛仔裤
stooped narrow chest 弯胸
stooping figure 驼背体型,弯腰屈背体型
stoop-shouldered 驼背的
stop button 定车按纽
stop latch spring （缝纫机）满线跳杆簧
stopper 绳索扣;拉绳(弹簧)扣;绳扣
stop spot 去渍药水
stopwatch 秒表
storage 贮藏
storage life 适用期,存放期,贮存限期
storage management 仓储管理
storage pocket 巨型口袋
storage roll 布卷
storage tube graphical display 存储管式图形显示器《计算机》
store 商店,店铺
store bought 现成的;从店里买来的
store bought suit 成品套装,现成套装
store brand 店属商标
store clothes 成衣,现售服装
store clothing 成品服装,商品销售的服装
stork fashion 孕妇款式
storm 防暴风雪服饰;防暴风雪靴
storm blue 蟹灰色
storm boots 风雪靴
stormbreaker 防风暴衣;风雪衣
storm coat 风雪大衣
storm collar 风雨领,防风立领

storm cuff	防风(松紧)袖头
storm flap	披胸布,暗门襟
storm gray	风暴灰
storm hat	风雨帽
storm patch	肩部风雪挡布
storm rubbers	风雪套鞋,防水橡皮鞋
storm serge	风雨粗哔叽,风雨薄哔叽(做女衣用)
storm strap	防风袖口束带,风雪束袢
storm suit	(险恶天气狩猎或钓鱼时穿着的)连帽两件套风雪装
storm tabs	防风袖袢,防风袖口
storm welt	鞋面滚条
storyboard	分镜头
stoting	接缝,织补破洞
stoting stitch	厚料拼接线迹,织补破洞线迹
stout	凸肚;大号尺寸
stout figure	胖体型,矮胖身材
stout form	矮胖体型
stove bottom	喇叭裤口
stoved shade	硫熏色
stove-pipe	瘦裤腿的,无折缝直筒裤腿的;烟囱裤,烟囱裤轮廓;大礼帽,高筒礼帽,丝质高礼帽
stove-pipe finish	大礼帽缎光整理,高级光泽整理
stove-pipe hat	缎制高顶大礼帽
stove-pipe neck	无颈线高领
stove-pipe pants	烟囱裤,烟囱形长裤,紧身筒裤
stove-pipes	瘦腿;紧身裤
stove-pipe satin	大礼帽缎
stowing	暗缝
stradella	斯特拉德拉提花羊毛围巾(法国制)
straight and bias cloth cutting machine	纵斜纹切布机
straight back	直背,直后式发型
straight back heel	后直跟
straight band collar	立领,直条领
straight bar hosiery machine	平袜机
straight bar knitting machine	钩针成型平机
straight bar linking machine	平式套口机,平式缝头机
straight barred end	直型加固缝端(锁钮孔)
straight bar rib machine	钩针平型罗纹机
straight belt	直腰带
straight binder	直线滚边器
straight B/L	收货人记名提单《贸易》
straight buttonhole	直纽孔,平眼锁纽孔
straight buttonhole machine	直线纽孔缝纽机,平眼锁眼机
straight buttonhole without taper bar	无套结直纽孔,无套结平眼锁纽孔
straight buttonhole with taper bar	有套结直纽孔,有套结平眼锁纽孔
straight buttonholing	平头扣眼
straight clothing	直筒型外套
straight coat	直筒型大衣
straight collar	西服领
straight/cross laser marker	(裁床定位用)直线/十字镭射灯
straight cuff(同 band cuff)	直筒袖口,条型袖口
straight cut	直线裁剪,直裁
straight cutting machine	直条切布机
straight edge	直尺,直规,标尺,校正尺
straight-edge seamer	直边缝纽机
straight-end buttonhole	平头眼,平眼,平头纽孔,直扣眼
straight English leather	英国直条仿麂皮针织物
straight eye	平头眼(纽孔),直钮孔
straight fit	紧身舒适型
straight gobelin stitch	直线哥白林网眼绣
straight hair	直发类发型
straight handle	直柄(伞)
straight hanging	直筒型西装轮廓线
straight hanging box jacket	直筒型宽松上衣
straight heel	直线跟
straight hem handkdrchief	窄边手帕
straight hemming	直缝边
straight hemming handkerchief	直边手帕,拉边手帕
straight hem stitch	直卷边线迹
straight hole	直纽孔,直线纽孔
straight jacket	囚犯约束衣,紧衣
straight joing	对合接,直接缝,无分支接头
straight joint	直接缝,对合接
straight knife	直裁刀,直刀
straight knife cutter	直刀裁剪机;直刀电剪
straight knife cutting machine	直刀裁剪机
straight knife machine	直刀裁剪机

straight lace 花纹网地花边
straight lailored 直裁的
straight laser marker 直线镭射灯
straight legs 直筒裤脚;直筒裤
straight-legs dart 直省,钉形省
straight-leg slacks 直筒裤,直脚裤
straight line 直线
straight linear fusing press 直线型压烫
straight line coat 直线条大衣
straight line stitching 直线缝
straight line system 传动线,流水线
straight lock stitch 直型锁式线迹
straight lock stitcher with automatic thread trimmer 自动剪线平缝机
straight neckline 一字领口,齐肩领口
straightness of stitches 线缝直线度
straight of the material 材料直向,织物经向,织物纵向
straight overcoat stitch 直线锁缝针迹,直线式覆盖刺绣针法
straight pants 直筒裤
straight pin 大头针
straight placket band 直开衩滚镶边
straight pleat 直裥
straight pocket 直线袋
straight quilting machine 直线绗缝机
straight ripple 横凸纹双面针织物
straight [round] edge 直[圆]边
straight ruffle 直型打褶皱边
straights 直筒裤
straight seam 直型缝,直线缝
straight seamer 缝纫机,直线缝缝纫机
straight sewer 缝纫机,直型缝缝纫机
straight shade 单色
straight sheath 收腰直筒裙
straight shoulder 直肩
straight-sided pocket 直边袋
straight silhouette 直筒型,直筒轮廓,直线型
straight skirt 细身裙,直筒裙,剑鞘形裙
straight sleeve 直筒袖
straight stitch 直型线迹;直线针法
straight stitched seam 直线缝,直型缝
straight stitcher 直线缝缝纫机,直型缝缝纫机
straight stitching machine 直线缝缝纫机
straight stitch type 直式线迹型
straight stretch stitch 直线伸缩线迹
straight swing machine 缝纫机,直型缝缝纫机

straight-tip (鞋)直尖式,一字式(鞋式);直线鞋头
straight toe-cap 鞋子直线式饰皮包头,直线式鞋头饰皮
straight top vest 吊带式平胸女背心
straight torso 直筒躯干(身材)
straight trimmer 裁缝剪刀
straight trousers 直筒裤
straight tucks 直线褶,直型塔克
straight veining stitch 直纹针法
straight waistband 直式腰带
straight wear 直线条服装
straight yoke 直线过肩,直线覆肩
straiken 苏格兰亚麻织物
strain cape 紧身斗篷
straining 绷紧
strait jacket 约束衣,紧衣
strait waistcoat 紧衣,约束衣
strand 一串珍珠;(皮革加工)抽刀
strange clothes 奇装异服
strap 软带(绳);皮带;布带,条带,吊带;塑料带;金属带;汗背心背带;袢;搭扣鞋
strap attaching 缝袢
strap bra pads 吊带胸垫
strap collar 窄条领,狭条领
strap leotard 背带式紧身衣
strapless 无肩带服装
strapless bodice 无肩带胸衣
strapless bra(ssiere) 无肩带(式)胸罩
strapless dress 无肩带(式)连衣裙
strapless evening gown 无肩带式女晚袍
strapless gown 无肩带裙装,无吊带裙装
strapless jumpsuit 无肩带连衫裤,女紧身连衫裤
strapless neck 无肩带的胸衣领
strapless neckline 无肩带领口,齐胸领口
strapless pants 无肩带胸衣的长裤(夜礼服)
strapless shirt 无肩带女式长衬衣
strapless slip 无肩带连身衬裙,无吊带衬裙(或套裙)
strapless top 无肩带弹力短上衣
strap neckline 窄条布形领口,条形领口,(有)肩带领口
strapped cuff 扣带袖口,束带袖口
strapping 布带子
strapping machine 打带机
strap pump 有带鞋,挂带袢布鞋
strappy sandal 搭扣便鞋,袢带凉鞋

strap seam 条饰缝,带形缝
strap shoes 搭扣鞋,袢带鞋(总称)
strap shoulder sleeve 肩章袖
strap-shoulder style 有肩带的服装款式
strap sleeve 吊带袖
strap stretch slacks 松紧带便裤;吊带弹性裤
strap work 绣花网眼花边,绣花网眼纱
strap wrist 腕口扣带
Strasbourg work 斯特拉斯堡刺绣
strass 铅玻璃
strass transfer 水晶烫贴
strategy of product location 产品定位
straw 草编织品,草帽;稻草色,浅黄色
strawberry 草莓红,紫红色;酒糟鼻
strawberry cream 草莓奶油色
strawberry pink 草莓玫瑰色
straw blond(e) 草莓红发人(尤指女子)
strawboard 草纸板;马粪纸;黄纸板
straw braid 草帽辫,麦秆辫
straw cap 无檐草帽
straw-colour 稻草色,淡黄色,麦秆色
straw dicer 麦秸草帽
straw hat 草帽
straw mesh 麦秆鞋
straw needle 疏缝针,假缝针
straw rain cape 蓑衣
straw sandal 草鞋
straw slippers 草编拖鞋
straw yellow 浅黄色,稻草色
streak 条纹,纹理,条痕,色线,闪光
streak dyeing 色柳,染色条花
streaks 色条花疵
streaky pattern 横条疵
stream days 运转日数,生产日数《管理》
streamer 飘带
streamline(form) 流线型,川流线
street dress 外出装,逛街装
street fashion 大众时装;街头时装
street length 逛街衣长(长至膝下1～3英寸)
street shoes 逛街鞋
street skirt 外出裙,逛街裙
street wear 外出装,上街装
strength (织物)强度
stress area 应力区(域)
stressed-skin structure 拼装结构
stress points 着力部位
stretch 弹力;伸展;拉伸;推;拔
stretchability 拉伸性

stretchable blind seam 弹性暗缝
stretchable fabric 弹性织物
stretch bandings 弹性带
stretch belt 松紧腰带
stretch blind hem 弹性暗卷边
stretch blind hem stitch 弹性暗卷边线迹
stretch body suit 弹力紧身装
stretch boots 弹力长筒靴
stretch bra[ssiere](同 one-size bra) 弹力胸罩
stretch briefs(同 garter-briefs) 吊袜弹力三角裤;吊袜弹力紧身裤
stretch crotch 拔裆
stretch-decorative stitch 伸缩装饰线迹
stretch denim 弹力坚固呢,弹力牛仔布
stretch denim pants 弹力牛仔裤
stretched stitch 伸缩性线迹
stretched vest 弹力背心
stretcher (撑开帽、鞋的)撑具
stretch fabric 弹力织物,弹力布,松紧织物
stretch flannel 弹力法兰绒
stretch front piece 推门(将平面衣片,推烫为立体衣片)
stretch garment 弹力服装
stretch girdle 松紧带
stretch hose 弹力袜
stretch hosiery 弹力袜
stretching 拉伸;拔(整烫工艺之一);拉幅;(皮革)拉皮
stretching crotch 拔裆
stretching machine 拉伸试验机
stretch in pressing 拔烫
stretch jeans 弹力牛仔裤
stretch jersey 弹力平针织物,弹力运动衫
stretch knit 弹性针织物
stretch mark 拉伸记号
stretch material 弹性缝料
stretch nylon socks 弹力锦纶袜
stretch nylon yarn 弹力耐纶丝
stretch overedge 伸缩性包缝
stretch overlock stitch 伸缩性包缝线迹
stretch pants 弹力裤
stretch polyester sewing thread 涤沦低弹丝缝纫线
stretch recovery 弹性回复
stretch ribbon 松紧带
stretch seam 弹性缝,伸缩缝
stretch seaming 弹力缝纫,伸缩缝纫
stretch-sensitiver fabtic 对张力敏感的

织物
stretch shirt 弹力女衬衣
stretch socks 弹力短袜,弹力袜
stretch stitch 伸缩[弹性]线迹
stretch stitching 伸缩线缝
stretch stitch sewing 伸缩线迹缝纫
stretch stocking 弹力袜,压力袜
stretch terry 弹力毛巾布,伸缩毛巾布
stretch-to fit garment 弹力贴身服装
stretch top 弹力贴身女衫
stretch toweling 弹力毛巾布
stretch twill 弹力斜纹布
stretch waistband 弹性腰带
stretch welt 松紧带边条
stretch wool fabric 弹性伸缩毛织物,弹力呢
stretch woven fabric 弹性织物
stretch yarn 弹力丝,弹力纱
stretch zigzag stitch 伸缩曲折形线迹
stretchy seam(同 stretch seam) 弹性缝,伸缩缝
striated fabric 条纹织物
stricot 斯特利考特呢
strict look 端庄款式
stride room 跨步余量
strides 裤子(英国)
strié 条痕,条花疵
strike back 渗胶
strike off 打样(配色样本),试印,印花打样
strike through 粘渗,渗胶(织物粘贴时黏合剂渗入织物)
striking suits 引人注目的成套衣服
string 线,带子,细绳
string bag 网袋,网线袋,细绳包
string bikini 用细带系的比基尼式泳装
string button 绳结扣
string color 棕灰色
stringer 半边拉链
string gloves 粗支纱网眼手套,绳状编织手套
string machine 剪带机
string necktie 丝带领带
strings 吊襻,色线头(织疵)
strings and tapes 绳带
string tie 警长领带;丝带领带;团长领带;西部领带;细绳领带,狭带领带;鞋带型领带;窄领结
string vest 条纹衬衣,条纹背心
string warp [lace] machine 网眼花边机

stringy selvedge 木耳边;松边
strip 条,带,细布条,细革片;(足球队员等穿的)彩条球衣
strip back (织物)背面擦损
strip band 针织滚条(用于低领、圆领、高领等的生产)
strip cloth 条子布
strip dyeing 衣片染色
stripe 条纹布;柳条布;条纹;色条;条纹囚衣;袖章;横档疵
stripe cloth 条子布
striped 抽条,柳条,条
striped cotton 条子布
striped crepe 柳条绉
striped crepe de Chine 条双绉
striped crepe twill 色条双绉,彩条双绉
striped defect 条花疵,经柳疵,横档疵
striped fabric 罗布;条子织物
striped fancy leno 涤棉间条纱罗
striped fancy suiting 条子花呢
striped habutae 条子电力纺
striped habutai 条子电力纺
striped hair cord 柳条麻纱
striped half hose 横条短筒袜
striped interlining woolens 条子驼绒
striped lining 条子衬里布
striped pajamas 条纹布睡衣裤
striped pattern card 条纹样卡
striped pyjamas 直条斜纹布
striped rayon voile 条子绡
stripe drill 条子斜纹布
striped scarf 彩条围巾
striped shadow velvet 条影绒
striped skunk 臭鼬,臭鼬毛皮
striped (spotted) material 带条料子
striped taffeta 条子塔夫绸
striped terry 条纹毛圈织物
striped towel 彩条毛巾
striped trousers 礼服用条纹西裤
striped voile 条子绡
striped voile broche 条子绡
striped worsted 精纺条纹毛织物
stripe fabric 条子织物
stripe fancy suiting 条花呢
stripe interlock fabric 抽条棉毛布
stripe jeans 条子斜纹布
stripe knit 条纹编织
stripe line 丝缕线,纹理线,布纹线(推档及排板的参考线)
stripe pattern 条纹图案,条纹花样

stripes 条纹织物,条纹布,条纹衫,警服等级条纹
stripe shirt 条子衬衫
stripe stitching machine 条纹缝编机
stripe voile broche 条花绡
stripiness （经编针织物）直条痕
strip opening 滚边开口
stripped-down model 落后型,朴实型
stripped seam 折边加固缝（缝后毛头分开,布条覆盖重缝）
stripping 条纹染色毛皮
strip test 垂直芯吸法;条样试验
stroke 笔划器《计算机》
stroke centerline 笔划中线《计算机》
stroke display 笔划显示《计算机》
stroke edge 笔划边缘《计算机》
stroke width 笔划宽度《计算机》
stroking 缝制皱裥
stroller 松散无腰带衣,松身式上衣
stroller jacket(同 sack jacket) 散步外套（半正式男礼服,套装式外套）
strollers 平底便鞋,轻便鞋
stroner 休闲毡帽;散步外套
strong blue 深蓝色,景泰蓝,靠蓝
strong butted seam 牢固对接线缝
strong color 强色,深色,浓色
strong(s) （色彩）浓、深、强烈
strong stitch 牢固线迹
strong tone 强烈色调
strontium tifanate 锶钛石
stroud 粗呢;粗呢衣服(旧时北美产)
stroud[ing] 精毯织物,粗毛毯大衣呢
structural design 织花花纹;结构设计,织花设计
structural diagram 结构图
structural drawing 结构图
structural element of knitted fabric 针织物结构单元
structural fabric 花式织物
structural pocket 结构袋
structural programming 结构化程序设计
structural surface effect 花式表面效应
structural type 结构类型
structure 结构
structure jacquard （针织）组织提花
structure line 结构线
structured fabric 花式织物
structured look 花式效应;装饰外观
struntain 窄编织带
St. Tropez skirt 圣·特罗比兹长裙（长至踝关节,多达八种不同面料制成）
stubbed beard 残茬短胡子
stucco 棕灰
stud 领扣,饰纽,袖扣,饰钉
stud button 饰纽
studded placket 纽扣开衩,纽扣开口
stud earring 后侧耳环,饰纽耳环
student bycocket 意大利学生帽
student cloth 学生呢
student coating 学生呢
student's uniform cloth 学生呢
student's wear 学生服
stud-fastening hood 拷纽可脱卸风帽
stud hole 纽孔
studio designer 影视服装设计师
stud-off sleeve 拷纽可脱卸袖
stud-on fishing leggings 拷纽可脱卸捕鱼裤
stud press 拷纽
studs 柱式穿孔耳环
study and fondle 观摩
stuff 原料;织品;精纺毛织物;衬里布;料子,呢绒;填料;素材
stuff cap 呢帽
stuffed seam 加厚缝
stuffed trunk-hose 桶形棉连袜裤
stuff goods 呢绒,毛织物
stuff gown 毛料礼服;旧时律师袍
stuffies 拖鞋,平底拖鞋
stuffing 填充,填料
stump 矮胖子
stump work 浮雕刺绣
sturdy fabric 质地坚牢的面料
sturdy shoes 强力鞋
sturdy texture （织物）质地坚实
S twist yarn S捻纱
ST yarn (self twist yarn) 自捻纱
style 式样,风格,款式;时尚;类型,类别;气派,风度,格调;品格;习性
style analysis chart 款式分析图
style armhole sleeve 变形袖
style book 服装款式样本,时装款式样本
style change 服装款式变化
style chasse [法]狩猎款式
style corsaire [法]海盗款式
styled armhole sleeve 变形袖
style design 风格设计;款式设计
style design drawing 服装款式设计图
style detail 款式细节
styled raglan 变化插肩袖

styled sleeve 变形袖
style effect 款式效果
style feature 服装款式特征
style landmark 款式标记线
style line 服装造型线,服装结构线
style lingerie 妇女内衣,装饰性内衣
style Louis Quatorze 路易十四风格
style Louis Quinze 路易十五风格
style Louis Treeize 路易十三风格
style No 品号(货号)
style number 款式号码,款号
style of clothing 服装样式
style of needle hole 针孔型
style pattern 服装款式纸样
style petite fille 少女服装款式
style piracy 剽窃设计
style planning 服装款式设计
styler 时装设计师
style show 时装表演
style tape 款式标带
style trim 款式点缀(装饰)
styling 款式,式样
styling device (服装)款式设计,款式方式
stylion 茎突点
stylish belt 时尚腰带
stylish button 时尚纽扣,时款纽扣
stylish clothes 流行服装,时装
stylish design 风格设计
stylish gament 流行服装,时装
stylish ladies' belt 时尚女装腰带
stylish necktie 时尚[款]领带
stylish silhouette 时装轮廓线条
stylist 服装设计师,服饰搭配师;产样设计师;发型师;造型师;流行规划专家
styliste [法]服装总设计师,服饰搭配师
stylistic form 超正宗款式,超传统款式
stylization 风格上的仿效;因袭
stylized flower (近代图案中)样式化花样
stylized print 抽象印花
stylized sketch (服装)效果图
stylon 胶绒织物
stylus 光笔,指示笔《计算机》
S-type hanging S形挂码
styrofoam 聚苯乙烯泡沫塑料
styrtop 高帮鞋
suba 匈牙利圆裁长羊皮袄
subahia 金钱,丝宽边格子织物
suba suit 潜水服
subclass 细类
subclass machine 系列缝纫机

subcompany 子公司,附属公司《贸易》
subcontract 托外加工《管理》
subdued 柔和光泽
subfus(c)k 深暗色衣着;(牛津大学师生穿的)黑礼服
subject matter 题材
subjective appraisal 主观评定,感官评定
subjective assessment 主观评定,感官评定
subjective brightness 亮度(英国色度用语)
subjective characteristic 主观特征,感官鉴定的特征
subjective comfort rating (SCR) 主观舒适度
subjective inspection 主观检验,感官检验
subjective judgement 主观判断
subjective wear trial of fabric comfort 织物舒适性主观评定
sub-line 辅助线,副线
sublistatic print paper 转移印花纸
sublimation transfer printing 升华转移印花
submarine armo(u)r 潜水服
submarine hose 潜水蹼,潜水袜
submission 协议
submit tender 投标
submodel 系列型,子型
subordination 从属
subroutine 子程序《计算机》
subsack (登山用的)辅助背囊
subsidiary company (同 subcompany) 子公司《贸易》
subsilk 丝光缝纫棉线
substance 纺织品质地
substernale 胸骨下点
substitutes 代用品
substrate 底布
substructure 用夹里和衬里直接成衣的方式
subteen (13岁以下女孩的)6～14号童服尺码
subtle brown tone 淡棕色调
subtle green 淡玉绿
subtle handle 滑爽手感
subtle perfume 淡香水;淡香水色
subtle tone 淡色调
subtractive colour 色彩暗度
subtractive mixture 减法混色
subtractive primary color 相减的基色

suburban coat	郊游外套,郊外上装
suburban wear	郊游服装
subzero clothing	严寒服
succatoons	苏卡通布,低档杂色棉布
succinite	琥珀色
sucker	泡泡纱
suclat	欧洲宽幅棉布
sucreton	松捻纱平纹棉布
sudanette	苏丹细布
sudation	发汗,出汗
suds	浓肥皂水
suds time	皂洗时间
suede	绒面革,仿麂布,人造麂皮,仿麂皮织物,小山羊皮,起毛皮革
suede cloth	仿麂皮织物,人造麂皮
suede collar jacket	羔皮领夹克
suede-covered button	仿麂皮包扣
sueded silk	仿麂皮绸
suede effect	麂皮效应,绒面效应
suede fabric	仿麂皮织物,人造麂皮
suede finish	仿麂皮整理
suede (fur)-covered button	仿麂皮(毛皮)包扣
suede gloves	仿麂皮手套,羊皮手套
suede imitation leather	仿麂皮绒,人造麂皮绒
suede jeans	仿麂皮牛仔裤
suede jeweler's pouch	羔皮珠宝商用袋,珠宝商用麂皮袋
suede knit	仿麂皮绒针织物
suede knitted fabric	仿麂皮绒针织物
suede leather	仿麂皮织物,软羔羊皮
suede leather jacket	羊皮夹克
suede-like fabric	人造麂皮
suede-like nonwovens	仿麂皮非织造布
suede mouton lamb	仿海獭羊皮
suede nap	麂皮绒(布)
suede nonwoven fabric	仿麂皮非织造布
suede shoes	绒面皮鞋;羊皮皮鞋;仿麂皮鞋
suede type synthetic leather	仿麂皮合成革
suedette	绒革仿制品,仿麂皮织物
suedine	仿麂皮织物,人造麂皮
suedine machine	仿麂皮起绒机
suedoise	法国哔叽
su embroidery fabric	苏绣
Sueo Irie	入江米男(日本时装设计师)
suffers	长至膝盖的紧身裤
suffolk lace	萨福克花边
suffragettes	女权运动
sufi	棉经丝纬织物,巴勒斯坦苏菲织物
sugar bag neckline	糖包领
sugar brown	砂糖棕,红糖色
sugar loaf hat	锥形帽,锥形糖帽
sulfur dyes	硫化染料
sulphur dyes	硫化染料
suit	成套衣服,套装;(一套)西装;常服;飞行衣;潜水服
suit bag	外衣口袋
suit box	套装纸盒
suit case	衣箱;手提箱,小型旅行箱
suitable colour	适应色
suitable concept	套装化概念
suitcase bag	提箱式手提包
suitcase lining	衣箱里子布
suit dress	套装式连裙装,套装式连衣裙
suit ensemble	全套装(由套装与同面料的外套组合而成)
suit hanger	套装衣架
suit hat	套装帽
suiting	西服料,套料;一身衣服,一套衣服;色织条子棉布,素色丝光棉布
suiting finish	棉织物仿亚麻布整理
suiting mixed	舒挺美
suitings	素色丝光棉布
suiting silk	呢,仿毛丝织物;质地丰满的丝织物
suit jacket	套装上衣
suit length	套头料
suit look	套装装式
suit mobility	(指尺寸大小的)服装规格机动性
suit of armo(u)r	整套盔甲
suit of clothes	西装;套装
suit of lights	斗牛士装
suit of sequins	斗牛士装
suit pattern	套装图案
suit shape	服装外形
suit sleeve	两片袖,套装袖
suit slip	上、下两色连身衬裙
suit style	套装风格
suit-suits	(相同衣料)正统套装
suit vest	套装式背心
suit with English drape	英式套装
Suizu costume	水族民俗服
suja knani	旁遮普花绸
sukiya	轻薄花卉绸
sukkerdon	印度东部细布
sulfur (yellow)	硫黄黄色,嫩黄色

sulphur dye 硫化染料
sulphur vat dyes 硫化还原染料
sulphur yellow 嫩黄色,硫黄色,带微绿的黄色
sultan 印度绸;洋红色
sultana 深紫红色
sulu 苏禄裙(斐济群岛男子围裙)
summarized chromatogram 简化色谱(图)
summer and spring colour 春夏色
summer blouse 夏季短上衣
summer cloth 夏令呢,奥尔良棉毛呢(英国名称)
summer clothing 夏季薄型外套,夏衣
summer coat 薄型外套,夏季外套
summer dark 夏天暗色
summer dress 夏季女服
summer fabric 夏季织物
summer green 夏绿色
summer hat 夏帽
summer knit 夏季针织服装
summer leather shoes 白皮鞋
summer night gauze 夏夜纱
summer night leno brocade 夏夜纱
summer school wear 夏季校服
summer shirt 夏季衬衫
summer silk 夏令绸
summer suit 夏季套装
summer tan 棕褐
summer tie 夏用领带
summer tuxedo 夏季晚便服
summer wear 夏装
summer-weight 轻薄滑爽、适合夏季穿的
summer-weight clothes 夏服
summer-weight shoes 夏鞋
summery dress 夏服
sumptuary laws 节约法令(限制个人服饰消费的法令)
sunback dress 背带式夏装
sunbeam 浅奶油色,草黄色
sun-belt climate 阳光地带气候
sunbonnet 阔边太阳帽,遮阳软帽
sun burn 日晒红,日晒黑,日灸棕(带红浅棕色)
sun burst 嵌有宝石的旭日针饰;浅旭日色,浅橘红色
sunburst pleat 辐射式褶裥,阳光褶襞(呈散射状细褶)
sunburst-pleated skirt 辐射式褶裥裙,撒开式女裙,散裥裙

sunburst tuck 放射状活褶
sun cloth 遮阳呢,遮阳布
sun clothes (露肩或露背的)太阳装
sunday best 星期日服装;礼拜服;节日盛装(泛指一个人最好的服装)
sunday clothes 节日服,礼拜服
sundays (在星期日或其他特殊场合穿的)最好的衣服,礼拜服,节日服,星期天外出服,(假日才穿的)最好衣服
sundown 阔边女帽;橙红色
sun dress 太阳裙,(前后低领口和无袖的)背心裙
sun fast 耐晒
sun flower(yellow) 向日葵色(黄色)
sun glasses 太阳镜,墨镜
sunglasses pocket 墨镜口袋
sun hat 太阳帽,阔边遮阳帽
sun helmet 木髓盔,遮阳盔,硬壳太阳帽
sunk cost 沉没成本
sun kiss 阳光色
Sunlight(color) 阳光色
sunlight fading 日光褪色,日晒变色
sunlight resistance 耐日光性
sunny yellow 日光黄
sun oil 日晒皮肤油
sun pleat 扇形褶裥,放射式褶裥
sunray pleats 日光褶,放射式褶裥
sun set 晚霞色,浅橘红色
sun set gold 晚霞金色
sun sets 太阳装
sun-shade (女用)阳伞;(复数)太阳镜;遮阳帽舌
sun shades 太阳眼镜,墨镜
sun shield 女用阳伞;遮阳帽舌
sunshine yellow 日光黄
sun shoes 尚夏得鞋
sunsuit 太阳装,日光浴装,夏季套装
suntan 日晒肤色化妆;棕黄色,土黄色
suntan-oil 日晒皮肤油
suntans 夏季土黄色装
sun-top 吊带式平胸女背心,无肩带胸衣
sun visor 女用阳伞;遮阳帽舌;遮阳帽檐
sun wear 日光衣,太阳装,日光服
Sun Yat Sen's uniform 中山装,孙逸仙制服
suo embroidery fabric 锁绣品
suonare 苏纳蕾发型
super 特级品,特优产品;特大号商品;特级;特大号
Super 新加坡条子棉平布

super casual	超级便装
supercilium	眉,眉毛
super classic	超正宗款式,超传统款式
super-coordination	超级组合装
super elegant clothing	超雅款式
super-excellent clean uniform	超静防尘服
superfine cloth	高级服装呢
superfine fancy suitings	(高级)单面花呢
superfine fiber	超细纤维,特细纤维
superfine glass fiber	超细玻璃纤维
superfine goods	特级货
superfine irreversible fancy suiting	单面花呢
superfine merino fancy suiting	超细美利奴花呢
superfines	特级商品
superimposed fabric	叠织织物
superimposed seam	叠缝
superimposed stitch	来去缝
superior brocade	妆花锦
superior goods	高档商品,优质商品
super jeans	超级牛仔裤
superlanered	超级重衣款式
superlative beauty	绝色
super layered	超级重衣款式(指衬衫加衬衫,毛衣加毛衣的穿衣款式)
super layered look	重衣款式,重衣风貌,超叠层型款式
super lemon	浓柠檬色
superlock seam	优质锁缝线迹
super loop	长毛圈织物
super major defects	主要疵点
superman look	超人款式
supermarket	超级市场
supermicro peach fabric	超细桃皮绒织物
supers	条子棉平布
super sell	特畅销;超级推销《销售》
super sleeve	大袖
super slim	超细裤脚
supersonic cleaner	超声波清洗器
supersonic noise	超声波噪声
supersonic wave sewing machine	超声波缝纫机
super stiffening finish	特硬手感整理
super stiff finish	特硬整理
super stitch	优质线迹
super thin but warm retention fabric	超薄保暖织物
super washing	石磨水洗
super wide look	超宽款式
super-stretch stitch	超弹性线迹,超伸缩线迹
supervisory instrument	检测装置,监视装置
superwash	超级耐洗
superwash finishing	羊毛超级耐洗加工
superzigzag decorative stitch	高级曲折形装饰线迹
supp-hose	(医用)护脚长袜
supple	柔软;柔软衣料;柔软衣料的服装
supple double-action stitch	辅助二步线迹
supple handle	柔软手感
supple look	柔软款式
supplementary dart	辅助省缝
supplementary measurement	补充尺寸
supplementary ornamental thread	辅助装饰线
supplementary press	辅助熨烫机
suppleness	柔软性
supplier	供应商
supply chain	供应链
supportasse	撑领架
support bandage	绷带
supported fabric	非织造复合织物
supporter	护身;三角带;丁字带;载体
support garment	护身服装
support girdle	护身绑肚
support hose	弹性护脚长袜
support hosiery	护腿袜,针织护腿套
supporting activity	(经营)辅助活动
supporting fabric	辅助织物
support legwear	针织护腿袜
support pants	护腿性裤子
support panty hose	护腿连裤女袜,护理性袜子
support socks	护理性短袜,护腿男用弹力中筒袜
support stockings	护腿长袜
support tights	护身紧身衣裤
suppression form	褶裥形式
suppression point	抑制点
suprasternale	胸骨上点
supukwenkin	围巾绸
surah	有光斜纹软绸,斜纹软绸
surah chevron	破斜纹绸
surah de laine	法国斜纹毛葛
surah ecossaise guadrille	格子斜纹绸
surah fantasie	法国印花绸
surah gros cote	粗棱条斜纹绸
surah silk	斜纹软绸

surat 印度苏拉特棉布
sur chemise 衬衫上衣
surcingle （古时教士长袍等的）腰带
surcingle belt 教士长袍腰带
sur clothing 上衣；外套
surcoat 外套,上衣,罩衣,中世纪女用背心式长外套,中世纪无袖铠甲袍
sure sewers 加固绱袖法
surepach 苏里巴契细布
surface blurn 表面燃烧
surface characteristic （织物）表面特征,表面风格
surface color 表面色
surface darning stitch 表面织补绣；表面缀纹绣
surface Design Association ［美］印染及图案设计协会
surface dyeing （表面染色的）浮色疵,浮色
surface interest 表面效应,外观影响；（织物）表面趣味；强调表面效果的衣料
surface modified fiber 表面改性纤维
surface placket 表门襟,外门襟
surface resistivity 表面电阻率
surface ruggedness of fabric 织物的粗壮两面
surface satin stitch 缎纹针迹
surface tension 表面张力
surfer look 冲浪发型,冲浪款式
surfers 冲浪裤
surfer swimsuit 冲浪游泳衣
surfer's shorts 冲浪运动短裤
surf green 竹青
surfing suit(同 wet suit) 冲浪服,潜水服
surfrider 冲浪裤
surf satin 塞夫重磅经面缎
surf trunks 男式冲浪裤
surgeon's gown 外科医生手术服
surgeon's mask 手术医生面罩
surgical bandage 外科［用］绷带
surgical boots 畸形矫正鞋,畸形矫正靴
surgical cap 手术帽
surgical cloth 外科纱布
surgical dressings 外科绷带
surgical gauze 外科纱布
surgical gloves 手术手套
surgical gown （外科）手术衣
surgical gowns and hats 手术衣帽
surgical gut 手术用缝线,肠线
surgical hose 外科整形袜子
surgical mask 手术（医生）面罩
surgical PBT bandage 医用高分子绷带
surgical sutures 医用缝合线,外科用缝线
surgical thread 外科用缝线
surgical wiper 外科［手术］用纱布
surjeteuse stitch （手套的）装饰线迹
suroit 宽后檐帽
surperfine fiber 超细纤维
surpied 绑扎鞋用皮带
surple 衣边,绣花衣边
surplice 斜叠襟,斜叠襟衣,白法衣,白袈裟,和尚袍,牧师法袍
surplice blouse 斜叠襟女衫
surplice cardigan 斜叠襟式毛衫,和服式开襟毛衫
surplice collar 葫芦领
surplice dress 斜叠襟服装
surplice front gown 斜胸襟长衫
surplice neckline 斜叠襟领口,白法衣领口
surplice sweater 斜叠襟式毛衫,法衣式毛线上衣
surplus and deficit 盈亏
surplus jacket 仿军服夹克
surrealism style 超现实主义风格
surrealism stylistic form 超现实主义款式
sur robe ［法］修尔装,修尔袍,重叠装
surtax 附加税
surtout ［法］瑟都帽；女式连帽斗篷,男式紧身长外衣
surveste 夹克外衣
survetement 运动员休息服（毛巾布制,保暖用）
survey report 检验报告
survital look 保温型款式,保温装
survival wear （适应于严酷自然环境的）防寒多用服
survival bag ［充气］救生袋
survival look 保温型款式；保温装（指滑雪、钓鱼等室外活动时穿用的服装）
survival time 生存时间
survival vest 救生背心
survival wear 防寒多用服
susces 印度轻磅塔夫绸
susetchen 本色薄软野蚕丝绸
susha 中国本色平纹绸
susi 苏西条格棉布,苏西柞丝条棉布
susi sufiyana 苏西线棉交织绸
susomawari 摆围
suspender 背带,吊带

suspender belt （长筒袜）吊袜带
suspender buckle 吊带扣；葫芦扣
suspender clip 吊袜带夹子，吊带夹
suspender girdle 绑肚
suspender jumper 背带裤
suspender pants 吊带裤，背带裤
suspenders ［英］吊袜带，吊带，背带
suspender shorts 背带短裤
suspender skirt 背带裙；吊带裙
suspender slacks 挂肩宽衣裤
suspender suit 背带式套装，背带服
suspender tape 吊带，背带
suspender trimming 悬挂装饰，绕肩饰带
suspension tack（同 swing tack） 三角针
suspensory 吊袜带，裤子背带
sussex lawn 苏赛克斯本色薄亚麻布
susurrus 丝鸣
sute machine 捨车缝（在布端稍微内侧的车缝，不折缝头而毛边不细碎）
suti poult 素绨
sutura lambdoidea 人字缝
suture 缝合；［外科手术用］缝合线；针脚
suture lambdoidea 人字缝
suture line 缝合线，袜跟缝
suture thread 袜跟线；缝合线
sutwan 匹染缎
SUUNTO 桑拓
suya jacquard crepe (de Chine) 苏亚绉
suyue rayon marguisette 素月纱
Suzhong twill damask 苏中花绫
Suzhou brocade 古香缎
Suzhou embroidery 苏绣
svelte 苗条身材
Svend 斯文德（法国女帽设计师）
swaddle 襁褓用长布条，绷带布
swaddling bands 襁褓带，襁褓，婴儿服
swaddling clothes 襁褓，婴儿服
swaddling clouts 襁褓，婴儿服
swaddlings 襁褓用长布条，绷带
swadeshi 印度斯瓦代希棉布
swag dress 低垂服装（用柔软衣料制作随身体曲线下垂）
swagger 漂亮的，时髦的；宽大女短大衣；毛呢运动帽
swagger cane（同 swagger stick） （军官等用的）轻便手杖
swagger clothes 时髦服装
swagger clothing 阔步大衣；潇洒外套
swagger coat 潇洒外套，宽大女短大衣
swagger hat 毛呢软帽，时髦帽

swagger jacket 宽大夹克，宽大上衣
swagger pouch 宽大轻便包
swagger stick 时髦手杖；（军官等用的）轻便手杖
swagger suit 套装，潇洒套装
Swakara （South West African Karakullamb）西南非洲产的卡拉库尔改良种羔羊的注册商标
swallow collar 刀领，燕子领
swallowtail 燕尾服
swallow-tailed clothing 晚礼服；燕尾服
swallow-tailed coat 燕尾服
swallow-tailed collar 燕尾领
swallowtails 燕尾服
swan bill busk 梨形胸衣，匙状胸衣
swanboy 双面天鹅绒
swanette 小天鹅防雨绸
swank 时髦，漂亮，优雅
swansdown 天鹅绒女式呢；纬面起绒厚棉布，软绒布；天鹅羽绒
swan shift 天鹅羽毛外套
swanskin 天鹅皮；柔软的斜纹织物；厚密法兰绒
swan velvet 漳绒，天鹅绒
SWAROVSKI 施华洛世奇
swatch 小块布样，样本，样品
SWATCH 斯沃奇
swatch-cutting machine 切布样机
swatching band 婴儿服
swatch kit 样本册
swathe 包装品；绷带，带子，包布
swathing cloth 绷带织物，襁褓带子织物
Swatow grass cloth 汕头夏布
swatow pineapple cloth 汕头菠萝叶纤维布
sway 斯凡；热敏变色纤维（商名，日本东丽）
swayback 凹背；臀部后翘
swazy 斯瓦齐彩格布
sweat 汗，汗水
sweat absorption finishing 吸汗整理
sweatband 帽内防汗带，帽内皮圈，吸汗带
sweat clothes 休闲服装，运动服装
sweat dress 汗衫料连衣裙
sweater 毛线衫，羊毛衫，运动衫，针织套头衫式衬衫，厚运动衫，背心式毛衣，针织开襟背心，针织套头V领背心
sweater blouse 线衫上衣
sweater brush 衣刷

sweater clothing 毛衫外套
sweater coat 毛衣外套,针织短外套
sweater dress 针织连衣裙,运动衫式连衣裙,编织连裙装,连裙毛衣
sweater-for-two 特大双领毛衣
sweater girl 穿紧身套衫的女郎,(穿紧身套衫)胸部丰满的女郎(尤指演员、模特儿等)
sweater hat 翻檐毛线帽
sweater heart neckline 鸡心领口,甜心领口
sweater jacket 针织外套,针织夹克
sweater-knit clothing 针织套衫
sweater knitting machine 毛衣编织机
sweater look 毛衣款式,运动衫风格,运动衫款式
sweater made of cotton yarn 棉线衫
sweater made of ramie yarn 麻线衫
sweater on sweater 重穿毛衣,重穿毛衫
sweater pants(同 warmup pants) 厚绒运动裤,毛线裤套装
sweater set 两件套毛衣,成对羊毛衫
sweater shirt 运动衫,针织套式衬衫
sweater socks 毛衣花纹短袜
sweater-strip 运动衫衣坯,卫生衫衣坯
sweater-strip machine 计件衣坯圆形针织机
sweater style frock 毛衫式长外衣
sweater suit 毛线套装,针织套装
sweater sweater 毛线衫
sweater-top dress 毛线上衣连裙装,针织上衣连裙装
sweater type construction 毛衫型结构(指无下层结构类型)
sweater vest(同 vest sweater) 背心式毛衣,针织套头V领背心,针织开襟背心
sweater with beadings 珠片毛衣
sweater with embroideries 绣花毛衣
sweater with patches 补花毛衣
sweater with reverse collar 翻领运动衫
sweatery sweater 毛线衫
sweat fastness 耐汗渍(色)牢度
sweath-bands 婴儿服装
sweatheart neckline 鸡心领口
sweating manikin 出汗人体模型,出汗假人(服装功能性评价用)
sweat jacket 针棉运动衫(用棉或化纤针织料制作的圆领长袖衫)
sweat pants(同 warmup pants) 宽松长运动裤,厚绒运动裤

sweat resistance 防汗;耐汗渍牢度
sweats 跑步套装(俚语)
sweat set (两件套)毛衣套装
sweat seventeen 豆蔻年华
sweat shields 腋下防汗布,腋下垫布
sweat shirt 汗衫,圆领长袖运动衫,训练衫,卫生衣
sweat shirt and pants 卫生衣裤
sweat shirt with elongated tail 长后摆衫
sweat shirt with hood 带帽卫生衣
sweat shop 血汗工厂(指工资低、劳动条件恶劣、劳动时间长的小工厂)《管理》
sweatshorts 竞走短裤,夏天短跑裤
sweat socks 吸汗袜,运动短袜
sweat suit 运动服,运动套装,运动衫裤
sweaty odour 汗臭
Swedish costume 瑞典民俗服,瑞典民族服饰
Swedish Crown 瑞典王冠
Swedish darning 瑞典织补刺绣
Swedish embroidery 瑞典刺绣
Swedish hat 瑞典帽
Swedish lace 瑞典花边(棱结花边,以饰带花为主)
Swedish work 手工瑞典窄带
sweep 下摆,衣裾;后摆;(衣裙等的)拖摆
sweep effect 下摆效果
sweeper's 胸褶裙
sweep line 下摆边,底边
sweep opening 下摆宽
sweet color 悦目颜色
sweet dress 甜美服装,舒适的服装,好看的服装
sweetheart neck 鸡心领,情侣领
sweetheart neckline 鸡心领口,情侣领口
sweethearts watch 情侣表
swell 一流的(超流行)
swelled edges 隆起边缘(离衣服边缘0.5~2cm处,缝一排线迹,装饰用)
swelling-resistant finish 抗膨化整理
swelling shrinkage 膨胀收缩
swept-back hair 后掠式发型
swept-up hair 掠起式发型
swim bra[ssiere] 泳衣胸罩
swim cap 游泳帽
swim dress 泳衣式裙衣,泳装式连衣裙
swim fin 脚蹼
swimmer 女泳衣
swimmers 廉价粗纺呢,(男)游泳裤
swimming [bathing] costume 游泳衣

swimming cap　游泳帽
swimming costume　游泳衣
swimming pants　游泳裤
swimming suit　泳装
swimming suits and trunks　游泳衣裤
swimming trunks　男游泳裤,游泳短裤
swimming wear　游泳衣,游泳服
swim pants　游泳裤
swimsuit　女游泳衣,游泳衣
swimsuit fabric　游泳衣料
swimsuit hook　泳装钩扣
swimsuit knit fabric　泳装针织料
swimtrunks　男游泳裤
swimwear　(总称)游泳衣,泳装
swing　(鞋底等的)曲线形轮廓
swing arm vertical cutting machine　摇臂裁剪机
swing back coat　大下摆(女式)大衣
swing cutting　联合裁剪
swinger bag　摆式拎包(提把在角上)
swing guide quilter　导向摆动绗缝机
swing hemmer　摆动卷边器
swinging arm　摆臂
swing jacket　高尔夫夹克
swing needle　缝纫机摆动针
swing-out binder　摆动式滚边器
swing-out bra(ssiere) and corset attachment　胸衣摆动附件
swing-out folder　摆动式折边器
swing-out hemmer　回转卷边器
swing rake plaiter　摆布料
swing room　休息室《管理》
swing shift　中班《管理》
swing skirt　摇曳裙,飘摆裙
swingy skirt　摇曳裙,飘摆裙
swing sleeve　浅口袖,后掠式袖
swing tack(同 suspension tack)　三角针
swing tag　吊牌
swing ticket　吊牌
swing top　高尔夫上衣,高尔夫短夹克,运动夹克
swing type cord welter attachment　摆动式嵌线滚边附件
swing[y] skirt　飘摆裙,摇曳裙
swirl skirt　蜗牛裙,漩涡裙
swish　(裙脚拖地而过时的)塞窣声;漂亮,时髦
swiss　薄纱
swiss applique　植绒薄纱
swiss bar　变密度编带

swiss batiste　巴蒂斯特高级薄纱
Swiss belt　瑞士宽腰带
swiss cambric　白色薄纱
swiss checks　漂白格子薄纱
Swiss costume　瑞士民俗服
Swiss crepe organdy　瑞士蝉翼绉
Swiss darning　瑞士织补法
Swiss embroidery　瑞士刺绣
swiss embroidery　细薄纱刺绣
Swissing　辊筒轧光工艺
Swiss lace　瑞士机绣花边;变密度编带
swiss mull　上浆薄细布
swiss muslin　薄纱,薄细布
swiss organdy　奥甘迪,蝉翼纱
swiss pongee　仿绸丝光棉布(英国用语)
swiss voile　轻薄透明纱
switch　马尾型假发;转手贸易
switch co-ordination　组合装,不同商标的单件服装组合
switched-off　不懂时髦的,过时的,老派的
switched-on　懂时髦的,时新的,新潮的
switching look　可变组合的式样
switch line　(服装的上下身)转换线
switch wig　摆动假发
swivel　挖花织物;摇臂,摇杆;旋转
swivel angular binder　转角滚边器
swivel arm　旋转臂
swivel bobbin　刺绣圆梭心
swivel cloth　挖花织物
swivel dot　挖花点子花纹,植绒点子花纹
swivel effect　挖花花纹
swivel fabric　挖花织物
swivel hemmer　旋转折边器
swivel ruffling attachment　旋转打褶附件
swivel weave　挖花织物
swizzing　辊筒轧光工艺
sword belt　剑带,刀带;佩刀剑带
sword guard　(刀枪柄部的)护手
Sybil Connolly　西比尔·康诺利(1921～,爱尔兰时装设计师,其著名设计有水平蘑菇形褶裥,手帕式亚麻晚礼服等)
syddo　粗仿硬衬呢(代替马尾衬)
sylvan green　森林绿
Sylvia Pedlar　西尔维亚·佩德拉(1901～1972,美国女内衣时装设计师,1929年创立爱丽丝内衣公司)
Sylvie Schurer　塞尔维·斯恰尔勒(德国时装设计师)

symbolic colours 象征色
symmetrical design 对称设计,对称花纹
symmetrical pattern 对称纹样
symmetry 对称,匀称
symmetry mark 对称符号
symphony 和谐,协调
synchro-flow 同步流行;单件产品渐进流水线
synchronized top feed 同步上送料
synchro system 同步系统
Syndicat 巴黎高级服装店工会
synergy 协同作用
synthetic detergent 合成洗涤剂
synthetic diamond 合成钻石,人造钻石
synthetic fabric 合成纤维织物;合纤布
synthetic fibre 合成纤维
synthetic (fibre) knitwear 合纤针织品
synthetic fibre yarn 合成纤维纱
synthetic gem 人造宝石
synthetic hook 尼龙弯钩
synthetic hook and pile fastener tape 尼龙搭扣带
synthetic leather 人造革,合成皮革
synthetic leather for heel piece 合成革鞋主跟
synthetic leather for shoe bottom liner 鞋内底合成革
synthetic leather for toe piece 鞋包头硬衬合成革
synthetic leather sandals 合成革凉鞋
synthetic resin 合成树脂
synthetic resin button 合成树脂纽扣
synthetic rubber 合成橡胶
synthetic rubber raincoat 人造胶雨衣
synthetic rutile 人造金红石
synthetics 合成纤维织物;化学合成物
synthetic sewing thread 化纤缝纫线
synthetic spinel 人造尖晶石
synthetic textile 合成纤维纺织品
synthetic thread 合成纤维线,合纤线
synthetic yarn 合成纤维纱;合纤纱
synthon fabric for outerwear 俄罗斯合纤外衣绸
sypers 16世纪英国轻薄织物
syrian crochet 钩边手帕花边织物
syrian embroidery 叙利亚刺绣
system 系统;体系;分类;方式
systematical costume 系列服装
systematical jeans 牛仔系列服装
systemization 系列化
system of human and machine 人机系统
szedria 与夹克配穿的滚边背心
szur coat 粗纺绒牧羊服

T

TA（technology assessment） 技术评定《管理》
tab 绳,衩;搭衩;(服饰用)小垂片;舌片;(帽)护耳,帽耳;鞋带末端的包头(英国);领角衬片;参谋军官的领章(英国)
tabard 塔巴德式外衣;(多作现代时髦女郎海滨浴装的)无袖(或短袖)外套,(中世纪穷人出门穿的)粗布外衣;(古代)骑士短披风;传令官无袖制服;铠甲罩袍,无袖圆领斗篷式上衣
tabaret 波纹毛葛,塔巴勒绸
tab attaching machine 钉标签缝纫机
tabbed closing 扣衩门襟
tabbinet 府绸,波纹塔夫绸,波纹毛葛
tabbis (波纹整理的)泰比斯绸
tabby 平纹;平纹织物;纳缝
tabby-back 平纹背起绒织物
tabby-back corduroy 平纹背灯芯绒
tabby velvet 平纹背平绒
tab closure 用悬垂片束的服装
tab collar 拉衩领,衩式立领,饰耳领
tab neckline 扣衩形领口
tab opening 贴边式开襟
tab placket 贴边式开襟
tab point (制鞋)接帮点
tab slider 拉链滑块,拉链头子
tabi socks (大拇趾单独分开的)日本式原底短袜
tabinet 府绸,波纹塔夫绸,波纹毛葛
tabis 法国波纹轧光平纹绸
table 桌,台,台板;工作台;案板
table board of sewing machine 缝纫机台板
table cloth check 桌布式格子织物
table cut glove 贴合手套
table down 顶面朝下
table drawer 台板抽头
table size 顶面大小,台板尺寸
tabless tab collar 无悬垂片的悬垂领
table stand 机架
tablet 图形输入板《计算机》
table technique 图表技术
table tennis wear 乒乓球装
table top 工作台板
tablier 旧时女服上的围裙式装饰;女子围裙
tablier de paysane 农民式罩衫
tablier skirt 围裙
taboosh with tassel 埃及流苏毡帽
taboret 绣花绷架
taborett 粗纺呢,印染窗帘布
taborine 英国塔博林波纹呢
tabouret 法国古老装饰用呢;绣框
tabs (1m 以下的)零头布,短码布;衣服滚边垂饰片
tabulation 算价《销售》
tache 搭扣,搭钩,钩扣
tachograph 转速表,速度记录器
tachometer 转数计,转速计
tachu 西藏锦缎
tack (为定样临时缝上的)粗缝,假缝,加固缝;黏着性,黏着力;平头钉,大头钉,图钉,鞋钉;打结
tack belt 钉带子
tack button 铆合扣
tack edge 粗缝边,假缝边
tacked herringbone stitch 打结人字绣
tack elastic 钉松紧带
tacker 打结机;加固机;用粗针脚缝纫的人;钉纽扣机
tack facing 粗缝贴边
tackiness 黏着性
tacking 粗缝,假缝,加固缝纫;打结;打铆钉扣;缝筒
tacking awls 加固锥
tacking clamp 加固夹
tacking cut (布)边沿小洞
tacking hook 加固钩
tacking length 加固长度
tacking machine 打结机,加固机,归拔机
tacking mark 假缝符号
tacking operation 加固操作
tacking or basting stitch 假缝线迹,疏缝线迹
tacking seam 加固缝
tacking sewing 加固缝纫
tacking shoulder strap 钉肩带
tacking stitch 加固线迹,假缝线迹,疏缝线迹
tacking the end 加固端部

tacking thread 加固线,粗缝线
tacking welts 钉带子
tackle box 工具箱式包
tackle twill 运动服斜纹布
tack-stitching 假缝,粗缝,加固缝
tack tear （涂层织物）脱粘性
tacky party 奇装异服竞赛会,俗艳服装会
tacky surface 发黏表面
tacky 发黏的；俗艳的
tact 触觉；美感,鉴赏力；节拍《管理》
tactel 塔克特（纤维或布料）；聚酰胺 bb 纤维（商名,英国 ICI）
tactile appraisal 手感评定
tactile impression 触感,质感
tactile interest 手感趣味,（材料）质感趣味
tactile sensation 触感,质感
tactile sensibility 触觉,触觉敏感性
tactility 触感,质感
tact system 流水作业（线）
tadpole éponge 粗纺毛圈织物
taekwondo wear 跆拳道服
taenia 头带（古罗马）
taffeta 绢；塔夫绸；仿塔夫绸衣料
taffeta alpaca 阿尔帕卡塔夫绸,羊驼毛塔夫绸
taffeta armure 花塔夫绸
taffeta brocade 花塔夫绸
taffeta broche 彩花绢纺
taffeta cameleon ［法］闪光塔夫绸
taffeta checked 格子塔夫绸
taffeta chiffon 细塔夫绸
taffeta coutil 白地彩色塔夫绸
taffeta faconne 花塔夫绢；提花塔夫绸
taffeta flannel 轻质条格法兰绒
taffeta fleuré 彩色小花卉纹塔夫绸
taffeta fuchun rayon 富春纺
taffeta glacé 光亮闪光塔夫绸
taffetaline 绢丝塔夫绸
taffeta lining 里子塔夫绸,塔夫绸里子
taffeta luster 光亮塔夫绸,平纹丝府绸
taffeta metallique 金属色彩塔夫绸
taffeta mousseline 精细塔夫绸
taffeta pipkin 平顶窄檐小女帽
taffeta plain 素塔夫绸
taffeta pointille 小花塔夫绸,小点花塔夫绸
taffeta ribbon 塔夫绸带,熔边切割带
taffetas arc-en-ciel ［法］虹彩塔夫绸

taffetas armure 花法国古老低级塔夫绸
taffetas boyaux 格子花纹塔夫绸
taffetas bright 高级素色塔夫绸
taffetas broché 挖花塔夫绸
taffetas changeant ［法］闪光塔夫绸
taffetas chiné 印经全丝塔夫绸
taffetas de florence 佛罗伦萨薄塔夫绸
taffetas d'herbe 草地塔夫绸,阿里达斯塔夫绸
taffetas fleuret 弗里雷特塔夫绸
taffetas glacé 光亮闪光塔夫绸
taffeta souplesse 软塔夫绸（匹染,柔软整理）
taffetas prismatique 虹彩塔夫绸
taffetas quadrille 格子塔夫绸
taffetas royal 多色条纹塔夫绸
taffetas tinsel check 格塔绢
taffeta tinsel check 格塔绢
taffeta uni 绢丝塔夫绸,单色塔夫绸
taffetine 丝棉塔夫绸,塔夫泰因里子绸,塔夫绸
taffetized fabric 仿塔夫绸棉织物
taffetized finish 仿塔夫绸整理
taffety 塔夫绸；绢
tafta 伊朗平纹绸
tag 标签；（服装上的）挂祥,垂饰；（鞋口跟部的）拉祥；（鞋带的）金属包头
tag cloth 标签布
tagging gun （装标签用）胶枪；标签枪
tagging needle 胶枪针嘴
Tagik ethnic costume 塔吉克族服饰
Tagilmust 撒哈拉男式包头巾
Taglioni coat （19世纪中期流行于欧美的）泰格里昂尼男式合身大衣
tag pin （打吊牌用）胶针
tag price 标价
tahj 圆锥形高帽
taikong burnt-out polyester sheer 太空绡
tail （衣服）后摆；燕尾服；辫子
tail bottom 燕尾式下摆
tail braid （女裙的）尾带
tail clothing 晚礼服；燕尾服
tailcoat（同 swallow-tailed coat） 燕尾服；男子夜礼服
tail cover gasket 后盖衬垫
tailing 印花色渗；染色头梢色差（印庇）
tailings 短码布（美国）
taille 腰身；女子紧身胸衣
taille flattante 宽大的腰围
taille haute 高腰

taille marquee　强调腰围
taille smock　腰围皱褶刺绣
tailleur　[法]套装;定制的服装(指男装套装、西装);男式女西服,夹克裙套;男裁缝;饰花,贴花
tailleur ateliers　做大衣、套装的工作间
tailleur Broadway　百老汇套装
tailleur masculin　男式套装
tail of shirt　衬衫后摆
tailor　裁制衣服;做裁缝;剪裁,制作;按男装式样裁制(女服);(指男服)裁缝师;成衣工;成衣商
tailorable　(衣料等)可以裁制成衣的
tailor basting　长斜针假缝,八字缝,定针
tailor chalk　裁缝(用)划粉
tailordom　裁缝
tailored　裁缝做的;缝制的;定做的;合身的;(女服)线条简单朴素且贴身的
tailored blouse(同 shirtwaist blouse)　男式女衬衫
tailored bow　(腰带上用的)蝴蝶结
tailored buttonhole　鸽眼纽孔,西装纽孔,凤眼纽孔
tailored clothing　西装型服装
tailored collar　西装领
tailored costume　定做服装
tailored fell seam　折边叠缝,折边平缝
tailored fibre　特制化学纤维
tailored garment　精做服装
tailored hat　定做的帽子
tailored jacket　西装式夹克,精做上衣
tailored jumpsuit　西装式连衫裤
tailored knit　针织西装料;定制针织衣
tailored lapel　精做驳领
tailored man's shirt　精做男衬衣
tailored notched collar　精做平驳领
tailored opening　袖口衩
tailored pajamas　西装式睡衣裤,精做睡衣裤
tailored pants　西装裤
tailored placket　法式袖口衩,半开襟袖口衩
tailored set-in sleeve　精做装袖
tailored shawl collar　精做丝瓜领
tailored shirt collar　西服衬衫领
tailored shoulder　西装型肩线
tailored skirt　定做的简洁裙装,西服裙
tailored skirt suit　西服裙套装
tailored sleeve　西装袖,两片袖
tailored slip　裙裥饰边的衬裙

tailored suit　精做西装,西式套装;西装式女套装
tailored toe　裁剪缝合袜头
tailored with turn-ups　翻边合体裤
tailoring　裁缝业,缝制;成衣;裁缝工艺,成衣工艺;成衣活
tailoring equipment　服装裁剪设备
tailoring garment　定做服装
tailoring machine　成衣机
tailoring process　成衣加工
tailoring shears　裁缝剪刀
tailoring technology　成衣加工技术
tailor kerchief　椭圆形花边头饰
tailor knot stitch　记号缝
tailor-made　定做服装;女服线条简单朴素且贴身的;定做的,特制的
tailor's button hole　圆头纽洞,圆头细孔,圆头纽孔
tailor's canvas　亚麻帆布
tailor's chair　(无脚靠背的)裁缝椅
tailor's chalk　(裁缝用)划粉
tailor's chalk holder　划粉套
tailor's clapper　拱型烫木
tailor's clippings　衣料小样;裁剪碎料
tailor's cushion　(熨烫用)烫模,垫子
tailor's cutter　成衣店的裁剪师
tailor's cuttings　裁剪碎料;零剪;单裁
tailor's donkey　馒头烫垫
tailor's goose　裁缝用大型熨斗
tailor's ham　烫垫;布馒头,烫衣胸部、裤臀部的托垫
tailorship　(裁缝)手工;裁缝业
tailor shop　裁缝铺,成衣店,西装店
tailor's loop　挂衣袢
tailor's points　厚刃剪刀
tailor's press board　熨衣板,马凳
tailor's roller　袖烫垫
tailor's sleeve(同 suit sleeve, two-piece sleeve)　套装袖,两片袖
tailor's square　裁缝用角尺,L型直尺
tailor's tack　纤缝,裁缝粗缝,线钉
tailor's tack stitch　记号缝
tailor's tape　裁衣用软尺,量身软尺
tailor's thimble　顶针,针箍
tailor's trimming　服装附件(总称)
tailor's twist　缝纫用粗丝线
tailor's wax　裁缝用蜡
tailor's wax chalk　裁衣用蜡制划粉
tailor's-to-the-trade　现成服装店(裁缝制作成衣直接出售)

tailor suit 男式女服套装
tailor tacker 钻孔锥,记号针
tailor tacking 绗缝,粗缝
tailor tacking mark(同 thread marking stitch) 裁剪假缝,缝纫粗线标记,钉线迹
tails 燕尾服
tails dress coat 燕尾服
tail style 尾状发型
taint 退色斑
Taishan doupioni faille 泰山绸
taj 伊斯兰教苦行僧头饰,圆锥形他其高帽
Tajikezu costume 塔吉克族民俗服
tajong 粗纺呢
taj toe 上翘小尖头式东方鞋头
Takada 肯安奏(高田贤三)
Takada Kenzo 高田贤三(1945～,日裔法国时装设计师)
Takao Ikeda 池田贵雄(日本时装设计师)
Takayuki Mori 森孝行(日本时装设计师)
take 拿;取;(衣服)改动
take in 改小(衣服);归拢;打褶
take measurements 量尺寸
take off 脱下(衣帽);改瘦(衣服)
Takeo Kikuchi 菊池武夫(日本时装设计师)
take on (服装式样)流行,风行
take (the) size 量尺码,量尺寸
take-up (缝纫机)挑线杆,松紧装置;衣服的褶裥,打裥;服装改短
take-up lever 挑线杆
take-up thread picker 挑线器
take-up thread trimmer 挑线剪线器
taking delivery 收货;收取支付
taking in 归拢(缝纫),归拢打裥缝纫
taking measurement 量身;采寸
taking off (衣服)改瘦
taking up (衣服)改短
takri 印度泰克里白棉布
talanche 法国粗麻毛织物
talari 印度泰拉里条子薄棉布
talaria 翼形凉鞋,神翼鞋
talcum powder 滑石粉,扑粉,爽身粉
taliped 畸形足者,畸形足的
talipes 畸形足,足畸形
talisman ring(同 friendship ring) 友谊戒指,祝福戒指,护身戒指
tall 大号衣服

tall and thin figure 细长体型
tallboy 高脚衣橱,双层衣橱
tall collar 高领
tall figure 高体型
tall hat 大礼帽,高顶礼帽
tallith 缨边下衣,缨边头巾,缨边肩巾
tall socks 高筒短袜
tally band 缀船名的水手帽帽檐缎带
tally ribbon 缀船名的水手帽帽檐缎带
Talma mantle (近代流行于欧美的)陶尔玛斗篷;塔尔玛半圆形长斗篷
talma 旧式大披肩;宽大短外衣
tam(同 tam o'shanter) 苏格兰便帽(大黑帽),(帽顶中央饰一绒球的)宽顶无檐圆帽
tambour 绣花绷圈,绷子刺绣品
tambour embroidery 链状针迹刺绣
tambour farthingale 英国式裙撑
tambour hook 绣花钩针
tambourine bag 鼓状提包,铃鼓包
tambourine brim 铃鼓形浅筒帽,铃鼓帽
tambourine hat 铃鼓帽
tambour lace 绷子刺绣花边
tambour muslin 绷子绣花用棉布
tambour needle 刺绣针
tambour pocket 滚边袋
tambour stitch 绷架刺绣花边,刺绣针迹
tambour work 绷子刺绣,绣花绷圈
tamein 条子棉布,条绸;缅甸妇女围裹服装
tamese 透明缎条纱
tametta 印度泰米泰手帕布
tamet woven 正反面相同的织物
tamin(y) 塔明有光薄呢
tamis 透明缎条纱
tamise [rep] 泰米斯薄呢
tamiska 低级精梳衣料
tammel 黑色平纹亚麻布
tammie 棉经毛纬平纹呢;透明缎条纱
tammy 棉经毛纬平纹呢;透明缎条纱
tam o'shanter 苏格兰便帽(大黑帽),(帽顶中央饰一绒球的)宽顶无檐圆帽
tamtine 绢丝塔夫绸
tan 棕黄色,黄褐色;晒成棕褐色的肤色;(制革的)鞣料
tandem 中级漂白亚麻平布
tandi shadow stripe 弹涤绸
Tanera 塔内腊聚酯纤维非织造人造革
Tanfibs 轻薄坦菲布
tang 印度手织平纹布

tanga	短裙,三角裤
tangerine	橘红色
tangible assets	坦吉平纹细薄棉布 有形资产(管理)
tangible goods trade	有形商品贸易
Tan Giudicelli	唐·朱迪塞利(法国时装设计师)
tangle	(饰于服装或头发上的)缠结
tangled yarn	网络丝
tanglelaced	水刺法非织造布
tanglelaced fabric	喷网法非织造布
tanjib	坦克平纹细薄棉布
tankini	潜游式泳装
tank suit	坦克套装(针织短袖衫配短裤);坦克车型泳装,(无袖无领)一件泳装,有肩带的女式泳装(美国),传统式连体女泳装;有肩垫浴袍
tank swimsuit	传统女式泳装(勺形领,连裁式肩带)
tank top	大圆领女背心,无袖圆领衫,吊带式女背心,运动背心,背心衫;连衣式游泳衣;连裁式肩带
tannage	鞣制;鞣革,鞣熟的皮革;鞣革厂,鞣皮厂
tanner	鞣皮工,制革工
tannery	鞣革厂,鞣皮厂
tanning	制革,鞣革
tanning leather	硝革过的皮料
tanny	中国宽幅布
tans	(尤指鞋的)棕黄色的穿着用品;棕黄色皮鞋
tantoor	斜利亚新娘长钉形头饰
tanzen	优质条子棉布;优质条纹绸
tanzib	坦吉平纹细薄棉布
tap	鞋掌;鞋跟铁片;踢跶舞鞋
tapa cloth	塔帕纤维布
tapalos	多色墨西哥女披肩、女围巾
tapatis	棉布
tape	狭幅织物;线带,织带,狭带,卷尺,带尺;贴边,镶边,牵条
tape advancing equipment	送带装置
tape and thread chopper	带子和缝线切断器
tape attaching	缝带
tape attaching on body	大身缝带
tape attaching on folded body	大身褶边缝带
tape attaching seam	缝带接缝
tape belt	吊袜带,吊袜宽腰带
tape binder	滚带器
tape binding	滚带
tape binding attachment	滚带器附件
tape brake	止带器
tape checks	英国坦普切克薄棉布
tape chopper	限带器
tape clipper	剪带器
tape cutter	剪带器
tape cutter guard	带刀的防护罩
tape cutting machine	裁带机,剪带机,条带式裁剪机
taped finish	带子滚边
tape distance	带距
taped seam	牵条缝,贴带缝,热风胶贴缝
tape edge	缝边
tape fabric	带状织物
tape fastener	带式扣合件
tape finished hem	镶带下摆
tape foot	缝带压脚
tape for apron	围裙带
tape for shoe rim	鞋口带
tape garter	吊袜带
tape guard	按扣牵带,用按扣固定的牵带
tape guide	引带器
tape heel seam	带子根部缝
tape holder attachment	托带附件
tape inserting presser foot	嵌条压脚
tape knife	条形裁刀;条刀
tape lace	狭条花边
tape line	软尺,带尺,卷尺
tape measure	卷尺,带尺,软尺
taper	锥形
tape rack	(缝纫机)卷带轴架
taper bar	锥形套结
taper bar tack	锥形套结
tapered	锥形的
tapered belt	锥形带
tapered bottom	窄裤脚
tapered cut	锥形裁剪,斜裁
tapered dress shirt	小腰身衬衫
tapered leg	小裤脚;锥形裤管;窄裤脚
tapered legged pants	小裤脚裤,锥形裤
tapered line	三角线形,锥形线
tapered pants	锥形裤
tapered shirt	紧身男衬衫;女式圆摆短袖长衬衫;女子连裤紧身衫
tapered shoes	锥形鞋
tapered silhouette	渐窄型轮廓,锥形轮廓
tapered skirt	锥裙
tapered slacks	小裤脚裤,锥形裤
tapered trousers	小裤脚裤,窄脚裤,锥

形裤
tape reinforcement 加固带
tapering hem 斜折边
tapes and buttons for ornament 装饰带扣
tapes for corsets 胸罩带
tape stay 牵条
tapestry 挂毯;花毯;绒绣;织锦;像景织物;墙毯色
tapestry satin 织锦缎
tape width 带宽
tape work 狭带钩编品
tape yarn 扁丝
taping armhole 敷袖隆牵条
taping attachment 镶边附件
taping back vent 敷背衩牵条
taping front edge 敷止口牵条
taping hood brim 绱帽檐
taping lapel roll line 敷驳口牵条
taping pocket opening 敷袋口牵条
taping waist line 敷腰口牵条
taping 镶边,带子贴边;敷牵条
tapir fur 白貘皮
tap panties 踢跶短内裤(脚口略带喇叭形)
tap pants 踢跶裤
tap shoes 踢跶舞鞋
tara ameni 泰拉阿曼尼羊绒披巾
tarboosh 土耳其帽(无边圆塔型)
tare 皮重
tare(weight) 皮重,包装重量
target-shooting sweater 打靶用毛衫
tariff 有效关税
tariff (of) sizes (服装)号比
tarnish 失泽,光泽变暗
tarnishing 白地沾色
tarpaulin (船员用的)油布帽;油布衣;防水布;油布
tarpaulin grey 油布灰
tarragon 龙高绿,浅暗黄绿色
tarred paper 防潮纸
tarsal bone 跗骨
tartan 格子花呢;格子织物;格子呢服装;塔登
tartan check 格子呢
tartan kilt (苏格兰高地男子穿的)格子呢折叠短裙
tartan plaid 格子花呢
tartan plaid jacket 格子花呢夹克
tartan stripe 苏格兰条纹
tartan tweed 格子粗花呢

Tartar sable 红貂皮,亚洲貂皮,鞑靼貂毛皮
tash 塔什锦缎(经用多色的丝绒,纬用金银线)
taslan[taslon] 塔斯纶(运动套装料)
Taspa fabric 塔斯帕织物
tasse 盔甲的腿甲,腿罩
Tasseau 三角围巾
tassel 穗,流苏,缨;别针,饰针
tassel band 穗带
tassel earring 流苏耳饰
tassel loafer 流苏乐福鞋
tassel necklace 缨穗项链,链须式长项链(多达12条短链)
tassel shoes 流苏状便鞋
tassel stitch 流苏针法
tassel-slip-on 流苏浅口鞋
tassel-tie 穗带,穗带末端蝴蝶结
tassel-top loafer(同 loafer) 平底便鞋
tasset (盔甲的)腿甲,腿罩
tassor 塔沙纬向条纹薄呢
taste 趣味;式样;风格;审美力,欣赏力
tasteful clothes 雅致的服装
taste level 鉴赏水平,审美水平;情趣标准
tat chotee 印度黄麻布
Tataerzu costume 塔塔尔族民俗服
tatamis 榻榻米凉鞋
Tatar ethnic costume 塔塔尔族服饰
Tatas 窄肩带
tatbeb (古埃及人穿的)凉鞋
tatters 零头布,碎布,碎呢;破衣服
tattersall 塔特萨尔花格呢
tattersall check 塔特萨尔格子花纹
tattersall (plaid) 塔特萨尔格子花纹
tattersall's check 塔特萨尔格子
tattersall vest 塔特萨尔小格子男背心(单排六纽,无领,四只有盖袋)
tattersall woollen check cloth 塔特萨尔格子粗呢
tatting 梭结花边;梭结法
tatting lace 梭结花边,枕头花边
tattletale gray 灰白色
tattoo 点青;文身
tattooing 文身刺花(术),文身花纹
tattoo look 文身风貌,文身款式
tattoo pantyhose 超薄型纹(裤)身连裤袜(袜身上间隔印有花卉图案花纹)
tattoos 文身花纹;文身;刺花
taunton 宽幅布

taupe 灰褐
taupe color 灰褐色,灰棕色
taurino 陶利诺羊毛牛毛混纺呢
tautness 抽紧,绷紧
taut stitching 拉推式缝纫法,可拉伸线缝
tavelle 窄花边带
tawdry 廉价而花哨的服饰
tawdry clothes 花哨俗气衣服
tawdry clothing 廉价而花哨的衣服
tawny 黄褐色,茶色
tax 税金
tax bearer 纳税人
tax collectors' uniform 税务服
tax dodging 偷税
tax free 免税
tax on business 营业税
tax on sales 销售税
tax on value added 增值税
tax payer 纳税人
tax rate 税率
tax reimbursement 退税
tax revenue 税收
T-back fitness top T形健美低胸上装,T形健美祖胸式上装
T-bandage T形绷带
T-bar T字系带鞋
T-bar shoes T形带袢鞋
T-blouse 针织圆领衫
TC (technical committee) 技术委员会;(Terylene cotton) 涤棉,涤棉织物
T/C blended yarn T/C纱;涤棉纱
TCCA (Textile Color Card Association) 纺织色卡协会
T/C cambric 涤/棉细纺
T/C coated poplin 涂层涤棉府绸
T/C dyed tussores 染色涤棉线绢
T/C embossed poplin 拷花涤棉府绸
T/C emerized poplin 磨毛涤棉府绸
T/C fabric 涤棉织物;T/C布,棉的确凉
T/C fancy suitings 涤棉花呢
T/C gloss finished poplin 油光涤棉府绸
tchamir 摩洛哥罩衫,圆领口摩洛哥衬衫
tchapan 土耳其男用长宽松袍
tcharchaf 土耳其女用非正式披肩
Tchembert 白细布,花细布
tcheutche 抗皱塔夫绸
T/C jacquard poplin with weft filament 涤棉纬长丝府绸
T/C knitted lining 涤棉针织里布
T cloth T牌重浆粗平布,英国T粗平布

T coat T形上衣
T/C pants 涤棉裤
T/C plain cloth 涤/棉平布
T/C poplin 涤/棉府绸
T/C poplin coated jacket 涤棉涂层夹克
T/C washer jacket 涤棉水洗布夹克
T-dress (长到足以当外衣穿的)短袖圆领衫
tea 茶色
tea apron 茶围裙
teaberry 酱果红
tea-cozy cap (茶壶保暖套似的)针织羊毛帽
tea drop 低级塔夫绸
tea gown (茶会时穿的)宽松的长袍,茶服,茶会女礼服
tea green 茶绿,茶青
tea jack 茶服
tea jacket 下午茶宽松夹克
teak (brown) 柚木棕(棕色)
teal 凫蓝色(绿光暗蓝色),蟹青色
teal blue 凫蓝,绿光暗蓝色
teal duck 水鸭色(绿光暗蓝色)
team plan 班组计划《管理》
team system 流水作业系统
teamed-up pants 与上装配色的裤子
teapot basket handbag 茶壶箩手提袋
teardrop (项链或耳环上的)宝石坠子;纬斜
teardrop bikini 水滴型胸罩的比基尼泳衣
teardrop bra 水滴型胸罩
teariness 纬斜
tearing goods 棉织物,亚麻织物
tearing strength 撕破强力
tear of toe 豁袜头
tea rose 茶玫红(淡橙红色)
tear resistance 抗撕裂性
tear strength 撕裂强度
tease 起绒,拉绒,起毛
teasel cloth 起绒织物
tea stain 茶渍
tebenna 半圆形头篷
technic data 技术数据
technical assistance 技术协助
technical back/technical face 异面织物
technical characteristics 技术性能
technical clothing 技术服装
technical commitee (TC) 技术委员会
technical face 织物工艺正面

technical feature 技术性能
technical innovation 技术革新
technicality 技术性,专门性
technicalization 技术化,专门化
technical ornament 工艺装饰
technical reverse 织物工艺反面
technical room 技术室
technical safety measures 安全技术措施
technical schedule 工艺规程
technical specification 技术规范,技术说明
technical standards 技术标准
technical terminology 技术用语,技术术语
technician 技师,技术员
technicism 技术性
technicolor 鲜艳颜色,五光十色
technics 工艺学;工艺
technics design 工艺设计
technique 技巧;技艺,技法,技术;手法
technique media 技法
techniques 工艺(学);织物下机后的表面修饰
technique theories 技法理论
technocracy 技术管理
techno-economic verification 技术经济论证《管理》
technofashion 科技装
technological change 工艺变革
technological cooperation 技术协作《管理》
technological design 工艺设计
technological flow 工艺流程
technological forecasting 技术预测《管理》
technological innovation 技术革新《管理》
technological level 工艺水平
technological process 工艺流程
technological progress 技术进步
technological requirements 工艺要求
technological transformation 技术改造
technologist 技术专家;工艺师;技术员,技师(美国)
technology 工艺规程;工程技术;术语;工艺学;工艺;技术
technology change 工艺变革
technology for making suit 西装制作技术
technology introduction 技术引进
technology management 工艺管理《管理》
tec-net 网眼薄纱;六角网眼花边
teddy 特迪式胸衣连裤,(流行于20世纪20年代)妇女连衫衬裤,连裤内衣
teddy bear 长绒毛织物;厚大衣,毛皮大衣;毛皮里子的高空飞行服;(流行于20世纪20年代的)妇女连衫衬裤;穿着花哨的人
teddybear cloth 长毛绒织物
teddybear fabric 长毛绒织物
Teddy-boy look (20世纪50年代英国男孩流行的)阿飞型风貌(男少年蓄长发,穿紧身裤、尖头鞋、挺领外套),阿飞男款式
Teddy-girl look 阿飞女郎型风貌(20世纪50年代流行,女少年穿紧身短裙,细高跟鞋以及高耸蜂窝头)
Teddy-girl style 阿飞女郎型,阿飞女款式
Teddy style 阿飞青年风貌,阿飞款式
Teddy Tinling 特迪·廷林(英国时装设计师)
Ted Lapidus 特德·拉比德斯(法国时装设计师)
teenage brassiere 少女胸罩
teenager clothes 青少年服装
teen boys' sizes (十几岁)男少年尺寸
teen bra[ssiere] 少女胸罩,浅胸杯胸罩
TEENMIX 天美意
teen sizes 少年尺寸,十几岁少年尺寸
teens wear 青少年服装
teens' wear 青少年服装
tee post sandal 夹趾凉鞋
Tee-s(同T-s) T恤衫,T字衫
tee shape(同T shape) T字形服装
tee shirt 圆领短袖汗衫;文化衫;T恤衫
tee shirt dress T形连衣裙,T恤衫式连衣裙
teeth 拉链牙
teething weave 经条灯芯绒
teflon tie 防污领带
tehband 泰班德素色条格布
tehlila 特丽拉红粗呢
tehsila 平纹彩条布
tekka (阿尔及利亚)羊毛裤带
tela 蒂拉布,地中海地区亚麻布
telegraphic transfer 电汇
telephoto lens 摄远镜头
telescope bag 缩叠式旅行袋
telescope box 套盒,套叠箱
telescope case 望远镜包
telescope-crown(hat) 望远镜帽
telescope hat 望远镜型帽
telescope sleeve 套筒形双重袖

telescopic umbrella 折(叠)伞
telescoping of processes 工序连续化
telextra 结子花线;雪花波拉呢
teliwalah 德利瓦什仿开司米披巾
telon 毛麻呢(法国麻经毛纬粗呢)
temdel 轧光靛蓝薄布
temei brocade 特美缎
temperament 气质;性情,性格;禀赋
temperature regulated iron 调温熨斗
temperature regulating fabric 调温织物
temperature within clothing 服装内温度
temperer 鞣革工人
template 模板,样板
template method 模板方法
temple 鬓角;眼镜脚;太阳穴
templet 模板,样板
templet-controlled sewing 模板控制缝纫
templet guide tail 模板导轨
temporary dart 纸上褶;临时省
temporary set 暂时定型
temporary slash 临时开口,辅助开刀
temporary stitch 临时线迹
temptronic steam iron 电子恒温熨斗
ten 10号尺码的衣着用品
tenacity 韧性,韧度,强度
tenant 租房商店(租用大街或大楼而设立的商店)《销售》
tenbunkoede 黑地花卉文绣纹山袖
tencel 天丝(木浆纤维素纤维)
tencel/ramie blended fabric 天丝苎麻混纺织物
tender 嫩的,柔和的,鲜艳的;投标,投标人
tender colour 柔和色
tenderer 投标商《管理》
tender for 投标,投标人
tender goods 发脆织物
tender green 嫩绿
tender look 娇嫩款式
tender lustre 光泽柔和
tenderness 柔和
tender spots 脆化斑点
tender tone 柔和色调
tender yellow 嫩黄色
tendon 腱
tendrils 卷鬘
tenera 塔纳勒
tenerif(f)e lace 轮形花纹花边
tenerif(f)e work 轮形花纹花边
ten footer 10英尺的超长领巾

ten-gallon hat 牛仔帽;西部帽,宽边高顶帽,美国牛仔毡帽
ten-gallon hat and hat band 美国牛仔毡帽
ten-gallon sombrero 牛仔帽,宽边高顶帽
tenjiku 粗平布
ten kinds of sichun brocade 十样锦
tennies 网球鞋(美国)
tennis 网球条纹
tennis ball cloth 网球呢
tennis bracelet 网球手镯(镶有钻石等的金属手镯)
tennis cap 网球帽
tennis cloth 色织压花布,色织起绒条子布,网球衣料
tennis costume 网球装
tennis dress 网球装
tennis flannel 条子法兰绒,网球服法兰绒
tennis flannels 法兰绒运动裤
tennis look 网球服型,网球服款式
tennis outfit 网球衣装
tennis shirt 网球衫
tennis shoes 网球鞋
tennis shorts 网球裤
tennis stripe 网球条纹(有规律的宽条花样)
tennis sweater 网球毛衫
tennis togs 网球衣
tennis wear 网球衣,网球衣裤
tennis whites 白色网球运动服
tennis wrapper 网球外套
ten panel skirt 十片裙
tens 10号尺码;10号尺码的手套(或鞋、袜等)
tensile strength 拉伸强度
tension 张力,拉力
tension complete 夹线器
tension disc 夹线盘
tensionless seam 无张力线缝
tension line 拉紧线条
tension nipper 夹线钳
tension pucker 张紧皱裥
tension puckering 张紧皱裥
tension regulator 纱线张力调整器
tension releasing 松线
tension releasing disc 松线盘
tension seat 夹线簧调节座
tension sensitive fabric 张力敏感织物
tension sensitive textile 张力敏感纺织品

tension spring 夹线簧
tension thread guard 拦线板
tensometer 张力器
tensor 张力器
ten-stick 狭编带,窄编带
tent 帐篷式连衣裙
tentative specifications 暂定(技术)规格
tentative standard 暂行标准
tentative test 暂行试验法
tent coat 大下摆(女式)大衣
tent dress 大下摆宽松式连衣裙,蓬松连衣裙;伞装,帐篷装,(塔型)天幕装
tenter 拉幅机
tenting 帆布;帐篷布
tent-shaped style 帐篷型款式
tent silhouette 伞形(服装)轮廓
tent stitch 点状线迹,双面线迹;双缝,针向平行线迹
tent topper 帐篷式宽松轻便短大衣
tenuguiji 手工印花平纹棉布
tepis 印度丝棉平纹布
terai 宽边毡帽(亚热带用)
teremp 马海长毛绒织物;椅子绒
teres major 大圆肌
teres minor 小圆肌
terindans 印度德林顿细布
terlice 法国德利斯色条细斜纹布
term 专门名词,名称,术语
term bill 期票
terminal use 最终用途,产品用途
terminology standard 术语标准
ternaux 德洛克斯开司米肩巾
terrace skirt 叠层式裙子
terra cotta ［意大利］赤褐色
terry 厚绒布;毛圈织物
terry astrakhan 充羔皮(毛圈)织物
terry blouse 毛巾衫
terry cloth 毛巾布;毛圈布,毛圈织物
terry cloth beach outfit 毛巾布海滨服
terry cloth jacket 毛巾布夹克
terry cloth shirt 毛巾布汗衫,毛巾布衬衫
terry handkerchief 毛巾手帕
terry hosiery knitted by double cylinder knitter 双针筒毛圈袜
terry knitted fabric 毛圈针织物
terry-loop goods 毛圈织物
terry-loop hosiery 毛圈袜
terry pile 毛巾绒
terry socks 毛圈短袜,毛巾短袜
terry towel 毛巾织物

terry towelling stitch 毛巾织物线迹
terry velvet 毛圈绸;毛圈天鹅绒
tertiary colour (由两种次色合成的)第三色
tertiary colours 再间色,三间色,三次色
tertiary fibre mix-fabric 三种纤维混纺织物
terylene 涤纶,特丽纶,的确良,聚酯纤维
terylene/cotton(T/C) 涤棉
terylene fancy suitings 纯涤纶花呢
test 试验;测试
testaceous 赤褐色的
test error 试验误差
test-fit 裁料前的试样
test garment 试衣
testing items 试验项目
testing method 测试方法
testing of shrinkage 试缩水率
testing procedure 试验程序,试验步骤
testing specimen 试样
test market 试销市场《销售》
test method 测试方法
test muslin pattern 坯布试样
test norm 试验标准
test of new products 新产品测试
test OK 正常,无故障
test-press 试烫
test sale 试销《销售》
tests for color fastness 色牢度试验
Tetoron 帝特纶(日本聚酯纤维商品名)
tetrachloroethylene 四氯乙烯(干洗剂)
tetrads 四色相(等间隔)配色
tewei brocatelle 特纬缎
tewke 衬布,帆布
tewly 红丝线
tex 特,特克斯;号(数)
textile 纺织品,织物;纺织原料;纺织
Textile/Apparel Linkage Council 纺织服装标准促进会
textile art 纺织艺术
Textileather 皮革(式)织物(商名)
Textile Care Allied Trades Association 美国纺织品洗涤用品协会
Textile-care labelling code using symbols 纺织品使用标识图形符号规则
textile caucus (美国两党参议员和代表组成的)纺织工业决策委员会
textile chemicals 纺织化学品
Textile Chemist and Colorist ［美］《纺织化学家与染色家》(月刊)

textile chemistry 纺织化学
textile cloth 纺织布料
Textile Clothing Technology Corporation 美国纺织服装工艺协会
Textile Colour Card Association(TCCA) 纺织色卡协会
textile commodity 纺织原料；纺织品
textile composites 纺织复合材料
textile converters 纺织计量换算表
Textile Council of Australia 澳大利亚纺织协会
textile crayon （裁剪用）划粉
textile design 纺织品设计,织物设计,意匠图
textile designer 纺织品设计师,织物设计师
Textile Designers Guild 纺织品设计协会
textile disposal ecology 纺织品处理生态学
Textile Distributors Association 纺织品批发商协会
textile dressing 电光整理,织物整理
textile ecology 纺织生态学
textile end product 纺织最终产品；纺织复制品
textile engineer 纺织工程师
textile engineering 纺织工程
textile fabric 织物,纺织品
textile factory 纺织厂
Textile Fiber Products Identification Act [美]纺织纤维制品鉴定条例
textile fiber 纺织纤维
textile film 纺织薄膜
textile finishing [treatment] 织物整理,织物后处理
textile for general use 民用纺织品
textile for health care 保健纺织品
textile for medical use 医疗用纺织品
textile glass 玻璃纤维
textile heat conductor 导热织物
textile human ecology 纺织品消费生态学
textile industry 纺织工业
Textile Information Retrieval program 纺织情报检索程序
Textile Information System 纺织情报系统
Textile Information Users Council 纺织信息协会
Textile Institute [英]纺织学会
textile integrity 织物集成性,织物完整性
Textile Journal of Australia 《澳大利亚纺织杂志》(期刊)
textile-like hand 纺织品手感,织物状手感
textile machinery 纺织机械
Textile Machinery Society of Japan 日本纺织机械学会
Textile Manufacturer and Knitting world 《纺织制造家和针织世界》(月刊)
textile material 纺织材料,纺织品
textile mill 纺织厂
Textile Month [英]《纺织月刊》(期刊)
textile of bast and leaf fiber 麻纺织品
textile plant 纺织厂
textile printing 织物印花
textile process 纺织加工
textile processibility 纺织品加工性
textile processing 纺织加工
textile processing reguirements 纺织加工规格
textile product 纺织品
Textile Progress [英]《纺织月刊》(期刊)
Textile Quality Control Association 纺织品质量控制协会
textile raw material 纺织原料
Textile Research Institute [美]纺织研究所
Textile Research Journal [美]《纺织研究杂志》(期刊)
textile semi-product 纺织半制品
textile speciality 纺织特品
textile standard 纺织标准
textile structural composite 纺织结构复合材料
textile stylist 纺织专业人员
textile surface structure 织物,布（泛指各类织物,包括非织造布）
Textile Surveillance Body 纺织品进出口监督机构
Textile Technical Federation of Canada 加拿大纺织技术协会
textile technologist 纺织工艺专家,纺织技术专家
textile technology 纺织科技,纺织工艺
Textile Technology Digest [美]《纺织工艺文摘》(期刊)
textile weave 织物组织
Textile world [美]《纺织世界》(月刊)
textilist 纺织专家
textural design 织物设计,织纹设计

textural effect 质感效果图案;(织物的)质感效应,起绒结构效应
texture 织物,织品;(织物)结构,质地;(材料的)纹理;(皮肤的)肌理
textured fabric 花式织物
textured filament (yarn) 变形丝
textured finishing 织物质地整理
textured hose 花式连裤袜
textured hosiery 花式袜
textured polyester damask satin 涤花缎
textured polyester thread 涤纶变形线
textured polyester twill 涤弹绫
textured stockings 花式长筒袜
textured stretch yarn 变形弹力纱
textured worsted 花式组织精纺呢
textured yarn (美)变形纱线(膨体纱),变形丝
textured yarn fabric 变形丝织物
texture mapping 纹理映射《计算机》
texture on texture 组合不同的素材而创造新感觉的技巧
texturity 不缩标签(保证缩率低于2%)
texturizing finish 织物变形整理
T-gauge T形尺;T形定型
Thadirbezi 土耳其细布
Thai costume 泰国民俗服
Thai silk 泰国丝绸
than 印度粗印花棉布
Thana silk 印度塔纳绸
thardwetch 波斯花丝缎
thd (thread) 线
T-head pin T字形大头针
Thea Porter 西亚·波特(1927~,英国时装设计师)
the arts and crafts movement 工艺美术
theater of fashion 时装舞台
theater suit 观剧女套服
theatrical costume 戏装,演出服装
theatrical costume designer 戏剧服装设计师
theatrical gauze 戏服纱罗
theatrical make-up 舞台化妆
the beatles style 甲壳虫乐队风格
thebois 印度梯博亚平布
the fifties (50s') look 50年代款式
the forties (40s') look 40年代款式
the fifties (50s') stylistic form 50年代款式
the Golden Thimble 金顶针奖
The Hosiery and Allied Trades Research Association 针织业研究协会
The Information Users Council 纺织信息用户委员会
the latest design 最新设计
the latest fashion 时装;(服装的)最新式样
the latest model 最新样式
the latest style 最新款式
thelion 乳头点
the long robe 法官服,教士服
the lowest quotation 最低报价
theme 主题,题目
theme shop 专营商店,专卖店《销售》
the mode 时装;时式
thenar 手掌;足底;鱼际(指大拇指根部掌上突出的肌肉)
the north face 北方面孔
the nude 裸体;裸体人像
Theo tie 无舌系带鞋
Theoni Aldridge 奥尔德里奇(美国著名戏剧服装设计师)
theories and history 史论
theories of Fine Arts 美术理论
theory 理论,学说
theory of colours 色彩理论,色彩学
theory of shadow 阴影理论
therbig 基本分解动作,基本动作要素
thermal-adaptable fabric 适温织物
thermal cloth 保暖衣
thermal discolouration 热褪色,热变色
thermal endurance 耐热性
thermal fabric 保暖织物
thermal gloves 保暖手套
thermal history 保暖随时间的变化
thermal insulation cloth 保温布
thermal insulation value 保温率(值)
thermal insulation wadding 保暖絮片
thermal knit fabric 保暖针织物
thermally bonded fabric 热黏结织物
thermally bonded nonwovens 热黏合非织造布
thermal manikin 热量人体模型,暖体人体模型,暖体假人(用于服装功能性评价,测量保温性能)
thermal neutrality 热适中
thermal property 热学性能,保暖性
thermal protective performance (TPP) 保热性能,防热性能
thermal resistance 热阻;耐热性
thermal retention 保暖性

thermals 保暖内衣,保暖服
thermal severity number 热传导系数
thermal shrinkage 热收缩
thermal stability 热稳定性,耐热性,耐热度
thermal storage fiber 蓄热纤维
thermal undergarment （绒里）保暖内衣
thermal underpants （绒里）保暖长衬裤
thermal underwear （绒里）保暖内衣
thermal wear 防寒保暖服
thermal woven 保暖机织物
thermation 乙烯自动纳缝机
thermo bonded fabric 热黏合非织造布
thermo bonding machine 热熔黏合机
thermo bonding powder 热黏合粉
thermo fusing press 热熔烫压机
thermobonded fabrics 热黏合非织造布
thermoboded nonwovens 热黏合非织造布
thermobonding nonwovens 热黏合非织造布
thermo-cementing edge folder 自动上胶折边机
thermo-dot bonded nonwovens 热轧点黏合非织造布
thermodurable textile 耐热织物
thermoelasticity 热弹性
thermoform 热压成形
thermofusible fabric 热熔性织物
thermofusible interlinings 热熔衬
thermofusion point bonding 热熔点黏合
thermo insulation value 保温率（值）
thermo-man 暖体假人仪,热人仪,暖体假人仪阻燃试验
thermoplastic 热塑性的;热塑性塑料
thermoplasticity 热塑性
thermoplastic toe puff applying machine 热塑套头成形机（制鞋用）
thermo printer 热转移印花机
thermo printing paper 热转移印花纸
thermoregulated textile 调温纺织品
thermoset 热固树脂;热固塑料
thermosetting 热定型
thermosol bonded fabric 热熔（黏合）衬布
thermostability 耐热性,热稳定性
thermostable plastic 耐热塑料
thermostatical control 恒温控制
thermotaxis 体温调节
the savage 野蛮人款式
the scoop 勺子发型
thesis 主题,命题

the sixties(60s') look 60 年代款式
the sixties(60s') stylistic form 60 年代款式
the spiv 懒人发型
the thirties(30s') look [style] 30 年代款式
the twenties(20s') look [style] 20 年代款式
the thirties(30s') star coif 30 年代明星发型
thew 肌肉,腱
The Women's Dress Reform Movement 妇女服装改革运动
the yuppi 雅皮
thiband 牛毛呢
thibaude 牛毛呢
Thiber 西伯斗篷织物
thibet cloth 仿驼毛呢;缩绒厚呢
thibet shawl 丝毛花色披肩（法国制）
thick 裤裆垫布;厚,浓,深,密
thick and sharp shoes 满族旗鞋,高底鞋
thick-and-thin places 厚薄段疵
thick-and-thin crepe 厚薄绉
thick-and-thin pattern 宽窄条交替花纹
thick-and-thin striped fabric 稀密条织物
thick elbow 粗大肘部
thick end 粗经疵
thick fabric 厚料
thick filling 粗纬（疵）
thick multiple band 厚型带
thickneedle 粗针
thickness 厚度,密度
thickness of work 缝纫厚度
thick raised knitted fabric 厚绒布
thick seam 厚缝,密缝
thick set (cord) 粗厚灯花绒
thick-strap sandal 阔带凉鞋
thick stripes 粗条痕
thick thread 粗线
thick waist 粗腰（围）
thick work 厚料,厚料缝纫
Thierry Mugler 蒂埃里·米勒(1946～,法国时装设计师)
thigh 大腿;股;大腿围;（裤）横裆
thigh bone 股骨
thigh boots （长及大腿的）股高长统靴
thigh girth 股围,腿围
thigh-high 大腿上部
thigh-high boots 股高长靴,时装长筒靴
thigh highs 高筒长袜

thigh length （衣服、靴子等）长及大腿的；大腿长
thigh line 横裆线
thigh narrowing 大腿部位收针
thigh size 腿围
thimble （缝纫用）顶针；针箍；针抵子
thin 浅色，淡色；稀薄的，细的，瘦的
thin arm 细臂
thin band 薄型带
thin bar 稀纬档
thin cloth 薄布，薄布料
thin elbow 细肘（部）
thin end 细经疵
thin fabric 薄料
thin filling 细纬疵
thing high boots 及股长靴
things artistic 艺术品
thin knitted cloth 薄针织料
thin material 薄料
thinness （织物）手感单薄
thinning scissors 削发剪
thin place 稀弄，薄段（织疵）
thin raised knitted fabric 针织薄绒布
thin spot 稀弄，薄段（织疵）
thin stripes 细条痕
thin work 薄料
thin yarn woven tape 薄线带
third class 三级，丙等；三等品
thirds 三等品
thirteen 13号尺码的衣服（或鞋、袜等）
thirties look 20世纪30年代西装型，20世纪30年代款式
thirties silhouette 20世纪30年代的轮廓造型
thirty 30号尺码的衣服（或鞋、袜等）
thobe 阿拉伯外衣
tholia 索莱阿帽
Thomas Tago Fisher 汤姆斯·菲希（瑞士时装设计师）
thong 平底人字拖鞋，人字凉鞋；凉鞋饰带，皮条饰带
thong belt 皮条饰带
thong brief 训练便服，有人字背带的运动服
thong hole 拉链带头空隙
thong sandal 平底人字拖鞋，人字凉鞋
thong type 夹带式
thoothpick toe 牙签型鞋头
thorax 胸廓，胸
thorn （花边）饰线针绣

thorn proof tweed 防刺粗呢
thorn stitch 棘状针法
thoroughbred look 英国传统款式
thought 思维，思潮，思想
thread（thd） 线，带，绳，袢；穿线；串珠；股线；纱线
thread a needle 穿针
thread bag 收口小布袋
thread bare 露底，露白；穿旧
thread bare fabric 磨损织物
thread belt loop 线带环
thread bobbin 线团
thread break 断线
thread breakage 断线
thread broken 断线
thread carrier 线袢，线链袢
thread cast off （梭子未钩进面线线环的）脱环
thread catcher 钩线器
thread chain 线链，线辫
thread chain cutter 线链切割器
thread chain guard 按扣线攀
1-thread chainstitch 单线链式线迹
thread clippers 纱剪
thread clips 线头剪；纱剪
thread colour （缝）线颜色
thread consumption 耗线量
thread cord 缝纫线
thread count （织物）经纬密
thread cutter 割线刀，剪线刀，剪线器
thread deflector （缝纫机）导线板
thread drawing finger 松线器
threaded herringbone stitch 绕线人字绣
threaded running stitch 平针穿线针法
thread end 线头
threader 穿线器
thread eye 线环；线袢；线挂耳
thread fabric 线经纱纬平纹棉布
thread feeder （缝纫机）梭头
thread gaberdine 线华达呢
thread guard （缝纫机）护线器；导线架；导线器；线耳
thread handling area 线通道
thread harness muslin 剪花薄纱
thread hook （缝纫机）导线钩
threading 穿线，上线
threading the machine 缝纫机上线
thread khaki drill 线卡其
thread knot 线结
thread lace 线制花边，手织花边，亚麻纱

花边
thread layer sewing-knitted nonwoven 纱线层缝编法非织造布
thread layer stitch-bonded fabric 纱线层缝编布
thread layer stitch-bonded nonwoven 纱线层缝编法非织造布
thread length 缝线长度
thread loop 线环
thread-loose 脱线
thread lubricator 润线器
thread lubricator oil 润线油
thread mark 线钉
thread marking 打线钉
thread marking stitch(同 tailor tacking mark) 缝纫粗线标记,钉线迹;裁剪假缝
threadness 露线度
thread nipper 摄子夹
thread oiler 润线器
3-thread overlock stitch 三线包缝线迹
5-thread overlock stitch 五线包缝线迹
thread passage 缝线通道,线道
thread path 缝线通道,线道
thread pickup 挑线器
thread poplin 线府绸
thread regulator 调线器
thread release 松线
thread residue 线头,线尾
thread retainer 三眼钩线
thread route 线路
threads 衣服(美俚);伦敦街头穿着时髦的年轻人(英俚)
thread saver 回线器
thread serge 线哔叽
thread-shank 线扣脚,纽扣下的绕脚
thread shrinkage 皱线
thread shrinkager 皱线装置
thread size 线号
thread slack 缝线松弛
thread slippage 脱线
thread slipping 脱线
thread socks 线袜
threads on card or board 线圈
threads on cones 宝塔线
thread space 脱线间隙
thread stand 纱线架,线架
thread stem 纽子的线柄,纽子的线柱
thread storage rack 储线架
thread strength 缝线强度
2-thread stretch stitch 双线弹性缝线迹

thread tail 线尾,线头
thread take-off 钩线
thread take-up 挑线,挑线杆
thread take-up cam 缝纫机挑线凸轮
thread take-up lever (cam type) 凸轮式挑线杆
thread take-up lever (link type) 连杆式挑线杆
thread take-up lever stroke 挑线杆行程
thread take-up spring 挑线簧
thread tension 缝线张力
thread tension device （缝纫机上的）夹线器
thread thickness 缝线厚度
thread-thrum sucking machine 吸线头机
thread tightener 拉线器
thread tracing 记号缝,线钉
thread trimmer 剪线器
thread trimmer and cleaner 修线员
thread trimming 剪线
thread tube 线管
thread under trimmer 下剪线器
thread webs 狭滚边,狭镶条
thread wiper 拭线器,拨线器
thready 线(织)
thready cloth 线纹呢
thready drill 线斜纹,线卡其
thready poplin(e) 线府绸
thready serge 线哔叽
thready twilled satin 线直贡
three-armhole dress 三袖窿服装
three attributes （色彩)三属性
three-bar fabric 三梳栉经编织物
three-bar warp knitted pile fabric 三梳栉经编绒头织物
three-button coat 三粒纽扣大衣
three button grouping 三粒组纽扣
three-button waistcoat 三扣背心
three-colored jacquard hosiery 三色提花袜
three-color knop yarn 三色结节纱
three-colors 三色,三原色
three-cord sewing 三线缝纫
three-cord-thread 三股棉线
three-cornered gauge 三用定规
three-cornered hat 三角帽
three-dimensional apparel computer-aided design system 三维服装计算机辅助设计系统
three-dimensional body-scanning system

三维人体扫描系统
three-dimensional braiding fabric 立体编结物
three-dimensional fabric 三维结构织物
three-dimensional fashioning 立体服装成形
three-dimensional form 三向造型,立体造型
three-dimensional garment knitting 立体织物编织
three-dimensional gloves 立体手套;自由式手套
three-dimensionality 立体性,立体感
three-dimensional knitting 三维针织
three-dimensional measurement of human body 人体立体测量
three-dimensional nonwoven 三维非织造
three-dimensional pleat pressing machine 立体打褶机
three-dimensional product 三维纺织产品
three-dimensional shape 三维形态
three-dimensional space 三度空间
three-dimensional structure 立体结构
three-dimensional textile 三维纺织品
three-dimensions (3-D) 三度,三维
three-dimension space 三度空间
three-faced rose 三小面玫瑰(宝石)
three-fold cotton thread 三股棉线
three-folded yarn 三股线
three-fold seam 三折缝
three-fold yarn 三股线
Three gun 三枪
three-harness twill 三页斜纹
three-hole button 三孔纽扣
three-in-one 三合一妇女胸衣(胸罩、紧身带及紧身褡三件合一)
three-in-one shirt dress 三合一衬衫式裙衫
three-in-one wool blended fancy suitings 毛三合一花呢
three-in-one wool fancy suiting 三合一毛花呢
three-leaf filling twill 三页纬面斜纹
three-leaf twill 三页斜纹
three-leaf warp twill 三页经面斜纹
three-less 三无西服(无后背,无前襟,无两肋)
three M 三M(时装的大、中、小三种尺寸)
three miss blister fabric 三列凸纹针织物

three needle flat bed sewing machine 三针平缝机
three-needle cover stitch 三线覆盖线迹
three-needle flat bed sewing machine 三针平缝机
three-needle frame 三针缝纫机
three-needle lock stitch 三针锁式线迹
three peak 三顶点(折叠)饰巾法
three-pet shoes 三变鞋
three-phase alternating current 三相交流电
three-pick terry cloth 三纬毛巾布
three-piece 三件一套的衣服;三件一套的;三片式披肩
three-piece cape 三片裁披风
three-piece dress 三件套装
three-piece raglan coat 三片套袖大衣
three-piece raglan sleeves 三片连肩袖
three-piece set knitted 针织三件套
three-piece skirt set 连裙三件套装
three-piece slacks set 连裤三件套装
three-piece split sleeve 三片式仿连肩袖
three-piece suit 三件套装;三件套西服
three-pile velvet 三重绒头丝绒
three-ply 三股,三层,三股线
three-ply cloth 三层布
three-ply yarn 三股线
three-point diamond 三分钻石
three-point perspective 三点透视《计算机》
three-primaries 三原色
three-primary colors 三原色
three-properties of color 色彩三要素,色彩三属性
three-quarter 七分长,四分之三长,七分袖;大半身像
three-quarter cape 七分圆披风,三夸特披肩
three-quarter goods (幅宽27英寸)三夸特织物
three-quarter hose 中筒袜,中筒罗口童袜
three-quarter length 七分长,四分之三长(指外套长度);(人像等)大半身像(展示身体四分之三的像)
three-quarter length vest 七分长背心
three-quarter sleeve 中袖,七分袖,半长袖,三夸特袖
three-quarter sleeve length 七分袖长
three-seamed raglan sleeve 三片插肩袖
three seam raglan sleeve 三线连肩袖

three season clothing　三季节(春、秋、冬)用外套
three season coat　春秋冬大衣,三季大衣,三季外套
three season jumper　三季工作夹克
three separates sets　三件套的套装
three-shaft twill　三页斜纹
three-sided stitch　土耳其抽绣
three space　三维空间
three star tie pin　高级领带别针
three step zigzag　三段曲折线迹,三针跳
three-stitch V　三针V形线迹
three-stitch zigzag　三针曲折线迹
three-thread double row feed　三线双行送料
three-thread overlock　三线包缝机
three-thread overlock machine　三线包缝机;三线锁边机;三线拷边机
three-thread overlock seam　三线包缝线缝
three-tone　三色调的
three to one skip stitch　三针跳针线迹
three-way bag　三用提包
three-way collar　三用领,三式领
three-way mirror　三向镜
threshold price　入门价格
thrift-shop dress　旧货店服装;廉价旧货店(美国)
thrift-shop　廉价旧货店(美国),旧货店服装
thrift store　旧货店服装;廉价旧货店(美国)
throat　喉头;颈前;鞋口
throat belt(同 dog collar necklace)　宽带贴颈短项圈
throat latch　喉衬,附于小领上的小悬垂片,上领小舌片
throat plate　(缝纫机)针板
throat tab　喉衬,上领小舌片
through and through　双面毛织物
through and through stitching　反复缝缀
through dart　通省
through pocket　插袋,挖袋
through stitching　全缝
throw　曲折缝(锯齿缝)宽度;围巾,肩巾,薄披巾
throw-away coveralls　用即弃型工作服
throw-away fabric　用即弃型织物
throw-away nappy　用即弃尿布,一次性尿布
throw-away panties　用即弃型裤,一次性裤子

throw-away type disposable　用即弃型非织造布
throw hair[look]　投射式发型
thrown silk　加捻丝线
throwover dress　斗篷
thumb　拇指;(手套的)拇指部分
thumb blue　(洗衣时上蓝用)靛蓝
thumb draw thread　手套大拇指拆线
thumb joint　拇指关节
thumb knot　一把结,筒子结
thumb pin　图钉
thumb screw　(缝纫机的)压脚螺丝;指旋螺丝,翼形螺丝
thumbstall　拇指套
thumb tack　图钉
thunder and lightening　雷电呢,牛津灰色呢
thunder jacket　(领内带帽)防雨夹克,雷雨夹克
THV(Total Hand Value)　综合手感值
tiansun brocade　天孙锦
TIANWANG　天王
tiaoyong　印经绸
tiara　(天主教教皇的)三重冕;妇女冕状头饰;古代波斯头巾
Tibet cloth　山羊毛羽纱,缩绒厚呢
Tibet ox hair overcoating　牦牛绒大衣呢
Tibet shawl　西藏丝毛披巾
Tibetan ethnic costume　藏族服饰
tibia　胫骨
tick　褥子,床垫
tick effect　鸟眼织纹
ticketing　色泽标签
ticket number　标号
ticket pocket　暗口袋,内袋,票袋,零钱袋
ticket pocket flap　标袋盖,票袋盖
ticket pocket seam　标袋缝,票袋缝
ticket-sewing machine　钉商标机
ticket tacker　钉商标机
ticket tacking　钉商标
ticking　细密条纹棉布,枕套及褥罩织物
ticking stripes　坚固蓝色窄条纹布
ticking work　被褥刺绣
tickle　痒,痒感
ticks　细密条纹棉布,枕套及褥罩织物
tick-tack effect　鸟眼织纹
tidal foam　潮水泡沫色
tiddly suit　(个人所有的)最好的一套衣服(英国)

tidy 小块装饰罩垫织物
tidying 装饰缝
tie 领带[结],蝴蝶结领结;系,扎,束紧;系结;打(结),结带
tie and dye 扎染
tie bar 领带棒(领带夹的一种);领带杆;领带夹
tie-back 松紧条痕
tie belt 垂尾饰带;打结腰带
tie bonnet 束紧帽带
tie-bow blouse 扎结领衬衫
tie center lining machine 领带衬里机
tie chain 领带饰链
tie clasp 领带杆[夹]
tie clip 领带杆[夹]
tie closing 系门门襟
tie closure 系门门襟
tie collar 围巾;领带领,打结领;领圈
tie cutter 环形电剪刀
tied closing 系带门襟
tied herring bone stitch 连线打结人字绣
tied loan 约束性货款
tied-in end (布面上)补头拖纱
tied-in float 单面提花针织物
tied-in sale 搭配销售《销售》
tie-dye 扎染织物;扎染图案;扎染方法
tie-dyed clothing 扎染服装
tie-dyed cotton cloth 扎染棉布
tie-dyed denim 扎染牛仔布
tie-dyed skirt 扎染裙
tie dyeing 扎染图案;扎染织物;(手工)扎染
tie-dyeing imitation 仿扎染
tie-dye look 扎染款式
tie ends 抽带尾端
tie fabric 领带织物
tie fasten 系带
tie holder 领带夹[扣]
tieing 缝线接头,结头
tie label 系标签
Tielocken 叠合前襟无扣束腰男雨衣,泰勒肯型男雨衣
Tielocken front 泰洛肯型前襟设计
tie match shirt 领带配合衬衫(指领带与衬衫同料、同花、同色)
tie material 领带绸;领带料
tie neckline 打结型领口
Tientsin twill 天津斜纹布
tie on 系上
tie-on cape 披肩罩

tie pack string 扎包绳
tie pin 领带夹,领带扣针,领带别针
tie press 领带颈扣
tier 层褶裥;一行褶裥;围裙,围诞(美国)
tiered dress 段层衣服(段与段之间有荷叶边或剪接等)
tiered flounce 多层裙襞;多层荷叶边裙子
tiered flounce skirt 宝塔荷叶边裙
tiered gown 多层式裙装
tiered look 层次风貌,层裥款式,层叠款式
tiered silhouette 层褶裥式,多层型轮廓
tiered skirt 宝塔裙,分层裙,多层裙,(竹)节裙
tiered sleeve 多层袖,塔层(式)袖
tiered with ruffles 荷叶层褶裥
tie ribbon in bow 打蝴蝶结(丝带)
tie-roll collar 打结领
tiers 多层褶裥
ties 结带;系带浅口鞋
tie scarf 系围巾
tie shirt collar 领带衬衫领
tie shoes 牛津系带鞋
tie shoe laces 系鞋带
tie-side style 单侧缚带式
tie silk 领带绸
tie space 领豁口
tie stitch 结子针迹(短针迹,离线端几厘米处打结)
tie tac(k) (有座台的)领带饰针,领带扣针
tie top cap 顶结帽
tie trim 结带饰
tie tuck 领带扣针
tie turner 翻领带机
tie up 服装系结(用布或绳、带等打结,以加强效果)
tie up look 打结款式
tie wig 后系缎带的假发,扎式假发
tiffany 丝绢,丝纱罗;上浆亚麻薄布
Tiffany mounting (珠宝的)蒂法尼式镶嵌,爪式镶嵌;爪式镶嵌底座
Tiffany setting (珠宝的)蒂法尼式镶嵌,爪式镶嵌;爪式镶嵌底座
tiffer 英国黑色圆顶礼帽
tiger 粗绒大衣呢;虎皮花样
tiger cashmere 仿开司米织物
tiger cat fur 豹猫毛皮
tiger cloth 粗呢,粗绒布

tiger fur 虎皮
tigering 修毛整理
tigerlily 卷丹红(红桔色)
tiger picture 虎皮花样
tiger skin plush 虎皮长毛绒
tight 紧身衣；紧身裤；连裤袜
tightass (衣服)紧贴臀部的
tight armhole 袖窿紧
tight band 紧身衣裤带
tight boots 窄靴
tight bottom stitch 紧底线迹
tight collar edge 领外口紧
tight course 抽紧横列
tight crotch 夹裆(横裆紧)
tight-fit 紧身，贴身；紧身式
tight-fitting cotton padded undercoat 紧身棉衣
tight-fitting cotton wadded undercoat 紧身棉衣
tight-fitting lined undercoat 紧身夹衣
tight-fitting quilted jacket 紧身棉衣,紧身绗缝棉夹克
tight-fitting silk wadded undercoat 丝棉紧身上衣
tight-fitting suit 紧身装
tight-fit trousers 紧身裤,细腿裤
tightish 紧身的,有点紧的
tight-laced 束紧腰围的；穿着束紧腰围外衣的
tight lapel 驳头外口紧
tight lower armhole 抬腋缝起绺
tight measure 紧身尺寸
tight neckline 紧领口,领卡脖
tight-needle stitch 密针线迹
tightness 身骨；紧密度,硬挺度；松紧感
tightness factor (织物)紧密系数
tightness of stitches 线迹密度
tight pants 紧身裤,健美裤
tight quilted jacket 紧身棉袄
tight row 紧密线圈横列；紧密针迹
tights 紧身衣裤(统称)；(运动员或舞蹈演员等穿的)紧身衣；(妇女穿的)紧身下装,连袜裤
tight selvage (织物)紧边
tight shirt 紧身衫；紧身裙
tight skirt 窄裙,紧身裙
tight sleeve 紧身袖,紧袖
tight slip 紧身衬裙(或套裙)
tight stitch 线迹过紧,紧密线圈
tight top collar 领面紧

tight top flap 袋盖面紧,袋盖反翘
tight torso 紧身胸衣
tight weave 密织布；紧密组织
tigre 虎斑花样
tigre cashmere 仿羊绒织物
tigrine 条子斜纹织物
Tilden sweater 蒂尔登毛衫,网球毛衫
tile 大礼帽；高筒礼帽；丝质高礼帽
tile beard 瓦形须
tile blue 瓦蓝(浅灰蓝色)
tile red 瓦红(橙色)
tillet 亚麻衬布,包布
tilt 原毛织物,篷帐
tilting helmet (比武用)头盔
tilt pocket 斜口袋
timberland 泰博兰德皮靴
Timberland 坦博兰德
timber wolf 森林狼毛皮
time exposure 定时曝光；"T"门曝光
time-lapse photography (照片)时间间隔图
timeless clothes 不随潮流的服装,无时代感的服装
Time Measurement Unit (TMU) 时间测量单元
time overlap 时间重叠,时间交错
time-place-occasion (TPO) (穿衣三要素的)时间—场所—场合
time-saving device 省时装置
time scale 时序表
time-schedule control 程序控制
time series 时间序列
time total look 永远流行
time value of funds 资金的时间价值
TIMEX 天美时
timiak 爱斯基摩人保暖衬衫
timing 定时；计时《管理》
timing adjustment device 定时调节装置
Tina Leser 蒂娜·莱莎(1910~1986,美国时装设计师)
tinampipi 大麻平纹薄布
tinct 加色,着色,染色；色泽,色调
tinctorial property 着色性能
tinctorial value 色值
tincture 颜色,色彩
Tinea Pellionalla L. 衣蛾(蛀虫)
tineid 蛀虫
tinge 淡色彩,淡色调
tingey 优质细密府绸,廷杰布
tingle 刺痛感

tin-pants 防水帆布裤
tinsel 金属饰片;金银线;金银丝交织物
tinsel check taffeta 格塔绢
tinsel embroidery 亮光绣,金银丝刺绣
tinsel fabric 金银线织物
tinsell 金银丝布
tinselled brocade 库锦
tinselled mixed tapestry satin 金银线织锦缎
tinselled striped velvet 嵌金银丝条纹丝绒,绿柳绒
tinsel ribbon 金银丝带
tinselry 镶有闪光金属丝的料子
tinsel union tapestry 诗轴绸
tinsuti 印度三股棉线布
tint 色泽,色彩,色度;浅色,淡色;染发剂
tintage 上色
tint mark 色点,色斑
tintometer 色调计,色辉计,比色仪
tiny button 微型纽扣
tiny check 小格子
tiny pocket 小口袋
tiny skirt 超短裙
tiny snip 小剪口
tip 针尖;尖端;顶端;鞋头;(如铁鞋掌等的)顶端附加物;帽坯顶;轻碰;尖状鞋掌
tipiti 韧皮纤维弹性褶裥织物
tipped fabric 绒尖染色绒织物
tippet (法官、教士用)披肩;(通常用整张狐皮或貂皮等制成的)女式披肩;教士黑色圣带;(旧时服装的)衣袖(或披肩等)下垂部分
tippet on elbow 长及肘线的披肩
tipping 包头缝
tipping cover 包头缝盖
tippy 针尖的全部
tip shearing 轻剪绒头
tip stretcher 帽顶扩展机
tip toe 脚趾尖,脚尖
tip top 指甲贴饰
tiraz 织有苏丹姓名的阿拉伯绸布
tire (古称)服装;(妇女)珠宝头饰
tiretaine 法国麻毛粗呢
tire woman 女服装员;女化妆师
tiring room 化妆室
Tirolean flower print 蒂罗尔花卉印花
tirolian(hat) 基罗利安帽
TISSOT 天梭

tissu 纺织品(法国名称);纺织材料
tissue 机织物;金银线花缎;薄纱,薄纱罗,薄绢
tissue and sheets for patients 伤患者用衣
tissue checks 彩色格子纱
tissue faille 薄罗缎
tissue figured cloth 大提花细薄布
tissue gauze 绞纱绸
tissue gingham 细薄方格织物,细薄彩条织物
tissue paper 薄纸,绵纸;白复印纸
tissue-paper-padded hanger 用软纸包裹的衣架
tissue taffeta 透明薄塔夫绸
tissu metallique 金属丝纱罗
tissu plume 鹅绒呢
tissuti 印度细花细平布
titan braid (军服用)泰坦粗毛饰带
titfer (黑色的)圆顶礼帽(英国)
titian (red) 橙红色;赤褐色
title block 标题栏,名称栏
title panel 标题栏,名称栏
TITONI 梅花
Tito Rossi 蒂托·罗西(意大利时装设计师)
T-line T形造型,T字形
TMU (Time Measurement Unit) 时间测量单元
toast 吐司黄,烤面包褐黄色
toasted almond 米棕
toasty-warm texture 保暖纺织品
tobacco(brown) 烟草棕
tobe 阿拉伯长内衣,阿拉伯裙;北非和中非的围裹式外衣
tobine 经浮花纹织物
toblier (女子的)围裙;(旧时女服上的)围裙式装饰
toboggan cap 长袜形帽,滑雪帽,绒线便帽
toby collar 有褶宽领
toc (欧洲农民)阔边毡帽;阿尔及利亚白头巾
tochirimen 日本棉绉
tocuyo 棕色棉棉布,棉床单坯布
tocuyo asargado 斜纹棉坯布
toddlers 幼儿衣服尺寸;儿童短衫
toddlers' dress 幼儿装
toddlers' sizes 婴幼儿尺码
toddler suit 幼儿套装
toddler's wear 幼儿装

toe	袜头;脚趾,足尖;(鞋、袜的)足尖部
toe-ankle chain	与脚趾相连的脚链
toe block	裤头加固
toe box	(鞋头与鞋衬里之间的)内包头
toe cap	鞋尖皮,鞋头,外包头,鞋前趾
toe closing (of hosiery)	缝袜头
toe closing machine	缝袜头机
toe-end	(鞋子)头部
toe guard	加固袜头
toe hosiery	袜头
toe-in figure	足尖内向体型
toe laster	(制鞋)绷尖机
toeless shoe	前空鞋
toe linking	袜头缝合
toe medallion	孔饰鞋头
toe nail	脚趾甲
toe narrowing	袜头收针
toe-out figure	足尖外向体型,八外字体型
toe piece	鞋尖皮;鞋头,外包头,包头硬衬,鞋前趾
toe plate	鞋底前掌铁
toe post sandal	足趾支柱凉鞋
toe puff	(鞋子)套头
toe ring	脚趾戒指
toe rubbers	女式合趾套鞋
toe shoes	芭蕾舞鞋,硬式芭蕾舞鞋
toe slippers	芭蕾鞋,芭蕾舞鞋
toe socks	分趾袜;脚趾分开的短袜
toe-spring	(鞋子)翘头
toe-tip	(鞋子)包头
toffee brown	中褐色
tog	上衣,外套,斗篷,披风,游泳衣,托格(服装等热阻单位,评价服装保温性能的热阻单位,1tog＝0.1m² · K/W)
toga	(古代罗马市民穿的)宽外袍;(行业,官职等)专用袍褂,官服,托加(古罗马男子白色宽袍);长方形印花棉布披肩
toga candida	白袍服(古罗马官吏候补者穿)
toga dress	不对称单肩礼服
toga nightgown	托加式睡袍(不对称单肩或传统上装,一侧或两侧的侧缝开到臀部)
toga picta	(古罗马)金色刺绣的红紫袍服
toga praetexta	(古罗马)紫红滚边白长袍
toga pulla	(古罗马)黑及茶色的丧服
toga pura	(古罗马一般男子穿的未经漂白的)羊毛袍服
togataru	红边腰带,紫边腰带
toga trabea	(古罗马)紫色祭典礼服
toga virilis	(古罗马)成年服
toggery	衣服,特殊衣服;[英]服装店,服装用品店
toggle	肘节套接;套索扣,牛角扣,套环,套索扣门襟
toggle arm	肘节臂
toggle button	棒形纽扣,套环纽扣
toggle closing	套索扣门襟
toggle closure	套索扣门襟
toggle coat	套索扣外套
toggled	套索扣
toggled piece	索结片
toggle fasten	套环式系扣
togs	特殊服装;衣服(口语)
toile	[法]单色古典图案;地布上的花边图案;织物;帆布;薄亚麻织物;(用平纹细布等廉价织物做的)试穿服装,样服
toile à chapeau	法国帽布
toile anglaise	蓝色印花棉布
toile à sac	法国袋布
toile à tamis	法国条子硬衬布
toile à veste	里子布
toile à voile	轻帆布
toile bizonne	本色亚麻织物
toile bleue	蓝色细亚麻布
toile brune	本色亚麻织物,本色棉织物
toile colbert	刺绣用棉布,刺绣用呢
toile d'alsace	高级阿尔萨斯薄织物(棉或亚麻女服料)
toile d'araignée	南杜花边;透孔毛织物
toile d'argent	金银线织物,银线绸
toile d'em ballage	法国包装布
toile d'embourrure	午纹粗亚麻布
toile d'or	金银线织物,金线绸
toile d'o range	坚固平布
toile d'ortie	荨麻织物
toile d'ourville	法国本色帆布
toile de coffré	法国优质亚麻布
toile de cotton	薄棉麻条子印花布
toile de fries	优质荷兰亚麻布
toile de halle	强力本色亚麻布
toile de Jouy	[法]花卉风景印花布
toile de laine	轻质平纹毛料
toile de lille	里尔色条亚麻布
toile de mulquinerie	细棉布
toile de religieuse	修女服用布,修女面纱
toile de rouen	鲁昂织物

toile de sac 法国粗袋布
toile de saxe 萨克平纹织物
toile des indes 漂白或印花棉布
toile de veste 衬里织物
toile de vichy 薄亚麻条纹布
toile douce 法国亚麻帆布
toile du nord 法国轧光条格布
toile écru 法国本色亚麻布
toile marseilles 马赛提花亚麻布
toile muslin 细帆布
toile normande 诺曼底提花亚麻布
toile ouvrée 法国透孔织物,浮松布
toile satinée 法国软薄棉布
toilet 梳妆,化妆;女服,服饰;盛装;装束;梳妆披巾;梳妆台,梳妆用具;盥洗室,卫生间
toilet article 梳妆用品
toilet case 化妆用包
toilet clothes 单面凸纹织物
toilet glass 梳妆镜
toileting 单面凸纹织物
toilet mirror 梳妆镜
toilet powder 扑粉;爽身粉
toilet quilt 单面凸纹织物
toiletries 化妆品;梳妆
toiletry 化妆品
toilet set 梳妆用具
toilet soap 香皂
toilet table 化妆台,梳妆台
toilette (正式)礼服;盛装;梳妆,化妆;正式礼服;时髦装束;绿色包衣布
toilet water 花露水
toilinet[te] 背心呢
toiliste [法]制样师
token number 号带
tol 小花纹窄幅布
tole (bag) 万用袋
tolerance 公差;容限,容许数,宽松度
tolerance time 耐受时间
toll 过境税
tollanette 18～19世纪背心呢;特色织横条布
toluylene red 甲苯红
Tom Brigance 汤姆·布林格斯(美国服装设计师)
Tom Fogerty 汤姆·福格蒂(美国时装设计师)
Tom Jones Shirt 汤姆·琼斯衬衫(宽松大身和肩势,宽松袖)
Tom Jones Sleeve 汤姆·琼斯袖(轻垂而宽松的袖子)
tomato puree 番茄酱红
tomato (red) 蕃茄红(橙红色)
tom-bons 阿富汗长棉裤
Tommy Hilfiger 汤米
tom-tom 锤打洗涤机
tom-tom machine 锤式缩绒机
tom-tom washer 锤式洗涤机
tonal 同色调的
tonal contrast 色调对比
tonal effect 同色调效果
tonality 色调,调子
ton anglais [法]英国式霜降色
Tonder lace 岑讷花边,梭结花边,汤达抽绣细薄布
tone 色调,色光;调子;旋律;风气;风度
tone color 色调
tone contrast 色调对比
tone harmony 色调调和,调色
tone illustration 色调图《计算机》
tone in tone 同色调配色
toneless color 沉闷色
tone on tone 同系配色(指同一色相不同色调的配色);同色但深浅搭配不同的服装
tone-on-tone check 同色深浅格纹
tone-on-tone dyeing 同色深浅效应的染色
tone-on-tone effect 同色深浅效果
tone-on-tone print 同色深浅印花
tone-on-tone stripe 同色深浅条纹
tone-shade effects 色彩渐变效应
tone up 使色调更明亮
tonga 染色围巾布
tonghe taffeta 同和纺
tongue 牛津鞋舌;(饰针、皮带搭扣等的)别针
tongued shoes 有鞋舌的鞋子
tongue hemmer 舌形卷边器
tongue lining 鞋舌里
tongue notch V形剪口
toning 调匀颜色
toning pattern 变花色花样,色调深浅花样
tonish 流行的,时髦的,时式的
Tonkinois [法]东京帽
tonlet 金属片喇叭裙
tonquin 白绸布
tonsorial artist 理发大师
tonsure (教士)秃顶帽;秃顶
tontise 彩色毛织物

tool 工具;器具
tool bag 工具袋
tool box 工具箱
tooled-leather bag 工具皮包
tooled-leather belt 雕花皮带
tool for designing and drafting 设计制版工具
tool of cutting and sewing 裁剪设备与工具
tool of smearing paste 刮浆刀
tool-worked eyelet 冲钉式小孔
to-ori nishiki 仿中国花缎
toosh 粗纺呢
toothless plastic slide fastener 无齿塑胶拉链扣
toothpick toe 牙签型鞋头
top (服装)上身,上装;袜口;高筒靴口;鞋面;正面电镀的金属纽扣;袒胸衣服;(羊毛)毛条;化纤条
top and bottom (织物)上下端;(服装)上下装调和
top and bottom binder 上下滚边压脚
top and bottom coverstitch 上下绷缝线迹
top and bottom feed 上下送料,上下推布
top and bottom thread trimmer 面线和底线修剪器
top arc 针编弧
top arm 上臂(围)
topaz 黄玉色,透明的浅黄褐色;黄水晶
top-bare 露肩连衣裙
top belt feed 皮带送料
top boots 长筒马靴,上部装饰靴,下翻式高筒靴;高筒靴
top buttonhole 衬衫最上的纽孔;领座上的纽眼
top cap 鞋尖
top centre (plait) 明门襟
top closing machine 缝袜头机
top clothing 上衣;上装;春秋外套
top coat 女工宽大衣,轻薄大衣
top coating 轻薄大衣料
top collar 上领;领面
top collar appears tight 领面紧
top collar fall 领高
top collar stand 领里口
top collar to appears tight 领面紧
top collar too loose 领面松
top collar too tight 领面紧
top cord (缝纫机的)上芯线
top covering 绷线

top covering machine 绷缝机
top covering ornamenting machine 绷缝装饰机
top covering vari-stitch machine 绷缝可变线迹机
top cover thread 顶部绷线
top cover width 上束腰围
top cuff 硬袖
top designer 前卫设计师
top dyeing 毛条染色
top edge 上缘;上边
top edge of hip pocket 后袋上缘
top edge of pocket 口袋上缘
top edge of pocket with stay and facing 贴边袋上缘,滚边
topee 遮阳帽,软木帽,通草帽,木髓盔帽,遮阳盔,兜帽
top fabric 轻薄织物
top fashion 最新流行款式,最新时装款式;最新时装
top feed 压脚送布,上送料,上推布
top flap 袋盖面
top flap appears tight 袋盖反翘
top fly 门襟,止口
top fuse 面熔衬
top fusing collar 硬领,熔衬领
top grain 头层革,面皮
top grip feed 夹具送料,上夹送料
top gripping feed 上夹送料
top hat 黑色大礼帽;(高顶)丝礼帽
top hat box 高顶大礼帽帽盒
top heavy figure 上重下轻体型
top hip 上臀
top hip line 腹围线
topi 遮阳帽,软木帽,通草帽,木髓盔帽,兜帽
topical color 表面色
topical finish 后整理
topical method 时新方法
topical printing 人物图形印花
topknot 头饰;顶髻
top lapel tight 驳头反翘
top lapel (to) appear tight 驳头[口]反翘
topless (服装)袒胸的,低胸的;穿袒胸衣的;不穿上装;袒胸式服装,低胸游泳衣,露肩服,上空装
topless bathing suit 低胸泳装,袒胸泳装
topless look 袒胸露臂风貌;低胸露臂服装款式
topless style 无上衣款式

topless suit 上空装;袒胸装
topless swim suit 低胸式 V 形背带泳装
topless umbrella 折(叠)伞
topless waitress style 低胸露臂的女侍服款式
top lid gasket 止盖衬垫
top lift (鞋后跟的)底层
top line (鞋子)沿口;上口线
top line of cuff 袖头上口线
top line of waistband 腰头上口线
top line of yoke 过肩上口线
top line stay 沿口皮
top making fleece 精纺用毛
top metering device 夹线计量装置
top mode 领先款式,最新式样,最流行式样
top note 头香
top of hat 帽顶
top of shoes 鞋面
top of trousers 裤头,裤腰
top operating speed 最大缝纫速度,最高操作速度
topper 大礼帽;妇女宽松短大衣;单上衣;套口机;缝袜头机,[裤子]蒸烫机
topper coat 宽松短女大衣
topper-one (立体熨烫用)人身机
topper suit 松身套装
top picker 上线钩
top piece (鞋子)掌面
topping 额发;(苏格兰帽子上的)羽饰;套口,套色,套染;套色府绸
topping printing 套色印花,罩印印花
1/2 (1/4, 1/8) top pocket 1/2 (1/4, 1/8) 斜袋
top puffed sleeve 高泡袖
top-quality garment 高档服装
top roller feed 上滚轮送料
tops 上衣;靴袜装饰边
topseile 蓝条粗棉布
topside 裤腰侧,腰头侧
topside sleeve 大袖片,大袖,外侧袖
top sleeve 大袖片,大袖,外侧袖
top sleeve lining[linen] 大袖里
top-stitch 明线迹,正面线迹,面缝线迹,饰缝线迹
top-stitch collar band bottom 正面缝领座底
top-stitch cuff 缉袖口
top-stitched collar 接领
top-stitched edge 面缝边缘

top-stitched finish 缉明线;缉止口
top-stitched hem 面缝边缘;缉明线下摆
top-stitched lapped seam 压缉缝,明搭接缝
top-stitched open seam 分坐缉缝
top-stitched patch pocket 压线贴袋
top-stitched seam 明缝,正面线迹压线缝合;缉止口
top-stitched waistband 明线腰头
top-stitch hem 正面缝底边
top-stitching 花式缝迹,明线缝,正面缝,缉明线
top-stitching front facing 面缝前贴边
top-stitching lapped seam 压缉缝
top-stitching under collar 缉领里,面缝领里
top-stitch mark 明线号
top-stitch pinking 锯齿形面缝
top-stitch seam 面缝
top-stitch under collar 缉领里
top stop (拉链的)头锁
top tension (缝纫)面线张力
top texture 织物正面
top thread 面线
top threading thread 上拨线
top-to-toe 从头到脚
top-weight 薄织物
top-weight fabric 薄型织物
toque 豆蔻帽,(女用)小圆帽,无边女帽,(16 世纪)羽饰丝绒帽,杜克帽,双层编结御寒帽,蓬巴杜垫;细布
toque machine 织帽机
torada 印度托拉达细布
torc (古代高卢人戴的)金属饰环;金属项圈,金属颈环
torchon 镶边花边,饰带花边
torchonette 稀松粗绉网眼棉布
torchon lace 粗梭结花边,饰带花边,镶花边,抹布花边
torchon lace machine 饰带花边机
Torcon 聚对苯硫醚纤维;PPS 纤维
toreador costume 斗牛士装
toreador hat 斗牛士帽
toreador jacket 斗牛士外套,斗牛士外套式女外套
toreador pants 斗牛士裤,紧身半长女运动裤
toreador suit(同 sequins) 斗牛士套装,斗牛士装
torn list 破边疵

torn selvedge 破边疵
torn size 裁片尺寸
torque 金丝项圈,项链;(古代)金属饰环
torsade （帽饰用）螺旋形流苏（或缎带）
torsette 无吊带紧身褡
torsion lace 扭扭[曲]花边
torsional elasticity 抗扭弹性
torsional strength 抗扭强度
torso （人体）躯干;无头和四肢的裸体模型;紧身上衣
torso blouse 合身长女衫(长至臀部)
torso dart 躯干褶,双向褶
torso dress 低腰紧身连裙装
tortoise shell 杂黄褐色;玳瑁壳;仿玳瑁
tortoise shell glasses 玳瑁边眼镜
tortoise shell pattern 龟甲花型
torsolette(同 bustier) 无吊带紧身褡
torso line 低腰纤细型
torso shape 低腰型
torso shirt 女紧身连衫裤
torso silhouette 低腰型[服装]
torso skirt 低腰裙
toshak 托沙克被褥布
total apparel system 综合成衣系统
total blindness 全色盲
total crotch length 全裆长
total crotch measurement 全裆长,全裆尺寸
total environment（art） 全环境艺术,全景艺术
total hand value （织物）综合风格值
total look 整体感,整体风貌,整体搭配风格
total posterior arm length 臂长
total quality control 全面质量管理
total weakness 全色弱
tote bag （女用）有带手袋,大手提袋,购物袋
tote hat 男用雨帽
tote 箱式提包
totemism 图腾
Totes 杜特斯靴（一种雨靴,商名）
tou 西藏高级粗呢
touanse 坚固缎
touch 触感,触觉,手感
touch and away skirt 合身裙
touch and close fastener 尼龙搭扣
touch up 润色（图画等）
touch-ups 轻烫
toughness 韧性,韧度

tough-ups 轻烫
toun look 都市风貌
toupee 遮秃假发,男用假发
tourangette 轻缩哔叽
touring cap 旅行帽
tourist coating 旅行大衣呢
tourist shoes 旅游鞋
tourmaline fiber 负离子纤维
tourmaline mink 高级淡色貂皮（大衣）
tourmaline 高级淡色貂皮
tournay 装饰用陶奈印花呢
tourneau 陀螺裙,锥形裙
tournoise 环型中空织物（供衣领、衣边等用）
tournure （旧时用以鼓起女裙后部的）衬垫,撑架;用衬垫的衣裙
tovaglias 意大利图瓦格利亚斯白亚麻布
Tovta stretch stitch 托伏泰伸缩线迹
tow 丝束;纤维束
tow cloth 手织亚麻布
tow clothing 双排扣外套
tow coat 纽袢外套
towel 手巾,毛巾
towel blanket 毛巾被,毛巾毯
towel cloth 毛巾布
towel fabric 毛巾织物
toweling 毛巾布
towel linen 巾类亚麻布
towel sheet 毛巾被
towel suiting 毛巾布料
tow linen 短亚麻织物
towel(l)ing 毛巾布,毛巾料
towel(l)ing blouse 毛巾衫
towel(l)ing embroidery 毛巾刺绣,绣花毛巾布
towel(l)ing garments 毛巾服装
towel(l)ing gown 毛巾长晨衣,毛巾浴衣
towel(l)ing pajamas 毛巾布睡衣裤
towel(l)ing skirt 毛巾裙
towel(l)ing socks 毛巾袜
towel(l)ing tea cloth 纯棉毛巾
tower 塔帽
towlia 印度手动织机毛巾布
town bag 上街手提包
town hat 上街帽
town look 都市风貌,上街款式
town shoes 外出鞋,街市鞋,上街鞋
town suit 外出套装,上街套装
town turban 穆斯林帽,头巾式无檐帽
town wear 上街装,外出服;（适合办公或

城市生活穿的)深色服装(总称)
tow yarn 短亚麻纱
toy 精纺毛织物,精纺毛与丝的交织物;蓝黑格子呢;(旧时苏格兰底层老年妇女戴的)垂肩头饰,苏格兰老妪帽,小饰物
toyama georgette 高级丝绉
toy hat 托约帽
toyo 帽身材料
toys clothes 玩偶衣服
T pin T字形大头针
TPO (time-place-occasion) 时间-场所-场合(穿衣三要素)
TPP (thermal protective performance) 保热性能,防热性能
trace drawing 描(好的)图
tracer (制图用)描绘工具;点线器,点线板
tracery 花边上的花纹凸边
tracing board 点线板
tracing braid 花边钩边用粗线;军服绦带
tracing cloth 透明布
tracing-off 描绘,复描
tracing paper 描图纸,复写纸(美国)
tracing thread 花边钩边用粗线
tracing wheel 插盘,擂盘;点线轮[器],描样手轮,描迹轮
track 钉鞋,跑鞋;田径短裤
tracking the launch 产品追踪
track shoes (田径)钉鞋,跑鞋
track shorts 田径运动短裤
4-tracks single jersey machine 四针道单面针织机
track stripe 轨道条纹
tracksuit 田径服,轻快套装,运动套装
tracksuit bottoms 田径运动裤
tracksuit top with zip 拉链田径运动衫
tractability 易处理,易控制
tractor feed 履带送料
trad 传统装
trade 贸易
trade block pattern 商用基准服装纸样
trade duty 贸易税
trade fair 展销会,展览会,博览会
trade group 贸易集团
trade law 贸易法
trademark 唛头,商标,牌号
trademark name 商标名称
trademark plate 商标牌
trade paper 行业报纸《销售》

trade price 同业售价;批发价《销售》
trade-quota 贸易配额
trade seam 明包缝,折伏缝
trade secret 行业秘密《销售》
trade term 行业用语,行话
trade wall 贸易壁垒
trading profit 营业利润,毛利《销售》
trading up 升格销售《销售》
traditional Chinese clothing 中式服装
traditional color 传统色
traditional costume 传统服装
traditional look 传统风貌;传统款式;美国东海岸流行款式
traditional pattern 传统花样;古典花样
traditional style 传统风格,传统型,美国传统男装风格
traditional stylistic form 美国东海岸传统款式
traditional tailoring 传统裁缝法
traditional tailoring technique 传统的服装制作工艺
traditional technology [workmanship] 传统工艺
traditional wall paper print 传统壁纸花样印花
traditional wool pattern 呢绒传统花样
traffic 运输
trail 贴胸线缝;裙裾,拖裾,花样设计拖曳部分;尾部
trailing sleeve 垂饰袖
trail shorts 露营短裤
train 衣裙,裙裾,拖裙
train bearer (仪式中)拖衣裙者
train blue 全毛针绣乔赛
train case (旅行时用以放梳妆用品等的)小旅行箱
trainer 训练鞋,(没有钉的)软运动鞋;跑鞋;训练衫,训练服
training belt 训练腰带
training pants 训练裤,运动短裤
training shirt 长袖运动衫,训练衫
training shoes 训练鞋,运动鞋
training wear 运动服,训练服
trait 胸省;特性,特色,性格;面部轮廓
traje charro 墨西哥骑马装
traje de corto 及腰骑马夹克
trame underlay 双抽丝线绣;双花饰线绣
tramline top-stitched cuff 缉明线的袖口
tramp (皮靴等的)鞋底铁片
tramped dornoch 苏格兰亚麻织物

trank 制手套用长方形皮革
transaction 成交
trans-casual 混合便装
transfer 转移;烫画(转移图案);变针
transfer flock printing 转移植绒印花
transfer knitted fabric 纱罗组织针织物,移圈针织物
transfer machine 移圈针织机
transfer of risks 风险转移
transfer paper 转移印花用纸
transfer press 转移印花压烫机
transfer printing 转移印花(烫画)
transfer printing machine 转移印花机
transfer stitch 转移线圈
transfer style socks 翻口短袜
transfigure 美化
transformation 女人假发
transformation dress 万用礼服
transhipment 转船装运《贸易》
transit dues 通过税
transition color 过渡色,中间色
transit trade 过境贸易
translucence 半透明,半透明性,半透明度
translucent thread 半透明缝线
transmissivity 可透性
transnational corporation 跨国公司
transnational fashion 超民族款式,超越民族流行款式
transparence 透明性,透明度,明晰度
transparency 透明性,透明度,明晰度
transparent button 透明扣
transparent clothing 透明服装
transparent color 透明色
transparent drawing paper 透明打样纸
transparent dress 透明装
transparent finishing 透明整理
transparent look 透明款式;透视风
transparent nylon socks 透明锦纶袜,透明尼龙袜
transparent plastic raincape 透明塑料雨斗篷
transparent printings 透明印花
transparent ruler 透明尺,透明方格尺
transparent sequins 装饰小圆片
transparent silk umbrella 油绸伞
transparent skin 透明皮
transparent thread 透明线
transparent velvet 乔其绒,透明丝绒,明立绒

transparent yellow 透明黄
transportation management 运输管理
transport bag 运输袋
transport conveyor 输送带
trans-season fashion 超季节款式,超越季节性服装款式
trans-sex 男女通用
trans-sexual fashion 男女通用款式,超越性别款式
transshipment 从第三国进口
trans-times fashion 超越时间的款式
transverse cutting machine 横向裁剪机
transverse seam 横缝
transverse shuttle 横摆梭
transverse tricot 横条针织物
trapeze dress 梯形裙装
trapeze line 梯形轮廓线,梯型
trapeze neckline 梯形领口
trapeze silhouette 梯形轮廓
trapezium nectline 梯形领口,不等边形领口
trapezius 斜方肌
trapping seasons 捕捉季节
trappings 服饰;礼服;标志服饰
trapunto 装饰垫衬绗缝;提花垫纬凸纹布;针绣花边;浮雕式缝饰
trapunto quilting 浮雕式缝饰
trashy look 粗杂感
travel accessory 旅行(饰物)配件
travel bag 旅行袋
travel clothing 旅行外套
travel coat 旅游外套,旅行装
traveling bag 旅行袋,旅行手提包
traveling cap 旅行帽
traveling carriage (绗缝机)花样靠模板
traveling case 手提箱,旅行箱
traveling cloak 旅行斗篷
traveling jacket 旅行上衣
traveling shoes 旅游鞋
traveling stitch (绗缝机)跟踪线
traveling wig 旅行假发
traveling wear 旅游装
travel rug 旅行毯
travers 横条纹,横条花
traverse net 六角网眼
traverse warp fabric 米兰尼斯经编织物
traverse warp frame 米兰尼斯经编机
Travis Banton 特拉维斯·班顿(1894～1958,美国时装设计师)
T/R blended yarn T/R 纱;涤纶人棉混

纺纱
T/R cloth　T/R布(涤纶人造棉混纺布)
tread　鞋底
treadle　(缝纫机)踏板
treadle operated machine　脚踏传动缝纫机
treadle plate　缝纫机踏板
treadle sewing machine　脚踏传动缝纫机
treadle stand　缝纫机架
treated diamond　处理钻石
treated fabric　已整理的织物,经处理的织物
treaty　合同
treble cloth　三层织物
treble feather stitch　三重羽状饰边绣
tree　(柱式)木架;鞋楦
tree bark　树皮绉
tree bark satin　树皮绉缎
tree bark stripe　树皮条纹
tree calf　树纹小牛皮(革)
tree structure　树结构《计算机》
trefoil　三叶饰
trellis　本色大麻帆布,花边的网眼地
trellis back stitch　格子花式针法
trellised pleat pressing machine　格花打褶机
trellis pattern　格子花样
trellis work　挖花刺绣
trellis work type design　格花刺绣花样
trembling cap　长尾圆锥帽,摇摆帽
tremella lace　银耳花边
tremont hat　尖冠窄檐的毡帽,特蒙帽
trench buckle　堑壕搭扣
trench clothing　堑壕外套;军用胶布夹雨衣
trench coat　风衣;(夹里可脱卸的)军服式雨衣,战壕雨衣;有腰带的双排纽男式雨衣(常有肩袢,袖袢及大口袋等);战壕外套;(有腰带的)双排纽外套(大衣)
trench dress　肩饰衬衫式连裙装(束腰带)
trencher cap　学士帽,学位帽,方顶帽
trencher hat　角形饰边女丝帽
trend　(流行)趋势;倾向;时尚;时新款式
trend color　流行色
trend set　(服装式样上)标新立异
trend setter　(服装式样)创新人,导向者
trendy　赶时髦的人,爱时髦的人
trendy clothes　时髦衣服
trentaine　法国特朗坦原色呢

tress　发辫;女长发;一绺(头发)
tressé　花式编带
tresse　一束长发;一绺(头发);发辫
tresses　(妇女或女孩)头发;披肩长发
trews　格子呢紧身短裤,格子呢女紧身裤
triacetate　三醋酯纤维
triacetate fiber　三醋酯纤维
triacetate rayon　三醋酯人造丝
triacetate staple fiber　三醋酯短纤维
triad pattern　(服装)复式纸样
triads　三色相(等间隔)配色
trial fitting　试穿
trial layout　试排料,一顺排
trial muslin　薄纱试衣布
trial running　试运转,试车
triangle　男三角裤;三角板,三角尺
triangle cape　三角形披风
triangle collar　三角领
triangle mould　(制鞋)三角模
triangle of product strategy　产品战略三角形
triangle ornament　三角形装饰
triangle poncho　三角斗篷,三角形穿头披巾
triangle scarf　三角围巾
triangle tacker　三角形加固缝缝纫机
triangular　三角形胸袋巾装饰法
triangular blanket stitch　三角形毛毯锁边绣
triangular bra[ssiere]　(游泳用)三角胸罩
triangular face shape　三角面型
triangular ruler　三角尺
triangular scale　三棱尺,比例尺
triangular-shaped bar　三角形套结
triangular tack stitch　三角打结线迹
triangulate　三角形化,三角剖分《计算机》
trianizing　锦纶针织物热定形
triarylmethane dyes　三芳甲烷染料
triaxial fabric　三轴向[平面]织物
triblatti　绸布面料
triblend fabric　三合一织物
tribute satin　花累缎
tribute silk　贡缎,贡绸
triceps　三头肌
trichlorofluoro methane　一氟三氯甲烷(干洗剂)
trichlorotrifluoro ethane　三氯三氟乙烷(干洗剂)
trichophyton mentagrophytes　须发癣菌
trichroism　三原色(现象)

trichromatic printing 三原色印花
trick 小饰件(美国)
trick wig 活动假发
tricky clothes 奇巧服装,(出奇的)巧妙服装(如披肩内侧有袖,上衣有三个以上袖子等)
tric look 古怪款式
tricolette 轻薄细针距针织物,针织网眼毛织物,粗毛针织物
Tricoline 特里科林双股棉线丝光府绸
tri-color 三原色,三基色;三色绶带;三色徽章
tricorn(e) (翻边的)三角帽
tricorn hat (翻边的)三角帽
tricot 经编针织物,斜纹毛织物,特里科经编织物,紧身芭蕾舞服
tricot cross 横条凸纹效应
tricot cut 一卷(或一匹)经编织物
tricot de berlin 平针棉织物
tricot de laine 绵羊毛针织物
tricot écossais 苏格兰钩编织物
tricot fabric 脱里考织物
tricot flannel 经编法兰绒
tricot goods 斜纹毛织物;经编针织物
tricot heat setting machine 经编织物热定形机
tricot hosiery 经编长袜
tricotine 条子细棉府绸;巧克丁;急斜纹精纺细毛哔叽;丝针织外衣织物;针织细绸;斜纹织物;精纺呢绒
tricot jersey 经编乔赛;经编平针织物
tricot knit corduroy 经编灯芯绒
tricot knitting machine 经编机,特里科经编机
tricot lace 经编花边
tricot long 纵向凸纹效应
tricot machine 经编机
tricot stockings 经编长袜
Tricot warp knitting machine 特里科经编机
tricot weft insertion machine 衬纬经编机
tricot wool cloth 多利科特呢
tri-dimensional 三维,三度
tri-dimentional effects finish 三维立体效应整理
trifocals 三光眼镜
trifolding umbrella 三折伞
trigonous hair style 三角发型
trijama 睡衣套装
trilam 压膜平纹涤纶布

trilby [hat] 软毡帽
trim (华丽)服装;修剪,修整,修光(衣帽等)装饰;饰边,镶滚;袖口滚边
trim cutted pieces 撇片
trim edge 镶边;修边
trimetric projection 三轴侧投影《计算机》
trim figure 苗条的身材
trim front edge 修剪止口
trim-master 修边att
trimmed edge 修光的边
trimming 绳带,饰带,装饰品;装饰物;镶嵌线;修剪;装饰,修饰,镶滚;剪屑
trimmer 修边机,修边器;剪刀,切刀;整修者
trimmers 裁缝剪刀
trimming braid 装饰带
trimming cut piece 修片
trimming cutter 修剪器
trimming edge 镶边
trimming front edge 修止口
trimming house 服饰附件店
trimming lace 缘饰花边
trimming machine 锁缝机,布边修剪机,滚边机
trimming margin 修边幅度
trimming material 镶边材料
trimming silk 花边绸
trimming tape 装饰带
trimming thread 剪线
trim paste 镶边
trim sleeve with leather 给衣袖装饰皮革
trinket 小饰品;廉价首饰
trinkhall 特林克哈尔金银丝花缎,金银丝绣花缎
trio 三件套
trio filling stitch 三叉式线迹
trios 三件套装
trios-quarts 七分长(指外套及袖子)
trios-quarters-fournis 半漂亚麻布
tripe 麻背毛绒织物;粗纺毛织物;棉绒,纬绒
triple check 三线格子纹
triple cloth 三层织物
tripled-feed 三重送料
triple-feed machine 三重送料机
triple-framed bag 三层包
triple interlock stitch machine 三线绷缝机
triple layered style 三重叠式

triple lock stitch 三重锁缝线迹
triple mirror 三面镜
triple needle 三针
triple-needle toe 针尖式鞋头
triple seam stretch stitch 三行缝弹性线迹,三重伸缩线缝
triple sheer crepe 半透明薄绉
triple sheer 半透明薄绸
triple shift 三班
triple stitch 三行线迹,三重线迹
triple stitching 三线缝纫;三线 线缝
triple straight stitch 三重直形线迹,三重直线固缝线迹
triple stripe 三线条纹
triple tucked blouse 三排裙裥上衣
triple voile 半透明薄绸
triplex 三层包
triple zigzag stitch 三重锯齿固缝线迹,三重曲折线迹
tripoline 丝经棉纬斜纹布
tripp 毛绒头织物
tri-suit "铁人三项"紧身运动服装
trita 精纺平纹呢
tritinum 股经绸
tritone 三色调
triumph toga (古罗马)凯旋袍
Triumph 戴安芬
tro (trochanterion) 股长上点,转子点
trochanter (臀部)大转子
trochanterion 股长上点,转子点
trolley (lace) 粗线条花边
trollopee 无裙撑宽松女晨袍
trolly lace 粗线轮廓梭结花边
trompe-l'œil [法]迷惑装
trompelneil [法]视幻觉花纹
trooper cap 骑警帽
tropal 衬里[用]织物,里子布
tropical 薄型外衣织物,夏令织物
tropical cloth 夏令织物,薄型织物
tropical color 热带色彩
tropical cotton 热带薄型棉布
tropical design 热带风情图案
tropical fancy suiting 薄花呢
tropical green 热带绿
tropical hat 凉帽
tropical island print 热带岛屿印花
tropical kit 热带服装
tropical look 热带款式,热带风貌
tropical suiting 薄西服料,薄型精纺呢,热季套装料,大众呢
tropical viyella 维耶勒棉毛薄呢
tropical wear 轻薄夏装,热带服装
tropical weight 薄型织物重量
tropical whipcord 薄型防雨马裤呢
tropical worsted 薄型精纺呢
tropics proof 耐热带气候的
trouse (爱尔兰人穿的)紧身格子呢裤
trouser 裤立裆,裤直裆
trouser backpart 裤子后片(幅)
trouser belt 裤带
trouser belt loop 裤(带)袢,腰袢
trouser bottoms 裤腿下部
trouser braid 裤子装饰带
trouser breech seam 后裤裆缝,裤子臀部缝
trouser buckle 活带扣,裤子搭扣
trouser button 裤扣
trouser clip (骑自行车时用的)裤脚夹
trouser cotton padded 棉裤
trouser crease line 裤烫迹线,裤子的挺缝线
trouser crotch seam 裤裆缝
trouser cuffs 裤脚口翻边
trouser curtain (腰头衬)雨水布
trousered 穿长裤的
trouser forepart 裤子前片(幅)
trouser front 裤子前片(幅)
trouser hanger 裤架
trouser hem 裤脚口
trouser hook 裤扣
trouser hook and eye 裤钩;裤头钩扣
trousering 裤料,西裤料
trousering in wool serge 毛哗叽裤料
trouser jeans 西裤式牛仔裤
trouser lace 金银编带(以前用来装饰英国陆军军官裤边)
trouser legger and general utility 裤管万能压平机
trouser leg press 裤管熨烫机
trouser legs 裤脚管
trouser leg seam 裤腿缝
trouser leg steam ironing machine 裤腿蒸烫机
trouser leg turned up (cuffed) 裤腿反折
trouser length 裤长
trouser outseam 裤中缝,裤外缝
trouser pannel serger 裤片包边机
trouser piece 裤片
trouser pleats 裤子打褶;裤前侧褶裥;腰褶裥

trouser pocket 裤袋
trouser pocket stitching 裤袋线缝
trouser press 熨烫成(或保作)裤管折线
trouser rise 裤直裆,裤立裆
trousers 裤子;长裤,西装裤;宽大长裤
trouser seam 裤子缝
trouser section 裤幅
trousers for adult 成人裤
trousers for kids 童裤
trouser side seam 裤侧缝
trouser-skirt 裙裤
trouser stretcher 伸裤器
trouser suit 裤套装
trousers waistband 裤带
trousers waist circumference 裤腰围
trousers without belt 无腰带西裤
trousers without crotch worn by men 套裤
trouser tape sewing machine 裤带缝纫机
trouser topper press 裤头熨烫机,裤腰压平机
trouser waist 裤腰
trouser waistband 裤腰带
trouser zip 裤子拉链
trouser zipper 裤拉链
trousse (装钢笔、眼镜套等用的)小袋
trousseau 嫁衣,嫁妆
truck stripe 铁路条纹,转向条纹
trucker's apron 小贩围裙(前面有大口袋)
true bias 正斜;正斜滚条
true blue 纯蓝色
true color 纯色
true lace 网眼花边
true lining 全夹
truelove knot 同心结
true navy 海军蓝
true percale 精梳密织床单布
trueran 棉涤纶,涤棉布,棉的确良
trueran bleached lawn 漂白涤棉细布
trueran drill 涤棉卡其;涤卡
trueran fabric 涤棉织物;涤棉布
trueran interlining 涤棉衬
trueran sheeting 涤棉平布
trueran shirt 的确良衬衫;涤棉衬衫
trueran thread 涤棉线
trueran trousers 涤棉裤
trueran yarn 涤棉纱
trueran yarn-dyed lawn 色织涤棉细纺
trueran yarn-dyed piqué 色织涤棉灯芯布
trueran yarn-dyed(printed)cambric 色织

(印花)涤棉细布
truerean yarn-dyed weft filament poplin 涤棉纬长丝府绸
true safety stitch machine 安全缝线迹缝纫机,双重安全缝纫机
true safety stitch 安全线迹
true selvage 光边
true stitch 双面刺绣线迹
true wool fiber 绒毛
trufflette 窄幅漂白亚麻布
truffle yarn 嵌芯花线;粗松螺旋花线
truing 描实
trumpet dress 紧身喇叭连衣裙
trumpet line 喇叭线型
trumpet pattern 喇叭花样
trumpet silhouette 喇叭型轮廓
trumpet skirt 牵牛花裙,喇叭裙
trumpet sleeve 喇叭袖
trunk 旅行衣箱,皮箱;男运动短裤;躯干
trunk breeches 南瓜裤
trunk hose (16～17世纪)桶形连袜裤
trunking cord 芯线
trunk-length hosiery 中筒女袜
trunks 男式短裤,大裤衩,抽带式宽松男短裤;游泳裤;运动短裤,拳击运动短裤
trunk show 展销
trunk sleeve 大炮状袖
trustworthy product 信得过产品《管理》
try 试,尝试
trying 假缝
trying on 试穿,试戴
try on 试穿(衣、鞋),试样
try-out run 试运转,试车
try square 曲尺,检验角尺
T-s (Tee-s)T恤,T字形
tsarouchia 绒球装饰的尖头鞋
tsarvuil 保加利亚凉鞋
TSC(total safety control) 全面安全管理《管理》
tschepken 库尔德夹克
T shape (tee shape) T字式服装
T-shape table T形台
T-shaped lapel (西装)T形翻领
T-shirt T恤衫,文化衫,圆领马球衫,短袖圆领运动衫
T-shirt bag T恤包
T-shirt dress(同 polo shirt dress, tee shirt dress) 加长T恤衫(盖肩袖,系腰饰带,丝网印花),T恤衫式连衣裙,T恤连裙装

T-shirt line　T恤线形
T-shirt pajamas　两件套T恤睡衣
tsin-tseon　薄软绸
tsitsith　蓝色流苏肩巾
T-square　T字尺,丁字尺
T-strap　(鞋子的)丁字形搭扣带;有T字形搭扣带的女鞋
T-strap pump　T形带浅口鞋,T形带碰拍鞋
T-strap sandal　T形带凉鞋
T-strap shoes　T字形搭扣带鞋,T字鞋
tsumugi　日本绵绸
tsuni　高绒头法兰绒
Tsutomu Sakagami　坂上勉(日本服装设计师)
TT look　紧身款式
T-topper　圆领衫
tuanse　缎子
Tu ethnic costume　土族服饰,土族民族服,土家族民族服
Tu nationality's costume　土族服饰
twanse　缎子
tub　矮胖子
tubbable silk(同 tub silk)　耐洗丝绸
tube border　管状滚边
tube bra　无带套头式弹力胸罩
tube dress　直筒装
tubed twine　晒衣绳
tube foot　管状压脚
tube line　筒形线形
tube skirt　筒裙,筒型裙
tube socks　无跟袜,筒形短袜
tube stockings　圆筒长袜
tube top　(妇女用)筒形弹力胸围;(弹性)套筒胸衣
tube trimming　镶边,滚边
tub fabric　耐洗织物
tub-fast　耐洗耐烫
tubing　管状织物
tubing turner　翻转布带器
tub silk(同 tubbable silk)　耐洗丝绸
tubular　管形套纽;无跟圆筒袜;圆筒形辅料《计算机》
tubular braiding machine　管状编带机
tubular cording　管状串带
tubular fabric　圆筒形织物
tubular hose　直筒袜
tubular hosiery　圆筒形针织物,圆筒针织品
tubular knit　筒形针织物

tubular knitted fabric　圆形针织物
tubular knitting machine　圆形针织机
tubular line　筒形轮廓
tubular-needle knitting machine　管针针织机
tubular plain stitch　圆筒形平针法;圆筒形平针织物
tubular ribbon　圆筒形饰带
tubular silhouette　筒形轮廓
tubular stocking　无缝圆袜
tubular stocking machine　圆袜机
tubular strip　针织圆带,圆筒形编织带
tubular style　筒形款式
tubular toggle　筒式索结绳扣
tubular welt　双吃线罗纹关边
tubular woven fabric　圆形织物
tuck　打褶,打裥搭克;缝褶,活褶,横裥织物;塔克;收针;折短,卷起;开花省
tuck and welt cloth　集圈-浮线织物
tuck-back selvedge　折人边
tuck cloth　绉纹织物
tucked　打裥
tucked hem　塔克滚边
tucked-in selvedge　折人边
tucked neckline　褶裥领口
tucked pleat　胖裥
tucked seam　打裥缝,褶裥缝,坐缉缝,活页接缝
tucked skirt　叠褶裙,褶裙
tucked sleeve　抽褶鼓肩袖
tucked yoke　打褶育克
tucker　(缝纫机)打裥装置;打裥者;打塔克员;(可装卸的)披肩;活动衣领
tuckeries　印度特喀利棉织物
tuck fabric　集圈针织物
tuck-float fabric　集圈-浮浅织物
tuck finger　打裥钩
tuck gauge　打裥定距规
tuck gloves　集圈手套
tuck in　塞进,塞入
tuck-in [blouse]　内上衣,下摆塞在裙子内的衬衫,塞腰衫
tucking　褶裥;胖花;集圈;缩绒工艺;横裥,打裥
tucking attachment　褶裥装置;集圈装置
tucking defect　花针疵;集圈线弧
tucking device　打裥装置
tucking gloves　集圈手套
tuck jersey　集圈花纹平针织物
tuck knee　袜膝段

tuck knitted fabric 集圈针织物
tuck lace 褶裥花边
tuck line 褶线,塔克线
tuck loop (针织物)花针疵;集圈线弧
tuck machine 压褶机
tuck pleats 缝褶裥,褶缝裥
tuck press 压褶机
tuck-rib fabric 集圈罗纹针织物
tuck ripple 集圈波纹组织
tucks 褶裥织物,褶绉织物
tuck seaming machine 褶缝缝纫机
tuck side 打裥边
tucks stain 折皱污迹
tuck stitch 半织,集圈组织
tuck up skirt 提起裙子
tuck up sleeve 卷起袖子
tuck weave 褶裥组织,起皱组织
tuck width 打裥幅度,褶宽
TUDOR 帝舵
tueedo 宴会准礼服
tuft (古代)金色帽缨子
tuftaffeta 绒条塔夫绸,重磅真丝塔夫绸
tufted bed-cover 簇绒床罩
tufted bed-spread 簇绒床罩
tufted blanket 簇绒毯
tufted fabric 栽绒织物,簇绒织物
tufting 栽绒,簇绒;簇绒法
tufting machine 栽绒机,簇绒机
tufting needle 簇绒针
tuill (一块)护股甲;(铠甲的)腰裙下摆
Tujia ethnic costume 土家族服饰
Tujiazu costume 土家族民俗服
tuke 衬布,帆布
tulip collar 郁金香式领子(形似郁金香,遮住下巴)
tulip design 郁金香图案
tulip dress 郁金香式裙装,狭下摆裹襟式裙装
tulip line 郁金香形轮廓;郁金香型
tulip skirt 郁金香式裙
tulip sleeve 郁金香式袖
tulipwood 鹅掌揪木色
tulle 面罩薄纱,薄纱;六角网眼经编织物
tulle baran 英国印花绒布
tulle embroidery 六角网眼刺绣
tulle grenadine 黑白网眼薄纱
tulle handkerchief 绢网手帕,六角网眼手帕
tulle lace 六角网眼花边
tulle [net] 网眼纱[布],[经编]六角网眼织物
tulle work 六角网眼刺绣
tumble-dry 滚筒式烘干
tumble drying 圆筒烘干
tumbler (衣服)干燥机;干衣机
tumbler drier (衣服等的)滚筒式烘干机
tumbler washer 转笼式洗衣机
tumbling machine 干衣机
tummy 肚子,胃
tune 和谐,协调;主旋律;粗绸布
tung shanssu silk 本山素绸
tunic 古罗马短袖束腰外衣,及膝外套;宽松罩衫;束腰外衣;(警察、士兵的)紧身短上衣,短祭袍,法衣
tunica 塔奈卡(古罗马男式内衣)
tunic and trousers 束腰外衣及长裤
tunic blouse 束腰女衬衫,束腰女外衣
tunic clothing 紧身短外套,束腰带外套
tunic coat 紧身短上衣
tunic dress (古希腊、古罗马)长袍,束腰裙装
tunic jumper 罩衫式背心裙(长至大腿)
tunicle 法衣;(天主教)祭服
tunic line 束腰外衣型;蒂尤尼克型
tunic pajamas 两件套罩衫式睡衣
tunic skirt 束腰裙,塔尼克裙,(军警用)紧身短裙
tunic suit 束腰套装,束腰外衣套装
tunic sweater 毛线长外套
tunic swimsuit 蒂尤尼克式两件套泳装,塔奈克式两件套泳装
tunic trousers 贴身型裤
tunic with V-neckline V领束腰外衣
tuning up hem 翻边缝
tunique grecque 希腊风格束腰短宽衣
Tunisian man's 突尼斯男夹克
Tunis crochet 突尼斯棱纹钩编花边
tunnel collar 隧道领
tunnel elastic folder 隧道式弹性折边器
tunnel elastic seam 隧道式弹性线缝
tunnel loops 隧道式腰带环(穿皮带用)
tunnel shank 柱状纽柄
tunnel waistline(同 drawstring waistline) 抽带腰节线
tunneling elastic 穿橡筋带
tunzeb 伊朗手织腾泽细薄布
tupoz 西纳梅麻布
tuppotlya 斯里兰卡图普特丽亚腰布
tuque 杜克帽,双层编结御寒帽,加拿大尖顶帽

turban 头巾式女帽；(妇女或儿童用)无檐帽，狭边帽；穆斯林头巾；女用头巾，(19世纪流行的)头巾装束的妇女发型
turban hat 头巾帽
turban scarf 头巾式围巾
turban shape 头巾，包头布；妇女发网；(妇孺)头巾式无檐帽
turc 丝经棉纬缎
turin 塔赖因布
Turk satin 土耳其缎
turk's head 头巾形饰结
Turkey blue 土耳其蓝(蓝色)
Turkey gauze 提花纱罗织物
Turkey red 土耳其红(鲜红)
Turkey reds 平纹土耳其红棉布
Turkey work 粗毛纱刺绣；东方风格挂毯；东方风格家具布
Turkin 粗纺呢
Turkish cap 土耳其帽
Turkish costume 土耳其民族服饰
Turkish embroidery 土耳其刺绣
Turkish lace 土耳其花边
Turkish pants 土耳其裤，膝下束紧宽松裤
Turkish point lace 奥亚赫花边；土耳其针绣花边
Turkish polonaise 爱尔兰式长裙
Turkish shawl 土耳其披巾
Turkish slippers 土耳其平跟软拖鞋
Turkish stitch 土耳其式线迹，土耳其式卷边线迹
Turkish towel 土耳其式毛巾，土耳其式浴巾
Turkish trousers 土耳其裤
Turkish tunic 土耳其式束腰外衣
Turkoman 土库曼装饰织物
turmeric 郁金姜黄(黄色)；黄色植物染料
turn 翻；翻新；翻折
turn-about 正反两面可穿的衣服
turn-about machine 往复式哗编机
turn back 翻转
turn back checks 素色宽边小格棉布
turn-back collar 翻领
turn-back cuff 翻袖口，翻折式袖口，卷袖口
turn-back hem 翻边，贴边
turn-back line 翻折线
turn back pocket 反盖式口袋
turnban 头巾，包头布；无边圆帽

turn belt loop 翻带祥，翻裤带袢
Turnbuli's blue 滕[部尔]氏蓝(深蓝色)
turn collar 翻领子
turn dam 土鲁丹细布，达卡细布
turn down 翻下
turn-down collar(同 rolled collar) 卷领，翻领
turn-down range 调节范围
turned cuff 翻袖
turned-back cuff 翻边袖口
turned down self-facing 折翻的贴边
turned in 收小
turned-in opening 连贴边的开口
turned out 放大
turned overedge 卷边；翻边(布疵)
turned overedge seam 翻转包缝
turned shoe construction 翻边鞋结构
turned twill 山形斜纹，人字斜纹
turned-under circular hem 折翻的弧形下摆，折翻的曲线下摆
turned-up hem (向上)翻折的下摆
turned up hemmer 朝上的卷边压脚
turned welt 双层平针袜口
turn flap 翻袋盖
turn hood 翻风帽
turn in(collar) band edge and stitch it 包底领
turning 翻袜口；翻转；拐角
turning a heel 旋转袜跟
turning cuff 翻袖口
turning fabric 翻转缝料
turning flap 翻袋盖
turning fly facing 翻门襟
turning fly shield 翻里襟
turning left fly 翻门襟
turning lining 翻里子
turning-off 缝口，套口，连圈
turning over 翻转
turning over collar 翻领
turning over cuff 翻袖头
turning over flap 翻袋盖
turning (over) sleeve 翻袖子
turning over waist band 翻腰带
turning right fly 翻里襟
turning sleeve 翻袖子
turning tab 翻小祥
turning up hem 翻边缝
turning waistband 翻腰头
turn in top collar edge and stitch it 包领面

turn left fly 翻门襟
turn lining 翻里子
turn out 翻出;装束,穿着;装备,设备;产量,产额
turn-out collar 大翻领
turnover 领子翻折
turn-over collar 翻领,翻折领
turn-over figure 两面相同的图案
turnover top 翻转袜口
turn pocket flap 翻袋盖
turn right fly 翻里襟
turn-sew-turn device 缝合-翻袜装置
turn shoes 翻口鞋
turn sleeve 翻袖子
turn tab 翻小祥
turn-under 折翻边
turn-under strip carrier 带状腰带袢,串带袢,蚂蝗袢
turn-up （衣服等的）翻折部分,翻边,裤脚反折部分;卷起;折起
turn-up bottom 卷边裤脚
turn-up collar 袖袂
turn-up cuff 翻袖口,卷袖口;双层袖头;裤卷脚
turn-up hem 反折边,反吊边
turn-ups 裤脚翻边;翻袖;卷袖
turn-up style 翻脚风格
turn-up trousers 翻边裤
turn waistband 翻腰头（或腰带）
turn woolens 翻新毛料衣服
turn wristband inside 袖口往里折
turquoise 宝蓝色,蓝绿色,绿松石色;绿松石;蟹青;细棱纹色里子布,线经纱纬轻质毛织物,色格精纺哔叽
turquoise blue 翠蓝,绿松石色,青绿色
turquoise green 绿松石绿,碧绿
turquòise green clear 浅碧绿
turquoise green deep 暗碧绿
turquoise green pale 淡碧绿
turtle collar 高翻领
turtle dove 斑鸠色
turtle green 龟绿色
turtle-neck 高翻领,圆翻领,双翻领;圆翻领毛衣,高翻领毛衣,高领绒衣;龟领,高而紧的衣领
turtle-neck collar 高翻领
turtle-neck convertible collar 翻领两用领（前中线装拉链,可拉成高领或呈Ｖ领）
turtle neckline 玳琅领口;龟颈形领口
turtle-neck pullover 高领套衫

turtle-neck sweater 高翻领毛衫,圆翻领毛线衫
turumagi 朝鲜男式围裹大衣
tuscany lace 方形网地满花手工花边
tuskin 素色粗纺呢
tusk pendant 牙形装饰
tussah 柞蚕丝（织物）
tussah bourette 大条丝绸
tussah/brass union industrial fabric 带电作业服服
tussah cloth 柞蚕丝绸,柞丝绸
tussah drapery fabric 柞丝纬装饰绸
tussah fabric 柞丝织物
tussah-filling kimono silk 柞绢和服绸
tussah herringbone twill 人字呢,条影呢
tussah linen-like cloth 亚麻布式柞丝绸
tussa home spun 大条丝绸
tussah/polyester union pongee 柞丝的确良绸
tussah pongee 柞丝绸,柞丝纺,本色柞府绸
tussah popline 柞丝粗棱茧绸
tussah shantung 柞丝山东绸
tussah silk 柞蚕丝;柞丝绸
tussah silk fabric 山东柞丝绸,茧绸
tussah silk sliver 柞蚕棉条
tussah twill faconne 柞花绫
tussah twill raye 柞丝彩条绸
tussah velvet 柞蚕丝绒
tussah waste 柞蚕挽手
tussah wool 仿柞丝呢
tussar silk 丝光棉布,柞蚕丝
tussore 柞蚕丝,丝光棉布
tussores 罗缎;柞蚕丝;野蚕丝
tutu 肥大蓬松裙,芭蕾舞短裙
tutulus [拉]圆锥帽
tutwork 计件工作《管理》
tux[edo] （男用）夜小礼服,无尾晚礼服,塔士多礼服;塔士多夜小礼服女外套,无腰带女外套,无纽扣外套;（男子）餐服（美国）;绢丝领
tuxedo blouse 塔士多女衫
tuxedo clothing 无腰带女外套
tuxedo coat 无腰带女外套,塔士多女外套,无纽扣外套
tuxedo collar 塔士多翻领,无尾礼服翻领
tuxedo girl style 塔士多女套裙款式
tuxedo jacket 便礼服上衣,塔士多夹克（大多是单排一粒纽扣）

tuxedo look　夜小礼服款式,塔士多款式
tuxedo pants　塔士多长裤(男式黑色长裤的两侧镶有缎子狭条,女式长裤在腰部打裥)
tuxedo pump　鞋面上有缎带结的低跟浅口皮鞋
tuxedo shirt　塔士多礼服衬衫,塔士多女衫,夜小礼服女衫
tuxedo shoes　塔士多鞋,礼服鞋
tuxedo suit　塔士多礼服,无尾晚礼服
tuxedo sweater　外衣式开襟无纽扣毛衣
tuxedo worsted　礼服呢
tuxedo worsted cloth　礼服呢
Tuzu costume　土族民俗服
TV fold　(装饰手绢)TV折叠法
twalle　人丝素绸
twanse　缎子
tweed　(粗)花呢;花呢衣服
tweed cap　粗花呢帽
tweed flannel　粗花呢;苏格兰呢
tweed hat　粗花呢帽
tweediness　粗花呢的品质
tweed jacket　粗花呢上装
tweed jersey　粗花呢运动衫;仿毛平针织物
tweed pullover　杂色粗呢套衫
tweeds　(粗)花呢服装
tweed skirt　粗毛花呢裙
tweed suit　女粗花呢套装,女便装
tweedy look　粗呢款式,都市运动服装款式,都市运动服装风貌
tweel　斜纹布;斜纹图案
tweezers　镊子;捏钳,镊子钳
twelve　12号尺码的衣服(或鞋、袜等)
twelve-color circle[wheel]　十二色相轮
twenties look　20世纪20年代服饰风貌
twenties silhouette　20世纪20年代轮廓造型
twenty　20号尺码的衣服(或鞋、袜等)
twenty-four-carat gold fabric　纯金饰花薄纱
twice stitch　双重线迹,双针线迹
Twig(g)y look　崔姬风貌,崔姬款式;特威盖风貌,特威盖款式
Twiggy stylistic form　特威盖款式
twilight mauve　朦胧紫红色
twilight purple　朦胧紫色
twill　绫,斜纹;斜纹布,斜纹织物;斜纹图案
twill-back　背面点花纹针织物

twill-backed　斜纹衬里
twill backing　(正面为另一组织的)斜纹背织物
twill-back velveteen　斜纹地纬绒织物
twill camlet　斜纹羽纱
twill cap　斜纹布帽
twill checkerboards　人字形斜纹格子布
twill checks　彩格单面斜纹织物
twill cloth　斜纹织物
twill coating　啥味呢;斜纹呢
twill crepe　斜纹绉
twill damask　斜纹锦缎,斜纹花缎
twill diaper　正反斜纹条子花纹
twill douppioni　斜纹双宫绸
twill(ed)　斜纹布条
twilled flannel　斜纹法兰绒
twilled fustian　斜纹绒布
twilled mat　斜纹方平织物
twilled satin　直贡缎;直贡呢,直贡
twilled shirting　细斜纹布
twilled swansdown　凸纹布,绒布
twilled tape　人字形斜纹带
twillet　服装,服饰
twillettes　厚斜纹工作服布
twill fabric　斜纹织物
twill-faced bedford cord　经面斜纹凸条布
twill-faced filling satteen　纬面棉缎
twill-filling figure　纬面斜纹花纹
twill fustian　斜纹绒布
twill habutai　斜纹绸,斜纹电力纺
twill jacket　卡其布夹克,斜纹布夹克
twill muslin　薄毛哔叽
twill of uniform line　双面斜纹
twill pants　斜纹布裤
twill pongee　斜纹绸
twills　斜纹织物;绫类织物;哔叽
twill satin　直贡缎
twill shawl　斜纹披巾(小亚细亚人包头用)
twill silk　斜纹绸
twill swansdown　凸纹布,绒布
twill tape　人字带,斜纹牵条
twill ticking　蓝白条子斜纹被套布
twill weave　斜纹
twill weave fabric　斜纹织物;斜纹布
twin belt　并排双皮带;成对皮带;双腰带
twine　麻线,绳子;股线
twine cloth　线呢,轧光线呢,仿亚麻衬衫棉布
twine cotton　棉股线

twine fringe 缨穗带子,流苏带子
twin fabric 组合使用织物
twin-head machine 双头缝纫机
twining off machine 缝袜头机,套口机
twin-knitter 双系统针织机
twin-lens reflex camera 双镜头反光照相机
twinneedle stitch 双针线迹
twinning (毛皮制作)分刀
twin-needle angular stitch machine 双针角形线迹缝纫机
twin-needle chain stitch 双针链式线迹
twin-needle covering stitch 双针绷缝线迹
twin-needle machine 双针缝组机
twin-needle needle feed lockstitch sewing machine 双针针送锁缝缝纫机
twin-needles 双针
twin-needle sewing 双针缝纫
twin-needle stitch 双针线迹
twin-plain hemmer 双平卷边器
twin-prong jeans button 双针工字纽
twin-riveting machine 双粒铆钉机(制革)
twin-seam 双线缝,双行缝
twin set 两件套款式,两件式服装;女式两件套毛衣(指一件连衣裙和一件开襟衫)
twin sweater 半开襟套衫
twin sweaters 成对式毛衣
twin sweater set 毛衫套装
twist 捻度;缝纫丝线
twisted back stitch 回针绕线绣
twisted bar stitch 棒状绕线绣
twisted chain stitch 扭形链缝,绕线链状绣
twisted insertion stitch 棒状绕线绣
twisted knot stitch 捻结线迹,绳结线迹
twisted pants leg 裤腿扭曲
twisted running stitch 绕线平针绣
twisted scarf 螺旋丝巾
twisted stitch fagoting 棒状绕线绣
twisted stripe 捻式条纹
twisted towel 螺旋毛巾
twisted turban 扭形头巾
twisted union yarn 棉毛合股纱;异种纱合股线
twist fabric 绞绕线圈针织物
twistless towel 无捻毛巾
twist look 扭腰舞款式
twist loop formation 线环形状扭转
twist multiplier 捻系数

twist on twist 同向加捻,同向复捻
twists 扭曲;紧捻纱线织物;绞绕针织物,双色双股毛织物
twist sleeve lining 袖里扭
twists per inch 每英寸捻度
twists silk 缝纫丝线
twist stitch 绕绣
twist style 扭曲发型
twist way Z捻,反手捻
twist yarn 捻线
two 2号尺码的衣服
two-and-three-needle double chain stitch 双针和三针双链线迹
two-and-three-thread overlock seam 双线和三线包缝
two-and-two check 双经双纬格子花纹
two-bar fabric 双梳栉经编织物
two-bar tricot 双梳栉经编织物
two-bar tricot machine 双梳栉经编机
two-button coat 两粒纽外套
two-button cuff 双扣袖头
two-button waistcoat 两扣背心
two-cloth 双幅织物,两幅织物
two cloth seaming machine 双缝料缝纫机
two-colored button 双色(金属)扣
two color decorative sewing 双色饰线
two-color decorative sewing 双色饰线
two colored jacquard hosiery 双色提花袜
two color effect 两色效应
two-color metal button 双色金属扣
two-cord sewing 双线缝纫
two dimensional defect 片状疵点,成片疵点
two-dimensions (2-D) 二度,二维
two-fabric 双幅织物,两幅织物
two-face 双面布
two-faced machine 双面针织机,双针床针织机
two-faced plush cloth 双面绒布
two-faced terry fabric 双面毛圈织物
two-feed knitter 双系统针织机,双吃线织袜机
two-folded yarn 双股线
two-gore skirt 两幅裙
two half-circle tacker 双半圆套结机
two-head embroidery machine 双头绣花机
two-hole button 两眼扣,双孔纽扣
two-in-one shir 双拼衬衫(两种不同衣料缝合而成的衬衫)

two jacket button single-breasted jacket 单襟式双扣上衣
two layer fabric 双层织物
two legs are uneven 两裤脚不齐
two medium-sized circles 双孔符号
two medium-sized perforation 双孔符号（纸样上表示织物折叠的折叠符号）
two-needle carding seam 双针凸纹缝
two-needle center seam 双针中央缝
two-needle chainstitch 双针链缝线迹
two-needle covering stitch 双针绷缝线迹
two-needle cylinder bed seam covering machine 双针筒式底板绷缝机
two-needle faggoting 双针饰缝
two-needle faggoting and chain scallop stitching 双针饰缝和链状锯齿线缝
two-needle five-thread safety stitch machine 双针五线安全缝缝纫机
two-needle flat-bed coverseam machine 双针平式绷缝机
two-needle flat-bed sewing machine 双针平缝机
two-needle four-thread mock safety stitch machine 双针四线装饰安全缝缝纫机
two-needle hemming 双针卷边
two-needle hemming machine 双针卷边机
two-needle high-speed lockstitch seamer 双针高速锁缝机
two-needle lockstitch 双针锁式线迹
two-needle lockstitch sewing machine 双针锁缝线迹缝纫机
two-needle low throw machine 双针下摆动缝纫机
two-needle machine 双针缝纫机
two-needle rub-down seam 双针防磨缝（缝在皮鞋后跟端部和侧面）
two-needle sewing machine 双针缝纫机
two-needle vamping stitch 双针鞋面补丁线迹
two-needle zipper unit 双针拉链缝纫机
two or three coloured jacquard hosiery 双、三色提花袜
two parallel stitches 双道平行线迹
two-part mold 分成两片的包纽模
two peak 双峰形,(折叠)饰巾法
two-piece 上下两件套
two-piece catsuit 尼龙女式紧身连衣裤
two-piece back interfacing 两片式后身衬
two-piece bathing suit （分胸罩和短裤的）两件式女泳装

two-piece catsuit 锦纶紧身女套装
two-piece collar 上下两片式盘领；双页领
two-piece costume 两件套服装
two-piece cuff 拼缝式克夫
two-piece dress 上下装,两件套女服
two-piece flap 两片裁的袋盖
two-piece frock 两截式外长衣
two-piece garment 两片式服装,两件套服装
two-piece gusset 两片式插角布
two-piece look 两片式款式,两件套款式,仿两件套式
two-piece outer garment 两件套外衣
two-piece pocket 两片袋
twopiecer 两件套服装；一套分成两件的服装
two-piece raglan sleeve 两片式插肩袖
two-piece raglan smock 两片连肩袖罩衫
two-pieces 上下两件套
two-piece set 两件套装
two-piece shirt collar 两片式衬衫领
two-piece silhouette 两片型轮廓
two-piece skirt 两片裙
two-piece sleeve（同 tailor's sleeve） 两片袖,套装袖
two-piece snowsuit 两件套保暖装
two-piece split sleeve 两片式仿连肩袖
two-piece suit 两件套装；两件套西服
two-piece swimsuit 两片式泳装
two-piece tailored sleeve 两片袖
two-pile velvet 双面天鹅绒
two-plex rib 双梳栉经编织物
two-ply alpaca crepe 双股羊驼毛绉
two-ply 叠织[起毛]织物
two-point diamond 两分[尖]钻石
two-point perspective 二点透视《计算机》
two-process hosiery manufacture 两步成形织袜法
two-ply 双股；双层起毛织物
two-ply yarn 双股线
two-seam regain sleeve 双线斜肩袖
two-season coat 春秋两用大衣
two-shuttle plush 双梭长毛绒
two-sided 双边十字绣
two-sided effect 正反面效应,阴阳面效应
two-sided goods 双面织物
two sided（Italian）cross stitch 双边（意利）十字绣
two-sided knit goods 双面针织物
two-sided stitch 双平针线迹

two-side knit goods 双面针织物
two-stage bidding 两段招标
two suiter 男用旅行箱
two super imposed fabric 双幅叠织的平面织物
two-thread chain stitch 双线锁缝线迹
two thread chain stitch joining seam and two thread overedge seam 双线链状接缝与包边缝
two-thread chain stitch machine 双线链式线迹机
two-thread double lockstitch 双线双锁迹;双线链式线迹
two-thread linking machine 双线套口机
two-thread operation 双线操作
two-thread overlock machine 双线包缝机
two-thread overlock seam 双线包缝
two-thread seaming 双线缝纫
two-tone color 同色不同色度的等合色
two-tone corespondent 双色鞋
two-tone effect 双色调效应
two-tone jacket 双色夹克
two-tone jumper 双色棒球服,双色运动夹克,衣身与袖子颜色不同的短夹克
two-tone pattern 双色调
two-tone pump 双色无带女鞋
two-tone shoes 双色鞋子
two-tone stripe 双色条纹
two-tone 双色调
two to one skip stitch 两针跳一针线迹
two-tuck 双褶
two water action （洗衣机）双向水流洗涤方式
two-way 正反可穿的,两面可穿的,两用的
two-way coat 两用外套
two-way collar 两用领
two-way design 双向花样
two-way face to face spreading 来回和合铺料法
two-way fashion 两用服装
two-way mixed tricot 两路进线交织经编织物
two-way print 双向印花

two-way raincoat 双面雨衣
two-way separating zip 双头拉链
two-way spreading 双向拉布
two-way stretchable knit 双向弹性针织物
two-way stretch foundation 轻薄弹力内衣
two-way stretch woven fabric 经纬双向拉伸机织物
two-way zipper 两头拉链,双头拉链
tyes 女用围裙
tygan 泰根仿革布
tying stitch 重结线迹
type 类型,型式;型号;典型;样本,样板
type approval 定型
type jumper 正统短夹克,领、袖、下摆三处用针织物的短夹克,三角型短上衣
type of body 体型
type of character 字体
Type of facial makeup 戏剧脸谱
type of figure 体型
types of goods 货品
type plate 型号牌,型号板
type sample 标样
type test 典型试验
typewriter cloth 打字带织物（也可用于羽绒上衣）
typewriter ribbon fabric 打字带织物（也可用于羽绒上衣）
typical form 典型式样
tyraline 品红,洋红
Tyrian purple 推罗紫色,红紫色,泰尔红紫（从海螺中获得的紫色染料）
Tyrian veil 泰尔面纱
tyrlind 法国蒂兰绸
Tyrolean design 罗尔图案
Tyrolean hat 蒂罗尔帽,登山软帽（锥形帽身和小帽檐的软毡帽）
Tyrolean jacket 蒂罗尔夹克（边缘和口袋有布条装饰,圆领,单排扣）
Tyrolean look 蒂罗尔风貌,蒂罗尔款式（吸取奥地利蒂罗尔地区民族风格设计的时装款式）
Tyrolean tape 蒂罗尔绣带,蒂罗尔饰带

U

U-2　优图
UAA（Undergarment Accessories Association）[美]内衣衣饰协会
U-back　U形后身
U-back(ed)　U形后身[的]
Ubangi necklace（同 Afro choker）　乌班吉贴颈项链；非洲型贴颈项链
Ubangi neckline　乌班吉领圈（特高贴颈型）
UBL（under bustline）　下胸围线，乳根围线
UCC(Uniform Code Council)　统一代码委员会
UD（ultra dull）全无光的（化纤）；(ultra-deep dyeing)超深染色性
uddrussa　乌兹别克斯坦彩条绸
ugly　（19世纪中叶流行的女帽上的）绸遮阳
uinare stylion　尺骨茎突点
Ukrainian costume　乌克兰民族服饰
Ukrainian peasant blouse　乌克兰农妇衫（长袖及臀衫加无袜短外套，前襟交叠）
ulang　精纺厚缎
ulna　尺骨
Ulster　阿尔斯特宽大衣（有扣带和贴带）；阿尔斯特长毛大衣呢（美国）
Ulster back　阿尔斯特外套（有后腰带）
Ulster（cloth）　阿尔斯特长毛大衣呢
Ulster clothing[coat]　阿尔斯特外套
Ulster collar　阿尔斯特大衣领；倒挂领，宽大翻领
Ulsterette　轻型阿尔斯特大衣
Ulster tuck　阿尔斯特活褶
ultra-dull（UD）全无光的（化纤）
ultra-fashionable　超时髦的，一时狂热流行的
ultra-fast computer　超速计算机
ultra-lucent look　超明亮款式
ultra-mini　超短式
ultra-fashionable　超时髦的，一时狂热流行的
ultra-feminine　超女性
ultra-feminine look　超女性款式
ultra-feminine style　超女性服装款式
ultra-fine corduroy　特细条灯芯绒

ultra-fine fibre　（0.1～1.0分特）超细纤维
ultra-high-speed sewing machine　超高速缝纫机
ultra-hi-heel　特高高跟鞋
ultra-marine　绀青色，青蓝色，群青色，佛青色
ultra-marine blue　群青，深蓝色，饱和蓝，绀青绿
ultra-matte stitch selector　超自动线迹选择器
ultra-mini skirt　超短裙
ultra-modern　超现代化的，超时髦的
ultrarapid cinematography　高速电影摄象机
ultra-red ray　红外线
ultra-sheer pantyhose　超薄连裤袜
ultra-sheer style　超薄型
ultra-short skirt　超短裙
ultrasonic apparel machinery　超声波成衣机械
ultrasonic dart　超声波短裥
ultrasonic eyelet　超声波纽孔
ultrasonic melt-joining　超声波熔接
ultrasonic quilting machine　超声纤缝机
ultrasonic sewing　超声波缝纫
ultrasonic thickness gauge　超声波厚度计
ultrasonic washing　超声波洗涤
ultrasuede　仿麂皮非织造布；纺麂皮可洗织物；仿麂皮衣服
ultrasuede garment　仿麂皮服装
ultrasuede yardage　仿麂皮的用料
ultraviolet light　紫外光（线）
ultraviolet rays　紫外线
ultraviolet resistance　防紫外线
ultraviolet resistant fiber　抗紫外线纤维
ulukh　手织云纹绸
Ulwan（shawl）　厄尔万手织羊绒披巾（克什米尔产）
umanori　（西装等上衣的）衩口，开衩；[法]小开的骑马缝
umber　茶褐，茶色，赭色，红棕
umbilical cord　（宇航员在太空工作时供氧用的）空间生命管线
umbilical region　近脐区
umbo　罗马花瓣状褶

umbrella	伞,阳伞,雨伞;伞形过肩,伞弧抵肩
umbrella brim	女工伞状帽檐
umbrella case	伞套
umbrella cloth	伞布
umbrella cover	伞面
umbrella drawers	伞形长衬裤
umbrella handle	伞柄
umbrella pleat	伞形褶襞[褶裥]
umbrella pleats skirt	伞褶裙
umbrella silk	伞绸
umbrella skirt	伞褶裙,伞形裙
umbrella style	伞状发型
umbrella tote	可放伞的大手提包
umbrella yoke	伞弧抵肩
Umbro	茵宝
umpen hat	搬运工帽
unbalance	不均衡,不平衡
unbalanced figure	不平衡体型
unbalanced legs	裤脚前后
unbalanced plain weave	不平衡平纹织物
unbalanced stripes	不均匀条纹
unbalanced weave	不平衡织物,两面组织不同的织物
unbecoming	(服饰等)不相配的,不合身的;难看的
unbelted corset	无带紧身衣
unbleached finish	本色棉织品
unbleached flannel	本色法兰绒,起毛后未经漂白的法兰绒
unbleached material	坯布
unbroken curve	连续曲线
unbutton	不扣上纽扣;解开纽扣
unbuttoned	纽扣解开的;没有纽扣的;无拘束的;随便的
unbuttoned shirt	敞开式衬衫
uncharacteristic style	无特征的服装款式
Uncle Sam suit	山姆大叔套装
uncolored cloth	本色布
unconstructed garment	不硬挺的服装,简做服装
unconstructed jacket	无衬里上衣,简做夹克
unconstructed suit	简易套装,无构造套装
unconstructed tailoring	简做缝制
uncontaminated air	洁净空气
uncorseted	未穿紧身褡的
uncostructed tailoring	(同 struc-turedtai-loring) 简做缝制
uncovered leg	光腿
uncreased legs	未烫缝圆裤腿
uncrushable	揉不皱的
uncuffed trouser leg	不翻折的裤脚
uncut brocade	金银线浮花锦缎
uncut dart	不剪裁的省道
uncut furskin	未切割的毛皮
uncut pile fabric	毛圈织物,非割圈式长毛绒针织物
uncut serge	未剪毛哔叽
uncut velvet	毛圈丝绒
uncut worsted	毛哔叽,精纺毛绒斜纹织物,轻缩绒精纺毛织物
undecorated	简朴的
underarm bag	夹式皮包,腋包
underarm (bust) dart	胁省,腋下省
underarm crutches	腋杖
underarm dart	胁省;袖底省
underarm dress zipper	连衣裙腋下拉链
underarm gore	袖笼 V 形三角布,腋下镶条
underarm gusset	(衣)腋下镶布
underarm handbag	臂夹式手袋,腋夹式手提袋
underarm length	内臂长
underarm point	袖下点,腋下点
underarm seam	袖下缝
underband	腰带里子
under belly	下腹部
under blouse	女衬衫,女衬衣
under bodice	紧身背心,紧身马甲
underbody	女衬里背心,衣里,束腰类衣着,短衬裙
underbody clothing	(古代)下裳
under-breeches	衬裤
underbrim	帽檐镶边
underbust	乳房下围的尺寸
underbust girth	乳下胸围
under bustline (UBL)	下胸围线,乳根围线
underbust seam	乳根缝
under cap	室内便帽
undercloth	俄罗斯衬衣布
underclothes	衬衣裤,内裤
underclothing	内衣裤;衬衣裤
undercoat	(穿在外衣内的)上衣;衬裙;(古代)里大衣;(毛皮的)下层绒毛
undercoated fabric	内涂层织物
undercoat fibre	(毛皮的)下层绒毛
undercollar	领里,领下片

undercollar band	领夹里
undercollar cloth	领衬布,垫布
undercollar felling machine	领里折缝机
undercollar felt	领底呢衬里;领底绒衬
undercollar seam	衣领下暗缝,领里暗缝
undercollar stand	领里座
under cover	内衣,贴身衣
under cuff	克夫里子,袖口里子
under-cutter	下剪线刀
underdeveloped limbs	发育不全的四肢
underdevelopment	显影不足
under drawers	内裤,衬裤
underdress	(女)内衣;衬裙
underdrying	烘燥不足,未烘干
underexposure	曝光不足
under flap	袋盖里
underfleece	(毛皮的)下层绒毛
under fly	(衣、裤的)里襟
under fold	暗褶裥
under [fold] line of collar	领下口
underfur	(毛皮的)下层绒毛
undergarment	内衣,衬衣,肚兜儿
underglaze red	釉底红,釉里红
underground fashion	反体制时装(美国现代青年的时装流派)
undergrowth	(动物的)下层浓密的绒毛;发育不全的
underhand appearance	淡色,浅色
underivative	独创的
underknot	安达领带结,结下花样领(领带花样在领结之下)
underlap	(衬衫等袖子)小衩边;小襟,里襟;垫衬
underlay	褶裥的下层;衬底,底基层;铺底层
underlay extension closing	遮盖式腰头,与门襟平齐式腰头
underline	下口线
underlinen	(麻或其他薄织物的)内衣,衬衣
underlinen of hood	兜帽下口线
underline of band	底领下口线
underline of collar	领下口(线)
underline of hood	(风)帽下口线
under line of waistband	腰头下口线
underline of yoke	过肩下口线
underlining	(衣服的)衬里,衬布
underlip	下唇
underneath side of a quilt	被里
under long pants	长裤的衬裤,内长裤
under looper	下弯针
under paint	画底色
underpants	内裤;衬裤;汗裤
underpinnings	(尤指妇女的)内衣,衬衣,贴身衣;人的双腿
underpocket	袋里片,口袋衬里
under pressing	半成品烫边,缝前熨烫
under-pressure	压力不足,负压
underproduction	生产不足;减产;产品供不应求
under proof	不合格的;低于标准的;在试验中
undersewing	反面缝纫
undershirt	汗背心,汗衫,贴身内衣
undershirt cloth	汗衫布
undershirt dress	汗衫装;连衣衬裙;内衣连衣裙
undershirt sweater	汗衫式套头毛衣
undershorts	男短衬裤;三角裤;男人及儿童用裤衩
undershot floats	跳花疵
underside	(织物的)反面;夹里,里子
underside finish	反面制作加工
underside sleeve	(袖身分为两片裁剪时的)小袖片
undersize	比一般尺寸小的;小尺寸的,小型的
underskirt	衬裙,内裙
underslacks	妇女下身缠布(外衣)
undersleeve	下袖,内袖,衬袖,小袖(片),袖下片,袖瘪肚
undersleeve cap curve	小袖深弧线
undersleeve depth line	小袖深弧线
undersleeve length	衬袖长,小袖长
undersleeve lining	小袖里
understandings	脚(俚语);腿;鞋子,靴子
understitched seam	落漏缝
understitching	暗定针
under stitching seam	暗定针接缝
understructure	(服装)里层结构
understructure material	服装里层结构材料
undertaker's cloth	寿衣呢,寿材呢
under tape conveying and trimming	下条带式输送切割
under taping	反贴边(线镶边)
underthings	女内衣裤
under thread	底线
under thread guide	(缝纫机)底线导线钩
under thread spool	(缝纫机)底线线轴

under thread tension 底线张力
under throat plate thread trimmer 针板下剪线
undertint 淡色,浅色,柔和色
undertoe 袜头底部
undertone 底彩;淡色,浅色,柔和色;暗色
undertrimmer 滚边机
undervest 贴身内衣;汗衫;汗背心
underwaist 衬里背心,儿童连裤背心
underwear （总称）内衣;贴身衣;衬衣
underwear linen 亚麻内衣服装布
underweight 重量不足;标准重量以下;未达标准重量的人
underwheel feed 下轮式送布
under width 幅宽不足
underwire body briefer 钢托式紧身全帮胸衣
underwire bra(ssiere) 钢托式胸罩
underwool 羊毛内衣;供做内衣用的羊毛;(毛皮的)内层绒毛
undies （女）内衣,妇女衬衫
undress 便服,军便服;休闲装;晨衣;脱衣;裸体（或半裸体）状态
undressed 未经加工的,未处理的;便服的,裸体的
undressed uniform 军便服
undressed worsted 缩绒精纺毛织物
undress skin 未鞣熟的兽皮
undress uniform 军便服;便装
undress wig 军便假发
undress worsted 缩绒精纺毛织物
undui ［俄］俄式软底毛靴
undulating hair 波浪形头发
undyed crease 未上色折皱(布疵)
U neck U形领
U neckline U形领口
une partie de campagne 田园境界
unequal 不均匀的,不对称的,不平衡的,不相等的
unequal rhythm 不匀等协调
unequal twill 单面斜纹,异面斜纹
unerea sed legs （未烫缝）圆裤脚
uneven 不匀,不匀称
uneven armhole 绱袖不圆顺
uneven basting 疏缝不匀
uneven breakline 驳口不直
uneven cloth 布面不匀(的)织物;布面不匀(织疵)
uneven craping 皱斑,皱疵,绉纹不匀,起绉不匀
uneven dyeing 染色不匀
uneven filling 横档,纬档,错纬(织疵)
uneven finishing 云斑
uneven flap edge 袋盖不顺直
uneven fly facing 门襟起绺
uneven front edge 止口不顺直
uneven gorge line 串口不直,串口线不直
uneven ground 地色深浅不匀
uneven hem 底边起绺
uneven hemline 参差裙摆线,裙摆线不圆顺
uneven leg hem 裤脚口不齐,裤下口不齐
uneven leg 吊脚,吊裤脚
uneven listing 色点斑疵
uneven loop 线圈不匀
uneven lustre 光泽不匀
uneven mark 印花不匀
uneven material feeding 送料不匀
uneven mercerization 丝光不匀
uneven milling 缩呢不匀,缩绒不匀
uneven napping （布面）起绒不匀
uneven neckline 领窝不平
unevenness 不匀率,不匀度,不匀性
uneven pressing 压烫不匀
uneven printing 印花不匀,印花斑
uneven printing of repeats 接版深浅
uneven quilting 绗棉起绺
uneven raising 起绒不匀,拉绒不匀,刮绒不匀
uneven rolling 轧痕,轧印
uneven scouring 练斑
uneven selvedge 布边不齐
uneven setting 定形不匀
uneven shearing 呢面剪毛不匀,绒面剪毛不匀
uneven shrinkage 缩率不匀
uneven skirt flare 裙浪不匀
uneven-sided twill 单面斜纹
uneven sleeve opening 袖口起绺
uneven stitch length 线迹长度不匀
uneven surface 布面不匀
uneven tension in zigzag stitch 曲折形线迹张力不匀
uneven waist seam 腰缝起皱
uneven weaving 云织(织疵)
uneven zipper 绱拉链起绺
unfading 不褪色的
unfast dyeing 染色不牢
unfavoring flap edge 袋盖止口反吐

unfinished 未整理的,未经整理的织物
unfinished products 在制品
unfinished worsted 轻微起绒整理的精纺毛织物
unfit 不合身
unfitted 不紧贴的,宽敞的
unflammability 不燃性
un-full shape 不宽松的形状,合体的形状
ungraceful 不优美的,不雅的,难看的
ungummed silk 精练丝织物
unhairing 生皮去毛
unhook 解开(衣服)钩扣
uni 平纹织物,单色织物
unicolor 单色
unidirectional cloth 经纱加固织物
Unidure 耐久性防皱整理(商名,用于黏纤及其混纺织物)
Unifast 耐洗防缩处理(商名,用于黏纤织物)
unified management 统一经营
uniform 制服;职业(工作)服;军服;制服式样
uniform cap 制服帽;号帽
uniform chromaticity scale diagram 均匀色品标尺图,均匀色度标尺图
uniform cloth 学生呢,制服呢
uniform coating 制服呢;制服料
Uniform Code Council (UCC) 统一代码委员会
uniform color space 均等色空间
uniform dress 制服
uniformless 无制服的,不穿制服的
uniform rhythm 均匀协调
uniform rules for collection 托收统一规则
uniform shirt 制服衬衫;军衬衫
uniform stitches 均匀线迹
unimaginative clothes 朴实无华的衣服
unimaterial fabric 纯纺织物
uninflammable 不易燃的
union 混纺织物,交织(棉)织物
union blanket 棉毛毯
union broadcloth 防雨厚毛毯;军用防雨披风
union cassimere 棉毛交织呢
union chemical fiber fabric 化纤交织呢
union cloth 大众呢,棉经毛纬再生起绒织物;混纺衣料
union damask 交织装饰布
union duck 交织帆布

union dyed fabric 染成同色的交织物
union dyeing 混纺织物染色
union dyes 混染染料
union fabric 混纺织物,交织织物
union feed 协调送布
union gray[grey] 混纺交织衬衫布
union huck towels 交织小花毛巾
union italian cloth 半毛五页缎
union linen 交织亚麻布(棉/亚麻交织)
union linen lawn (女衣及内衣用)细亚麻交织布
Union of Needletrades, Industrial and Textile Employees 纺织从业人员联盟
union serge 棉经毛纬哔叽
union shirt 中袖女式衬衫
union silk for necktie 宝带绸
union suit 连衫裤,连裤内衣
union towel 交织毛巾
union twist yarn 混纺线
union yarn 混纺纱
uniped 独脚的人,独腿的人
unipod (照相机等的)独脚架
unique style 独特风格
unique flower yajing bourette 独花鸭江绸
unisex 中性服装;男女通用服装;(服饰、发式等的)不分男女的倾向(或状态)
unisex clothes 男女通用的服装
unisex fashions 中性服装
unisex garment 男女通用的服装
unisex hair style 男女不分的发式,中性发性
unisex jogging clothes 男女通用的跑步装
unisex jogging suit 男女通用的跑步装
unisex look 中性风格,无性别款式
unisex pants(同 **drawstring pants**) 束带紧腰裤;前开襟男女通用裤
unisex shirt 男女通用的衬衫
unisex shop 出售男女通用服装的商店
unison feed 同步进料
unit (机械等的)部件,元件,组件,构件,装置;(计数)单位,单元;品目
unitard 强力紧身衣,弹力全身紧身衣(通常为体操运动员、舞蹈演员所穿,也有作生活服装)
unit construction 块式成衣,组合式成衣
unit dress 组合时装,配套时装
unite cloth 团结布
UNITED COLORS OF BENETTON 贝纳通童装
United States Association of Importers of

Textiles and Apparel 美国纺织品进口商协会
unit pattern 单元花样
unit price 单价(销售)
uni-twist (合股线与单纱相同的)单向加捻;单向捻度
unit work plan 模块式制衣工作程序图,组合式制衣工程程序图
unity 和谐,协调;总体效果;总体布局;统一
universal agent 全权代理人
universal blind stitch 万向暗缝,无规律多方向暗缝
universal compasses 万能圆规,通用圆规
universal feed 多方向送布
universal foot 通用压脚,万能压脚
universal pleat pressing machine 通用褶裥机
universal presser foot 万能压脚,通用压脚
universal quilter 多功能缝纫机,多功能绗缝机,万能绗缝机,万能缝纫机
universal sewing machine 多功能缝纫机,万能缝纫机;通用缝纫机
universal stitch 多方向线迹,无规律多方向线迹
universal thread tension 通用线张力
universal upper feed 多方向上送料
universal zigzag sewing machine 多方向曲折缝缝纫机
universe look shoes 宇宙鞋
universe wadding 太空棉
university coat 学院外套
university gown 大学礼服,大学外氅(一般在举行典礼时穿)
unknit 拆散针织物
unlace 解开带力
unlevel dyeing 染色不匀
unlevel shade 不匀色泽
unlimited fancy stitch 无限制花式线迹
unlimited pattern area 无限制花型范围
unlimited price 不限价
unlined 单衣,无衬里的服装;无衬里的;无皱纹的,无线条的
unlined coat 单外套,单大衣,无夹里外衣
unlined dress 单衣
unlined garment 单服装,无夹里服装
unlined jacket 单上衣;无夹里上衣,无里夹克
unloading 卸货

unmatchable color 不相配的颜色
unmatched collar 绱领偏斜
unmatched crotch cross 下档十字缝错位
unmatched front fly 门里襟长短不一
unmatched shoes 不配对的鞋子
unmatched suit 异料服装,用不同布料做成的套装
unmeet bound pocket mouth 袋口裂
unmeet front edge 止口豁
unmeet pleat 裙裥豁开
unmeet vent 背衩豁
unmentionables 裤子(古代)
unmerchantable 不可出售的,不适合市场销售的;销不掉的《销售》
unmounted sleeve 与大身相连的袖子(如连袖、半连袖、套肩袖、和服袖)
unnapped 未起绒的
unnatural blended fabric (化纤为主的)化纤混纺织物
unnatural hand 非天然纤维手感
unneatened seam 毛边缝
unoiled yarn 不上油的纱线
unpick 拆去缝线(或针脚)
unpin 取下别针,为取下别针脱掉衣服
unplanked hat body 未缩绒帽身,未缩绒帽坯
unpressed pleat 无压褶,不定形的柔和细褶,免烫的细裙裥,软褶,泡泡褶
unpressed pleated skirt 不定形褶裙,软褶裙,(免烫)活裥裙子
unrelated colour 非相关色
unrobe 脱长袍,脱衣
unroving (套口后)拆袖头,脱散
unsal 二上二下重磅棉织物
unseamed waistline 无剪接腰线,无接缝腰线(腰部无任何横向剪接)
unselvedged edge 无包边折边
unsharp 轮廓不清
unshoed feet 光脚
unshrinkable finish 防缩整理
unshrinking 防缩整理
unsiructured tailoring 简做缝制
unskilled labo(u)r 非熟练工人
unskilled work 非技术工作
unsolid dyeing 极光疵;染色不透
unstitched stitch stripe 仿刺绣条纹
unstructured suit 简做套装
unstructured tailoring (同 uncon-structed tailoring) 简做缝制
unstudied charm 自然美

unsuits 便服,无构造服,无构造套装,柔型套装,简便套装
unsure sewers 一般缩袖法,腋下定点缩袖法
unsymmetrical leg 裤脚不对称
unsymmetry 不对称,不匀称,不调和
untapped leather 未鞣的皮革
untie 解开;松开
untie one's tie 解开领带
untie scarf 松开围巾
untwisted yarn towel 无捻毛巾
unwashable fabric 不宜水洗的织物(只适于干洗)
unwashable suits 不宜水洗的服装(只适于干洗)
unwearable style 陈旧式样
unwoven fabric 非织造织物,无纺织物,不织布,非织造布
unzip 拉开拉链
up-and-down effect (织物)从上到下的花纹效果
up-and-down print 单向印花
upbeat color 生气勃勃的色彩,活泼(的)色彩
upclassing 提高等级,提高质量
up-date 现代派
updated 最新式的
up-dated look 最新式风貌;最新型,当代型
updo 阿波杜发型,顶卷发式
upgarments 上装,上衣
upgrade (售出货物时)以次充好《销售》
up hair 高耸发型
upholsterer's welt seam 家饰缝
upholstery brocade 装饰花缎
upholstery cord 装饰带
upholstery fabric 家具布
upholstery jacquard silk 装饰花绸
upholstery material 室内装饰布料
upholstery needle 弧形针
upholstery silk 装饰用绸
upholstery twill silk for picture mounting 裱画绫
upkeep 保养,维修
uplift 乳罩,胸罩
uplift bra 提升胸罩
uplifted bosom 隆起胸部
upline of armhole 袖窿翘高线
upline of waist 后翘线
upper 上衣,上装,衫;鞋帮,鞋面,布鞋罩;腰带;[复]绑腿,腿套

upper abdomen 上腹
upper and lower feed 上下送料
upper and lower feed lockstitch sewing machine 上下送料锁式线迹缝纫机
upper and under throat plate thread trimmer 针板上下剪线
upper arm 上臂
upper arm circumference 上臂最大围
upper arm girth 上臂围
upper arm line 上臂围线
upper back 上背部
upper blind looper 暗缝上弯针
upperbustline (UBL) 上胸围线
upper cloth 上缝布,上缝料
upper clothes 外衣
upper collar 领面,领上片
upper end of waistband 腰头上口
upper feed 上送料,面送料
upper garment 上装,上衣,衫
upper knife 上切片
upper leather 面革
upper limb 上肢
upper limb bone 上肢骨
upper lip 上唇
upper looper 上弯针
upper looper thread 上弯针线
upper looper thread eyelet 上弯针过线钩
upper outer garments 上装
upper pocket 上口袋
upper posterior arm length 肘长
uppers 上排假牙
upper sleeve 大袖片
upper stocks 男式筒形裤
upper tape conveying and trimming 上条带输送切割
upper tension 上线张力
upper-thigh length 至大腿长度
upper thread (缝纫机的)面线,上线
upper thread hook 上弯针线钩
upper thread(ing) 缝纫机用面线,缝纫机上线
upper thread tension 面线张力
upper thread tension regulator (缝纫机头上的)面线调节器,夹线器
upper throat plate thread trimmer 针板上剪线
upper torso 上躯干
upper wheel feed 上轮式送布
upright cloth cutter 立式裁剪机
upright flap pocket 立式有盖口袋

upright pile 立绒
upright pile ladies cloth 立绒女式呢
upright stance 笔直(站立)姿势
upright stitch 直立针法
upright twill 急斜纹
up-scale fashion area 高档时尚区
upshoes 室内练舞鞋,室内体操鞋
up style 高耸发型,顶卷发式,向上卷发型,阿波杜发型
up-swept hair style 两叶翻翘式发型
upswept Rococo coiffure 洛可可式高发髻
up-to-date 最新式的;现代派
up-to-date style 时兴式样
up-to-neck 直开领(开至颈根,如学生装领)
upturn 袖口,裤脚翻边
upturned lotus-petal design 仰莲瓣纹
upturn hemmer foot 上翻卷边压脚
upward fold 上卷边
uracus (古埃及王冠上的)圣蛇像
urba 厄巴丝花缎
urban cowboy style 都市牧童风格,都市牛仔风格
urban jeans 都市牛仔裤
urban look 城市款式
urban wear 都市服装
urchin cut (女子的)"顽童"超短发式
urea button 电木扣;尿素扣
uretan form 泡状合成纤维布(用作西装衬布、里子布等)
urethane coated fabric 氨基甲酸酯涂层织物
urethane elastic fibre 聚氨基甲酸酯弹性纤维,氨纶
urethane foam 聚氨基甲酸酯泡沫塑料
urethane sole (制鞋)聚氨酯鞋底
urine transfer connector (太空服中的)输尿连接管
urmark 厄马克绣花马鞍呢
Ursrainian costume 乌克兰民俗服
urticant 发痒的,刺痒的
usable life 适用期;使用寿命
usable width [织物]除边幅宽
US army stylistic form 军队款式
U-scissors (U形)纱剪;线头剪
use press cloth 隔布压烫
use value 使用价值
used clothes 旧衣服
used goods 旧货,旧纺织品
use-life 使用寿命
user interface 用户界面,用户接口《计算机》
use-surface 织物正面,毯面
use-yellowing 使用泛黄
usha 高腰连衣裙;高腰迷你裙(起源于印度,满花土布,饰边)
ustar 乌斯塔里子布(印度制,有平纹,斜纹两种)
usual proof 通常证明
usugin(u) 薄纱,日本薄绸
usuji sofu 棉平纹床单布
utiliscope 工业电视装置
utilitarian cloth 无色彩织物,实用织物
utilitarian stitch 实用线迹
utilities 实用程序,工具程序《计算机》
utility boots 特种实用靴(用于警察或林业工作人员)
utility cloth 实用织物
utility clothes 实用服装
utility coat 实用外套,功能性外套
utility factor of machine time 设备时间利用率
utility look jeans 实用款式牛仔裤
utility percale sheeting 实用中支高密被单布
utility pocket 多用途口袋
utility press 通用熨烫机;万能熨烫机
utility program 实用程序
utility routine 实用程序
utility stitch 实用线迹
utility theory 效用理论
U-tip U形鞋头,U字形鞋头
U-tip tassel 带流苏的U字形鞋头
utrecht velvet 乌德勒支丝绒
uttariya 波斯式外衣,印度式外衣
UV-cut proceeded fabric 防紫外线加工织物
UV-no textile 防紫外线系列纺织品
UV-protection apparel textile 防紫外线服用纺织品
Uygur cap (新疆)维吾尔族小圆帽
Uygur ethnic costume 维吾尔族服饰
Uygur nationality's costume 维吾尔族服饰
Uygurzu cap (新疆)维吾尔族帽
Uygurzu costume 维吾尔族民俗服
Uzbek nationality's costume 乌孜别克族服饰
Uzbekzu costume (同 Uzbegzu costume)乌孜别克(族)民俗服
uzel 法国优质亚麻布
uzen silk satin 羽前耐洗丝缎
uzen washable satin 羽前耐洗丝缎

V

V (vertex)　头顶点;(violet)紫色;V形领口
vacancy look　休假款式
vacancy wear　休假服装
vacuum blowing table　吹吸熨烫台
vacuum board　抽湿台,吸风烫台
vacuum cutting table　真空裁床[裁剪台]
vacuum drum washing range　真空转鼓水洗机
vacuum dryer　真空洪干机
vacuum dyeing　真空染色
vacuum hydroextractor　真空脱水机
vacuum ironing table　(熨烫)抽湿台
vacuum table　抽湿台
vacuum transfer printing　真空转移印花
vacuum wet absorbing machine　真空抽湿机
vagabond hat　便帽,流浪汉帽(帽檐朝下,与便装一起穿戴,西方流行汉用)
vaglan　插肩,插肩大衣
vair　栗鼠皮,松鼠皮,灰白色松鼠皮,青白相间毛皮
val　瓦郎西安花边
valance　拉布,床沿挂布,桌帷;凸纹亚麻背心料
valencia　巴伦西亚斜纹薄呢;凸纹背心料
valencia vesting　凸纹背心料
valenciennes　[法]扁平梭结花边
Valenciennes lace　瓦朗谢讷花边,网眼梭织花边
Valenciennes batardes　瓦朗谢讷花边,网眼梭织花边
Valenciennes brabant　瓦朗谢讷花边,网眼梭织花边
valentia　背心呢,棉毛丝交织格子织物
Valentina　瓦伦蒂娜(美国时装设计师)
Valentino　瓦伦蒂诺;华伦天奴
Valentino Garabani　瓦伦蒂诺·加拉巴尼(意大利时装设计师)
valentino Garavani　瓦伦蒂诺·格拉瓦尼(1932~,法国高级时装设计师)
valet　衣物架,衣帽架(美国)
valeur　色193
valgus　(膝、足、髋等的)外翻;膝外翻的人;腿呈罗圈形的人(古代)

valid　有效的
validated license　有效许可证
validity　有效期
valid period　有效期
valise　背囊;手提包;手提旅行箱
valitin　凡立丁,精纺呢绒
val lace　瓦朗西安花边(六角网眼梭结花边)
vallancy　瓦伦西假发
valle cypre　瓦勒西伯尔绉纱,波洛尼亚丝绉纱(丧服用)
valuables　(首饰等)贵重物品
value　值;大小;价值;能力;浓淡色度,色明度;调子
value added criterion　增值标准
value added tax　增值税
value engineering　价值工程
value of import　进口额
value pattern　色彩花样
value scale　明度阶段
valures　缎纹精纺呢
vambrace　前臂铠甲,全臂铠甲
vamp　靴(鞋)面,前帮;前帮皮,靴(鞋)面皮;补丁,补片;补缀物
vamp crimping machine　鞋面定形机
vamp fashion　妖艳装,诱惑装
vamping stitch　补钉线迹,鞋面补缀线迹
vamp insole reinforced nalling machine　鞋面中底加固打针机
vamp lace stapling machine　鞋面条花钉花机
vamp lining　前帮衬里
vamp pattern stapling machine　鞋面饰花钉花机
vamp point　鞋楦表面中心线与跗围线的交点
vamp steaming heater　鞋面蒸湿加温机
vamp style　荡妇风格
Vandyke　锯齿形饰边,扇形饰边,范达克领;花边手帕
Vandyke beard　范达克尖削型胡须;(下巴上的)短尖髯
vandyke braid　锯齿纹狭棉带
Vandyke brown　深褐色,铁棕
Vandyke collar　(17世纪)范达克领(锯齿

边宽衣领)
Vandyke edge 范达克月牙花边,带子的荷叶边
Vandyke hat 范达克帽,伦勃朗帽
Vandyke stitch 范达克针法,经编缎纹组织
vane 轧染床单
vanilla 香草黄,淡杏黄色,香草冰淇淋色
vanished cloth 漆布
vanishing cream 雪花膏,消散面霜
vanishing stripes 断续条子花纹
vanity (梳妆用)小手提包,小手袋;梳妆盒
vanity bag (梳妆用)小手提包,小手袋,梳妆包
vanity box 梳妆盒,化妆盒
vanity case 梳妆盒,化妆盒
vanity set 一套化妆品
vapeur 松棉布,雾气细布
vapor 汽雾色
vapor barrier garment 不透气服装
vapor blue 蒸气蓝色
vapor permeability 透汽性
vareuse 粗纺羊毛罩衫
variable color fabric 变色织物
variable feed 可变送料,可调式送料
variable hand 手感不一
variable machine speeds 可变(可调)缝纫机速度
variable strip knitter 花式条纹针织机
variable top feed 可变式上送料,可调式上送料
varicolor 杂色,五颜六色
varicolored fabric 杂色织物
varied zigzag 可变曲折缝
variegated silk 彩绸
variegated [silk] umbrella 花绸伞,花伞
variegation 彩色,杂色
variety 品种,多样性;变化
Variknit single cylinder knitting machine 维鲁尼特单面针织机
vari-overlock 多种包缝
vari-pattern socks 多色提花花袜,闪色花袜
vari-stitch 多种线迹
vari-tuck machine 多种打褶机
vari-tuck method 多种打褶法
vari-tuck smock machine 多种打褶刺绣机
varnished cloth 漆布

varnished cloth garment 漆布服装
varnished composition leather 假漆皮
varsity sweater 字母毛衣
varus (膝、足、髋等的)内翻
vary-form curve rule 曲线尺
vaseline 凡士林;淡黄色
vasocons triction 血管收缩
vasodilation 血管扩张
vasquine 松针纤维松软布;女式外套;华丽衬裙
vassar rose 深品红到暗红的色彩;深品红到暗红色的纺织品
vastus externus 股外肌
vastus internus 股内肌
vaszon 亚麻帆布
vat bromo-indigo 还原溴靛蓝
vat direct black 还原直接黑
vat dye 还原染料
V-bed flat machine V形横机
V-bed rib machine V形罗纹横机
V-belf (缝纫机上)V形带
VC (voiuntary chain) 自由联锁店《销售》
V-cut 毛皮的)V形切割
V-cut yoke V形抵肩,V形过肩
vector 矢量
vee 三角形褶裥,三角形缺口;V形的
vee neck V字形领
vee neckline V字形领口
veev machine 圆纬衬经机
vegetable down 木棉
vegetable fiber 植物纤维
vegetable flannel 松叶纤维织物,粗亚麻内衣布
vegetable hair 植物纤维(填料)
vegetable tannage leather 植物鞣革
vegetable wool 人造羊毛,黏胶短纤维仿毛棉布
Veicro strap 尼龙搭扣带
Veicro tape 尼龙搭扣带
veil 面纱,面罩,面网,披纱
veiline 卷绒冬大衣呢
veiling 面纱料,蒙面纱,纱罗衣料
veiling lace 纱罗花边
vein 纹理,条纹,纹路,破洞疵,裂缝疵
veining 纱罗式条子花纹;花边镶边,夹缝饰边
veiours soleil 闪光丝绒,光亮丝绒
Velazquez silhouette 委拉斯凯兹型轮廓(1599~1660年西班牙贵妇装,上衣紧合而裙子宽大,且强调臀围的宽横效

果)
velcro closing 尼龙搭扣门襟
velcro closure 尼龙搭扣门襟
velcro(fastener) 尼龙搭扣带,维可牢(商名)
velcro pad construction 尼龙搭扣结构
velcro strap 尼龙搭扣带;魔术贴
velcro tape 尼龙搭扣带;魔术贴
veld(t) schoen (南非人穿的)生皮短筒靴
velline 卷绒冬大衣呢
velludo 经立绒织物,经起绒织物;麻绒
vellum 充牛皮纸棉织物;透明描图布
vellum bookcloth 仿牛皮面布
vellum cloth 双面轧纹薄布,透明描图布
vellum paper 牛皮纸
Vellveteen-like fabric 纬平绒织物
velocimeter 测速计
velocity 速度,速率
velour 天鹅绒,丝绒,立绒,棉绒;拉绒织物;维罗呢,(制帽用)兔绒皮(海狸绒皮)
velour an sabre 经起绒丝绒
velour cap 丝绒帽
velour finish (粗纺呢)起绒整理
velour hat 兔皮帽,丝绒帽
velour overcoating 起绒大衣呢
velours 丝绒,天鹅绒,立绒,棉绒;拉绒织物;维罗呢;(制帽用)毡
velours à deux poils 双股线线天鹅绒
velours à double corps 双色绒条丝绒
velours à la reine 皇后绒
velours albigeois 阿尔比窄丝绒条纹织物
velours antique 古香丝绒
velours antique écossais 苏格兰格子波纹府绸,古典格子丝绒
velours à ramages 缎地天鹅绒,锦缎绒
velours biseautés 斜面丝绒饰带
velours bombes 驼背丝绒
velours bosselle 波浪丝绒
velours broché 挖花丝绒
velours broderie 仿西塞莱天鹅绒
velours cameleon 闪光丝绒
velours cannele 高低棱条丝绒
velours chiffon 雪纺丝绒
velours chine 彩经绒,印经丝绒
velours cisele 西塞莱天鹅绒,凸绒刻花天鹅绒
velours col 衣领绒
velours contre-semple 倒毛花丝绒

velours couche 丝平绒
velours crepe-de-chine 双绉丝绒
velours d'angleterre 英国纬毛圈丝绒
velours de cotton 棉平绒,棉立绒
velours de France 法国彩色拷花丝绒
velours de guex 贝格丝绒,居安丝绒
velours de harlem 荷兰阿雷姆厚棉绒,厚重平绒
velours de hollande 丝天鹅绒
velours de laine 弹性法兰绒
velours de paris 巴黎绢丝绒
velours d'italie 横棱纹绒
velours d'oran 异经条纹绒
velours double (有闪光效应的)单双股绒经丝绒
velours du nord 棉背丝绒
velours d'utrecht 棉地拷花丝绒
velours ecossais 法国格子丝绒
velours ecrase 异向平绒
velours embosse 叠花丝绒,拷花丝绒
velours envers satin 缎背丝绒,缎背丝绒带
velours epingle 毛圈丝绒
velours figure 花丝绒
velours Florentin 法国弗洛朗坦丝绒
velours francais 双色经纬丝绒
velours frappe 拷花丝绒
velours frise 毛圈丝绒
velours frise uni 纯色毛圈绒
velours gandin 缎地丝绒
velours gaufre 拷花丝绒
velours glace 闪光丝绒
velours gregoire 印经画景丝绒
velours gros-grain 粗横棱绒
velours mirror 镜面丝绒
velours ombre 色彩丝绒
velours ottoman 奥特曼丝绒,双面交替宽条丝绒
velours overcoating 拉绒大衣呢
velours panne 平绒
velours paon 平绒
velours par chaine 经线丝绒
velours ras 法国毛圈绒,毛圈或毛圈绒
velours ras d'Angleterre 英式毛圈横棱丝绒
velours raz 法国毛圈绒,毛圈或毛圈绒
velours rayes 条子丝绒
velours renaissance 复兴丝绒
velours russe 棱纹彩花丝绒,俄罗斯彩色斜凸条丝绒

velours sans pareil 织纹天鹅绒
velours sculpte 叠花丝绒
velours shirt 半开襟拉链衫
velours simple 闪光单双股线经丝绒
velours soleil 光亮丝绒
velours surface uneven 绒面高低不平
velours towel 绒面毛巾,割绒毛巾
velours toweling 单面剪绒毛巾
velours travers 色横条丝绒,纬条毛巾
veloursurface overcoating 绒面花式大衣呢
velours venetian 威尼斯丝绒
velour toweling 剪绒毛巾布
velouté 天鹅绒,丝绒
veloutine 绒面呢;(衣里用)斜纹绒头薄呢
velpel (制帽用)长毛丝绒
velreteen-like 仿平绒
velure 丝绒,天鹅绒;仿天鹅绒织物;天鹅绒刷子
velutine 短绒头棉平绒
velutum 天鹅绒,丝绒
velventine 纬绒
velveret 粗天鹅绒,印花棉绒
velvet 天鹅绒,丝绒,立绒,经绒,平绒;天鹅绒连衣裙,丝绒连衣裙
velvet and velveteen 平绒
velvet article 绒头织物
velvet bath towel 绒面浴巾
velvet board (供绒毛织物整烫的)针(毯)烫垫
velvet brocade 花丝绒,锦地绒;天鹅绒织锦
velvet carpet 天鹅绒地毯
velvet cord 凸条灯芯绒
velvet costume 丝绒戏服
velvet cover 绒面
velvet crepe 丝缕绉
velvet ecrase 锦装绒
velveteen 平绒;棉绒;纬绒;棉绒衣服(尤指裤)
velveteen-like fabric 仿平绒,仿平绒织物
velveteen plush 长毛棉绒
velveteen rayon 绒花绸
velveteen ribbon 棉绒带
velveteens 棉绒连衣裙,棉绒裤子
velvet fabric 丝绒,天鹅绒;丝绒织物,经绒织物
velvet-figured 花丝绒
velvet flat embossing calender 丝绒拷花机
velvet flower 绒花
velvet foot (缝纫机)丝绒压脚
velvet for toy 玩具绒
velvet georgette 乔其绒,绒面透明纱
velveting 丝绒织物,精细绒头织物
velvet knitted fabric 天鹅绒针织物
velvet lain 平绒
velvet lining 绒里
velveton 仿丝绒织物;(制鞋用)纬起绒棉布
velvet-on-velvet 叠花丝绒
velvet picture 画景丝绒,印经丝绒
velvet pile tapestry carpet 印经绒头花毯
velvet-plain 平绒
velvet ribbon 丝绒带
velvet satin 丝绒花缎,缎地花丝绒
velvets pile 割绒,绒毛织物
velvet streamer 丝绒飘带,天鹅绒细长饰带
velvet stuff 天鹅绒织物
velvette 天鹅绒,丝绒,立绒,经绒;天鹅绒连衣裙,丝绒连衣裙
velvet towel 绒面毛巾
velvet trim 天鹅绒饰边
velvet velludo 麻绒,经立绒织物,经起绒织物
velvet vest 丝绒背心
velvet weft knitted fabric 天鹅绒针织物
velvet wool 丝绒毛料,羊毛天鹅绒
velvet work 土耳其风格刺绣
velvety look 丝绒状外观
velvety stuff 天鹅绒织物
vendeuse (指时装店的)女售货员,女店员《销售》
vendibility 可销售性;市场价值《销售》
vendue 拍卖
Venetian 直贡呢,贡呢,棉直贡,威尼斯缩绒呢,威尼斯精纺细呢
Venetian blind pleats 百叶褶,单向褶
Venetian blue 威尼斯蓝
Venetian chalk 裁剪用粉笔
Venetian cloak 威尼斯披肩
Venetian cloth 威尼斯缎纹织物
Venetian crepe 威尼斯绉
Venetian embroidery 威尼斯刺绣(透孔刺绣与凸花刺绣组合而成)
Venetian fabric 威尼斯织物
Venetian green 威尼斯绿
Venetian guipure 针绣挖花花边

Venetian hem stitch 威尼斯卷边线迹,条带卷边线迹
Venetian lace 威尼斯花边(挖花花边与针绣花边)
Venetian ladder work 威尼斯梯形刺绣
Venetian pearl 威尼斯珍珠
Venetian pink 威尼斯淡粉红
venetian point lace 威尼斯针绣花边
Venetian raised point 威尼斯凸纹针绣花边
Venetian red 威尼斯红(红光棕色)
Venetian rose point 网地花边;威尼斯玫瑰花纹
Venetians 威尼斯式裤,膝下束紧的宽大裤子
Venetian sleeve 威尼斯袖
Venetian velvet 威尼斯丝绒,色织丝绒
Venezuela lace 委内瑞拉花边,抽绣花边
venise 大花亚麻锦缎台布
venise à réseau 威尼斯针绣花边
venise point lace 威尼斯针绣花边
ven shan silk 文尚葛
vent (衣服)开缝,开衩,衩口,背衩;骑马缝
vent allowance 开衩宽放量
vent knit 浮线组织图案
vent line 衩位线;开衩线
vent seam 开缝,开衩缝
ventaglio 优质网眼针绣花边
ventail (中世纪)护面具
venter 腹
ventilated clothing 通风服
ventilated collar 通风领(有小孔的白色硬立领)
ventilated rigid protector (橄榄球)通风刚性保护器
ventilated shoes 通风鞋
ventilating insole 透气性袜底,透气性鞋垫
ventilating net underwear 通风网状内衣
ventile 文泰尔防雨布
ventile fabric 防雨[布]织物,透气织物
ventile material 文泰尔材料
ventilette 文提列特网眼织物
2-vent style 两衩式
3-vent style 三衩式
venustra 棉经黏胶丝纬巴里纱
Vera Maxwell 维拉·马克斯韦尔(1903~,美国时装设计师)
Vera Wang 薇拉·王

verdancy 翠绿,嫩绿,铜绿
verdant green 嫩绿色
verdigris color 铜绿色
verdigris green 铜绿色,铜锈
verdour 风景挂毯
verdugado 西班牙拱形衬裙
verdugal(l)e 西班牙拱形衬裙
verdugo 西班牙拱形衬裙
verdure 青绿色,青翠色
verification 核验,证实,校验,鉴定
verisimilitude 逼真
vermeil 朱红色,鲜红色
vermeil vermilion 朱红
vermicelli 窄编带,波纹线迹绗缝法
vermil(l)ion 朱红色,鲜红色
vermin-proof 防虫的
verona serge 棉毛交织薄哔叽
veronica 有耶稣面像的汗布,印有耶稣面像的织物(或衣服);婆婆纳色(青莲色)
Veronica Lake (hair style) 维罗妮卡·莱克发型
versatile application 多用途,多能应用
versatile laying-up machine 多能展叠料机
versatile machine 多能机
versatile sewing machine 多功能缝纫机
versatility 多样性
versatility sewing 多功能缝纫
versicolour 闪色,虹彩;多色,杂色
version 型,形式,版本
vert (纹章的)绿色
vert pomme [法]青苹果黄绿色
vertebra 椎骨,脊椎
vertebral column 脊柱
vertex (v) 头顶点
vertical 立式,竖边
vertical-axis sewing machine 垂直轴缝纫机
vertical border machine 立式饰边缝纫机
vertical cutting 垂直剪切
vertical cutting machine 直刀裁断机
vertical cutting mechanism 垂直切割机构
vertical dart (垂)直省
vertical division 垂直分割
vertical electric cutting machine 立式电动裁剪机
vertical grain 直丝绺
vertical hemming stitch 垂直缝边针法,直缝边针迹,直折缝
vertical hook 立式钩梭

vertical knife cutting machine 立式刀裁剪机
vertical leno 直罗
vertical line 垂直轮廓,垂直线,直条疵
vertical mark 直角符号
vertical multi-finisher 垂直多功能熨烫机
vertical pants finisher 垂直式裤子熨烫机;裤型烫机
vertical plane 垂直面
vertical pocket 垂直口袋
vertical pointed twill 经向山形斜纹
vertical seam 垂直缝
vertical section 纵断面
vertical stitching 垂直线缝
vertical streak 纵条痕(织疵)
vertical stripe 纵条痕
vertical style line 垂直款式线
vertical trimmer 立式修剪机
vertical type chain cutter 立式线辫切刀
vertiver green 香根草绿
vertomat 自动整烫机
vertugadin [法]裙撑
vertugado skirt with aro 带骨架的裙撑
vertugale 裙撑
verve 活力热情;神韵
very berry 纯酱果色
very grape 纯葡萄色
very important person jeans (VIP) 名家牛仔裤,名人牛仔裤
very slightly imperfect (V.S.I) 微瑕;二号花
very very slightly imperfect (V.V.S.I) 极微瑕;一号花
vessel 船舶
vest 汗衫,内衣,衬衣;背心,马甲,防护衣;(女服)胸前V形装饰布;(教士的)法衣;祭服;(古代)衣服,服装,长袍,外衣
vest button 背心纽扣
vest de chasse 猎装夹克
vest de judo-ka (日式)柔道上衣
vested suit 有马甲的套装
vestee (女服前胸)V形饰布;背心形胸衣
vestiary 衣帽间,藏衣室;(教堂等的)法衣存放室
vestido 手织羊毛长背心(至腿部,有饰摆)
vesting 马甲料;背心料
vest in jacket style 夹克式背心
vest look 背心款式

vestment 制服;法衣;祭服;官服,礼服
vestment cloth 中世纪金丝绣丝绸,织成动物(田园)图祭、用浮雕法织造的丝织物
vestments 衣服,服装
vestock (牧师戴的)长绸领带
veston 西服便装;[法]西服上装
veston-chemise 夹克式衬衫;衬衫式夹克
vest pocket 背心口袋,表袋
vestry room (教堂的)法衣圣器储藏室
vest suit 背心套装
vest sweater 背心式毛衣,针织开襟背心,针织套头V领背心
vesture 衣服,服装;罩衣,罩袍
vest with hood 带帽背心
vest with poppers 按扣背心
Vibram 伐柏拉姆牌登山靴橡胶底;伐柏拉姆牌橡胶底登山靴;耐用合成橡胶商标
vibrant colour 鲜明色,富有活力的色彩
vibrant green 明亮绿
vibrating presser [foot] 摆动压脚
vibration free 防震,无震动
vibrant yellow 鲜黄色,菜花黄
vichy 色织格子布,大方格布,双色纱棉布,维希呢(黑白小格的衣料)
VICHY 薇姿
vicogne 仿骆马绒织物
Victoria 维多利亚绸,维多利亚布,维多利亚双重斜纹呢
Victoria crepe 维多利亚绉布
Victoria lawn 维多利亚细节
Victorian 维多利亚风格;维多利亚式服饰
Victoriana 维多利亚印花
Victorian blouse 维多利亚女衫
Victorian collar (同 choker collar) 贴颈领,硬高领
Victorian fashion (1837～1901)维多利亚风格
Victorianism 维多利亚时代的服饰
Victorian lawn 维多利亚细布
Victorian mood 维多利亚样式
Victorian style 维多利亚风格
Victorian-styled collar 维多利亚式领
Victoria shawl 维多利亚提花披肩
Victoria silk 无丝鸣丝毛织物
Victoria sleeve 维多利亚袖
Victoria's secret 维多利亚的秘密
victorine (女式)毛皮披肩,毛皮围脖

Victor Stiebel 维克托·斯蒂贝尔(英国时装设计师)
victory stripes 胜利斜纹布(蓝白或棕白条子,蓝色纬)
vicugna 小羊驼毛服装,骆马绒服装;小羊驼毛织物,仿小羊驼毛织物,仿骆马绒织物
vicuna 骆马绒服装,小羊驼毛服装;小羊驼毛织物,仿小羊驼毛织物,仿骆马绒织物
vicuna fancy suitings 骆马绒花呢;维柯纳花呢
vicuna finish 仿骆马绒整理
vicuna fur 美洲驼毛皮
vicuna overcoating 骆马毛大衣呢
vicuna suiting 骆马绒花呢
vicuna wool fabric 骆马绒毛织物
vieley cloth 色纱绐布
Vienna clothing 维也纳外套(燕尾式后摆)
Vienna coat 维也纳外套
vigans [法]维冈粗呢
vignette 渐晕[图案],晕映
vignette print 渐晕印花花纹
vignetting 渐晕,光阑阻
vignetting effect 渐晕效应
vigogne [法]驼马绒;仿驼马绒织物;棉毛纱
vigogne yarn 棉毛纱,充棉毛纱
vigomemom 法国天然彩毛混纺呢
vigorous and firm 雄浑
vigoureux printing 毛条印花
viking lamb 海盗羔羊皮
Vilene 维莱恩衣衬
vine 蔓草纹样
vine black 葡萄黑
vineyard green 葡萄园绿
vintage 怀旧
vintage clothes 老式服装,过时服装
vintage fashion 古典服装
vintage look 古代服饰风貌
vinyl 乙烯基织物;聚乙烯基薄膜
vinyl baby pants 聚氯乙烯塑胶婴儿尿裤
vinyl boots 乙烯面人造靴
vinyl coating 维尼纶涂层的紧身短上衣
vinyle-cotton fine fabric 维棉细布
vinylidene chloride 偏二氯乙烯纤维
vinyl leather 乙烯人造革
vinyl leather raincoat 聚氯乙烯塑胶雨衣
vinyl rain cape 聚氯乙烯塑胶雨披

vinyl rain coat 聚氯乙烯塑胶雨衣
vinyl rain cover 聚氯乙烯塑胶雨衣
vinyl sandals 塑料凉鞋
vinyl shell coat 聚氯乙烯塑胶里衬雨衣
vinyl shoes 塑料鞋;乙烯面人造革鞋
vinyl slippers 塑料拖鞋
vinyl sponge leather 聚氯乙烯涂料海绵布
vinylon 维尼纶
vinylon/cotton blended yarn 维棉纱
vinylon/cotton yarn-dyed check 色织维棉条格布
vinylon stretch sewing thread 维纶缝纫线
vinylon thread 维纶线
vinylon yarn 维纶纱
viola 菜色,青紫色
violaceous 紫罗兰色
violet 紫色,紫罗兰色;紫色衣服,紫色衣料
violet ash 淡白紫
violet azalea 紫杜鹃色
violet black 黑紫,墨紫
violet bloom 紫罗兰粉霜色
violet cream (化妆用)紫罗兰香型乳霜
violet deep 暗紫色
violet light 鲜紫色
violet mink 紫貂皮
violet ray 紫色光线;紫外线
violet tulip 紫郁金香,雪青
violin bodice 小提琴形紧身胸衣
Vionnet 维奥内(法国时装设计师)
VIP jeans (very important person jeans) 名人牛仔裤,名家牛仔裤
virago sleeve 藕节袖
virdine green 艳黄绿色
virescence 开始呈现的绿色
virginia cloth 廉价粗麻布
virginie 弗吉尼亚斜纹绸,弗吉尼亚花哔叽
virgin wool 初剪羊毛,新羊毛
viridescent 淡绿色的,带绿色的
viridian 碧绿,翠绿,深翠绿
viridine yellow 浓黄绿色
viridis 松石绿
viridity 碧绿,翠绿;新鲜,活力,雄浑有力
virtuousity 艺术鉴赏;精湛技巧
visage 脸,面容,面貌,外表
visagist(e) 美容师,化装师
visca braid 维斯卡编带

VISCAP 维斯凯
viscose 黏胶纤维,黏纤
viscose/acrylic jacquard decorative fabric 黏腈大提花装饰织物
viscose/cotton blended yarn 黏棉纱
viscose/cotton plain cloth 黏棉平布
visco-elastic body 黏弹体
visco-elasticity 黏弹性
viscose fancy 黏胶花式绸,依群绸
viscose fiber 黏胶纤维
viscose filament 黏胶长丝
viscose filament yarn 黏胶长丝
viscose plain cloth 黏胶纤维平布
viscose/polyamide gabardeen 黏锦华达呢
viscose rayon 黏胶纤维,黏胶人造丝
viscose rayon taffeta 黏胶丝塔夫绸,黏纤塔夫绸
viscose/rayon tow 黏胶丝束
viscose/rayon yarn 黏胶人造丝
viscose rayon yarn embroidery thread 人造丝绣花线
viscose serge 黏胶纤维哔叽
viscose staple fibre 黏胶短纤维
viscose staple fiber spun yarn 黏胶纤维纱线
viscose top 黏胶毛条
viscose/vinylon plain cloth 黏维平布
visibility 可见度
visible defect 外观疵点
visible fleecy fabric 平针衬垫针织物
visible light 可见光
visible outline 外形线
visible pleat 明褶裥
visible spectrum 可见光谱
visible tuck 明裥
vision 视觉
vision electronic recording apparatus 电子录像机
visite 无袖斗篷拜访服
visiting dress 拜访裙装,访问服
vison 貂皮
visor (帽子上的)遮阳,帽檐,帽舌;(头盔上的)面甲,面罩,护face;假面具
visored cap 大盖帽
visual aid 直观教具
visual appearance 外观特征
visual artist 观赏艺术家,视觉艺术家
visual arts 视觉艺术,观赏艺术
visual assessment 目测鉴定
visual communication design 视觉传达设计
visual difference 视差
visual discrimination 目测鉴别
visual education 直观教育
visual effect 视觉效应
visual evaluation 目光鉴定,目光评价
visual examination 外观检验,目测检验
visual feature 视觉特征
visual inspection 目测,肉眼检查
visual measurement 目测
visual merchandising 展示销售(强调利用视觉的陈列效果进行推销)《销售》
visual perception 视觉
visual persistence 视觉暂留
visual sensation 视觉
visual standard 目测标样
visual test 肉眼检验
vital characteristics 主要特性,重要特性
vitality of company 企业活力
vital measuring 量身材
vitamin(e) color 维生素色(指维生素片剂的鲜嫩颜色)
vitaminised cloth 维生素布,维他命布
vitelline 蛋黄色
vitmoutiers 本色粗松亚麻布
vitrail light (钻石)紫光
vitrail medium (钻石)七彩
vitrees 漂白亚麻布,法国大麻船帆布
vitreous luster 玻璃光泽
vivid (色、光)强烈的,鲜明的,鲜艳的
vivid blue 碧蓝,深湖蓝
vivid chartreuse 嫩黄绿
vivid color 鲜艳明亮色泽
vivid green 鲜艳绿,鲜绿色,嫩绿色
vividness 鲜艳度,鲜明度
vivid tone 鲜艳色调
Vivienne Tam 谭玉燕
Vivienne westwood 维维恩·韦斯特伍德(1941~,英国时装设计师)
vixy 维克塞
viyella 毛棉混纺法兰绒
viyella flannel 维耶勒法兰绒
vizard 帽子遮阳,帽舌;面罩,面甲,脸盔,护面;假面具;伪装
vizor (头盔上的)面甲,面罩,护面;(帽子上的)遮阳,帽舌;假面具
vladimir 起毛薄呢
vleiger 高领女装
V line style V形服
V look V形款式

V-neck blouse　V形领上衣
V-neck(line)　V形领口；V字领
V-neck pullover　尖领套衫；V领套衫
V-neck pullover sweater　V形领套头毛衫
V-neck sweater　V形领毛衫，夹领毛衫
V-neck woolen sweater　V字领羊毛开衫
V. N. mixed gabardine (viscose-nylon mixed gabardine)　黏锦华达呢
vocabulary in garment design　服装设计用语
vocational career apparel　普通工作服
vocational career apparel fabric　职业装织物
vocational training　职业训练
vogue　时尚，流行，时髦，潮流；流行物
VOGUE　《时髦》,《时尚》(杂志)
VOGUE BAMBINI　《流行童服》(意大利)
VOGUE HOMME　《时髦男装》(法国)
void cheque　作废支票
void content　孔隙率,含孔隙量
voided fabric　烂花织物
voided velvet　花式丝绒
voigt model　伏伊特模型
voile　巴里纱,玻璃纱,薄纱,绡
voile blouse　巴厘纱衬衫
voile broche　绡
voile de laine　透明毛薄纱,透明毛巴里纱
voile gabardine　稀薄轧别丁,稀薄华达呢
voile marquisette　全丝薄纱罗
voile ribbon　薄纱带
voilette　[法]女帽短面纱；仿巴里纱；芙阿兰特花边,细网眼地机制花边
voile yarn　巴里纱用线
voiron　精细家用亚麻布
volant　宝塔式荷叶边裙
volet　伏立披纱；伏立旗

volley-ball wear　排球装
volley-ball shoes　排球鞋
voltex sewing-knitting machine　伏尔特克斯缝编机
volume　量感
volume color　空间色
volume fashions　大众化流行款式
volume novelty　膨体花式纱线
volume of imports　进口量
volume production　大批量生产《管理》
volume zone　最佳价格带《销售》
voluminous wig　多圈假发
voluminous yarn　膨体纱
voluntary chain(VC)　自愿联营批发站《销售》
vortex spun yarn　涡流纺纱
voucher　传票
vraie　手织带
vrai réseau　花边底网
vrai valenciennes　丝绣花边
vray　手织带
V-shape look　V形服装款式
V-shape neckline　V形领口
V-shape silhouette　V形轮廓
V-shaped space　领嘴
V-shaped yoke　V形过肩
V style　V形款式
V-throat　V字鞋
V-type knitting machine　V型横机,V型畦编机
vulcanized rubber　硫化橡胶,橡皮
vulcanized shoes　硫化鞋
vulcanizing press　模压《制鞋》
VV　前后V形领口
V zone　(西装的)领开部分,V形敞开领部分

W

W 腰围（waist）；女子尺码（women's size）
Wacoal 华歌儿
wad （用填料）填塞；填絮；填块，填料；软衬料
wadded bedford cords 垫经凸条纹布，垫经灯芯布
wadded cloth 衬垫经织物，衬垫纬织物
wadded clothes 棉衣；填棉服装
wadded double cloth 衬垫双层织物
wadding 填絮，絮料，絮片，软填料；衬料；衬垫
wadding sheet 定形棉，合纤絮片
wadding threads yarn-dyed fabric 填芯织物
waders （涉水、捕鱼等时用的）高筒防水胶靴；钓鱼靴
wadmal 瓦德麦尔呢，重厚拉毛粗纺毛织物
wadstena lace 瑞士农民梭结花边
Wa ethnic costume 佤族服饰
waffle backing 泡沫层衬垫，蜂窝衬垫
waffle cloth 蜂窝纹布，方格纹织物
waffle pique 蜂窝纹布；上等蜂窝组织棉织物
waffle shirring 方格式碎褶（以松紧线做方格式压缝而成的效果）
waffle stompers 宽底旅行鞋
waffle surface 凹凸表面，蜂窝形表面
wafuku 日本和服
wage level 工资水平《管理》
wage pattern 工资标准《管理》
wage rate 工资率《管理》
wage scale （行业或地区等的）工资等级、工资级别《管理》
wahrendorp 未漂粗麻布，漂白细麻布
waidemar 瓦尔德马优质棉绒布
waif check 华夫格
waif satin brocade 华夫锦
waikiki [shirt] 夏威夷衬衫，阿罗哈衬衫
waist 腰，腰部；腰围（W）；上身部分；腰部分；背心；紧身胸衣，乳袼；儿童内衣；（鞋子）腰窝；（衣裤的）腰身
waist adornment 佩饰
waist area 腰区
waist around 腰围
waist back length 背长；后腰节长
waist bag 腰袋
waistband 腰带，裤带，裙带；裤（裙）腰，腰头
waistband braid 腰头饰带
waistband carrier 腰袢，腰头蚂蝗袢
waistband closure 腰带搭扣
waistband crepe 腰带绉
waistband extension 腰带嘴，腰头探出
waistband for pyjamas 睡衣腰带
waist banding 绱裤腰头
waistband interfacing 腰里（腰头里子）
waistband interlining 腰头衬里
waistband lining 腰里，腰头衬里，腰头夹里
waistband machine 绱腰头机
waistband neekline 腰头下口线
waistband roll line 腰头上口线
waistband seam 腰带缝
waistband seam opening press 腰头缝分开熨烫机
waistband sewing attachment 缝腰带（裤带、裙带）附件
waistband sewing machine 腰带缝纫机
waistband stiffener 腰带硬衬
waistband width 腰带阔，裙带阔，裤带阔，裙腰阔，裤腰阔
waist belt 腰带，裤带，裙带
waist bolster 裙环垫圈，腰臀垫圈
waist cinch(er) 法国腰带；束腰宽带
waist circumference 腰围
waist cloth 围腰布
waistcoat 马甲，西服背心紧身内衣；（18世纪的）女骑装马甲；（16～17世纪的）男衬身服
waistcoat belt 背心腰带
waistcoat blouse 背心式女短上衣
waistcoating 西装背心，马甲，背心面料
waistcoat in jacket style 夹克式背心
waistcoat paletot 女式及膝长外套
waist coat pocket（同 watch pocket, vest pocket）表袋，背心口袋
waist cut seam 腰省缝
waist dart 腰褶，腰省
waist-deep 齐腰深（的）

waist depth briefs 齐腰三角裤
waist drawstring （衣服）腰部拉绳
waist edge （鞋子）腰窝边
waisted dress 高腰裙
waisted heel 腰形鞋跟,计时砂漏式鞋跟
waist extended （松紧）腰围拉度;腰围拉开计
waist girth 腰围;胕围
waist girth line 腰围线,掐腰线
waist-high 齐腰高
waist-hip length 腰臀长
waist-hip ratio adjustmen （孕妇服）腰臀比调节器
waisting 女用仿男式衬衫料,背心料子
waist-length 腰长,腰节长
waist-length hood jacket 齐腰连帽夹克
waist-length jacket 齐腰长夹克
waist level (W.L.) 腰围线
waist line (W.L.) 腰节;腰围;收腰线,围线,腰节线;（女服的）腰身部分;裙腰缝线
waistline casing 腰线处串带管
waistline dart 腰褶;裤腰省
waistline seam 腰缝
waist mark 显示腰围
waist measurement 腰身
waist nipper 夹腰,紧身腹带;束腰式紧身衣
waist ornament 腰饰
waist petticoat 无上身衬裙
waist-plated last （底面仅腰裆部位安装铁板的）鞋楦
waist pleat 裤腰褶裥
waist pocket 腰袋
waist press 腰围（里）压烫机
waist relaxed 平腰围,缩后腰围
waist rib 腰罗纹,下摆罗纹
waist-seam 腰缝,腰围缝合线
waist-seam jacket 腰围缝线夹克
waist-seam line 裙腰缝线,裤腰缝线
waist slip 短衬裙
waist stay 腰部牵带,腰部固定带
waist stretch 伸长腰围,未缩腰围
waist support (er) 护腰
waist-suppressed 紧腰身的,腰身收缩的
waist tab （上衣、背心的）腰扣襻
waist tag 腰卡（裤腰上的纸牌）
waist-to-crotch area 腰线到裤裆的区域
waist-to-hip 直裆,立裆
waist-to-hip area 腰至臀部的区域
waist vest 马甲

waist watcher 尼龙搭扣式紧腰弹力带
waist welt pocket （西服背心）下口袋
waist yoke 抵腰
waiter's 侍者服
waiter's uniform 服务员制服
waiter's wear 服务员服装
waiting time 等待时间《管理》
waive duty 免收税款
walachain embroidery 瓦拉几安刺绣
wale 条,线圈纵行;纹路清晰斜纹;凸条纹
wale deflection 线圈纵行歪斜
walenki ［俄］俄式毡靴
wale streak 经柳疵,直条花疵
walkers （低跟）轻便鞋;男用外穿短裤;（幼儿）学步鞋
walking boots 散步靴;便鞋
walking coat 散步服
walking costume 轻便服
walking dress 散步服,轻便服,外出服
walking figure 行走姿式
walking foot 双送压脚
walking [frock] suit 散步套装
walking gloves 散步手套
walking jacket 散步服
walking pleat 倒褶裥,助行褶
walking shoes 步行鞋
walking shorts 步行短裤,男用外穿短裤
walking stick 拐杖;散步手杖
walking stitch 步行线迹
walking suit 散步套装,女子散步裙装（七分长外套和毛皮饰带的粗花呢直裙）
walkjanker 瓦尔克防寒外套（奥地利山岳地区穿用）
walking frock suit 散步福乐套装
walk test 踩踏试验,行走试验
wallaby 小袋鼠式靴,袋状靴;小型袋鼠毛皮
Wallace Beery knit shirt 华莱士·比里针织衫,罗纹圆领半开襟针织衫
wallachian embroidery 线圈针迹刺绣
wall carpet 壁毯
wall clock 挂钟
wall cloth 贴墙布
walled toe 墙式鞋夹
wallet （放钞票等的）皮夹子;皮制小工具装,钩鱼袋;（香客等的）行囊;旅行袋
wall furnishing fabric 壁饰织物
wall linen 亚麻贴墙布
wall paper 墙纸
wall paper print 墙纸花样;墙纸印花

Wall Street 华尔街风格
wam(m)us 连腰带开襟绒线衫
walnut 胡桃色
walnut brown 胡桃壳棕(浅棕色)
walnut husk 胡桃壳色
walrus 海象皮革;矮胖子
Walter Albini 瓦尔特·阿尔比尼(意大利时装设计师)
Waltex loom 瓦尔特克斯管针经编机
Waltex loom knitting machine 瓦尔特克斯管针针织机
waltz-length 中腿肚长度
waltz-length gown 中长袍
waltz-length nightgown 中长睡袍
wampum belt 印地安饰带
wampum peag 白贝壳串珠
Wan 纨
wan blue 暗蓝色
wanshou brocaded velvet 万寿绒
wanshou crepe satin brocade 万寿缎
wanshow satin brocade 万寿缎
wanzi tapestry velvet 万紫绒
wappen 袋口刺绣,装饰纹章,衣料制徽章
war 纨
war bonnet 羽毛头饰
war color 卡其色;保护色
war look 战时风貌,战时款式,军装款式
wardrobe (个人的)全部服装;(为某季节或某种活动用的)全部服装;(剧团的)全部服装,行头;戏装保管室,衣服保管库;挂衣箱,衣柜,衣橱
wardrobe case 挂衣箱
wardrobe dealer 旧衣商《销售》
wardrobe glass 衣柜镜;穿衣镜
wardrobe master 戏装男保管员
wardrobe mirror 衣柜镜;穿衣镜
wardrobe mistress 戏装女保管员
wardrobe trunk 柜式衣箱
warehouse 仓库
warehouse coat 工作服,防护服,仓库工作服
warehouse receipt 仓库收据
ware room 商品陈列室;商品贮藏室
warfare fabric 军用织物
warm 保暖的东西;罩衫;头布;军用双排纽扣短大衣
warm air comb 热吹风梳
warm apricot 橙杏色
warm based colour 暖基色
warm clothes 保暖服装

warm cobalt 暖钴色
warm (cold) tone 暖(冷)色调
warm color 暖色
warm color scheme 暖色配色
warmer 保暖衣物;暖颈套
warm hue 暖色调
warming up period 预热时间
warm iron 中等温度熨烫
warmness 暖感
warmness of hamdle 温暖手感
warm outwear 防寒服
warm pants 暖裤(比热裤稍长的短裤)
warm rinse 温水漂洗
warmth retaining property 保暖性能
warmth retention property 保暖性能
warmth underwear 保暖内衣
warm tone 暖色调
warm-up jeans 保暖型牛仔裤
warm-up jumpers and pants 运动服;运动衫裤
warm-up pants(同 sweat pants) 厚绒运动裤;(滑雪用)衬里套裤
warm-up (suit) 保暖服;(做运动准备活动时穿的)运动服
warm wear 防寒服
warm weather fabric 热天织物
warning color 警戒色
warning coloration 涂有警戒色
warning mark 警告性标志
warp (织物的)经纱;罩衫;头布
warp and weft insertion knitted fabric 衬经衬纬针织物
warp around 围卷裙
warp backed cloth 经二重织物
warp count (织物的)经密
warp density 经密
warp effect 经面花纹,经面效应
warp faced fabric 经面织物
warp faced twill 经面斜纹
warp falling 缺经,断经(织疵),经吊痕
warp fault 错经(织疵)
warp figured fabric 经纹织物
warp finings 梭织花边曲折线迹
warp flat 双经(织疵)
warp float 经向跳花(织疵)
warp flush sateen 经缎,经面棉缎
warp flush twill 经面斜纹
warp gauze weave 直罗
warp grain 经向纹理,经向丝缕
warp holding place 松紧条痕(织疵),经吊痕

warp ikat 扎染经纱布
warp-insertion weft knitted fabric 衬经纬编针织物
warp-knit fabric 经编针织物
warp(-knit) lace 经编花边
warp-knitted bath towel 经编浴巾
warp-knitted biaxial fabric 经编双轴向织物
warp-knitted blanket 经编毯,经编毛毯
warp-knitted branching tubular fabric 经编筒状分枝织物
warp-knitted brushed 经编绒布
warp-knitted brushed or raised fabric 经编绒布
warp-knitted camleteen 经编骆驼绒
warp-knitted cauterizing fabric 经编烂花织物
warp-knitted check fabric 经编方格织物
warp-knitted check or marquisette net fabric 经编格子网眼织物
warp-knitted clipped fancy fabric 经编剪割花纹织物
warp-knitted cloth 经编呢
warp-knitted corduroy fabric 经编灯芯条织物,经编灯芯绒织物
warp-knitted cot-press fabric 经编花压织物
warp-knitted couterising fabric 经编烂花织物
warp-knitted curtain with weft-in-serted yarn 经编全幅衬纬窗帘织物
warp-knitted double-faced fabric 经编双面织物
warp-knitted double-face terry fabric 经编双面毛圈织物
warp-knitted double-needle bar velour 经编双针床丝绒织物
warp-knitted duo-elastic fabric 经编双向弹力织物
warp-knitted elastic bandage 经编弹力绷带
warp-knitted embroidered fabric 经编绣纹织物
warp-knitted eyelet fabric 经编网眼织物
warp-knitted fabric 经编针织物
warp-knitted glaces 经编闪光缎
warp-knitted high pile fabric 经编长毛绒
warp-knitted imitation fu 经编人造毛皮
warp-knitted jacquard bed cover 经编提花床罩
warp-knitted jacquard fabric 经编提花织物
warp-knitted jacquard terry fabric 经编提花毛巾织物
warp-knitted jacquard velour 经编贾卡丝绒织物
warp-knitted knotless net 经编无结网
warp-knitted lace fabric 经编花边织物
warp-knitted laid-in fabric 经编衬纬织物
warp-knitted man-made fur 经编人造毛皮
warp-knitted mesh fabric 经编网眼针织物
warp-knitted miss-press fabric 经编缺压织物
warp-knitted mosquite net fabric 经编蚊帐织物
warp-knitted multi-axial fabric 经编多轴向织物
warp-knitted multibar decoration fabric 经编多梳栉装饰织物
warp-knitted napped fabric 经编起绒织物
warp-knitted net 网孔经编织物
warp-knitted net or eyelet fabric 经编网眼织物
warp-knitted net singlet fabric 经编网眼汗衫布
warp-knitted overfeed fabric 经编超喂圈毛织物
warp-knitted packing sack 经编包装袋
warp-knitted partial laid-in fabric 局部衬纬经编织物
warp-knitted pile fabric 经编毛绒织物
warp-knitted plaited fabric 经编褶裥织物
warp-knitted press fabric 经编压纱织物
warp-knitted queen's cord 经编昆士织物
warp-knitted raincoat fabric 经编雨衣布
warp-knitted scarf fabric 经编头巾织物
warp knitted sharkskin fabric 经编雪克斯金织物
warp-knitted sheared terry fabric 经编毛圈剪绒织物
warp-knitted silk fabric 经编真丝绸
warp-knitted single-face terry fabric 经编单面毛圈织物
warp-knitted singlet fabric 经编衬衣织物
warp-knitted stockings 经编长筒女袜
warp-knitted space fabric 经编间隔织物

warp-knitted stretch fabric	经编弹力织物
warp-knitted suede fabric	经编仿麂皮织物
warp-knitted table cloth	经编台布
warp-knitted terry fabric	经编毛圈织物,经编毛巾布
warp-knitted towel	经编毛巾
warp-knitted tubular fabric	经编筒状织物
warp-knitted tulle net fabric	经编六角网眼织物
warp-knitted twill fabric	经编斜纹织物
warp-knitted velvet	经编天鹅绒,经编立绒
warp-knitted velvet fabric	经编丝绒织物
warp-knitted warp fabric	经编围巾织物
warp-knitted yarn-dyed fabric	经编色织物
warp-knitting cloth	经编呢
warp-knitting machine	经编针织机
warp lace	经编花边
warp lace machine	经编花边机
warp look	卷缠款式
warp loom	经编机
warp machine knitter	经编机
warp neck	叠领
warp net	经编花边
warp net frame	花边织机
warp ondulé	经向波纹织物
warp pants	卷裹裤(宽腰带打结)
warp pile astrakhan	经起毛仿羔皮织物
warp pile fabric	经起绒织物
warp pile velvet	经编天鹅绒
warp pique	经向凸纹布
warp plush	经长毛绒织物
warp printed cretonne	印经装饰布
warp printed fabric	印经布,印经织物
warp printed taffeta	印经塔夫绸
warp printing	经纱印花
warp rep	经棱织物
warp rib top machine	绣花罗纹袜口机
warp sateen	经缎,经面棉缎
warp satin	经面缎纹,经面缎织物
warp soleil	经浮光亮塔夫绸
warp stitch	经纹线迹,链状线迹
warp stockings	经编长筒袜
warp stretch woven fabric	经向拉伸织物
warp striped fabric	经条呢
warp-to-filling seam	斜跨经纬接缝
warp topper	宽松短大衣
warp-to-warp seam	横跨经纱接缝
warp twill	经面斜纹
warp twist	Z 捻,反手捻;经纱捻度
warp velvet	经平绒,丝天鹅绒
warp wadded double cloth	垫经双层织物
warp-way twist	Z 捻,反手捻
warp yarn	经纱
warsaw-Oxford rules	华沙-牛津规则
wartime style	战时服装风格
waschi	中世纪红绸
wash	(服装)洗水,洗涤;洗涤物;洗涤剂;洗衣房,洗衣店;水洗;耐洗
washability	耐洗性
washable	可洗的,耐洗的
washable silk	耐洗丝绸,洗水丝绸,树脂整理丝绸
wash and use [wear]	免烫,洗可穿
wash and wear	免烫,洗可穿;洗可穿服装
wash and wear cycle	洗可穿周期
wash and wear finish	洗可穿整理,免烫整理
wash and wear finishing	永久定形熨烫,PP 加工
wash and wear hairstyle	永久卷曲发型
wash and wear properties	洗可穿性能
wash and wear suit	洗可穿,免烫;洗可穿服装
washateria	自助洗衣店
wash-bag	梳妆袋
wash blonde	白网眼棉花边
wash board	洗衣板,搓板
wash boards	(针织布)洗后扭斜,(洗衣板状)凹凸不平
wash brush	洗衣刷
wash care label	洗标,洗涤标志
wash cleaning	水洗
wash cloth	方巾,浴巾,毛巾
wash clothes	洗衣
wash cotton	耐洗棉布
wash drawing	(以线条为轮廓的)淡彩画;淡彩画法,单色画法
washed denim	水洗牛仔布
washed fabric	水洗布
washed-out jeans	褪色型牛仔裤
washed retro look	水洗做旧
washed silk	水洗绸
washed wool	水洗毛
washer	洗衣机;洗衣工;洗水布;水洗槽;廉价粗仿呢;垫圈

washer cloth 水洗绉,水洗绉织物
wash-crease resistant 防洗皱
washerette 自助洗衣店
washer extractor 洗衣脱水机
washer for knitted fabric 针织物洗涤机
washer mark 洗呢斑
washer standard of costume 服装洗涤标志
washer wrinkle fabric 水洗布(有绉纹)
washer wrinkle 水洗整理织物;洗涤折痕
washes 洗涤液;洗涤废水
wash fabric 耐洗织物
wash fast 耐洗的
wash fastness 耐洗牢度
wash-finishing fabric 水洗织物
wash glove 沐浴擦背袋
wash goods 耐洗衣服;耐洗织物
wash house 洗衣间,洗衣店
washi 桑叶纤维非织造布
washing 洗水;洗涤;洗涤剂;洗涤物;洗净
washing cap 水洗帽
washing cloth 水洗布
washing equipment 水洗设备
washing filler 水洗棉
washing instruction 洗涤说明;洗涤法
washing label 洗水唛;洗涤商标;洗涤说明
washing machine 洗衣机,水洗机
washing machine for woolen yarn 绒线洗涤机
washing powder 洗衣粉
washing quality is not good 水洗不良
washing resistance 耐洗性
washings 洗液
washing shrinkage (织物)缩水率
washing silk 耐洗丝绸
washing treatment 水洗处理,洗涤处理
wash inside out (以保护正面的)反面洗涤
wash leather 油鞣革;揩拭用麂皮
wash out 洗去;洗掉
wash-out jeans 退色型牛仔裤
wash-proof finish 耐洗整理
wash rag 毛巾,洗涤用布
wash removability 洗污能力
wash satin 耐洗缎
wash separately 洗涤分开
wash silk 耐洗绸缎,耐洗丝绸
wash tank 水洗槽

wash test 耐洗牢度试验
wash-waisted 蜂腰的,细腰的;束腰的;紧胸的
wash-wear 免烫,洗可穿
wash-wear rating 洗可穿等级
wash well 耐洗
wash-white 廉价粗纺呢
wash whites 白色薄法兰绒
waspie 妇女紧身胸衣,妇女紧身胸带
wasp-waist(ed) silhouette 束腰型轮廓,蜂腰型轮廓
waspwaist 细腰,蜂腰
wasp-waisted 细腰的,蜂腰的;束腰的;紧胸的
wasp-waisted dress 紧身女礼服
wasp-waisted suit 束腰式套装
waste 回丝,废料
waste cloth 废纺纱织物;零碎布料,布头布尾
waste cotton 废棉
waste duck 回纱帆布(用于便服),再生纬纱帆布
waste fabric 废布碎料
waste plains 低级平布,纬起绒棉平布
waste silk 废丝,绢丝
waste silk blend spinning fabric 绢丝混纺织物
waste silk yarn 绢丝
waste yarn 废纺纱
watch 手表,挂表
watch band 表带
watch bracelet 表镯
watch cap 巡夜水兵帽,值班站岗风帽
watch chain (挂表的)表链
watch cloak 巡夜外套,站岗防寒大衣
watch coat 巡夜外套,站岗防寒大衣
watch strap 表带
watchet 14世纪英国浅蓝毛织物
watchet blue 浅蓝的,淡蓝的;淡蓝色的布
watchet kersey 英国粗纺斜纹呢
watch guard 表带,表链
watchman plaid 巡夜者格子纹
watch pocket (同 vest pocket, waist coat pocket) 表袋
watch ribbon 表缎带
water 水色
water absorbability 吸水性,吸湿性
water bean 睡莲色
water blue 清水蓝,浅暗绿蓝色

water boots　水靴
water bottles belt　水壶带
water buffalo hide　水牛皮
water color　水彩画；水彩颜料；水彩画技艺
water content　含水量,含水率
water-cooled clothing　水冷服
water-cooled suit　水冷服（装）
water-cooling jacket　水泠却夹套
water damage　水渍
watered gauze　香云纱
watered poplin　波纹无葛,波纹府绸
watered silk　波及绸
water-fall　披挂长卷发
water-fall back　波浪裙
water-fall frill　大瀑布式花边
water-fall front skirt　垂瀑前饰裙
water-fall necktie　男式瀑布形大领巾
water fastness　耐水牢度
water grass green　水草绿
water green　水绿（浅暗黄绿色）
water impermeability　不透水性
watering　波纹,云纹；云纹绸；松板印（疵）；云斑（疵）；衣料下水（预缩水）
watering cloth　波纹布,云纹布
water-jet cutter　高速喷水裁剪器
water-jet entangled nonwoven　水刺法非织造布
water-jet nowoven　水刺法非织造布,射流针刺法非织造布
water-jet woven fabric　喷水织机织物
water-laden fabric　湿透的织物,吸满水的织物
water lily　睡莲色
watermark　水渍,水印
water marked finish moire finish　波纹轧光整理
water marked tabby　波纹塔夫绸
water opal　水蛋子石
water permeability　渗透性,渗水性
water pimpernel　水绿色
waterprint design　水纹图案
waterprint effect　水纹图案
water-proof　防水（性）；防水布；雨衣,防水服,防水衣
water(-proof) boots　防水靴；雨靴
water-proof breathable cloth　防水透气布,透气性防雨布
water-proof breathable fabric　防水透气织物

water-proof but breathable fabric　可呼吸织物
water-proof but permeable damp fabric　防水透湿织物
water-proof cloak　防水披风
water-proof cloth　防水布；防水织物
water-proof clothing　防水服
water-proof coat　雨衣
water-proof coating　防水涂层
water-proof duck　防水帆布
water-proof fabric　防水织物；防水布
water-proof finish　防水整理
water-proofing property　防水性
water-proof & moisture permeating fabric　防雨透湿织物
water-proof shorts　防水短裤
water-proof/windproof fabric　防水、防风织物
water rat　河鼠毛皮
water repellency　拒水性
water repellent finish　拒水整理
water repellent finished fabric　拒水整理织物
water resistance　抗水性
water skiing suit　滑水服
water ski shoes　滑水运动鞋
water-soluble fiber　水溶性纤维
water-soluble interlining　水溶衬
water spot　水渍,水花,水滴
water spotting　水渍（变色）试验
water sprayer　喷水器
water stain　（熨烫的）水渍,水花,水渍印痕
water stain in ironing　水花（熨烫产生）
water-tight clothing　防水衣,防水雨衣
water vapor transmission（WVT）　透水（蒸）汽性
water vapour permeability　透水气性,透湿性
water-wash cloth　水洗布
water wave　水烫波浪式（发型）；水烫法
water wave hair　水烫波浪式发型
water white　水白色
watery blue　浅蓝色
watte　填料,填衬
Watteau　（18世纪初风行的）华托裙；法兰西裙
Watteau back　华托背（女式礼服从领口到下摆有宽裥的不束腰背部）
Watteau bodice　（华托式）紧身胸衣

Watteau hat 华托帽(浅帽身,后部上卷,有花饰)
Watteau pleat 盒状形裙裥,华托褶(箱形褶裥)
Watteau sacque 华托外套(后背有数个盒状褶,前身合贴的驮篮式外套)
Watteau with hat pin (带别针并饰口缎带及驼鸟羽毛)扁平草帽,扁平毡帽
wattelin 棉毛针织布
waulking 脚踏缩呢
wave 波,波浪形;云斑,云纹;仿羔皮粗纺呢
wave braid 波纹织带,波纹编带,波形饰带
wave crepe knitting 水纹绉
waved stitch 波状线迹,波形线迹
waved twill 锯齿形斜纹,波形斜纹
waved welt 波形凹凸织物
wave filling stitch 波形贴线绣
wave knife 波纹形裁刀
wave length 波长
wave pattern jacquard crepe 浪花绉
waver 烫发师;卷发器
wave set 卷发液
wave-shade defect 松板印(疵)
wave stitch 锯齿形针迹,波形缝
wavy cloth 厚薄段织物,不平整织物,波纹起毛织物
wavy design 波浪形纹样,波纹花样
wavy face 波浪形布面
wavy hair 卷发
wavy pongee 波浪绸
wavy selvedge 松(布)边,木耳(布)边
wavy shantung 波浪绸
wavy stitch 波形线迹
wax 蜡
wax cloth 蜡布,油布
wax coating 蜡涂层
wax(ed) end 蜡缝线
waxed thread (制鞋)蜡线
waxed thread faille 蜡线绨
wax finish 上蜡整理
waxing 上蜡加工
waxing finishing 上蜡整理,上光整理
wax like handle 蜡状手感
wax printings 蜡防印花
wax resist printed fabric 蜡防花布
wax resist printings 蜡防印花
wax resist printing fabric 蜡染织物
waxy 蜡色

wax yellow 蜡黄色
waxy touch 蜡状手感
3-way blindstitch drapery hem 三折暗缝线迹装饰卷边
2-way collar 敞领
3-way lockstitch 三折锁缝线迹
3-way lockstitch drapery hemmer 三折锁缝线迹装饰卷边器
ways 习俗;风度;作风;癖性
Wazu costume 佤族民族服
WCRA(wed crease recovery angle) 湿折皱回复角
weak acid cyanine 弱酸性深蓝
weak acid dye 弱酸性染料
weak color 浅色,弱色
weak twist 弱捻
wear 穿,戴,佩;服装,穿戴物;时装;流行样式;穿破,磨损
wearability 服用性能,穿着性能,耐磨损性
wearable junk 可穿用的破旧衣物
wearables 服装,衣服,衣着类
wear away (使)磨损,磨薄
wear behavio(u)r 服用性能
wear black 穿丧服
wear denim 穿牛仔服装
wear-comfort 穿着舒适性
wear-comfort of clothing 服装舒适性
wear-crease resistant 耐穿着起绉的
wearer trial 穿着试验
wear down (使)磨损
wear gloves 戴手套
wear grey 穿灰衣服
wearing apparel (总称)服装;衣服
wear[ing] comfort 穿着舒适性
wearing of cloth 布的耐磨性,织物服用性
wearing parts 易损件
wearing performance 穿着性能
wearing property 服用性能
wearings (古语)衣服,服装
wearing sleeve 悬饰袖
wearing test (服装和布料)耐磨测试
wearing textile 服(装)用纺织品,衣用纺织品
wearing value 耐穿性
wear leather shoes 穿皮鞋
wear level 服装穿着洗涤周期(次数)
wear-life [织物的]穿着寿命,可服用期
wear off 磨去;磨损掉

wear out 穿破,穿坏,用旧
wear-prone parts 易损零件
wear-proof 耐用的,防磨损的
wear property 服用性能
wear-refurbishing cycle 服装穿着洗涤周期;耐磨循环次数
wear-resistance 耐服用性,耐磨性
wear ring 戴戒指
wear-service condition 服装穿着状况
wear spectacles 戴眼镜
wear stockings 穿袜子
wear sunbonnet 戴太阳帽
wear test 穿着试验
wear testing 穿着试验,耐磨试验
wear triat 穿着试验
wear uniform 穿制服
wear watch 戴手表
wear well 耐穿;耐用
wear wool 穿毛料衣服
weasel (老式长薄型)熨斗,烙铁;黄鼠狼毛皮,鼬鼠毛皮
weasel fur 黄鼠狼毛皮,鼬鼠毛皮
weatherability 耐气候性
weather-all 晴雨兼用外衣,晴雨伞
weather-all cloth 防雨帆布
weather-all clothing 风雨衣;晴雨大衣
weather-all coat 晴雨大衣
weather blouse 防雨外套
weather blouson 防雨外套
weather cloth 防雨帆布,耐气候织物,篷帆布,晴雨两用布料
weather coat 风雨衣,风雪大衣,晴雨大衣
weathered piece 风溃布匹
weather fastness 耐气候性,耐气候坚牢度
weathering 风化,老化,风蚀
weathering jacket 晴雨夹克衫
weathering quality 耐气候性,耐风蚀性
weathering test 耐气候试验
weather jacket 晴雨夹克衫
weather mark 风印,风溃
weather-o-meter 耐气候性试验仪
weather-proof and winter clothing 防风雨保暖服装
weather-proof clothing 风衣,雨衣
weather-resistant finish 耐气候整理
weather stain 风溃
weave 织物;织法,编法;编织式样
weave analysis 织物分析

weave cutting 飘拂发型
weave design 织纹设计
weave-knit fabric 织编织物
weave structure 织物结构
weaving 机织;织造;织布
weaving defects 织疵
weaving-knitting fabric 织编物
weaving yarn 织造用纱
web 带子;织物;一匹布,一卷布;雪鞋;纤网,纤维网
webbed belt 行军腰带
webbing (textile) 狭机织物;带子;带状织物
web knit 网状针织物
web-like nonwoven fabric 纤维网状非织造布
web stitch-bonding machine without yarn 无纱线纤网型缝编机
wedding band (同 wedding ring) 结婚戒指(美国)
wedding-band collar 婚礼腰带形领
wedding-band earring 婚礼宽轮式金耳环
wedding dress 婚纱服,结婚礼服
wedding dress lace 礼服花边
wedding dress gown 婚纱服
wedding favour (古代结婚时佩带的)白色缎带花结
wedding garment 结婚礼服
wedding garter 婚礼服装饰吊袜带
wedding gown 结婚礼服
wedding hat 婚礼帽
wedding ring (同 wedding band) 结婚戒指
wedding ring box 结婚戒指盒
wedding ring velvet 软薄丝绒,雪纺丝绒
wedding shoes 结婚鞋
wedding trio 三件套结婚戒指
wedding veil 婚礼面纱;结婚披纱,新婚面纱
wedding white 婚礼服白色
wede 丧服,黑纱
wedge V形,楔形;腰省;(鞋子的)坡跟;V形短发,楔形短发;(毛皮加工)开皮器
wedge cut V形发型,楔形发型
wedge dart 钉子省,楔形省
wedge dress V形装,楔形装
wedge hairstyle V形发型,楔形发型
wedge heel 坡跟,楔形后跟,大插跟;坡跟鞋,楔形后跟鞋

wedge inset cuff　有V形嵌饰布的袖口
wedge-point needle　楔头缝针
wedge sandals　坡跟凉鞋,楔(形鞋)跟凉鞋
wedge-shaped silhouette　楔形轮廓
wedge shoes　楔形跟鞋
wedge silhouette　楔形轮廓
wedge sleeve　楔形袖
wedge sole　坡跟鞋底,楔形鞋底
wedge trousers　楔形裤,马裤
wedgie　坡跟(女)鞋,楔形后跟(女)鞋
wedgies　女式坡跟鞋(商标名),楔形后跟女鞋
Wedgwood blue　韦奇伍德蓝色(中间色调钝感蓝色,有淡蓝灰和深蓝紫两种)
Wedgwood cameo pin　韦奇伍德陶瓷别针
Wedgwood print　韦奇伍德印花布(仿韦奇伍德彩底白浮雕纹瓷器效果)
weed　服丧黑纱
weeds　(寡妇穿的)丧服,服丧黑纱;(古代)服装,衣服
weekend bag　周末旅行袋
weekend case　周末旅行袋
weekender　周末旅行女套装;周末旅行袋
weekend suit　周末套装,郊游服套装
weeper　黑丧服,黑纱,服丧佩带物;(帽子上的)下垂物
weepers　两腮的长须
weft　(织物)纬纱
weft backed cloth　纬二重织物
weft bow　弓纬(布疵)
weft cord　博坦尼毛纬面斜纹精纺呢
weft count　纬密
weft crackiness　稀密路(布疵)
weft density　纬密
weft effect　纬面花纹
weft faced fabric　纬面织物
weft faced satin　纬面缎纹
weft-face fabric　纬面织物
weft face twill　纬面斜纹
weft filament fabrice　纬长丝织物
weft filament mixed fabric　纬长丝织物
weft float　浮纬,纬跳纱(布疵)
weft grain　纬纱方向
weft insertion raschel machine　拉舍尔衬纬经编机
weft knitted fabric　纬编针织物
weft knitted pile fabric　纬编长毛绒针织物
weft knitted sofa-cover fabric　纬编针织沙发布
weft-knitted velour　纬编针织天鹅绒
weft knitting　纬编
weft kntting fabric　纬编针织物
weft knitting machine　纬编针织机
weftless fabric　无纬布,无纬织物
weftless needled felt　无纬刺刺毛毯
weft loop　缩纬(布疵);纬编线圈
weft nests　糙纬(绸缎布疵)
weft pile astrakhan　纬起毛仿羔皮织物
weft pile fabric　纬起毛织物,纬无圈型缝织物
weft plain knit fabric　纬平针织物
weft plain-knitted fabric　纬平针织物
weft plush　纬长毛绒织物
weft rib　纬向棱纹织物,纬重平织物
weft sateen　纬缎,纬面棉缎
weft satin　纬面缎纹织物
weft shiner　亮纬,紧纬(布疵)
weft soleil　浮纬光亮塔夫绸
weft stitched weave　纬编缝花织物
weft streaks　罗纹档(绸缎织疵)
weft stretch woven fabric　纬向拉伸织物
weft stripe　纬向条纹
weft tight　紧纬(布疵)
weft-way twist　S捻,顺手捻
weft-way yarn　S捻纱,顺手纱
weft yarn　纬纱
wei-cheng silk suitings　丝直贡呢;纬成丝呢
weighing commission　过磅费《贸易》
weight　(布料的)重量;(压布用的)压铁;体重
weight belt　(一种平衡人在水中浮力的)重量带
weight deduction　(碱)减量处理
weighted cloth　加重织物
weighted coefficient　加权系数
weighted fabric　加重织物
weighted finish fabric　增重整理织物
weighted mean　加权平均
weighted mean skin temperature　加权平均皮肤温度
weighted silk　增重丝绸
weighted tape　(包有小铅盘的)加重布带
weight giving finish　增重整理
weighting cotton fabric　重浆棉布
weighting finished fabric　增重整理织物
weighting finishing　(织物)加重整理
weighting scale　体重计

weight jacket 厚夹克衫;(训练用的)负重衣,砂背心
weight lifter's wear 举重服
weight lifting shoes 举重鞋
weight lifting suit 举重服
weight linear meter 每米长重量
weight list [memo] 重量单
weight loss treatment (碱)减量整理
weight-minus finished fabric 减量整理织物
weight per square meter (布料)每平方米(克)重
weight per yard (布料某个幅宽的)每码重量
weights 压铁
weight ton 重量吨
weijin brocatelle 微锦绸
weird dress 奇装异服
weiwen satin brocade 微纹缎
Weiwuerzu costume 维吾尔族民俗服
wéi yāo 腰巾;围腰
welding 熔合缝制法
welding technical textile 熔接[黏合]织物
welfless fabric 无纬织物
well-built garment 式样优美的衣服
well-covered (人)丰满的;肥胖的
well-covered cloth 布面丰满的织物
well-cut 裁制得式样很好的
well-defined shoulder line 轮廓清晰的肩线
well fashion 流行,流行服装
well fit (服装)合身的
welline 威利纳波状拉绒织物,威利纳波状结子大衣呢
Wellington 威灵顿防雨呢,英国斜纹军服呢;威灵顿靴
Wellington (boots) 威灵顿靴,威灵顿长筒高帮靴,威灵顿军用橡胶跟皮靴
Wellington style (19世纪初)威灵顿男装风格
well-lit dressing room 明亮试衣室
well-proportional figure 匀称体型
well proportioncd form 匀称体形
wellset figure 健美的体型
well suited 成套
well-turned form 女子的优美体型
well-turned leg 匀称的腿
Welsh flannel 威尔士手织法兰绒,松织毛法兰绒
Welsh margetson 威尔士绵羊毛薄呢

welt 扎袜口,起口边;浮线;贴边,滚边,嵌革,嵌线;嵌条;松背凹凸织物,(制鞋)沿条
welt allowance 压缝量
welt button hole 滚(扣)眼
welt foot 折边压脚
welt hook 扎袜口针
welting 扎袜口,套袜口;贴边,滚边;贴边料,滚边料;嵌线
welting cord 贴边带,滚边带
welting machine 缝袖口机,扎袜口机,罗纹边机
welting seam 折边缝,贴边缝;嵌线缝
welting tape 滚边带
welt machine 折边机
welt pocket (同 singlewelt pocket, stand pocket) 开缝口袋,单嵌线袋,一字嵌袋,单贴边袋,立式口袋,挖袋
welt pocket flap 开缝袋袋盖
welt pocket with lip 有袋盖,开缝口袋
welt pocket without lip 无袋盖开缝口袋
welt ripple 浮线波纹组织
welt seam 贴边缝,折边缝,嵌线缝,暗包缝,伏卧缝
welt-seam opening 伏卧缝开口
welt-seam placketing 伏卧缝开口
welt stitch 折边线迹;浮线组织
welt strap 袜口饰带
welt tying-up 自动扎袜口装置
welt yarn 袖口线;袜口线
wendy Dagworthy 温迪·达格华兹(英国时装设计师)
wengi twill [taffeta] faconne 文绮绸
wenming faconne 文明绸
wen-shang(plain)faiwld 素文尚葛
weskit 西装背心,马甲,紧身背心
weskit dress 背心裙装
western 美国西部牛仔风格
western belt (美国)西部皮带
western boots (美国)西部牛仔长靴
western cowboy hat 美国西部牧童帽,美国西部牛仔帽
western cuff 西部牛仔风格明线缝袖头
western dozen 宽幅毛呢
western dress 西装
western dress shirt (美国)西部竞技衬衫
western flap pocket 牛仔型盖口袋
western formal(同 western suit) 西部夜礼服
western Han 西汉

western hat　（美国）西部牛仔帽
western house coat　西部牛仔骑马外套
westernism　西方习俗,西方特色
western jacket　（美国）西部夹克,长流苏装饰的皮夹克
western look　（美国）西部款式,西部风貌,西部牛仔式样
western pants（同 frontier pants）　（美国）西部牛仔裤;拓荒裤;度假裤
western pocket　美国西部牛仔型裤装口袋
western (scout) hat　牛仔帽;西部帽
western shirt　美国西部衬衫,牛仔衬衫
western shirt yoke　美国西部牛仔衬衫覆肩
western shorts　（美国）西部短裤
western style　西方风格;西式
western style bed sheet　西式床单
western style clothes　西服,西式服装
western stylistic form　西部款式
western suit　（美国）西部套装,西部夜礼服,西装
western tie　西部式领带,牛仔领带
western touch　（美国）西部味,西部格调
western yoke　牛仔式育克
west of England fabric　英国西部毛织物
west point　西点华达呢
wet　潮湿
wet absorption　吸湿
wet blowing　湿蒸呢
wet cleaning　水洗
wet cloth　湿光布,表面有淋湿感的衣料
wet crease recovery angle（WCRA）　湿折皱回复角
wet cut　湿理发
wet decatizing　湿蒸呢
wet finish　湿整理
wet-formed fabric　湿法非织造布
wet ironing　湿熨烫
wet-laid nonwoven　湿法成网非织造布
wet look　（织物等的）淋湿光泽外观;淋湿感款式
wet on dry　湿罩干印花
wet on wet　湿罩湿印花
wet-salting　兽皮盐腌法
wet setting　湿热定型
wet spun yarn　湿纺纱
wet suit（同 surfing suit）　潜水服;冲浪服
wettability　可湿性;吸湿度,润湿度
wetting property　润湿性

wet transfer printing　湿转移印花
weyd　丧服黑纱
whale leather　鲸鱼皮
whang (leather)　生牛皮,皮带皮
wheat　小麦色,淡黄色
wheaten　灰黄色
wheat sheaf stitch　捆扎绣,麦束绣
wheat stitch　表穗状线迹,麦穗缝
wheat straw knitted fan　麦秸编织扇
wheel　轮子
wheel farthingale　轮状裙撑环
wheel feed　轮式送料
wheel horse　实干家《管理》
wheel stitch　车轮线迹
whetstone　磨石,油石
whip　锁边,包缝,织补
whipcord serge　粗斜纹哔叽
whipcord　缎纹卡其;马裤呢（精纺呢绒）
whip net　网状纱罗织物
whipper　八角钉
whipped back stitch　回针绕线绣
whipped blanket stitch　绕线毯子锁边绣
whipped buttonhole stitch　绕线锁孔绣
whipped chain stitch　绕线链状绣
whipped outline stitch　绕线包梗绣,绕线轮廓线
whipped running stitch　绕线平针绣
whipped stem stitch　绕线包梗绣;绕线轮廓线
whipper　绕边者,素描或样式画家
whipping　包缝,锁边,缝袜头,套口;扣眼锁缝
whipping stitch　顺编线迹
whip stitch　锁缝线迹,顺编线迹,搭接缝
whip stitching　顺编线迹
whip stitching with gimp　顺编粗线辫带密针缝纫
whirling　涡纹整理
whisk　女式花边大翻领,扫帚式花边立领
whisk broom　衣刷
whisk (collar)　威斯克领,扫帚领
whiskerage　胡须式样,须式
whiskers　连鬓胡子,颊须,髯
whiskey　威士忌酒色
whispy curl　威士沛卷发
white　白色;(肤色)白皙的
white and black overcoating　雪花大衣呢
white asparagus　芦笋白
white-back denim　白背劳动布(蓝色经,本色纬),白背牛仔布

white-back duck 劳动布
white balance 白色平衡(光照术语)
white belly sheep 白肚羊皮
white belt (柔道)白腰带,柔道白带
white blue 影青
white(bow)tie 白蝶结领带,白色蝶形领结
white bucks 白皮牛津鞋(鞋面与鞋舌相连)
white calico 漂白棉细布
white cap 白帽;戴白帽者
white cap gray 白浪灰
white cashmere 白羊绒
white-clad wear 白色服
white clothing 素服
white coat 白海豹毛皮
white collar 白领服饰
white core 露白(新织物有意做成磨损露白的陈旧外观)
white dextrin 白糊精
white discharge printings 拔白印花,雕白印花,白色防染印花
white-dotted cotton 白点印花(棉)布
white dozens 漂白平纹衬衫布
white drill 漂白卡其布
white dyeing 增白,加白;上蓝
white elastic-waist gym pants 白色宽紧束腰体操短裤
white embroidery 白色刺绣(用白线绣在白布上)
white flannels 法兰绒便裤
white fox 白狐毛皮,白狐皮
white garment 白衣,白大褂
white gloves (法官戴的)白手套
white gold 人造白金
white goods 漂白织物
white goose down 白鹅绒
white grament 白衣,白大褂
white hat 白帆布水兵帽,白帽
white jute 白麻
white kid gloves 白色羔羊皮手套
white lining 白亚麻布衬里衣
white look 白色款式
white mink 白貂皮,白色水貂皮
white mull 白色麦尔纱
whitened fabric 增白织物
whiteness 白(色)度
whiteness meter 白度仪
whiteness retention 白度保持力
whiteness standard 白度标准

whiteness value 白度值
whitening 增白
white-on-white 白织提花织物
white point 白色点,中性点,中和点,非彩色点(色彩学用语)
white print (白底的)蓝图;正像复印品
white resist 白色防染
white resist printing 白色防染印花
whites 白色衣服,白制服
white seal 白海豹皮
white selvage melton 白边麦尔登
white selvedge melton 白边麦尔登
white sheet 白色忏悔服
white sheeting 漂白布
white shirt 白衬衫
white silk 白绢,素绢
white slip 白色凹凸织物窄边
white smoke 白烟色
white squirrel 银鼠皮
white standard 白色标准
white stick (盲人用)白色手杖
white style 白色款式
white swan 白天鹅色
white tie 白领带,白领结;配白领结的男士晚礼服
white tie and tails 燕尾服
white vest (与燕尾服配穿的)男背心
white vinegar (洗丝织物用)白醋
white waistcoat 白色马甲
white wash 皮肤增白剂
white wax 白蜡
white weasel fur 白鼠皮,白鼬毛皮
white work 白织物上的白色刺绣
white yarn gloves 白纱手套
whitish 淡白,带白
whitney 惠特尼波纹起绒大衣呢
whitney finish 惠特尼波纹起绒整理
whitney-long 惠特尼径向波纹大衣呢
whitney point blankets 惠特尼羊毛毯
whittle 粗厚呢;白色大披肩,怀多披巾
whizzer 离心干燥机;脱水机
whole back 全背式(上衣背面为整块衣料)
whole cloth (未裁剪过的)整幅布
whole colored 同色的;单色的
whole cut 整帮(制鞋)
whole falls 连体男裤前门襟
whole plated last 全底面镀铁板的鞋楦
whole skin 整张皮
whole stitch 十字绣

whole stone 全钻	幅领
whole-cut vamp pump 无缝船鞋	wide-square neckline 大开口方领口
whole-length statue 全身像	wide stitch 宽线迹
whole-stitch 梭结花边的平纹线迹	wide tuck 宽褶
whorl design 圆涡纹	wide vee neckline 宽 V 形领口
wick chromatogram 灯芯色谱(图)	wide V-neckline 宽 V 形领口
wicker 柳条	wide wale 宽条凸纹呢,宽棱条布
wicking ability 吸咐能力	wide wale cloth 粗条纹呢
wicking action 芯吸作用	wide wale corduroy 粗条灯芯绒
wide-awake 宽檐男毡帽	wide wale serge 粗条纹哔叽
wide back 宽后背	wide welt pocket 单嵌线袋,一字嵌袋
wide belt 宽腰带,阔皮带	wide wrist 宽腕
wide brim hat 宽檐帽	widow's bonnet 寡妇(软)帽
wide-brimmed cap 宽边便帽	widow's hood 寡妇头巾
wide-brimmed felt hat 宽边呢帽,骑士帽	widow's lawn 英国包头细布
wide-brimmed hat 宽边帽	widow's peak 寡妇(软)帽,女式心形小
wide-brimmed straw hat 宽边草帽	帽,慰兜帽
wide cloth 宽幅布	widow's silk 丧服用绸
wide collar 宽展领,宽幅领,大八字领	widow's weeds 寡妇穿的全黑丧服
wide cording 宽嵌线	width 宽度;(布料)幅宽;(某宽度的)一
wide cowl neckline 宽大垂褶领口	块料子;排料图的最大宽度;肥度(制
wide crew-necked shirt 海员宽翻领衬衫	鞋)
wide dart 宽省	width across chest 胸宽
wide duck 宽幅帆布,阔幅帆布	width across shoulder 肩宽
wide-edge sole 宽边底	width barrier 横档疵
wide fabric 阔幅布,阔幅织物	width between selvedge(s) 布身宽度
wide flare 喇叭裤脚	width ease 宽度放松量
wide hip 宽臀	width line of back crotch curve 后裆宽线
wide lapel 宽驳头;大翻领	width of bottom 裤脚宽
wide lapel suit 宽驳领西装	width of chest 胸宽
wide-legged boots 宽筒靴	width of cuff 袖口宽
wide-legged pants 宽腿裤	width of front 门襟宽
wide legs 宽裤脚	width of knee (裤子的)膝部宽
wide-mest crackle 冰裂纹	width of leg 横裆
wide number duck 宽幅编号帆布	width of thigh 裤裆宽,裤子的大腿部宽
wide-open zigzag stitch 宽的 Z 形线迹	width of zigzag stitch 曲折线迹宽
wide pointed collar 大尖角领	width over 阔幅
wide(long, short) pointed collar 大尖角领	width tolerance 幅宽公差
wide polo belt 宽马球护腰带	width under 狭幅
wide rose 野玫瑰色	widthway 布幅;横向
wider pointed collar 大尖角领	wig 假发,(法官戴的白色卷曲)假发套
wider shoulder 宽肩	wigan 棉衬,威根织物;本色厚实平布,无
wide sheeting 宽幅平布,阔幅平布	光上浆衬里布
wide shepherdess brimmed hat 牧羊女宽	wig block 假发,假发套
檐帽	wig cap 假发帽,假发套
wide shoulder 宽肩	wiggin 威杰织物
wide special shirring foot 宽面特定打折	wig hat 钩编软帽
压脚	wiglet 局部假发,假发束
widespread collar 大八字领,宽展领,阔	wig powder 发粉,假发粉
	wig sewing 假发缝纫

wild belt 宽腰带
wild leather 动物原皮革
wild lime 嫩绿
wild look 粗犷款式,粗野风貌
wild make-up 粗野化妆法
wild mink 野生貂皮
wild orchid 野兰花色
wild rose 野玫瑰红色
wild silk 野蚕丝,柞蚕丝
wild silk pongee 柞蚕丝和棉交织茧绸
wildbore 女服呢,英国粗纺呢
wildcat fur 猞猁毛皮,野猫毛皮,豹猫毛皮
wildcat(1eather) 山猫毛皮
wilderness boots 荒野靴
wilderness life 荒野生活
wildlife 野生动物
Willi Smith 威尔·史密斯(1948～1987,美国时装设计师)
Willi Wear 威利·韦尔(美国时装设计师)
William Morris design 威廉·摩里斯式设计
Willian Penn hat 威廉·佩恩帽(适中帽檐,高圆顶)
Willie Smith 威尔·史密斯(美国时装设计师)
willow 柳色;芦苇草和棉交织的威罗布
willow calf 搓纹小牛面革
willow grain (皮革)褶皱
willow green 柳绿色
wilton cloth 威尔顿粗纺呢
wiltshire lace 威尔特郡花边
wimple (妇女)披巾,肩巾,褶裥(苏格兰);修女用肩巾
wincey 棉毛绒布,上等绒布
winceyette 色织棉法兰绒
wind blown bob 风吹发型
wind bonnet 挡风护发罩帽
windbreaker (皮或呢制)防风外衣,风衣,防风运动服
windbreaker cloth 防风布,防风厚呢
windbreaker suit 风衣套装
windcheater 防风衣,防风夹克衫
wind chill chart 风寒图
wind clothing 风衣
wind cuff 卷袖口,翻袖口,喇叭袖口
winder (缝纫机上的)绕线器,挑线器
wind hosen 防风裤
winding bobbin 绕梭芯

winding bobbin thread 绕底线
winding jacket 防风夹克,防风上衣
winding jumper 防风夹克,防风上衣
wind jacket 防风夹克,防风上装
wind jumper 防风夹克,防风上装
window 橱窗;窗口《计算机》
window buttonhole (窗式)嵌线纽孔
window curtain 窗帘
window display 橱窗展示,橱窗陈列
window dressing 橱窗[服饰]布置
windowpane 窗格花纹
windowpane check 窗格花纹,棋盘图案,方格花
windowpane hose 窗格纹花袜
windowpane method (缝制嵌线纽孔的)窗玻璃法,缝制嵌线纽孔的窗格法
windproof fabric 防风织物
wind proof garment 防风服
windproof jacket 风雪衫
wind rose 深粉红
windsor 温莎花色簇绒织物
Windsor brilliant 温莎耐洗光亮棉衣料
Windsor check 温莎格子呢
Windsor collar 温莎领,大八字领
Windsor duck 温莎印花帆布
Windsor knot 温莎领结,温莎大型领结
Windsor look 温莎款式
Windsor louisine 温莎优质耐洗印花棉衣料
Windsor Red 温莎红
Windsor tie 温莎领带,(松散蝴蝶结式)阔领带
Windsor uniform 温莎宫殿礼服
wind surfing textiles 冲浪运动用纺织品
wind tunnel technique 风道技术
wind up 卷绕式(服装)
wind up an account 结算
wineberry 白里叶莓色,红光紫,茄皮紫
wine(colour) 红葡萄酒色,深红色,紫红色
wine red 酒红(深红色),紫红
wine yellow 酒黄(浅灰黄色)
wing broque 布洛格鞋
wing collar 燕子领,翼形领
wing-collared shirt 翼形衬衫
winged collar 翼领,燕子领
winged cuff 翼形袖口,翼形克夫
winged-deer design 翼鹿纹
winged sleeve 翼形袖
wing look 翼状款式

wing needle 蝶形针
wing ruffle sleeve 翼形荷叶袖,褶皱翼形袖
wings 肩翼;(女童子军佩带的)绿色翼章
wing sleeve 翼形袖
wing style 翼形发型
wing tie 翼形领带,蝴蝶领结
wing tip 翼形鞋头;(男式皮鞋的)翼形盖饰;翼形盖饰男皮鞋
wing-tip collar 翼形领,燕子领
wing-tip heel 翼梢跟
wing-tip oxford 翼形饰孔牛津鞋
wing-tipped collar 翼领,燕子领
wing-tip shoes 翼形鞋
winker 睫毛;眼睛;眼睑
winkers 男式高领;护目眼镜
winkle pickers 尖头皮鞋
winnie (美国服装行业一年一度颁发的)时装设计优胜奖
winoey 棉经毛纬温诺伊厚呢
winsey 棉毛绒布,上等绒布
winter boots 冬季长筒靴
winter buckskin (冬季用)鹿皮呢
winter cap 冬帽,防寒帽
winter clothing 寒衣,御寒衣
winter coat 冬用外套
winter commodity 冬季商品
winter cotton 棉织物冬装
winter fabric 冬季织物
winter green 冬绿色
winter lace 冬季花边,紧密花边,密织带
winter overcoat 冬大衣
winter pastel look 冬季清淡款式,冬季淡而柔和色调款式
winter pear 冬梨色
winter sky 冬天天蓝色
winter sportswear 冬季运动服装
winter uniform 冬服
winter wardrobe 冬装
winter warm but summer cool 冬暖夏凉
winter warm but summer cool fabric 冬暖夏凉织物
winter wear 冬服
winter weight fabric (冬季用)厚重织物
winter white 冬天白色,灰白色,粉白色
winter white look 冬天白色款式,冬天白色风貌;白色冬装
wintry color 冷色
wintry hair 皓发
wipes 揩布

wire basting machine 金属线疏缝机,金属线粗缝机
wire braiding machine 钩针编织机
wire brush (毛皮加工用)钢丝刷
wire cloth 金属丝织物
wire covering machine 金属丝包缝机
wired bra(ssiere) 金属丝型胸罩
wire fabric 金属丝织物,金属线织物
wire frame 线框图
wire frame model 线框图模型
wire hanger (金属)丝衣架
wire-mesh mask 击剑面罩
wire mesh garment 金属丝网服装
wire mesh mask 击剑面罩
wire printer 针式打印机《计算机》
wires 穿耳孔的耳环(总称)
wire shank 金属纽柄
wire shank button 金属柄纽扣
wire velvet 天鹅绒
wiriness 手感粗硬
wiry fabric 金属网织物
wishing cap 如意帽
wist stitch 绳状线迹,凸纹线迹
wistaria 紫藤色(浅紫色)
wisteria 紫藤色(浅紫色)
witchoura 波兰狼皮大衣
with back fullness skirt 后褶裙
withered leaf 枯叶色
with godets skirt 大叶形裙
with nap 倒顺毛
with nap layout 有倒顺毛的排料
with nap yardage (倒顺毛)起绒织物的尺码
with nature 接近自然
witney 威特尼毛毯;双面起绒哔叽
witney point blanket 威特尼厚绒毛毯
witney unio blanket 威特尼混纺厚绒毛毯
witzchoura 毛皮里长大衣,威乔拉外套
W.L. ①(waist level)腰围线;②(waist line)腰节;腰围;腰围线,腰节线;(女性的)腰身部分;裙腰缝线
W.L.P. mixed worsted suiting (wool-linen-polyester mixed worsted suiting) 毛麻涤薄花呢
woaded blue 菘蓝色
wolf 狼皮,狼毛皮
wolfskin 狼皮;狼皮制品
wolverine 狼獾毛皮,貂熊毛皮
womens 妇女服装大型尺寸(常指38号

至52号的尺寸）
woman's beauty 女性美
women's clothes 女式服装
women's design 女装设计
woman's dress 连衣裙
Women's Dress Reform Movement （19世纪后半叶美国）妇女服装改革运动
women's extra large size（WX）女式特大号
woman's fasten 女式纽扣开口,右压左式纽扣开口
women's figure 女子体型
women's half size 胖妇女尺寸
women's overcoat cloth 俄罗斯女用大衣呢
women's pleat skirt 百裥裙
women's Quarpel raincaot 女式夸佩尔雨衣
women's red-skirt 石榴裙
women's shorts 女式短裤
women's size 妇女尺码,女式尺码
women's slacks 女裤
women's suit 女套装;女西装;女式常服
women's sweater 女毛衫
women's tensioned slacks 女紧身裤
woman's upper outer garment 女上装
woman's walking boots 女士散步靴
women's wear 女装;女服,女式服装
Women's Wear Daily（W.W.D.）［美］《妇女时装日报》
women's wear design 女装设计
women's wear standard 女装标准
woman's worksuit 女式工装
wonderbra 理想胸罩,提升胸罩
wood 木色
wood bead 木［制］珠
woodbine 忍冬色
wood buckle 木扣
wood button 木纽扣
woodchuck fur 美洲旱獭毛皮,花白旱獭毛皮
wood drawer 羊毛长内裤
wooden accessories 木饰物
wooden button 木饰扣
wooden cane 木手杖
wooden clothes pin （衣服用）木饰针
wooden last 木鞋楦
wooden mallet （熨烫用）木凳,木台板
wooden mask 木面罩
wooden ornament 木制饰物

wooden peg 木衣夹
wooden shoe 木屐（总称）
wooden sole 木鞋底
wooden stick 木手杖
wooden walking stick 木手杖
wood fabric 19世纪木纤维为纬纱的织物
wood hanger 木制衣架
woodstock 鹿皮骑士手套
woof （织物的）纬纱;布,织物,机织物
wool 羊毛;毛线,绒线;羊毛织物;毛衣;毛料衣服;（黑人的）浓密短卷发
wool/acrylic blended fabric 毛腈混纺呢绒
wool/acrylic blended yarn 毛腈纱
wool/acrylic lady's dress 毛腈女衣呢
wool/acrylic lady's dress cloth 毛腈女衣呢
wool auction system 羊毛拍卖制度
wool backed cloth 毛背织物
wool backed satin 毛背丝缎
wool bat 羊毛絮
wool batiste 全毛细薄呢
wool blanket 厚垫呢
wool blend 羊毛混纺物
wool blended fabric 混纺毛织物
wool blended tussah square 丝毛呢
wool blended tussah tweed 丝毛呢
wool blend jersey 混纺毛衫,羊毛混纺运动衫
wool blend mark 混毛标志
wool boucle 包喜呢
wool broadcloth 全毛绒面呢
wool bunting 旗纱（稀松平纹毛织物）
wool canvas 毛衬
wool cap 毛绒帽,羊毛帽
wool checked flyback skirt （后身中间暗裥的）方格绒裙
wool chenille cloth 毛雪尼尔
wool chiffon （极细薄的）雪纺呢
wool cloth for shoes 俄罗斯鞋用呢
wool clothing 呢绒大衣
wool coat 毛呢大衣
wool coating 呢料
wool crash 手织苏格兰式粗呢
wool crepe 全毛绉呢,绉纹呢
wool denim 毛牛仔布
wool drawer 羊毛长内裤
wool dress 毛料连衣裙
wool dyed 羊毛染色

wooled jacket 呢(皮)夹克
wool embroidery thread 毛绣花线
woolen 毛织物,毛料,呢绒;粗纺毛织物,粗梳毛织物
woolen/acrylic sweater 毛腈混纺衫
woolen and worsted interweave fancy suiting 精粗交织花呢
woolen blanket 毛毯
woolen cap 毛线帽;羊毛帽;呢绒帽
woolen check 格子呢
woolen chiffon velvet 全毛雪纺薄绒
woolen cloth 粗纺毛织物;呢绒
woolen coat 毛呢大衣
woolen cord 粗纺经向灯芯绒
woolen crepe 全毛绉呢
woolen double cloth 双层毛呢
woolenet(te) 薄呢
woolen fabric 粗纺毛织物
woolen finishing 粗纺毛织物整理
woolen fleece 长毛大衣呢
woolen garment 呢绒服装,毛呢服装
woolen gauze 巴雷格毛纱罗,毛纱罗
woolen gloves 羊毛手套
woolen goods 呢绒,粗纺毛织物
woolen grenadine 丝经毛纬纱罗
woolen hand knitted sweater 棒针毛衫
woolen jacket 毛呢夹克
woolen jersey 毛针织物,隐密细条平纹粗花呢
woolen knitted garment 羊毛针织服装
woolen knitwear 毛针织品
woolen knotted carpet 羊毛栽绒地毯
woolen lady's cloth 女大衣呢
woolen material 粗纺毛料
woolen necktie 毛呢领带
woolen needlepoint tapestry 绒绣
woolen overcoating 大衣呢
woolen piece 呢
woolen plush 长毛绒,粗纺长毛绒
woolen pressed felt 传统羊毛毡
woolen raised scarf 羊毛起绒长巾
woolen satin 缎纹呢
woolen scarf 羊毛围巾
woolen serge 毛哔叽
woolen sheer 轻薄型毛料(夏服用)
woolen shirt 毛料衬衫
woolen skirt 毛呢裙
woolen sleeveless pullover 毛背心
woolen stockings 毛袜
woolen suiting 粗纺毛料,粗花呢

woolen sweater 绒线衫,羊毛衫
woolen sweater and trousers 羊毛衫裤
woolen textile 毛纺织品
woolen tricot 经编毛织物
woolen trousers 毛呢裤,料子裤
woolen type (W type) 毛型
woolen underwear 羊毛内衣
woolen union 混纺或交织毛织物
woolen work cloth 劳动呢
woolen yarn 粗纺毛纱
woolen yarn knitted underverst 毛针织内衣
wool fabric 毛织物,呢绒
woolfell 羊毛皮
wool felt hat 毛毡帽
wool fibre 羊毛纤维,毛纤维
wool gabardine cap 毛华达呢帽
wool garment 呢绒服装
wool hair 绵羊绒
wool hat 羊毛帽,毛线帽,毡帽
wool hat body 呢帽坯
wool hosiery 羊毛袜
wool imitation 仿羊毛
Wool Industries Research Association 羊毛工业研究协会
wool interfacing fabric 毛衬垫织物
wool jersey 毛线衣,轻薄毛针织衣
Woolknit Associates 毛针织品协会
wool knit bed jacket 床上外披短毛衫
wool knit petticoat 毛针织衬裙
wool-knitted tie 毛针织领带
wool knit undercoat 毛线衣
wool knitwear 羊毛衫
wool labeling 毛织物质量标签(美国)
wool lace 羊毛花边
wool-lama cloth 金银毛花呢
woolled skin 仿绵羊皮
woollen 毛织物,毛料,呢绒;粗纺毛织物,粗梳毛织物
woollen and worsted union fancy suitings 精粗交织花呢
woollen blended 毛粗纺混纺(交织)织物
woollen cap 呢帽,羊毛帽
woollen chiffon velvet 全毛雪纺薄绒
woollen cloth 粗纺毛织物
woollen cord 粗纺灯芯绒
woollen crepe 全毛绉呢
woollen double cloth 双层毛呢
woollen fabric 粗纺呢绒;粗纺毛织物
woollen fleece 长毛大衣呢,顺毛大衣呢

woollen gauze 毛纱罗
woollen gloves 羊毛手套
woollen goods 呢绒,粗纺毛织物
woollen hat 毛线帽
woollen jersey 毛针织物;隐密细条平纹粗呢
woollen knitted garment 羊毛针织服装
woollen knitting yarn 毛线,绒线
woollen labour cloth 劳动呢
woollen ladies cloth 粗纺女式呢
woollen ladies cloth for outerwear 俄罗斯粗梳女式外衣呢
woollen lady's cloth 粗纺女式呢
woollen necktie 毛料领带
woollen nylon 弹力尼龙,仿毛尼龙
woollen overcoating 大衣呢
woollen plush 长毛绒
woollen rags 碎呢,呢片
woollen raised scarf 羊毛起绒长巾
woollensatin 缎纹呢
woollen scarf 羊毛围巾
woollen-spun yarn 粗纺毛纱
woollen stockings 毛袜
woollen sweater 绒线衫,羊毛衫
woollen thread 粗纺毛线
woollen tricot 经编毛织物
woollen type (W type) 毛型
woollen underwear 羊毛内衣
woollen union 混纺或交织毛织物
woollen work cloth 劳动呢
woollen yarn 粗纺毛线
woollen yarn knitted undervest 毛针织内衣
woollie 毛线衣,羊毛内衣
wool-like fabric 毛型织物,仿毛织物,凉爽呢
wool-like fabric with low-elastic polyester yarn 低弹涤纶丝仿毛织物
wool-like fancy suiting with lowelastic filament 纯涤纶低弹长丝仿毛花呢
wool-like finishing 仿毛整理
wool-like handle 毛型手感
wool lined 绒里(手套)
wool lined hunting pants 衬绒猎装裤
wool lined jacket 衬绒袄,衬绒夹克
wool lined robe 衬绒长袍
wool/linen blended fabric 毛麻混纺织物
wool linked hunting pants 衬绒猎装裤
woolly (常用复数)毛线衣;羊毛内衣
woolly aspect 毛型感

woolly finish 毛型整理
woolly nylon 仿毛耐纶,仿毛尼龙布料
wool-rich materials 羊毛为主的混纺织物
Wool Manufacturer's Council 毛纺织品制造商协会
wool mark （国际羊毛局）纯羊毛标志,羊毛合格标记《贸易》
wool material 纯毛料子
wool mixed damask 达马斯克;西洋缎子
wool mixed jeans 羊毛牛仔布
wool-mix thread 混纺毛线
wool muffler 羊毛围巾
wool/nylon blended yarn 毛锦纱
wool/nylon fabric 羊毛锦纶混纺呢绒
wool one-side gabardeen 全毛单面华达呢
wool pleated plaid skirt 长方格羊毛褶裥裙
wool/polyester blended yarn 毛涤纱
wool/polyester cheviot 毛涤啥味呢
wool/polyester fabric 毛涤织物,毛涤薄花呢,毛的确良
wool/polyester fancy suitings 涤毛花呢
wool/polyester gabardeen 毛涤华达呢
wool/polyester herringbone 毛涤海力蒙
wool/polyester palace 毛涤派力斯
wool/polyester single face fancy suitings 毛涤单面花呢
wool/polyester tricotine 毛涤巧克丁
wool/polyester tropical 凉爽呢
wool/polyester tropical suitings 毛涤混纺薄花呢,凉爽呢,毛的确良
wool/polyester valitin 毛涤凡立丁
wool/polyester venetian 毛涤直贡呢
wool/polyester whipcord 毛涤马裤呢
wool/polyester worsted flannel 毛涤啥味呢
wool/polyester/viscose suiting 毛、涤、黏三合一衣料
wool poplin 精纺平纹呢,精纺纬重平呢
wool product 羊毛制品
Wool Products Labeling Act (W.P.L.A.) 羊毛制品标记条例《贸易》
wool/ramie fancy suiting 毛麻花呢
wool/ramie/polyester mixed worsted suitings 毛麻涤薄花呢
wool-rayon blended muffler 毛黏混纺围巾
wool-rich and polyester fancy suiting 高比例毛涤花呢
wool-rich blends 羊毛为主的混纺织物

wool-rich materials 羊毛为主的混纺织物
Wool Record and Textile World [英]《羊毛记录和纺织世界》(月刊)
Wool Research Organization of New Zealand 新西兰羊毛研究组织
wool sateen 横贡呢
wool satin 毛缎,直贡呢
wool/satin-backed gabardeen 全毛缎背华达呢
Wool Science Review [英]《羊毛科学评论》(月刊)
wool seersucker 毛泡泡纱
wool serge 全毛哔叽,毛哔叽
woolsey 棉经毛纬交织物,麻经毛纬交织物
wool shawl 羊毛围巾
wool sheer 轻薄毛织物
wool shirting 精纺和时纺
wool shorts 羊毛短裤
wool skin 羊毛皮
wool/spandex woven fabric 毛氨纶机织物
wool sport shirt 羊毛运动衫
wool suiting 毛料,呢子
wool taffeta 毛塔夫(紧密光滑的毛织物),密织粗纺毛织物
wool tie 毛料领带,毛领带
wool touch 毛型手感
wool tow 毛条
wool trimming in the length 羊毛长装饰带
wool tweed skirt 杂色粗呢裙
wool type (W type) 毛型(指化纤)
wool type fiber 毛型纤维
wool type viscose staple fiber 人造毛
wool velvet 海虎绒,弹性法兰绒
wool venetian 毛直贡呢;威尼斯毛呢
wool vest 羊毛背心
wool/viscose blended yarn 毛黏纱
wool/viscose cheviot 毛黏啥味呢
wool/viscose fabric 毛黏混纺呢绒
wool/viscose fancy suitings 毛黏花呢
wool/viscose flannel 毛黏混纺法兰绒
wool/viscose gabardeen 毛黏华达呢
wool/viscose melton 毛黏(混纺)麦尔登
wool/viscose overcoating 毛黏(混纺)大衣呢
wool/viscose (polyester) fancy suiting 毛黏(涤)花呢
wool/viscose/ramie blend cloth 毛/黏/麻粗花呢
wool/viscose serge 毛黏哔叽
wool/viscose tartan overcoating 毛黏格子大衣呢
wool/viscose whipcord 毛黏马裤呢
wool voile 毛巴里纱
wool white 羊毛白
wool work 绒绣,梳毛纱刺绣,绒线绣
wool worsted check ensemble 格子毛绒套装
wool worsted sweater 精梳羊毛套衫
woolworth shoe 伍尔活斯鞋(商标)
wool yarn 毛纱
work 产品,制品,织物;工艺品;工艺;刺绣,针线活,缝补
workability 可加工性,加工性能
work aid 附件,缝纫附件
work apron 工作围裙
work bag 工具袋,针线袋
work basket 针线筐
work boots 工作靴,劳动靴
work box 针线盒,工具箱
work capacity 生产能力,工作量《管理》
work clamp 布料夹
work clamp foot 压脚
work clearance 工作间隙
work clothes 工作服
work-clothes button 滑带扣环
work clothing 工作服
work clothing fabric 工作服织物
work coat 工作大衣
work cycle 工作周期
workdays for a bale of yarn 每件纱扯用工
workdays for ten-thousand meters standard fabric 折标准万米布用工
work dress 工作服
worked buttonholes 手锁纽孔
worker's overshirt (宽松的)劳动衬衫
work factor system 工作因素制《管理》
work gloves 工作手套
work guide (缝纫机)导布器
work holder 缝料夹持器
workhouse sheeting 博尔顿斜纹粗布
work in 归拢;织入
working boots 工作靴
working cap 工作帽
working clothes 工作服
working dart 纸上褶,纸上省(纸样设计)
working drawing 图样,工作图,效果图

working dress 工作服
working girl 女工
working gloves 工作手套,劳工手套
working hat 工作帽
working in of fullness 缝线隆起
working instruction 工作规程
working jacket 工作上衣
working life 适用期;使用寿命
working look 工装风貌,工装款式
working pants 工作裤
working shirt 工作衬衫,作业衬衫
working sketch 草图,款式设计图
working space 工作空间
working specification 操作规程
working speed 工作速度
working station 工位
working surface 工作台面
working trousers 工作裤,作业裤
working uniform 工作服
working wardrobe 工作衣柜
working wear 工作服,劳保服装
working wear look 工作服款式,工作服风貌
workman's cap 工人便帽
workmanship 技艺;工艺;技巧;手艺;做工;工艺品
work master 熟练工人,老师傅;领班
workmen's overal 工装裤
work of rupture 断裂功,断脱功(终值断裂功)
work-oriented post design 根据制品设计
workout bra[ssiere] 训练胸罩
work out corner 翻衣角
workout suit (同 aerobic ensemble, exercise suit) 训练服
work pants 工作裤
workpiece 工作件,缝件
workpiece thickness 缝件厚度
work plate 送布板,送料板
work place layout 作业面设计,作业面安排
work room 工作室,打样间
work schedule 作业计划
work sheet 工作单,加工单
work shirt 工作衫
work shoes 工作鞋
workshop 车间,工作室
workshop section 工段
work shop system 工场制
work space 工作空间

work table 工作台
work trousers 工作裤,作业裤
workwear 工作服,工作服式样的服装
workwear line 工作服型
workwear look 工作服款式,工作服风貌
work week 工作周
workwoman 女工
world brand 世界名牌
world coordinate system (WCS) 世界坐标系《计算机》
world Patents Index (WPI) 国际专利索引
world time clock 世界时钟
World Trade Organization (WTO) 世界贸易组织
World War I helmet 第一次世界大战帽盔
World War II helmet 第二次世界大战帽盔
worm mark 虫渍,污渍
worn clothes 旧衣服,穿过的衣服
worn garment 旧衣服,穿过的衣服
worn-out look 破旧型款式,破烂款式
worry beads 念珠;定心珠
worsted 毛线,精纺毛纱,精纺毛线;精纺毛料;精纺毛料服装,精纺套服;精纺毛织物,精梳毛织物
worsted blended or interwoven fabric 毛精纺混纺(交织)织物
worsted cheviot 精纺缩绒粗呢
worsted cloth 精纺毛织物,精梳毛织物
worsted covert cloth 芝麻呢
worsted crepe 精纺绉呢
worsted damask 毛织花缎
worsted-edge sateen 有羊毛边纱的棉横贡
worsted fabric 精纺毛织物,精梳毛织物
worsted fabric for outerwear 俄罗斯精梳女式外衣呢
worsted fancy suiting 精纺花呢
worsted finshing 精纺毛织物整理
worsted flannel 精纺法兰绒,啥味呢
worsted gabardine 精纺轧别丁,精纺华达呢
worsted goods 精纺毛织物,精纺毛织品
worsted homespun 钢花呢
worsted lady's dress 精纺女衣呢
worsted lasting 精纺厚实毛织物
worsted loose-structure fabric 精纺松结构织物
worsted material 精纺毛料

worsted melton 精纺麦尔登
worsted merino 精纺棉毛混纺呢
worsted organdie cloth 精毛和时分
worsted pile fabric 精纺长毛绒织物
worsted plain fabric 精纺平绒毛织物
worsted rays 精纺厚重毛呢
worsted serge 精纺毛哔叽
worsted shetland 设得兰精纺毛织物
worsted socks 毛线袜,绒线袜
worsted suiting 精纺毛料
worsted union 精纺交织毛织物
worsted venetian 贡呢,精纺缎纹呢
worsted wool tweed 精纺粗支花呢
worsted yarn 精纺毛纱
worumbo 混纺毛呢
would-be welt pocket 假嵌线袋
wound dressing 绷带
wound goods 布卷
woven and knitted fabric 机织针织物
woven asbestos cloth 石棉布
woven design 织物花纹
woven elastic webbing 机织弹性带
woven fabric 梭织物,梭织布,机织物
woven felt 机织毡
woven fur-like plush 机织仿兽皮长毛绒
woven garment 梭织服装
woven glass fabric 玻璃纤维织物,玻璃纤维布
woven handkerchief 机织手帕,织造手帕
woven hemp sash 麻织宽腰带
woven interlining 机织衣衬,有纺衬,梭织衬
woven jacquard carpet 机制提花地毯
woven label 织唛
woven lace 机织花边,手编花边,梭织花边
woven like 仿梭织物外观
woven looks 类机织物外观
woven patch 织章
woven picture 像景织物
woven polypropylene-bag 聚丙烯编织袋
woven ribbon 机织带
woven seersucker 机织泡泡纱,织成的泡泡纱(非化学处理形成的)
woven stretch fabric 弹力织物
woven stripe handkerchief 织条手帕
woven stripe pajamas 柳条睡衣,织条睡衣
woven tape 织带
woven ticking 色织被褥布,提花被褥布

woven tube 管状类织物
woven twill ticking 色织斜纹床单
woven wear 机织服装,梭织服装
woven welts 横梭织物
WPI (World Patents Index) 国际专利索引
wramp 拉姆泼
wrangler 西部牛仔服饰
wrangler jacket (美国)西部夹克
wrangler shade 中蓝色牛仔裤
wranglers 蓝哥牛仔裤
wrap 包缠;吊线;围巾,外套;包身布;覆盖物;围裹式服装,叠襟式服装
wrap and tie 叠襟系带式服装
wrap-around (同 wrap-round) 围裹式服装
wrap-around coat (同 wrap coat) 围裹式大衣,裹襟式大衣
wrap-around glasses 箍带式太阳眼镜
wrap-around knot 缠扎结
wrap-around robe 围裹式长袍
wrap-around skirt 围裹裙,卷腰裙,浪花裙
wrap-around sweater 围裹式毛衣
wrap blouse 围裹式罩衫,裹襟式罩衫
wrap circular-knitting machine 衬经圆形针织机,吊线圆形针织机
wrap closing 叠襟
wrap clothing 宽胸式外套;卷裹外套
wrap coat 宽胸外套,围裹式外套,裹襟式大衣,围裹式大衣
wrap collar 叠领
wrap cuff links 围裹式袖口带
wrap dress 围裹式裙装,裹襟式裙装
wrap hosiery machine 吊线袜机
wrap jacket 围裹式夹克便装
wrap jumper 围裹扎结式上衣
wrap look 围裹型款式,围裹式风貌
wrap machine 吊线针织机;经纱提花针织机;添纱针织机
wrap neckline 围裹式领口,披肩式领口
wrap over 围裹式斜襟衣;裹身叠合式衣裙;(衣服或裙子的)褶子;(衣服上身的)部分叠合
wrap over bateau 堆叠式船形领
wrap over front 暗门襟
wrap over skirt 裹叠式裙
wrappage 围巾,头巾,毯子;宽松女服,女子便服;包裹物,包装材料
wrap pants 卷裹裤,围裹裤

wrap pattern socks	吊线花袜,绣花短袜
wrapped and tied	叠襟系带式的
wrapped around button	包纽,钉绕脚纽
wrapped cardigan	和服式无扣开衫
wrapped form	包缆式,螺旋缠绕式
wrapped leggings	绑腿布
wrapped seam	包卷缝
wrapped skirt	围裹裙,包裙
wrapped top	布条裹身扎结的上衣
wrapped turban	缠裹式头巾帽
wrapped yarn	包线,包芯花式线,包缠花式线
wrapper	(妇女室内穿用的)宽大长衣,浴衣,晨衣,睡衣,化妆衣;童装长斗篷;裹袍;包装材料;家居长袍
wrapping	绑腿,裹腿护腿,包装;(常用复数)用于包裹材料
wrapping buckle	布包带扣
wrapping button	布包扣
wrapping button shank	绕扣柄
wrapping cloth	包装布;包裹布
wrapping paper	包装纸;烫缝时的衬纸
wrapping paper cushion	熨烫衬纸垫
wrapping paper stripe	熨烫用衬纸条
wrapping plate	包装板
wrappings	包装材料;包装纸;包装布
wraprascal	(19世纪)拉布拉斯卡洛外套,男式瑞丝卡宽绰型外套
wrap rib top machine	绣花罗纹袜口机
wrap-round	围裹式衣服(指裙)
wrap-round skirt	家用病人裙子
wraps	围巾,头巾;披肩;毯子;手帕;外衣
wrap skirt	围裹裙,宽搭门裙
wraps-on-socks	吊线袜子
wrap style	围裹式,叠襟式,和服式(用腰带等方式扎住,无纽扣)
wrap sweater	围裹式毛线上衣,无扣开衫
wrap tennis skirt	缠绕式网球裙
wrap topper	女宽松短大衣,宽短大衣
wreath	花冠,花环
wrestle costume	摔跤服
wring dry	绞干
wrinkle	腕,(衣服、手套等的)腕部;皱,皱褶,皱纹,折皱
wrinkle fabric	折皱布
wrinkle fly shield	里襟起绉
wrinkle-free	免熨烫,不起皱,抗皱,防皱
wrinkle-free batch oven	抗皱处理烘箱
wrinkle-free clothing	免熨烫服;抗皱服装
wrinkle-free cotton clothes	全棉免烫服装
wrinkle-free equipment	抗皱设备
wrinkle-free finishing	无皱整理免烫整理
wrinkle free pants	免熨烫裤;防皱裤
wrinkle-free shirt press	抗皱衬衫熨烫机
wrinkle-free trousers press	抗皱裤熨烫机
wrinkle front crotch	小裆起绉
wrinkle line	皱纹线条
wrinkle lower armhole	腋窝起绉
wrinkle mark	皱痕,接缝痕
wrinkle recovery	折皱回复性
wrinkle resistant	防皱,耐折皱性
wrinkle resistant finish	防皱整理
wrinkles at collar band facing	底领里起皱
wrinkles at front rise	小裆不平(前裆起绉)
wrinkles at hem	底边起绉;卷边起绉
wrinkles at lapel	驳头起皱
wrinkles at shoulder	塌肩(肩斜度小)
wrinkles at sleeve opening	袖口起绉
wrinkles at top collar	领面松
wrinkles at top fly	门襟起绉
wrinkles at top lapel	驳口起皱
wrinkles at waistband facing	腰缝起皱
wrinkles at zipfly	绱拉链处起绉
wrinkling	起皱,起绉
wrinkly cloth	绉面织物
wrist	手腕;(袖子、手套的)腕部
wristband	腕带,腕套;袖口;表带;门襟;翻口皮
wrist bone	腕骨
wrist circumference	腕围
wrist effect	(仿麂皮)指痕(绒毛方向发生变化而留下的指印)
wrist girth	腕围
wrist-length	及腕长度
wrist-length sleeve	及腕袖
wrist line	腕线
wristlet	袖头,腕套,腕带;手镯
wrist point	腕点
wrist sleeve length	九分袖长
wrist sling	套腕带
wrist strap	金属扣宽皮腕带,腕部扣带
wrist support(er)	护腕
wrist thickness	腕厚
wrist watch	挂表,手表
wrist width	腕宽
write for purchase	承购
write number on carton	填箱号

wrong button placement 纽眼位置不对	WTCA 世界贸易中心协会
wrong colour 错色,颜色配错	WTO 世界贸易组织
wrong craping 皱疵,皱斑	W type (woollen type, wool type) 毛型
wrong draw-in 错序	Wu silk 五丝
wrong narrowing mark 收针花错	WVT (water vapor transmission) 透水(蒸)汽性
wrong side (WS) 织物反面,织物背面,内面	w-wash 洗旧整理,不磨烂整理
wrong side of cloth 布料的反面	W.W.D. (Women's Wear Daily) (美国)妇女时装日报
wrong side marking 反面标记	WX (women's extra large size) 女式特大号
wrong size assortment 尺寸搭配错	
wrong striping 夹色错	wylie coat 保暖内衣,羊毛内衣,法兰绒内衣
wrong waist shaping 腰部不平的服装	
wrought 手织花纹织物;全成形的	wyvis 威维斯
wrought hosiery 成形针织物	

X

X (King size) 特大号,特长号
xanthic 黄色的,带黄色的
xanthochroous （皮肤和头发）呈淡黄色的,带黄色的
xanthomelanous 有橄榄色(或黄色)皮肤和黑头发的
xanthous 黄色的,浅黄色的;有黄棕色毛发的;黄色人种的,蒙古人种的
Xerga 哔叽;斜纹毛毯
xi 舄,中国古代贵族大礼时穿的鞋子;缔（古代织物）
xǐ 屣
xiang embroidery fabric 湘绣品
xiangle crepe(de chine) 香乐绉
xingfu brocade 幸福缎
xíng téng 绑腿
Xinjiang carpet 新疆地毯
xiubei matelasse 袖背绸
Xibe ethnic costume 锡伯族服饰
Xibozu costume 锡伯族民俗服

Xiehe 斜褐(古代)
xihu gauze 西湖纱
xihu jacquard crepon 西湖呢
xihu taffeta faconne 西湖绢
xining wool 西宁毛
Xinwha crepe 新华呢
XL (extra large) 特大号;加大码
X-leg 叉形腿
X-ligne button X号纽扣
X-line X型(强调肩宽、腰细、裙摆扩展的造型)
X-ray dress 透明装
x-ray proofing fabric 防X射线织物
X-shape X形;交叉形;X型
X-shape tack X形加固缝
X-sizing 特大号
Xstyle X型款式
Xuchow crepe 湖绉
XXL (extra extra large) 超特大号

Y

y（yard） 码
Y 偏瘦体型（服装体型分类代号）；(yellow)黄色,黄色服装,黄色料子
yacht cap 快艇帽
yacht cloth 快艇呢,浅色薄法兰绒
yachting cap 游[快]艇帽,海军帽
yachting coat 男式快艇外套（双排四扣,快艇俱乐部铜扣）
yachting costume 快艇服饰
yachting parka 快艇派克防寒服
yachting shoes 游艇鞋
yacht line sneaker 快艇用运动鞋（鞋底上部有绳纹,鞋面有条纹）
yacht parka 游艇防寒大衣,游艇派克大衣,快艇派克防寒服
yacht sneakers 快艇（帆布）运动鞋（帆布鞋）
Yajiang bourette(tussah pongee) 鸭江绸
Yajiang checked bourette 条格鸭江绸
Yajiang tussah pongee 鸭江绸（传统柞蚕丝织物）
yak 牦牛绒毛花边,精纺毛纱花边
yak hair 牦牛毛[绒]
yak hair knitwear 牦牛绒衫
yak hair overcoating 牦牛绒大衣呢
yak lace 牦牛毛花边,亚克粗梭结花边,羊毛钩编花边
yak overcoating 牦牛绒大衣呢
yak rukha 织物表面
yaktara 牦牛毛平纹织物
yakut woman's apron and coat 雅库特女毛皮衣裙
Yalan 雅兰
Yale 耶鲁衬衫布,牛津布
Yale blue 耶鲁蓝（美国耶鲁大学制服的颜色）,绿蓝色
yam 山药色
yamamai 日本山蚕丝
yamamai silk 山蚕丝（产于日本）,天蚕丝
yangnian crepe jacquard 羊年绉
Yankee look 扬基款式
yantsou 烟台丝绸
Yao ethnic costume 瑶族服饰
Yaozu costume 瑶族民俗服
yard（yd 或 y） 码

yardage 码数（长度）
yardage goods 匹头;按码出售的织物
yardage including loss 包括裁剪废料的用布量
yardage roll 布卷,成卷的布
yard-dyed flannelette cheeked 色织彩格绒布
yard goods 按码出售的织物;匹头
yardgoods knitting 匹头针织物,按码计算的针织物
yardgoods machine 匹头布针织机
yard measure 码尺（直尺或卷尺）
yardstick 码尺（直尺）
yardwand 码尺（直尺）
yarkee necktie 美式领带
yarmalke （犹太人）圆形小便帽
yarmelke （犹太人）圆形小便帽
yarmulka （犹太）亚莫克便帽,圆形小便帽;男式无檐便帽
yarmulke （犹太）亚莫克便帽,圆形小便帽;男式无檐便帽
yarn 纱,纱线,丝
yarn bobbin 纱线筒子;木纱团
yarn-bonded fabric 纱线黏合织物
yarn brake 纱线张力器,纱线制动器
yarn coloured article 色织物,色织布,原纱染色织物
yarn combination （不同纱线）交织
yarn count 纱支
yarn cutting （缝纫）断线
yarn darner 粗型缝补针（规格 14～18）
yarn dyed 色织
yarn-dyed bed sheet 色织被单布
yarn-dyed broken filament fabric 雪花呢
yarn-dyed bulky acrylic tweed 色织腈纶膨体粗花呢
yarn-dyed bulky acrylic upholstery fabric 色织膨体腈纶装饰织物
yarn-dyed burnt-out cloth 色织烂花布
yarn-dyed checks 全棉向阳格布
yarn-dyed combed cotton high count poplin 色织纯棉精梳高支府绸
yarn-dyed corduroy 色织灯芯绒
yarn-dyed cotton shoe canvas 色织纯棉双层鞋面帆布

yarn-dyed cotton/arylic mixed upholstery fabric 色织棉腈交织装饰织物
yarn-dyed crepe 色织绉纱
yarn-dyed crepon 色织高档中长织物
yarn-dyed denim 牛仔布,靛蓝劳动布,坚固呢
yarn-dyed embossed velvet 色织拷花绒
yarn-dyed fabric 色织物,色织布,原纱染色织物
yarn-dyed fabric with broken filament 色织断丝织物
yarn-dyed fabric with embroidery effect 色织浮纹仿绣织物
yarn-dyed flannelette checked 色织彩格绒布
yarn-dyed gingham 色织自由条布
yarn-dyed glittering poplin 色织复合丝闪光绸
yarn-dyed half-thread poplin 色织半线府绸
yarn-dyed imitation silk and linen fancy suiting 色织涤纶竹节丝仿丝麻花呢
yarn-dyed interlock fabric 色织棉毛布
yarn-dyed jacquard wool-like fancy suiting 大提花中长纬低弹花呢
yarn-dyed lawn with pin-waled 色织涤棉牙签条
yarn-dyed light-weight seersucker 色织薄型弹力绉
yarn-dyed linen 色织亚麻布
yarn-dyed looped cloth 色织起圈织物
yarn-dyed medium-weight wool-like fancy suiting 色织中长厚花呢
yarn-dyed ombre fabric 色织凤尾布
yarn-dyed oxford 色织牛津布
yarn-dyed polyester/cotton boucle fancy suiting 色织涤棉结子花呢
yarn-dyed polyester/cotton jacquard poplin 涤棉提花府绸
yarn-dyed polyester/cotton lawn 色织涤棉细纺
yarn-dyed polyester/cotton sarong 色织涤棉裙布
yarn-dyed polyester/cotton thick and thin stripe fabric 色织涤棉稀密条纹织物
yarn-dyed polyester/cotton tussores 色织涤棉线绢
yarn-dyed polyester/cotton/viscose furnishing fabric jacquard 色织涤/棉/黏大提花装饰布

yarn-dyed polyester interlacing yarn fancy suiting 色织低弹中长网络花呢
yarn-dyed polyester/viscose cvalry 色织中长马裤呢
yarn-dyed polyester/viscose harris 涤黏海力斯
yarn-dyed polyester/viscose hopsack 色织中长板司呢
yarn-dyed polyester/viscose panama 色涤黏巴拿马花呢
yarn-dyed polyester/viscose valitin 色织中长凡立丁
yarn-dyed polynosic lawn 色织富强纤维细纺
yarn-dyed polypropylene air textured yarn tweed 色织丙纶吹捻丝粗花呢
yarn-dyed poplin 色织府绸
yarn-dyed preshrunk denim 色织防缩坚固呢(劳动布)
yarn-dyed raising gingham 色织格绒布
yarn-dyed resin finished chambrag 色织纯棉树脂青年布
yarn-dyed satin 熟织丝缎
yarn-dyed seersucker 色织泡泡纱
yarn-dyed silk satin 熟丝缎
yarn-dyed suitings 色织花呢
yarn-dyed tarpaulin for chiidish car 童车篷帆布
yarn-dyed tree bark like crepe 色织树皮绉
yarn-dyed vinylon/cotton gingham 色织维棉条格布
yarn-dyed vinylon/cotton mixed fancy suiting 色织维棉交织花呢
yarn-dyed viscose jacquard furnishing fabric 色织黏胶大提花装饰布
yarn-dyed viscose/polyester satin brocade 色织涤黏大提花织锦
yarn-dyed wool-like fancy suiting 色织中长花呢
yarn-dyed wool-like herringbone 色织中长海力蒙
yarn-dyed wool-like palace 色织派力司
yarn-dyed wool-like semi-finish serge 色织中长哈味呢
yarn-dyed wool-like tropical suiting 色织中长薄型花呢
yarn-dyed wool-like two-layer tweed 色织仿毛双层粗厚花呢
yarn-dyeing 纱线染色

yarn-dyeing fabric 色织布
yarn fault 纱疵
yarn finish 纱线整理
yarn for raised fabric 起绒用纱
yarn in hanks 绞纱
yarn linear density 纱线的线密度
yarn number 纱线支数,纱支
yarn on cones 筒子纱;筒纱
yarn printed fabric 印纱织物
yarn severance 缝纫时底线断头
yarn size 纱线细度,长丝细度,纱支
yarns per inch 每英寸纱线根数
yarn stockings 纱袜
yarn trapper (缝纫机)夹线器;夹纱器
yarn twist 纱线捻度
yarn variation 横路疵
yarn woven fabric 纱织物
yashimagh (伊拉克)亚希玛戈浮纹布
yas(h)mac 亚瑟玛面纱,(穆斯林妇女)双层面纱,浮纹面纱
yashmak (穆斯林妇女)双面纱;浮纹面纱
yaw 织物薄段(疵)
yazma 三角印花大头巾,素色三角大头巾
Y coat Y形外套,Y形大衣
yd (yard) 码
yearly output (Y.O.) 年产量《管理》
year round coat 四季穿大衣
year round fashion 四季穿流行女服组合,四季流行女服
year round suit 全年可穿套装,四季套装
year round type 全年可穿服装款式,全年通用式
year-round suit 全年通用套装,四季套装
yeddo crepe 中原印花织物
yeddy girl style (女)阿飞款式
yehtansuchow 烟台茧绸
yelek 土耳其上衣
yelin jacquard faille 椰林葛
yelken-bez 厚篷帆布
yellow (Y) 黄色;黄色服装;黄色料子
yellow-based red 黄底红色
yellow brown 黄棕
yellow cast 黄色光
yellow cream 黄奶油色,嫩黄
yellow deep 暗花色,暗黄,深黄
yellow earth 土黄
yellower [色深]偏黄
yellow green (YG) 黄绿色

yellow-gold 金黄色
yellow gum silk 黄生丝
yellowing 泛黄,发黄
yellowing resistance 抗泛黄性
yellowish brown 黄光棕,土黄,黄棕色的
yellowish ironing 烫焦(疵点)
yellowish orange (yO) 黄光橙色
yellowish pink 肉色
yellowish red (yR) 黄光红色
yellow jacket 黄马褂
yellow jersey 黄色紧身运动套衫
yellow lapis lazuli 黄色蓝宝
yellow mid 中黄,旗黄色
yellow ocher 土黄,赭石黄(浅暗桔黄色)
yellow olive 橄榄黄
yellow orange 橙黄
yellow red (YR) 红黄赤
yellow soap 洗衣肥皂,家用肥皂
yellow steam filature 黄厂丝
yellow system 黄色类
yellow ultramarine 佛黄色
yellow-undertone complexion 淡黄色皮肤
yellow wish green (yG) 黄光绿
yellow wish grey 黄光灰
yemeni 也门头巾
yerges 雅尔其粗厚毛毡,粗毛马鞍毡
yeri 和服刺绣领
yé-yé 耶耶派服式
Y-fronts 男用Y形短裤,弹力内裤
YG (yellow green) 黄绿色
yG (yellowish green) 黄光绿
Y-heel Y形袜跟
yield point 屈服点
Yi ethnic costume 彝族服饰
yifeng tussah flake 异风绸
yihao faille 一号绨
yishbizh 棱条毛毯
yi shing kung silk 南京中山绸
yistlo 平纹毛毯
yiwen poplin brocade 意纹绸
yixin jacquard sand crepe 意新绸
yixin poplin faconne 一心绸
yizu costume 彝族民俗服
Ykk zipper (日本名牌)Ykk拉链
Y-line Y型,Y形线形
Y.O. 黄光橙色(yellowish orange);年产量(yearly output)《管理》
yobes 礼服
yoga wear 瑜伽服
Yogi Bear tee-shirt "瑜伽熊"短袖圆领衫

Yohji Yamamoto 山本耀司(日本时装设计师)
yoke 育克,过肩,覆肩;抵肩,抵腰;针织机筒子架,横机三角座滑架;裤或裙的前后翘;裙腰拼接布
yoke and gathers 过肩和抽褶
yoke and sleeve in one 过肩式连肩袖,育克式连肩袖
yoke-attaching machine 过肩自动缝合机
yoke attachment 过肩附件
yoke belt 抵腰带
yoke blouse 育克罩衫
yoke bone 颧骨
yoke collar 抵肩领,育克领,过肩领
yoke-front 前过肩,前抵肩
yoke garment 有过肩的服装,有育克的服装,有覆肩的服装
yokel color 乡土色彩
yoke length [width] 过肩长[宽]
yoke line 过肩线,育克线,覆肩线
yoke seam 过肩线缝,育克线缝
yoke skirt 育克裙(在臀围部位剪接的裙子),抵腰裙,剪接裙
yoke sleeve 过肩式连肩袖,过肩袖,育克式连肩袖
yoke slide insert 衬垫,过肩衬垫
yoke under line 过肩下口线
yoke width 过肩宽
yoke width line 过肩宽线
yoke with pocket 带肩饰口袋的育克
Yokohama crepe 横滨绉,紧密无光丝绉
yolk yellow 蛋黄色,深黄色
YONEX 尤尼克斯
yonghal lace 友哈耳花边
yorks 裤腿绑带
Yorkshire flannel 约克夏法兰线(全毛),约克夏棉毛混纺呢
york tan glove 有分割的皮革手套
York wrapper 约克袍(19世纪早期欧洲妇女晨礼服)
yoryu crepe 柳条绉
yoryu georgette 柳条乔其绉
Yoshie Inaba 稻叶贺惠(日本时装设计师)
Yoshiyuki Konishi 小西良幸(日本时装设计师)
Yotumo (5~13岁穿)童和服
youghal lace 爱尔兰编带
young 纵向弹力针织物,经向弹力针织物

young adult 青年(指22~25岁的人)
young boy's jumpsuit 儿童短外套
young boy's suit 小男孩装
young fashion 青年款式,青年流行装
young junior/teen size 中学生尺码,十几岁少年尺码
youngmen's clothing 青年服
youngmen's coat 青年装上衣
young wear 青年服
young wheat 嫩麦色
youth full adult 年轻成人
youth's range (童鞋)过渡尺寸(日本)
yoyo 交替拉伸和收缩织物
ypres 粗方网眼梭结花边;精纺细呢
Ypres lace 伊珀尔花边
YR (yellow red) 红黄赤
yR (yellowish red) 黄光红色
yranny boots 祖母靴
Ysatis 伊塞迪丝
Y-shaped suit Y型西装
Y-stitch Y字线迹
yu 绤(中国古代织物)
yuchausa 驼毛呢
Yue embroidery fabric 粤绣品,广绣品
yuenching suchienyong 黑丝绒
Yues Saint Laurent 伊夫·圣·洛朗(1936~ ,法国时装设计师)
YUE-SAI 羽西
Yugoslavian costume 南斯拉夫民俗服
Yugoslavian embroidery 南斯拉夫刺绣
Yugur ethnic cotume 裕固族服饰
Yuguzu costume 裕固族民俗服
yukata [日]浴衣,和服式轻便衣
Yuki Torii 鸟居由纪(日本时装设计师)
Yukiko Hanai 花井幸子(日本时装设计师)
Yukio Kobaya 小林由纪雄(日本时装设计师)
yuling 强捻厚斜纹织物
yun brocade 云锦
yunfeng lame 云风锦
yunt 蓝棉布
Yuppie (young urban professlonais) 雅皮士,(城市)少壮职业人士
yutun 丝经毛纬驼毛呢
yuxiang crepe georaette 玉香绉
yuyun voile 宇云绡
yuzen birodo 友禅法兰绒
Yves Saint Laurent (YSL) 伊夫·圣·洛朗(法国时装设计师)

Z

zacarilla　19世纪细薄布
zaffer blue　钴蓝色
zaftig　（女子）体态丰满而富有曲线美的，身段漂亮的；胸脯高突的，乳房丰满的
zamarra　［西］（牧民穿）羊皮上衣，羊皮外套
zambelotto　马海毛呢
Zamdra Rhoades　赞德拉·罗兹（1942～，英国时装设计师，主要创新有：在柔软的面料上进行手工筛网印花，毛皮大衣，锯齿形线条，大块泼墨般溅洒式图案等）
zanella　扎纳拉伞布，伞用斜纹织物
zangoz khata　莎车粗驼毛织物
Zangzu costume　藏族民俗服
zante lace　（希腊）纤绣挖花花边
zanza silk　盛泽纺，复兴纺（中国制）
zanzibar cloth　印度原色棉织物；双经双纬杂色条子衬衫布
zanzibars　桑给巴尔布
zarasas　标准印花布
zarbaf　轻薄金丝花缎
zazou　［法］爵士乐迷款式
zebra　有黑白相间条纹的；斑马条纹
zebra leather　斑马皮
zebra splicing　无缝袜粗线条加固
zebra stripe　粗条纹，色地等宽条纹，斑马条纹
zeck　条子斜纹衬里（作外衣、雨衣衬里用）
zelan finish　（防水防污的）泽伦整理
zemana　泽马纳双层织物
zenana　18世纪轻薄凸纹布
zenana（cloth）　（女用）薄质衣料
zendado　头巾，披肩
zenith（blue）　天顶蓝（淡紫光蓝色）
zephirities　泽菲里蒂斯薄纱（绸）
zephyr　薄衫；轻薄色织席纹呢；细薄织物，轻薄色织直条平布，轻罗；高级细软毛线；藕灰
zephyr blue　薄纱蓝
zephyr cloth　薄纱绸，轻罗绸
zephyr finishing　轻薄（织物）整理
zephyr flannel　（丝毛混纺）轻薄法兰绒
zephyr gingham　高级轻薄方格平布，轻薄格子布
zephyrlite　［美］泽弗来特帆布
zephyr shawl　轻薄棉毛交织围巾
zephyr shirting　薄纱衬衫料
zephyr silk barege　丝毛薄呢
zephyr worsted　轻软精纺毛线
zephyr yarn　轻软精纺毛线
zerak　深蓝衬衫棉布
zerape　彩色毛呢披肩
zerbase　伊朗泽贝斯织锦
zero defect　无缺陷
zero needing　零需求
zero twist　无捻，零捻
zero twist filament　无捻丝
zha-bo　垂胸领饰；胸饰领；垂胸饰领衬衫
zhangzhou（brocaded）velvet　漳绒，天鹅绒
zhangzhou velvet satin　漳缎
zhao jia　罩甲（中国古代外褂）
zhe　帻（中国古代男子头巾）
zhee lay　假胸片女衫
zhi cheng　绒，织绒，织成（中国古代丝织花纹服料）
zhi shen　直身（中国古代长袍）
zhi shun　只孙，质孙（中国古代一色衣）
zhishan brocade　织闪缎
zhivago blouse　齐瓦戈女衫
zhivago coat　齐瓦戈外套
zhivago collar（同cossack collar）　齐瓦戈领；哥萨克领
zhivago dress　齐瓦戈衬衫式裙装
zhivago shirt　齐瓦戈衬衫
zhong-jing headdress　忠靖巾（中国古代帽饰）
zhongshan clothing　中山装
zhongshan coat　中山装上衣
zhongshan coat pocket　中山装袋；老虎袋，吊袋
zhongshan（coat）collar　中山服领，中山装领
zhongshan suit　中山装
zhongshan suit pocket　吊袋，老虎袋
zhu　纻（中国古代织物名称）
zhuang brocade　壮锦
zhuang ethnic costume　壮族服饰

zhuang hua silk 妆花
zhuangzu costume 壮族民俗服
zhu-li crepe 珠丽纹
zhu-si silk 纻丝
zibel(1)ine 齐贝林有光长绒呢；紫貂毛皮；黑貂皮；紫貂绒；齐贝林大衣呢；绒结子毛纱
zigzag 之字形；Z形；人字形；曲折形；锯齿形；跳针；锯齿状花样；交错
zigzag and satin stitch 曲折和缎纹形线迹
zigzag attachment 曲折缝附件
zigzag cable chain stitch 锯齿锚链线迹
zigzag chain stitch Z形线迹,锁缝线迹
zigzag circle tacker 曲折圆形加固缝纫机
zigzag connection 交错连接
zigzag coral stitch 锯齿珊瑚线迹
zigzag decorative stitch 曲折装饰线迹
zigzag design 人字纹
zigzag double chain stitch 曲折双线链式线迹
zigzag double locked chainstitch 曲折双锁链式线迹
zigzag finger tip control 曲折缝指头控制器
zigzag flat bed sewing machine 曲折缝平底缝纫机
zigzag gathered band trimming Z形抽褶饰边
zigzagged seam 锯齿缝线迹
zigzagger 曲折缝缝纫机
zigzag heart tacker 曲折心形加固缝纫机
zigzag hem stitch 曲折缝卷边线迹
zigzag high-speed seamer 高速曲折缝纫机
zigzag inside pocket 锯齿形里[内]袋
zigzag lock stitch 曲折锁式线迹
zigzag machine 曲折缝缝纫机,锯齿形锁缝缝纫机
zigzag neck 锯齿领
zigzag necklace[neckline] 锯齿形领口
zigzag neckline 锯齿形领口
zigzag-overstitch 曲折包缝线迹
zigzag pattern Z形图案
zigzag rectangular tacker 长方形绕缝加固机
zigzag rope-stitch Z形绳纹线迹
zigzag ruler 曲尺
zigzag scissors 锯齿剪刀,花齿剪
zigzag seam 曲折形线缝,锯齿形线迹,之字形线缝
zigzag semi-circle tacker 曲折半圆加固机
zigzag sewing 锯齿缝,曲折缝；曲折缝纫
zigzag sewing foot 曲折缝压脚
zigzag sewing machine 之字针迹缝纫机，曲折缝缝纫机；花针机
zigzag sewing operation 曲折缝缝纫操作
zigzag square tacker 方形线缝加固机
zigzag stitch 花针法；锯齿形线迹,工形线迹,Z形线迹,曲折线迹；弯线迹（疵点）
zigzag stitching 曲折线缝,齿状线缝,缝之字线
zigzag stitching machine 曲折缝缝纫机，之字针迹缝纫机,花针机
zigzag tacks 曲折形加固缝
zigzag twills 人字呢,山形斜纹织物
zigzag type hook 曲形钩
zigzag width 曲折缝宽
zigzag wire Z形金属丝（鱼骨撑的替代品）
zimarra 宽松女袍,传教士长袍
zinbei 和尚衫(和服的一种)；日本和服
zinc(color) 锌灰；锌色
zinc finish 锌辊轧花整理,锌辊轧光整理
zinc white 锌白,锌氧粉
zindai 人体模型
zinken 人造丝
zinnia 百日草色
zip 拉链；拉[扣]拉链
zip and stud 拉链纽扣双开口
zip-bag 拉链包,拉链袋
zip fastener 拉链
zip-fly 拉链门襟
zip foot （缝纫机）拉链压脚
zip-front 前开拉链,前襟拉链
zip-front jacket 拉链衫,拉链夹克衫,拉链短上衣
zip front pullover 前拉链套衫
zip guards 拉链垫襟
zip insertion 装拉链
zip-in inner-lining 拉链脱卸式夹里
zip-in liner 拉链脱卸式夹里
zip-in lining 拉链脱卸式夹里
zip-in sleeve 拉链脱卸袖
zip-in/zip-out lining 活里
zip length 拉链式裂解的链长
zip-lined coat 装拉链衬里大衣
zip necktie 拉链领带
zip-off coat (同 duo-length coat) 双长度

大衣(至中腿肚或全长,由拉链控制长度,常为毛皮大衣,由拉链控制长度的大衣)
zip-off sleeve 拉链脱卸袖
zip-out inner-lining 拉链脱卸式夹里
zip-out liner 拉链脱卸式夹里
zip-out lining 拉链脱卸式夹里
zipped cuff 拉链袖口
zipper 拉链
zipper adhesive tape 拉链黏性牵条
zipper assembler 缝拉链的人;缝拉链机,装拉链机
zipper bag 拉链袋;拉链包
zipper cardigan 拉链开襟衫
zipper clamp 拉链夹头
zipper clanging 换拉链
zipper closing pinwale corduroy jacket 直条灯芯绒拉链夹克
zippered closing 拉链门襟
zippered foundation 拉链式束腰胸衣
zippered girdle 拉链式紧身褡
zippered pants 拉链裤
zippered pocket 拉链口袋
zipper fly stay 拉链基布
zipper foot 缝拉链压脚
zipper front jacket 拉链夹克
zipper front pullover 拉链套衫
zipper front sport shirt 拉链运动衫
zipper front vest 拉链背心
zipper inserting 装拉链,缝拉链
zipper inserting machine 缝拉链机
zipper insertion 装拉链
zipper opening 拉链式开口
zipper placket 拉链式开口
zipper pocket 拉链袋
zipper pull 拉链拉头
zipper puller 拉链头;链头
zipper pull label 拉链头织唛
zipper pullover 拉链套衫
zipper shield 拉链暗门襟
zipper slider 拉链头;链头
zipper strength 拉链强力,拉开强力
zipper tail 拉链尾部(比实用拉链长的部分)
zipper tape 拉链(齿)带;拉链(齿)布边
zipper teeth 拉链齿;链齿
zipper underlay 拉链里襟,拉链垫襟
zipper with double sliders 双头拉链
zipper with semi-autolock slider 半自动头拉链

zip pocket 拉链口袋
zip pocket open 拉开口袋拉链
zip setting 装拉链
zip slider 拉链滑锁
zip suit 拉链套装(拉链分隔衣裤)
zip tape 拉链布带
zip teeth 拉链齿
zip through front 前装拉链;拉链开襟
zip with guard 有护垫的拉链
zipper-sealing mechanism 拉链密闭机构
zippy coat 拉链衫
zircon blue 浅蓝色
zircon 锆石
zirjoumeh 波斯妇女居家连衣裙
Zizi Jeanmaire 齐齐·让梅蕾(法国时装设计师)
zizzy tie 俗艳的领带
Z-lay Z捻
Z-leg Z形边脚
zoccoli 意式木鞋
zod 腰带;黄道十二宫图
zodiac 腰带;黄道十二宫图
zodiac bag 织有黄道十二宫的手提布包
zodiac necklace 雕有黄道十二宫的垂饰项链
zoguri silk 日本坯绸
zoisite 黝帘石
zona 崇娜带,佐纳皮饰带
zone 带子,腰带;色区
zoneless 不佩带子的,不佩腰带的
zoom-back 拉摄,拉镜头
zoom colour 直升色
zoom finder 变焦距取景器
zoom lens 变焦距镜头
zoom ratio 变焦比
zoomorphic design 兽形花纹(图案),动物形花纹
zoom-up 推摄,推镜头
zoo pattern 动物园花样
zoo-suit 宽肥套装(上装宽大,裤肥,裤口收紧)
Zoot look 佐特风貌(一种上衣及膝而裤子狭窄的服式,流行于20世纪40年代)
Zoot shirt (与佐特套装配穿的)艳色衬衫
Zoot [suit] 佐特套装(上衣肥大、肩宽而长、裤口狭窄、收紧)
Zootsuiter 穿阻特服者(赶时髦,着廉价服装的人)
zori [日本]草履;平底人字拖鞋

zoster （古希腊用的）带,束带,腰带
Zouave 祖阿芙制服;短至膝上的灯笼裤;（妇女）绣花短上衣
Zouave [jacket] 祖阿芙型（绣花）女短上衣
Zouave pants 祖阿芙步兵裤;半长灯笼裤
Zouave style 祖阿芙款式
zrikas 二上二下斜纹呢
Z twist yarn Z捻纱,左手捻纱
zucchetto （天主教戴）室内圆形小便帽（神父用黑色,主教用紫色,红衣主教用红色,教皇用白色）,教士便帽,教士圆顶小帽
zug stitch 卒格针法
zukin 印花方形薄头巾
zulu cloth 密织斜纹刺绣底布;祖卢格子布,平纹红或蓝底彩格棉布（英国出口非洲）,绣花底布
zunnar 教徒腰带
Zuo-gi silk-like fabric 佐其
zuohua twill faconne 柞花绫
zygoma 颧骨
zygomatic bone 颧骨

附录 服装服饰常用缩写词

AABPDF　Allied Association of Bleachers, Printers, Dyers and Finishers　纺织品漂印染整工人联合会

AAC　Association for Anti Contamination　污染防治协会

AACA　American Apparel Contractors Association　美发国服装协会

AAE　American Association of Engineers　美国工程师协会

AAEI　American Association of Exporters and Importers　美国进出口商协会

AAFA　American apparel & Footwear Association　美国服装与鞋袜协会

AAMA　American Apparel Manufacturers Associaton　美国制衣商协会

AATCC　American Association of Textile Chemists and Colorists　美国纺织化学家和染色家协会

AATT　American Association of Textile Technology　美国纺织工艺协会

ABCM　Association of British Chemical Manufacturers　英国化学品制造商协会

ABF　Adhesive Bonded Fabric　黏合织物, 黏合法非织造布

ABLC　Association of British Launderers & Cleaners　英国洗衣工和清洗工协会

ABS　American Bureau of Standards　美国标准局

a/c, acc.　account　账单

ACA　American Cotton Association　美国棉业协会

ACAM　Australian Confederation of Apparel Manufacturers　澳大利亚服装制造商协会

ACF　Activated Carbon Fiber　活性碳纤维

ACFIF　Asian Chemical Fiber Industries Federation　亚洲化学纤维工业联合会

ACMA　①American Cotton Manufacturers Association　美国棉纺织品制造商协会, 美国棉纺织同业协会; ②Athletic Clothing Manufacturers' Association　运动服制造商协会(英国)

ACMI　American Cotton Manufacturers' Institute　美国棉业协会

Acod.　Acodenmy　学院,学会

ACPTC　Association of college Professors of Textiles and Clothing　纺织服装大学教授协会

ACSMA　American Cloak and Suit Manufacturers Association　美国外套和套装制造商协会

ACT　Acetate Cloth Tape　醋酯布带

ACTM　Asspcoation of Cotton Textile Merchants　棉纺织品商业协会

ACTMA　Associated Cotton Textiles Manufacturers of Australia　澳大利亚棉纺织品制造商联合会

ACTWU　Amalgamated Clothing and Textile Workers Union　纺织与服装工人联合会

ACWA　Amalgamated Clothing Workexs of American　美国服装工人工会

Add.　address　地址

ADM　Affiliated Dress Manufacturers　服装制造商联盟(美国)

ADMI　American Dye Manufactures Institute　美国染料制造商协会

AEJI　Association of European Jute Industries　欧洲黄麻工业协会

AEPC　Apparel Export Prmotion Council　服装出口促进理事会

AFA　American Flock Association　美国植绒协会

AFIA　Apparel and Fashion Industry's Association　服装工业协会

AFMA　①American Fiber Manufacturers Association　美国纤维制造商协会; ②American Fur Merchants' Association　美国毛皮商协会

AFP　Association of Flock Processors　绒毛生产商协会

AFTAC　American Fiber, Textile, Apparel Coalition　美国纤维、纺织品、服装联盟

agt.　agent　代理商;代理人

AH　armhole　袖隆

AHSCA　American Home Sewing & Craft Association　美国家庭缝纫和手工艺协会

AIC ①American Institute of Chemists 美国化学工作者学会；②Association Internationale Colour 国际色彩协会（法国）

AICTC Associazione Italiana di Chimica Tessile e Coloristca 意大利纺织化学家和染色家协会

AIDS Association of Interior Decoration Specialists 室内装饰专家协会（美国）

AIL American Institute of Laundering 美国洗涤研究所

AIP Association International Polynosigues 国际波里诺西克协会，国际高湿模量黏胶纤维协会

AJSM Association of Jute Spinners and Manufacturers 黄麻制造商协会（英国）

AKFM Association of Knitted Fabrics Manufacturers 针织品生产者协会

AKIC Australian Knitting Industries Council 澳大利亚针织工业理事会

AKTA Apparel, knitting and Textile Alliance 服装针织和纺织业同盟（英国）

AL ①alginate fiber 海藻纤维，藻酸纤维；②artificial leather 人造革；③artificial life [simulation] 人工寿命[模拟]（纺织品、服装）

ALCA American Leather Chemists' Association 美国皮革化学家协会

ALCOTEXA Alexandria Cotton Exporters Association 亚历山大棉花出口商协会（埃及）

ALD American Laundry Digest 《美国洗涤文摘》（月刊）

ALG alginate 海藻酸盐，海藻纤维

ALMT Association of London Master Tailor 伦敦成衣商协会

ALRA Australian Leather Research Assocition 澳大利亚皮革研究协会

AM animalized cotton 动物质化棉；羊毛化棉

a.m. ante meridiem 上午

AMA Adhesive Manufacturers Association 黏合剂制造厂商协会

AMCC Apparel Manufacturers' Council of Canada 加拿大服装厂商联合会

AMMA American Millinery Manufacturers Association 美国女帽制造商协会

amt. amount 总计，合计，总额

ATEC American Textiles Export Co. 美国纺织品出口公司

ATEX American Textile Partnership 美国纺织合作协会

ANEC Asian Nonwovens Exhibition and Conference 亚洲非织造布展览与会议

ANFI Association of Nonwoven Fabrics Industry 非织造布工业协会

ANG Amarican Needlepoint Guild 美国刺绣协会

ANIC Asia Nonwovens Industry Conference 亚洲非织造布工业会议

ann. ①annals 年表，年鉴；②annual 年刊

Ann. Rep. Annual Report 年报，年鉴

anti-G suit anti-Gravitation suit 抗超重飞行衣

AOPM apparel order processing module 服装按指令加工模块

AP. arm point 袖圈

APFC American Printed Fabrics Council 美国印花协会

approx. approximately 约计

Apr. april 四月

Ad anthraguinone dye 蒽醌染料

AQL acceptable quality level 合格质量标准，正品标准

A.R. all risks 一切险，综合险

a/r. all round 共计

ARITT Arrangement Regarding International Trade in Textiles 国际纺织品贸易协定

ARN Apparel Research Network 服装研究网

art. article 物品，商品，条款

Art. silk artificial silk 人造丝

ARTA American Reusable Textile Association 美国再生纺织品协会

ARTP Association of Reclaimed Textile Processors 再生纺织品加工者协会

A.S. arm size 肘围

asb asbestos 石棉

ASC America Silk Council 美国蚕丝理事会

ASDC Associate of Society of Dyers and Colorist 染色工作者学会会员（英国）

ASF acrylic staple fibers 丙烯腈系短纤维

Asiatex Asia Textile Machinery Exhibition 亚洲纺织机械展览会

ASKT American Society of Knitting

Technologists 美国针织工艺师学会
ASME American Society of Mechanical Engineers 美国机械工程师学会
ASN average sample number 平均样本数
Asoc., assn., assocn. association 协会，团体
ASQC American Society for Quality Control 美国质量控制协会
ASSF IBRE Associazione Italiana Produttori Fiber Chimiche 意大利化学纤维生产厂协会
ASVT apparel spatial visualization test 服装体态同测
ASY acrylic spun yarn 丙烯腈纤维纱
ASYM Association of Synthetic Yarn Manufactures 合成纤维厂联合会
AT ①after tax 税后；②Air jet texturing 空气变形纱
ATC ① Agreement on Textile and Clothing 纺织品与服装协定（世界贸易组织）；② Asia Textile Conference 亚洲纺织会议
ATCA Alliance of Textile Care Associations 纺织品维护协会联盟（美国）
ATES advanced technology exposure suit 高技术防护服
ATI Asbestos Textile Institute 石棉纺织品学会（美国）
ATIE Association of Textile Industrial Engineers 纺织工业工程师协会（美国）
ATIRA Ahmedabad Textile Industry's Research Association 艾哈迈达巴德纺织工业研究协会
ATMA ① Alabama Textile Manufacturers Association 阿拉巴马州纺织品生产厂商协会（美国）；②American Textile Machinery Association 美国纺织机械协会
ATME-I American Textile Machinery Exhibition-International 美国国际纺织机械展览会
ATMI American Textile Manufacturers Institute 美国纺织厂商协会
ATY air textured yarn 喷气［法］变形丝
AUA Allied Underwear Association 内衣业协会联盟（美国）
Aug. august 八月
av. average 平均
aw ①actual weight 实际重量；②all width 全宽，全幅宽
a.W. all wool 全毛；纯毛
a/w., aw actual weight 实际重量
AWC ①American Wool Council 美国羊毛协会；② Australian Wool Corporation 澳大利亚羊毛协会
AWEX Australian Wool Exchange 澳大利亚羊毛交易所
AWIC Australian Wool Industry Conference 澳大利亚羊毛工业会议
AWRC Australian Wool Research Commission 澳大利亚羊毛研究委员会
AWTA Australian Wool Testing Authority 澳大利亚羊毛测试中心，澳大利亚羊毛检验局
AYSA American Yarn Spinner's Association 美国纱厂协会
B ①bale 包，捆；②bust 胸围；③button 按纽，纽扣；④ bottom 脚口，下摆；⑤blue 蓝
b bright 明亮的
B.A.wool Buenos Aires wool 布宜诺斯艾利斯羊毛
B.C. biceps circumference 上臂围；袖长
B.D. ①bust depth 胸高；②back depth 后腋深；③battle dress 战地服装；野战服
B.L. ①back length 后长；②bust line 胸围线
B.N. back neck 后领围
B.N.P. back neck point 后颈点
B.O. branch office 分公司
B.P. bust point 胸高点；乳峰点
B.R. ①back rise 后（直）裆；②body rise 股上
B.S.L. back shoulder line 后肩线
B.T. bust top 乳围
B.W. back width 背宽
B/L, b/l bill of lading 提货单
BA Belt Association 带类织物协会
BACMM British Association of Clothing Machine Manufacturers 英国服装机器制造商协会
BATC British Apparel and Textile Confederation 英国服装与纺织品联盟
BATEXPO Bangladesh Apparel and Textile Exposition 孟加拉国服装与纺织品展览会
BBAA Bride and Bridesmaid Apparel As-

sociation 新娘装和女傧相装协会
BBSI British Boot and Shoe Institution 英国靴鞋学会
BC bill for collection 托收汇票
BCC ①British Color Council 英国颜料染料委员会；②Buffalo Color Corp. 布法罗染料公司（美国）
BCDTA British Chemical and Dyestuffs Traders' Association 英国化学品及染料商协会
BCEA British Carpet Export Association 英国地毯出口协会
BCF bulked continuous filament 膨化变形长丝
BCF Nylon bulked continuous filament nylon 膨化变形耐纶长丝
BCGA British Cotton Growing Association 英国植棉协会
BCIRA British Cotton Industry Research Association 英国棉纺织工业研究协会
BCITA British Carpet Industry Technical Assciation 英国地毯工业技术协会
BCMA British Color Makers Association 英国颜料制造商协会
BCWA British Cotton Waste Association 英国废棉协会
BDA Bradford Dyes Association 英国布雷德染色工作者协会
BFA ①Bleachers'and Finishers' Association 漂整工作者协会（英国）；②British Fabric Association 英国织物协会
BFF bonded-fiber fabric 黏合纤维非织造布
BG. blue green 蓝绿
bG. blueish green 蓝调绿,带蓝色绿
BGMEA Bangladesh Garments Manufacturers and Exporters Association 孟加拉服装制造商和出口商协会
bh. buttonhole 纽孔;扣眼
BHMA British Hosiery Manufacturers' Association 英国针织品制造商协会
BIC Belgian International Carpets 比利时国际地毯博览会
BIFF British Industry Fashion Fair, Birmingham 英国国际时装博览会（伯明翰）
BJA Burlap and Jute Association 黄麻和黄麻织物协会（美国）,美国黄麻及麻布协会

BJEC Bangladesh Jute Export Company 孟加拉国黄麻出口公司
BJFB British Jensey Fabric 英国针织物研究会
BJMA Bangladesh Jute Mills Association 孟加拉国黄麻厂协会
BKCEC British Knitting and Clothing Export Council 英国针织品与服装出口理事会
BKEC British Knit Export Committee 英国针织品出口委员会
bl bale 包;捆
BL bust line, breast line, bust level 胸围线
BLA Belgian Linen Association 比利时亚麻协会
BLC British Leather Confederation 英国皮革同盟
BLF British Lace Federation 英国网眼织物联合会
BLMRA British Leather Manufacture Research Association 英国皮鞋制造研究协会
BLRA British Launderers' Research Association 英国洗衣业研究协会
bls. bales 包;捆
BMF boron metal fiber 硼金属纤维
BMFF, BMMFF British Man-made Fibers Federation 英国化学纤维联合会
BMS ①body measurement system ［三维］量身系统,人体尺寸测量系统；②British Mohair Spinners 英国马海毛纺织工联合会
BNMA British Nonwovens Manufacturers' Association 英国非织造布制造商协会
BOC. bank of china 中国银行
BPS body protection system 人体防护系列化服装产品（高尔夫球诸全天候服装）
BPTA British Polyolefin Textiles Association 英国聚烯烃纺织品协会
BR bills receivable 应收票据
brd braid 编织;编织物
brlp burlap 打包麻布
BRMA Braided Rug Manufacturers' Association 编织地毯制造商协会（美）
BRMF British Rainwear Manufacturers' Federation 英国雨衣生产商联合会
BRRA British Rayon Research Association

英国人造丝研究协会
BSAC Brotherhood of Shoe and Allied Craftmen 制衣业及同业工人兄弟会
BT ①back tacker 固缝机；②before tax 税前；③bright triangle 有光三角形截面（丝）
BTC bachelor of textile chemistry 纺织化学学士
BTBA Britainish Textile By-products Association 英国纺织副产品协会
BTC Britainish Textile Confederation 英国纺织联合会
BTD bachelor of textile dyeing 织物染色学士，织物染色工艺学士
BTE bachelor of textile engineering 纺织工程学学士
BTMA ①Britain Textile Machinery Association 英国纺织机械协会；②British Towel Manufacturers' Association 英国毛巾制造商协会
BTRA Bombay Textile Research Association 印度孟买纺织研究协会
BWC British Wool Confederation 英国羊毛联合会
BWMB British Wool Marketing Board 英国羊毛营销部，英国羊毛贸易部
BWP back waistline point 腰围线后中点
BWTA Boston Wool Trade Association 美国波士顿羊毛贸易协会
bx. boxes 箱；盒（复数）
C ①case 箱；②chest 胸围；③cord 帘线，绳；④cotton 棉
CAC Computer-aided coloring 计算机辅助测色配色
CACF continuous activated carbon fiber 连续活化碳纤维
CAD computer aided design 计算机辅助设计
C.A.D. cash against documents 凭单据付款
CAD/CAM computer aided design/computer aided manufacture 计算机辅助设计与制造
CADC color analysis display computer 颜色分析显示计算机
CADD computer-aided design and drafting 计算机辅助织物设计与制图
CAFCA Coty American Fashion Critics Award 科蒂美国时装评论奖
CAI ①Career Apparel Institute 职业服装学会（美国）；②color alteration index 色变指数；③Crochet Association International 国际钩针编织协会
CAL Computer Aided layout 计算机辅助排料
CAM. computer aided manufacture 计算机辅助生产
CAMA Children's Apparel Manufacturers Association 童装制造商协会（加）
Canad. Text. J. Canadian Textile Journal 《加拿大纺织杂志》（月刊）
C. and P. coat and pants 男式上衣和短裤
CAPIB Clothing and Allied Products Industry Board 服装工业及服装产品委员会（英国）
CAR Clemson Apparel Reseach 克莱姆森［大学］服装研究［中心］
cat. catalogue 样品目录
CATCC Canadian Association of Textile Colorists and Chemists 加拿大纺织染色家与化学家协会
CATD Computer Aided Textile Design 计算机辅助纺织品设计
CATMAA Cotton and Allied Textile Manufacturers Association of Australia 澳大利亚棉业与棉纺织厂商联合会
CB garment chemical-biological protective garment 防生化服装
C.B. centre back 后中缝
CBC carpet backing cloth 地毯底布
CBCF carbon bonded carbon fiber 碳黏合碳纤维
CBCS chemical-biological combat suit 防生化军服
CBD. cash before delivery 付款交货
CBF cellulose butylate fiber 纤维素丁酸酯纤维，丁酸纤维素纤维
CBN-waist Center back neck point-waist 后颈点至腰
CBWT Confederation of British Wool Textiles 英国毛纺织业联盟
CC ①centric covercore 包芯纱，皮芯纱；②Clothing Center, London 服装中心（伦敦）；③color code 色码，色标
CCAI Chamber of Commerce of the Apparel Industry 服装业商会（美国）
CCFA China Chemical Fibers Association 中国化学纤维协会

CCI ①Canadian Carpet Institute 加拿大地毯学会；②Cotton Corporation of India 印度棉花公司；③Cotton Council International 国际棉花理事会(美国)
CCM computer color matching 计算机配色，计算机拼色
CCMI Cashmere & Camel Hair Manufacturers Institute 开司末与驼绒制造商协会
CCOIC China Chamber of International Commerce 中国商会
CCS computer color searching 电脑色样检索
CCV color contrast value 色对比度
C.C.V.O. combined certificate of value and origin 估价和原产地联合证明书
cd cord 帘线；绳
C/D. certificate of delivery 交货证明书
C.&D. collected and delivered 货款两清
CDE chemical defense ensemble 整套防化服
CDG Costume Designers Guild 服装设计师同业公会(美国)
CDTRA Committee of Directors of Textile Research Associations 纺织研究协会会长委员会(英国)
C&E clothing and equipment 服装与设备
CEC Clothing Export Council of Great Britain 英国服装出口理事会
CECF Chinese Export Commodi-ties Fair 中国出口商品交易会
CEPC Carpet Export Promotion Council 地毯出口促进理事会(印度)
cert. certificate 证书；执照
CETIH Centre d'Etudes Technique des Industrie de l'Habillement 服装工业技术研究中心(法国)
CF ①carbon fiber 碳纤维；②center front 前(片)中心线；③continuous filament 连续长丝；④cover fabric 织物紧度，覆盖系数
C.F. centre front 前中缝
C.F., c.f., c.&F. cost and freight 离岸加运费价格
C/F, c/f. continuous filament 连续长丝，长丝
C&F. cost and freight price 成本加运费价格

CFA Contract Flooring Association 配套地毯料协会(英国)
CFB Cotton & Flax Bureau 棉麻局
CFDA Council of Fashion Designers of America 美国时装设计师理事会
CFFA Chemical Fabric & Film Association 化学织物和薄膜协会(美国)
C-fiber carbon fiber 碳[素]纤维
CFL center front line 前[片]中心线
CFS container freight station 集装箱货运站
C.H. custom house 海关
chg. change 费用
chq. cheque 支票
CI ①Clothing Institute 服装研究所(英国)；②color Index 颜色指数，比色指数；③Cotton incorporated 棉业联合会
CI, C.I. colour index 颜色指数；比色指数
C/I certificate of insurance 保险证明书
CIA ①Cotton Industries Association 棉纺织工业协会(加拿大)；②Cotton Insurance Association 棉花保险协会
CIC Cotton Institute of Conada 加拿大棉花学会
CIF, C.I.F. cost insurance and freight 到岸价
CIF, C.I.F., c.i.f. cost, insurance and freight 到岸价格，包括成本、保险费和运费的价格
CIFC cost, insurance, freight and commission 到岸价格加佣金价
CIFE cost, insurance, freight and exchange 到岸价格加汇费价
CIFW cost, insurance, freight and war risks 到岸价格加战争险价
CILC Confederation Internationale du Lin et du Chanvre 国际亚麻和大麻协会
CIM computer integrated manu-facturing 计算机综合制造
CIRFS Comité International de la Rayonne et des Fibres Synthétiques 国际人造纤维和合成纤维委员会(法国)
CITA Committee for the Implementation of Textile Agreements 纺织品协议执行委员会(美国)
CITDA Computer Integrated Textile Design Association 计算机综合织物设计协会

CITME China International Textile Machinery Exhibition 中国国际纺织机械展览会
cl. cloth 布,织物
C.L. coat length 衣长
CLF chlorofiber 含氯纤维(聚氯乙烯系纤维)
CLS China Leather society 中国皮革学会
Clthg. clothing 服装
cm. centimetre 厘米,公分
C/M certificate of manufacture 制造商证明书
CMA Clothing Manufacturers Association of the USA 美国服装制造商协会
CMC Color Measurement Committee 颜色度量委员会
CMD cellulose modal 高湿模量黏胶纤维(莫代尔纤维)的国际代码
CMFA Carpet Manufacturers Federation of Australia 澳大利亚地毯制造商联合会
CMFF Cotton and Man-made Fibers Federation 棉花和化学纤维联合会(澳大利亚)
CMP computer match prediction 计算机配色预测
CMT cutting, making, trimming 裁剪、缝制、整烫
CNITA China Nonwovens and Industrial Textiles Association 中国非织造布及工业织物协会
CNTA China Nonwovens Technical Association 中国非织造布技术协会
CNTIEC China National Textiles Import and Export Corporation 中国纺织品进出口总公司
Co., co., coy. company 公司
C/O certificate of origin 产地证明书
c/o, c.o. care of 代收,转交
C.O.D., c.o.d. cash on delivery 货到付款
COMIC Colorant Mixture Computer 模拟式配色电子计算机(配色系统,商名,美国戴维和赫曼丁格)
comm. commission 佣金
cont. contract 合同
cords. corduroy trousers 灯芯绒裤子
Corp. corporation 公司
C.O.S. cash on shipment 装船付款

COSC China Ocean Shipping Company 中国远洋运输总公司
cot. web. cotton webbing 棉布带,棉线带
cott. cotton 棉花,棉纱,棉织物
COTTD Center for the Organization for Textile Technology Development 中国纺织技术开发组织中心
C.&P. coat and pants 男式上衣和短裤
CPA ①Calico Printer's Association 棉布印花工作者协会;② chemical protective apparel 防化服
CPAI Canvas Products Association International 国际帆布制品协会
CPO jacket chief petty officer jacket 美国海军士官上装
CPO shirt chief petty officer shirt 美国海军士官衬衫
CQRS Cotton Quality Research Station 美国棉花质量研究所
CRAFFM Consortium for Research in Apparel, Fiber and Textile Manufacturing 服装、纤维和纺织品生产研究会(美国)
CRF Crease resistant finish 防皱整理
CRI Carpet and Rug Institute, USA 美国地毯协会
CRILC Canadian Research Institute of Launders and Cleaners 加拿大洗衣业与干洗业研究所
C.P.L. collar point length 领尖长
C.P.W. collar point width 领尖宽
c/s cases 箱(复数)
CSB Central Silk Board 中央蚕丝委员会(印度)
CSC China Silk Corporation 中国丝绸公司
CSI Commission Sericole International 国际蚕丝业委员会
CSMFRA Cotton, Silk and Man-made Fibers Research Association 棉、丝及化学纤维研究会
CSMIEC China Silk Materials Import and Export Corp 中国丝绸进出口公司
CSRC Canadian Silk and Rayon Committee 加拿大丝绸和人造丝委员会
CSY core-spun yarn 包芯纱,包芯线
CT carat 克拉

CTA ①China Textile Academy 中国纺织科学研究院；②Cotton Textile Agreement 棉织品协定
CTAT Center for Textile and Apparel Technology 纺织和服装工艺中心
CTC Clothing Technology Center 服装工艺中心
CTDA Custom Tailors and Designers Association of America 美国定制服装裁剪师与设计师协会
CTI ①Canadian Textiles Institute 加拿大纺织学会；②Cotton Textile Institute 棉织品学会
CTMG China Textile Machinery Group 中国纺织机械集团
CTMTC China Textile Machinery and Technology Import and Export Corporation 中国纺织机械和技术进出口总公司
CTR Council for Textile Recycling 再生纺织品协会(美国)
CTRC China Textile Resources Corporation 中国纺织原材料公司
CTRL Cotton Technological Research Laboratory 棉花技术研究实验所(印度)
CTRS Central Tasar Research Station 蚕丝中心研究所(印度)
CTTL Canadian Textile Testing Laboratories 加拿大纺织品试验实验室
CTWBA Cape Town Wool Buyers' Association 开普敦羊毛买主协会(南非)
CVC chief value of cotton 棉为主(50%或以上)的混纺织物；以棉为主的涤棉混纺织物
C.W. cuff width 袖口宽
CWC Central Wool Committee 中央羊毛委员会(澳大利亚)
CYCA Craft Yarn Council of America 美国手工编织委员会
d. dull 浊的
D/A documents against acceptance 承兑交单
D and T double and twist 双色螺旋花式线
D.B. double breast 双排钮
DC Denim Council 劳动布协会(美国)
DCLI Dry Cleaners & Launderers Institute 干洗工和洗衣工协会(加拿大)
DCS data communication system 数据传输系统
D.D. delightful dressing 欢愉的装饰
D/D. demand draft 票汇
Dec. december 十二月
den., dens density 密度
des. design 设计
destn. destination 抵达地
3D fabric Three dimensional faric 三维织物
D.F. flat duck double filling flat duck 双纬帆布
DF flat duck double filling flat duck 双纬帆布
DFA Dyers' and Finishers' Association 染色工作者和整理工作者协会
dg. dark grayish 暗灰调的,暗灰色调
diam. diameter 直径
DJ dinner jacket (没有燕尾的)晚礼服
D&K damge and kept 认赔
dk. dark 暗的
dm. decimetre 分米
d/o delivery order 提货单
DOC data on children's steepwear 儿童睡衣阻燃试验数据
DOD drop on demand 喷墨印花
dol. dollar 元
doz. dozen 一打
DP finishing durable press finishing 耐久压烫整理
dp. deep 深的
D/P ①document against payment 付款交单；② D/P after sight documents against payment after sight 远期付款交单；③ D/P sight documents against payment at sight 即期付款交单
DPC dimensional pattern concept (服装)整体设计放码系统
DPL Delta pine land cotton 岱字棉,岱塔派棉
Dr. doctor 博士
DSA Dress Salesmen's Association 服装销售商协会(美国)
D.,T. denim double and twist denim 双色纬纱劳动布;双色纬纱粗斜纹布
d.t. delivery time 交货时间
DTD doctor of textile dyeing 纺织染色博士
DTF Domestic Textile Federation 家用纺织品联合会(英国)
DTS doctor of textile science 纺织科学

博士
DTT doctor of textile technology 纺织工艺博士
D.T yarn double and twist yarn 双色螺旋花线；混色仿螺旋花线
dup. duplicate 副本
e ①elasticity 弹性；②elongation 伸长
E. east 东(方)
EA environment auditing 环境审核
EANOTR European Association of National Organizations of Textile Retailers 欧洲纺织品零售商全国性组织联盟
EATP European Association for Textile Polyolefins 欧洲聚烯烃纤维协会
EC European Community 欧洲共同体
E.C. elbow circumference 肘围
ECDP easy cation-dyeable polyester 阳离子染料易染型聚酯
ECE European Colorfastness Establishment 欧洲染色坚牢度协会
ECFC European Continental Fastness Convention 欧洲大陆染色坚牢度公约
ECGC Empire Cotton Growing Corporation 帝国植棉公司(英国)
ECGO Egyptian Cotton General Organisation 埃及原棉联合会
E-CT electro-conductive textile 导电纺织品
EDANA European Disposable & Nonwovens Association 欧洲用即弃和非织造布制造商协会(比利时)
EDDP easy disperse dyeable polyester 分散染料易染聚酯纤维
EFFI Educational Foundation for the Fashion Industry 时装工业教育基金会(美国)
EFMI Elastic Fabric Manufacturers' Institute 弹力织物制造商协会(美国)
EFTA European Free Trade Association 欧洲自由贸易联合会
EFTC European Free Trade Committee 欧洲自由贸易委员会
EH controller electric hemming controller 电动卷边控制器
EHA electric hemming apparatus 电动卷边装置,缝线防偏装置
EI electric iron 电熨斗
EITP environmentally improved textile products 生态纺织品,绿色纺织品
EL ①export license 出口许可证；②environment labels 环境标志
E.L. ①elbow length 肘长；②elbow line 肘线
Embr. embroidery 绣花,刺绣,绣品
EMPB Embroidery Manufacturers Promotion Board 刺绣品制造商促进委员会(美国)
EMS environment management systems 环境管理体系
enc., encl. enclosure 附件
E.&O.E. errors and omissions excepted 差错待查
e.o.m. end of month 月底
e.o.s. end of season 季末
E.P. elbow point 肘点
e.p. ever press 永久熨烫
EPE environment performance evaluation 环境绩效评估
eq. equal 等于
etc. et cetera 等等
EVS extravehicular suit 宇航服(在航天航外活动的成套服装)
ex example 例如
exam. examination 检验
exp. export 出口
fab., fab fabric 织物,布
FBCM Federation of British Carpet Manufacturers 英国地毯制造商联合会
fbr, fbr. fibre 纤维
FCI Fellow of the Clothing Institute 服装学会会员(英国)
Feb. February 二月
F.D. front depth 前腋深
FFF Furnising Fabrics Federation 装潢用织物联合会(英国)
f/glass fibre glass 玻璃丝；玻璃纤维
FH fabric handle 织物手感
FHTF Frankfurt Heimtextil Trade Fair 法兰克福纺织品交易会
FI flammability index 耐燃烧性指数
fib fiber 纤维
fil. filament 长丝,单纤维
FIMT Federation International of Master Tailors 国际高级西装师联盟
FIT Fashion Institute of Technology 时装工艺学院(纽约时装学院)
FITEI Federation of Indian Textile Engineering Industry 印度纺织工程工业联合会
fl flake yarn 雪花线；竹节花式线

F.L. front length 前长
FMC Felt Manufacturers Council 毛毡制造商理事会(美国)
F.N front neck 前领圈
F.N.F. flying needle frame 管针经编机
F.N.P. front neck point 前颈点
FOB,F.O.B.,fob,f.o.b. free on board 船上交货;离岸价格
f.o.c. free of change 免费
FOR free on rail 火车上交货价
F.P.A. free from particular areraqe 平安险
fr. franc 法郎
F.R. front rise 前(直)裆;
F.R. fabrics flame retardant fabrics 阻燃织物
Fri. Friday 星期五
frt. freight 运费
F.S. fist size 手头围
ft. foot,feet 英尺
FTF Fiber Trade Federation 纤维业联合会(英国)
FW fabric weight 织物重量
F.W. ①front width 前胸宽;②full weight 全重,毛重
FWWMR fire, water, weather and mildew resistance 防火、拒水、耐气候和防霉[整理法]
FY filament yarn 长丝(纱)
g. grayish 灰调的
G. green 绿
g.,gm gram(s) 克
G.A general average 共同海损
GATT General Agreement on Tariffs and Trade 关税及贸易总协定
gB. greenish blue 绿调蓝
GCA Greenwood Cotton Association 格林伍德棉业协会(美国)
GCAD garment computer aided design 服装计算机辅助设计
GCV gross calorific value 总热值
gds. goods 货物
GFC ①glass fiber cloth 玻璃纤维布; ②graphite fiber composites 石墨纤维复合材料
G-fiber graphite fiber 石墨纤维
gfpt grams-force per tex 克力/特(纤维强度单位)
GFRABS glass fiber reinforced ABS 玻璃纤维增强 ABS

GFY glass filament yarn 玻璃长丝纱
GG gray goods 坯布,本色布
GGL garment grain leather 正面服装革
GK golf knickers 短裤
GPCF General purpose Pitch-based Carbon Fiber 通用沥青基碳纤维
gr. ①gross 总额,总量,罗(＝12打); ②gram(s) 克
GRF germ resistant fabric 抗菌织物
GSAC Garment Salesman Association of Canada 加拿大服装营销商协会
GSP generalised system of preferences 普遍优惠制(简称普惠制)
G string gee string （印第安人或歌女、舞女等当三角裤用的)G 带,遮羞布,腰布
G suit Gravity suit 宇航员用重力防护服,抗超重飞行衣
G.W.,gr.wt. gross weight 毛重
GWBA Geelong Wool Brokers' Association 基朗羊毛经纪人协会(澳大利亚)
gY. greenish yellow 绿调黄
GZU-SF Guangzhou Spring Fair 广州春季交易会
H ①hairweight 毫特克斯;②hip 臀围;③hue 色相
H. hip 臀围
HATRA the Hosiery and Allied Trades Research Association [英]针织业研究协会
HBT herringbone twill 人字形斜纹;山形斜纹;人字形斜纹布
HE high elasticity 高弹性
HESC Hand Evaluation and Standardization Committee [日]手感评定和标准化委员会
HFA Hard Fibers Association 韧皮纤维协会
HFAK hollow fiber artifical kidney 中空纤维人造肾
HIA Handkerchief Industry Association [美]手帕工业协会
HKA Hand knitting Association [美]手工编织协会
HKGMA Hosiery and Knit Goods Manufacturers Association 针编织品生产商协会
HL home laundry 家庭洗涤
H.L. ①hip line 臀围线;②head length

头长
HLCC Home Laundering Consultative Council [英]家庭洗涤咨询理事会
Hmd hemmed 缝好边的,镶好边的,折边的
HMGF high modulus glass fiber 高模量玻璃纤维
HMHS high modulus high strength 高模高强
H.O. head(or home) office 总公司,总部
HOY highly oriented yarn 高取向丝,高定向丝
HPCF high performance carbon fiber 高性能碳纤维
HPY high-power yarn 高收缩丝,高应力丝
hr(s). hour(s) 小时
HS fiber high-shrinkage fiber 高收缩纤维
H.S. head size 头围
HSY high shrinkage yarn 高收缩[率]丝
HT carbon fiber high tenacity carbon fiber 高强力碳纤维
HT yarn high tenacity yarn 高强力丝
ht. height 高度
HTA Household Textiles Association 家用纺织品协会
HTR,H.T.R. high tenacity rayon 高强力黏胶人造丝
HTS ①high temperature setting 高温热定形;②high temperature stability 高温稳定性;③Home Textile Show 家用纺织品展览会
HV hand value 织物风格值
HWL heat-fast,water-fast,light-fast 耐热,耐洗,耐晒(织物)
HWM rayon high wetmodulus rayon 高湿模量黏胶纤维
HWNA Hosiery Wholesalers' National Association [美]全国袜子批发商协会
HWS hot water stability 耐热水性
HWWC hand wash with care 手工精洗
I institute 学会;研究所;学院
I. inseam 内长
IAA Intimate Apparel Associates 贴身内衣行业协会
IACD International Association of Clothing Designer 国际服装设计师协会
IAGM International Association of Garment Manufacturers 国际服装制造商

协会
IAJAM Industrial Association of Juvenile Apparel Manufacturers [美]青少年服装制造商协会
IATCL International Association for Textile Care Labelling 国际纺织品洗整标识协会
IB identification bracelet 鉴别镯
IBM inclined bed machine 斜台缝纫机
IBSMMF International Bureau for the standardization of Man-made Fiber 国际人造纤维标准局
ICA International Colour Authority 国际色彩局
ICAC International Cotton Advisory Committee 国际棉业咨询委员会
ICB intenational competitive bid-ding 竞争性招标
ICC International Chamber of Commerce 国际商会
I C.C. institute cargo clauses 协会货物条款
ICCA Infants' and Children's Coat Association (美)婴儿和儿童罩衣协会
ICCO International Carpet Classification Organization 国际地毯分类组织
ICF International Carpet Fair,Harrogate 国际地毯博览会(英国哈罗盖特)
ICFF Information Council on Fabric Flammability 织物耐燃性情报理事会
ICMF Indian Cotton Mills' Federation 印度棉纺织厂联合会
ICPA International Cotton Producers' Association 国际棉花生产者协会
IDC International Drycleaners Congress 国际干洗工作者会议
IDCRC International Dry Cleaning Research Committee 国际干洗研究委员会
IE industrial engineering 工业管理学
i.e. id est (拉丁语)那就是;即
IFAI Industrial Fabrics Association International [美]国际产业用纺织品协会
IFC International Fashion Council 国际流行[服装]评议委员会
IFCATI International Federation of Cotton and Allied Textile Industries 国际棉花与棉纺织工业联合会
IFF International Fashion Fair 国际时

装博览会

IFFE International Fashion Fabrics Exhibition 国际时装织物展览会

IFI International Fabricare Institute 国际纺织品洗涤研究院

IFJ International Fiber Journal ［美］《国际纤维杂志》(期刊)

IFS Indian Fiber Society 印度纤维学会

IGF India Garment Fair 印度服装博览会

IHE International Hosiery Exposition 国际针织品博览会

IIC International Institute for Cotton 国际棉业协会

IIGF India International Garment Fair 印度国际服装博览会

IIL Institute of Industrial Launderers ［美］工业洗衣工学会

IKAE International Knitting Arts Exhibition Tokyo 国际针织艺术展览会(东京)

IKF International Knitwear Fair 国际针织品博览会

IKL imitation kid leather 仿小山羊革

IL import license 进口许可证

I.L. inside length 股下,下档长

ILA International Laundry Association 国际洗涤协会

ILGWU International Ladies Garment Workers Union 国际女装生产者协会

ILPC International Linen Promotion Commission 国际亚麻促进会

IMA International Mohair Association 国际马海毛协会

IMBWEX International Mens'& Boys' Wear Exhibition, London 国际男装及男童服装展览会(英国伦敦)

imp import 进口

in. inch(es) 英寸

INB International Nonwovens Bulletin 《国际非织造布通讯》(期刊)

ince, ins. insurance 保险

ind indent 订单

INDA International Nonwovens and Disposables Association 国际非织造布和用即弃物品协会

Indian J Fiber & Fext Res Indian Journal of Fiber & Textile Research 《印度纤维和纺织学报》(期刊)

Indian Text J Indian Textile Journal 《印度纺织杂志》(期刊)

Industrie Text. L'ndustrie Textile 《纺织工业》(月刊)

INFEX International Floorcovering Exhibition, Brighton 国际地毯展览(英国布赖顿)

inorfib. inorganic fibre 无机纤维

inorfil. inorganic filament 无机长丝

inst. ①institute 学会,协会,学院；②institution 学会；③instant 本月

Int Dyer International Dyer and Textile Printer ［英］《国际染印漂整工作者》(期刊)

int. interest 利息

INTERCOLOR International Commission for colour in Fashion and Textile 国际时装纺织品流行色委员会

Internat. Text. Bull.: Weaving International Textile Bulletin: weaving ［瑞士］《国际纺织通报—织造》(季刊)

inv. invoice 发票

IOL International Old Lacers 国际带子制作者和收藏者协会

IPE individual protection ensemble 个人防护服(整套服装)

IRSA International Raw Silk Association 国际蚕丝协会

IRSFC International Rayon and Synthetic Fibers Committee 国际再生纤维和合成纤维委员会

IRTAC Impoter-Retailer Textile Advisory Committee ［美］纺织品进口及销售咨询委员会

ISC International Sericultural Commission 国际蚕丝业委员会

ISCC Inter-Society Color Council, USA 美国色彩协会

ISCL International Symposium for Care Labelling Textiles 国际纺织品维护标识研讨会

ISIFM International Society of Industrial Fabric Manufacturers 国际产业用纺织品生产者协会

ISIYM International Society of Industrial Yarn manufacturers 国际工业用纱线制造商协会

ISO. IOS International Organization for Standardization 国际标准化组织

ISPA International Sleep Products Association 国际床品协会

ISS International Sportswear Salon, Monaco [摩]国际运动服展览会

ISTCL International Scientific and Technical Committee on Laundering 国际洗涤科学与技术委员会

ISTDA Institutional and Service Textile Distributors Association [美]制服用纺织品批发商协会

ITA ①Indonesia Textile Association 印度尼西亚纺织协会；② International Textile Agreement 国际纺织品协定

ITAA International Textile and Apparel Association 国际纺织服装协会

ITAE International Textile Accessory Exhibition, Istanbul [土]国际纺织附件展览会(伊斯坦布尔)

ITB International Textile Bulletin [瑞士]《国际纺织通报》(季刊)

ITC International Textile Club 国际纺织联合会

ITCLC International Textile Care Labelling Code 国际纺织洗整标志

ITF ①Institut für Textil-und Faserforchung [德]纺织品及纤维研究所；②Institut Textile de France [法]纺织研究院

ITGLWF International Textile Garment and Leather Workers Federation 国际纺织服装和皮革工人联合会

ITGME International Textile and Garment Machinery Exhibition 国际纺织及服装机械展览会

ITGWF International Textile and Garment workers' Federation 国际纺织和服装工人联合会

ITI International Textile Institute [英]国际纺织学会

ITMA ①International Textile Manufactures Association 国际纺织业联合会；②International Textile Marketing Association 国际纺织品销售协会

ITME International Textile Machinery Exhibition 国际纺织机械展览会

ITMF International Textile Manufacturers Federation 国际纺织生产者联盟

ITO International Trade Organization 国际贸易组织

ITS Institute of Textile Science [加]纺织科学研究所

ITT Institute of Textile Technology [美]纺织[技术]研究院

IWF Irish Wool Federation 爱尔兰羊毛联合会

IWS International Wool Secretariat 国际羊毛局，国际羊毛咨询处

IWT Institute for Wool Technology [澳]毛纺工艺研究所

IWTO International Wool Textile Organization 国际羊毛纺织品组织

JAC Joint Advisory Council for the Carpet Industry 地毯工业联合咨询委员会

JAFET Japanese Association for the Functional Evaluation of Textiles 日本纺织品功能评定协会

Jan. January 一月

JASMA Japan Sewing Machinery Manufacturers Association 日本缝纫机制造商协会

JCAT Japanese Color Aptitude Test 日本颜色适应性试验

JCF Journal of Coated Fabrics [美]《涂层织物杂志》(期刊)

JCFA Japan Chemical Fibers Association 日本化学纤维协会

JCI Jute Corporation of India Ltd. 印度黄麻公司

JCTA Japan Cotton Traders' Association 日本棉花贸易商协会

JECC Joint Egyptian Cotton Committee [美]联合埃及棉花委员会

JECMA Japan Export Clothing Makers' Association 日本出口服装制作商协会

JFCA Japan Fashion Color Association 日本流行色协会

JFF Junior Fashion Fair, London 少年时装博览会(伦敦)

JIAME Japan International Apparel Machinery Exhibition 日本国际服装机械展览会

JIT Just-in-time 最少库存

jkt. jacket 短上衣，夹克衫

JMFU Japan men's Fashion Unity 日本男装协会

JRI Jute Research Institute 孟加拉黄麻研究所

JRSA Japan Raw Silk Association 日本蚕丝协会

JSA ①Japan Silk Association 日本丝业

协会；②Japan Spinners' Association 日本纺纱工协会

J. Soc. Dyers Col Journal of the Society of Dyers and Colorists ［英］《染色协会会志》（月刊）

JTI Journal of the Textile Institute ［英］《纺织学会会志》（期刊）

JTN Japan Textile News 《日本纺织新闻》（月刊）

JTR Japan Textile Research 《日本纺织研究》（期刊）

JTRL Jute Technological Research Laboratories ［印］黄麻工艺研究所

Jul. July 七月

Jun. June 六月

JUSTIS Japan-United States Textile Information Service 日美纺织情报所

JWTC Japan Weathering Testing Center 日本耐气候牢度试验中心

KAE Knitting Arts Exhibition 针织艺术品展览会

KC K.C. Knit coat 针织短外衣

KCA Karachi Cotton Association ［巴］卡拉奇棉业协会

KCFA Korean Chemical Fiber Association 韩国化纤协会

KEA Knitwear Employers' Association ［美］针织业雇主协会

KES Kawabata Evaluation System 川端织物手感评价体系

KES-F handle tester Kawabata Evaluation System-F handle tester 川端织物风格仪

KF Knitwear Fair, London 针织品博览会（伦敦）

kg. kilogram 公斤

KIF Knitting Industries' Federation ［英］针织工业联合会

K.L. Knee line 膝围线

KMMA Knitting Machine Manufacturers' Association ［英］针织机制造商协会

Knitt. Internat Knitting International 《国际针织》（月刊）

Knitt. Times Knitting Times 《针织时代》（月刊）

KTA Knitted Textile Association ［英］针织品协会

KTDF Knitted Textile Dyers' Federation ［英］针织染色业联合会

L. ①length 衣（裤，裙等）长；②large 大号

l. ①light 浅的；②left 左

lb. libra （一）磅

LBC Leather Bureau of Canada 加拿大皮革局

LBD little black dress 基本型黑色裙装

LBL leather boot lace 皮靴带

L/C, l/c letter of credit 信用证；信用

LCA Liverpool Cotton Association ［英］利物浦棉业协会

L cotton Lambert cotton 苏丹兰伯特棉

Ld, Ltd. limited 有限股份（公司）

LDCMMA Laundry and Dry Cleaners Machinery Manufacturers Association 洗衣与干洗机机械制造厂协会

LDCO laundry and dry cleaning operations 洗衣和干洗作业

LDT long duration test 耐久性试验

LENG, lg., lgth. Length 长度

LF laminated fabric 层压织物

LFE London Fashion Exhibition ［英］伦敦时装展览会

LFF London Fashion Fair ［英］伦敦时装博览会

lg. light grayish 明灰调的

L/G banker's letter of guarantee 银行保证书，银行保函

l/g letter of guarantee 保函

LGF long glass fiber 长玻璃纤维

LIA ①Lace Importers Association ［美］花边进口商协会；②Leather Industries of America 美国皮革业联合会

LIYFF London International Youth Fashion Fair ［英］伦敦国际青少年时装博览会

Lj life jacket 救生衣

L-joint lap joint 搭接

LM looping machine 套口机；缝线头机

lng lining；衬里；内衬

LNM latch needle machine 舌针针织机

LOA length over all 全长；总长

LPFS London Public Fur Sales 伦敦毛皮拍卖行

LPW Lombardi Pattern Warp machine 朗伯迪花式圆形针织机

LRA Lace Research Association ［英］花边研究协会

LSAA linen Supply Association America 美国亚麻布供应商协会

LTA ①Linen Trade Association ［美］

亚麻行业协会；②Longterm agreement 棉制品长期合作协定
LTTH London Textile Testing House 伦敦纺织试验所
LWMMA London Wholesale Millinery Manufacturers' Association 伦敦女帽制造商协会
LWTU lightweight thermal underwear 轻质保暖内衣
m. metre(s) 公尺，米
Mar. march 三月
Mass TAC Massachusetts Textile and Apparel Council 马萨诸塞州纺织服装委员会
mat. Material 原料
max. maximum 最大
m/c machine 机器
MCA ①Manchester Cotton Association [英]曼彻斯特棉业协会；②Memphis Cotton Exchange [美]孟菲斯棉花交易所；③Mohair Council of America 美国马海毛协会
MCC Moscow Chamber of Cotton 莫斯科棉花管理委员会
Melliand Textilber Intern Melliand Textiberichte International [德]《梅利安德纺织学杂志，国际版》（期刊）
MF ①metal fiber 金属纤维；②microfilament 微丝，微细纤维；③mineral fiber 矿物纤维
MFA ①Men's Fashion Association 男式时装协会；②Multi Fiber Arrangement 纺织品多边贸易协定，多边纤维贸易协定
M/F bicomponent fiber matrix fibril bicomponent fiber 基质型双组分纤维，天星型双组分纤维
M-fibers modified viscose fiber 变性黏胶纤维
MFN most favoured nation 最惠国
MFPC Man-made Fibers Producers Committee [英]化学纤维生产者委员会
M/F type bicomponet fiber matrix-fibril bicomponent fiber 基质原体型双组分纤维，海岛型双组分纤维
MFU Japan men's Fashion Unity 日本男式时装协会
MGF metallized glass fiber 镀金属玻璃纤维

M.H. middle hip 中臀围
M high modulus fiber 高模量纤维
M.H.L. middle hip line 中臀围线
MIECS multiple independent embroidery control system 多路独立绣花控制系统
min. minimum 最小
MIS management information system 经营管理信息处理系统
Mitsubishi R Mitsubishi Rayon Co.Ltd. [日]三菱人造丝公司
mk. mark 商标；标志
mkt. market 市场
mm. mllimetre 毫米
MMAA Merchandise Market Apparel Association [美]商品市场服装协会
M., med. medium 中号，中等
MMF Man-made fiber 人造纤维，化学纤维
MMFPA Man-made Fiber Producers Association [美]化学纤维生产者协会
MMVF man-made Vitreous fiber 透明化学纤维，人造透明纤维
Mod.Text. Modern textiles [美]《现代纺织品》（月刊）
MODTEPS modular toxic environment protective suit 有毒环境下使用的组装式防护服
Mon. monday 星期一
mooh. mohair 马海毛，马海毛织物
mos. months 月（复数）
MR moisture regain 回潮率
Mr. Mister 先生
Mrs. Mistress 夫人，太太
Ms. Miss 小姐
M/T mail transfer 信汇
MTB Melliand Textilberichte [德]《梅利安德纺织学杂志》（期刊）
MTN multilateral trade negotiations 多边贸易谈判
M.V. wool Montevideo wool 蒙特维迪奥美利奴短羊毛
MWA Melbourne woolbrokers' Association 墨尔本羊毛经纪人协会
MWK multiaxial warp knit 多轴向经编针织物
MWRA Men's Wear Retailer's Association [美]男式服装零售商协会
N nylon 尼龙；锦纶
N. ①neck 颈；领；②North 北（方）

NABM ① National Association of Bedding Manufacturers [美]全国床上用品制造商协会；② National Association of Blouse Manufacturers [美]全国女衬衫制造商协会

NADFD National Association of Decorative Fabric Distributors [美]装饰织物经销商协会

NAFAD National Association of Fashion and Accessory Designers [美]全国服装服饰设计师协会

NAFTA ① North American Free Trade Agreement 北美自由贸易协定；② North Atlantic Free Trade Area 北大西洋自由贸易区

NAGM National Association of Glove Manufacturers [美]全国手套制造商协会

NAHM National Association of Hosiery Manufacturers [美]袜类生产者协会

NAIEHS National Association of Importers and Exporters of Hides and Skin [美]全国皮革进出口商协会

NALGM National Association of Leather Glove Manufactureres [美]全国皮手套制造商协会

NAMSB National Association of Men's Sport-wear Buyers [美]全国男式运动服装采购商协会

NAN National Academy of Needlearts [美]全国针锈协会

NASA look National Aeronautics and Space Administration look 宇航服款式,宇航服风貌

NASMD National Association of Sewing Machine Dealers [美]全国缝纫机销售商协会

NASWM National Association of Scottish Woolen Manufacturers [英]苏格兰全国毛织物制造商协会

NATAW National Association of Textile & Apparel Wholesalers [美]全国纺织品及服装批发商协会

NATMM National Association of Textile Machinery Manufacturers [美]全国纺织机械制造商协会

NATO North Atlantic Treaty Organization 北大西洋公约组织

NATS National Association of Textile Supervisors [美]全国纺织专家协会

NAUM National Association of Uniform Manufacturers [美]全国制服制造商协会

NAWCM National Association of Wiping Cloth Manufacturers [美]全国揩布制造商协会

NAWM National Association of Wool Manufacturers [美]全国羊毛加工厂商协会

NAWWO National Association of Woolen and Worsted Overseers [美]全国精纺和粗仿毛织物监管人员协会

NBA National Button Association 全国纽扣协会

NBBDA National Burlap Bag Dealers' Association [美]全国麻袋商协会

NBC nuclear, biological, chemical [suits] 防核、生物及化学服装(成套衣服)

NBF National Bedding Federation [美]全国床上用品联合会

NBS National Bureau of Standards [美]国家标准局

NCA Neighborhood Cleaners Association 社区洗衣协会

NCCA National Cotton Council of America 美国国家棉花委员会

NCF Nylon coated fabric 锦纶胶布,锦纶涂层织物,尼龙胶布,尼龙涂布织物

NCGA National Cotton Ginners' Association [美]全国轧棉业协会

NCPA National Cotton Products Association [美]全国棉制品协会

NCRC Nonwovens Cooperative Research Center 非织造布合作研究中心

NCRPM National Committee on Radiation Protection and Measurement [美]国家辐射防护和检测委员会

NCS ① national colour system 天然色系统；② Natural Color System 自然颜色系统

NCSU North Carolina State University [美]北卡罗来纳州立大学

NCTE National Council for Textile Education [美]全国纺织教育委员会

NCTMA North Carolina Textile Manufacturers Association [美]北卡罗来纳纺织生产者协会

NCTRU Navy Clothing and Textile Research Unit [美]海军服装与纺织研究所

n.c.v. no commercial value 无商业价值

NCWA National Children's Wear Association [英]全国童装协会

NDMA National Dress Manufacturers Association [美]全国服装制造商协会

NFC/PFC needled and pile floor carpet 针刺地毯/绒头地毯

NFDC National Federation of Dyers and Cleaners [英]全国染色和干洗工作者联合会

NFG Narrow Fabric Group [英]窄幅织物集团

NFI Narrow Fabric Institute [美]窄幅织物研究所

NFMA Needleloom Felt Manufacturers' Association [英]针刺机制毛毯制造商协会

NFPA National Flexible Packaging Association [美]全国柔性包装材料协会

NFY nylon filament yarn 尼龙长丝纱

NGMA National Garment Manufacturers' Association [加]全国服装制造商联合会

N.H. neck hole 领孔,领口,领圈

NICD National Institute of Cleaning and Dyeing [美]全国洗染学会

NID National Institute of Drycleaning 全国干洗学会

NIE National Institute of Environment [美]国家环境研究所

NIFT National Institute of Fashion Technology [印]国家时装工艺学院

NIH National Institute of Health [美]国家健康研究所

NITMA Northern India Textile Mills Association 北印度纺织厂协会

NITPA Northern India Textile Processors' Association 北印度纺织品加工者协会

NKMA National Knitwear Manufacturers Association 美国针织品生产者协会

NKSA National Knitwear and Sportswear Association 美国针织品和运动衣协会

N.L. ①neck length 领长;②neck line 领口线,领围线

NNA ①National Neckwear Association [美]全国颈部装饰品协会;②National Needlework Association [美]全国针绣协会

NNWF needled nonwoven fabric 针刺非织造布

NO.,no. number 数;号码

NOSA National Outerwear and Sportswear Association [美]全国外衣和运动衣协会

Nov. November 十一月

NOY non-sizing yarn 免浆丝,交络丝

NP ①natural pigment 天然颜料;②natural polymer 天然聚合物

N.P. neck point 颈点,肩顶

NPPon Rayon Nippon Rayon Co. Ltd. 日本人造丝公司

N.R. neck rib 领高

NS nylon strap 锦纶带

N.S. neck size 颈围

n.s. new style 时髦式样

NSGA National Sporting Goods Association [美]全国运动制品协会

NSI National Shoe Institute [美]全国制鞋学会

N.S.P. neck shoulder point 颈肩点

NSSA National Skirt and Sportswear Association [美]全国女裙和运动服协会

NSY ①non-sizing yarn 免浆丝,交络丝;②non-stretch yarn 非伸缩丝,非弹力丝

nt.wt.,n.wt. net weight 净重

NTA Northern Textile Association [美]北方纺织协会

NTBS non tariff barriers 非关税壁垒

NTC ①National Textile Ceonter [美]国家纺织中心;②National Textile Corporation [印]国家纺织公司

NTCF nylon tire cord fabric 锦纶轮胎帘子布

NTDCTI Nonwovens Technology Deveiopment Center of Textile Industry 纺织工业非织造布技术发展中心

NTIS National Technical Information Service [美]国家技术情报服务处

NTP normal temperature and pressure 常温常压

NTPG National Textile Processors Guild [美]全国纺织加工协会

NUF mat Nonwoven unidirectional fiber glass mat 非织造单向玻璃纤维垫

NUFLAT National Union of Footwear, Leather and Allied Trades [英]全国鞋袜、皮革及同业联合会

N.W.,nt.wt net weight 净重

NWGA National Wool Growers' Association 全国羊毛生产者协会
N.W.L. neck waist length 背长
NWMC National Wool Marketing Corporation [美]全国羊毛销售公司
NWTA National Wool Trade Association [美]全国羊毛同业协会
NWTEC National Wool Textile Export Corporation [英]全国毛纺织品出口公司
NYCE New York Cotton Exchange 纽约棉花交易所
NYCSA New York Coat & Suit Association [美]纽约外套与套装协会
NYFD New York Fashion Designers [美]纽约时装设计师协会
N.Y. New York (美)纽约
NZWA New Zealand Woolbrokers' Association 新西兰羊毛经纪人协会
NZWB New Zealand Wool Board 新西兰羊毛管理局
NZWDC New Zealand Wool Disposal Commission 新西兰羊毛处理委员会
NZWIRI New Zealand Wool Industries Research Institute 新西兰毛纺织工业研究所
O. orange 橙
Oct. October 十月
OE yarn open-end yarn 自由端纱,气流纱
OECS open end cover spun [yarn] 气流纺包缠花式纱
Oeko-Tex International Association for Research & Testing in the Field of Textile Ecology 国际纺织生态领域研究及检验协会;生态纺织品
OF ①olefin fiber 烯烃类纤维;②optical fiber 光学纤维,光导纤维
OLCCM on-line color continuity monitor 在线色连续监控系统
OR gown operating gown 外科手术衣
OR,or.,orn. orange 橙色的
OS oversize 超大尺码
O.S. outside seam 外长
OTC Organization for Trade Cooperation 贸易合作组织
OTEXA Office of Textile and Apparel 美国纺织服装管理处
owf on weight of fabric 按织物重量(计算)

OWM Open-Weave material 稀松组织织物
Oxf. Oxford (英)牛津
oz. ounce 盎司;英两,两
P ①page 页;②polynosic 波里诺西克,富强纤维
P. ①pale 浅色;淡的;②purple 紫色,紫色布;紫袍
PA interlining polyamide interlining 聚酰胺黏合衬
PA wool Punta Arenas Wool 蓬塔阿雷纳斯羊毛
PA P.A. Polyamide fiber 聚酰胺纤维
PAIF polyamide-imide fiber 聚酰胺酰亚胺纤维
PALF pineapple leaf fibers 菠萝纤维
PANCF polyacrylonitrile carbon fiber 聚丙烯腈系碳纤维
PANOF polyacrylonitrile oxidized fiber 聚丙烯腈预氧化纤维
pat. patent 专利,专利权,专利件
pat.,patt. pattern 花纹组织,图案,花样,式样,样板
PB. purple blue 蓝紫
PBA poly-p-benzamide fiber 聚对苯甲酰胺纤维
PBI fiber polybenzimidazole fiber 聚苯并咪唑纤维(耐高温纤维)
PBI hollow fiber polybenzimidazole hollow fiber 聚苯并咪唑中空纤维
PBO fiber poly-p-phenylene benzobisthiazole fiber 聚对亚苯基并双噻唑纤维(新型高强度合成纤维)
PBS fabric point-bonded staple fabric 点黏合热熔非织造布
PBS point-bonded staple 点黏合纤维(用于非织造布)
PBSM post bed sewing machine 柱式缝纫机
PBSP polar bear skin plush 北极熊皮长毛绒
PBT fiber polybutylene terephthalate fiber 聚对苯二甲酸丁二酯纤维(弹性聚酯纤维)
PC,pc blend polyester and cotton blend 涤棉混纺织物
pc. ①piecc 件;匹;②price 定价;价格
P.C. percent 百分之
P/C,P-C,pc polyester/cotton,polyester-cotton 涤棉混纺(织物)

PCC pretema Color Computer 普雷特马配色计算机(商名,瑞士)
PCF Plastic coated fabric 塑料涂覆织物
PCG plains Cotton Growers 得克萨斯州棉花生产加工协会
PCS fiber polycarbosilane fiber 聚碳硅烷纤维
pcs.,ps. pieces 件;个(复数)
PCT&S Philadelphia College of Textile and science [美]费城纺织及科学院
pd paid 付讫
PE fiber polyester fiber 聚酯纤维
PEB poly-p-ethyleneoxy benzoate 聚对苯甲酸亚乙氧基酯纤维
PECS Vest Protective Environmentai Control System Vest 环保控制系统背心(利用变相材料制作的轻薄,透气凉爽背心,美军防核、防生化,在高温条件下使用的配套服)
PEF polyethylene fiber 聚乙烯纤维
PES interlining polyester interlining 聚酯衬
PF ①performance factor 性能因素;②perspiration fastness 耐汗渍[色]牢度;③plant fiber 植物纤维;④pressed felt 压制毡
P-fabric paper-like nonwoven fabric 仿纸非织造布
PFF polyester filament fabric 聚酯长丝织物,涤纶长丝织物
PFMA Pressed Felt Manufacturers'Association[英]压制毡制造商协会
PFY polyester filament yarn 涤纶长丝,聚酯长丝
PGM pattern grading machine 放样机
PGS propective glove system 防护手套体系
PHST packaging, handling, storage and transportation 包装、装卸、存储与运输
PIT Provincial Institute of Textiles [美]地方纺织学会
PJMA Pakistan Jute Mills Association 巴基斯坦黄麻工厂协会
pk pink 桃红色,粉红色
pkg package 卷装,包装,装箱
PL pearl leather 珍珠革
PLA fiber polylactic acid fiber 聚乳酸纤维
p.m. past meridian 下午

pntg. printing 印花(工艺)
pp pages 页(复数)
PP ①performance parameter 性能参数;②permanent press 耐久压烫,耐久定形
PP fiber polypropylene fiber 聚丙烯纤维
PP finish[ing] permanent press finish[ing] 耐久压烫整理
PPFY polypropylene filament yarn 聚丙烯长丝
PPL physical property of leather 皮革物理性能
PPM product portfolio management 产品系列平衡法
PPSF polypropylene staple fiber 聚丙烯短纤维
PPT[fider] polyphenylene triazole[fiber] 聚亚苯基三唑[纤维]
PPTA fiber polyphenylene terephthalamide fiber 聚对苯二甲酰对苯二胺纤维
pR. purplish red 紫调红
Prof. Professor 教授
prox. proximo 下月
P.S. palm size 掌围
P.S.,p.s. postscript 再启;又及
PSA pressure suit assembly [全套]增压服(宇航)
PSF polyester staple fiber 聚酯短纤维
PTD fiber poly-1,3,4-thiadiazole fiber PTD 纤维(聚-1,3,4-噻二唑纤维)
PTFE fiber polytetrafluoroethlene fiber 聚四氟乙烯纤维
P.T.O.,p.t.o. please turn over 请看背面,见下页
PTO fiber polyterephthaloyl oxalicbisamidrazone fiber 聚对苯二甲酰草酸双脒腙纤维
PTT fiber polytrimethylene terephthalate fiber 聚对苯二甲酸丙二醇酯纤维(新型聚酯纤维,可用常压沸染法)
PTY polyester texturised yarn 聚酯变形丝,涤纶变形丝
PU elastomer filament polyurethane elastomer filament 聚氨基甲酸酯弹性体长丝(弹力丝)
PU leather polyurethane leather 聚氨基甲酸酯合成革,PU 革
PU synthetic leather polyurethane synthetic leather 聚氨酯合成革

PUF polyurethane fiber 聚氨基甲酸酯纤维

PVAF polyvinyl alcohol fiber 聚乙烯醇纤维

PVC leather polyvinyl chloride leather PVC 合成革,聚氯乙烯合成革

PVCCF polyvinyl-chloride-coated fabric 聚氯乙烯涂层织物

P.W. point width 乳间宽,乳中

PWF pure woolen felt 纯羊毛毡

Q fiber Q 纤维(聚-2,6-萘二酸乙酯纤维,商名,日本)

QAAS quality assurance acceptance standarde 质量保证验收标准

QC quality control 质量管理,品质控制

QE quality evaluation 质量鉴定

QEST quality evaluation systems test 质量鉴定系统试验

QF quartz fiber 石英纤维

QI,Q.I. quality index 质量指标,质量指数

QIC quality inspection criteria 质量检查标准

QM quality management 质量管理

QR zipper quick release zipper 快速拆卸拉链

qr. quarter 四分之一,季度

QSC quick style change[system]快速花样设计转换[系统],快速产品更新[系统]

QT qualification test 质量鉴定试验

QTPT qualification test and proof test 合格试验与验证试验

QTY,qt. quantity 数量

qua. quality 品质,质量,性质

QVR quality verification report 质量验证报告

QVT quality verification test 质量验证试验

R rayon 人造纤维,人造丝

R. ①red 红;②registered trademark 注册商标

r. right 右

RACE rapid automatic checkout eguipment 快速自动检测设备

rd. road 路

rep. representative 代表

RIF resin-impregnated fabric 树脂浸渍织物

rO. reddish orange 红调橙

RP. red purple 红紫

RPFMA Rubber and Plastic Footwear Mamufacturers' Association [英]胶鞋及塑料鞋厂商协会

R.S. right side 正面

RSCC Raw Silk Classification Committee 生丝分级委员会

RTP Review of Textile Progress[英]《纺织进展评论》(期刊)

RTQC real-time quality control 实时质量控制

R.T.W ready to wear 成衣

RY rayon 人造丝,人造纤维

rY. reddish yellow 红调黄

s count 纱支

S staple 纤维;短纤维

s. ①Sleeve 袖长;②shoulder 肩宽;③South 南(方);④Small 小号,小码;⑤strong 强烈的

S.A. Seventh Avenue 第七大街

SAA Supima Association of America 美国西南长绒棉纤维协会

SAFLINC Sundries and Apparel Findings Linkage Council[美]服装联盟

SAHM semi-automatic hose machine 半自动织袜机

Sat. saturday 星期六

SAWTRI South African Wool Textile Research Institute 南非毛纺织研究所

S.B. ①single breast 单排纽;②slack bottom 裤脚口

SC Society of Cotton 棉花学会

S.C. stand collar 领座

SCA Southern Cotton Association [美]南部棉花协会

S-CECF Spring Chinese Export Commodities Fair 春季中国商品交易会,春交会

SCGA Southern Cotton Ginners Association[美]南部棉花轧花协会

SCTMA South Carolina Textile Manufactures Association [美]南卡罗来纳纺织工业协会

SD ①Service dress 制服,军便服;②Synthetic detergent 合成洗涤剂

S.D. scye depth 腋深

s.d. sports dress 运动服

SDA Surface Design Association [美]印染及图案设计协会

Sep. September 九月

SF ① Sable fur 黑貂毛皮；② Satin finish 缎光整理；③ Stocking frame 织袜机
SF staple fibre 短纤维
S.F. flat duck single filling flat duck 单纬帆布
SFMI Soft Fiber Manufacturers' Institute [美]韧皮纤维制造者学会
Silk Rayon Ind Silk and Rayon lndustries of India 《印度丝绸工业》(期刊)
SIS Salon International Sportswear, Monte Carlo [摩]蒙特卡洛国际运动服博览会
SITNME Shamghai Intermational Techtextiles, Nonwovens and Machinery Exhibition 上海国际技术纺织品,非织布及纺织机械展览会
SITRA South lndia Textile Research Association 印度南方纺织研究协会
SKA Scottish Knitwear Association 苏格兰针织业协会
SL Synthetic linings 合成衬里材料
S.L. ① sleeve length 袖长；② side length 裙长,裤长
SLEMA Schiffli Lace and Embroidery Manufacturers Association [美]席弗里花边和刺绣品制造商协会
SLS snow leopard skin 雪豹皮
SL slider self locking slider 自锁拉链头
SM, S.M. sewing machine 缝纫机
SMTA Sewing Machine Trade Association [美]缝纫机同业协会
S/N, s/n shipping note 装船通知
SNLSM single needle lock stitch machine 单针锁缝机
S.N.P. ①shoulder neck point 肩颈点；②side of neck point 旁颈点
S/O shipping order 装货单
sp. specimen 样品
S.P. shoulder point 肩点
sq. square 平方
S.S. sleeve slope 肩斜
SSM sack sewing machine 缝袋机
S.S.P. shoulder sleeve point 肩袖点
S.T. sleeve top 袖山
st. stitch 针脚；缝线
STA Southern Textile Association [美]南部纺织品协会
std. standard 标准
Sun. sunday 星期天

S.W. shoulder width 肩宽
SWAK Spinning and Weaving Association Korea 韩国棉纺织协会
t. ton (一)吨
TABI Textile and Apparel Business Intelligence [美]纺织与服装企业信息协会
TAC Trade Advisory Council 贸易咨询委员会
TAIAC Textile and Appavel Industry Advisory Council [澳]纺织及服装工业咨询委员会
TALC Textile Apparel Linkage Council 纺织服装标准促进会
tapy tapestry 花毯；挂毯；织花壁毯
tar tarpaulin 防水帆布；雨衣
TAT Textile Association of Thailand 泰国纺织协会
TC Terylene cotton 涤棉织物
TC 248 欧洲标准委员会纺织技术委员会
TC 38 国际标准组织纺织技术委员会
T/C terylene/cotton 棉涤纶；涤棉
TCA Textile Council of Australia 澳大利亚纺织协会
TCB Textile and Clothing Board [加]纺织服装委员会
TCC & ADR Textie Chemist and Colorist & American Dyestuff RePorter [美]《纺织化学家和染色家及美国染料报道》(期刊)
TCCA Textile Color Card Association 纺织色卡协会
T/C cambric Terylene Cotton Cambric 涤棉细纺
T/C drill Terylene Cotton drill 涤棉卡其
T/C fancy suiting Terylene Cotton fancy Suiting 涤棉花呢
T/C plain cloth Terylene Cotton Plain cloth 涤棉平布
T/C poplin Terylene Cotton poplin 涤棉府绸
T/C rope for jacguard card Terylene Cotton rope for jacguard card 涤棉纹板绳
TDA Textile Distributors Associaation [美]纺织品批发商协会
TDG Textile Designers Guild 纺织品设计协会
TEC total easy care 完全免烫,完全易

保养[羊毛]
Tech Textiles Int　Technical Textiles Intenational　[英]《国际技术纺织品》(期刊)
TECK　total easy care knitwear　全免烫针织品
Tel.,tel.　telephone　电话
Text Horizons　Textile Horizons　[英]《纺织天地》(期刊)
Text Inst Ind　Textile Institute and Industry　[英]《纺织学会与工业》(期刊)
Text Mfr　Textile Manufacturer　[英]《纺织品厂商》(期刊)
Text Mfr Knitt World　Textile Manufacturer & Knitting World　[英]《纺织制造家和针织世界》(期刊)
Text Month　Textile Month　[英]《纺织月刊》(期刊)
Text Outlook Intl　Textile Outlook International　[英]《国际纺织展望》(期刊)
Text Prog　Textile Progress　[英]《纺织进展》(期刊)
Text Res J　Textile Research Journal　[美]《纺织研究杂志》(期刊)
Text World　Textile World　[美]《纺织世界》(月刊)
Text.　textile　纺织的;纺织品;纺织原料
Textile Ind　Textile Industries　[美]《纺织工业》(期刊)
Textile Inst Ind　Textile Institute and Industry　[英]《纺织学会与工业》(期刊)
Textile Mus　Textile Museum　纺织博物馆
TFA　①Textile Fiber Association　[美]纺织纤维协会;②Textile Finishers Association　[英]纺织品整理工作者协会
TFPIA　Textile Fiber Products Identification Act　[美]纺织纤维制品鉴定条例
TFTA　Textile Finishing Trades Association　[英]纺织品整理行业联盟
tg.　telegram　电报
Thur(s)　thursday　星期四
T.I.　①Textile Industry　纺织工业;②Textile Institute　[英]纺织学会
TIRP　Textile Information Retrieval program　纺织情报检索程序
TIT　Technological Institute of Textile　[印]纺织工程学会
TITUS　Textile Information Treatment Users Service　国际纺织信息中心
TIUC　Textile Information Users Council　纺织信息协会
TJA　Textile Journal of Australia　《澳大利亚纺织杂志》(期刊)
T.L.　trousers length　裤长
TMA　Tape Manufacturers' Association　[英]窄幅织物制造商协会
TMAMA　Textile Machinetury and Accessory Manufacturers' Association　[英]纺织机械和附件制造商协会
TMDS　total material design system　整体材料设计系统(服装面料)
TMI　textile machinery Industry　纺织机械工业
TMMA　Textile Machinery Manufacturers' Association'India　印度纺织机械制造商协会
TMSJ　Textile Machinery Society of Japan　日本纺织机械学会
TPM　①trigger price mechanism　启动价格制;②total productive maintenance　全员生产维修
TQC　total quality control　全面质量管理
TQB　Textile Quota Board　纺织品配额委员会
TQCA　Textile Quality Control Association　[美]纺织品质量管理协会
TQM　total quality management　全面质量管理,综合质量管理
TQS　total quality system　全面质量系统
T.R.　trouser rise　裤(直)裆
tr.　tare　皮重
TRI　Textile Research Institute　[美]纺织研究所
TRJ　Textile Research Journal　[美]《纺织研究杂志》(期刊)
T.S.　thigh size　腿围
TSA　①Textile Salesmen's Association　[美]纺织品推销商协会;②Textile Society of Australia　澳大利亚纺织学会
TSB　Textile Surveillance Body　纺织品进出口监督机构
TSC　Textile Society of Canada　加拿大纺织学会
TSS　Toyota Sewing system　丰田[公司]制衣生产方法

T/T　telegraphic transfer　电汇
TTD　Textile Technology Digest　[美]《纺织工艺文摘》(期刊)
TTF　①Taiwan Textile Fair　中国台湾纺织品博览会；②Taiwan Textile Federation　台湾纺织联合会
TTFC　Textile Technical Federation of Canada　加拿大纺织技术协会
TTI　Textile Technology Inc.　[美]纺织工艺公司
TTMA　Tufted Textile Manufacturers' Association　[美]簇绒纺织品制造商协会
Tu., Tues.　Tuesday　星期二
TV, T.V.　television　电视
TW　①Textile World　[美]《纺织世界》(月刊)；②tricot warper　经编整经机
TWA　Textile Wholesalers Association　[加]纺织品批发商协会
UAA　Undergarment Accessories Association　[美]内衣衣饰协会
UBL　Under bustline　乳根围线，下胸围线
UGWA　United Garment Workers America　美国服装工人协会
UI　Underwear Institute　[美]内衣协会
UICWA　United lnfants' and Children's Wear Association　[美]婴儿装和儿童装协会联合会
UKFA　United Kingdom Fellmongers' Association　英国毛皮商协会
UKML　United Knitwear Manufactrers' League　[美]针织品制造商总会
ult.　ultimo　上月
USA-ITA　United States Association of Importers of Textile And Apparel　美国纺织品进口商协会
USS, U.S.St　United States Standrrd　美国标准
v.　vivid　鲜艳的
via　by way of　经由
VICS　Voluntary Interindustry Communications Standards Committee　美国纺织标准联络委员会
VIP jeans　Very important person jeans　名人牛仔裤，名家牛仔裤
VN mixed gabardine　Viscose-nylon mixed gabardine　黏锦华达呢
vol.　volume　体积
VS. cotton　Vsibo cotton　伍西布棉；

V.S.I　very slightly imperfect　二号花(钻石)；微瑕
V.V.S.I　very very slightly imperfect　一号花(钻石)；极微瑕花；极微瑕
V'vet　Velvet　天鹅绒；丝绒
W.　waist　裤(裙)腰；腰节；腰围
W., w.　①west　西(方)；②weight　重量；③width　宽度
W.A.　Width average　宽度平均值
WCMFGB　Wholesale Clothing Manufacturers' Federation of Great Britain　英国服装批发厂商联合会
Wcy　wincey　上等绒布；棉毛绒布
WDL　Worn denim look　牛仔布磨旧外观，牛仔布穿旧外观
Wed.　Wednesday　星期三
w.f.　wrinklefree　无皱纹
WFBMA　Woven Fabric Belting Manufacturers' Association　[美]编织腰带制造商协会
WFFA　Women's Fashion Fabric Association　[美]女式时装织物协会
wg.s.　wrong side　织物反面；织物背面
wh.　white　白色的
WHIA　Woolen Hosierg Institute of America　美国毛织品学会
Wirk. Strik. Tech　Wirkerei und Strikerei Technic　[德]《针织技术》《刊物》
WIWK　Weft-inserted warp knit　衬纬经编针织物
wk.　week　星期
W.L., WL, wl　waist line　腰围线；腰节线
W.M.A.　Wabbing Mamfacturers' Association　织带制造商协会
w/p　water proof　防水
W/R　warehouse receipt　仓库
WLP　mixed worsted suiting　wool-linen-polyester mixed worsted suiting　毛麻涤薄花呢
WMC　Wool Manufacturers' Council　[美]毛纺织品制造商协会
Wool Rec　Wool Record　[英]《羊毛纪事》(期刊)
WP　weatherproof　防风雨的；不受气候影响的
WP, WP., WPG　waterproofing　防水的；不透水的
WPLA　Wool Products Labeling Act　羊毛制品商标法

WR water repellency 拒水性
wstd. worsted 精梳毛纺的;精纺的
wt. weight 重量,质量
WTA ①Wholesale Textile Association [英]纺织批发商协会;②World Textiles Agreement 世界纺织品协定
WTCA World Trade Centers Association 世界贸易中心协会
WTO World Trade Organization 世界贸易组织
wt per ft, wt per ft. weight per foot 每英尺重量
W.T.R.C. Wool Textile Research Council 毛纺研究委员会
W.W., W&W wash and wear 洗可穿,免烫
WWB water proof, wind proof, and breathable fabric 可呼吸织物,防水透湿防风织物
WWEPC Wool and Woolens Export Promotion Council 羊毛和粗纺毛织物出口促进协会
WX women's extra large size 女式服装特大号
XL extra outsize, extra large （衣服等)特大号,超大的
XOS extra outsize （衣服等）特大号
XXL extra extra large 超特大号
y ①yard(s) 码;②year(s) 年
Y. yellow 黄
yd. yard(s) 码
YG. yellowgreen 黄绿
yG. yellowish green 黄调绿
YO. yellowish orange 黄调橙
yr(s) year(s) 年
yR. yellowish red 黄调红